Global Biodiversity

Status of the Earth's Living Resources

A Report compiled by
the
World Conservation Monitoring Centre

Editor: Brian Groombridge

WORLD CONSERVATION MONITORING CENTRE

in collaboration with
The Natural History Museum, London
and in association with
IUCN - The World Conservation Union
UNEP - United Nations Environment Programme
WWF - World Wide Fund for Nature
and the
World Resources Institute

THE NATURAL HISTORY MUSEUM UNEP WWF

With project sponsorship from
Overseas Development Administration, UK
and additional support from
The Ministry of Foreign Affairs, The Netherlands
The Ministry of the Environment, Denmark
and
The World Bank

SPRINGER-SCIENCE+BUSINESS MEDIA, B.V.

This report is a contribution to GEMS - The Global Environment Monitoring System

First edition 1992

Originally published by Chapman & Hall in 1992
Softcover reprint of the hardcover 1st edition 1992

Reproduced from camera-ready copy prepared by WCMC.

Recycled paper supplied by Robert Horne Paper Co Ltd.

ISBN 978-94-010-5012-8   ISBN 978-94-011-2282-5 (eBook)
DOI 10.1007/978-94-011-2282-5

Citation: World Conservation Monitoring Centre (1992) *Global Biodiversity: Status of the Earth's living resources*. Chapman & Hall, London. xx + 594pp.

Also available from IUCN Publications Services Unit, 181a Huntingdon Road, Cambridge,CB3 0DJ

Cover Photos
Mugger, *Crocodylus palustris:* Brian Groombridge
Guzmania lingulata: D. Muleax
Fish market, Indonesia: Tom Moss/WWF Photo library
Henri Pittier National Park, Venezuela: Paul Goriup

The designations of geographical entities in this book, and the presentation of the material, do not imply the expression of any opinion whatsoever on the part of WCMC or its sponsoring organisations concerning the legal status of any country, territory, or area, or of its authorities, or concerning the delimitation of its frontiers or boundaries. In recent years geopolitical entities have become more fluid and this is not reflected consistently in the tables which were completed at different times and use data of different ages.

A catalogue record for this book is available from the British Library

Contents

How to Use this Book viii

World Conservation Monitoring Centre ix

Acknowledgements x

Preface xii

Biodiversity - An Overview xiii

Part 1. BIOLOGICAL DIVERSITY

SYSTEMATICS AND DIVERSITY

1. GENETIC DIVERSITY 1

The nature and origin of genetic variation 1
Measuring genetic variation 1
The interpretation of variation 3
The environment and the distribution of genetic variation 3
Conclusion 6

2. SYSTEMATICS AND DIVERSITY 7

Biological classification 7
Phylogenetic relationships and their estimation 7
From hierarchy to classification 7
Taxonomic nomenclature and its regulation 9
Major features of the hierarchy of life 10
Systematics and the measurement of biodiversity 11

3. SPECIES CONCEPTS 13

What is a species? 13
Early species concepts 13
Evolutionary theory and polytypic species 13
The biological species concept 14
The phylogenetic species concept 15
Species in practice 15
Conclusion 16

4. SPECIES INVENTORY 17

Current status 17
Prediction from the existing partial inventory 20
Other approaches to predicting patterns 23
Uncharted realms of species richness 26
Sampling the hyper-diverse but poorly known 31
New data on tropical insects and what they convey 32
Prospects for improved species richness estimates 36

SPECIES DIVERSITY

5. SPECIES DIVERSITY: AN INTRODUCTION 40

A brief history of diversity 40
Measuring biological diversity 41
The global distribution of species richness 43
Species and energy 45
Explanations and hypotheses 46

6. MICROORGANISMS 47

Taxonomic scope 47
Assessment of diversity 47
Species concepts in microorganisms 47
Extent of genetic diversity 49
Regions and habitats of maximum diversity 49
Role of microorganisms in biodiversity maintenance 50
Role of microorganisms in biosphere functions 51
Potential contribution of microorganisms to sustainable development 51
The need for diversity amongst microorganisms 52
Ex situ conservation of microorganisms 52
The taxonomic challenge 52

7. LOWER PLANT DIVERSITY 55

Bryophytes 55
Lichens 55
Algae 57

8. HIGHER PLANT DIVERSITY 64

The groups of higher plants 64
The distribution of higher plants 65

9. NEMATODES 88

Nematode diversity 88
The ecological importance of nematodes 90

10. DEEP-SEA INVERTEBRATES 93

Deep-sea communities 93
Ocean trenches 95
Hydrothermal vents 98
Cold seeps 100

11. SOIL MACROFAUNA 103

Soil and soil fauna 103
Patterns of species richness 105
General patterns of diversity 113

12. FISHES 116

The diversity of fishes 116
Freshwater fishes: species richness and endemism 116
Subterranean fishes 121
Coral reef fishes 122

13. HIGHER VERTEBRATES 136

The groups of higher vertebrates 136
The distribution of higher vertebrates 137

14. ISLAND SPECIES 147

Plants on oceanic islands 148
Land snails 149

15. CENTRES OF SPECIES DIVERSITY 154

Introduction 154
Methods of determining areas of conservation priority 154
Centres of plant diversity 157
Centres of avian endemism 161

SPECIES LOSS

16. SPECIES EXTINCTION 192

How species become vulnerable to extinction 192
A brief history of extinctions 196
Extinctions in recent history 198
Current and future extinction rates 202

17. THREATENED SPECIES 234

What is a threatened species? 234
Globally threatened animals 236
Aquatic habitats 242
Threatened species on islands: plants 244
Threatened species on islands: birds 245

HABITATS AND ECOSYSTEMS

18. GLOBAL HABITAT CLASSIFICATION 248

Ecosystem mapping 248
Estimating rates of change of ecosystems 250

19. BIODIVERSITY AND GLOBAL CLIMATE CHANGE 254

Modelling global climate change 254
Effects of climate change on vegetation zones and biodiversity 254

20. TROPICAL MOIST FORESTS 256

What are tropical moist forests? 256
The global area of tropical moist forest 257
Factors involved in changes in forest cover 264
Management practices in tropical forests 267
Case studies 269

21. GRASSLANDS 280

The world area of grassland 280
Origins and floral diversity of grasslands 280
The 20th century impact on grasslands 284
Case studies 288

22. WETLANDS 293

Global extent and distribution of wetlands 293
Regional extent and distribution of wetlands 295
Values and threats 297
Loss of wetlands 298
Conservation of wetlands 302

23. CORAL REEFS 307

Occurrence of reefs 307
Values and threats 307

24. MANGROVES 324

The mangrove habitat 324
Value of mangroves 324
Threats to mangrove habitats 324

Part 2. USES AND VALUES OF BIODIVERSITY

USES OF BIOLOGICAL RESOURCES

25. PLANT USE 331

Food plants 331
Timber 342
Rattans 350
Medicinal plants 350
Ornamental plants 353

26. ANIMAL USE 359

Introduction 359
Food: terrestrial animals 359
Food: fisheries 365
Non-food uses 374
Domestic livestock 389

VALUING BIODIVERSITY

27. BIODIVERSITY AND ECONOMICS 407

Valuing the environment 407
Loss of biodiversity as an economic process 409
Current uses of diverse resources 411
Community use of wildlife resources 411
Ecotourism 413
Existence values 415
The valuation of diverse ecosystems 417
The value of tropical forests 417
The value of wetlands 421
Preserving future options 425
The value of diversity in providing insurance: crop yields 426
Sources of yield variability 429
Crop insurance: the response to increased agricultural risk 430
The value of agricultural genetic diversity 432
The value of biodiversity in the production of pharmaceuticals 434

Part 3. CONSERVATION AND MANAGEMENT OF BIODIVERSITY

NATIONAL POLICIES AND INSTRUMENTS

28. NATIONAL LEGISLATION 441

The protection of wild flora 441
The protection of wild fauna 442
Limitations of species legislation 444
The protection of natural habitats 444

29. PROTECTED AREAS 447

National protected area systems 447
International protected area systems 459

INTERNATIONAL POLICIES AND INSTRUMENTS

30. MULTILATERAL TREATIES 479

Multilateral treaties 479

31. INTERNATIONAL POLICY AND LEGAL ASSISTANCE 490

Funding 490
International obligations: protected areas 494
Intellectual property rights for biotechnology 495
Regulated trading in wildlife products 499
Regional seas programme 501

32. INTERNATIONAL AID 506

International development assistance 506
Bilateral development assistance 507
Multilateral development assistance 509
International assistance in forest management 514
The tropical forestry action plan (TFAP) 514
The international tropical timber agreement (ITTA) 518
Debt purchase 522

33. MANAGEMENT OF INTERNATIONAL RESOURCES 528

International fisheries management commissions 528
Antarctica: the evolution of an international resource management regime 534

BIODIVERSITY CONVENTION

34. CURRENT PRACTICES IN CONSERVATION 543

In situ conservation of threatened plant species 545
In situ conservation of crops and wild relatives of crops 546
Institutions involved in *ex situ* conservation of plants 549
Techniques for *ex situ* plant conservation 557
In situ conservation of animals 561
Ex situ conservation of animals 563
Ex situ conservation of animal genetic resources 571
Ex situ conservation of microbial diversity 571

34. CURRENT PRACTICES IN CONSERVATION 543

Background 576
The biodiversity convention 576
The biodiversity country studies and unmet financial needs 579
Future data needs: networking and global monitoring 581

GLOSSARY 585

How to Use this Book

An extensive review of global biodiversity obviously generates substantial quantities of data with the concomitant problem of how best to present this mass of material. *Global Biodiversity* is intended to be a source-book of information and analysis rather than be read cover to cover, so assisting the reader find his/her way around the book is essential.

The primary means of accessing this wealth of information is through the Contents list (page iii). This is therefore very detailed and serves some of the function of an index (which it has not been practical to include because of excess length). **The reader is urged to browse the Contents** before dipping into the text.

The book is divided into three Parts, each of which opens with a brief overview of its structure and contents. The Parts are then divided into ten Sections that group together Chapters that address a common theme. This structure is outlined below as a guide to the overall organisation of the book.

Part 1. Biological Diversity

- Systematics and diversity
- Species diversity
- Species loss
- Habitats and ecosystems

Part 2. Uses and Values of Biodiversity

- Uses of biological resources
- Valuing biodiversity

Part 3. Conservation and Management of Biodiversity

- National policies and instruments
- International policies and instruments
- Current practices in conservation
- Biodiversity Convention

Individual Chapters are divided thematically by major sub-headings, and these are fully listed in the Contents, which is therefore the key entry point for access to the information.

As far as possible, plain English has been used rather than scientific terminology, but when the use of obscure technical language has been unavoidable a definition has been provided in the Glossary.

World Conservation Monitoring Centre

The Earth's biological diversity and other natural resources provide many economic, social and aesthetic benefits to mankind. Effective programmes for sustainable human development must, therefore, incorporate conservation objectives. Responsible institutions and individuals need access to a service that provides factual information on conservation issues in a timely, focused and professional way.

This service is provided by the **World Conservation Monitoring Centre,** Cambridge, UK. Established in 1988 as a company limited by guarantee with charitable status, WCMC is managed as a joint-venture between the three partners in the *World Conservation Strategy* and its successor *Caring For The Earth:* IUCN - The World Conservation Union, UNEP - United Nations Environment Programme, and WWF - World Wide Fund for Nature. Its mission is to provide information on the status, security, management and utilisation of the world's biological diversity to support conservation and sustainable development.

To implement this mission, WCMC maintains substantial databases on the status and distribution of plant and animal species of conservation and development interest; habitats of conservation concern, particularly tropical forests, coral reefs and wetlands; the global network of national parks and protected areas; and the international trade in wildlife species and their derivative products. Much of this information is managed with Geographic Information Systems, and is supported by an extensive bibliography of published and "grey" literature. WCMC is also involved in providing support for the expansion of national data management and monitoring capabilities in developing countries, and in developing communication networks for the flow of information.

WCMC contributes its biodiversity data to GEMS - the Global Environment Monitoring System, co-ordinated by UNEP. GEMS is a collective programme of the world community to acquire, through global monitoring, and assessment, the data that are needed for the rational management of the environment. GEMS is an element of the United Nations Earthwatch programme.

WCMC Biodiversity Report Team

John McComb	Project Manager
Dr Brian Groombridge	Editor and Research Co-ordinator
Esther Byford	Production Supervisor
Crawford Allan	Research
John Howland	Research
Dr Chris Magin	Research
Helen Smith	Research
Veronica Greenwood	Production
Lindsay Simpson	Production

Consultant Assistant Editors

Martin Jenkins (general)
Timothy M. Swanson (economics and policy)
Hugh Synge (plants)

WCMC Staff who contributed to the compilation and review process:

Mike Adam, Clare Billington, Simon Blyth, Gillian Bunting, John Caldwell, Lorraine Collins, Dr Mark Collins, Mary Cordiner, Helen Corrigan, Robert Cubey, John Easy, Jeremy Harrison, Tim Inskipp, Dr Timothy Johnson, Beverley Lewis, Dr Richard Luxmoore, Lesley McGuffog, Sheila Millar, Dr Ronald I. Miller, James R. Paine, Dr Robin Pellew, Corinna Ravilious, Jonathan Rhind, Sarah Skinner, Jo Taylor, Dr Kerry Walter.

Acknowledgements

The production of this Report has been based largely upon the substantial biodiversity databases that WCMC manages, supplemented by a major world-wide data gathering and standardisation programme. This information is drawn from an extensive network of scientists, research workers, park managers, wildlife authorities, conservation bodies and government organisations. WCMC would like to express its thanks for the contributions of these individuals and agencies, too numerous to mention individually, without whose support we would not be able to operate.

WCMC particularly recognises with gratitude the fundamental contribution of both the IUCN Species Survival Commission (SSC) and the Commission of National Parks and Protected Areas (CNPPA), whose commitment to WCMC over the years has enabled the Centre to expand its databases. Their data have been used extensively in the Report. WCMC also acknowledges the support of the IUCN Environmental Law Centre, whose legal data and expertise have contributed significantly.

In compiling the Report itself, WCMC particularly acknowledges the major contributions of the following people whose names appear at the end of the relevant chapter (unattributed chapters were provided by WCMC staff):
Dr John Akeroyd, Bruce Aylward, Dr Keith Banister, Dr Gordon Brent Ingram, Dr B.N.K. Davis, Victoria Drake, Alan Eddy, B.C. Eversham, Alix Flavelle, Shirra Freedman, D.J. Galloway, Sarah Gammage, Dr Stephen J.G. Hall, Dr P.N. Halpin, Peter Hammond, David Hanrahan, Dr Caroline Harcourt, Prof. D.L. Hawksworth, Richard J. Hornby, Nigel Howard, Martin Jenkins, David M. John, Sam Johnston, A.S. Jolliffe, E.A. Leadlay, Dr Rik Leemans, Mike Maunder, Sara Oldfield, Greg Rose, Timothy M. Swanson, Hugh Synge, Richard Thomas, Ian Tittley, Susan M. Wells, Dr P.S. Wyse Jackson, Dr R.I. Vane-Wright.

The first four chapters were contributed by staff of The Natural History Museum, London. WCMC is especially grateful for their assistance, and for the efforts of John Peake (Associate Director, Scientific Development) in facilitating this collaboration.

In addition, WCMC would like to express thanks to the following who have contributed to the compilation of this Report in a variety of ways:
Dr Dennis Adams, Suraya Affiff, Dr J.Crinan Alexander, M. Altieri, Dr Chris Andrews, Dr Martin Angel, Prof. Peter Ashton, Margerita Astralaga. Dr Paul Bamps, Dr John Beard, Dr S. Beck, Dr Henk Beentje, Dr Colin J. Bibby, Dr Mike Bingham, Dr William Block, Dr Robert Boden, Dr A. Bogan, Dr Attila Borhidi, Dr Philippe Bouchet, Prof. Loutfy Boulos, Dr David Bramwell, Dr F.J. Breteler, Dr Harold Brookfield, Dr Dick Brummitt, David Brunner, Dr Francoise Burhenne-Guilmin, Victor Bullen, Dr Yvonne A. Byron, Dr J. Cardiel, Dr Jan Cerovsky, Jim Chapman, Dr Arthur D. Chapman, Dr A. Cleef, P. Colman, Dr R. Cowie, Dr Quentin Cronk, Mike J. Crosby, James Culverwell II, Michael Dadd, Dr Patricia Davila, Stephen D. Davis, D.G. Debouck, Dr Jean-Jacques de Granville, Dr Robert DeFilipps, A. Delsaerdt, Nelly Diego, Dr C. Dodson, Dr John Dransfield, Dr J. Duivenvoorden, M. Dulude, Dr K. Emberton, Lynne Farrell, Prof. Philip M. Fearnside, Dr Richard Felger, Rosa M. Fonseca, Dr F.R. Fosberg, Dr F. Friedmann, Dr Ib Friis, Dr John D. Gage, Dr F. Galena, Dr Rodrigo Gamez, Dr Sam M. Gan III, Dr Martin Gardner, N. Gardner, Dr Steve Gartlan, Dr Alwin Gentry, Dr David Given, Prof. César Gómez Campo, Dr Roger Good, Dr R. Gopalan, Dr Frederick Grassle, Peter Green, O. Griffiths, Liz Guerin, Prof. Nimal Gunatilleke, Dr M. Hadfield, Dr S. Halloy, Dr Ole Hamann, Dr Alan Hamilton, Dr Stephen Harris, Melanie F. Heath, Dr A.N. Henry, Dr Derral Herbst, Prof. Vernon Heywood, Craig Hilton-Taylor, Dr A. Hoffmann, Dr Martin W. Holdgate, E. Hoyt, Dr Otto Huber, Prof. Dr Gordon Brent Ingram, Dr Frank Ingwersen, Dr Walter Ivantsoff, Prof. K. Iwatsuki, Dr Peter Wyse Jackson, Dr J. Jérémie, Prof. Robert Johns, Dr Marshall Johnston, Dr M. Jorgensen, Dr Calestous Juma, Prof. Horng Jye-Su, Dr Ruth Kiew, Dr T. Killeen, Prof. V. Krassilov, Dr John Lambshead, Prof. Elias Landolt, Dr R. Lara, Dr John Leigh, Dr David Lellinger, Christine Leon, Blanca León, Dr E. Lleras, Dr Paul V. Loiselle, Adrian J. Long, Francisco Lorea, Dr Rosemary Lowe-MacConnell, Lucio Lozado, Prof. Grenville Lucas, Olga Herrera-MacBryde, Dr Kathy MacKinnon, Jane MacKnight, Lynne Maclennan, Dr Domingo Madulid, Mike Maunder, Dr Niall McCarten,

Dr Bill McDonald, Bob McDowall, Jeffrey A. McNeely, Dr Tim Messick, Robert Mill, Dr Kenton R. Miller, Dr Tony Miller, Danya Miskov, Lino Monroy, P. Mooney, Dr Norman Moore, Prof. P. Morat, Dr P.B. Mordan, Dr Scott Mori, Dr Larry Morse, Michael Moser, Fred Naggs, Dr David Neill, Dr B. Nelson, Dr Dan Nicholson, Dr Hans Nooteboom, Dr Rosa Ortiz, Dr Maria Tereza Jorge Padua, Dr Christopher Page, Dr W. Palacios, Dr Mark Perry, Prof. Ghillean T. Prance, Dr M. Prashanth, Robert Prescott-Allen, Han Qunli, Dr L. Ramella, Dr Orlando Rangel, Dr Peter Raven, Dr Tony Rebelo, Marcia Ricci, J. Robertson-Vernhes, Joyce Rushton, Dr B.D. Sharma, Samar Singh, Dr D.K. Singh, Dr Mark Skinner, Joel Smith, D. Smits, Dr Sy Sohmer, C. Sperling, Alison Stattersfield, Dr George Staples, Dr G. Stephens, Wendy Strahm, Dr Tod Stuessy, Prof. Dr H. Sukopp, Dr R.W. Sussman, Glen Swindlehurst, Lesley Taylor, Dr Simon Thirgood, Dr Duncan Thomas, Dr F. Thompson, Dr Jim Thorsell, Dr Mats Thulin, Dr S. Tillier, Simon Tonge, Dr Shigeru Tsuda, Dr Verena Tunnicliffe, Dr C. Ulloa, Dr E. Vajravelu, Dr Vu Van Dung, Dr Leo Vanhecke, Jane Villa-Lobos, Dr C. Villamil, Dr J.-F. Villiers, Dr David Wagner, Dr Warren H. Wagner, Dr H. Waldén. Richard Warner, Dr Tom Wendt, Julie S. Wenslow, Dr Dagmar Werner, Dr Gerry Werren, Dr Tony Whitten, J.T. Williams, Julia Willison, Dr David S. Woodruff, Dr Richard Wunderlin, Prof. Wang Xianpu, Dr K.R. Young, Prof. Yang Zhouhuai,

The authors are grateful for the assistance provided by the librarians of the Monks Wood Experimental Station, the University of Cambridge, the Natural History Museum (General, Zoological and Entomological) and the British Antarctic Survey.

Finally, WCMC recognises with gratitude the substantial financial contributions made by our sponsors listed on the title page. Their confidence in our ability to complete such an ambitious project is appreciated. In particular, WCMC thanks the Overseas Development Administration, UK and especially David Turner, Ian Haines and Mark Lowcock, together with the Ministry of Foreign Affairs, The Netherlands, particularly Ton van der Zon and Egbert Pelinck. The Ministry of the Environment, Denmark, through the endorsement of Veit Koester, also contributed, whilst the World Bank, through Mohan Munasinghe has distributed copies into the developing world. WCMC expresses its sincere gratitude these organisations and individuals.

Preface

We Need Your Data for Future Editions of this Report

In your hands you now hold the most comprehensive review of global biodiversity ever compiled. It represents the product of numerous scientists, consultants and research institutes each of whom has generously contributed data or assistance to the compilation of this Report, together with the substantial information holdings that WCMC already manages. Yet so vast and diverse are the Earth's living resources - the genes, species and ecosystems that comprise the planet's biotic wealth - and the threats that these resources now face, that this massive effort has barely scratched the surface.

To build the information store on which this Report is based, **we need your help**. WCMC will continue to expand its global biodiversity database and intends to republish the Report every two years. This volume is therefore **the first of a proposed series** that will document changes to the status, utilisation and management of the world's biological resources. We need your contribution to fuel this expansion. We are embarked upon a long-term process, the aim of which is to mobilize the substantial amounts of data available throughout the world to encourage a more enlightened conservation practice. Your piece of the jigsaw puzzle may fit into the overall picture we are trying to create. If you are able to contribute data to expand this Report, we urgently want to hear from you - don't quibble with its deficiencies which inevitably are numerous; instead be more constructive by contributing your specialist knowledge to this global conservation effort. We plan to distribute the database itself later this year in machine-readable format, and it is not too late to include your information.

The need for reliable quantitative information about the impact of people upon nature has never been greater. Good intelligence is the key to good decisions, whether about priorities, policies or investments. We need to develop data gathering and monitoring capabilities at the local and country levels, particularly in the developing world, and to build networks for the early-warning of new threats to biodiversity. The realisation of these needs is encompassed in *Agenda 21 of the UN Conference on Environment and Development*, in the *Biodiversity Convention*, and in the *Global Biodiversity Strategy*, but the basic common factor for the implementation of all these initiatives is good information. WCMC will make available its information to support these global enterprises, but to be really effective, we need your data and your participation.

This process of expanding the global database through the networking of national centres must be linked directly into the *Biodiversity Convention*. Despite the delays and frustrations in its negotiation, which are discussed in Chapter 35, the Convention could provide a potent mechanism for implementing global conservation and sustainable use of biodiversity. Assuming a Convention is eventually agreed, its effectiveness will depend upon its access to reliable up-to-date scientific information. WCMC will mobilize its substantial data holdings to support the Convention: information will be its life-blood and WCMC will act as the catalyst for its operation by providing a massive blood transfusion.

Robin Pellew
Director

24 April 1992

World Conservation Monitoring Centre
219 Huntingdon Road
Cambridge
CB3 0DL
UK

BIODIVERSITY: AN OVERVIEW

This introduction is intended to map out in general terms some of the principal themes to be encountered in the field of biological diversity. It will provide a context for the remainder of the report, in which many of these themes are further developed.

WHAT IS BIODIVERSITY?

The word `biodiversity' is a contraction of *biological diversity*. Diversity is a concept which refers to the range of variation or differences among some set of entities; *biological* diversity thus refers to variety within the living world. The term `biodiversity' is indeed commonly used to describe the number, variety and variability of living organisms. This very broad usage, embracing many different parameters, is essentially a synonym of `Life on Earth'.

Management requires measurement, and measures of diversity only become possible when some quantitative value can be ascribed to them and these values can be compared. It is thus necessary to try and disentangle some of the separate elements of which biodiversity is composed.

It has become a widespread practice to define biodiversity in terms of *genes*, *species* and *ecosystems*, corresponding to three fundamental and hierarchically-related levels of biological organisation.

Genetic diversity

This represents the heritable variation within and between populations of organisms. Ultimately, this resides in variations in the sequence of the four base-pairs which, as components of nucleic acids, constitute the genetic code.

New genetic variation arises in individuals by gene and chromosome mutations, and in organisms with sexual reproduction can be spread through the population by recombination. It has been estimated that in humans and fruit flies alike, the number of possible combinations of different forms of each gene sequence exceeds the number of atoms in the universe. Other kinds of genetic diversity can be identified at all levels of organisation, including the amount of DNA per cell, and chromosome structure and number.

This pool of genetic variation present within an interbreeding population is acted upon by selection. Differential survival results in changes of the frequency of genes within this pool, and this is equivalent to population evolution. The significance of genetic variation is thus clear: it enables both natural evolutionary change and artificial selective breeding to occur.

Only a small fraction (often less than 1%) of the genetic material of higher organisms is outwardly expressed in the form and function of the organism; the purpose of the remaining DNA and the significance of any variation within it is unclear.

Each of the estimated 10^9 different genes distributed across the world's biota does not make an identical contribution to overall genetic diversity. In particular, those genes which control fundamental biochemical processes are strongly conserved across different taxa and generally show little variation, although such variation that does exist may exert a strong effect on the viability of the organism; the converse is true of other genes. Further, an astonishing amount of molecular variation in the mammalian immune system, for example, is possible on the basis of a small number of inherited genes.

Species diversity

Perhaps because the living world is most widely considered in terms of species, biodiversity is very commonly used as a synonym of *species diversity*, in particular of `species richness', which is the number of species in a site or habitat. Discussion of global biodiversity is typically presented in terms of global numbers of species in different taxonomic groups. An estimated 1.7 million species have been described to date; estimates for the total number of species existing on earth at present vary from five million to nearly 100 million. A conservative working estimate suggests there might be around 12.5 million. In terms of species number alone, life on earth appears to consist essentially of insects and microorganisms.

The species level is generally regarded as the most natural one at which to consider whole-organism diversity. Species are also the primary focus of evolutionary mechanisms, and the origination and extinction of species are the principal agents in governing biological diversity in most senses in which the latter can be defined. On the other hand, species cannot be recognised and enumerated by systematists with total precision, and the concept of what a species is differs considerably between groups of organisms.

Further, a straightforward count of the number of species only provides a partial indication of biological diversity, for implicit within the term is the concept of degree or extent of variation; that is, organisms which differ widely from each other in some respect by definition contribute more to overall diversity than those which are very similar.

The more different a species is from any other species (as indicated, for example, by an isolated position within the taxonomic hierarchy), then the greater its contribution to any overall measure of global biological diversity. Thus, the two species of Tuatara (genus *Sphenodon*) in New Zealand, which are the only extant members of the reptile order Rhynchocephalia, are more important in this sense than members of some highly speciose family of lizards.

Developing this argument, a site with many different higher taxa present can be said to possess more *taxonomic diversity* than another with fewer higher taxa but many more species. Marine habitats frequently have more different phyla but fewer species than terrestrial habitats; i.e. higher taxonomic diversity but lower species diversity. Measures under development endeavour to incorporate quantification of the evolutionary uniqueness of species.

The ecological importance of a species can have a direct effect on community structure, and thus on overall biological diversity. For example, a species of tropical rain forest tree which supports an endemic invertebrate fauna of a hundred species evidently makes a greater contribution to the maintenance of global biological diversity than a European alpine plant which may have no other species wholly dependent on it.

Ecosystem diversity

The quantitative assessment of diversity at the ecosystem, habitat or community level remains problematic. Whilst it is possible to define what is in principle meant by genetic and species diversity, and to produce various measures thereof, there is no unique definition and classification of ecosystems at the global level, and it is thus difficult in practice to assess ecosystem diversity other than on a local or regional basis and then only largely in terms of vegetation. Ecosystems further differ from genes and species in that they explicitly include abiotic components, being partly determined by soil parent material and climate.

Ecosystem diversity is often evaluated through measures of the diversity of the component species. This may involve assessment of the relative abundance of different species as well as consideration of the types of species. In the first instance, the more equally abundant different species are, then in general the more diverse that area or habitat is considered to be. In the second instance, weight is given to the numbers of species in different size classes, at different trophic levels, or in different taxonomic groups. Thus a hypothetical ecosystem which consisted only of several species of plants, would be less diverse than one with the same number of species but which included animal herbivores and predators. As different weightings can be given to these different factors when estimating the diversity of particular areas, there is no one authoritative index for measuring diversity. This obviously has important implications for the ranking of different areas.

Biodiversity: its meaning and measurement

The differences between these conceptual perspectives on the meaning of *biodiversity*, and the associated semantic problems, are not trivial. Management intended to maintain one facet of biodiversity will not necessarily maintain another. For example, a timber extraction programme which is designed to conserve biodiversity in the sense of site species richness may well reduce biodiversity measured as genetic variation within the tree species harvested. Clearly, the maintenance of different facets of biodiversity will require different management strategies and resources, and will meet different human needs.

Even if complete knowledge of particular areas could be assumed, and standard definitions of diversity be derived, the ranking of such areas in terms of their importance with respect to biological diversity remains problematic. Much depends on the scale that is being used. Thus, the question of what contribution a given area makes to *global* biological diversity is very different from the question of what contribution it makes to local, national or regional biological diversity. This is because, even using a relatively simplified measure, any given area contributes to biological diversity in at least two different ways - through its richness in numbers of species and through the endemism (or geographical uniqueness) of these species. The relative importance of these two factors will inevitably change at different geographical scales, and sites of high regional importance may have little significance at a global level. Neither of these factors include any explicit assessment of genetic diversity.

Although the word *biodiversity* has already gained wide currency in the absence of a clear and unique meaning, greater precision will be required of its users in order that policy and programmes can be more efficiently defined in the future.

BIODIVERSITY: CHANGES IN TIME AND SPACE

Changes over time

The fossil record of life in geological time is very incomplete. There is marked variation between higher taxa and between species in different ecosystems in the extent to which individuals are susceptible to preservation and to subsequent discovery. Chance factors have played a large part, and interpretation by palaeontologists of the available material is beset by differences of opinion. Thus, the record is relatively good for shallow-water hard-bodied marine invertebrates, but poor for most other groups, such as plants in moist tropical uplands.

Two salient points appear well-substantiated. Firstly, *taxonomic diversity*, as measured by the number of recognised phyla of organisms, was greater in Cambrian times than in any later period. Secondly, and keeping in mind the difficulty of disentangling artifacts of the record from the underlying pattern, it appears that *species diversity* and number of families have undergone a net increase between the Cambrian and the Pleistocene epoch, although interrupted by isolated phases of mass extinction (few of which are reflected in the fossil record of plants).

Changes in space

In general, species diversity in natural habitats is high in warm areas and decreases with increasing latitude and altitude. On land, diversity is also usually higher in areas of high rainfall and lower in drier areas. The richest areas are undoubtedly tropical moist forests. If current estimates of the number of species (mainly insects) comprising the microfauna of tropical moist forests are credible, then these areas, which cover perhaps 7% of the world's surface area, may well contain over 90% of all species. If the diversity of larger organisms only is considered, then coral reefs and, for plants at least, areas with Mediterranean climate in South Africa and Western Australia, may be as diverse. Gross genetic diversity and ecosystem diversity will, by definition, tend to be positively correlated with species diversity (although there are indications that some tropical species show more genetic diversity than related temperate species, and some habitat generalists more than habitat specialists).

The reasons for the large-scale geographic variation in species diversity, and in particular for the very high species diversity of tropical moist forests, are not fully understood and involve two interconnected questions: the *origin* of diversity through the evolution of species and the *maintenance* of diversity. Both these involve consideration of the present and historic (in a geological or evolutionary sense) conditions prevailing in particular areas, principally climatic but also edaphic and topographic. Climatically benign conditions (warmth, moisture and relative aseasonality) over long periods of time appear to be particularly important.

It is often assumed that areas with so-called climax ecosystems will be more diverse than areas at earlier successional stages. However, an area with a mosaic of systems at different successional stages will probably be more diverse than the same area at climax provided that each system occupies a sufficiently large area of its own. In many instances, human activities artificially maintain ecosystems at lower successional stages. In areas that have been under human influence for extended periods, notably in temperate regions, maintenance of existing levels of diversity may involve the maintenance of at least partially man-made landscapes and ecosystems, mixed with adequately sized areas of natural climax ecosystems.

Loss of biodiversity

The loss of biological diversity may take many forms but at its most fundamental and irreversible it involves the extinction of species.

Over geological time, all species have a finite span of existence. Species extinction is therefore a natural process which occurs without the intervention of man. However, it is beyond question that extinctions caused directly or indirectly by man are occurring at a rate which far exceeds any reasonable estimates of background extinction rates, and which, to the extent that it is correlated with habitat perturbation, must be increasing.

Unfortunately, quantifying rates of species extinction, both at present and historically, is difficult and predicting future rates with precision is impossible.

Documenting definite species extinctions is only realistic under a relatively limited set of circumstances, where a described species is readily visible and has a well-defined range which can be surveyed repeatedly. Unsurprisingly, most documented extinctions are of species that are easy to record (e.g. land snails, birds) and inhabit sites which can be relatively easily inventoried (e.g. oceanic islands). The large number of extinct species on oceanic islands is not solely an artefact of recording, because island species are generally more prone to extinction as a result of human actions.

Rather than being derived from observed extinctions, therefore, quoted global extinction rates are derived from extrapolations of measured and predicted rates of habitat loss, and estimates of species richness in different habitats. These two estimates are interpreted in the light of a principle derived from island biogeography which states that the size of an area and of its species complement tend to have a predictable relationship; fewer species are able to persist in a number of small habitat fragments than in the original unfragmented habitat, and this can result in the extinction of species.

Even on best available present knowledge, these estimates involve large degrees of uncertainty, and predictions of current and future extinction rates should be interpreted with very considerable caution. Pursuit of increased accuracy in the estimation of global extinction rates, however, whilst of great concern, is not a crucial activity; it is more important to recognise in general terms the extent to which populations and species which are not monitored are likely to be subject to fragmentation and extinction.

Loss of biodiversity in the form of crop varieties and livestock breeds is of near zero significance in terms of overall global diversity, but genetic erosion in these populations is of particular human concern in so far as it has implications for food supply and the sustainability of locally-adapted agricultural practices. For domesticated populations, loss of wild relatives of crop or timber plants is of special concern for the same reason. These genetic resources may not only underlie the productivity of local agricultural systems but also, when incorporated in breeding programmes, provide the foundation of traits (disease resistance, nutritional value, hardiness, etc.) of global importance in intensive systems and which will assume even greater importance in the context of future climate change.

Erosion of diversity in crop gene pools is difficult to demonstrate quantitatively, but tends to be indirectly assessed in terms of the increasing proportion of world cropland planted to high yielding, but genetically uniform, varieties.

The causes of loss of biological diversity

Species may be exterminated by man through a series of effects and agencies. These may be divided into two broad categories: direct (hunting, collection and persecution), and indirect (habitat destruction and modification).

Overhunting is perhaps the most obvious direct cause of extinction in animals, as it has affected several large and well-known species. In terms of overall loss of biodiversity, however, it is undoubtedly far less important than the indirect causes of habitat modification and loss. Nevertheless, as it self-evidently selectively affects species which are or have been considered a harvestable resource, it has important implications for the management of natural resources.

Genetic diversity, as represented by genetic differences between discrete populations within wild species, is liable to reduction as a result of the same factors affecting species. The genetic diversity represented by populations of crop plants or livestock is liable to reduction as a result of mass production; the desired economies of scale demand high levels of uniformity.

Virtually any form of sustained human activity results in some modification of the natural environment. This modification will affect the relative abundance of species and in extreme cases may lead to extinction. This may result from the habitat being made unsuitable for the species (for example, clear-felling of forests or severe pollution of rivers), or through the habitat becoming fragmented. The latter has the effect of dividing previously contiguous populations of species into small sub-populations. If these are sufficiently small, then chance processes lead to raised probabilities of extinction within a relatively short time.

A major, though at present largely unpredictable, change in natural environments is likely to occur within the next century as a result of large-scale changes in global climate and weather patterns. There is a high probability that these will cause greatly elevated extinction rates, although their exact effects are at present unknown.

MAINTAINING BIOLOGICAL DIVERSITY

The maintenance of biological diversity at all levels is fundamentally the maintenance of viable populations of species or identifiable populations. This can be carried out either on site or off site. Some integrated management programmes have begun to link these basically dissimilar approaches.

In situ conservation

The maintenance of a significant proportion of the world's biological diversity at present only appears feasible by maintaining organisms in their wild state and within their existing range. This is generally preferable to other courses of action because it allows for continuing adaptation of wild populations by natural evolutionary processes and, in principle, for current utilisation practices to continue (although these often require enhanced management).

Ex situ conservation

Viable populations of many organisms can be maintained in cultivation or in captivity. Plants may also be maintained in seed banks and germplasm collections; similar techniques are under development for animals (storage of embryos, eggs, sperm) but are more problematic. In any event, *ex situ* conservation is clearly only feasible at present for a small percentage of organisms. It is extremely costly in the case of most animals, and while it would in principle be possible to conserve a very large proportion of higher plants *ex situ*, this would still amount to a small percentage of the world's organisms. It often involves a loss of genetic diversity through founder effects and the high probability of inbreeding.

WHY CONSERVE BIOLOGICAL DIVERSITY ?

This question can be asked from a number of different perspectives, all conditioned by a variety of cultural and economic factors. The various answers given, arguing for the maintenance of biological diversity, have tended to become increasingly confused. Different goals have different implications for the elements and extent of biological diversity that must be maintained. Among these goals are the following:

- the present and potential *use* of elements of biodiversity as biological resources
- the maintenance of the biosphere in a state supportive of human life
- the maintenance of biological diversity *per se*, in particular of all presently living species.

Biological diversity as a resource

It is evident that a certain level of biological diversity is necessary to provide the material basis of human life: at one level to maintain the biosphere as a functioning system and, at another, to provide the basic materials for agriculture and other utilitarian needs.

Food

The most important direct use of other species is as food. Although a relatively large number of plant species, perhaps a few thousand, have been used as foodstuffs, and a greater number are believed to be edible, only a small percentage of these are nutritionally significant on a global level, and only very few of these have been intensively managed on a commercial scale. Similarly, very many animal species are eaten (mostly fishes), but only a very small percentage are globally of nutritional significance. A few dozen species, mostly mammals, are managed in some kind of husbandry system, and a handful of these are globally significant.

It is clear that successful cultivation of agricultural crops on a large scale requires a suite of other organisms (chiefly soil microorganisms and, in a few cases, pollinators) but these probably amount to a statistically insignificant percentage of global biological diversity. Highly productive agricultural systems also require the virtual absence of some elements of biological diversity (pest species) from given sites.

Whilst relatively little diversity is currently used in commercial food production, the very high probability of global climate change, predicted to result in large-scale shifts in natural vegetation and in agricultural systems, has focused attention on the need for conservation of plant genetic resources in order to maintain crop productivity under different climatic regimes. This 'insurance value' of diversity is also evident in contemporary conditions, where increased genetic uniformity is correlated with increased crop yield variation.

Pharmaceuticals

Medicinal drugs derived from natural sources make an important global contribution to health care. An estimated 80% of people in less-developed countries rely on traditional medicines for primary health care; this shows no signs of decline despite availability of western medicine. Some 120 chemicals extracted in pure form from around 90 species are used in medicines throughout the world. Many of these cannot be manufactured synthetically: the cardiac stimulant digitoxin, the most widely used cardiotonic in western medicine, is extracted direct from dried *Digitalis* (foxglove); synthetic vincristine, used to treat childhood leukaemia is only 20% as efficacious as the natural product derived from *Catharanthus roseus* (Rosy Periwinkle).

As with agriculture, and excluding traditional medicines, at present only a very small percentage of the world's biodiversity contributes on a global scale to health care. Many argue that technological advances within the pharmaceutical industry, and in particular those involving the design and manufacture of synthetic drugs, will mean that this contribution is more likely to fall than rise. However, natural diversity might be increasingly valued for the 'blueprints' it provides for new synthetic drugs.

Other material values of biological diversity

Many natural or semi-natural ecosystems, some of which may be of high biological diversity, are of considerable benefit to man. Examples are:

- the role of forests in watershed regulation and stabilisation of soils in erosion-prone areas
- the role of mangroves in coastal zone stabilisation and as nursery areas for fisheries species
- the role of coral reefs in supporting important subsistence fisheries
- the role of natural ecosystems protected as national parks in generating income from wildlife tourism.

In general, however, these values are only indirectly related to biological diversity. That is, a certain level of species richness is required for these functions but there is not necessarily a direct correlation between the value of the ecosystem and its diversity, nor in all cases do a particular set of species have to be present. Thus, mangrove ecosystems are generally of far lower diversity than adjacent lowland terrestrial forests but in resource terms are likely to be of comparable value. The savannas of east and southern Africa, which are of great importance in generating revenues from tourism, are less diverse than the moist forests in these countries which have far less potential for tourism.

The precautionary principle

While it is evident that at present a relatively small proportion of the world's biological diversity is actively exploited by man, other elements of biological diversity may be important for different reasons:

- they have values which are unused or unknown at present but which could enhance the material well-being of mankind if these values were discovered and exploited
- they may become useful or vital at some time in the future owing to changing circumstance.

These factors support a precautionary line in maintaining biological diversity - that is, actually or potentially useful resources should not be lost simply because we do not know about or value them at present. However, although this precautionary argument has wide applicability it has limited force. It is based on estimates of the potential value of a given element of biological diversity which must be balanced against the actual cost of maintaining it or refraining from destroying it. Thus, unless a given element is identified as vital, it must have a finite value and there must therefore come a point at which the projected costs required to maintain it will outweigh any probable benefits. The fact that these costs and benefits are rarely if ever precisely quantifiable means that such calculations will involve the estimation of probabilities and risks.

Conclusions on resource values

Experience and general ecological theory indicate that no single species is indispensable in maintaining basic ecological processes on a global scale and that, in general terms, the rarer a species is, the less likely it is to play an important ecological role on even a local level. In other words, every species has a finite resource value and, although in some cases this value may be very high, in the case of increasingly rare species it tends to zero.

Similarly, with respect to species which may be directly useful to man, chiefly as food and pharmaceuticals, the vast majority of species can be said with high probability to have little potential. Experience enables us to identify those groups of taxa where there is a higher probability of value (e.g. wild relatives of crop species, and certain plant families for pharmaceuticals).

General conclusions to be drawn from the above discussion may be that *considering species only as material resources*, it would be more cost-effective to:

- maintain systems and areas rich in species than those poor in species
- maintain those known to be useful, or regarded as having a high probability of being useful, than to maintain other species.

These conclusions indicate that resource values of biodiversity, and in particular the cost-benefit approach to conservation, do not of themselves provide justification for the wide-ranging approach to biodiversity conservation that many seek to pursue. Such arguments must have limited applicability and limited force, and considerable caution must be exercised when citing them, especially when extrapolating from the particular (the rationale for maintaining particular species or a certain level of biological diversity) to the general (that all biological diversity is inherently valuable as a resource and must therefore be preserved).

Biodiversity and the biosphere

Human activities are affecting the biosphere on a global scale. It is important in the present context to establish the extent to which losses in biological diversity may contribute to these changes in having an impact on man.

One of the most obvious of such global changes is the perturbation of the carbon cycle, leading to a steady increase in atmospheric CO_2 levels. This will probably have far-reaching, although at present unpredictable, effects on global climate patterns which may in turn have serious consequences for human welfare.

A significant part of this is ascribable to industrial processes, especially the burning of fossil hydrocarbon fuels for energy generation. However, it is believed that alteration of existing natural or semi-natural ecosystems is also important. In particular the large-scale destruction of tropical moist forests is implicated, both in contributing to atmospheric CO_2 through burning and in decreasing the carbon-fixing potential of the biosphere. The high risk of serious consequences for humans of global climate changes is itself a strong argument for decreasing rates of forest clearance. It must, however, be stressed that this argument applies to tropical moist forest as 'forest', rather than as 'a highly diverse ecosystem'. Diversity is important only to the extent that it contributes to the system functioning as a carbon sink and the argument applies equally to other systems with a similarly high capacity for carbon fixation, such as tropical freshwater swamps, although these are far less diverse than tropical moist forest. In more general terms, there appears to be no direct or obvious link between the importance of an ecosystem in maintaining essential global ecological processes and its diversity, although more research is required.

Non-resource values of biological diversity

It is evident that resource-based arguments for the maintenance of biological diversity have very considerable but finite force; therefore any fundamental justification for striving to maintain *all* currently existing biological diversity must lie outside the realm of material resource values. Such justification usually devolves onto two principles - ethics and aesthetics - which themselves lie outside the realm of science.

Ethics

For some cultures, ethical beliefs provide the strongest grounds for maintaining biological diversity, and indeed in some eastern countries much of the remaining diversity in densely populated areas can be attributed directly to religious practices. However, without recourse to an absolutist moral code, it is difficult to argue compellingly for an ethical imperative for the maintenance of *all* existing biological diversity. Whilst the killing of any living organism may be morally unacceptable to some people, there are problems in extending this argument to the conservation of biological diversity. At an extreme level, any individual organism that is not genetically identical to another represents a facet of this diversity, and a strict ethical argument would proscribe its destruction. It may be understandable to object to the killing of an elephant on moral grounds, but is it any less moral to eat wheat, which is grown from genetically diverse seeds, than to eat potatoes, most of which are grown from genetically identical clones? Similarly, there are difficulties in demonstrating that a species, which is to some extent a human construct, has any greater 'right' to existence as an entity than any one of the individuals of which it is comprised.

Neverthess, the fact remains that ethics provides a powerful argument against the destruction of biological diversity. In practice, this argument is often contingent on other grounds, particularly the precautionary principle. For example, it may be considered immoral to destroy something which is now, or may be in the future, regarded as valuable to others. This is embodied in the 'stewardship' argument. The principle of inter-generational responsibility underpins the ethical case for conservation in the developed world, although it may be of little practical relevance to a desperate farmer faced with the reality of survival in a developing country.

Aesthetics

Arguments for the maintenance of biological diversity for its aesthetic appeal are compelling but have limited force, as they must be dependent on relative aesthetic judgements. Such judgements could presumably discard some organisms (those not visible, for example) as not worthy of being maintained. They are also unlikely to hold sway in the face of counter arguments that certainly exist for the destruction in the wild of harmful organisms, such as malarial *Plasmodium* species. Further, because genetic diversity is not susceptible to aesthetic appreciation, aesthetic criteria can be applied only to species and ecosystem aspects of biodiversity.

Regardless of individual aesthetic judgements, it is undoubtedly the case that humans very strongly favour variety in most areas of their experience. This need is particularly evident in the realm of the natural world. That is, diversity itself, and biological diversity in particular, is held in some poorly-definable but fundamental sense to be a highly desirable phenomenon. This is no mere notion, but a need that is very deeply felt, and a fundamental part of the spiritual life of many people. It is not important that the reasons for this cannot be fully articulated; the need is strongly manifest and should have force in determining action.

Overall, while it is evident that neither ethical nor aesthetic arguments provide of themselves sufficient grounds for attempting to maintain all existing biological diversity, a more general and pragmatic approach recognises that different but equally valid arguments (resource values, precautionary values, ethics and aesthetics, and simple self-interest) apply in different cases, and between them provide an overwhelmingly powerful case for biodiversity conservation.

PART 1

BIOLOGICAL DIVERSITY

Part 1 introduces some of the principal elements comprising biological diversity. Where appropriate, it discusses the ways in which they are measured, their patterns of distribution in space and the changes they have undergone over time, and notes their ecological importance. The main emphasis is on diversity at the species level. The chapters in Part 1 are grouped into four sections.

The first section (Chapters 1-4) is concerned broadly with the science of systematics as the primary approach to biodiversity. The opening three chapters cover: genetic diversity among species and populations, the scope and practice of systematics, and the meanings of the word 'species'. Although most debate about biodiversity has been in terms of species, it is important to recognise that the 'species' is not a standard unit; the way species are defined differs between groups and between taxonomists. The fourth chapter deals in considerable detail with the complex topic of global species numbers: how many species have been named and how many species probably exist but are as yet unknown and undescribed? There is considerable uncertainty about the number of valid described species, and extreme uncertainty about the global species total: conservative working estimates suggest 1.7 million described species and 12.5 million in total (estimates of the latter range up to 100 million).

The second section (Chapters 5-15) presents a review of biodiversity at the species level. Chapter 5 provides a general introduction to the subject of species diversity, while Chapters 6 to 14 present a series of case studies of different taxonomic or ecological groups. Many of the data sets presented here are entirely new. No attempt has been made systematically to cover all organisms in a consistent manner. The groups included and the kinds of data presented have to a great extent been dictated by the availability of information and expertise, although we have tried to cover some groups and communities that are less familiar, or highly diverse, or both. Species richness of tropical forest insects is discussed at length in Chapter 4, along with an outline of sampling procedures which could result in much-improved data on their distribution. Groups that have not received detailed review will be considered in future editions of this report.

In this section, Chapter 8, on ferns, gymnosperms and flowering plants, and Chapter 13, on vertebrates (excluding fishes), include large data tables which attempt to give an estimate, for each major group, of the total number of species in each country of the world, and an estimate of the number endemic (restricted) to each country. Chapter 12 includes data tables of freshwater fish species number and endemism in rivers and lakes.

This section closes with a discussion (Chapter 15) of some of the ways in which data on species distribution can be analysed to identify sites or areas which are particularly rich in species or contain a high proportion of endemic species. Conservation of these areas will be particularly important in efforts to maintain global biodiversity. This approach is illustrated by data derived from two global-level projects dealing with plants and with birds.

The third section contains two chapters which deal with trends in species diversity over time. Chapter 16 introduces the phenomenon of extinction; while extinctions in palaeontological time are discussed, the main emphasis is on historical and recent extinction, and the problems of predicting current and future rates of species loss. An attempt to list the animal species known to have become extinct since 1600 is included in this chapter. Chapter 17 discusses species threatened with extinction, in particular those which have been assigned to one of the IUCN threatened species categories. It covers the taxonomic, habitat and geographic distribution of species listed by IUCN as threatened, and discusses the factors leading to population decline.

The fourth and final section moves on to look at the habitat and ecosystem level of biodiversity. The opening chapter (18) introduces the theme of global community classifications, and notes some of the conceptual and practical difficulties which hinder their construction. Chapter 19 briefly outlines evidence for global climate change, and its predicted impact on protected areas. Both these chapters are illustrated by full colour maps.

Chapters 20 to 24 in turn cover five ecosystem types: tropical rain forest, grassland, wetlands, coral reefs and mangrove forest. A selection of systems which are species-rich or under particular threat have been included; no attempt has been made systematically to review all ecosystems (others will be included in future editions). Chapter 20, on tropical forests, discusses in some detail the various attempts that have been made to estimate the rate at which this habitat is being modified, and the difficulties inherent in such estimation. This should be read in conjunction with Chapter 4, on species inventory, and Chapter 16, which in discussing estimates of current and future rates of extinction, notes that no precise quantitative link can be made between species number in tropical forests, rates of forest loss, and rates of species extinction.

1. GENETIC DIVERSITY

This section introduces concepts from genetics necessary for an understanding of the generation and maintenance of biological diversity.

THE NATURE AND ORIGIN OF GENETIC VARIATION

Genes are the blueprints that make us and all the other organisms around us what we are. They consist of a discrete segment of deoxyribonucleic acid (DNA), a linear molecule composed of sequences of four different nucleotide bases. From the seemingly simple code contained in the sequence of these four bases of DNA comes the overwhelming complexity and diversity of the living world.

Living organisms can be divided very broadly into *eukaryotes*, in which the cell nucleus is bounded by a membrane, contains a number of organelles, and has its DNA combined with proteins to form chromosomes, and *prokaryotes*, in which these features are lacking. All higher organisms are eukaryotes; bacteria are prokaryotes.

Bacteria generally have a single copy of each of their genes located on a single piece of DNA and usually they tend to reproduce asexually, that is without the coming together of genetic information from another individual. Sometimes bacteria obtain some or all of the genetic material from other individuals in a process analogous to sexual reproduction in animals and plants. Thus, concepts of species developed principally with reference to higher organisms do not apply exactly to bacteria. Work is just beginning to characterise the nature and extent of genetic variation in a few bacteria. Given the huge diversity that has evolved over three billion years it is not surprising that bacteria appear to be a very complex group.

Genes are arranged linearly along the DNA and in most eukaryote organisms there are something like 50,000 of them. The actual quantity of DNA in each cell of different species of eukaryotes varies over three orders of magnitude (Fig. 1.1). Much of this DNA is not coding for anything and it is still an active area of research to understand what, if anything, all this apparently 'extra' DNA is doing. Our ignorance of its function, however, does not stop it from being useful for answering some kinds of questions, as discussed below. In most of the organisms we can see with the naked eye (animals and plants) the DNA of a cell is divided among a number of chromosomes. Humans have 23 different chromosomes. These chromosomes generally exist in two copies within each cell of the body and the organism is then said to be diploid; thus humans have a total of 46 chromosomes per cell. For the majority of organisms, which have sexual reproduction, one of these copies comes from the mother and the other from the father. Sex in genetic terms is just this, the coming together of genetic information from separate individuals. In this way genetic differences from different individuals may be combined in their offspring to produce new combinations upon which evolutionary processes can work. Asexually reproducing organisms must wait for the occurrence of different mutations in the same lineage to achieve these new combinations of genes.

Mutations are changes in the DNA. They occur in many ways. Mutations produce variation and variation is the raw material of evolution. The same gene can exist in a number of variants and these variants are called ***alleles***. If the two copies of a particular gene possessed by an individual are different alleles, the individual is said to be heterozygous at that gene. If the two copies are the same allele the individual is homozygous at that gene. A population of a species that has more than one allelic form of a particular gene is said to be polymorphic for that gene. If there are two alleles for a gene there are two possible homozygotes and one heterozygote. If there are three alleles, there are three homozygotes and three heterozygotes. For four alleles there are four homozygotes and six heterozygotes, and so on. Now consider the possibilities when we look at two polymorphic genes, and three, and on to the thousands that are polymorphic in most outbreeding organisms.

The number of possible combinations is vast - much larger than the number of individuals making up a species. This is the variation that the evolutionary process works on, and that provides the production attributes which agricultural development seeks to incorporate into crop varieties and livestock breeds.

The material below considers what is known of the implications of all this variation, how it changes and spreads, and the effects of human activities on genetic diversity and evolutionary processes.

MEASURING GENETIC VARIATION

Measurements of genetic variation are useful for studies of two broad classes of problems. One of these is the testing of theories about the nature of the forces acting on genetic variants - the nuts and bolts of evolution. There is a large body of mathematical and statistical theory about the genetics of populations, the basis of which was formulated by 1930. Only now, with the advent of DNA technology, do we have sufficiently powerful tools to begin rigorously testing these theories and their more recent elaborations. The other class of problems uses measures of genetic variation as a tool for understanding relationships among organisms and the diversity within and divergence between them.

There are necessarily important connections between the two sets of problems. Indeed, the central debate in evolutionary genetics is about whether most of the genetic variation seen in natural populations is maintained by natural selection or is neutral and therefore is subject only to the laws of chance. The issues at stake in this debate are crucial to the understanding of the mechanisms of the evolutionary process but they are not so important in the very practical matters of assessing differences between individuals, populations and species that are our main concerns here.

Allozymes

The first widely applicable technique for measuring genetic variation does so at one remove from the DNA itself. This

Figure 1.1 Range of DNA content in eukaryote organisms

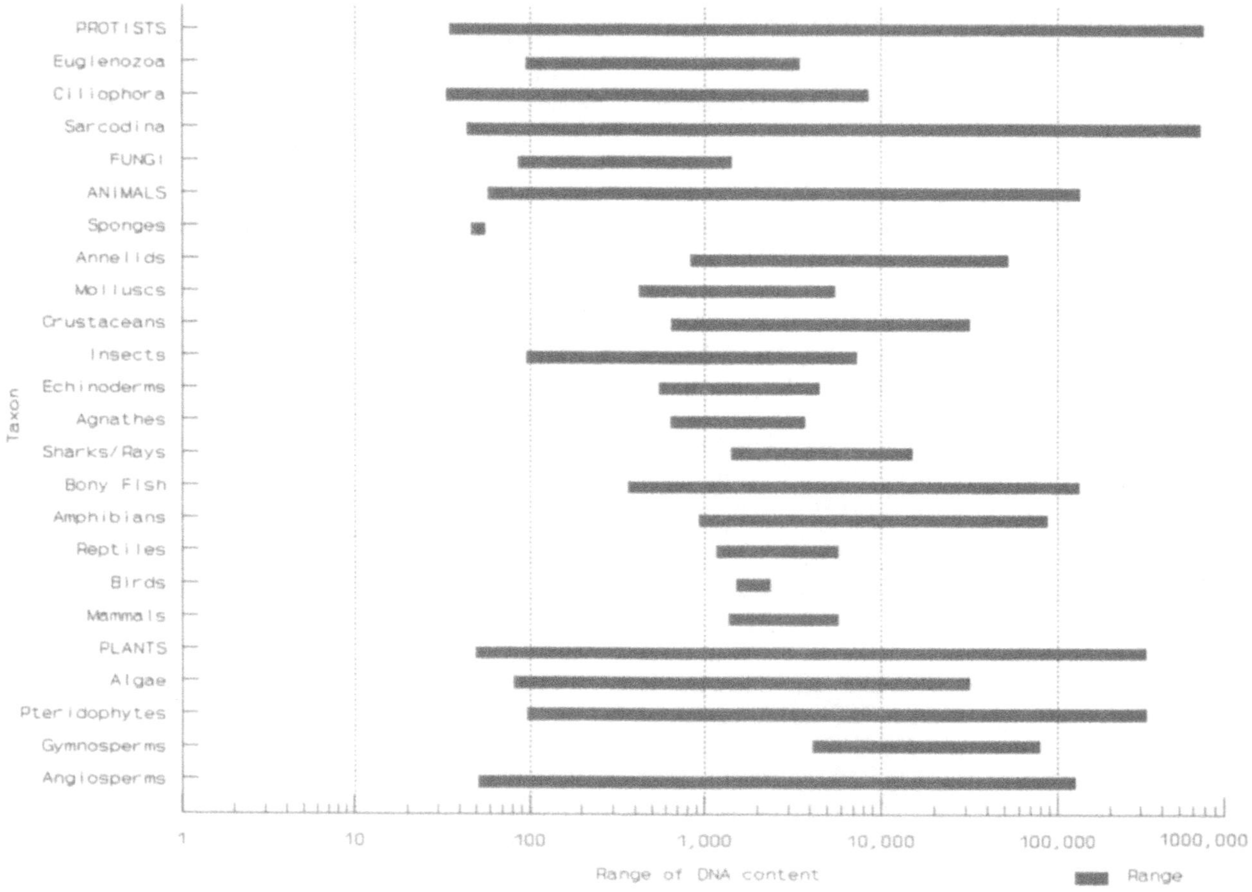

Source: from data tabulated by Li, W. and Graur, D. 1991. *Fundamentals of Molecular Evolution*. Sinauer Associates Inc., Sunderland, Mass.

technique is protein electrophoresis and it depends on the differences in electrical charge between variants of specific enzymes (allozymes) coded for by DNA. These charge variants migrate at different rates in gels subjected to an electric field and can therefore be differentiated from one another. It was this method that first revealed that on average 20-30% of the proteins of most organisms exist in more than one allelic form. This level of variation was not expected and the search for adequate explanations for it has been a major force in evolutionary genetics for more than 20 years.

By measuring the frequencies of different variants in groups of individuals sampled from different areas we can quantify the amount of variation within and between individuals and thereby get a picture of the geographic structure of the species in genetic terms. Not all changes in DNA result in a charge change which allows variants to be separated on a gel but this method still provides a good approximation to changes at the DNA level, at least within species and between closely related species. At greater taxonomic distances the probability of two different variants showing the same mobility in the gel system becomes high enough for the method to break down. Nevertheless, analysis of allozyme frequencies still has an important role to play in the study of intraspecific variation and the bulk of available data on genetic variation comes from studies of frequencies of electrophoretic variants of a number of enzymes.

Very recently development of DNA technology has provided us with the means to sample genetic variation directly at the DNA level. At present, however, these methods are more expensive and generally more difficult to perform than allozyme techniques. Therefore there are not yet the large amounts of data on within-species variation available from allozyme studies. This situation is changing rapidly as the necessary technology becomes more widely available and less expensive. The following sections give brief descriptions of the main techniques of use in phylogenetic and population genetic studies.

Because different parts of the DNA evolve at different rates we can choose to study particular segments to answer particular questions. Some genes change very slowly and can be used to study relationships among groups of organisms which diverged from one another hundreds or even thousands of millions of years ago. Other regions of DNA change at such a rapid rate that every individual in a population, except for identical twins and other such clones, is distinct. Still other regions of DNA show intermediate levels of variability which are useful for studies of variation within and between populations of a species, or of variation between closely related species.

Restriction fragment polymorphisms (RFLPs)

The DNA-based techniques most widely used for studies of

within- and between-population variation make use of the properties of enzymes derived from various species of bacteria which use them to protect themselves from infection by viruses by cutting (restricting) invading viral DNA. These restriction enzymes are very specific in the DNA sequence they recognise and cut, and they form the backbone of the technology of DNA manipulation. If DNA from an individual is extracted and cut with a restriction enzyme and the resulting fragments separated by length in an electrophoretic gel a pattern is obtained. Another individual may have a change in its DNA which produces an additional site recognised by the enzyme, or it might have changed in such a way that a recognition site has disappeared, thereby changing the pattern of restriction fragments seen on a gel. By repeating this process with other individuals and restriction enzymes, patterns of variation can be seen and analyzed to estimate the amount of variation in the DNA sequences among the individuals. These restriction fragment length polymorphisms (RFLPs) are very useful for determining the geographic structure of populations. By measuring the frequencies of different patterns in populations of a species we can estimate the amount of gene flow or genetic cohesion among the populations.

DNA sequencing and the polymerase chain reaction

Another more powerful (and more expensive) method of assessing genetic variation is to sequence a portion of the DNA itself. With the advent of the polymerase chain reaction technique (PCR), which can be used to make millions of copies of a particular region of DNA, it is now possible to obtain DNA sequence data from a wider variety of organisms much more quickly than was possible previously. The exquisite sensitivity of the PCR permits the amplification of a sequence from minute amounts of starting material - as little as a single cell. This has very important implications for obtaining data from very small organisms which contain too little tissue to use with RFLPs, and from larger organisms without having to kill or otherwise injure them. A minute drop of blood or a hair root or a feather are now adequate material for DNA sequence-based work. This has obvious importance in dealing with rare and endangered species.

THE INTERPRETATION OF VARIATION

Different measures of variation can be used to investigate relationships ranging from very distant groups, such as phyla, to closely related individuals within a population.

Often an understanding of relationships among closely related individuals is necessary for understanding behaviour and evolutionary processes within a species. Similarly, with breeding programmes for endangered species it is important to know the degree of genetic relatedness of individuals so that deleterious effects from inbreeding can be minimised. The technique of genetic fingerprinting can provide this information. Fingerprinting makes use of a common but peculiar group of DNA sequences known as minisatellites. These are dispersed throughout the genome and consist of tandemly repeated copies of short sequence units. High levels of variation in the numbers of these repeated units are exploited in fingerprinting to identify close relatives.

Biologists have long wanted to know if the genetic differences between species were of a different sort from the differences between individuals within a species. The answer appears to be that interspecific differences are not different in kind from intraspecific variation. Animal species usually differ at a large number of genes; single mutations are seldom, if ever, responsible for speciation events. The genetics of speciation is not discussed here although information on genetic distance between species-level populations in selected vertebrate genera, derived from methods outlined above, is show in Fig. 1.2.

In an outbreeding species every individual has a unique combination of alleles and the shuffling of genes that occurs in sexual reproduction insures that every future individual will be unique as well. If every individual is unique, what use are genetic data in making decisions about conservation problems? This question gets us to the heart of some fundamental problems in biology. Our knowledge of how a genotype is translated into a phenotype, a body, is very sketchy and this is an area of major research effort in biology. Genetic criteria for uniqueness and justification for conservation are not simple problems. In the sections below we will outline some of the issues, the problems and prospects for the use of genetic data in conservation.

THE ENVIRONMENT AND THE DISTRIBUTION OF GENETIC VARIATION

The earth is not a homogeneous place. This obvious fact has profound implications for the ways in which organisms live and evolve and is very probably responsible for much of the diversity of life around us. Limitations of the extent of particular habitats and differences in the ways in which organisms get their livings contribute in part to the large differences in the amounts and distributions of genetic variation which we observe. The following sections describe some of the basics of population genetics theory.

Gene flow and range expansion

One organism's minor inconvenience to free movement can be another's insurmountable barrier. These barriers can be physical, as for an animal which cannot cross a small stream, or behavioural, as for a small rodent which refuses to cross a small gap between patches of forest, or a plant reliant on a particular species of animal for pollination or dispersal of its seeds. Behavioural traits can have a large influence on the distribution of variation within a species. Even organisms which range over vast areas of ocean can have very different genetic population structures as a result of behavioural differences. An example of the extremes possible are the North American Eel which inhabits streams along 4000km of coastline and the Humpback Whales of the North Pacific and North Atlantic Oceans. The eels migrate to the Sargasso Sea to reproduce as one massive population and as a consequence the individuals inhabiting streams show no geographic differentiation. Other fish species inhabiting the same streams, but which do not leave their home streams to spawn, show substantial genetic differentiation. Humpback Whales on the other hand show genetically distinct subpopulations within ocean basins despite their ability to roam over huge distances. This differentiation is apparently the result of female traditions in migratory destinations. The eels then show a very high rate of gene flow while the humpbacks have a low rate of gene flow among subpopulations, despite ranging over comparable areas.

These differences in rates of gene flow and population structuring have major effects on the course of evolution. A few broad generalisations are possible, though subject to all sorts of *caveats* in particular situations. Species inhabiting large geographic areas and showing high rates of gene flow show very little or no local differentiation. Conversely, species with low rates of gene flow are often divided into distinct populations. At least some of this distinctness represents adaptation to the local environment. Adaptations of this sort are familiar to us all in varieties of crop plants and domesticated animals which, as a result of artificial selection by humans, perform better in particular climates and agricultural regimes. Natural selection can work in a similar way in producing populations with adaptations to local conditions.

The Earth has only very recently (in geological and evolutionary terms) emerged from an ice age. This and other events in the planet's history have had, and continue to have, major effects on the nature and distribution of living things. Much of the northern hemisphere was under thick ice 10,000 years ago. Most of that ice is now gone and in its place is forest, prairie, lakes and tundra, all teeming with life which has managed to colonise these newly available habitats. Natural processes of change are still visibly occurring in these regions, suggesting that populations inhabiting them are not likely to be in genetic equilibrium. This means that patterns and amounts of genetic variation reflect historical factors as well as the present-day situation.

Genetically effective population size

The number of individuals we can count in a population at any given time can be a surprisingly deceptive measure of the size of that population in genetic terms. At one extreme are organisms with vegetative or asexual reproduction such as aspen where we can stand in a forest surrounded by genetically identical individuals and a large area can be populated by only a handful of clones. A number of other factors commonly found in nature tend to reduce the genetically effective size of populations below that of the observed census size. Organisms with limited dispersal abilities tend to mate with individuals who are more closely related to themselves than the average for the population at large. This inbreeding reduces the overall genetic variation of the population relative to what it would have been if individuals mated at random across the whole population. Variation in number of offspring produced by different individuals in a population produces the same effect. If some individuals have many offspring while others have few or none the genetic variation of the population is reduced relative-to what it would have been if everyone had the same number of offspring. Similarly, populations which fluctuate in size or pass through a bottleneck of small population size can also show reduced genetic variation relative to that expected, all else being equal. Population geneticists have developed mathematical formulae to take account of these complicating factors in order to express population sizes of different organisms in comparable terms - the genetically effective population size. All these factors, and others too, can be operating and indicate the complexities of understanding genetic population structure of natural populations. We now have the tools with which to study this structure. Much remains to be done before we can hope to have a deep understanding of the structures of natural populations.

One of the most dramatic examples of the potential discord between our visual impression of a species and its genetic reality is the Cheetah. This cat was until quite recently widely distributed throughout Africa and Asia. It has undergone a severe reduction in its range and numbers but is still found in widely separated areas of Africa. Recent surveys of genetic variation in the Cheetah have found almost no variation - individuals from widely separated parts of the species range are genetically almost identical. These results indicate a severe population bottleneck and subsequent inbreeding. Cheetahs both in the wild and in captive populations show pronounced effects of inbreeding not seen in other wide ranging carnivores. This inbreeding shows itself in reproductive difficulties such as very low numbers of sperm, many with morphological aberrations, and high susceptibility to epizootic diseases resulting from very low amounts of genetic diversity in their immune systems. The bottleneck responsible for these difficulties may well have been due to events following the retreat of the last ice sheet thousands of years ago. Human assaults on the Cheetah's range and numbers have certainly not aided its recovery from the effects of this bottleneck. Similar effects of inbreeding resulting from recent population bottlenecks are seen in relictual populations of lions in the Gir Forest Sanctuary of western India and in the Ngorongoro Crater in the Serengeti of Kenya.

Outbreeding depression is the converse of inbreeding depression. If individuals have differentiated genetically over their range, the mating of individuals from different parts of that range can result in deleterious effects. This is presumably because genes from one area do not necessarily work harmoniously with genes from another area. The experimental difficulties involved in trying to understand these effects are great but we do have observations attesting to their existence in a number of plant species. Several experiments have demonstrated an 'optimal outcrossing distance', that is fertilization by pollen from distances greater than the optimum results in reduced fitness, just as fertilization by pollen from individuals close by can result in inbreeding depression. There are a few dramatic examples of outbreeding depression in animal populations. In Czechoslovakia, Turkish and Nubian Ibex were mixed with the local Tatra Mountain Ibex and the hybrids were so poorly adapted that the entire population went extinct.

The genetic effects of habitat alteration and fragmentation

Human activities cause genetic changes in species by altering their population structures. Disruption of dispersal and migration routes and reduction of population sizes are the most obvious factors. As with natural processes, effects of particular activities vary depending on the species considered, some may be affected virtually not at all while

Figure 1.2 Means and ranges of genetic distance between species in selected vertebrate genera

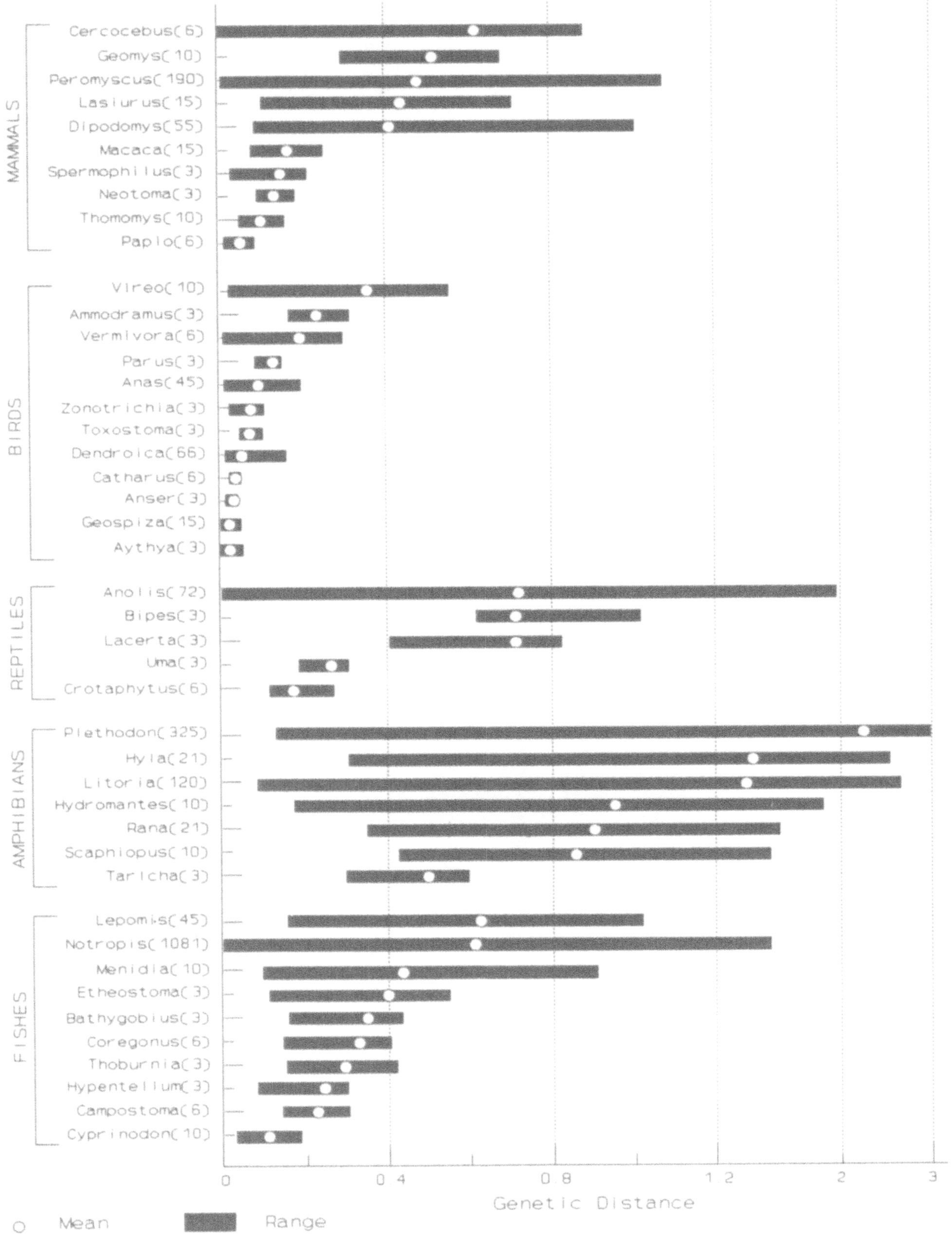

Note: The numbers of pairwise comparisons of species are in parentheses.
Source: Avise, J.C. and Aguardro, C.F. 1982. A comparative summary of genetic distances in the vertebrates. In: Hecht, M.K., Wallace, B. and Prance, G.T. (Eds), *Evolutionary Biology*. Volume 15. Plenum Press, New York.

others may be devastated. For example, many tropical forest trees occur at very low densities over wide areas and rely on particular species of insects or birds for pollination. Fragmentation of the forest results in very small numbers of individuals in each patch. If their pollinators are unable to cross the gaps between patches, severe inbreeding or failure to reproduce can result. So even if accidents of nature do not remove these rare individuals from isolated patches of forest, they are genetically speaking dead, despite appearances to the contrary. Other species in the same isolated patches of forest may maintain large genetically effective population sizes, either by being present in higher densities within patches or by having better dispersal abilities between patches, or both. Of course, not all species are adversely affected by habitat fragmentation, especially those dependent on 'edge' habitats such as where forests and open country meet. Species that thrive in these circumstances range from animals usually perceived as desirable like White-tailed Deer, to the vectors of a number of the most devastating human parasites and diseases.

The effects of small population size depend on the breeding system of the species and the duration of the bottleneck. If population size expands rapidly immediately after a bottleneck, relatively very little genetic variation will be lost. If the bottleneck lasts for many generations or recovery is very slow a great deal of variation can be lost. Of course a population which remains at a very small size for an extended period is very likely to go extinct as a result of demographic accidents, probably before deleterious genetic effects manifest themselves.

If habitat fragmentation eliminates gene flow between parts of a species' range these newly isolated populations have independent evolutionary futures. What this means for the long-term future for a species is difficult to predict. It is certainly time to put some serious effort into trying to find out.

CONCLUSION

The genetic diversity inherent in most species provides the raw material to respond rapidly to changed circumstances. This response may not always be adequate and it may not be in the best interests of humans, as when agricultural pests and human pathogens develop resistance to our control measures. Change is, of course, the normal state of affairs in the living world. What makes our present situation unique is the rapidity and scale of the change. Our fragmentation and destruction of habitats constitutes a massive uncontrolled experiment in ecology and genetics. We are beginning to understand in outline what needs to be done to mitigate at least some of the negative effects of this experiment. Knowledge of the population structures, i.e. the distribution and amount of genetic variation, of a wide range of organisms is necessary, as is a much deeper understanding of the biological significance of different sorts of variation.

References

Avise, J.C. and Aguardro, C.F. 1982. A comparative summary of genetic distances in the vertebrates. In: Hecht, M.K., Wallace, B. and Prance, G.T. (Eds), *Evolutionary Biology*. Volume 15. Plenum Press, New York.

Li, W.-H. and Graur, D. 1991. *Fundamentals of Molecular Evolution*. Sinauer Associates Inc., Sunderland, Mass.

Contributed by Richard Thomas, Molecular Biology Unit, The Natural History Museum (London).

2. SYSTEMATICS AND DIVERSITY

This chapter provides a short introduction to systematics: the branch of biological science responsible for recognising, comparing, classifying and naming the millions of different sorts of organisms that exist. As such, systematics provides the basic framework for the whole of biology, and is the fundamental discipline of biodiversity. The work can be divided into a number of activities, including classification, identification and nomenclature. These are often grouped as *taxonomy*, broadly defined as the classification and naming of organisms. This chapter gives the background for Chapter 3, which discusses some key theoretical and practical problems arising from the concept of *the species*.

BIOLOGICAL CLASSIFICATION

The ultimate task of systematics is to document and understand the extent and significance of biological diversity. Within this framework, taxonomy performs four basic functions: differentiation (recognition of taxa), identification (universal diagnosis of taxa), symbolisation (application of universal names), and comparison (relative relationships of taxa). Vernacular or folk taxonomies provide limited local systems for the first three but have little to tell us about the last.

Individuals and characters are the most basic units of biological classification. On the basis of features held in common (attributes or characters), individuals can be grouped together into a large number of different classes. These classes are of two kinds (often regarded as sharply distinct, although in reality they form a continuum). On the one hand, individual organisms can be divided into such groups as freshwater, marine, terrestrial, planktonic, nocturnal, pollinators, etc. Alternatively, they can be placed into taxonomic categories of species, genera, families, orders and so on. The former are regarded as *artificial* classes, constructed only to serve a particular purpose, whereas the latter are seen, ideally, as *natural* groups.

Natural groups comprise individuals with a very large number of attributes in common, whereas individuals belonging to artificial groups have relatively few shared characters. Thus the essential difference between, for example, 'marine animals' and Mammalia is that individuals of the latter class have far more in common than those of the former. A natural group, being based on a large number of characters, can be used for a far wider range of generalisations and predictions than an artificial group.

Artificial and general classifications are not restricted to biology. Biology, however, has a unique theory of its own, the theory of organic evolution. Ideas about evolution can be divided into a general theory of descent with modification and special theories about the processes of that descent (natural selection, neutral theory, etc.). Modern systematists consider that the general theory of evolution not only provides a compelling justification for seeking one natural, general classification for living organisms but also suggests the basis on which that classification can be most securely founded: the hierarchic pattern of the ancestor-descendant sequence, or *phylogenetic relationships*.

PHYLOGENETIC RELATIONSHIPS AND THEIR ESTIMATION

In the past, many biologists have denied that we have access to sufficient or appropriate information to determine the phylogenetic relationships of organisms. In the last 25 years, however, spectacular advances in such areas as molecular biology threaten to overwhelm us with suitable data. Moreover, during this same period great advances have also occurred in the theory of systematics and methods of data analysis.

In an absolute sense, being part of remote history, phylogenetic relationships cannot be known. What is done instead is to estimate the most basic feature of the ancestor-descendant sequence, the pattern of branching points or nodes. Relationships are defined in terms of common ancestry. If two species are considered to have a common ancestor which they do not share with a third species, then the first two are considered to be more closely related to each other than either is to the third. This represents the fundamental three-taxon problem, basic to all phylogenetic (or *cladistic*) analysis.

Cladistic analysis rests on three basic assumptions: features shared by organisms (homologies) form a hierarchic pattern; the hierarchic pattern can be expressed by branching diagrams (cladograms); and the nodes in a cladogram symbolise the homologies shared by the organisms subtended by that node (groups). Where data are in conflict (as they usually are, to a greater or lesser extent), parsimony is used to find the best supported or most efficient solution.

Cladistics differs from other methods of classification because, based on these principles, only special resemblances are used as evidence of relationship or group membership. This is in sharp contrast to methods such as phenetics, in which all resemblances, including character absences, are regarded as equally informative. Some of the principles involved here are illustrated in Fig. 2.1. Cladistics has been at the centre of heated debate, but is now widely acknowledged to be the best way of approximating the branching patterns of phylogenetic history.

FROM HIERARCHY TO CLASSIFICATION

Once a justified hierarchy of phylogenetic relationships has been established, what relationship should exist between the hierarchy and classification? Organisms are divided into kingdoms (animals, plants, etc.), kingdoms into phyla (Arthropoda, Chordata), phyla into classes (Crustacea, Mammalia), classes into orders (Decapoda, Rodentia), families (Cancridae, Muridae), genera (*Cancer*, *Rattus*) and species (*Rattus norvegicus*, *Rattus rattus*). Each group contains the entirety of one or more groups at a lower level. The categories most often used are shown in Table 2.1, and see Fig. 2.2. Multiple membership of categories is not permitted (thus an organism cannot belong to two or more orders, genera or species at once, with the possible exception of hybrids).

Figure 2.1 Establishing the phylogenetic hierarchy

Taxa	Characters 0	1	2	3	4	5	6	7
MAN	-	G	L	S	D	G	V	L
ALLIGATOR	M	E	L	S	D	Q	V	L
TUNA FISH	-	-	-	-	-	A	V	L
SHARK	-	-	-	-	-	T	V	N

Notes: Establishing the phylogenetic hierarchy - overall similarity or special resemblance? In this simple example of a four taxon problem, the characters are the amino acids (guanine, lucine, serine etc) found to vary at eight homologous positions in the amino acid sequence of myoglobin A in man (taxon A), an alligator (B), a tuna fish (C) and a shark (D) (note that the positions 0-4 in man and alligator have no equivalent in tune or shark, and that position 0 is also represented in man). If both presences and absences are counted as equally informative, the branching diagram on the left most efficiently summarises the data, but if only presences (special resemblances) are counted, the cladogram on the right is best. The first solution is that of the phenetic school of classification, which would continue to accept the 'fish' as a natural group. The second solution is that of the cladistic school of classification, which would wish to recognise that, in terms of recency of common ancestry, there is good evidence that the tuna fish is more closely related to man and alligator than it is to the early-diverging shark. As a result, the fish is seen to be a paraphyletic group, of little or no value on natural (phylogenetic) classification (see also text; based on Patterson, 1980).

Groups which, on the evidence of shared unique characters (special resemblances), are considered to contain all the living descendants of a common ancestor are called monophyletic groups; the mammals are an example. Use of characters which have evolved more than once leads to the formation of *polyphyletic* groups (groups of organisms which have multiple origins, such as placing birds and turtles together because they have beaks). If groups are formed on the basis of unspecialised or non-unique characters, such as reptiles (which can only be recognised collectively as members of the amniote vertebrates that are not mammals or birds), these are termed *paraphyletic* groups.

These distinctions are important because they relate to a continuing debate over the relationship between genealogical hierarchy (as discovered by cladistic analysis of taxonomic characters) and formal classification. Most taxonomists agree that polyphyletic groups once recognised should be abandoned (although some remain in use, such as lophophores, a false grouping of the animal phyla Phoronida, Ectoprocta and Brachiopoda). But many paraphyletic groups continue to be very widely used, such as the invertebrates (Metazoa minus Chordata), fish (Chordata minus Tetrapoda) and Reptilia (Amniota minus birds and mammals).

Nevertheless, Darwin's view that our classifications should correspond to genealogies is becoming more and more widely accepted. In the last 2-3 decades much progress has been made in discovering the phylogenetic relationships of organisms but far more needs to be done. In what follows it is therefore necessary to appreciate the ideal of hierarchic classification based on phylogenetic relationships and the compromise that most existing classifications still represent.

Table 2.1 The taxonomic hierarchy

KINGDOM
DIVISION (Botany) or PHYLUM (Zoology)
CLASS
ORDER
FAMILY
GENUS
SPECIES
  Subspecies
    Variety (Botany)
    Form (Botany)

Note: The categories of the taxonomic hierarchy in descending order of rank and inclusiveness. There are a few additional less commonly-used categories, (subphylum, superfamily, tribe, etc.).

Figure 2.2 Basic principles of classification

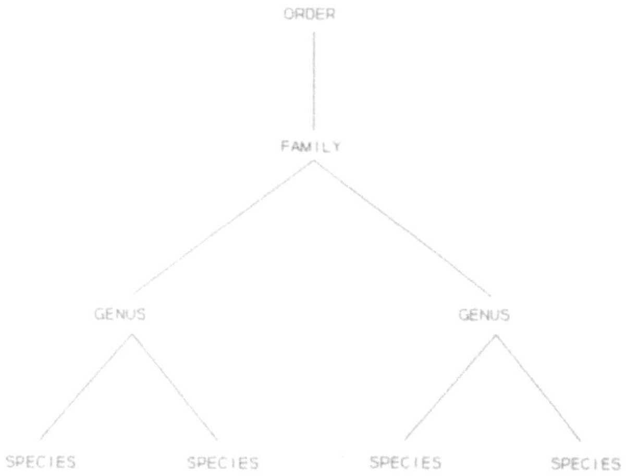

Notes: Diagram to illustrate the basic principles for turning a scheme of phylogenetic relationship into a classification (based on Goodrich, 1919, *The Evolution of Living Organisms*). According to Goodrich, "individuals are grouped into species, species into genera, these again into families, orders, classes and phyla, divisions of increasing size and importance ... the only "fixed points" in a phylogenetic system of classification are the points of bifurcation, where one branch diverges from another ... it is here that our divisions should be made." Goodrich's formulation of the problem remains appropriate today, except that to avoid mis-matches in the ranking of higher categories, division should proceed from top downward, not from the terminals (species) upwards. Simple principles of this sort allow us to translate the phylogenetic hierarchy into a classification hierarchy, although some compromises and exceptions are still widely accepted in practice, notably through continued recognition of paraphyletic groups such as invertebrates, fish and reptiles (see text), and problems created by hybridization (particularly in plant taxonomy).

TAXONOMIC NOMENCLATURE AND ITS REGULATION

A separate problem from classification, but often confused with it, is taxonomic nomenclature. The ultimate goal of scientific nomenclature is a universal system of unambiguous names for all recognised taxa. Scientific names are basic to all biology, and biodiversity is no exception. In particular, their exact significance has important implications for conventions, red lists, export controls, licences or any other legal instruments used to manage biological diversity.

Taxonomic nomenclature is controlled by voluntary application of internationally agreed rules or codes. Separate codes apply to the animal kingdom, plants (including fungi), and bacteria. In this section the operation of the zoological code will be outlined and the other codes briefly compared by noting a few of their differences.

International Code of Zoological Nomenclature

The formation and application of names at the rank of species (including subspecies), genus and family are regulated by the *International Code of Zoological Nomenclature* (the Code), and by the use of type specimens. Cases in dispute are settled through submissions to the International Commission on Zoological Nomenclature (ICZN). Names in use below the rank of subspecies (for polymorphic forms, seasonal variations, hybrids etc.) and above the rank of superfamily (orders, classes, phyla etc.) lie outside the scope of the zoological code, and are simply regulated through usage. This might appear unsatisfactory, but in practice it gives rise to few difficulties. The major problems occur with the names of species and, to a lesser extent, genera and families.

The zoological code depends on two operational principles - *availability* and *priority* - and also governs the formation of names. To be considered nomenclaturally valid, a species name must be introduced in combination with a generic name, and in Latinised form. The species name follows the generic, never takes a capital, and is usually printed, as with the generic name, in italics (e.g. *Homo sapiens*, *Rattus norvegicus*, *Papilio machaon*). If a species is considered divisible into two or more taxonomically distinct subspecies, formal trinomens can be introduced. The subspecies including the population originally described is designated by tautonymy (*Papilio machaon machaon*); other subspecies receive distinguishing third names (*Papilio machaon britannicus*).

Availability

For a name of a subspecies, species, genus or family to be recognised within zoological nomenclature, a number of requirements must be met. If all these are satisfied, the name is said to be *available*; if not, the name is considered *unavailable* for the purposes of nomenclature. For a species name these requirements normally include: a statement that the name is proposed for a newly recognised species or subspecies; an indication of how the new taxon differs from other, related species; and proposal of the name in Latinised binominal form (i.e. the new species name must be proposed in combination with a generic name). These are some of the basic ingredients of the *description*, which must be properly published, in printed form.

Priority

The second basic principle is priority. If what is currently considered a single species, genus or family has received two or more available names independently, how would you choose between them? The basic principle of *priority* simply directs that, wherever possible or practical, the oldest or *senior* available name must be used. Binominal nomenclature for animals was first consistently introduced in the 10th edition of Linnaeus's *Systema Naturae*, published in 1758, and this gives a baseline for priority. For zoological nomenclature it is therefore unnecessary to consider names published in any work before 1758 (with the exception of a single work on spiders published in 1757).

Name, author and date

The two principles of availability and priority come together in the original published description. It is for this reason that, when a name is mentioned formally (as in a catalogue), the original author of the name and year-date of publication should also be mentioned; thus: *Papilio machaon* Linnaeus, 1758.

Types and their function

Species and other taxa are concepts about the organisation of the natural world, whereas names are artefacts, symbols intended to designate those concepts. As taxonomic concepts change, difficulties arise with the application of existing names. One of the commonest problems occurs when there are more names available than taxa to be designated. Which old names apply to which newly circumscribed taxa? Objectivity in the application of names is achieved by the use of *type specimens*.

The code strongly recommends that in original descriptions the author selects a particular specimen as the type (strictly *holotype*) and ensures that it is clearly so labelled and preserved in a permanent place (normally a museum) so that it can be studied again in the future. What is the purpose of such types? It is quite commonly supposed, by those unfamiliar with biological nomenclature, that the type specimen represents some sort of 'standard' (typical) for defining the species, perhaps analogous to the standard metre or standard kilogram used to calibrate rulers or weights. Nothing could be further from the truth. The type specimen is simply the name-bearer - it is the specimen to which the original name is attached. In cases of doubt over identification with a particular species concept, if you can decide to which concept the type specimen fits, then the name automatically follows. Where more than one name is found to apply, then priority will normally determine which one is to be used; the other names are *synonyms*.

Why do names change?

Everyone who makes regular use of biological classifications soon becomes aware that 'official' names can change. The instability of scientific names is irritating and, as conservation and wildlife trade legislation becomes more complex, can lead to real difficulties. Some systematists, embarrassed that instability gives taxonomy a bad name, have proposed that a stabilised 'official list' should be created (for one of the latest rounds of discussion, see Hawksworth, 1991).

Changes in nomenclature occur for two basic reasons: problems with names and their application (homonymy, synonymy, and misidentification, as normally decided by interpretation of the international code), and revisions of the system of classification necessary to reflect new scientific discoveries about taxa and their natural relationships. Frequently these problems are compounded. While responsible efforts to avoid 'unnecessary' changes brought about by slavish application of the code are to be encouraged (because taxonomy is a science to which nomenclature ought to be subservient), it is futile to imagine that some fixed, permanently stable list of names can be drawn up.

To insist on fixity would be far more damaging to biological science than to accept the minor irritation that, as our understanding of natural classification changes and steadily improves, it is necessary to adjust nomenclature accordingly. However, there are situations where automatic application of the code can lead to changes considered so unacceptable that the normal rulings of the code are best set aside. Such cases are submitted to *The International Commission on Zoological Nomenclature*, an international panel of experts in animal nomenclature whose role is to decide on the best action in such cases, and then publish their decisions through the *Bulletin of Zoological Nomenclature*.

International Code of Botanical Nomenclature

This code governs the names of fungi as well as green plants. The ICBN operates in a broadly similar way to the zoological code, but differs in many details. One obvious difference is the 'double citation' whereby, if there has been any change in taxonomic assignment or rank of a taxon since its original proposal, the name is formally to be cited with the original author's name in parentheses, followed by the name of the taxonomist who proposed the change.
Thus the plant known in English as the scentless mayweed was named by Linnaeus as *Matricaria inodora*. Later, it was moved by Schultz-Bipontinus to a separate genus, *Tripleurospermum*. This is the accepted name today, and its authority is formally quoted as *Tripleurospermum inodorum* (L.) Sch-Bip. Another difference is that tautonyms are not permitted for species names. Thus a name like *Bison bison*, acceptable under the zoological code, would not be acceptable in botany. (Tautonymous names below the rank of species do, however, occur in botany, being created automatically when plant species are first named; these so-called *antonyms* apply to varieties and subspecies.) Unlike zoological nomenclature, to establish a valid botanical name it is essential that the original description includes a Latin diagnosis.

Cultivars are specifically the subject of an additional code, the International Code of Nomenclature for Cultivated Plants. Because of biological and other peculiarities, a number of special provisions also apply to fungi, lichens, plant hybrids and certain other groupings. One example is that the name of a lichen is taken to apply to the fungal part, should it be necessary to consider priority over the application of names to its constituent algal or fungal elements. At a more fundamental level, there are subtle but important differences between the botanical and zoological codes regarding availability and the significance of types. Changes in the botanical code, and appeals against the strict application of its provisions, must be directed to the Nomenclature Section of an International Botanical Congress, for decision in plenary session.

Codes for the nomenclature of bacteria, actinomycetes, and viruses

Names for bacteria and actinomycetes are controlled by the ICNB, the *International Code of Nomenclature for Bacteria*, itself controlled by the International Committee for Systematic Bacteriology. In some respects the bacterial code is similar to the botanical code (e.g. double citation) but there are many differences in detail. A particularly important development occurred recently when the nomenclatural starting date for all bacteria was revised to 1 January 1980, to coincide with publication of the *Approved List of Bacterial Names* (Skerman, McGovern and Sneath, 1980).

The names of viruses present exceptional difficulties, and no international or standard system has been followed. During the 1966 International Congress for Microbiology the problem was addressed by an International Committee on Nomenclature of Viruses (ICNV). This produced a report, *Classification and Nomenclature of Viruses* (Wildy, 1971), including recommendations for rules. Since then the ICNV has become the International Committee on Taxonomy of Viruses (ICTV), revising and re-revising the rules and recommendations of Wildy's report. An almost complete statement is to be found in Matthews' (1979) report, *Classification and Nomenclature of Viruses*, the nearest approach yet to an international code for viral nomenclature.

MAJOR FEATURES OF THE HIERARCHY OF LIFE

The evidence of molecular biology, notably the universality of the genetic code, strongly favours the idea that all modern life on Earth is monophyletic.

Ernst Haeckel (1866) was amongst the first to recognise the enormous diversity of bacteria and other unicellular organisms, separating many of these life forms (together with many others that would no longer be included) into a major group, the Protista, equal in rank to the plants and animals. This group is no longer formally recognised; some 'protists' are currently classified amongst the prokaryotes. This basal, paraphyletic assemblage comprises the eubacteria (for which there is good evidence of monophyly) and archaebacteria (which may or may not form a natural group). The prokaryotes represent an evolutionary grade in which DNA is not organised within a nuclear envelope.

The higher organisms, the eukaryotes, form a clade characterised by possession of a double nuclear membrane. The eukaryote clade includes the three major groups of macro-organisms, the green plants, fungi and animals, together with many unicellular and other simple organisms now often referred to as 'protists'. The protists include the myxomycetes (slime moulds), protozoans and various groups of algae, including green algae (chlorophytes), chromists (chrysophytes, golden brown algae etc.), and rhodophytes or red algae; the chlorophytes form a monophyletic group with the green plants (Bremer, 1985).

Plants
A major group, comprising the green algae and the land plant kingdom, can be recognised as a natural group. Of three primary divisions, the Chlorophyta (green algae) comprise a complex paraphyletic group from within which the land plants (embryophytes) have arisen. The most basal groups of land plants are the liverworts and hornworts, and then the mosses. The next level of organisation is represented by the tracheopytes (characterised by the possession of vascular tissue), including lycopods, horsetails and ferns. Beyond this level are the seed plants (spermatophytes), including cycads, *Ginkgo*, conifers, a group comprised of *Ephedra, Gnetum* and *Welwitschia*, and finally the flowering plants (angiosperms). The angiosperms are a vast and complex assemblage, traditionally divided into the monocotyledons (probably monophyletic) and the dicotyledons (paraphyletic).

Fungi
The fungi form a major kingdom, divisible into the Oomycetes and the true fungi, the Eumycota. According to Tehler (1988), the true fungi (identifiable as a natural group on the basis of 25S RNA and chitin cell walls) can be divided into four divisions, one of which includes the Dicaryomycotina. The dicaryomycetes are themselves divided into three classes: the Ascomycotina (moulds, yeasts), Protobasidiomycotina, and Basidiomycotina (smuts, rusts, bracket fungi, mushrooms, toadstools). A number of poorly-known fungal groups probably do not fit into this scheme, but the 'fungi-imperfecti' (Deuteromycotina) are an unnatural assemblage of forms (including many moulds) unknown in their sexual stage, most of which are believed to be non-sexual stages of ascomycetes and basidiomycetes.

Animals
The higher, multicellular animals (Mesozoa and Metazoa) are usually regarded as monophyletic, the principal basal members being the mesozoans and poriferans (sponges), followed by coelenterates (jelly fish and cnidarians) and platyhelminths (flatworms). The molluscs, arthropods (including insects), echinoderms (starfish, sea urchins etc.) and vertebrates are conventionally grouped together at the apex of the animal hierarchy.

In conclusion, although some major features are discernible, our knowledge of the hierarchical pattern of life, even at this most general level, appears very limited. However, new molecular evidence, such as the 18S rRNA data studied by Lake (1990) and others, holds the promise of yielding far greater understanding. Margulis and Schwartz (1988) should be consulted for further information on all the recognised phyla of organisms, their biology, relationships and taxonomy.

SYSTEMATICS AND THE MEASUREMENT OF BIODIVERSITY

Ecologists have measured diversity either by estimating species richness (number of species) in an area, or by one or more indexes combining species richness and relative abundance within an area. Some attempts have also been made to measure change in species richness (species turnover) between areas. These solutions to the problem of measuring biodiversity are limited because species richness takes no account of the differences between species in relation to their place in the natural hierarchy, and because relative abundance is not a fixed property of species, varying widely from time to time and place to place. Furthermore, in many environments most taxa are virtually or even completely unknown.

For some time conservationists have called for a measurement of diversity more clearly related to overall genetic difference. For example, regarding the problem of differential extinction, IUCN/UNEP/WWF (1980) noted that "the size of the potential genetic loss is related to the taxonomic hierarchy because ... different positions in this hierarchy reflect greater or lesser degrees of genetic difference ... the current taxonomic hierarchy provides the only convenient rule of thumb for determining the relative size of a potential loss of genetic material."

Measurements of diversity are now being proposed that either attempt to measure genetic difference directly, or indirectly through use of the taxonomic (cladistic) hierarchy (Williams *et al.*, 1991; Faith, in press). Apart from scientific debate still not fully resolved, the latter approach is more practical because we already have a "rule of thumb" taxonomic hierarchy (which is being steadily improved through the application of cladistic analysis, notably to molecular data), whereas reliable estimates of overall genetic differences between taxa are virtually non-existent.

Based on the shared and unshared nodes between taxa (equivalent to position in the taxonomic hierarchy), a number of taxonomic diversity indices have now been developed. Of these, the most distinct are root weight, higher taxon richness and taxonomic dispersion. The first places highest individual value on taxa which separate closest to the root of the cladogram and comprise only one or relatively few species; in effect this gives high weighting to relict groups. Higher taxon richness favours taxa according to their rank and number of included species. Dispersion, the most complex of the measures proposed so far (Williams *et al.*, 1991), endeavours to select an even spread of taxa across the hierarchy, sampling a mixture of high, low and intermediate ranking groups. See Fig. 2.3 for illustration of these concepts.

For a given group these measures, together with simple species richness if desired, can be used to compare the biotic diversity of any number of sites. The measures can also be expressed as percentages. Thus a site with viable populations of all species in a group would have a diversity score of 100%, while a site without any species of the group in question would score zero. In reality, of course, most sites have only a selection of species, and so receive various intermediate scores.

Such assessments allow us to compare all sites with each other, and rank them individually from highest to lowest

Figure 2.3 Measures of biodiversity

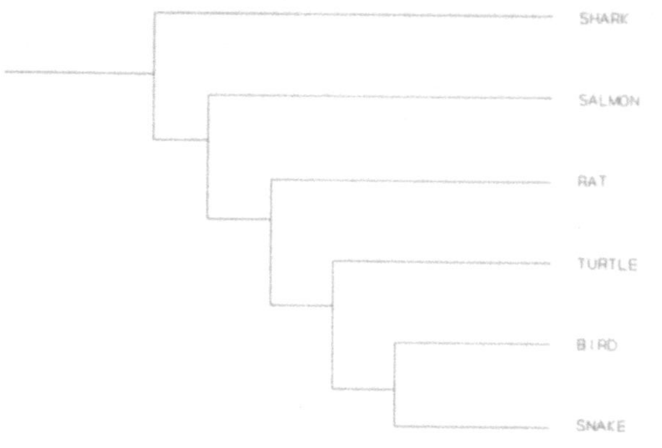

Notes: The practical need for measures of biodiversity. Assume there is a small zoo keeping six species of vertebrates: a shark, a bony fish (salmon), a rat, a turtle, a bird and a snake but only half can be maintained in future (each one costs the same). If the objective is to display as 'good' a sample of biodiversity as possible, which three should be selected? Accepting the phylogenetic relationships in the diagram, species richness offers no help - all 20 possible choices are the same. Taxic diversity measures will help us choose, but the result will be dependent on which index we use. Root-weight selects shark, bony fish and rat. Higher taxon richness selects a shark, bony fish plus one of the remainder. Taxic dispersion chooses shark and rat plus bird or snake. Dispersion is probably the criterion that corresponds most closely to an intuitive notion of diversity.

diversity. However, if we then take some action on this (such as conserving a particular site), the same measures are unlikely to be directly comparable for making a second decision (such as choosing a second conservation site). This is because, in most real situations at least, there will be considerable overlap in the presence of species at particular sites.

In a seminal work on the measurement of diversity, Whittaker (1972) introduced the concepts of alpha, beta and gamma diversity. The measurements just described, giving diversity values for single sites, are examples of alpha diversity. The beta and gamma diversity concepts relate to changes in diversity between sites at local (beta) and geographical (gamma) scales. An essential part of these relational concepts is the idea of species turnover - the degree to which species present at one site are replaced by others at different sites. For use in assessing the relative value of multiple sites for the conservation of biodiversity, the idea of species turnover is translated into the principle of *complementarity*, implemented in combination with a taxonomic diversity index. This is returned to in Chapter 15.

References

Bremer, K. 1985. Summary of green plant phylogeny and classification. *Cladistics* 1:369-385.

Faith, D. (in press). Conservation evaluation and phylogenetic diversity. *Biological conservation.*

Goodrich, E.S. 1919. *The Evolution of Living Organisms.* Jack and Nelson, London.

Haeckel, E. 1866. *Generelle Morphologie der Organismen,* 2. Berlin.

Hawksworth, D.L. (Ed.) 1991. Improving the stability of names: needs and options. Koeltz Scientific Books, Koenigstein. (*Regnum Vegetabile* 123).

International Code of Nomenclature for Bacteria. 1975. American Society for Microbiology, Washington.

International Code of Botanical Nomenclature. 1988. International Association for Plant Taxonomy (Europe).

International Code of Nomenclature for Cultivated Plants. 1980. International Commission for the Nomenclature of Cultivated Plants, IUBS.

International Code of Zoological Nomenclature. 1985. International Trust for Zoological Nomenclature, London.

IUCN/UNEP/WWF 1980. *World Conservation Strategy: living resource conservation for sustainable development.* Gland, Switzerland.

Lake, J.A. 1990. Origin of the Metazoa. *Proceedings of the National Academy of Science USA* 87:763-766.

Margulis, L. and Schwartz, K.V. 1988. *Five Kingdoms: an illustrated guide to the phyla of life on earth.* W.H. Freeman, New York.

Matthews, R.E.F. 1979. Classification and nomenclature of viruses. Third report of the International Committee on Taxonomy and Viruses. *Intervirology* 12:131-296.

Patterson, C. 1980. Cladistics. *Biologist* 27:234-240.

Skerman, V.D.B., McGowern, V. and Sneath, P.H.A. (Eds) 1980. Approved list of bacterial names. *International Journal of Systematic Bacteriology* 30:225-420.

Tehler, A. 1988. A cladistic outline of the Eumycota. *Cladistics* 4:227-277.

Whittaker, R.H. 1972. Evolution and measurement of species diversity *Taxon* 21:213-251

Wildy, P. 1971. Classification and nomenclature of viruses. In: Melnick, J.C. (Ed.), *Monographs in Virology* 5. London: Academic Press, London.

Williams, P.H., Humphries, C.J. and Vane-Wright, R.I. 1991. Measuring biodiversity: taxonomic relatedness for conservation priorities. *Australian Systematic Botany*, 4:665-679.

Williams, P.H. (unpublished). Afrotropical antelopes - priority areas for biodiversity. Progress report to the Natural History Museum, London, WCMC and IUCN-SSC.

Abridged from a document written by R.I. Vane-Wright, Biodiversity Programme, The Natural History Museum (London).

3. SPECIES CONCEPTS

An understanding of the species concept is basic to an understanding of biological diversity because species are almost universally used as the units in which diversity is measured.

WHAT IS A SPECIES?

This simple question has troubled biologists for more than two centuries. Although accepted so widely as a 'natural', basic or fundamental unit, many conflicting definitions of species have been coined, and agreement is still lacking. The range of definitions reflects, to a large degree, the differing interests and differing theories of individual scientists about the origin of diversity itself - literally from Genesis to Darwin and DNA. This process has not stopped, continuing for example with the debate over the importance of neutralism or selectionism in the evolutionary process. Furthermore, many scientists have entered the debate from practical knowledge of particular groups of animals or plants. As there are major differences in the biology of different groups, with consequent variations in the patterns and processes of species formation, it is hardly surprising that species and species concepts are heterogeneous both in theory and practice.

One of the most fundamental aspects of the problem is variation. Most if not all animals and plants show variation, every individual often being demonstrably unique. Within a population variation can be continuous (such as height or weight) or discontinuous (such as sex or handedness), environmental in origin (such as human language) or genetic (such as blood group). Variation can also be seen in time between successive generations (seasonal variation), and in space across allopatric populations (geographical variation: clines, demes, races and subspecies).

The species problem is, in part, a history of how biologists have tried to manage this problem of variation. In particular, how can we classify variable organisms into discrete groups, tempered by knowledge of the existence, origin and maintenance of that variation? Modern species concepts divide into two main groups, those concerned with process and those concerned with pattern. We thus need to examine the processes of segregation, isolation and recognition responsible for the differentiation and cohesion of populations, and the patterns we perceive through comparison of the products of those differentiation processes.

EARLY SPECIES CONCEPTS

The word 'species' literally means outward or visible form. Conspicuous natural species have long been recognised by people of many local cultures. With the emergence of natural science in the 17th and 18th centuries, attempts were made to catalogue the whole of biological diversity, in all its manifestations and variations. Early approaches to dealing with the species problem were influenced by two very different philosophical views, essentialism and nominalism. In practice, however, both were usually abandoned in the face of increasing empirical knowledge of the life cycles of organisms and how they reproduce.

According to the typological species concept, based on essentialist principles, which was widely adopted during much of the 18th and 19th centuries, every organism corresponds to some idealised plan. The task of the taxonomist involved recognising each fundamental design, and describing, diagnosing or divining the essential features of those designs or 'types', so that individual organisms could be assigned to them.

In practice this often led to arbitrary divisions. Very different plants or animals were often lumped together because they shared certain 'essential' features; this was particularly evident amongst higher taxa, such as Linnaeus's group *Vermes*. By the same token, what we would now recognise as different forms of one and the same animal or plant were often separated because they conformed to different idealised types - in its most extreme manifestation, in many higher taxa two different sexes exist which according to this view could be classified as separate species, plainly a nonsensical view.

The most extreme opposing view states that only individuals exist in nature. Taxonomic groups are seen as man-made abstractions allowing us the convenience of being able to refer to large numbers of individuals collectively, and nothing more. They have no objective or independent basis but are merely convenient 'pigeon-holes' for dividing up or handling diversity.

Few scientists now accept that this nominalist approach is applicable to species, but it is still widely considered to apply to higher taxa. Most cladists and other taxonomists concerned with natural classification deny nominalism at all levels - the kingdom is seen as 'real' as the species (Loevtrup, 1987). Some cladists, however, deny reality to the species level, seeing species as only something in the process of becoming, while higher taxa are considered permanent real entities. With such deep divisions in the philosophical views of taxonomists, it is hardly surprising that there is still no agreement over the species concept.

EVOLUTIONARY THEORY AND POLYTYPIC SPECIES

Evolution and genetics

Following the emergence of Darwinism in the 1860s, and the general acceptance of the theory of evolution, the typological approach began to be questioned. Darwin himself suggested that "our classifications will become, so far as possible, genealogies". To Darwin, species were no different from other taxa (a view currently advocated by Nelson, 1989), and he expressed relief at being freed from "the vain search for the undiscovered and undiscoverable essence of the term species". Darwin, however, had no reliable theory of inheritance. With the development of genetics and population biology, including statistics, scientists began to develop rational explanations for the origin and inheritance of variation, and apply this understanding to a radically different view of the nature of taxa - and species in particular.

Polytypic species

One of the first major impacts of population thinking on taxonomy was the concept of polytypic species. According to this idea, many widespread species-level taxa show more or less discontinuous geographical variation describable by the use of trinomens, or subspecies. Previously such variation was recognised haphazardly by the occasional naming of 'varieties' or, alternatively, by the description of increasingly large numbers of allopatric species, many of which differed only in details of coloration or other superficial characters. Such patterns were seen to reflect both common ancestry and local adaptation, as species were thought to spread from their geographical places of origin and differentiate under the influence of natural selection. Subspecies were seen virtually as 'species in the making'. This approach, including the trinominal nomenclature (genus, species and subspecies), was preadapted to become the basis of an influential new vision of the species.

THE BIOLOGICAL SPECIES CONCEPT

The biological species concept is particularly associated with the work of three zoologists, Theodozius Dobzhansky, Julian Huxley and Ernst Mayr. This view concentrates not on logical classes or plans but on the idea of the species as a process, a closed reproductive community or breeding system. According to Mayr (1969), *species are groups of interbreeding [or potentially interbreeding] natural populations that are reproductively isolated from other such groups*. The basic idea of a biological species is that of a 'pool' of genes available for re-combination through sexual reproduction, but not with genes belonging to other gene pools, from which they are 'protected' by a variety of recognition and isolation mechanisms (behavioural, physiological, genetical, etc.). Thus the biological species to which a given individual belongs is determined by the limits of the populations with which it interbreeds, or potentially interbreeds.

The biological species concept, or some variant of it, is probably the most widely accepted view of the species held by biologists today. Extreme versions of the concept, such as Huxley's (1940) definition of species as "distinct self-perpetuating units with an objective existence in nature, and therefore on a different theoretical footing from genera or families or other higher categories" approach the *evolutionary species concept*, in which species are seen as the fundamental units of evolution (rather than haphazard by-products of it).

Recently, certain proponents of the biological species concept have split into two 'camps': those supportive of the idea that species distinctness is mainly brought about and maintained by selection for isolating mechanisms (isolation concept), and those emphasising greater importance for inherent mate-recognition systems in this role (the recognition concept). The debate has led to further proposals, such the *cohesion concept*. According to Templeton (1989) this idea draws on all three major variants of the biological species (the evolutionary, isolation and recognition concepts), and defines species as the "most inclusive population of individuals having the potential for phenotypic cohesion through intrinsic cohesion mechanisms". The intrinsic mechanisms relate to gene flow and ecological equivalence.

All variations of the biological species concept suffer from a number of practical shortcomings and limitations. They are inapplicable to the very large number of animals and plants that reproduce with only irregular genetic recombination, or without it altogether (asexual or agamospecies). In sexually reproducing species the limits of genetic re-combination are rarely known and have to be inferred from indirect evidence, and there is further uncertainty regarding species limits when the concept is applied over wide geographical ranges or over time.

Superspecies and syngameons

As already noted, the biological species concept was developed by zoologists from the idea of grouping allopatric (not overlapping geographically), modestly differentiated races or subspecies into single *polytypic species*. This system was elaborated to include a further concept, that of the *superspecies*, consisting of assemblages of more strongly differentiated groups of populations, or *semispecies*. Semispecies have geographically non-overlapping but contiguous (parapatric) distributions, permitting gene exchange at their boundaries. Most significantly, they are seen as ecological equivalents, and thus unable to coexist as stable, fully differentiated species.

Following Turesson, botanists have long recognised a related concept, the *syngameon*, whereby groups of sympatric (geographically overlapping) semispecies coexist. Gene flow may be slight or extensive, and their continued existence depends on ecological vicariance, occupying stable and distinct local habitats (such as contiguous forest and open formations). If such a patchy environment is destroyed and replaced by a different ecosystem, the separate semispecies usually fuse through hybridisation.

The advent of genetic fingerprinting techniques has now permitted zoologists to appreciate that gene flow between more or less closely related but perfectly 'good' sympatric species of animals may be commonplace. One of the most recent discoveries of this kind is reported by Templeton (1991), who quotes work showing that significant gene exchange can occur between *Bison* and certain species of *Bos* (domestic cattle). This example demonstrates that species sufficiently distinct to have been placed in different genera can have this type of relationship, empirically violating the most basic tenet of the biological species concept, the separateness of gene pools.

Tokogenetic and phylogenetic relationships

In order to understand continuing disagreements over the significance and definition of species, it is necessary to appreciate that two quite separate goals are being pursued. Species serve as the basis for describing and cataloguing the elements of biodiversity, and in our attempts to discover the historical relationships of those diverse elements. Species are also widely regarded as fundamental units of evolution, being both the products of speciation and the things which are thought to speciate. Thus the single word, species, serves the needs of systematics (discovery of empirical patterns) and the needs of population biology (formulation of process theories). Once the existence of these two separate goals is acknowledged, it becomes easier to make sense of the multiplicity of species concepts, many of which represent only differences of emphasis within the two major divisions.

Another way to think about this problem is to consider two major sorts of genetic relationships: those between individuals (tokogenetic, or blood relationships) and those between taxa (phylogenetic, or historical relationships). What is truly unique about species may simply be that they lie at the junction of both types of relationship (Nixon and Wheeler, 1990). Higher taxa, and their interrelations, represent a fixed, historical past. Below the species, at the level of demes and populations, all is change, with mutation and genetic recombination affecting every new life cycle, every generation of individuals. Species exist at a dynamic limit between the two, with tokogenetic processes maintaining cohesion yet allowing change, while historical accidents fragment species into separate phylogenetic lineages. Such ideas form the basis of yet another species concept, that of phylogenetic species.

THE PHYLOGENETIC SPECIES CONCEPT

According to this view, species are irreducible clusters of organisms diagnosably distinct from other such clusters, and within which there are parental networks of ancestry and descent. Nixon and Wheeler (1990) have defined the concept as "the smallest aggregation of populations (sexual reproduction) or lineages (asexual reproduction) diagnosable by a unique combination of character states in comparable individuals".

This view of species places the emphasis not on reproductive process but on the most general aspect of taxonomic diversification, that of *differentiation*. In some cases differentiation results in reproductive isolation but in many cases it does not. Thus the existence of reproductive isolation is evidence of diagnostic characters but new characters which become fixed within a population do not necessarily affect reproductive isolation.

An inherent danger in such a view is that, by *reductio ad absurdum*, every population, stage, morph or even individual organism could be elevated to separate species status. For this type of definition to be operational it would also be essential to emphasise the critical importance of reproductive community, or *cohesion*, more or less in Templeton's sense. Even then, a consequence of applying the phylogenetic species concept, compared with the biological species concept, would be a very large increase in the number of species recognised (Nelson and Platnick, 1981).

SPECIES IN PRACTICE

Empirical consequences of different concepts

Cracraft (1989) has provided some examples of the striking differences that can arise in evolutionary and taxonomic conclusions, dependent on the species concept applied. Cracraft's examples all concern parapatric birds of debatable specific or subspecific status, with evidence of hybridisation in contact zones. His cladistic analyses suggest that many biologically defined 'subspecies' that hybridise on contact are less closely related to each other by descent than they are to other, full 'species' with disjunct distributions.

Thus, as accepted under the phylogenetic species concept, species separable on phylogenetic criteria may be interfertile, while polytypic species recognised on biological (interbreeding) criteria may not be the 'units of evolution'. At the practical level, these alternative approaches give rise to major differences in the classification and status given to populations and groups of populations. As already noted, the phylogenetic concept or approach leads to the recognition of far more species (and fewer subspecies) than the biological species concept. In terms of formal classification, it lacks the major practical advantage of trinomens - we would tend to lose sight of the wood for the trees.

Subspecies

Many species of geographically variable and conspicuous organisms, such as birds, have been subdivided into numerous subspecies. Butterflies, for example, are thought to comprise about 17,500 full species, but the number of currently recognised subspecies approaches 100,000. Many of these subspecific taxa (particularly those from small islands or isolated mountains) are fully diagnosable - that is, virtually every individual can be reliably identified to subspecies, regardless of knowledge of where it was found. Such subspecies would qualify as species under a phylogenetic species concept.

On the other hand, this is not true for all so-called subspecies, notably many of those described from large islands or continental areas. In many of these cases subspecies are only recognised on a statistical basis, so that individuals cannot be reliably diagnosed, and only identified with the aid of knowing where they came from. Typically, this represents the phenomenon of clinal geographic variation. At the extreme, the most distant populations in long clines may be so distinct that in areas of overlap they may behave as separate biological species and be fully diagnosable locally (rassenkreis and ring species: Mayr, 1963). Even in less extreme situations, the opposite ends of a cline may be more strikingly distinct than related, fully diagnosable subspecies, or even full species.

The implications of this are that for the assessment of biodiversity there is no easy answer to 'the subspecies problem' any more than there is to the species problem. Species status bears no direct or simple relationship to degree of phenetic differentiation, or to any measure such as genetic distance. Species (and subspecies) are determined by relational properties, not by absolute criteria, be they essences, reproductive mechanisms or distance measures.

The state of the science

At the broadest scale, we know very few organisms well enough to consider the subtle, albeit highly significant, interpretations that such insights as the phylogenetic species concept or the syngameon might lead us to consider. In particular, the vast majority of named species are known only from morphology and limited knowledge of their geographical distributions. For these species we know virtually nothing about their individual breeding systems, gene flow, ecology or even, in most cases, their cladistic relationships. Such species are often referred to as morphospecies.

The present state of taxonomy, carried out by different scientists working at different times to different theories and philosophies and on imperfectly known groups of widely differing size, taxonomic apparency and life-cycle characteristics, ensures that species currently recognised are not comparable entities. Following the successive rise of population biology and phylogenetic systematics, there is some prospect if not of harmonising species concepts at least of clarifying what is meant by a particular scientist in a particular context.

CONCLUSION

For the present we have to manage with a very imperfect and inconsistent system of classification, even at the supposedly fundamental level of species. In practice we have not advanced much beyond the position outlined long ago, that a species is what a competent systematist says it is (Regan, 1926). Although much can and should be done to improve this state of affairs, a lack of certainty should be accepted as inherent to the subject.

However, this strong limitation on the use of species as comparable units is all too often forgotten when species numbers are handled in aggregate, as with many practical conservation issues or theoretical discussions of biodiversity. Conclusions reached on this basis run a risk of being inaccurate, spurious or even completely misleading. If species, instead of being treated like independent and equivalent units of diversity, are placed in their proper relational context of the entire hierarchical classification, some of the problems caused by this limitation can be avoided.

References

Cracraft, J. 1989. Speciation and its ontology: the empirical consequences of alternative species concepts for understanding patterns and processes of differentiation. In: Otte, D. and Endler, J.A. (Eds), *Speciation and its Consequences*. Sinauer, Sunderland, Mass. Pp.28-59.

Huxley, J.S. 1940. Introductory: towards the new systematics. In: Huxley, J. (Ed.), *The New Systematics*. Oxford University Press, London. Pp.1-46.

Loevtrup, S. 1987. On species and other taxa. *Cladistics* 3:157-177.

Mayr, E. 1963. *Animal Species and Evolution*. Harvard University Press, Cambridge, Mass.

Mayr, E. 1969. *Principles of Systematic Zoology*. McGraw-Hill, New York.

Nelson, G. 1989. Species and taxa: systematics and evolution. In: Otte, D. and Endler, J.A. (Eds), *Speciation and its Consequences*. Sinauer, Sunderland, Mass. Pp.60-81.

Nelson, G. and Platnick, N. 1981. *Systematics and biogeography: cladistics and vicariance*. Columbia University Press, New York.

Nixon, K.C. and Wheeler, Q.D. 1990. An amplification of the phylogenetic species concept. *Cladistics* 6:211-223.

Regan, C.T. 1926. Organic evolution. *Report of the British Association for the Advancement of Science* 1925:75-86.

Templeton, A.R. 1989. The meaning of species and speciation: a genetic perspective. In: Otte, D. and Endler, J.A. (Eds), *Speciation and its Consequences*. Sinauer, Sunderland, Mass. Pp.3-27.

Templeton, A.R. 1991. Genetics and conservation biology. In: Seitz, A. and Loeschcke, V. (Eds), *Species Conservation: a population-biological approach*. Birkhauser, Basel. Pp.15-29.

Text written by R.I. Vane-Wright, Biodiversity Programme, The Natural History Museum (London).

4. SPECIES INVENTORY

The objective of this section is to explore how far global biodiversity may have been accounted for by taxonomic description, emphasising diversity at the species level. This is done with reference to the total number of species currently recognised (itself very imprecisely known) and the degree to which we can estimate the completeness of taxonomic knowledge.

Existing knowledge of geographical and other variation in species richness provides a useful starting point, but this knowledge is heavily biased. Unfortunately, our understanding of the best-known taxonomic groups and best-known parts of the world remains an insufficient basis for predicting more general patterns, or for rigorously testing explanations for such patterns as have been identified. Any estimation that may be made of the overall extent of global species richness remains staggeringly imprecise. Even so, for at least multi-celled animals and green plants, and perhaps for all eukaryotes (i.e. all of life except for microorganisms such as bacteria) it is possible to predicate useful lower and (with less confidence) upper limits to the extent of regional and global species richness of the major groups. Also, it is now reasonably clear just what the major gaps in our understanding are, so that we have a good idea of which new data are needed to improve on present estimates. Work in progress that involves intensive sampling of species-rich groups (e.g. insects) in especially species-rich areas (e.g. moist tropical forests) promises to provide a much more reliable picture of major global species richness patterns and a more reliable basis for estimating the number of species with which we share the planet.

It must be emphasised that data discussed in this section that may be pertinent to species richness estimates should not, in the current poor state of knowledge, be applied directly to estimations of possible species extinction rates via loss or degradation of habitat. Existing data on range sizes, patchiness of distribution and population structure of the poorly-known organisms discussed here are such that no direct connection between numbers of species present at one site or in one region and the threat posed to the continued existence of any one of those species by the loss of a given area of habitat can be made.

CURRENT STATUS

Here we consider how many extant species of organisms have already been described and assess at what rate the existing inventory is growing and improving.

The number of described species

The number of species which have been described and the number currently regarded as valid are not precisely known for many groups of organisms. For the best known groups, all of which are relatively small (e.g. birds with 9,881 species, Sibley and Monroe, 1990), catalogues and counts are very complete. Variations in published figures are largely because of differences in whether certain taxa are regarded as 'good' species or not. Accurate figures for currently recognised species are also available for some groups (e.g. bacteria with 3,058 recognised species as of 1991) in which it can be assumed a major proportion remains undescribed. Much improved counts have recently become available for some substantially larger groups, such as the vascular plants (260,000 species in total) and fungi (70,000 species). Counts for animal groups with many described species, as Tables 4.1 and 4.2 illustrate, mostly remain much less precise. Disparities between the various figures very recently furnished for individual groups such as the molluscs, annelids and platyhelminths (Table 4.1) and Diptera (Table 4.2) are particularly striking. On investigation, only some of the apparent discrepancies turn out to be because of differences in the year up to which counts had been made; others appear to result from confusion between the number of *nominal* species (i.e. all species that have ever received a separate name no matter what their current status) and the often much lower number of species recognised as valid, as well as from simple miscalculation or oversight. Some of the largest groups (e.g. the insect orders Coleoptera and Diptera) are, in fact, relatively well catalogued, but animal taxonomists have tended to place little emphasis on providing accurate tallies of described species that are regarded as valid at any particular point in time. Largely as a result, figures for the biota as a whole that have been published in recent years vary considerably, from around 1.4 million to more than 1.8 million. This imprecision is far exceeded by that involved in the attempts to estimate *total* species richness (including as yet undiscovered and undescribed species) discussed below, but it is in some respects more surprising. Estimates for numbers of currently recognised and described species are given here, mostly rounded to the nearest five thousand, for the groups that make the largest contributions (see Table 4.3 and Fig. 4.5) - but without any pretence to high accuracy. In arriving at these figures relevant specialist opinion, as well as the most recent literature, was taken into account. Including all of the smaller groups not listed in Table 4.3, the overall figure reached is approximately 1.7 million. A more accurate count is likely to produce a somewhat higher figure.

Deficiencies of the existing database

The evidently low priority accorded by taxonomists to keeping track of how many species have been described stems in large measure from the knowledge that the biological significance of these data is slight. For all but the best known groups, if a species count is a measure of anything it is of taxonomic effort expended, and this is clearly seen to be arbitrary by most biological criteria. Even in terms of the taxa that ostensibly have been dealt with by the descriptive process much uncertainty exists, as catalogues of described species, however carefully compiled, include the results of poor taxonomy as well as good. When careful reassessments (taxonomic revisions) are made, it is common to find that a relatively high proportion of previously recognised 'species' are not, in fact, distinct. To put it in taxonomists' jargon, most parts of the existing inventory contain substantial amounts of unrecognised or at least unreported synonymy. In some groups, further imprecision arises from a fundamental lack of agreement as to just what constitutes a species.

Table 4.1 Estimated numbers of described extant species in major animal groups

	Mayr *et al.* (1953)	Barnes (1989)	May (1988)	May (1990)	Brusca & Brusca (1990)
'Protozoa'	-	-	260,000	32,000	35,000
Porifera	4,500	5,000	10,000	-	9,000
Cnidaria	9,000	9,000	10,000	9,600	9,000
Platyhelminthes	6,000	12,700	-	-	20,000
Rotifera	1,500	1,500	-	-	1,800
Nematoda	10,000	12,000	1,000,000 ?	-	12,000
Ectoprocta	3,300	4,000	4,000	-	4,500
Echinodermata	4,000	6,000	6,000	6,000	6,000
Urochordata	1,600	1,250	-	1,600	3,000
Vertebrata	37,790	49,933	43,300	42,900	47,000
Chelicerata	35,000	68,000	63,000	-	65,000
Crustacea	25,000	42,000	39,000	-	32,000
'Myriapods'	13,000	10,500	-	-	13,120
Hexapods	850,000	751,012	1,000,000 ?	790,000	827,175 +
Mollusca	80,000	50,000	100,000	45,000	100,000 +
Annelida	7,000	8,700	15,000	-	15,000

Notes: 'Protozoa', a paraphyletic group, is used in the 'traditional' zoological sense. The 'Myriapods' consist of the Chilopoda (centipedes) and Diplopoda (millipedes) together. Apart from two exceptionally high figures - those for Protozoa and Nematoda - provided by May (1988), who presumably intended these as estimates of *actual* rather than described species, most estimates, even the highest of a range for any given group, are probably conservative. The extent of the great variation in totals for some groups is inexplicable. For example, while Brusca and Brusca (1990) suggest 100,000+ as a likely figure for described species of molluscs a totalling of the figures given for the individual mollusc classes by the same authors provides a total of around 50,000.

Table 4.2 Number of described species in the four major insect orders

	Southwood (1978)	Arnett (1985)	May (1988)	Brusca & Brusca (1990)
Coleoptera	350,000	290,000	300,000	300,000+
Diptera	120,000	98,500	85,000	150,000
Hymenoptera	100,000	103,000	110,000	125,000
Lepidoptera	120,000	112,000	110,000	120,000

Notes: Some recent estimates of the number of described species in the four major insect orders. An accurate figure for Hymenoptera is probably not very different from any of the fairly consistent estimates shown, while one for Diptera probably lies towards the middle of the very wide range indicated here. The estimates for Coleoptera and Lepidoptera, on the other hand, are probably all far too low.

The existing inventory of described species may, as discussed below, provide a poor basis for estimating the true extent of global species richness. However, where extrapolative methods are adopted that do involve the use of described species counts, the accuracy with which the counts have been made will often have a substantial influence on results. For example, let us assume that our approach to estimating global species richness is to (1) estimate what proportion of the biota belongs to a particular group, (2) estimate what proportion of species in that group has already been described, and (3) use these estimates and the number of described species in the group to calculate a total for all groups. Should the group chosen be the Coleoptera our estimates might be that this group contains (say) 20% of all living species, and that (say) only one in every five or even ten Coleoptera species has been described. Recent published estimates for the number of described species of Coleoptera, like those for most other large groups, are extremely variable. Several put the figure at around 300,000 species (Table 4.4), although thorough counts of a sample of coleopterous families suggest that 400,000 is a likely minimum. If the figure of 300,000 does, in fact, represent an underestimate of the order of 100,000 species its use in the calculations outlined above would lead to underestimation of the biota as a whole by as much as five million.

Current rates of growth

Current rates of description of new species and other taxa and how these rates vary from group to group can tell us a good deal about how the task of inventorying biotic diversity is proceeding. Whether or not rates of description have any value for predicting just how much of the task remains to be done is another matter, considered below. Numbers of newly described species recorded in the *Zoological Record* for a range of animal groups, for each year between 1979 and 1988, are given in Table 4.5. The

Table 4.3 Numbers of species in the groups of organisms likely to include in excess of 100,000 species (plus vertebrates)

	DESCRIBED SPECIES	ESTIMATED SPECIES HIGHEST FIGURE	WORKING FIGURE	
Viruses	5,000	500,000+	500,000	Sy
Bacteria	4,000	3,000,000+	400,000	Ma/Te/Sy
Fungi	70,000	1,500,000+	1,000,000	Te/Sy
Protozoans	40,000	100,000+	200,000	Ma/Te/Sy
Algae	40,000	10,000,000+	200,000	Ma
Plants (Embryophytes)	250,000	500,000+	300,000	Te
Vertebrates	45,000	50,000+	50,000	Ma/Te
Nematodes	15,000	1,000,000+	500,000	Ma/Te/Sy
Molluscs	70,000	180,000+	200,000	Ma/Te
Crustaceans	40,000	150,000+	150,000	Ma
Arachnids	75,000	1,000,000+	750,000	Te
Insects	950,000	100,000,000+	8,000,000	Te

Notes: The figures for described species (mostly given to the nearest 5,000) were arrived at by consulting relevant specialists as well as by critically reviewing the literature. The 'highest figure' estimates for existing species, many of them frankly speculative, are the highest encountered during a survey of recent literature. The 'working figure' estimates are conservative. The figure for bacteria has been arbitrarily 'capped' at 100 undescribed to 1 described species on the grounds that projections involving more than two orders of magnitude are inherently unsafe. The biggest question marks lie over the true numbers of species of viruses, bacteria and algae. Substantial upward revisions from the working figures for these groups *may* prove justified with time. The figures for fungi, protozoans and nematodes are also insecurely based. Note that the Fungi and Protozoa are used in the 'traditional' sense, while Bacteria includes cyanobacteria. The figures for 'insects' include all hexapods, and that for described insect species assumes totals of 400,000 for Coleoptera, 150,000 for Lepidoptera, 130,000 for Hymenoptera and 120,000 for Diptera. The final column in the table gives an indication of where the major proportion of species in each group is concentrated. All groups listed include at least some symbionts (Sy), obligately associated as parasites, mutualists or commensals with other organisms, and all (except for viruses) have at least some free-living representatives in marine (Ma), terrestrial (Te) and freshwater systems. Despite high *local* species richness in some groups, the overall contribution of freshwater species to group totals is relatively small, unsurprising in view of the fact that freshwater covers well below 1% of the earth's surface.

most striking aspect of these figures is the extremely low variation between years. Indeed, the yearly overall totals (for all groups included in the *Zoological Record*) for the years 1979-1988 show a standard deviation of less than one-twentieth of the mean annual figure. Description rates for many of the individual groups listed in Table 4.5 are almost equally invariant through this decade.

Description rates can tell us roughly how quickly various parts of the taxonomic inventory are growing. As the description of new species in most groups is accompanied by continuing reappraisal of the existing inventory, the rate of description of new species is rarely precisely the same as the rate of increase in the number of recognised species. The description rate may be substantially higher; in some of the larger groups of insects, for example, the current rate at which previously described species disappear into synonymy is around one-quarter to one-third the rate at which new species are described. Unfortunately, for some groups, newly recognised synonymies are not systematically reported or recorded in abstracting journals, so that the extent of the disparity between rates of description and rates of growth in number of recognised species is extremely difficult to assess.

How current description rates for a range of groups compare with earlier rates is indicated in Table 4.6. In relation to averages for the post-Linnean period, and with the exception of groups such as birds where few new species are being discovered, current rates are uniformly high. For some groups, such as nematodes, current rates are as high as or even higher than they have ever been, but the average figures conceal the fact that the period of maximum species description for a number of groups is well in the past. Detailed information on how taxonomic activity, as reflected in description rates, has varied through time is available in a range of reviews dealing with individual groups, and a summary of this has been provided by Simon (1983). The picture for groups such as the birds is predictable, with most species described early on and half of the present day total of 9,000 or so recognised species having been reached by 1843 (see Fig. 4.1). After a sometimes rather slow start, description of new species in some groups has otherwise proceeded at a fairly steady rate, while in others there has been a marked decline after a peak of activity which in many instances falls towards the end of the 19th century. Groups in which the maximum activity is taking place now include some of the better-known as well as those, such as nematodes and fungi, in which only a small fraction of species is likely to have been described so far.

Description rates or growth rates do not, of course, necessarily provide a good measure of taxonomic effort expended or of how effective this is (see below). A rough and ready way of determining how such effort is being applied is to look at publication rates (Barnes, 1989, May 1988). Most instructive, perhaps, is to compare publication rates for various groups with their size, both in terms of currently recognised species and also projected overall species totals. Publication data from the *Zoological Record* quoted by May (1988) reveal, not surprisingly, that the

Table 4.4 Number of species in various families of beetles (Coleoptera)

	DESCRIBED SPECIES			ACTUAL SPECIES	
	Arnett (1967)	Lawrence (1982)	All Sources (1990)	Arnett (1967)	
Byrrhidae	154	c. 300	319	300	TEMP
Derodontidae	10	19	20	19	TEMP
Discolomidae	30	c. 400	443	50	TROP
Dryopidae	178	c. 200	253	300	trop
Elmidae	263	c. 700	1,170	350	trop
Limnichidae	67	c. 200	297	80	trop
Lymexylidae	37	c. 50	64	100	trop
These families in total	739	c. 1,869	2,566	1,195	
All Coleoptera	219,409	340,500	?	290,199	

Notes: Number of described species in various families of beetles (Coleoptera) compared with one taxonomist's estimates (Arnett, 1967) for actual numbers of existing species in these families. Three sets of figures for numbers of described species are provided. Arnett's (1985) estimates of described species were based on catalogues published between 1910 and 1915 while Lawrence's (1982) estimates were based on more up-to-date sources. The 1990 figures are based on direct counts from the most recent catalogues available supplemented by *Zoological Record* entries for subsequently described species and new synonymies to June 1990. A small part of the increase in numbers of Elmidae recognised in 1990 is due to species transferred from Dryopidae. If the 7 families for which the 1990 total of described species is 2,566 were representative of the Coleoptera as a whole in terms of growth since the 1910-1915 period (Arnett's figures) *total* described beetle species regarded as valid up-dated to 1990 should be 761,845 (i.e. 219,409 x 2566/739). Using the more up-to-date estimate (Lawrence, 1982) as starting point the 1990 total for Coleoptera should be 467,481 (i.e. 340,500 x 2566/1869). In fact, recent species description rates for some of the families in the table (e.g. Elmidae and Limnichidae) are likely to be well above average for the Coleoptera as a whole, so both extrapolations may produce overestimates. No accurate count has been made for the whole Order but the actual number of described Coleoptera species regarded as valid as of 1990 is probably in the region of 400,000. The table illustrates the way in which cautious taxonomists are strongly influenced by the number of species already known when predicting the number that might actually exist. For relatively well-known beetle groups, largely those in which most species occur in temperate regions, this may not lead to drastic underestimation, but clearly may do so in the case of less well- known groups in which most species are tropical. TEMP = family shows a strong bias away from the Tropics; TROP = family shows a strong bias towards the Tropics; trop = family with a weaker tropical bias.

ratio of papers published to number of species already described has been high in recent years for vertebrates, varying from around two papers per species in mammals to one paper for every two or three species in fish. Leaving aside these well-known groups, the number of papers published per recognised species per year is more or less *inversely* correlated with the size of the group in terms of *described* species. For example, in all major groups including fewer than 50,000 described species the ratio of publications to species is below 1:50 and often below 1:10, whereas ratios are generally much higher, exceeding 1:100 in the case of the Coleoptera, in groups containing a greater number of described species.

Despite this evident bias in taxonomic attention against groups containing many described species, actual growth rates, i.e. number of new species described in relation to the number already described (see Table 4.6), currently vary remarkably little.

PREDICTION FROM THE EXISTING PARTIAL INVENTORY

Inherent limitations

The total number of species so far described, of course, gives us some idea of the minimum extent of global species richness, while knowledge of how described species in the better-known groups are distributed gives us an impression of the way in which species richness is allocated between regions, ecosystems, etc. However, there is every indication that described species do not account for the major portion of the world's species and, more importantly, that described species represent a very biased sample. Is it then possible to use these figures for described species in any way at all as a basis for projecting actual world totals?

If we examine the available data on absolute rates of description and how these have varied through time, and on how the overall taxonomic effort is and has been apportioned between the major groups of organisms, we might justifiably conclude that they tell us very little about the size of the descriptive task that remains. However, a careful scrutiny of these data can help to demonstrate the biases that exist in the inventory as it stands, and may help to reveal what underlies them. This may be helpful in making initial judgements as to where the major sources of unexplored diversity are to be found. It may also prove useful in assessing the likely accuracy of species richness estimates, such as those of taxonomic specialists considered below, that are difficult to evaluate in any other way.

Perhaps the most obvious among biases in how taxonomic effort is applied are those that stem from everyday human interests and preoccupations. Large organisms, those that are considered particularly attractive (flowering plants, butterflies, etc.) or appealing in some other way, those most closely resembling humans themselves (vertebrates, especially mammals), and those that have a direct impact on human affairs, usually as pests of one kind or another, are all favoured objects of study and description. By association, certain other groups such as fleas and lice - because they are parasites of birds and mammals - may also receive a relatively large share of attention.

Table 4.5 Number of new species listed in the *Zoological Record* 1979-1988

	1979	1980	1981	1982	1983	1984	1985	1986	1987	1988
Protozoa	463	309	353	280	272	374	432	385	315	377
Platyhelminthes	293	357	341	290	307	340	359	336	297	243
Nematoda	407	389	364	422	383	354	365	367	331	258
Annelida	234	114	215	186	127	208	200	149	161	131
Mollusca	338	355	391	239	344	316	419	412	392	419
'Other' invertebrates	180	214	341	174	303	353	187	252	195	339
Crustacea	645	638	763	736	695	599	700	843	708	660
Arachnida	1,610	1,523	1,199	1,182	1,145	1,504	1,353	1,185	1,488	1,307
'Other' arthropods	98	57	120	154	119	123	55	94	146	122
Hemiptera	1,277	1,115	1,191	981	1,275	1,038	1,072	1,162	903	1,016
Lepidoptera	675	445	506	580	631	685	658	915	623	699
Diptera	1,019	1,015	1,130	899	973	928	1,115	1,095	1,303	1,000
Hymenoptera	1,167	1,134	1,086	1,031	1,215	1,496	857	1,184	1,084	1,705
Coleoptera	2,116	2,804	2,243	2,454	1,960	2,259	2,220	2,843	2,130	2,051
Other Insecta	725	824	992	878	1,331	1,030	836	822	1,110	703
Pisces	183	241	273	240	260	223	220	204	234	229
Other Chordata	146	170	134	186	177	168	230	117	191	138

Source: *Zoological Record Online*, data search organised and carried out by BIOSIS UK.
Notes: Number of new species listed in the *Zoological Record* 1979 to 1988 (Vols 116-125), showing the remarkable constancy of description rates in the larger animal groups. The figures given are for newly described species. While extinct groups such as the Trilobita are excluded, fossil species (mostly relatively few in number) of extant groups (e.g. Mollusca, Insecta, Pisces) are included in the counts. As new synonymies are not accounted for the figures do not provide a precise measure of growth in numbers of described species regarded as valid. The 'other' invertebrates category includes all extant non- arthropod groups not otherwise listed. The 'other' arthropod category includes all taxa listed in Section 12 of the *Zoological Record*; in major part these are Chilopoda (centipedes) and Diplopoda (millipedes). The other Chordates category includes reptiles, mammals, amphibia and birds.

Other strong biases have more to do with taxonomic taste and fashion, and the ease with which the organisms are found, collected, studied and preserved as specimens.

Organisms that can be studied without complex procedures or expensive equipment, that are not too small, that exhibit distinctive characteristics and can be readily sorted tend to be dealt with preferentially. These various biases have influenced the historical pattern of description, starting with Linnaeus himself, who described very few small organisms, and confirmed by Gaston (1991b) who showed a clear relationship between body size and date of description in the approximately 4,000 species of British beetles.

It is also evident that most areas distant from the centres of human population as well as more obviously inaccessible regions such as the ocean depths are rather poorly inventoried. It should also be noted that most taxonomists, however much they may travel, remain based in the cities of the north temperate zone. Recent patterns of description for birds and mammals suggest that the few species that remain to be discovered in such well-known groups will almost certainly turn out to be tropical. Despite clear indications that the greater part of global species richness is to be found in the tropics, it is evident (no precise counts are available) that, in terms of *described* species, those of tropical origin are considerably outnumbered by those from temperate and boreal regions. An analysis of recent description data for insects (Gaston, unpublished) reveals that, for this speciose group at least, the bias against tropical species is still not being redressed. In some groups, the rate at which new species are described may be determined, at least in part, by what material is immediately available for study, which in turn depends on its general accessibility in nature. However, this can scarcely be the major influence on description rates in those groups (e.g. the larger insect orders) for which the world's museums already contain hundreds of thousands of undescribed species.

Time-series of species descriptions

Changes in the rate at which new species have been and are being described have been used to make an estimation of the likely future growth of each of the major groups of organisms. Various statistical procedures can be used to identify trends and project description rates forwards (May, 1990), with different techniques sometimes producing very different results (Frank and Curtis, 1979; Simon, 1983). Not surprisingly perhaps, attempts to use trends in description rates to predict global species richness of major contributors (i.e. excluding groups such as birds, mammals, etc.) or of the biota as a whole have been singularly unsuccessful (see Erwin, 1991). First of all, as a glance at Table 4.5 reveals, very few trends in description rates can be observed over the short term. Longer term trends, where they are evident, are often erratic but, except for a few small groups, commonly involve an increase in the pace of description over time. Species description in the least apparent and least tractable groups often gets off to a slow start but once this gets going, description rates may be more or less monotonic.

For those few groups in which species description is

Table 4.6 Current species description rates for various animal groups and for fungi

	SPECIES DESCRIBED PER ANNUM (1978-1987)	'GROWTH' RATE PER ANNUM (1978-1987)	CURRENT RATE/ OVERALL RATE	PROPORTION OF SPECIES DESCRIBED TO DATE
Vertebrates	367	0.82	1.90	High
Birds	5	0.05	0.13	Very high
Mammals	26	0.59	1.37	"
Amphibians and Reptiles	105	1.17	2.72	High
Fish	231	1.22	2.83	"
Molluscs	366	0.52	1.22	Moderate
Sponges	50	0.56	1.30	"
Cnidarians	57	0.63	1.48	"
Platyhelminths	316	1.58	3.68	"
Ectoprocts	58	1.29	3.00	"
Annelids	173	1.15	2.57	"
Protozoans	356	0.88	2.00	Moderate/low
Crustaceans	699	1.74	3.91	"
Insects	7,222	0.76	1.77	Low
Lepidoptera	642	0.43	1.00	Moderate/high
Coleoptera	2,308	0.57	1.34	Low
Diptera	1,048	0.87	2.03	Low/very low
Hymenoptera	1,196	0.92	2.14	"
Arachnids	1,350	1.80	4.19	"
Fungi	1,700	2.43	5.67	Very low
Nematodes	364	2.43	5.65	"

Notes: Current species description rates for various animal groups and for fungi, expressed as number of species described per annum (mean of years 1978-1987) (Column 1), compared with number of already described species (figures from column 1 (x 100) divided by number of described species) (Column 2), and with an *approximation* to average description rates for the whole of the period since 1758 (mean number of species described per annum 1978-1987 divided by mean number of species *currently recognised* as valid described per annum between 1758 and the present) (Column 3). The fourth column provides an indication of the proportion of each group that is likely to have been described so far; very high = c. 90% or more already described; high = c. 50-90%; moderate = c. 20-50%; low = c. 10-20%; very low = less than 10%

nearing completion, description rates may be expected to have some predictive value. Even here, however, they are likely to tell us what we already know, and may be distinctly misleading. Growth curves based on time-series of species descriptions are a convenient way of portraying the relevant data. The time-series for the very well-known groups such as birds generally forms a classic S-shaped growth curve (see Fig. 4.1, but note that here the curve is not S-shaped as the vertical axis is on a logarithmic scale). Any other form of curve indicates that the group in question is unlikely to be almost completely inventoried, but may tell us little else. We may note that birds are exceptional among relatively high-ranking taxa in that the description of new species has slowed to a trickle. Even mammal species (see Table 4.6) are still being described at a rather high rate that gives no clear sign of an asymptote, although it is fairly certain that the number of species awaiting discovery is relatively low. Some slowing down of the rate, finally, is perhaps indicated by figures for the last decade or so, with an average of 37 mammal species described per year from 1978 to 1982 and 20.5 per year from 1983 to 1988.

There are various ways of gathering the data for time-series graphs, and these can have significant effects on our ability to predict. For example, the number of species recognised within a group at any one time can be assessed from contemporary taxonomic works. However, over the past 120 years, there have been major changes in how species status is evaluated, which can make the figures non-comparable and the curves uncertain.

A better curve is usually obtained by making a cumulative graph of the dates of first description of all currently recognised species. In the case of the 120 or so species of crows (Fig. 4.2A), for example, we see a truncated version of the S-shaped curve, with nearly 10% of all currently recognised crow species having been described within the decade 1758-1767. With the curve for crows flat for the last quarter century it would be a bold person who would predict a rise to even 130 species, let alone a higher figure. However, such dramatic shifts can occur, even in relatively well-known groups.

An example is provided by a group of blue butterflies (Fig. 4.2C), subject of a recent major revision (Eliot and Kawazoe, 1983). Linnaeus knew only one species, the familiar European Holly Blue *Celastrina argiolus*, and the time-series is very slow up until the decade ending 1877. After that it goes through a rapid growth-phase, and then flattens at about the same time as the curve for the crows. Since 1967, however, there has been a new burst of species description, bringing the current total to a point at least

Figure 4.1 Discovery curves for species from 1758 to 1970

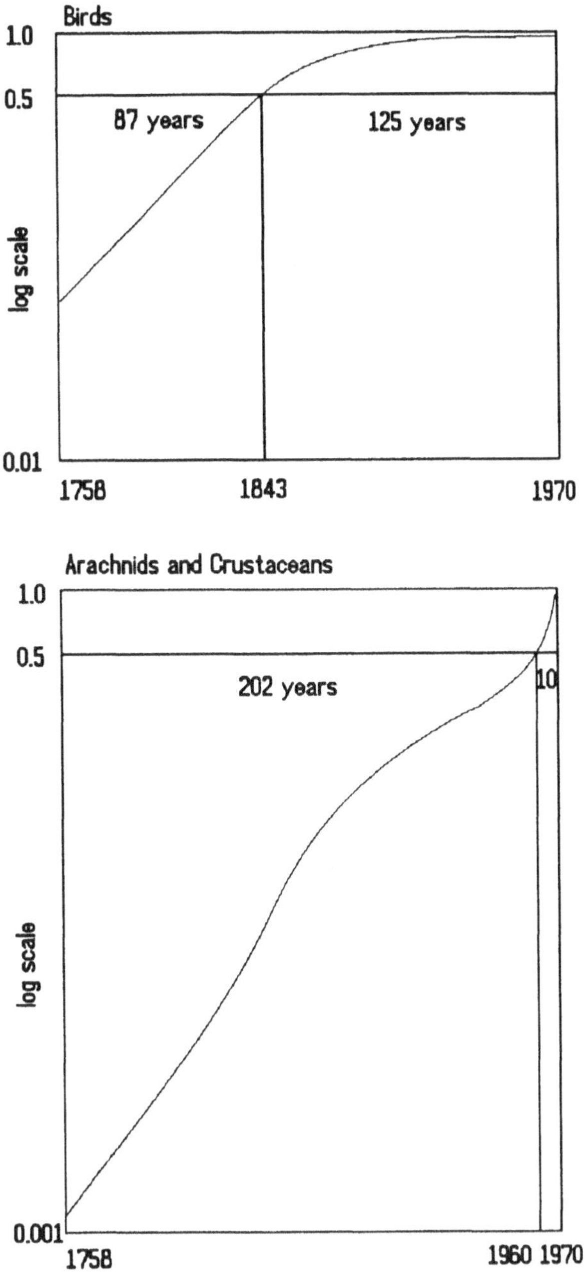

Source: Following May (1990) after Simon (1983).
Notes: Numbers of known species (expressed as a fraction of those known in 1970 on a logarithmic scale) are plotted against time. The vertical and horizontal lines show the points at which half of the 1970 totals had been reached. Although a trickle of new species of birds continues to be described the shape of the curve for birds as a whole resembles that for crows. The curve for Arachnida + Crustacea (i.e. the majority of non-insect arthropods) shows that up to 1970 description of new species had an ever-increasing pace, with the 1960 total doubled by 1970. Description of new species in these groups now proceeds at a steady rate of some 2,000 per annum (see Table 4.5).

33% higher than the plateau level. The explanation here is not poor taxonomy or a change in species concept, but a combination of exceptionally painstaking work coupled with vigorous collecting in previously inaccessible parts of Southeast Asia, where these butterflies form many island or mountain endemics.

OTHER APPROACHES TO PREDICTING PATTERNS

Estimates by taxonomic specialists

The opinions of taxonomists specialising in particular groups of organisms have traditionally played a considerable part in the formulation of views on the extent and pattern of species richness at every scale. Indeed, the preliminary tentative working figures for global species richness of the major groups used in this section have inevitably been influenced by the opinions and estimates of relevant taxonomists. However, the simple approach of collating views based on the specialist knowledge of the taxonomic community has not been systematically pursued, a major exception being the recent essay by Gaston (1991a) to assemble and interpret a cross-section of taxonomists' opinions concerning likely global insect species richness.

The approach adopted by Gaston has the merit of involving a large number of data points so that no one estimate has an overriding effect on the overall result. In addition, the sources are experienced taxonomists whose work generally involves exposure to at least part of the richness of species located in poorly studied regions. This said, it is likely that the way in which taxonomists actually arrive at their conclusions is quite varied, may be distinctly idiosyncratic and tends to the conservative. Indeed, the generally rather poor track record for such estimates suggests a possible correlation between the degree to which any given taxonomist has been exposed to relevant data (e.g. representative samples from many areas, including some of the richest) and the extent to which he or she is prepared to extrapolate beyond the relatively sure ground of already described species. Very early estimates by such as John Ray who, in the late 17th century, considered that the insects of the world as a whole might amount to some 10,000-20,000 species, may lend some support to this view.

To the extent that taxonomists work largely with what happens to come their way, it is likely that the collections they examine do not fully represent the richness to be found in less-known regions of the world, such as the tropics. In making their assessments of overall species richness it is also likely that they make some use, however unsystematically, of described to undescribed species ratios (see also below) in the small groups with which they are most familiar. If the group already contains (say) 100 nominal species, and the taxonomist in question is aware that 10 of these are not 'good' but is also aware of a further 60 undescribed species, the new provisional total for the group will be 150 species, representing an increase of 50%. The value of this figure for generalising will, of course, depend very much on how typical the sample group is and how well the available material represents its true size.

Nevertheless, if accepted for what they are, and if we accept also that recent estimates by taxonomists are based, in comparison with their predecessors, on a relatively extensive (if still fragmentary) coverage of the world, the surely conservative figures produced by the cautious and pragmatic approach may have considerable value as *minimum* estimates.

Gaston's conclusions have attracted strong criticism (Erwin,

1991), the main focus of which is that the reliability of results obtained in this way is impossible to judge; taxonomists' estimates represent opinions that have been arrived at in ways that we cannot know. The arguments for and against have broadened to include the merits of other approaches as well as the usefulness of collated opinion, providing an area of active debate (see Gaston, 1992).

First principles and empirical relationships

The broad understanding we have of how life evolved and how species interact could be used to estimate, from first principles, how many species are likely to be found in a given region or in the world as a whole (May, 1988). General rules concerning: body size relations, commonness and rarity, range sizes, and the relationship between species numbers and area have all been used to suggest explanations for observed species richness patterns and why there are so many (or so few) species overall. Understandably, only tentative use has been made of rules of this type for actually predicting major species richness patterns for poorly-known groups. Any real test of their predictive power in these areas awaits the provision of many more data concerning the exceptionally diverse but little-known groups than are available at the moment. This applies, for example, to the empirical rules, derived mainly from the larger terrestrial animals, that describe the way in which species numbers increase with decreasing size. Using only described species these rules begin to break down at body lengths of below about 1cm. Arbitrary extrapolation to smaller size classes (down to lengths of about 0.2mm) that are poorly represented among described species produces an estimated global total for terrestrial animals of around 10 million species (May, 1988).

The use that may be made of other empirical relations that concern the structure of food webs, and the numbers of parasitic or other symbiotic species that are typically associated with individual host species, has also been well reviewed by May (1988, 1990). While rules concerning the number of levels in food webs are sufficiently well established to form the basis for relatively reliable generalisation, the same cannot be said for the numbers of species and overall numbers of links involved in webs of various types. Species richness patterns involving parasite, parasitoid or (less often) predator species and their hosts or prey have received much attention. In well-known regions such as the British Isles it is possible to calculate the approximate number of potential host species for a given group of, for example, parasites and relate this to the overall number of species of these parasites that are present. If we take British vascular plants (2,089 species) and the insects that directly exploit them (assuming this to be around 25% of the British total or c. 5,500 species) as an example, we can derive a ratio, in this case of around 2.6 (associated insect species) to 1 (plant species). This type of simple relationship tells us very little, of course, about host-specificity. Nevertheless, the question of host-specificity levels, rather than any empirical relationship between the number of hosts and the number of associated parasites, has received some attention as a possible means of predicting overall numbers of parasite species. The difficulties involved in evaluating such patchy host-specificity data as exist and using them for extrapolative purposes are great (see May, 1990 for discussion).

Keeping to vascular plants and their associates as the example, we see that the most useful data on how many species may be effectively specialised to one host come from detailed single species studies. Intensive studies, whether of oak trees or passion vines (see May, 1990), may help to reveal something of the processes underlying the way in which these plants are exploited, while at the same time elucidating a series of contrasting patterns. However, they cannot be expected to provide what is required for any prediction of overall numbers of plant associated species. The simple questions for which answers are needed here are: how many species depend on the *average* plant species throughout its range, and how many species depend on the same average plant species in one place at one time? For practical purposes it is also advantageous if these data can be related to sampling phenomena, so that it is known what proportion of the associated species present at one place are obtained in a particular type of sample.

As far as species associated with green plants are concerned there are indications that patterns vary with moisture, latitudinal and other gradients. Host-specificity levels may also tend to be lower where plant species richness is especially high, particularly when the plants in question are trees, as in tropical moist forests. Indeed, there are strong suggestions that the general architecture of forests may be a better predictor of the number of small animal species and fungi present in a given area than is the number of different vascular plant species that occur.

Taxon to taxon and region to region relationships

Using some aspect or aspects of the diversity profile of a well-known group such as birds or mammals as a reference point, a variety of simple extrapolations to other less well-known groups may be made. We may use butterflies, a well-known group, as an example. Of the roughly 22,000 species of insects to be found in Britain some 67 are butterflies. The number of described species of butterflies in the world is fairly accurately known at around 17,500, the true figure almost certainly not exceeding 20,000 or so. If the ratio of butterfly species to all insect species is the same globally as it is in Britain then the world insect species total should lie at around 22,000 x 17,500/67, that is 5.75 million.

A more involved extrapolation may be made by taking tropical to extratropical ratios as the point of departure. For both birds and mammals, for example, there are roughly two to three times as many tropical as non-tropical species. To extrapolate successfully from this we need to have a good estimate of the proportion of described species that are from extra-tropical areas in the more significant of the less well-known groups, coupled with a good estimate as to the proportion of extra-tropical species that have been described. In practice our estimates for the first are unlikely to be very accurate and for the second unreliable. However,

Figure 4.2 Time series of first descriptions of currently recognised species in decades from the time of Linnaeus (1758) to 1987

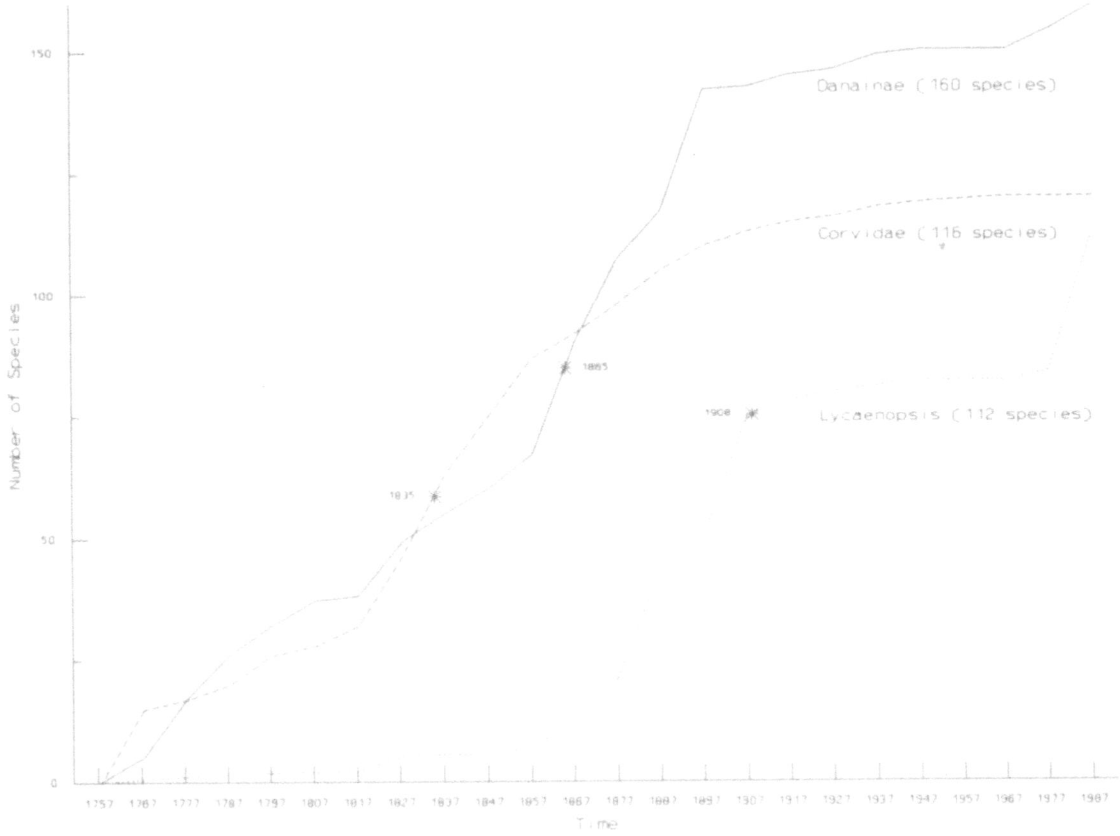

Source: Data for crows (Corvidae) based on Goodwin, 1986, *Crows of the World*, London: BM(NH), that for milkweed butterflies (Danainae) on Ackery and Vane-Wright, 1984, *Milkweed Butterflies*, London: BM(NH), and that for the *Lycaenopsis* group of blue butterflies (Lycaenidae on Eliot and Kawazoe, 1983, *Blue Butterflies of the Lycaenopsis Group*. London: BM(NH).

Notes: Asterisks on each curve indicate the points at which half of the 1987 totals had been reached. All three of the groups depicted are 'well-known' with few if any species left to be discovered and described. The rate at which new species of crows (Corvidae) were recognised and described declined steadily from the mid 1800s so that 90% were known by around 1880. Description of milkweed butterflies (Danainae) followed a largely similar pattern, with 90% of the apparently settled total achieved by 1937 and maintained for the next three decades also reached by around 1880. However, intensive studies over the past two decades have led to a further (and unpredicted) small burst of description. After a much slower start, the *Lycaenopsis* group of blue butterflies (*Lycaenopsis*) also reached a seemingly stable plateau (by around 1920). As with the milkweeds an unpredicted burst of description, although in this instance a much larger one, has characterised the last decade or so.

again taking insects as the example, if we take one million as the rough number of described species, and assume that (1) roughly 60% of described insect species are from temperate and boreal regions, and (2) 40% of extra-tropical species have been described, then ratios of two or three tropical species to one extratropical species give us world insect species totals in the range 4.5-6 million. There are few suitable data points to use for microorganisms and some of the other groups such as nematodes and mites discussed below, even for north temperate sites and regions, but extrapolations based on the pattern of species richness in vascular plants, various vertebrate groups and on butterflies all produce roughly the same kinds of answers for the remainder of the biota, including the insects.

All such calculations, of course, depend on how similar bird, mammal, butterfly or other patterns used in calculations are to those found in the much richer but less well-known groups. If we were, in fact, confident that patterns found in groups such as birds were universal we would be close to achieving reasonable understanding of the global picture. But just how 'typical' are these well-studied groups with respect to species richness patterns, including their local species richness in tropical as opposed to temperate areas, and the rates at which species accumulate as the area considered is enlarged? We know enough to be clear that latitudinal gradients of species richness are not the same in all major groups (although species richness does generally increase dramatically with reducing latitude). Turnover rates also vary substantially from group to group, although evidence presently available (mostly of course for well-known groups) fails to reveal any clear correlation between these rates and size or other significant biological attributes that might suggest large average differences between (say) mammals and small invertebrates. A more rapid turnover in tropical as opposed to temperate regions does, however, seem to be indicated by the evidence, and various explanations for this have been advanced.

However, in the absence of data that might be used for more direct approaches to calculating species richness in the largest and most poorly-known groups, simple extrapolations from well-known groups are likely to provide us with the most securely based, if very conservative, estimates attainable at present. To do distinctly better it will be necessary to identify clearly which of the poorly-known groups might eventually make a major contribution to the taxonomic inventory (see below), and gather fresh relevant data by direct sampling from nature.

The relationship between the number of described and undescribed species in any group requires comment. The usefulness of this relationship as a means of predicting the number of species in a group depends on the extent to which representative samples are available and the accuracy with which the proportion of species that are undescribed can be ascertained. In practice, the latter is generally time-consuming and difficult, if not impossible. Unfortunately, where most feasible (e.g. in very small groups and groups in which most species have already been described), the results obtained will tend to be uninformative. Where the approach is potentially most valuable (e.g. very speciose groups in which 75% or more of the species remain undescribed), it is most difficult to apply. Here, there is a premium on accuracy, but this can only be achieved by someone who has close familiarity with all of the described species that might be present in the sample. Nevertheless, the effort may be worth making for groups likely to make a major contribution to global species richness. Any indication as to whether undescribed species are, for example, around three times as numerous (i.e. 75% undescribed) or (say) 19 times (i.e. 95% undescribed) as numerous as described species would be of considerable value.

UNCHARTED REALMS OF SPECIES RICHNESS

Here we turn away from the existing taxonomic inventory and knowledge of species richness patterns in well-known groups to consider directly where the major part of as yet unassessed species richness might lie. For which ecosystems, taxonomic or other groups are there indications of great unassessed species richness? Is it possible to pinpoint the areas that it is essential to take into account if global totals are to be roughly estimated? Included in the discussion are the principal among the biological 'new frontiers' that have attracted attention in recent years. Evidence or the presumption that local species richness is at least sometimes high provides the first hint that a taxonomic group or a type of community might make a large contribution to the global species total. However, in sifting the stronger indications out from less telling anecdotes or the merely hyperbolic, it is helpful to remember that high local species richness, although necessary, by no means provides a sufficient demonstration that the group in question makes a particularly large contribution overall.

The marine realm

The oceans, occupying over two-thirds of the Earth's surface, have been described by Colinvaux (1980) as making up "a vast desert, desperately short of nutrients and with living things spread most thinly through them". This blunt description, dismal as it may seem, nonetheless provides an effective summary of what is known of marine productivity, turnover time and biomass. Average biomass (per unit area) in the seas has been estimated to be of the order of one thousandth that on dry land while marine productivity (again per unit area) is about one-fifth of the average for terrestrial systems (Valiela, 1984). In absolute terms it has been calculated, for example, that the world's seas produce some 92,000 million tons of plant tissue *per annum*, as against 272,000 million tons for dry land plants. Although new data may necessitate some revision of figures of this type they are unlikely to change the general picture. Against this background it may be unsurprising that there are few data to suggest that the oceans contribute more than a small fraction to the world total of species, at least of multicellular animals and plants. In contrast, the marine realm makes an exceptional contribution to biotic diversity at higher levels (all major eukaryote groups are represented and more than 80% of all phyla are restricted to the seas).

Of all currently described species it has been estimated that somewhat less than 15% are marine. The views of relevant taxonomists (see Barnes, 1989, etc.), supported by the generally rather high proportion of described species in samples taken from poorly studied areas, suggest that fairly high percentages of the marine 'macrofauna' (mostly species of molluscs, crustaceans and polychaete worms) and multicellular algae are already known. The position with regard to smaller organisms, including nematodes and protists, is very much less certain. Moderately high species richness at the local level can be found in some inshore communities where productivity is high, those of tropical reef systems providing good and well documented examples. However, total areas occupied by these rich communities are small and many of the species have fairly large ranges; thus local species richness of the apparently relatively well-described littoral and shallow water marine communities is not reflected in especially high regional or global described species totals.

Although the ranges occupied by most marine organisms are poorly understood, patterns observed in the better-known groups suggest that turnover of species, the rate at which species numbers increase with increasing area, may be generally lower in the seas, perhaps especially in the open oceans and the ocean depths. Unlike the continents the oceans are contiguous; also the deep sea appears to have few areas sufficiently isolated for boundaries to be defined and thus few limits to dispersal which, even for small sediment-dwelling animals, may be through planktonic larvae. Although volumetrically great, the seas are also architecturally not very varied. As noted above, systematists working on most marine groups (see Barnes, 1989) appear reluctant to suggest that large numbers remain to be described and, compared to terrestrial arthropods, for example, this may well be true for such groups as Echinodermata, the larger Mollusca and Crustacea, etc., as well as fishes.

The deep sea is one of the more remarkable biological 'new frontiers' that has become evident in the past few decades (see Grassle, 1989, 1991; Grassle *et al.*, 1991). Although some parts of the deep sea floor are apparently poor in species, high local species richness of macrofauna in deep sea sediments appears to be the rule over the fairly large areas that have now been investigated in the Gulf of Mexico, the West Atlantic (Grassle, 1991) and elsewhere. This is manifest mostly among polychaete annelids, certain groups of Crustacea and, to a lesser extent, molluscs. Low productivity, sediment patchiness and ease of immigration are among the factors suggested to explain this diversity. Distinct depth and sediment type assemblages have also been shown to occur, but there is little indication in the macrofauna of high turnover across all spatial scales. Indeed, the major part of local species richness seems to be exhibited at a very small scale, so that the majority of species to be found at one site are obtained by very few samples. The smaller organisms or meiofauna of deep ocean sediments often equal the macrofauna in biomass and are present in much greater abundance, the major component being nematodes. However, whether meiofaunal species richness equals or possibly exceeds that of the macrofauna remains to be established. Although relevant data may be forthcoming from studies in progress, as yet how nematode species of deep ocean sediments accumulate as we move from site to site is more or less unknown.

Although these new data on the deep sea, coupled with recent discoveries of a whole new realm of protistan, bacterial and other picoplankton suggest that total marine biotic diversity could be considerably greater than previously assumed, evidence to support the contention that this richness rivals that found in tropical forests, except perhaps at the smallest of scales (i.e. the range below $1m^2$) is wanting. New data on both pelagic and benthic microorganisms and the deep sea meiofauna may yet confound this view, but the evidence so far suggests that the oceans, including their poorly explored depths, contribute less to total global species richness, by an order of magnitude or more, than do moist tropical forests.

Parasites

Parasite loads for a few large animals (mostly vertebrates) and some green plants may be high, involving many parasites that are specific to a single host or a narrow range of host species. However, the overall numbers of large animal and large vascular plant symbionts, unless there are many more unknown than we suppose, are insufficient in themselves to make a very large contribution to global species richness. In contrast, very little is known concerning loads and levels of host-specificity with respect to the microorganisms, small nematodes, mites and others that are associated as parasites with members of the most species-rich groups, such as terrestrial arthropods. In relatively well-known areas such as the British Isles the recorded numbers of such parasites are low, but even here it is not unusual for small invertebrate animals to turn out on close examination to possess previously unknown parasites. Clearly, if there are many such undetected parasite species, their numbers could lead to a considerable inflation of global species figures. For example, if each insect species has, on average, one completely specific associated parasite or other symbiont this would entail at least doubling estimates of insect species to obtain a minimum figure for overall global species richness. As yet there is little evidence that this may be necessary, as where a range of insects and other small potential host species have been relatively well studied, large numbers of host-specific parasites have not been found. We may note that such negative results (absence of parasites) often go unremarked and unreported. There is also an inevitable general tendency for host ranges to be underestimated. In addition, it is reasonable to assume that the sometimes high parasite loads observed in widely distributed pest species are not, in fact, typical, and furnish a poor basis for extrapolation. We should also not be too eager to generalise from the situation in large vertebrates and vascular plants whose size and bodily complexity furnish many potential niches for exploitation. The great majority of organisms, small in size, clearly offer very different opportunities to potential parasites. On first principles, levels of parasitism may be expected to vary very widely, depending not only on the size of the host but also its defences and its population structure. Potential hosts that are very hard to find will generally have few obligate parasites.

Fungi and microorganisms

Although far fewer species have been described than of green plants it has long been considered likely that the fungi (using the term in its traditional non-phylogenetic sense) might eventually prove to be the most species-rich of all groups, insects excepted. Interestingly, at a time when only a few thousand species of fungi had been described, some 19th century mycologists early on recognised the likelihood that some hundreds of thousands might actually exist. However, with around 70,000 described species now recognised, we are still not in a position to say much more than this about the size of the group. In the absence of good data on tropical fungal communities, on latitudinal or other gradients in diversity, and how the numbers of fungus species accumulate as we move from one spatial scale to another, any estimates of overall fungus species richness can only be tentative.

In a thorough review of the significance and possible magnitude of fungal diversity, Hawksworth (1991, and see this report) has settled on 1.5 million as a conservative estimate for the world's species of fungi. This figure was arrived at by taking into account several types of evidence, but finds its most firm basis in the relationship between the number of species of fungi known to occur in the British Isles and the number of British species of vascular plants. The list of fungus species recorded from the British Isles currently stands at around 12,000. Taking a figure of 2,089 (i.e. garden species, etc. excluded) for British vascular plant species, we arrive at an approximately 6:1 ratio in favour of the fungi. Applying this ratio to a conservative global figure for vascular plant species of 270,000 yields a global total for fungi of around 1.6 million species.

As already discussed above, the reliability of extrapolations made in this way depends on the extent to which species richness patterns are shared, in this instance between fungi and vascular plants. At least some fungus species have extremely large ranges; should average range size in fungi be significantly greater than the average in vascular plants, some lowering of the 1.5 million figure for fungi would be in order. Similarly, should fungi exhibit a less steep latitudinal gradient in species richness than that found in vascular plants, this should also point to a lower figure. Data on tropical fungi remain extremely scant, but we may note that the rather low proportions of undescribed species found in recent tropical collections as yet provide no indication of especially great tropical diversity.

Taking a cautious approach similar to that adopted here

towards other poorly-known groups, a minimum figure for global fungus species might be put at around half a million. An alternative, less cautious but well-supported, approach is presented in Chapter 6 of this book. The arbitrary 'working figure' of one million incorporated in Table 4.3 represents a compromise between this and the 1.5 million estimate given by Hawksworth (1991).

Microorganisms, including the smaller fungi, algae and 'protozoans', as well as bacteria and viruses, present the greatest challenge to any serious attempt to assess the overall scale of global species richness. The great genetic diversity and general significance of microorganisms is highlighted in Chapter 6, where the problem of applying to them the species concepts that are more or less consistently used for many larger organisms is also discussed.

What is clearly an immense diversity of very small organisms, perhaps especially bacteria, viruses and unicellular algae, remains largely unaccounted for by the existing taxonomic inventory. However, whether the diversity of these organisms, often lacking sexual processes and many of them clonal, is best expressed in terms of the number of phenetic groups recognised as species is a moot point. The comparability of, for example, viral 'species' and those of multi-cellular organisms, in which sexual reproduction predominates, is very questionable. Virtually nothing is known of any latitudinal or other gradients of diversity that microorganisms might exhibit while, even in temperate regions, at no scale is species richness well documented. Probable range sizes are also known for very few species, but very small organisms (and those with very small dispersal stages, such as fungal spores) are known, in some instances, to have very broad if not cosmopolitan distributions. Coupled with a generous measure of caution in extrapolating too far from the known, all of these considerations are reflected in the arbitrary 'working figures' for species richness of microorganism groups given in Table 4.3.

Nematodes, mites and insects

Despite a considerable increase in resources devoted to nematode taxonomy over the past few decades and a commensurate surge in the rate of description of new nematode taxa, this group of worms probably still remains the least well inventoried group of metazoan animals. Although relatively early attention had been devoted to some of the larger and, in human terms, more significant parasitic species, up until 1860 only 80 species of plant, soil and freshwater species had been described. This compares with an annual rate of around 140 species of the same groups described in the 1960s and the present overall description rate (including parasitic and marine taxa) of more than 300 species per annum. The current total of described species is very uncertain but has been estimated to stand at around 15,000.

Nematodes

Indications that nematode species richness may be of an extremely high order stem more than anything from the abundance of free-living forms (a few millions of individuals may be present in 1km^2 of suitable soil or mud) and the great number of free-living species that may be found in samples taken from a very small area. Two hundred or more species have been reported from samples of just a few cm^3 of coastal mud.

While parasitic species totals may prove to be significantly high (see above), and free-living terrestrial and freshwater species also very numerous (Poinar, 1983), recent work on estuarine, shallow-water and deep-sea sediment nematodes suggests that the marine realm (see above) could make an even greater contribution to a total count of the world's nematodes. However, how high levels of species richness at the smallest scales bear on the question of the overall number of nematode species remains unclear. Good data on species turnover in both terrestrial and marine nematode assemblages are conspicuously lacking, as is any indication that assemblages of tropical nematodes are especially rich. In the absence of any direct indication of massive unaccounted for species richness at larger scales a somewhat cautious approach to estimating the likely overall number of nematode species is probably advisable. However, it would be surprising if this number were not at least some hundreds of thousands.

As in the case of protists and other microorganisms the taxonomic study of nematodes is made difficult by uncertainties with regard to the application of species concepts. Many species are entirely uniparental or contain some uniparental populations. Apart from their frequently very small size, the sorting to species of nematode samples is often hampered by a very low incidence of diagnostic males. At best, species recognition is beset by many difficulties and may, in some instances, remain frankly subjective.

Mites

In the case of mites (Acari) there are fewer problems with interpreting species limits but, as with nematodes, the number (around 30,000 or so) of described species clearly represents only a small proportion of the actual total. Knowledge of tropical mite faunas in particular is very scant, lagging well behind that of other arachnids, including spiders. Reliable quantitative sample data that give anything more than a hint of what mite species richness might be at any site in the tropics appear to be unavailable. However, it may be reasonable to expect that free-living terrestrial mites, although flightless and differing from insects in various other respects, do roughly follow patterns, in terms of coexistence, range sizes, turnover, etc., already tentatively established for certain insect groups. If so, and despite the fact that we have a less complete knowledge of temperate mites than, say, of beetles, it is difficult to envisage a world total of less than a few hundred thousand species. Suggestions that the global number of mite species is in the region of one million or even higher may prove defensible once good data for tropical sites are forthcoming.

Insects

There is abundant evidence to suggest that insects exhibit high species richness at most scales (i.e. from a few m^2 to ecosystems) except perhaps the very smallest. The number of already recognised and described species - around one million - is sufficient to establish that insects comprise a substantial portion of the world's species. Most insect groups are taxonomically tractable and the rate at which the process of inventorying advances depends largely on the level of resources devoted to the task. Samples containing many species can often be fairly rapidly as well as reliably sorted, and this makes several major insect groups suitable for a range of species richness studies, even when most of the species being examined are undescribed. Some of the ways in which data from samples of tropical insects may be used to tackle the problem of assessing insect global species richness are discussed below. We may note, however, that attaining any reasonably accurate idea of what proportion of species in total are insects is less easy. This is likely to depend as much on achieving advances in estimating the diversity of microorganisms and other poorly understood groups as on better data for the insects themselves.

Tropical forest canopies: the height of tropical diversity?

Tropical forests have long been known to harbour a great richness of life and, although they cover only 6% of the earth's land surface, it has been widely supposed that they may contain as many species of organisms as, or even more than, the rest of the world together. One part of these forests, the world of the tree tops, has tended to evade close inspection by biologists but, with the development over the past two decades of new methods for studying forest canopy organisms, notably (but not only) the use of insecticide fogging techniques, canopy communities even in tall tropical forests have become much more accessible (Erwin, 1990).

There is now sufficient information to indicate that local species richness of many of the insect and other arthropod groups that have been the main focus of recent attention are very high in tropical forest canopies, much higher (often by a factor of 10 or more at the level of a single tree) than in temperate forests. It is equally clear that not only are a high proportion of the species undescribed (this is the case for all strata in moist tropical forests) but a proportion of them are not or are only exceptionally found at lower levels. Data have now been gathered that give some idea of the usual sort of numbers of species of at least some of the more important insect groups (notably Coleoptera and Hemiptera) to be found in various neotropical and palaeotropical forest canopies at the level of individual trees and small quadrats (e.g. 12 x 12m), up to about the one hectare level.

Fewer data are available to allow confident estimation of canopy species numbers at a larger scale within relatively uniform tropical forest. Indications are that much of the patchiness in the canopy is at or below the one hectare level and that samples from adjacent hectares are about as different in species composition as samples taken several kilometres apart. The picture that is beginning to emerge is of a mosaic less defined by tree species than by a variety of other factors, including the condition of each tree, and the patchwork distribution of resources, including epiphytes, that manifests itself at a much smaller scale than an individual tree canopy. Some data are available to show that adjacent but radically different forest types have very different canopy faunas but inadequate sampling does not allow any even remotely accurate estimation as yet of the extent of 'turnover' in moving from one forest type to another, or whether this is higher or lower than species turnover in the forest's lower strata.

In sum, quite enough is known to indicate that high local species richness (although not of all groups) and considerable patchiness at quite a small scale are typical of tropical forest canopy arthropod communities. How large a contribution canopy-dwelling species or species that are present in canopy samples (not exactly the same thing) make to overall arthropod species richness at one site is less clear. The contribution made by canopy species to faunas at regional and other scales is even less well understood, despite claims that the canopy is where maximum tropical biodiversity occurs (Erwin, 1990).

Against this background, it is rather surprising that speculations as to the number of species of arthropods that might be found overall in the canopies of tropical forests (Erwin, 1982, etc.) have come to occupy centre stage in recent general discussion (May, 1988, 1990; Stork, 1988; etc.) of the possible magnitude of the global species inventory. At the same time, and stemming from the view that tropical forest canopies harbour an unparalleled diversity of life, suggestions that the global species total for terrestrial arthropods alone may be as high as 50 or even 100 million have also been widely reported, and have found expression in a number of reports concerned with the conservation of biotic diversity (Wolf, 1987; Reid and Miller, 1989; National Science Board, 1989; etc.). The attention paid to these suggestions perhaps justifies a closer look at data that may give some hints as to the likely richness of tropical forest canopy arthropod assemblages.

Tropical forest canopies: reassessment of the evidence

Critical examination of the available data (many of them still unpublished) might usefully begin with some evaluation of how fully the richness of canopy arthropod assemblages is reflected in samples that are routinely studied. Most of the significant data points come from insecticide fogging studies. The proportion of species that might be expected to be obtained by this technique has been the subject of some discussion (Adis *et al.*, 1984; Erwin, 1990; Stork, 1991; etc.), but without firm conclusions being reached. However, restricting attention to adult stages only, we know that some species that mine or burrow within living or dead plant or fungal tissue and some of the fauna of suspended litter and soil are poorly collected by fogging, as are certain arthropods that are firmly attached (e.g. scale-insects) to leaf surfaces, along with an uncertain proportion of the larger species of some groups that may escape capture by flight. On the other hand, species that are present as 'tourists', most of them presumably resting on exposed surfaces or in flight, seem to be well sampled locally. Characteristically, their pattern of occurrence in the canopy is patchy and unpredictable, with the result that tourist species accumulate steadily as sample size is increased. A good number of groups (e.g. ladybirds, ants, adult psyllid bugs, etc.) seem to be sufficiently well sampled by fogging that results give an accurate impression of the relative and even absolute abundance of individual species, as well as a good account of which species are present.

However, canopy samples obtained by means other than the application of insecticides reveal that a proportion of true canopy species are not or are not readily taken by fogging. The most telling evidence for this comes from studies

(Hammond, 1990; Hammond and Stork, unpublished) where canopy fogging has been carried out in tandem with additional extensive sampling of both canopy and lower forest strata by other means. In such instances we find a certain number of species well represented in, for example, baited traps or interception traps placed in the canopy, but absent from traps of the same type operated at ground level as well as from fogging samples.

Ignoring the proportion of species (probably rather small) that are not well sampled by the technique, how much fogging is necessary to give a reliable picture of the size of a local canopy arthropod community, and how are its components distributed? A number of studies in both temperate and tropical countries suggest that, with an appropriate pattern of sampling (including adequate seasonal coverage) relatively few trees or quadrats may be needed. Particularly good evidence on this point is emerging from the results of a fogging programme carried out in a relatively uniform tract of lowland tropical forest in Sulawesi (Hammond and Stork, unpublished). In this study a number of samples, covering all seasons, were taken from each of 20 different 12 x 12m quadrats distributed through a 500ha study area. A strong indication that a representative sample of the canopy insects present in the study area was obtained is furnished by the rate at which species accumulated with sampling effort (see Fig. 4.3).

How near are we to determining the proportion of all arthropod species present in a given tropical forest that are likely to be taken by canopy fogging, and is this more or less a constant? If canopy samples are to be used as a means of directly estimating overall species richness of a forest, either locally or at a larger scale, it is clearly vital that the relationship between numbers of species present in the canopy and the number of species found overall be roughly understood. If canopy samples are to be used for comparing local species richness directly it would obviously be helpful if proportions varied little from one place to another. Finally, if global figures for arthropod species richness are to be derived from canopy fogging data (see below) these will be on a particularly sure basis if the number of species present in canopy samples is a very high as well as constant and a known proportion of the whole. That this is the case, for neotropical forests at least, has been asserted by Erwin (1991) who in earlier work (1982) suggested that canopy arthropod communities were at least twice as rich overall as those of the forest strata below. Working from first principles, this sort of relationship might seem unlikely. Most of the production of living tissue in a forest starts off in the canopy, but most of this - fallen leaves, fruit and wood, insect, bird and other excrement, and whole fallen trees - ends up forming a rich mosaic of resources on the forest floor. Not surprisingly, the abundance and biomass of arthropods is greatly skewed in favour of the lowest levels in a forest. Strictly comparable figures for both canopy and forest floor are not available, deriving as they do from fogging samples for the canopy (undersampling internal and concealed feeders, etc.) and a range of different 'standing crop' methods for the forest floor. For example, in neotropical forests investigated by Adis and Schubart (1985), disregarding the Collembola and mites which made up 60-80% of the individuals in soil/litter samples, an average of around 30 times as many arthropods were found, per m^2, in the soil/litter layer as in the canopy. Methods used in studies such as this are known to undersample small arthropods, mites and springtails in particular, because of poor extraction from soil and other substrates, and also ignore or underplay the large contribution made by significant but patchily distributed resources such as carrion, fallen fruit, large fungus fruiting bodies and decaying wood.

Both baited traps and those not involving attractants (e.g. Malaise traps and window traps) collect far fewer individuals and species at canopy level than on the ground. This is a common finding of studies in several countries. Some tropical studies (e.g. Hammond, 1990), for example, show a relationship of around three species of Coleoptera in ground-level Malaise trap samples to one for the same trapping effort in the canopy. A much higher ground to canopy ratio is characteristic for some other groups (e.g. Hymenoptera) and higher ratios all round are generally found in catches from interception or other traps that do not favour plant-climbing species.

Apart from temperate forests where the overall proportion of species present at a site that can be found in the canopy probably rarely exceeds 20%, the most compelling evidence for much lower local species richness in the canopy than at other levels comes from the Sulawesi study already mentioned (Hammond, 1990), where as complete an inventory as possible was made of the Coleoptera and some other insect groups found in the 500ha study area. The extensive canopy fogging that formed part of the sampling and inventorying programme produced around 30% of the beetle species found in total, and around 20% of those conservatively estimated actually to occur in the study area.

More than three-quarters of the species taken by fogging in the Sulawesi study were also present in samples of various types taken at ground level. Analysis of their pattern of occurrence in all ground and canopy-level samples suggests that many of these were present in the canopy only as 'tourists', and that overall less than two-thirds of species found in the canopy belong to the canopy fauna proper, either as 'specialists' (species largely restricted to the canopy) or 'generalists' (species found regularly both in the canopy and at lower levels). Making allowance for canopy species not obtained by fogging, canopy species proper amount to at most 20% of the area's species, of which no more than half (i.e. probably less than 10% of the total fauna) may be regarded as canopy specialists.

Results from other palaeotropical and from neotropical sites suggest that although canopy insect species richness in tropical moist forests is somewhat variable, it is not exceptionally low at the Sulawesi site. Somewhat higher levels of local species richness might be expected, however, in canopies that contain more tree species and forests in which canopy, understorey and ground layers are more clearly demarcated. Data available for temperate forests suggests relatively weak stratification, a very small canopy specialist component and a 'typical' overall canopy to ground arthropod species ratio of around 1:10 or more. Variation is to be expected in tropical forests, with the lowest ground to canopy ratios most likely to be found where the ground component is relatively small (e.g. dry

Figure 4.3 Accumulation of beetle species in canopy samples

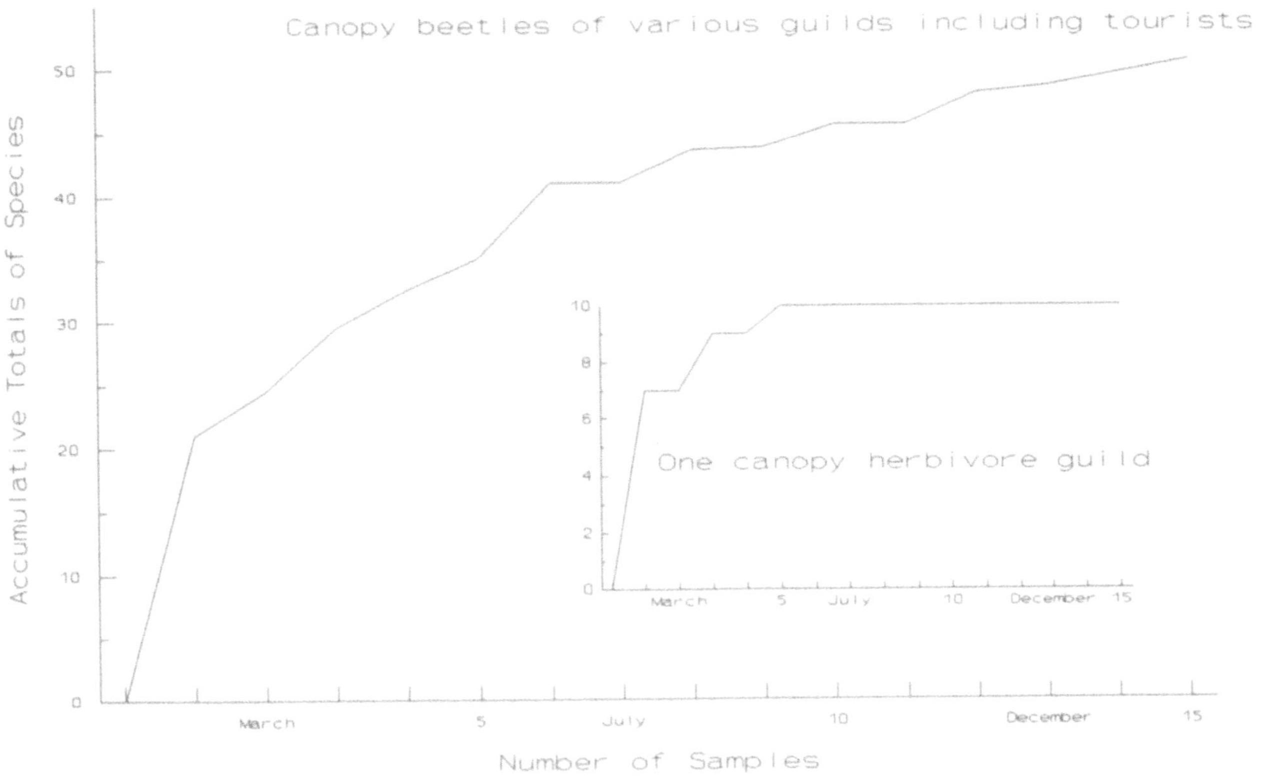

Notes: Beetle species in canopy fogging samples from a single tropical site and how these accumulate with increased sampling effort. The upper curve is for a 'representative' selection of 51 species (out of 900 beetle species in the total sample) comprised of 23 'regular' canopy species and 28 that are present in the canopy as 'tourists'. It shows a steady decrease in increments with sampling effort but no distinct flattening. The inset curve is for some of the species - the 10 members of a herbivore guild (broad-nosed leaf-chewing weevils that are all either canopy specialists or generalist species regularly feeding in the canopy) - included in the upper curve. This shows how, with a dataset restricted to canopy species proper, the species accumulate much more rapidly, in this case reaching a plateau after 5 (out of 15) samples had been taken.

forests) or, conversely, where the canopy component is high as a consequence of great stratification, as may be the case in some of the tallest closed-canopy moist forests. The ratios found for Coleoptera in Sulawesi (about one in five species belong to the canopy fauna proper, about one in ten species are canopy specialists) may not be modal for tropical forests, but further results are needed before any firm view on what 'typical' ratios are can be taken.

Despite the large numbers of arthropod and other species to be found in tropical forest canopies, there are few data providing any clear support for the view that the upper levels of tropical forests are truly the "heart of biotic diversity". If anywhere, it would seem more likely that this is to be found on and under the forest floor.

SAMPLING THE HYPER-DIVERSE BUT POORLY KNOWN

Knowledge of large organisms and some temperate regions provide an inadequate basis on which to extrapolate with any confidence to groups and areas that are poorly-known. Well-established species richness patterns exhibited by groups such as birds are, of course, a useful starting point in attempts to gauge better the species richness of less well-known groups, but there is every reason to suppose that they provide no more than general guidance. Well-known organisms are a biased sample of the biota. Apart from being mostly large, they may also be unrepresentative in many other ways. Vascular plants, for example, may be much less dependent on surface moisture levels than many small animals. Butterflies, unlike the majority of insects, are all essentially herbivorous.

Taxonomic groups, functional groups and ecosystems that might be expected to make the largest contributions to global species richness have been briefly surveyed above. For some of these, there are strong indications of considerable diversity that is as yet unaccounted for by the taxonomic inventory, while for others the hints are more vague. Many more data for these poorly-known groups and areas of the world are needed for the magnitude of their contributions to biodiversity to be even roughly assessed. How some of these data might be gathered and applied to species richness estimates is discussed below.

What, where and how

Almost any new data on species richness patterns in the groups discussed in the previous section are likely to prove useful, but the pace at which our understanding of these patterns improves will depend heavily on which data we choose to gather first, and on how economical and effective the methods are that we adopt.

The questions of what and where to sample and how best to gather sample data to improve our knowledge of major species richness patterns has been well reviewed in a recent report (Solbrig, 1991) where the need to focus efforts on high diversity groups and ecosystems is highlighted. More precise proposals with regard to the choice of sites for intensive study and the choice of indicator or focal groups (see below) have been advanced by di Castri *et al.* (in press). Clearly, there is an urgent need for better data on all of the hyper-diverse groups: insects, nematodes, fungi, bacteria, etc.. However, it is equally clear that we cannot expect progress to be made at an even rate on all fronts. The point of departure varies from group to group, as does the ease and reliability with which good sample data may be obtained. Some groups are distinctly more tractable than others, in the sense that large samples may be rapidly and reliably sorted to species.

To make the most of the considerable effort involved in gathering species richness data for groups of any size, two complementary approaches are necessary. The *intensive* approach entails in-depth studies, inevitably feasible for large groups at only a few sites, aimed at establishing the number of species present as precisely as possible. If coupled with appropriate quantitative sampling, the process of intensively inventorying a single site may be exploited to identify and calibrate methods that are needed for studies of a more *extensive* type. Thus, complete site inventories are needed to furnish the 'knowns' against which sampling methods can be calibrated and more extensive sample data compared. The actual methods used for inventorying will, of course, vary from group to group, habitat to habitat, and biome to biome.

The current emphasis on terrestrial arthropods in biodiversity research is perhaps to be explained as much by the general amenability of these animals to study as by the likely size of their contribution to the global species inventory.

In extensive studies of hyper-diverse groups it may often prove necessary to deal with just part of the group rather than treat it in its entirety. In such instances the 'indicator' group or groups chosen need to be as 'representative' as possible. It is also helpful if, in species terms, they constitute a more or less unvarying proportion of the group as a whole.

Where to look first if we aim to advance rapidly our knowledge of species richness patterns in the ultra-diverse groups is fairly clear. In the marine realm there is an evident need for many more data from the ocean depths. For terrestrial organisms in general the most urgent requirement is for more data from the moist tropics. Despite their undoubted richness, tropical forests remain the least well studied of major terrestrial ecosystems.

Kinds of extrapolation

Extrapolation of one sort or another is likely to be employed at every stage in the process of assembling and interpreting species richness data on poorly studied groups of organisms or regions. Although all extrapolative procedures involve the same assumption: that a ratio obtaining in a known situation holds in an unknown one, some kinds of extrapolation may, in practice, be seen to be more trustworthy than others.

For the purposes of this discussion, perhaps the most important distinction to make is between ratios that are extrapolated from one site to another and those that are used to extrapolate across spatial scales. Some of the different kinds of ratio that may be extrapolated from site to site have already been mentioned above while discussing the intensive/extensive approach to obtaining species richness data. Most commonly, when dealing with sites of a generally similar type, ratios used will be those relating less complete (sample/focal group) data to more complete (inventory/larger group) data. Here, the reliability of extrapolation will depend in part on how extensively the ratio has been calibrated, but also of relevance is the notion of comparing like with like. For example, a ratio that has been shown to obtain at a series of sites in the moist tropics might well be considered unlikely to hold at temperate sites.

It goes almost without saying that species richness data for poorly-known groups that we may wish to use as the basis for extrapolation will generally relate to single sites, as few reliable data for larger areas are available. If we start with single site data and wish to extrapolate to species richness of such groups at the regional or global level, we face a dilemma, as the ratios needed can only come from the few very well-known groups of organisms in which species number relationships across spatial scales are more or less established. Such ratios, derived as they are from groups which in the main may be expected to have quite different species turnover rates, should be used only with the greatest caution.

It is, of course, possible to extrapolate directly from species richness data for a single site or even a single sample to species richness at the ecosystem, regional or global level. Naturally enough, approaches that offer the possibility of moving from sample or site figures to global figures in a single step are tempting to use. However, given its inevitably speculative nature, extrapolation in this way is probably best avoided. The limitations of methods that involve empirical species richness relationships between very different groups of organisms (e.g. vascular plants and insects, butterflies and nematodes), host specificity levels, and proportions of species remaining undescribed have already been discussed. In some instances, ratios made use of (e.g. host: parasite species numbers) are likely to be extremely poorly calibrated. In most cases, the extrapolation from site to region or globe involves the essentially unsafe (and often unstated) assumption that the relationships used scale evenly (see May, 1990).

NEW DATA ON TROPICAL INSECTS AND WHAT THEY CONVEY

It is widely assumed that insect species outnumber all others. The belief is not without some foundation, as more than half of all described species are insects, and it is evident that at least several times as many remain undescribed. Ultimately, however, the question of the size of the contribution that insects make to the global species inventory is not to be settled by data on the insects themselves. A much improved understanding of microorganismal diversity and a better appreciation of species richness in groups such as the fungi and nematodes is needed for the insect contribution to be seen in perspective. This said, the insect part of the equation is a matter of obvious interest, particularly if we concede that an approximate answer to the question of how many insect species there are is within reach.

In comparison with other speciose groups such as nematodes or mites, knowledge of tropical insects is relatively advanced. Although the actual evidence remains fragmentary and anecdotal in the main, it has long been recognised that the tropics, and moist tropical forests in particular, contain far greater numbers of species than extra-tropical regions. Arguably, therefore, a reasonably accurate estimate of the number of tropical insect species would provide a good indication of the scale of insect species richness overall. For some of the smaller and best-known insect groups, such as butterflies and dragonflies, tropical species richness patterns are, in fact, rather well understood. The same cannot be said of the largest insect groups, although enough is known concerning a range of family-level taxa to suggest that the proportional representation of these groups (Coleoptera, Diptera and Hymenoptera) in the tropics may differ significantly from that in well-studied parts of the temperate regions.

New quantitative data, including a number not yet referred to in print, are beginning to both broaden and give greater precision to our understanding of tropical insect species richness and how it is distributed. However, few hard data on the number of species of any of the major insect groups to be found at individual tropical sites have yet emerged. Only for the very best-known groups, such as butterflies, is there any sound appreciation of turnover rates and the relationship between single site and regional species richness.

In spite of these difficulties, two datasets concerning the number of species of major insect groups present in large samples taken at moist tropical sites have already been used (Erwin, 1982; Hodkinson and Casson, 1991) to generate estimates for total tropical and also global insect (or arthropod) species richness. The estimates produced from these now widely quoted studies, both of them involving explicit assumptions, but with regard to ratios of very different kinds, are strikingly divergent, with Hodkinson and Casson arriving at a figure of around two million for insects globally and Erwin at a figure of 30 million for arthropods in the tropics alone. If correct, the first figure implies that around half of all insect species have already been described, while the second would suggest that undescribed insect species outnumber those described by a factor of 30 or more. However, not too much significance need be read into the discrepancy between the results, as both approaches entail the use of ratios that are essentially uncalibrated. Recognising this, Erwin's (1982) original calculations have been tentatively reworked by others (e.g. Stork, 1988; May, 1990), illustrating well how ostensibly reasonable but different assumptions will produce widely varying results from the same chain of reasoning. The same applies, if with less force, to Hodkinson and Casson's calculations (see below).

Hodkinson and Casson use a single data point - the number of species of bugs (Hemiptera *sensu lato*) in samples from the Dumoga area of N. Sulawesi, Indonesia. They suggest that the bug samples studied "contained a significantly high proportion of the species present", but there is good reason to suppose that the recorded total of 1,690 species represents a considerable underestimate. However, for the first of the two separate calculations employed by Hodkinson and Casson, the extent to which their data accurately reflect the size and composition of the bug fauna of their study area is not directly relevant. They begin by estimating the ratio of undescribed to described species in the Dumoga sample of bugs and then, treating this as a subsample of the world bug fauna, extrapolate directly to a global figure for the group. Only two considerations are of significance here: the accuracy of the undescribed to described ratio for Dumoga bugs, and whether the Dumoga sample is in fact representative in global terms. On the second count, we lack the data to make any reasonable judgement, but with regard to the first it is clear that the estimates on which the ratio is based, as might be expected, are in no way precise. In fact, the figure of 62.5 for the percentage of species undescribed could well turn out to be rather conservative.

The second line of attack adopted by Hodkinson and Casson begins with the number of undescribed species of Hemiptera (see discussion above) considered to occur in the Dumoga area (i.e. 62.5% of 1,690 = 1,056) and the ostensibly empirical relationship between this and the number of tree species found there, estimated to be around 500. Direct extrapolation to the tropics as a whole (with an estimated 50,000 tree species) yields a figure of 105,600 undescribed tropical bug species. Added to the 81,700 species of bugs already described, this furnishes a total of 187,300, no allowance being made for undescribed extratropical species. It should be noted that the relationship presumed to exist between the numbers of bug species and numbers of tree species present in a given area includes the hidden assumption that this scales evenly, that is to say that an area containing, for example, 5,000 tree species may be expected to contain 10 times as many (rather than 5 or 20 times as many) bug species as an area with 500 tree species. This problem of scaling is as relevant to empirical relationships of the type considered here as it is to those based on host-specificity (see discussion in May, 1990).

For both sets of calculations Hodkinson and Casson scale up to global insect species overall by using figures of 7.5% or 10% for the proportion of the world's insects that are Hemiptera. The first of these figures represents the proportion of described insects that are Hemiptera, more reasonably put at around 8.5%, and the second is the proportion of insect species in Bornean canopy fogging samples that are bugs. Both are probably over-estimates. Bugs, like several other mainly plant-associated groups, are known to be over-represented in fogging samples; for a number of reasons, including their taxonomic apparency, it may be reasonable to assume that bugs are proportionately better described than the insects as a whole. Taking a figure of 5% (rather than 7.5% or 10%) as the proportion of insects that are bugs and applying this to the revised bug estimates produced above, we see that it is possible to reach figures for world insects in the range 6.5 to 11 million rather than the two million or so that Hodkinson and Casson conclude with.

The ostensible basis for the estimate of 30 million tropical arthropods obtained by Erwin (1982) is an interesting study of the beetles (of some 1,200 species) obtained by fogging the canopies of 19 individual trees of the neotropical species *Luhea seemannii* (Erwin and Scott, 1980). In the light of how little is known of insect species:tree species relationships, this might seem an unlikely source for an estimate of tropical arthropod species richness. However, closer examination of the chain of reasoning adopted by Erwin reveals that the data obtained from the field on *Luhea* insects play a relatively minor part in the calculations. Of much greater significance in terms of the results are two major assumptions that are unrelated to the field data. The first of these, and one which we are far from being in a position to test concerns average levels of host-specificity in tree-dwelling tropical insects (see also May, 1990). The second assumption, one that, at least at the local level, is much easier to test, concerns the proportion of tropical forest species that are to be found in the canopy. Other factors involved in Erwin's chain of argument, including the proportion of canopy arthropods that are beetles, and the number of species of tropical trees, are less problematic, as the figures used may reasonably be expected to be of the right general order. It should be added that further implicit rather than explicit assumptions that relate to problems of scaling (see discussion in May, 1990) are involved.

The role played by the estimate of 163 for the number of beetle species specialised on the average species of tropical tree in Erwin's estimate is crucial. Essentially, it is this that generates the very high figure for tropical insect species richness that eventually emerges from his chain of calculations. Unfortunately, although there are good reasons to suppose that the degree of host-specificity exhibited by tropical canopy insects is generally low, there are few data that give even a hint as to what actual levels of host-specificity might be. More importantly, and as has already been noted, the use of host-specificity data for species richness calculations is beset with problems (see discussion in May, 1990). Even in the British Isles, where the host ranges and preferences of canopy-dwelling insects are relatively well documented, specificity data are far too imprecise to be used for any calculation of the number of tree-associated insect species.

Bearing these limitations in mind, reworking of Erwin's calculations may be viewed as of little practical value. However, it should be noted that truly staggering numbers are generated if the ratio of tropical canopy beetle species to tropical beetle species overall is revised in the light of findings discussed above. If the 1:4 or so canopy to total ratio found to obtain in Sulawesi is substituted for Erwin's 2:3, but all else in Erwin's chain of calculations is left as it is, we arrive at an estimate for tropical forest arthropods alone of around 100 million, rather than 30 million. If we should conclude, reasonably enough in view of what is known of tropical canopy insects, that beetles are typically less than 40% of canopy arthropod species, let us say 25% (see Stork, 1987), the estimate for tropical arthropods rises again to approaching 200 million.

Some of the relationships used by Erwin are important ones for almost any kind of estimates of global insect species richness that we might envisage, and some of these, for example the proportion of tropical forest beetles that are to be found in the canopy, are also amenable to test. However, this is far from true for the key relationship that Erwin employs, concerning numbers of beetle species that are effectively specialised on individual species of tree. In fact, it would seem likely that only when we know most of the answers that we are actually seeking, i.e. the number of species of insects to be found in the tropics and how many of them are found in the canopy, will we be in a position to start gaining some idea of how many are exclusively associated with the average tropical tree species.

The methods of estimating tropical insect species richness used by Erwin on the one hand and Hodkinson and Casson on the other have been discussed in some detail here with the intention of stressing the problems involved in the short-cut approach. Any extrapolatory route, from sample or inventory data to a summary for the tropics as a whole, that avoids the explicit use of ratios concerning relative species richness at different spatial scales is bound to be tempting. However, if the alternative is to invoke relationships that cannot be calibrated, the temptation is perhaps best avoided.

The valuable datasets (Casson, 1988; Erwin and Scott, 1980) on which the Hodkinson and Casson and Erwin estimates discussed above were based are just a part of a whole crop of new data that have recently become available for tropical insects. Although most results pertain to rather narrow taxonomic groupings, they are nevertheless leading to a steady improvement in our overall understanding of such questions as altitudinal gradients in species richness, species turnover at small spatial scales, and the contribution made by elevational assemblages and pronouncedly different but adjacent forest types to species richness at the level of the 'extended site'.

Data of a particularly extensive type have come from one recent large study based on an area of moist tropical forest in northern Sulawesi, Indonesia. The work of analysing results is still in progress, but many data concerning local species richness of beetles (Hammond, 1990) have already become available. The full dataset for beetles includes the results of quantitative sampling by a variety of means through all seasons of one year, as well as an inventory of species found within the principal study area (500ha of relatively uniform lowland forest). Valuable if less comprehensive data for several other insect groups, e.g. Hemiptera (Casson, 1988) and Hymenoptera (Noyes, 1989) are also available. The data from this study offer the possibility, for the first time, of (1) establishing a figure for overall local species richness of some major insect groups at a tropical moist forest site, (2) assessing what proportion of species is found in the canopy as opposed to lower layers (see above), and (3) of calibrating a range of sampling methods against knowns (total inventory results) in a tropical forest setting. Finally, the detailed sample data and inventory provide a comprehensive enough picture of the assemblage of insects present that, with sufficient general knowledge of their biology, it is possible to assess the proportions that belong to different functional groups, and that are associated with particular microhabitats and the various forest strata. The biases of various sampling methods with respect to these and other characteristics, such as body size and taxonomic group membership, may also be determined.

The findings of most direct relevance to overall tropical insect species richness to emerge so far from this study are:

- Species richness of Coleoptera at this tropical site, at scales of 1ha up to around 500ha is some five times greater than the average for a range of temperate forest sites. The species richness of Hemiptera, in relation to temperate sites, may be of the same general order, while that of Lepidoptera and Hymenoptera is also higher than in temperate forests, but by a less certain factor (probably between two and four)
- The numbers of species of some major insect groups and of insects overall that are found in the canopy are low compared with numbers found at ground level
- For an equivalent intensity and pattern of sampling, some of the sampling methods used obtain the same proportion of species present as they do at comparable sites in temperate regions (see Fig. 4.4).

In the long term, the last of these findings may turn out to be the most significant. Following calibration against the Sulawesi site inventory, simple 'sampling packages' that have already been shown reliably to reflect local species richness of Coleoptera and/or other major insect groups at 'known' temperate sites, might reasonably be expected to provide a good indication of species richness at other moist tropical sites. In fact, a number of trials of these sampling packages at a range of sites in the Indo-Australian and New World tropics have now been made. Assuming that the results being obtained (Hammond, unpublished) are reliable, they suggest ratios for the number of Coleoptera species between the tropical sites investigated and average temperate forests, that vary, except for one small tropical island with substantially lower beetle species richness, from around 3:1 to 8:1.

New data on the overall species richness of major groups at single well-defined sites make an obvious contribution to our general understanding of the pattern of insect species richness in the tropics. Furthermore, if accurate, they provide us with the essential base-line from which improved estimates of tropical insect species richness might eventually grow. For the moment, our poor understanding of species turnover in the tropics means that we have little to go on, if we wish to use single site data for extrapolation to regional or global figures.

Of course, starting with the ratio of five beetle species at a moist tropical site to one at a temperate site, crude extrapolation to a global insect species total is possible, but to do this it is necessary to make a series of major assumptions, not the least of which concern the proportion of insect species that are beetles and, as we have noted, species turnover rates. For a start, we may repeat the simple extrapolation made earlier on, based on the assumption that we are already able roughly to estimate the number of extratropical beetle species. If we take 400,000 as the number of described beetle species, and assume (no good count is available) that roughly 50% of described beetle species are from extratropical regions, and make an educated guess that around 50% of extratropical species have been described, an overall ratio of five tropical beetle species to one extratropical species yields a world total for beetles of 2.4 million, of which two million are tropical and 0.4 million extratropical species. If we then take the proportion of insect species that are beetles (see below) to be 33%, the figure we reach for insect species globally is around 7.2 million. This, of course, involves the dubious assumption that the tropical to temperate ratio scales evenly from site upwards, in both tropical and extratropical regions. Assuming much higher species turnover rates in the tropics, but bearing in mind that the extratropical component includes contributions from broad latitudinal bands in both southern and northern hemispheres, a tropical turnover 'factor' may be brought into play. If we take this to be (say) 1.3 and apply it to the calculation already made our figure for insect species worldwide is 9.4 million.

Figure 4.4 Beetle species richness: tropical *vs* temperate

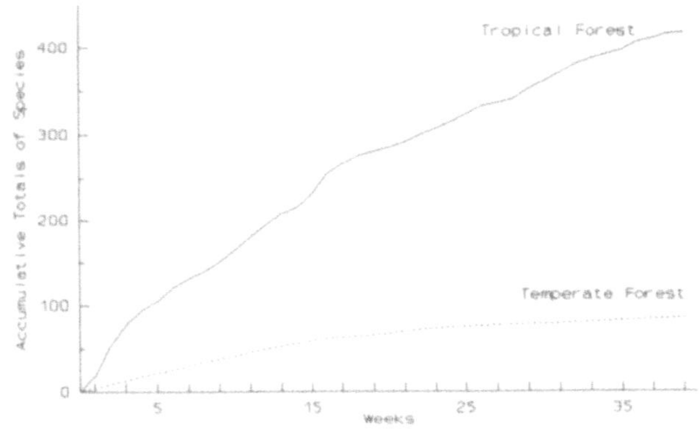

Notes: Comparison of beetle species richness in comparable sets of samples from single tropical and temperate sites, showing a relationship of around 5 to 1. The graph depicts accumulative numbers of species over time collected by representative single Malaise traps of modest size (see Hammond, 1990). Tropical data are for moist lowland forest in N Sulawesi and temperate data for mixed deciduous woodland in southern Britain (Hammond, unpublished). Traps chosen for illustration are those producing total beetle species nearest to the means of 412 per trap for Sulawesi (9 traps) and 83 for Britain (5 traps).

An alternative approach is to take the beetle species total for the Sulawesi site, and scale up directly to a figure for the tropics as a whole, using available data on tropical species turnover for relatively well-known groups as a rough guide. Using information patched together from many groups, including the best-known families of beetles themselves, an extrapolation may be made from the Sulawesi site inventory of 6,000 or so beetle species to 28,000 for the northern part of Sulawesi, to 70,000 (Sulawesi as a whole), 700,000 (Asian tropics) and finally 1.8 million beetle species for the entire moist tropics. Using the same figure for extratropical beetles as before, we reach a global beetle species figure of 2.3 million beetle species and, assuming (as before) that beetles comprise 33% of the global insect species inventory, 6.9 million insect species worldwide.

Finally, we might compare these results with those obtained by a Hodkinson and Casson type approach to the Sulawesi beetle data. In fact, no estimate is available for the proportion of species undescribed in the sample as a whole,

Figure 4.5 Major groups of organisms: described species as proportions of the global total

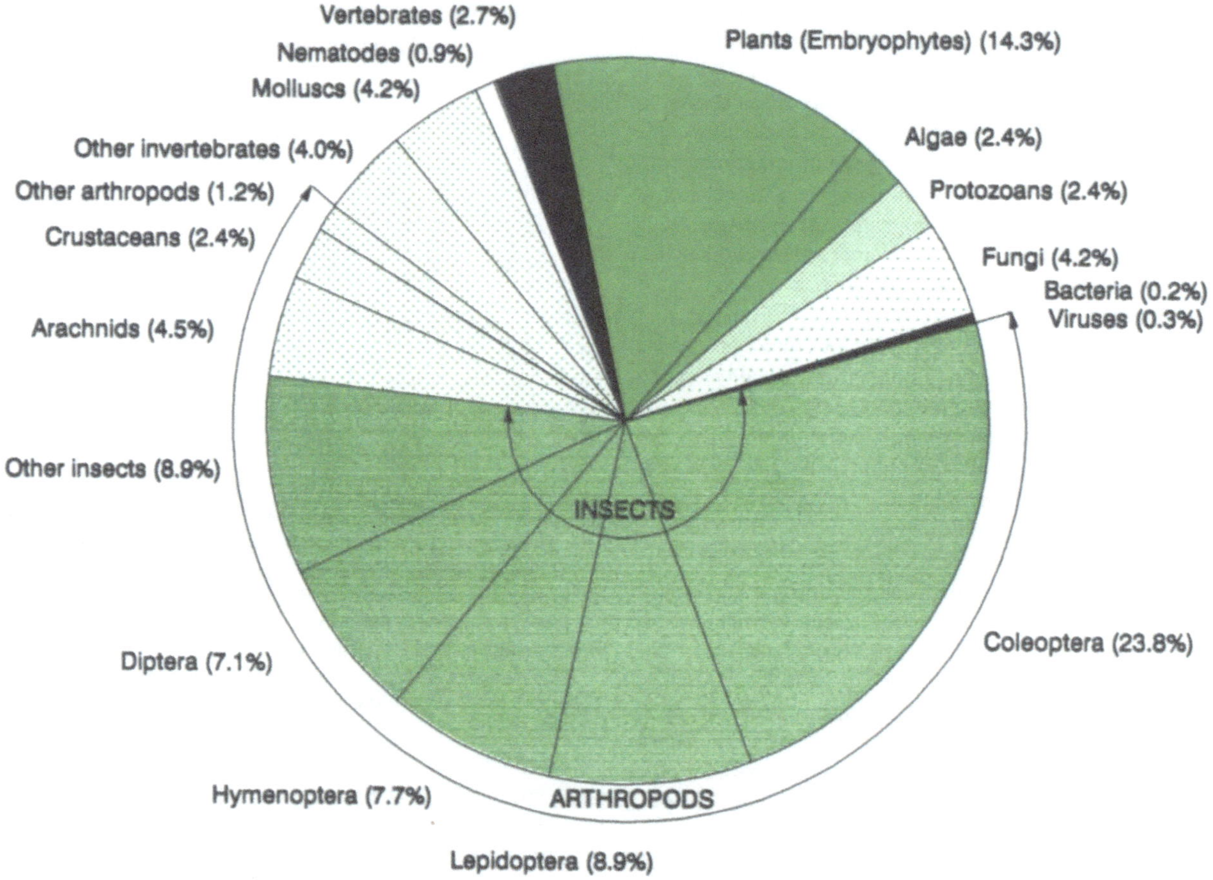

Notes: Proportions of major groups of organisms in terms of described species (estimated to total approximately 1.7 million). Groups included in the pie-chart are those considered likely to contain in excess of 100,000 species when as yet undescribed species are taken into account, along with vertebrates for comparison. Numbers of described species used in this diagram are those given in Table 4.3, with the exception of plants for which an earlier lower estimate of 240,000 was used.

but assuming this (on the basis of a small and probably unrepresentative sample) to be 75%, we generate a world figure of 1.6 million species for Coleoptera and, using the 33% formula from above, one of approaching five million for insect species worldwide.

Of course, all of the more significant ratios used in these simple calculations derive, at best, from informed guesses, but they are not simply plucked from the air. First-hand experience of how heterogeneity manifests itself at very small scales at tropical sites, and a feel for the extent of the contribution made by the different elements (e.g. elevational assemblages and different forest types) involved at more 'extended' sites may provide particularly useful guidance. Knowledge of vicariance patterns, especially as they differ between the three major tropical regions, may also be of considerable assistance. Finally, an awareness of the biases of various sampling methods, and the many factors that influence how well sampled and studied particular groups are likely to be, will be of great help when attempting to grasp the significance of fragmentary data.

The more important ratios used, those concerning the proportional representation of the major insect groups in terms of species and tropical to extratropical relationships, in the simple extrapolations made above were derived by patching together small fragments of data from many sources. The conclusions reached and assumptions used in reaching them cannot be detailed here, but it should be mentioned that higher tropical to temperate ratios were assumed for Coleoptera, Hemiptera and Lepidoptera, as opposed to Diptera and Hymenoptera (see Gaston, 1991a). The relative species richness of what seem certain to be the three largest insect groups was based on separate assessments of their possible overall species richnesses in both tropical and extratropical regions. 'Working figures' arrived at for the percentage of insects overall that are Coleoptera, Hymenoptera and Diptera in extratropical regions were 25%, 30% and 30% respectively, while those for the tropics were 35%, 27% and 20%, yielding (if we assume a 5:1 tropical to extratropical ratio for beetle species) overall working figures of 33% Coleoptera, 27.5% Hymenoptera and around 22% Diptera.

PROSPECTS FOR IMPROVED SPECIES RICHNESS ESTIMATES

Currently available estimates of species richness for all but the best-known groups such as birds, and best-known regions such as northern Europe, all involve substantial margins of error. By simple extrapolation from the well-known, only a very rough idea may be gained of how many species exist overall. The many uncertainties, especially with respect to microorganisms, make an upper

Figure 4.6 Major groups of organisms: possibly-existing species as proportions of the global total

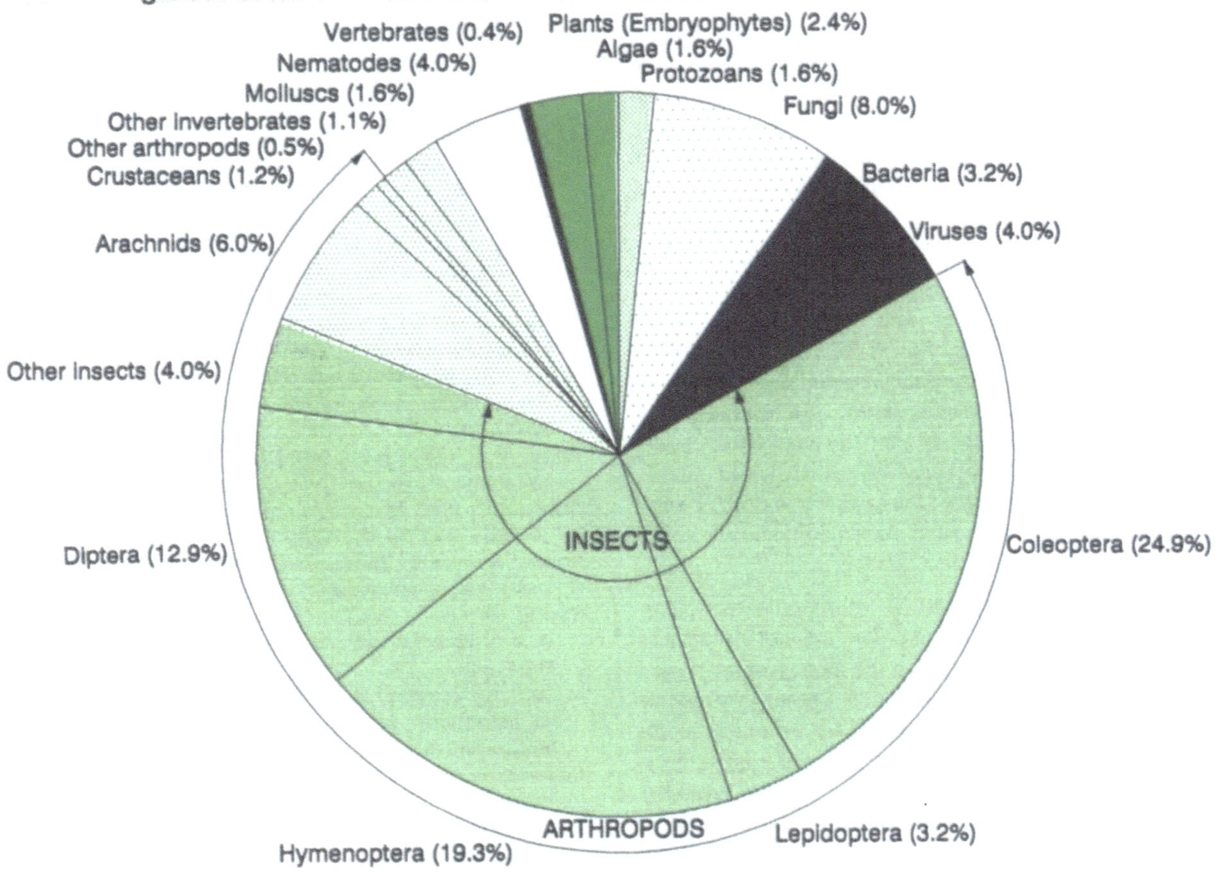

Notes: Possible proportions of major groups of organisms based on conservative estimates (see Table 4.3) providing a total for all groups of approximately 12.5 million species. All groups considered likely to contain in excess of 100,000 species are picked out in the pie-chart, along with vertebrates for comparison.

bound to the size of the global species inventory particularly difficult to establish. Despite numerous indications that this could be very great, claims that extant species number many tens of millions or even more can not be supported, for the moment, by any firm evidence. However, a lower bound to the global figure is much easier to set, and the available data, some of them discussed above, suggests that this might safely be put at a level considerably higher than the current described species total (approaching two million), perhaps at around eight million. The 'working figure' adopted here of 12.5 million species for the biota as a whole (see Fig. 4.6), arrived at by examining the data for each major group separately, is an avowedly conservative one.

In a situation where the most species-rich groups are at the same time the least known, an unwillingness to take into account anything but incontrovertible evidence is always likely to result in underestimation, as the record of early attempts to estimate the scale of global species richness well illustrates. Nevertheless, if we are to have any confidence in species richness estimates, there is no real alternative to working forwards by steadily enlarging the area of knowns. While new observations concerning little-known taxonomic groups and poorly explored habitats continually alert us to additional possibilities of as yet unassessed species richness, it would be naive to make too much of each and every anecdote.

To speed up the rate at which our understanding of species richness patterns and the overall dimensions of global biodiversity grows, it will be necessary to identify key questions and, if feasible, turn our attention first to them. The most obvious general line of attack is to focus efforts on the groups of organisms and parts of the globe that seem most likely to make the greatest overall contribution to the species inventory. New and pertinent data are needed for all of the ultra-diverse groups, but quicker and more substantial returns for efforts made are to be expected from some groups rather than others. If the main emphasis of this section has been on terrestrial arthropods, it is not because these animals (however numerous their species) can supply all of the answers, but rather because answers to key questions concerning their patterns of species richness are seen to be distinctly and not too distantly attainable.

Perhaps the greatest need is for good sample data on microorganisms and fungi. Because of seasonality and difficulties in detecting and/or culturing small species, exhaustive inventories may not be achievable but, in line with recent recommendations, these should be attempted at representative sites in the major biomes. If tropical to temperate species richness ratios are to be established for these groups, there will be a need to develop sampling methods and protocols that allow reliable comparison between sites without a complete inventory being taken.

For nematodes there is a pressing need for data on the species richness of free-living forms in both marine and terrestrial environments, including the moist tropics. Sample data that allow some estimation of species turnover at least at relatively small scales (i.e. in the m^2 to km^2 range) are a particular need, while any results concerning the less easily addressed problem of turnover at larger scales would be of great value. As in the case of microorganisms, advances in both the theory and practice of species recognition and discrimination will be needed if data gathered are to be truly informative.

In the case of terrestrial arthropods, the more tractable groups that are also large and 'representative' (e.g. Coleoptera) may be expected to receive considerable attention. For some of these groups, sampling programmes at various tropical and other sites, are already well advanced. While there is a need for the analysis of results already obtained to be speeded up, this should not be allowed to stand in the way of the application of the best of the methods so far developed at many additional sites.

For some of the major terrestrial arthropod groups, e.g. Diptera and Acari (mites), data on the numbers of species to be found at any one location in the moist tropics remain extremely limited and largely anecdotal. Reasonably reliable estimates of the species richness of these groups at single tropical sites are eminently attainable, and the acquisition of the appropriate datasets is a particular priority. Another clear need is for a better understanding of the proportional representation, in species terms, of the major terrestrial arthropod groups at single sites, and how this varies from region to region.

A separate agenda of research is needed for the investigation of species richness patterns in the marine realm. Here, data from the ocean depths remain too fragmentary for any confident estimation of the contribution that this 'new frontier' might make to marine or overall global species richness. There is a particularly urgent need for results that give some idea of species turnover in deep-ocean sediment assemblages, especially at the larger spatial scales.

Attention has been directed in this section almost entirely towards species, which for sound theoretical as well as operational reasons are often considered "central to the concept of biodiversity" (Reid and Miller, 1989). However, it should be stressed that a species count falls far short of any full assessment of biotic diversity, which expresses itself at a number of levels, from genes to ecosystems (Solbrig, 1991).

Our perception of the full dimensions of biotic diversity remains very hazy, but there is much of an immediate nature that can be done and is being done to remedy the situation. Indeed, there is every reason to suppose that advances in our understanding of some significant species richness patterns will be made very rapidly. Of course, we shall not get to know, even approximately, how many other species we share the planet with overnight, but we may reasonably expect our global species estimates to be made with steadily increasing confidence and precision.

References

Ackery and Vane-Wright. 1984. *Milkweed Butterflies*. BM(NH), London

Adis, J., Lubin, Y.D. and Montgomery, G.G. 1984. Arthropods from the canopy of inundated and terra firma forests near Manaus, Brazil, with critical considerations of the Pyrethrum-fogging technique. *Studies on Neotropical Fauna and the Environment* 19:223-236.

Adis, J. and Schubart, H.O.R. 1985. Ecological research on arthropods in central Amazonian forest ecosystems with recommendations for study procedures. In: Cooley, J.H. and Golley, F.B. (Eds), *Trends in Ecological Research for the 1980s*. NATO Conference Series, Series 1: Ecology. Plenum Press, London. Pp.111-144.

Arnett, R.H. 1967. Present and future systematics of the Coleoptera in North America. *Annals of the Entomological Society of America* 60:162-170.

Arnett, R.H. 1985. *American insects: handbook of the insects of America north of Mexico*. Van Nostrand Reinhold, New York.

Barnes, R.D. 1989. Diversity of organisms: how much do we know? *American Zoologist* 29: 1075-1084.

Brusca, R.C. and Brusca, G.J. 1990. *Invertebrates*. Sinauer, Sunderland, Massachusetts.

Casson, D. 1988. Studies on the Hemiptera communities of Dumoga-Bone National Park, Sulawesi. M.Phil. Thesis. Liverpool Polytechnic.

di Castri, F., Vernhes, J.R. and Younes, T. (in press). A proposal for an international network on inventorying and monitoring of biodiversity. *Biology International, Special Issue* 27.

Colinvaux, P. 1980. *Why Big Fierce Animals Are Rare*. Pelican Books, London.

Eliot, J.N. and Kawazoé, A. 1983. *Blue Butterflies of the Lycaenopsis Group*. BM(NH) London.

Erwin, T.L. 1982. Tropical forests: their richness in Coleoptera and other arthropod species. *Coleopterists' Bulletin* 36:74-75.

Erwin, T.L. 1990. Canopy arthropod biodiversity: a chronology of sampling techniques and results. *Revista Peruana de Entomología* 32:71-77.

Erwin, T.L. 1991. How many species are there? Revisited. *Conservation Biology* 5:1-4.

Erwin, T.L. and Scott, J.C. 1980. Seasonal and size patterns, trophic structure and richness of Coleoptera in the tropical arboreal ecosystem: the fauna of the tree *Luehea seemannii* Triana and Planch in the Canal Zone in Panama. *Coleopterists' Bulletin* 34:305-322.

Frank, J.H. and Curtis, G.A. 1979. Trend lines and the number of species of Staphylinidae. *Coleopterists' Bulletin* 33:133-149.

Gaston, K.J. 1991a. The magnitude of global insect species richness. *Conservation Biology* 5:283-296.

Gaston, K.J. 1991b. Body size and probability of description; the beetle fauna of Britain. *Ecological Entomology* 16:505-508.

Gaston, K.J. 1992. Estimates of the near-imponderable: a reply to Erwin. *Conservation Biology* 5:564-566.

Goodwin. 1986. *Crows of the World*. BM(NH), London.

Grassle, J.F. 1989. Species diversity in deep-sea communities. *TREE* 4:12-15.

Grassle, J.F. 1991. Deep-sea benthic biodiversity. *Bioscience* 41:464-469.

Grassle, J.F., Laserre, P., McIntyre, A.D. and Ray, C.G. 1991. Marine biodiversity and ecosystem function. *Biology International, Special Issue* 23:i-iv, 1-19. IUBS, Paris.

Hammond, P.M. 1990. Insect abundance and diversity in the Dumoga-Bone National Park, N. Sulawesi, with special reference to the beetle fauna of lowland rain forest in the Toraut region. In: Knight, W.J. and Holloway, J.D. (Eds), *Insects and the Rain Forests of South East Asia* (Wallacea). Royal Entomological Society, London. Pp.197-254.

Hawksworth, D.L. 1991. The fungal dimension of biodiversity: magnitude, significance and conservation. *Mycological Research* 95:641-655.

Hodkinson, I.D. and Casson, D. 1991. A lesser predilection for bugs: Hemiptera (Insecta) diversity in tropical rain forests. *Biological Journal of the Linnean Society of London* 43:101-109.

Lawrence, J.F. 1982. Coleoptera. In: Parker, S.P. (Ed.), *Synopsis and Classification of Living Organisms*. McGraw-Hill, New York. Pp. 482-553.

May, R.M. 1988. How many species are there on earth? *Science* 241:1441-1449.

May, R.M. 1990. How many species? *Philosophical Transactions of the Royal Society* B330:293-304.

Mayr, E., Linsley, E.G. and Usinger, R.L 1953. *Method and principles of systematic zoology*. McGraw-Hill, New York.

National Science Board 1989. Loss of Biological Diversity: a global crisis requiring international solutions. National Science Foundation, Washington, DC.

Noyes, J.S. 1989. The diversity of Hymenoptera in the tropics with special reference to Parasitica in Sulawesi. *Ecological Entomology* 14:197-207.

Poinar, G.O. 1983. *The Natural History of Nematodes*. Prentice Hall, Englewood Cliffs, NJ.

Reid, W.V. and Miller, K.R. 1989. *Keeping options alive*. The scientific basis for conserving biodiversity. World Resources Institute, Washington, DC.

Sibley, C.G. and Monroe, B.L. Jr 1990. *Distribution and Taxonomy of Birds of the World*. Yale University Press, Yale.

Simon, H.R. 1983. Research and publication trends in systematic zoology. Ph.D. thesis. The City University, London.

Solbrig, O. (Ed.) 1991. From genes to ecosystems: a research agenda for biodiversity. Report of an IUBS-SCOPE-UNESCO workshop, Harvard Forest, Petersham, Ma. USA, June 27-July 1, 1991. IUBS, Cambridge, Mass.

Stork, N.E. 1987. Guild structure of arthropods from Bornean rain forest trees. *Ecological Entomology* 12:69-80.

Stork, N.E. 1988. Insect diversity: facts, fiction and speculation. *Biological Journal of the Linnean Society of London* 35:321-337.

Stork, N.E. 1991. The composition of the arthropod fauna of Bornean lowland rain forest trees. *Journal of Tropical Ecology* 7:161-180.

Southwood, T.R.E. 1978. The components of diversity. In Mound, L.A. and Waloff, N. (Eds), *Diversity of Insect Faunas*. Symposia of the Royal Entomological Society of London. 9. Blackwell Scientific Publications, Oxford.

Valiela, I. 1984. *Marine Ecological Processes*. Springer Verlag, New York.

Wolf, E.C. 1987. On the brink of extinction: conserving the diversity of life. [Worldwatch Paper No. 78]. Worldwatch Institute, Washington, DC.

Abridged from a document provided by Peter Hammond, Environmental Quality Programme, The Natural History Museum (London).

Data presented in Table 4.5 retrieved from Zoological *Record Online by BIOSIS, UK.*

5. SPECIES DIVERSITY: AN INTRODUCTION

A BRIEF HISTORY OF DIVERSITY

Knowledge of the history of diversity through geological time is based on analysis of the fossil record. Because the fossil record gives only a very incomplete and highly biased view of the past history of life on earth, the reconstruction of that history has been, and continues to be, the subject of great debate. It is generally accepted that the fossil record can give a reasonable insight into past diversity in terms of taxonomic richness, particularly at higher taxonomic levels. However, it is far more difficult to derive other, more ecologically based, measures of diversity from it, as these require the reconstruction of palaeoenvironments, a far more contentious exercise than palaeotaxonomy.

While detailed patterns of taxonomic richness through the earth's history remain debatable, the overall outline is generally accepted. There are believed to have been relatively few species in total during the Palaeozoic and early Mesozoic; since then, that is for the past hundred million years, diversity has increased markedly. This recent diversification has passed through one major extinction event, at the Cretaceous-Tertiary boundary, and probably two minor events since then (see Chapter 16). Apart from these, the diversification appears to have continued more or less unabated, with the world apparently reaching its highest ever level of species richness during the Pliocene and Pleistocene, when climatic change and the advent of organised human activity finally halted the process. Significantly, however, diversity at higher taxonomic levels does not conform with this pattern, as evinced by the far higher number of animal phyla present in the early Cambrian than today (see below).

The early history of Life - the Precambrian

Recent consensus suggests that cellular life on the planet (in the form of procaryotes, at least some of which were probably very similar to living cyanobacteria) originated sometime between 3,900 and 3,400 million years ago (Mya). The origin of the earliest eucaryotes has proved difficult to establish, but it is generally accepted that the Precambrian microfossils known as 'acritarchs', which are recorded as far back as 1400 Mya, are almost certainly the cysts of marine algae and the earliest known eucaryotes. If this analysis is correct, then life on earth consisted only of procaryotes for at least 2,000 million years, or well over half its history. There is sufficient morphological variation in the fossil remains to permit some analysis of changes in diversity of these presumed early procaryotes in the late Proterozoic era. Vidal and Knoll (1983) have hypothesised a gradual increase in diversity from 1400 Mya to 750 Mya, when there was a peak of around 30 taxa in the fossil record, followed almost immediately by a sharp drop to around 10 taxa, possibly owing to a period of glaciation. After this there is an exponential increase in diversity, corresponding with the start of the Phanerozoic era.

The early Phanerozoic

For many years it was assumed that metazoans (multicellular organisms with internal organs) originated in the Cambrian era at the base of the Phanerozoic. This is now known not to be the case, as a wide range of fossil metazoans is now known from well before this time, including recognisable arthropods and possibly echinoderms. Most fossils from this time, however, appear completely unrelated to extant forms, and consist mainly of enigmatic frond- and disc-shaped soft-bodied animals: the so-called Ediacaran fauna.

The lower Cambrian marks a dramatic change from this early fauna, with the sudden appearance in the fossil record of a wide range of metazoans, many with calcareous skeletons. It is generally accepted that this represents a genuine explosion of diversity which took place over only a few million years, and is not an artefact of the fossil record. The lower Cambrian thus represents the most important period of high-level diversification in the history of animal life on earth. Very many phyla may have existed at this time, no more than five of which have origins traceable to before the Cambrian-Precambrian boundary. These include every well-skeletalised animal phylum living today (with the possible exception of the Bryozoa), indicating that virtually no new animal phyla have appeared during the many subsequent evolutionary radiations. Perhaps most significantly, no new animal phyla appeared with the colonisation of land, some 50-100 million years after the Cambrian radiation.

The Cambrian appears to have represented not only a peak of diversification but perhaps also a peak of higher order taxonomic diversity, as suggested by the presence of many more animal phyla than the 35 or so now extant.

Changes in diversity of marine animal taxa through the Phanerozoic

Although the number of phyla has decreased markedly since the Cambrian, diversity at all lower taxonomic levels has either increased overall or in a few cases remained more or less level.

The number of orders (of marine animals) present in the fossil record climbed steadily through the Cambrian and Ordovician, levelling off towards the end of the Ordovician to a figure of between 125 and 140, which has been maintained throughout the Phanerozoic.

The diversity of families represented in the fossil record shows a similar pattern of increase through the Cambrian and Ordovician, levelling off at around 500, a figure which was maintained until the late Permian mass-extinction (see Chapter 16). This extinction event resulted in the loss of around 300 families; subsequent to this, family diversity has increased to the modern level, with a number of temporary reversals in the form of the series of extinction events outlined in Chapter 16.

The trend in number of species in the fossil record is even more extreme. From the early Cambrian until the mid-Cretaceous, the number of marine species remained low; since then, that is in the past 100 million years, it has probably increased by a factor of 10.

Diversity patterns in terrestrial animals

Colonisation of land by animals has occurred many times; although the oldest body fossils of terrestrial animals date from the early Devonian, it is generally accepted that the primary period of land invasion by animals was the Silurian.

The overwhelming number of described extant species of terrestrial animals are insects and arachnids. The fossil record for both these groups is generally scanty.

Some attempt has been made, however, to chart changes in insect diversity at the generic level. Insects first appear in the fossil record in the Carboniferous. The number of genera then increased through much of the Palaeozoic and first part of the Mesozoic, interrupted by a sharp drop coinciding with the late Permian mass extinction, and then levelling off during the late Triassic. Diversity then doubled during the Cenozoic or Tertiary, coincident with the radiation of the angiosperms.

The fossil record of terrestrial vertebrates is much better, particularly that of tetrapods. The bird record is much less substantial than that for other groups, probably because their light skeletons have been less frequently preserved. Terrestrial vertebrates first appear in the fossil record in the late Devonian. Diversity remained relatively low during the Palaeozoic, with around 50 families, and actually declined overall during the early Mesozoic. From the mid-Cretaceous the number of families started to increase rapidly, reaching a Recent peak of around 340. Diversity of genera follows this overall pattern in a more exaggerated form. These trends are shown in Fig. 5.1.

Figure 5.1 Fossil diversity: terrestrial vertebrates

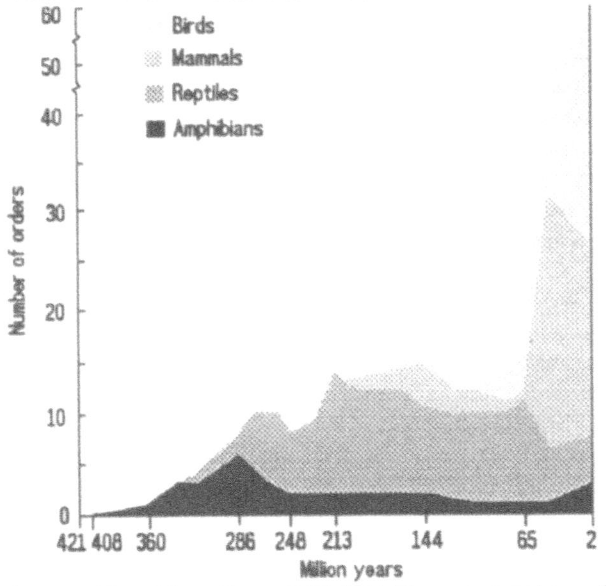

Source: Adapted from Signor, P.W. 1990. The geological history of diversity. *Annual Review of Ecology and Systematics* 21.
Note: Diversity is here measured in terms of number of taxonomic orders present.

Diversity patterns in vascular plants

It is generally accepted that vascular terrestrial plants first arose in the Silurian, although some palaeobotanists argue for a Late Ordovician origin. Diversity increased during the Silurian, and then more rapidly during the Devonian, owing to the first appearance of seed-bearing plants, leading to a peak of over 40 genera during the late Devonian. Diversity then declined slightly, but started to increase markedly during the Carboniferous, with at least 200 species recorded by the mid Carboniferous. Following this, diversity increased only slowly until the end of the Permian. There was a minor decrease in diversity at the end of the Permian, coinciding with or preceding the mass extinction of animal species, followed by a rapid rebound to previous levels. Diversity then continued increasing slowly, reaching around 250 species in the early Cretaceous. Starting at the mid-Cretaceous, diversity began increasing at an accelerating pace.

This overall pattern masks important changes with time in the composition of the flora, most notably in the relative importance of the three main groups of tracheophytes: the pteridophytes, gymnosperms and angiosperms. The Silurian and early Devonian are marked by a radiation of primitive pteridophytes. During the Carboniferous, more advanced pteridophytes and gymnosperms developed and underwent extensive diversification. Following the late Permian extinction event, pteridophytes were largely replaced (although ferns remain abundant) by gymnosperms which became the dominant group until the mid-Cretaceous. The dramatic increase in plant diversity since then is entirely due to the radiation of the angiosperms which first appeared in the lower Cretaceous. These trends are shown in Fig. 5.2.

Figure 5.2 Fossil diversity: terrestrial plants

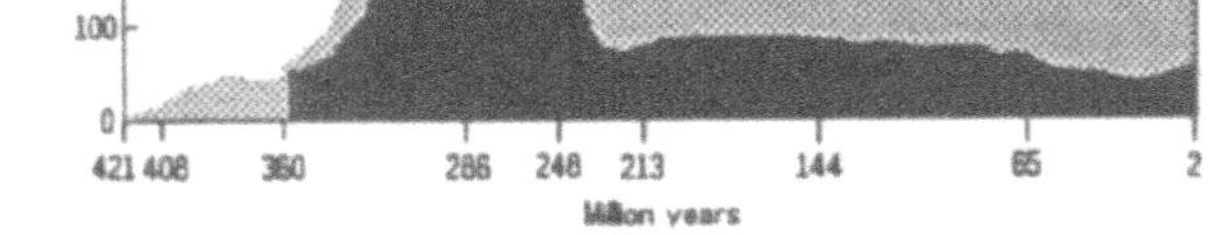

Source: Adapted from Signor, P.W. 1990. The geological history of diversity. *Annual Review of Ecology and Systematics* 21.
Note: Diversity is here measured in terms of number of species present.

MEASURING BIOLOGICAL DIVERSITY

A central problem in the maintenance of biological diversity is an assessment of the relative importance, in terms of diversity, of different areas, habitats or ecosystems. Only by understanding this can priorities in conservation efforts be usefully assigned. However, this importance can be assessed in different, though related, ways. The first, and most obvious, makes reference to its 'intrinsic' diversity, so that an area with higher diversity is deemed more important than one with lower diversity. The second attempts an assessment of the contribution any given area makes to theoverall diversity of a given geographic region, such as a country, continent or, ultimately, to the world overall. From this perspective, some areas with lower intrinsic diversity may be more important than others with higher diversity. This will be discussed further below; see also Chapters 2 and 15.

Assessments of diversity pose considerable problems, both practical and theoretical. In the first instance, the concept of diversity in an ecological context has to be made clear.

Local biological diversity

Species richness

Biological diversity measures for particular areas, habitats or ecosystems are often largely reduced to a straightforward measure of species richness. In its most ideal form this would consist of a complete catalogue of all species occurring in the area under consideration. In practice this is clearly unrealistic outside very small areas which will be of only limited interest in a global context. Even with small sites, a complete enumeration of all species will be impossible to carry out if micro-organisms are included.

Species richness measures will therefore in practice be based on samples. Such samples could consist of complete catalogues of all species in a particular, generally taxonomic, group (e.g. all birds, all ferns) or may consists of measures of species density (i.e. all the species in a sample plot of standard area) or of numerical species richness, defined as the number of species per specified number of individuals or biomass.

Although straightforward measures of species richness may convey relatively little ecologically important information, in practice because they are the most easily derived, they are perhaps the most useful index for comparisons of biological diversity on a large scale.

Species abundance

From an ecological viewpoint, simple species richness indices have limited value. More meaningful measures of diversity take into account the relative abundance of the species concerned. In general, the more equally abundant the species in the area or ecosystem under consideration are, the more diverse it is considered to be. A number of models have been developed which derive diversity indices from measures of species abundance. As different mathematical and biological assumptions are made in these models, they will often generate different diversity measures from the same sets of data. Thus there is no one authoritative index for measuring diversity.

Taxic diversity

Furthermore, weight can also be given to the relative abundance of species in various categories, for example in different size classes, at different trophic levels, in different taxonomic groups, or with different growth forms. Thus a hypothetical ecosystem which consisted only of several species of primary producers, such as photosynthesising plants, would be less diverse than one with the same number of species but which included herbivores and predators. Similarly, an ecosystem with representatives from four different phyla would be more diverse than one with representative of only two.

Based on cladistic analysis, a number of taxonomic diversity indices have now been developed. Some of these give higher weight to so-called relict groups, that is taxonomic groups not closely related to other living groups and consisting of few species; others favour higher taxonomic groups with large numbers of species. The most complex measure so far developed is taxonomic dispersion, which endeavours to select an even spread of taxa in any given group.

Comparisons of different areas

Once a measure of diversity has been decided upon, it should be possible to compare the diversity of different areas. Such comparisons may not, however, be straightforward.

Diversity measures for ecological entities such as communities, habitats and ecosystems make the assumption that these entities are not site-specific, that is that they occur in essentially the same form over a wide area or in a number of different places. In practice, species composition and species abundance are very rarely constant either in space or in time; thus the existence of communities or ecosystems definable by species composition is seriously questioned by many ecologists. This therefore undermines the extent to which diversity measures derived from particular sites can be used as a basis for generalisation. Nevertheless, these ecological concepts still retain considerable force, even if they cannot be rigorously defined, and much discussion of biological diversity is couched in terms of comparisons between different habitats and ecosystems.

Species/area relationships

The relative diversity of different sites will often depend on the scale at which diversity is measured. Thus $1m^2$ of semi-natural European chalk grassland will contain many more plant species than $1m^2$ of lowland Amazonian rain forest whereas for an area of, say, $1km^2$ or more this will be reversed. This is because as an area is sampled the number of species recorded increases with the size of the area, but this rate of increase varies from area to area.

A wide range of observations has demonstrated that, as a general rule, the number of species recorded in an area increases with the size of the area, and that this increase tends to follow a predictable pattern, known as the Arrhenius relationship, whereby:

$$\log S = c + z\log A$$

where S = number of species, A = area and c and z are constants.

The slope of the relationship (z in the equation above) varies considerably between surveys, although is generally between 0.15 and 0.40, and some surveys do not fit the relationship at all. This relationship is shown graphically in Fig. 5.3.

Figure 5.3 A typical species-area plot

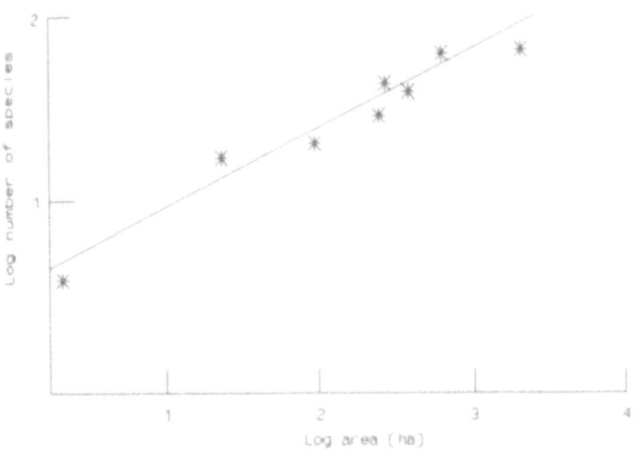

Note: The data are plotted on logarithmic axes resulting in a straight line graph, the slope of which (z) indicates the rate at which species number changes in step with changing area.

The most widely quoted generalisation from this finding is that a ten-fold reduction in an area (i.e. loss of 90% of habitat) will result in the loss of from *c.* 30% (with $z = 0.15$) to *c.* 60% (with $z = 0.40$) of the species present. This is often reduced to the rule-of-thumb that a ten-fold decrease in area leads to a loss of half the species present.

The causes of the species-area relationship appear to be relatively straightforward, and involve a combination of sampling effects and environmental heterogeneity. On a small scale, the increase in number of species with an increase in area is probably overwhelmingly a result of the former: that is, put very crudely, a given habitat in which species are randomly distributed will become increasingly more completely sampled as the area sampled increases. At larger scales, environmental heterogeneity will be more important: that is, as the area sampled increases, so different habitats with different species in them will be included in the sample.

Diversity at different scales

Thus the overall diversity of any given area will be a reflection both of the range of habitats it includes and the diversity of the component habitats. The greater the differences between the various component habitats in terms of species composition, then the greater the overall diversity will be. The differences between habitats are referred to as beta (β) diversity, while the diversity within a site or habitat is alpha (α) diversity. Thus an area with a wide range of dissimilar habitats will have a high β-diversity, even if each of its constituent habitats may have low α-diversity. Differences in site diversity over large areas, such as continents, are sometimes referred to as gamma (γ) diversity.

An area with relatively low species diversity may therefore still make an important contribution to the overall diversity of the larger region it is found in if it contains a significant number of species which do not occur elsewhere (endemics). Oceanic islands (see Chapter 14) and continental montane regions are examples of geographical entities which typically have comparatively low species diversity but high rates of endemism.

Assessing the relative importance of areas with high species diversity and low rates of endemism compared to areas with lower rates of diversity and high endemism remains an intractable problem. Attempts have been made to circumvent this by using somewhat different approaches, such as Critical Faunal Analyses, but these also generally do not generate unequivocal results (Chapter 15).

THE GLOBAL DISTRIBUTION OF SPECIES RICHNESS

Analysis of worldwide trends in biological diversity almost always treats this in terms of species richness, as this is the only indicator of diversity for which anything approaching adequate data is available on a global scale. Biological diversity is not evenly distributed around the globe.

Latitudinal gradients

The single most obvious pattern in the global distribution of species is that overall species richness increases with decreasing latitude. At its crudest this means that there are far more species per unit area and in total in the tropics than there are in temperate regions and far more species in temperate regions than there are in polar regions.

Not only does this apply as an overall general rule, it also holds within the great majority of higher taxa (at order level or higher), and within most equivalent habitats, although the most obvious and frequently cited are forests and shallow-water marine benthic communities, with, respectively, tropical moist forests and coral reefs being renowned for their remarkably high levels of species diversity.

This overall pattern masks a large number of minor trends where species richness in particular taxonomic groups or in particular habitats may show no significant latitudinal variation, or may actually decrease with decreasing latitude; nevertheless it remains a phenomenon of overwhelming biogeographical importance.

As well as latitude, changes in diversity can also be correlated with a many other variables, some of which are discussed briefly below. For some of these it is not easy to establish a significant relationship because there are often confounding variables, and because there are too few comparable datasets.

The maps in Fig. 5.4 demonstrate broad gradients in species richness in frogs (left) and trees (right) in the Americas (data extracted from Duellman, 1988 and Gentry, 1988). For these groups in this part of the world, climatic factors appear to play a large part in determining such gradients.

Elevational gradients

In terrestrial ecosystems, diversity generally decreases with increasing altitude. This phenomenon is most apparent at extremes of altitude, with highest regions at all latitudes having very low species diversity. There are fewer examples showing gradients of species richness with altitude, although amongst vertebrates this has been demonstrated for bird species in New Guinea (Kikkawa and

Figure 5.4 Gradients in species richness: frogs and trees in the Americas

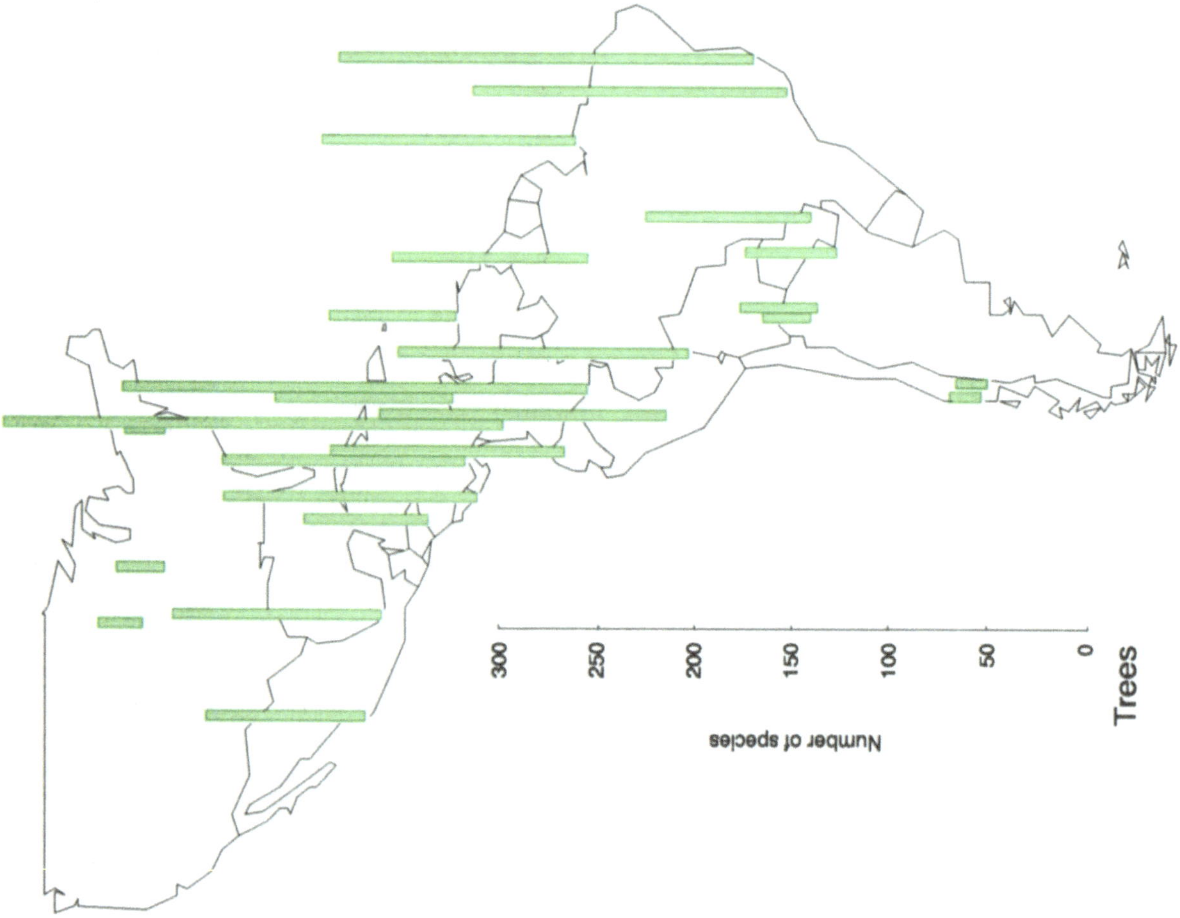

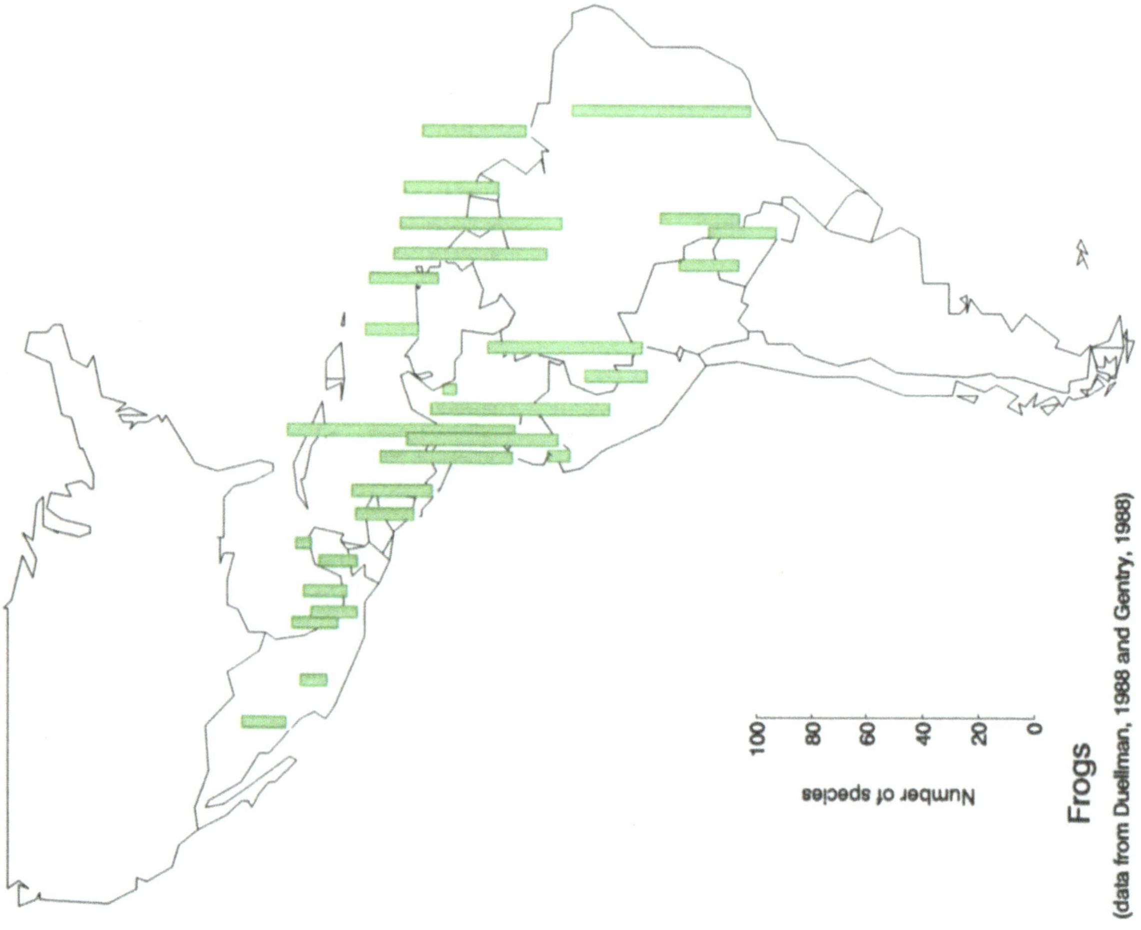

(data from Duellman, 1988 and Gentry, 1988)

Williams, 1971 cited in Brown) and on the Amazonian slope of the Andes in Peru (Terborgh, 1977). Gentry (1988) demonstrates it for woody plants in tropical forests, although notes that the data for upland sites are very incomplete. Suggestions have been made that, in tropical forests at least, diversity may be higher at mid-altitudes than in lower areas. However, there appear to be no substantiating data for this 'mid-altitude bulge' as a general phenomenon, although it has been noted in particular cases such as a desert mountain in Arizona where diversity at lower and higher altitudes is believed limited by aridity and low temperate respectively (Brown, 1988).

The decrease in straightforward species numbers with increasing altitude may in part be a reflection of species-area relationships, as available area generally decreases with increasing altitude, and number of species is closely related to area. Measurement of species numbers in standard-sized plots, such as those of Gentry (1988) take account of this, demonstrating that the relationship between altitude and species diversity is real, although not necessarily discounting the role that decreased available area may play in causing this phenomenon. It should also be noted that ß diversity will often be higher in areas of varied topography because of increased environmental heterogeneity.

Precipitation gradients

Precipitation is generally believed to be an important factor governing terrestrial diversity. However, the relationship between precipitation and diversity is not straightforward, and it seems that seasonality in precipitation may be as important as absolute amount. As with altitude, the relationship between precipitation and diversity is most apparent at one extreme, as highly arid environments are well-known to be much less diverse than less arid, or more mesic, environments at similar altitudes and latitudes. There are, however, apparently few quantifiable data to demonstrate this. Gentry (1988) in his study of forest diversity, demonstrated a strong correlation between plant species richness and absolute annual precipitation. However, he notes that this correlation may not apply at all in the Palaeotropics, and that there were strong indications that the length and severity of the dry season were more important than absolute annual rainfall. In the Neotropics, there is a strong relationship between annual rainfall and strength of the dry season, which is much less marked in the Palaeotropics. The importance of seasonality was borne out by a preliminary study of a Brazilian site with a relatively low, evenly-distributed annual rainfall, which showed a much higher species diversity than would be expected from total rainfall measures alone. Moreover, there appears to be a marked tailing-off of increasing diversity with increasing rainfall at high rainfalls, with little or no increase in diversity once rainfall exceeded 4,000-4,500mm per year.

However, it should be noted that the limits on diversity may in fact represent a limitation of sampling technique: in the two most diverse sites sampled (in areas of year-round rainfall of 3,000-4,000mm p.a.), diversity was so high in the plots sampled (in one site 300 species $\geq$ 10cm diameter out of 606 individual plants in one plot), that it seems likely that only by increasing the size of the survey plots would any further trends be discerned (Gentry, 1988).

Nutrient levels

Although there are few studies of global trends in diversity and soil nutrients, the relationship between plant community richness and tropical soil nutrient levels has been the subject of considerable interest. The data that are available indicate that the relationship may not be straightforward. Studies in Southeast Asia indicate that diversity may be highest at intermediary levels of nutrition, with a decrease at higher levels, while in the Neotropics diversity generally seems to increase with increasing nutrient levels, being most strongly correlated with Potassium (K) levels. This overall trend is apparently also shown by a variety of other organisms, including bats, birds and butterflies. In general, however, diversity in tropical forest ecosystems seems much less strongly dependent on nutrient levels than other factors, notably latitude, altitude and precipitation (Gentry, 1988).

The relationship between nutrient levels and diversity in other ecosystems is also complex: declines in diversity with increasing nutrient levels of temperate freshwater habitats (eutrophication) and grasslands are well-documented,but it is difficult to draw general conclusions from these (Brown, 1988).

Salinity gradients

In aquatic ecosystems, salinity appears to act as a strong 'normalising' factor on diversity. Thus, in coastal areas, diversity almost invariably declines when salinity deviates from 'normal' sea water (i.e. 35 ppt), while in freshwaters diversity decreases when salinity increases above *c.* 2 ppt; this results in a bimodal distribution of diversity with increasing salinity (Brown, 1988).

Islands

The study of diversity on islands, both real and theoretical, has been an important factor in much of biogeography and conservation biology. In particular the equilibrium theories of island biogeography elaborated by MacArthur and Wilson (1967) have had an important influence on both disciplines. More recently discussion in this, as in many other areas of ecology, has tended to move away from assumptions of equilibrium to more realistic, but far more complex, non-equilibrium theories.

SPECIES AND ENERGY

The relationship between diversity and productivity has been the subject of long-standing debate in ecology. Recent studies have indicated that available energy is strongly correlated with species diversity on a large-scale, at least in terrestrial ecosystems. A study of North American tree species (Currie and Paquin, 1987) demonstrated that realised annual evapotranspiration, a measure of available energy, statistically explained 76% of the variation in species richness across the continent. Such recent studies have shown that diversity gradients in tree species are more closely related to indices of climatic productivity than to other geographical parameters, including latitude (Adams, 1989). These results could be used to predict accurately tree species richness patterns in Great Britain and Ireland. Preliminary analysis of the diversity of terrestrial vertebrates in North America apparently yielded very similar results.

EXPLANATIONS AND HYPOTHESES

The explanation of geographic and temporal variation in species diversity is one of the central problems of biology. It has also proved one of the most intractable. The problem has generated an enormous amount of literature in which many different hypotheses have been proposed to attempt to account for it; these hypotheses often operate at different levels of explanation and much confusion has arisen as a result. It is beyond the scope of this report to attempt a thorough review of the subject, although, ultimately, an understanding of the importance of biological diversity should rest on an understanding of how and why it has the form that it does.

It is self-evident that, ultimately, all non-random patterns in species diversity must depend on past or present variations in the physical environment. How such variations result in the patterns observed is often far from clear. It is evident, however, that any complete explanation must involve both historical events and current ecological processes - the former implicit in any explanation of the *origin* of diversity, the latter in explanations of its *maintenance*, these being two separate, although intimately linked, problems. The relative importance of these two factors in determining present patterns is still a subject of considerable debate.

References

Adams, J.M. 1989. Species diversity and productivity of trees. *Plants today* Nov.-Dec. 183-187.

Brown, J.H. 1988. Species diversity. In: Myers, A.A. and Gillet, P.S. (Eds), *Analytical Biogeography*. Chapman and Hall, London.

Currie, D.J. and Paquin, V. 1987. Large-scale biogeographical patterns of species richness of trees. *Nature* 329:326-327.

Duellman, W.E. 1988. Patterns of species diversity in anuran amphibians in the American tropcis. *Annals of the Missouri Botanical Garden* 75:70-104.

Gentry, A.H. 1988. Changes in plant community diversity and floristic composition of environmental and geographical gradients. *Annals of the Missouri Botanical Garden* 75:1-34.

Kikkawa, J. and Williams, E.E. 1971. Altitudinal distribution of land birds in New Guinea. *Search* 2:64-69.

MacArthur, R.H. and Wilson, E.O. 1967. *The Theory of Island Biogeography*. Princeton University Press, Princeton.

Signor, P.W. 1990. The geological history of diversity. *Annual Review of Ecology and Systematics* 21:509-539.

Terborgh, J. 1977. Bird species diversity on an Andean elevational gradient. *Ecology* 58:1007-1019.

Vidal, G. and Knoll, A.H. 1982. Radiations and extinctions of plankton in the late Proterozoic and early Cambrian. *Nature* 297:57-60.

Chapter contributed by Martin Jenkins.

6. MICROORGANISMS

TAXONOMIC SCOPE

This section provides an overview of the phylogenetically extremely diverse groups collectively regarded as 'microorganisms'. This term is misleading as by no means all are microscopic. The definition accepted here is: organisms which *either* belong to phyla many members of which cannot be seen by the unaided eye, *or* where microscopic examination, and in many cases growth in pure culture, is essential for identification (Hawksworth, 1992). Some of the themes touched on here with specific reference to microorganisms are developed from a broader perspective elsewhere in the report (Chapter 4). The glossary should be consulted for definitions of certain terms.

The classification of the various microorganism groups at the rank of kingdom, and both below and above that level, is currently in a state of flux. For the purposes of this contribution, the terms algae, bacteria, fungi, protozoa, and viruses are treated in their traditional non-phylogenetic sense, with some minor modifications (Table 6.1). However, as the macroalgae (charophytes and seaweeds) and the lichen-forming fungi (lichens) are discussed elsewhere in this publication (Chapter 7), these non-taxonomic groupings are given only brief mention here.

ASSESSMENT OF DIVERSITY

The diversity of microorganisms in terms of the numbers of species currently known, and those estimated to occur in the world, was considered by leading specialists in the various groups at an IUBS/IUMS workshop in 1991 (see below; Hawksworth and Colwell, 1992 and in prep.). While the total number of known species is reliably estimated at 159,000 (Table 6.1), considerable difficulty arises in the estimation of those which remain undescribed. Nevertheless, the conclusion that less than 5%, and probably less than 3%, of the world's microorganisms have been described is not expected to be unduly pessimistic.

Algae

While the number of recognised algal species can be asserted with some confidence, the estimated world figure of 350,000 now proposed has large error margins - indeed it has been hinted that the chromophyte algae alone might eventually prove to comprise either 100,000 or up to 10 million species, the diatoms being the most speciose (Andersen, in press). The terrestrial algal species, especially those on bark and rocks, and minute ocean species have received particularly scant attention. Further, the marine picoplankton, which can make up to 25% of the phytoplankton biomass in polar waters, were first recognised only in 1980.

Bacteria

The number of bacterial species accepted in the *Approved List of Bacterial Names* was 3,058 in July 1991 (Trüper, 1992); the figure of 4,000 in Table 6.1 has been increased to allow for cyanobacteria. Perceptions of the true number of bacteria in the world have changed dramatically during the last 5-10 years. It has become increasingly evident, primarily as a result of the application of molecular techniques (Liesack and Stackebrandt, 1992), that there are enormous numbers of as yet uncultured bacteria to be found in soils, deep sea sediments, as mutualists in protozoans and other organisms and, most importantly, in the digestive tracts and pockets of a wide variety of animals - including most insects (Trüper, in press). It has been suggested that one genus of wall-less bacteria inhabiting insect guts, the mollicute *Spiroplasma*, may prove to be the largest genus on Earth with well over one million species (Whitcomb and Hackett, 1989).

Fungi

The number of fungi estimated to occur in the world has recently been conservatively estimated at 1.5 million species (Hawksworth, 1991a). This figure contrasts markedly with the 70,000 now described - that figure has been increased from the 69,000 cited by Hawksworth (*op. cit.*) to allow for fungi newly published since 1990. The 1.5 million figure is conservative as in the calculations leading to it: (1) a modest world estimate of vascular plants was employed, (2) no special allowance was made for fungi to be expected on the large numbers of insects now postulated, (3) the UK vascular plant:fungus ratio of 1:6 used must be an underestimate as additional fungi continue to be found in that country, and (4) no provision was made for any proportionately increased numbers in the tropics or polar regions. Whether an upward revision of the 1.5 million figure is defensible must await in-depth studies of particular tropical sites. See Chapter 4 for an alternative view.

Protozoa

Corliss (1991) estimated the number of known non-fossil protozoan species at 40,000. No calculated predictions of the number of world species have been prepared, but many groups, such as the heterotrophic heterokonts in soil, have scarcely been investigated. The total world estimated number of 100,000 used here could prove to be a gross understatement.

Viruses

No comprehensive catalogue of the world's known viruses currently exists, but it is expected that about 5,000 will be recognised in a compilation being planned by the International Committee on the Taxonomy of Viruses for publication in 1993. The estimate of 500,000 species presented here reflects the substantial numbers of new viruses to be expected on yet unstudied non-crop plants, and especially insects. Also scarcely investigated are viruses only recently recognised as frequent in marine plankton, plasmids in fungi, and phages on bacteria - not least on 'unculturable' bacteria.

SPECIES CONCEPTS IN MICROORGANISMS

Comparisons of species numbers between microorganisms and macroorganisms, and indeed also between the different microorganism groups, are complicated by variations in species concepts. While the idea of the 'biological species' is not without appeal to microbiologists, in practice in the majority of cases it is not readily applicable. This difficulty arises both because sexual processes are absent or difficult

Table 6.1 Estimates of the number of described species and possible undescribed species of microorganisms

GROUP	DESCRIBED SPECIES	ESTIMATED SPECIES	PER CENT KNOWN
Algae	40,000	350,000	11.0
Bacteria (incl. cyanobacteria and 'unculturables')	4,000	3,000,000	0.1
Fungi (incl. yeasts, lichen-forming fungi, slime moulds, and oomycetes)	70,000	1,500,000	5.0
Protozoa (proctoctists, excl. algae and oomycete fungi)	40,000	100,000	40.0
Viruses (incl. plasmids, phages, etc.)	5,000	500,000	1.0
TOTAL	159,000	5,450,000	3.0

Source: Based primarily on data in Hawksworth, D.L. and Colwell, R.R., (Eds) (in prep.). Biodiversity amongst microorganisms and its significance. *Biodiversity and Conservation* 1.

to detect in many microorganism groups, and further when they are known to occur it is often impractical to determine breeding groups. In practice microbiologists tend to be pragmatic, recognising as 'species' specimens or strains with a high degree of morphological, biochemical, or molecular similarity and which produce replicating lineages. The scale of characters used is inversely proportionate to the size and number of morphological characters. In the bacteria and yeasts suites of assimilation, substrate utilisation, and cultural attributes are extensively used, while these feature to a much lesser extent in algae, filamentous fungi, and protozoa.

Stress is invariably placed on the recognition of marked discontinuities in several characters, but emphasising those features which are relevant in human terms - for example the ability to cause diseases in particular animals or plants, to form toxins, to conduct economically important fermentations, or to produce desired chemical products.

Clones, the progeny derived from a single cell and which do not exhibit any genomic variation or recombination, are to be found in all microorganism groups. However, these are not always easy to recognise, and the practice has been to accept as species clones fulfilling the requirements of distinctness normally associated with that rank. Clones are frequently opportunistic organisms well-adapted to particular ecological niches; in the case of the conidial fungi, clones are derived from a part of the life-cycle of sexually reproducing species in which the sexual stage has sometimes been entirely lost.

Particular aspects of the use of species concepts in the different microorganism groups are considered further below:

Algae
The biological species concept is theoretically usable in those algae which are entirely sexual or have such stages in their life cycles, but experimental verification is rarely practical as many species cannot be readily grown in pure culture. Its application in practice has thus been extremely limited. In large groups such as the diatoms and coccoliths, while sexual stages are known or expected to occur, in reality morphological species concepts have to be used, increasingly employing characters only visible by the Scanning Electron Microscope (SEM). In eight algal classes, however, sexual reproduction is entirely unknown. Chemical characteristics are extensively used as aids to species differentiation in certain groups (Kessler, 1985). Mating complex studies and molecular approaches are also increasingly being used. The latter approaches are illustrating that considerably diverse taxa are sometimes grouped in the same genus or species, although the converse situation is also known. An overview of species concepts in algae is provided in Andersen (1992).

Bacteria
As sexual differentiation does not occur in bacteria and most reproduction is asexual, and further, as recombination between different strains is difficult to detect, species concepts in bacteria have largely been based on overall similarities. Since the early 1960s, numerical taxonomic studies utilising 50 to several hundred biochemical and cultural tests have played a major role in defining bacterial species. Similarity coefficients are computed, and phenetic groups formed at about the 80-85% similarity level are generally taken as equivalent to species (Austin and Priest, 1986; Sneath, 1989). The advent of molecular techniques has enabled species concepts derived from phenetic methods to be reassessed. DNA homologies of 20-50% are found between species in the same genus, and 60-70% between subspecies within the same species (Johnson, 1989). The International Commission on Systematic Bacteriology recommends that a minimal DNA homology of 70% be required for species-level treatment (Wayne, 1987).

Fungi
A consequence of the wealth of morphological characters in fungi is that species continue to be mostly distinguished by marked discontinuities between those features. The assimilative and predominant phase is haploid, and most

fungi are either sexual or derived from ancestors that were so. Despite the considerable literature on speciation in fungi (Burnett, 1983), the delimitation of populations from a biological standpoint remains in its infancy. As particular examples are studied in depth, it is becoming increasingly apparent that several discrete reproductively isolated groups are not uncommonly present within single morphospecies (Brasier, 1986). Mycologists have been reluctant to recognise such groups at the rank of species, but this can be expected to change where particular groups also have other important features such as pathogenicity to different crops. While a wide range of biochemical and molecular techniques are currently being employed in the fungi (Hawksworth and Bridge, 1988), the application of many of these is limited to the 20% of the known species which can be grown in pure culture. Where DNA homology studies have been conducted, notably in yeasts and certain economically significant genera such as *Aspergillus*, the differences between morphological species tend to be in the 20-50% range (Kurtzman, 1985), as they are in bacteria.

Protozoa
In contradistinction to the fungi, many protozoan species are diploid. In numerous groups information on life-cycles and sexuality are still lacking, rendering it difficult to apply a biological species concept. Clonal protozoans are, however, often described as species, while in contrast, as in the case of fungi, morphologically defined species may be found on more critical analysis to consist of a number of discrete gene pools.

Viruses
While some biologists are reluctant to recognise viruses as 'living', that they are functional biological entities is inescapable. They possess genomes, replicate, evolve, occupy specific ecological niches, and exhibit intrinsic variability. Ultrastructure, serological tests, physical and chemical structure and features, and the ability to infect particular hosts are used in species separation. The species concept in virology has been analysed by Regenmortel (1990). He took a pragmatic stance and defined a virus species as a polythetic class of viruses constituting a replicating lineage and occupying a particular ecological niche. This definition has the attraction of being applicable both to groups which are able to undergo recombination and those which are clonal.

EXTENT OF GENETIC DIVERSITY

The extent of genetic diversity exhibited by microorganism groups is vast in comparison to that of macroorganisms. This conclusion was to be expected bearing in mind that the earliest bacteria probably arose around 3.5 billion years ago on an Earth formed only one billion years earlier, whereas the first land plants, for example, did not emerge until about 0.4 billion years ago; i.e. microorganisms have had nine times as long to diverge as land plants.

This diversity is illustrated to some extent in terms of the numbers of phyla recognised, but most forcefully at the molecular level. Of the 95 phyla accepted by Margulis and Schwartz (1988), 52 belong to the microorganisms as defined here (*less* the virus groups not considered by those authors). More significantly, the study of 16S-like rRNAs in prokaryotes led to the suggestion that they should be split into two separate groups, Archaebacteria and Eubacteria, and that these were roughly equivalent to the Eukaryotes. Recognising that most biologists would be unwilling to accept plants and animals as belonging to the same kingdom, the higher rank of "domain" has been applied to these three groups, i.e. the domains Archaea, Bacteria, and Eucarya (Woese *et al.*, 1990). Studies with the gut protozoan *Giardia lamblia*, however, have further demonstrated that at least some eukaryotic microorganisms are much more remote from each other than had hitherto been assumed; for example, on the basis of 16S-like rRNAs, the crustacean *Artemia salina* and *Homo sapiens* are ten times closer to each other than either are to *Giardia* (Sogin, 1991).

The extent of genetic diversity now demonstrated between the higher ranks of microorganisms is reflected also at the species level. Both the genetic diversity within single microbial species, and that between several species referred to the same genus, can also be vast in comparison with macroorganism groups. This is especially true at the DNA homology level where 20-50% similarities are regularly encountered between species (see above), whereas primate 'species' may still be regarded as distinct although sharing 90+% DNA homology.

One consequence of the considerable genetic diversity within microbial species is that in certain microorganism groups infraspecific categories are utilised to an extent not otherwise seen outside the higher vertebrates. These include subspecies, *pathovars*, 'special forms', and *serotypes*. In addition, complex race notations have been developed within particular species of major medical or plant pathogenic importance. This tradition has developed as a pragmatic response to the need to label populations to a finer degree because of the different effects they have on humans or their crops.

From this discussion it will be apparent that if identical DNA homology criteria were used for species separations in both macro- and microorganisms, the numbers of known and estimated species in Table 6.1 would have to be inflated by not less than an order of magnitude.

REGIONS AND HABITATS OF MAXIMUM DIVERSITY

The variety of ecological niches exploited by the major groups of macro- and microorganisms is directly related to their geological age; ecology recapitulates phylogeny (Price, 1988). The greatest niche breadth is consequently seen in the bacteria, and then, in declining sequence, in the algae and protozoa, fungi, animals, and plants.

While there is every reason to suppose that regions and habitats with a maximum diversity of macroorganisms will also be particularly rich in microorganisms - a consequence of the larger numbers of host-specific parasites, mutualists, and saprobes to be expected - there are additional habitats of no importance for macroorganisms which are important for the conservation of microorganism diversity.

Amongst the bacteria are species able to grow in extreme

saline substrata or at high sugar (low water activity) concentrations, ones which thrive at high concentrations of heavy metals, sulphur, or other generally toxic compounds, major groups restricted to anaerobic situations, and ones able to tolerate or even thrive at extremely high (e.g. *Thermotoga* lives at 90°C) or low temperatures (e.g. at or below freezing point in the Antarctic).

Trüper (in press) identified the following environments as ones dominated by microorganisms or ones which are strongly influenced or stabilised by them:

- hypersaline neutral and alkaline lakes (salt lakes and soda lakes), e.g. East African rift valley lakes, the Dead Sea
- hot springs (hydrotherms, fumaroles, solfatoras) which have not been disturbed
- natural leaching environments (acid crater lakes, acid mine waters)
- peat mosses, permafrost tundra, cypress and mangrove swamps
- stratified (meromictic) lakes
- hot deserts (sand and rocks) which have not been disturbed)
- bare lichen-encrusted rock areas (with associated bacteria and fungi), in all climatic regions
- estuaries (salt marshes, mud flats, beaches)
- deep sea environments (hydrothermal vents, hypothermal zones, manganese nodule areas).

Extreme environments also continue to be a particularly rich source of previously unknown microorganisms belonging to diverse groups. Even though not all the species are known, it is evident that due accord needs to be given to extreme environments when drawing up international, national, or regional plans for the establishment of protected areas.

As a consequence of the antiquity of the groups, there is a tendency for microorganisms to have much broader geographic ranges than macroorganisms. Biogeographic studies, except in the case of lichens and macroalgae, are rarely undertaken. However, there is no reason to suppose that while there are a considerable number of almost cosmopolitan species, many others do not have geographically restricted ranges. This is certainly true for the fungi, but current perceptions of distributions on a global scale are skewed by inadequate sampling. Mycologists, for example, would take in their stride the discovery of a species previously known only from Europe in an undisturbed habitat in Australia, whereas a similar event would cause amazement among workers in most other groups.

Conversely, detailed biogeographic analyses from the world level down to national mapping programmes, clearly demonstrate that in the fungi numerous species are narrowly restricted geographically. Studies on the numbers of species of particular families and genera of fungi in different geographic regions can potentially lead to the recognition of centres of maximum diversity, as demonstrated for certain ascomycete groups by Pirozynski and Weresub (1979). A shortage of authoritative inventories and surveys currently limits the utilisation of such approaches in site-selection.

ROLE OF MICROORGANISMS IN BIODIVERSITY MAINTENANCE

Microorganisms have played a major role in the evolution and diversification of macroorganisms. They contributed key organelles such as mitochondria and chloroplasts to eukaryotic cells, and as mutualists are either involved in nutrient-supply or perform other biochemical processes on which they depend (Margulis and Fester, 1991). Bacteria, fungi, and protozoa in the guts of insects and herbivorous mammals perform crucial roles in their digestive processes, particularly in the breakdown of celluloses and lignins, and without which they could not exist (Smith and Douglas, 1987). About 85% of the Earth's vascular plants form mycorrhizas with fungi. This life-style is often obligate in nature, the mycorrhizas being crucial to the absorption of growth-limiting nutrients (Read, 1991). The very existence of many macroorganisms is consequently dependent on the continued availability of the mutualistic microorganisms they require.

In the marine environment, up to 80% of the biomass and productivity in open waters is contributed by ultraplanktonic algae (Andersen, 1992). Further, dinoflagellates form mutualisms with coelenterate stony corals, and the outer ridges of major reefs taking the full force of the oceans are formed by crustose coralline algae cementing detritus together (Round, 1981). In the absence of these mutualistic microorganisms, coral reef ecosystems simply could not exist (Smith and Douglas, 1987). Without the coral mutualists one of the most biologically diverse habitats on Earth would never have been formed.

At the ecosystem function level, food networks of all life on Earth are ultimately dependent on microorganisms. This holds for terrestrial and marine ecosystems (Andersen, 1992; Grassle *et al.*, 1991; Price, 1988), yet ecologists and conservationists only exceptionally take it into account.

The greatest biomass in soil, on the basis of current evidence, is that of the microorganisms, especially the fungi (Lee, 1991; Lynch and Hobbie, 1988). These play a variety of roles related to the maintenance of soil structure and composition both through the biodegradation and incorporation of dead plant and animal remains, and by extra-cellular fungal polysaccharides which bind soil particles together, thus increasing soil aggregation and stability (Lal, 1991).

Microorganisms also contribute to the maintenance of ecosystem structure through natural biocontrol. Plant pathogenic microorganisms can limit plants that would otherwise expand explosively in the absence of their co-existing pathogens. Similarly, entomogenous microorganisms can limit the populations of insects that would otherwise become major pests (e.g. defoliants) of trees or other plants. In these two cases, if their targets have crucial ecological roles, the loss of the containing microorganism would lead to major changes in the ecosystem.

ROLE OF MICROORGANISMS IN BIOSPHERE FUNCTIONS

Bacteria shaped the early atmosphere of Earth, the start of life coinciding with a fall in carbon dioxide and an increase in methane at around 3.8 billion years ago. The photosynthetic cyanobacteria were subsequently instrumental in producing oxygen, and microorganisms on land would have increasingly removed carbon dioxide from the early atmosphere in rock weathering (Lovelock, 1988). In the absence of these activities there would have been no macroorganisms or humans. Microorganisms continue to play a major role in the maintenance of the biosphere and global ecology through the various biogeochemical cycles. They perform unique and indispensable roles in the circulation of matter in the world (Stolz *et al.*, 1989). The principal biogeochemical cycles with which they are involved are:

Carbon

It has been estimated that about 40% of the carbon fixed by photosynthesis on the Earth is carried out by algae and cyanobacteria, especially those in oceans and seas. Bacteria also fix atmospheric carbon dioxide anaerobically and in methanogenesis. Methanogenic archaean bacteria generate about 58% of the Earth's methane. Conversely, wood-decay fungi are instrumental in releasing around 85 billion tonnes of carbon (as carbon dioxide) into the atmosphere each year. Ruminant gut microbial populations also produce methane, and other methyl gases are produced by fungi during wood decay. The tissues of microorganisms further have roles as carbon sinks, and their removal of carbon from the atmosphere in rock weathering is an on-going process.

Nitrogen

The Earth's nitrogen cycle is dependent on bacteria (including cyanobacteria) for nitrogen fixation, the oxidation of ammonia, nitrification, and nitrate reduction. The magnitude of the amounts involved is staggering: each year bacteria fix 240 Tg of nitrogen, release 210 Tg of nitrogen by denitrification, and release 75 Tg of ammonia (Trüper, 1992).

Sulphur

The sulphur cycle on Earth is dependent on sulphur-reducing bacteria for the reduction of sulphate into hydrogen sulphide, on purple and green photosynthetic bacteria for the oxidation of sulphides to sulphur, and sulphur oxidising bacteria for the conversion of sulphur to sulphates. Bacteria are also involved in the biogenesis of dimethylsulphide, a substance of particular relevance as a greenhouse gas and postulated as performing an equilibrating function for the planet (Lovelock, 1988).

Minerals

Microorganisms of various types, including algae, bacteria, fungi, and protozoa, are important in the production of a wide range of biogenic minerals, notably in the processes of rock weathering. These include diverse kinds of carbonates, phosphates, oxalates, sulphates, silicates, sulphides, and further oxides of iron and manganese (Krumbein, 1983; Leadbeater and Riding, 1986; Stolz *et al.*, 1989).

POTENTIAL CONTRIBUTION OF MICROORGANISMS TO SUSTAINABLE DEVELOPMENT

Microorganisms have the potential to contribute to sustainable development in multifarious ways (Hawksworth, 1991*b*; Persley, 1990). Production on existing agricultural land may be increased through:

- the selection and introduction of the most efficacious nitrogen-fixing *Rhizobium* strains into legume crops
- the enhancement of natural nitrogen fixation by the application of cyanobacterial inocula, either directly or through mutualists (e.g. improvement of cyanobacteria of *Azolla* for use in rice-fields)
- the use of bacteria and fungi as biocontrol agents for insect pests, plant pathogens, disease vectors, and noxious weeds
- the mass production of the most efficacious mycorrhizal (and in the future almost certainly also beneficial endophytic) fungi for inoculation into seeds or seedlings on or prior to planting.

Genes from bacteria and fungi with useful properties, for example the production of an insecticidal metabolite or enzyme, can be cloned and inserted into the genome of a crop plant by an increasing range of methods. Indeed, a plasmid in the crown gall bacterium *Agrobacterium tumefaciens* is well-established as a practical mechanism by which genes from any source can now be engineered into over 20 major world crops.

A wide array of pharmaceutical and other industrial products are already obtained from microorganisms grown under factory conditions. These include, for example, organic acids, vitamins, antibiotics, anti-inflammatory drugs, immunoregulators (e.g. cyclosporin from a saprobic fungus which is now routinely used in human transplant surgery), food colourings, fragrances, and food preservatives. The discovery and studies of the actions of naturally occurring compounds can also lead to semi-synthetic drugs of great potential, as in the case of ivermectin first used against helminths parasitic on livestock but now also employed in humans against onchocerciasis (river blindness).

In addition, cellulosic and lignosic wastes from agricultural and industrial sources can be biodegraded by microorganisms and converted to animal feedstuffs.

Waste-water treatments using anaerobic bacteria and filter-feeding ciliates reduce pressure on freshwater supplies. Microorganisms are crucial to the functioning of sewage filter-beds. Bacteria can also be employed in the removal of toxic chemicals, especially heavy metals, from liquid waste; any valuable metals can be recovered for reuse. The bioremediation of major oil spills at sea can be achieved by applying nitrogen fertilizers which encourage the naturally present hydrocarbon degrading microorganisms to proliferate.

Biogas (methane) production from a variety of agricultural and other wastes for use as fuel is dependent on anaerobic bacteria. This has the potential to reduce the pressure on forests by providing an alternative energy supply.

An expanded range of sources of food for humans can be derived both from the mass-production of certain algae and filamentous fungi (e.g. the *Fusarium graminearum* strain in 'Quorn'), and through the increased use of waste materials for the commercial production of a wide range of edible macrofungi.

The design and development of technologies to increase the utilisation of microorganisms for human benefit therefore merit interpretation as activities integral to the formulation of long-term sustainable development programmes.

THE NEED FOR DIVERSITY AMONGST MICROORGANISMS

While sufficient diversity of microorganisms to enable the various functions necessary for ecosystem maintenance and the operation of biogeochemical cycles is clearly crucial, the extent to which individual species are important is less certain. In monitoring microorganisms with reference to the conservation of biodiversity in macroorganisms, the maintenance of functional groups rather than individual species can be presumed to be limiting - except where a particular microorganism is a keystone species.

There has been considerable debate as to the significance of functional redundancy in ecosystem function and maintenance (Solbrig, 1991). The presence of a wide variety of species able to perform similar roles is unquestionably beneficial as it provides an ecosystem with increased resilience to perturbations. For example, in the case of ectomycorrhizas of temperate and boreal forests, the ability of a tree to form associations with a variety of fungi (over 100 in the case of *Betula*) enables that tree to grow satisfactorily even if only a few of the candidate mycorrhizal fungi are present in a particular soil. Further, if the mycorrhizal species are differentially sensitive to pollutants, the tree can continue provided at least some of those fungi can tolerate the ambient pollution levels. In the event that too many species from a functional group are eliminated, at some point an ecosystem will start to break down irretrievably. In this regard, the implications for trees of the recently reported widespread and dramatic losses of ectomycorrhizal fungi in Europe are of particular concern (Jaenike, 1991).

Single microorganisms can also function as keystone species crucial to the maintenance of particular ecosystems. This applies to marine environments such as coral reefs, kelp forests formed by *Macrocystis* in temperate waters, and lichen-dominated deserts, heaths and rocks. Microorganisms are most important as keystone organisms when they function as mutualistic symbionts in organisms that dominate an ecosystem, and in low productivity/high diversity systems (Solbrig, 1991). Examples include dinoflagellates in corals, endomycorrhizal fungi in tropical forest trees, and nitrogen-fixing bacteria in tree roots.

Individual microorganisms which are major parasites can also function as keystone species through natural biocontrol processes. For instance, trypanosomes in East Africa keep cattle out of wide areas and so may limit soil degradation.

The present state of ignorance of the biology, ecology, and biochemical activities of so many microorganisms is comparable to that of their role in food-webs (cf. above). It is consequently often difficult or impossible to assert whether a particular microorganism is functionally redundant or a keystone species. Thus, while the presence of a variety of lignosic wood decay fungi might at first be assumed to be a case of functional redundancy, in practice the species of wood attacked can be restricted, and in most instances the specific enzymes being formed are unknown. Several species of fungi with different but complementary properties may need to work simultaneously or successionally in the decay of a single log. Furthermore, one or more of the decay fungi in that log might be a source of digestive enzymes for an insect of ecological importance in that ecosystem (Martin, 1987).

EX SITU CONSERVATION OF MICROORGANISMS

A wide range of techniques is available for the preservation of microorganism strains, freeze-drying (lyophilisation) and storage in liquid nitrogen (cryopreservation) being the most efficacious for long-term storage. Although not all microorganisms can yet be preserved by such methods, the development of programmable coolers and cryomicroscopy is enabling protocols to be devised for the successful cryopreservation of organisms previously considered recalcitrant. Even where species cannot be grown in pure culture, host tissue including them (e.g. plant leaves infected with rust fungi) or samples of the substrate itself (e.g. soil) can be conserved by cryopreservation. A survey of the existing technology is provided by Kirsop and Doyle (1991), and the World Federation for Culture Collections (1990) has issued guidelines for the establishment and operation of such collections.

Further information on *ex situ* culture collection is provided in Part 3.

THE TAXONOMIC CHALLENGE

Studies on the biodiversity and roles of almost all microorganism groups are frustrated by an inadequate taxonomic base. Not only are there vast numbers of species yet to be described, there are few modern monographs, keys, and other readily available aids, and disproportionately few taxonomists so that assistance with identifications is difficult to obtain. This issue requires priority attention at national, regional, and international levels. It is clearly unrealistic for most countries even to contemplate the provision of comprehensive microorganism identification services. However, attention could be focused on strengthening existing centres of expertise, developing north-south and south-south linkages, establishing networks of centres and specialists, and endeavouring to ensure that research agendas are complementary and collaborative.

Action to improve the knowledge base

An IUBS/SCOPE workshop on Ecosystem Function of Biological Diversity held in Washington DC in June 1989 recognised that the issue of microbial diversity and its function had been neglected and was in urgent need of attention; the workshop recommended that IUBS and IUMS

(International Union of Microbiological Societies) establish a cooperative programme to address this problem (Di Castri and Younès, 1990).

An IUBS/IUMS workshop on *Biodiversity amongst Microorganisms and its Relevance* was therefore convened in Amsterdam 7-8 September 1991. Representatives of relevant international scientific organisations concerned with different groups of microorganisms, together with other specialists, presented overviews of the current knowledge base (Hawksworth and Colwell, in prep.). A 14-point action statement, MICROBIAL DIVERSITY 21, was drawn up detailing the remedial work necessary to raise to an appropriate level our knowledge of the biodiversity of microorganisms and its relevance. The various action points are currently being developed and costed, but it must be recognised that substantial international resources will be required to implement this programme at the level necessary for it to realise its objectives. This Chapter draws heavily on the presentations and discussions which took place during the IUBS/IUMS workshop and the proceedings (Hawksworth and Colwell, in prep.) should be consulted for further information on many of the topics discussed here.

References

Andersen, R.A. (in press). The diversity of eukaryotic algae. *Biodiversity and Conservation* 1.

Austin, B. and Priest, F. 1986. *Modern Bacterial Taxonomy*. Van Nostrand Reinhold, Wokingham. 145pp.

Brasier, C.M. 1986. The dynamics of fungal speciation. In: Rayner, A.D.M., Brasier, C.M. and Moore, D. (Eds), *Evolutionary Biology of the Fungi*. Cambridge University Press, Cambridge. Pp.231-260.

Burnett, J.H. 1983. Speciation in fungi. *Transactions of the British Mycological Society* 81:1-14.

Corliss, J.O. 1991. Introduction to the protozoa. In: Harrison, F.W. and Corliss, J.O. (Eds), *Microscopic Anatomy of Invertebrates*, 1. Wiley-Liss, New York. Pp.1-12.

Di Castri, F. and Younès, T. 1990. Ecosystem function of biological diversity. *Biology International, Special Issue* 22:1-20.

Grassle, J.F., Lasserre, P., McIntyre, A.D. and Ray, G.C. 1991. Marine biodiversity and ecosystem function. *Biology International, Special Issue* 23:1-19.

Hawksworth, D.L. 1991a. The fungal dimension of biodiversity: magnitude, significance, and conservation. *Mycological Research* 95:641-655.

Hawksworth, D.L. (Ed.) 1991b. *The Biodiversity of Microorganisms and Invertebrates: its role in sustainable agriculture*. CAB International, Wallingford. 302pp.

Hawksworth, D.L. (in press). Biodiversity in microorganisms and its role in ecosystem function. In: Solbrig, O.T. and van Emden, H.A. (Eds), *Biological Diversity and Global Change*. Springer Verlag, New York.

Hawksworth, D.L. and Bridge, P.D. 1988. Recent and future developments in techniques of value in the systematics of fungi. *Mycosystema* 1:5-19.

Hawksworth, D.L. and Colwell, R.R. 1992 Biodiversity amongst microorganisms and its relevance. *Biology International* 24:11-15.

Hawksworth, D.L. and Colwell, R.R., (Eds) (in prep.). Biodiversity amongst microorganisms and its significance. *Biodiversity and Conservation* 1.

Jaenike, J. 1991. Mass extinction of European fungi. *Trends in Ecology and Evolution* 6:174-175.

Johnson, J.L. 1989. Nucleic acids in bacterial classification. In: Holt, J.G. (Ed.), *Bergey's Manual of Systematic Bacteriology*, 4. Williams and Wilkins, Baltimore. Pp.2306-2308.

Kessler, E. 1985 ["1984"]. A general review on the contribution of chemotaxonomy to the systematics of green algae. In: Irvine, D.E.G. and John, D.M. (Eds), *Systematics of the Green Algae*. Academic Press, London. Pp.391-407.

Kirsop, B.E. and Doyle, A. (Eds) 1991. *Maintenance of Microorganisms*, 2nd edn. Academic Press, London. 308pp.

Krumbein, W.E. (Ed.) 1983. *Microbial Geochemistry*. Blackwell Scientific Publications, Oxford. 330pp.

Kurtzman, C.P. 1985. Molecular taxonomy of fungi. In: Bennett, J.W. and Lasure, L.L. (Eds), *Gene Manipulations in Fungi*. Academic Press, Orlando. Pp.35-63.

Lal, R. 1991. Soil conservation and biodiversity. In: Hawksworth, D.L. (Ed.), *The Biodiversity of Microorganisms and Invertebrates: its role in sustainable agriculture*. CAB International, Wallingford. Pp.89-104.

Leadbeater, S.C. and Riding, R. (Eds) 1986. *Biomineralization in Lower Plants and Animals*. Clarendon Press, Oxford. 401pp.

Lee, K.E. 1991. The diversity of soil organisms. In: Hawksworth, D.L. (Ed.), *The Biodiversity of Microorganisms and Invertebrates: its role in sustainable agriculture*. CAB International, Wallingford. Pp.73-87.

Liesack, W. and Stackebrandt, E. (in press). Unculturable microbes detected by molecular sequences and probes. *Biodiversity and Conservation* 1.

Lovelock, J.M. 1988. *The Ages of Gaia*. Oxford University Press, Oxford. 252pp.

Lynch, J.M. and Hobbie, J.E. (Eds) 1988. *Microorganisms in Action: concepts and applications in microbial ecology*, 2nd edn. Blackwell Scientific Publications, Oxford.

Margulis, L. and Fester, R. (Eds) 1991. *Symbiosis as a Source of Evolutionary Innovation*. Massachusetts Institute of Technology Press, Cambridge, Mass. 454pp.

Margulis, L. and Schwartz, K.V. 1988. *Five Kingdoms. An illustrated guide to the phyla of life on Earth*, 2nd edn. W.H. Freeman, New York. 376pp.

Martin, M.M. 1987. *Invertebrate-Microbial Interactions. Ingested fungal enzymes in arthropod biology*. Comstock Publishing Associates, Ithaca. 148pp.

Persley, G.J. (Ed.) 1990. *Agricultural Biotechnology: opportunities for international development*. CAB International, Wallingford. 495pp.

Pirozynski, K.A. and Weresub, L.K. 1979. A biogeographic view of the history of ascomycetes and the development of pleomorphism. In: Kendrick, [W.] B. (Ed.), *The Whole Fungus*, 1. National Museums of Canada, Ottawa. Pp.93-123.

Price, P.W. 1988. An overview of organismal interactions in ecosystems in evolutionary and ecological time. *Agriculture, Ecosystems and Environment* 24:369-377.

Read, D.J. 1991. Mycorrhizas in ecosystems - nature's response to the "Law of the Minimum". In: Hawksworth, D.L., (Ed.), *Frontiers in Mycology*. CAB International, Wallingford. Pp.101-130.

Regenmortel, M.H.V. van 1990. Virus species, a much overlooked but essential concept in virus classification. *Intervirology* 31:241-254.

Round, F.E. 1981. *The Ecology of Algae*. Cambridge University Press, New York. [Not seen.]

Smith, D.C. and Douglas, A.E. 1987. *The Biology of Symbiosis*. Edward Arnold, London. 302pp.

Sneath, P.H.A. 1989. Numerical taxonomy. In: Holt, J.C. (Ed.), *Bergey's Manual of Systematic Bacteriology*, 4. Williams and Wilkins, Baltimore. Pp.2303-2305.

Sogin, M.L. 1991. The phylogenetic significance of sequence diversity and length variations in eukaryotic small subunit ribosomal RNA coding regions. In: Warren, L. and Koprowski, H. (Eds), *New Perspectives on Evolution*. Wiley-Liss, New York. Pp.175-188.

Solbrig, O.T. (Ed.) 1991. *From Genes to Ecosystems: a research agenda for biodiversity*. International Union of Biological Sciences, Cambridge, Mass. 124pp.

Stolz, J.F., Botkin, D.B. and Dastoor, M.N. 1989. The integral biosphere. In: Rambler, M.B., Margulis, L. and Fester, R. (Eds), *Global Ecology*. Academic Press, San Diego. Pp.31-49.

Takishima, Y., Shimura, J., Udagawa, Y. and Sugawara, H. 1989. *Guide to World Data Center on Microorganisms with a List of Culture Collections in the World*. World Data Center on Microorganisms, Saitama. 249pp.

Trüper, H.G. (in press). The prokaryotes, an overview with respect to biodiversity and environmental importance. *Biodiversity and Conservation* 1.

Wayne, L.G. (Ed.) 1987. Report of the *ad hoc* committee on reconciliation of approaches to bacterial systematics. *International Journal of Systematic Bacteriology* 37:463-464.

Whitcomb, R.F. and Hackett, K.J. 1989. Why are there so many species of mollicutes? An essay on prokaryote diversity. In: Knutson, L. and Stoner, A.K. (Eds), *Biotic Diversity and Germplasm Preservation: Global Imperatives*. Kluwer Academic Publishers, Dordrecht. Pp.205-240.

Woese, C.R., Kandler, O. and Wheelis, M.L. 1990. Towards a natural system of organisms: proposal for the domains Archaea, Bacteria, and Eucarya. *Proceedings of the National Academy of Sciences, USA* 87:4567-4579.

World Federation for Culture Collections 1990. *Guidelines for the Establishment and Operation of Collections of Cultures of Microorganisms*. World Federation for Culture Collections, Campinas. 16pp.

This section was prepared by Professor D.L. Hawksworth, International Mycological Institute, UK.

7. LOWER PLANT DIVERSITY

The term 'lower plants' is a convenient but imprecise label for a disparate group of plants and plant-like organisms which are defined primarily by their lack of vascular tissue (the transport system for water and nutrients within higher plants). Under this heading we here discuss bryophytes, lichens and larger algae. Many authorities would only include the first of these among the 'true' plants (defined as those developing from an embryo; see Chapter 8). The lichens are composite organisms, not true plants, discussed here for convenience.

BRYOPHYTES

The bryophytes comprise some 14,000 species, consisting of 8,000 mosses and 6,000 liverworts. This is a very diverse group of plants containing several classes that are only distantly related. These classes, and their main subdivisions (orders) vary in their evolutionary history and geographical points of origin, and hence vary also in their current regions of maximum abundance and diversity. On a global scale, therefore, a more accurate assessment of areas of biodiversity should rely more on numbers of taxa within major taxonomic divisions of the bryophytes than on the oversimplified picture derived from crude summations of the whole group. Nevertheless, as with other plants, certain areas of the world are recognised as being particularly rich in bryophyte species, usually (but by no means always) the same areas where mosses and/or liverworts form more than 50% of the active biomass. In general terms, although bryophytes occur almost throughout the world, the majority of taxa are distributed in areas of high oceanicity, i.e. with cool or temperate, consistently moist climates. Their maximum diversity is to be found in regions where such conditions have persisted over geological time, and where tectonic factors have brought about an amalgamation of several regional floras. The regions of high species richness are noted in Table 7.1.

In contrast to many groups of mosses, liverworts generally (with the exception of the highly adapted Marchantiales) show little adaptation to desiccation, either physiologically or by reduction from perennial to annual growth cycles. In general, therefore, liverworts reach their maximum diversity and only achieve dominance in highly oceanic regions. There are fewer recognised genera than in the mosses but this is offset to a degree by the much larger numbers of species in some of them (e.g. *Frullania*, with up to 400 species; *Plagiochila*, with about 500). For convenience, the Hornworts (Anthocerotae) are included here with the liverworts.

Both mosses and liverworts (and hornworts) consist of major divisions into orders and families that may have widely different habitat preferences. There are too many such divisions to detail here, but the more significant groups (orders and families) are listed in Tables 7.2 and 7.3 to provide a reasonably representative picture.

Table 7.1 Regions of high bryophyte diversity

REGION	SPECIES (approximate)
Indo-Australian archipelago (esp. New Guinea, Sulawesi and Borneo)	3,000
South America (temperate, montane)	3,000
S Australasia (esp. Tasmania and New Zealand)	2,400
N America (Pacific, subarctic)	2,000
NE Asia (Pacific, subarctic)	2,000
Himalayas	2,000
E. Africa (and adjacent islands)	2,000
Europe (Atlantic areas, incl. British Isles)	1,800
British Isles	1,000

LICHENS

Lichens are composite organisms consisting of a usually dominant fungal partner in symbiosis with one or more photosynthetic partners, the resulting composite, organised structures behaving as independent entities. The fungal partner (mycobiont) is, in most cases, an ascomycete, rarely a basidiomycete, while the autotrophic partner (photobiont) may be a green alga or a cyanobacterium. Lichen photobionts come from a small number of genera most of which occur widely in nature while lichen mycobionts are exclusively lichen-forming and are taxonomically diverse, many coming from orders that also have non lichen forming taxa. The lichen symbiosis is one of the most successful known in nature. Of the 46 orders in the Ascomycotina some 16 have lichenised taxa to a greater or lesser degree, and out of some 238 families, 81 consist entirely of lichens or at least have some lichenised taxa. Lichenisation is a polyphyletic process that has occurred at many different times.

Currently, the consensus of known lichenised taxa world wide varies from 13,500 to 17,000. On the basis of recent monographic revision of a number of widespread lichen genera, and the collection of lichens from areas of the world previously unknown or little known lichenologically, it is safe to assume that a realistic world total for lichens will be closer to 17,000 and possibly even to 20,000. It seems probable that at present we know 50-70% of the world's lichens, though future discoveries of short-lived, fast-growing lichens on leaves and on bryophytes, and of Southern Hemisphere microlichens, could substantially alter this estimate (Galloway, 1992).

Although for higher plants the tropics are regarded as major sites of biodiversity, much less is known about tropical lichens whose biodiversity tends to be richest in canopy vegetation, which is still very poorly sampled in many tropical areas. Temperate areas of the world, on the other hand, with their wide variations of habitat, geology and climate are known to be major sites of lichen diversity. Of great importance are the temperate rainforests of the Southern Hemisphere, especially South America, New Zealand, Tasmania, south eastern Australia and the highlands of the tropical Pacific islands.

Table 7.2 Selected orders and families of mosses

FAMILIES	GENERA	DISTRIBUTION AND ECOLOGY
Sphagnales		
1	1	Cosmopolitan. About 80 species. Maximum diversity in cool oceanic regions of N Hemisphere: c. 40 spp in W Europe, similar in N America; 13 in SE Asia and Pacific; 15 in tropical S America; 13 in E. Africa; < 6 in Australasia? Terrestrial, mainly calcifuge.
Polytrichales		
2	21	Old and diverse group with regional endemism and widely differing areas of diversity. About 200 species.
Bryales		
85 families incl:		Most mosses; about 7,000 species.
Dicranaceae	45	Cosmopolitan; tropical montane and high latitude; greatest diversity probably in W Europe, N America and NE Asia. Mainly calcifuge.
Pottiaceae	>70	Cosmopolitan; most diverse in, Mediterranean or continental climates, extending to semi-deserts; principally in temperate to subarctic N Hemisphere but strongly represented in Australasia and Africa. Xerophytic on soil and rocks, rarely epiphytic, many annuals.
Calymperaceae	12	Lowland tropics and subtropics. Greatest diversity in SE Asia and W Pacific; absent from cool temperate regions.
Grimmiaceae	12	Mostly subarctic and alpine. Highest diversity in W Europe and N America.
Bryaceae	20	Cosmopolitan; most numerous in cool temperate to polar regions; ecologically important in polar deserts.
Orthotrichaceae	21	Ecologically very important in the epiphytic biome of the montane tropics. Major diversity in W Pacific (esp. New Guinea) with other areas in Australasia, S America, E Africa. Mainly photophilic epiphytes.
Spiridentaceae	2	Endemic to W Pacific (esp. New Guinea). Epiphytes.
Hypnodendraceae	2	Endemic to W Pacific and Australasia. Epiphytes and lignicoles.
Amblystegiaceae	21	Mainly cool temperate to arctic with high diversity in NW Europe, N America and NE Asia. Hygrophilous and subaquatic.
Pterobryaceae	30	Tropical montane rainforests, especially abundant in SE Asia and W Pacific where many genera endemic. Frondose epiphytes.
Meteoriaceae	19	Tropical montane rainforests, especially SE Asia and W Pacific. Pendulous epiphytes.
Hookeriaceae	27	Greatest diversity in the humid tropics, especially S America and SE Asia with significant endemism in both. Hygrophilous.
Sematophyllaceae	49	Temperate to tropical montane. Maximum diversity in SE Asia/W Pacific (esp. Indonesia, Papua New Guinea) and tropical America. Mainly acidophilous, lignicolous and epiphytic.
Brachytheciaceae	c 30	Temperate to arctic. Greatest diversity in N America, NW Europe and NE Asia. Mainly ground-dwelling.
Hypnaceae	40	Cosmopolitan but with strong regional speciation in all of the areas mentioned above.

Other regions of important local, lichen biodiversity are the unique coastal fog lichen communities (*nebeloasen*) found in northern Chile, Peru, Baja California and Namibia, where members of the family Roccellaceae are particularly well-developed.

Islands also often show high lichen biodiversity in comparison with large continental areas, not only islands surrounded by water, but biogeographical islands (i.e. areas of habitat or climate diversity such as rock outcrops, mountains or ranges in an otherwise uniform forest or grassland landscape). Lichens are particularly successful pioneer colonisers, and so are important components of vegetation in many harsh environments of the world, such as alpine and polar regions, in hot and cold deserts, and in often toxic, mineralised environments.

Comparative figures of lichen diversity for a number of areas are presented in Table 7.4. The information is derived from published accounts of varying age and reliability, most of which are recorded in the bibliography of Hawksworth and Ahti (1990), and from unpublished data.

Table 7.3 Selected orders and families of liverworts (including hornworts)

FAMILIES	GENERA	DISTRIBUTION AND ECOLOGY
Calobryales		
2	2	Living fossils with disjunct distributions, all in moist temperate habitats.
Treubiales		
1	2	An ancient group best represented in the W Pacific.
Jungermanniales		
39 families incl:		The 'leafy liverworts'.
Herbertaceae	4	Confined to oceanic regions, mainly tropical and subtropical montane but 2 spp. in W Europe.
Lepidoziaceae	24	Greatest richness in W Pacific and temperate S Hemisphere. Mainly humicolous and lignicolous.
Lophoziaceae	18	Widespread in cool oceanic regions: W Europe, NW North America, NE Asia, S America, Australasia.
Jungermanniaceae	11	Cool temperate to subarctic regions. Terricolous and strongly hygrophilous.
Gymnomitriaceae	3	Arctic and alpine preferences: main diversity in the cool N Hemisphere, especially W Europe.
Schistochilaceae	2	Almost confined to montane forests around the W Pacific.
Lophocoleaceae	15	Cosmopolitan. Mainly ground-dwelling.
Plagiochilaceae	6	*Plagiochila* is most important genus with large numbers of species in tropical montane rainforests. Very strongly represented in SE Asia and W Pacific.
Radulaceae	1	Cosmopolitan, but greatest diversity in rainforest vegetation: SE Asia, S America.
Lejeuneaceae	c. 70	Extremely diverse and important family, especially within the tropics; greatest diversity in W Pacific, Indonesia and tropical America. In Europe almost confined to the Atlantic seaboard. Mainly corticolous and epiphyllous
Frullaniaceae	3	*Frullania*, with over 400 species, is the most important genus. Greatest diversity in the montane tropics of SE Asia and America. Mainly strongly photophilic epiphytes but also epilithic etc.
Metzgeriales		
5 families incl:		Thalloid liverworts, strongly hygrophilous.
Dilaenaceae	11	Comparatively few species, more or less evenly distributed. Soil-dwelling, mainly riparian
Aneuraceae	2	Montane forests; SE Asia, S America, W Europe and N America. Hygrophilous, mainly on soil, rocks and rotting wood
Metzgeriaceae	1	Widely distributed.
Marchantiales		Thalloid liverworts
	30	Mediterranean type climate: W Mediterranean, S Africa, India. Xerophytic tendencies. Includes some 'weedy' cosmopolitan species.
Anthocerotales		Hornworts
	5	Widespread. Epiphylls maximum diversity in the W Pacific. Terrestrial hygrophytes, epiphytes and epiphylls.

ALGAE

Chlorophyta (Green Algae)

The class Chlorophyta is cosmopolitan in distribution and occurs in marine and brackish water, freshwater, and terrestrial environments. It comprises approximately 1,040 species in 170 genera and contains eight orders (Silva, 1982) some of which have restricted geographical distributions. Many of the larger Chlorophyta are restricted to either marine or freshwater conditions; a few are sufficiently tolerant to be found in both environments. Table 7.5 lists the orders and constituent families of the larger Chlorophyta (excluding unicellular forms) and indicates their broad geographical distributions and salinity tolerances.

The largest family, the Cladophoraceae, occurs globally and

Table 7.4 Lichen diversity

REGION	GENERA	TAXA
USA and Canada	401	3409
Australia	299	2499
France	181	2200
Sweden and Norway	216	2142
West Indies	173	1751
United Kingdom	250	1600
New Zealand	243	1162
India	163	1150
Mexico	130	997
Philippines	137	974
Argentina	122	942
Sardinia	178	901
Hawaii	104	750
Tasmania	173	655
East Africa (macrolichens)	79	639
Central America	120	635
Guianas	165	600
New Guinea	137	537
Galápagos	80	196
Juan Fernández	31	194
Ecuador	160	?

in a wide range of salinities; the next largest family, the Codiaceae, is restricted to the marine environment and does not occur in the colder waters of the polar regions. Temperature-dependent distribution is clearly seen at both order and family levels. The Acrosiphoniaceae is restricted to colder waters in contrast to the Siphonocladales, Caulerpaceae and Udoteaceae' which occur only in tropical and subtropical waters.

Of the selected floras assessed for species diversity (Table 7.8), the North Atlantic, the tropical/subtropical western Atlantic, and the Japanese region of the Pacific are the most species-diverse. Although the green seaweed flora of southern Australia is not so species-diverse, it probably contains the highest number of endemics (46% of the total). Particularly impoverished floras are those of the tropical west coast of Africa (e.g. Gambia to Angola) and the west coast of South America (e.g. Colombia, Peru), areas where there are major cold water upwellings. Other impoverished floras include those of small isolated islands (e.g. Macquarie Island) and the polar regions.

Phaeophyceae (Brown Algae)

The Phaeophyceae are global in distribution, occurring in polar, temperate and tropical zones. The brown algae are principally marine plants, with only very few species in freshwater. The class contains about 265 genera and in excess of 1,500 species arranged in 14 orders (Wynne, 1982, Table 7.6).

Table 7.5 Orders and families of larger green algae

	FAMILIES	GEN	SPP	DISTRIBUTION	ECOLOGY
Ulotrichales	1	16	80	Global	Mostly freshwater
Ulvales	5	12			
Capsosiphonaceae		1	3		Brackish water
Percursariaceae		1	1		Brackish water
Ulvaceae		6	86	Global	Mostly seawater/brackish water
Monostromaceae		3	12		seawater/brackish water
Prasiolales	1	3	40	Global	Seawater/brackish water/freshwater
Acrosiphoniales	2	4		Cold waters	Seawater/brackish water
Codiolaceae		16		Cold waters	Seawater/brackish water
Acrosiphoniaceae		2	3	Cold waters	Seawater/brackish water
Cladophorales	2	20			
Cladophorales		14	300	Global	Seawater
Anadyomenaceae		6	26	Tropical/subtropical	Seawater
Siphonocladales	2	12		Tropical	Seawater
Siphonocladaceae		8	48	Tropical	Seawater
Valoniaceae		4	35	Tropical	Seawater
Bryopsidales	6	24			
Bryopsidaceae		6	60	Global	Seawater
Ostroebiaceae		1	6	Shells	Seawater
Dichotomosiphonaceae		1	1		
Caulerpaceae*		1	75	Tropical/subtropical	Seawater
Udoteaceae		13	87	Tropical/subtropical	Seawater
Codiaceae**		1	100	Global	Seawater
Dasycladaceae***	1	11	50	Tropical/subtropical	Seawater

Notes: * Temperate in southern hemisphere, ** one relict genus in inland brackish water, *** not polar waters. GEN = genera, SPP = species.

Table 7.6 Orders and families of brown algae

	FAMILIES	GENERA	SPECIES	DISTRIBUTION	ECOLOGY
Ectocarpales	3			Global	Marine
Ectocarpaceae		29		Global	Marine
Ralfsiaceae		17			Marine
Sorocarpaceae		2	2	N Atlantic	Marine
Chordariales	10				
Myrionemataceae		11			Marine
Elachistaceae		5		Global	Marine
Corynophloeaceae		5			Marine
Spermatochnaceae		5			Marine
Acrotrichaceae	1	1		N Atlantic	Marine
Chordariaceae		29		Global	Marine
Ischigiaceae		1	2	Limited	Marine
Chordariopsidaceae		1		S Africa	Marine
Notheiaceae		1	1	Australasia	Marine
Splachnidiaceae		1		S Africa	Marine
Cutleriales	1	3		Warm waters	Marine
Tilopteridales	1	2		N Atlantic	Marine
Sphacelariales	4			Global	Marine
Sphacelariaceae		5		Global	Marine
Stypocaulaceae		4	10	Global	Marine
Cladostephaceae		1		N Atlantic/Australasia	Marine
Choristocarpaceae		2		N Atlantic/Mediterranean	Marine
Dictyotales	1	16		Tropical/subtropical*	Marine
Sporochnales	1			Warm waters	Marine
Desmarestiales	2			Cold waters	Marine
Desmarestiaceae		3		Cold waters	Marine
Arthrocladiaceae		1		N Atlantic/Mediterranean	Marine
Dictyosiphonales	7				Marine
Myriotrichaceae		1		N Atlantic/Mediterranean	Marine
Giraudiaceae		1	1	N Atlantic/Mediterranean	Marine
Striariaceae		9			Marine
Delameriaceae		4			Marine
Punctiariaceae		17		Temperate	Marine
Chnoosporaceae		1	2	Tropical/subtropical	Marine
Dictyosiphonaceae		2		N Hemisphere	Marine
Scytosiphonales	1	8			Marine
Laminariales	4			Temperate/polar	Marine
Chordaceae		1	2	N Atlantic	Marine
Laminariaceae		15		Temperate/polar	Marine
Lessoniaceae		8		NE Pacific/S Hemisphere	Marine
Alariaceae		7		Temperate/polar	Marine
Fucales	6				Marine
Fucaceae		7		N Hemisphere	Marine Brackish
Himanthaliaceae		1	1	NE Atlantic	Marine
Hormoseiraceae		1	1	Australasia	Marine
Phyllosporaceae		6		Australasia	Marine
Sargassaceae		6		Tropical/temperate	Marine
Cystoseiraceae		16		Tropical/temperate	Marine
Durvilleales	1	1	4	Australasia/Antarctic	Marine
Ascoseirales	1	1	1	Antarctic	Marine

The more primitive orders (Ectocarpales, Chordariales) are global in distribution, although some of the constituent families, particularly the smaller ones (e.g., Sorocarpaceae) are geographically restricted. The small orders Cutleriales, Dictyotales and Tilopteridales are limited, respectively, to warmer waters, the tropics and subtropics, and the North Atlantic, while the Durvilleales and Ascoseirales occur only in Australasia and Antarctica. The kelps (order Laminariales) are disjunctly distributed in temperate waters of both northern and southern hemispheres. In the most

Table 7.7 Orders and families of red algae

	FAMILIES	GENERA	SPECIES	DISTRIBUTION	ECOLOGY
Bangiophycideae					
Porphyridiales	4				
Porphyridiaceae		7		Global	Marine/freshwater
Goniotrichaceae		3			Marine/brackish/freshwater
Phragmonemataceae		4			Freshwater
Bangiales	3				Marine/freshwater
Erythropeltidaceae		6			Marine
Bangiaceae		?			Marine/brackish/freshwater
Boldiaceae		1	1		
Compsopogonales	1	1	12	Tropical/subtropical	
Rhodochaeteles	1	1	1		Marine
Florideophycideae					
Nemaliales	13			Global	Marine/brackish/freshwater
Acrochaetiaceae		1		Global	Marine/brackish/freshwater
Batrachospermaceae		3			Freshwater
Lemaneaceae		1			Freshwater
Thoreaceae		2			Freshwater
Helminthocladiaceae		3			Marine
Nemaliaceae		4			Marine
Dermatonemaceae		2			Marine
Chaetangiaceae		7			Marine
Naccariaceae		3			Marine
Bonnemaisoniaceae		5			Marine
Gelidiaceae		2			Marine
Gelidiellaceae		1			Marine
Wurdemanniaceae		1	1		Marine
Cryptonemiales	13			Global	Marine
Weeksiaceae		3			Marine
Dumontiaceae		6			Marine
Choreocolacaceae		2	3		Marine
Cryptonemiaceae		20		Global	Marine
Corynomorphaceae		1	1		Marine
Pseudoanemoniaceae		2		S Hemisphere	Marine
Kallymeniaceae		10	70	Global	Marine
Endocladiaceae		2			Marine
Crossocarpaceae		4	6		Marine
Gloiosiphoniaceae		2			Marine
Tichocarpceae		1	1	NW Pacific	Marine
Pterocladiophilaceae		1	1		Marine
Peyssonneliaceae			?		Marine
Corallinales	1	30	400	Global	Marine
Hildenbrandiales	1	1			Marine/brackish/freshwater
Gigartinales	27				Marine
Gymnophleaceae		4	30		Marine
Gracilariaceae		?			Marine
Sebdeniaceae		1	2		Marine
Calosiphoniaceae		2	3		Marine
Petrocelidaceae		2			Marine
Phyllophoraceae		4			Marine
Gigartinaceae		4			Marine
Chondriellaceae		1	1	Juan Fernandez	Marine
Polyideaceae		2			Marine
Nizymeniaceae		2	3	S Australia	Marine
Rhizophyllidaceae		3	12		Marine
Acrotylaceae		5	5	S Hemisphere	Marine
Plocamiaceae		2			Marine
Phacelocarpaceae		2		S Hemisphere	Marine
Sarcodiaceae		4			Marine
Furcellariaceae		3	5		Marine
Solieriaceae		?			Marine
Hypneaceae		3			Marine
Rissoellaceae		1	1	Mediterranean	Marine

Table 7.7 Orders and families of red algae (continued)

	FAMILIES	GENERA	SPECIES	DISTRIBUTION	ECOLOGY
Rhabdoniaceae		5			Marine
Cubiculosporaceae		1	1		Marine
Rhodophyllidaceae		3			Marine
Mychodeaceae		1	11	S Australia	Marine
Dicranemaceae		4		S Australia	Marine
Ahnfeltiales	1	1			Marine
Rhodymeniales	2				Marine
Rhodymeniaceae		30			Marine
Champiaceae		6			Marine
Palmariales	1	4			Marine
Ceramiales	4			Global	Marine
Ceramiaceae		100		Global	Marine/brackish*/freshwater*
Delesseriaceae		100	300	Global	Marine/freshwater*
Dasyaceae		12	100		Marine
Rhodomelaceae		100	500		Marine/brackish*

Note: * A few species only.

advanced order, the Fucales, the family Cystoseiraceae occurs widely in the tropical and temperate zones, whereas the family Sargassaceae is mostly confined to the tropical waters. Other families have more circumscribed distributions, the Hormoseiraceae and Phyllosporaceae occurring only in Australasia; the Fucaceae is restricted to the northern hemisphere and the monotypic Himanthaliaceae is endemic to the north-eastern Atlantic.

Brown algae attain greatest species richness in the Japanese region of the Pacific, the North Atlantic and, to a lesser extent, southern Australia. The last region, however, is probably highest in endemics, with 18% of genera and 54% of species endemic while only 26% of the flora comprises widespread species. Species-depauperate floras are, as in the Chlorophyta, those in cold-water upwelling areas (e.g. Angola, Colombia and Peru), on isolated islands (Ascension, St Helens), or a combination of both (e.g. Macquarie Island).

Rhodophyta (Red Algae)

The Rhodophyta is the largest of the three main seaweed groups, with over 555 genera (Dixon, 1982, Table 7.7); it contains more species than the Chlorophyta (Green) and Phaeophyceae (Brown) together. The Rhodophyta is divided into two subclasses, the subclass Bangiophycideae, the smaller of the two, occurs throughout the world in marine, brackish and freshwater environments. The subclass Florideophycideae comprises eight orders and is predominantly marine. None of the orders is clearly circumscribed geographically; some small families (e.g. Mychodeaceae, Phacelocarpaceae) are restricted in occurrence to Australia or the southern hemisphere generally. The larger families are widely distributed.

As with the Green and Brown Algae, the most species-rich floras are those of the Japanese Pacific region, the tropical and subtropical western Atlantic, and the North Atlantic (including temperate and arctic regions). Other rich floras of Red Algae are those of California and Chile. Although precise data are not available for southern Australia, it is probably also species-rich, with 75% of species and 30% of genera endemic to the area. Species-poor floras are those referred to previously on the tropical west coasts of Africa and South America where there are cold-water upwellings. Red algae floras decrease in species abundance in cool waters.

General remarks on marine algal floras

The most species-rich algal flora assessed is that of the Japanese region of the Pacific (1,503 species, Table 7.8). The North Atlantic and tropical and subtropical western Atlantic are also species-rich, with over 1,000 species recorded.

In the North Atlantic, eastern and western seaboards differ in diversity. The western (American) coastline is relatively species-poor; 65% of the North Atlantic flora is restricted to the eastern (European) coast, 35% is common to both coasts, and only 5% restricted to the American coast. A reduction in species also occurs from south to north, with the Arctic flora the least diverse and characterised by hardy, cosmopolitan species and very low endemism. The flora of the British Isles is relatively species-rich (over 700) exceeding that north western Pacific America and one of the richest of the 18 floras assessed. The flora of the Eastern Mediterranean is probably much richer than indicated by the 430 species listed for Aegean Greece. The latter flora is fairly high in endemics (20% of the total species, while 28% of all species are Mediterranean-Atlantic in distribution).

The flora of the tropical and subtropical regions of the Atlantic contrast with those in higher latitudes to the north in having the most species-diverse area on the western (American) side, where 1,058 species are recorded. On the eastern (tropical African) side the number is only about 300. Similar comparison of the Chlorophyta (green algae) and Phaeophyta (brown algae) floras (groups having better data for subtropical and tropical Africa) shows 253 species of green algae from the west and 153 from the east, and 150 species of brown algae from the west and 125 from the east. The flora of tropical west Africa contains 56% of species that also occur in the Indian Ocean and 58% of species in the Pacific Ocean.

Moderately rich floras are those of Chile (temperate), North-west America (temperate), California (subtropical) and tropical East Africa. The flora of southern Australia probably also falls into this group, but is probably much higher in endemics.

Species-poor floras generally occur in polar waters, and on isolated islands - the further from the nearest landmass the poorer the flora (e.g. St Helena with only 68 species). Potential for endemism exists in water masses isolated from the main oceans, such as the Mediterranean Sea (which only has very small water exchange with the Atlantic Ocean), the Black Sea (for similar reasons), and the Caspian Sea (now completely isolated and brackish, but retaining an impoverished seaweed flora).

An important characteristic of many tropical and subtropical regions is the occurrence of coral reefs; algae are a major constituent of these long-stable ecosystems and the sheltered lagoons they protect. Coral reefs support a unique and generally diverse algal flora that includes many crustose coralline algae whose numbers are likely to increase with further study. Mangrove areas are also restricted to the tropics and subtropics and support a well-defined and interesting algal vegetation, contrasting with that of saltmarshes in the temperate zones, which are generally more species-poor. Sandy coastlines are floristically depauperate areas and often form barriers to seaweed dispersal. Some anthropomorphic changes to the coastline involving creation of additional habitats have locally enhanced species diversity; pollution, in contrast, has reduced species diversity, especially in lagoons, mangrove areas and coral reefs. In the latter, pollution-tolerant weedy species appear to outcompete and replace pollution-sensitive species. Land reclamation, rice-paddies and salt-pan development have led to the loss of algal habitat in many coastal areas in the tropics.

Table 7.8 Diversity of marine algal (seaweed) floras

FLORA	CHLOROPHYTA		PHAEOPHYCEAE		RHODOPHYTA		TOTAL	
	GENERA	SPECIES	GENERA	SPECIES	GENERA	SPECIES	GENERA	SPECIES
Japan	60	234	108	379	267	900	475	1503
N Atlantic	67	253	127	324	193	539	387	1116
W Atlantic	64	253	63	150	134	655	321	1058
Chile	31	131	60	140	150	480	241	751
California	22	72	69	137	186	459	279	668
E Africa	30	159	29	118	121	366	180	643
NW America	51	117	66	143	161	373	273	635
Antarctica		88		118		357		563
S Africa								547
E Mediterranean	30	73	49	90	131	267	210	430
Viet Nam		115		86		223		424
Red Sea	30	92	34	118	79	173	143	383
Tropical W Africa	19	59	22	42	88	198	129	299
Angola	8	34	18	22	71	140	97	196
Peru		29		20		107		156
Colombia	12	23	13	21	46	79	71	123
Macquarie I	12	15	25	28	46	60	81	103
St Helena	10	13	9	10	34	45	53	68
Ascension I	9	14	11	15	16	23	35	52
S Australia	39	119	104	231				
Tropical/subtropical								

Charophyta (Stoneworts)

The charophytes or stoneworts are a very distinctive group of macrophytic green algae that occur from Spitzbergen in the north (80°N) to the Kerguelen Islands in the south (c. 50°S). A few are restricted to brackish water but the large majority are widely distributed in such freshwater habitats as ponds, lakes, ditches, temporary pools, streams, rivers and swamps. The six extant genera are placed in two tribes: the Chareae - *Chara*, *Lamprothamnium*, *Nitellopsis* and *Lychnothamnus*; and the Nitelleae - *Nitella* and *Tolypella*. It is difficult to undertake a biogeographic analysis based on charophyte species because of the current uncertainty surrounding taxonomic limits and the ranking of infrageneric taxa. In Wood's world monograph on the group (Wood, 1965) the infrageneric taxa were divided into sections, species, subspecies, varieties and forms. He considered morphologically similar monoecious and dioecious taxa to be 'species pairs' and combined them. Other charologists do not accept Wood's views on merging monoecious and dioecious taxa and continue to regard them as distinct.

Unlike most other groups of freshwater algae sufficient regional information exists on the distribution of charophytes to allow for global analysis. In carrying out such an analysis Khan and Sarma (1985) used Wood's classification but did not recognise the merging of monoecious and dioecious taxa. They included charophytes described after 1965 and taxa reduced by Wood to synonymy but subsequently shown to be distinct. Khan and Sarma recognised 440 taxa, of which 274 were known from only one region or continent ('endemics'). For assessing the geographical distribution of taxa eight broad zones (regions/continents) were recognised: North America, South America, Africa, Europe, Asia (including Japan but excluding India), India, Pacific Island region, and Australia. Antarctica was not included as it is the only continent for which charophytes have yet to be reported.

Table 7.9 Stonewort diversity

REGION	GENERA	SPECIES	ENDEMIC SPECIES
North America	4	114	50
Asia	5	122	48
Africa	4	116	42
Europe	6	91	41
Australia	5	62	25
South America	5	89	25
India	6	125	23
Pacific Region	4	72	19
World	6	440	274 *

Source: Khan, M. and Sarma, Y.S.R.K. 1984. Cytogeography and Cytosystematics of Charophyta. In: Irvine, D.E.G. and John, D.M. (Eds), *Systematics of the Green Algae*. Academic Press, London and Orlando.

Note: * Majority of the remainder (c. 166) have a restricted distribution (normally two or three regions/continents) and about seven are to be regarded as cosmopolitan. Fewer than a dozen taxa have been published since 1985 and most are from underworked regions (e.g. South America; Asia, especially China).

It is impossible to determine to what extent tabulated estimates are significant or simply reflect collecting. Europe is one of the most intensively collected regions and so the lower numbers reported are likely to represent a real difference in diversity. The general unsatisfactory state of the taxonomy will continue to hamper biogeographical analysis.

Charophytes form extensive and sometimes diverse associations in marl rich water bodies and are especially sensitive to nutrient enrichment or eutrophication. In some countries they have become dramatically less common and more restricted in distribution as a result of nutrient enrichment primarily from agricultural sources. This would seem to be the main threat to these algae along with the general loss of aquatic habitats through land reclamation. Brackish-water lagoons is an example of a habitat under threat in many countries and one the genus *Lamprothamnion* is almost wholly confined to it. In the British Isles this is the only charophyte protected by government legislation although several freshwater species may also be under threat and have been recommended for protection. The conservation status of charophytes is difficult to determine without considerably more information on habitat requirements.

Other groups of algae

Comments on the diversity and global distribution of most groups of microalgae are not possible because of inadequate knowledge of the algal floras of the world. A reasonable coverage exists for a few regions but only for fairly well-defined algal groups such as the desmids (Division Chlorophyta, Order Desmidiales) and the diatoms (Division Bacillariophyta). Only a few attempts to analyse and interpret regional distribution patterns go so far as to consider the wider distribution of individual taxa. Doubt is often attached to the reliability of published lists so that the findings of regional comparisons need to be treated with caution. Frequently, 'regional endemics' have had to be reduced to synonymy because the describing authors failed to take adequate account of the taxonomic literature covering other regions. Sometimes the converse is true, and endemics are not recognised because they are incorrectly attributed to an extant taxon using identification guides written for another region. If progress is to be made it is essential to have sounder species concepts, more accurate identification, and considerably more information on the algal floras of under-collected parts of the world.

References

Dixon, P.S. 1982. Rhodophycota. In: Parker, S.P. (Ed.), *Classification of Living Organisms*. McGraw Hill, New York. Pp.62-79.

Galloway, D.J. 1992. A lichenological perspective. Biodiversity and Conservation: submitted September 1991.

Hawskworth, D.L. and Ahti, T. 1990. A bibliographic guide to the lichen floras of the world, 2nd edn. *Lichenologist* 22:1-78.

John, D.M. 1986. The algal flora: its analysis and biogeography. In: John, D.M., *The Inland Waters of Tropical West Africa*. E. Schweizerbart'sche, Stuttgart. Pp.133-160.

Khan, M. and Sarma, Y.S.R.K. 1984. Cytogeography and Cytosystematics of Charophyta. In: Irvine, D.E.G. and John, D.M. (Eds), *Systematics of the Green Algae*. Academic Press, London and Orlando. Pp.303-330.

Silva P.C. 1982. Chlorophycota. In: Parker, S.P. (Ed.), *Classification of Living Organisms*. McGraw Hill, New York. Pp.133-161.

Wood, R.D. 1965. In: Wood, R.D. and Imahori, K. (Eds), *A Revision of the Characeae*, Part I. Cramer, Weinheim.

Wynne, M.J. 1982. Phaeophyceae. In: Parker, S.P. (Ed.), *Classification of Living Organisms*. McGraw Hill, New York. Pp.115-125.

Chapter abridged from material contributed by the following staff of the Department of Botany, The Natural History Museum. (London):

Alan Eddy (Bryophytes); D.J. Galloway (Lichens); David M. John (Algae); Ian Tittley (Green Algae).

8. HIGHER PLANT DIVERSITY

The higher plants, characterised by vascular tissue and reproducing either by spores, cones, or flowers, dominate the world's flora and vegetation. Along with the bryophytes (Chapter 7), they develop from an embryo resulting from the sexual fusion of cells. They consist of three groups:

- the pteridophytes or ferns and fern allies, such as clubmosses, horsetails, quillworts and whiskferns
- the gymnosperms, mainly the conifers and cycads
- the angiosperms or flowering plants.

THE GROUPS OF HIGHER PLANTS

Pteridophytes

Estimates of the total number of ferns and their allies vary between 10,000 and 13,000 species but is probably close to 12,000, the majority of which are native to the moist tropics.

The so-called 'fern allies' probably do not form a natural group but rather represent the end points of several distinct evolutionary lineages. Like the true ferns, they reproduce by spores. The earliest known vascular land plants belong to this group. These psilophytes (Psilophyta), which dominated the landscape during the Silurian and Devonian around 400 million years ago (Mya), are all but extinct; they are only represented by two relict genera - *Psilotum* (tropics) and *Tmesipterus* (Australia, New Zealand, South Pacific). *Psilotum* is extremely primitive, lacking both roots and leaves.

Today, the lycopods (Lycopodiophyta) are represented by only five relict genera (*Isoetes, Lycopodium, Phylloglossum, Selaginella*, and *Stilites*), but their fossil record extends back to the Carboniferous (c. 300 Mya), when they formed the dominant vegetation. These extinct forms grew to 40m high and had a stem diameter of 2m; their remains form part of the coal reserves we rely on today.

The horsetails and scouring rushes (Sphenophyta) are another ancient group, and are also all but extinct. They are represented by a single genus, *Equisetum*, containing some 15 species found throughout the world, but especially well represented in North temperate bogs.

The true ferns (Pteridophyta or Filicophyta) are much more diverse than are the fern allies. They show great range of form, from the tiny, delicate filmy ferns (Hymenophyllaceae) to tropical tree-ferns (Cyatheaceae and Dicksoniaceae) more than 15m tall; leaves vary in length from 5mm to 10m. Ferns are cosmopolitan in distribution but are scarce in arid zones and occur in greatest numbers in the moist tropics, where they often grow epiphytically. It has been estimated that 12.5% of the world's fern species are to be found in Papua New Guinea (Johns and Bellamy, 1979), and 10% in India (Dixit 1984). Some species have a very wide distribution, notably Bracken *Pteridium aquilinum*, which is found throughout the temperate zones and over much of the tropics, while other species are extremely limited in their distribution.

Gymnosperms

The gymnosperms are trees (or occasionally shrubs) whose seeds lack the covering characteristic of the flowering plants. They include some 500 species of conifer, 100 species of cycad, and a few other small but scientifically fascinating families. They first appear in the fossil record in the Carboniferous (c. 300 Mya) as the so-called 'seed ferns' (which were not true ferns at all, but intermediates between ferns and gymnosperms). Gymnosperms dominated the earth until the rise of the flowering plants.

Conifers occur worldwide, but they reach their greatest diversity of species and genera in parts of Oceania and on the margins of the Pacific Ocean. They are the softwoods of commerce and are widely grown for timber and ornament. A conifer from the western USA, the Giant Sequoia *Sequoia sempervirens* is the tallest tree in the world, reaching a height of 110m; another conifer from western USA, the Bristlecone Pine *Pinus aristata* is thought to include the oldest living individual trees on earth, some being 4,900 years of age. The largest genera are the pines *Pinus*, firs *Abies*, and spruces *Picea*, which form extensive, economically important forests in the boreal zone of Eurasia and North America and in the mountains of the northern hemisphere. The podocarps *Podocarpus* are widespread in tropical and subtropical forests of the southern hemisphere. Locally, other genera are prominent, such as kauri pines *Agathis* (exploited for resin) in wet forests from Malesia to New Zealand, and Chinese Fir *Cunninghamia lanceolata*, the major timber tree of South and West China.

Cycads, palm-like tropical trees, occur mostly in Central and South America, South Africa, and from Southeast Asia to Australasia. They include the Sago-palms *Cycas*, an ancient group which originated at least 240 Mya and are thus of considerable scientific interest. Many of them are highly restricted in their distribution and are of great conservation concern.

Other gymnosperms include the famous maidenhair tree *Ginkgo biloba*, an isolated, ancient relict species native to China, the yews *Taxus* (source of the promising drug taxol) and their allies; joint-pines *Ephedra*, leafless 'switch plants' of scrub and semi-desert, *Gnetum*, mostly lianes of moist tropical forests, and the remarkable *Welwitschia bainesii*, which looks like a great woody turnip bearing only two huge, strap-shaped leaves and a cluster of either male or female cones, restricted to the coastal fog-belt of the Namib desert of Angola and Namibia. As a general rule, however, Africa has a very poor gymnosperm flora.

Angiosperms

The flowering plants, or Angiosperms, are an extremely diverse group of plants, containing some 250,000 species (see Table 8.2). From their first appearance in the fossil record around 135 million years ago, they evolved quickly and have come to dominate all other land plants, except in certain habitats (such as the boreal region, in which gymnosperms dominate). Most of our food comes from angiosperms, as do many spices, drugs, poisons, fibres, building materials. Many angiosperms are much utilised for their valuable timber (see Part 2).

Angiosperms are seed-producing plants that bear flowers that are often insect- or bird-pollinated. The plants range in size from 1mm (*Wolffia* spp.) to over 100m tall (*Eucalyptus regans* from Tasmania). The flowers can reach over 1m across (*Rafflesia arnoldii* from Sumatra and Borneo).

Estimates of the number of flowering plant species vary between 240,000 and 750,000, but most botanists accept 250,000 species as the best figure. These species are grouped into some 17,000 genera. Despite an enormous diversity of growth form and floral structure, the number of flowering plant families recognised is relatively small. It has varied over the years from 200 to over 600, but there is now general agreement on a basic 300-400 'core' families of flowering plants. Many of these families, such as Compositae (daisy and dandelion family) and Cruciferae (cabbage family) are natural units, and can be recognised without too much difficulty by the non-botanist, while others are characterised by more technical features not easily discernible by the layman.

Families vary greatly in the number of species they contain: on the one hand there are massive families like Orchidaceae (orchid family) with 25,000-35,000 species and Leguminosae (pea and bean family) with about 14,500 species (see Table 8.2). In fact, only 31 families contain 62% of known flowering plant species. At the other extreme are the 36 families with a single species, such as the Adoxaceae, the family of the well known North European woodland flower, Moschatel *Adoxa moschatellina*.

The grouping of these families into higher taxonomic levels such as orders and subclasses is somewhat more problematical, reflecting uncertainty about the fundamental evolutionary relationships between families. A commonly used scheme (after Cronquist, 1981) is presented in Table 8.2.

THE DISTRIBUTION OF HIGHER PLANTS

Higher plants occur in virtually all ecosystems of the world, even in the sea, but their distribution is very uneven. Two-thirds of the world's flowering plants are tropical, emphasising the great importance of plant conservation in the tropics. Many large or economically important families such as Annonaceae (custard-apple family), Lauraceae (cinnamon family), Moraceae (fig family), Dipterocarpaceae (dipterocarp family), Ebenaceae (ebony family) and Meliaceae (mahogany family) are almost entirely restricted to the tropics. This contrasts with the distribution of those who study plants, for specialists in plant taxonomy work mostly in Europe or the USA. The richest continent for plants, and still the least explored botanically, is South America, home to perhaps as much as one-third of the world's higher plants.

Table 8.1 gives an assessment of the numbers of species of higher plants in various regions of the world. Some of the figures, however, are provisional estimates that need to be treated with caution. It must be emphasised also that the species concept used varies from one region to another, which means that any comparison of the numbers of plants between regions must be done with care.

In particular, the differences in species richness between the regions of the world shown in Table 8.1 may be somewhat exaggerated. The species concept commonly used in Latin America, for example, tends to recognise more species, based on characters visible in the field, than the taxonomy of botanists working on the Malesian region. South America is still the continent with the most plants, but the differences between this region and tropical Asia or Africa may in time be found to be less than suggested. For example, estimates of the size of the flora of Colombia, a territory with high levels of species diversity and endemism, fell over a ten-year period from 45,000 (Prance, 1977) to 35,000 (Forero 1988).

A degree of convergence is apparent. In 1985, IUCN cited figures of 20,000 species in North America and 11,300 in Europe (Davis *et al.*, 1986). In Table 8.1 the estimate for North America has dropped to 17,000, following revised estimates by the Flora of North America workers, while that for Europe has risen to 12,500, following predictions based on the many species added to the recently revised first volume of *Flora Europaea*. It is fair to assume that North America does have more plants than Europe, but further convergence between the two figures is likely.

These changes in numbers of species do not result strictly from extinctions or the evolution of new species, although both of these processes are happening. In most cases, they result from decisions of botanists as to the delimitation of individual species. Many species in a flora are not clearly defined entities, as is, for example, the Gingko tree *Ginkgo biloba*, but are members of a complex group of species between which differences may be small. This is particularly true of some tropical and Mediterranean floras, where many species are extremely difficult to identify in the field. At the same time, collaboration between botanists who study the floras of different continents (facilitated by modern information technology and electronic data retrieval systems) is helping to rationalise and standardise the classification of plants that have in the past been treated as distinct species in different regions. Opinions will naturally vary as to the use of the rank of species, subspecies or merely variety.

Individual botanists tend to study either the plants of a particular country or the members of a particular family. Consequently, few data are available as to the numbers of species in individual habitats. Nevertheless, some general points can be made. Tropical forests, especially moist forests, are of enormous importance as habitats for plants. The species diversity of these forests, alongside fossil evidence, has led many botanists to argue that the flowering plants evolved in tropical forests, although it is more likely that they represent a 'museum' of evolution (Stebbins, 1974). Probably half or slightly under half of all higher plant species are restricted in the wild to tropical forests, a proportion that may be a little lower than that of animals because of the exceptional plant richness of Mediterranean ecosystems, a richness that is not reflected in faunal

Table 8.1 Distribution of higher plants by continents

Latin America (Mexico through S America)	85,000 [1]
Tropical & Subtropical Africa	40,000 - 45,000
North Africa	10,000 [2]
Tropical Africa	21,000 [3]
Southern Africa	21,000 [4]
Tropical & Subtropical Asia	50,000 [5]
India	15,000 [6]
Malesia	30,000 [7]
China	30,000 [8]
Australia	15,000 [6]
Caribbean	
Pacific	
North America	17,000 [9]
Europe	12,500 [10]

Sources: [1] Gentry, A.H. 1982. Neotropical floristic diversity: phytogeographical connections between Central and South America, Pleistocene climatic fluctuations, or an accident of the Andean orogeny? *Annals of the Missouri Botanical Garden* 69:557-593. [2] Based on figures for the size of country floras given in Quezel, P. 1985. Definition of the Mediterranean region and the origin of its flora. In: Gómez-Campo, C. (Ed.), *Plant Conservation in the Mediterranean Area*. Junk. P.17. [3] Estimate by A.L. Stork, quoted by Peter Raven, pers. comm., 1991. [4] Cowling, R.M. *et al.* 1989. Patterns of plant species diversity in southern Africa. In: Huntley, B.J. (Ed.), *Biotic Diversity in Southern Africa: concepts and conservation*. Oxford, Cape Town. [5] From Raven, P.H. 1987. The scope of the plant conservation problem worldwide. In: Bramwell, D. *et al.* (Eds), *Botanic Gardens and the World Conservation Strategy*. Academic Press. Pp.19-29. [6] From Davis, S. *et al.* 1986. *Plants in Danger: What do we know?* IUCN, Cambridge and Switzerland. [7] M.M.J. van Balgooy, Leiden, *in litt.* to J.R. Akeroyd, August 1991. [8] Prof. Wang Siyu, Beijing, *in litt.* to J.R. Akeroyd, October 1991. [9] Nancy Morin, pers. comm. via Peter Raven, 1991. [10] Estimate by J.R. Akeroyd, based on *Flora Europaea*, 1964-80, and the revision of Volume 1, in press.

Note: 'Malesia' consists of the nations of Malaysia, Brunei, Indonesia, Philippines and Papua New Guinea.

diversity. It is estimated that the Mediterranean basin has a flora of 25,000 species of higher plants (Quezel, 1985), a high proportion of which are endemic. The other regions of the world with a Mediterranean climate - the Cape Province of South Africa, SW Australia, California, and Central Chile - are also rich in endemics.

Patterns of plant distribution

Typical of most, but not all, groups of organisms, the diversity of higher plants increases as one moves from the poles to the equator. Plant species diversity, however, varies markedly on smaller scales. Between 40 and 100 tree species may occur on one hectare of tropical moist forest in Latin America, compared to 10-30 per hectare in forests in eastern North America. In a study done near Iquitos, Peru, Gentry found approximately 300 tree species per hectare with trunks greater than 10cm in diameter (Gentry, 1988).

Myers (1990) has estimated that 18 places on earth (termed 'Hot-Spots') support nearly 50,000 endemic plant species - about 20% of the world's total flora - but comprise only 0.5% of the earth's surface. These 18 places, which range widely in scale, are as follows: Atlantic coast of Brazil, California Floristic Province, Cape Floristic Province, Central Chile, Colombian Choco, Eastern Arc forests of Tanzania, Eastern Himalayas, Côte d'Ivoire, Madagascar, New Caledonia, Northern Borneo, Peninsular Malaysia, Philippines, South Western Australia, Sri Lanka, Western Amazonia uplands, Western Ecuador, and the Western Ghats. This and other approaches to distinguishing areas of high diversity are discussed further in Chapter 15.

Although the hot-spots *sensu* Myers are not defined by habitat, they can be considered in such terms. Six units - the Atlantic coast of Brazil, the Colombian Choco, Northern Borneo, Peninsular Malaysia, the Philippines and the Western Amazonia uplands - are areas of which the natural vegetation cover (now severely degraded) is almost entirely tropical rain forest, a large proportion of it lowland forest. Two more units - the Eastern Arc forests of Tanzania and the Western Ghats in India - represent areas of tropical montane forest. The vegetation of Western Ecuador is essentially a mixture of both (Gentry, 1991). Madagascar, Côte d'Ivoire and Sri Lanka each have a range of habitats but those with by far the richest floras are the tropical moist forests. The Eastern Himalayas are a region of subtropical to warm-temperate forests, and New Caledonia has a wide range of tropical habitats (Schneckenburger, 1991). The four other units - the California and Cape Floristic Provinces, Central Chile and SW Australia -are regions of predominantly Mediterranean vegetation.

Geopolitical distribution of plant diversity

Table 8.3 is a new compilation of higher plant richness and endemism assessed on a territorial basis. The associated figures are based on selected data from this table, and illustrate the approximate percentage of country floras composed of single-country endemic species (Fig. 8.1) and the relative species richness of different countries. The 25 most species-rich countries are represented in Fig. 8.2 and countries grouped by continent in Figs. 8.3-8.8 (note that graph scales differ between continents).

It should be noted that these data reflect the size and topographic complexity of the countries represented, in addition to diversity per unit area as a function of climatic and other factors. Nevertheless, the figures do confirm the great floristic richness of the regions of moist tropical forest. Territories that lie along the equatorial zone of moist trade winds can have enormous numbers of species, especially in South America: Venezuela has 15,000-25,000, Colombia has 35,000, Brazil may have as many as 55,000 flowering plant species. African countries show a similar high level of diversity, although numbers of species are not as great as in South America, perhaps because of prehistoric climatic fluctuation. Cameroon has an estimated 8,000 flowering plant species, Gabon 6,000-7,000 and Tanzania 10,000. Floras in SW Asia are intermediate in size between those of Africa and South America: there are an estimated 20,000 flowering plant species in Indonesia and 12,000 in both Malaysia and Thailand.

Amongst the richest floras are those of larger oceanic islands in tropical and warm-temperate latitudes. Cuba has a flora of 6,499 higher plant species, 3,233 of them endemic; Japan has 5,372 species, some 2,000 of them endemic; New Caledonia has 3,094 species, 2,480 of them endemic; New Zealand has 2,371 species, 1,942 of them endemic. The richest island flora is probably that of Madagascar, estimated at up to 10,000 species, with perhaps as many as 8,000 endemics. These include eight endemic families of flowering plants, most notably the spiny, rather cactus-like Didiereaceae that are a major constituent of the vegetation in the drier parts of the island.

Smaller oceanic islands, even in the tropics, have small floras due to the problems of long-distance dispersal for plants, but the low total number of species frequently includes a large endemic element. Mauritius, including Réunion, has a native flora of 878 higher plant species, of which 329 are endemic; Socotra has 788 flowering plants, 268 of which are endemic; St Helena has a native flora of just 89 species, but 74 of these are endemic. Even some of the very tiny atoll territories in Oceania usually have one or a few endemic higher plants.

Drier tropical and subtropical regions, on the other hand, have relatively poor levels of floral diversity when assessed purely on a numerical basis. Most of the arid sub-Saharan territories of the Sahel belt have smaller floras than have many countries in N. Europe: for example, Burkino Faso (1,100 higher plant species), Chad (1,600 species), Mali (1,741 species) and Niger (1,178 species). These territories have but a tiny number of endemics, perhaps no more than a dozen between them. That is not to say that the Sahel flora is not important, for it contains potentially valuable drought-resistant and economic plants. They certainly show a good deal less floristic diversity than the territories of the Mediterranean region (noted above). Several of the territories that border its shores have very high floral diversity: Greece has 4,900 flowering plants, 742 of them endemic; Spain about the same number, 941 of them endemic; and Turkey 8,472 with 2,651 endemics. These figure compare favourably with those from many tropical territories, although they also reflect more thorough levels of floristic exploration.

References

Airy Shaw, H.K. (ed.). *A Dictionary of the Flowering Plants and Ferns. Eighth Edition.* Cambridge Univ. Press. 1245 pp.

Cowling, R.M. *et al.* 1989. Patterns of plant species diversity in southern Africa. In: Huntley, B.J. (Ed.), *Biotic Diversity in Southern Africa: concepts and conservation.* Oxford, Cape Town.

Cronquist, A. 1981. *An Integrated System of Classification of Flowering Plants.* Columbia University Press, NY.

Davis, S. *et al.* 1986. *Plants in Danger: What do we know?* IUCN, Cambridge and Switzerland.

Dixit, R.D. 1984. *A Census of the Indian Pteridophytes.* Botanical Survey of India, New Delhi.

Forero, E. 1988. Botanical exploration and phytogeography of Colombia: past, present and future. *Taxon* 37:561-566.

Gentry, A.H. 1982. Neotropical floristic diversity: phytogeographical connections between Central and South America, Pleistocene climatic fluctuation, or an accident of the Andean orogeny? *Annals of the Missouri Botanical Garden* 69:557-593.

Gentry, A.H. 1988. Tree species richness of upper Amazonian forests. *Proceedings of the National Academy of Sciences* 85:156-159.

Gentry, A.H. 1991. Biological extinction in western Ecuador. *Annals of the Missouri Botanical Garden* 78:273-295.

Johns, R.J. and Bellamy, A. 1979. *The Ferns and Fern Allies of Papua New Guinea.* Papua New Guinea Forestry College.

Myers, N. 1990. The biodiversity challenge: expanded Hot-Spots analysis. *The Environmentalist* 10(4):243-255.

Prance, G.T. 1977. Floristic inventory of the tropics: where do we stand? *Annals of the Missouri Botanical Garden* 64:659-684.

Quezel, P. 1985. Definition of the Mediterranean region and the origin of its flora. In: Gómez-Campo, C. (Ed.), *Plant Conservation in the Mediterranean Area.* Junk. P.17.

Raven, P.H. 1987. The scope of the plant conservation problem worldwide. In: Bramwell, D. *et al.* (Eds), *Botanic Gardens and the World Conservation Strategy.* Academic Press. Pp.19-29.

Schneckenburger, S. 1991. *Neukaledonien. Pflanzenwelt einer Pazifikinsel.* Palmengarten Sonderheft 16. Palmengarten, Frankfurt.

Stebbins, G.L. 1974. *Flowering Plants. Evolution above the species level.* Edward Arnold. Pp.165-170.

Based on a document written by John Akeroyd and Hugh Synge.

Table 8.2 Vascular plants: a summary of systematic diversity

MAJOR GROUP (CLASS) SUBCLASS ORDER FAMILY	GENERA	SPECIES	DISTRIBUTION
Pteridophytes			
Lycopodiaceae	4	587	cosmopolitan
Selaginellaceae	1	725	mainly tropical, with some temperate species
Isoetaceae	1-2	77-80	temperate and tropical (aquatic)
Equisetaceae	1	22	cosmopolitan, except Australasia
Psilotaceae	2	3-10	tropical and subtropical
True ferns			
Ophioglossaceae	3	81	temperate with some tropical
Marattiaceae	4	204	mostly Old World tropical; some New World tropical
Osmundaceae	3	18	temperate and tropical
Plagiogyriaceae	1	36	mostly Old World tropical; some New World tropical
Schizaeaceae	5	143	pantropical
Adiantaceae	38	712	pantropical; subtropical; warm temperate
Parkeriaceae	1	4	pantropical
Vittariaceae	9	113	pantropical
Pteridaceae	7	259	pantropical
Marsileaceae	3	67	temperate and tropical
Hymenophyllaceae	5	600	pantropical
Hymenophyllopsidaceae	1	8	northern South America
Stromatopteridaceae	1	1	New Caledonia
Matoniaceae	2	4	Malesia
Gleicheniaceae	2	140	pantropical
Cheiropleuriaceae	1	1	tropical Asia and Malesia
Dipteridaceae	1	8	tropical Asia; Malesia; Australia; Fiji
Polypodiaceae	40	1,068	pantropical; subtropical; some temperate
Metaxyaceae	1	1	pantropical
Loxsomataceae	2	4	New World tropical; New Zealand
Thyrsopteridaceae	2	6	pantropical
Dicksoniaceae	3	41	pantropical
Lophosoriaceae	1	1	New World tropical
Cyatheaceae	4	623	pantropical
Thelypteridaceae	30	1,000	pantropical; some in subtropical and temperate
Dennstaedtiaceae	18	486	pantropical
Aspleniaceae	14	711	pantropical; subtropical; some temperate
Woodsiaceae	18	705	pantropical; some temperate
Tectariaceae	19	431	pantropical
Dryopteridaceae	20	464	temperate; tropical
Lomariopsidaceae	8	615	pantropical
Davalliaceae	6	218	pantropical
Blechnaceae	8	238	pantropical
Salviniaceae	1	10	pantropical; subtropical; a few temperate
Azollaceae	1	6	pantropical; subtropical; some temperate
Gymnosperms - Cycads			
Zamiaceae	8	80	tropical and subtropical
Cycadaceae	1	20	Madagascar; eastern and Southeast Asia; Indomalaysia; Australia; Polynesia
Stangeriaceae	1	1	South Africa
Boweniaceae			
Gymnosperms - Conifers			
Pinaceae	10	250	Northern Hemisphere, south to Sumatra, Java, Central America and West Indies
Taxaceae	5	20	Northern Hemisphere, south to Celebes and Mexico; one species in New Caledonia
Taxodiaceae	10	16	eastern Asia; Tasmania; North America
Cupressaceae	19	130	cosmopolitan

Table 8.2 Vascular plants: a summary of systematic diversity

MAJOR GROUP (CLASS) / SUBCLASS / ORDER / FAMILY	GENERA	SPECIES	DISTRIBUTION
Araucariaceae	2	38	Southern Hemisphere (excluding Africa) to Indochina and the Philippines
Cephalotaxaceae	1	7	eastern Himalayas to Japan
Phyllocladaceae	1	7	Malaysia; Tasmania; New Zealand
Podocarpaceae	6	125	mostly Southern Hemisphere, extending north to Japan, Central America, and West Indies
Gymnosperms - Ginkgo			
Ginkgoaceae	1	1	eastern China
Gymnosperms - Gnetophytes			
Ephedraceae	1	40	warm temperate North and South America; warm temperate Eurasia
Gnetaceae	1	30	tropical (Indomalaya; Fiji; northern tropical South America; western tropical Africa)
Welwitschiaceae	1	1	southwestern Africa
Angiosperms - Dicots			
Magnoliidae			
Magnoliales			
Winteraceae	9	100	primarily islands of southwestern Pacific
Degeneriaceae	1	1	Fiji
Himantandraceae	1	1-3	New Guinea; Molucca Is.; northeastern Australia
Eupomatiaceae	1	2	New Guinea and eastern Australia
Austrobaileyaceae	1	1	northeastern Australia
Magnoliaceae	12	220	widespread, especially Northern Hemisphere
Lactoridaceae	1	1	San Juan Islands (Chile)
Annonaceae	130	2,300	mainly tropical
Myristicaceae	15	300	tropical
Canellaceae	6	20	tropical Africa; Madagascar; South America
Laurales			
Amborellaceae	1	1	New Caledonia
Trimeniaceae	2	5	New Guinea; New Caledonia; Fiji; southeastern Australia
Monimiaceae	30-35	450	tropical and subtropical, especially Southern Hemisphere
Gomortegaceae	1	1	central Chile
Calycanthaceae	3	5	China; North America
Idiospermaceae	1	1	northern Australia
Lauraceae	30-50	2,000	tropical and subtropical
Hernandiaceae	4	60	tropical
Piperales			
Chloranthaceae	5	75	tropical and subtropical
Saururaceae	5	7	eastern Asia; eastern and western North America
Piperaceae	10	1,400-2,000	tropical
Aristolochiales			
Aristolochiaceae	8-10	600	mainly tropical
Illiciales			
Illiciaceae	1	40	Southeast Asia; southeastern United States; Caribbean; Mexico
Schisandraceae	2	50	tropical and temperate eastern Asia; southeastern United States
Nymphaeales			
Nelumbonaceae	1	2	warm Asia and Australia; eastern United States
Nymphaeaceae	5	50	cosmopolitan distribution
Barclayaceae	1	4	tropical Southeast Asia to New Guinea
Cabombaceae	2	8	tropical and warm temperate
Ceratophyllaceae	1	6	cosmopolitan
Ranunculales			

Table 8.2 Vascular plants: a summary of systematic diversity

MAJOR GROUP (CLASS) / SUBCLASS / ORDER / FAMILY	GENERA	SPECIES	DISTRIBUTION
Ranunculaceae	50	2,000	widespread, especially North temperate and boreal
Circaeasteraceae	2	2	Southeast Asia
Berberidaceae	13	650	widespread, especially temperate Northern Hemisphere
Sargentodoxaceae	1	1	China, Laos, Vietnam
Lardizabalaceae	8	30	Himalayas to Southeast Asia; Chile
Menispermaceae	70	400	tropical and subtropical
Coriariaceae	1	5	disjunct in tropical America, Europe, Asia
Sabiaceae	3	60	Southeast Asia; tropical America
Papaverales			
Papaveraceae	25	200	temperate & tropical Northern Hemisphere
Fumariaceae	19	400	mainly North temperate; also South Africa
Hamamelidae			
Trochodendrales			
Tetracentraceae	1	1	Nepal; central and southeastern China; Burma
Trochodendraceae	1	1	Korea, Japan to Taiwan
Hamamelidales			
Cercidiphyllaceae	1	2	China; Japan
Eupteleaceae	1	2	Japan, China, Assam
Platanaceae	1	6-7	eastern Mediterranean to Himalayas; Mexico to Canada
Hamamelidaceae	26	100	widespread, especially eastern Asia
Myrothamnaceae	1	2	Africa, Madagascar
Daphniphyllales			
Daphniphyllaceae	1	35	Asia and Malay Archipelago
Didymelales			
Didymelaceae	1	2	Madagascar
Eucommiales			
Eucommiaceae	1	1	montane forests of western China
Urticales			
Ulmaceae	18	150	widespread, especially Northern Hemisphere
Barbeyaceae	1	1	northeastern Africa and adjacent Arabia
Cannabaceae	2	3	North temperate
Moraceae	40	1,000	tropical and subtropical
Cecropiaceae	6	276	tropical
Urticaceae	45	700	tropical and subtropical
Leitneriales			
Leitneriaceae	1	1	southeastern United States
Juglandales			
Juglandaceae	7-8	60	widespread in Northern Hemisphere and into South America
Rhoipteleaceae	1	1	southwestern China and North Vietnam
Myricales			
Myricaceae	3	50	mostly temperate and subtropical
Fagales			
Balanopaceae	1	9	Southwest Pacific, especially New Caledonia
Fagaceae	6-8	800	cosmopolitan, except tropical and South Africa
Betulaceae	6	120	mainly temperate and cool Northern Hemisphere
Casuarinales			
Casuarinaceae	1	50	Australia, Pacific islands, Asia
Caryophyllidae			
Caryophyllales			
Phytolaccaceae	18	125	tropical and subtropical
Achatocarpaceae	2	8	warm North America; Central America; South America

Table 8.2 Vascular plants: a summary of systematic diversity

MAJOR GROUP (CLASS) SUBCLASS ORDER FAMILY	GENERA	SPECIES	DISTRIBUTION
Nyctaginaceae	30	300	tropical and subtropical, especially New World
Aizoaceae	12	2,500	South Africa; Australia
Didiereaceae	4	11	Madagascar
Cactaceae	30-200	1,000-2,000	American deserts
Chenopodiaceae	100	1,500	cosmopolitan, especially deserts and semideserts
Amaranthaceae	65	900	tropical and subtropical
Portulacaceae	20	500	cosmopolitan, especially western North America and Andes
Basellaceae	4	15-20	tropical and subtropical, mostly New World
Molluginaceae	13	100	tropical and subtropical, especially Africa
Caryophyllaceae	75	2,000	widespread, especially North America
Polygonales			
Polygonaceae	30	1,000	mainly temperate Northern Hemisphere
Plumbaginales			
Plumbaginaceae	12	400	widespread, especially Mediterranean
Dilleniidae			
Dilleniales			
Dilleniaceae	10	350	tropical and subtropical, especially Australia
Paeoniaceae	1	30	Eurasia, especially temperate eastern Asia
Theales			
Ochnaceae	30	400	tropical, especially Brazil
Sphaerosepalaceae	2	14	Madagascar
Sarcolaenaceae	10	30	Madagascar
Dipterocarpaceae	16	600	tropical, especially rain forests of Malaysia
Caryocaraceae	2	23	tropical America, especially Amazon basin
Theaceae	40	600	tropical and subtropical
Actinidiaceae	3	300	tropical and subtropical
Scytopetalaceae	5	20	tropical western Africa
Pentaphylacaceae	1	1	southern China to Malay peninsula and Sumatra
Tetrameristaceae	2	2	Malaysia; southern Venezuela (Guayana Highlands)
Pellicieraceae	1	1	Costa Rica, Panama, Columbia
Oncothecaceae	1	1	New Caledonia
Marcgraviaceae	5	100	tropical America
Quiinaceae	4	40	tropical America, especially Amazon basin
Elatinaceae	2	40	tropical and subtropical
Paracryphiaceae	1	1	New Caledonia
Medusagynaceae	1	1	Seychelles
Guttiferae (= Clusiaceae)	50	1,200	moist tropical and North temperate
Malvales			
Elaeocarpaceae	10	400	tropical and subtropical
Tiliaceae	50	450	tropical and subtropical
Sterculiaceae	65	1,000	tropical and subtropical
Bombacaceae	20-30	200	tropical, especially Central and South America
Malvaceae	75	1,000-1,500	cosmopolitan, especially tropical
Lecythidales			
Lecythidaceae	20	400	tropical, especially rain forests of South America
Nepenthales			
Sarraceniaceae	3	15	easter and northwestern United States; northern South America
Nepenthaceae	1	75	East Indies to Madagascar; to northern Australia and Southeast Asia
Droseraceae	4	100	temperate and tropical
Violales			

Table 8.2 Vascular plants: a summary of systematic diversity

MAJOR GROUP (CLASS) SUBCLASS ORDER FAMILY	GENERA	SPECIES	DISTRIBUTION
Flacourtiaceae	85	800	tropical
Bixaceae	3	15	tropical
Peridiscaceae	2	2	tropical South America
Cistaceae	8	200	mostly in temperate and warm temperate
Huaceae	2	3	tropical Africa
Lacistemataceae	2	20	tropical America
Scyphostegiaceae	1	1	Borneo
Stachyuraceae	1	5-6	Himalayan region to Japan
Violaceae	16	800	cosmopolitan
Tamaricaceae	4-5	100	Eurasia and Africa, especially Mediterranean region
Frankeniaceae	3	80	cosmopolitan, especially Mediterranean region
Dioncophyllaceae	3	3	rain forests of tropical Africa
Ancistrocladaceae	1	15-20	Southeast Asia; India; tropical Africa
Turneraceae	8	120	tropical and subtropical America and Africa; Madagascar
Malesherbiaceae	1-2	25	Andes from Chile to Peru
Passifloraceae	16	650	tropical and warm temperate, especially tropical America and Africa
Caricaceae	4	30	tropical and subtropical America; Africa
Achariaceae	3	3	South Africa
Fouquieriaceae	1	11	arid parts of Mexico and southwestern United States
Hoplestigmataceae	1	2	western tropical Africa
Cucurbitaceae	90	700	tropical and subtropical; rarely temperate or cool temperate
Datiscaceae	3	4	Malesia; Asia; western North America
Begoniaceae	3-5	1,020	tropical, especially northern South America
Loasaceae	14	200	temperate and tropical North and South America
Salicales			
Salicaceae	2	340	mostly North temperate; also Australia and Malay Archipelago
Capparales			
Tovariaceae	1	2	tropical America
Capparaceae	45	800	tropical and subtropical
Cruciferae (= Brassicaceae)	350	3,000	cool temperate or warm temperate Northern and Southern Hemisphere
Moringaceae	1	10	xeric Africa; Madagascar; India
Resedaceae	6	70	Northern Hemisphere, mostly Old World, especially Mediterranean
Batales			
Gyrostemonaceae	5	17	Australia
Bataceae	1	2	tropical and subtropical America; Galapagos; Hawaii; New Guinea and northeastern Australia
Ericales			
Cyrillaceae	3	14	northern South America; Central America; West Indies; southeastern United States
Clethraceae	1	65	tropical America; southeastern United States; Southeast Asia; East Indies
Grubbiaceae	1	3	South Africa (Cape Province)
Empetraceae	3	5	cold Northern Hemisphere; southern South America; eastern United States; Europe
Epacridaceae	30	400	mostly Australia, New Zealand, and East Indies
Ericaceae	125	3,500	temperate, cool and subtropical regions; montane tropical
Pyrolaceae	4	45	Northern Hemisphere, especially temperate and boreal

Table 8.2 Vascular plants: a summary of systematic diversity

MAJOR GROUP (CLASS) SUBCLASS ORDER FAMILY	GENERA	SPECIES	DISTRIBUTION
Monotropaceae		10	12
Diapensiales			
Diapensiaceae	6	18	arctic & North temperate; south to Himalayas
Ebenales			
Sapotaceae	70	800	tropical
Ebenaceae	5	450	tropical and subtropical
Styracaceae	10	150	widely disjunct in both hemispheres
Lissocarpaceae	1	2	tropical South America
Symplocaceae	1	300-400	tropical and subtropical America; southern and eastern Asia; Australia; East Indies
Primulales			
Theophrastaceae	4	100	mostly New World tropical
Myrsinaceae	30	1,000	tropical and subtropical New and Old World; also temperate Old World
Primulaceae	30	1,000	mostly temperate and cold Northern Hemisphere; montane tropical
Rosidae			
Rosales			
Brunelliaceae	1	50	tropical America
Connaraceae	16-24	300-400	tropical, especially Old World
Eucryphiaceae	1	6	eastern Australia; Tasmania; Chile
Cunoniaceae	25	350	Southern Hemisphere, especially Australia, New Guinea and New Caledonia; also Mexico and West Indies
Davidsoniaceae	1	1	northeastern Australia
Dialypetalanthaceae	1	1	Brazil
Pittosporaceae	9	200	tropical and warm temperate Old World, especially Australia
Byblidaceae	2	4	Australia and South Africa
Hydrangeaceae	17	170	temperate and subtropical Northern Hemisphere; southeastern Asia and Malesia
Columelliaceae	1	4	Andes, from Colombia to Bolivia
Grossulariaceae	25	350	cosmopolitan
Greyiaceae	1	3	South Africa
Bruniaceae	12	75	South Africa and Natal
Anisophylleaceae	4	40	tropical or subtropical forests, mostly Africa and Indomalaysia; South America
Alseuosmiaceae	3	12	New Zealand and New Caledonia
Crassulaceae	25	900	cosmopolitan, except Australia and Polynesia
Cephalotaceae	1	1	southwestern Australia
Saxifragaceae	40	700	cosmopolitan, especially temperate and cold Northern Hemisphere
Rosaceae	100	3,000	cosmopolitan, especially temperate and subtropical Northern Hemisphere
Neuradaceae	3	10	deserts in Africa, across Middle East to India
Crossosomataceae	3	10	arid western United States and adjacent Mexico
Chrysobalanaceae	17	450	pantropical, especially New World
Surianaceae	4	6	Australia and tropical maritime
Rhabdodendraceae	1	3	tropical South America
Fabales			
Leguminosae (= Fabaceae)	590	14,200	cosmopolitan, especially tropical and subtropical
Proteales			
Elaeagnaceae	3	50	temperate and subtropical Northern Hemisphere, to tropical Asia and northern Australia

Table 8.2 Vascular plants: a summary of systematic diversity

MAJOR GROUP (CLASS) / SUBCLASS / ORDER / FAMILY	GENERA	SPECIES	DISTRIBUTION
Proteaceae	75	1,000	tropical and subtropical, especially warmer Southern Hemisphere
Podostemales			
Podostemaceae	40	200	mostly tropical, especially Asia and America
Haloragales			
Haloragaceae	8	100	cosmopolitan, especially Southern Hemisphere
Gunneraceae	1	50	Southern Hemisphere to southern Mexico
Myrtales			
Sonneratiaceae	2	10	Old World tropical
Lythraceae	24	500	mainly tropical; also temperate
Penaeaceae	7	20	Cape Province (South Africa)
Crypteroniaceae	1	4	India, Philippines, Malay Archipelago
Thymelaeaceae	50	500	cosmopolitan
Trapaceae	1	15	tropical and subtropical Africa and Eurasia
Myrtaceae	140	3,000	tropical and subtropical; temperate Australia
Punicaceae	1	2	Balkans to northern India; Socotra
Onagraceae	17	675	temperate and subtropical, especially New World
Oliniaceae	1	8	tropical and southern Africa; St Helena
Melastomataceae	200	4,000	tropical and subtropical, especially South America
Combretaceae	20	400	tropical and subtropical, especially Africa
Rhizophorales			
Rhizophoraceae	14	100	tropical and subtropical
Cornales			
Alangiaceae	1	20	eastern and tropical Asia; eastern Australia; Pacific islands; Madagascar; western Africa
Nyssaceae	3	7-8	eastern North America; eastern Asia; Pacific islands; China
Cornaceae	11	100	North temperate; irregularly tropical and South temperate
Garryaceae	1	13	western North and Central America, from Washington to Panama
Santalales			
Medusandraceae	1	1	rainforests of tropical western Africa
Olacaceae	24-30	250	tropical and subtropical
Dipentodontaceae	1	1	southern China and Burma
Opiliaceae	9	50	tropical and subtropical
Santalaceae	35	400	nearly cosmopolitan, especially arid climates
Misodendraceae	1	10	temperate South America
Loranthaceae	60-70	700	mostly tropical and subtropical
Viscaceae	7-8	350	cosmopolitan, especially tropical
Eremolepidaceae	3	12	tropical America
Balanophoraceae	19	45	tropical and subtropical
Rafflesiales			
Hydnoraceae	2	10	drier parts of Africa, Madagascar
Mitrastemonaceae	1	2	Borneo and Sumatra to Indochina and Japan; Mexico and Central America
Rafflesiaceae	7	50	tropical and subtropical
Celastrales			
Geissolomataceae	1	1	South Africa (Cape Province)
Celastraceae	50	800	pantropical, some in temperate regions
Hippocrateaceae	2-13	300	tropical
Salvadoraceae	3	12	Africa; Madagascar; India; Sri Lanka; Southeast Asia
Stackhousiaceae	3	20-25	Australia and New Zealand; southwestern Pacific
Aquifoliaceae	4	320-420	more or less cosmopolitan
Icacinaceae	50	400	pantropical

Table 8.2 Vascular plants: a summary of systematic diversity

MAJOR GROUP (CLASS) SUBCLASS ORDER FAMILY	GENERA	SPECIES	DISTRIBUTION
Aextoxicaceae	1	1	Chile
Cardiopteridaceae	1	3	Asia to New Guinea and Australia
Corynocarpaceae	1	5	New Zealand; northeastern Australia; New Guinea
Dichapetalaceae	3	235	pantropical, mainly Africa
Euphorbiales			
Buxaceae	5	60	nearly cosmopolitan
Simmondsiaceae	1	1	western United States and Mexico
Pandaceae	3	26	Africa, Asia, New Guinea
Euphorbiaceae	300	7,500	cosmopolitan, especially tropical and subtropical
Rhamnales			
Rhamnaceae	55	900	cosmopolitan, especially tropical and subtropical
Leeaceae	1	70	pantropical
Vitaceae	11	700	tropical and subtropical; a few in temperate regions
Linales			
Erythroxylaceae	4	200	pantropical, especially New World
Humiriaceae	8	50	mainly tropical South America, with one species in Africa
Ixonanthaceae	5	30	pantropical
Hugoniaceae	7	60	tropical
Linaceae	6	220	widespread, especially temperate and subtropical
Polygalales			
Malpighiaceae	60	1,200	tropical and subtropical, especially South America
Vochysiaceae	7	200	mostly tropical America, 1 in Africa
Trigoniaceae	3	26	subtropical in moist lowland forests
Tremandraceae	3	28	Australia and Tasmania
Polygalaceae	12	750	nearly cosmopolitan
Xanthophyllaceae	1	40	Indomalaysian region
Krameriaceae	1	15	Argentina and Chile, mainly in dry regions
Sapindales			
Staphyleaceae	5	50	Americas, Eurasia, Malay Archipelago
Melianthaceae	2	8-36	Africa
Bretschneideraceae	1	1	mountains of western and southwestern China
Akaniaceae	1	1	eastern Australia
Sapindaceae	140	1,500	tropical and subtropical; some in temperate regions
Hippocastanaceae	2	16	North America to northern South America; Europe; Southeast Asia
Aceraceae	2	112	temperate and subtropical, especially Malesia; China
Burseraceae	16-20	600	pantropical, especially tropical America and Northeast Africa
Anacardiaceae	60-80	600	mainly pantropical, some in temperate regions
Julianiaceae	2	5	tropical America (Central America, Peru)
Simaroubaceae	25	150	pantropical, some in warm temperate regions
Cneoraceae	1	3	Mediterranean, Canary Is., Cuba
Meliaceae	51	550	tropical and subtropical; some in temperate regions
Rutaceae	150	1,500	nearly cosmopolitan, especially South Africa and Australia
Zygophyllaceae	30	250	mostly arid tropical and subtropical, sometimes in saline habitats
Geraniales			

Table 8.2 Vascular plants: a summary of systematic diversity

MAJOR GROUP (CLASS) SUBCLASS ORDER FAMILY	GENERA	SPECIES	DISTRIBUTION
Oxalidaceae	7-8	900	tropical and subtropical; some in temperate regions
Geraniaceae	11	700	temperate and warm temperate regions; some tropical
Limnanthaceae	2	11	temperate North America
Tropaeolaceae	3	92	Mexico to Chile (in mountains), Patagonia
Balsaminaceae	2	450	tropical Asia and Africa, some in temperate regions; India to Java
Apiales			
Araliaceae	70	700	tropical and subtropical; some in temperate regions
Umbelliferae (= Apiaceae)	300	3,000	nearly cosmopolitan, especially North temperate regions and tropical mountains
Asteridae			
Gentianales			
Loganiaceae	20	500	tropical and subtropical; relatively few species in temperate regions
Retziaceae	1	1	Cape Province of South Africa
Gentianaceae	75	1,000	cosmopolitan, especially temperate and subtropical regions and tropical mountains
Saccifoliaceae	1	1	southern Venezuela
Apocynaceae	200	2,000	tropical and subtropical; relatively few species in temperate regions
Asclepiadaceae	250	2,000	tropicals and subtropical, especially Africa, with relatively few species in temperate regions
Solanales			
Nolanaceae	2	66	northern Chile and southern Peru, often along the seashore
Duckeodendraceae	1	1	Amazon basin of Brazil
Solanaceae	85	2,800	nearly cosmopolitan, especially tropical South America
Convolvulaceae	50	1,500	nearly cosmopolitan, especially tropical and subtropical
Cuscutaceae	1	150	nearly cosmopolitan, especially warmer parts of New World
Menyanthaceae	5	30-35	cosmopolitan
Polemoniaceae	18	300	North temperate (Eurasia, Alaska to western South America), especially temperate North America
Hydrophyllaceae	20	250	wide-ranging, especially dry western United States
Lamiales			
Lennoaceae	3	4-5	New World from southwestern United States to Colombia and Venezuela
Boraginaceae	100	2,000	cosmopolitan, especially western North America and Mediterranean region; east into Asia
Verbenaceae	100	2,600	pantropical, with only a few species in temperate regions
Labiatae (= Lamiaceae)	200	3,200	cosmopolitan, especially Mediterranean region and into central Asia
Callitrichales			
Hippuridaceae	1	1	temperate and boreal Northern Hemisphere; Australia; southern South America
Callitrichaceae	1	35	nearly cosmopolitan
Hydrostachyaceae	1	20	Madagascar; tropical and southern Africa
Plantaginales			
Plantaginaceae	3	254	cosmopolitan
Scrophulariales			
Buddlejaceae	10	150	mainly tropical and subtropical

Table 8.2 Vascular plants: a summary of systematic diversity

MAJOR GROUP (CLASS) / SUBCLASS / ORDER / FAMILY	GENERA	SPECIES	DISTRIBUTION
Oleaceae	30	600	nearly cosmopolitan, especially Asia and Malesia
Scrophulariaceae	190	4,000	cosmopolitan, especially temperate regions and tropical mountains
Globulariaceae	10	300	Africa; Madagascar; Europe; western Asia
Myoporaceae	3-4	125	Australia; Asia; Pacific islands; West Indies; northern South America
Orobanchaceae		17	150
Gesneriaceae	120	2,500	pantropical, with a few species in temperate regions
Acanthaceae	250	2,500	tropical, with only a few species in temperate regions
Pedaliaceae	20	80	mostly tropical, especially along seacoast or in arid regions, with only a few species in temperate climates
Bignoniaceae	100	800	mainly tropical, especially tropical America
Mendonciaceae	2-4	60	South America; tropical Africa; Madagascar
Lentibulariaceae	5	200	cosmopolitan
Campanulales			
Pentaphragmataceae	1	30	Southeast Asia and nearby Pacific islands
Sphenocleaceae	1	2	pantropical; western Africa
Campanulaceae	70	2,000	cosmopolitan
Stylidiaceae	5	155	Australasia; south and Southeast Asia; southernmost South America
Donatiaceae	1	2	southern South America; New Zealand; Tasmania
Brunoniaceae	1	1	Australia
Goodeniaceae	14	300	primarily Australia; also New Zealand, Japan, and tropical and subtropical Old and New World
Rubiales			
Rubiaceae	450	6,500	cosmopolitan, especially tropical and subtropical
Theligonaceae	1	3	temperate eastern Asia to Mediterranean region and Canary Islands
Dipsacales			
Caprifoliaceae	15	400	mostly North temperate and boreal regions; also tropical mountains
Adoxaceae	1	1	circumboreal
Valerianaceae	13	300	nearly cosmopolitan, especially North temperate regions and Andes
Dipsacaceae	10	270	Eurasia and Africa, especially Mediterranean region
Calycerales			
Calyceraceae	6	60	Central and South America
Asterales			
Compositae (= Asteraceae)	1,100	20,000	cosmopolitan, especially temperate and subtropical regions
Angiosperms - Monocots			
Alismatidae			
Alismatales			
Butomaceae	1	1	temperate Eurasia
Limnocharitaceae	3	7-12	tropical and subtropical
Alismataceae	12	75	cosmopolitan, especially Northern Hemisphere
Hydrocharitales			
Hydrocharitaceae	15	15 100	cosmopolitan
Najadales			
Aponogetonaceae	1	40	Old World tropical to South Africa
Scheuchzeriaceae	1	1	cool Northern Hemisphere

Table 8.2 Vascular plants: a summary of systematic diversity

MAJOR GROUP (CLASS) / SUBCLASS / ORDER / FAMILY	GENERA	SPECIES	DISTRIBUTION
Juncaginaceae	5	20	temperate and cold Northern and Southern Hemisphere
Potamogetonaceae	1	100	cosmopolitan
Ruppiaceae	1	1-2	temperate and subtropical
Najadaceae	1	35	cosmopolitan
Zannichelliaceae	4	7-8	cosmopolitan
Posidoniaceae	1	3	Mediterranean, Australia
Cymodoceaceae	5	18	tropical and subtropical seacoasts
Zosteraceae	3	18	subarctic, temperate, subtropical seacoasts
Triuridales			
Triuridaceae	7	70	tropical and subtropical
Petrosaviaceae	1	2	southern China and southern Japan to Malay Peninsula and Borneo
Arecidae			
Arecales			
Palmae (= Arecaceae)	200	3,000	tropical and warm temperate
Cyclanthales			
Cyclanthaceae	11	180	tropical America
Pandanales			
Pandanaceae	3	682-782	Old World, especially tropical (Malesia)
Arales			
Araceae	110	1,800	mostly tropical and subtropical
Lemnaceae	6	31	cosmopolitan
Commelinidae			
Commelinales			
Rapateaceae	16	100	tropical South America, with one species in tropical western Africa
Xyridaceae	4	200	tropical and subtropical; a few species in temperate region
Mayacaceae	1	4	tropical western Africa; tropical and warm temperate America
Commelinaceae	50	700	tropical and subtropical
Eriocaulales			
Eriocaulaceae	13	1,200	tropical and subtropical, with a few species in temperate regions
Restionales			
Flagellariaceae	1	3	Old World tropical
Restionaceae	30	400	widely distributed in Southern Hemisphere, especially Australia and South Africa
Joinvilleaceae	1	2	Pacific Islands
Centrolepidaceae	4	35	Australia; Southeast Asia; Pacific Islands; southernmost South America; mostly in nutrient-poor soils
Juncales			
Juncaceae	8	300	temperate or cold regions, or montane tropical
Thurniaceae	1	3	Amazon basin and Guayana
Cyperales			
Cyperaceae	70	4,000	cosmopolitan, most abundant in temperate regions
Gramineae (= Poaceae)	500	8,000	cosmopolitan, especially tropical and North temperate semi-arid regions with seasonal rainfall
Hydatellales			
Hydatellaceae	2	7	Australia, New Zealand, Tasmania
Typhales			
Sparganiaceae	1	13	chiefly North temperate regions, to Australia and New Zealand
Typhaceae	1	10	cosmopolitan

Table 8.2 Vascular plants: a summary of systematic diversity

MAJOR GROUP (CLASS) SUBCLASS ORDER FAMILY	GENERA	SPECIES	DISTRIBUTION
Zingiberidae			
Bromeliales			
Bromeliaceae	45	2,000	New World, except one species in western tropical Africa
Zingiberales			
Strelitziaceae	3	7	tropical
Heliconiaceae	1	100	tropical and subtropical South and Central America; one species widespread in southwestern Pacific islands
Musaceae	2	42	tropical and subtropical Old World
Lowiaceae	1	6	southern China; Malay Peninsula; Pacific islands
Zingiberaceae	47	1,000	tropical regions, especially southern and Southeast Asia
Costaceae	4	150	pantropical, especially New World
Cannaceae	1	50	tropical and subtropical New World
Marantaceae	30	400	pantropical, especially New World
Liliidae			
Liliales			
Philydraceae	4	5	Australia; western Pacific islands to Japan and mainland Southeast Asia
Pontederiaceae	9	30	tropical and subtropical; into North temperate regions
Haemodoraceae	16	100	mostly Southern Hemisphere, but reaching northern United States
Cyanastraceae	1	7	forests of tropical Africa
Liliaceae	280	4,000	widespread, especially dry, temperate to subtropical regions
Iridaceae	80	1,500	cosmopolitan, especially South Africa
Velloziaceae	6	250	South America; Africa; Madagascar; southern Arabia
Aloeaceae	5	700	Africa, Madagascar, Arabia, nearby islands; especially South Africa
Agavaceae	18	600	warm, mostly arid regions of New and Old Worlds; a few in distinctly temperate climates
Xanthorrhoeaceae	9	55	Australia; Tasmania; New Guinea; New Caledonia
Hanguanaceae	1	1-2	Malesia; Sri Lanka
Taccaceae	1	10	pantropical, especially Southwest Asia and Polynesia
Stemonaceae	3	30	eastern Asia; Malesia; northern Australia; southeastern United States
Smilacaceae	12	330	tropical and subtropical, especially Southern Hemisphere; also in parts of North temperate region
Dioscoreaceae	6	630	tropical and subtropical, with a few species in North temperate region
Orchidales			
Burmanniaceae	20	130	pantropical, with a few species in temperate regions
Geosiridaceae	1	1	Madagascar and other Indian Ocean islands
Corsiaceae	2	9	New Guinea, Chile
Orchidaceae	800-1,000	25,000-35,000	cosmopolitan

Sources: Flowering plant information modified from Cronquist, A. 1981. *An Integrated System of Classification of Flowering Plants*. Columbia Univ. Press. 1262 pp.; other information from Airy Shaw, H.K. (ed.). *A Dictionary of the Flowering Plants and Ferns. Eighth Edition.* Cambridge Univ. Press. 1245 pp.; and other sources.

Table 8.3 Species richness and endemism: higher plants

	FLOWERING PLANTS	GYMNO-SPERMS	FERNS	NUMBER OF ENDEMICS	% ENDEMISM	ESTIMATE/ COUNT	COMPLETION	DATE
ASIA								
Afghanistan	3,500	–	–	–	[30–35%]	e2	2	1989–91
Bahrain	195	1	1	0	0.0	c	1	1991
Bangladesh	5,000	–	–	–	–	e2	2	1972
Bhutan	5,446	22	–	50–100	1.4	e1	3	1991
British Indian Ocean Territory	100	1	–	0	0.0	e1	1	1971
Brunei	3,000	28	–	7	0.2	e2	5	1990
Cambodia	–	–	–	–	–	–	–	–
China	30,000	200	2,000	18,000	55.9	e2	3	1991
Cyprus	1,650	12	20	88	5.2	c	1	1977–85
Hong Kong	1,800	4	180	25	1.3	e2	2	1978–91
India	15,000	–	1,000	5,000	31.3	e2	2	1983–84
Indonesia	20,000	–	2,500	15,000	66.7	e3	4	1991
Iran, Islamic Rep	6,500	33	–	–	[30–35%]	e2	1	1989–91
Iraq	2,914	7	16	190	6.5	c	1	1966–86
Israel	2,294	8	15	155	6.7	c	1	1982–84
Japan	4,700	42	630	2,000	37.2	c	1	1987
Jordan	2,200	6	6	–	–	c	2	1982–85
Korea, Dem People's Rep	{2,898	–	–	107	{14.0	{c	–	{1976–83
Korea, Rep	{2,898	–	–	224	{14.0	{c	–	{1976–83
Kuwait	234	1	1	0	0.0	c	1	1991
Laos	–	–	–	–	–	–	–	–
Lebanon	2,000	12	40	–	[10%]	e3	2	1984–91
Malaysia	12,000	–	500	–	–	e3	3	1991
Maldives	260	2	15	5	1.8	c	1	1983
Mongolia	2,272	–	–	229	10.1	c	1	1984
Myanmar	7,000	–	–	1,071	15.3	e2	4	1961
Nepal	6,500	23	450	315	4.5	c	2	1978–82
Oman	1,018	3	14	74	7.1	c	1	1991
Pakistan	4,917	21	–	372	7.5	e1	2	1986
Philippines	8,000	31	900	3,500	39.3	e2	3	1982–91
Qatar	220	1	0	0	0.0	c	1	1991
Saudi Arabia	1,729	8	22	34	1.9	c	2	1991
Singapore	2,000	2	166	1	0.1	e1	1	1989–91
Sri Lanka	2,900	–	314	900	28.0	c	2	1982–83
Syria	2,000	12	40	–	[10%]	e3	2	1984–91
Taiwan	2,983	20	565	–	[25%]	c	1	1982–91
Thailand	12,000	25	600	–	–	e2	3	1979–85
Turkey	8,472	22	85	2,651	30.9	c	1	1988
United Arab Emirates	340	2	5	0	–	c	1	1991
Viet Nam	–	–	–	–	–	–	–	–
Yemen, People's Dem Rep[1]	1,373	3	41	58	4.1	c	2	1991
Yemen, Arab Rep[1]	959	1	14	77	7.9	c	1	1991
USSR[2]								
	22,000	74	207	–	–	e1	2	1991
EUROPE								
Albania	2,965	21	45	24	0.8	c	2	1980–88
Andorra	980	6	26	0	0.0	e1	1	1981
Austria	2,850–3,050	12	66	35	1.2	e1	1	1978–91
Belgium	1,250–1,550	2	50	1	0.1	e1	1	1978–83
Bulgaria	3,505	15	52	320	9.0	c	1	1991
Czechoslovakia	2,507	11	72	62	2.4	c	1	1991
Denmark	1,000–1,400	2	50	1	0.1	e1	1	1984–91
Faeroe Islands	236	1	25	1	0.4	c	1	1991
Finland	1,040	4	58	0	0.0	c	1	1988
France	4,500	20	110	133	2.9	c	1	1991
Germany	2,600	10	72	6	0.2	e1	1	1984–91
Greece	4,900	21	71	742	14.9	e1	2	1989
Hungary	2,148	8	58	38	1.7	c	1	1991
Iceland	340	1	36	1	0.3	e1	1	1984–91
Ireland	892	2	56	0	0.0	c	1	1991
Italy	5,463	29	106	712	12.7	c	1	1982
Liechtenstein	1,400	10	–	0	0.0	e2	2	1977
Luxembourg	1,200	4	42	0	0.0	e1	1	1984–91
Malta	900	3	11	5	0.5	e1	1	1984
Monaco	–	4	18	0	0.0	–	–	1973
Netherlands	1,170	3	48	0	0.0	c	1	1991
Norway	1,550–1,750	4	61	1	0.1	e1	1	1978–91
Poland	2,200–2,400	10	62	3	0.1	e1	1	1978–91
Portugal[3]	2,400–2,600	8	65	150	5.8	e1	1	1978–91
Romania	3,000–3,350	11	62	41	1.3	e2	2	1977–78
San Marino	–	–	–	0	0.0	–	–	1991
Spain[4]	4,916	18	114	941	18.6	c	2	1984–91
Sweden	1,550–1,750	4	60	1	0.1	e1	1	1978–91
Switzerland	2,927	16	87	1	0.1	c	1	1989
United Kingdom	1,550	3	70	16	1.0	e1	1	1991
Vatican City	–	–	–	0	0.0	–	–	1991
Yugoslavia	5,250	23	78	137	2.6	e2	2	1978–91

Table 8.3 Species richness and endemism: higher plants (continued)

	FLOWERING PLANTS	GYMNO-SPERMS	FERNS	NUMBER OF ENDEMICS	% ENDEMISM	ESTIMATE/ COUNT	COMPLETION	DATE
NORTH AND CENTRAL AMERICA								
Anguilla	321	–	0	1	0.3	c	–	1991
Antigua and Barbuda	766	1	33	–	[0.7%]	c	1	1938–91
Aruba	460	–	–	25	5.4	c	–	1991
Bahamas	1,172	3	43	115	9.4	c	1	1982–91
Barbados	542	–	30	5	0.8	c	2	1984–91
Belize	2,500–3,000	10	134	150	5.2	e2	2	1989–91
Bermuda	147	0	20	15	9.0	c	1	1991
Canada	2,920	33	65	147	4.9	c	1	1967–91
Cayman Islands	518	1	20	19	3.4	c	1	1984
Costa Rica	10,000–12,000	9	1,000	1,800	15.0	e2	3	1989–91
Cuba	5,996	23	495	3,229	49.6	c	2	1991
Dominica	1,127	1	197	11	0.8	c	1	1991
Dominican Republic	{5,000	{7	{650	{1,800	–	c	{2	1984–91
El Salvador	2,500	8	400	17	0.6	e2	3	1989–91
Greenland (Denmark)	497	1	31	0	0.0	c	1	1978
Grenada	919	1	148	4	0.4	e3	2	1979–91
Guadeloupe	{1670	1	261	26	1.6	c	1	1991
Guatemala	8,000	29	{652	1,171	13.5	e2	3	1989–91
Haiti	{5,000	{7	{650	{1,800	31.8	c	{2	1984–91
Honduras	5,000	30	325	148	2.8	e2	3	1978–91
Jamaica	2,746	4	558	906	27.4	c	2	1991
Martinique	{1670	1	259	30	1.9	c	1	1979–91
Mexico	20,000–30,000	71	1,000	3,624	13.9	e2	3	1984–91
Montserrat	554	–	117	2	0.3	c	1	1991
Netherlands Antilles	–	–	–	–	–	–	–	–
Nicaragua	7,000	–	500	57	0.8	e2	3	1989–91
Panama	9,000	12	577	1,222	12.7	e2	3	1989–91
Puerto Rico	2,128	1	364	235	9.4	e2	2	1982–91
St Kitts and Nevis	533	–	122	–	–	c	1	1979
St Lucia	909	–	118	11	1.1	c	–	1991
St Vincent and the Grenadines	1,000	1	165	–	–	e3	3	1979–91
Trinidad and Tobago	2,132	–	289	226	9.3	e2	1	1981–91
Turks and Caicos Islands	440	1	7	9	2.0	c	1	1982
United States	18,956	113	404	4,036	20.7	c	2	1978–91
Virgin Islands (British)	–	–	–	–	–	–	–	–
Virgin Islands (US)	–	–	–	–	–	–	–	–
SOUTH AMERICA								
Argentina	9,000	13	359	–	[25–30%]	e1	2	1984–91
Bolivia	15,000–18,000	–	–	–	–	e3	5	1989
Brazil	55,000	–	–	–	–	e3	4	1979
Chile	4,750–5,500	17	150	2,698	51.1	e1	2	1983–91
Colombia	35,000	–	–	1,500	4.3	e2	4	1989
Ecuador	16,500–20,000	–	1,100	4,000	20.7	e2	4	1986–91
French Guiana	5,000	–	318	–	–	e1	2	1991
Guyana	6,000	–	–	–	–	e3	2	1991
Paraguay	7,000–8,000	–	–	–	–	e3	4	1985
Peru	13,000	11	1,000	–	–	e2	4	1984–91
Suriname	4,500	2	293	–	–	e1	2	1978–91
Uruguay	–	2	81	–	–	c	–	1991
Venezuela	15,000–25,000	14	1,059	8,000	38.0	e3	4	1979–91
OCEANIA								
American Samoa	328	0	125	10	2.2	c	1	1991
Australia	15,000	–	–	–	[80%]	c	3	1990
Cook Islands	184	0	100	3	1.1	c	1	1991
Fiji	1,307	11	310	812	49.9	c	1	1991
French Polynesia	–	–	–	–	–	–	–	–
Guam	330	–	–	–	[69%]	–	1	1970
Kiribati	60	0	–	2	3.3	e2	1	1973–74
Marshall Islands	100	1	10	4	3.6	e2	2	1960–82
Micronesia, Federated States of	–	–	–	–	–	–	–	–
Nauru	50	0	4	1	1.9	e2	2	1982
New Caledonia	2,750	44	300	2,480	80.2	c	2	1991
New Zealand	2,160	22	189	1,942	81.9	c	1	1991
Niue	150	0	28	1	0.6	c	1	1991
North Marianas Islands	250	1	64	81	25.7	e2	3	1978–82
Palau	–	–	–	–	–	–	–	–
Papua New Guinea	10,000	44	1,500	–	[55%]	e2	4	1979–91
Pitcairn Islands	56	0	20	14	18.4	c	2	1960–83
Solomon Islands	2,780	22	370	30	0.9	e1	3	1991
Tokelau	26	0	6	0	0.0	c	1	1991
Tonga	360	1	102	25	5.4	c	1	1991
Tuvalu	–	–	–	–	–	–	–	–
Vanuatu	1,000	–	–	50	5.0	e1	1	1978
Wallis and Futuna Islands	250	–	–	5	2.0	e1	1	1983
Western Samoa	493	0	200	117	16.9	c	1	1991

Table 8.3 Species richness and endemism: higher plants (continued)

	FLOWERING PLANTS	GYMNO-SPERMS	FERNS	NUMBER OF ENDEMICS	% ENDEMISM	ESTIMATE/ COUNT	COMPLETION	DATE
ANTARCTICA								
Antarctica	41	0	11	11	21.2	c	1	1990
Falkland Islands (Malvinas)	146	0	19	14	8.5	c	1	1991
French Southern Territories	30	0	20	11	22.0	c	1	1990
AFRICA								
Algeria	3,100	18	46	250	7.9	e1	2	1975–84
Angola	5,000	–	185	1,260	24.3	e2	3	1991
Benin	{3050	{1	{200	0	0.0	c	2	1991
Botswana	–	0	15	17	–	c	3	1970–78
Burkina Faso	1,100	0	–	0	0.0	e3	3	1954–85
Burundi	2,500	–	–	–	–	e2	2	1991
Cameroon	8,000	3	257	156	1.9	e2	3	1964–83
Cape Verde	740	0	34	86	11.1	c	1	1985
Central African Rep	3,600	2	–	100	2.8	e3	3	1958
Chad	1,600	–	–	–	–	e1	1	1991
Comoros	660	1	60	136	18.9	c	2	1917
Congo	4,350	7	–	–	[5–10%]	e2	2	1988–91
Cote d'Ivoire	3,517	0	143	62	1.7	c	2	1985
Djibouti	635	2	4	2	0.3	c	1	1989
Egypt	2,066	4	6	70	3.4	c	1	1974–84
Equatorial Guinea	3,000	0	250	66	2.0	e3	4	1991
Ethiopia	6,000–7,000	3	100	600–1400	15.1	e2	4	1989
Gabon	6,000–7,000	1	150	–	[5–10%]	e2	4	1991
Gambia	966	0	8	0	0.0	c	1	1991
Ghana	3,600	1	124	43	1.2	e2	2	1991
Guinea	3,000	0	–	88	2.9	e3	3	1991
Guinea–Bissau	1,000	0	–	12	1.2	e2	2	1991
Kenya	6,000	6	500	265	4.1	e2	2	1984
Lesotho	1,576	0	15	2	0.1	c	1	1971–75
Liberia	2,200	0	–	103	4.7	e3	4	1991
Libya	1,800	10	15	134	7.3	c	1	1975–84
Madagascar	8,000–10,000	5	500	5,000–8,000	68.4	e3	4	1987
Malawi	3,600	4	161	49	1.3	e2	2	1970–75
Mali	1,741	0	–	11	0.6	c	1	1991
Mauritania	1,100	0	–	–	–	e2	1	1976
Mauritius	700	0	178	329	37.5	e1	2	1978–91
Mayotte	–	–	–	–	–	–	–	–
Morocco	3,600	19	56	600–650	17.0	e1	2	1975–84
Mozambique, People's Rep	5,500	9	183	219	3.8	e1	2	1960–70
Namibia	3,128	1	45	–	–	c	1	1976
Niger	1,170	0	8	0	0.0	c	1	1983
Nigeria	4,614	1	100	205	4.3	e2	2	1991
Reunion	750	0	240	175	17.7	e1	1	1991
Rwanda	2,288	2	–	26	1.1	c	2	1978–88
Saint Helena	50	0	24	59	79.7	c	1	1991
Sao Tome and Principe	744	1	150	134	15.0	c	1	1973
Senegal	2,062	0	24	26	1.2	c	1	1973
Seychelles	1,139	1	500	250	15.2	c	2	1989–91
Sierra Leone	1,700–2,480	0	–	74	3.5	e2	1	1962–91
Somalia	3,000	2	26	500	16.5	e2	3	1991
South Africa	2,300	40	380	–	[70–80%]	e2	1	1984
Sudan	3,132	5	–	50	1.6	c	2	1952–56
Swaziland	2,636	8	71	4	0.1	c	1	1983
Tanzania	10,000	8	–	1,122	11.2	e2	1	1968
Togo	{3050	{1	{200	0	0.0	c	2	1991
Tunisia	2,150	10	36	–	–	e1	1	1976–84
Uganda	5,000	6	400	30	0.6	e2	2	1984
Western Sahara	330	–	–	–	–	e2	2	1976
Zaire	11,000	7	–	3,200	29.1	e2	5	1991
Zambia	4,600	1	146	211	4.4	e2	3	1960–70
Zimbabwe	4,200	6	234	95	2.1	e2	2	1970–75

Notes: { Indicates figure is a combined total with another country. This applies to both Korean nations (flowering plants); Guadeloupe and Martinique (flowering plants); Guatemala and Belize (ferns) Benin and Togo (all plants); Dominican Rep. and Haiti (data for Hispaniola only). % endemism: calculated from data unless in square brackets. Estimate/count: c count; e1 approximate count; e2 extrapolation; e3 estimate on basis of any available information and comparable floras. Completion: Percentage of flora still to be described. 1: <5%(+/–known). 2: 5–10%. 3: 10–15%. 4: 15–20%. 5: >20%. Date: date of information. – no data available. [1] no data available for the new combined Yemen Republic. [2] USSR: covers the former Union of Soviet Socialist Republics [3] Portugal: data include the Azores. [4] Spain: data do not include the Canary Islands

Table compiled for WCMC by John Akeroyd.

Figure 8.1 Percent endemism of country floras

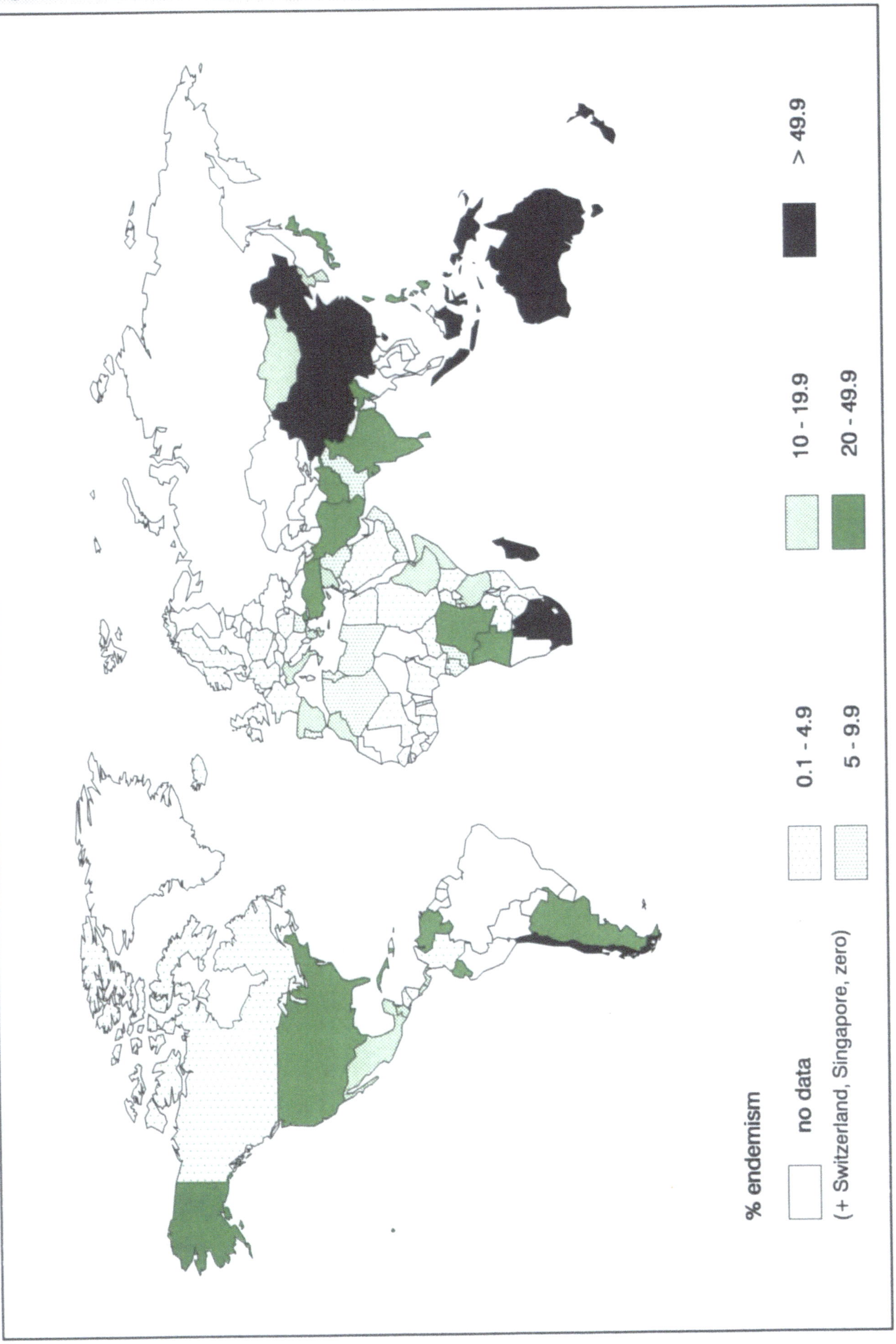

Figure 8.2 The 25 most plant-rich countries

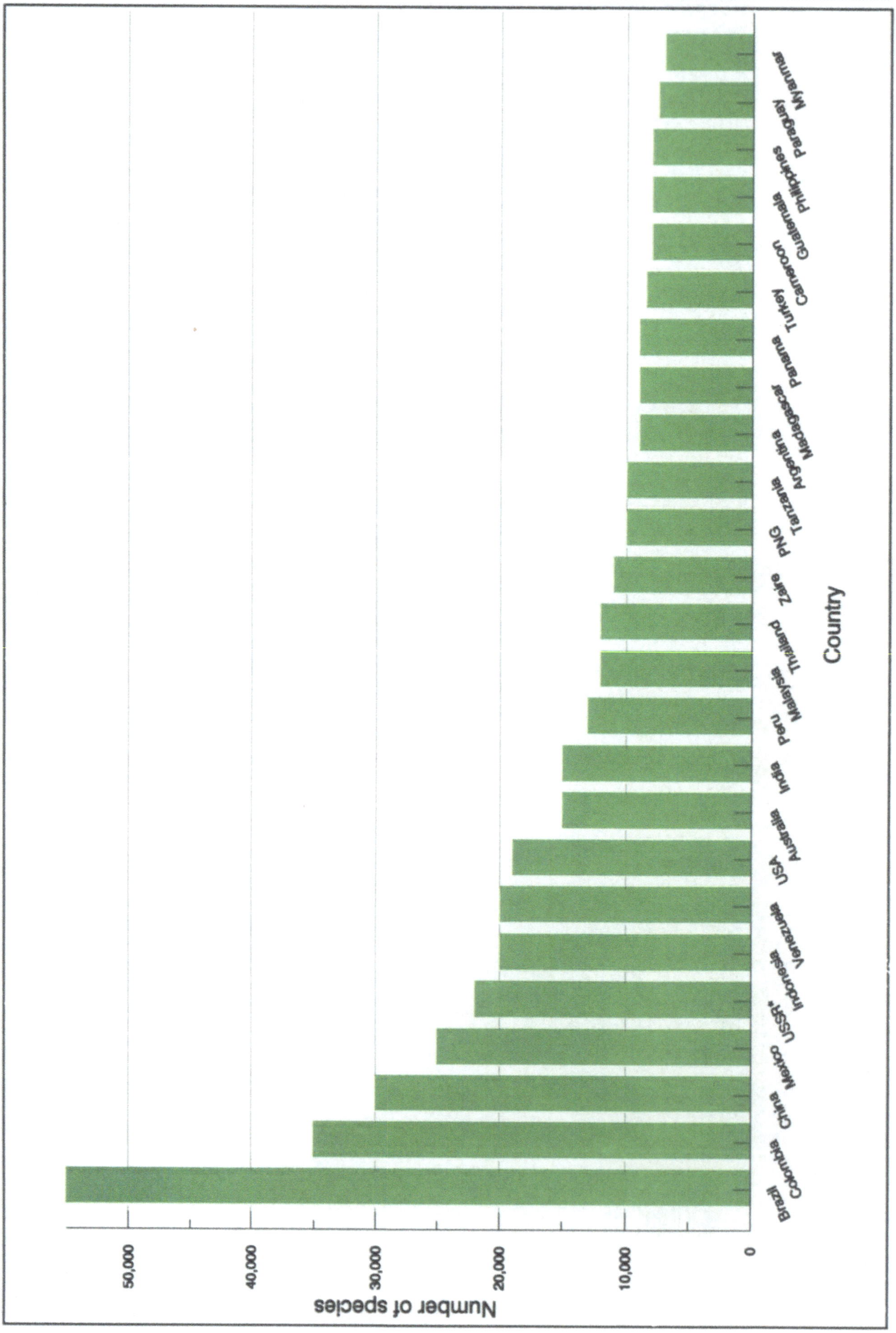

Figure 8.3 Flowering plant richness: Asia and 'USSR'

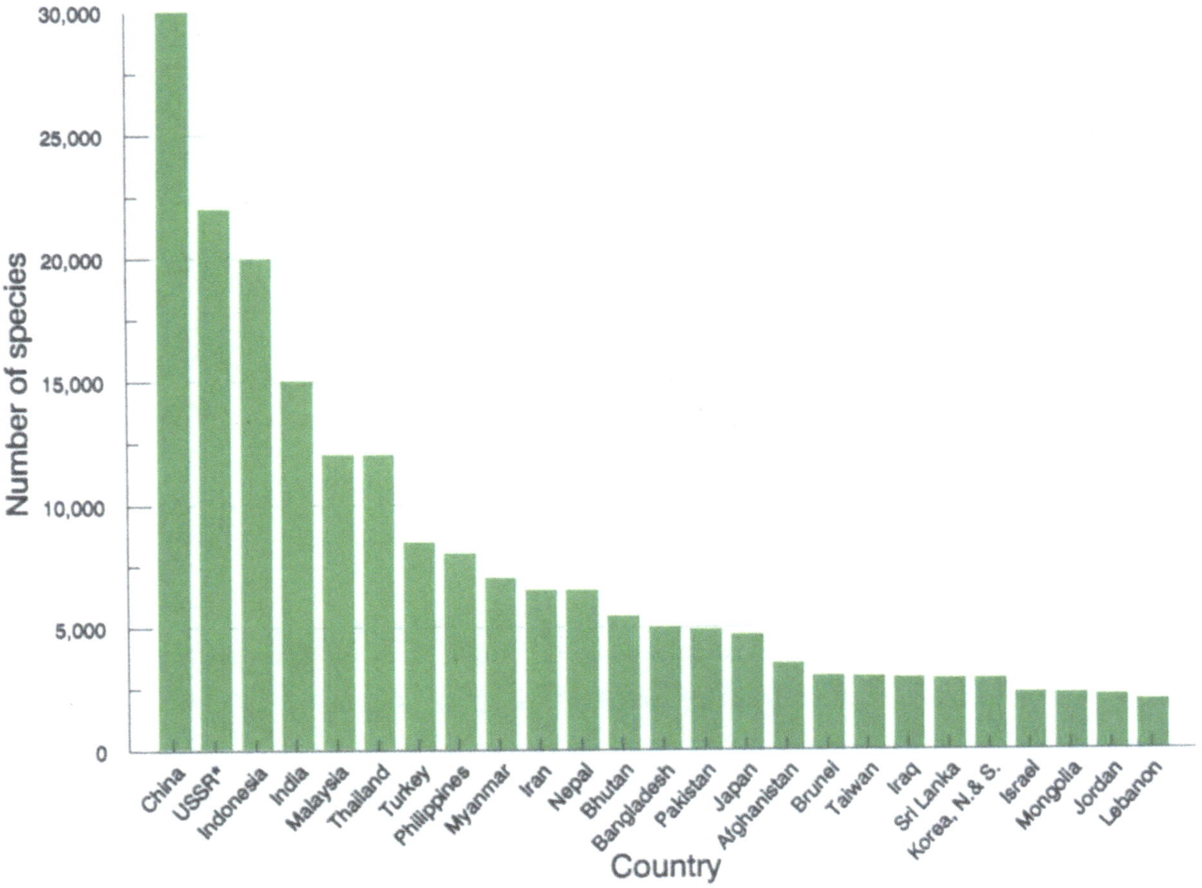

Figure 8.4 Flowering plant richness: Europe

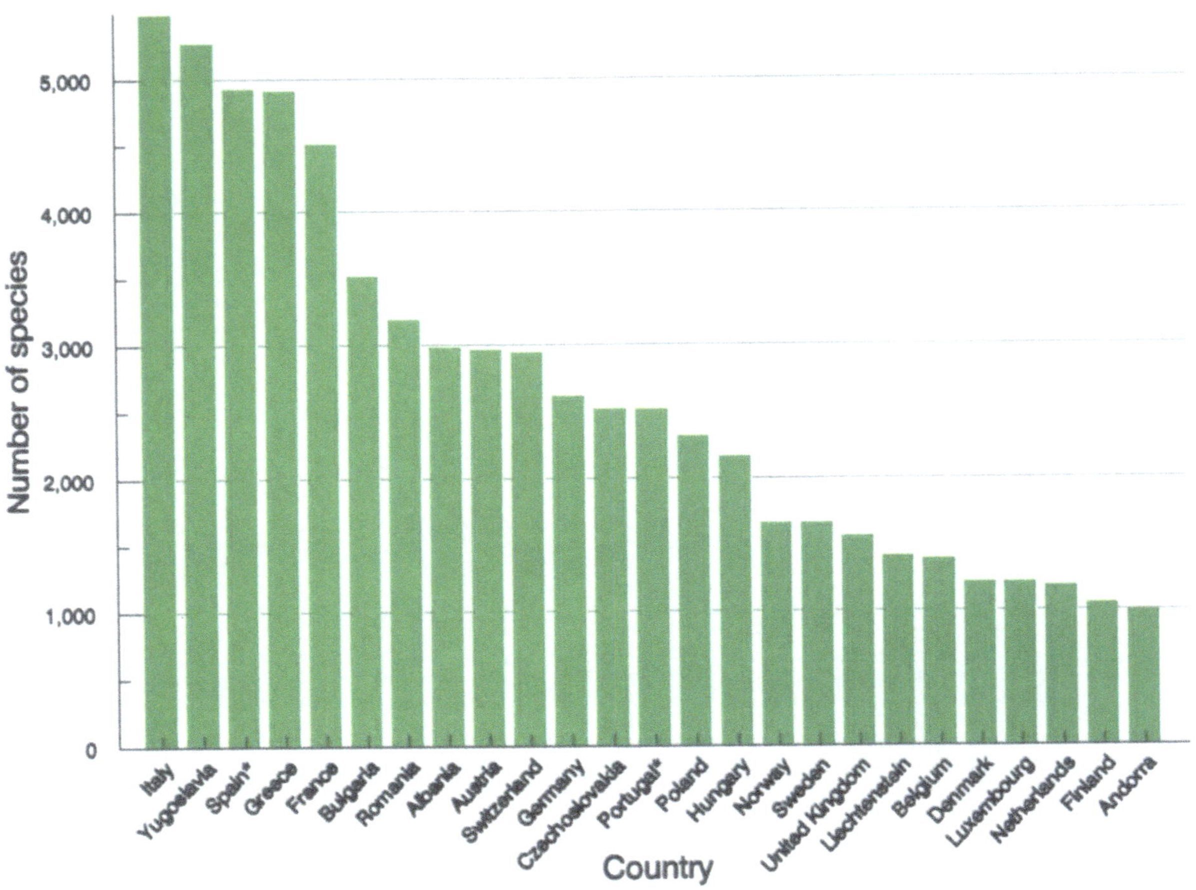

Figure 8.5 Flowering plant richness: North and Central America

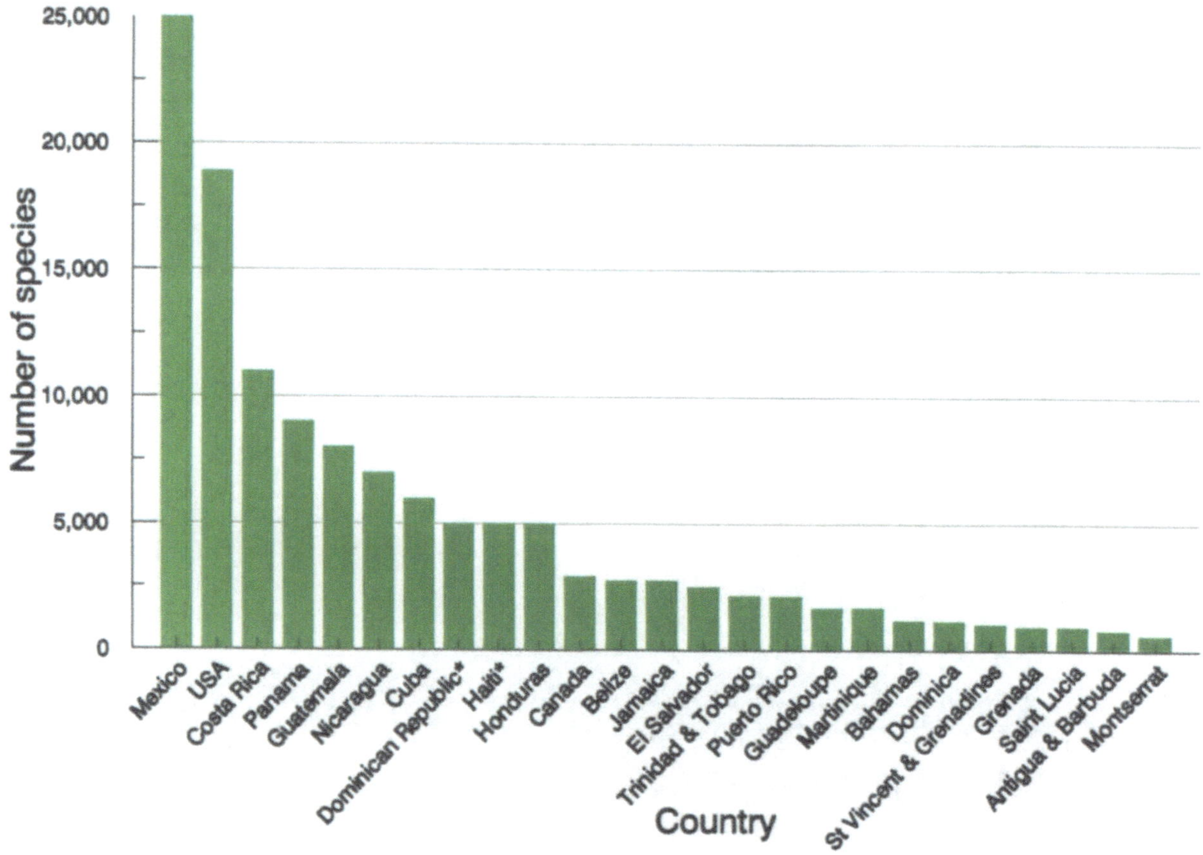

Figure 8.6 Flowering plant richness: South America and Antarctica

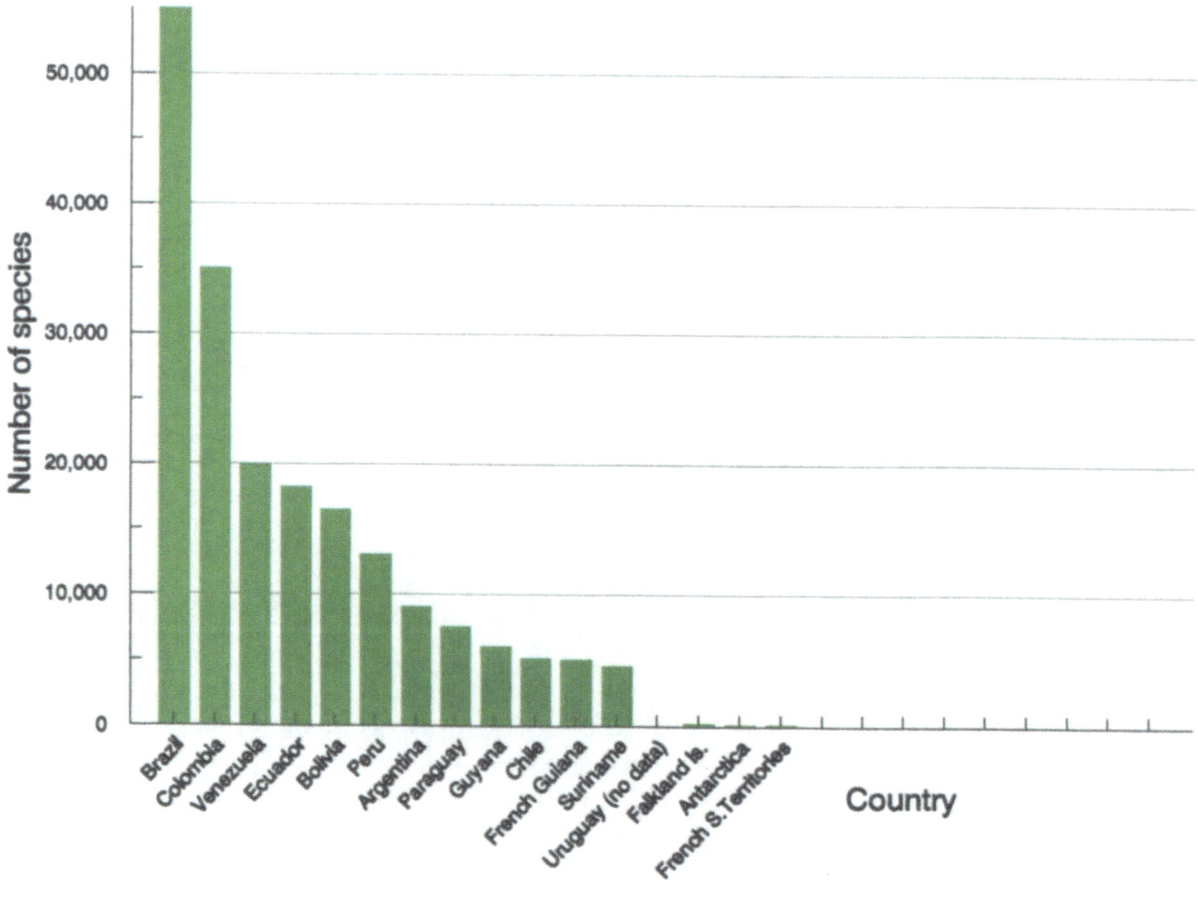

Figure 8.7 Flowering plant richness: Oceania including Australia

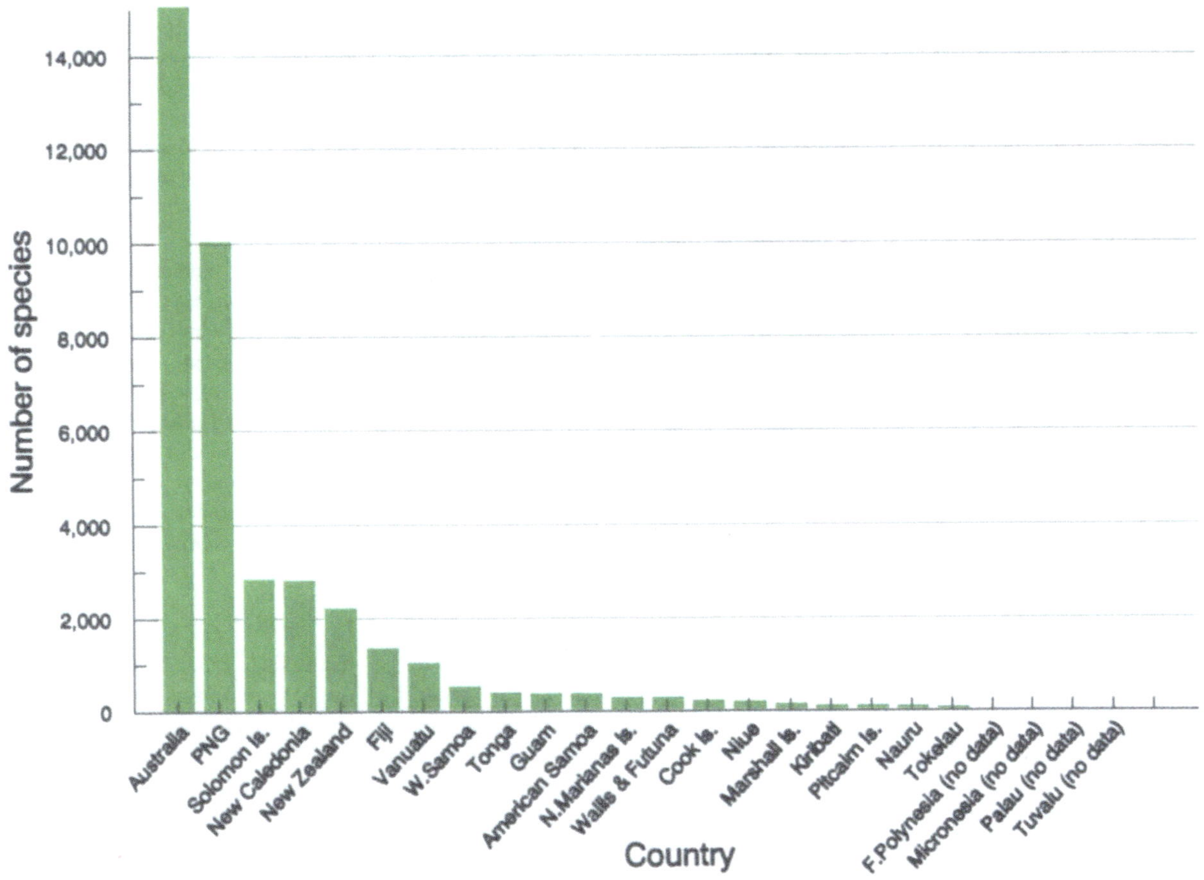

Figure 8.8 Flowering plant richness: Africa and Madagascar

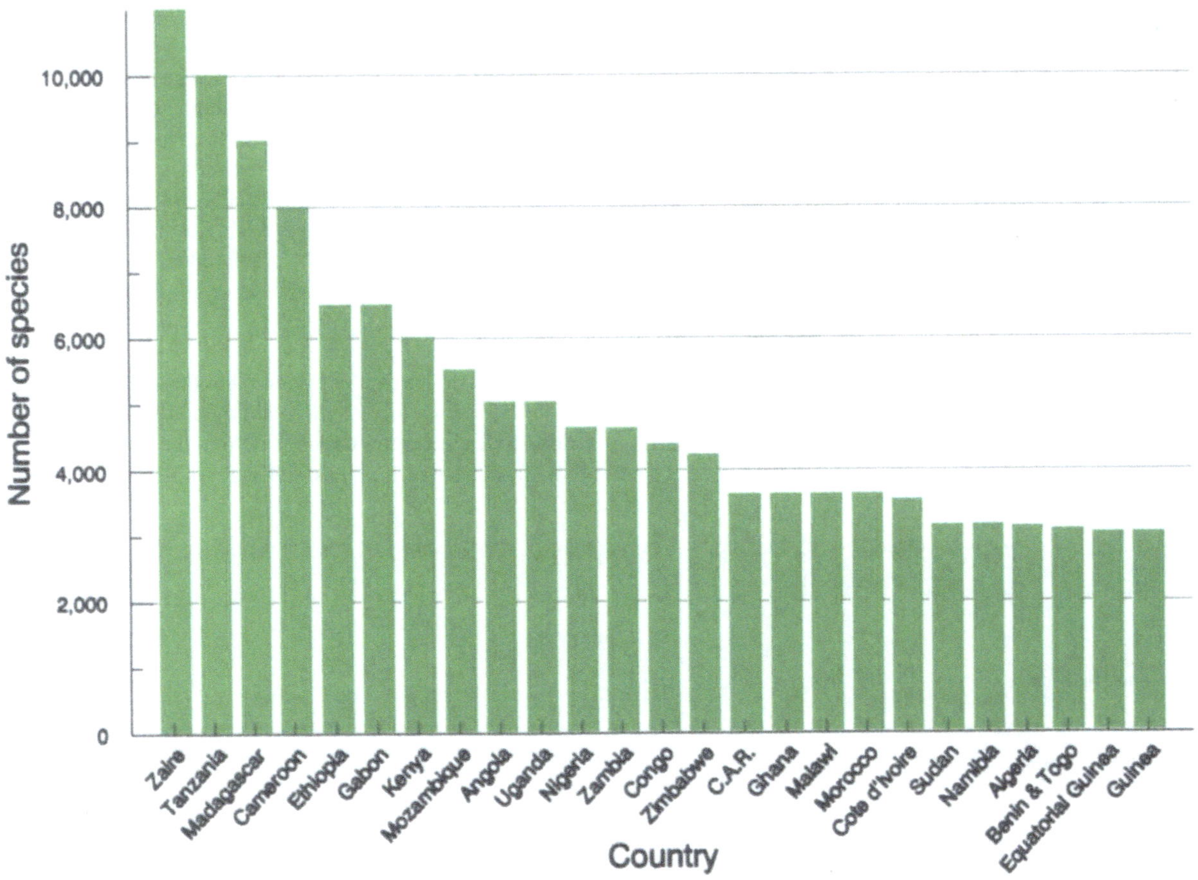

9. NEMATODES

The phylum Nematoda includes a very large number of very small worm-like animals which have a great impact on humans, often directly deleterious, as with many parasitic forms, but also with an important role in decomposition and nutrient cycling. The group contains a large number of described species, but the true proportion of the world's species that are nematodes is suspected of being very large indeed. This section is intended to introduce some features of nematodes important in the context of biological diversity.

NEMATODE DIVERSITY

More than 15,000 species have been described and the total number of species has been estimated at between 500,000 species (Poinar, 1983) and around one million (J. Lambshead, pers comm.). Nematodes show a wide range of life histories, from the entirely free-living to almost totally parasitic in plants and many kinds of animals. The parasitic forms which afflict humans, domesticated animals and plants are among the best-studied species. Anderson (1984) showed that approximately a third of known nematode genera are parasitic on vertebrates (Table 9.1).

Of the non-parasitic forms, those feeding on micro-organisms (especially bacteria) can be described as *microbotrophic*, and those that feed on multicellular metazoan organisms are described as *predaceous*. All others are described as parasitic on plants and fungi, invertebrates or vertebrates (Poinar, 1983).

Nematodes are usually long and cylindrical in shape (giving rise to the common name 'roundworms') and their cuticle is of a type of secreted collagen thought to be peculiar to nematodes. Uniquely, muscle-nerve links arise during development from the muscle not the nerve, as is usually the case (Barnes, 1980). Nematodes have a relatively complicated reproductive system and lack dispersive larvae. These features might be implicated in the high species richness of the group (J. Lambshead, pers. comm.). Body length varies enormously. One of the smallest known marine nematodes, *Greeffiella minutum*, is only 82μm long; however, the largest nematode known, *Placentonema gigantissima*, which is parasitic in the placenta of the sperm whale has been recorded at over 8m (Poinar, 1983).

Taxonomic procedures are difficult because of the small size of many nematode species. There have been several major taxonomic reviews over the last few decades. Classification is almost entirely based on morphological characteristics visible under a compound microscope (Poinar, 1983). Many species, especially those with parasitic relationships with other organisms, cannot be kept in culture and thus are not amenable to biochemical or genetic study. Scientists of different disciplines frequently work independently of each other, resulting in confusing taxonomic revisions.

Estimates of the total number of nematode species vary greatly, current figures ranging from 500,000 to around one million. Recent work on species diversity in the meiofauna of deep-sea benthic samples has found very high diversity in each sample. However, taxonomic problems and the sheer number of organisms involved means that the species similarity between samples is still unresolved (J. Lambshead, pers. comm.). If many of these samples constitute separate species, nematodes may approach or even exceed the insects in species richness.

Microbotrophic nematodes

The microbotrophic nematodes, especially some marine forms, are generally thought to represent the most primitive organisms in the phylum, although there is an alternative hypothesis that extant microbotrophs are secondarily derived from parasitic forms (Poinar, 1983). It is difficult to elucidate the evolutionary history of a group which leaves few fossil remains but it is thought that microbotrophic nematodes were probably well represented in the Cambrian period, c. 600 million years ago.

Microbotrophic nematodes are one of the most widespread and abundant animal groups known. Wherever a suitable food source exists they are found, even under extreme conditions such as hot sulphur springs or polar ice. Because of their relatively small size (although some grow to over 10mm, most cannot be seen with the naked eye) they tend to go unnoticed even though present in great numbers. For example, about 90,000 nematodes of several different species have been found in a single rotting apple in an orchard and about 50,000 nematodes of at least eight different species have been reported from a single fig (Barnes, 1980).

These nematodes can be divided into three groups - marine, freshwater and terrestrial - although even the so called terrestrial species are dependent upon the water film around soil particles and in interstitial spaces. Those species which

Table 9.1 Approximate numbers of nematode families and genera known from different habitats

HABITAT	FAMILIES	GENERA
Marine and freshwater	41	730
Soil	64	429
Plant (parasitic)	26	166
Invertebrate (parasitic)	42	187
Vertebrate (parasitic)	83	759
TOTALS	256	2271

Source: Anderson, R.V. 1984. The origins of zooparasitic nematodes. *Canadian Journal of Zoology*, 62:317-28.

live in environments with only a periodic water supply, such as deserts, survive mostly as inactive larvae and only emerge when water is present.

Marine species live in bottom sediments of many habitats from sandy shores and salt-marshes to ocean trenches and have been reported in numbers ranging from 100,000 to 10 million individuals per m² (Poinar, 1983). Thus they are the most important metazoan element of the meiofauna in all samples. Samples reported by Nicholas (1984) taken at various depths down to about 400m show a range in number of species from 3 to 125 per site and a range in densities of 110,000 to 5,261,000 animals per m². These samples were derived from sediments, algae, shells and rocks, where bacteria and other micro-organisms flourish. In one study of deep-sea nematodes, examination of 216 individuals yielded a total of 148 species (J. Lambshead, pers. comm.).

Several groups of nematodes live in fresh and brackish waters, and transitional zones. Many of these species tolerate rapid fluctuations in salinity. As in marine habitats, the animals are usually present in the sediment, although they may occasionally swim freely. The most dense nematode faunas are associated with a reasonable oxygen supply and sediment with a high organic content. Lakes have a very variable fauna which probably depends upon their physical attributes, such as isolation and thermal stratification. Shallow marginal waters may be quite rich, probably sharing some species with wet terrestrial habitats. However, deeper waters seem to be species-poor unlike marine systems. A notable exception to this is Lake Baikal, where, as among other animal groups, considerable speciation has occurred and endemism appears to be high (Nicholas, 1984).

In the soil the distinction between microbotrophic and parasitic nematodes becomes very blurred in certain taxa. All kinds of soils support large nematode communities (see Table 9.2 below) and the richest tend to be where there is plenty of organic matter, fine plant roots, etc. The interactions with plant roots and other organisms, such as fungi, are extremely complex and difficult to assess.

It is thought that parasitism has arisen independently in several nematode taxa, and certainly the microbotrophic forms illustrate a great variety of interactions which could be considered as stages in the evolution of parasitism. For example, there are many examples of phoretic relationships with invertebrates. These range from larval stages attaching externally to mobile hosts who carry them to the next food source, to larval stages which live within a host apparently without harming it, but which cannot escape to continue their life cycle until the host dies of natural causes. Many of these relationships are very finely tuned to the life cycle of a specific carrier whilst others use a variety of suitable invertebrates. Not all relationships benefit the nematode alone: in some cases the carrier may also feed upon the nematodes. Relationships with plants may be equally complex, as nematodes may often feed upon the bacteria on and in decaying roots. However, some species are suspected of spreading disease to increase their food resource or of being able to feed upon living plant tissue as an alternative to bacteria. Even within one species, different forms may show different degrees of interaction, making rigid definitions impossible.

Predaceous nematodes

Predaceous nematodes are found in all habitats but are most abundant in terrestrial systems. All eat a few to many multicellular organisms in the course of their development, although bacteria, ciliates and organic particles may also be eaten. Little is known about prey-specificity in nature, as most studies, by necessity, have been carried out under laboratory conditions. However, some extremely common groups include other nematodes as prey items and may be potential biological control agents for nematode pests of plants. For instance, a single nematode of the family Mononchidae has been observed to kill over 1,000 nematodes in a three-month period and estimates of density suggest that up to 300 million mononchid nematodes might be contained in an acre of soil (Poinar, 1983). However, observations also suggest that almost any invertebrate of the correct size may be eaten and prey location is a chance affair.

Little is known of the aquatic predaceous nematodes. However, observations which suggest that some marine forms may be able to penetrate foraminiferan tests to get at the body inside are of considerable interest as borings similar to those attributed to these nematodes have been seen in fossilized foraminiferan tests from the Holocene and Cretaceous periods (Poinar, 1983).

Table 9.2 Abundance and biomass of soil nematode fauna from different types of ecosystem

ECOSYSTEM	ABUNDANCE x 1000m⁻²		BIOMASS*	
	MEAN	RANGE	MEAN	RANGE
Tundra	3,490	800-10,000	1,350	265-4,130
Coniferous forest	3,330	1,125-15,000	510	180-1,696
Eucalyptus forest	5,467	4,040-7,449	1,423	770-2,050
Deciduous forest	6,270	255-29,800	2,760	75-15,200
Temperate grassland	9,190	2,432-30,000	3,800	650-17,800
Fen, bog, heathland	1,660	330-3,900	660	350-900
Desert	760	423-1,100	410	125-700
Tropical forest	1,700	1,500-1,900	-	-

Source: Sohlenius, B. 1980. Abundance, biomass and contribution to energy flow by soil nematodes in terrestrial ecosystems. *Oikos* 34:186-94.
Note: * biomass is measured here in mg weight per m².

Table 9.3 Distribution of nematode genera among groups of vertebrates

	FISH	AMPHIBIANS	REPTILES	BIRDS	MAMMALS	HOST SPECIES PER NEMATODE GENUS
Fish	<u>62</u>	6	7	4	1	250
Amphibians		<u>20</u>	22	2	1	50
Reptiles			<u>62</u>	4	4	60
Birds				<u>113</u>	17	60
Mammals					<u>387</u>	8
TOTAL GENERA	80	51	99	140	410	

Source: Modified from Inglis, W.G. 1965. Patterns of evolution in parasitic nematodes. In: Taylor, A.E.R. (Ed.), *Evolution of Parasites*. Blackwell Scientific Publishers, Oxford, UK and Poinar, G.O. 1983. *The Natural History of Nematodes*. Prentice-Hall Inc., New Jersey, USA.
Note: Numbers underlined indicate genera exclusive to each vertebrate group.

Parasitic nematodes

Plant parasitic nematodes have been found in most species of terrestrial plants, all over the world. Many are polyphagous and consequently a plant species may be attacked by a wide range of nematode species. For instance, Poinar (1983) lists 36 nematode species in 15 genera which have been identified parasitising potatoes and six species in four genera from wild chicory. Most fungi also suffer from nematode attacks, some species being serious pests in mushroom culturing operations. Plant parasites are apparently much less common in aquatic habitats, and relatively few species are known from seaweeds and marine fungi.

Parasitic nematodes are similarly widespread in both invertebrate and vertebrate hosts and have evolved some remarkably complex life cycles. The greatest number of invertebrate parasites known are in the insects and some of these have been studied in great depth in the hope of developing successful biological control methods. As with the plant parasites many nematode species can attack a wide range of insect hosts. Others are highly specialised and adapted to the life cycle of one particular host. Of the former group, two nematode families include genera which have evolved mutualistic relationships with a single bacterium genus, which is unkown in a free living state. These nematodes introduce bacterial cells into a host insect which dies soon after becoming infected. The bacteria then grow on the body and the nematodes feed on the bacteria, ensuring some are carried to infect a new host. Insects of ten different orders are known to be attacked by these species (Poinar, 1983).

Vertebrate parasites are equally widespread and, here again, some may utilise a whole range of hosts whilst others are extremely host specific. Many have developed complicated methods of dispersal which may involve invertebrates (or occasionally other vertebrates) as intermediate hosts. Some of the world's most debilitating diseases are spread by this method, such as onchocerciasis (river blindness).

Nematode parasites tend to become more specialised in more developed vertebrate groups. The majority of nematode genera are confined to a single vertebrate genus. Of those that do have a wider host range a few can utilise different classes, but most are restricted to similar animals. Table 9.3 illustrates the higher diversification of nematode parasites in the higher vertebrate groups and the greater specifity associated with this.

THE ECOLOGICAL IMPORTANCE OF NEMATODES

Free living nematodes are vital components of ecosystems. Although not themselves decomposers, many feed on the primary decomposers, the bacteria and fungi, which break down complex organic molecules and thus make these nutrients available in the food chain again. They are therefore elementary in the decomposition cycle. The predaceous species are also important consumers near the

Table 9.4 Estimated crop losses due to *Meloidogyne* species in tropical regions

CENTRAL AMERICA AND CARIBBEAN Crop	% loss	SOUTH AMERICA Crop	% loss	BRAZIL Crop	% loss	WEST AFRICA Crop	% loss	SOUTHEAST ASIA Crop	% loss
Tomato	38	Cucumber	33	Tomato	25	Tomato	46	Tomato	24
Chayote	38	Tomato	27	Coffee	24	Cowpea	43	Melon	18
Guava	35	Bean (common)	24	Soybean	23	Okra	42	Bean (common)	18
Pumpkin	22	Watermelon	23	Cotton	17	Carrot	38	Eggplant	17
Bean (common)	16	Pepper	22	Papaya	15	Pigeon pea	35	Black pepper	16
Yam	16	Eggplant	20	Yam	15	Melon	33	Chinese Pechay	16
Mean % loss (all crops)	15		15		13		25		11

Source: Adapted from Sasser, J.N. 1979. Economic importance of *Meloidogyne* in tropical countries. In: Lamberti, F. and Taylor, C.E. (Eds), *Root-knot Nematodes (Meloidogyne species)*. Academic Press, London, UK.
Note: Only the six worst-affected crops are shown in each case.

base of food webs, feeding on unicellular algal primary producers and smaller metazoans.

Nematodes are most often studied in their destructive capacity, as pests of agricultural crops and as parasites of livestock and humans. However, there is also potential for biological control applications, against a wide range of insect pests and against other nematode species. Free-living nematodes have been used as models for various experiments on the functioning of ecosystems and as indicators of environmental health, such as water pollution. Other potential and actual uses include nematodes as indicators of the quality of terrestrial soils, freshwater and marine sediments (van der Wal and de Goede, 1988), and in a whole range of biological research projects (see Nicholas, 1984, for examples).

There are many aspects to the problems of nematode association with crops. For instance, nematode species which are useful in controlling pathogenic root fungi in one situation may in another destroy mycorrhizal fungi, necessary for good plant growth. Similarly, some nematodes which feed harmlessly or even usefully on bacteria most of the time may also be able to move into plant roots, either to eat healthy plant tissue directly or to infect them to provide more food for their bacteria. This change may depend on environmental conditions, for instance the soil drying out, and may produce a sudden reaction in the crop which superficially resembles water stress. These cases are the cause of some debate and considerable research. However, other nematodes are without doubt serious crop destroyers. Poinar (1983) quotes estimates which suggest that 7-15% of the annual crop production of the USA is destroyed by nematodes. Table 9.4 shows the estimated yield losses in several tropical regions due to species *Meloidogyne*, one of the most destructive nematode genera. Only the six worst-affected crops in each region are shown here in detail but in the original table Sasser (1979) gives figures for up to 21 crops in each region. The most destructive species in each case is *M. incognita*, followed by *M. javanica*, *M. arenaria* and *M. hapla*.

Some of these problems have arisen as a result of crop monoculture which reduces natural control systems that normally keep such pests within acceptable limits. Various methods of control are possible, including timed planting to miss the most active cycle of the parasite, crop rotations which can include crops poisonous to the nematodes, and flooding. Another form of natural control which has received considerable attention in recent years entails use of fungi that are predaceous or parasitic upon nematodes. At least one of these former, a trap-forming deuteromycete in the genus *Arthrobotrys*, is commercially available (Poinar, 1983) and is effective in tomato fields and greenhouses against *Meloidogyne* species. Various other fungi have been tested with varying results and other fungi which apparently produce nemotoxins are also being studied.

On the other hand, control by nematodes of fungal plant diseases and weeds have been investigated. For example, an encysting plant parasite *Paranguina picridis* has been used with some success in the USSR to control knapweed (Poinar, 1983). Predaceous nematodes have also been considered as control agents for ectotrophic root parasites, especially other nematodes, and microbotrophic nematodes for control against certain infective bacteria.

In contrast to plant parasites, the invertebrate parasites are rarely a problem to man (except where plants or higher animals are also part of the life-cycle). In fact many have great potential for control of pest insects. In particular, certain nematodes have been intensively studied for possible mosquito control and others which parasitise water snails may be able to control schistosome-bearing snails. Insect pests of crops and livestock are also targeted by research programmes; several examples which have been tried are shown in Table 9.5, adapted from Poinar, 1983.

Table 9.5 Examples of nematode species investigated as biological control agents

FAMILY	SPECIES	INSECT PEST	LOCATION	HABITAT
Mermithidae *culicivorax*	*Romanomermis*	mosquitoes	North America, Taiwan, Europe, Africa, Oceania, Central America, Thailand	Ponds, ditches, lakes.
Diplogasteridae *uniformis*	*Pristionchus*	Colorado beetle	Poland	Soil
Steinernematidae *glaseri*	*Neoaplectana*	Japanese beetle	Eastern USA	Soil
Heterorhabditidae *bacteriophora*	*Heterorhabditis* (click beetles)	*Agriotes* spp.	Italy	Soil
Neotylenchidae *siricidicola*	*Deladenus* (wood wasp)	*Sirex noctilio*	Australia	Trees
Allantonematidae *autumnalis*	*Heterotylenchus* (face fly)	*Musca autumnalis*	North America	Dung
Sphaerulariidae	*Tripius sciarae*	Sciarid flies	England (greenhouse)	Soil

Source: Adapted from Poinar, G.O. 1983. *The Natural History of Nematodes*. Prentice-Hall Inc., New Jersey, USA.

Table 9.6 Estimates of nematode infections in man (in millions)

DISEASE	NEMATODE/S	AFRICA (excl. 'USSR')	ASIA & SOUTH AMERICA	CENTRAL	OCEANIA AMERICA	NORTH (excl. 'USSR')	EUROPE	'USSR'
Ascariasis	*Ascaris lumbricoides*	159	931	104	1	5	39	30
Hookworms	(various)	132	685	104	2	3	2	4
Human pinworm	*Enterobius vermicularis*	24	136	40	1	29	75	48
Trichuriasis	*Trichuris trichiura*	76	433	94	1	1	41	41
Trichinosis	*Trichinella spiralis*	1		3		35	5	2
Others		9	49	21	<1	1	1	3
Elephantiasis	*Wuchereria bancrofti* and *Brugia malayi*	59	300	22	2			
Other filariae		178	57	39				

Source: Peters, W. 1978. Comments and discussion II. In: Taylor, A.E.R. and Muller, R. (Eds), *The Relevance of Parasitology to Human Welfare Today*. Blackwell Scientific Publications, Oxford, UK.

Nematode parasites of vertebrates are an enormous drain upon human resources, both in the effects on domestic animal species and on human life directly. The World Health Organization produces estimates for the numbers of people afflicted with the major parasitic diseases. Poinar (1983) gives figures for four of these for 1977-78: hookworm disease, onchocerciasis, ascariasis and trichuriasis in Africa, Asia and Latin America. Of these the first two cause the greatest number of deaths each year: 50-60 thousand deaths among 7 million to 900 million people with hookworm disease, and 20-50 thousand deaths out of 30 million estimated cases of onchocerciasis. A different presentation of similar data is given in Table 9.6, adapted from Peters (1978). These estimates are apparently based on data collected in the 1940s although Peters suggests they adequately represent the current situation.

The monetary costs caused by livestock disease are also immense, in terms of prevention, treatment, animals lost and human time. Where these parasites are also transmittable to humans, such as several of those affecting pigs, precautions against infection are also costly and time-consuming. Thus, unlike the possible benefits from free living and plant parasitic nematodes, and the considerable potential in invertebrate parasites, there are no obvious uses of vertebrate parasites with benefit to humans.

References

Anderson, R.V. 1984. The origins of zooparasitic nematodes. *Canadian Journal of Zoology* 62:317-28.

Barnes, R.D. 1980. *Invertebrate Zoology*, 4th edn. Holt-Saunders Tokyo, Japan. 1,089pp.

Inglis, W.G. 1965. Patterns of evolution in parasitic nematodes. In: Taylor, A.E.R. (Ed.), *Evolution of Parasites*. Blackwell Scientific Publishers, Oxford, UK. Pp.79-124.

Nicholas, W.L. 1984. *The Biology of Free-living Nematodes*, 2nd edn. Clarendon Press, Oxford, UK. 251pp.

Peters, W. 1978. Comments and discussion II. In: Taylor, A.E.R. and Muller, R. (Eds), *The Relevance of Parasitology to Human Welfare Today*. Blackwell Scientific Publications, Oxford, UK. Pp.25-40.

Poinar, G.O. 1983. *The Natural History of Nematodes*. Prentice-Hall Inc., New Jersey, USA. 323pp.

Sasser, J.N. 1979. Economic importance of *Meloidogyne* in tropical countries. In: Lamberti, F. and Taylor, C.E. (Eds), *Root-knot Nematodes (Meloidogyne species)*. Academic Press, London, UK. Pp.359-374.

Sohlenius, B. 1980. Abundance, biomass and contribution to energy flow by soil nematodes in terrestrial ecosystems. *Oikos* 34:186-94.

Wal, A.F. van der and Goede, R.G.M. de (Eds) 1988. *Nematodes in Natural Systems*. Report of a workshop held at the Dept. of Nematology, Agricultural University, Wageningen, The Netherlands, 16-18 December 1987. Mededeling 199.

10. DEEP-SEA INVERTEBRATES

DEEP-SEA COMMUNITIES

Until the mid-1960s it was believed that oceanic diversity was concentrated in shallow water around coasts and declined with both depth and distance from land as food resources became more remote. The first reports of unexpectedly high species diversity in bottom living communities arrived in 1967 with samples collected using a new technique: the epibenthic sled (Hessler and Sanders, 1967). Although many were initially sceptical of the conclusions, the deep-sea environment has been an active area of research and is now known to support communities rich in species, high in endemism and often ecologically unique. In terms of species numbers alone, the marine environment provides a relatively minor proportion of the global total.

Approximately 71% of the Earth's surface is covered by sea, and about 51% of its surface by ocean over 3,000m in depth. Deep-sea communities are thus prevalent over a major proportion of the planet. All deep-sea habitat is in the aphotic zone, well below the distance sunlight can penetrate. Community structures and food webs are therefore very different from those found on land and in the shallower parts of seas in that, except in the specialist case of hydrothermal vents (described below), there is no primary production and all life relies on organic material from other parts of the ocean. As deeper and deeper levels are reached biomass falls exponentially (Rowe, 1983). This was misinterpreted as being synonymous with falling species diversity (Grassle, 1991). Because, despite their enormous volume, the deep oceans appear to be relatively simple ecosystems, there was little reason to imagine that they should make any significant contribution to overall global species diversity. That species diversity in the benthic community should rise with increasing depth was therefore a major discovery.

The benthic samples taken by Hessler and Sanders (1967) and later workers have revealed a hitherto unexpectedly high species richness. This discovery has prompted speculation that the deep sea is a site of prolific speciation and, as one of the most stable and ancient environments on Earth, perhaps the origin of certain higher-level taxa (Gage and Tyler, 1991). Several ideas have been postulated to explain this high diversity but it would appear that a combination of factors is important. Grassle (1991) suggests four major influences:

- the relative lack of environmental extremes such as those of temperature, salinity, low oxygen and major disturbances
- patchy food resources
- local disturbances and structures caused by animal activities
- a large area with few barriers to dispersal.

The first three of these are equivalent to the processes thought by some to be fundamental to the high species diversity in tropical terrestrial and shallow water ecosystems. Environmental stability allows the development of high species diversity with many highly specialised species. This is supported by observations in deep-sea areas that do not have long-term environmental stability, such as trenches and areas of strong bottom currents; these usually have a much reduced species diversity although their faunas may be of interest in other ways (Thorne-Miller and Catena, 1991).

The patchiness of food availability and local disturbance can be compared to the importance of gap appearances in the canopy of tropical forests, both involving small scale habitat diversity within a larger homogenous area and the maintenance of a mosaic of disequilibrium populations (Grassle, 1989). Most organisms which live in the deep sea are totally dependent on organic detritus falling from euphotic zones. This is largely of planktonic and faecal origin but larger masses such as pieces of wood, carcasses and algal mats are also of importance. Local disturbances such as feeding activities and burrowing and mound-building by polychaete worms also ensure local topographic variations which provide a variety of microhabitats. Weak bottom currents allow particulate organic matter to concentrate in hollows and lees.

The large area of the deep ocean zone, coupled with the above factors, results in a very large species pool with wide dispersion potential. Grassle (1991) estimates that if the currently observed species-area relationship is extrapolated the total species pool may be in the order of 10 million. Although, as discussed below, there are many problems with predictions of this type, even this figure may be conservative.

Faunal composition

Studies of the benthic species assemblages of different regions are still in their infancy. The major difficulty is obtaining quantitative samples, since the depths involved are far greater than a diver can go. Much of the work which has been carried out has not been coordinated, leading to different sieve sizes for sampling, different collection techniques and different assessments of biomass (Rowe, 1983). This makes comparisons between sites difficult. In addition, taxonomic problems in certain taxa have meant that, while it may be possible to have a species count from any one sample, it is not possible to say what the similarity is *between* samples. Nearly half the species in each new sample may be undescribed (Grassle, 1989), and there may be few taxonomists working on any one group, raising problems of species identification. Even in well sampled areas, sample sizes are small compared to the regions they are supposed to represent, and it is uncertain to what extent results can be extrapolated. However, the rate of discovery of new species and the proportion of species currently known from only one sample both indicate that a great number remain to be discovered (Grassle, 1991).

Benthic fauna is usually classified into size classes, increasing from the nanobiota, through the meiofauna, the macrofauna and finally to the megafauna. A problem with this type of classification is it splits natural taxonomic groups and even age classes of the same species. Many workers prefer to classify all the members of certain taxa into the size class which best represents the group; for example, all nematodes are often considered as meiofauna.

Small size and taxonomic problems mean that few comparative data are currently available on meiofaunal diversity. The major taxonomic groups in this size class are the nematodes and foraminiferans (protozoa). Other important taxa in this size class include the harpacticoid copepods and ostracods.

More information is available for the macrofauna, which has been more extensively studied than other size classes. This is typically dominated by polychaetes (up to 75% numerically), peracarid crustaceans (including cumaceans, tanaids, isopods and amphipods) and a variety of smaller molluscs (Gage and Tyler, 1991), but most other phyla are also represented. Figures from Grassle (1991) (see Table 10.1 and Fig. 10.1) demonstrate the species, family and phylum composition of a typical sample; (however, note these samples were taken at bathyal rather than abyssal depths - see below). Wolff (1977) also provides examples of the taxonomic composition of the macrofauna (and megafauna using the taxa listed here) in a number of regions (Table 10.2 and Fig. 10.2). However, the sampling methods are not the same in each area so these results may not be comparable.

Table 10.1 Diversity in benthic samples *

GROUP	NO. OF FAMILIES	NO. OF SPECIES
Annelida	49	385
Arthropoda	40	185
Mollusca	43	106
Echinodermata	13	39
Nemertina	1	22
Cnidaria	10	19
Sipuncula	3	15
Pogonophora	5	13
Hemichordata	1	4
Echiura	2	4
Priapulida	1	2
Brachiopoda	1	2
Ectoprocta	1	1
Chordata	1	1
TOTAL	171	798

Source: After Grassle, J.F. 1991. Deep-sea benthic biodiversity. *Bioscience* 41(7).
Note: * Sea-bed samples from 1,500m to 2,500m depth off New Jersey, north-east Atlantic.

Table 10.2 Composition of benthic macrofauna (percentage of total species present)

LOCATION SAMPLE TYPE DEPTHS (m)	TRENCHES* TRAWL 6,000-10,000	CENT.N PACIFIC A.D. 5,600	NW ATLANTIC A.D. 4,400-5,000	NW ATLANTIC E.S. 4,700
Polychaetes	7	55	55	8
Peracarid crustacea (total)	5	24	33	32
Tanaidacea	<1	18	19	1
Isopoda	4	6	12	18
Amphipoda	<1	0	2	5
Bivalvia	19	7	4	47
Echinodermata (total)	57	1	1	2
Ophiuroidea	2	<1	<1	2
Holothuroidea	54	<1	0	0
Others	11	11	8	11
Number of individuals	21,589	287	681	3,737

Source: After Wolff, T. 1977. Diversity and faunal composition of the deep-sea benthos. *Nature* 267:780-785.
Note: * Trenches = Kurile-Kamchatka, Japan, Kermadec and Java A.D. = Anchor Dredge; E.S. = Epibenthic Sledge.

In the megafauna, echinoderms of several classes are often the dominant mobile (or errant) life forms on or in association with the sea bottom. Their distribution may be very uneven, they may sometimes occur in great numbers on patches of detritus fallout and some scavenging forms may be found in large congregations at bait. Giant scavenging amphipods, growing up to about 18cm in length, are also characteristic in many areas. However, the high mobility of these animals means they are rarely caught in trawls and have been less well studied than less active animals. Other arthropods include a variety of sea spiders (Pycnogonida) and decapods of several families (both errant and sessile). Errant animals of several other taxa occur, including polychaetes, hemichordates, cephalopods and fish. Sessile animals generally occur on any suitable surfaces. Sponges (Porifera), especially the glass sponges, are widely distributed and coelenterates (Cnidaria) are also well represented, dominated by anthozoans. Other taxa include bryozoa and brachyopoda.

Distribution of deep ocean biodiversity

There are general trends in species richness with respect to depth in benthic communities. The picture is very incomplete, however, so conclusions must be tentative. Rex (1983) examined data from four major taxonomic groups (polychaetes, gastropods, protobranchs and cumaceans) along a depth gradient down to 5,000m. All showed maximum diversity between 2,000m and 3,000m. The three of these taxa which had been sampled with an epibenthic sled rather than an anchor dredge also showed a higher diversity between 4,000m and 5,000m than between 0m and 1,000m (polychaetes were the exception). However, whether this is an artefact of the difference in sampling technique or a true difference between the taxa is unclear.

The assemblages of different depth zones have differing patterns of geographical distributions. Abyssal species appear to have the most widespread distributions (Angel, 1991), probably because there are fewest barriers to larval dispersal. For example, in the Polychaeta, which is one of the less cosmopolitan groups, 78% of all North Atlantic abyssal species are found in both the East and West Atlantic, compared to 58% of the bathyal species (Gage and Tyler 1991). Hadal, or ultra-abyssal, communities again have more disjointed distributions as only about 1% of the Earth's surface is covered with water of such depth. Plain communities at these depths are poorly sampled but of particular interest are trench faunas (although not all trenches reach hadal depths). These are dealt with separately below.

Latitudinal patterns are even less well studied. In pelagic communities there is a general trend for the number of species to increase from the polar to the tropical regions. Buzas and Culver (1991), also report a definite latitudinal gradient in the foraminiferans of open ocean sediments, typically ranging from 10-30 species in a few millimetres of sediment at high latitudes to 50-70 species in tropical latitudes. Whether this pattern is representative of the benthos as a whole is unclear.

OCEAN TRENCHES

Physical evolution and properties

Ocean trenches are formed as a consequence of plate tectonic processes where sectors of expanding ocean floor pushes upon an unyielding continental mass or island arc, resulting in the crust buckling downwards (subducting) and being destroyed within the hot interior of the Earth. As oceanic crust ages and cools, it becomes denser and stiffer, resulting in a steeper angle of subduction and a deepening trench. Fig. 10.3 shows the locations of the principal known trenches, those occurring along the western edge of the Pacific being both the deepest, and geologically the oldest. Seismically, ocean trenches are highly active, as subduction is an erratic rather than a smooth process. This results in an unstable and unpredictable habitat compared to the relative environmental stability of the adjacent abyssal plains (Angel, 1982).

Being generally close to land masses, ocean trenches tend to have relatively high rates of sedimentation, a significant amount of which is of organic origin and an important available food source for trench communities. Several trenches also underlie highly productive cold water upwelling zones, the organic fallout from which contributes greatly to their richness. The water within trenches generally originates from the surrounding bottom water, which is derived from cold surface water at high polar latitudes and is relatively well oxygenated (Angel, 1982).

Endemism, diversity and biomass

Trenches tend to be isolated linear systems. This, combined with their high seismic activity, would suggest that faunas low in species diversity but relatively high in numbers of endemic species should be found. These would be expected to show strong affinities at generic and family levels to other trenches in the same system, having all originated from the same parental species inhabiting the surrounding

Table 10.3 Endemism among hadal species

GROUP	TOTAL NO. OF HADAL SPECIES >6000m	NO. OF SPP. EXCLUSIVELY AT DEPTHS >6000m	% ENDEMIC HADAL SPECIES
Cumacea	3	3	100.0
Harpacticoida	2	2	100.0
Ostracoda	2	2	100.0
Crinoidea	9	8	88.9
Gastropoda	16	14	87.5
Pogonophora	26	22	84.6
Amphipoda	17	14	82.4
Tanaidacea	19	15	78.9
Isopoda	49	37	75.5
Porifera	12	9	75.0
Coelenterata	12	9	75.0
Pisces	4	3	75.0
Bivalvia	26	17	65.4
Holothurioidea	22	14	63.6
Echiurida	8	5	62.5
Ophiuroidea	5	3	60.0
Asteroidea	12	6	50.0
Others	5	2	40.0
Polychaeta	32	12	37.5
Cirripedia	3	1	33.3
Pycnogonida	3	1	33.3
Foraminifera	126	35	27.8
Sipunculida	4	0	0.0
Total spp.	417	234	56.1
Excl. Foraminifera	291	199	68.4

Source: Wolff, T. 1970. The concept of the hadal or ultra-abyssal fauna. *Deep-Sea Research* 17:983-1003.

Figure 10.1 Species and family diversity in sea-bottom samples

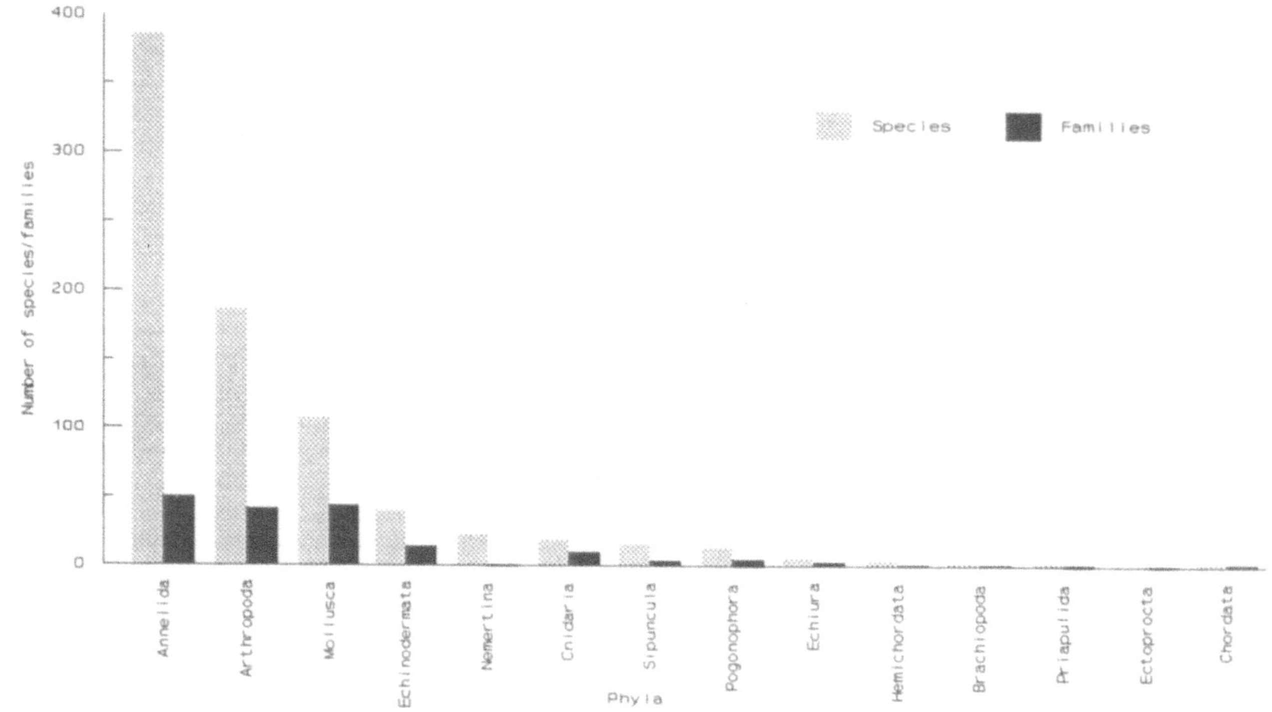

Source: Grassle, J.F. 1991. Deep-sea benthic biodiversity. *Bioscience* 41(7).
Note: Samples taken at 1,500-2,500m depth off New Jersey, USA.

Figure 10.2 Composition of benthic macrofauna

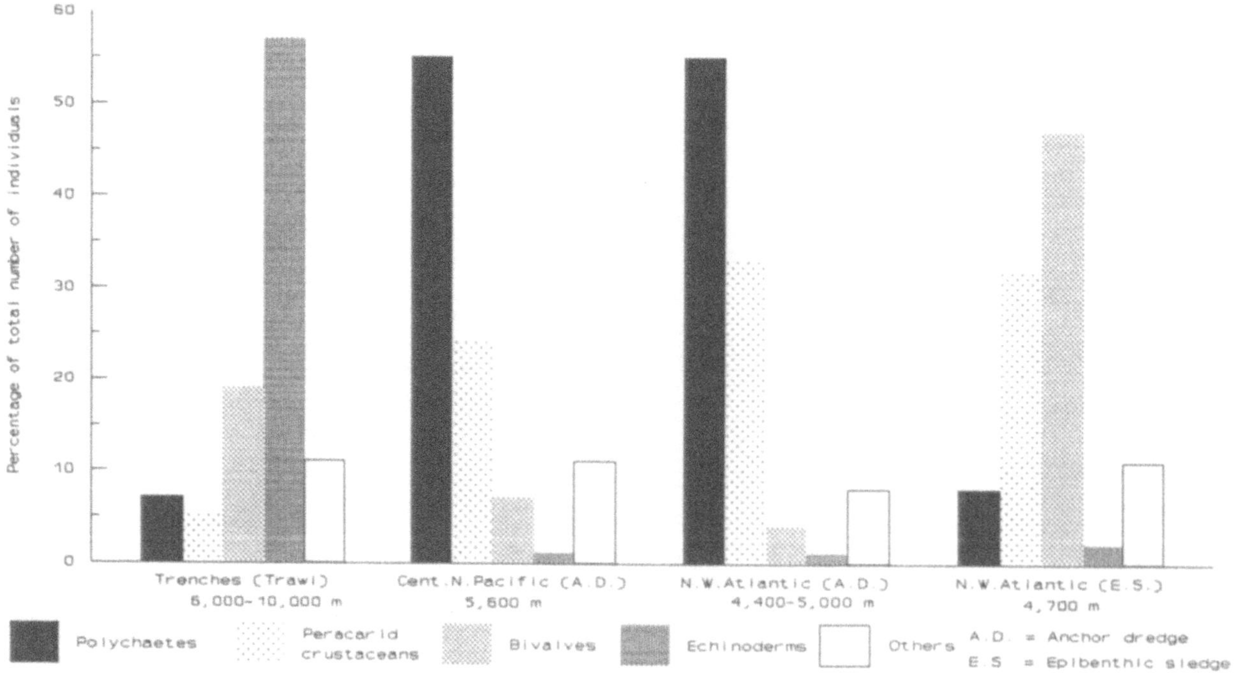

Source: Wolff, T. 1977. Diversity and faunal composition of the deep-sea benthos. *Nature* 267:780-785.

Figure 10.3 Distribution of the main ocean trenches

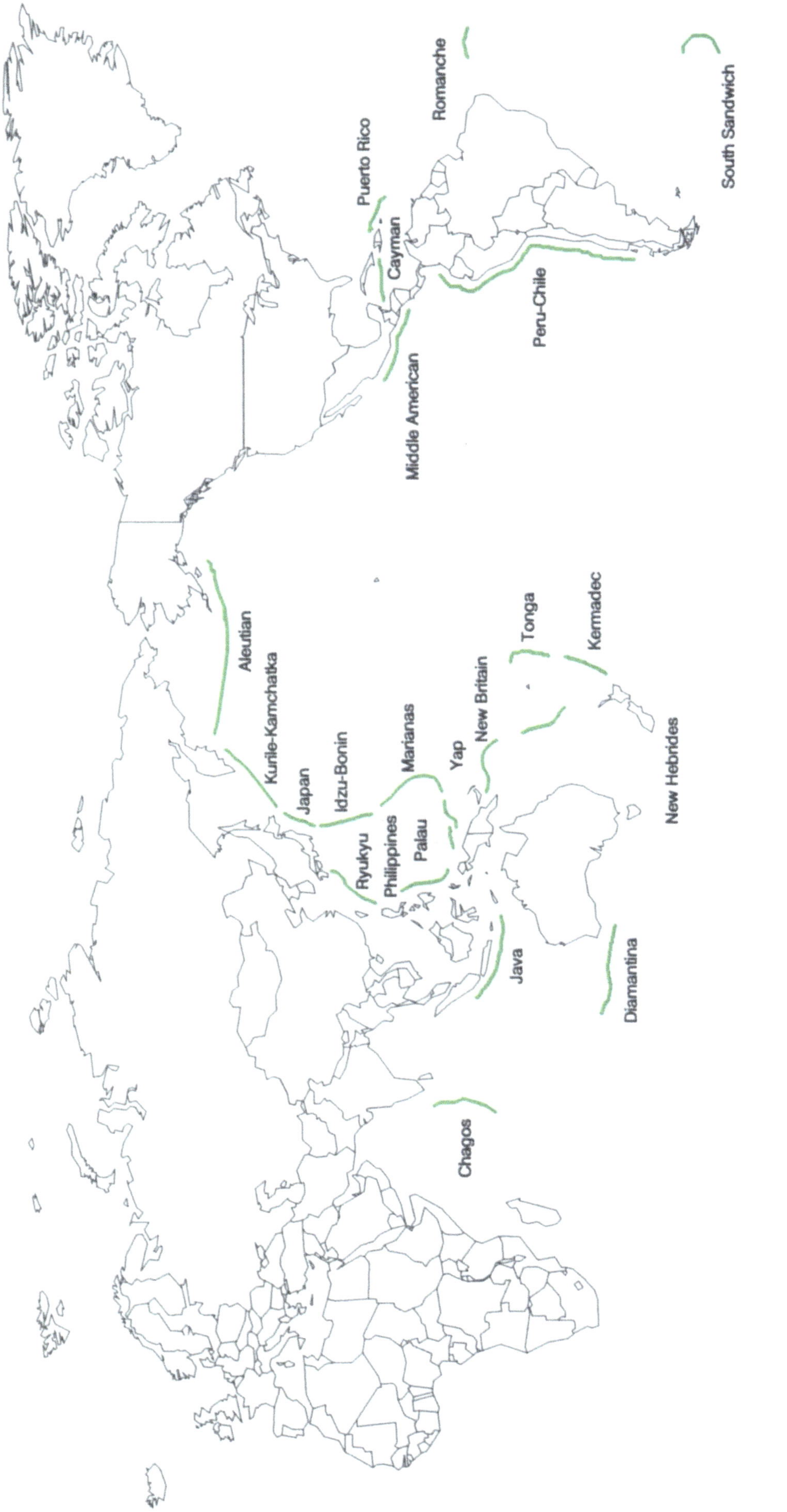

(modified from Angel, 1982)

abyssal plains (Angel, 1982). In the few trenches studied, these hypotheses would appear to be true; however, as with other aspects of deep-sea diversity, generalisations remain tentative.

Angel (1982) quotes Professor G.M. Belyaev (from which the following paragraph of information is taken). Between 50% and 90% of the fauna of each ocean trench is endemic, compared to the overall endemism of hadal, or ultra-abyssal faunas, which is in the region of 57-60% (for example, see data given by Wolff, 1970, Table 10.3). There are some 25 known endemic hadal genera, representing some 10-25% of the total number of genera occurring in the hadal zone, and two known endemic hadal families; the Galatheanthemidae (Actinaria) and Gigantapseudidae (Crustacea). The latter family contains a single species: *Gigantapseudes adactylus*. The greatest number of endemic species known from a single trench is a sample of 200 from the Kurile-Kamchatka Trench; this may be compared with 10 endemic species known from the Ryukyu and Marianas Trenches. The Banda Trench has the lowest recorded proportion of species endemism (33%), and is probably the youngest trench geologically. In total, representatives of 33 classes, 150 families and about 240 genera are known from hadal depths.

As noted, high seismic activity may tend to produce low species diversity. Rapid sedimentation may have a similar effect. For example, the Aleutian Trench and the Japan Trench have relatively low macrofaunal diversities, attributable to frequent catastrophic slumping of canyon wall sediment (Grassle, 1989).

In general, comparative data are sparse because of the variety of collection techniques employed. The composition of trench faunas is unusual (compared to abyssal faunas) in that they tend to be dominated by deposit-feeders (Angel, 1982) and show a higher percentage of species of amphipods, polychaetes, bivalves, echiurids and holothurians, and a lower percentage of sea stars, echinoids, sipunculids and brittle-stars, and especially non-actinian and scyphozoan coelenterates, bryozoans, cumaceans and fishes, than in the surrounding abyss. Decapod crustaceans are completely absent (Gage and Tyler, 1991).

Trenches appear to have a higher biomass than adjacent shallower areas, although within the trenches themselves the stocks of macrofauna decrease with depth at a rate similar to the general declining pattern. The higher biomass in trenches is probably a reflection of the net accumulation of sediment from the adjacent shallow continental margins (Rowe, 1983), as the amounts of available nutrients have a profound effect on trench faunas; $8.8g/m^2$ of living organisms have been assessed from the nutrient-rich South Sandwich Trench and $3.44g/m^2$ from the Kurile-Kamchatka Trench, compared to $0.008g/m^2$ from the nutrient-poor Marianas and Tonga Trenches (Angel, 1982).

HYDROTHERMAL VENTS

Hydrothermal vent communities were first discovered in 1977, at a depth of 2,500m on the Galápagos Rift. They are now known to be associated with almost all known areas of tectonic activity at various depths (see Fig. 10.4). These include: along the East Pacific Rise off Mexico, in the Guaymas Basin in the Gulf of California, on the Juan de Fuca Ridge off Washington State, in subduction areas off Oregon and Japan, on the Mid-Atlantic Ridge at 26°N, in the Mariana Trough near the Mariana Trench, and in the Lau and North Fiji Basins to the west and east of Fiji (Gage and Tyler, 1991). These tectonic regions include ocean-floor spreading centres, subduction and fracture zones, and back-arc basins (Gage and Tyler, 1991). Cold bottom-water permeates through fissures in the ocean floor close to ocean-floor spreading centres, becomes heated at great depths in the Earth's crust and finds its way back to the surface through hydrothermal vents. The temperature of vent water varies greatly, from around 23°C in the Galápagos vents, to around 350°C in the vents of the East Pacific Rise, and they may be rich in metalliferous brines and sulphide ions (Angel, 1982). Although the vent water may be at a high temperature, the majority of species live out of the main flow at temperatures of around 2°C, the ambient temperature of deep-sea water.

Although vent communities are often separated from one another by gaps of a kilometre or so, they can be up to 100km apart. They have yet to be found in certain areas of known hydrothermal activity, such as the Red Sea (Grassle, 1986). Hydrothermal vents and their associated communities are relatively short-lived at any particular site, probably only being active for between several years and several decades. This has been suggested by discoveries of 'dead' vents (visible from the remains of white shells which dissolve away completely in about 15 years) and by growth measurements of individual organisms (indicating very rapid growth to maturity at a large size) (Gage and Tyler, 1991). However, active hydrothermal centres appear to move relatively slowly, thus allowing dispersal of vent organisms.

Areas of tectonic activity are connected over most of the earth's surface, and although this network is in a dynamic state, new areas are linked to old and so vent communities could be part of a unique ecosystem at least 200 million years old (Grassle, 1985). Studies on variation in vent species, comparing those in the main network and those isolated in remote parts of the system, provide important opportunities for evolutionary and genetic studies. Vent species are also of interest in that they flourish in the dark at high pressures and low temperatures (Grassle, 1986), which previously had been thought to inhibit productivity.

Hydrothermal vent communities are unique in that they are supported by a non-photosynthetic source of organic carbon, i.e. chemosynthetic primary production. The enriched hydrothermal fluid supports large numbers of bacteria (predominantly *Thiomicrospira* species) which form dense bacterial 'mats', and are capable of deriving energy from reduced compounds such as hydrogen sulphide (Grassle, 1986, Gage and Tyler, 1991). Many of the vent species filter-feed on these bacteria, whilst others rely on symbiotic sulphur bacteria for energy (Angel, 1982).

Endemism, diversity and biomass

The overall species diversity at vents is low compared with other deep-sea soft-sediment areas (Grassle, 1986), but endemism is high. More than 20 new families or sub-

Figure 10.4 Hydrothermal vent and cold seep communities

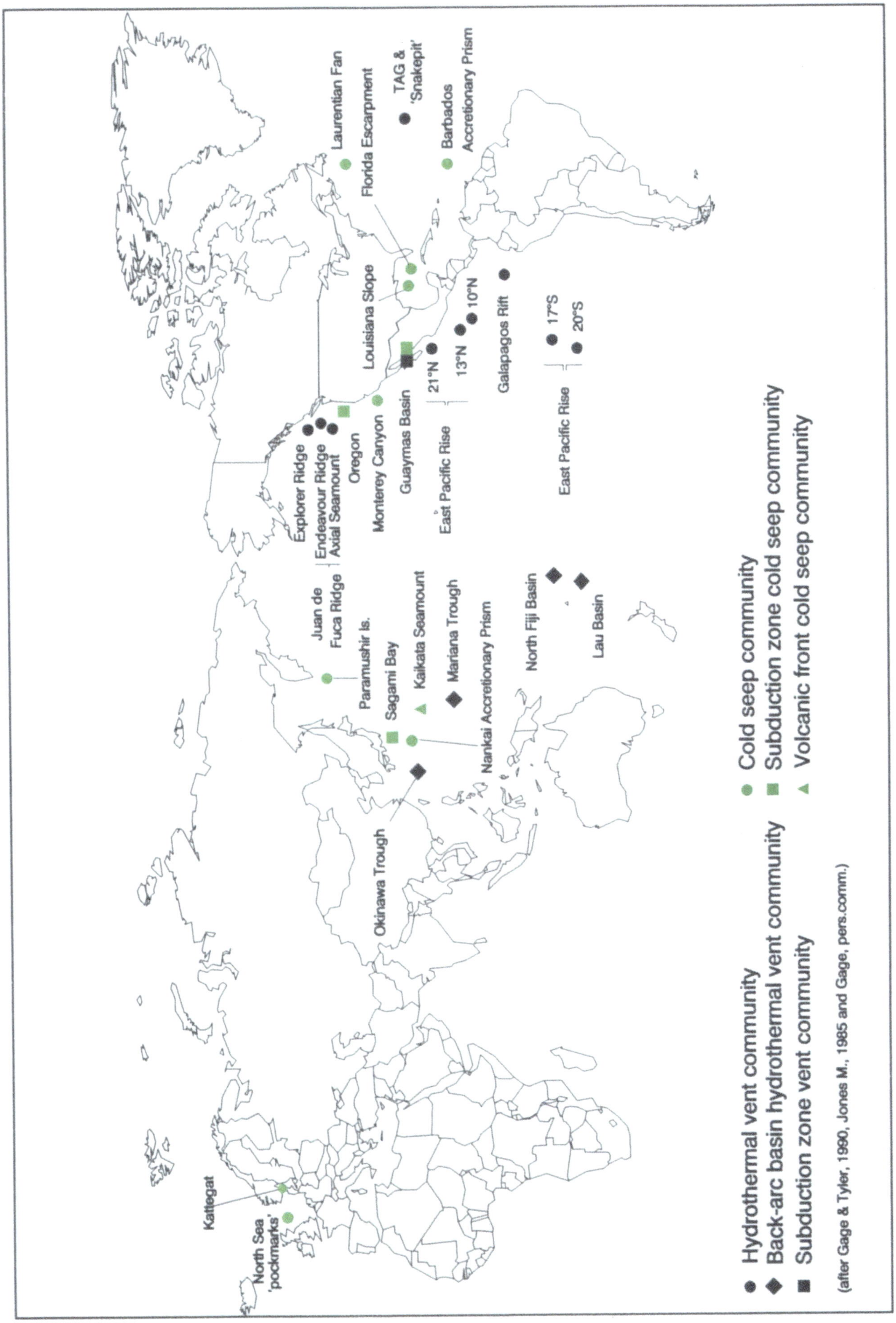

families, 50 new genera and nearly 160 new species have been recorded from vent environments, including brine and cold seep communities (discussed below) (Grassle, 1989; Gage and Tyler, 1991). Examples of these new taxa are given in Table 10.4.

In biogeographic terms, vents can be regarded as ephemeral, biogeographic islands. With increasing spatial separation the species composition can vary considerably, with some species being replaced by closely related forms. Differences in subsurface flux of hydrothermal fluids and in vent configuration can result in large differences in faunal composition over short distances within or between vent fields. Geochemical differences between vent communities may also result in faunal dissimilarities, between the Galápagos and East Pacific Rise vents, for example (Grassle, 1986). However, the major features of the fauna at each vent site are consistent, whilst none of the species seems to be ubiquitous. The larvae of many vent species appear to have relatively poor dispersal abilities (Grassle, 1986), and this could contribute to maintenance of high endemism.

The biomass of vent communities is usually high compared to other areas of similar depth, and varies according to water temperatures and chemistry, reaching 8.5kg wet weight per m^2 at lower temperature vents, and averaging 2-4kg wet weight per m^2 at the hottest vents (200-360°C) (Gage and Tyler, 1991). Dense colonies of tube-worms, clams, mussels and limpets typically constitute the major proportions of biomass. Swarms of the probably vent-specific copepod species *Isaacsicalanus paucisetus* reached densities of 920 individuals m^{-2} and a dry weight biomass of 133 mg^{-2} at one site. Microbial production at low-temperature vents (10°C) is thought to be two or three times that of photosynthetic production at the surface in the same region (Gage and Tyler, 1991).

Features of some major vent regions are noted below.

Galápagos Spreading Centre

This consists of 12 known active populated vents and three 'dead' vents along a 30km section of ridge-crest. The two large bivalves *Calyptogena magnifica* and *Bathymodiolus thermophilus*, and vestimentiferan worms (especially the tube-dwelling *Riftia pachyptila*) are the most distinctive species of these hydrothermal vents (Grassle, 1986).

Eastern Pacific Rise

These hydrothermal vents support a similar fauna to the Galápagos Spreading Centre, including the same two bivalve species (which can occur in enormous densities - the biomass of *B. thermophilus* may exceed 10kg/m^2), and *Riftia pachyptila*. More than 30 species of limpet-like gastropod have been recorded (mostly as yet undescribed), and mussels, shrimp, anemone and limpet species (Gage and Tyler, 1991). The spreading rate of 11-12cm/year is greater than that of the Galápagos spreading centre (Grassle, 1986).

Mid-Atlantic Ridge

The active hydrothermal vents discovered on this ridge are characterised by the presence of two species of caridean shrimp belonging to the new family Bresiliidae. These occur in great numbers, along with mats of bacteria. Compared to the eastern Pacific, the vent faunas are less varied; bivalve mussels appear to be uncommon, and tubeworms absent (Gage and Tyler, 1991).

Mariana Trough

This back-arc spreading centre borders the subduction zone of the Mariana Trench. It is isolated from the main mid-ocean ridge system. The vent-fauna is very different from those of the eastern Pacific, and is dominated by a sessile barnacle (the most primitive living barnacle species known), limpets and anemones. The giant bivalves of the eastern Pacific are replaced by a large, hairy-shelled gastropod (Gage and Tyler, 1991).

Shallow-water hydrothermal vents

Vents at depths of less than 20m have been described off the Palos Verdes Peninsula, California. They support a diverse assemblage of colourless chemosynthetic bacteria similar to those of deep-sea vent sites, which form mats around the vent openings. The mats provide nourishment for the mollusc *Haliotis cracherodii* (Kleinschmidt and Tschauder, 1985), commonly known as black abalone.

COLD SEEPS

Cold sulphide and methane-enriched groundwater seeps occur near the base of the porous limestone of the Florida Escarpment, as well as in the Gulf of Mexico (Fig. 10.4). The seeps support a dense faunal community associated with a covering or mat of bacteria on the sediment surface. These communities are strikingly similar in taxonomic composition to the hydrothermal vents of the east Pacific, a fact which points to a common origin and evolutionary history for both community types (Hecker, 1985). The community consists of large mussels and the vestimentiferan worm *Escarpia laminata*, as well as galatheid crabs, serpulid worms, anemones, soft corals, brittle stars, gastropods and shrimps. Mussel densities appear to be linked to methane levels in the water, whilst tubeworm density may be correlated with the hydrocarbon loading of the sediment (Gage and Tyler, 1991).

Tectonic subduction zone seeps

Subduction seeps are more diffuse and lower in temperature than hydrothermal vent seeps, and are rich in dissolved methane. They are known to occur off Oregon, where the fauna includes species of *Lamellibrachia* and large vesicomyid bivalves, and in the Guaymas Basin in the Gulf of California, where thick bacterial mats cover the sulphide and hydrocarbon-coated sediment. The cold Japanese subduction zone seeps occur at a depth of 1,000m in Sagami Bay near Tokyo and in the subduction zones of the trenches off the east coast of Japan. The communities vary, but include dense benthic assemblages dominated by *Calyptogena* clams associated with a stone crab *Paralomis* sp., sepulid worms, sea anemones, galatheid crabs, swimming holothurians and amphipods (Gage and Tyler, 1991).

Other colonised deep-sea seepage sites include a cold seep to the east of Barbados dominated by the mussel *Bathymodiolus*, vesicomyid bivalves and vestimentiferan

Table 10.4 Examples of new taxa found at vents and seeps

PHYLUM	CLASS	ORDER	FAMILY	GENUS	SPECIES	WHERE FOUND	NOTES
(Methanogenic Bacteria)				*Methanococcus*	*jannaschii**	21°N	First known thermophilic methanogen. Prolific growth rate at 85°C (optimum).
(Chemautotrophic Bacteria)				*Thiomicrospira*	*crunogena**		
Cnidaria	Anthozoa			?	?		Several undescribed anemone species often common near vents.
	Hydrozoa	Siphonophora	Rhodaliidae	*Thermopalia**	*taraxaca*	Galapagos Rift, 13°N	Galapagos 'dandelion'. Similar animals seen at other vents, 21°N and Juan de Fuca Ridge.
Mollusca	Bivalvia		Vesicomyidae	*Calyptogena*	*magnifica**	Fairly widespread	One of the characteristic vent species with symbiotic bacteria in gills.
	Gastropoda	Archeogastropoda	Trochidae	*Melanodrymia*	*aurantiaca**	21°N	
			Neomphalidae*	*Neomphalus*	*fretterae*		These five limpet families have been found only at hydrothermal vents. Some species seem to be endemic to a single vent, some are more wide–spread. Relict species from the Mesozoic and late Paleozoic.
			4(+) undescribed	?	c. 21 undescribed	Various locations	
Vestimentifera (Pogonophora)			Lamellibrachiidae*	*Lamellibrachia*	*barhami*	Seep sites off Oregon and California	
				Lamellibrachia	*luymesi*	Off Guyana coast	Known from single specimen only.
			Riftiidae*	*Riftia*	*pachyptila*	Galapagos Rift and East Pacific Rise off Mexico	
			Ridgeiidae*	*Ridgeia*	2 species	Juan de Fuca Ridge	
			Tevniidae*	*Tevnia*	*jerichonana*	13°N	Known as 'Jericho worms' Other species probably also in this genus.
				Oasisia	*alvinae*	21°N	
			Escarpiidae*	*Escarpia*	2 species	Florida escarpment (1 sp.) and California Borderlands (1 sp.)	
Annelida	Polychaeta		Ampharetidae	*Alvinella**	*pompejana*	13°N & 21°N	Known as 'Pompeii worms'. Live near hottest water.
				*Paralvinella**	*grasslei*	Fairly widespread	Hot water.
				Amphisamytha	*galapagensis**	All five Pacific Ocean vent areas studied so far.	
			Polynoidae	*Lepidonotopodium*	2 undescribed	Galapagos & East Pacific Rise	
	Hirudinea		?	*Bathybdella**	*sawyeri*	Galapagos vents	Probably parasitic on fish.
Arthropoda	Copepoda		Spinocalanidae	*Isaacsicalanus**	*paucisetus*	21°N	Low temperature vents.
			Dirivultidae*	*Dirivultus*	*dentareus*	Southern California	Associated with vestimentiferans.
	Cirripedia		Scalpellidae	*Neolepas**	*zevinae*		Only known survivor of Mesozoic family.
	Malacostraca	Leptostraca	?	*Dahlella**	*caldariensis*	13°N & 21°N, Galapagos	
		Isopoda	Epicaridea[1]	*Thermaloniscus**	*cotylophorus*	13°N	Described from single larva.
		Decapoda	Bythograeoidea*[2]	*Bythograea*	*thermydron*	Quite widespread	Decapods are the most obvious crustaceans present. These are the most unusual examples.
				Bythograea	*microps*	Quite widespread	
				Cyanograea	*praedator*	13°N & 21°N	Preys upon *Alvinella pompejana*.
			Bresiliidae	*Alvinocaris**	*lusca*	Galapagos vents	
			Galatheidae	*Munidopsis*	*lentigo**	21°N	
	Arachnida	Acarina	Halicaridae	*Copidognathus*	*papillatus**	Galapagos vents	Predatory mite associated with mussel detritus.
Hemichordata	Enteropneusta		Saxipendiidae*	*Saxipendium*	*coronatum*	Galapagos vents	Known as 'Spaghetti worms'. The posterior end drapes over rocks, anterior drifts free.
Chordata	Pisces		Zoarcidae	*Thermaces**	*cerberus*	Eastern Pacific vents	An eelpout. Probably endemic.
				*Thermaces**	*andersoni*	Eastern Pacific vents	An eelpout. Probably endemic.

Source: compiled from information given in Grassle, 1986
Notes: * indicates the highest taxonomic level of newly–described taxa from this environment in each case. [1] suborder. [2] superfamily.

worms. Dense communities, probably of recent origin, have also been discovered on the Laurentian Fan on the south-east Canadian continental margin. These include vesicomyid and thyasirid bivalves, gastropods, pogonophorans, galatheid crabs and bacterial mats. There is evidence that, like hydrothermal vents, cold seeps are ephemeral and cyclic. However, many species are now known to occur in the deep sea in a variety of similar sulphur and other compound-reducing habitats, and have been found to occur around such temporary habitats as oil-rich whale carcasses, which may be important 'stepping-stones' for chemosynthetic-dependent deep-sea animals (Gage and Tyler, 1991).

References

Angel, M.V. 1982. Ocean trench conservation. *Commission on Ecology Papers* No. 1. IUCN.

Angel, M.V. 1991. Biodiversity in the deep ocean. A working document for ODA. Unpublished MS.

Buzas, M.A. and Culver, S.J. 1991. Species diversity and dispersal of benthic foraminifera. *Bioscience* 41(7):483-489.

Gage, J.G. and Tyler, P.A. 1991. *Deep-sea Biology: A natural history of organisms at the deep-sea floor.* Cambridge University Press.

Grassle, J.F. 1985. Hydrothermal vent animals: distribution and biology. *Science* Vol.229.

Grassle, J.F. 1986. The ecology of deep-sea hydrothermal vent communities. *Advances in Marine Biology* Vol. 23. Academic Press.

Grassle, J.F. 1989. Species diversity in deep-sea communities. *TREE* 4(1).

Grassle, J.F. 1991. Deep-sea benthic biodiversity. *Bioscience* 41(7).

Hecker, B. 1985. Fauna from a cold sulphur-seep in the Gulf of Mexico: comparison with hydrothermal vent communities and evolutionary implications. *Biological Society of Washington Bulletin* 6:465-473.

Hessler, R.R. and Sanders, H.L. 1967. Faunal diversity in the deep-sea. *Deep-Sea Research* 14:65-78.

Kleinschmidt, M. and Tschauder, R. 1985. Shallow-water hydrothermal vent systems off the Palos Verdes Peninsula, Los Angeles County, California. *Biological Society of Washington Bulletin* No.6.

Rex, M.A. 1983. Geographic patterns of species diversity in the deep-sea benthos. In: Rowe, G.T. (Ed.), *Deep-Sea Biology*. Volume 8, *The Sea*. John Wiley and Sons, New York. Pp.453-472.

Rowe, G.T. 1983. Biomass and production of the deep-sea macrobenthos. In: Rowe, G.T. (Ed.), *Deep-Sea Biology*. Volume 8, *The Sea*. John Wiley and Sons, New York. Pp.453-472.

Thorne-Miller, B. and Catena, J. 1991. *The Living Ocean: understanding and protecting marine biodiversity*. The Oceanic Society of Friends of the Earth, Washington DC.

Wolff, T. 1970. The concept of the hadal or ultra-abyssal fauna. *Deep-Sea Research* 17:983-1003.

Wolff, T. 1977. Diversity and faunal composition of the deep-sea benthos. *Nature* 267:780-785.

11. SOIL MACROFAUNA

Compared to conspicuously diverse habitats such as tropical rain forest or coral reefs, the soil as a habitat in its own right, with its own rich fauna and flora, is often overlooked. It supports, however, a wide array of diverse animals, with representatives from every major phylum in the animal kingdom except the coelenterates and the echinoderms (Wallwork, 1976).

SOIL AND SOIL FAUNA

The soil habitat is not a uniform environment. Examination of a vertical section of the profile of a mature soil will often reveal several layers reflecting its past history and development. This sequence, from the organic litter layer on the soil surface to the parent material below, can be divided into four main horizons (Eisenbeis and Wichard, 1985):

O-horizon: organic upper layer of plant debris lying on the surface of the mineral soil

A-horizon: upper, fine mineral soil permeated by organic material

B-horizon: weathered, rough mineral soil coloured by small deposits of humus

C-horizon: original, unweathered material.

The organic layer (O-horizon) can often be further subdivided into three sub-layers: the leaf litter layer, the fermentation layer and the humus layer, in a downward succession (Wallwork, 1976). The actual depth that the O-horizon attains is dependent on the rate of input from the covering vegetation and the rate of decomposition.

The term 'soil fauna' can be used to encompass a large number of animal species, including any which spend a proportion of their life cycle in the soil, on the soil surface, or in the leaf litter. The soil fauna contains numerous life forms adapted to a great variety of microhabitats. In an attempt to clarify the distribution of organisms within the soil, Kevan (1962) proposed three categories, in terms of their respective adaptations to life in the soil:

Euedaphon: inhabitants of the mineral soil, e.g. most earthworms, all Symphyla, many mites
Hemiedaphon: inhabitants of the litter and fermentation layer, such as many woodlice and millipedes
Epedaphon: inhabitants of the soil surface, such as most ground-beetles and scorpions.

These categories are widely used, although some later authors (e.g. Eisenbeis and Wichard, 1985) have modified the definitions. Any given taxonomic group may include species in more than one of the above categories, as well as species which are not considered soil fauna.

Table 11.1 shows all taxonomic groups to be considered in this context, with the term soil fauna being defined as narrowly as practicable. Taxonomic level varies from phylum to family. Taxonomic sequence follows Barnes (1984).

Patterns of soil-fauna research

Although research is being conducted on the key soil groups, much of it is limited to individual genera or species rather than whole orders; and it covers only a few of the major habitats. Thus, it is very difficult to build up a picture of the total fauna of a region using literature primarily on soil research.

The taxonomic precision of the primary literature also varies through time. The 1940s and 1950s saw a peak of species-level identifications by ecologists. Since the 1960s, ecological and taxonomic interests have developed, so few soil ecologists now provide species lists in their papers. Data are now more often presented at order level, with emphasis on biomass and productivity rather than on species assemblages. Very recently, there have been moves to revive taxonomic competence among ecologists (Erzinclioglu, 1989; Dempster, 1991).

Other types of literature, such as general guides to animal groups, identification keys, and taxonomic monographs, provide useful information, but many are dated, thus reducing the accuracy of their assessment of species totals for a region. Many are also only the result of brief collecting expeditions and so can only be considered preliminary markers of the possible species richness.

As our knowledge of the soil fauna and habitat expands, so our appreciation of its faunal diversity increases, sometimes ten-fold. The estimated world total of Pseudoscorpiones recently rose from 1,300 (Levi *et al.*, 1968) to 3,000 (Davies *et al.*, 1985), and of Collembola from 1,500-2,000 (Wallace and Mackeras, 1970) to 10,000-20,000 (Greenslade and Greenslade, 1983).

This is in line with the trend shown by invertebrate diversity estimates in general. The degree to which current figures for soil biodiversity may be relied upon is geographically patchy: some areas have comprehensive and up-to-date lists for most groups, and these have been relatively stable for several decades despite an increasing pace of ecological and biogeographic research (e.g. Britain); a few others, such as Australia, are attempting to produce comprehensive overviews; but in most countries, the literature is becoming narrower and less easily used.

Ecological functioning and importance of soil fauna

The soil is basic to most terrestrial ecosystems, and the health and functioning of the soil relies heavily on the activities of soil fauna. The initial formation of soil, for instance, at the end of a glaciation, depends greatly on detritivores to help in cycling of nutrients and humus formation. The accumulation of the latter is responsible for the development of the soil through time. The role of soil invertebrates in these pioneer phases must be considerable: several groups of invertebrates are known, from the fossil record, to have colonised newly exposed areas well in advance of the vascular flora (Buckland and Coope, 1991).

The soil fauna is also a major vector of microorganism and

Table 11.1 Taxonomic distribution of soil macrofauna

	PROPORTION OF GROUP WHICH ARE TERRESTRIAL	PROPORTION OF TERRESTRIAL SPP. LIVING IN THE SOIL	EXTENT TO WHICH SOIL SPP. UTILISE THE SOIL
Platyhelminthes:			
Tricladida	●	●●●●	●●●●
Nemertea	●	●●●●	●●●●
Nematoda	●●●	●●●	●●●●
Annelida:			
Oligochaeta*	●●●	●●●	●●●
Mollusca:			
Gastropoda	●●	●●	●/●●●
Crustacea:			
Isopoda*		●●	●●●●●●
Amphipoda	●	●●●●	●●●●
Decapoda	●	●●	●
Chelicerata: Arachnida:			
Scorpiones*	●●●●	●●●●	●●●
Pseudoscorpiones*	●●●●	●●●	●●●
Uropygi	●●●●	●●●●	●●●
Amblypygi	●●●●	●●●●	●●●
Palpigradi	●●●●	●●●●	●●●●
Ricinulei	●●●●	●●●●	●●●●
Solifugae	●●●●	●●●●	●●
Opiliones*	●●●●	●●●	●●/●●●
Araneae	●●●	●●	●
Acari*:			
Mesostigmata	●●●●	●●●	●●●
Prostigmata	●●	●●	●●●
Astigmata	●●●●	●	●●●
Cryptostigmata	●●●●	●●●	●●●
Onychophora	●●●●	●●●	●
Uniramia:			
Diplopoda*	●●●●	●●●	●●●
Pauropoda	●●●●	●●●●	●●●●
Chilopoda*	●●●●	●●●	●●●
Symphyla	●●●●	●●●●	●●●●
Diplura	●●●●	●●●●	●●●●
Collembola*	●●●	●●●	●●●
Protura	●●●●	●●●●	●●●●
Thysanura	●●●●	●●	●●●
Embioptera*	●●●●	●●●●	●●●
Orthoptera*:			
Gryllotalpidae	●●●●	●●●●	●●●
Tridactylidae	●●●●	●●●●	●●●
Cylindrachetidae	●●●●	●●●●	●●●
Tetrigidae	●●●●	●●●●	●●●
Dermaptera*	●●●●	●●●	●●/●●●
Isoptera*	●●●●	●●●	●●●●
Blattaria*	●●●●	●●●	●●
Psocoptera	●●●●	●●	●
Thysanoptera	●●●●	●●	●●
Homoptera	●●●	●●	●●●
Coleoptera:			
Carabidae*	●●●	●●●	●●/●●●
Staphylinidae*	●●●	●●●	●●/●●●
Tenebrionidae	●●●●	●●●	●●●
Scarabaeoidea	●●●●	●●●	●●
Elateroidea	●●●●	●●	●●
Cantharaoidea	●●●●	●●	●●
Hymenoptera:			
Formicidae*	●●●●	●●●	●/●●/●●●
Megaloptera	●	●●	●●
Diptera	●●	●●	●●

Notes: * indicates taxa considered key soil groups for which adequate biogeographic and taxonomic information has been located and which are therefore considered in detail in this review. Nematodes are discussed in Chapter 9. Columns 2 and 3 are coded as follows: ●●●● all species; ●●● most species; ●● some species; ● few species. Column 4 is coded as follows: ●●●● obligate soil-dwellers; ●●● usually soil-dwelling, but may at times climb vegetation etc.; ●● temporarily present, normally for a particular part of their life cycle (e.g. Diptera larvae); ● regular users of the soil (i.e. for foraging) but able to spend much or all of their life in other microhabitats.

cryptogam propagules (Gerson and Seaward, 1977; McCarthy and Healy, 1978). Many soil groups include decomposers which are important in the breakdown and recycling of nutrients throughout mature ecosystems. Most temperate soils differ from soils at lower latitudes in having a greater 'standing crop' of plant detritus, owing to the lower decomposition rates, and tend to be deeper and with a more elaborate profile, partly due to seasonality of precipitation and the effects of frosts.

All of these processes, where soil invertebrates function as pioneers and as facilitators of cycling in later seral stages, are important in the rehabilitation of damaged ecosystems where, for example, the vegetation and/or top-soil has been lost. Earthworms have been shown to aid the development of vegetation in derelict industrial sites; monitoring of their population levels can therefore be used as an indicator of the recovery of the habitat (Davis, 1986).

In most terrestrial habitats, the soil fauna is also important for niche creation: that is, the activities of some groups provide niches for other soil animals. Most importantly, the actions of earthworms are largely responsible for the structure of many soils, and their burrows allow access to deeper parts of the soil not normally penetrable by other groups; they thus provide retreats in the face of predation or desiccation.

The soil macrofauna, in synergy with microorganisms, also acts as a major link between the soil and non-soil habitats. The role of these groups, and in catalysing the processes of nutrient cycles, releasing minerals for uptake by vascular plants, as well as providing a physical soil structure which strongly influences the development of plant communities, is crucial to the final appearance of the vegetation.

Many non-soil animals such as birds and mammals feed on soil fauna regularly. For some groups, such as shrews (Soricidae), hedgehogs (Erinaceidae), some wading birds (Charadriiformes) and many reptiles and amphibians, soil fauna may make up the bulk of their diet, at least during part of the year.

PATTERNS OF SPECIES RICHNESS

This section contains a systematic account of the biogeographic patterns of the key soil groups prefaced by notes on their biology. The higher level classification of groups is not necessarily the same as followed elsewhere in this volume. The reference list for data cited in the tables below is available on request from WCMC.

Phylum Annelida

Sub-class Oligochaeta

The Oligochaeta is divided into two main soil-dwelling groups, the earthworms (Lumbricina) and the potworms (family Enchytraeidae). Oligochaetes are soft-bodied segmented worms adapted to burrowing in the soil; they include the only truly terrestrial annelids and are ecologically a very important group.

Oligochaeta: Lumbricina

Earthworms are largely absent in highly acidic soils, such as peatlands and heathlands: few species can tolerate a pH lower than 4.0 (Wallwork, 1976). In base-rich soils, however, they often constitute a high proportion of the total animal biomass. Members of the family Lumbricidae dominate the fauna in north temperate regions, and range in size from less than 1cm to 35cm. Several other families occur in warm temperate and tropical countries, the best known being the Megascolecidae, some species of which can exceed 3m.

Populations of 2.4-7.2 million earthworms per hectare have been reported from rich permanent grassland habitats in Britain (Cloudsley-Thompson and Sankey, 1968). They are of considerable importance in soil processes (Sims and Gerard, 1985). In addition to the benefits in agricultural soils, earthworms are fundamental to the production of the soil structure within which many other soil invertebrates can live (Lavelle, 1983). Several important groups of soil fauna are able to penetrate deep into the soil, and thereby survive during dry weather, solely because of the network of fine passages created by earthworms.

Oligochaeta: Enchytraeidae

The potworms comprise the terrestrial members of a family many of whose members are freshwater or marine. They are rarely more than 25mm in length and can tolerate acid conditions much better than lumbricids; large populations, of the order of thousands per m^2, may be found in more acid soils of oak woodlands and moorland peats (Wallwork, 1976). Few other groups are as successful in colonising the rather sterile and water-logged soils of bogs; only nematodes and Diplura (Eversham, unpublished) thrive equally well in these conditions.

Areas which were covered by ice during the last glaciation or which supported only tundra vegetation for several millennia lost almost all their earthworms; post-glacial recolonisation appears to have been restricted to a few highly mobile, eurytopic species. Thus, Britain and north-west Europe support a small fauna of only 10-30 species and the earthworms of natural habitats are only slightly more diverse than those of improved agricultural soils. Countries bordering the Mediterranean have many more species, with particular concentrations in the Iberian Peninsula and Italy; even the French fauna reaches 97 species (Lavelle, 1983). In such areas, agricultural improvement displaces the more stenotopic species. A similar pattern is found in North America. Of nearly 400 lumbricid species recognised, only 5% occur in the northern areas which were overlain by ice sheets. A whole superfamily, the Crilodriloidea, is now confined to a small area of the southern USA (Sims and Gerard, 1985).

There is limited evidence that the centres of species richness in the southern hemisphere are now being threatened by the introduction of north temperate species, which are associated with agricultural soils but may be able to out-compete the indigenous fauna (Ljungström, 1972).

Other aspects of family level distribution throw light on much more ancient geomorphological history. A consequence of the two effects (relict Gondwana distributions and Pleistocene glacial defaunation) is seen, for instance, in the much richer earthworm faunas of southern hemisphere islands like New Zealand (192 species; Lee, 1959) compared with that of a similar-sized landmass such as Britain (28 species), which should for reasons of island biogeographic theory be expected to acquire species much more readily from its nearby continent. The impoverished fauna of Iceland (8 species; Lavelle, 1983) is an even more extreme example of this; it is likely that the whole of this fauna is recently introduced by man. The position of Japan in relation to the Eurasian landmass is reflected in its relatively rich fauna (75 species; Easton, 1981).

Table 11.2 Soil species: Oligochaeta

TOTAL SPECIES	1,200
Lumbricina	
New Zealand	192
France	97
Japan	75
Oregon (USA)	267
UK	25
Little Carpathians (East Europe)	22
Denmark	19
Sweden	13
Washington (USA)	137
Iceland	8
Enchytraeidae	
North America	143
Europe	111
Little Carpathians (East Europe)	24

Phylum Crustacea

Almost all the terrestrial Crustacea belong to the order Isopoda, the familiar woodlice, slaters or sowbugs. A very few members of the mainly aquatic Amphipoda and land crabs have also adopted a terrestrial existence.

Class Malacostraca: Isopoda

Some genera rarely venture up to the soil surface, whereas others spend most of their existence among leaf litter and grass roots, and a few forage regularly among herbaceous vegetation or even in the lower branches of trees (Sutton, 1980). In all these cases, the major part of woodlouse diet is probably dead plant matter, though some species have been observed browsing on the living foliage of trees. Isopods have developed a wide range of behavioural adaptations to avoid desiccation.

Eurasian Isopoda appear to have a strong centre of diversity around the Mediterranean - especially in Spain, Italy and North Africa; it has been suggested that the fauna in these areas is even more diverse than that of tropical sub-Saharan Africa (P.T. Harding, pers. comm.). A few north-west European species have been widely spread by man, and make up a large proportion of the common synanthropic woodlice of North America and other temperate regions, and a few species are now almost cosmopolitan, e.g. *Cylisticus convexus* (Harding and Sutton, 1985). Although superficially amongst the better-known soil macrofauna, some woodlice are very small and cryptic, and occupy narrow and obscure niches: for example, the coastal shingle-bank fauna of northern Europe is only just being discovered. In well-worked regions, there is clear evidence of this niche specialisation, as well as landscape-scale differentiation (Harding *et al.*, 1991).

Some species associated with ancient natural habitats are now threatened by agricultural change and other human modifications of the landscape, such as the clearance or replanting of ancient woodland, the drainage of wetlands (e.g. *Ligidium hypnorum* in England), and general disturbance of coastal habitats (e.g. *Armadillidium album* throughout its range).

Table 11.3 Soil species: Isopoda

France	156
North America	100
Germany	60
UK	42
Holland	35
Little Carpathians (East Europe)	28

Phylum Chelicerata

Class Arachnida

Second only to the insects among the arthropods in terms of species numbers, the arachnids are an ancient and mainly terrestrial group.

Arachnida: Scorpiones

A morphologically rather uniform group of nocturnal predators with modern species ranging from 1.3cm to 18cm in length, but some Carboniferous species attained a length of 86cm (Barnes, 1980). Although usually thought of as typical of arid regions, there are many species which require a humid environment and occur in tropical moist forests. Most species occur in warm regions, but a few occur near the snow-line in mountains, and a single species occurs as far north as Canada (Levi *et al.*, 1968).

Table 11.4 Soil species: Scorpiones

TOTAL SPECIES	1,000 +
Africa (S)	159
Iran	36
USA	20-30
Kenya (N)	26
Arabia	23
Tanzania/Kenya	23
Africa (W)	17
Iraq	15
Israel	15
Australia (W)	11
Syria	11
Turkey	11
Egypt	9
Israel (N)	8
Trinidad	7
Libya	6
Madagascar	6
Kenya (S)	6
Jordan	5
Tobago	3

It is possible that the ranges of some temperate species are currently not entirely climatically determined, but reflect incomplete recolonisation after the glaciation. From the literature it appears that scorpions have a surprisingly even distribution of species richness in the warmer regions of the world. There is no evidence of any areas of marked radiation, which may be characteristic of an ancient and morphologically conservative group. The only exception is the apparent radiation of the rich southern African fauna, which parallels the high diversity of certain plant groups, especially *Erica* and *Protea* (Good, 1964). An additional factor in producing the even distribution of species, with few areas of very high diversity, may be their mode of life: they are bare-ground active hunters of large invertebrates and small vertebrates, and consequently occupy a broad niche space which cannot easily be partitioned between species, even if individual population density is high. In this respect, they provide an interesting contrast with the Carabidae, another group of surface-active generalist predators as discussed below.

Arachnida: Pseudoscorpiones

These small arachnids, the largest being only 8mm long and most only 2-3mm, superficially resemble scorpions. Most species live in leaf litter, and require a high humidity; in suitable woodlands, very high densities may be attained, with over 500 per m^2 commonly recorded, and peaks of over 900 per m^2 reported (Gabbutt, 1967). The efficient dispersal of species, particularly those with narrow microsite requirements (nests, barns, decomposing vegetation), is enhanced by phoresy - attaching themselves to other arthropods, especially Diptera and occasionally Coleoptera, and remaining attached until the host reaches another patch of suitable habitat. The maximum diversity of pseudoscorpions is widely believed to be in the tropics (Wallwork, 1976), but the available data are very patchy, with no comprehensive regional reviews, even in generally well documented areas such as Europe and North America; almost all the literature focuses on individual genera or species. The classification at species and family level is still in a state of flux. There is some evidence of microsite specialisation at the landscape scale (Legg and Jones, 1988), but few sites support a particularly rich range of pseudoscorpions.

Table 11.5 Soil species: Pseudoscorpiones

TOTAL SPECIES	3,000
North America	200
South Africa	109
Australia	99
Brazil	40
UK	26

Arachnida: Opiliones

The harvestmen or harvest-spiders have an average body length of 5-10mm, but the largest tropical species reach 20mm with a leg length of 160mm (Barnes, 1980). Harvestmen are abundant in leaf-litter and low vegetation in most habitats. In some tropical forests, the predatory force of harvestmen is thought to exceed that of the spiders (Dalingwater, 1983), although they will also scavenge on dead animals, and will eat a very wide range of organic matter, including fruit. Many species forage on the trunks and branches of trees when adult, but even these species tend to spend most of their juvenile life in litter or grass roots. Many species are nocturnal, as an adaptation to avoiding desiccation; the more resistant species are able to be active by day (Todd, 1949).

There are two ecological/systematic divisions in the harvestmen which show contradictory distribution trends, although the overall pattern is of higher diversity in the tropics. The actively predatory Laniatores are almost exclusively tropical and can be regionally diverse, e.g. 581 species in South America (Lawrence, 1931). The other main group, the scavenging Palpatores, show the reverse trend, with high diversity in the temperate regions: the known South American fauna contains a mere 29 Palpatores, while Europe possesses 215 species, compared to the meagre 14 native south-European Laniatores, of which only eight occur in central/northern Europe (Lawrence, 1931; Martens, 1978).

Harvestmen are more sensitive to desiccation than most arachnids, so are ill-adapted to a desert environment. This may explain the low diversity in Australia, for instance (where there is a rich spider fauna), compared with the rich Opiliones fauna in moist tropical forests, where they may be able to out-compete the spiders (Dalingwater, 1983).

Table 11.6 Soil species: Opiliones

TOTAL SPECIES	3,500
South America	581
Europe	232
Africa (excluding S Africa)	201
New Zealand	170
Europe (N of Mediterranean)	110
North America	104
South Africa	90
Madagascar	69
China	60
Australia	38
UK	23
Holland	21

Arachnida: Acari (Oribatei)

There are seven major groups of mites and ticks in this huge order, but only mites of three suborders occur predominantly in soil. They have a worldwide distribution. The Cryptostigmata or Oribatei are generally saprophagous, some feeding directly on decomposing litter fragments while others eat the fungi and bacteria which coat the litter. The Mesostigmata also include some saprophages, but many species are predatory. Prostigmatid mites are very varied in form and habit and include many non-soil-dwelling species. Of these three suborders the Cryptostigmata or Oribatei are by far the best known, in terms both of taxonomy and ecology; the prostigmatid and mesostigmatid soil-mites have received less attention and are less confined to the soil.

Most Oribatei are less than 1mm long, some very much smaller. High population densities can occur, with figures of 130,000 per m^2 being unexceptional. They occur in a wide range of soil and litter microhabitats the world over

(Luxton, in prep.) and play an important part in litter decomposition, both directly (those which feed directly on litter will consume about 20% of their body weight in litter each day), and indirectly (they stimulate microbial action in the litter). As the single most important fungivorous group in the soil, up to 50% of microfungal grazing and spore dispersal is attributable to oribatids (Eisenbeis and Wichard, 1985).

The available numerical data suggest that temperate soils support a more diverse cryptostigmatic mite fauna than the tropics, but this is almost certainly an artefact of sampling. The British Isles, with 300 species, would appear to have the richest concentration of any area, but also has the only up-to-date checklist (Luxton, in prep.). The available world literature concentrates almost entirely on generic and species taxonomy, or the fauna of very small sampling areas within atypical habitats. Oribatids are such an important group within the soil, being geographically and biotopically ubiquitous, that their overall biodiversity pattern will be of great interest when sufficient comparable data have accumulated.

Table 11.7 Soil species: Acari (Oribatei)

TOTAL SPECIES	7,000
UK	300
USSR (European)	278 +
Bulgaria	250
Japan	170
Arctic	144
Little Carpathians (East Europe)	129
Canada (N)	106
Peru	91
Ghana	52
Alaska	10
India	(8)

Phylum Uniramia

Class Chilopoda

This class comprises the centipedes, an important group of elongate, swift and agile predators which play a considerable part in most soil ecosystems. The class may be divided into four orders, representing the four main lines of morphological adaptation: the Geophilomorpha, the most subterranean group of centipedes, rarely seen on the surface; Lithobiomorpha; Scolopendromorpha, including the largest of all centipedes, some reaching almost 30cm in length; Scutigeromorpha, the majority of which live in dry, rocky habitats, hunting among rocks and scree. Several families of centipede are better represented at lower latitudes, the Scutigeromorpha in particular being confined to warm-temperate and tropical regions, though a few species occur inside human habitations further north. The trend in family distribution appears, from the very limited data, to be reflected in species richness too; but the accessible literature on centipedes is fragmentary, and even the most thoroughly researched areas such as northern Europe still have many areas of taxonomic confusion.

Table 11.8 Soil species: Chilopoda

TOTAL SPECIES	3,000
Peru	74
Germany	60
Transvaal	47
Natal-Zululand	42
UK	41
Holland	35
Canada (N)	29-31
Africa (SW)	27
Congo	10
Bermuda	7
Tunisia	7
Cyprus	6
Peru (NE)	(6)
Arctic	3

Class Diplopoda

The millipedes live in litter, under bark, and in the soil, being active in the open only after dark. Some are cave-dwelling, and several species live commensally in the nests of ants. All millipedes are predominantly saprophages, feeding on dead leaves, fallen logs and branches of trees, though some may also occasionally browse on mosses, lichens, algae or even living vascular plants. They often occur at high densities, and can be the main shredders of leaf litter in woodland soils that are too acid to support a rich earthworm population (Blower, 1985).

Documented diversity is rather low in most areas, including tropical Africa, but there is a high figure for North America. This suggests a Nearctic warm-temperate peak of diversity, enhanced by the absence of east-west geographical barriers in the Americas (where the main mountain ranges run north-south). This may have permitted much greater northward spread of taxa than in Eurasia (where there are major physical barriers - Pyrenees, Alps, Himalayas etc. - running east-west, and restricted post-glacial recolonisation of the region).

Table 11.9 Soil species: Diplopoda

TOTAL SPECIES	7,000
Central America and Mexico	750+
North America	749
France	250
Natal-Zululand	188
Germany	160
Peru (NE)	78
Transvaal	69
Congo	67
Madeira	53
UK (1958)	52
Holland	45
Denmark	39
Little Carpathians (East Europe)	31
Africa (SW)	17
Bermuda	8
Cyprus	6
Tunisia	(1)

The four-fold difference in recorded diversity between Britain (41 species) and Natal-Zululand (188 species) may be partly owing to the glacial effect; but millipedes are considered to be largely woodland/forest animals. Southern Africa has supported much more extensive woodlands with a stable history, throughout the Quaternary. The very low diversity in the Arctic probably reflects the low primary productivity, and thus the limited vegetable detritus for millipedes to consume. An extreme example of island speciation in soil fauna because of natural barriers may be found on Madeira, where 25 of the 53 species are now considered endemic. Ecological segregation in those regions which have been adequately studied tends to be on a macrohabitat scale, with grassland, woodland or sand-dune species, for instance, rather than intensive multispecies resource partitioning within a single habitat (Blower, 1985).

Class Oligoentomata: Collembola

The collembolans or springtails are small apterygotes (primitive insect-like hexapods; Dohle, 1988), seldom greater than 5mm long.

Like the mites, they have a cosmopolitan distribution, ranging from the seashore to high mountain-tops, and from the equator to the poles. Similarly, they can occur in very high densities, the smaller species reaching hundreds per cm^2 in ideal conditions. Species vary in their desiccation tolerance, so that different microsites in a habitat will support different species. The majority of Collembola are saprophages, feeding on decomposing plant and animal debris, although a few are predators, and others are small-scale pests of crops, notably the Lucerne Flea *Sminthurus viridis*. Because of their enormous densities and ubiquity, springtails are a crucial food-source for many small soil predators, including pseudoscorpions, and some staphylinid and carabid beetles.

Like the Oribatei, the Collembola appear to show a trend to higher diversity in temperate regions than in the tropics. Again, data quality may be suspect, but appears to be considerably higher and more uniform for Collembola than for Oribatei. A possible explanation of this trend proposed by Rapoport (1982) is that temperate soils are richer in nutrients and organic matter, as well as being more elaborately structured.

Comparing similarly-sized land masses with broadly similar climate reveals a constancy of collembolan fauna: Britain (300 species), Japan (241 species) and New Zealand (293 species) (Chinery, 1973; Rapoport, 1982). However, it is almost certain that the majority of species have yet to be found: for instance, Wallace and Mackeras (1970) could refer to only 215 described species, whereas Greenslade and Greenslade (1983) estimated there were 1,000-2,000 Australian species. The figures for mainland North America, lower than for Britain, are likely to be a sampling artefact.

Little has been published on patterns of collembolan endemism, but many species and genera have wide geographic ranges, implying some effective mechanism for long-distance dispersal, possibly wind-blown or rain-blown eggs. The available figures for the Tasmanian fauna contrast the native forest fauna, where up to 40% of species are endemics, with that of managed grassland, where only 1-2% of species are endemic and the majority are cosmopolitan (Greenslade and New, 1991). This clearly suggests that the Collembola will be highly sensitive to human impacts on natural and semi-natural vegetation; unfortunately, little research has been done elsewhere in the world.

Table 11.10 Soil species: Collembola

TOTAL SPECIES	10,000-20,000
Australia	1,000-2,000
UK	300
USSR	300
California (USA)	150
Little Carpathians (East Europe)	143
Peru	97
Arctic	91
Iceland	58
Philippines	37
Sudan	24

Class Pterygota: Dermaptera

The earwigs are a distinctive order of medium-sized insects allied to the Orthoptera. Although often hiding among litter or in the soil during the day, many species forage nocturnally among vegetation, flying readily and climbing trees (Imms, 1957). They are included here as soil fauna because almost all return to the soil to breed. Most species are thought to be omnivorous (Marshall and Haes, 1988).

Earwigs are essentially tropical and subtropical in distribution. Most species are sedentary, so individual species tend to have rather small geographic ranges, and consequently the fauna of each region contains a high proportion of endemics. The African fauna has been more intensively studied than others. Central Africa appears to be an important centre of diversity, particularly for the more primitive families of earwigs; it contains about 30% of the known world species of the ancient Carcinophoridae, for example, but only 18% of the more advanced Labiidae (Brindle, 1973). Literature on other tropical regions is sparse, although the Indian subcontinent appears, like Africa, to hold important concentrations of species.

Table 11.11 Soil species: Dermaptera

TOTAL SPECIES	1,200
Africa	298
India	185
Australia	60
USSR	26
USSR (European)	17
California (USA)	10
UK	5
Iceland	1

Only a very few earwig species are truly cosmopolitan, although their lifestyle makes them susceptible to accidental transport through commerce. Many such casual translocations lead only to temporary establishment, but if the climate is suitable a species may become more widespread. For example, the Indo-Australian species *Marava arachidis* is now well established in Africa and the Americas, but occurs only sporadically in Britain and northern Europe, usually in warehouses of imported organic materials (Brindle, 1973; Marshall and Haes, 1988). The sole cosmopolitan temperate species, the European *Forficula auricularia* is the common garden earwig in North America and elsewhere, though in the tropics it occurs mainly in montane areas. It has been implicated in the demise of three endemic earwigs of the genus *Anisolabis* in Hawaii (Howarth and Ramsey, 1991).

Pterygota: Embioptera

This primitive order comprises small to medium-sized soft-bodied cylindrical insects, commonly known as web-spinners. The females of most species are believed to be predominantly herbivores, while the males' diet may include other insects and soil arthropods.

Web-spinners are essentially tropical animals. The small numbers of European species are confined to the south, their northern limits being the Crimea, Bulgaria and the shores of the Mediterranean, although a few species occur further inland in Spain. The American fauna totals over 70 species, of which three are introduced and the rest are endemic (Ross, 1944). The highest concentration of species is probably in Australia (65 species (Ross, 1970)).

Overall, web-spinner species occur in widely-scattered, isolated areas, the group distribution being highly discontinuous.

Table 11.12 Soil species: Embioptera

TOTAL SPECIES	100
Australia	65
South America	44
Europe and Mediterranean	24
Central America	15
USA	12
Europe (S)	5
USSR	2
California (USA)	3
USSR (European)	1

Pterygota: Orthoptera

Four families of Orthoptera are largely soil-dwelling: the Gryllotalpidae, Tridactylidae, Cylindrachetidae and Tetrigidae. Many other species of grasshoppers and crickets spend some of their time among leaf litter and/or lay their eggs in the soil but are not included here because a significant part of their life-cycle takes place away from the soil.

The literature on the three soil-dwelling groups of orthopteroids is partial and fragmented. It is thus difficult to draw global conclusions at this stage.

Orthoptera: Gryllotalpidae

The mole-crickets are a small and specialised family of large, bulky insects which construct burrows mainly for feeding. Although found mostly in natural grasslands, they occasionally reach pest status by attacking root crops, especially (in temperate areas) potatoes (E.C.M. Haes, pers. comm.). Most species can fly, and can therefore colonise new areas. They are rare in cool-temperate regions and more diverse in warm-temperate ones.

Many species are phenotypically very similar, but there may be genetically-isolated cryptospecies awaiting recognition. Those species already described are fairly uniformly distributed between the main biogeographic regions, with no marked concentrations apparent from the literature. A few species are occasionally transported by man, mainly among root-crops; and the commonest Eurasian species, *Gryllotalpa gryllotalpa* has been introduced into North America.

Table 11.13 Soil species: Gryllotalpidae

TOTAL SPECIES	50
Australia	7
USSR	3
USSR (European)	2
UK	1

Orthoptera: Tridactylidae and Cylindrachetidae

The pigmy mole-crickets are not closely related to Gryllotalpidae, but have converged on the same lifestyle and acquired the same modifications of body form. They are relatively small - less than 10mm long - and live in damp sandy soils usually close to water.

The Tridactylidae are widely scattered in warm-temperate and subtropical regions, whereas the Cylindrachetidae are confined to Australia, New Guinea and Patagonia (Imms, 1957); the latter probably indicative of the family's early evolutionary origins on Gondwanaland.

Table 11.14 Soil species: Tridactylidae and Cylindrachetidae

TOTAL SPECIES	50
Australia	4
USSR	4
USSR (European)	3

Orthoptera: Tetrigidae

The groundhoppers or grouse-locusts are relatively small, usually less than 20mm. Most are found in damp microsites, such as river or pond margins. The eggs are often drought-resistant, enabling species to occupy seasonally-wet habitats (Hartley, 1962). Most species are unable to fly.

The Tetrigidae is a large group, with many described species. They appear, from the limited figures available, to be best represented in warmer regions, the Australian fauna being among the largest, though quite high concentrations have been described in some cool temperate areas. However, their taxonomy is still being clarified, and the ecological distinctions between closely-related species are only just beginning to be determined, even in western Europe (Devriese, 1990).

Table 11.15 Soil species: Tetrigidae

TOTAL SPECIES	700
Australia	70
Europe, Asia and N Africa	50
USSR	14
USSR (European)	9
UK	3

Class Pterygota: Blattaria

Small (e.g. temperate *Ectobius*, 5-7mm) to large (e.g. tropical Blaberidae, up to 15cm) insects. Most of the world fauna lives in low vegetation or on the ground, probably as scavengers; dense populations can occur in the litter layer of warm forests, and some species occur in caves. A handful of cosmopolitan species are pests and can be very abundant in domestic situations.

Cockroaches are characteristic of tropical moist forests, which support the largest diversity. Australia, for instance, has 439 species, most found in the native forests. Comparing two areas of roughly equal size, the British Isles support only three species, all in the genus *Ectobius*, whereas the West Indies are home to 156 species, including representatives of all the major families. There are several cosmopolitan species spread by man and now established in most countries. For this reason, published checklists, particularly in colder regions, often overestimate the indigenous fauna by including aliens which are restricted to heated domestic premises, e.g. Britain has three native and 23 casual or introduced species, of which five are well established (Marshall and Haes, 1988); the whole of the USSR has 41 native and 12 alien species.

Table 11.16 Soil species: Blattaria

TOTAL SPECIES	3,500
Australia	439
Africa (W)	300
West Indies	156
USA	55
USSR	50+
California (USA)	5-6
UK	3

Pterygota: Isoptera

The termites or 'white ants' are one of two main groups of soil-dwelling social insects (the others being the ants, Hymenoptera: Formicidae) whose colonies consist of a complex caste system, in which four main types can be recognised: the queen(s), workers, soldiers, and alate sexuals. The most primitive types are wood-boring and feeding, making no external modification to the decaying timber in which they live; such forms generally lack the worker caste. Certain genera may become pests by boring in domestic timbers. The remaining families are more exclusively soil-dwelling, some simply excavating galleries underground with little surface protrusion, while others construct large termite-mounds or termitaria which extend the nest many metres above the soil surface and form a conspicuous feature of the landscape in African and Australian scrub-grasslands. Many species feed in the same manner as earthworms, ingesting the soil detritus, microfungi and bacteria, or upon the roots of grasses and other plants. Others cultivate elaborate 'fungus gardens' on compost pre-prepared from vegetable matter. The majority of these more advanced species do not forage beyond the confines of the nest, unlike social Hymenoptera. The actual impact of termites on tropical ecosystems is still being evaluated (e.g. Collins, 1980, 1983, 1989).

In addition to their direct contribution to biodiversity, termites are important in providing niches for an extensive cohabiting fauna in their nests, ranging from commensals to symbionts, parasites and specialist predators.

Termites occur widely outside the polar and cold-temperate regions, except in the Palaearctic. The Ethiopian region appears to possess the richest diversity of genera as well as species, and is thus probably the most important centre of termite evolution (Bouillon, 1970). It contains the largest proportions of endemics. High numbers of species are also found in South America and the oriental region.

Broad patterns of temperature explain much of the variation in termite diversity. In the northern hemisphere, a strong correlation between diversity and latitude has been found (Sutton and Collins, 1991), though this may be a slight over-simplification: the correlation would be far less clear using southern-hemisphere data, because of the rich termite fauna of Australia, which extends beyond the Tropic of Capricorn.

Table 11.17 Soil species: Isoptera

TOTAL SPECIES	2,000
Ethiopian Region	570
South America	499
Oriental Region	434
Australia	182
Congo and Cameroon	78
Thailand	74
Palaearctic Region	41
Myanmar	39
Pakistan (W)	30
California (USA)	15
Mexico(W)	15
New Zealand	11
USSR	4+
Europe	2

Pterygota: Hymenoptera (Formicidae)

The ants are morphologically conservative but behaviourally diverse social insects with an elaborate caste system. Their nests vary from a few individuals in a space of less than 1cm^3 contained inside a dead twig (e.g. *Leptothorax*) to huge soil-based mounds with hundreds of thousands of foraging workers, which may be the dominant predatory force in whole forests (Brian, 1977). The diversity of individual size and feeding ecology allows many species to coexist in an area, and to partition resources, thereby avoiding competition (Davidson, 1978). The majority of ant nests are situated either within the mineral soil, or in the litter layer; although with deserved reputations as predators, many species also consume large volumes of plant material, especially seeds. Quite a high proportion of ant species have complex interactions with other ants. Like termites, ants also interact elaborately with other invertebrates, thereby increasing invertebrate diversity, through providing a range of additional niches within their nests; the range of symbiotic, inquiline, commensal, scavenging, parasitic and predatory lifestyles closely parallels those found within termite nests.

The ants are a large and diverse group, with most species in tropical regions, and a sharp decline toward the cool-temperate. Even on a small scale, in Europe and North America, there is a clearly marked latitudinal decline in diversity (Cushman and Lawton, in press); for instance, France has 180 species, whereas Britain has only 46 including introductions. The Palaearctic and Nearctic faunas are roughly equal in total diversity and pattern of species richness, their post-glacial colonisation apparently being unaffected by the topographic differences between the continents described under Diplopoda. This may be because the winged queens of ants are highly mobile, and so could travel long distances and recolonise virgin habitats as they became available with the retreat of the ice-sheet. This could also be the reason why Britain (46 species) has twice as many species as New Zealand (23) despite the fact that the total fauna of Oceania is much richer than that of Europe: the isolation of New Zealand is too great for uncontrolled flight to convey large numbers of species.

Ants have been the focus of much ecological research and speculation over the past 40 years. It has recently been observed that in Europe and temperate North America there is a latitudinal cline in individual mean size, with larger ant species in the boreal forest and many more tiny species around the Mediterranean/southern USA (Cushman and Lawton, in press). Further explanations of regional biodiversity have been related to vegetation patterns (Greenslade and New, 1991), and Australian work has also shown a high species turnover (beta diversity) across the continent.

Table 11.18 Soil species: Hymenoptera (Formicidae)

TOTAL SPECIES	10,000
Neotropical Region	2,233
Australia	1,100
North America (+USA)	585
USA	400+
California (USA)	200+
France	180
Sweden	61
Denmark	49
Finland	47
Norway	46
UK	46
New Zealand	23

Pterygota: Coleoptera (Carabidae)

The ground-beetles and tiger-beetles may be the largest of all families in terms of total species; over 40,000 species are described (Erwin *et al.*, 1979). They are ecologically very wide-ranging, in diet varying from obligate herbivore and detritivore to highly specialised predator. Their size ranges from less than 2mm to several centimetres, and they occupy almost all habitats from permanently waterlogged soils to the driest deserts. Although a proportion of forest species forage in the canopy, and rest under bark, the great majority are closely linked to soil and litter. Ground-beetles can reach high diversity in small habitat patches because of the variety of ways in which they can divide up the food resource, microsites, and time (different species being diurnal, nocturnal or crepuscular) (Greenslade, 1963).

With over 40,000 described species, the Carabidae are potentially valuable in analysing patterns of soil fauna distribution. Unfortunately, many areas still lack comprehensive reviews of their fauna, so the available literature remains patchy. However, the high diversity reported from the main tropical landmasses is probably a genuine effect; these areas did not suffer the extremes of recent glaciations, and the long periods of stability may have allowed local speciation to occur.

One of the most striking examples of intensive local speciation is provided by the tiger-beetles (sub-family Cicindelinae) in India, where there are 150 species in the genus *Cicindela*. The explanation of this high diversity is probably a complex of past dispersal, ecological isolation (largely through local climatic effects) and habitat specialisation (Pearson and Ghorpade, 1989). This contrasts with the low diversity of other surface-dwelling generalist predators such as scorpions.

There is a rich boreo-montane fauna in the northern hemisphere: carabids make up a large proportion of most European early post-glacial fossil deposits (Atkinson, Briffa and Coope, 1986), and this highly mobile element is equally important in North America - hence the rich Canadian/Alaskan fauna (850 species, Lindroth, 1969). The comparison of Britain (350 species) with New Zealand (538 species, Hudson, 1934) probably reflects local speciation on the oceanic island: over 90% of New Zealand's terrestrial arthropods are endemic (Howarth and Ramsey, 1991). In contrast, Britain has in effect only been partially recolonised from mainland Europe because of the breach of the land bridge to Europe by the English Channel, and has only a single 'endemic' carabid, *Tachys edmondsi* (Lindroth, 1974).

Table 11.19 Soil species: Coleoptera (Carabidae)

TOTAL SPECIES	40,000
Neotropical Region	5,000
North America	2,500
Australia	1,613
California (USA)	800
New Zealand	538
UK	350
Iraq	176

Although tropical forests support a very rich carabid fauna, arid grasslands are less rich; this is in part because of their replacement by the more drought-adapted Tenebrionidae. For example, whereas Britain has a mere 44 tenebrionids in a beetle fauna of over 3,000 species, Morocco has 711 species, which amounts to 15% of the total fauna (Kocher, 1958).

Pterygota: Coleoptera (Staphylinidae)
The rove-beetles range in size from less than 1mm to several centimetres. Many species are predatory, but others feed on decaying organic matter - vegetation, dung or animal corpses. A number of species occur in ants nests, some commensally or scavenging, others partially predatory on the ant brood, but often providing the ants with a sweet secretion in return. As a group, the Staphylinidae are an important predatory force in moist temperate habitats (Hammond, in prep.), perhaps rather less so in the tropics. Many species are difficult to identify, and they are therefore often excluded from surveys.
The rove-beetles are less well-known than the carabids, but the existing numerical data reveal several patterns among the temperate fauna. Most noticeably, the staphylinids outnumber the carabids in each documented area in the northern hemisphere, whereas in the southern, the reverse is true. One possible explanation for this is that rove-beetles are more prone to flying and were thus able to continue colonising new areas despite rising sea-level after the last Ice Age. The lower diversity in the southern hemisphere is harder to explain, and data are too few to evaluate with confidence; in some cases (e.g. Australia, with only 650 species) the generally more arid climate may limit the Staphylinidae.

Table 11.20 Soil species: Coleoptera (Staphylinidae)

TOTAL SPECIES	27,000
North America	2,800
California (USA)	1,000
UK	1,000
Australia	650
West Indies	468
Morocco	423
New Zealand	216

GENERAL PATTERNS OF DIVERSITY

This preliminary study has shown that the different groups of soil macrofauna function ecologically in very different ways and that most trends in distribution will be group-specific. The soil fauna is such a diverse group that the distributional trends within, for example, scorpions may run counter to those of the Collembola. In a more detailed study, it may thus be better to consider the major groups separately: the differences between soil groups may be greater than those between soil and non-soil members of the same group.

It would thus be an over-simplification to look for a single pattern of soil faunal biodiversity. That said, there are some indications of global pattern which hint at concentrations of species very different from those found in most plant and animal groups.

The usual trend towards higher diversity in the tropics compared with temperate regions is certainly apparent in some soil groups such as the scorpions, solifugids and Orthoptera. However, the limited information available for others, such as the Collembola, appears to show the reverse: temperate faunas may be more diverse than tropical ones. A possible explanation lies in the difference between the profiles of the two soils; tropical soils do not possess the depth or varied horizons seen in temperate ones. This is because of efficient re-cycling processes producing a low organic content, and lack of thermal seasonality (Rapoport, 1982). Both of these factors reduce the niche space and habitat quality of the soil, and consequently the soil-fauna diversity that it can support. A more fundamental difference is revealed when the respective ages of the soils are considered. The older tropical soils, such as those in Australia and Africa, are strongly leached and weathered, while the temperate soils, such as those in northern Europe, possess large areas of unweathered rock left by the retreating ice-caps of the last glaciation. The latter therefore have a higher mineral content, and a steady release of inorganic nutrients, which enhances the fertility of the soil.

It is premature to identify centres of diversity and endemism with any confidence although a few areas on present evidence stand out. The faunas of South Africa, Australia, and the Mediterranean Basin are richer than the average in most groups. That of New Zealand shows a higher degree of endemism than other similar-sized areas, and is species-rich in some groups such as the Carabidae. In many, the South American fauna is too poorly described in the literature to allow detailed comparison, but the few available figures suggest it is very rich in many groups.

Explanations of patterns of diversity depend on several different effects, which may be contradictory. For example, post-glacial history may have led to an impoverished fauna in large parts of the northern hemisphere, yet it is also responsible for the elaborate soil structure and landscape mosaic seen in many areas of Europe and North America, which enhance diversity. These two effects are jointly responsible for the Mediterranean species concentrations in several groups: during the glaciation, large numbers of species appear to have survived in Mediterranean refugia, and failed to recolonise the rest of northern Europe during the post-glacial. At the same time, the seasonality of the climate round the Mediterranean helps to diversify the soil habitat, enabling many more species to co-exist.

One factor underlying patterns of diversity which is more theoretical and harder to verify derives from the ecology of the groups. Some generalist predators such as scorpions and solifugids may have such broad niches that rather few species can coexist in an area, although the regional diversity in such groups can be high if the individual species have small ranges, and species complementing occurs on a smaller scale than usual.

Several recent estimates have suggested that the true diversity of soil fauna, in common with most invertebrates, may be ten times or more than the number of described species (Erwin, 1982; May, 1988).

References

Atkinson, T.C., Briffa, K.R. and Coope, G.R. 1986. Reconstruction of late glacial climates from Coleoptera using the mutual climate range method. In: Berger, W.H. and Labeyrie, L.D., *The Book of Abstracts and Reports from the Conference on: abrupt climate change* (NATO/NSF). University of California, Los Angeles. Pp.56-59.

Barnes, R.D. 1980. *Invertebrate Zoology*. Saunders, Philadelphia.

Barnes, R.S.K. (Ed.) 1984. *A Synoptic Classification of Living Organisms*. Blackwell, Oxford.

Blower, J.G. 1985. *Millipedes*. Synopses of the British Fauna New Series 35. Brill/Backhuys, London.

Bouillon, A. 1970. Termites of the Ethiopian region. In: Krishna, K. and Weesner, F.M. (Eds), *Biology of Termites*, Vol. II. Academic Press, New York. Pp.153-280.

Brian, M.V. 1977. *Ants*. Collins, London.

Brindle, A. 1973. The Dermaptera of Africa. Part 1. *Annales du Musée Royal de l'Afrique Centrale* Sér. 8 (Zoologie) 205:1-335.

Buckland, P.C. and Coope, G.R. 1991. *A Bibliography of Quaternary Entomology*. Collis, Sheffield.

Chinery, M. 1973. *A Field Guide to the Insects of Britain and Northern Europe*. Collins, London.

Cloudsley-Thompson, J.L. and Sankey, J.H.P. 1968. *Land Invertebrates*. Methuen, London.

Collins, N.M. 1980. The effects of logging on termites (Isoptera) diversity and decomposition processes in lowland dipterocarp forests. *Tropical Ecology and Development*:113-121.

Collins, N.M. 1983. Termite populations and their role in litter removal in Malaysian forests. In: Sutton, S.L., Whitmore, S.C. and Chadwick, A.C. (Eds), *Tropical Rain Forest: ecology and management*. Blackwell, Oxford. Pp.311-325.

Collins, N.M. 1989. Termites. In: Lieth, H. and Werger, M.J.A. (Eds), *Tropical Rain Forest Ecosystems. Ecosystems of the world*, vol. 14b. Elsevier, Amsterdam.

Cushman, J.H. and Lawton, J.H. (in press). Latitudinal variation in body size and species richness: patterns in the structure of European ant assemblages. *American Naturalist*.

Dalingwater, J. 1983. IX International Congress of Arachnology, Panama, August 1983. *Newsletter of the British Arachnological Society* 38:2-4.

Davidson, D.W. 1978. Size variability in the worker caste of a soil insect (*Veromessor pergandei* Mayr) as a function of the competitive environment. *American Naturalist* 112:523-532.

Davies, V.T., Harvey, M.S. and Main, B.Y. 1985. *Volume 3: Arachnida; Mygalomorphae, Araneomorphae in part, Pseudoscorpionida, Amblypygi and Palpigradi*. Bureau of Flora and Fauna, Canberra. Zoological Catalogue of Australia. Australian Government Publishing Service, Canberra.

Davis, B.N.K. 1986. Colonization of newly created habitats by plants and animals. *Journal of Environmental Management* 22:361-371.

Dempster, J.P. 1991. Opening remarks. In: Collins, N.M. and Thomas, J.A. (Eds), *The Conservation of Insects and their Habitats*. Academic Press, London. Pp.143-153.

Devriese, H. 1990. Herziene Tetrigidae-tabel. *Saltabel* 3:29-30.

Dohle, W. 1988. *Myriapoda and the Ancestry of Insects*. Manchester Polytechnic/British Myriapod Group, Manchester.

Easton, E.G. 1981. Japanese earthworms: a synopsis of the Megadrile species (Oligochaeta). *Bulletin of the British Museum (Natural History) Zoology* 40:33-65.

Eisenbeis, G. and Wichard, W. 1985. *Atlas on the Biology of Soil Arthropods*. Springer-Verlag, London.

Erwin, T.L. 1982. Tropical forests: their richness in Coleoptera and other arthropod species. *Coleopterists' Bulletin* 36(1):74-75.

Erwin, T.L., Ball, G.E., Whitehead, D.R. and Halpern, A.L. 1979. *Carabid Beetles: their evolution, natural history and classification*. Junk, The Hague.

Erzinclioglu, Y.Z. 1989. Campaign for real zoology. *New Scientist* 123(1676):62-63.

Eversham, B.C. 1982. A conspectus of European Diplura. (privately circulated).

Eversham, B.C. and Arnold, H.R. (in press). Introductions and their place in British wildlife. In: Harding, P.T. (Ed.), *Biological Recording of Changes in British Wildlife*.

Gabbutt, P.D. 1967. Quantitative sampling of the pseudoscorpion *Chthonius ischnosceles* (Hermann) from beech litter. *Journal of Zoology, London* 151:469-478.

Gerson, U. and Seaward, M.R.D. 1977. Lichen-invertebrate interactions. In: Seaward, M.R.D., *Lichen Ecology*. Academic Press, London.

Good, R. 1964. *The Geography of the Flowering Plants*. Longman, London.

Greenslade, P. and New, T.R. 1991. Australia: conservation of a continental insect fauna. In: Collins, N.M. and Thomas, J.A. (Eds), *The Conservation of Insects and their Habitats*. Academic Press, London. Pp.33-70.

Greenslade, P.J.M. 1963. Daily rhythms of locomotor activity in some Carabidae (Coleoptera). *Entomologia Experimentalis et Applicata* 6:171-180.

Greenslade, P.J.M. and Greenslade, P. 1983. Ecology of soil invertebrates. In: *Soils: an Australian viewpoint*. Division of Soils. CSIRO. Melbourne/Academic Press, London. Ch.40:645-669.

Hammond, P.M. (in prep.). *Provisional Atlas of the Rove Beetles (Coleoptera: Staphylinidae, Omaliinae) of the British Isles*. Biological Records Centre, Huntingdon.

Harding, P.T., Rushton, S.P., Eyre, M.D. and Sutton, S.L. 1991. Multivariate analysis of British data on the distribution and ecology of terrestrial Isopoda. In: Anon., *The Biology of Terrestrial Isopods III*. Université de Poitiers, Poitiers. Pp.65-72.

Harding, P.T. and Sutton, S.L. 1985. *Woodlice in Britain and Ireland: distribution and habitat*. Institute of Terrestrial Ecology, Abbots Ripton.

Hartley, J.C. 1962. The egg of *Tetrix* (Tetrigidae, Orthoptera), with a discussion of the probable significance of the anterior horn. *Quarterly Journal of Microscopical Science* 103:253-259.

Howarth, F.G. and Ramsey, G.W. 1991. The conservation of island insects and their habitats. In: Collins, N.M. and Thomas, J.A. (Eds), *The Conservation of Insects and their Habitats*. Academic Press, London. Pp.71-107.

Imms, A.D. 1957. *A General Textbook of Entomology*, 9th edn. Methuen, London

Kevan, D.K. McE. 1962. *Soil Animals*. Witherby, London.

Kocher, L. 1958. *Catalogue commente des Coléoptères du Maroc. Fascicule 2: Hydrocanthares, Palpicornes, Brachelytres. Fascicule 6: Tenebrionides*. Ministère de l'Instruction publique et des Beaux-arts, Rabat.

Lavelle, P. 1983. The structure of earthworm communities. In: Satchell, J.E. (Ed.), *Earthworm Ecology: from Darwin to vermiculture*. Chapman and Hall, London. Pp.449-466.

Lawrence, R.F. 1931. The harvest-spiders (Opiliones) of South Africa. *Annals of the South African Museum* 29(2):342-508.

Lee, K.E. 1959. The earthworm fauna of New Zealand. *Bulletin New Zealand DSIR* 130:1-486.

Legg, G. and Jones, R.B. 1988. *Pseudoscorpions*. Synopses of the British Fauna New Series 40. Brill, London.

Levi, H.W., Levi, L.R. and Zim, H.S. 1968. *Spiders and their Kin*. Golden Press, New York. 160pp.

Lindroth, C.H. 1961-1969. The ground beetles (Carabidae excl. Cicindelidae) of Canada and Alaska. Parts 1-6. *Opusc. Entomol.* Suppl. 20:1-200, Part 2 (1961): 24:201-408, Part 3 (1963): 29:409-648, Part 4 (1966): 33:649-944, Part 5 (1969): 34:945-1192, Part 6 (1969): 35:i-xviii, Part 1 (1969).

Lindroth, C.H. 1974. Coleoptera Carabidae. *Handbook for the Identification of British Insects* 4(2):1-148.

Ljungström, P.O. 1972. Taxonomical and ecological notes on the earthworm genus *Udeina* and a requiem for the South African acanthodrilines. *Pedobiologia* 12:100-110.

Luxton, M. (in prep.). *Provisional Atlas of the Moss Mites of the British Isles (Arachnida, Oribatida)*. Biological Records Centre, Huntingdon.

McCarthy, P.M. and Healy, J.A. 1978. Dispersal of lichen propagules by slugs. *Lichenologist* 10:131-134.

Marshall, J.E. and Haes, E.C.M. 1988. *Grasshoppers and Allied Insects of Great Britain and Ireland*. Harley Books, Colchester.

Martens, J. 1978. Weberknechte, Opiliones. *Die Tierwelt Deutschlands* 64. 464pp.

May, R.M. 1988. How many species are there on earth? *Science* 241:1441-1449.

Pearson, D.L. and Ghorpade, K. 1989. Geographical distribution and ecological history of tiger beetles (Coleoptera: Cicindeldae) of the Indian subcontinent. *Journal of Biogeography* 16(4):333-344.

Rapoport, E.H. 1982. *Areography. Geographical Strategies of Species.* Pergamon Press, Oxford.

Ross, E.S. 1944. A revision of the Embioptera or Web-Spinners of the New World. *Proceedings of the United States National Museum* 94:401-504.

Ross, E.S. 1970. Embioptera. In: *The Insects of Australia. A textbook for students and research workers*. CSIRO. Carlton, Victoria (Melbourne University Press).

Sims, R.W. and Gerard, B.M. 1985. *Earthworms*. Synopses of the British Fauna New Series 31. Brill/Backhuys, London.

Sutton, S.L. 1980. *Invertebrate Types: woodlice*. Pergamon Press, Oxford. 144pp.

Sutton, S.L. and Collins, N.M. 1991. Insects and tropical forest conservation. In: Collins, N.M. and Thomas, J.A. (Eds), *The Conservation of Insects and their Habitats*. Academic Press, London. Pp.71-107.

Todd, V. 1949. The habits and ecology of the British harvestmen (Arachnida: Opiliones) with special reference to those of the Oxford district. *Journal of Animal Ecology* 18:204-229.

Wallace, M.M.H. and Mackeras, I.M. 1970. The Entognathous hexapods. In: *The Insects of Australia. A textbook for research workers*. CSIRO. Carlton, Victoria (Melbourne University Press).

Wallwork, J.A. 1976. *The Distribution and Diversity of Soil Fauna.* Academic Press, London.

This is a condensed version of a consultancy report prepared by B.C. Eversham, A.S. Jolliffe and B.N.K. Davis of Monks Wood Experimental Station, a unit of the Institute of Terrestrial Ecology (Natural Environment Research Council). The full unpublished report is entitled: Soil fauna biodiversity: a preliminary global review. Project T13061A1.

12. FISHES

THE DIVERSITY OF FISHES

Fishes make up the most abundant class of vertebrates, both in terms of numbers of species and of individuals. They exhibit enormous diversity in size, shape, biology, and in the habitats they occupy. They are also the least known of vertebrates. It is clear, however, that the group of animals popularly termed *fishes* is defined by the retention of primitive vertebrate features (aquatic, gills, fins, 'cold-blooded') and the extant groups include several rather distantly-related evolutionary lineages. The first jawed vertebrates, around 500 million years ago, were fishes, and the first tetrapod land vertebrates arose from among the fishes around 400 million years ago.

There are in excess of 22,000 described species of fish. Vertebrates as a whole comprise around 43,000 species; thus, approximately half of all described vertebrates are fishes. Given that some 200 new species of fish have been described annually in recent years, probably well over half of all vertebrate species are fishes.

The great majority is comprised of bony fishes, mainly teleosts (advanced jawed fishes); in addition, there are around 800 species of cartilaginous fish (sharks, rays, chimaeras) and 70 jawless fish (lampreys and hagfishes).

Fishes range in size from around 1cm (as shown by a Philippines Goby *Pandaka pygmaea*, which is about 1.2cm in adult length, and another in the Indian Ocean, about 1cm) to the Whale Shark *Rhincodon typus*, which attains 15m. Some fish, typified by eels, are long and slender, others are globular; some are almost colourless, others are brilliantly coloured; some are fast and graceful, others sedentary.

They occupy almost every kind of aquatic habitat, ranging from sub-zero waters under the Antarctic icecap to near-boiling hot springs, and in water that is almost pure or highly saline. Many occupy the lightless ocean depths, a few dozen inhabit lightless cave systems (and some have lost both eyes and skin pigment).

Liquid water in lakes and rivers totals around 126,000km^3, equivalent to 0.0093% of the total volume of liquid water in the world. The oceans comprise about 1,320,000,000km^3, or 97% of the total. More than 8,400 fish species, or about 40% of all fishes, live in freshwater. There is thus around 100,000 km^3 of water for each marine species but a mere 15km^3 for each freshwater species: a difference of several orders of magnitude.

It has been calculated that some pelagic marine species may attain population levels of 10^{12} individuals, although a more typical value might be 10^9. The mean value for freshwater species has been estimated to range down to 10^6. Given the different water volume available per species, this represents a possible ten-fold decrease in water volume per individual in freshwater over marine species. This is not inconsistent with the greater net primary productivity per unit area, and greater plant biomass, in freshwater as compared with marine habitats.

Fishes provide the major world source of food derived from wild animals. Whether assessed in terms of tonnage traded or proportion of total dietary protein, fishes are a global resource of the first magnitude. Although the tropics generally have far higher species richness and endemism than temperate or arctic regions, and include 50% and 30% of the world's open water and continental shelf water, respectively, tropical fisheries contribute only about 16% of world fish production (Longhurst and Pauly, 1987).

Table 12.2, modified from Nelson (1984), lists the orders of extant fishes. Most orders are geographically very widespread, with representatives in the Atlantic, Indian and Pacific Oceans and/or on most continents: those with less wide distributions are noted in the table. Also listed are the numbers of families, genera and species in each order, with estimates of the number of species in marine and freshwater habitats.

We have made no attempt to deal comprehensively with the biodiversity of fishes, but have concentrated on aspects of species diversity, and include below material dealing with species richness and endemism in freshwaters, and notes on subterranean and coral reef fishes.

FRESHWATER FISHES: SPECIES RICHNESS AND ENDEMISM

Estimates have been made of species richness on major landmasses (Table 12.1), and detailed information is now available for a few families, but Tables 12.6 to 12.10 below are a first preliminary attempt to collate data on species richness and endemism of indigenous freshwater fishes on a global scale. Summary data on rivers and lakes are represented graphically in Figs 12.2 and 12.3.

Table 12.1 Freshwater fishes: species richness by continents

South America	2200
Africa	1800
Asia	1500 [1]
North America	950
Central America	354
Europe	250
Australia	170
New Zealand	27 [2]

Sources: Estimates cited in Nelson, J.S. 1984. *Fishes of the World*, 2nd edn. John Wiley and Son, New York.

Notes: [1] Estimate probably should be much higher, (Nelson, 1984). [2] Mostly diadromous.

In contrast to practice in other parts of this book, data on species diversity are presented in terms of water bodies rather than country units. River systems, for example, frequently cross several country boundaries or themselves constitute the boundary, making a country approach to data compilation more difficult and biologically less meaningful.

Introduced species are excluded from the counts wherever possible, as are subspecies (although some information sources are too imprecise to allow this in all cases; these

Table 12.2 Living fishes: a summary of systematic diversity

ORDER	COMMON NAME	DISTRIBUTION[1]	FAMILES	GENERA	MARINE SPECIES	FRESHWATER SPECIES[2]	TOTAL SPECIES
Myxiniformes	Hagfishes	temperate	1	6	32	0	32
Petromyzontiformes	Lampreys	cool temperate	1	6	9	32	41
Chimaeriformes	Chimaeras	Atlantic, Pacific	3	6	30	0	30
Hexanchiformes	Frill and Cow Sharks		2	4	5	0	5
Heterodontiformes	Bullhead Sharks	tropical Indo-Pacific	1	1	8	0	8
Lamniformes	Sharks		7	65	239	0	239
Squaliformes	Sharks		3	21	87	0	87
Rajiformes	Skates, Rays, Sawfishes		9	54	410	14	424
Ceratodontiformes	Lungfish	Australia (Qld)	1	1	0	1	1
Lepidosireniformes	Lungfish	Africa, Brazil, Paraguay	2	2	0	5	5
Coelacanthiformes	Coelacanth	S Africa, Comoros	1	1	1	0	1
Polypteriformes	Bichirs	tropical Africa	1	2	0	11	11
Acipenseriformes	Sturgeons and Paddlefishes	N Hemisphere	2	6	10	15	25
Lepisosteiformes	Gars	N and Cent. America, Cuba	1	1	0	7	7
Amiiformes	Bowfin	east N America	1	1	0	1	1
Osteoglossiformes	Bonytongues, Elephantfishes, Knifefishes		6	26	0	206	206
Elopiformes	Tarpons, Bonefishes, etc.		3	4	11	0	11
Notacanthiformes	Halosaurs and Spiny Eels	deep sea	3	6	25	0	25
Anguilliformes	Eels and Gulpers		19	147	597	0	597
Clupeiformes	Herrings, Anchovies, etc.		4	68	305	26	331
Gonorynchiformes	Milkfish, etc.	Africa, Indo-Pacific	4	7	2	25	27
Cypriniformes	Minnows, Carps, Loaches, Suckers, etc.		6	256	0	2,422	2,422
Characiformes	Characins, etc.		10	252	0	1,335	1,335
Siluriformes	Catfishes		31	400	56	2,155	2,211
Gymnotiformes	Knifefishes	Cent. and S America	6	23	0	55	55
Salmoniformes	Pikes, Smelts, Salmonids, Galaxiids, etc.		15	90	225	95	320
Stomiiformes	Lightfishes, Viperfishes, etc.		9	53	248	0	248
Aulopiformes	Greeneyes, Lizardfishes, etc.		12	40	188	0	188
Myctophiformes	Lanternfishes, etc.		2	35	241	0	241

Table 12.2 Living fishes: a summary of systematic diversity (continued)

ORDER	COMMON NAME	DISTRIBUTION[1]	FAMILES	GENERA	MARINE SPECIES	FRESHWATER SPECIES[2]	TOTAL SPECIES
Percopsiformes	Trout-perches, Cavefishes	N America	3	6	0	9	9
Gadiformes	Cods, Hakes, etc.		7	76	413	1	414
Ophidiiformes	Cusk-eels, Pearlfishes, etc.		4	86	290	4	294
Batrachoidiformes	Toadfishes		1	19	59	5	64
Lophiiformes	Anglerfishes		16	64	265	0	265
Gobiesociformes	Clingfishes and Singleslits		2	36	112	2	114
Cyprinodontiformes	Killifishes, Goodeids, Guppies, etc.		13	120	170	675	845
Atheriniformes	Silversides, etc.		5	48	150	85	235
Lampriformes	Oarfishes, Threadtails, etc.		11	20	39	0	39
Beryciformes	Squirrelfishes, etc.		14	38	164	0	164
Zeiformes	Dories, etc.		6	21	36	0	36
Gasterosteiformes	Sticklebacks, etc.	N Hemisphere	3	8	8	2	10
Indostomiformes		Myanmar (Lake Indawgyi)	1	1	0	1	1
Pegasiformes	Seamoths	Indo-W Pacific	1	1	5	0	5
Syngnathiformes	Pipefishes, Seahorses, etc.		6	63	254	3	257
Dactylopteriformes	Flying Gurnards		1	4	4	0	4
Synbranchiformes	Swamp-eels		1	4	4	11	15
Scorpaeniformes	Scorpion-fishes, Sculpins, etc.		20	269	1,070	90	1,160
Perciformes	Cichlids, Perches, Snappers, Butter-fly Fishes,		150	1,367	6,684	1,107	7,791
Pleuronectiformes	Flatfishes		6	117	535	3	538
Tetraodontiformes	Triggerfish, Puffers		8	92	321	8	329
TOTALS			445	4,044	13,321	8,411	21,732

Source: based on a table in Nelson, J.S. 1984. *Fishes of the World*, 2nd edn. John Wiley and Son, New York, with additional information from the same source.
Notes: [1] Most orders have a very wide distribution, often with representatives in the Atlantic, Indian and Pacific Oceans, or on most landmasses; an entry is made in this column only if the order is less widespread. [2] An estimate of the number of species always confined to freshwaters (and inland lakes even if saline); includes all primary division families, most secondary division families and many peripheral families (see text).

exceptions are recorded in the notes). In general, river systems or lakes were included only when estimates of both the total and the endemic fish fauna were available. Several very important rivers for which no useful data could be traced (e.g. the Ganges, Irrawaddy, Sikiang, etc.) have been excluded. For some of these, information does exist, but is too outdated or incomplete to use.

Faunal knowledge

The quality and extent of faunal studies vary widely from country to country. The geography of countries, which are artificial constructs, often bears little relationship to the geography of water bodies, which are natural. Because the great majority of faunal inventories are on a sub-national basis, substantial gaps in the coverage of a multinational river system frequently result. The probability of such incompleteness should be borne in mind when using the figures given below. In general, data quality is highest in developed countries, where species richness is lowest, and faunal inventory will probably remain least satisfactory in systems that cross several developing country boundaries (e.g. the Mekong), where species richness is undoubtedly high. It is of some concern that ichthyologists do not know with precision how many species of fish exist in multinational rivers, and often have but sparse knowledge of the fauna of rivers where fishes are an important human food resource. The ability to provide appropriate management remains correspondingly impoverished.

Taxonomic knowledge

Difficulties concerned with the taxonomic status of fishes arise from scientific disagreement or ignorance. Freshwater fish species vary in morphology throughout their range as a result of genetic, dietary and other factors. A specimen from one region may therefore have received a different taxonomic name from specimens from another population of the same species in another region. This result of parochial taxonomic research can be corrected when a taxonomist has access to a suitably large sample on which to work. A similar problem occurs when different taxonomic status is given to the same species by different authors so that the same biological species can appear under different names in different faunal lists. When uncritical overviews have been undertaken the same species can appear as two or more nominal species in the same list, thereby artificially inflating the number of species. Without the time to compare specimens and refer to the original descriptions, the number of species (both widespread and endemic) quoted in this document is based on a reasonable interpretation of the published literature and reference to people actively working in particular fields.

What are 'freshwater fishes'?

Freshwater fishes are customarily categorised as *primary* or *secondary*. This categorisation is essentially ecological not taxonomic, although based on families. Primary freshwater fishes, in this usage, are those families with little salt tolerance (stenohaline) and therefore confined to fresh waters. This has meant that the sea is a barrier and their current distribution is a result of physiographical events. The families Cyprinidae, Characidae and Cobitidae are examples of this group. Secondary freshwater fish families contain species which mostly live in fresh water but have some degree of salt tolerance and can cross salt waters. The Cichlidae, for example, are in this category. There is an accepted third category, the *peripheral* fishes, containing families that do not conform to either of the other two categories. Some may spend most of their life in fresh waters; others live in brackish waters. Marine families with representatives in fresh waters are also grouped here along with some anadromous or catadromous fish. The usefulness of categorising fish on the basis of salt tolerance had been challenged by Rosen (1974) who thought that fish should be regarded as continental or oceanic.

A pragmatic approach has been taken to determining which species to include in these estimates. In general, if most members of the species live in the sea, isolated freshwater populations are not included below. Some arbitrary decisions have been made. Anguillid eels have been excluded on the grounds that they breed in the sea, can move overland, and no populations isolated in fresh water are known.

Endemism

In normal biological usage, an endemic species is one confined to some given area, which may be defined as a site, a country, a continent or, in this compilation, a discrete river system or lake. If, however, a list of fish is constructed on the basis of river systems the full picture of very localised species will not emerge. For example, most riverine endemic fish species live in head-waters and often in very short stretches of river. The geophysical process of head-water capture has frequently resulted in one highly localised species living in two or more much larger river systems yet, in reality, being confined to a world distribution of just a few square miles. It has not been possible for the present document to compile adequate data on species which are local or regional endemics but are not endemic to one particular river system. They are not included below and it should be borne in mind that there are many more species of fishes with extremely localised distributions than is apparent here.

Species richness in rivers and lakes

The number of fish species present in subtropical and tropical rivers is highly correlated with the area of the river basin; temperate rivers show a similar pattern although the number of species rises more steeply with increasing basin area in tropical systems than in higher latitudes (Welcomme, 1979). The relationship appears to break down at high latitudes, where some tundra rivers are very extensive but have few fish species. Data gathered by Welcomme (1990) and Daget and Economidis (1975) are tabulated (Table 12.3) and shown graphically (Fig. 12.1).

Lake area is in general positively correlated with species richness, but a variety of additional factors may be involved. On a global scale, surface area and latitude together account for about one-third of the overall variation in species number (Barbour and Brown, 1974). For a sample of 14 lakes in North America, these factors accounted for most of the variation in species number, the

Figure 12.1 Number of fish species and river basin area

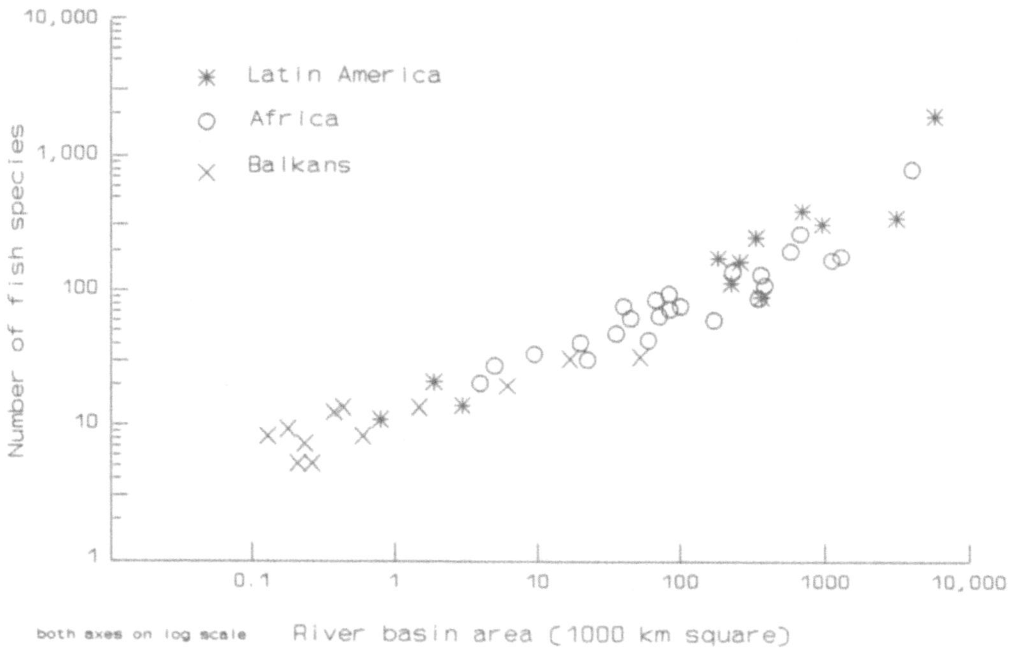

Source: Data in Welcomme, R.L. 1990. Status of fisheries in South American rivers. *Interciencia* 15(6):337-345; and Daget, J. and Economidis, P.S. 1975. Richesse spécifique de l'ichtyofaune de Macédoine orientale et de Thrace occidentale (Grèce). *Bulletin du Muséum National d'Histoire Naturelle*, 3e série, no 346, écologie générale 27:81-84.

Table 12.3 Numbers of fish species and river basin area

RIVER	NUMBER OF SPECIES[1]	BASIN AREA (km²)	RIVER	NUMBER OF SPECIES[1]	BASIN AREA (km²)
Latin America			Sassandra	71	84,140
Sucio	11	794	Bandama	75	100,000
Paz	21	1,884	Cunene	59	169,824
San Tiguel	14	2,985	White Nile	136	229,087
Paraguay	178	181,970	Senegal	86	342,768
Uruguay	115	223,872	Kasai	129	357,273
Magdalena	166	256,622	Volta	107	378,443
Negro	254	331,131	Chari	195	575,440
Parnaiba	90	362,000	Ubangui	263	668,344
Madeira	398	691,831	Niger	166	1,100,000
Orinoco	318	950,000	Zambezi	178	1,280,000
Parana, La Plata	355	3,100,000	Zaire	790	3,968,000
Amazon	2000	5,711,000	**Balkans**		
Africa			Aspropotamos	8	129
Me	20	3,981	Laspopotamos	9	180
Bouda	27	5,012	Loutos	5	211
Bia	33	9,441	Marmaras	7	235
Shire	40	19,953	Potamos	5	265
Cavally	30	22,387	Bospos	12	376
Sokoto	47	35,481	Kossithnos	13	435
Oueme	75	39,811	Kompsatos	8	600
Kafue	61	44,668	Filiouris	13	1,490
Ruaha	42	59,566	Nestos	19	6,178
Tana	84	66,834	Strymon	30	17,035
Comoe	63	70,795	Evros	31	52,788
Gambia	93	83,176			

Source: Data in Welcomme, R.L. 1990. Status of fisheries in South American rivers. *Interciencia* 15(6):337-345; and Daget, J. and Economidis, P.S. 1975. Richesse spécifique de l'ichtyofaune de Macédoine orientale et de Thrace occidentale (Grèce). *Bulletin du Muséum National d'Histoire Naturelle*, 3e série, no 346, écologie générale 27:81-84.

Note: [1] Total fish number given above will differ in some instances from numbers given in the main set of tables (12.6-12.10) because of different original data sources.

Table 12.4 Number of fish species, lake area and latitude

LAKE	NUMBER OF SPECIES[1]	SURFACE AREA (km²)	LATITUDE
Africa			
Albert	46	5346	1.7°N
Bangweulu	68	2072	11.1°S
Chad	93	17500	13.0°N
Chilwa	13	673	15.3°S
Edward	53	2150	0.5°S
Kivu	17	2370	2.0°S
Malawi	245	28490	12.0°S
Mweru	88	4413	9.0°S
Nabugabo	24	30	0.6°S
Rudolf	37	9065	3.5°N
Rukwa	22	3302	8.0°S
Tana	18	3626	12.0°N
Tanganyika	214	32893	6.0°S
Victoria	177	69484	1.0°S
Canada			
Athabasca	21	7154	59.2°N
Big Trout	24	616	53.8°N
Great Bear	12	31153	66.0°N
Great Slave	26	27195	61.4°N
Keller	13	406	63.9°N
Kootenay	19	399	49.5°N
La Ronge	19	1425	55.0°N
Opeongo	22	60	45.7°N
Great Britain			
Loch Lomond	15	71	56.1°N
Windermere	9	15	54.3°N
Guatemala			
Peten	23	98	17.0°N
Yzabal	48	684	15.5°N
Italy			
Maggiore	21	212	46.0°N
Japan			
Biwa	46	676	35.2°N
Mexico			
Chapala	14	1080	20.2°N
Pátzcuaro	7	111	19.6°N
Zirahuén	5	8	19.4°N
Nicaragua-Costa Rica			
Nicaragua	40	8264	11.5°N
Peru-Bolivia			
Titicaca	18	9065	16.0°S
Philippines			
Lanao	20	357	7.9°N
USSR			
Aral Sea	17	64500	45.0°N
Baikal	50	31500	54.0°N
Balkhash	5	18500	46.0°N
Beloe	22	1125	60.2°N
Black Sea	156	423488	43.0°N
Caspian Sea	74	436	42.0°N
Gusinoe	13	165	51.2°N
Issyk Kul	11	6206	42.0°N
Ladoga	48	18400	61.0°N
Leprindo	14	24	56.5°N
Onega	28	10340	61.5°N
Pestovo	17	2	58.3°N
Sea of Azov	17	38000	46.0°N
Seliger	21	221	57.2°N
Taimyr	13	4650	74.5°N
Teletskoe	14	231	51.6°N
USA			
Black	10	5	34.7°N
Canandaigua	37	41	42.8°N
Cayuga	60	171	42.7°N
Erie	113	25719	42.2°N
Huron	99	59596	44.5°N
Jones	13	1	34.7°N
Keuka	30	44	42.5°N
Michigan	114	58016	44.0°N
Ontario	112	19477	43.5°N
Otisco	17	10	42.8°N
Owasco	10	85	42.8°N
Salters	14	1	34.7°N
Seneca	39	174	42.6°N
Singletary	14	3	34.6°N
Skaneateles	14	54	42.8°N
Superior	67	82414	47.5°N
Waccamaw	36	36	34.3°N
Walnut	30	1	42.6°N
White	19	5	34.6°N
Yugoslavia-Albania			
Ohrid	17	347	41.0°N

Source: Data in Barbour, C.D. and Brown, J.H. 1974. Fish species diversity in lakes. *The American Naturalist* 108 (962):473-489.
Note: [1] Total fish number given above differs in some instances from numbers given in the main set of tables (12.6-12.10) because of different original data sources.

very strong effect of latitude probably a reflection of climatic severity and isolation from colonisation sources. In contrast, in a sample of 14 lakes in tropical Africa, surface area, depth and conductivity were the primary factors involved (increasing depth in a sense represents an increased area available to non-pelagic fishes). Select data used by Barbour and Brown are given in Table 12.4.

SUBTERRANEAN FISHES

At least 47 species of fishes are either cave-adapted or have cave-adapted populations. These highly localised populations are widely distributed across the globe, from about 38°N southward to the Tropic of Capricorn. Information on these fishes is given here in order to illustrate a facet of vertebrate biodiversity that is little-known although of great intrinsic interest and of scientific value in illustrating aspects of the evolutionary process.

Taxonomic and distributional data are summarised in Table 12.11 and site localities mapped in Fig. 12.4.

These 47 cave species represent seven orders and 13 families. Although frequently called cave fishes, this is not wholly accurate as some live in honeycombed rocks (aquafers) in which there are not necessarily any caves that can be entered by humans. Indeed, some species are only known from artesian wells that have penetrated these aquafers. It is therefore better to refer to these fishes as 'subterranean', 'cave-adapted' or *troglobionts*.

Characteristic of such species is a marked trend toward

eyelessness, lack of pigment and low metabolic rate. It is interesting to note that similar physical characteristics have evolved in some freshwater species confined to rapids and torrents in Africa and South America. These torrenticolous species have presumably lost their eyes and body pigment as a result of a lack of light in their habitat under stones and rocks in turbid rapids.

Of the 13 families which include cave-adapted fishes, nine are among the primary freshwater group. Indeed, the Homalopteridae, Ictaluridae, Pimelodidae, Trichomycteridae, Cyprinidae, Cobitidae and Amblyopsidae have particularly narrow salinity requirements. The families Ophidiidae, Synbranchidae and Eleotridae are primarily marine. The subterranean members of these families live near the coast in caves where they have been trapped in some cases, by land uplift. In all cases, cave species form a very small minority of the species in their respective families.

Population sizes are generally unknown. *Nemacheilus smithi*, for example, is known from just one specimen. Only two cave-adapted forms have been bred in captivity. An eyeless population of *Astyanax fasciatus* is on widespread sale as the 'blind cave tetra'. The blind form of this species breeds true, yet if mated with the above ground (epigean) form, as happens in nature, a complete range between eyeless and fully-eyed, and depigmented and fully pigmented, forms will result. Most laboratory based behavioural studies have been conducted on this species (e.g. Wilkens, 1971). Some observations on *Phreatichthys* were made by Ercolini and Berti (1975).

Studies on subterranean fishes in the wild are lacking. There is some evidence that breeding is seasonal and related to the influx of water into the subterranean environment. The young of *Caecobarbus geertsi* are only found after the rainy season (M. Poll, pers. comm.).

Because of the conspicuous superficial differences between a subterranean (hypogean) species and its epigean relatives it had been considered normal practice to allocate a hypogean species to a different genus. This action is now considered to be phylogenetically unjustified (Roberts and Stewart, 1976; Banister, 1984) and published nomenclatural changes are used in the species list below.

Not all cave fishes show the same degree of non-development of eyes or pigment. Some have very small eyes (are microphthalmic) or have eyes covered with skin, some are lightly pigmented. Such species can be regarded as not yet fully cave-adapted. The acquisition of extreme cave morphology implies the passage of time and this notion has been used by some authors (e.g. Wilkens, 1982) to argue that the fully cave-adapted species have been in their environment longer than those that are partially adapted. This argument involves the questionable assumption that evolutionary rates are the same in all species. These arguments also do not take account of the evidence for neoteny in cave fishes (Gould, 1977; Banister, 1984).

The subterranean fishes are of particular scientific value in exemplifying dramatic evolutionary phenomena. Within seven orders and 12 families of fishes there are 46 examples of parallel evolution occurring in similar environments. These evolutionary microcosms are often now under threat. The waters in which these species live and have evolved are a final sump for water soluble chemicals used on land. In the regions where subterranean fish live, water is often at a premium for human consumption and tapped for that purpose (the only habitat of *Satan eurystomus* is also the water supply for San Antonio, Texas).

CORAL REEF FISHES

Coral reef fishes are those associated with coralline structures. Many of these species can also occur in habitats other than coral reefs and in regions outside the geographic range of reef-building corals (Sale, 1980). Coral reefs are tropical, shallow water ecosystems, largely restricted to the area between the latitudes 30°N and 30°S (see Chapter 23). These complex systems are highly productive, a result of efficient recycling, high nutrient retention, and a structure which provides habitat for a great range of organisms (UNEP/IUCN 1988a,b,c).

Central parts of the Indo-West Pacific contain the highest number of reef fish species (Ehrlich, 1975), and richness decreases with increasing distance from this core area. Sale (1980) considers that this general pattern cannot be accounted for entirely by ecological hypotheses based upon latitudinal gradients in diversity, but may be due to historical factors. The origin and maintenance of high diversity is subject to debate. One view is that high diversity is sustained on reefs because of resource partitioning between species, fish assemblages being equilibrium communities (Dale, 1978; Robertson and Lassig, 1980; Smith and Tyler, 1972). An opposing view is that these communities are non-equilibrium unstable systems, and that species abundance is determined through independent differential responses to unpredictable environmental changes (Sale, 1977, 1978, 1980;, 1978; Sale and Williams, 1982).

Most reef fish species are relatively rare in terms of individuals in the community. Thus, at Toliara (south-west Madagascar) only about 25% (136) of the total number of fish species present were ranked as abundant (Harmelin-Vivien, 1989). Many families of coral reef fishes have a circum-tropical distribution, although there are pronounced differences at species level; the number of reef fish species within a single zoogeographic region varies between 100s and 1,000s. Most families in tropical seas include species that occur in the coral reef fauna, and some families are almost entirely restricted to reefs, such as Chaetodontidae, Scaridae, and Labridae. Within the demersal component (feeding on benthic organisms), the families Acanthuridae, Balistidae, Belennidae, Holocentridae, Ostraciodontidae, Pomacentridae (damselfish) and Serranidae tend to dominate. Principal pelagic families associated with reefs, other than the top predators such as Carangidae, *Sphyraena* and sharks, include Atherinidae (silversides), Pomacentridae and small lutjanids such as *Caesio* and its relatives (Longhurst and Pauly, 1987).

Small-sized species tend to predominate, although the range is from 2-3cm for some *Eviota* species to over 5m for some sharks. Fish distribution is highly heterogeneous within a particular geomorphological reef zone because of stochastic processes involved in fish larvae settlement (Gladfelter *et al.*, 1980). Complexity in reef structure contributes to species richness among reef fish by providing a wider variety of niches. On a local scale, fish community structure varies markedly between reef flat and outer reef slope; these zones are subject to different environmental factors affecting egg type, size-class categories, and feeding ecology. Other zones, including boulder tract, seagrass beds and deep outer flagstone all harbour characteristic fish assemblages.

There is a strong positive correlation between coral and fish species richness at given sites, although this is less evident on a small scale within reef zones (Table 12.5). It has also been suggested that there is a positive correlation between the degree of live coral cover and species richness and abundance of reef fishes (Bell and Galzin, 1984). In addition, the presence of dietary specialist fish species is often related to specific coral growth forms; for example, the exclusive coral feeders in the Chaetodontidae are positively correlated with the abundance of tall-branched coral colonies (Bouchon-Navarro *et al.*, 1985).

Table 12.5 Numbers of reef fishes and coral species

CORAL REEF SITE	NUMBER OF FISH SPECIES	NUMBER OF CORAL SPECIES
Great Barrier Reef (Australia)	2,000	500
New Caledonia	1,000	300
French Polynesia	800	168
Heron Island (Great Barrier)	750	139
Society Islands	633	120
Toliara (Madagascar)	552	147
Aqaba	400	150
Moorea (Society Is)	280	48
St Gilles (Réunion)	258	120
Tutia Reef (Tanzania)	192	52
Tadjoura (Djibouti)	180	65
Baie Possession (Réunion)	109	54
Kuwait	85	23
Hermitage (Réunion)	81	30

Source: Data from Harmelin-Vivien, M.L. 1989. Reef fish community structure: an Indo-Pacific comparison. In: Harmelin-Vivien, M.L. and Bourlière, F. (Eds), *Vertebrates in Complex Tropical Systems*. Springer-Verlag, New York.

References

Banister, K.E. 1984. A subterranean population of *Garra bareimiae* (Teleostei: Cyprinidae) from Oman, with comments on the concept of regressive evolution. *Journal of Natural History* 18:927-938.

Barbour, C.D. and Brown, J.H. 1974. Fish species diversity in lakes. *The American Naturalist* 108 (962):473-489.

Bell, J.D. and Galzin, R. 1984. The influence of live coral cover on coral reef fishes communities. *Marine Ecology Progress Series* 15(3):265-274.

Bouchon-Navarro, Y., Bouchon, C., and Harmelin-Vivien, M.L. 1985. Impact of coral degradation on a chaetodontid fish assemblage (Moorea, French Polynesia). *Proceedings of 5th International Coral Reef Symposium* 5:427-432.

Daget, J. and Economidis, P.S. 1975. Richesse spécifique de l'ichtyofaune de Macédoine orientale et de Thrace occidentale (Grèce). *Bulletin du Muséum National d'Histoire Naturelle*. 3e série, no 346, écologie générale 27:81-84.

Dale, G. 1978. Money-in-the bank: a model for coral reef fish coexistence. *Environmental Biology of Fishes* 3(1):103-108.

Ehrlich, P.R. 1975. The population ecology of coral reef fishes. *Annual Review of Ecology and Systematics* 6:211-247.

Ercolini, A. and Berti, B. 1975. Light sensitivity experiments and morphological studies on the blind phreatic fish *Phreatichthys andruzzi* Vinciguerra from Somalia. *Monitore Zoologico Italiano* (NS) Suppl. 6:29-43.

Gladfelter, W.B., Ogden, J.C. and Gladfelter, E.H. 1980. Similarity and diversity among patch reef fish communities: a comparison between tropical western Atlantic (Virgin Islands) and tropical central Pacific (Marshall Islands) patch reefs. *Ecology* 61(5):1156-1168.

Gould, S.J. 1977. *Ontogeny and Phylogeny*. Belknap Press of Harvard University Press. ix + 501pp.

Harmelin-Vivien, M.L. 1989. Reef fish community structure: an Indo-Pacific comparison. In: Harmelin-Vivien, M.L. and Bourlière, F. (Eds), *Vertebrates in Complex Tropical Systems*. Springer-Verlag, New York. Pp.21-60.

Longhurst, A.R. and Pauly, D. 1987. *Ecology of Tropical Oceans*. Academic Press Inc., San Diego, London.

Nelson, J.S. 1984. *Fishes of the World*, 2nd edn. John Wiley and Son, New York.

Roberts, T.R. and Stewart, D.J. 1976. An ecological and systematic survey of fishes in the rapids of the lower Zaire or Congo river. *Bulletin of the Museum of Comparative Zoology* 147(6):239-317.

Robertson, D.R. and Lassig, B. 1980. Spatial distribution patterns and coexistence of a group of territorial damselfishes from the Great Barrier Reef. *Bulletin of Marine Science* 30:187-203.

Rosen, D.E. 1974. Phylogeny and zoogeography of salmoniform fishes and relationships of *Lepidogalaxias salmondroides*. *Bulletin of the American Museum of Natural History* 153 2:265-326.

Sale, P.F. 1977. Maintenance of high diversity in coral reef fish communities. *American Naturalist* 111:337-359.

Sale, P.F. 1978. Coexistence of coral reef fishes: a lottery for living space. *Environmental Biology of Fishes* 3(1): 85-102.

Sale, P.F. 1980. The ecology of fishes on coral reefs. *Oceanography and Marine Biology Annual Review* 18:367-421.

Sale, P.F. and Williams, D.Mc.B. 1982. Community structure of coral reef fishes: are the patterns more than those expected by chance? *American Naturalist* 120:121-127.

Smith, C.L. and Tyler, J.C. 1972. Space resource sharing in a coral reef fish community. *Bulletin of the Natural History Museum Los Angeles Science Bulletin* 14:125-170.

UNEP/IUCN 1988a. *Coral Reefs of the World. Volume 1: Atlantic and Eastern Pacific*. UNEP Regional Seas Directories and Bibliographies. IUCN, Gland, Switzerland and Cambridge, UK/UNEP, Nairobi, Kenya. 373pp., 38 maps.

UNEP/IUCN 1988b. *Coral Reefs of the World. Volume 2: Indian Ocean, Red Sea and Gulf*. UNEP Regional Seas Directories and Bibliographies. IUCN, Gland, Switzerland and Cambridge, UK/UNEP, Nairobi, Kenya. 389pp., 36 maps.

UNEP/IUCN 1988c. *Coral Reefs of the World. Volume 3: Central and Western Pacific*. UNEP Regional Seas Directories and Bibliographies. IUCN, Gland, Switzerland and Cambridge, UK/UNEP, Nairobi, Kenya. 329pp., 30 maps.

Welcomme, R.L. 1979. *Fisheries ecology of floodplain rivers*. Longman, London and New York.

Welcomme, R.L. 1990. Status of fisheries in South American rivers. *Interciencia* 15(6):337-345.

Wilkens, H. 1971. Genetic interpretation of regressive evolutionary processes: studies on hybrid eyes of two *Astyanax* cave populations (Characidae, Pisces). *Evolution* 25:530-544.

Wilkens, H. 1982. Regressive evolution and phylogenetic age: the history of the colonization of freshwater of Yucatan by fish and Crustacea. *Bulletin of the Association of Mexican Cave Studies* 8:237-244 and *Bulletin of the Texas Memorial Museum* 28:237-244.

Chapter abridged from a consultancy report by Keith Banister, with additional material by WCMC staff (general introduction, reef fishes).

Table 12.6 Freshwater Fishes: Eurasia

	No of species	No of endemics
Baltic Sea basin		
Neva	40	0
Dvina	43	0
Vistula	43	0
Black Sea basin		
Danube	67	5
Dniepr	58	0
Dniestr	58	0
Kuban	46	0
Don	56	0
Crimea	14	0
Sakayra basin (Turkey)	40	0
Caspian Sea basin		
Volga	61	0
Ural	48	0
Terek	37	0
Kura & Araxes	47	0
Sefid & Atrek	*c* 22	*c* 6
Aral Sea basin		
Amu-Darya	44	1?
Syr Darya	46	3
Issy-Kul Lake basin	11	3
Lake Balkash basin	12	3
Tarim basin [1]	14	1
White Sea drainage		
N Dvina	25	0
Pechora	22	0
Arctic Ocean basin		
Ob	43	0
Yenisei (excl. Lake Baikal)	42	2
Lake Baikal	50	23
Lena	43	0
Kolmya	29	0
Bering Sea drainage		
Anadyr	20	1
Kamchatka	15	0
Pacific Ocean drainage		
Amur	*c* 90	6?
Yalu	74	*c* 4
Hong Ha (Red River may include brackish water species)	*c* 180	77
North Vietnam rivers	203	*c* 7
South Vietnam rivers [2]	255	very few
Mekong	*c* 500	?
Malayan Peninsula [3]	183	3 -83
Tasek Bera swamp (Malaysia)	95	0
Gombak (Malaysia)	28	0
Japan		
Overall [4]	*c* 120	*c* 53
Lake Biwa (Japan) [5]	< 63	8
Philippines		
Lake Lanao (Mindanao) [8]	*c* 24	*c* 18
Indonesia		
Kapuas (Kalimantan) [6]	*c* 250	*c* 35
Java	*c* 100	*c* 6
Lake Poso (Sulawesi) [9]	10	8
Papua New Guinea		
Fly river [7]	103	17
Northern rivers	*c* 84	*c* 36
Sri Lanka		
	54	<7?
Indian Ocean drainages		
Mae Khong [10]	215	15
Lake Indawgyi	43	2
Lake Lortak	13	0
Lake Inlé	28	7-8
Nepal (rivers Arun, Trisuli Mardi-Kola, all Ganges head-waters)	101	0
Indus	147	22
(Kabul, Chamkani-Kurram, Zhob Gowmal, all Indus head-waters)	45	10?
Tigris & Euphrates	62	5?
Endorheic basins: Mongolia		
Ugiy Nuur	7	0
Biger Nuur	8	0
Boon Tsagaan Nuur	2	0
Endorheic basins: China		
Upland lakes of Yunnan [11]	65	44
Er-Hai (Yunnan) [12]	6	5
Endorheic basins: Afghanistan-Iran		
Helmand-Sistan basin	27	1 - 5?
Hari-Tedzhen	12	0
Murgals	15	0
Lake Reza Iyeh (Urmin) [13]	14	5
Arabian Peninsula and Levant		
Oman mountains	3	2
Red Sea, Gulf of Aden and Wadi Hadramut systems	8	7
Rub al Khali drainage	3	3
Jordan river drainage	24	12
Azraq Oasis [14]	1	1
Western and southern Europe		
(excluding the river basins considered elsewhere)		
Lake Ohrid [15]	17	3
Europe [16]	76	10

The rivers are arranged roughly clockwise from the Baltic Sea drainage. Major lakes are included along with their tributaries, as most literature treats the ichthyofauna on a regional or basin basis. Within an Eurasian context this treatment is biologically rational, as many of the lake fishes are anadromous.

Regrettably, no reliable data could be found for several major river systems in this region; these include the Hwang Ho, Sikiang, Irrawaddy, Ganges, and rivers of peninsular India (notably the Godaveri, Cauvery and Narmada).

Notes

1 *In toto* there are 16 species endemic to the Aral Sea basin.
2 Of these species, about 80 are brackish water inhabitants or largely marine.
3 The freshwater fishes of Peninsular Malaysia are divided into 3 faunal zones: the northwest, the northeast and central, and the south zone. The numbers of freshwater fishes in each division are given respectively as 47, 98 and 58. The number of species listed in various articles as endemic varies extremely widely. Some reliance can be placed on the total number of primary freshwater fish, at least as to the order of magnitude, but very little on the number of endemic species.
4 This figure probably includes some euryhaline species.
5 The total figure includes introductions and subspecies.
6 The number of endemic species will be higher if the immediately adjacent rivers were included (Roberts 1989).
7 The endemic tally would be 47 if one or more other rivers from central-southern New Guinea were included (Roberts 1978).
8 This lake is famed for a reported endemic species-flock of cyprinids. However, since 1962 when alien species were introduced the indigenous fauna has become extinct (see Kornfield & Carpenter 1984; Reid 1980). Furthermore, many of the original specimens collected by Herre that led to the idea of the Lanao species flock were destroyed during the Japanese invasion in World War II. The number of species on other islands varies widely but has not been the subject of detailed listings.
9 It is not known if 2 of these species still survive. Introductions are likely to be responsible for their possible extirpation (Kottelat 1990).
10 This figure apparently includes about 40 brackish water and introduced species. The endemics are loaches and homalopterids from head-water streams.
11 These figures include subspecies as well as 2 endemic genera. It also seems that endemic in this context means very limited distribution but in more than one water body (Li 1982).
12 All these species are cyprinids.
13 The fauna of this region, especially of Lake Urmin, is very badly in need of re-examination.
14 There have been very many instances in historical times of translocation of fishes within and into this area that well over half the fishes now living in the system are not indigenous. These have not been included above.
15 One of the 'endemics' occurs in immediately adjacent lakes.
16 The European fish fauna is richest in the west and becomes increasingly depauperate towards the Mediterranean, Atlantic and North Sea coasts. This trend is even more marked in the off-shore islands which were separated from continental Europe at the end of last ice age, before the full complement of the refugia fauna had moved westwards. Only the most widespread or euryhaline forms live in Ireland, for example. The endemic species live in Dalmatia (1), Greece (3), Spain (3), North Italy south of the Alps (1), Italian rivers draining into the northwest Adriatic (1) and the Rhone (1). The European fishes have been much studied but rarely in a global context, and the significance of minute differences has been given greater importance than is probably justified. Only recently has a trend started to look at European fishes in an Eurasian context, which probably will affect the classification of the fishes quite considerably.

Table 12.7 Freshwater Fishes: North America

	Total species	Endemic species
Far north		
Hudson Bay drainage [1]	101	0
Ungava Bay watershed [2]	18	0
Arctic archipelago [3] (no primary freshwater fish)	8	0
St Lawrence River	98	1
Newfoundland rivers (no primary freshwater fish)	20	0
Labrador rivers (2 primary freshwater fish)	26	0
Northern Appalachian rivers		
Delaware to Nova Scotia [4] (about 85 primary fresh-water fish)	150	1
Central Appalachian Atlantic drainages		
Edisto	55	0
Santee	90	5
Peedee	76	1
Waccamaw	51	2
Cape Fear	71	1
Neuse	70	0
Tar	66	0
Roanoke	87	7
James	70	3
York	49	0
Rappahanock	50	0
Potomac	65	1
Susquehanna	61	1
Central Appalachian western drainages		
(Ohio system headwaters)		
Allegheney	92	0
Muskingum	111	0
Monongahela	89	0
Little Kanawha	72	0
Kanawha: below falls	90	0
Kanawha: above falls	49	6
Guyandotte	67	0
Big Sandy	94	0
Southeastern USA		
Ogeechee	58	0
Savannah	75	0
Apalachicota drainage	86	7
Choctawhatchee	74	0
Perdido	57-64	0
Mobile Bay drainage	157	*c* 40
Kissimmee river (and Lake Okeechobee)	37	0
Suwannee (and Withlacoochee)	43	0
Mississippi-Missouri [5]	*c* 260	*c* 72
Rio Grande Basin	121	69
California Coastal to Oregon		
(and internal basins)		
Colorado [6]	30	18
Sacramento system [7]	38	6

	Total species	Endemic species
Far north		
Death Valley system	8	6
North central basins	4	2
Lahontan basin	13	5
Bonneville basin	19	8
Oregon lakes	15	3
Klamath river	28	6
N California - Oregon rivers	29	3
Cascadia		
(the Columbia system north to Stikine)		
Columbia	45	13
Fraser	39	0
Skeena	32	0
Nass	27	0
Stikine	27	0
Yukon and Mackenzie basins		
Peace	24	0
Mackenzie river	34	0
Yukon river [8]	33	0
Lakes		
Superior	44	0
Erie [9]	99	0
Ontario	95	0
Michigan[10]	78-130	0
Huron	86	0
Pontachartrain	76	1
Lahontan - see Lahontan basin		
Tahoe - see Oregon lakes		
Great Slave lake and tributaries	36	0

Notes

1. The many recent introductions are excluded here. The drainage covers a wide range of climate zones and most of the species are in the south of the region and are probably recent (post-glacial) migrants.
2. This figure includes freshwater species with some degree of euryhalinity.
3. Fish have only occupied this area for 14,000 years. Much of it is ice-covered in winter.
4. The unique endemic is a rare anadromous coregonid found only in the fresh waters of the southern tip of Nova Scotia.
5. Of these endemics 56 come from the Cumberland, Tennessee and Arkansas drainages, ie, a very small part of the system.
6. The number of species and endemic species could change substantially at any time as there is disagreement about the specific or subspecific status of some forms, as well as known problems with hybridization.
7. One of the Sacramento endemics has been widely introduced elsewhere and the number of endemics would have been much higher if small, adjacent, but quite separate rivers had been included here.
8. Migratory forms are included in the Yukon figures.
9. One endemic subspecies now extinct.
10. Larger figure includes tributaries.

Table 12.8 Freshwater Fishes: Central and South America

	Total species	Endemic species
Mexico		
Santiago	17	2
Lerma	35	15
Morelia	14	1
Patzcuaro	10	2
Zirahuén	8	0
San Juanico	6	2
Valle de Mexico [1]	5	1
Puebla plateau	4	2
Atonilco	7	0
Ameca	20	7
Magdalena	8	1
Armeria	11	3
Coahuayana	9	3
Balsas	27	8
Papagayo	4	0
Varde Atoyac	9	1
Panuco	75	22
Gulf coast	21	1
Papaloapan	57	9
Nicaragua		
Lake Nicaragua [2]	*c* 39	0
Pacific slopes	12	0
Atlantic slopes	32	0
Lake Xiloa	12	0
Lake Managua	26	1
South America		
Trinidad	36	5
Magdalena	166	?
Maracaibo	108	31
Caribe	48	6
Lago de Valencia	35	4
Orinoco [3]	318	88
Amazon [4]	*c* 2000	*c* 1800
Rio Negro (Amazon)	436	35
Lake Titicaca	20	14
Trans-Andean region	390	*c* 100
La Plata [5]	*c* 550	*c* 110
Uruguay	*c* 160	*c* 35

Notes

1. In this region 3 additional former endemic species have recently become extinct.
2. The zero for the number of endemic species in the lake does not reflect the fact that it contains species of extremely limited distribution which variously occur in associated water bodies.
3. So far as can be ascertained, these figures include subspecies and probably also include some not strictly freshwater fish.
4. This figure is extremely imprecise. Most published figures vary widely, and the relevance of detailed studies at one locality to the fauna of the appropriate part of the subsystem is in doubt.
5. The total number of species in this river includes an unknown number of euryhaline species. The Parana, above the Guayra falls, has a high proportion of endemics in its fauna which is depauperate when compared to the rest of the system.

The entities listed in the table include rivers, lakes, and one island (Trinidad).

Table 12.9 Freshwater Fishes: Australia and New Zealand

	Total species	Endemic species
Australia [1]		
	3	5
	c 110	c 105
New Zealand [2]	c 30	27

Notes

[1] It is very difficult to categorize the Australian fishes in the same way as in other parts of the world. Strictly speaking, primary freshwater fish number just 3, of which 2 are endemic. The total number of species living all or the major part of their lives in fresh water is about 150. Of these, about 110 seem to be confined to fresh waters, even if they are capable of living in sea water. A further difficulty is that many of the 'fresh' waters are remarkably saline, especially in the desert regions. The great majority of fishes are confined to the short, peripheral, coastal rivers. All the 110 or so species had marine ancestors and many have marine close relatives; they are either physically confined to non-marine waters or are supposed to inhabit and breed in the freshwater parts of rivers. However, this figure could easily vary by 15% either way.

[2] Similar problems occur in evaluating the status of New Zealand fishes, except that there are no primary freshwater fishes there.

Table 12.10 Freshwater Fishes: Africa

	Total species	Endemic species
Atlantic drainage		
Senegal [1]	83	3
Gambia	79	0
Tominé	36	1
Koukouré	>44	5
Great Scarcie (Kolenté)	23	0
Sassandra	65	2
Bandama	77	1-2
Komoé	74	1
Volta [2]	132	8
Mono	39	0
Ouemé	62	1
Niger	149	13
Mungo-Meme	27	5
Rio Muni	81	>6
Zaïre [3]	c 700	c 500
Cunene [4]	55	2?
Orange-Vaal	16	5
Cape drainage		
Rivers of the great escarpment and eastern plateau [5]	13	10
Indian Ocean Drainage		
Olifants river	10	5
Limpopo	49	2
Zambezi	122	c 25
Great Ruaha [6]	>36	3
Tana		>2
Mediterranean drainage		
Nile	115	26
Tunisian rivers	6	1
Internal drainage rivers		
Omo (Lake Turkana	20	1
Chari (Chad) [7]	c 162	c 25?
Malagarazi (Tanganyika) [8]	>14	1
Ruzizi (Tanganyika) [9]	92	29
Cubango (Okavango) [10]		
Natural lakes		
Afrera (Guilietti)	2	1
Albert	46	9
Bangwelu [11]	86	0
Barombi-Mbo [12]	17	12
Chad [13]	93	1-30
Chilwa [14]	13-18	1
Edward-George	c 55	c 35
Eyasi complex [15]	1	1
Jipe	4?	2
Kivu [16]	17-33	6-7
Malawi [17]	>250	>230
		c 338
		c 1000
Mweru [18]	85	0
Nabugabo [19]	24	5
Natron and Magadi [20]	1	1
Rukwa	<17	1
Tanganyika [21]	>250	>230
Tsana (Tana) [22]	c 20	1?
Tumba [23]	>100	1
Turkana (Rudolf) [24]	48	10
Upemba lakes [25]	c 130	1
Victoria (including Kyoga) [26]	>250	>225
Zwai [27]	<20	3
Madagascar [28]		
	c 40	38

Notes

[1] The 3 endemic species are only found in small headwater streams.

[2] There is a large number of small rivers draining south from the Guinea highlands which hold many species restricted to several rivers in that region. In the original species descriptions the localities are given but cannot be put into context as the total fauna of these rivers has not been described. The high level of regional endemicity is not reflected in this table. Only the larger rivers (Tominé to Volta) have been studied in sufficient detail to make an adequately reliable entry.

[3] The Zaïre figures are a consensus of the most recent estimates. Over the last few years reduction in the number of nominal species by synonymisation has roughly equalled the descriptions of new taxa. The given figure has been based on the collections made at relatively few sites within the vast river network. (See, in particular, Banister, 1986: 215-216.)

[4] Comments made in Note [2] apply equally to the small rivers of Angola between the Quanza and the Cunene. For example, Ansorge, made a collection of fish in the early years of this century close to Lucalla railway station on the Lucalla river. His collection contained 25 species, of which 11 were unique to that site. No more recent records of fish collections from that region have been located.

[5] There is a very high level of endemicity in this localized Cape fauna. The main named rivers are the Berg, Breder and Buffalo rivers. The indigenous fauna is not speciose but now there are many introductions, to the detriment of the local fauna.

[6] This figure is based on a pre-impoundment survey in just one part of the Rufigi system.

[7] Although 25 is the most commonly cited number of endemics in the Chari-Logone system, it seems likely that, at best, many are sub-species. The basin fauna consists largely of widespread Nilotic fishes with a contribution of Niger-Benue faunal elements. The Chari-Benue watershed is extremely low and the systems connect during periods of heavy rain.

[8] The Malagarazi is a swampy river flowing sluggishly westward across a plain to Lake Tanganyika. Its poorly known fauna is Zaïrean in origin as the present Malagarazi is a now isolated former part of the Zaïre system.

[9] The Ruzizi is the main inflow to Lake Tanganyika, yet it is only about 12,500 years old. At that time the water level in Lake Kivu rose to such an extent that it overflowed southwards and the Ruzizi was formed. The upper and lower reaches of the river have different faunas and different hydrological conditions: Upper reach - Total 27. Endemic 7; Lower reach - Total 65, Endemic 20; Common to both - Total 13, Endemic 6. There are difficulties in evaluating the fauna of the lower reaches because of fish movements between the Ruzizi and the lake. The lake cichlids, however, rarely penetrate far up the river. In addition to the species enumerated above, there are 3 endemic species in streams flowing from the west into Lake Tanganyika. The streams are not meaningfully named.

[10] The fauna of this endorheic river is essentially that of the Zambezi (*q.v.*). However, its upper reaches and headwaters are very poorly known.

[11] The 86 species include those that live in the surrounding interconnecting small lakes, creeks and marshes. Lake Bangweulu does not have clearly defined limits. Poll (1957) stated that 17 species live in the main lake.

[12] The total number includes the species that inhabit the feeder streams and may occur in the lake itself at the feeder inflows. Of the 12 endemics, 11 are cichlids.

[13] Chad is a rapidly dissociating lake in a shallow basin. Formerly it was much more extensive. The total number of species is that of the entire basin. Only one species is endemic to the nucleus of the lake, but 25-30 are endemic to the entire basin.

[14] This lake periodically dries up. The fish take refuge in residual pools or in feeder streams when this happens or when the conductivity gets too high. The higher figure for the total number of species includes those that normally live in the feeder streams, but all must live together at times of desiccation.

[15] This is a series of 4 small lakes on the Tanzanian shield, Eyasi, Kitangiri, Manyara and Singida, that are the remnants of a former, much larger shallow lake.

[16] In the earlier literature, no distinction was made between an occurrence in a feeder stream and in the lake itself. The figures given above are, respectively, for the lake basin and the lake, but at least one of the basin species occurs in the lake but only at the mouth of feeder streams.

[17] Lake Malawi illustrates the uncertainties involved in compilation of this list. The three lines of species numbers above demonstrate the difference between published figures, current knowledge, and a probable future number when the lake fauna is well known. The top line is the published estimate. The second line is the current number of species described or known to be in press and to be published within the next year or so. The third estimate is based on information from Prof J. Stauffer (Pennsylvania State University): "Additionally, there are at least 200 entities which most authorities working in the lake recognize as valid species, but for which no species descriptions exist. Many of these are known by common names in the aquarium trade. Based on the number of undescribed species which occur in the trawl samples and the fact that little is known about the fishes inhabiting the Mozambique coast, I estimate that there are at least 1000 species which inhabit Lake Malawi. Approximately 95% of the total fish fauna is endemic to the lake." (*in litt.* to K. Banister 25 June 1991.) Whichever number is most correct, only 38 species are not cichlids.

[18] This lake lies on a shallow watershed between the Zambezi and Zaïre systems and contains fish from both systems.

[19] The formation of this lake, an offshoot of Lake Victoria, has been dated at 4,000 years BP. All the endemics are cichlids.

[20] These are relict, highly alkaline lakes, formerly parts of a larger lake.

[21] There is a much higher percentage of non-cichlid endemics than in the other rift valley lakes and a much higher number of families with endemic representatives.

[22] The alleged one endemic is the loach *Nemacheilus abyssinicus*. There is considerable suspicion that the unique specimen was accidentally translocated from a collection of middle eastern fishes into the Degen collection of fish from Lake Tsana and inadvertently described by Boulenger (1902) as indigenous to that lake.

[23] Although definable as a lake, it is a zone of permanent inundation up to 10 metres deep.

[24] Of the 48 species, 36 are exclusively in the lake. The other 12 occur only in the Omo River inflow.

[25] The Upemba lakes lie in the Kamalondo depression and are a shifting series of permanent, shallow, eutrophic lakes that are in varying contact with the Lualaba river. Of necessity, the number of the species has to include those also present in the Lualaba.

[26] Giving a reliable number of Lake Victoria species is very difficult as two contradictory factors are involved. First, there are an unknown number of yet undescribed cichlid species in museum collections. Second, the recent introduction of the predatory Nile perch (*Lates niloticus*) into the lake has apparently caused the extirpation of some species. The fauna of Lake Victoria is in a state of flux and the figures must be treated correspondingly.

[27] Unusually, none of the endemics are cichlids. They are cyprinids and probably spend some time in the lake as well as in the Maki river. The lake is drying out and there is no recent information on the fish fauna.

[28] All are 'secondary' freshwater fishes; see Introduction.

References

Banister, K.E. 1986. Fish of the Zaïre system. In: Davies, B.R. and Walker, K.F. (Eds), *The Ecology of River Systems*. Dr W. Junk, Dordrecht, Netherlands. Pp.215-224.

Boulenger, G.A. 1902. Descriptions of new fishes from the collection made by Degen, E. in Abyssinia. *Annals and Magazine of Natural History (Series 7)* 10(60):421-439.

Kornfield, I. and Carpenter, K.E. 1984. Cyprinids of Lake Lanao, Philippines: taxonomic validity, evolutionary rates and speciation scenarios. In: Echelle, A.A. and Kornfield, I. (Eds), *Evolution of Species Flocks*. University of Maine at Orono Press, Orono. Pp.69-84.

Kottelat, M. 1990. Synopsis of the endangered Buntingi (Osteichthyes: Adrianichthyidae and Oryziidae) of Lake Poso, Central Sulawesi, Indonesia, with a new reproductive guild and descriptions of three new species. *Ichthyological Exploration in Fresh Waters* 1(1):49-67.

Li Shusen 1982. Fish fauna and its differentiation in the upland lakes of Yunnan. *Acta Zoologica Sinica* 28(2):169-176 [In Chinese with English summary].

Poll, M. 1957. Les genres des poissons d'eau douce de l'Afrique. *Annales du Musée Royal du Congo Belge*. Tervuren, Sciences zoologiques 54:1-191.

Reid, G.Mc.G. 1980. "Explosive speciation" of carps in Lake Lanao (Philippines) - fact or fancy? *Systematic Zoology* 29:314-316.

Roberts, T.R. 1978. An ichthyological survey of the Fly river in Papua New Guinea with descriptions of new species. *Smithsonian Contributions to Zoology* 281:1-72.

Roberts, T.R. 1989. The freshwater fishes of western Borneo (Kalimantan Barat, Indonesia). *California Academy of Sciences* 14:1-120.

Tables prepared for WCMC by Keith Banister.

Figure 12.2 Freshwater river fishes: species richness and endemism

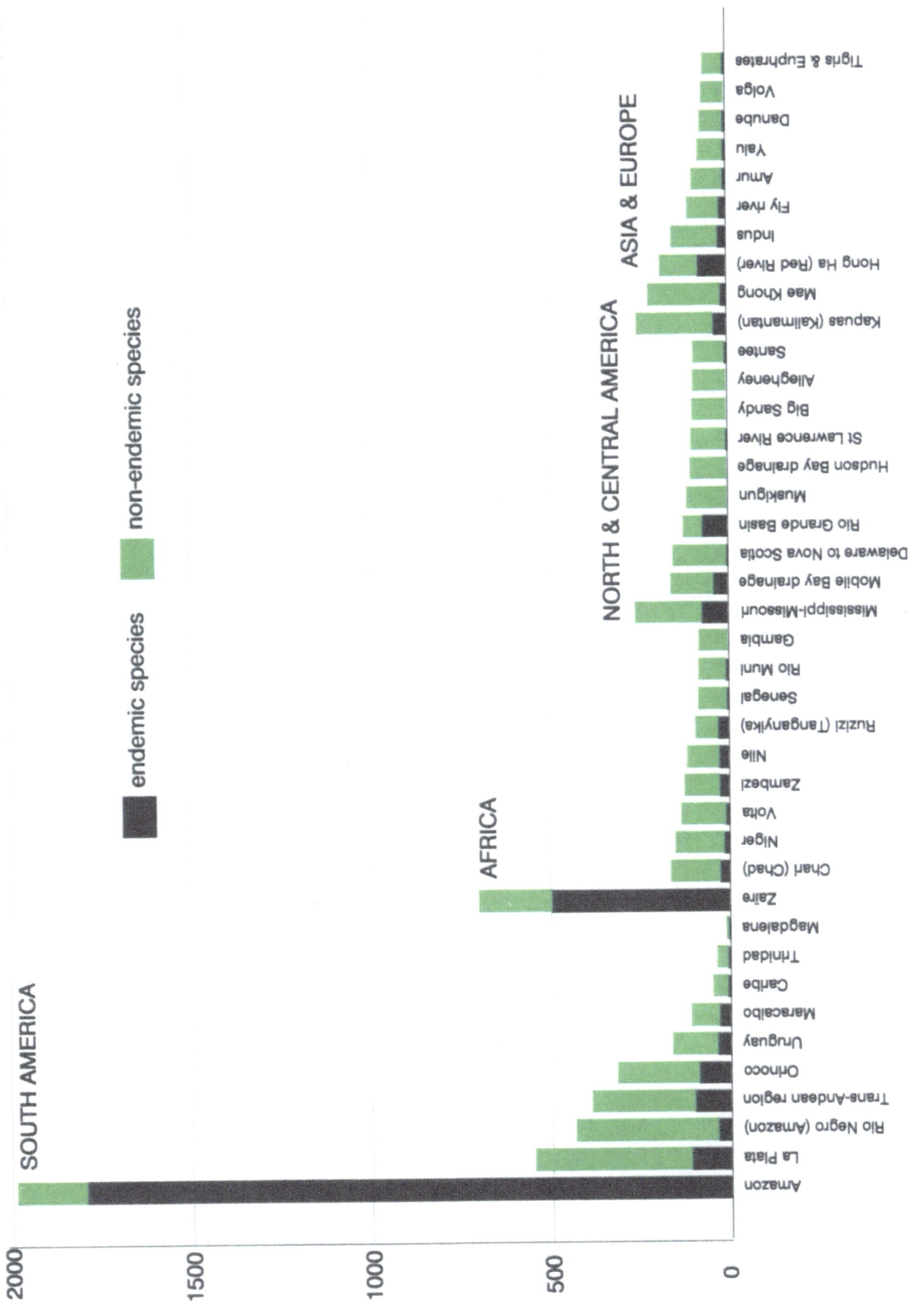

Figure 12.3 Freshwater lake fishes: species richness and endemism

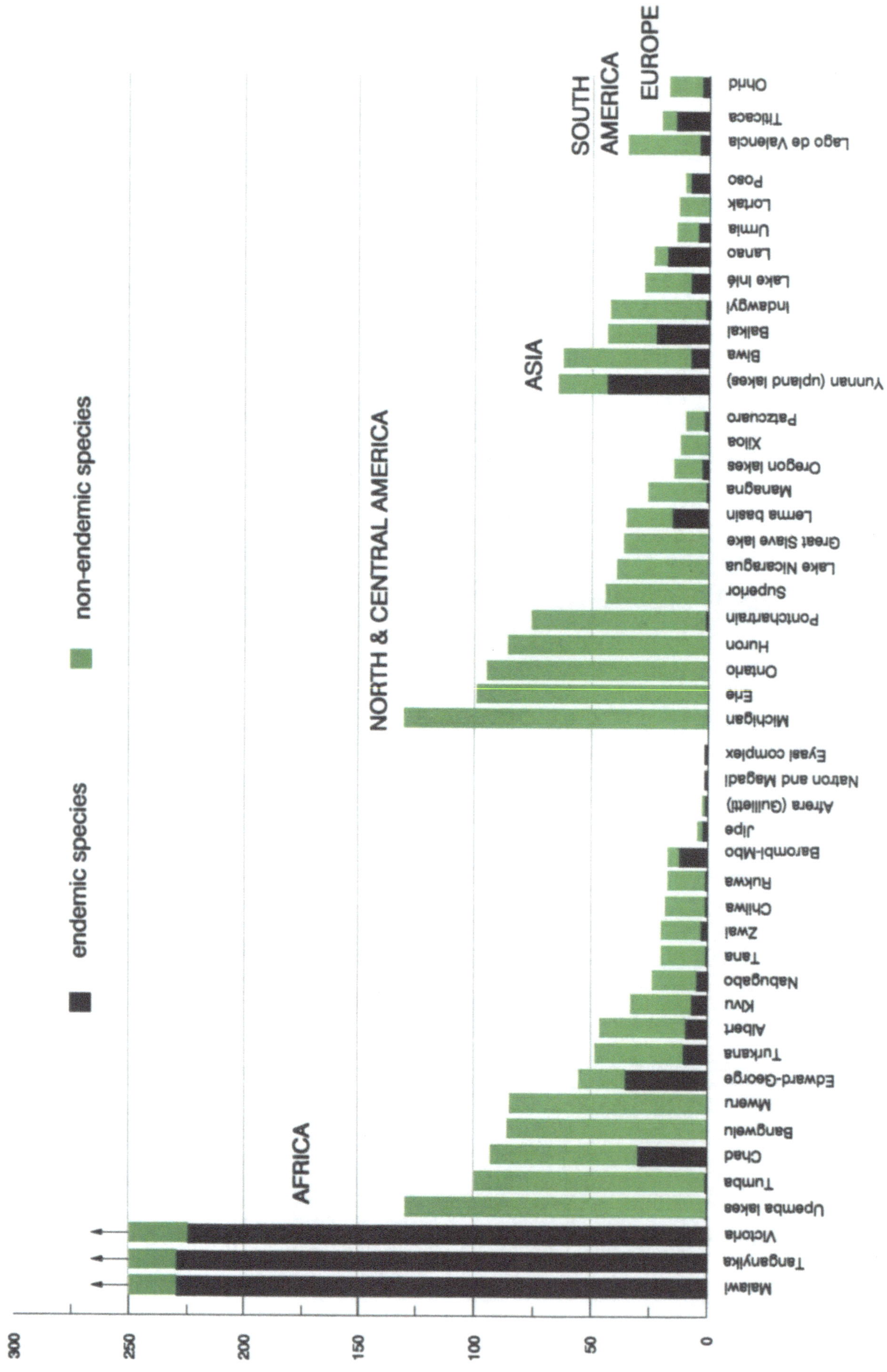

Table 12.11 Subterranean Fishes

FULL SPECIES

MAP NO	COUNTRY	SPECIES	ORDER: FAMILY	LOCATION	NOTE
1	AUSTRALIA	*Milyeringia veritas*	Perciformes: Eleotridae	Yardie Creek Wells, NE part of the NW Cape of Australia, 12km from Vlamingh Head lighthouse	Appears to be relatively common unlike *Ophisternon candidum* with which it occurs. Occasionally seen swimming slowly at the surface of sinkholes, possibly attracted there by insects that have fallen in. In summer can tolerate a water temperature of 30°C
2		*Ophisternon candidum*	Synbranchiformes: Synbranchidae	(as *Milyeringia veritas*)	Originally described as *Annomatophasma candidum*. Only 4 specimens known
3	BRAZIL	*Stygichthys typhlops*	Characiformes: Characidae	Jaiba, Minas Geraïs State	
4		*Ancistrus cryptophthalmus*	Siluriformes:Loricariidae	Cave at Passa Três, Rio Sao Vicente system, Sao Domingos, Goias	Species based on ca 20 examples from different parts of cave system. Has variable degrees of eye reduction. At the most extreme the eyes are highly reduced and in adults completely covered by bony plates
5		*Caecorhamdella brasilensis*	Siluriformes: Pimelodidae	Sao Paulo province	
6		*Phreatobius cisternatum*	Siluriformes: Trichomycteridae	Marajo Island (mouth of the Amazon), north of Soure	
7	CHINA	*Nemacheilus gejiuensis*	Cypriniformes: Cobitidae	Underground river at Bajianjing, near Qifang of Gejiu, Yunnan	
8		*Typhlobarbus nudiventris*	Cypriniformes: Cyprinidae	*circa* 24°N, 102°E	
9	CUBA	*Lucifuga simile*	Ophidiiformes: Ophidiidae	North-west Cuba	
10		*Lucifuga subterranea*	Ophidiiformes: Ophidiidae	Subterranean waters from Guira de Melena to Canas	Differences between *L. simile* and *L. teresinarum* and the widespread *L. subterranea* are small. A fourth species, *L. spelaeotes*, has been found in marine sink holes and caves in the Bahamas
11		*Lucifuga teresinarum*	Ophidiiformes: Ophidiidae	Two caves at Artemisa, Havana Province	Species founded on 2 male specimens
12		*Stygicola dentata*	Ophidiiformes: Ophidiidae	Subterranean waters from Jovellanos and Alacranes to Canas	
13	INDIA	*Horaglanis krishnai*	Siluriformes: Clariidae	A well at Kottayam, Kerala State	
14	IRAN	*Nemacheilus smithi*	Cypriniformes: Cobitidae	Baq-e-Loveh pool, Ab-i-Serum Valley, near Tang-e-haft railway station, Zagros Mountains	The pool is a seepage outlet. Although several feet deep, there is no connection through which a human can pass to the water-bearing strata inside the mountains. It is surmised that the aquafer supporting this species (and its fellow *Nemacheilus smithi*) is honeycombed rock
15		*Iranocypris typhlops*	Cypriniformes: Cyprinidae	(as *Nemacheilus smithi*)	
16	IRAQ	*Caecocypris basimi*	Cypriniformes: Cyprinidae	A sinkhole at the Sheik Hadid shrine, near Haditha (*circa* 34°4'N, 42°24'E)	A monotypic genus unlike any other cyprinid. Much rarer than the other species in the same subterranean system (*Typhlogarra widdowsoni*)
17		*Typhlogarra widdowsoni*	Cypriniformes: Cyprinidae	(as Caecocypris basimi)	

Table 12.11 Subterranean Fishes

FULL SPECIES (continued)

MAP NO	COUNTRY	SPECIES	ORDER: FAMILY	LOCATION	NOTE
18	LIBERIA	*Typhlosynbranchus boueti*	Synbranchiformes: Synbranchidae	A spring in a marsh about 2.5 miles from the sea near Monrovia	
19	MADAGASCAR	*Typheleotris madagascarensis*	Perciformes: Eleotridae	Sinkhole at Mitoho on the Mahalofy plateau	
20		*Typheleotris pauliani*	Perciformes: Eleotridae	Cave at Andranomaly, Andolambezo region, near Marombe SW Madagascar	The eleotrids are mostly marine and brackish water fishes. There is also another, undescribed species from Madagascar, (K, Banister, pers. comm.)
21	MALAYSIA	*Sundoreonectes tiomanensis*	Cypriniformes:Cobitidae	A granitic cave in Gunung Kajang (elevation 2900 feet), Tioman island, Malaysia. The 2 known specimens came from pools about 4 inches deep. The water was cold to the touch and the air temperature was about 15°C. Co-habiting the pools were tadpoles and freshwater crabs	Two specimens only were collected of the dozen seen. Reportedly most closely related to *S. obesus* from east and south Kalimantan and Sarawak. No loaches were found in any other caves in the region
22	MEXICO	*Prietella phreatophila*	Siluriformes: Ictaluridae	Well at foothills of Sierra Santa Rosa, Coahuila State	Closely related to the genus *Noturus* (Lundberg, 1982)
23		*Typhliasina pearsi*	Ophidiiformes: Ophidiidae	Balaam Cauche Cave, near Chichen Itza, Yucatan	
24		*Rhamdia reddelli*	Siluriformes: Pimelodidae	Cueva del Nacimiento del Rio San Antonio, Oaxaca, 9km SW of Acatlán	The main passage of the cave extends to a deep lake containing catfish and crayfish. Catfishes occur in the deeper, ponded parts of side streams. There is also a rich invertebrate fauna
25		*Synbranchus infernalis*	Synbranchiformes: Synbranchidae	Cave at Hoctun between Merida and Chichen Itza	
26	NAMIBIA	*Clarias cavernicola*	Siluriformes: Clariidae	Aigamas Cave, North Otavi	This species is depigmented. The eyes are covered with skin or absent
27	OMAN	*Garra dunsirei*	Cypriniformes: Cyprinidae	Pools in side caves at the bottom of a large sinkhole at Tawi Atair (17°06'N, 54°34'E) in the Jabal Qara (variously Jabal Samhan) mountains, Dhofar	A microphthalmic species isolated geographically from its relatives. The closest natural population of other *Garra* species is more than 400 miles away. The side caves are over 200m down from the lip of the sinkhole
28	SOMALIA	*Uegitglanis zammaranoi*	Siluriformes: Clariidae	Caves near the Uegit and Webi Shebeli rivers, west of Mogadishu	
29		*Barbopsis devecchi*	Cypriniformes: Cyprinidae	Wells in NE Somalia near Eil (Nogal system) and Scusiban (Luth and Darror systems)	Includes *Eilichthys micorphthalmus* and *Barbopsis stefaninii* as synonyms. This species is microphthalmic and faintly pigmented. Appears to have no close relatives among the African cyprinids and its phylogentic position remains obscure
30		*Phreatichthys andruzzi*	Cypriniformes: Cyprinidae	Bud-Bud, Somalia, 270km NW of Mogadishu, 4°11'N, 46°30'E	The subterranean waters are described as "warm and saline", this is significant because cyprinids are very rarely salt-tolerant

Table 12.11 Subterranean Fishes

FULL SPECIES (continued)

MAP NO	COUNTRY	SPECIES	ORDER: FAMILY	LOCATION	NOTE
31	THAILAND	*Nemacheilus troglocataphractus*	Cypriniformes: Cobitidae	Tham Sai Yok Noi (= Tham Nam Tok), 3km NNW of Nam Tok (14°15'N, 99°04'E)	Kottelat (1990) provisionally attributed a further 3 specimens from an imprecisely located cave, possibly in the Sai Yok karstic region, to this species.
32		*Schistura jarutanini*	Cypriniformes:Cobitidae	Amphoe Sri Sawat, Tham Ba Dan, Kanchanaburi Province	Known only from the 22 specimens comprising the type series. Eyes reduced or absent. Surprisingly, it is well pigmented. No information is available on its habitat (apparently an underground stream in the Mae Nam Kwae Yai basin)
33		*Schistura oedipus*	Cypriniformes: Cobitidae	Mae Hong Son Province	
34		*Homaloptera thamicola*	Cypriniformes: Homalopteridae	Mae Hong Son Province	This species is based on one 3cm specimen from a ledge on an outflow stream from a cave, about 600m from the cave entrance. If the species be valid it is the only known homalopterid showing cave-adapted features
35	UNITED STATES	*Amblyopsis rosea*	Percopsiformes: Amblyopsidae	Sarioxie, Missouri (Ozark uplift region)	
36		*Amblyopsis spelea*	Percopsiformes: Amblyopsidae	Mammoth Cave, Kentucky and southern Indiana	This species has a very low reproduction rate and broods the eggs in its mouth. Vandel (1965) noted that *Amblyopsis spelea* can live for 2 years without food, so low is its metabolic rate
37		*Speoplatyrhinus poulsoni*	Percopsiformes: Amblyopsidae	Lentic subterranean waters on north bank of Tennessee river, west of Florence, Alabama	This rare and highly cave-adapted species has been the subject of a model recovery plan (Cooper, 1982)
38		*Typhlichthys subterraneus*	Percopsiformes: Amblyopsidae	Subterranean waters in 2 disjunct ranges: Ozark plateau of southern Missouri and northeast Arkansas; the Cumberland and Interior low plateaux of northern Alabama, central Tennessee and Kentucky and southern Indiana	
39		*Satan eurystomus*	Siluriformes: Ictaluridae	Artesian wells near San Antonio, Texas. The well is over 300m deep and relies on interstitial water from aquafers in the Edwards limestone formation. This is the sole source of drinking water for the town of San Antonio.	
40		*Trogloglanis pattersoni*	Siluriformes: Ictaluridae	As for *Satan eurystomus*	
41	'USSR'	*Troglocobitis starotsini*	Cypriniformes: Cobitidae	Subterranean waters of Kugitangtau mountains, Turkmenia	
42	ZAIRE	*Caecobarbus geertsi*	Cypriniformes: Cyprinidae	Caves at Thysville (*circa* 5°S, 15°E)	This species, not closely related to any extant epigean species of the region, lives in 2 cave systems isolated from each other. Collection for the aquarium trade is a serious threat

Table 12.11 Subterranean Fishes

CAVE-DWELLING POPULATIONS OF SURFACE SPECIES

MAP NO	COUNTRY	SPECIES	ORDER: FAMILY	RANGE	NOTE
43	BRAZIL	*Pimelodella lateristriga*	Siluriformes: Pimelodidae	Eastern Brazil to Argentina	Includes *Pimelodella kronei* from the Areias and Bombas caves, Sao Paulo
44	MEXICO	*Astyanax fasciatus*	Characiformes: Characidae	Southeastern North America; Central America.	Includes *Anoptichthys jordani, A. hubbsi* and *A. alvarezi*
45	OMAN	*Garra barreimiae*	Cypriniformes: Cyprinidae	Oman	A depigmented, eyeless population occurs in the caves 9km ESE of Al-Hamra (23°05'N, 57°21'E) on the Jabal Akhdar Mountains. The normal, eyed, epigean form is found at many localities in the mountain range (Banister 1984)
46	TRINIDAD	*Rhamdia quelen*	Siluriformes: Pimelodidae	Northern and eastern South America and offshore islands	Includes *Caecorhamdia urichi* from the Guacharo Cave
47	UNITED STATES	*Gronias nigrilabris*	Siluriformes: Ictaluridae	Cave at Connestoga, South Pennsylvania	Probably abnormal examples of the widespread bullhead *Ictalurus melas*

Table based on data provided by Keith Banister.

Figure 12.4 Subterranean fishes

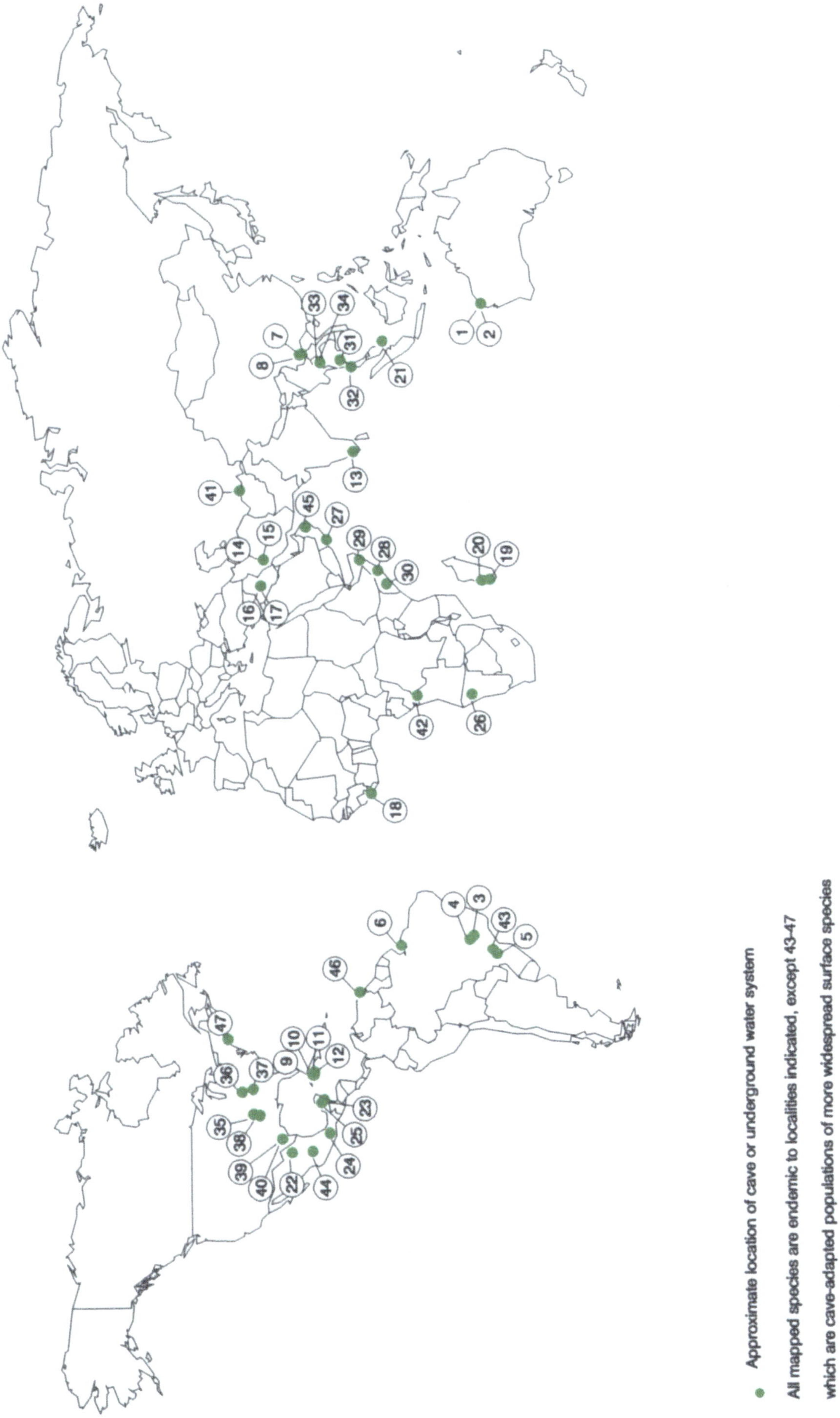

13. HIGHER VERTEBRATES

The vertebrates, with around 43,000 known species, make up a very minor proportion of the global total of some 1.7 million described species but, by virtue of their size, adaptions and ecological role, often exert a major effect on the structure of communities and habitats.

Vertebrates (Craniata or Vertebrata) make up the principal sub-phylum of the three included in the phylum Chordata. The vertebrates share the common characteristic of a hard endoskeleton (interior skeleton) with a backbone. There are seven classes (Parker, 1982), of which three, namely the Cephalaspidomorphi or Agnatha (lampreys and hagfishes), the Chondrichythes (sharks, rays, and skates) and Osteichthyes or Teleostomi (the bony fishes) are commonly referred to collectively as fishes. They are the extant members of diverse early vertebrate lineages, no longer recognised as a monophyletic taxonomic group. Fishes comprise nearly half of all extant vertebrate species. Data on species richness and endemism in freshwater fishes are presented in Chapter 12. The remaining four classes - Amphibia (amphibians), Reptilia (reptiles), Aves (birds) and Mammalia (mammals) - are often referred to as the higher vertebrates or tetrapods.

THE GROUPS OF HIGHER VERTEBRATES

Amphibians

No single characteristic uniquely defines the amphibians. All are ectotherms, using external environmental sources of energy to regulate body temperature, and have highly permeable skin and pedicellate teeth. Most, but not all, have a dual life-cycle, being aquatic as larvae and terrestrial or semi-terrestrial as adults. Amphibians are the only four-limbed animals in which metamorphosis, the abrupt transformation from larvae to adult, occurs.

There are currently in excess of 4,000 described species of amphibian (Frost 1983), divided into approximately 400 genera, 34 families and three orders (Halliday *et al.*, 1986). The three living orders are the Urodela or Caudata (salamanders, newts and their allies); the Anura (frogs and toads); and the Gymnophiona (the caecilians). Amphibians are ecologically less versatile than other higher vertebrate groups, in general being dependent on adequately high temperatures, moist conditions, and the availability of water for breeding and larval development.

Reptiles

The most obvious feature of reptiles is their covering of dry, horny scales, formed by localised thickenings of the keratin layer of the epidermis. Other characteristics include air-breathing, ectothermy - the dependence on external sources of heat to maintain a rather variable blood temperature - simple unspecialised 'homodont' teeth, and reproduction (normally on land) via the production of shelled eggs or live young.

Approximately 6,550 species of living reptile have been described, classified into about 905 genera, 48 families and four orders (Halliday *et al.*, 1986). These comprise the Chelonia (tortoises, turtles and terrapins), the Crocodylia (crocodiles and alligators), the Rhynchocephalia (which contains two species of tuatara *Sphenodon*), and the Squamata. This last group is divided into three suborders: Sauria (lizards); Serpentes (snakes) and Amphisbaenia (worm lizards) (many taxonomists recognise each of these as an order, rather than suborder). Unlike most amphibians, most reptiles are truly terrestrial and do not require environments rich in water; a number of species have also been able to adapt to marine habitats. The most species-rich, abundant and widely distributed reptile groups are the lizards and snakes.

Birds

All birds are, like mammals, endotherms, but their distinctive characteristic is that they possess feathers. Feathers are an evolutionary modification of reptilian scales which initially probably simply served a thermoregulatory function but now, in conjunction with the development of the forelimbs into wings, allow most birds the power of flight. Birds have lost all teeth from their bills, and reproduce by laying hard-shelled eggs.

Sibley and Monroe (1990) recognise 9,672 species of bird, organised into 2,057 genera, 144 families, and 23 orders. Birds are therefore the most diverse terrestrial vertebrate group. They have adapted to all the major habitats of the world, including equatorial forests, hot deserts, and the high Arctic and Antarctic.

Mammals

Mammals are animals whose bodies are insulated by hair (often in the form of a thick pelt or fur), which nurse their infants with milk produced from mammary glands, and which share a unique jaw articulation between the dentary (the main bone of the lower jaw) and the squamosal bone of the skull. Present-day mammals are 'heterodont', i.e. their teeth are specialised to fulfil different functions, and endothermic, i.e. their internal body temperatures are maintained by energy generated from metabolic processes within the body.

Corbet and Hill (1991) list 4,327 recognised mammal species, arranged into approximately 1,000 genera, 135 families, 18 orders and two subclasses (Macdonald, 1984). The division into subclasses reflects a separation which occurred almost 200 million years ago between the egg-laying Prototheria (the only survivors of which are three Monotremes: the platypus and two echidnas) and the Theria which bear live young. The live-bearing mammals diverged around 90 million years ago into the groups now recognised as marsupials (infraclass Metatheria) and the placental mammals (infraclass Eutheria). As a class, mammals are extremely versatile and have adapted to almost all terrestrial and aquatic habitats. Monotremes are only found in Australasia and marsupials are confined mainly to Australasia and the Neotropics, but placental mammals have spread throughout the globe, including the polar regions.

THE DISTRIBUTION OF HIGHER VERTEBRATES

Patterns of higher vertebrate distribution

As with many other organisms, species richness of land vertebrates tends to increase at lower latitudes. Amphibians, for example, are generally absent at very high latitudes (although one salamander species, *Hynobius keyserlingii*, ranges as far north as the Arctic circle 66.5°N) and species richness in most groups increases progressively towards the equator. Trends along moisture and altitudinal gradients are superimposed upon the latitudinal trend. For example, in North America, the greatest numbers of species are found in areas of high rainfall, principally in the south-eastern USA and secondarily in the north-west. Amphibian species diversity generally declines with altitude, so that a transect along the Equator from the Amazon basin to the crest of the Andes reveals a gradual reduction from 81 species at 340m to only four species above 3,500m (Duellman and Trueb, 1985). Reptiles are extremely sensitive to cold conditions and species diversity is very low in very high latitudes. Reptile species diversity increases towards the subtropics and tropics, to which some groups, such as the Crocodylia, are completely confined. Similarly, the diversity of birds and mammals increases towards the Equator.

The geopolitical distribution of higher vertebrates

Table 13.1 is a new compilation of data on species richness and endemism in vertebrates other than fishes, assessed on a geopolitical basis. This table is intended to complement the parallel compendium of flowering plant data (Table 8.3) earlier in this book.

Figs 13.1-13.8 show select data from Table 13.1 in graphic form. In this set of figures we have focused on single-country endemic species of mammals, birds and amphibians; the data are less complete for reptiles. We do not yet have a full data set for total country numbers, and here show (Fig. 13.2) the countries with most mammal species. Figs 13.3-13.8 represent the same countries shown in the higher plant graphs (Chapter 8).

Content and format

The table attempts to give realistic estimates of:

- the total number of species of mammals, birds, reptiles and amphibians present in each country of the world
- the number of species in each group that is endemic to each country.

'Endemic' in this context means that the species distribution is entirely within the political boundaries of a given country; they are single-country endemics, as opposed to site or area endemics.

It is important to note that for the purposes of this table, islands are included with their parent country (unless separately listed). Thus, the Galápagos are included with Ecuador, Hawaii with USA, the Canary Islands with Spain, and so on. Some apparent anomalies in the estimates are a result of this political aggregation; for example, the UK has 13 endemic birds listed, but 12 of these are from overseas territories (Henderson, Inaccessible, St Helena, S Georgia and S Sandwich Is, Tristan da Cunha). This affects the bird data in particular for a small number of countries.

Criteria for inclusion

Certain conventions have been followed wherever possible.

- Marine cetaceans, sea turtles and sea snakes are excluded. However, in a very few cases, especially where data have been taken from non-primary sources, we have been unable to establish whether cetaceans, for example, have been included or not.

- Data for birds include regular breeding species and exclude non-breeding migrants, occasional visitors and vagrants. It was felt that this would give a more consistent basis for comparison, and would avoid, for example, the problems involved in enumerating vagrants. Data available for some countries have not allowed us to make these exclusions and the figures will be correspondingly inflated - for example, an estimate of the birds of a Sahara-Sahel country will be low if only regular breeding species are counted, but more than twice as large if winter migrants and vagrants are included.
- Species known to be recently extirpated from or recently introduced to a country have been excluded.

Data quality

The estimates will become increasingly accurate as more and better data become available. Errors arise principally because of inadequate species inventory within countries and continual flux in the taxonomic status given to different population groups.

Species inventory based on field survey work is to varying degrees incomplete. Knowledge of the fauna of many developing countries is based largely on old and taxonomically outdated literature, often from colonial times. Taxonomic work results in continuing changes in nomenclature and the delimitation of species boundaries; populations recognised by one authority as belonging to one species will often be assigned to one or more other species by another taxonomist.

A further complication arises from the fact that animal distribution is dynamic not static; the geographical limits of species change over time, either as a slow advance or retreat of populations at the edge of a species range or as a more rapid population collapse or colonisation event (the latter perhaps most evident with bird populations).

We have made no systematic attempt to survey the primary literature for taxonomic changes that post-date the published works consulted. In general, the number of species reported in older literature to occur in any given country will have been both reduced by synonymy and enlarged by the description of new species.

These factors mean that a substantial margin of error is associated with all these data. It has not been possible to make a rigorous assessment of the extent to which estimates from several sources for a given parameter differ, but informal comparisons suggest a margin of plus or minus

10% is quite common and greater variation is not uncommon.

In the 'endemic species' columns we have attempted to minimise problems arising from taxonomic differences by deriving estimates for each group for almost all countries from a single consistent source. These sources are marked with an asterisk in the list below. In a few cases, later estimates based on new fieldwork have been incorporated.

The 'total species' columns include data from a variety of sources. These include published country or regional faunal monographs and the WCMC species database (itself based upon the former category of sources, but not complete for all vertebrate classes for all countries of the world). The extent of variety among these data sources, in terms of data quality, publication date and place of origin, will have led to a corresponding variety in data quality among the figures provided.

References

* Corbet, G.B. and Hill, J.E. 1991. *A World List of Mammalian Species*, 3rd edn. Natural History Museum Publications and Oxford University Press.

Duellman, W.E. and Trueb, L. 1985. *Biology of Amphibians*. McGraw-Hill, London, New York.

Frost, D.R. 1983. *Amphibian Species of the World. A taxonomic and geographical reference*. Allen Press Inc. and the Association of Systematics Collections, Lawrence, USA.

Halliday, T., Adler, K. and O'Toole, C. 1986 (Eds). *The Encyclopaedia of Reptiles and Insects*. Unwin, London, UK.

Macdonald, D. 1984 (Ed.). *The Encyclopaedia of Mammals*. Unwin, London, UK.

Parker, S.P. 1982. *Synopsis and Classification of Living Organisms*. McGraw-Hill, London, New York.

* Peters, J.A., Donoso-Barros, R. and Orejas-Miranda, B. 1986. *Catalogue of the Neotropical Squamata*. (Part I *Snakes*, Part II *Lizards and Amphisbaenians*). Smithsonian Institution.

Porter, K.R. 1972. *Herpetology*. Saunders Company, Philadelphia, London and Toronto.

* Schwartz, A. and Henderson, R.W. 1991. *Amphibians and Reptiles of the West Indies: descriptions, distributions, and natural history*. University of Florida Press.

* Sibley, C.G. and Monroe, B.L. 1990. *Distribution and Taxonomy of Birds of the World*. Yale University Press, New Haven and London.

* Welch, K.R.G. 1982. *Herpetology of Africa: a checklist and bibliography of the orders Amphisbaenia, Sauria and Serpentes*. Robert E. Krieger Publishing, Malabar, Florida.

Table 13.1 Species richness and endemism: higher vertebrates

	MAMMALS		BIRDS		REPTILES		AMPHIBIANS	
	Species known	Endemic species	Species known	Endemic species	Species known	Endemic species	Species known	Endemic species
ASIA								
Afghanistan	123	0	456	0	103	–	6	1
Bahrain	–	0	–	0	25	0	–	0
Bangladesh	109	0	354	0	119	–	19	0
Bhutan	109	0	448	0	19	–	24	0
British Indian Ocean Territory	–	0	–	0	–	–	–	0
Brunei	155	0	359	0	44	–	76	0
Cambodia	117	0	305	0	82	–	28	0
China	394	62	1100	63	282	–	190	131
Cyprus	21	0	80	2	23	1	4	0
Hong Kong	38	0	107	0	61	0	23	2
India	317	38	969	69	389	156	206	110
Indonesia	515	165	1519	258	511	150	270	100
Iran, Islamic Rep	140	4	–	1	164	3	11	5
Iraq	81	1	145	1	81	–	6	0
Israel	–	2	169	0	–	–	–	0
Japan	90	29	>250	20	63	28	52	35
Jordan	–	0	132	0	–	–	–	0
Korea, Dem People's Rep	–	0	–	0	19	1	13	0
Korea, Rep	49	0	–	0	18	–	13	1
Kuwait	–	0	27	0	29	0	2	0
Laos	173	0	481	1	66	–	37	1
Lebanon	52	0	124	0	–	–	–	0
Malaysia	264	14	501	4	268	–	158	39
Maldives	–	0	24	0	–	–	–	0
Mongolia	–	6	–	0	–	–	–	0
Myanmar	300	8	?867	4	203	29	75	9
Nepal	167	1	629	1	80	–	36	7
Oman	46	3	–	0	64	11	–	0
Pakistan	151	3	476	0	143	22	17	2
Philippines	166	90	395	172	193	131	63	44
Qatar	–	0	–	0	17	0	–	0
Saudi Arabia	–	1	59	0	84	5	–	0
Singapore	57	1	118	0	–	–	–	0
Sri Lanka	86	12	221	20	144	75	39	19
Syria	–	0	165	0	–	–	–	0
Taiwan	62	13	160	15	67	20	26	6
Thailand	251	5	616	2	298	39	107	13
Turkey	116	0	284	0	102	5	18	2
United Arab Emirates	–	0	–	0	37	1	–	0
Viet Nam	273	5	638	12	180	–	80	26
Yemen	–	1	–	8	77	25	–	1
USSR*								
	276	55	–	13	168	–	37	2
EUROPE								
Albania	68	0	215	0	31	0	13	0
Andorra	–	0	104	0	–	0	–	0
Austria	83	0	227	0	14	0	20	0
Belgium	58	0	180	0	8	0	17	0
Bulgaria	81	0	242	0	33	0	17	0
Czechoslovakia	81	0	227	0	12	0	19	0
Denmark	43	0	185	0	5	0	14	0
Faeroe Islands	–	0	75	0	0	0	0	0
Finland	60	0	230	0	5	0	5	0
France	93	0	267	9	32	0	32	3
Germany	76	0	237	0	12	0	20	0
Greece	95	2	244	0	51	4	15	1
Hungary	72	0	203	0	15	0	17	0
Iceland	11	0	80	0	0	0	0	0
Ireland	25	0	141	0	1	0	3	0
Italy	90	2	254	0	40	1	34	10
Liechtenstein	64	0	134	0	7	0	10	0
Luxembourg	55	0	130	0	7	0	14	0
Malta	22	0	28	0	8	1	1	0
Monaco	–	0	–	0	6	0	3	0
Netherlands	55	0	187	0	7	0	16	0
Norway	54	0	235	0	5	0	5	0
Poland	85	0	224	0	9	0	18	0
Portugal	63	1	214	2	29	1	17	0
Romania	84	0	249	0	25	0	19	0
San Marino	–	0	–	0	–	0	–	0
Spain	82	4	275	6	53	13	25	2
Sweden	60	0	249	0	6	0	13	0
Switzerland	75	0	201	0	14	0	18	0
United Kingdom	50	0	219	13	8	0	7	0
Vatican City	–	0	–	0	–	0	–	0
Yugoslavia	95	2	245	0	41	2	23	0

Table 13.1 Species richness and endemism: higher vertebrates (continued)

	MAMMALS		BIRDS		REPTILES		AMPHIBIANS	
	Species known	Endemic species	Species known	Endemic species	Species known	Endemic species	Species known	Endemic species
NORTH AND CENTRAL AMERICA								
Anguilla	5	0	–	0	–	1	–	0
Antigua and Barbuda	7	0	–	0	9	4	2	0
Aruba	–	0	–	0	10	2	1	0
Bahamas	12	2	88	3	24	16	5	0
Barbados	6	0	24	0	–	3	–	0
Belize	125	0	528	0	107	2	–	0
Bermuda	–	0	–	1	–	–	–	0
Canada	139	4	426	3	41	0	40	0
Cayman Islands	8	0	45	0	–	6	–	0
Costa Rica	205	8	848	6	214	17	162	34
Cuba	31	15	159	22	100	79	41	36
Dominica	12	1	59	2	13	2	2	0
Dominican Republic	20	0	125	0	–	22	–	15
El Salvador	135	1	?450	0	73	4	23	0
Greenland (Denmark)	–	0	–	0	–	0	–	0
Grenada	14	0	50	1	12	1	3	0
Guadeloupe	10	2	–	1	–	2	–	2
Guatemala	184	4	480	0	231	19	88	25
Haiti	20	0	–	0	–	29	–	17
Honduras	173	1	–	1	152	11	56	9
Jamaica	22	3	159	25	–	25	–	18
Martinique	9	0	53	1	–	3	–	0
Mexico	439	136	961	88	717	368	284	169
Montserrat	8	0	43	1	–	2	–	0
Netherlands Antilles	–	0	–	0	–	4	–	0
Nicaragua	–	2	–	0	161	6	59	2
Panama	?218	11	?922	6	?226	18	164	22
Puerto Rico	13	0	94	11	46	20	22	14
St Kitts and Nevis	7	0	40	0	9	0	3	1
St Lucia	8	0	51	4	15	5	4	0
St Vincent and the Grenadines	9	0	108	2	16	3	4	0
Trinidad and Tobago	100	1	258	1	–	2	–	2
Turks and Caicos Islands	–	0	184	0	–	5	–	0
United States	346	93	650	69	–	–	–	122
Virgin Islands (British)	–	0	–	0	–	3	–	1
Virgin Islands (US)	–	0	–	0	–	4	–	1
SOUTH AMERICA								
Argentina	258	47	–	21	–	63	123	37
Bolivia	280	7	1257	15	250	11	110	14
Brazil	394	68	1573	191	468	172	502	294
Chile	91	11	432	15	78	33	39	25
Colombia	359	22	1721	73	383	104	407	141
Ecuador	271	21	1435	37	337	100	343	136
French Guiana	152	1	–	1	–	1	–	2
Guyana	193	0	–	0	–	2	–	10
Paraguay	156	3	?650	0	120	4	85	4
Peru	344	46	1705	106	298	96	241	86
Suriname	187	2	–	0	–	0	–	7
Uruguay	81	0	–	0	–	1	–	2
Venezuela	288	11	1308	45	–	55	–	76
OCEANIA								
American Samoa	3	0	38	0	11	–	0	0
Australia	282	210	571	351	700	616	180	169
Cook Islands	–	0	28	7	–	–	0	0
Fiji	4	1	87	25	25	9	2	2
French Polynesia	0	0	67	25	–	–	0	0
Guam	–	0	23	3	10	1	0	0
Kiribati	–	0	15	1	–	–	0	0
Marshall Islands	–	0	18	0	7	1	0	0
Micronesia, Federated States of	–	3	47	18	–	–	0	0
Nauru	–	0	9	1	–	–	0	0
New Caledonia	7	3	116	20	32	23	0	0
New Zealand	–	3	285	74	40	40	3	3
Niue	1	0	16	0	4	0	0	0
North Marianas Islands	–	0	31	3	–	–	0	0
Palau	–	0	48	10	22	3	1	1
Papua New Guinea	242	49	578	54	249	–	183	100
Pitcairn Islands	0	0	19	0	–	–	0	0
Solomon Islands	47	18	163	72	57	9	15	2
Tokelau	0	0	65	0	7	0	0	0
Tonga	1	0	39	2	6	0	0	0
Tuvalu	–	0	9	0	–	–	0	0
Vanuatu	12	2	84	10	22	4	0	0
Wallis and Futuna Islands	–	0	14	0	–	–	0	0
Western Samoa	3	1	44	8	8	0	0	0

Table 13.1 Species richness and endemism: higher vertebrates (continued)

	MAMMALS		BIRDS		REPTILES		AMPHIBIANS	
	Species known	Endemic species	Species known	Endemic species	Species known	Endemic species	Species known	Endemic species
ANTARCTICA								
Antarctica	–	0	–	1	0	0	0	0
Falkland Islands (Malvinas)	–	0	63	1	0	0	0	0
French Southern Territories	–	0	–	1	0	0	0	0
AFRICA								
Algeria	92	1	192	1	–	3	–	0
Angola	276	4	872	12	–	18	–	23
Benin	188	0	630	0	–	1	–	0
Botswana	154	0	569	0	143	2	36	1
Burkina Faso	147	1	497	0	–	3	–	0
Burundi	107	0	633	0	–	–	–	2
Cameroon	297	10	848	11	–	19	–	65
Cape Verde	–	0	36	4	12	10	0	0
Central African Rep	209	2	668	0	–	–	–	0
Chad	134	0	496	0	–	1	–	0
Comoros	12	2	99	9	22	3	–	0
Congo	200	1	500	0	–	1	–	1
Cote d'Ivoire	230	2	683	0	–	2	–	2
Djibouti	–	0	311	0	–	–	–	0
Egypt	102	4	132	0	83	1	6	0
Equatorial Guinea	184	1	392	3	–	4	–	2
Ethiopia	255	26	836	26	–	6	–	30
Gabon	190	3	617	0	–	3	–	4
Gambia	108	0	489	0	–	1	–	0
Ghana	222	0	721	1	–	1	–	4
Guinea	190	1	529	0	–	3	–	4
Guinea–Bissau	108	0	376	0	–	2	–	1
Kenya	309	10	1067	7	187	15	88	10
Lesotho	33	0	288	0	–	2	–	1
Liberia	193	1	590	2	62	2	38	4
Libya	76	4	80	0	–	1	–	0
Madagascar	105	67	250	97	252	231	144	142
Malawi	195	0	630	0	124	6	69	1
Mali	137	0	647	0	16	2	–	1
Mauritania	61	1	49	0	–	1	–	0
Mauritius	–	1	102	10	–	2	2	0
Mayotte	–	0	–	0	15	1	–	0
Morocco	105	5	209	0	–	8	–	2
Mozambique, People's Rep	179	2	666	0	–	5	62	2
Namibia	154	2	640	1	–	25	32	2
Niger	131	0	473	0	–	–	–	0
Nigeria	274	2	831	2	>100	7	>60	1
Reunion	2	0	33	0	–	3	–	0
Rwanda	151	0	669	0	–	1	–	0
Saint Helena	–	0	–	0	–	–	–	0
Sao Tome and Principe	8	2	124	24	16	6	9	9
Senegal	155	1	625	0	–	1	–	1
Seychelles	–	1	126	9	15	13	12	11
Sierra Leone	147	0	614	0	–	1	–	2
Somalia	171	8	639	11	193	66	27	3
South Africa	247	27	774	7	299	76	95	36
Sudan	267	7	938	0	–	6	–	2
Swaziland	47	0	381	0	106	0	39	0
Tanzania	306	12	1016	13	245	48	121	40
Togo	196	1	630	0	–	1	–	3
Tunisia	78	1	173	0	–	1	–	0
Uganda	315	4	989	3	119	2	44	0
Western Sahara	15	1	60	0	–	–	–	0
Zaire	415	25	1086	23	–	33	–	53
Zambia	229	3	732	0	–	2	83	1
Zimbabwe	196	2	635	0	153	2	120	3

Notes: See text for general conventions adopted and sources for endemics data. Dependent islands are included with the parent territory. ? may include marine species where data refers to mammals or reptiles, or may include non–breeding species where data refers to birds. – no data. USSR*: covers the former Union of Soviet Socialist Republics.

Figure 13.1 Higher vertebrates: the 25 most endemic-rich countries

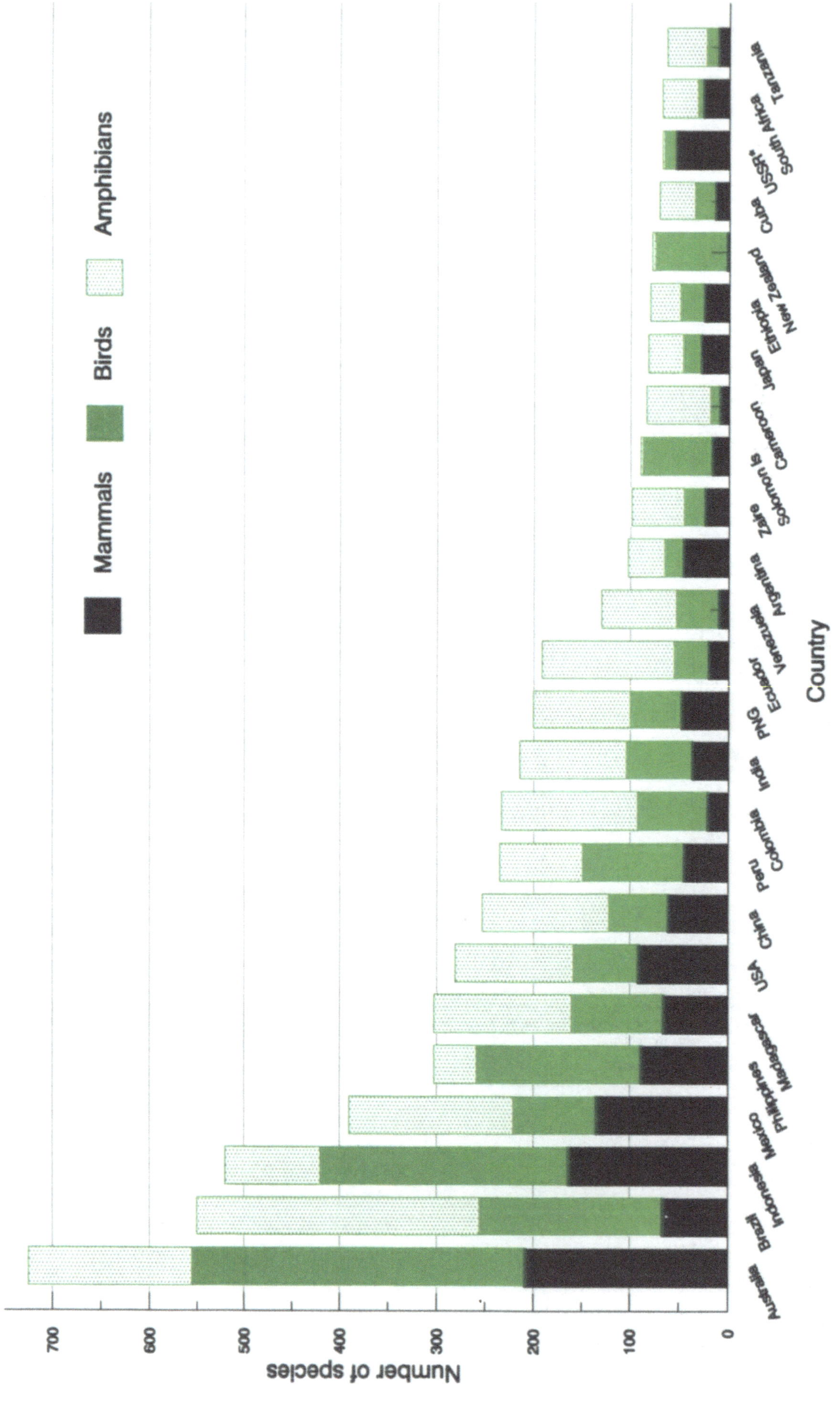

Figure 13.2 Mammal richness and endemism: major countries

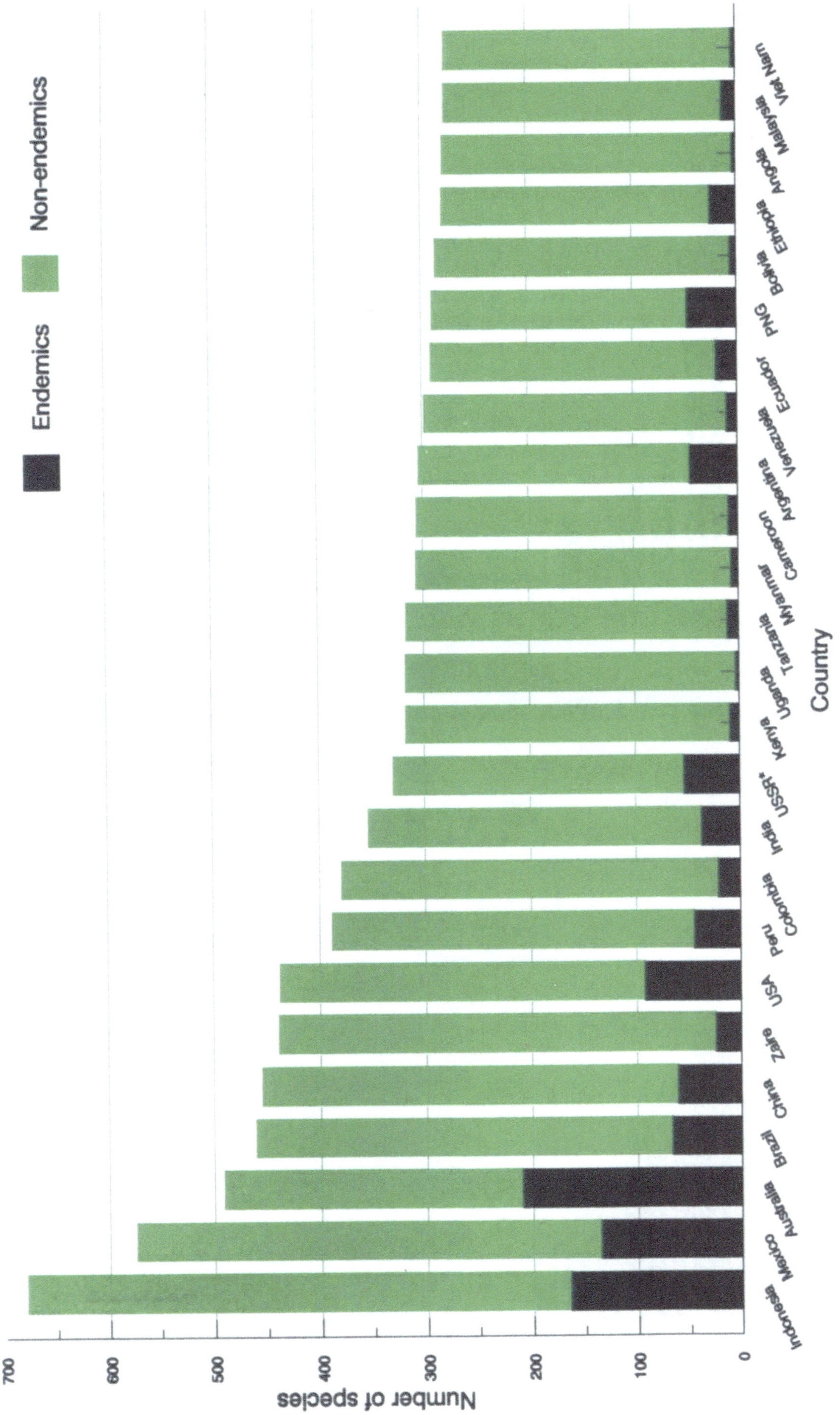

Figure 13.3 Higher vertebrate endemism: Asia

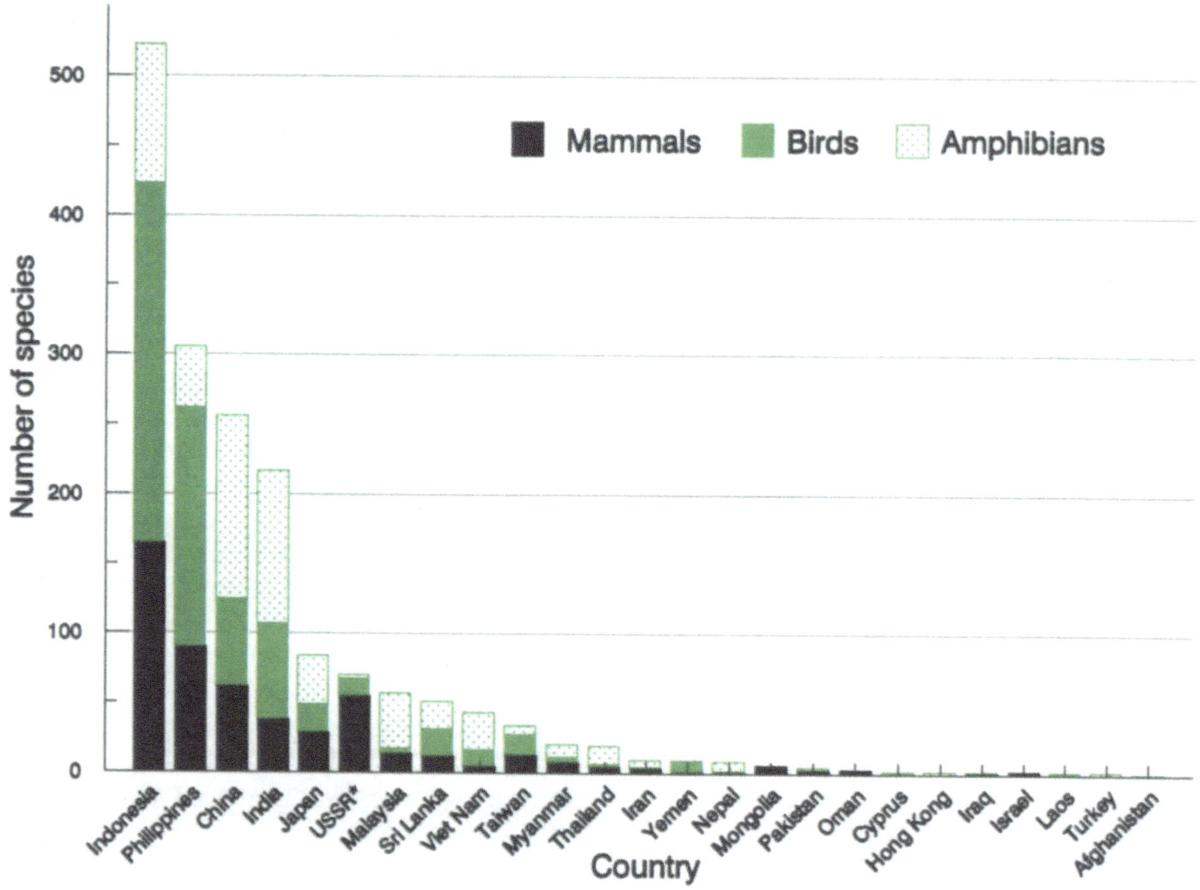

Figure 13.4 Higher vertebrate endemism: Europe

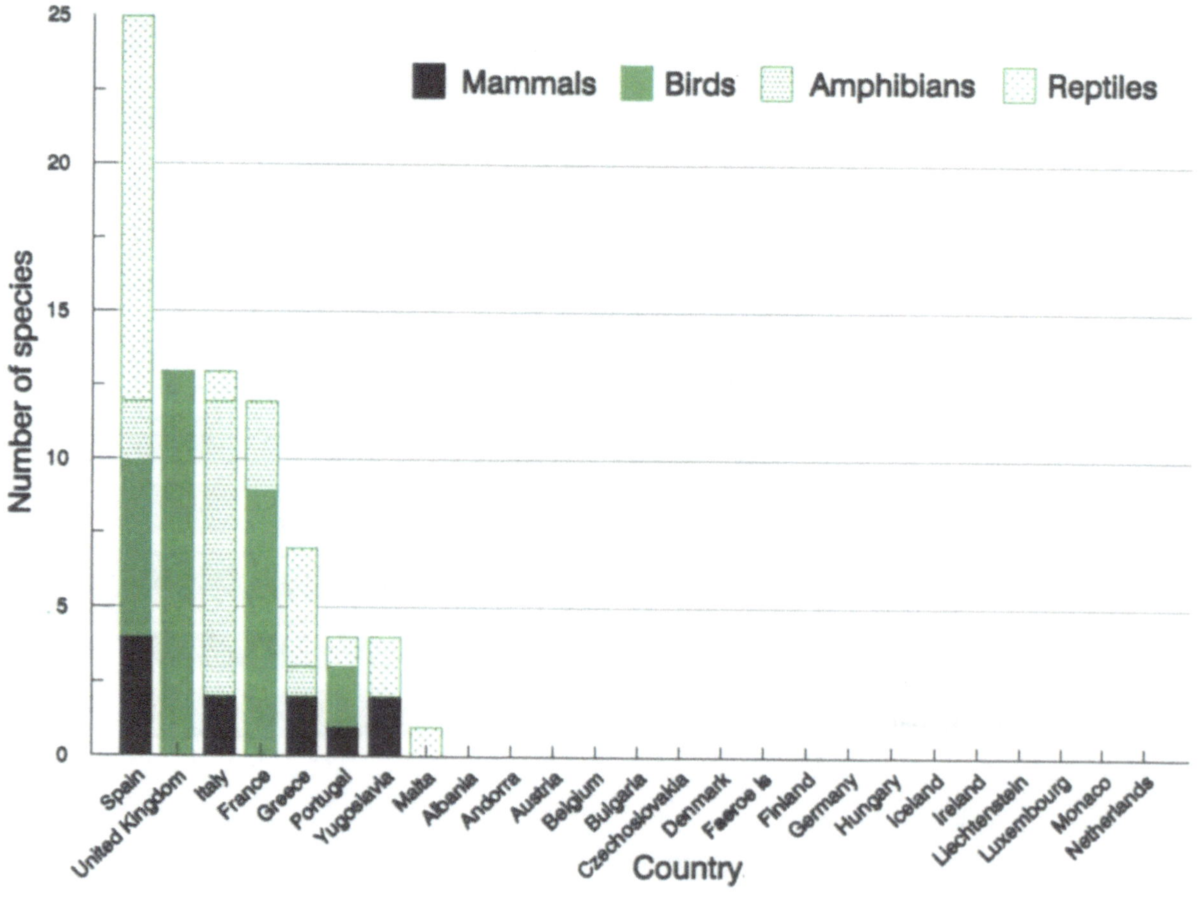

Figure 13.5 Higher vertebrate endemism: North and Central America

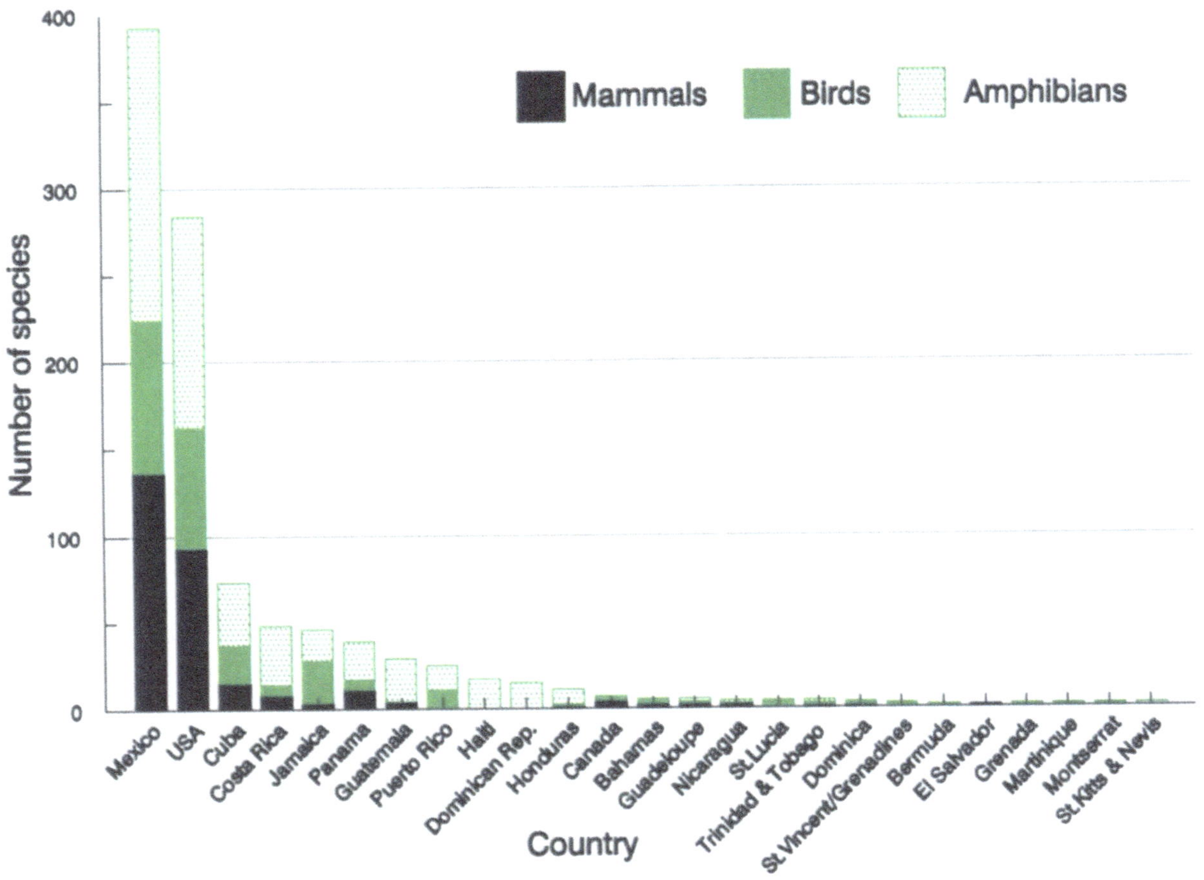

Figure 13.6 Higher vertebrate endemism: South America and Antarctica

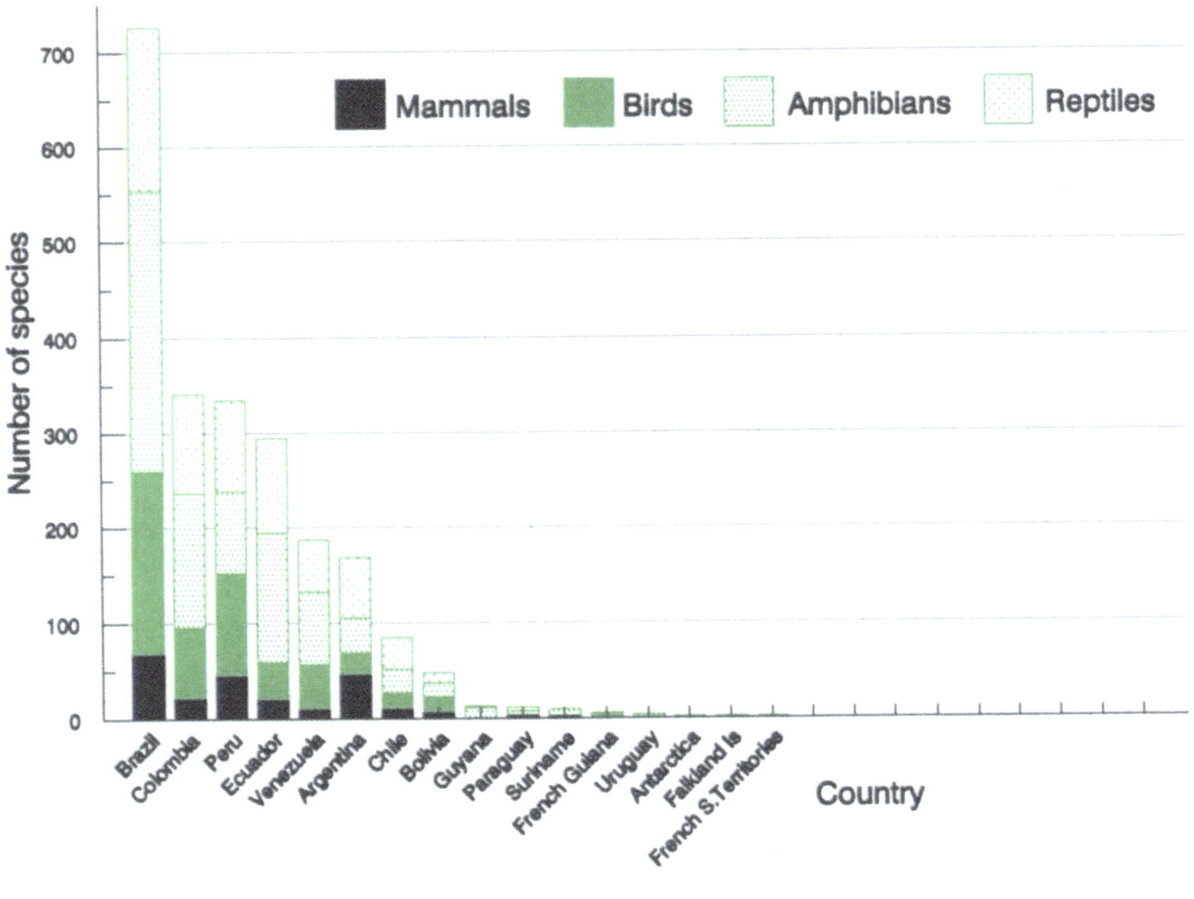

Figure 13.7 Higher vertebrate endemism: Oceania including Australia

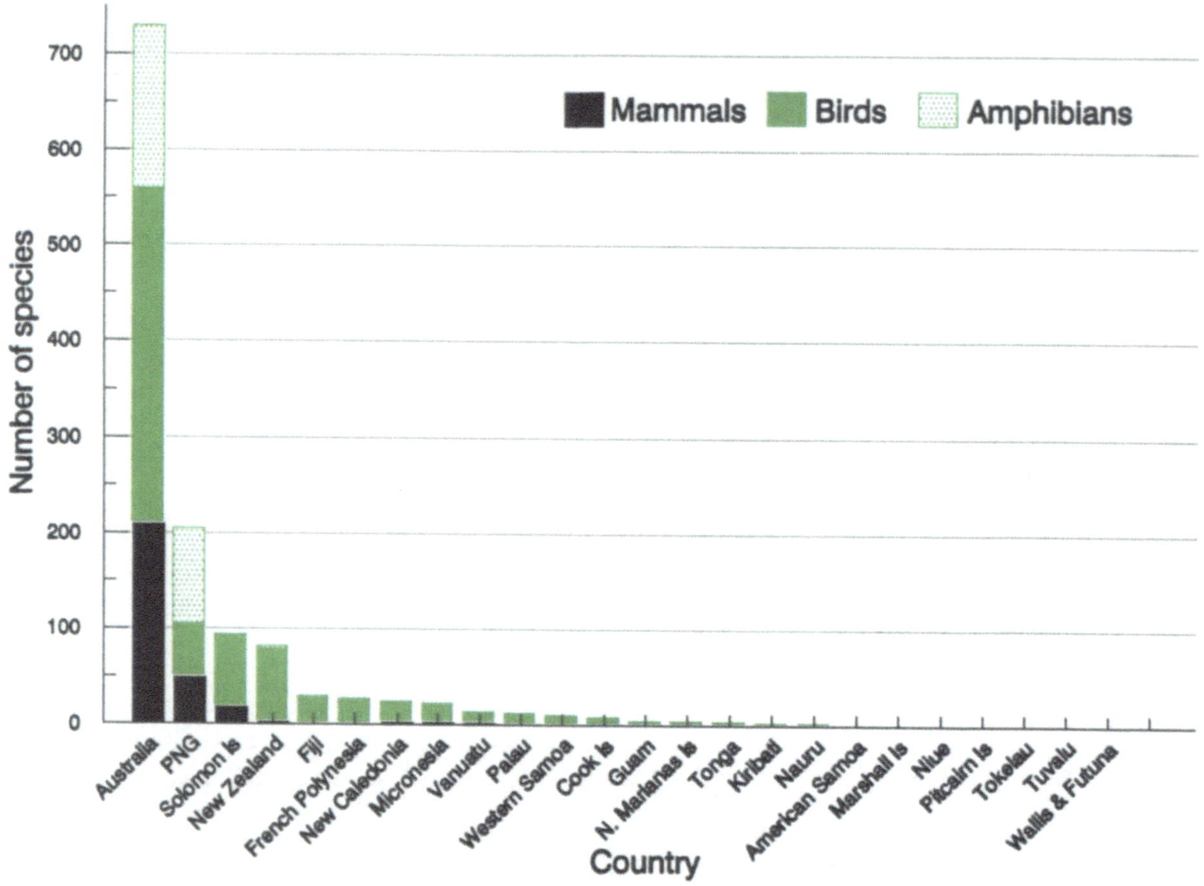

Figure 13.8 Higher vertebrate endemism: Africa and Madagascar

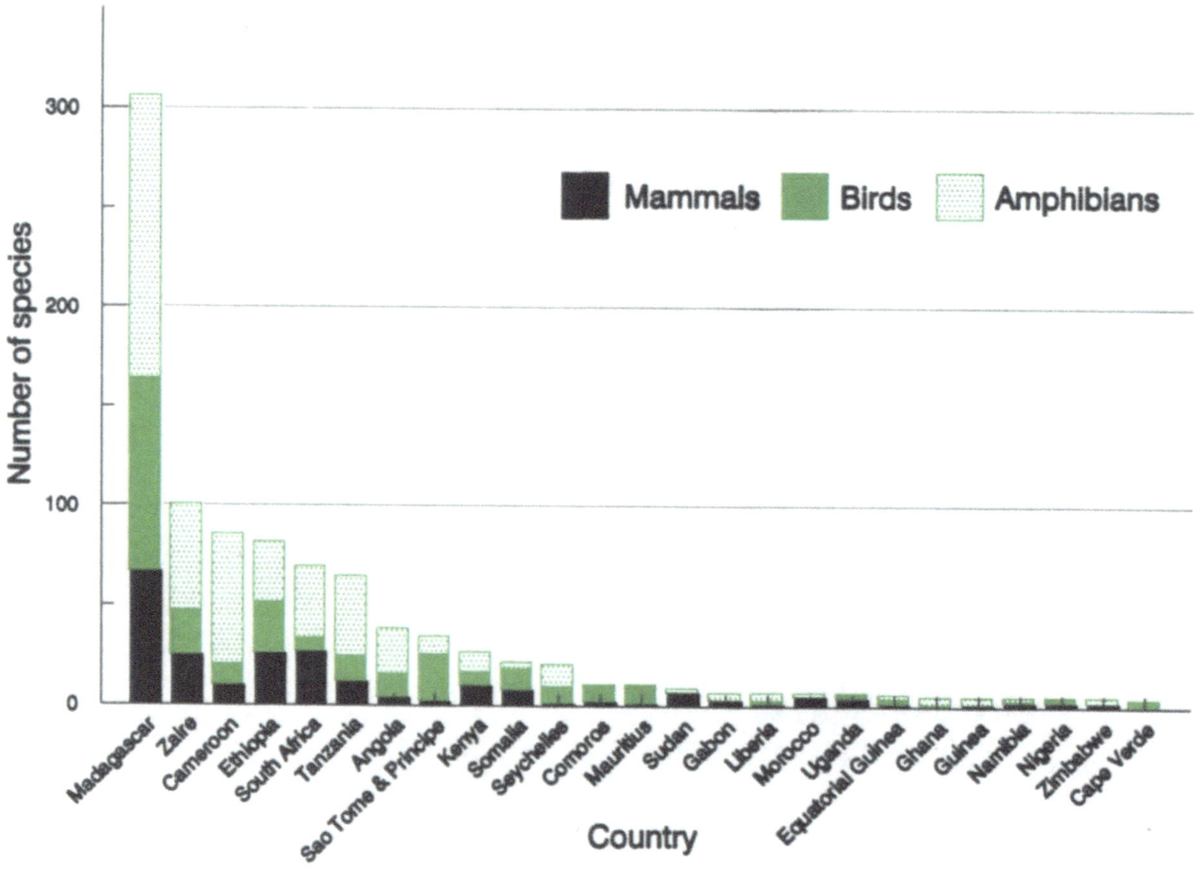

14. ISLAND SPECIES

Islands frequently have distinctive and often unique assemblages of species. In general they have lower species diversity than equivalent continental areas, but tend to have elevated numbers of endemic species. The number of species in a particular taxonomic group on a given island and the proportion of these which are endemic appears to depend on a wide variety of factors, both historical and ecological. Among these are the degree of isolation, age, size, topography and climate of the island and the biological characteristics of the taxonomic groups concerned, in particular their vagility (the ease with which they disperse). Historical accident also appears to play a large part in patterns of species occurrence on islands.

Island endemics tend to be of two types: relict species which appear to have been more widespread in the past and species which have evolved in isolation on the island concerned. Relict species are generally confined to islands which were previously part of larger land masses but which have been isolated through processes of continental drift or changes in sea level. Madagascar and New Caledonia are examples of this, although, because of its size and long period of isolation, Madagascar is perhaps more accurately regarded as an island continent than an oceanic island. In contrast, many island species are believed to represent the results of adapative radiation *in situ* following accidental colonisation by individuals. The biotic composition of isolated, oceanic islands which have never been part of larger land-masses (and are generally volcanic in origin) is largely a result of this process. The taxa represented on these islands are those which have (or whose ancestors had) the capacity for long-range dispersal. Thus, at a very general level, oceanic islands may have good representation of, and high levels of endemism in, plants, birds and some invertebrate groups, such as land snails and some insects, while having low diversity of groups such as non-volant mammals and amphibians.

Once islands have been colonised, other factors play an important role in determining subsequent patterns of evolution and speciation. Species which are highly vagile tend not to speciate and diversify - this applies to, for example, most groups of sea-birds and to strandline vegetation. Species in these groups tend to have very wide distributions, so that, for example, most tropical and sub-tropical Pacific islands have essentially the same, small number of species forming their shoreline vegetation. In contrast, groups such as the rails (Rallidae), pigeons (Columbiformes) and tortoises (Testudinidae) which are essentially terrestrial but which have the capacity for long-range dispersal will tend to form separate species on islands or island groups which they successfully colonise. The degree of speciation which occurs on islands subsequent to colonization appears to be highly dependent on habitat diversity, which is itself dependant on the size, topography and climate of the island. Thus, low-lying oceanic islands, such as coral atolls, tend to have low diversity and low rates of endemism for most groups, while montane (generally volcanic) islands tend to have much higher species diversity and rates of endemism. As with continental ecosystems, other factors being equal, species diversity increases with decreasing latitude.

Island - especially oceanic island - biotas tend to share similar features, such as gigantism in plants and reptiles, dwarfism in large mammals (although most examples of this are extinct) and flightlessness in birds. These may arise from the disharmonic colonisation of islands and the subsequent evolution of plants and animals in isolation (Bramwell, 1979). Of particular importance to conservation are those factors which appear to lead to an increasing extinction-pronenessamongst island species (discussed more fully in Chapter 16). These are largely related to the evolution of island species generally in the absence of large terrestrial 'predators' - for plants these being grazing mammals, for animals these being carnivores. This helps explain the often catastrophic effect of the introduction of animals such as rats, rabbits, goats, pigs and cats on native island biotas.

This report discusses two important island groups - plants and land snails - in some detail. Available data on these two groups has been collated in Tables 14.1 and 14.3.

It is impractical, in a global approach, to treat each island individually, but appropriate to consider them in groups. In this report, we have mainly followed the classification of islands into 147 units made by the International Working Group on Taxonomic Databases for Plant Sciences (TDWG) (Hollis and Brummitt, in press).

For plants, coverage of true oceanic islands is reasonably complete. Most important continental shelf islands other than those of the Sunda Shelf and New Guinea have also been included.

Where complete datasets are available for particular islands, regression analysis shows a moderately close relationship between numbers of endemic plants and snails (see Fig.14.1) but no clear relationship between snails and birds or between plants and birds (bird data not shown).

Figure 14.1 Islands: relationship between plant and snail endemism

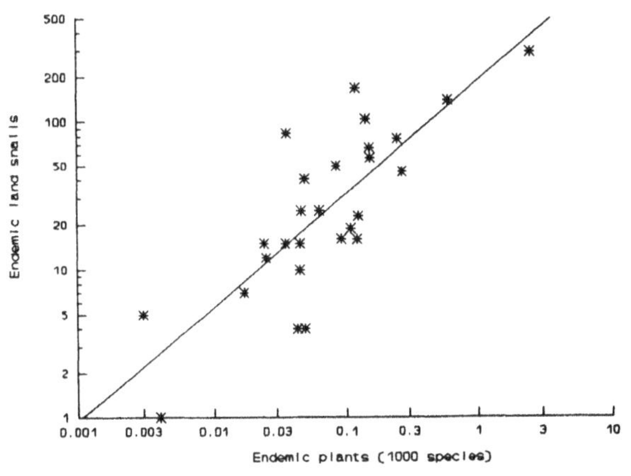

PLANTS ON OCEANIC ISLANDS

The number of endemic species and the proportion of the flora that is endemic varies considerably from island to island and appears to depend on a number of the factors outlined above. On some island groups, like the Hawaiian Islands, the flora can be described as consisting mainly of 'endemics and aliens'; here the endemic species form 89% of the native flora (Wagner *et al.*, 1990). However, on other islands, such as those of the Caribbean, the endemics form only a small element in a diverse flora of predominantly widespread continental species.

The extent to which island endemic floras consist of relict species tends to be a matter of speculation. Greuter (1979) suggests that about half the flora of Crete, for example, is of the relict element. Palaeontologists have found fossils of some Canarian endemics in southern Europe and south Russia; these species include the famous Dragon Tree (*Dracaena draco*), and the dominant species of the Canarian laurel forests, at present a vegetation type now only found in parts of the Canaries, Madeira and to a lesser extent the Azores. The implication is that this remarkable type of forest, now endangered in much of its range, once covered much of the Mediterranean Basin in the Miocene Period, up to 20 million years ago (Bramwell and Bramwell, 1974).

The relict species include an extraordinary array of endemic monotypic genera and even families. Monotypic families (i.e. families with only one species each) on islands include Lactoridaceae (*Lactoris fernandeziana*) on Juan Fernández, Dirachmaceae (*Dirachma socotrana*) on Socotra, and Degeneriaceae (*Degeneria vitiensis*) on Fiji. All are threatened species and, in consequence, threatened families.

In contrast, many of the endemics have evolved in isolation on islands. In the Canary Islands, for example, adaptive radiation of colonists has led to over 30 endemic species in each of the genera *Echium* (Vipers Bugloss), *Limonium* (Sea Lavender) and *Aeonium*. The most outstanding example of diversification and adaptive radiation in the plant world is the Hawaiian Islands, where some genera, such as *Cyanea* and *Cyrtandra*, have over 50 endemic species. Wagner *et al.* (1990) propose that 469 Hawaiian species, in 20 large genera, evolved from only 26-32 different colonists, clearly showing the scope of the evolutionary capacity of isolated islands. The species that result from adaptive radiation tend to be difficult to classify, often with much hybridisation between the various species. As with the Galápagos finches, which helped Darwin develop the theory of evolution and natural selection, these series of evolving and evolved island endemics are of great importance to science.

One of the most extraordinary features of island plants is the phenomenon of gigantism. A group of plants that is otherwise herbaceous and often weedy is represented on some islands as tall shrubs or trees. For example, the endemic species of Vipers Bugloss (*Echium*) and the Sea Lavenders (*Limonium*) in the Canary Islands include woody shrubs with stems several metres high. Some of the most remarkable examples are the tree daisies (Compositae) on St Helena in the Atlantic Ocean and on the Juan Fernández islands off Chile.

In assessing the importance of islands for conservation of the world's plants, the best single measure is simply the number of species endemic to the island or island group. In virtually all cases, for plants, estimates of some kind are available, varying from counts made from detailed floristic analyses to estimates by knowledgeable botanists. This is one of the few datasets on biodiversity, at least for plants, that is complete to a reasonable standard of accuracy worldwide.

Table 14.1 lists the islands and island groups of the world of less than 120,000km^2 in size in declining order of endemic plant species (covering flowering plants, gymnosperms and ferns). This gives a rough guide to the importance of each for botanical conservation. Of the greatest importance are those three islands with over 1,000 endemic plant species each - Cuba with 3,233, New Caledonia with 2,480 and Hispaniola (the Dominican Republic and Haiti) with 1,800. The Hawaiian Islands were previously included but the first comprehensive and complete account of the plants has reduced the number of endemics to below the thousand.

The number of endemics is an effective measure of the importance of individual islands or island groups for plant conservation worldwide. However, this very simple approach is less appropriate where a significant part of the endemic flora is shared between two or more of the island groups used. For isolated islands or island groups like St Helena, Juan Fernández and the Hawaiian Islands, the number of endemics shared with other island groups is very small. But in the Lesser Antilles (the Leeward and Windward Islands) in the Caribbean the shared endemics form a considerable proportion of the endemic flora as a whole. This is partly a consequence of the geographical classification used; because many of the islands are individual nation states, the classification tends to treat each individual island as a single unit, rather than to cluster them together, as with, for example, the Galápagos Islands or Canary Islands. It is partly a consequence of the geography and biology of the islands; the islands tend to be close together, and have similar climates and land forms. Also, because of their proximity to the Greater Antilles (Puerto Rico, Cuba, Hispaniola and Jamaica), and to Central and South America, there are numerous shared species both within the Lesser Antillean chain and between various islands of the chain and neighbouring areas.

The completion of the *Flora of the Lesser Antilles* (Howard, 1989) has permitted an analysis of the endemics of this region. The results are given in Table 14.2, below, and in the accompanying map (Figure 14.2). The table shows the number of plant endemics with different patterns of distribution recorded in the *Flora*. As can be seen, only 107 of the 327 species endemic to the Lesser Antilles as a whole are endemic to single TDWG units (and so are included in Table 14.2). The highest number of endemics for any island is 25 on Guadeloupe, which is also home to a further 111 Lesser Antillean endemics.

The map shows the distributions of 190 of the 327 endemics. Most of the combinations of islands that had only one or two endemics were omitted from the map, as were all combinations of over five islands, as being too complex

Table 14.1 Oceanic islands in declining order of endemic plant species

ISLAND	NO. OF ENDEMIC PLANTS	DATE OF INFORMATION	ISLAND	NO. OF ENDEMIC PLANTS	DATE OF INFORMATION
Cuba	3233	1991	American Samoa	27	1982
New Caledonia	2480	1991	Virgin Is (US)	27 [1]	1974
Hispaniola	1800 [1, a]	1984	Sardinia	26	1991
Jamaica	894	1988	Guadeloupe	25	1974-89
Taiwan	892 [5]	1982-91	Tonga	25	1991
Hawaii	850 [2]	1990	Martinique	24	1974-89
Fiji	700 [5]	1984	Tuamotu Is	20 [4,5]	1931-5
Canary Is	593 [4]	1990	Pitcairn Is	19 [3]	1983
Caroline Is	293 [3,4]	1979, 82	St Vincent	19	1974-89
Socotra	267	1991	Netherlands Antilles	7-19	?
Mauritius	246 [b]	1991	Cayman Is	18 [1]	1984
Puerto Rico	234	1982	Annobon	17	1973
Trinidad-Tobago	215 [1]	1981	Christmas I	17	1980s
Ogasawara-Shoto	152	1978	Coco, Isla del	15 [5]	1966
Vanuatu	150	1975	Bermuda	14	1991
Galápagos Is	148	1980s	Guam	14	1991
Andaman Is	144	1989	Dominica	12	1974-89
Tubuai Is	140 [1,5]	1984	Falkland Is	12	1991
Comoros	136	1917	Gambier	11 [1]	1974
Juan Fernández	123	1991	St Lucia	11	1974-89
Réunion	120 [6],	1991	Ascension I	10	1991
Madeira	118	1980s	Kazan Retto	9	1991
Bahamas	112	1982	Turks and Caicos Is	9	1982
São Tome	108	1944	Auckland Is	6 [1]	1985
Marquesas Is	105	1931-35	Easter I	6	1990
Cape Verde	92	1974-79	Antigua-Barbuda	5 [1,5]	1938
Cyprus	90	1977-91	Maldives	5	1961
Lord Howe I	84	1991	Malta	5	1991
Northern Marianas	81 [3,4]	1979, 82	Wallis and Futuna	5 [1]	1977
Nicobar Is	72 [c]	1989	Antipodean Is	4	1981
Balearic Is	70	1991	Grenada	4	1974-89
Seychelles	63 [d]	1991	Selvagens	4	1980s
Western Samoa	57 [7]	?	Barbados	3	1974-89
Azores	49	1980s	Campbell Is	3 [1]	1961
Bioko	49	1978	Cook Is	3	1991
St Helena	46	1991	Macquarie I	3 [1]	1960
Corsica	45	1991	Montserrat	2	1974-89
Rodrigues	45	1991	St Kitts-Nevis	2	1974-89
Aldabra	43 [4]	1980	St Martin-St Barthélémy	2	1974-89
Sicily	41	1991	Marion and Prince Edward Is	1-2	1989
Tristan da Cunha	40	1965, 81	Anguilla	1	1974-89
Chatham Is	36	1991	Antigua-Barbuda	1	1974-89
Norfolk I	36	1991	Kerguelen Is	1 [1]	1975
Príncipe	35	1944	Nauru	1	?
Solomon Is	30	1991	Netherlands Leeward Is	1	1974-89

Sources: Compiled from numerous sources. See Davis, S. *et al.* 1986. *Plants in Danger: what do we know?* for many pre-1986 references.
Notes: [1] Omits ferns; [2] Omits ferns and gymnosperms; [3] Omits monocotyledons; [4] Includes subspecies and varieties; [5] Estimated from a given percentage of endemism; [6] A slight underestimate as omits endemic species treated as infraspecific level in the WCMC plants database; [7] Certainly an underestimate; [a] Covers Haiti and Dominican Republic; [b] An underestimate as omits full treatment for families not yet covered in the Flore des Mascareignes; [c] Probably an underestimate as only for Great Nicobar Island; [d] Omits the coralline islands, which are listed under Aldabra.

to display graphically. However, the map does show the broad pattern of plant endemism in the region, and provides a convincing argument for a regional approach.

LAND SNAILS

"If we take the whole globe, more species of land shells are found on the islands than on the continents."

Alfred Russel Wallace 1892. Island Life, 2nd edn. Macmillan, London. 563pp.

Global distribution of snail diversity

Recent estimates of world land snail species richness suggest a total of between 30,000 and 35,000 species (Solem, 1984).

Table 14.2 Plant endemism in the Lesser Antilles

NO OF SPECIES	NO. OF TDWG UNITS	NO. OF SPECIES MAPPED
107 occur in	1 unit	107
55 occur in	2 units	48
47 occur in	3 units	20
40 occur in	4 units	15
18 occur in	5 units	8
60 occur in	> 5 units	0
TOTAL ENDEMICS 327		190

Source: Howard, R.A. 1974-89. *Flora of the Lesser Antilles*. 6 vols. Endemics counted by Hugh Synge, 1991.
Notes: Geographical units used: (from north to south) Anguilla, St Martin-St Barthélémy, Netherlands Leewards (Saba and St Eustatius), St Kitts-Nevis, Barbuda-Antigua, Montserrat, Guadeloupe (including Marie Galante, Les Saintes and Le Désirade), Dominica, Martinique, St Lucia, St Vincent, Barbados, Grenada. The TDWG classification divides the Grenadine islands between St Vincent and Grenada, and so records for 'The Grenadines' in the Flora have been disregarded.

Figure 14.2 Plant endemism in the Lesser Antilles

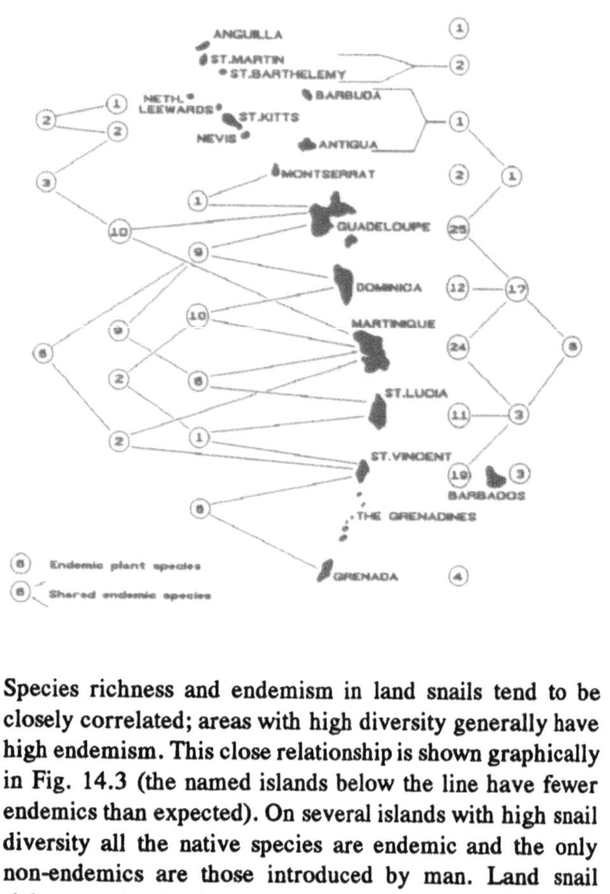

Species richness and endemism in land snails tend to be closely correlated; areas with high diversity generally have high endemism. This close relationship is shown graphically in Fig. 14.3 (the named islands below the line have fewer endemics than expected). On several islands with high snail diversity all the native species are endemic and the only non-endemics are those introduced by man. Land snail richness and endemism are distributed very unevenly around the world, and tend to be highest on islands and in mountains.

A major problem in discussing mollusc richness and endemism is the lack of information for several regions of the world, notably Asia, the Neotropics and the Nearctic; some continental tropical areas are particularly under-recorded and new data could significantly change the current picture of land snail diversity patterns.

Although islands often have highly diverse habitats, not all islands have rich snail faunas. Work in Melanesia (Peake, 1969), and on the Greek islands in the Aegean Sea, suggests that there is a direct correlation between island size and snail species richness. Other work in the Pacific suggests that this relationship is not always a simple one, and Solem (1973) (also Peake, 1981) concluded that highest diversities are found on islands about 15-40km^2 in area and with an elevation of over 400m. Altitude is thus an important factor, and atolls, for example, do not have high snail richness or endemism.

There is some indication that isolation is also an important factor. The island with the greatest number of species is Rapa, one of the smallest and most remote islands in French Polynesia. The location with the highest known snail species richness (i.e. greatest number of species per unit area) is Manukau Peninsula in North Island, New Zealand, where 82 species have been found in a small area.

There is some evidence that although islands often have remarkably high diversity and abundance (in the absence of human impact), their snail faunas are often not 'saturated' and additional snail species could survive. Evidence for this is seen from work in Madeira and on the Greek Islands, where humans have introduced species but the numbers of endemic species have stayed the same (Solem, 1984).

Correlation of land snail diversity with other species

Patterns of land snail diversity and endemism are generally considered not to correlate strongly with those for other groups of animals, particularly higher vertebrates. Available data for islands show a marked positive correlation between numbers of endemic plant species and endemic molluscs (Fig. 14.1), but not between molluscs and birds. There is a lack of data on mollusc faunas of tropical continental areas, and it is thus difficult to make more general statements.

Solem (1984) draws attention to the following islands as known or believed to be important for snails:

- Reasonably well studied large snail faunas on the small high islands of Micronesia, Melanesia, Polynesia, Indonesia, Philippines, Mascarenes, Antilles, Madeira.
- Surveys or studies under way suggest important snail faunas in Japan, Oahu, Tahiti, New Caledonia, New

Table 14.3 Land snails: species richness and endemism on islands

	TOTAL SPECIES	ENDEMIC SPECIES	% ENDEMICS
ATLANTIC			
Atlantic (Macaronesian) Islands			
Azores	98	41	41.8
Canary Is	181	141	77.9
Cape Verde Is	37	16	43.2
Madeira	237	171	88
Selvagens	1	1	100
Mid-Atlantic Islands			
Annobon (Pagalu)	9	7	77.7
Bioko (Fernando Po)	6	c.4	c.66.6
Principe	26	15	57.7
Sao Tomé	26	19	73
St Helena	c. 31	c. 25	c. 80
South Atlantic			
Falkland Is	1	0	0
Northern European Islands			
Faeroe Is	20	0	0
Iceland	35	0	0
Svalbard	0	0	0
MEDITERRANEAN			
Corsica	c. 100	c. 10	c. 10
Cyclades	88	>20	c. 23
Malta	c. 46	c. 7	c. 15
Pityuse Is	36	4	11
Sardinia	–	21	–
INDIAN OCEAN			
Aldabra	c. 9	c. 4	c. 44
Adamans and Nicobars	81	75	93
Anjouan	58	–	–
Comoros (inc. Mayotte)	136	–	–
Grand Comore	37	–	–
Ile Europa	6	0–3	0–50
Mascarene Is	145	127	87.6
Mayotte	90–95	32–41	29–39
Mauritius	109	77	70.6
Moheli	18	–	–
Réunion	40	16	40
Rodrigues	25	15	60
Seychelles	c. 57	c. 24–26	c. 44
Socotra	49	46	94
Sri Lanka	c. 265	–	c.95
Madagascar	380	361	95
CARIBBEAN			
Barbuda	10	0	0
Barbados	37	c.5	c.7
Cuba	c. 600	–	–
Guadeloupe	53	9	17
Jamaica	400–450	–	80–95
Martinique	37	15	c. 40
St Bethelemy	–	0	–
St Martin	c.36	0	0
Saba	14	0	0
Puerto Rico	>85	–	–
Mona	12	6	50
PACIFIC			
Eastern			
Japan	492	c. 487	99
Southwestern			
Fiji	60	–	–
Viti Levu	58	–	–
Lakemba	22	–	–
Karoni	20	–	–
Mothe	13	–	–
New Caledonia	300	c. 299	99
Tutuila	–	8	–
Upolu	44	–	–
Solomon Is	200–270	–	–
Tikapia	16	7	44
Vanuatu	58	57	98
Wallis	15	0	0
Futuna	21	c. 2	c. 5
South-Central			
Henderson	c. 18	3	c. 16
Tahiti	80	c. 72	90
Rapa	>105	>105	100?
North and North-Central			
Hawaiian Is	c. 1000	c. 1000	c. 99.9
Oahu	395	c. 387	98
Kauai	70–80	71	99
Maui	167	–	–
Lanai	54	–	–
Molokai	126	–	–
Hawaii	128	–	–
Pacific Islands off Central & South America			
Galápagos	c. 90	>66	c. 73
Juan Fernández Is	23	23	100
Australia and New Zealand			
New Zealand	c. 1000	–	–
Kermadec Is	c. 20	c. 20	–
Lord Howe I	c. 85	c. 50	c. 60
Norfolk I	84	c. 84	100

Source: table provided by Susan M. Wells (IUCN/SSC Mollusc Specialist Group)
Notes: c. approximated figure. > figure is minimum estimate.

Figure 14.3 Island snails: relationship between species richness and endemism

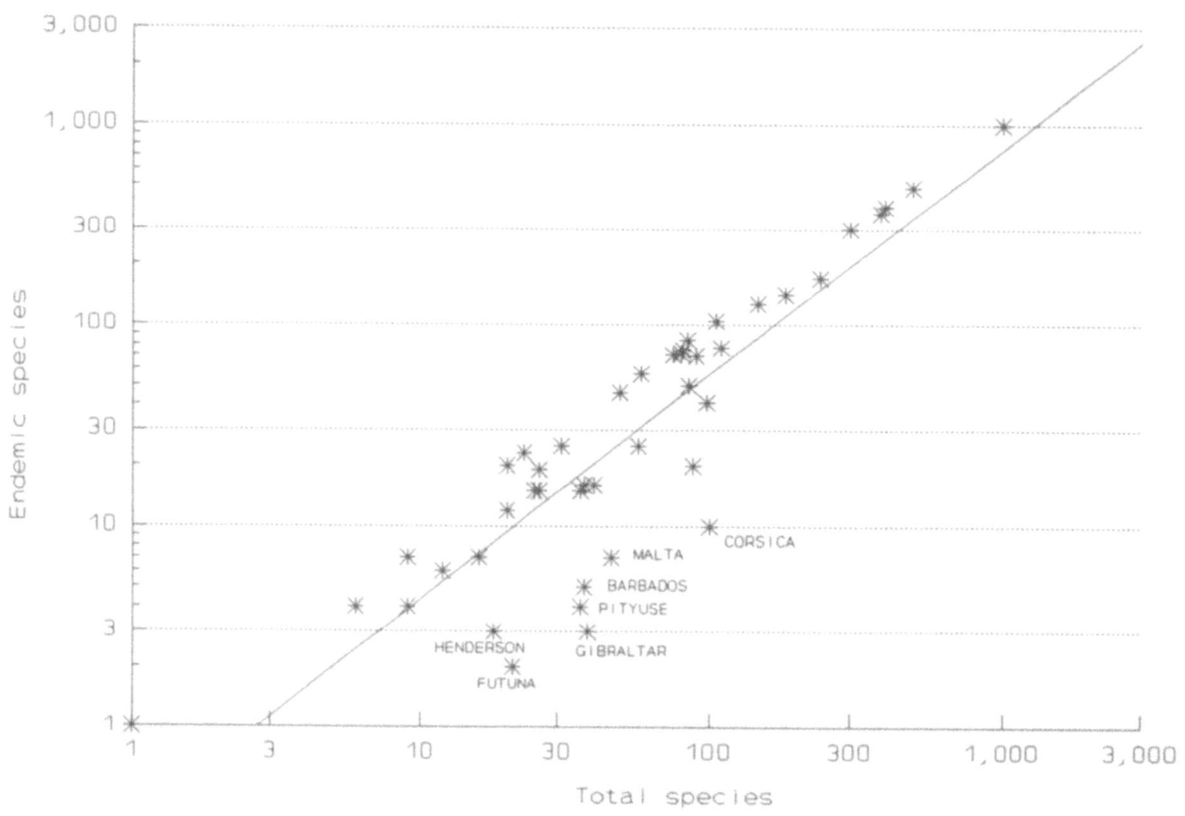

Zealand, Madagascar, Madeira.
- Poor information available but almost certainly important islands: Hispaniola, Cuba, Jamaica, New Guinea.

Some of these areas, particularly the small high islands, do not have particularly high diversities of vertebrates.

Ecology of snails and diversity patterns

Snails that have colonised islands and subsequently speciated tend to be those that are good at dispersal and thus tolerant of stress: the key factors are the presence of a shell to resist desiccation (few slugs are found on islands), and ovoviviparity. On most islands which have high snail diversity, snails are largely confined to the interiors and more mountainous regions and are often forest species restricted to primary forest.

Viable populations of certain snail species appear to be able to exist in very small areas over very long periods of time; this must contribute to maintenance of high species richness. Factors favourable to land snail speciation and the persistence of diverse faunas are: (1) a stable and moderate water supply providing a moist habitat (without either torrential downpours or arid periods), (2) deep litter, (3) a topography of gullies along streams sheltered from prevailing winds, (4) lack of disturbance by man, (5) small-scale vegetation changes e.g. as a result of climatic variation, (6) little predation. Such criteria are found on many volcanic islands and in mountains.

Environmental conditions that are not optimal for snails include: (1) certain types of forest such as rain and monsoon, which may have little litter, an overabundance of rain, acidic soils and seasonal climates; (2) grassland (which may however provide local conditions leading to high abundance); (3) deserts (except where there are mountain refugia).

Threats and extinctions

Known extinctions of island land snails are listed in Chapter 16. Solem's work in the Pacific (Solem, 1976, 1983) gives some idea of the rates of extinction that may be taking place. The endodontoid snails (Families Endodontidae, Charopidae, Punctidae) are tiny tropical snails, only a few millimetres in diameter and are the most diverse group in the Pacific where over 600 species have been described. Over 100 may have become extinct this century; they are mainly ground dwellers in primary forest and are threatened by habitat loss and introduced ants (that prey on the eggs).

Other important island families are entirely or largely arboreal, such as the Partulidae. This family is restricted to the Pacific and comprises about 120 species, most of which are probably threatened. Most is known about the *Partula* of the Society Islands, where they are threatened particularly by the introduced carnivorous snail *Euglandina rosea*. Many populations of achatinelline snails in Hawaii have been lost because of over-collecting and habitat modification; these species are rendered highly vulnerable to extinction because of very low lifetime fecundity (6-24) (Hadfield, 1986). Tillier (*in litt.*, 10 Sept. 1991) says that from his experience (Caribbean, New Caledonia) the island land snails most at risk are those in dry lowland forests which may be lost to cattle grazing or development more rapidly than upland forest.

In New Zealand at least, and probably elsewhere, the native snails are totally dependent on native plant associations for survival. In this country the rate of extinction is apparently fast outstripping the rate of description of undescribed species, many of which are 'spot' endemics, restricted to tiny alpine localities or areas of limestone outcrop (Climo *et al.*, 1986).

References

Bramwell, D. 1979. Introduction. In: Bramwell, D. (Ed.), *Plants and Islands*. Academic Press. Pp.1-10.

Bramwell, D. and Bramwell, Z. 1974. *Wild Flowers of the Canary Islands*. Stanley Thornes (Publishers), London.

Climo, F.M., Roscoe, D.J. and Walker, K.J. 1986. Research on land snails in New Zealand. *WRLG Research Review* No. 9. Wildlife Research Liaison Group, Wellington, NZ. 28pp.

Greuter, W. 1979. The origin and evolution of island floras as exemplified by the Aegean Archipelago. In: Bramwell, D. (Ed.), *Plants and Islands*. Academic Press. Pp.87-106.

Hadfield, M.G. 1986. Extinction in Hawaiian achatinelline snails. *Malacologia* 27(1):67-81.

Hollis, S. and Brummitt, R.K. (in press). *World Geographical Scheme for Recording Plant Distributions*. International Working Group on Taxonomic Databases for Plant Sciences (TDWG) and Hunt Botanical Library, Pittsburg.

Howard, R.A. 1974-89. *Flora of the Lesser Antilles*. 6 vols. Endemics counted by Hugh Synge, 1991.

Peake, J.F. 1969. Patterns in the distribution of Melanesian land Mollusca. *Philosophical Transactions of the Royal Society B* 255:285-306.

Peake, J.F. 1981. The land snails of islands - a dispersalist's view point. In: Forey, P.L (Ed.), *The Evolving Biosphere*. British Museum (Natural History), Cambridge University Press. Chapter 19.

Solem, A. 1973. Island size and species diversity in Pacific island land snails. *Malacologia* 14:307-400.

Solem, A. 1976. Endodontoid land snails from Pacific Islands (Mollusca: Pulmonata: Sigmurethra). Part I. Family Endodontidae. Field Museum of Natural History, Chicago. 508pp.

Solem, A. 1983. Endodontoid land snails from Pacific Islands (Mollusca: Pulmonata: Sigmurethra). Part II. Families Punctidae and Charopidae, zoogeography. Field Museum of Natural History, Chicago. 336pp.

Solem, A. 1984. A world model of land snail diversity and abundance. In: Solem, A. and Bruggen, A.C. van (Eds), *Worldwide Snails*. E.J. Brill/W. Backhuys, Leiden. Chapter 1, pp.6-22.

Wagner, W.L. *et al.* 1990. *Manual of the Flowering Plants of Hawai'i*. 2 vols. University of Hawaii Press, Bishop Museum Press.

Chapter based on plant account provided by Hugh Synge and snail account supplied by Susan M. Wells (and the IUCN/SSC Mollusc Specialist Group.)

15. CENTRES OF SPECIES DIVERSITY

INTRODUCTION

A principal goal of conservation activity is to ensure the long-term survival of as many species as possible. Traditionally, most resources available have been allocated to single 'flagship' species, either through *in situ* measures or through *ex situ* captive breeding efforts. Often these are large, charismatic species which generate considerable public interest. Habitat destruction and modification are the most important factors now affecting species survival and although conservation initiatives focused on single species may protect a particular organism's habitat, and by extension a host of other associated species, they do not necessarily conserve those habitats which contain the most species.

Biodiversity is not distributed uniformly across the globe: some habitats, particularly tropical forests among terrestrial systems, possess a greater number or density of species than others. Thus a 13.7km^2 area of the La Selva Forest Reserve in Costa Rica contains almost 1,500 plant species, more than the total found in the 243,500km^2 of Great Britain, while Ecuador harbours more than 1,300 bird species, or almost twice as many as the USA and Canada combined (Myers, 1988). Given the budgetary constraints on conservation and the competing demands of other forms of land-use, some system is necessary for identifying the areas in which a certain allocation of effort will maximise species survival. It is widely accepted that the identification and prioritisation of important centres of biodiversity are necessary at both the national and the global scale. A number of methods by which such areas could be determined have been suggested.

METHODS OF DETERMINING AREAS OF CONSERVATION PRIORITY

Overall species diversity

The simplest method of suggesting target areas for conservation action is to identify countries with the highest number of species (greatest species richness). For example, Mittermeier (1988) and Mittermeier and Werner (1990) recognised that a very small number of countries situated mainly in the tropics possess a large fraction of the world's species diversity, and introduced the concept of 'Megadiversity Countries' which, they suggested, merit special international attention. McNeely *et al.* (1990) used country species lists of vertebrates, swallowtail butterflies, and higher plants to identify 12 such megadiversity countries: Mexico, Colombia, Ecuador, Peru, Brazil, Zaire, Madagascar, China, India, Malaysia, Indonesia and Australia. Together these countries hold up to 70% of the world's species diversity in these groups. This approach is relatively simple in that it involves species inventory within a given geopolitical boundary; it also recognises that conservation action is managed at the country level. One drawback to this approach, however, is that it fails to take into account the *uniqueness* of the fauna and flora of a country or region. There may be considerable overlap in species composition between different regions with high species numbers, particularly if they are situated close to one another geographically. Taking mammal species in two of the megadiversity countries listed above as an example, 271 species of mammal (excluding Cetacea) have been recorded from Ecuador and 344 from neighbouring Peru, but 208 of these are common to both countries. In addition, high diversity regions may contain large numbers of very widely distributed species which are currently neither threatened nor otherwise of special conservation concern.

Endemic species diversity

An alternative approach is to identify areas with the greatest numbers of 'endemic' or 'restricted-range' species. An endemic species is one restricted to some given area, which might be a mountain top, a river, a country or continent. In this context, the assessment is often based on single-country endemics, or on some small identifiable region within a country. At the global level these are areas of high conservation priority because if unique species are lost they can never be replaced. Although not biologically meaningful, the choice of country boundaries for assessing endemicity is of great practical significance because conservation action is usually administered at the national level.

An important study that attempted to use endemic plant species to identify areas of global conservation concern was that of Myers (1988). Focusing on tropical forests, Myers identified 10 regions or 'Hot Spots' that are characterised by high concentrations of endemic species and are experiencing unusually rapid rates of habitat modification or loss (Table 15.1). These 10 areas cover only 292,000km^2, or 0.2% of the Earth's land surface, and comprise 3.5% of the remaining primary forest. Together, however, they harbour 34,400 endemic plant species (27% of all tropical forest species and 13% of all plant species worldwide).

In a subsequent publication, Myers (1990) identified a further eight terrestrial hot spots, four in tropical forest areas and four in Mediterranean-type areas (Table 15.1). Together these contain 15,555 endemic plant species, or 6% of the world's total, in 454,400km^2 or 0.3% of the world's land area. This second selection of eight areas are therefore not nearly as rich in endemic species as the first 10, containing only 45% as many plant species in an area one and a half times as large. In total these 18 sites contain approximately 49,955 endemic plant species, or 20% of the world's plant species, in just 746,400km^2, or 0.5% of the Earth's land surface.

Despite its limitations (e.g. the difficulty of quantifying threats to the existing habitat, and the paucity of distributional information available for many of the world's plant species), Myers' work is an important step towards determining areas where conservation requirements are greatest and where the potential benefits from conservation measures would be maximised.

From the wider conservation perspective, the question of interest is whether levels of endemism in one taxon are correlated with those in others. If endemism follows similar patterns for different taxa, then conservation measures

Table 15.1 Numbers of endemic species present in 18 'Hot Spots'

REGION	HIGHER PLANTS	MAMMALS	REPTILES	AMPHIBIANS	SWALLOWTAIL BUTTERFLIES
Cape Region (South Africa)	6,000[2]	15	43	23	0
Upland western Amazonia	5,000[1]	-	-	c. 70	-
Atlantic coastal Brazil	5,000[1]	40	92	168	7
Madagascar	4,900[1]	86	234	142	11
Philippines	3,700[1]	98	120	41	23
Borneo (north)	3,500[1]	42	69	47	4
Eastern Himalaya	3,500[1]	-	20	25	-
SW Australia	2,830[2]	10	25	22	0
Western Ecuador	2,500[1]	9	-	-	2
Colombian Chocó	2,500[1]	8	137	111	0
Peninsular Malaysia	2,400[1]	4	25	7	0
Californian floristic province	2,140[2]	15	15	16	0
Western Ghats (India)	1,600[2]	7	91	84	5
Central Chile	1,450[2]	-	-	-	-
New Caledonia	1,400[1]	2	21	0	2
Eastern Arc Mts (Tanzania)	535[2]	20	-	49	3
SW Sri Lanka	500[2]	4	-	-	2
SW Côte d'Ivoire	200[2]	3	-	2	0
TOTAL	49,955	375	892	737	59

Sources: For plants, Myers (1988[1], 1990[2]); for animals, miscellaneous sources (WCMC).
Notes: - indicates no data yet available. All regions are classed floristically as tropical forest, with the exceptions of four regions which have Mediterranean-type floras, i.e. Cape Region South Africa, SW Australia, Californian floristic province and Central Chile.

focused in areas of high endemism will generate enhanced returns in terms of overall biodiversity conservation. Myers' botanical hot spots are undoubtedly good sites to conserve endemic plants, and they often contain high numbers of endemics among other groups. There are exceptions, however, and the strength of such relationships remains to be investigated. Area, size, scale, and the biogeography of different taxa will be among the important variables.

Bibby *et al.* (1992) examined available data for other groups to compare with bird data, and showed that endemism at least among larger vertebrates is often, though not always, related. Countries with high numbers of endemics in one vertebrate group often also have high numbers of endemics among other vertebrates (see Table 15.2). Statistically, numbers of mammals and birds, and of mammals and reptiles, correlate quite closely. Country size is probably an important factor underlying these correlations: larger countries tend to have larger numbers of species and also larger numbers of endemic species of each taxon.

Even if associations do exist between levels of endemicity in different taxa, care must be exercised in their interpretation and application since correlations are merely generalisations. For example, Table 15.1 shows that while there may be some broad similarities amongst endemic species numbers in different vertebrate and plant taxa, there are significant discrepancies. Thus, although the Colombian Chocó has high numbers of endemic reptiles and amphibians (137 and 111 respectively) it has relatively few endemic mammals (8); and the Cape Region of South Africa, which has the highest number of endemic plant species (6,300) has only 15 endemic mammals. Overall conservation priorities should therefore be based on a synthesis of detailed analyses of different taxonomic groups, not an analysis of the pattern of endemicity in just one taxon.

Critical faunas analysis

Whether simple species richness or levels of endemism are initially used to assess the biological importance of sites, the concept of 'complementarity' and its application in 'critical faunas analysis', first introduced by Ackery and Vane-Wright (1984), is increasingly used to determine conservation priorities objectively. In this approach the entire set of taxa within the group under consideration, e.g. single-country endemic amphibians, constitutes the 'complement'. The single most important site for conservation is that at which the greatest proportion of the complement is represented. The portion of the complement not included is called the 'residual complement'. The priority for second site selection can be determined by identifying the site that adds the greatest proportion of the residual complement to the initial choice. The process can be continued in a step-wise sequence until all sites have been considered and allocated a priority. The advantage of this process is that it produces an objective and optimised selection sequence, against which the performance of any other (sub-optimal) sequence can be judged for its relative efficiency in representing total biodiversity.

Table 15.2 Countries rich in endemic land vertebrates

COUNTRY RANK ORDER	ENDEMIC TAXON							
	MAMMALS		BIRDS		REPTILES		AMPHIBIANS	
1	Australia	210	Indonesia	356	Australia	605	Brazil	293
2	Indonesia	165	Australia	349	Mexico	368	Mexico	169
3	Mexico	136	Brazil	176	Madagascar	231	Australia	160
4	USA	93	Philippines	172	Brazil	178	Madagascar	142
5	Philippines	90	Peru	106	India	156	Ecuador	136
6	Brazil	70	Madagascar	97	Indonesia	150	Colombia	130
7	Madagascar	67	Mexico	88	Philippines	131	India	110
8	China	62	New Zealand	74	Colombia	106	Indonesia	100
9	USSR	55	Solomon Islands	72	Ecuador	100	Peru	87
10	PNG	49	India	69	Peru	95	Venezuela	76
11	Argentina	47	Colombia	58	Cuba	79	Cameroon	65
12	Peru	46	Venezuela	45	South Africa	76	Zaire	53

Source: WCMC database.

Collins and Morris (1985) performed a critical faunas analysis at the country level, examining endemicity in swallowtail butterflies. They found that if the five countries with the highest numbers of *endemic* swallowtail species enacted conservation plans to protect swallowtails, then 54% of the world's *total* number of swallowtail species would be conserved. If the next five countries were included, the total protected would rise to 68%. Increments decreased as further blocks of five countries were added, with 15, 20, 25, 30, 35, 40 and 45 countries respectively including 77, 90, 93, 95, 96, 97 and 99% of the world's swallowtails.

This type of analysis can be used to direct international and national attention to faunistically important countries, states or provinces. Local knowledge must however remain the basis for more detailed conservation planning, in order to identify precise centres of species richness and importance within a country, and to plan a system of protection around those centres.

Whilst earlier studies were based on species numbers alone, more sophisticated studies of this kind are now being developed which attempt to take into account species turnover between sites, not only in a simple numerical sense but by use of some taxic diversity index. Taxonomic dispersion is the most complex but perhaps intuitively most attractive of these, in that, given a hypothesis of the evolutionary relationships among members of a group, it attempts to select an even spread of taxa across the hierarchy (see Chapter 2).

Table 15.3 shows one application of this procedure, to determine the priority sequence of African protected areas for the conservation of antelopes. Serengeti National Park (Tanzania) is the richest single site, holding breeding populations of 24% of all African antelope species. The highest incremental change occurs with the addition of Kafue National Park (Zambia): together the two parks hold 38%. The addition of two further reserves, Haut Dodo Faunal Reserve (Côte d'Ivoire) and Ouadi Rimé-Ouadi Achim Faunal Reserve (Chad) brings the representation of African antelope species diversity to over 56% in just four protected areas.

In critical faunas analysis, if all the species in the world in the taxon under consideration are to be conserved, and if all species are treated as taxonomically equal, then *a priori* endemics are accorded a high value in the prioritisation sequence. Thus Ackery and Vane-Wright (1984) found that in order to conserve all 158 species of milkweed butterflies (Lepidoptera: Danainae) a total of 31 sites or 'critical faunas' needed protection. Site selection was made starting with the site that contained the highest number of endemics - in this case Sulawesi. Of these 31 sites, 24 were sufficient to protect all the narrow endemics, and a further seven were sufficient to complete the list. In practice, even if the conservation of 100% of the Earth's biodiversity is the goal, some species will of necessity be neglected. The critical faunas approach may not always offer a sufficiently flexible strategy for planning conservation at the global level (Vane-Wright *et al.*, 1991).

Conclusion

Although species are normally used as the basis for critical faunas evaluation or distributional analysis, other taxonomic groupings such as genus or family can be used instead. Different forms of weighting system can also be introduced, so that for instance a species in a monotypic genus, such as the Giant Panda *Ailuropoda melanoleuca* might be allotted a higher conservation priority than a species with many congeners. New measures of biodiversity are now being developed which can take into account the genetic distinctiveness of species based on the relative position of species and other taxa in the classification hierarchy. For example, Vane-Wright *et al.* (1991) suggest using a 'taxic diversity measure' based on the information content of cladistic hypotheses (indicating the branching pattern of evolution), which would provide a measure of taxonomic distinctiveness. Bibby *et al.* (1992) apply a simple method of assigning taxonomic uniqueness to endemic species based

Table 15.3 Biodiversity scores for Afrotropical antelopes

STEP NO.	DIVERSITY INCREMENT %	DIVERSITY CUMULATIVE %	CONSERVATION AREA NAME	COUNTRY
1	23.95	23.95	Serengeti NP	Tanzania
2	13.70	37.65	Kafue NP	Zambia
3	9.99	47.64	Haut Dodo FR	Côte d'Ivoire
4	9.32	56.96	O. Rime-O. Achim FR	Chad
5	4.85	61.81	Yangudi Rassa NP	Ethiopia
6	4.71	66.52	Odzala NP	Congo
7	5.27	71.79	W. Pretorius GR	S Africa
8	3.50	75.29	Manovo-G-St Floris NP	C African Rep
9	2.81	78.10	De Hoop NR	S Africa
10	2.82	80.91	Gorongosa NP	Mozambique

Note: Part of the optimised priority area sequence of protected areas in terms of their potential for conservation of African antelopes, based on the taxonomic dispersion measure and complementarity. Serengeti National Park (Tanzania) is the richest single site, holding breeding populations of species accounting for 24% of African antelope diversity. The highest incremental addition occurs in Kafue National Park (Zambia); in combination the two total 38%. The addition of two further reserves (one in Côte d'Ivoire; one in Chad) brings the representation of African antelope taxonomic diversity to over 56%. (Based on data from East (1988, 1989, 1990) and Gentry (in press) and analysis of Williams (unpublished report).)

on the diversity of the genus and family to which the species belongs.

The kinds of technique outlined above are useful tools which enable conservation biologists to prioritise sites and allocate scarce resources. Care must be taken to base overall global conservation priorities on a number of taxa, which ideally should be well-represented throughout the world. It should, however, be remembered that the identification of areas of high diversity is but the first step in determining effective conservation plans. The size and heterogeneity of the sites under consideration also have serious implications for conservation biology through their effects on minimum viable population sizes, stochastic ecological effects, etc. Planners must seek to conserve multiple populations whenever possible, to allow for chance local extinctions. Progress in designing the protection of a functional ecological system or set of systems has recently been made in Australia (e.g. Margules, 1989), where wildlife services are developing step-wise analyses intended to take these kinds of factors into account.

Two major projects have developed and refined approaches to the systematic identification of centres of species endemism or diversity at the global level. The IUCN Plant Conservation office is identifying centres of plant diversity, and the International Countil for Bird Preservation (ICBP) has identified centres of endemism among restricted-range birds. The approaches and major findings of these two projects are detailed below. Simple visual comparison of the two world maps relating to these projects (Figs 15.1 and 15.2) shows much broad correspondence between the sites, although there are differences in detail (e.g. more botanically diverse areas identified in Mediterranean regions). The sites concerned are noted in Tables 15.6 (plants) and 15.7 (birds).

CENTRES OF PLANT DIVERSITY

The IUCN Plant Conservation Programme is at present carrying out a project to identify the several hundred major Centres of Plant Diversity (CPD). These are defined as places particularly rich in plant life which would if protected safeguard the majority of wild plants in the world. The book IUCN is preparing with the help of collaborators worldwide will provide detailed data sheets on some 250 selected areas. It will also document the many benefits, economic and scientific, that conservation of these areas would bring and will outline the potential value of each for sustainable development.

IUCN has defined the CPD 'sites' as of three types:

- botanically rich sites that can be defined geographically (e.g. Mt Kinabalu in Borneo)
- geographically defined regions with high species diversity and/or endemism (such as the Atlas Mountains, or the Cordillera Bética in Spain)
- vegetation types and floristic provinces that are exceptionally rich in plant species (such as the Amazon rain forests and the South-West Botanical Province of Western Australia).

The formal criteria for inclusion of sites in the Centres of Plant Diversity project specify that each must have one or both of the following two characteristics:

- the area is evidently species-rich, even though the number of species present may not be accurately known
- the area is known to contain a large number of species endemic to it.

The following characteristics are also considered in the selection: a) the site contains an important gene pool of plants of value to man or plants that are potentially useful; b) the site contains a diverse range of habitat types; c) the site contains a significant proportion of species adapted to special edaphic conditions; d) the site is threatened or under imminent threat of large-scale devastation.

The selection is therefore based on botanical importance rather than on degree of threat. A site that could be considered safe one year could be severely endangered the next. This is particularly likely in the tropics where pressures on land continue to increase.

The site selection process involved extensive consultations with experts in all major regions. In Africa, China, India, North and South America, this has resulted in Workshops at which data on lists of proposed CPD sites have been reviewed, and the final site selection made. For the Central Asian region, the final selection has not yet been made.

Figure 15.1 Centres of plant diversity: the world

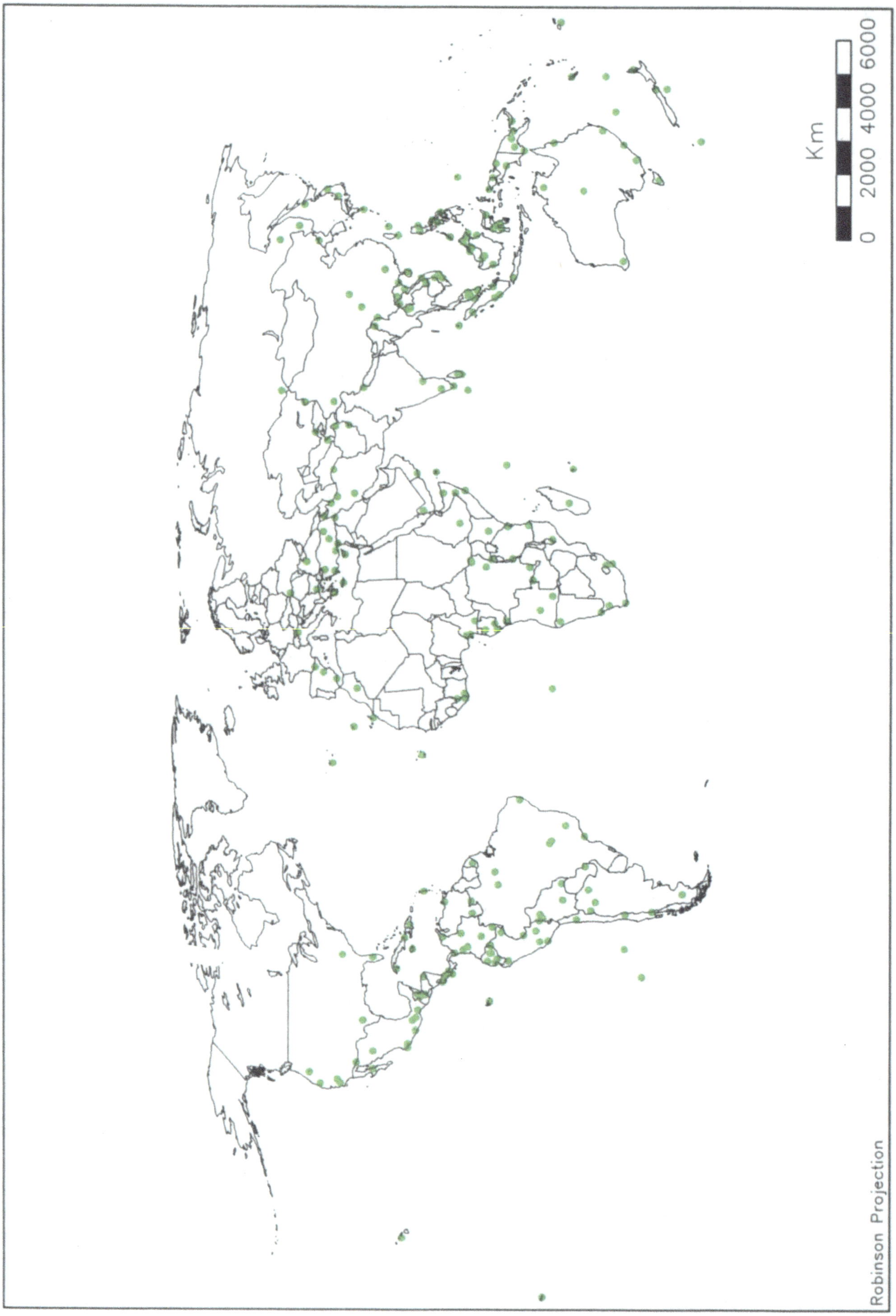

Figure 15.2 Endemic bird areas: the world

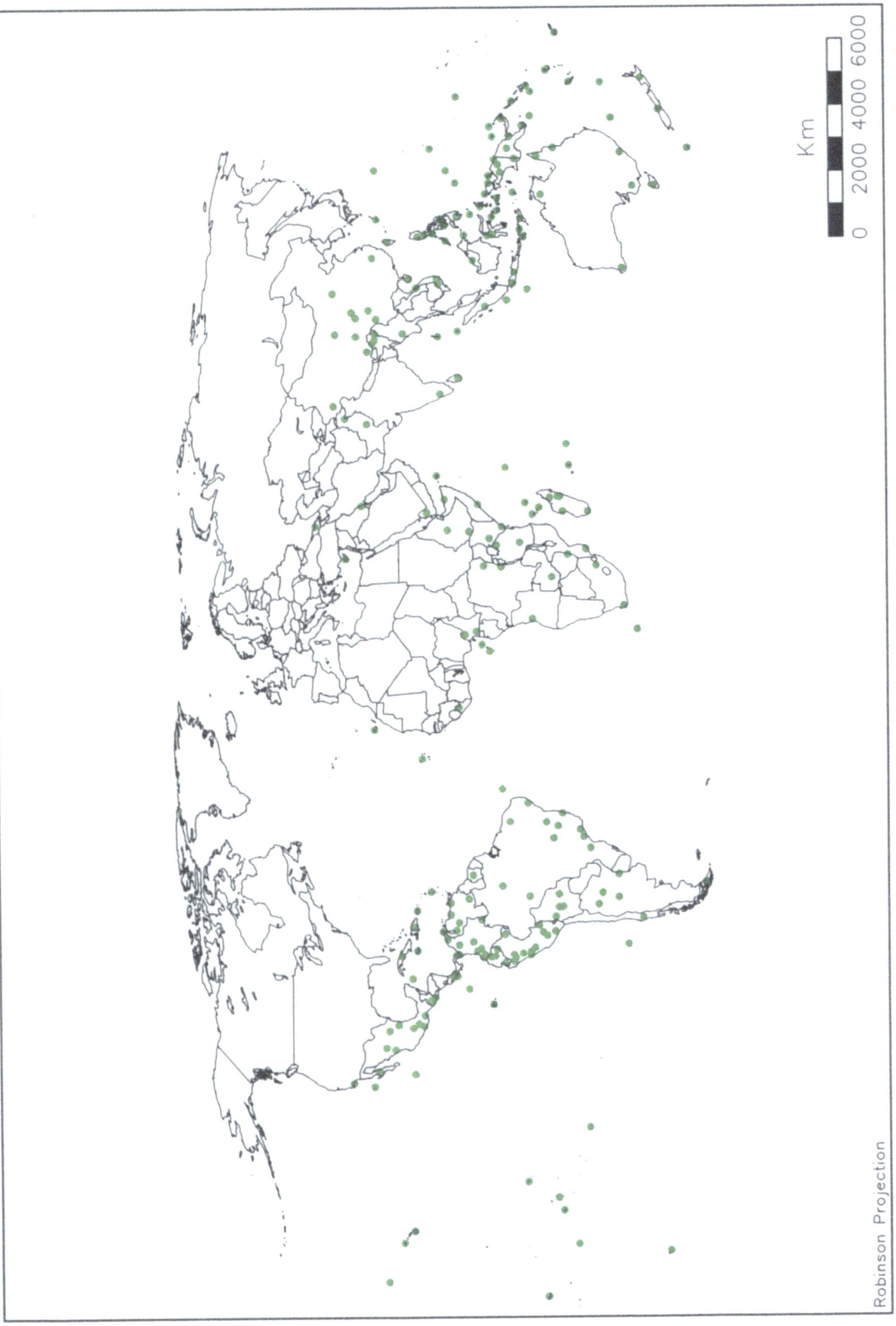

The difficulty in selecting sites varies greatly from one part of the world to another. In some regions the selection is easy. In West Africa, for example, it has long been known that the famous Taï Forest National Park is the only large portion of rain forest in Côte d'Ivoire still intact; with over 150 plants endemic to the park, the Taï is an obvious candidate for inclusion. Often, especially in Africa, the Centres of Plant Diversity are mountains, like Mt Nimba where the borders of Guinea, Liberia and Ivory Coast meet, Mt Mulanje in Malawi, and the Aïr Mountains in the Sahara. Such mountains have a wide range of diverse plant communities but are often delimited by low-diversity habitat making identification of sites relatively easy.

In other areas the selection is much more difficult. The islands of Borneo and New Guinea, for example, contain the largest floras in Asia. Virtually all the habitats are rich in plants, but floral diversity varies from place to place in very complex ways. As a result, it is very hard to specify which parts of Kalimantan, if protected, would include the most plant species. In Irian Jaya, botanical knowledge is not yet sufficient to say with any degree of confidence which areas are richest in plant species.

In some regions, the selection of the sites that need to be protected cannot be made on botanical criteria alone. For example, the Atlantic forests of Brazil are reduced to 2-5% of their original extent, and they have a quite different complement of species to the much larger Amazonian forests. To save their flora, as many as possible of the surviving remnants should be protected. Where such remnants provide two similar sites, with similar complements of species, socio-economic considerations rather than botanical ones will influence the decision as to which sites might be protected. In such cases the CPD project will identify the whole vegetation type - in this case the Atlantic forests of Brazil - and not recommend detailed protection strategies for the various sites within that region.

In addition to data sheets on the selected sites, the CPD publication will contain Regional Overviews which will describe the general patterns of vegetation and plant distribution. Opinions will naturally vary as to the exact choice of sites for coverage at international level, and in order to avoid implications that only the 250 or so sites outlined should be protected, the Regional Overviews will also contain lists of other sites for botanical conservation, many of a lesser priority but important nonetheless.

The 'Centres' concept is particularly appropriate for plant conservation because it focuses on the plant-rich tropics. As the map of sites (Fig. 15.1) shows, most of the 241 sites selected so far are in the tropics, where it is not usually possible to identify threatened plant species individually. Botanists can, however, say which areas are rich in plants and which are not without knowing the status of every single species. Thus, while identifying threatened species provides a practical approach to planning plant conservation in most temperate countries, and on most islands, identifying Centres of Plant Diversity is the best approach in most of the tropics.

It is as yet unknown to what extent the sites identified as Centres of Plant Diversity can also be described as centres of diversity for animals; it is intended to investigate this during later stages of the project.

All the 241 Centres of Plant Diversity selected so far are listed in Table 15.6. As the data on degree of protection show, many are already protected areas, such as Bwindi (Impenetrable) Forest (Uganda), the wet tropics of Queensland (Australia) and the Sinharaja Forest (Sri Lanka). In virtually all cases, however, more conservation work is needed to ensure the full complement of plants survives intact. Below we show the areas selected for Africa and Peninsular Malaysia, showing the application of the approach in regions of very different size.

Centres of Plant Diversity: Africa

White (1983) recognises 17 major phytogeographic divisions (phytochoria) for mainland Africa. Of these, seven are classed as Regional Centres of Endemism, each having more than 50% of its species confined to it and a total of more than 1,000 species endemic to it. Two more phytochoria (Afromontane and Afroalpine) are patchily distributed on mountains. The remainder are termed transition zones, having low species endemism and, in some cases, very impoverished floras. These main divisions, which cover vast areas, were used as the starting point for selecting sites for the CPD project. In general, floristically-rich phytochoria have been allocated more Data Sheet sites than those with impoverished floras; however, some regions with very high endemism, such as the Cape, will be treated as one 'super-site'.

Salient features of White's phytochoria are outlined below, and indicated in Fig. 15.3. The CPD sites are shown in the same map, superimposed on these regional divisions and details presented in Table 15.4. For comparison, areas of bird endemism identified by ICBP are mapped in Fig. 15.4 and detailed in Table 15.5.

Guineo-Congolian (A)
8,000-12,000 vascular plant species; endemism very high, 80%. The tropical rain forest in west and central Africa. Western block (Guinea) floristically distinct from central block, mostly cleared or threatened. Gulf of Guinea islands, especially São Tome, also have high endemism. Central block (Congo) has two main centres of plant diversity: west (especially Gabon - the most species-rich rain forest in Africa, and Cameroon), and east (especially Zaire). In Cameroon, forests nearer coast richer (e.g. Korup), extending into south-east Nigeria (e.g. Oban). In Zaire, forests near coast (e.g. Mayombe) reported to be floristically distinct, threatened; forests on east side (e.g. Maiko, Kahuzi-Biega, Ituri, probably Itombwe) appear to be richer than those in centre.

Zambezian (B)
8,500 vascular plant species; high endemism, 54%. Miombo, mopane and chipya woodland. Most diverse area is Haut Shaba, Zaire (including Kundelungu). Zambia: richest miombo is in wetter area near Zaire border (extension of Haut Shaba). Angola: Huila Plateau rich in endemics, Itigi thicket near Tanzania/Zambia border also rich. Local endemics in Zaire on metalliferous soils and on serpentine in Zimbabwe (e.g. Great Dyke) need protection.

Sudanian (C)
2,750 vascular plant species, most widely distributed; regional endemism low, 35%. Woodland (mainly *Isoberlinia*, *Khaya*).

Somalia-Masai (D)
2,500 vascular plant species; 50% regional endemism. *Acacia*, *Commiphora* woodland. Rather homogenous; Somalia the richest country.

Cape (E)
Fynbos, with 8,600 vascular plant species; high endemism, 60-68%. Extraordinarily rich in species and endemics. Many important areas. Invasive species a major problem.

Karoo-Namib (F)
6,000 vascular plant species; 35-40% regional endemism. Dwarf succulent shrubland. Unparalleled diversity of succulents. Important centres in north (Gariep centre, including the Richtersveld) and south (southern Namibia and western Cape Province, South Africa).

Mediterranean (G)
4,000 vascular plant species; endemism low, 20%. Evergreen oak forest, macchia, maquis. High Atlas the most outstanding area botanically, many species and endemics.

Afromontane (H)
4,000 vascular plant species; endemism very high, 75%. Forests, afroalpine vegetation. Afroalpine vegetation (above forest limit) especially rich in local endemics. Eastern Arc mountains in Tanzania and south-east Kenya have many species absent from central Africa, especially in submontane forest. Richest mountains in central Africa uncertain, but possibly East Kivu (e.g. Itombwe) and Bwindi. All African mountain forests especially important for watershed protection.

Indian Ocean coastal (M & O)
Comprising Zanzibar-Inhambane regional mosaic in north, and Tongaland-Pondoland in south. Both with 3,000 vascular plant species and low endemism, 15-20%. Most important are the coastal forest remnants, floristically similar to the Guineo-Congolian, but with c. 40% of species endemic to coastal belt, many with very restricted distributions. In Kenya, c. 50 forest patches, most very small. In Tanzania, number of sites uncertain, need more fieldwork to determine which areas are key; includes Rondo Plateau.

CENTRES OF AVIAN ENDEMISM

The International Council for Bird Preservation, in its Biodiversity Project, has undertaken a major data collation and analysis project to identify areas supporting aggregations of restricted range endemic birds. This project has served two functions: first, it applies rigorous scientific criteria for identifying areas of high conservation value for birds; and second, it reviews the information on patterns of endemism in other taxonomic groups so that the value of birds as biodiversity indicators can be assessed. The major results of the project are now published in Bibby *et al.* (1992).

Locality records were gathered for species with breeding ranges below 50,000km^2 (about the size of Sri Lanka, Costa Rica or Denmark). Remarkably, there are 2,608 species or 27% of the world's birds with such small ranges. In all, some 55,000 separate locality records of birds were accurately geo-referenced and mapped with the aid of a Geographic Information System.

Species of restricted range tend to occur together, for instance on islands or in isolated areas of a particular habitat, such as tropical montane forest. Boundaries of these natural groupings of species have been identified (designated as Endemic Bird Areas or EBAs). They number 221 and embrace 2,480 species, which is the vast majority of all restricted range birds. Both the numbers of species involved and the number of EBAs divide roughly equally between continental areas and islands.

The tropics, with 76% of all Endemic Bird Areas, are the most important zone and there are very few at north temperate latitudes (Fig. 15.2). Indonesia is by far the most important country, with 411 restricted range species of which 339 are confined to the country. Peru, Brazil, Colombia, Papua New Guinea, Ecuador, Venezuela, the Philippines, Mexico and the Solomon Islands all have more than 100.

Table 15.7 shows the political affiliation, altitudinal range and habitats, and richness in restricted range birds of each EBA. The size of EBAs varies considerably, from the Northwestern Hawaiian Islands (5km^2) to the Guianas (170,000km^2). However, over 30% of EBAs have areas of less than 10,000km^2 and are therefore considerably smaller than the maximum range size allowed for any one species. Island EBAs are generally smaller than continental EBAs. For instance, 29% of island EBAs are smaller than 1,000km^2, whereas no continental EBAs are this small. The extent of EBAs in Africa, Middle East and Europe are shown in Fig. 15.4 and sites are detailed in Table 15.5.

The number of restricted range bird species contained within EBAs also varies, from the minimum of two used to define an EBA to 67 in the Solomon Islands EBA. A large number, 757 (29%), of these birds are threatened, and they constitute 77% of all threatened birds. Most EBAs (85%) have one or more threatened restricted range bird species (see Table 15.5 for Africa and adjacent areas). The principal habitat used by birds in the EBAs is forest (69% of restricted range species) with smaller numbers using scrub (12%). Other habitats such as grasslands are poorly represented, largely because species in these habitats are generally more widespread.

Figure 15.3 Centres of plant diversity: Africa

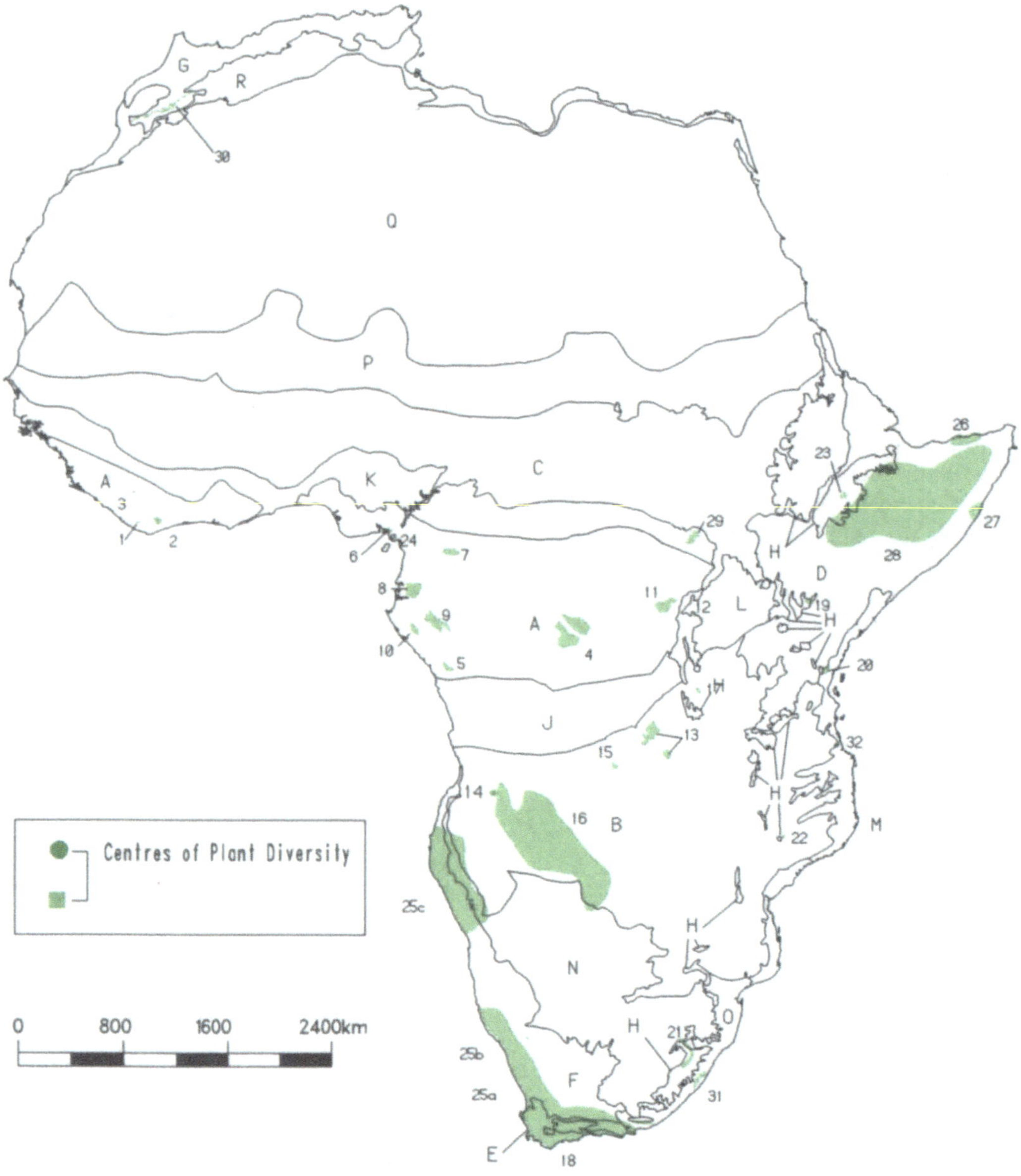

Notes: Data Sheet sites are shown superimposed on the main phytochoria (after White, 1983). Letter codes denote the following: (A) Guineo-Congolian regional centre of endemism. (B) Zambezian regional centre of endemism. (C) Sudanian regional centre of endemism. (D) Somalia-Masai regional centre of endemism. (E) Cape regional centre of endemism. (F) Karoo-Namib regional centre of endemism. (G) Mediterranean centre of endemism. (H) Afromontane archipelago-like regional centre of endemism. (J) Guinea-Congolia/Zambezia regional transition zone. (K) Guinea-Congolia/Sudania regional transition zone. (L) Lake Victoria regional mosaic. (M) Zanzibar-Inhambane regional mosaic. (P) Sahel regional transition zone. (O) Tongaland-Pondoland regional mosaic. (Q) Sahara regional transition zone. (R) Mediterranean/Sahara regional transition zone.

Figure 15.4 Endemic bird areas: Africa, Middle East, Europe

(Source: ICBP)
Robinson projection

Table 15.4 Centres of Plant Diversity: sites in continental Africa

SITE NO.	SITE NAME	COUNTRY	NO. OF PLANT SPECIES
1	Sapo Forest	Liberia	
2	Taï Forest	Ivory Coast	
3	Mt Nimba	Guinea, Ivory Coast, Liberia	>2,000
4	Salonga National Park	Zaire	
5	Mayombe-Cabinda	Congo, Cabinda, Zaire	
6	Korup-Oban	Cameroon, Nigeria	3,500
7	Dja	Cameroon	2,000
8	Crystal Mountains	Gabon	>3,000
9	Massif du Chaillu	Gabon	>3,000
10	Massif de Doudou	Gabon	>1,000
11	Maiko	Zaire	
12	Bwindi (Impenetrable) Forest	Uganda	1,000 taxa
13	Kundelungu/Upembe	Zaire	
14	Huila Plateau	Angola	
15	Zambesi Source Area	Zambia	
16	Okavango-Kwando	Angola, Namibia, Botswana	
17	Mbali-Mahali Hills	Tanzania	
18	Cape Floristic Province	South Africa	8,600
19	Mt Kenya	Kenya	800
20	Eastern Arc Mts: Usambaras	Tanzania	1,921 taxa*
21	High Drakensberg	South Africa	
22	Mt Mulanje	Malawi	>800
23	Bale Mts	Ethiopia	>1,000
24	Mt Cameroon	Cameroon	1,200
25a	Karoo-Namib region	South Africa	5,000 taxa
25b	Gariep Centre	South Africa, Namibia	
25c	Brandberg-Kaokoveld	Angola, Namibia	
26	Cal Madow	Somalia	>1,000
27	Hobyo	Somalia	<1,000
28	Limestone bush/woodland, Ogaden	Ethiopia, Kenya, Somalia	
29	Garamba	Zaire	
30	High Atlas mts	Morocco	
31	Pondoland Plateau	South Africa	
32	Rondo Plateau	Tanzania	

Note: Figures refer to the number of vascular plant species (if known) estimated to occur in the area. * denotes number of vascular plants so far recorded.

Importance for other taxonomic groups

The review of other taxonomic groups suggests that the EBAs are also of great importance for mammals, reptiles, amphibians, molluscs, insects and plants. However, there are gaps in data on these other groups, and additional data on these could significantly change the conservation evaluation of some EBAs. It seems likely, also, that in other groups different scales of endemism may occur (more fine-grained for various invertebrate groups and some plant taxa, for instance), and it is worth noting that entirely different approaches are needed to deal with non-terrestrial endemism.

Evaluation for importance and threats

EBAs were evaluated on biological importance and threat. The biological importance index reflects richness in restricted range species per unit area, modified to allow for taxonomic uniqueness of the species involved. On this basis, ICBP assigns EBAs to three categories. EBAs which are also significant centres of endemism for at least two other taxonomic groups are upgraded by one category. This has the overall effect of increasing the priority of EBAs which are important for other groups: uniform quantitative data on other groups are urgently needed to refine these priorities. Threats to EBAs are evaluated on the proportions of restricted-range species threatened, and the extent of coverage by the protected areas system (see Table 15.5).

Conclusions

Bibby *et al.* (1992) conclude that 20% of all bird species are confined to just 2% of the world's land surface. The total area of all 221 EBAs accounts for 4.5% of the land surface. Since more widely ranging species also occur in these EBAs, the proportion of the world's birds that could be conserved if EBAs were secured would greatly exceed the 27% whose ranges are highly restricted.

Further work and follow-up

ICBP aims to promote the conservation of all 221 EBAs, in collaboration with other international and national

Table 15.5 Endemic bird areas of Africa, the Middle East and Europe

AREA NAME	SIZE (km²)	SPP. CONFINED T	SPP. CONFINED N	SPP. CONFINED Tot.	SPP. OCCURRING T	SPP. OCCURRING N	SPP. OCCURRING Tot.	SPP. R	PA (%)
Sites in continental Africa									
Upper Guinea forests	113,000	4	1	5	5	1	6	5.5	7
Cameroon mountains	7,300	8	4	26	9	4	28	27.0	7
Cameroon and Gabon lowlands	40,000	4	-	5	5	-	6	5.5	22
Angola	14,000	5	4	14	6	4	15	14.5	4
North-east Somalia	41,000	2	-	5	2	-	5	5.0	0
Central Ethiopian highlands	37,000	2	1	4	2	1	4	4.0	0
South Ethiopian highlands	15,000	4	-	4	4	-	4	4.0	0
Central Somalian coast	1,200	1	1	2	1	1	2	2.0	0
East Zairean lowlands	49,000	4	1	5	4	1	5	5.0	3
Albertine Rift Mountains	44,000	8	4	37	10	4	40	38.5	12
Kenyan mountains	46,000	2	-	6	2	-	7	6.5	6
Serengeti	47,000	-	1	3	-	1	3	3.0	43
Kenyan and Tanzanian coastal forests	8,800	5	-	7	6	-	8	7.5	7
Eastern Arc Mountains	39,000	11	2	26	13	2	30	28.0	11
South Zambia	47,000	1	1	2	1	1	2	2.0	14
East Zimbabwean mountains	4,900	-	2	2	1	2	4	3.0	7
South-east African coast	43,000	-	2	2	-	3	3	2.5	3
South-east African grasslands	60,000	2	-	2	2	-	2	2.0	0
Cape region	24,000	-	1	3	-	1	4	3.5	50
Sites outside continental Africa									
Canary Islands and Madeira	8,100	6	-	9	6	-	10	8.5	30
Cape Verde Islands	4,000	1	-	4	1	-	4	4.0	9
Principe	140	-	-	6	3	-	12	8.8	0
Sao Tome	860	6	1	15	9	1	21	16.8	0
Tristan da Cunha Islands	200	5	-	6	5	-	6	6.0	0
Caucasus	64,000	-	1	2	-	1	2	2.0	9
Cyprus	9,300	-	-	2	-	-	2	2.0	0
Iraq marshes	40,000	-	-	2	-	-	2	2.0	0
Arabian mountains	59,000	1	-	7	1	-	7	7.0	1
Socotra	3,500	1	-	6	1	-	6	6.0	0
Granite Seychelles	240	6	-	10	6	-	10	10.0	3
Aldabra	160	1	1	2	1	1	4	2.7	4
Comoro Islands	1,900	4	-	10	4	-	13	11.2	0
Mayotte	360	1	-	3	1	-	6	4.2	0
West Madagascan dry forest	30,000	1	-	3	2	-	4	3.5	3
East Madagascan humid forests	112,000	13	1	17	14	1	18	17.5	5
Central Madagascan lakes	2,000	2	-	2	2	-	2	2.0	0
West Madagascan coastal wetlands	5,000	2	-	2	2	-	2	2.0	11
South Madagascan Didiera scrub	30,000	2	1	8	2	1	8	8.0	3
Reunion	2,500	1	-	3	1	-	7	5.0	2
Mauritius	1,900	6	-	6	6	-	10	8.0	2
Rodrigues	100	2	-	2	2	-	2	2.0	0

Key: T=threatened; N=near threatened; SPP. R=species richness; PA=coverage by protected areas.

organisations. The list of priority areas should enable a wide range of organisations to develop both regional and local programmes to help implement measures to prevent mass species extinctions. These measures will range from establishment and management of protected areas to the sustainable use of natural resources in the centres of endemism, and will require political and economic collaboration at all levels. There is a need to strengthen local data on birds and other taxa, and within EBAs, to study habitats which are vital for the survival of restricted range species. Key sites must be identified within EBAs for the targeting of conservation resources.

References

Ackery, P.R. and Vane-Wright, R.I. 1984. *Milkweed Butterflies.* British Museum (Natural History), London.

Collins, N.M. and Morris, M.G. 1985. *Threatened Swallowtail Butterflies of the World. The IUCN Red Data Book.* IUCN, Cambridge, UK and Gland, Switzerland. vii+401pp.+8pls.

Bibby, C.J., Crosby, M.J., Heath, M.F., Johnson, T.H., Long, A.J., Stattersfield, A.J. and Thirgood, S.J. 1992. *Putting Biodiversity on the Map: global priorities for conservation*. ICBP, Cambridge, UK.

East, R. (Ed.) 1988. *Antelopes. Global survey and regional actions plans. Part 1. East and northeast Africa*. IUCN, Gland.

East, R. (Ed.) 1989. *Antelopes. Global survey and regional actions plans. Part 2. South and south-central Africa*. IUCN, Gland.

East, R. (Ed.) 1990. *Antelopes. Global survey and regional actions plans. Part 3. West and central Africa*. IUCN, Gland.

Gentry, A. (in press). The subfamilies and tribes of the family Bovidae.

Margules, C.R. 1989. Introduction to some Australian developments in conservation evaluation. *Biological Conservation* 50:1-11.

McNeely, J.A., Miller, K.R., Reid, W.V., Mittermeier, R.A. and Werner, T.B. 1990. *Conserving the World's Biological Diversity*. IUCN, Gland, Switzerland.

Mittermeier, R.A. 1988. Primate diversity and the tropical forest: case studies from Brazil and Madagascar and the importance of the megadiversity countries. In: Wilson, E.O. and Peter, F.M. (Eds), *Biodiversity*. National Academic Press, Washington, DC. Pp.145-154.

Mittermeier, R.A. and Werner, T.B. 1990. Wealth of plants and animals unites 'megadiversity' countries. *Tropicus*:4(1):1,4-5.

Myers, N. 1988. Threatened biotas: 'hot spots' in tropical forests. *The Environmentalist* 8(3):187-208.

Myers, N. 1990. The biodiversity challenge: expanded hot-spots analysis. *The Environmentalist* 10:243-256.

Vane-Wright, R.I., Humphries, C.J. and Williams, P.H. 1991. What to protect? - systematics and the agonies of choice. *Biological Conservation* 55:235-254.

White, F. 1983. *The Vegetation Map of Africa. A descriptive memoir to accompany the Unesco/AETFAT/UNSO Vegetation Map of Africa*.

Williams, P.H. (unpublished). Afrotropical antelopes - priority areas for biodiversity. Progress report to The Natural History Museum, London, WCMC and IUCN-SSC.

Text, table and maps on plant diversity supplied by IUCN Centres of Plant Diversity Project. Text, table and maps on bird diversity provided by ICBP Biodiversity Project. Additional material from R.I. Vane-Wright, Biodiversity Programme, The Natural History Museum (London).

Table 15.6 The centres of plant diversity: summary data on sites selected for data sheet treatment

SITE	TYPE	SIZE (km^2)	ALTITUDE	FLORA	EXAMPLES OF USEFUL PLANTS	VEGETATION	PROTECTED AREAS	THREATS	ASSESSMENT
AFRICA									
Cameroon									
Mount Cameroon	S	1100	0-4006m	Iquitos 1200	Timber trees, especially African mahogany, medicinal plants	Lowland rain forest to afromontane forest, scrub, subalpine grassland	Mount Etinde Reserve ($200km^2$), Bambuko Forest Reserve ($220km^2$), Limbe Bot. Garden	Agricultural encroachment, fire, logging; potentially grazing	Threatened; most seriously on lower slopes of east & north; montane forest at risk from fire
Dja	S	8100	2-500m	2000	Useful timbers	Dense rain forest, semi-deciduous forest	Reserve naturelle		
Cameroon/Nigeria									
Korup/Oban Hills	S	6500	?-1079m	3500		Lowland tropical evergreen rain forest	Korup NP (Cameroon), Cross Rivers NP (Nigeria) covers all area		
Ethiopia									
Bale Mountains	F	4400	1500-4377m	>1000	Wild & semi-wild arabica coffee	Humid montane forest, juniper/hagenia/erica forests, woodlands, swamps, alpine vegetation	Bale Mountains NP ($2200km^2$)	Logging, grazing	At risk at most altitudes
Gabon									
Crystal Mountains	S	9000	0-911m	>3000	Timber trees (e.g. Okoumé), semi-wild oil palms, raphia, medicinal plants	Mainly tropical lowland & hill rain forest	None	No serious threats at present; logging is a potential threat	Not protected; at risk
Massif de Chaillu	S	12000	600-1000m	>3000	Timber trees (e.g. Okoumé, Caesalpinioideae)	Mainly tropical hill rain forest	None	Logging, shifting cultivation	Not protected; threatened
Massif de Doudou	S	1750	100-800m	>1000	Timber trees (e.g. Caesalpinioideae)	Tropical lowland and hill rain forest		Formerly logging	
Guinea/Ivory Coast/Liberia									
Mount Nimba	S		?-1752m	>2000		Lowland & transitional rain forest, grassland	Strict Nature Reserve (Guinea/Ivory Coast)	Mining of iron ore (Liberia/Guinea)	Greatly damaged by mining
Kenya									
Mount Kenya	S	1500	2000-5199m	800	Timber trees, fruit trees, medicinal plants	Montane moist & dry forest, bamboo, woodland, giant heaths, moorland	Above 3100m, NP; lower down, FR	Felling, monocultures on the lower slopes	
Malawi									
Mount Mulanje	S	500	650-3000m	>800	Mulanje cedar, Brachystegia, bamboos, fruit trees	Woodland, evergreen forest, montane grassland, high-altitude scrub, rupicolous communities	FR	Deforestation, illegal logging of cedars, uncontrolled collection of fuelwood, invasive pines	Severely threatened; action urgently needed

Table 15.6 The centres of plant diversity: summary data on sites selected for data sheet treatment

SITE	TYPE	SIZE (km^2)	ALTITUDE	FLORA	EXAMPLES OF USEFUL PLANTS	VEGETATION	PROTECTED AREAS	THREATS	ASSESSMENT
AFRICA (continued)									
Somalia									
Cal Madow	S	9600	0-2400m	>1000	Frankincense, myrrh,	Dry montane forest,	Daalo Forest Reserve,	Logging, grazing	No protection in practice,
Hobyo area	S	3000	0-440m	<1000		Deciduous bushland & woodland, dune vegetation	Proposed game reserve	Grazing	No protection, but threats to flora probably not severe
Somalia/Ethiopia/Kenya									
Limestone/bush woodland, Ogaden	V	200,000	200-1500m	2000	Frankincense, myrrh, yeheb nut	Deciduous woodland, bushland	None	Grazing	No protection, extremely vulnerable to over-grazing. Threatened.
South Africa									
Cape Floristic Region	F	90000	0-2325m	8600	Ornamentals (e.g. bulbs, succulents, proteas, ericas)	Fynbos, shrubland, montane forest	Many reserves covering 19% of region, almost all in mountains	Agriculture, urbanization, fire, introduced species	Most mountain fynbos effectively protected; lowlands much less so
South Africa/Namibia									
Karoo-Namib region	F	111212	0-1907m	4000-5000	Ornamentals (e.g. bulbs, succulents), food and medicinals	Succulent shrubland (veld) with associated annuals	NP (270km^2) & a few other reserves covering >2% of the region in total	Overgrazing, agriculture, mining, plant collecting, invasive species, urban development	Inadequate coverage of protected areas; threatened
Tanzania									
East Usambaras	S	231 of forest	150-1500m	1921 taxa*	African violet, timber & pole species	Lowland semi-deciduous & evergreen submontane forests	25 Catchment Forest Reserves	Logging, pole-cutting, clearance of forest for agriculture, invasive species	Reserves intact; more extensive forest outside threatened & declining
Uganda									
Bwindi (Impenetrable) Forest	S	321	1160-2610m	1000 taxa	Timber trees, bamboos, medicinal plants	Moist evergreen submontane & montane forest	Forest Reserve/Wildlife Sanctuary	High local population density, leading to removal of timber & fuelwood, grazing	Isolated forest very threatened despite protection status and conservation projects
Zambia									
Zambezi Source Area	F	8000	1100-1550m		Timber trees, melliferous plants	Riverine, swamp & dry evergreen forest; miombo woodland, bushland, savanna	25% gazetted as Protected Forest Area	Clearance for agriculture, refugees	At risk from inappropriate land management practices
ATLANTIC OCEAN									
St Helena	F	122	0-825m	60	Endemic ebony, gumwood	Tree fern thicket, semi-desert, scrub, woodland, severely degraded	3 Forest Reserves (31ha), 1 Nature Reserve (7ha), whole island proposed as BR	Invasive plants, over-grazing by domestic & feral animals	Rescue & rehabilitation programme underway, but remaining native vegetation still severely threatened

Table 15.6 The centres of plant diversity: summary data on sites selected for data sheet treatment

SITE	TYPE	SIZE (km^2)	ALTITUDE	FLORA	EXAMPLES OF USEFUL PLANTS	VEGETATION	PROTECTED AREAS	THREATS	ASSESSMENT
AUSTRALIA/NEW ZEALAND									
Australia									
Alpine Region, S.E. Australia	F	30000	200-2200m	700	Timber trees (especially eucalypts)	Grassland, woodland, shrubland, forest, alpine vegetation	NPs/BR (16,000km^2); other parks 2000km^2	Tourism, grazing, hydro-electric facilities, fire, feral animals, exotic plants	Generally well protected; reasonably secure
Border Ranges, NSW/Queensland	F	600	0-1360m	>850	Timber trees (e.g. Red Cedar, Rose Mahogany), ornamental plants	Various types of rain forest, eucalypt forest, woodland, shrubland, heath	NP (majority of area), state forest, WHS (NSW part)	Clearance for grazing, rural development, visitor pressure, fire, invasive plants	Generally well protected; proposal to add Queensland part to WHS will improve protection
Kakadu-Alligator Rivers Region, N.Territory	F	30000	0-370m	1340	Many plants used by aborigines	Tropical sclerophyll forest, woodland, rain forest, swamp forest, mangrove, saltmarsh	NP (19,815km^2, most of which is also WHS)	Tourism, invasive weeds	WHS area well protected; proposal to extend WHS
Lord Howe/ Norfolk Island	F	15 (LH) 39 (NI)	0-866m 0-319m	393	Ornamental plants (e.g. Kentia palms, Norfolk Island pine)	Evergreen rain forest, palm forest, "mossy" forest, pine & hardwood forests	WHS covers all Lord Howe; NP of 4.6km^2 on Norfolk I.	Tourism, cattle grazing, introduced plants on Norfolk Island	Lord Howe well protected; both areas secure
South-West Botanical Province, W.Australia	F	309840	0-400m	5500	Timber trees, ornamental plants	Eucalypt forest, woodland, mallee, scrub	NP & reserves (34,716km^2), state forest (17,710km^2); total reserves cover 17% of area	Root rot fungus, wild fires	Reserves well protected but coverage inadequate; no further clearing likely
Sydney Sandstone Region, NSW	F	24000	0-1300m	2000	Australian Red Cedar & other hardwoods, ornamental plants	Forest, woodland, shrubland, grassland, coastal dune & swamp, mangrove	NPs & reserves cover 11,615km^2	Urban & industrial development, tourism	Generally well protected
Wet Tropics of Queensland	F	11000	0-1622m	1160	Timber trees, ornamental plants, native food plants	Various types of rain forest, woodland, mangrove	WHS (8900km^2), c. 50 NPs (1830km^2), remainder mainly Crown Land	Tourism, lowland forest clearance, telecommunication facilities, feral animals	Well protected now as WHS
Western Tasmanian Wilderness	F	16300	0-1617m	800	Timbers, especially Huon Pine, King Billy Pine, ornamental plants	Cool temperate rain & eucalypt forests, alpine vegetation, sub-alpine scrub, moorland, grassland	WHS (13,800km^2) incl. NPs & 1 BR, most of WHS also has State Reserve status	Tourism, fire, *Phytophthora* root fungus	WHS well protected. North-west forests unprotected & at risk, but recommended as reserve
New Zealand									
Campbell Island	F	113	0-558m	228*	Insignificant	Scrub, herbfield, feldmark	NR	Some invasive weeds	Reasonably safe; isolation causes some management problems

Table 15.6 The centres of plant diversity: summary data on sites selected for data sheet treatment

SITE	TYPE	SIZE (km²)	ALTITUDE	FLORA	EXAMPLES OF USEFUL PLANTS	VEGETATION	PROTECTED AREAS	THREATS	ASSESSMENT
AUSTRALIA/NEW ZEALAND (continued)									
Chatham Island	F	970	0-290m	320*	Many species used in Maori culture, ornamental plants	Evergreen cool-temperate forest, peatlands, lagoons, coastal communities	c. 12 (covering ??) with others under negotiation	Clearance for agriculture, feral animals, invasive plants, fire	Degradation persists with many important sites unprotected
North Auckland	F		0-600m		Timber trees, ornamental plants, many species used in traditional culture	Moist temperate evergreen lowland forest, swamps, coastal communities	>200 protected areas, incl. Waipoua State Forest Sanctuary (91km²)	Clearance for agriculture, feral animals, weeds, tourism, plantations	Despite many protected areas, many endemics insufficiently protected & at risk
North-west Nelson	F		0-1800m	1200	Timber trees, ornamental plants, plants used in traditional culture	Moist temperate to montane forest, wetlands, alpine vegetation, grassland	NP (covering ??), Forest reserve (covering ??), >40 other reserves	Logging, mining, development plans	
CARIBBEAN									
Cuba									
Oriente	F	18000	0-1994m	>3000	Pines, podocarps, palms, potential ornamentals	Seasonal evergreen forest, montane & submontane rain forests, pine forest, semi-evergreen scrub	4 NPs covering ??km²	Removal of timber & fuelwood, mining, tourism	Reasonably safe due to rugged terrain being unsuitable for agriculture
Pinar del Rio	F	1150	0-692m	>1500	Pine, oak, cycad	Seasonal forests, succulent & thorn scrub	NP (??km²)	Removal of fuelwood, tourism	Lowlands at risk; vegetation on rugged terrain & cliffs reasonably safe
Dominica									
Morne Trois Pitons NP	S	70	600-1383m	500	Craft materials; potential ornamentals	Rain forest, montane forest, elfin woodland, secondary palm brakes, volcanic lakes	NP	Tourism, water supply development	Protected, but inadequate management resources
Jamaica									
Cockpit country	F	430	320-790m	1500	Few remaining timber trees, a few potential ornamentals	Tall seasonal subtropical forest, scrub thicket	None	Cutting of poles, charcoal production, slash & burn cultivation	Severely threatened
CENTRAL ASIA									
Armenia/Turkey/Iran									
Armenian Uplands/Arid Transcaucasus	F		500-2500m		Cereal relatives, dyes, oils, tannins, ornamentals	Desert & semi-desert, mountain steppes, juniper & oak forests	Numerous small reserves	Quarrying, felling	
Azerbaijan/Iran									
Girkan forests, Talysh Mts	V	30	500-600m		Ornamentals (e.g. lilies, shrubs), fruit trees, medicinal plants	Mixed broadleaved & oak forests	Girkan Reserve covers 29km²	Tree felling	Well protected in Azerbaijan

Table 15.6 The centres of plant diversity: summary data on sites selected for data sheet treatment

SITE	TYPE	SIZE (km^2)	ALTITUDE	FLORA	EXAMPLES OF USEFUL PLANTS	VEGETATION	PROTECTED AREAS	THREATS	ASSESSMENT
CENTRAL ASIA (continued)									
Kazakhstan Karatau	F	28000	?-2100m	1666 taxa*	>200 species have been used for food, medicines, industry, ornamental plants	Desert, mountain steppes, valley forests	Berkara (31km^2), Djambulsky (64km^2)	Mining, clearance for agriculture, overgrazing	
Kazakhstan/China Khrebet Dzhungarskiy Alatau	F	24000	? - 4463m	2168 taxa*	300 species have been used for food, medicines, industrial uses & for their omanamental value	Lepsinsky Reserve (258km^2)	Clearance for agriculture, grazing, over-exploitation of medicinal plants	Lower slopes particularly threatened	
Russia Primorskye Kray	F	680000		2000	>150 medicinal plants, southern regions included in Vavilov centre	Spruce forests, mixed spruce-broadleaved forests, meadows, steppes	60 protected areas, incl. numerous Strict Nature reserves & 100 wildlife reserves, BR (3470km^2)		
Russia/Kazakhstan/Mongolia Altai Mountain Range	F	100000	1000-4506m	2500	Ornamental plants, pines	Steppe, spruce & fir forests, alpine vegetation	Altaysky (A) Reserve (8637km^2), Makrokolsky (M) Reserve (714km^2)	Forest clearance, fire, tourism, plant collecting, dam construction	Coverage of existing reserves inadequate, more being planned
Russia/Mongolia Baikal region	F	500000	500-2500m	2300 taxa	Medicinal plants, ornamental plants, pines	Coniferous forests, scrub, alpine meadows, steppe, swamps	Baikal Reserve (1657km^2), Barguzin Reserve (2632km^2), proposed NP	Logging, grazing pressure, industrial development, tourism, plant collecting	Proposed NP will protect most of interesting parts of Baikal shore
Tadzhikistan Zeravshan River basin & Samarkand Mts	F	20000	?-5000m	1900 taxa	c. 200 species are used for food, medicines, industry, ornamental plants	Desert-steppe communities, wormwood scrub, alpine meadows	3 reserves covering 392km^2	Overgrazing, clearance for agriculture, tree felling, fire	Reserve coverage inadequate
Tadzhikistan/Uzbekistan Varzob River basin & Gissarsky Ridge, Pamir-Altai mts	F	2100	800-4000m	1455	c. 150 species have been used for food, medicines, in industry, ornamentals	Deciduous & evergreen forests, alpine grass meadows, savannas	2 Nature Reserves (700km^2)	Mining, clearance for cultivation, pollution from aluminium smelting	Lower slopes & basin particularly threatened
Turkmeniya/Iran Kopetdag	F	50000	?-2800m	1766 taxa*	c. 200 species have been used for food, medicines, industry, ornamental plants	Semi-desert grassland steppes	Kopetdag Reserve (498km^2), Sunt Khasardag Reserve (297km^2)	Overgrazing, clearance for agriculture	

Table 15.6 The centres of plant diversity: summary data on sites selected for data sheet treatment

SITE	TYPE	SIZE (km^2)	ALTITUDE	FLORA	EXAMPLES OF USEFUL PLANTS	VEGETATION	PROTECTED AREAS	THREATS	ASSESSMENT
CENTRAL ASIA (continued)									
Ukraine									
S.Crimea/West Transcaucasus	F		0-1200m		Medicinal plants, tannins, oils	Oak, juniper & pine forests	Numerous reserves, incl. Yalta Mt Forest Reserve ($142km^2$)	Road construction, urban development, quarrying, agriculture, tourism	Threatened
Ukrainian Carpathians	F	3700	180-2061m	>2000	Wild relatives of cherry, apple, pear, c. 300 medicinal plants, crop relatives	Oak, beech & spruce forests, scrub, alpine vegetation	Carpathian Forest Reserve ($127km^2$), Carpathian NP	Clearance for agriculture, overgrazing, tourism	Threatened
CHINA									
Peoples Republic of China									
Changbai Mt region, Jilin	F	30000	300-2691m	2385*	Medicinal plants, timber trees	Coniferous & broadleaved forests, alpine vegetation	BR for $2000km^2$	Logging, colonization, illegal collection of medicinal plants, tourism	Area lacks effective management; at risk
Gaoligong Mt, Nu Jiang River & Biluo Snow Mt, Yunnan	F	10000	1000-5128m	2500-3000	Medicinal plants, pines, bamboos, house plants	Evergreen broadleaved & sub-alpine coniferous forests, alpine vegetation, savanna	2 Nature Reserves covering $5000km^2$	Clearance for cultivation, logging, illegal collection of medicinal plants	Part safe; part severely threatened
Hainan tropical forest region	V	33920	0-1867m	4200-4500	Timber trees, medicinal plants, rattans, wild litchi relative	Seasonal rain forest, montane seasonal rain forest, mangroves, savanna	51 reserves covering $1500km^2$	Clearance for cultivation, logging; illegal collection of medicinal plants	Severely threatened
Min Jiang River basin, Hengduan Mts, Sichuan	F	20000	600-6250m	3500-4000	Medicinal plants, timber trees, bamboos	Broadleaved evergreen & montane forests, alpine scrub & meadows, savanna	Protected area of $2000km^2$	Logging, illegal collection of medicinal plants	Area lacks effective management; at risk
Limestone region, S.W. Zhuang	F	20000	100-1300m	2500-3000	>1000 species used locally; timbers, medicines, bamboo, rattan, ornamentals	Seasonal rain forest, montane evergreen broadleaved & limestone evergreen forests	11 Natural Reserves covering $4000km^2$	Clearance for cultivation, logging, colonization, illegal collection of medicinal plants	Protected areas need effective management; severely threatened
Nan Ling Mt region	F	40000	200-2145m	3000-3500	Medicinal plants, timber trees, pines, bamboos	Broadleaved evergreen, montane & mixed broadleaved forests	17 Nature Reserves covering $5000km^2$	Clearance for cultivation, logging, illegal collection of medicinal plants	Part safe; part severely threatened
Taibai Mt region, Shaanxi	F	2379	500-3767m	1800-2000	Timber trees, medicinal plants	Broadleaved deciduous & sub-alpine coniferous forests, alpine vegetation	Nature Reserve covers $540km^2$	Logging, illegal collection of medicinal plants	Part safe; part severely threatened

Table 15.6 The centres of plant diversity: summary data on sites selected for data sheet treatment

SITE	TYPE	SIZE (km^2)	ALTITUDE	FLORA	EXAMPLES OF USEFUL PLANTS	VEGETATION	PROTECTED AREAS	THREATS	ASSESSMENT
CHINA (continued)									
Xishuangbanna region, Yunnan	F	19690	500-2429m	4000-4500	Timber trees, medicinal plants, bamboos, rattans	Seasonal rain forest, montane evergreen broadleaved forest	5 Nature Reserves covering 2400km^2	Clearance for cultivation, logging, colonization, fire, illegal collection of medicinal plants	Severely threatened
Taiwan									
Kenting National Park	S	177 (land area)	0-526m	>1350	Timber trees, medicinal plants, legumes, ornamental plants	Evergreen broadleaved rain forest, semi-deciduous & littoral forests, grassland	NP	Tourism, grazing, plant collecting, military facilities	Generally well protected, some parts under threat
EAST ASIA/INDOCHINA									
Japan									
Iriomotejima	S	322	0-470m	1217*	Ornamental plants	Subtropical rain forest, warm temperate evergreen forest	NP	Logging, road construction, agriculture, tourism	National park, but effective controls needed. At risk.
Yakushima	S	540	0-1935m	1342*	Ornamentals (e.g. rhododendrons)	Warm temperate to cold temperate forests with some subtropical elements	NP	Visitor pressure; formerly logging due to road building	Reasonably secure (?)
Vietnam									
Bach Ma - Hai Van	S	600	0-1450m	2500	200 timber trees, 108 medicinals, ornamentals, fibres, rattans, edible fruits	Lowland evergreen forest, tropical montane evergreen forest	220km^2 as Bach Ma NP	High local population density, felling of timber trees, fuelwood cutting	Threatened
Cat Tien	S	1376	60-754m	2500	200 timber trees, 120 medicinals, rattans, bamboos, orchids	Lowland evergreen & semi-evergreen forests, freshwater swamps	379km^2 as Nam Cat Tien NP, Rhino Reserve	Illegal logging, over-exploitation of rattans & resins	Reasonably secure
Cuc Phuong	S	300	100-637m	1800*	Timber trees, medicinal plants, rattans, bamboos, ornamental plants	Forest on limestone, lowland evergreen forest	222km^2 as Cuc Phuong NP	Illegal felling of timber trees	Reasonably secure
Lang-bian Plateau	S	400	1400-2167m	2000	100 medicinal plants, ornamental plants (especially orchids), resin trees	Pine forest, tropical montane evergreen forest	46km^2 as nature reserve	Felling for timber & charcoal production, over-exploitation of resins & medicinals	Severely threatened
Mount Fan-Si-Pan	S	2000	1000-3142m	3000	Timber trees, medicinal plants, essential oils	Tropical & subtropical montane deciduous forests	200km^2 as proposed Hoanglien Son Nature Reserve	Shifting cultivation, clearance for agriculture, over-exploitation of non-wood products	No management plan; at risk

Table 15.6 The centres of plant diversity: summary data on sites selected for data sheet treatment

SITE	TYPE	SIZE (km^2)	ALTITUDE	FLORA	EXAMPLES OF USEFUL PLANTS	VEGETATION	PROTECTED AREAS	THREATS	ASSESSMENT
EAST ASIA/INDOCHINA (continued)									
Yok Don	F	650	200-482m	1500	150 timber trees, tannins, resin trees, ornamental plants	Dry dipterocarp forest, lowland semi-evergreen forest, riverine forest	582km^2 as Yok Don NP	Shifting cultivation, felling of timber trees, over-exploitation of non-wood products	Threatened
EUROPE									
Bulgaria/Yugoslavia									
Balkan (Sar, Rila, Pirin) Mts	F	5000	500-2900m	3000	Medicinal plants, ornamentals	Beech & mixed coniferous forests, rock & cliff communities	2 NPs (390km^2), 4 BRs (30km^2)	Tourism, changes in land tenure	Extensive & expanding areas under protection; some threats arising from new political order
Cyprus									
Troodos Mts	S	1000	1000-2130m	1200	Medicinal herbs, ornamentals	Evergreen scrub, some evergreen oak & coniferous forests, rock & cliff communities	Some areas of forest protected	Tourism & skiing development, road building, fire	Conflict between tourism and conservation, but forests well protected
Greece									
Crete	F	8700	0-2450m	1608*	Crop relatives, culinary herbs, ornamentals	Evergreen scrub, rock & cliff vegetation	1 NP (48km^2), a few other small reserves	Tourism, light industry, agriculture, plant collecting	Protected area coverage inadequate, but threats less acute than in many other areas
Mountains of S. & C. Greece	F	13400	1000-2500m	4000	Culinary herbs, chickpea relative	Conifer & deciduous forests; rock, cliff & scree vegetation	4 NPs	Fire, excessive grazing, tourism, mining, botanical collecting	Threats to flora less severe than at other sites; conservation measures taken
INDIAN OCEAN									
Mauritius/Réunion									
Mascarene Is	F	4484	0-3070m	>970 taxa	Timber trees (e.g. black ebony), palm hearts, coffee relatives, medicinal plants	Tropical/subtropical coastal & dry lowland forest to moist montane forest, high altitude heath	Mauritius: numerous reserves (2.4% of area), new national park; Réunion: 1 reserve (0.7km^2); Rodrigues: 2 small reserves	Invasive introduced plants, introduded animals, development pressure, increasing human population	Réunion: inadequate reserve system; Mauritius & Rodrigues: good reserve network but flora severely threatened. Overall - flora endangered, but good opportunities still exist to save what remains
Sri Lanka									
Knuckles "Conservation Area"	S	217	430-1920m	>1000	Timbers, medicinal plants, bamboos, fruit tree relatives, spices, ornamental plants	Lowland semi-evergreen to montane evergreen forests, grasslands	No legal protection	Clearance for cardomom, settlements & agriculture, mining	Area below 1200m under threat

Table 15.6 The centres of plant diversity: summary data on sites selected for data sheet treatment

SITE	TYPE	SIZE (km^2)	ALTITUDE	FLORA	EXAMPLES OF USEFUL PLANTS	VEGETATION	PROTECTED AREAS	THREATS	ASSESSMENT
INDIAN OCEAN (continued)									
Peak Wilderness/ Horton Plains	S	224/32	500-2238m	>1000	Timbers, medicinal plants, bamboos, fruit tree relatives, spices, ornamental plants	Lowland, submontane & montane wet evergreen forests, montane grassland	Sanctuary & NP	Religious tourism, fuelwood cutting, mining, agricultural expansion, fire	Reasonably secure, but boundaries at risk
Sinharaja Forest	S	89	300-1100m	>600	Timber trees, rattans, fruit tree relatives, medicinal plants, ornamentals	Lowland tropical evergreen rain forest	National Heritage Wilderness Area & WHS; plans to increase area protected to 113km^2	Population pressure, agricultural encroachment	Southern part still under threat, otherwise threats declining & reasonably secure
INDIAN SUBCONTINENT									
India									
Agastyamalai Hills	F	2000	67-1868m	2000	Medicinal herbs, timber trees, bamboos, rattans, crop relatives	Tropical dry to wet forests	3 Wild Life Sanctuaries protect c. 870km^2 of forest; BR proposed	Clearance for hydro-electric projects, plantations, tourism	
Nallamalais		6840	200-950m	750	Medicinal plants, crop relatives	Scrub forests, dry & moist deciduous forests	Wild Life Sanctuary covering 1200km^2 of forest	Forest clearance, bamboo cutting for paper, fire	
Namdapha	S	7000	300-4500m	3000	Wild relatives of banana, citrus, pepper; timber trees, ornamental plants	Tropical evergreen & semi-evergreen rain forests, alpine vegetation	NP covers 1985km^2; proposed BR with core of c. 2500km^2	Shifting cultivation, refugees & other settlers, illegal timber felling	
Nanda Devi	S	2000	?-7816m	800	Medicinal species, ornamental plants	Coniferous, birch & rhododendron forests, alpine vegetation	NP (630km^2), WHS	Almost nil	Effectively secure
Nilgiri Hills	F	5520	150-2000m	3240	Medicinal plants, timber trees, crop relatives	Wide range of tropical evergreen to deciduous forests	Proposed as BR; Silent Valley NP covers 89.5km^2	Forest clearance for timber, plantations, roads, development projects	
MIDDLE AMERICA									
Costa Rica									
Atlantic Highland-to-Lowland Forest	S	450	30-2906m	800 trees	Monstera, vanilla, cocoa, palms, timber trees	Tropical wet forest, swamp forest, to lower montane rain forest	NP, Biological Station, BR	Squatters, illegal logging, agriculture, grazing	Reasonably secure
Osa Peninsula	S	22500	0-745m	>500 trees	Timber trees	Tropical wet & tropical rain forest	418km^2 as Corcovado NP	Mining, logging, colonizationPark threatened	
Costa Rica/Panama									
Talamanca Mts	S	320km long	100-3563m		Timber trees, medicinal plants	Lowland tropical wet forest to cloud forest & paramo	International Park, BR, WHS (Costa Rica: 1939km^2; Panama: 2000km^2)	Colonization, agriculture, cattle ranching, fire	Threatened

Table 15.6 The centres of plant diversity: summary data on sites selected for data sheet treatment

SITE	TYPE	SIZE (km^2)	ALTITUDE	FLORA	EXAMPLES OF USEFUL PLANTS	VEGETATION	PROTECTED AREAS	THREATS	ASSESSMENT
MIDDLE AMERICA (continued)									
Guatemala									
Petén	F	35854	200-1000m		Mahogany, edible palms, Parlour Palm house plants, medicinal plants	27,900km^2 of semi-evergreen seasonal forest, savanna, wetlands	Maya BR (14000km^2)	Logging, road-building, colonization, grazing, slash & burn agriculture, oil exploration	Several protected areas, but region under threat
Honduras									
Rio Plátano Biosphere Reserve	S	5250	0-1326m		Timber trees	Mostly humid tropical & subtropical forests; also mangrove, swamp & elfin forests, pine savanna	BR, WHS	Logging, shifting agriculture, cattle grazing, potential road	Threatened
Mexico									
Cañón de Zopilote	F	4383	600-3100m		Softwood & hardwood trees, medicinal plants	Deciduous tropical forest, oak & coniferous forests, cloud forests	1 state park	Logging	Inadequate protection and coverage; under threat
Chamela Biological Station	S	1.6	50-500m	800	Valuable timber trees (e.g. rosewood, linum vitae)	Tropical deciduous & semi-deciduous forests	Protected area, managed by Universidad Nacional Autónoma de México	Illegal encroachment	Reasonably secure, but at risk from isolation. A new reserve adjoins it to the west & south
Lacandon Rain Forest	F	13000	400-1500m		Timber, fruit & spice trees	Tropical rain forest, montane rain forest,semi-deciduous tropical forest	Montes Azules BR for 3312km^2; 6000km^2 as FR, not yet implemented	Colonization, logging, cattle, road-building, hydro-power, oil drilling, development plans	>50% of forest lost, much of rest going at an accelerating rate; severely threatened
Sierra de Manantlán	S	1400	400-2860m	2070*	Wild perennial maize, wild frijol, oaks, pines	Tropical deciduous forest to cloud forest, or firs, or pines & oaks	BR	Logging, fire, grazing	About 30% in good condition
Sierra de Juárez	F	1700	100-3250m	2000	Timber trees (e.g. pines, firs, oaks), ornamentals (e.g. orchids, tree ferns)	Cloud forest, fir forest, mixed pine & oak forest, pine forest, oak forest	None	Logging, agriculture, colonization	Threatened
Tehuacán Valley	F	Valley 170 x 40km	700-1700m		Cacti & other ornamental plants	Predominantly dry scrub to early-deciduous forest	Small botanic garden	Overgrazing by goats, farming, plant collecting	Inadequate protection; under threat
Uxpanapa-Chimalapa	F	7700	100-2250m		Timber (including true mahogany) & fruit trees, edible palms, medicinals	Evergreen & semi-evergreen rain forest, montane rain forest		Deforestation from logging, colonization, agriculture, grazing, dams & roads	Threatened

Table 15.6 The centres of plant diversity: summary data on sites selected for data sheet treatment

SITE	TYPE	SIZE (km^2)	ALTITUDE	FLORA	EXAMPLES OF USEFUL PLANTS	VEGETATION	PROTECTED AREAS	THREATS	ASSESSMENT
MIDDLE AMERICA (continued)									
Panama									
Darién NP	S	5970	0-1900m		Cativo & other valuable timbers, medicinal plants	Tropical moist & tropical rain forests	NP, WHS	Logging, mining, proposed Pan-American Highway	Threatened
NORTH AMERICA									
Mexico/U.S.A.									
Apachian/ Madrean region	F	600 x 300km	500-3500m	3500	60-80 wild relatives of crop plants, >300 food plants, >450 medicinals	Coniferous, oak-coniferous, tropical deciduous forests, savanna, chapparal, thorn scrub	NPs, National Monument, National Forests, TNC reserves (U.S.A.); forest & private reserves (Mexico)	Logging, erosion, agricultural encroachment, over-grazing	
U.S.A.									
California Floristic Province	F	324000	0-4400m	3500	Timber trees & other forest products, wild grape, ornamental plants, medicinal plants	Coniferous & mixed evergreen forests, oak woodlands, chaparral, coastal scrub, grassland	Parks & reserves protect c. 11% of land area, mostly montane	Population pressure, urban & agricultural development, mining, dams, overgrazing, exotic weeds, off-road vehicle recreation	Highlands reasonably well protected; lowland habitats threatened & inadequately protected
Central Highlands, Florida	F	150km long x few km wide	20-94m		Few timber trees, food plants, potential insecticides	Xerophytic pine & oak scrub, palmetto flatwoods, wetlands	State, federal & private reserves, state & local parks	Citrus plantations, urban, tourist & recreational developments	Inadequate protected area coverage; threatened
Edwards Plateau, Texas	F	100000	100-1000m	2300	Forage species, juniper oil, few timber trees	Semi-arid temperate semi-evergreen forest, grassland, semi-desert scrub	State & municipal parks, several reserves cover <0.05% total area	Clearance for agriculture, grazing, dams, urbanization, introduced species	Protected area coverage very inadequate; threatened
Klamath-Siskiyou region	F	55000	0-2750m	3500	Timber trees, beargrass, medicinal plants	Coniferous & mixed evergreen forests, prairies, savanna, subalpine meadows	Wilderness areas, national monuments, federal, state & private reserves	Logging, agriculture, urbanization, mining, dams, tourism	Most is relatively safe due to remoteness, but logging & mining are considerable threats
Vernal pools, California	V	20000	0-600m (-2500m)	>200	*Limnanthes* is potential sperm whale oil substitute, ornamental plants	Annual herbs, wet grassland, aquatics, oak & pine forests	Numerous small pools protected, but more resrves needed	Agriculture, grazing, urban development, mining	Inadequate protected area coverage; severely threatened
U.S.A./Canada									
Serpentine flora	V	4550	0-2200m		Flax & sunflower relatives adapted to low nutrient soils	Grasslands, chaparral, montane woodland	Frenzel Creek Research Natural Area (2.7km²), TNC Ring Mountain Preserve (25 ha)	Mining, logging, off-road vehicles, urbanization	Inadequate protected area coverage; threatened

Table 15.6 The centres of plant diversity: summary data on sites selected for data sheet treatment

SITE	TYPE	SIZE (km^2)	ALTITUDE	FLORA	EXAMPLES OF USEFUL PLANTS	VEGETATION	PROTECTED AREAS	THREATS	ASSESSMENT
PACIFIC									
Chile									
Juan Fernández	F	100	0-916m	362*	Some potential horticultural species	Mainly forested	NP	Feral animals, invasive plants	Flora acutely threatened; WWF-sponsored rescue programme underway
Ecuador									
Galápagos Islands	F	7900	0-1707m	596* taxa	Tomato & cotton relatives, native timber trees	Wide range of generally arid vegetation types, Scalesia forests (much reduced) on higher islands	96.7% of land area as NP	Feral animals, invasive plants, over-exploitation of native woody species, increasing human population, fire	Many plants still at risk despite conservation measures
Fiji	F	18235	0-1323m	1518*	Timber trees, medicinal plants, culturally important plants	Tropical rain, dry & montane forests, grassland, scrub, coastal vegetation	16 protected areas ($65km^2$) incl. 8 Nature Reserves ($62km^2$, mainly rain forest), 1 NP ($2.4km^2$, coastal), several other small forest parks	Logging, clearance for agriculture, feral animals, invasive plants, population pressure, tourism, mining	Threatened; reserves & funding inadequate, numerous proposals never implemented. Need involvement of local population
New Caledonia									
Grande Terre	F	16890	0-1628m	3250	Timber trees, ornamental plants, medicinal plants	Humid evergreen & sclerophyll forests, maquis, mangroves, marshes	9% covered by protected areas (incl. 14 special botanical reserves & 1 Strict Nature Reserve)	Fire, mining, clearance for agriculture & grazing, urbanization, invasive plants	Coverage of protected areas inadequate. Sclerophyll & calcareous forests particularly threatened
U.S.A.									
Hawaiian Islands	F	16641	0-4205m	>1100	Native plants used by indigenous people for all their needs; ornamentals, hardwood trees	Wide range of vegetation types	Many protected areas but not all plant-rich communities covered	Development, introduced plants & animals, fire	Possibly the most endangered island flora in the world; severely threatened
SOUTH AMERICA									
Argentina									
Anconquija	S	3000	300-5500m	1700	Timber trees, crop relatives (e.g. beans), wild tree tomato & pawpaw, medicinal & ornamental plants	Amazonian winter dry rain forest & chaco monsoon forests to temperate cloud forest, páramo, high-altitude vegetation	Several protected areas covering $>340km^2$, proposed NP of $3000km^2$	Logging, clearance for agriculture, fire, grazing, plant collecting, road building, dams	Varying degrees of threat; some protected areas reasonably secure, remote areas self-protected, lower forests of Tipa-Pacará severely threatened

Table 15.6 The centres of plant diversity: summary data on sites selected for data sheet treatment

SITE	TYPE	SIZE (km^2)	ALTITUDE	FLORA	EXAMPLES OF USEFUL PLANTS	VEGETATION	PROTECTED AREAS	THREATS	ASSESSMENT
SOUTH AMERICA (continued)									
Bolivia									
Apolobamba-Alto Madidi	F	250000	300-6000m	8000	Timber trees, medicinal plants, ornamental plants	Mainly humid evergreen tropical montane forest, dry ridge tops, cloud forest, Andean high grassland	Proposed NP	Logging, road building, colonization, petroleum exploration, gold mining	Mostly intact, but without protection; under threat
Catariri Rio Grande	F	150000	1200-3100m	500-1000	Wild potatoes, edible palms, timber trees	Mainly deciduous forest, some semi-evergreen to evergreen forests, grassland	None	High local population density, deforestation from agriculture, grazing, fuelwood cutting	Last remnant of inter-Andean dry forest; threatened
Llanos de Mojos (Beni)	F	200000	130-250m	1000-1500	Forage grasses, legumes, timber trees, palms, pineapple relatives	Savanna dominates, mosaic of lowland vegetation from aquatic communities to evergreen forests	3 Wildlife Refuges (2700km^2), 1 Natural Reserve (2750km^2)	Fire, over-grazing, road building, logging	Key forest areas severely threatened; small formerly protected areas lost by change of ownership
Serranías de Chiquitos y Sunsas	S	7000	200-1200m		Timber trees, forage grasses	Tropical semi-deciduous forest, cerrado, campo rupestre	Santa Cruz La Vieja NP (171km^2)	Mechanized agriculture, road construction	Largest area of undisturbed semi-deciduous forest on continent; severely threatened
Bolivia/Paraguay/Argentina									
Gran Chaco	F	1.1 million	100-500 (-2400)m	1200	Timber trees, medicinals, dyes, ornamentals, forage plants, fibres, oils	Mostly xeric deciduous forest, some savannas & grasslands	5 NPs, 1 Natural Reserve (Argentina & Paraguay)	Timber cutting, grazing, fire, road-building, oil exploration	Inadequately protected & threatened; extensions & new protected areas recommended
Brazil									
Atlantic Moist Forests	V	Original: 1 million km^2	0-3000m		Jacaranda, rosewood, pau brasil & other valuable timbers	Moist forest (<5% remaining)	A range of NPs, national biological reserves, ecological stations & other protected areas, covering ??km^2	Deforestation, in particular for crops, cattle & logging; collection of fuelwood	The most endangered tropical moist forests in the Wetern Hemisphere. Only ?? % protected
Distrito Féderal	S	500	850-1260m	2500	Timber, fruit trees, medicinal plants, ornamental plants	Cerrado	NP, BR (??km^2)	Population pressure, invasive species, fire	Inadequate funding and management of existing reserves; threatened
Padre Bernardo	S	30	600-800m		Timber trees	Deciduous & semi-deciduous forests on limestone	None	Deforestation due to agriculture, grazing, logging (?), charcoal production, fire	Severely threatened
Serra do Espinhaco	S	1000	800-1800m	3000	Ornamental plants, medicinal plants	Campo rupestre	NP (338km^2)	Grazing, erosion, fire, charcoal production, plant collecting	Protected area not yet established & inadequately funded; threatened

Table 15.6 The centres of plant diversity: summary data on sites selected for data sheet treatment

SITE	TYPE	SIZE (km^2)	ALTITUDE	FLORA	EXAMPLES OF USEFUL PLANTS	VEGETATION	PROTECTED AREAS	THREATS	ASSESSMENT
SOUTH AMERICA (continued)									
Chile									
Mediterranean Deciduous Forests	V	200	700-2200m	1500	Timber from *Nothofagus*	Deciduous *Nothofagus* forest, submontane & montane	NP (La Campana, 80km^2), 3 proposed national reserves (c. 130km^2)	Mining, fire, removal of fuelwood, grazing	Partially protected
Atacama Desert	F				Cacti, medicinal plants			Mining, grazing, road building, cactus cutting for wood	
Temperate Rain Forests	V		0-1500m	1000	Timber trees, incl. Araucaria, Alerce, Southern Beeches, Tepa, Tineo	Temperate rain forests	Many NPs & other reserved areas	Timber cutting, agriculture, grazing	Coverage & staffing of existing reserves inadequate
Colombia									
Caquetá River's Mid-branch Watershed	F	12000	200-340m	3000 in sample plots	Timber trees, medicinal plants, fibres, dyes	Tropical rain forest	Río Cahuinarí NP (5750km^2)	Potential threat: encroachment by settlers	Almost all intact, 50% protected, reasonably secure
Sierra Nevada de Santa Marta	S	6000	0-5770m			Humid tropical forest, montane tropical rain forest, cloud forest, paramo	NP of 3830km^2	Colonization, agriculture, cattle ranching	
Colombia/Panama/Ecuador									
Chocó	F/V	9000	?-1900m	8000-9000	Timber trees, fruit trees, ornamentals	Mainly lowland tropical rain forest, montane forest	(see WWF-US)	Logging, colonization, agriculture, grazing, mining	Large areas of forest under threat
Ecuador									
Jatún Sacha	S	5	400m	2000	Many medicinal plants, fruit trees	Lowland tropical wet forest, transitional to premontane wet forest	Private reserve; proposed purchase of 4.3km^2 extension	Colonization, subsistence agriculture	At risk
Cerro Blanco	S	20	0-430m	>1000	Timber species	Tropical dry & moist forests, mangroves	Designated a "Bosque Protector" by Ecuadorian Government	Colonization, at present by homeless, long-range threat of housing development	
Western Ecuador: Moist to Pluvial Forests	V	3690	0-900m	5000	Timber trees (e.g. caoba), fruit trees, ornamentals	Moist forest: <1500km^2 (4% of original); wet forest: 90km^2 (0.8%); pluvial forest: c. 2100km^2 (10%)	Samples in 3 private reserves (c. 2.8km^2), 2 government reserves & several "Bosques Protectores"	Deforestation from logging, conversion to oil palm plantations, grazing, road building, colonization	Severely threatened & one of the most endangered floras in the world; most "Bosques Protectores" cut over

Table 15.6 The centres of plant diversity: summary data on sites selected for data sheet treatment

SITE	TYPE	SIZE (km^2)	ALTITUDE	FLORA	EXAMPLES OF USEFUL PLANTS	VEGETATION	PROTECTED AREAS	THREATS	ASSESSMENT
SOUTH AMERICA (continued)									
Sumaco	S	4000	500-3700m	4000	Medicinal plants, crop relatives, culturally important plants	Lowland tropical forest to wet paramo	Protection forest	Deforestation from road building, colonization, subsistence agriculture, pastures	Inadequately protected; threatened
Yasuní National Park	S	6800	200-350m		Wild rubber, vegetable ivory palm, valuable hardwoods, medicinal plants	Tropical moist forest	NP & BR, proposed extension of $1600km^2$	Oil exploration, logging, colonization	Severely threatened
French Guiana									
Saül region	S	1336	200-760m	1800	Timber trees, edible palms, rosewood oil, medicinal plants	Mostly lowland moist forest, swamp forest, montane forest	Proposed NP	Slash & burn agriculture, charcoal production, firewood cutting, gold mining, potential road	Not seriously threatened at present; repeated efforts to protect area since 1975
Paraguay									
Mbaracayú	S	600	140-450m		Valuable timbers, fruit trees, edible palms	Semi-evergreen subtropical moist forest	None	Logging & agriculture	Protected, but management plan yet to be developed
Peru									
Cerros de Amotape	S	913	200-1400m		Timber trees	Dry forest	NP, BR	Timber felling, firewood collection, fire, grazing, desertification	At risk
Eastern slopes of Peruvian Andes	F	250000	400-3500m	7000-10000	Timber trees, ornamentals	Tropical & subtropical premontane & montane forests	3 functioning NPs (8% of area), several reserves & sanctuaries (77% of area)	Deforestation from colonization, roads, agriculture, logging, narcotics	NPs reasonably safe; protection and management of other reserves inadequate
Huancabamba	S	3000	1000-3500m		Timber trees, ornamentals, medicinal plants	Montane cloud forest	None	Logging, road-building, colonization	Isolated forest; severely threatened
Iquitos	S		120-140m		120 fruits, timber trees, medicinal plants (incl. curare), fibres	Evergreen rain forest	None	Clearance along rivers for settlements & agriculture, loss of dispersal agents	Unprotected; some habitats threatened
Manú National Park (Cocha Cashu)	S	15320	300->4000m		Spanish cedar, mahogany, cocoa relatives, edible palms & fruits	Mostly evergreen tropical forest	NP, BR	Agriculture & grazing, potential oil exploration & road building	Inadequate funding resulting in lack of trained staff; highlands particularly threatened

Table 15.6 The centres of plant diversity: summary data on sites selected for data sheet treatment

SITE	TYPE	SIZE (km^2)	ALTITUDE	FLORA	EXAMPLES OF USEFUL PLANTS	VEGETATION	PROTECTED AREAS	THREATS	ASSESSMENT
SOUTH AMERICA (continued)									
Peru/Chile									
Lomas	V		0-1000m	>1000	Forage species	Islands of low montane desert scrub & thorny steppe	1 national reserve (Lachay Nature Reserve: 50.7km^2)	Grazing, collection of firewood, pollution from mining, urbanization	Reserve coverage inadequate; severely threatened
Venezuela									
Coastal Cordillera	F	45000	0-2765m	>5000	Quinine & other medicinal plants, quality hardwoods	Montane/submontane semi-deciduous & evergreen forests, cloud forest, coastal scrub, mangroves	11 NPs, 5 Natural Monuments protecting large parts of remaining vegetation	Population pressure, deforestation, colonization, agriculture, roads	Protected areas reasonably secure, but coverage inadequate
Pantepui region	S/F	7000	1300-3015m	3000	Ornamentals (e.g. bromeliads, carnivorous plants), undoubtedly rich in potentially useful species	Montane forests, tepui scrub, pioneer communities on cliffs & rocky areas	National park covers entire region	No serious threats, but potential threat from increased tourism & over-collecting	Safe at present, but potentially at risk without more effective controls
S.E.ASIA									
Indonesia (Irian Jaya)									
Gunung Lorentz National Park (proposed)	S	21500	0-4884m			Lowland to montane rain forests, mangroves, bogs, swamps, heaths, grasslands, alpine vegetation	NR; proposals for NP status	Mining, logging, petroleum exploration, road building, colonization	
Indonesia (Java)									
Gede-Pangrango National Park (proposed)	S	150	1000-3019m		Many medicinal plants	Mostly montane and submontane rain forest, alun-alun meadows	NR & BR with proposed NP status	Agricultural encroachment, tourism, plant collecting, fuelwood collection	Encroachment around boundaries; at risk
Indonesia (Kalimantan)									
Bukit Raya/ B. Baka	S	7705	100-2278m	4000	Timber trees (especially dipterocarps), fruit trees, illipe nuts, rattans	Lowland tropical rain forest, swamp forest, montane forest, "mossy forest", ericaceous scrub	Bukit Raya NR (1100km^2), Bukit Baka NR (705km^2); proposed extension (5900km^2)	Logging, road construction, shifting cultivation	Encroachment in west; at risk
Indonesia (Sulawesi)									
Dumoga-Bone National Park (proposed)	S	3300	200-1968m		Timber trees, rattans	Tropical lowland semi-evergreen rain forest, montane forest, forest on limestone	Game Reserve and Nature Reserve complex proposed as NP	Shifting agriculture, rising local human population, mining	Effectively managed but boundaries at risk

Table 15.6 The centres of plant diversity: summary data on sites selected for data sheet treatment

SITE	TYPE	SIZE (km^2)	ALTITUDE	FLORA	EXAMPLES OF USEFUL PLANTS	VEGETATION	PROTECTED AREAS	THREATS	ASSESSMENT
S.E.ASIA (continued)									
Lore Lindu National Park (proposed)	S	2310	200-2610m		Wild relatives of fruit trees (e.g. durians, bananas, jackfruit), macadamia nuts, timber trees, rattans	Mostly tropical montane rian forest, small area of lowland rain forest	Proposed NP & WHS status for Game Reserves & Protection Forest; BR	Agricultural encroachment, over-exploitation for forest products, proposed hydro-electric scheme	Effectively managed but boundary at risk
Mt Rantemario (Gunung Latimojong)	S		100-3440m			Montane rain forests	Protection Forest covering $300km^2$ proposed as NR	Few, but encroachment from shifting agriculture up to 1,700m	Boundaries & lower slopes severely threatened; rest safe
Indonesia (Sumatra)									
Gunung Leuser National Park (proposed)	S	7927	0 - 3466m		Timber trees (especially dipterocarps), fruit trees, ornamentals	Lowland dipterocarp rain forest, montane & subalpine forests, freshwater swamp forest, marshes	NP (not yet legally gazetted)	Encroachment from settlements, agriculture, illegal logging	Threatened, especially in the lowlands; increasing local population pressure, inadequate funding
Indonesia/Malaysia									
Gunung Bentuang dan Karimun/ Lanjak Entimau/ Batang Ai	S	8142	300-1960m		Timber trees, especially dipterocarps, illipe nuts, rattans, fruit tree relatives, gharu	Lowland, evergreen rain forest, heath, swamp & montane forests	Proposed NR ($6000km^2$), Wildlife Sanctuary ($1688km^2$), proposed NP & extensions	Agricultural encroachment, logging	Boundaries & lower slopes at risk
Malaysia									
East Sabah Lowland Dipterocarp Forest	V	10000	0-4220m		Timber trees, rattans, fruit trees	Tropical lowland evergreen rain forest			
Endau Rompin, Pen Malaysia	S	2024	?-1500m		Wild banana relative, rattans, dipterocarps, 52 medicinal herbs	Mainly tropical lowland rain & hill dipterocarp forest	Proposed NP; $920km^2$ as State Park	Oil palm plantations, logging on fringes, proposed new town	Partly protected. At risk
Kinabalu Park, Sabah	S	780	1000-4105m	4500	Timber trees, ornamental plants, such as pitcher plants, orchids	Mostly montane rain forest, some tropical lowland rain forest, ultrabasic forest, alpine vegetation	State Park	Forest clearance for shifting cultivation, copper mining, tourism	Boundaries threatened
Lambir Hills, Sarawak	S	70	30-467m	684 trees; c.60 palms	Timber trees, incl. 69 dipterocarps, mango & durian relatives, rattans	Lowland mixed dipterocarp forest, heath forest, scrub	NP	Logging, clearance for agriculture	Encroachment around boundaries; threatened
Limestone flora, Pen. Malaysia	F	269	?-713m	1216	Ornamentals, especially orchids, begonias, palms gesneriads	Limestone forest, scrub, lithophytic vegetation	A few outcrops protected in Taman Negara, Templer Park, & at Batu Caves	Quarrying, mining, encroachment from agriculture, fire, tourism	Many outcrops severely threatened, some at risk, a few safe

Table 15.6 The centres of plant diversity: summary data on sites selected for data sheet treatment

SITE	TYPE	SIZE (km^2)	ALTITUDE	FLORA	EXAMPLES OF USEFUL PLANTS	VEGETATION	PROTECTED AREAS	THREATS	ASSESSMENT
S.E.ASIA (continued)									
Malaysia/Brunei									
Gunung Mulu NP/ Labi Hills/Batu Patam/Sg Ingei	S	529 (Mulu)	30-2376m	4000	Timber trees, especially dipterocarps, fruit trees, sago palm, rattans	Lowland mixed dipterocarp to montane forests on sandstones, limestones & shales, heath forests	NP (Mulu), Protected Forest (Medalam, 125km^2), Primary Conservation Area (Sg Ingei, 185km^2)	Mulu: logging around perimeter, shifting cultivation along some rivers, potential threat from road construction	Mostly safe at present, but would be at risk if plans to build road went ahead; buffer zones need to be implemented
Philippines									
Mt Isarog	S	101	?-1966m		Almaciga, rattans, timber trees, bamboo	Lowland dipterocarp forest, montane forest	NP	Illegal logging	Threatened
Mt Pulog & Mt Tabayoc	S		?-2929m		Pines, oaks	Montane forest, pine forest, oak forest, grassland, bamboo scrub	NP (115km^2)	Cultivation, fire, tourism	Mt Pulog under threat
Palanan Wilderness Area (Sierra Madre Mts)	S	1872	?-1672m		Timber trees, rattans	Lowland dipterocarp forest, lower montane forest, ultrabasic & limestone forests	Wilderness Area, Forest Reserve	Illegal logging, shifting cultivation	Mostly safe
Palawan	F	14896	0-2085m	> 2000	Timber trees, rattans, almaciga, fruit trees, orchids, nipa palm	Lowland evergreen & semi-deciduous forests, ultrabasic & limestone forests	MAB Reserve; 63 reserves covering 542km^2	Illegal logging, mining, shifting cultivation, tourism	Poorly protected; threatened
Sibuyan Island	F	449	0-2052m		Timber trees, palms, pandans	Lowland dipterocarp forest, lower montane forest	Provincial Watershed, proposed NP	Illegal logging, over-exploitation of rattan	Mt Giting-Giting is under threat
Southern Samar	F	13080	?-1000m		Timber trees, rattans, pandans, palms	Ultrabasic & limestone forests, lowland dipterocarp forest, mangroves	NP (area?)	Illegal logging, shifting cultivation	Park under threat
Papua New Guinea									
Huon Peninsula, Bangeta Huon Terraces	S	20000	0-3500m	>5000	Hardwoods, fruit trees, ornamental plants (e.g. ferns, orchids)	Lowland tropical rain forest to subalpine forest, grasslands, mangroves	Proposed NP	No immediate threats; logging in lowlands & some traditional gardening	Reasonably safe at present
Bismarck Falls, Ramu, Mt Otto, Mt Wilhelm	S	6000	0-4300m	5000-6000		Lowland swamp & rain forest to alpine vegetation	Provincial Park, proposed NP	Population pressure, logging, agriculture	

Table 15.6 The centres of plant diversity: summary data on sites selected for data sheet treatment

SITE	TYPE	SIZE (km^2)	ALTITUDE	FLORA	EXAMPLES OF USEFUL PLANTS	VEGETATION	PROTECTED AREAS	THREATS	ASSESSMENT
S.E.ASIA (continued)									
Mt Bosavi	S	30000	2000-3000m	>3000	Traditional food & medicinal plants	Lowland & lower montane vegetation, lowland dipterocarp forest		Logging in dipterocarp forests, clearance for agriculture	
Wassi-Kussa	S	50000	0-100m	>3000	Traditional food & medicinal plants	Savanna, monsoon forest, freshwater swamps, mangroves	None	Introduced deer, possible logging, clearance for gardening, fires	At risk
S.W.ASIA/MIDDLE EAST									
Oman/Yemen Dhofar Fog Oasis	F	30000	0-2100m	900	Frankincense, traditional food, fibre & medicinal plants	Mainly dry deciduous shrubland, montane evergreen shrubland, semi-desert grassland	1 small bird sanctuary, otherwise none	Population pressure, over-grazing, cutting of wood for fuel & timber	Severely threatened; IUCN proposals for reserves in Oman not yet implemented
Saudi Arabia/Yemen S.W. Arabian Highlands	F	70000	200-3760m	2000	Qat (stimulant), coffee, barley, wheat & sorghum relatives, myrrh	Deciduous & evergreen bushland & thicket, Juniper woodland	Asir NP (4150km^2) in Saudi Arabia, proposed areas in Yemen	Uncontrolled cutting of wood for fuel, timber & charcoal; over-grazing, erosion	No effective protection over most of area, remaining woodland severely threatened
Turkey Anti-Taurus Mts	F	60000	700-3600m	>3200	Medicinal plants, dyes, onion relatives	Steppe, montane steppe, alpine vegetation	NP (428km^2), 1 proposed game reserve	Dam construction, re-afforestation projects, rock climbing leading to erosion	Urgent action needed to protect more of the region; threatened
Isaurian, Lycaonian & Cicilician Taurus	F		0-3524m	>2500	Fig, pomegranate, nuts, dune stabilizers	Cilician fir & cedar forests, scrub, thorn-cushion plants, scree vegetation	NP (very small), bird sanctuary, game reserve (30km^2)	Road building, over-grazing, clearance for agriculture, tourism	Many important ecosystems unprotected; threatened
Mountains of S.E. Turkey, N. Iraq, W. Iran	F	147332	1400-4135m	>2500	Pears, almonds, hawthorns, gum tragacanth	Oak forest, alpine thorn cushion scrub, montane steppe, alpine grassland, scree vegetation	Iran: NP (4636km^2); Turkey: 1 forest recreation area, 1 game reserve	Influx of refugees, re-afforestation projects, potential dam & irrigation schemes	Inadequate coverage of reserve system, but threats to flora less severe than elsewhere
N.E. Anatolia	F	33200	0-3932m	>2460	Cherries, hazelnuts, timber trees	Coastal humid subtropical forest to fir forest, rhododendron scrub, scree vegetation	1 NP (? area)	Illegal logging, clearance for agriculture, export of wild bulbs	None of the Little Caucasus is protected; at risk but not as seriously threatened as other areas

Table 15.6 The centres of plant diversity: summary data on sites selected for data sheet treatment

SITE	TYPE	SIZE (km^2)	ALTITUDE	FLORA	EXAMPLES OF USEFUL PLANTS	VEGETATION	PROTECTED AREAS	THREATS	ASSESSMENT
S.W.ASIA/MIDDLE EAST									
S.W. Anatolia	F	87500	0-3070m	>3365	Cypress, cedar, Oriental sweet gum, medicinal herbs	Mediterranean forest & scrub, cedar forest	Some NPs, BR (84km^2), 4 biogenetic reserves (181km^2)	Tourism, over-grazing by goats, large-scale export of wild bulbs	Coastal areas severely threatened
Yemen									
Socotra	F	3625	0-1519m	815*	Dragon's Blood, other resins, gums, aloes	Semi-desert, dry deciduous shrubland, montane semi-evergreen thicket, secondary grassland	None	Overgrazing by goats, cutting of wood for fuel; potential new development projects	No protection but traditional practices have prevented serious exploitation so far

Notes: Data for the table were extracted from existing Data Sheet drafts and from information generously provided by a number of contributors ahead of the full Data Sheet accounts. **Type:** A letter code categorising each of the areas selected for Data Sheet treatment. The following codes are used: S - site, where the area is a discrete geographical unit, and where the whole area needs to be conserved; F - distinct floristic province, often covering a very wide area, or centre of plant diversity and/or endemism covering a whole region. Effective conservation of the flora often requires a network of reserves to be established, as in many cases it would be impractical to protect the entire province or area; V - vegetation type. As in 'F', effective conservation often requires representative samples to be protected. **Area** Usually to nearest 10km^2. **Altitude** Altitude range in metres. **Flora** Unless otherwise stated, the number of indigenous vascular plant species present, or an estimate based on current botanical knowledge of that (or similar) sites. Where the exact number is known, indicated by using an asterisk (*) after the number. Otherwise an estimate is included to the nearest 100 species (or for some large tropical sites, to the nearest 500-1,000 species). **Examples of Useful Plants** Important plants or major groups are listed, using vernacular names of species and commodity groups. **Vegetation** The major vegetation formations present. **Protected Areas** For centres of plant diversity that are fully included within protected areas, the category of protected area(s) is given. Abbreviations used: NP = National Park; BR = Biosphere Reserve; WHS = World Heritage Site. Where part of the site falls within protected area(s), the area is given. **Threats** Only the main threats, with the most important ones first, are listed. **Assessment** A summary of the conservation status of the area. A keyword summary is included to indicate the degree of threat. Categories include: Severely Threatened, Threatened, At Risk, Reasonably Secure, Safe. It must be stressed that the table is at present in draft form only, and that additional data would be welcomed for any of the sites with which you are familiar. Any updates or additional data should be sent to the IUCN Plant Conservation Office, Descanso House, 199 Kew Road, Richmond, Surrey TW9 3BW, UK.

Table prepared by IUCN Centres of Plant Diversity Project.

Table 15.7 Endemic bird areas of the world

NAME	POLITICAL UNIT(S)	ALTITUDE (m)	HABITAT(S)	SIZE (km²)	SPP. R.
AFRICA, THE MIDDLE EAST AND EUROPE					
Canary Islands and Madeira	Spain, Portugal	250-2,000	forest, rocky	8,100	8.5
Cape Verde Islands	Cape Verde	0-160	rocky, mixed	4,000	4.0
Upper Guinea forests	Côte D'Ivoire, Ghana, Guinea, Liberia, Sierra Leone	100-1,400	forest	113,000	5.5
Cameroon mountains	Cameroon, Equatorial Guinea, Nigeria	700-2,900	forest	7,300	27.0
Cameroon and Gabon lowlands	Cameroon, Gabon, Equatorial Guinea, Nigeria	0-800	forest	40,000	5.5
Principe	Sao Tome and Principe	0-1,000	forest, mixed	140	8.8
Sao Tome	Sao Tome and Principe	0-2,000	forest	860	16.8
Angola	Angola	0-1,500	forest, mixed	14,000	14.5
Tristan da Cunha Islands	St Helena	0-300	grassland, mixed	200	6.0
Caucasus	USSR, Turkey	1,500-4,000	rocky, mixed	64,000	2.0
Cyprus	Cyprus	0-1,900	forest, scrub	9,300	2.0
Iraq marshes	Iraq, Iran	0-100	wetland, mixed	40,000	2.0
Arabian mountains	Saudi Arabia, Yemen	1,800-3,200	scrub, mixed	59,000	7.0
Socotra	Yemen	0-1,400	scrub, grassland	3,500	6.0
North-east Somalia	Somalia	300-2,100	rocky, mixed	41,000	5.0
Central Ethiopian highlands	Ethiopia	1,300-3,100	rocky, scrub	37,000	4.0
South Ethiopian highlands	Ethiopia	1,275-2,300	scrub, mixed	15,000	4.0
Central Somalian coast	Somalia	0-100	desert, grassland	1,200	2.0
East Zairean lowlands	Uganda, Zaire	700-1,500	forest	49,000	5.0
Albertine Rift Mountains	Burundi, Rwanda, Uganda, Zaire	1,000-3,200	forest	44,000	38.5
Kenyan mountains	Kenya, Tanzania	1,100-3,700	forest, mixed	46,000	6.5
Serengeti	Kenya, Tanzania	1,100-2,100	savanna	47,000	3.0
Kenyan and Tanzanian coastal forests	Kenya, Tanzania	0-500	forest	8,800	7.5
Eastern Arc Mountains	Malawi, Mozambique, Tanzania	750-3,000	forest	39,000	28.0
South Zambia	Botswana, Zambia, Zimbabwe	600-1,000	forest, savanna	47,000	2.0
East Zimbabwean mountains	Mozambique, Zimbabwe	1,200-2,400	forest	4,900	3.0
South-east African coast	Mozambique, South Africa	0-100	forest, scrub	43,000	2.5
South-east African grasslands	Lesotho, South Africa	1,700-2,200	grassland	60,000	2.0
Cape region	South Africa	0-1,000	forest, mixed	24,000	3.5
Granite Seychelles	Seychelles	0-900	forest	240	10.0
Aldabra	Seychelles	0-8	forest	160	2.7
Comoro Islands	Comoros	400-2,600	forest	1,900	11.2
Mayotte	Comoros	0-1,700	forest	360	4.2
West Madagascan dry forest	Madagascar	0-800	forest	30,000	3.5
East Madagascan humid forests	Madagascar	0-2,300	forest	112,000	17.5
Central Madagascan lakes	Madagascar	750-1,500	wetland	2,000	2.0
West Madagascan coastal wetlands	Madagascar	0-100	wetland, forest	5,000	2.0
South Madagascan Didiera scrub	Madagascar	0-200	scrub, forest	30,000	8.0
Reunion	Reunion	200-2,300	forest	2,500	5.0
Mauritius	Mauritania	300-800	forest, scrub	1,900	8.0
Rodrigues	Reunion	0-390	forest, scrub	100	2.0

Table 15.7 Endemic bird areas of the world

West China	China	900-1,300	desert, scrub	15,000	2.0
Western Himalayas	Afghanistan, India, Nepal, Pakistan	1,600-3,600	forest	33,000	9.0
Indus valley	India, Pakistan	0-200	wetland, scrub	37,000	2.0
Western Ghats	India	0-2,450	forest	28,000	16.0
Sri Lanka	Sri Lanka	0-2,260	forest	36,000	23.0
Tibetan valleys	China	3,600-4,600	scrub, rocky	7,900	2.0
South Tibet	China	2,700-5,000	scrub, forest	18,000	2.0
Eastern Himalayas	Bhutan, China, India, Myanmar, Nepal	900-4,000	forest	70,000	22.8
Assam plains	Bangladesh,India	0-1,000	wetland, grassland	43,000	3.8
Tirap Frontier	India, Myanmar	500-1,800	scrub, mixed	14,000	2.0
Qinghai mountains	China	1,800-5,100	rocky, mixed	22,000	3.0
Central Sichuan mountains	China	1,500-3,600	forest	30,000	9.8
West Sichuan mountains	China	2,700-4,900	forest, mixed	24,000	3.0
South Chinese forests	China	300-1,900	forest	11,000	4.0
Yunnan mountains	China, Myanmar	1,500-3,650	forest	26,000	3.3
Burmese plains	Myanmar	0-1,000	scrub, agricultural	16,000	2.0
Andaman Islands	India	0-700	forest	8,200	10.0
Nicobar Islands	India	0-600	forest	2,000	7.0
Annamese lowlands	Laos, Viet Nam	0-1,500	forest	12,000	5.3
Hainan	China	500-1,800	forest	12,000	2.8
Da Lat Plateau	Viet Nam	900-2,300	forest	7,400	5.1
Cochinchina	Viet Nam	0-1,200	forest	15,000	2.5
Shanxi mountains	China	2,000-2,800	forest	14,000	2.0
Fujian mountains	China	200-2,000	forest	45,000	3.8
Taiwan	Taiwan	300-3,300	forest	36,000	15.3
Nansei Shoto Islands	Japan	0-500	forest, mixed	4,500	9.3
Ogasawara Islands	Japan	100-400	forest, mixed	100	1.5
SOUTH-EAST ASIAN ISLANDS AND AUSTRALIA					
Luzon mountains	Philippines	350-2,800	forest	36,000	11.8
Luzon lowlands and foothills	Philippines	0-1,300	forest	12,000	14.8
Mindoro	Philippines	0-1,500	forest	10,000	7.3
Negros and Panay	Philippines	0-1,300	forest	26,000	10.6
Cebu	Philippines	0-1,300	forest	5,100	1.0
Palawan	Philippines	0-1,000	forest	14,000	17.8
Samar, Leyte, Bohol and Mindanao lowlands	Philippines	0-1,500	forest	66,000	15.6
Mindanao mountains	Philippines	700-3,000	forest	32,000	21.6
Sulu Archipelago, excluding Basilan	Philippines	0-790	forest	1,700	4.5
Bornean mountains	Indonesia, Malaysia	0-3,000	forest	27,000	27.8
Sumatra and Peninsular Malaysia	Indonesia	600-3,000	forest	53,000	25.2
Enggano	Indonesia	0-150	forest, mixed	370	2.0
Javan and Balinese mountains	Indonesia	800-3,000	forest	18,000	23.9
Javan and Balinese lowlands	Indonesia	200-800	forest, scrub	16,000	4.0
Flores and associated islands	Indonesia	0-2,300	forest	36,000	23.2
Sumba	Indonesia	0-1,400	forest	11,000	11.2
Timor and associated islands	Indonesia	0-2,600	forest	26,000	26.7
Tanimbar and associated islands	Indonesia	0-1,750	forest	5,600	31.1
Talaud and Sangir Islands	Indonesia	0-1,700	forest	1,600	7.1
Sulawesi mountains	Indonesia	500-3,000	forest	24,000	30.8
Sulawesi lowlands	Indonesia	0-2,000	forest	24,000	16.9
Banggai and Sula Islands	Indonesia	0-2,300	forest	6,900	11.9
Buru	Indonesia	0-1,750	forest	8,000	18.4

Table 15.7 Endemic bird areas of the world

NAME	POLITICAL UNIT(S)	ALTITUDE (m)	HABITAT(S)	SIZE (km^2)	SPP.R
SOUTH-EAST ASIAN ISLANDS AND AUSTRALIA (continued)					
Seram	Indonesia	0-1,750	forest	14,000	19.3
Halmahera	Indonesia	0-1,750	forest	29,000	32.5
West Papuan Islands and Vogelkop lowlands	Indonesia	0-900	forest	14,000	11.6
Vogelkop mountains	Indonesia	600-3,000	forest	26,000	13.3
Geelvink Bay Islands	Indonesia	0-700	forest, mixed	3,200	9.6
North New Guinean mountains	Indonesia, Papua New Guinea	600-2,200	forest	11,000	4.5
North New Guinean lowlands	Indonesia, Papua New Guinea	0-900	forest	32,000	7.1
Adelbert and Huon mountains	Papua New Guinea	500-3,500	forest	19,000	8.3
Central New Guinean high mountains	Indonesia, Papua New Guinea	2,700-4,600	forest, mixed	6,800	13.5
Central New Guinean mid mountains	Indonesia, Papua New Guinea	500-3,800	forest	98,000	32.8
Trans-Fly and Upper Fly	Indonesia, Papua New Guinea	0-1,000	forest, wetland	64,000	6.6
Christmas Island	Christmas Island	0-350	forest	140	2.0
Kimberley and the Top End	Australia	0-1,700	rocky, mixed	105,000	13.0
Cape York	Australia	0-500	mixed	43,000	4.3
Atherton region	Australia	0-1,700	forest	28,000	14.8
South-west Australia	Australia	0-500	mixed	115,000	13.5
Murray-Darling region and adjoining coast	Australia	0-500	scrub, mixed	98,000	5.5
South-east Australia	Australia	0-1,200	forest	85,000	9.5
Tasmania	Australia	0-1,600	forest, mixed	68,000	15.0
NORTH AND CENTRAL AMERICA					
California	USA	0-550	forest, scrub	30,000	3.0
Guadalupe Island	Mexico	0-1,300	mixed	280	2.0
Baja California	Mexico	0-1,000	mixed	17,000	3.0
Sierra Madre Occidental	Mexico, USA	1,200-3,050	forest	36,000	3.5
North-west Mexican Pacific slope	Mexico, USA	0-1,000	forest, scrub	14,000	9.0
Sierra Madre Oriental	Mexico, USA	1,800-3,500	forest	16,000	2.0
North-east Mexican Gulf slope	Mexico, USA	0-1,000	mixed	77,000	4.0
Central Mexican marshes	Mexico	1,500-2,500	wetland	10,000	2.0
Yucatan Peninsula	Belize, Guatemala, Honduras, Mexico	0-300	forest, scrub	138,000	15.1
Revillagigedo Islands	Mexico	0-300	scrub, forest	280	5.0
Central Mexican highlands	Mexico	900-3,500	scrub, forest	41,000	15.8
Sierra Madre del Sur	Mexico	300-2,000	forest, scrub	18,000	6.8
Isthmus de Tehuantepec	Mexico	0-1,000	scrub, forest	7,700	2.3
North Mesoamerican highlands	Belize,El Salvador, Guatemala, Honduras, Mexico, Nicaragua	600-3,000	forest	68,000	21.0
North Mesoamerican Pacific slope	Mexico, Guatemala, El Salvador, Nicaragua, Honduras	0-1,050	forest, scrub	15,000	3.0
South Central American Caribbean slope	Costa Rica, Guatemala, Nicaragua, Panama	0-1,200	forest	25,000	8.0
South Central American Pacific slope	Costa Rica, Panama	0-1,500	forest	24,000	13.0
Costa Rican and Panamanian highlands	Costa Rica, Nicaragua, Panama	600-3,350	forest	27,000	52.5
North Choco and Darien lowlands	Colombia, Costa Rica, Panama	0-1,000	forest	14,000	10.0
Darien highlands	Colombia, Panama	600-1,600	forest	4,200	13.5
Cocos Isles	Costa Rica	0-700	forest, scrub	47	3.0
Cuba and the Bahamas	Bahamas, Cuba, Turks and Caicos Is	0-2,000	forest, scrub	93,000	21.3
Jamaica	Jamaica	0-2,200	forest, scrub	11,000	30.3

Table 15.7 Endemic bird areas of the world

NAME	POLITICAL UNIT(S)	ALTITUDE (m)	HABITAT(S)	SIZE (km²)	SPP.R.
NORTH AND CENTRAL AMERICA (continued)					
Hispaniola	Dominican Republic, Haiti	0-3,000	forest, scrub	76,000	26.1
Puerto Rico	Puerto Rico	0-1,200	forest, mixed	9,000	17.9
East Caribbean	Antigua and Barbuda, Anguilla, Netherlands Antilles, Barbados, Dominica, Grenada, Guadeloupe,St Kitts-Nevis, St Lucia, Martinique, Montserrat, St Vincent and Grenadines, British Virgin Is, Virgin Is (US)	0-1,500	forest, scrub	6,600	30.3
SOUTH AMERICA					
North Choco and Darien lowlands	Colombia, Costa Rica, Panama	0-1,000	forest	14,000	10.0
Darien highlands	Colombia, Panama	600-1,600	forest	4,200	13.5
Guianas	Brazil, French Guiana, Guyana, Suriname	0-1,100	forest	174,000	11.5
Tepuis	Brazil, Guyana, Venezuela	500-2,800	forest	35,000	39.0
Cordillera de Caripe and Paria Peninsula	Venezuela	700-2,500	forest	4,000	8.6
North Venezuelan mountains	Venezuela	750-2,400	forest	7,100	11.3
Venezuelan llanos	Colombia, Venezuela	0-1,100	savanna, mixed	57,000	2.0
Merida mountains	Venezuela	750-4,000	forest	18,000	18.0
Guajiran lowlands	Colombia, Venezuela	0-600	scrub, forest	36,000	10.5
Santa Marta Mountains	Colombia	750-4,600	forest	5,400	17.6
Nechi lowlands	Colombia	0-1,500	forest	28,000	8.5
Eastern Andes of Colombia	Colombia, Venezuela	900-5,200	forest, wetland	67,000	22.2
Upper Rio Negro white sand forests	Colombia, Venezuela	100-500	forest	10,000	11.5
Cauca valley	Colombia	600-2,700	forest	19,000	7.5
Magdalena valley	Colombia	200-2,700	forest	29,000	8.0
Choco	Colombia, Ecuador	0-1,200	forest	59,000	17.0
Western Andes of Colombia and Ecuador	Colombia, Ecuador	500-3,300	forest	27,000	37.7
Galapagos Islands	Ecuador	0-1,300	scrub, forest	8,000	23.0
Central Andes of Colombia and Ecuador	Colombia, Ecuador	2,100-5,200	forest, mixed	37,000	15.8
Eastern Andes of Ecuador	Colombia, Ecuador, Peru	400-2,000	forest	24,000	13.0
Napo lowlands	Brazil, Ecuador, Peru	100-600	forest	129,000	8.0
Ecuadorian dry forests	Ecuador, Peru	0-2,000	forest, scrub	57,000	47.5
North Peruvian cloudforests	Ecuador, Peru	1,500-3,200	forest	9,200	6.0
Maranon valley	Peru	200-2,400	forest, scrub	11,000	11.0
North-east Peruvian riverine forests	Peru, Ecuador	100-450	forest	11,000	2.0
East cordilleran ridgetop forests	Ecuador, Peru	1,000-2,400	forest	8,900	6.5
East Peruvian cordilleras	Peru	1,900-3,700	forest	44,000	22.0
North Peruvian coast	Ecuador, Peru	0-500	scrub, mixed	31,000	5.5
Western Andes of Peru	Peru	1,800-4,300	scrub, forest	59,000	20.0
Junin grasslands	Peru	3,700-5,000	wetland, grassland	17,000	4.0
Eastern Andes of Peru	Peru	700-1,600	forest	11,000	6.5
South-east Peruvian lowlands	Brazil, Peru	100-400	forest	155,000	14.0
South-east Peruvian Andes	Peru	2,500-4,300	forest, scrub	13,000	11.0
South Peruvian Pacific slope	Chile, Peru	0-3,000	scrub, mixed	76,000	8.0
Upper Bolivian yungas	Bolivia, Peru	1,800-3,700	forest	19,000	15.0
Lower Bolivian yungas	Bolivia, Peru	700-2,400	forest	38,000	20.5
Bolivian Andes	Argentina, Bolivia, Peru	1,400-4,600	scrub, forest	32,000	13.0
East Bolivian lowlands	Bolivia, Brazil	200-750	forest, grassland	93,000	7.0
North Argentinian Andes	Argentina	2,000-4,000	scrub, mixed	17,000	6.0

Table 15.7 Endemic bird areas of the world

NAME	POLITICAL UNIT(S)	ALTITUDE (m)	HABITAT(S)	SIZE (km²)	SPP.R.
SOUTH AMERICA (continued)					
Argentinian grasslands	Argentina	100-500	scrub, wetland	34,000	3.0
Argentinian cordilleras	Argentina	1,600-2,900	grassland, mixed	10,000	3.0
Juan Fernandez Islands	Chile	0-1,300	forest, scrub	180	3.0
Central Chile	Argentina, Chile	0-1,600	forest, scrub	137,000	10.0
Tierra del Fuego and the Falklands	Argentina, Chile, Falklands	0-1,200	grassland, wetland	119,000	9.0
Central Amazonian Brazil	Brazil	0-300	forest	35,000	16.0
West Amazonian Brazil	Brazil	0-400	forest	30,000	2.0
Fernando de Noronha	Brazil	0	forest, scrub	26	2.0
North-east Brazilian caatinga	Brazil	0-900	forest, scrub	100,000	9.5
Alagoan Atlantic slope	Brazil	0-1,000	forest	30,000	13.7
Bahian deciduous forests	Brazil	250-900	forest	8,000	2.5
Minas Gerais deciduous forests	Brazil	300-500	forest	10,000	2.0
Serra do Espinaco	Brazil	700-1,600	grassland, scrub	30,000	5.5
Bahian and Espirito Santo Atlantic slope	Brazil	0-600	forest	40,000	10.7
South-east Brazilian lowland to foothills	Argentina, Brazil, Paraguay	0-1,500	forest	50,000	44.2
South-east Brazilian mountains	Brazil	500-2,200	forest	65,000	22.5
South-east Brazilian Araucaria forest	Argentina, Brazil, Paraguay	0-1,000	forest	30,000	4.0
Entre Rios wet grasslands	Argentina, Uruguay	0-200	wetland	25,000	3.0
PACIFIC ISLANDS					
Mariana Islands	Guam, N Marianas	0-950	forest, mixed	1,000	10.8
Yap	Micronesia	0-180	mixed, forest	56	4.2
Palau Islands	Palau	0-240	forest, mixed	500	12.0
Micronesian Islands	Micronesia	0-800	forest, mixed	580	13.1
Admiralty Islands	Papua New Guinea	0-700	forest, mixed	1,900	8.6
St Matthias Islands	Papua New Guinea	0-650	forest, mixed	300	4.5
New Britain and New Ireland	Papua New Guinea	0-2,200	forest, mixed	46,000	41.4
D'Entrecasteaux and Solomon Sea Islands	Papua New Guinea	0-2,200	forest, mixed	3,400	2.9
Louisiade Archipelago	Papua New Guinea	0-800	forest	1,300	5.7
Solomon Islands	Papua New Guinea, Solomon Is	0-2,500	forest	32,000	51.3
San Cristobal	Solomon Is	0-2,000	forest	3,300	19.0
Rennell Island	Solomon Is	0-110	forest, mixed	840	7.3
Vanuatu and the Santa Cruz Islands	Solomon Is, Vanuatu	0-1,800	forest, mixed	16,000	20.4
New Caledonia and the Loyalty Islands	New Caledonia	0-1,600	forest, mixed	19,000	23.3
Samoan Islands	American Samoa, Samoa	0-1,200	forest, mixed	3,000	11.7
Fijian Islands	Fiji	0-1,200	forest, mixed	18,000	27.8
Norfolk Island	Australia	0-320	forest, mixed	35	4.0
Lord Howe Island	Australia	0-760+	forest, scrub	17	2.0
New Caledonia North Island	New Zealand	0-2,000	forest	41,000	1.5
South Island	New Zealand	0-2,500	forest, mixed	43,000	8.5
Auckland Islands	New Zealand	0-600+	grassland	570	1.0
New Caledonia Islands	New Zealand	0-270+	mixed	960	5.0
Northwestern Hawaiian Islands	USA	0-300	scrub, mixed	5	4.0
Hawaiian Islands	USA	0-3,100	forest	6,200	20.5
Hawaii	USA	0-3,100	forest	10,000	11.5
Marquesas Islands	French Polynesia	0-1,200	forest	1,000	10.5
Society Islands	French Polynesia	0-1,700	forest, mixed	400	5.3
Tuamotu Archipelago	French Polynesia	0-110	forest, plantation	690	7.3
Lower New Caledonia Islands	French Polynesia	0-700	forest, mixed	190	6.6
Pitcairn Islands	Pitcairn	0-33	forest	36	3.5

Note: Spp. R. is an index of the numbers of restricted-range species occurring in each EBA taking into account the sharing of species between EBAs.

Table supplied by ICBP

16. SPECIES EXTINCTION

Species extinction is a natural process. The fossil record suggests that all species have a finite lifespan and that the vast majority of species that have ever existed are now extinct, with extinct species outnumbering living species by a factor of perhaps a thousand to one.

Species become extinct when all individuals die without producing progeny. They disappear in a different sense when a species lineage is transformed over evolutionary time, or divides into two or more separate lineages (so-called pseudo-extinction). The relative frequency of true extinction and pseudo-extinction in evolutionary history is unknown, although the former's great importance is demonstrated by the disappearance of entire, and once highly diverse, lineages such as trilobites and ammonites.

HOW SPECIES BECOME VULNERABLE TO EXTINCTION

Two broad categories of process are believed to affect the dynamics of populations, and provide the fundamental mechanisms of species extinction:

- deterministic processes (or cause and effect relationships) e.g. glaciation or direct human interventions such as deforestation
- stochastic processes (chance or random events), which may act independently or influence variation in deterministic processes.

The magnitude of the effects of these processes depends on the size and degree of genetic connectedness of populations. Four types of stochastic processes can be distinguished (Shaffer, 1987): *demographic uncertainty* (resulting from random events in the survival and reproduction of individuals); *environmental uncertainty* (due to unpredictable changes in weather, food supply, disease, and the populations of competitors, predators, or parasites); *natural catastrophes* (floods, fires or droughts); and *genetic uncertainty* (random changes in genetic make-up, to which several factors contribute).

Models of the effects of stochastic processes suggest that:

- demographic uncertainty is only a hazard for relatively small populations (numbering tens or hundreds of individuals)
- there is no critical population size that once reached guarantees a high level of long-term security from environmental uncertainty
- progressively larger increases in population size yield diminishing returns in persistence times for a given catastrophic event.

When demographic and environmental uncertainty interact, their effects compound each other, so that in a variable environment any loss in population size proportionally increases the chance of population extinction. Thus, to be reasonably certain of conserving a species for a significant length of time, one must preserve either very large population sizes (hundreds to millions of individuals or more, depending on the biology of the species) or numerous populations (Schaffer, 1987).

The isolation of populations

The 'equilibrium theory' of island biogeography developed by MacArthur and Wilson (1963 and 1967) is an extension of the species-area relationship (see Chapter 5). Whilst originally used to model species richness and turnover on real islands, it has subsequently been used to predict changes in species number in isolated habitat islands.

The area of an island sets an upper limit to the maximum population size of each species. Since small populations are inherently more prone to extinction than large (for reasons discussed above), extinction rates tend to be inversely proportional to island area. Successful colonisation by new species is not affected so much by area as by the degree of isolation of the island: islands near to the mainland or to other islands are colonised at higher rates than those farther away. Increased isolation of populations not only reduces the incidence of colonisation by new species, but decreases the probability that immigrants of an existing species will arrive. Over time, an equilibrium is eventually reached on any island at which the loss of species through extinction is balanced by the arrival and colonisation of new species.

A later modification of the theory incorporates the 'rescue effect' (Brown and Kodric-Brown, 1977). The immigration of new, unrelated individuals can play an important role in maintaining an isolated population, because their demographic and genetic contributions tend to increase its size and genetic fitness, thereby reducing the possibility that it will become extinct. The significance of the rescue effect is that fewer immigrants are needed to rescue an existing population than to successfully found a new one.

Island biogeographic theory has far-reaching implications for conservation biology. Rates of habitat modification are currently so high that virtually all natural terrestrial habitats and protected areas are destined to become ecological 'islands' in surrounding 'oceans' of habitat much altered by human activity. Not only is the total area of many natural habitats rapidly decreasing, but those large natural habitat islands that now exist are being fragmented into archipelagos of habitat islands. This process of fragmentation and isolation is predicted to lead directly and indirectly to accelerated species extinctions at both the local and global scales.

Consequences of insularisation

The combination of short- and long-term insularisation effects is predicted to reduce the number of species to a lower equilibrium. A study of understorey birds in fragments of tropical forest ranging from 0.1 to 571ha in the Usambara Mountains, Tanzania, found just this result (Newmark, 1991). Since separation, smaller forest fragments have lost more bird species than larger areas, and more isolated fragments have lost more species than those close to a source of potential colonists. Similarly, Klein (1989) observed communities of dung and carrion beetles (subfamily Scarabaeinae) in fragmented habitat patches of different sizes in the Amazon rain forest of Brazil. He found that forest fragments had lower species richness, an

(1989) observed communities of dung and carrion beetles (subfamily Scarabaeinae) in fragmented habitat patches of different sizes in the Amazon rain forest of Brazil. He found that forest fragments had lower species richness, an increased proportion of rare species, and sparser populations in comparison with continuous undisturbed forest. These differences were more pronounced in small fragments (< 1ha) than large.

Many researchers, however, are now convinced that calculation of rates of species loss in habitat islands or reserves using the species-area relationship is unjustified as a basis for detailed conservation recommendations. Boeckeln and Gotelli (1984) argue that the models developed ignore species identity, habitat heterogeneity and population sizes, and have such wide margins of error that they have low explanatory power and give unreliable estimates. For example, Soulé *et al.* (1979) predicted on the basis of a simple species-area model that the Serengeti National Park will lose 50% of its large mammals (some 15 ungulate species) in the first 250 years of isolation, while Western and Ssemakula (1981) attempted to incorporate habitat diversity data and predicted that only one species will be lost. Zimmerman and Bierregard (1986) argue that beyond the ecological truism that species richness increases with area, the equilibrium theory of biogeography has revealed little that is of "real value for planning real reserves in real places". In designing reserves to protect Central Amazonian forest frogs, Zimmerman and Bierregard consider that critical breeding habitat and places that contain quality habitat at high density must be found before the reserve size question is addressed. In general, biologists need empirical studies that directly measure the effects of habitat fragmentation on specific groups (Klein, 1989).

Ecological correlates of vulnerability to extinction

There is considerable evidence that the number of species in an isolated habitat will decrease over time, although the probable rates of such extinctions (and whether the equilibrium theory of island biogeography can be used to predict these) are in dispute. The crucial issue for conservationists now is whether those species which are most at risk from extinction following habitat fragmentation can be predicted from a knowledge of their biology and ecology. At least nine ecological or life history traits (some of which may actually be highly correlated with each other) have been proposed as factors determining an animal species sensitivity to fragmentation (Karr, 1991; Laurance, 1991):

Rarity
Several studies have found that the abundance of a species prior to habitat fragmentation is a significant predictor of extinction. For example, Newmark (1991) found that after fragmentation, rare understorey bird species occupied fewer forest fragments per species than common ones. This is only to be expected, since fewer individuals of a rare species than a common species are likely to occur in habitat fragments, and the mechanisms of extinction mean that small populations are inherently more likely to become extinct than large.

Dispersal ability
If animals are capable of migrating between fragments or between 'mainland' areas and fragments, the effects of small population size may be partly or even greatly mitigated by the arrival of 'rescuers'. Species that are good dispersers may therefore be less prone to extinction in fragmented habitats than poor dispersers.

Degree of specialisation
Ecological specialists often exploit resources which are patchily distributed in space and time, and therefore tend to be rare. Specialists may also be vulnerable to successional changes in fragments and to the collapse of coevolved mutualisms or food webs.

Niche location
Species adapted to, or able to tolerate, conditions at the interface between different types of habitats may be less affected by fragmentation than others. For example, forest edge species may actually benefit from habitat fragmentation.

Population variability
Species with relatively stable populations are less vulnerable than species with pronounced population fluctuations, since they are less likely to decline below some critical threshold from which recovery becomes unlikely.

Trophic status
Animals occupying high trophic levels usually have small populations: e.g. insectivores are far fewer in number than their insect prey and, as noted above, rarer species are more vulnerable to extinction.

Adult survival rate
Species with naturally low adult survival rates may be more likely to become extinct, as Karr (1991) has proposed for island birds on Barro Colarado Island, Panama.

Longevity
Long-lived animals are less vulnerable to extinction than short-lived.

Intrinsic rate of population increase
Populations which can expand rapidly are more likely to recover after population declines than those which cannot.

Laurance (1991) has, however, studied extinction proneness among 16 species of non-flying land mammals in fragmented rain forest in Queensland, Australia. Seven traits were examined: body size, longevity, fecundity, trophic level, dietary specialisation, natural abundance in continuous rain forest, and 'matrix' abundance (abundance of the species in modified habitats surrounding original fragments). Of these, matrix abundance was the best predictor of vulnerability. Once its effects were removed, partial correlations showed no other significant predictors of extinction proneness.

Laurance therefore suggests that tolerance of modified habitats is important in determining survival in fragmented habitats. Species that were able to exploit modified habitats tended to remain stable or even to increase in number in fragments, whereas those that avoided these habitats tended to disappear. The most vulnerable species even avoided using the corridors of secondary growth forest that existed along streams, a finding which highlights the importance of maintaining corridors of primary vegetation to act as pathways for dispersing individuals between patches of habitat.

Viable populations, genetic variation and extinction

Increasingly the attention of conservationists has become focused on the management and preservation of isolated small populations confined to habitat islands, usually in protected areas. An essential requirement is to ascertain how many individuals of a species should be conserved in order to ensure its survival in a particular area. There are two approaches to estimating such a Minimum Viable Population (MVP) size, the demographic and the genetic. The process of applying a demographic or genetic MVP model to a particular species or population, and proposing the management interventions that should be undertaken to increase its chances of survival, is known as Population Viability Analysis (PVA).

In the demographic approach, estimates of a population's average growth rate (which is in part determined by the species's body size), the variance in this growth rate attributable to environmental fluctuations, and the population's maximum size, are used in mathematical models to calculate its expected persistence time to extinction. There are two main factors that need to be considered: the population size, and the length of time it requires to be preserved. Normally, an MVP is taken to be that size of population that has a 95% probability of persistence for x number of years, where for consistency x is usually taken as either 100 or 1,000.

Clearly, there is no such thing as a standard MVP that can be applied to all species. Belovsky (1987) has calculated that over a range of body masses from 10g (the size of a European Common Shrew *Sorex araneus*) to 10^6g (the size of a Black Rhinoceros *Diceros bicornis*), MVP sizes for mammals range from hundreds to millions. The Minimum Area Required (MAR) to support these populations ranges from tens to millions of square kilometres. As body mass increases, MVP size decreases, but larger mammals require proportionately larger ranges. MARs are larger for carnivores than for herbivores and larger for tropical than for temperate species.

MVPs can also be examined from a genetic perspective, in which not only the number of individuals surviving but their genetic variation or heterozygosity are considered important. In the long term, this genetic variation is necessary for evolution by natural selection to occur, and is required for adaptation to potential future changes in the environment. In the short term, heterozygosity is positively correlated with fitness, including survival, disease resistance, growth and developmental rate and stability (Allendorf and Leary, 1986).

Genetic variation can quickly be lost through breeding with closely-related individuals (inbreeding) which leads to low levels of heterozygosity and lowered offspring fitness, a phenomenon known as inbreeding depression (Falconer, 1981). The most likely explanation is that new mutations, which are almost always harmful, can accumulate in a species genome providing they are fully or partially recessive and are not therefore expressed. Inbreeding increases the probability that the effects of these harmful genes will be expressed.

Franklin (1980) has proposed that in the short term an effective population size of 50 is the MVP required to guard against the negative effects of inbreeding for a population of large mammals with no immigration or introduction of unrelated stock. Populations of this size will nevertheless eventually become inbred over time, to a degree directly related to the generation interval (randomly-breeding populations of 50 mice will become more inbred in a decade than 50 elephants will in a century). In the long term an effective population size of 500, corresponding to a real population size of several times this number, has been suggested as a suitable genetic MVP for large mammals, since in a population of this size rates of mutation will renew genetic variation as quickly as it is lost by inbreeding and genetic drift (Franklin, 1980; Lande and Barrowclough, 1987).

Although biologists have suggested the figures quoted above as useful first estimates of MVP sizes, in both the demographic and the genetic approach the actual numerical value arrived at depends not only on the criteria chosen to define the MVP (e.g. the number of years the population is required to persist) but also on the values of the parameters used in the model. These values cannot always be assigned in an objective manner. Thus even for one particular species there is no single number that is universally valid, and this reservation is doubly true when different species are compared. Each situation is unique and should be considered separately. For example, a species that exhibits a boom and crash population cycle will require a larger MVP than one which inhabits a stable environment and whose population is relatively stable.

Both MVPs and PVAs have now been applied to a variety of species. Examples include: large mammals such as the Sumatran Rhinoceros *Didermocerus sumatrensis* and the Florida Panther *Felis concolor coryi*; and birds such as the Bali Starling *Leucopsar rothschildi*, Caribbean parrot species, and Asian Hornbills.

Analysis of the existing worldwide protected areas system indicates that few if any large mammal species will be adequately conserved with the current scale of ecosystem coverage, as most protected populations are too small to constitute MVPs (Grumbine, 1990).

The fact that a population has declined in number to below the theoretically determined MVP does not automatically mean that its situation should be considered hopeless. Some species, such as the Northern Elephant Seal *Mirounga angustirostris* (Bonner and Selander, 1974) and captive populations of Golden Hamster *Mesocricetus auratus* have survived through population bottlenecks of just a few individuals, following which numbers have increased to substantial levels. Eventually, if a large population is re-established, genetic variation may be regenerated by mutation, thus restoring the potential for adaptive evolution.

These examples, however, may be the exceptions rather than the rule. Other species that have declined to such low levels may have vanished altogether. Even if populations do recover numerically from a bottleneck, inbreeding and consequent loss of heterozygosity may cause noticeable declines in fitness effectively prejudicing the species long-term chances of survival. For example, O'Brien *et al.* (1985) found high rates of juvenile mortality, incidence of sperm abnormalities, and susceptibility to disease in several populations of Cheetah *Acinonyx jubatus*, and attributed this to the low level of genetic variation found in all Cheetah populations examined.

Perhaps the most compelling evidence to date of the negative consequences of population bottlenecks comes from a study of Lion *Panthera leo* in Ngorongoro Crater and the neighbouring Serengeti Plains in Tanzania. In 1962 the relatively isolated Lion population in the Crater dropped from around 70 individuals to 10 as a result of an outbreak of biting flies *Stomoxys calcitrans*. The population has since recovered to its pre-plague levels. Packer *et al.* (1991) have found that compared to the larger outbred population of Serengeti Lions, those in Ngorongoro suffer high levels of sperm abnormality. Their reproductive performance has also diminished over the years since the bottleneck, and both effects are apparently correlated with the lower levels of heterozygosity in the Ngorongoro population.

Metapopulation theory

The MVP models discussed so far have considered all individuals as belonging to a single isolated population, which is rarely the case in the real world. In practice most species are patchily distributed, and are best regarded as a population of subpopulations, or a *metapopulation*, in which subpopulations are geographically isolated but interconnected by patterns of gene flow, extinction and recolonisation. Thus, studies over a 25-year period by Erhlich and colleagues of a purported single population of Checkerspot Butterfly *Euphydryas editha bayensis* in the Jasper Ridge Preserve (USA) demonstrated that although the population occupied three nearly contiguous habitat patches, it actually consisted of three demographic units whose sizes fluctuated independently in response to annual changes in rainfall. One of these units became extinct, was re-established by immigration, and became extinct again several years later (Wilcox and Murphy, 1985). Relaxing the single population assumption of the MVP model so that immigrants can be received from neighbouring populations will lengthen the projected persistence times.

Habitat heterogeneity and the existence of many subpopulations are an important element of population dynamics, and have profound implications for conservation biology. Pulliam (1988) introduced a simple model of metapopulation dynamics incorporating density-dependent immigration as the linking factor between source and sink populations in severely fragmented habitats. In his model, a limited number of reproductively successful 'source' subpopulations produce an excess of offspring over and above the number that the habitat can absorb. The surplus individuals migrate to other less favourable areas, occupied by 'sink' subpopulations which would be doomed to extinction without persistent immigration.

Supporting evidence for the source-sink metapopulation theory is available from a number of field studies; for example, King and Mewaldt (1987) found that isolated montane populations of White-crowned Sparrows *Zonotrichia albicollis* were unable to persist without periodic immigration.

Metapopulation theory should help biologists determine which populations are priorities for conservation. The importance of identifying and preserving source populations and habitats is obvious: without them the metapopulation cannot persist. However, the presence of breeding individuals at a particular site does not necessarily indicate that it is suitable for the species in the long term, since it could still be a sink habitat. In general, source populations will not only have higher annual reproduction rates than annual mortality rates but will also have more stable populations than sink populations. In the case of long-lived species the identification of source populations will therefore necessitate continuous, long-term monitoring. In addition to the identification and protection of demographic source populations, the conservation of buffer habitats and marginal subpopulations should also be a part of comprehensive conservation plans, and the long-term status of even apparently secure metapopulations should be carefully monitored.

Conclusion

Current models of the extinction process and estimates of habitat loss, principally tropical forest, predict that species extinctions are occurring at very high rates on both a local and global scale. The primary cause is habitat modification and fragmentation by human activities. This process not only decreases overall population sizes of many species but splits previously continuous populations into smaller isolated sub-populations. Deterministic and stochastic effects mean that small populations are more susceptible to extinction than large. Conservation biologists have enlisted the help of various theories and models to try and predict how many species, and which ones, will be lost. It is possible to make reasonable predictions of which species will be most adversely affected by habitat fragmentation.

The species-area relationship is not now thought to be a good predictor of species loss in habitat fragments, but has implications for the design and positioning of reserves. With a realisation that ecosystems are often best preserved by concentrating on keystone species, efforts have switched to conducting population viability analyses for selected species in an attempt to estimate the minimum viable population sizes that must be conserved to ensure their long-term survival. MVPs can be examined from either the demographic or genetic perspective - both approaches give estimates of a similar order of magnitude. A shortcoming of MVP estimates is that they consider only a single population. The incorporation of metapopulation theory should improve the accuracy and utility of these models, and allow the identification of the most important subpopulations, facilitating the determination of conservation priorities.

A BRIEF HISTORY OF EXTINCTIONS

Knowledge of extinction patterns through geological time is based on analysis of the fossil record, which represents a small and highly biased sample of the taxa that have existed - it may represent only one in every 20,000 species that has existed. The best preserved group consists of marine animals, chiefly invertebrates, with durable, highly mineralised exoskeletons. Caution has to be exercised in extrapolating from this group to others, particularly plants, as they may show different patterns of extinction.

Mass extinction events in marine organisms

The fossil record indicates that overall extinction rates have not been constant over time (Fig. 16.1). Around 60% of extinctions have occurred in a number of relatively short episodes. The earliest period for which there is evidence of a major loss of diversity is during the late Precambrian, around 700 Mya (million years ago) although the Precambrian fossil record is too incomplete to allow detailed analysis.

The fossil record for the Phanerozoic (i.e. from the Cambrian to the present, see Fig. 16.2) is much more detailed. During this time there have been five major 'mass extinction events'. These events took place late in each of the Ordovician, Devonian, Permian, Triassic and Cretaceous periods. By far the most severe was in the late Permian (245 Mya). At that time, the number of families of marine animals recorded in the fossil record declined by 54% and the number of genera by 78-84%. Extrapolation from these figures indicates that species diversity may have dropped by as much as 96%. The second most severe mass extinction, at the end of the Ordovician (440 Mya), resulted in the loss of 22% of families of marine taxa, a slightly greater figure than the late Devonian and late Triassic events (21% and 20% respectively). The late Cretaceous event was the least important, resulting in the loss of around 15% of marine families.

Figure 16.1 Extinction events in marine organisms

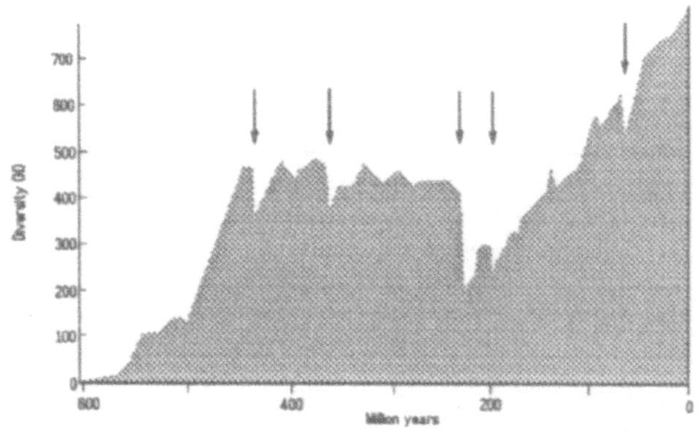

Source: Modified from Erwin, D.H., Valentine, J.W. and Sepkoski, J.J. 1987. A comparative study of diversification events: the early Palaeozoic versus the Mesozoic. *Evolution* 41(6).

Note: The curve plots diversity of marine animal families and indicates five major extinction phases.

Figure 16.2 The geological time scale

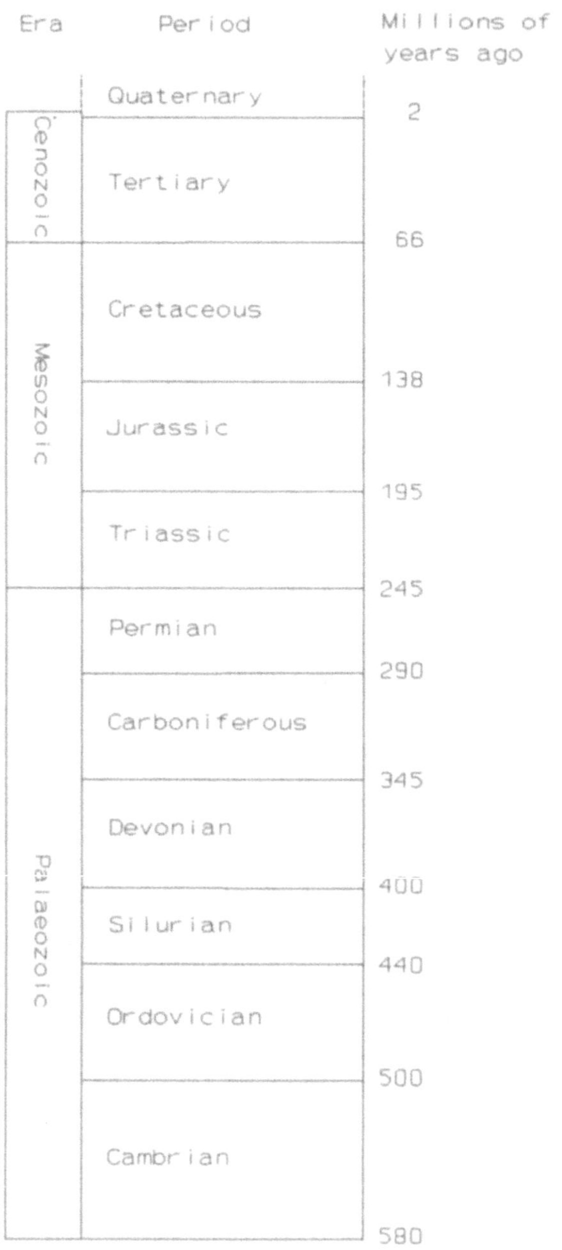

Note: Dates are approximate; scale covers the Phanerozoic only.

The causes and timespans of these events have been the subject of much debate and study. It is now widely accepted that the late Permian mass extinction was a long-term event, lasting for 5-8 million years. It appears to have been associated with geologically-rapid global physical changes (including the formation of the supercontinent Pangea), climate change, and extensive, tectonically-induced marine transgression and increased volcanic activity. There is no direct evidence of a single, catastrophic event such as impact by an extra-terrestrial body, although this cannot be ruled out as a contributory factor in the event. Interpretation of the late Triassic event is hampered by the absence of a good stratigraphic record; some indications suggest this was also a protracted period of extinction, although this is uncertain. The late Devonian extinction also appears to have spanned a considerable length of time, with elevated extinction rates throughout much of the middle and late Devonian. However, this extinction phase probably consisted of a series of discrete shorter extinction events rather than one protracted episode.

In contrast to these, the late Ordovician and late Cretaceous extinctions are thought to have taken place over a much shorter period. The late Ordovician event appears to be correlated with global glaciation 439 Mya (the Hirnantian glaciation) with three separate episodes of extinction spread over only 500,000 years.

The late Cretaceous extinction is probably the best known, but in terms of overall loss of diversity is also the least important. There is some evidence that this extinction event was associated with an extra-terrestrial impact, although this remains controversial.

As well as these major mass extinction events, a large number of less dramatic, but still significant, episodes can be identified from the marine fossil record. It has been argued that those following the late Permian extinction event have a periodicity of 26-28 million years, indicating some underlying unifying cause, although this remains unproven. It is notable that these more minor events account in total for more extinctions than the five major events outlined above.

Mass extinctions in vertebrates

The vertebrate fossil record, especially for terrestrial tetrapods, is much less amenable to analysis of extinction rates than the invertebrate record chiefly because it is less complete and less diverse. However, studies indicate that tetrapods have been subject to at least six mass extinction events since their appearance in the late Devonian, while fishes have experienced eight such events since their recorded origin in the Silurian. Some of these events coincide with each other and with those recorded for marine invertebrates; in particular, the five major mass extinction events outlined above are paralleled by losses in vertebrate diversity. The most significant is the late Permian event, which is the largest recorded extinction both for fishes (44% of families disappearing from the fossil record) and tetrapods (58% of families disappearing). The late Cretaceous event was more significant for tetrapods than for other groups, with 36 of the 89 families in the fossil record disappearing at this time. These families were, however, virtually confined to three major groups which suffered complete extirpation - the dinosaurs, plesiosaurs and pterosaurs. Most other major vertebrate taxa were almost completely unaffected.

Evidence for correlation between the more minor extinction events in vertebrates and the postulated periodic extinctions in marine invertebrates is currently poor.

Extinctions in vascular plants

In general, the plant fossil record does not clearly show the same sudden mass-extinction events seen in the animal record. Part of the explanation for this may lie in the nature of the plant fossil record itself and in the difficulties in interpreting it, but there also seem likely to be genuine differences between plants and animals in patterns of species origination and extinction. Plant extinction rates (based on analysis of families and genera) do vary with time, but in general, periods of elevated plant extinction appear to be more protracted than animal extinction events and do not usually coincide with them. It is argued that these periods may be more to do with competitive displacement by more developed plant forms, or with gradual climatic change, than with any sudden catastrophic events (Knoll, 1984).

The major exception to this is the end-Cretaceous catastrophe, which appears to have had a major influence on the structure and composition of terrestrial vegetation and on the survival of species. Data from fossil leaves suggest that perhaps 75% of late Cretaceous species became extinct, although data from fossil pollens indicate a lower though still significant level of extinction. During the Tertiary there are two other periods of widespread enhanced extinction rates, during the late Eocene and from the late Miocene to the Quaternary, although in the latter, extinction of taxa at generic level and above appears to have been mainly regional rather than global.

Background extinction rates

A corollary of the finding that the majority of extinctions recorded in the fossil record have taken place over relatively short time periods (geologically speaking) is that extinction rates for the remainder of the Phanerozoic have been low.

The average lifespan of species in the fossil record is around four million years which would give, at a very gross estimate, a background extinction rate of four species each year out of a total number of species of around 10 million. However, it can be argued that the fossil record is heavily biased towards successful, often geographically wide-ranging, species which undoubtedly have a far longer than average persistence time. Most species will therefore survive for less than four million years, and real extinction rates at any given time will be correspondingly higher. Nevertheless, even if background extinction rates were ten times higher than this, extinctions amongst the 4,000 or so living mammals would be expected to occur at a rate of around one every 400 years, and amongst birds at one every 200 years.

It is indisputable that the extinction rate in recent times has been far higher than this and that man has been the overwhelming cause. It is also widely accepted that mankind is in danger of precipitating further extinctions on a scale and at a rate at least comparable with those of the major extinction events in the distant past.

Extinctions and the spread of mankind

Documenting man's impact on the world's biota, and in particular quantifying species extinctions induced by man, is difficult for a variety of reasons, associated with: identifying species, especially those known only from subfossil or fossil remains; unequivocally demonstrating that extinction has occurred; and establishing a causal link between man's activities and extinction of the species in question.

Man may have first had a significant impact on the survival of other species during the late Pleistocene. Humans spread into Europe and Asia about one million years ago but slow advances in culture and technology seem to have restricted the impact on the fauna of these regions. However, man's arrival on previously isolated continents, around 50,000 years ago in the case of Australia and 11,000 years ago for North and South America, seems to coincide with large-scale extinctions in certain taxa. The exact timings are unclear and hence the cause and effect in each case are open to debate. However, Australia lost nearly all its species of very large mammals, giant snakes and reptiles, and nearly half its large flightless birds around this time. Similarly, North America lost 73% and South America 80% of their genera of large mammals at around the time of the arrival of the first humans. In these cases there is more direct evidence to link the events, although climatic upheavals at around the same time could also be implicated.

EXTINCTIONS IN RECENT HISTORY

The European Age of Expansion in the 15th and 16th centuries initiated another wave of extinctions. Indeed it has often been assumed that all, or at least the great majority, of modern man-induced extinctions date from this period. However, this may well be based more on the fact that a dramatic increase in documentation of natural phenomena, in large measure induced by the great voyages of discovery, also dates from this time.

It is now known that in some parts of the world a significant number of extinctions occurred before the arrival of Europeans. The Polynesians, who colonised the Hawaiian Islands in the 4th and 5th centuries AD, appear to have been responsible for exterminating around 50 of the 100 or so species of endemic land birds in the period between their arrival and that of the Europeans in the late 18th century. A similar impact seems to have been felt in New Zealand, which was colonised some 500 years later than Hawaii. Here an entire avian megafauna, consisting of members of the family Anomalopterygidae (the Moas) was apparently exterminated, also by the end of the 18th century. As with the late Pleistocene extinctions, there has been some controversy over the extent to which humans were responsible; however there is now a broad consensus that man was indeed responsible, probably through a combination of direct hunting and large-scale habitat destruction through burning.

Although most information from this period relates to avian extinctions, there is evidence that other groups, particularly mammals, had been similarly affected. On Madagascar, in addition to 6-12 ratites, including the Giant Elephantbird *Aepyornis maximus* (the largest bird ever recorded), at least 14 lemur species, most of them larger than any surviving species, have become extinct within the last 1,500 years, as have two giant tortoises. In the Caribbean, at least two ground sloths in the family Megalonychidae, several large rodents and three insectivores in the family Nesophontidae survived into the period of Amerindian settlement, but had become extinct before Europeans arrived at the end of the 15th century. The case for man being solely responsible for these extinctions is more equivocal than it is for New Zealand. However, on balance this appears to remain the most likely explanation, although it is possible that, on Madagascar at least, climate change leading to progressive desiccation of the environment also played a part.

While documentation has improved considerably since the 15th and 16th centuries, it still remains far from complete. This applies even to the best known groups, namely birds and mammals; for most lower vertebrates and virtually all invertebrates knowledge of extinction rates remains extremely scanty.

The main problem for documentation is that the majority of the world's species, especially tropical invertebrates, have not been scientifically named. A significant percentage of these may well become extinct before they have ever been collected and described. Of described taxa, numbering around 1.1 million animal species and around 270,000 vascular plants, accurate information on status and abundance is available for only a tiny proportion. The vast majority of the world's species, even in the best-known groups such as mammals and birds, are not subject to systematic monitoring and species may be locally or completely extirpated before their plight becomes known.

In general, it can only be stated with any confidence that a taxon is extinct when unsuccessful attempts have been made to locate it, or when it has not been sighted for several decades. Animal species thought to have become extinct, using this criterion and expert opinion, are listed in Table 16.1. Even here it is often difficult to demonstrate unequivocally that a species has become extinct and consequently several species are marked as possibly still being extant. Many species may persist unrecorded (albeit often in very low numbers) despite intensive efforts to locate them. This is borne out by the periodic reappearance of 'Lazarus taxa', after many years or decades of presumed extinction. Plants (Table 16.2), some of which produce seeds that can lie dormant and undetected for many years before germination, present particular monitoring problems.

Historical records of extinctions may thus be expected to be heavily biased, both taxonomically and geographically. Taxonomically, information on snails, particularly terrestrial species, birds and mammals is good, while that for most other groups is poor. Geographically, information on Europe and North America (including Hawaii) is much better than that for the rest of the world, although relatively few species extinctions have been recorded in Europe in recent times. Figures 16.6-16.10, taken from Table 16.6, illustrate these biases.

These biases make analysis of extinction patterns problematic. However, certain generalised patterns do

Table 16.1 Summaries of animal extinctions on islands and continents

	MOLLUSCS	BIRDS	MAMMALS	OTHER	TOTAL
ISLANDS	**151**	**104**	**34**	**74**	**363**
% of islands total	41.6	28.7	9.4	20.4	100
% of grand total	31.2	21.5	7	15.3	75
CONTINENTS	**40**	**11**	**24**	**46**	**121**
% of continents total	33.1	9.1	20	38	100
% of grand total	8.3	2.3	5	9.5	25
TOTALS	**191**	**115**	**58**	**120**	**484**
% of total on islands	79	90.4	59	61.7	75
% of TOTALS	39.5	23.8	12	24.8	100

emerge. The most important of these is the preponderance of extinctions on islands over those in continental areas (Table 16.1). Exactly 75% of recorded animal extinctions since 1600 have been on islands. For the three groups with best information, the proportion of island extinctions varies from 90% for birds to 58% for mammals, with molluscs intermediate at 80%. Of the continental extinctions, at least 66% can be classified as aquatic species (this includes amphibians and insects with aquatic larval stages but excludes birds such as ducks and grebes). Most striking, perhaps, is the very small number of extinctions recorded to date in continental tropical forest ecosystems, which are precisely the areas where mass extinction phenomena are predicted to be taking place at present (see below).

There appear to be several reasons for the elevated extinction level amongst island species. Most straightforwardly, island species, especially those confined to single islands, tend to have very restricted and completely circumscribed ranges: they consist effectively of single populations. Adverse factors are thus likely to affect the entire species and bring about its extinction. In contrast, continental species tend to occupy larger ranges existing as meta-populations, with a number of more-or-less isolated subpopulations. It is likely that some of these subpopulations will not be affected by a given adverse factor. Thus, the species itself will survive even if a number of subpopulations are extirpated. These concepts apply both to real islands and ecological islands, that is, areas of habitat separated from other such areas by inhospitable environments which act as an effective barrier to dispersal. In this context, aquatic species in isolated inland waters behave similarly to terrestrial species on isolated oceanic islands, which helps to explain the significantly elevated number of extinctions amongst continental freshwater species.

Many island species are innately vulnerable to extinction because of their biology. Species on islands have often evolved in the absence of terrestrial predators and may therefore be highly vulnerable to introduced predators. Tameness, flightlessness and reduced reproductive rates characterise many island birds and appear to have been major contributory factors in their extinction, through predation by humans or introduced species. Similarly, many island land snails, such as the Hawaiian *Achatinella* and French Polynesian *Partula* species, have low reproductive rates and, apparently, no defences against introduced predators, most notably the snail *Euglandina*. The elevated species extinction rates on islands can also be ascribed to taxonomic practices, as there has been a tendency for island populations to be designated as full species when they may more reasonably be regarded as subspecies of species on adjacent islands or on the mainland.

Causes of Extinction

A brief analysis of the 'Possible causes' column of Table 16.4 shows that introduced animals and direct habitat destruction by man have been major factors involved in these extinctions, being implicated in 17% and 16% respectively (see Fig. 16.3). These are equivalent to 39% and 36% if only those extinctions for which causes are assigned are counted. Hunting and deliberate extermination also contribute significantly (23% of extinctions with known cause). For a large number of animals, no information on cause of extinction was known.

Figure 16.3 Causes of animal extinction

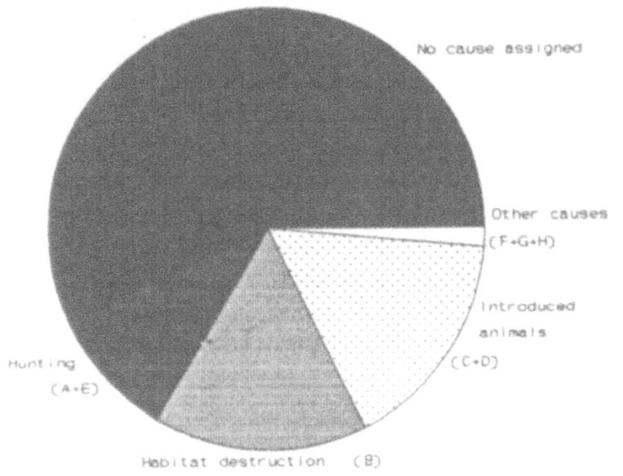

Notes: These figures were compiled by giving each species a score of 1 in the appropriate category if there was only one cause, 0.5 in each for two, etc. Where there were multiple causes C/D was counted as one part, C and D as two parts.

Time Series

Figures 16.4 and 16.5 and Table 16.2 present a breakdown of recorded extinctions in 30-year intervals from the year

Table 16.2 Time series of animal extinctions on islands and continents

	ISLANDS					CONTINENTS					COMBINED
	MOLLUSCS	BIRDS	MAMMALS	OTHER	TOTAL	MOLLUSCS	BIRDS	MAMMALS	OTHER	TOTAL	TOTAL
TOTALS	151	104	34	74	303	40	11	24	46	121	484
1600–1629	0	2	0	1	3	0	0	0	0	0	3
1630–1659	0	4	0	1	5	0	0	0	0	0	5
1660–1689	0	9	0	0	9	0	0	0	0	0	9
1690–1719	0	5	0	2	7	0	0	0	0	0	7
1720–1749	0	4	0	0	4	0	0	0	0	0	4
1750–1779	0	10	1	0	11	0	0	0	0	0	11
1780–1809	0	2	0	4	6	0	0	1	0	1	7
1810–1839	0	8	0	1	9	0	1	2	0	3	12
1840–1869	2	9	2	3	16	0	1	1	1	3	19
1870–1899	67	16	3	4	90	0	1	6	1	8	98
1900–1929	11	19	3	18	51	6	4	3	7	20	71
1930–1959	37	10	2	6	55	25	2	7	15	49	104
1960–	9	5	3	7	24	4	2	2	12	20	44
No date	25	1	20	27	73	5	0	2	10	17	90

Note: these summaries do not take into account 4 species (2 birds, 1 mammal and 1 'other') which are not assignable to either island or continent.

1600. These data should be interpreted cautiously. In only a few cases are the extinction dates reasonably certain; more often they are approximate to within one or two decades. In other cases, they are simply the date when the species was last recorded, and it is unknown how accurately they reflect the actual date of extinction (assuming the species is truly extinct). The uncertainties are most marked for species in areas which have only been occasionally surveyed (e.g. land snails on many tropical islands), and create difficulty in interpreting trends in extinction rates.

Of the individual taxa presented, island birds are the best documented group. There is no consistent trend over the full 400 years; peaks occur in the mid-17th and mid-18th centuries, and there is a clearer increase for the early 19th century until the 1930s. The apparent fluctuations for the first 200 years may represent real effects from introduced species, hunting and habitat modification associated with increasing levels of human settlement. Continental bird extinctions and the entire mammal data set are numerically smaller and thus harder to interpret. Of the 14 dated mammalian extinctions on islands, 13 have taken place since 1840. Most of the 20 undated mammalian extinctions (chiefly Caribbean rodents and insectivores) are believed to have taken place before the middle of the 19th century, showing little indication of a marked overall trend.

Information on mollusc extinctions was not available prior to the mid-19th century, and although high numbers of extinctions are documented for island molluscs in two 30-year periods, uncertainty in the dates again confuses interpretation.

Two trends are apparent in the time-series data for all taxa: first, that documented island extinctions began almost two centuries earlier than continental extinctions; second, that both island and continental extinctions have increased rapidly from early or mid-19th century to the mid-20th century. This increase has been more pronounced for continental species, although the island extinctions exceed continental ones numerically in all periods. The late 19th century for islands has the highest rate of all periods, reflecting a high contribution for mollusc extinctions on islands during this period.

The apparent decline in rate for both continental and islands for 1960-1989 is probably attributable to two causes; one is the expected time-lag in recording extinctions from 1960 onwards. As noted above, extinction is normally only attributed when a species has not been recorded over a significant time span. For some purposes, such as the designation of 'Extinct' under the Convention on Trade in Endangered Species of Wild Fauna and Flora (CITES), this time period is arbitrarily taken as 50 years. By this criterion, therefore, no species would be accepted as having become extinct since 1960 as 50 years would not have elapsed since its last being recorded. A more realistic and flexible approach has been adopted here, on the grounds that some species recorded since 1960 are regarded with a high degree of certainty to have become extinct, while conversely many species not observed by specialists in the wild for over 50 years are almost certainly still extant. Nevertheless, the general principle holds that the longer a species has not been recorded the more likely it is to be regarded as extinct, and vice-versa. A significant number of species are therefore likely to have become extinct recently without being recorded as such.

A second, more positive contributory factor to explain the apparent recent decline in extinction rates is the great increase in conservation action over the past 30 years. During this time, attention has focused largely on saving well-known species under imminent threat of extinction; most efforts to preserve these species havesucceeded, at least in the short or medium term. Several projects have taken the last wild individuals into captivity to build up populations until environmental conditions and populations are suitable for re-introduction to the wild (Tables 16.7 and 16.8). Thus well-documented species most vulnerable to

Figure 16.4 Time series of animal extinctions on islands and continents: selected taxa

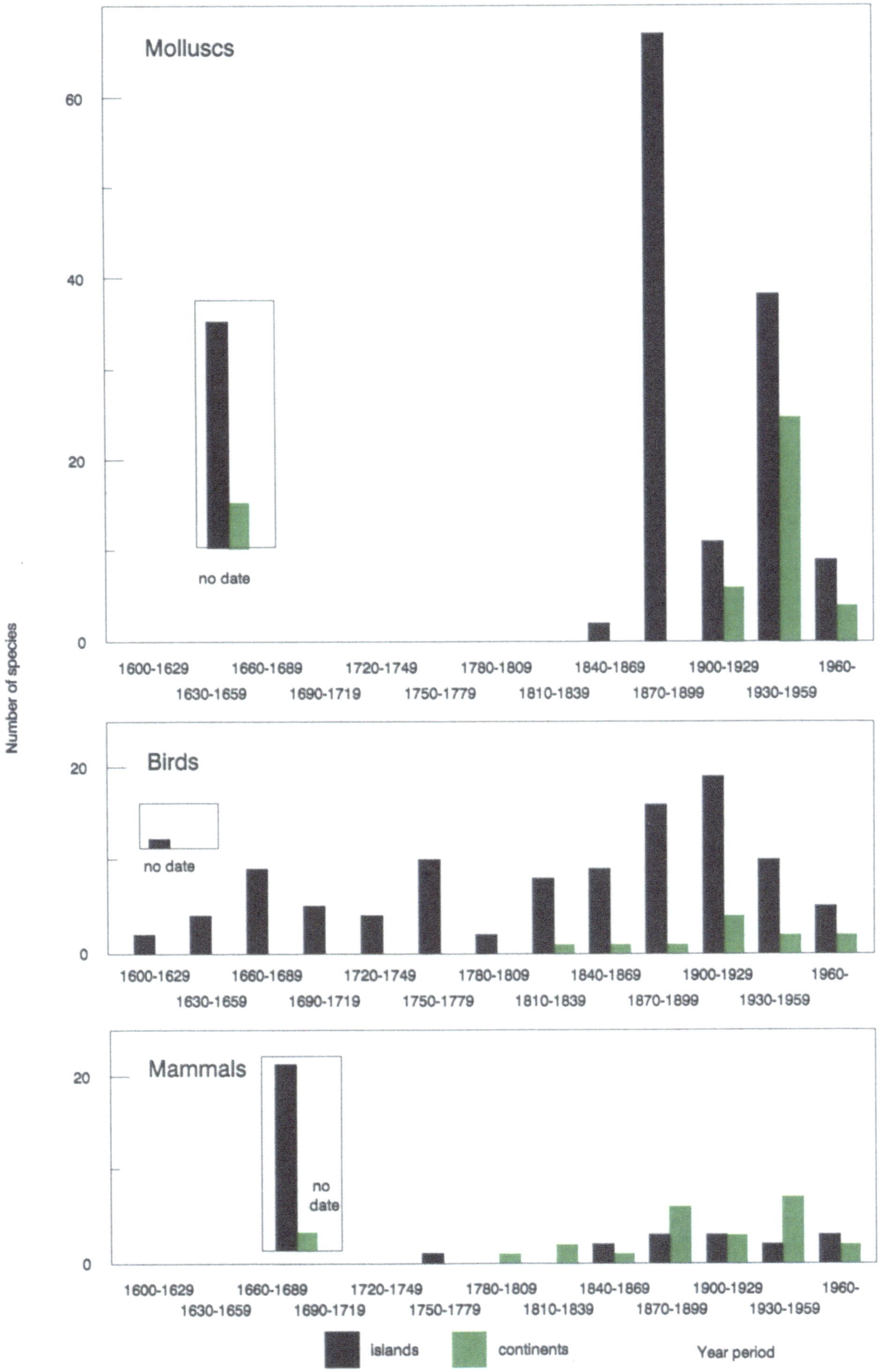

Figure 16.5 Time series of animal extinctions on islands and continents: all taxa

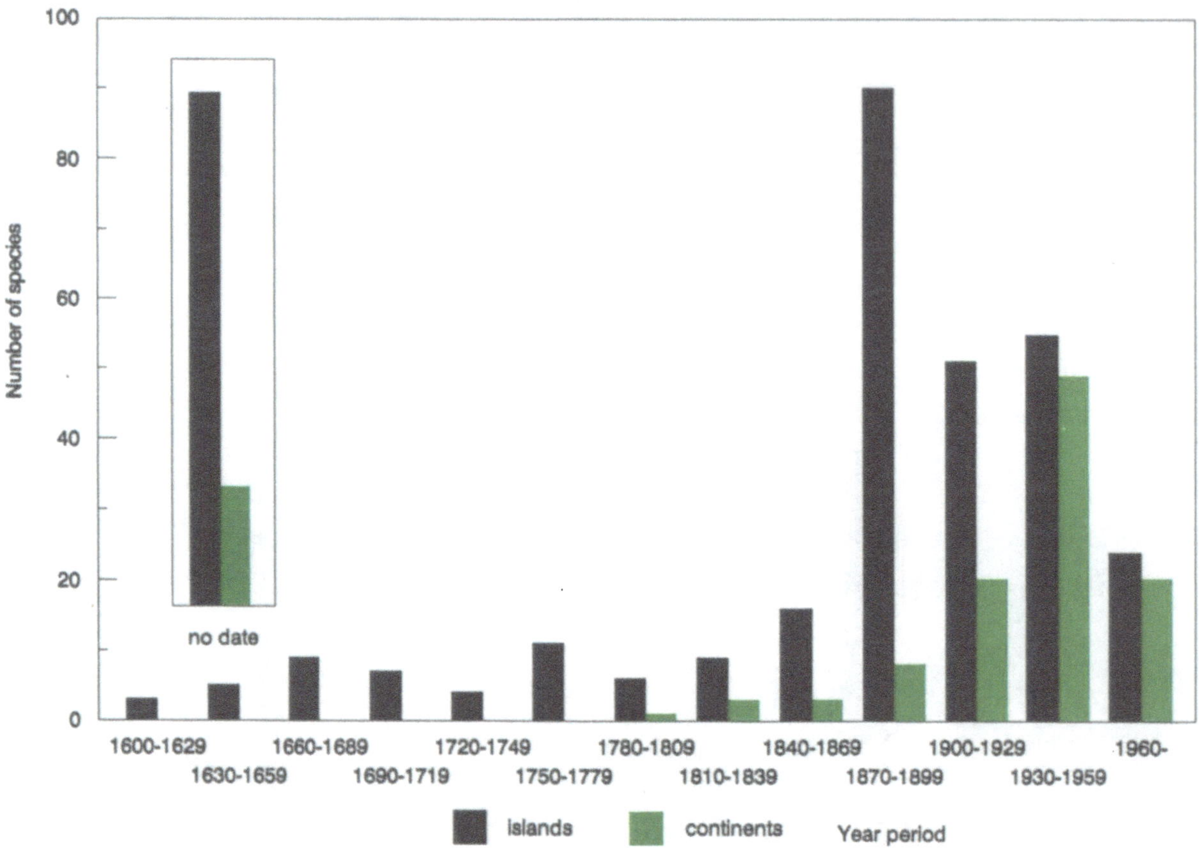

extinction during the past 30 years have often not become so, as a result of direct manipulative intervention. As noted above, it seems probable that significant numbers of undocumented continental species will have become extinct during this time.

CURRENT AND FUTURE EXTINCTION RATES

Habitat destruction, modification, and fragmentation are widely recognised as the most serious current threats to biological diversity, and the primary cause of recent extinctions. Estimates for present and projected global extinction rates have not been based on observed or recorded species extinctions, but rather on extrapolations from estimates of habitat loss coupled with assumptions derived from biogeography, relating numbers of species to area of habitat. A range of estimates are given in Table 16.3

In practice, most predictions of global extinction rates have been based on estimates of species richness in tropical forests, combined with estimates of actual and projected deforestation rates. Equating global species extinction with tropical forest species extinction has been justified by the recognition that the vast majority of terrestrial species occur in tropical moist forests.

The extrapolations from estimates of habitat loss are coupled with biogeographic assumptions using the species-area (Arrhenius) relation ($\log S = c + z\log A$) where S = number of species, A = area and c and z are constants (see Chapter 5). Values for z used are between 0.15 and 0.40. The most widely quoted generalisation is that a ten-fold reduction in area (i.e. loss of 90% of habitat) results in the loss of half the species present (30% with $z = 0.15$; 60% with $z = 0.40$).

Recent estimates based on these assumptions include those of Ehrlich and Wilson (1991) and Reid and Miller (1989). The former, on the basis of a 1.8% loss of rain forest per year, and using 'conservative' estimates from biogeographic theory (i.e. low z values), estimate a loss of 2-3% of rain forest species per decade. Reid and Miller, using z values of 0.15-0.40 and the assumption that forest loss is 1-2 times that projected by FAO for the period 1980-85, derive a similar figure of 2-5% loss per decade. This translates into a loss of some 5-15% by the year 2020, assuming rates of forest loss continue to increase.

Reid (1992) has refined the analysis somewhat, applying figures for forest area and rates of loss separately to Latin America, Africa and Asia, and accounting for observed differences in species diversity between the three regions. Using z values of 0.15-0.35 he concludes that global loss of closed-forest species will be of the order of 1-5% per decade, or 2-8% in total between 1990 and 2015. Reid stresses (and this applies to other estimates of species loss) that this is the number of species 'committed' to eventual extinction as a result of forest loss, not the number which will actually become extinct during that time - in many cases, there will be a delay between reduction in area of habitat and the extinction of species dependent on that habitat, especially for longer-living species.

Table 16.3 Estimated rates of extinction

ESTIMATE	% GLOBAL LOSS PER DECADE	METHOD OF ESTIMATION	REFERENCE
One million species between 1975 and 2000	4	Extrapolation of past exponentially increasing trend	Myers (1979)
15-20% of species between 1980 and 2000	8-11	Estimated species-area curve; forest loss based on Global 2000 projections	Lovejoy (1980)
12% of plant species in neotropics. 15% of bird species in Amazon basin	-	Species-area curve (z=0.25)	Simberloff (1986)
2000 plant species per year in tropics and subtropics	8	Loss of half the species in area likely to be deforested by 2015	Raven (1987)
25% of species between 1985 and 2015	9	As above	Raven (1988a.b)
At least 7% of plant species	7	Half of species lost over next decade in 10 'hot spots' covering 3.5% of forest area	Myers (1988)
0.2-0.3% per year	2-3	Half of rain forest species assumed lost in tropical rain forests to be local endemics and becoming extinct with forest loss	Wilson (1988, 1989)
5-15% forest species by 2020	2-5	Species-area curve ($0.15 < z < 0.35$); forest loss assumed twice rate projected by FAO for 1980-85	Reid and Miller (1989)
2-8% loss between 1990 and 2015	1-5	Species-area curve ($0.15 < z < 0.35$); range includes current rate of forest loss and 50% increase	Reid (1992)

Source: Reid, W.V. 1992. How many species will there be? In: Whitmore, T.C. and Sayer, J.A. (Eds), *Tropical Deforestation and Species Extinction*, Chapman Hall, London, UK.
Notes: See original source for additional notes referring to this table and reference citations.

Estimates such as these are often combined with estimates of species numbers in tropical rain forests to provide figures for numbers of species disappearing daily, yearly or each decade. Figures of 100,000 species lost per year (based on estimates of 20 million tropical forest species) are frequently quoted. The vast majority of the hypothesised extinctions would occur among undescribed arthropods because these comprise the majority of the total number of species estimated to occur in tropical forest.

Earlier estimates, some based on similar biogeographic assumptions and others using different models, gave even higher projected rates of extinction, with figures of 20-50% species loss by the end of the century (Myers, 1979; Ehrlich and Ehrlich, 1981). In the light of the more recent estimates based on increased sophistication of the model, these earlier predictions now look exaggerated.

Problems with the model

Both the theoretical assumptions and the figures used in deriving estimates from the species-area model are open to question.

The principal assumption underlying the model is that species richness and habitat destruction within tropical forests are distributed evenly. This is not the case, as richness is known to vary considerably between different areas of tropical moist forest at all scales of comparison (see Chapter 4). Many ecologists and taxonomists would agree that, given the inadequate data available on the poorly-known groups which make up most of the world's total complement of species, no realistic assessment can be made of the extent to which reduction of an area of forest habitat will affect the species present.

Areas also differ greatly in the number of species confined to them (i.e. endemics). Self-evidently, the complete destruction of even a small area with a large number of endemics will contribute more to global extinction than the destruction of the same-sized area with few or no local endemics, even if the latter is richer in species. Thus, if habitat destruction preferentially takes place in areas with large numbers of endemics it will lead to extinction rates higher than those estimated from mean species-area relationships, while if it is concentrated in areas with few endemics, the reverse will be the case.

Figures for rates of habitat destruction are also open to question (see Chapter 20). Calculations tend to take figures for forest conversion as equivalent to forest loss, that is complete destruction of forest and replacement by habitats in which none of the original biota can survive. In reality, forest conversion covers a range of conditions, from selective logging which may have relatively little impact on species composition, through small-scale patch-work clearing for agriculture, to clear-felling of extensive areas. Forest conversion thus covers a range of degrees of degradation, with only the most extreme resulting in complete elimination of all species from a particular area. This will tend therefore to reduce the estimates for extinction rates. In addition, projections of extinction rates are based on an assumption that deforestation rates will remain constant. This is evidently not the case. It is widely agreed that rates of forest conversion are increasing, and will continue to increase until easily accessible areas which are not legally protected have been cleared, following which they will decrease.

Furthermore, the estimate from a straightforward global species-area curve does not take into account the presumed 'residual' extinctions which will occur through remaining forest becoming fragmented: on the basis of island biogeographic theory it is argued that these fragments will suffer elevated rates of extinction through stochastic processes. Already many species may be committed to extinction in that without direct human intervention, their residual numbers are non-viable. The list of threatened species in Table 17.1 show 140 species of mammals as endangered and likely to become extinct in the near future unless the threat to their survival is alleviated: this is more than twice the total number of mammals that has gone extinct over the last four hundred years. Instead of concentrating on extinctions, it is important to monitor the status and threats to a wide array of species if global trends of species diversity are to be assessed.

Finally, estimates of extinction rates do not - and cannot - take into account the impact of unpredictable large-scale changes in environmental conditions, such as global climate change, which is likely to have a profound influence upon species survival.

Conclusion

There are many unsatisfactory assumptions underlying current estimates of global extinction rates, and the resulting numerical values are fraught with imprecision. Alternative models, possibly based on a greater understanding of the ecological or life history traits correlated to extinction proneness, would be highly instructive in either confirming current estimates or refining them by avoiding some of the major short-comings in the species - area method. However in the absence of such alternatives, conclusions from the different studies using the current model must be examined, even if the methodology is known to be flawed. In large measure, these agree about the accelerating rates of species extinctions arising from the continued loss of tropical forests. The most recent refinement of the estimates (Reid, 1992) predicts that at current rates of deforestation, we will commit some 2-8% of the planet's species to extinction in the next 25 years.

However, what is equally clear is that quantifying the precise rate of extinction is of no greater relevance to conservation practice than is determining a precise figure for the number of species on earth. Policymakers and the public may like to assess the magnitude of the extinction crisis, and thus the priority to be given to the issue, on the basis of an absolute rate, but investment of time and effort in refining such predictions contributes little to tackling the root causes of the problem. Indeed, obsession with an absolute extinction rate may give an unrealistically optimistic impression in that no allowance is made for the genetic impoverishment of the multitude of species brought to the verge of extinction through the progressive loss of discrete sub-populations.

Rather than focus on refining extinction rates, we need to develop the capability to identify areas or localities of high species endemism and diversity (see Chapter 15), and ensure that these sites are placed under a system of conservation management that maintains their ecological integrity before they are perturbated by logging, mining or forest clearance. Such proactive conservation practice could stem the tide of the accelerating species extinction crisis.

References

Allendorf, F.W. and Leary, R.F. 1986. Heterozygosity and fitness in natural populations of animals. In: Soulé, M.E. (Ed.), *Conservation Biology: the science of scarcity and diversity*. Sinauer Associates, Inc., Mass. Pp.57-76.

Belovsky, G.E. 1987. Extinction models and mammalian persistence. In: Soulé, M.E. (Ed.), *Viable Populations for Conservation*. Cambridge University Press, Cambridge, New York. Pp.35-58.

Boeckeln, W.J. and Gotelli, N.J. 1984. Island biogeographic theory and conservation practice: species-area or specious-area relationships? *Biological Conservation* 29:63-80.

Bonner, M.L. and Selander, R.K. 1974. Elephant seals: genetic variation and near extinction. *Science* 184:908-909.

Brown, J. H. and Kodric-Brown, A. (1977). Turnover rates in insular biogeography: effect of immigration on extinction. *Ecology* 58, 445-449.

Ehrlich, P.R. and Ehrlich, A.H. 1981. *Extinction: the causes and consequences of the disappearance of species*. Random House, New York.

Ehrlich, P.R. and Wilson, E.O. 1991. Biodiversity studies: science and policy. *Science* 253:758-762.

Erwin, D.H., Valentine, J.W. and Sepkoski, J.J. 1987. A comparative study of diversification events: the early Palaeozoic versus the Mesozoic. *Evolution* 41(6).

Falconer, D.S. 1981. *Introduction to Quantitative Genetics*, 2nd edition. Longman, London.

Franklin, I.R. 1980. Evolutionary change in small populations. In: Soulé, M.E. and Wilcox, B.A. (Eds), *Conservation Biology: an evolutionary-ecological perspective*. Sinauer Associates, Inc., Mass. Pp.135-149.

Grumbine, R.E. 1990. Viable populations, reserve size, and federal lands management: a critique. *Conservation Biology* 4(2):127-134.

Karr, J.R. 1991. Avian survival rates and the extinction process on Barro Colarado Island, Panama. *Conservation Biology* 4(4):391-397.

King, J.R. and Mewaldt, L.R. 1987. The summer biology of an unstable insular population of White-crowned Sparrows in Oregon. *Condor* 89:549-565.

Klein, B.C. 1989. Effects of forest fragmentation on dung and carrion beetle communities in central Amazonia. *Ecology* 70:1715-1752.

Knoll, A.H. 1984. Patterns of extinction in the fossil record of vascular plants. In: Nitecki, M.H. (Ed.), *Extinctions*. University of Chicago Press, Chicago, IL. Pp.22-68.

Lande, R. and Barrowclough, G.F. 1987. Effective population size, genetic variation, and their use in population management. In: Soulé, M.E. (Ed.), *Viable Populations for Conservation*.

Cambridge University Press, Cambridge, New York. Pp.87-124

Laurance, W.F. 1991. Ecological correlates of extinction proneness in Australian tropical rain forest mammals. *Conservation Biology* 5(1):79-89.

MacArthur, R.H. and Wilson, E.O. 1963. An equilibrium theory of insular zoogeography. *Evolution* 17:373-387.

MacArthur, R.H. and Wilson, E.O. 1967. *The Theory of Island Biogeography*. Princeton University Press, Princeton, N.J.

Myers, N. 1979. *The Sinking Ark: a new look at the problem of disappearing species.* Pergamon Press, Oxford, UK.

Newmark, W.D. 1991. Tropical forest fragmentation and the local extinction of understory birds in the eastern Usambara Mountains, Tanzania. *Conservation Biology* 5(1):67-78.

O'Brien, S.J., Roelke, M.E., Marker, L., Newman, C.A., Winkler, D., Meltzer, D., Colly, L., Evermann, J.F., Bush, M. and Wildt, D.E. 1985. Genetic basis for species vulnerability in the cheetah. *Science* 227:1428-1434.

Packer, C., Pusey, A.E., Rowley, H., Gilbert, D.A., Martenson, J. and O'Brien, S.J. 1991. Case study of a population bottleneck: lions of the Ngorongoro Crater. *Conservation Biology* 5(2):219-230.

Pulliam, H.R. 1988. Sources, sinks and population regulation. *American Naturalist* 132:652-661.

Reid, W.V. 1992. How many species will there be? In: Whitmore, T.C. and Sayer, J.A. (Eds), *Tropical Deforestation and Species Extinction*. Chapman Hall, London, UK. Pp.55-73.

Reid, W.V. and Miller, K.R. 1989. *Keeping Options Alive: the scientific basis for conserving biodiversity*. World Resources Institute, Washington, DC.

Schaffer, M. 1987. Minimum viable populations: coping with uncertainty. In: Soulé, M.E. (Ed.), *Viable Populations for Conservation*. Cambridge University Press, Cambridge, New York. Pp.70-86.

Soulé, M.E., Wilcox, B.A. and Holtby, C. 1979. Benign neglect: a model of faunal collapse in the game reserves of East Africa. *Biological Conservation* 15:259-272.

Western, D. and Ssemakula, J. 1981. The future of savannah ecosystems: ecological islands or faunal enclaves? *African Journal of Ecology* 19:7-19.

Wilcox, B.A. and Murphy, D.D. 1985. Conservation strategy: the effects of fragmentation on extinction. *American Naturalist* 125:879-887.

Zimmerman, B.L. and Bierregard, R.O. 1986. Relevance of the equilibrium theory of island biogeography and species-area relations to conservation with a case from Amazonia. *Journal of Biogeography* 13:137-143.

Based on text prepared by Martin Jenkins with additions by WCMC staff

Table 16.4 Animal species extinct since circa 1600

SPECIES	ENGLISH NAME	DISTRIBUTION	LAST RECORDED	POSSIBLE CAUSE
CORALS ETC. (CNIDARIA)				
Order MILLEPORINA				
Family Milleporidae				
Millepora sp.		Panama	1983	
MOLLUSCS				
Order ARCHAEOGASTROPODA				
Family Acmaeidae				
Lottia alveus	Eelgrass Limpet	USA		B
Order MESOGASTROPODA				
Family Hydrobiidae				
Bythiospeum pfeifferi		Austria		
Clappia umbilicata	Umbilicate Pebblesnail	USA		
Ohridohauffenia drimica		Yugoslavia	1980s	
Family Pleuroceridae				
Elimia clausa	Closed Elimia	USA		B
Elimia fusiformis	Fusiform Elimia	USA		B
Elimia hartmaniana	High-spired Elimia	USA		B
Elimia impressa	Constricted Elimia	USA		B
Elimia jonesi	Hearty Elimia	USA		B
Elimia laeta	Ribbed Elimia	USA		B
Elimia pilsbryi	Rough-lined Elimia	USA		B
Elimia pupaeformis	Pupa Elimia	USA		B
Elimia pygmaea	Pygmy Elimia	USA		B
Elimia varians	Puzzle Elimia	USA		B
Gyrotoma incisa	Excised Slitshell	USA	1924	
Gyrotoma lewisii	Striate Slitshell	USA	1924	
Gyrotoma pagoda	Pagoda Slitshell	USA	1924	
Gyrotoma pumila	Ribbed Slitshell	USA	1924	
Gyrotoma pyramidata	Pyramid Slitshell	USA	1924	
Gyrotoma walkeri	Round Slitshell	USA	1924	
Leptoxis clipeata	Agate Rocksnail	USA		
Leptoxis formanii	Interrupted Rocksnail	USA		
Leptoxis ligata	Rotund Rocksnail	USA		
Leptoxis lirata	Lirate Rocksnail	USA		
Leptoxis occultata	Bigmouth Rocksnail	USA		
Leptoxis showalterii	Coosa Rocksnail	USA		
Leptoxis vittata	Striped Rocksnail	USA		
Family Pomatiasidae				
Tropidophora carinata		Mauritius	1881	B
Order STYLOMMATOPHORA				
Family Endodontidae				
Discus guerinianus		Madeira (Portugal)	1870s	
Kondoconcha othnius		Rapa (F. Polynesia)	1934	
Libera subcavernula		Raratonga (Cook Is)	1880s	
Libera tumuloides		Raratonga (Cook Is)	1880s	
Mautodonta acuticosta		Raiatea (F. Polynesia)	1880s	
Mautodonta boraborensis		Borabora (F. Polynesia)	1880s	
Mautodonta ceuthma		Raivavae (F. Polynesia)	1880s	
Mautodonta consimilis		Raiatea (F. Polynesia)	1880s	
Mautodonta consobrina		Huahine (F. Polynesia)	1880s	
Mautodonta maupiensis		Maupiti (F. Polynesia)	1880s	
Mautodonta parvidens		Society Is (F. Polynesia)	1880s	
Mautodonta punctiperforata		Moorea (F. Polynesia)	1880s	
Mautodonta saintjohni		Borabora (F. Polynesia)	1880s	
Mautodonta subtilis		Huahine (F. Polynesia)	1880s	
Mautodonta unilamellata		Raratonga (Cook Is)	1880s	
Mautodonta zebrina		Raratonga (Cook Is)	1880s	
• *Opanara altiapica*		Rapa (F. Polynesia)	1934	
• *Opanara areaensis*		Rapa (F. Polynesia)	1934	
• *Opanara bitridentata*		Rapa (F. Polynesia)	1934	
• *Opanara caliculata*		Rapa (F. Polynesia)	1934	
• *Opanara depasoapicata*		Rapa (F. Polynesia)	1934	
• *Opanara duplicidentata*		Rapa (F. Polynesia)	1934	
• *Opanara fosbergi*		Rapa (F. Polynesia)	1934	
• *Opanara megomphala*		Rapa (F. Polynesia)	1934	
• *Opanara perahuensis*		Rapa (F. Polynesia)	1934	
• *Orangia cooki*		Rapa (F. Polynesia)	1934	
• *Orangia maituatensis*		Rapa (F. Polynesia)	1934	
• *Orangia sporadica*		Rapa (F. Polynesia)	1934	
* *Pilula cycloria*		Mauritius		
• *Rhysoconcha atanuiensis*		Rapa (F. Polynesia)	1934	
• *Rhysoconcha variumbilicata*		Rapa (F. Polynesia)	1934	
• *Ruatara koarana*		Rapa (F. Polynesia)	1934	
• *Ruatara oparica*		Rapa (F. Polynesia)	1934	
Taipidon anceyana		Hiva Oa (F. Polynesia)	1880s	
Taipidon marquesana		Nuku Hiva (F. Polynesia)	1880s	
Taipidon octolamellata		Hiva Oa (F. Polynesia)	1880s	
Thaumatodon multilamellatus		Raratonga (Cook Is)	1880s	
Family Bulimulidae				
Amphibulima patula		Guadeloupe		B
Bulimulus duncanus		Galapagos (Ecuador)	late 1800s	H
Leuchocharis loyaltyensis		New Caledonia	1900s	B
Leuchocharis porphyrocheila		New Caledonia	1900s	B

Table 16.4 Animal species extinct since circa 1600 (continued)

SPECIES	ENGLISH NAME	DISTRIBUTION	LAST RECORDED	POSSIBLE CAUSE
MOLLUSCS (continued)				
Family Charopidae				
Helenoconcha leptalea		St Helena	1870s	
Helenoconcha minutissima		St Helena	1870s	
Helenoconcha polyodon		St Helena	1870s	
Helenoconcha pseustes		St Helena	1870s	
Helenoconcha sexdentata		St Helena	1870s	
Helenodiscus bilamellata		St Helena	1870s	
Helenodiscus vernoni		St Helena	1870s	
Pseudohelenoconcha dianae		St Helena	1870s	
Pseudohelenoconcha laetissima		St Helena	1870s	
Pseudohelenoconcha persoluta		St Helena	1870s	
Pseudohelenoconcha spurca		St Helena	1870s	
Sinployea canalis		Raratonga (Cook Is)	1872	
Sinployea decorticata		Raratonga (Cook Is)	1872	
Sinployea harveyensis		Raratonga (Cook Is)	1872	
Sinployea otareae		Raratonga (Cook Is)	1872	
Sinployea planospira		Raratonga (Cook Is)	1872	
Sinployea proxima		Raratonga (Cook Is)	1872	
Sinployea rudis		Raratonga (Cook Is)	1872	
Sinployea tenuicostata		Raratonga (Cook Is)	1872	
Sinployea youngi		Raratonga (Cook Is)	1872	
Family Achatinellidae				
Achatinella abbreviata		Hawaii (USA)	1963	A,C,D
Achatinella buddii		Hawaii (USA)	early 1900s	A,C,D
Achatinella caesia		Hawaii (USA)	early 1900s	A,C,D
Achatinella casta		Hawaii (USA)		A,C,D
Achatinella decora		Hawaii (USA)	early 1900s	A,C,D
Achatinella elegans		Hawaii (USA)	1952	A,C,D
Achatinella juddii		Hawaii (USA)	1958	A,C,D
Achatinella juncea		Hawaii (USA)		A,C,D
Achatinella lehuiensis		Hawaii (USA)	1922	A,C,D
Achatinella papyracea		Hawaii (USA)	1945	A,C,D
Achatinella rosea		Hawaii (USA)	1949	A,C,D
Achatinella spaldingi		Hawaii (USA)	1938	A,C,D
Achatinella stewarti		Hawaii (USA)	1961	A,C,D
Achatinella thaanumi		Hawaii (USA)	1900s	A,C,D
Achatinella valida		Hawaii (USA)	1951	A,C,D
Achatinella vittata		Hawaii (USA)	1953	A,C,D
x *Elasmias jauffreti*		Rodrigues (Mauritius)		
x *Elasmias sp.*		Mauritius		
Partulina crassa		Hawaii (USA)	1914	C?
Partulina montagui		Hawaii (USA)	1913	C?
Family Partulidae				
Partula exigua	Moorean Viviparous Tree Snail	Moorea (F. Polynesia)	1977	C?
Partula filosa	Tahiti Viviparous Tree Snail	Tahiti (F. Polynesia)		
Partula producta	Tahiti Viviparous Tree Snail	Tahiti (F. Polynesia)		
Partula salifana		Guam		
Samoana abbreviata		American Samoa	1940	B
Family Amastridae				
Carelia anceophila		Hawaii (USA)	1930	B,C,D
Carelia bicolor		Hawaii (USA)	1970	B,C,D
Carelia cumingiana		Hawaii (USA)	1930	B,C,D
Carelia glossema		Hawaii (USA)	1930	B,C,D
Carelia kalalauensis		Hawaii (USA)	1945/47	B,C,D
Carelia knudseni		Hawaii (USA)	1930	B,C,D
Carelia olivacea		Hawaii (USA)	1930	B,C,D
Carelia paradoxa		Hawaii (USA)	1930	B,C,D
Carelia periscelis		Hawaii (USA)	1930	B,C,D
Carelia tenebrosa		Hawaii (USA)	1930	B,C,D
Carelia turricula		Hawaii (USA)	1930	B,C,D
Family Vertiginidae				
Campolaemus perexilis		St Helena	1870s	
Nesopupa turtoni		St Helena	1870s	
Family Pupillidae				
x *Gibbulinopsis sp.*		Rodrigues (Mauritius)		
Leiostyla abbreviata		Madeira (Portugal)	1870s	
Leiostyla cassida		Madeira (Portugal)	1870s	
Leiostyla concinna		Madeira (Portugal)	1870s	
Leiostyla gibba		Madeira (Portugal)	1870s	
Leiostyla laevigata		Madeira (Portugal)	1870s	
Leiostyla lamellosa		Madeira (Portugal)	1870s	
Leiostyla simulator		Madeira (Portugal)	1870s	
Pupa obliquicostata		St Helena	1870s	
Family Helixarionidae				
Colparion madgei		Rodrigues (Mauritius)	1938	B
Ctenoglypta newtoni		Mauritius	1871	B
x *Ctenophila planorbina*		Mauritius		
Diastole matafaoi		American Samoa	1940	?D
x *Erepta thirouxi*		Mauritius		
x *Erepta sp.*		Mauritius		
Pachystyla rufozonata		Mauritius	1869	B
x *Plegma bewsheri*		Rodrigues (Mauritius)		
x *Plegma duponti*		Mauritius		
x *Plegma sp.*		Mauritius		
Family Ferussaciidae				
Cecilioides eulima		Madeira (Portugal)	1870s	
Family Subulinidae				
Chilonopsis blofeldi		St Helena	1870s	
Chilonopsis exulatus		St Helena	1870s	
Chilonopsis helena		St Helena	1870s	

Table 16.4 Animal species extinct since circa 1600 (continued)

SPECIES	ENGLISH NAME	DISTRIBUTION	LAST RECORDED	POSSIBLE CAUSE
MOLLUSCS (continued)				
Chilonopsis melanoides		St Helena	1870s	
Chilonopsis nonpareil		St Helena	1870s	
Chilonopsis subplicatus		St Helena	1870s	
Chilonopsis subtruncatus		St Helena	1870s	
Chilonopsis turtoni		St Helena	1870s	
Family Helicidae				
Discula lyelliana		Madeira (Portugal)	1870s	
Discula tetrica		Madeira (Portugal)	1870s	
Geomitra delphinuloides		Madeira (Portugal)	1870s	
Lemniscia galeata		Madeira (Portugal)	1870s	
Pseudocampylaea lowei		Madeira (Portugal)	late 19th C	
Family Streptaxidae				
Edentulina thomasetti		Seychelles	1908	
Gibbus lyonetianus		Mauritius	1905	B
Gonidomus newtoni		Mauritius	1867	B
x *Gonospira cirneensis*		Mauritius		
x *Gonospira heliodes*		Mauritius		
x *Gonospira majusculus*		Mauritius		
Imperturbata violescens?		Seychelles		
Family Assimineidae				
x *Omphalotropis plicosa*		Mauritius	1878	B
x *Omphalotropis caldwelli*		Mauritius		
x *Omphalotropis dupontiana*		Mauritius		
x *Omphalotropis maxima*		Mauritius		
x *Omphalotropis multilirata*		Mauritius		
x *Omphalotropis sp.*		Mauritius		
Family Pomatiasidae				
x *Tropidophora bewsheri*		Rodrigues (Mauritius)		
x *Tropidophora bipartita*		Rodrigues (Mauritius)		
x *Tropidophora deflorata*		Réunion		
x *Tropidophora lienardi*		Mauritius		
x *Tropidophora mauritiana*		Mauritius		
Order UNIONOIDA				
Family Unionidae				
Alasmidonta mccordi	Coosa Elktoe	USA		
Alasmidonta wrightiana	Ochlacknee Arc-mussel	USA		
Epioblasma arcaeformis	Sugarspoon	USA	1940s	B
Epioblasma biemarginata	Angled Riffleshell	USA	1960s	B
Epioblasma flexuosa	Leafshell	USA	1940s	B
Epioblasma haysiana	Acornshell	USA		
Epioblasma lenior	Narrow Catspaw	USA	1965	B
Epioblasma lewisii	Forkshell	USA	1964	B
Epioblasma personata	Round Combshell	USA	1930	B
Epioblasma propinqua	Tennessee Riffleshell	USA	1930	B
Epioblasma sampsonii	Wabask Riffleshell	USA	1950s/60s	B
Epioblasma stewardsoni	Cumberland Leafshell	USA	1930	B
Medionidus mcglameriae	Tombigbee Moccasinshell	USA		
Order AMPHIPODA				
Family Crangonyctidae				
Stygobromus hayi	Hay's Spring Scud	USA	1957	
Stygobromus lucifugus	Rubious Cave Amphipod	USA		
Order DECAPODA				
Family Astacidae				
Pacifastacus nigrescens	Sooty Crayfish	USA	1860s	
Family Atyidae				
Syncaris pasadenas	Pasadena Freshwater Shrimp	USA	1933	
Order EPHEMEROPTERA				
Family Siphlonuridae				
Acanthometropus pecatonia	Pecatonica River Mayfly	USA	1927	
Family Ephemeridae				
Pentagenia robusta	Robust Burrowing Mayfly	USA		
Order ORTHOPTERA				
Family Tettigoniidae				
Neduba extincta	Antioch Dunes Shieldback Katydid	USA	1937	
Order PHASMATOPTERA				
Family Phasmatidae				
Dryococelus australis	Lord Howe Island Stick-insect	Lord Howe I (Australia)	1969	
Order DERMAPTERA				
Family Labiduridae				
* *Labidura herculeana*	St Helena Earwig	St Helena	1967	
Order PLECOPTERA				
Family Chloroperlidae				
Alloperla roberti	Robert's Stonefly	USA		

Table 16.4 Animal species extinct since circa 1600 (continued)

SPECIES	ENGLISH NAME	DISTRIBUTION	LAST RECORDED	POSSIBLE CAUSE
INSECTS (continued)				
Order HOMOPTERA				
Family Pseudococcidae				
Clavicoccus erinaceus		Hawaii (USA)		
Phyllococcus oahuensis		Hawaii (USA)		
Order COLEOPTERA				
Family Cerambycidae				
Xylotoles costatus	Pitt Island Longhorn Borer	Chatham I (NZ)	1930s	B,C
Family Curculionidae				
Dryophthorus distinguendus		Hawaii (USA)		
Dryotribus mimeticus		Hawaii (USA)		
Hadramphus tuberculatus		New Zealand	1910	C
Macrancylus linearis		Hawaii (USA)		
Oedemasylus laysanensis		Hawaii (USA)		
Pentarthrum blackburnii		Hawaii (USA)		
Rhyncogonus bryani		Hawaii (USA)		
Family Carabidae				
* *Aplothorax burchelli*		St Helena	1967?	
* *Mecodema punctellum*		Stephens I (NZ)		B,G
Order DIPTERA				
Family Tabanidae				
Stonemyia volutina	Volutine Stoneyian Tabanid Fly	USA		
Family Dolichopodidae				
Campsicnemus mirabilis		Hawaii (USA)		
Family Drosophilidae				
Drosophila lanaiensis		Hawaii (USA)		
Order TRICHOPTERA				
Family Rhyacophilidae				
Rhyacophila amabilis	Castle Lake Caddis–fly	USA		
Family Hydropsychidae				
Hydropsyche tobiasi	Tobias' Caddis–fly	Germany	1920s	
Family Leptoceridae				
Triaenodes phalacris	Athens Caddis–fly	USA		
Triaenodes tridonata	Three–tooth Caddis–fly	USA		
Order LEPIDOPTERA				
Family Zygaenidae				
Levuana iridescens	Levuana Moth	Fiji	1929	E
Family Lycaenidae				
Glaucopsyche xerces	Xerces Blue	USA	early 1940s	
Family Libytheidae				
Libythea cinyras		Mauritius	1865	
Family Nymphalidae				
Euthalia malapana		Taiwan		
Family Pyralidae				
Genophantis leahi		Hawaii (USA)	early 1900s	
Hedylepta asaphombra		Hawaii (USA)	1970s	
Hedylepta coninuatalis		Hawaii (USA)	1958	
Hedylepta epicentra		Hawaii (USA)	early 1900s	
+ *Hedylepta euryprora*		Hawaii (USA)		E
+ *Hedylepta fullawayi*		Hawaii (USA)		E
Hedylepta laysanensis		Hawaii (USA)		
+ *Hedylepta meyricki*		Hawaii (USA)		E
+ *Hedylepta musicola*		Hawaii (USA)		E
Hedylepta telegrapha		Hawaii (USA)		
Oeobia sp.		Hawaii (USA)	1911	
Family Geometridae				
Scotorhythra nesiotes		Hawaii (USA)	early 1900s	
Scotorhythra megalophylla		Hawaii (USA)	early 1900s	
Scotorhythra paratactis		Hawaii (USA)	early 1900s	
Tritocleis microphylla		Hawaii (USA)	1890s	
Family Sphingidae				
Mandura blackburni		Hawaii (USA)	1960s	
Family Noctuidae				
Agrotis crinigera	Poco Noctuid Moth	Hawaii (USA)	1926	E
Agrotis fasciata	Midway Noctuid Moth	Hawaii (USA)		
Agrotis kerri		Hawaii (USA)	1923	
Agrotis laysanensis		Hawaii (USA)	1911	
Agrotis photophila		Hawaii (USA)		
Agrotis procellaris		Hawaii (USA)	pre–1900	
Helicoverpa confusa		Hawaii (USA)	post–1927	
Helicoverpa minuta	Minute Noctuid Moth	Hawaii (USA)	pre–1911	
Hypena laysanensis	Laysan Dropseed Noctuid Moth	Hawaii (USA)	1911	
+ *Hypena newelli*		Hawaii (USA)		
+ *Hypena plagiota*		Hawaii (USA)		
+ *Hypena senicula*		Hawaii (USA)		
Peridroma porphyrea		Hawaii (USA)		
Order HYMENOPTERA				
Family Colletidae				
Nesoprosopis angustula	Lanai Yellow–faced Bee	Hawaii (USA)		
Nesoprosopis blackburni	Blackburn's Yellow–faced Bee	Hawaii (USA)		
Nesoprosopis connectens	Connected Yellow–faced Bee	Hawaii (USA)		

Table 16.4 Animal species extinct since circa 1600 (continued)

SPECIES	ENGLISH NAME	DISTRIBUTION	LAST RECORDED	POSSIBLE CAUSE
FISHES				
Order PETROMYZONTIFORMES				
Family Petromyzontidae				
Lampetra minima	Miller Lake Lamprey	USA	1953	E
Order CYPRINIFORMES				
Family Cyprinidae				
Evarra bustamantei		Mexico	1970	B
Evarra eigenmanni		Mexico	1970	B
Evarra tlahuacensis		Mexico	1970	B
Gila crassicauda	Thicktail Chub	USA	1957	B,C/D
Lepidomeda altivelis	Pahranagat Spinedace	USA	1940	C/D
Notropis amecae	Ameca Shiner	Mexico	1970	C/D
Notropis aulidion	Durango Shiner	Mexico	1965	C/D
Notropis orca	Phantom Shiner	Mexico, USA	1975	B,C/D
Pogonichthys ciscoides	Clear Lake Splittail	USA	1970	B,C/D
Rhinichthys deaconi	Las Vegas Dace	USA	1955	B
Stypodon signifer	Stumptooth Minnow	Mexico	1930	B
Family Catostomidae				
Chasmistes muriei	Snake River Sucker	USA	1928	B
Lagochila lacera	Harelip Sucker	USA	1910	G
Order SALMONIFORMES				
Family Retropinnidae				
* *Prototroctes oxyrhynchus*	New Zealand Grayling	New Zealand	1920s	B,D,H
Family Salmonidae				
Coregonus alpenae	Longjaw Cisco	USA, Canada	1978	A,C
Coregonus johannae	Deepwater Cisco	USA, Canada	1955	A,C/D
Salvelinus agassizi	Silver Trout	USA	1930	A,C/D
Order CYPRINODONTIFORMES				
Family Fundulidae				
Fundulus albolineatus	Whiteline Topminnow	USA	1900	B,C/D
Family Poeciliidae				
Gambusia amistadensis	Amistad Gambusia	USA	1973	B
* *Gambusia georgei*	San Marcos Gambusia	USA	1983	B,C/D
* *Priapella bonita*	Guayacon Ojiazul	Mexico		
Family Goodeidae				
Characodon garmani	Parras Characodon	Mexico	1900	?B
Empetrichthys merriami	Ash Meadows Killifish	USA	1953	B,C/D
Family Cyprinodontidae				
Cyprinodon latifasciatus	Perrito de Parras	Mexico	1930	B
Cyprinodon sp.	Monkey Spring Pupfish	USA	1971	B,C/D
Cyprinodon sp.		Mexico		
Cyprinodon sp.		Mexico		
Order SCORPAENIFORMES				
Family Cottidae				
Cottus echinatus	Utah Lake Sculpin	USA	1928	B,C/D
AMPHIBIANS				
Order ANURA				
Family Discoglossidae				
Discoglossus nigriventer	Israel Painted Frog	Israel	1940	B
Rana fisheri	Relict Leopard Frog	USA	1960	B
REPTILES				
Order TESTUDINES				
Family Testudinidae				
Cylindraspis borbonica		Réunion	1800	
Cylindraspis indica		Réunion	1800	A
Cylindraspis inepta		Mauritius	early 18th C	A,C/D
Cylindraspis peltastes		Rodrigues (Mauritius)	1800	A,B,C/D
Cylindraspis triserrata		Mauritius	early 18th C	A,C/D
Cylindraspis vosmaeri		Rodrigues (Mauritius)	1800	A,C/D
Order SAURIA				
Family Gekkonidae				
Hoplodactylus delcourti		New Zealand (?)	mid 19th C?	
Phelsuma edwardnewtoni	Newton's Day Gecko	Rodrigues (Mauritius)	1917	C
Phelsuma gigas	Giant Day Gecko	Rodrigues (Mauritius)	end 19th C	C
Family Iguanidae				
Leiocephalus eremitus		Navassa I (USA)	1900	C
Leiocephalus herminieri		Martinique	1830s	–
Family Teiidae				
Ameiva cineracea		Guadeloupe	early 20th C	–
* *Ameiva major*	Martinique Giant Ameiva	Martinique		C?
Family Anguidae				
Celestus occiduus	Jamaican Giant Galliwasp	Jamaica	1840	C?
Family Scincidae				
# *Leiolopisma mauritiana*		Mauritius	1600	C
Macroscincus coctei	Cape Verde Giant Skink	Cape Verde	early 20th C	A?
* *Tiliqua adelaidensis*	Adelaide Pigmy Bluetongue	Australia	1959	B,C

Table 16.4 Animal species extinct since circa 1600 (continued)

SPECIES	ENGLISH NAME	DISTRIBUTION	LAST RECORDED	POSSIBLE CAUSE
REPTILES (continued)				
Order SERPENTES				
Family Boidae				
* *Bolyeria multocarinata*		Round I (Mauritius)	1975	
Family Typhlopidae				
Typhlops cariei		Mauritius	17th C	C
Family Colubridae				
* *Alsophis ater*	Jamaican Tree Snake	Jamaica	1950	A,C
Alsophis sancticrucis	St Croix Racer	Virgin Is (US)	20th C	A,C
* *Liophis cursor*	Martinique Racer	Martinique	1963	C
* *Liophis perfuscus*	Barbados Racer	Barbados	mid 20th C?	C
BIRDS				
Order STRUTHIONIFORMES				
Family Dromaiidae				
Dromaius diemenianus	Kangaroo Island Emu	Kangaroo I (Australia)	1803	B
Family Aepyornithidae				
Aepyornis maximus	Great Elephantbird	Madagascar	1650	A,B
Family Anomalopterygidae				
Dinornis torosus	Brawny Great Moa	New Zealand	1670	A,B
Eurapteryx gravis	Burly Lesser Moa	New Zealand	1640	A,B
Megalapteryx didinus	South Island Tokoweka	New Zealand	1785	A,B
Order GALLIFORMES				
Family Phasianidae				
Coturnix novaezelandiae	New Zealand Quail	New Zealand	1875	F
Ophrysia superciliosa	Himalayan Mountain Quail	India	1868	A
Order ANSERIFORMES				
Family Anatidae				
Alopochen mauritianus	Mauritian Shelduck	Mauritius	1698	–
Anas theodori	Mauritian Duck	Mauritius, ?Réunion	1696	–
Camptorhynchus labradorius	Labrador Duck	Canada, USA	1878	A,B
Cygnus sumnerensis	Chatham Island Swan	Chatham I (NZ)	1590–1690	–
Mergus australis	Auckland Island Merganser	New Zealand	1905	A,B,C
* *Rhodonessa caryophyllacea*	Pink–headed Duck	India, Nepal	1935	A
Sheldgoose sp.		Réunion	1674	–
Order CORACIIFORMES				
Family Alcedinidae				
Halcyon miyakoensis	Ryukyu Kingfisher	Nansei–shoto (Japan)	1841	–
Order CUCULIFORMES				
Family Cuculidae				
* *Coua delalandei*	Snail–eating Coua	Madagascar	1930	A,B,C/D
Order PSITTACIFORMES				
Family Psittacidae				
Anodorhynchus glaucus	Glaucous Macaw	Brazil, Uruguay	1955	
Ara tricolor	Cuban Red Macaw	Cuba	1885	A,E
Charmosyna diadema	New Caledonia Lorikeet	New Caledonia	1860	B
Conuropsis carolinensis	Carolina Parakeet	USA	1914	E
Cyanoramphus ulietanus	Raiatea Parakeet	Raiatea (F. Polynesia)	1773	–
Cyanoramphus zealandicus	Black–fronted Parakeet	Tahiti (F. Polynesia)	1844	B
'Lophopsittacus' bensoni	Mauritius Grey Parrot	Mauritius	1765	C/D
Lophopsittacus mauritianus	Mauritius Parrot	Mauritius	1675	A,C
Mascarinus mascarinus	Mascarene Parrot	Réunion	1775 (1834 in captivity)	B
'Necropsittacus' rodericanus	Rodrigues Parrot	Rodrigues (Mauritius)	1761	A,C/D
Nestor productus	Norfolk Island Kaka	Phillip I (Australia)	1851	A,E
Psittacula exsul	Rodrigues Ring–necked Parakeet	Rodrigues (Mauritius)	1876	B
Psittacula wardi	Seychelles Alexandrine Parrot	Seychelles	1870	A,B
Order TROCHILIFORMES				
Family Trochilidae				
Chlorostilbon bracei	New Providence Hummingbird	Bahamas	1877	
Family Caprimulgidae				
* *Siphonorhis americanus*	Jamaica Least Pauraque	Jamaica	1859	C
Order STRIGIFORMES				
Family Strigidae				
Athene blewitti	Forest Owlet	India	1914	
'Athene' murivora	Rodrigues Little Owl	Rodrigues (Mauritius)	1726	B
?Sauzieri sp.	Mauritian Owl	Mauritius		
* *Sceloglaux albifacies*	Laughing Owl	New Zealand	1914	B,C
'Scops' commersoni	Mauritian Owl	Mauritius	1836	
Family Aegothelidae				
* *Aegotheles savesi*	New Caledonia Owlet–frogmouth	New Caledonia	1880	–
Order COLUMBIFORMES				
Family Raphidae				
'Ornithaptera' solitaria	Réunion Solitaire	Réunion	1710–1715	A
Pezophaps solitarius	Rodrigues Solitaire	Rodrigues (Mauritius)	1765	A
Raphus cucullatus	Dodo	Mauritius	1665	A,C,D

Table 16.4 Animal species extinct since circa 1600 (continued)

SPECIES	ENGLISH NAME	DISTRIBUTION	LAST RECORDED	POSSIBLE CAUSE
BIRDS (continued)				
Family Columbidae				
Alextroenas nitidissima	Pigeon Hollandais	Mauritius	1835	A,C
'Alextroenas' rodericana	Rodrigues Pigeon	Rodrigues (Mauritius)	1726	C/D
Columba jouyi	Ryukyu Wood Pigeon	Nansei–shoto (Japan)	1936	B
Columba versicolor	Bonin Wood Pigeon	Ogasawara–shoto (Japan)	1889	C
Ectopistes migratorius	Passenger Pigeon	USA	1914	A,B
* *Microgoura meeki*	Solomon Island Crowned–pigeon	Choiseul (Solomon Is)	1904	C
* *Ptilinopus mercierii*	Marquesas Fruit–dove	Marquesas Is (F. Polynesia)	1922	C/D
Order GRUIFORMES				
Family Rallidae				
Aphanapteryx bonasia	Red Rail	Mauritius	1700	A,C/D
Aphanapteryx leguati	Rodrigues Rail	Rodrigues (Mauritius)	1761	–
Atlantisia elpenor	Ascension Flightless Crake	Ascension I (UK)	1656	G(A,C)
Fulica newtoni	Mascarene Coot	Mauritius, Réunion	1693	–
Gallinula nesiotis	Tristan Moorhen	Tristan da Cunha (UK)	1875–1900	C
Gallinula pacifica	Samoan Woodhen	Savaii (Western Samoa)	1908–1926	C
Gallirallus pacificus	Tahiti Rail	French Polynesia	1773–4	
Nesoclopeus woodfordi	Woodford's Rail	Bougainville (Papua New Guinea)	1936	–
Porphyrio albus	Lord Howe Purple Gallinule	Lord Howe I (Australia)	1834	A
Porzana monasa	Kosrae Crake	Federated States of Micronesia	1827	C
Porzana palmeri	Laysan Rail	Hawaii (USA)	1944	C,D
Porzana sandwichensis	Hawaiian Rail	Hawaii (USA)	1898	C
Rallus dieffenbachii	Chatham Island Banded Rail	Chatham I (NZ)	1840	B,C
Rallus modestus	Chatham Island Rail	Chatham I (NZ)	1900	D
Rallus wakensis	Wake Island Rail	Wake I (USA)	1945	A
* *Tricholimnas lafresnayanus*	New Caledonia Rail	New Caledonia	1904	–
Order CICONIIFORMES				
Family Scolopacidae				
Prosobonia leucoptera	Tahitian Sandpiper	Tahiti, Moorea (F. Polynesia)	1773	D
Family Charadriidae				
Haematopus meadewaldoi	Canarian Black Oystercatcher	Canary Is (Spain)	1913	G
Vanellus macropterus	Javanese Wattled Lapwing	Java (Indonesia)	1940	A,B
Family Laridae				
Alca impennis	Great Auk	Canada, Iceland, Faeroes UK, 'USSR', Greenland	1844	A
Family Falconidae				
Falco sp.		Réunion	1674	–
Polyborus lutosus	Guadalupe Caracara	Guadalupe (Mexico)	1900	A,D,E
Family Podicipedidae				
Podiceps andinus	Colombian Grebe	Colombia	1977	–
Podilymbus gigas	Atitlan Grebe	Guatemala	1980–1986/7	A,D
Tachybaptus rufolarvatus	Lake Alaotra Grebe	Madagascar		
Family Phalacrocoracidae				
Phalacrocorax perspicillatus	Spectacled Cormorant	Bering Straits ('USSR')	1852	A
Family Ardeidae				
Ixobrychus novaezelandia	New Zealand Little Bittern	New Zealand	1900	–
Nycticorax mauritianus	Mauritius Night–heron	Mauritius	by 1700	–
Nycticorax megacephalus	Rodrigues Night–heron	Rodrigues (Mauritius)	1761	–
Nycticorax sp.		Réunion	by 1700	–
Family Threskiornithidae				
Borbonibis latipes	Reunion Flightless Ibis	Réunion	1773	–
Family Ciconiidae				
Ciconia sp.		Réunion	1674	–
Family Procellariidae				
* *Oceanodroma macrodactyla*	Guadalupe Storm–petrel	Guadalupe (Mexico)	1912–1922	C
Pterodroma sp.		Rodrigues (Mauritius)	1726	–
Order PASSERIFORMES				
Family Acanthisittidae				
Xenicus longipes	Bush Wren	New Zealand	1972	B,C
Xenicus lyalli	Stephens Island Wren	Stephens I (NZ)	1874	C
Family Pycnonotidae				
Hypsipetes sp.		Rodrigues (Mauritius)	1600s?	–
Family Muscicapidae				
Acrocephalus familiaris	Laysan Millerbird	Hawaii (USA)	1912–1923	B,D
Eutrichomyias rowleyi	Caerulean Paradise–flycatch	Sangihe (Indonesia)	1978	B
Myiagra freycineti	Guam Broadbill	Guam	1983	
* *Turnagra capensis*	Piopio	New Zealand	1955	B,C
Turdus ravidus	Grand Cayman Thrush	Cayman Is	1938	B
Zoothera terrestris	Kittlitz's Thrush	Ogasawara–shoto (Japan)	1928	C
Babbler sp.		Rodrigues (Mauritius)	1600s?	–
Family Dicaeidae				
Dicaeum quadricolor	Four–coloured Flowerpecker	Cebu (Philippines)	1906	B
Family Zosteropidae				
Zosterops strenua	Lord Howe White–eye	Lord Howe I (Australia)	1928	A,B,C/D
Family Meliphagidae				
Chaetoptila angustipluma	Kioea	Hawaii (USA)	1860	B
Moho apicalis	Oahu Oo	Hawaii (USA)	1837	A,B,C/D
* *Moho nobilis*	Hawaii Oo	Hawaii (USA)	1934	A,B,C/D

Table 16.4 Animal species extinct since circa 1600 (continued)

SPECIES	ENGLISH NAME	DISTRIBUTION	LAST RECORDED	POSSIBLE CAUSE
BIRDS (continued)				
Ciridops anna	Ula-ai-hawane	Hawaii (USA)	1892	–
Drepanis funerea	Black Mamo	Hawaii (USA)	1907	–
Drepanis pacifica	Hawaii Mamo	Hawaii (USA)	1899	A,B
* *Hemignathus obscurus*	Akialoa	Hawaii (USA)	1960	
Hemignathus sagittirostris	Greater Amakihi	Hawaii (USA)	1900	B
* *Paroreomyza flammea*	Kakawihie or Molokai Creeper	Hawaii (USA)	1963	–
Psittirostra kona	Kona Grosbeak	Hawaii (USA)	1894	–
Rhodacanthis flaviceps	Lesser Koa-finch	Hawaii (USA)	1891	–
Rhodacanthis palmeri	Greater Koa-finch	Hawaii (USA)	1896	–
Family Icteridae				
Quiscalus palustris	Slender-billed Grackle	Mexico	1910	B
Family Ploceidae				
Foudra sp.	Reunion Fody	Réunion	1671	
Family Fringillidae				
Chaunoproctus ferreorostris	Bonin Grosbeak	Ogasawara-shoto (Japan)	1890	B,C/D
Spiza townsendi	Townsend's Finch	USA	1833	
Family Sturnidae				
Aplonis corvina	Kosrae Mountain Starling	Kosrae (Fed. States Micronesia)	1828	C
Aplonis fusca	Norfolk Island Starling	Norfolk I (Australia)	1925	–
Aplonis mavornata	Mysterious Starling	Cook Is	1825	C/D
* *Aplonis pelzelni*	Pohnpei Mountain Starling	Pohnpei (Fed. States Micronesia)	1956	B
Fregilupus varius	Réunion Starling	Réunion	1850–1860	B,C/D
Necrospar rodericanus	Rodrigues Starling	Rodrigues (Mauritius)	1726	–
Family Callaeidae				
Heteralocha acutirostris	Huia	New Zealand	1907	A,B,C/D
MAMMALS				
Order MARSUPIALIA				
Family Macropodidae				
* *Caloprymnus campestris*	Desert Rat-kangaroo	Australia	1935	A,B,C
+ *Lagorchestes asomatus*	Central Hare-wallaby	Australia	1931	
Lagorchestes leporides	Eastern Hare-wallaby	Australia	1890	
Macropus greyi	Toolache Wallaby	Australia	1927	C
Onychogalea lunata	Crescent Nailtail Wallaby	Australia	1964	C,D
Potorous platyops	Broad-faced Potoroo	Australia	1875	C
Family Peramelidae				
Chaeropus ecaudatus	Pig-footed Bandicoot	Australia	1907	C,D
Perameles eremiana	Desert Bandicoot	Australia	1935	
Family Thylacomyidae				
Macrotis leucura	Lesser Bilby	Australia	1931	A,C
Family Thylacinidae				
Thylacinus cynocephalus	Thylacine	Tasmania (Australia)	1934	E
Order CHIROPTERA				
Family Pteropodidae				
Acerodon lucifer	Panay Giant Fruit Bat	Philippines	1888	
Dobsonia chapmani	Chapman's Bare-backed Flying Fox	Philippines	1964	
Pteropus pilosus	Palau Flying Fox	Palau	19th C	
Pteropus subniger	Lesser Mascarene Flying Fox	Mauritius, Réunion		
Pteropus tokudae	Guam Flying Fox	Guam	1968	
Family Molossidae				
Mystacina robusta	New Zealand Lesser Short-tailed Bat	New Zealand	1960s	
Order INSECTIVORA				
Family Nesophontidae				
# *Nesophontes hypomicrus*	Atalaye Nesophontes	Haiti, Dominican Republic		C
# *Nesophontes micrus*	Western Cuban Nesophontes	Cuba		C
# *Nesophontes paramicrus*	St Michel Nesophontes	Haiti, Dominican Republic		C
# *Nesophontes zamicrus*	Haitian Nesophontes	Haiti, Dominican Republic		C
# *Nesophontes sp.*		Cayman Is		
Order LAGOMORPHA				
Family Ochotonidae				
Prolagus sardus	Sardinian Pika	Corsica (France), Sardinia (Italy)	18th C	
Family Leporidae				
* *Sylvilagus insonus*	Omilteme Cottontail	Mexico		
Order RODENTIA				
Family Arvicolidae				
Pitymys bavaricus	Bavarian Pine Vole	Germany		
Family Capromyidae				
# *Capromys sp.*		Cayman Is		
# *Geocapromys colombianus*		Cuba		
Geocapromys thoractus		Little Swan I (Honduras)	1950s	
# *Geocapromys sp.*		Cayman Is		
# *Isolobodon portoricensis*		Haiti, Dominican Republic		
# *Plagiodontia velozi*		Haiti, Dominican Republic		
Family Cricetidae				
Megalomys desmarestii	Martinique Rice Rat	Martinique	1902	
Megalomys luciae	St Lucia Rice Rat	Saint Lucia	19th C	
Megaloryzomys curioi		Galapagos (Ecuador)		
Megaloryzomys sp.		Galapagos (Ecuador)		
Nesoryzomys darwini	Santa Cruz Rice Rat	Galapagos (Ecuador)		
Nesoryzomys sp.		Galapagos (Ecuador)		

Table 16.4 Animal species extinct since circa 1600 (continued)

SPECIES	ENGLISH NAME	DISTRIBUTION	LAST RECORDED	POSSIBLE CAUSE
MAMMALS (continued)				
Oryzomys victus	St Vincent Rice Rat	Saint Vincent	1897	
* *Peromyscus pembertoni*	Pemberton's Deer Mouse	Mexico		
Family Echimyidae				
# *Boromys offella*		Cuba		
# *Boromys torrei*		Cuba		
# *Brotomys voratus*		Haiti, Dominican Republic		
Family Muridae				
Conilurus albipes	Rabbit-eared Tree-rat	Australia	1875	
* *Crateromys paulus*	Ilin Bushy-tailed Cloud-rat	Philippines		
Leporillus apicalis	Lesser Stick-nest Rat	Australia	1933	
* *Notomys amplus*	Short-tailed Hopping-mouse	Australia	1894	
* *Notomys longicaudatus*	Long-tailed Hopping-mouse	Australia	1901	
+ *Notomys macrotis*	Big-eared Hopping-mouse	Australia	pre-1850	
+ *Notomys mordax*	Darling Downs Hopping-mouse	Australia	pre-1846	
+ *Pseudomys fieldi*	Alice Springs Mouse	Australia	1895	
+ *Pseudomys gouldi*	Gould's Mouse	Australia	1930	
Rattus macleari	Maclear's Rat	Christmas I (Australia)	1908	
Rattus nativitatis	Bulldog Rat	Christmas I (Australia)	1908	
Order CARNIVORA				
Family Canidae				
Dusicyon australis	Falkland Island Wolf	Falklands Is	1876	E
Family Procyonidae				
+ *Procyon gloveralleni*	Barbados Racoon	Barbados		
Order PINNIPEDIA				
Family Phocidae				
Monachus tropicalis	Caribbean Monk Seal	Caribbean	1962	A
Order SIRENIA				
Family Dugongidae				
Hydrodamalis gigas	Steller's Sea Cow	Bering Straits ('USSR')	1768	A
Order PERISSODACTYLA				
Family Equidae				
Equus quagga	Quagga	South Africa	1883	A,E
Order ARTIODACTYLA				
Family Bovidae				
Gazella rufina	Red Gazelle	Algeria?	19th C	A
Hippotragus leucophaeus	Bluebuck	South Africa	1800	E
Family Cervidae				
Cervus schomburgki	Schomburgk's Deer	Thailand	1932	A

Key: * indicates species generally regarded as extinct but for which there may still be some chance of survival. + indicates taxa which may be conspecific with extant forms. # indicates species known from post-Columbian (i.e. post 1500) deposits in the Caribbean; some may have become extinct before 1600. • indicates species last recorded from Rapa in 1934, and which were considered likely to become rapidly extinct. x indicates species recorded from subfossil deposits in the Mascarenes which are considered very likely to have become extinct following settlement in 1723 although may possibly have become extinct earlier.

Possible causes column': A Hunting (includes for food, skin, sport, live trade, feathers); B Direct habitat alteration by man; C Introduced predators (e.g. cats, rats, mustelids, mongooses, snails, monkeys); (C/D predators or others not specified); D Other introduced animals (e.g. goats, rabbit, pigs); E Destroyed as a pest species; F Introduced disease; G Indirect effects; H Natural Causes; – causes uncertain.

Note: The proceedings of a symposium entitled *St Helena Natural Treasury* (Edited by P. Pearce-Kelly and Q.C.B. Cronk, published by the Zoological Society of London, 1990) were procured too late to include data in these lists. An additional eight extinct endemic bird species are listed from that island, six of which should be included in our analysis. They are thought to have become extinct as a result of the human discovery of the island in 1502, and should therefore be included in the same sort of category as those species recovered from post-Columbian deposits in the Caribbean (i.e. those marked #). The report would also seem to indicate that it may be premature to declare the two insects *Labidura herculeana* and *Aplothorax burchelli* extinct, and they should perhaps be excluded from this list at present. The effect these additions and changes have on the graphs and maps should be borne in mind, especially the increase in early island bird extinctions.

Source: compiled from multiple sources; details available from WCMC. Most bird data compiled by A Stattersfield, and kindly made available by the International Council for Bird Preservation. Mollusc data assembled by Sue Wells with the assistance of members of the SSC Mollusc Specialist Group and other malacologists.

Table 16.5 Extinct higher plant taxa*

MAJOR GROUP (DIVISION) / FAMILY / TAXON	COMMON NAME	HISTORIC RANGE
Fern Allies		
Lycopodiaceae		
Huperzia nutans Brackenr.		United States - Hawaii
Selaginellaceae		
Selaginella orizabensis Hieron.		Mexico - Veracruz
Isoetaceae		
Isoetes dixitii Shende		India - Maharashtra State
Isoetes sampathkumarnii L.N. Rao		India - Karnataka State
True Ferns		
Aspidiaceae		
Diplazium laffanianum (Baker) C.Chr.		Bermuda
Dryopteris speluncae (L.) Underwood		Bermuda
Lastreopsis wattii (Beddome) Tagawa		India - Manipur State
Aspleniaceae		
Asplenium fragile K. Presl var. *insularis* C. Morton		United States - Hawaii
Asplenium leucostegioides Baker		United States - Hawaii
Diellia manii		United States - Hawaii
Diellia unisora Wagner		United States - Hawaii
Blechnaceae		
Doodia lyoni Degener		United States - Hawaii
Marsileaceae		
Marsilea paradoxa Diels		Australia - Western Australia
Ophioglossaceae		
Botrychium subbifoliatum Brackenr.	makou	United States - Hawaii
Thelypteridaceae		
Christella altissima Holttum		South Africa - Natal
Thelypteris macilenta E. St. John	Edward's maiden fern	United States - Florida
Gymnosperms		
Zamiaceae		
Encephalartos woodii Sander		South Africa - Natal
Zamia monticola Chamberlain		Mexico
Dicots		
Acanthaceae		
Dicliptera abuensis Blatter		India - Rajasthan State
Dicliptera falcata (Lam.) Bosser & Heine		Mauritius
Hypoestes inconspicua Balf. f.		Mauritius - Rodrigues
Hypoestes rodriguesiana Balf. f.		Mauritius - Rodrigues
Hypoestes serpens R. Br.		Mauritius
Justicia brachystachya Thouars ex Schultz		Mauritius
Justicia eranthemoides F. Muell.		Australia - New South Wales
Justica psychotrioides Thouars ex Schultz		Mauritius
Aizoaceae		
Gibbaeum esterhuyseniae L. Bolus		South Africa - Cape Province
Trianthema cypseloides (Fenzl) Benth.		Australia - New South Wales
Amaranthaceae		
Achyranthes atollensis St. John		United States - Hawaii
Achyranthes mutica A. Gray ex H. Mann		United States - Hawaii
Amaranthus mentegazzianus Passer.		Argentina
Blutaparon rigidum (Robinson & Greenman) Mears		Ecuador - Galapagos
Ptilotus caespitulosus F. Muell.		Australia - Western Australia
Ptilotus extenuatus Benl		Australia - New South Wales
Ptilotus fasciculatus Fitzg.		Australia - Western Australia
Ptilotus pyramidatus (Moq.) F. Muell.		Australia - Western Australia
Anacardiaceae		
Buchanania mangoides F. Muell.		Australia - Queensland
Aquifoliaceae		
Ilex ternatiflora (C. Wright) R.A. Howard		Cuba
Asclepiadaceae		
Caralluma arenicola N.E. Brown		South Africa - Cape Province
Marsdenia coronata Benth.		Australia - Queensland
Marsdenia tubulosa F. Muell.		Australia - NSW - Lord Howe Island
Matelea balbisii (Dcne.) Woods.	Balbis' milkvine	United States - Arizona
Matelea radiata Correll	Falfurrias Anglepod	United States - Texas
Begoniaceae		
Begonia cowellii Nash		Cuba
Begonia opuliflora Putz.		Panama
Boraginaceae		
Cryptantha aperta (Eastw.) Payson	Grand Junction cat's-eye	United States - Colorado
Cryptantha insolita (J.F. Macbr.) Payson	unusual cat's-eye	United States - Nevada
Heliotropium muticum Domin		Australia - Western Australia

Table 16.5 Extinct higher plant taxa*

MAJOR GROUP (DIVISION) / FAMILY / TAXON	COMMON NAME	HISTORIC RANGE
Heliotropium pannifolium Burchell ex Hemsley		St Helena
Lindelofia angustifolia (Schrenk) A. Brand.		former Union of Soviet Socialist Republics
Myosotis petiolata Hook.f. var. *pottsiana* L. Moore		New Zealand - North Island
Onosma affine Hausskn. ex H. Rield		Turkey
Onosma discedens Hausskn. ex Bornm.		Turkey
Plagiobothrys diffusus (Greene) I.M. Johnston	San Francisco popcornflower	United States - California
Plagiobothrys lamprocarpus (Piper) I.M. Johnston	popcornflower	United States - Oregon
Plagiobothrys orthostatus J. Black		Australia - South Australia
Bruniaceae		
Staavia trichotoma (Thunb.) Pillans		South Africa - Cape Province
Thamnea depressa Oliver		South Africa - Cape Province
Thamnea uniflora Solander ex Brongn.		South Africa - Cape Province
Cactaceae		
Hylocereus cubensis Britton & Rose		Cuba
Leptocereus wrightii Leon		Cuba
Lobivia vatteri Krainz		Argentina
Opuntia lindheimeri Engelmann var. *linguiformis* (Griffiths) L. Benson		United States - Texas
Pyrrhocactus aricensis Ritt.		Chile
Pyrrhocactus longirama Ritt.		Chile
Pyrrhocactus nuda Ritt.		Chile
Pyrrhocactus occultus Ritt.		Chile
Campanulaceae		
Campanula oligosperma Damboldt		Turkey
Clermontia multiflora Hillebrand		United States - Hawaii
Cyanea arborea (H. Mann) Hillebrand var. *arborea*		United States - Hawaii
Cyanea asplenifolia (H. Mann) Hillebrand	spleenwort-leaved cyanea	United States - Hawaii
Cyanea comata Hillebrand		United States - Hawaii
Cyanea dunbarii Rock		United States - Hawaii
Cyanea giffardii Rock		United States - Hawaii
Cyanea glabra (F. Wimmer) St. John	smooth cyanea	United States - Hawaii
Cyanea grimesiana Gaudich. ssp. *cylindrocalyx* (Rock) Lammers		United States - Hawaii
Cyanea linearifolia Rock		United States - Hawaii
Cyanea longissima (Rock) St. John		United States - Hawaii
Cyanea obtusa (A. Gray) Hillebrand		United States - Hawaii
Cyanea pohaku Lammers		United States - Hawaii
Cyanea procera Hillebrand		United States - Hawaii
Cyanea profuga C. Forbes		United States - Hawaii
Cyanea pycnocarpa (Hillebrand) F.E. Wimmer		United States - Hawaii
Cyanea quercifolia (Hillebrand) F.E. Wimmer var. *quercifolia*		United States - Hawaii
Cyanea recta (Wawra) Hillebrand		United States - Hawaii
Cyanea scabra Hillebrand var. *longissima* Rock		United States - Hawaii
Cyanea undulata C. Forbes		United States - Hawaii
Delissea fallax Hillebrand		United States - Hawaii
Delissea laciniata Hillebrand var. *laciniata*	cut-leaf delissea	United States - Hawaii
Delissea lauliiana Lammers		United States - Hawaii
Delissea parviflora Hillebrand	small-flowered delissea	United States - Hawaii
Delissea rivularis (Rock) F.E. Wimmer		United States - Hawaii
Delissea sinuata Hillebrand ssp. *lanaiensis* (Rock) Lammers		United States - Hawaii
Delissea sinuata Hillebrand var. *sinuata*	wavy-leaf delissea	United States - Hawaii
Delissea undulata Gaudich.	undulata delissea	United States - Hawaii
Lobelia monostachya (Rock) Lammers		United States - Hawaii
Lobelia remyi Rock		United States - Hawaii
Rollandia parvifolia C. Forbes		United States - Hawaii
Rollandia purpurellifolia Rock		United States - Hawaii
Wahlenbergia burchellii A.DC.		St Helena
Wahlenbergia roxburghii A.DC.		St Helena
Wahlenbergia saxifragoides V. Brehm.		South Africa - Cape Province
Caryophyllaceae		
Alsinidendron viscosum (H. Mann) Sherff		United States - Hawaii
Schiedea amplexicaulis H. Mann		United States - Hawaii
Schiedea helleri Sherff		United States - Hawaii
Schiedea implexa (Hillebrand) Sherff		United States - Hawaii
Schiedea spergulina A. Gray var. *leiopoda* Sherff		United States - Hawaii
Schiedea stellarioides H. Mann var. *stellarioides*	laulihilihi; kawelu; ma'oli'oli	United States - Hawaii
Silene cryptopetala Hillebrand		United States - Hawaii
Silene oligotricha Huber-Mor.		Turkey
Silene rectiramea Robinson		United States - Arizona

Table 16.5 Extinct higher plant taxa*

MAJOR GROUP (DIVISION) / FAMILY / TAXON	COMMON NAME	HISTORIC RANGE
Stellaria elatinoides Hook. f.		New Zealand
Celastraceae		
Hexaspora pubescens C. White		Australia - Queensland
Maytenus lineata C. Wright		Cuba
Chenopodiaceae		
Hemichroa mesembryanthema F. Muell.		Australia - South Australia
Sclerolaena ramsayae (Willis) A.J. Scott		Australia - Victoria
Suaeda duripes I.M. Johnston	hardtoe seepweed	United States - Texas
Compositae		
Abrotanella rhynchocarpa Balf. f.		Mauritius - Rodrigues
Acanthocladium dockeri F. Muell.		Australia - New South Wales, South Australia
Argyroxiphium virescens Hillebrand var. *virescens*	greensword	United States - Hawaii
Artemisia insipida Vill.		France
Brachycome muelleri Sonder		Australia - South Australia
Calocephalus globosus M. Scott & Hutch.		Australia - Western Australia
Cirsium toyoshimae Koidz.		Japan
Commidendrum rotundifolium (Roxb.) DC.		St Helena
Crepidiastrum ameristophyllum (Koidz.) Nakai		Japan - Ogasawara-Shoto
Crepidiastrum grandicollum (Koidz.) Nakai		Japan - Ogasawara-Shoto
Erigeron perglaber Blake		United States - Arizona
Felicia annectens (Harvey) Grau		South Africa - Cape Province
Helianthus praetermissus E. Watson		United States - New Mexico
Helichrysum oligochaetum F. Muell.		Australia - Western Australia
Helichrysum selaginoides (Sonder & F. Muell.) Benth.		Australia - Tasmania
Helichrysum spiceri F. Muell.		Australia - Tasmania
Helipterum guilfoylei Ewart		Australia - Western Australia
Hemizonia mohavensis Keck	Mojave tarweed; Mojave tarplant	United States - California
Leptorhynchos gatesii (Williamson) J.H. Willis		Australia - Victoria
Lipochaeta bryanii Sherff		United States - Hawaii
Lipochaeta ovata R. Gardner		United States - Hawaii
Lipochaeta perdita Sherff	ko'oko'olau; nehe	United States - Hawaii
Marasmodes undulata Compton		South Africa - Cape Province
Olearia arida Pritzel		Australia - South Australia, Western Australia
Olearia flocktoniae Maiden & E. Betcke		Australia - New South Wales
Olearia oliganthema Benth.		Australia - New South Wales
Osteospermum hirsutum Thunb.		South Africa - Cape Province
Perityle inyoensis (Ferris) A. Powell	Inyo laphamia	United States - California
Perityle villosa (Blake) Shinn.	Hanaupah laphamia	United States - California
Senecio behrianus Sonder & F. Muell.		Australia - New South Wales, South Australia, Victoria
Senecio georgianus DC.		Australia - South Australia, Victoria, Western Australia
Senecio laticostatus Belcher		Australia - Victoria
Senecio sandwicensis Less.		United States - Hawaii
Solidago porteri Small	Porter's goldenrod	United States - Georgia, North Carolina
Tetramolopium arenarium (A. Gray) Hillebrand var. *arenarium*		United States - Hawaii
Tetramolopium arenarium (A. Gray) Hillebrand var. *confertum* Sherff		United States - Hawaii
Tetramolopium arenarium (A. Gray) Hillebrand ssp. *laxum* Lowrey		United States - Hawaii
Tetramolopium capillare (Gaudich.) H. St. John		United States - Hawaii
Tetramolopium consanguineum (A. Gray) Hillebrand ssp. *consanguineum*		United States - Hawaii
Tetramolopium conyzoides (A. Gray) Hillebrand		United States - Hawaii
Tetramolopium lepidotum Less. ssp. *arbusculum* (A. Gray) T.K. Lowrey		United States - Hawaii
Tetramolopium tenerrimum (Less.) Nees		United States - Hawaii
Tracyina rostrata Blake	showy indian clover	United States - California
Vernonia africana (Sonder) Druce		South Africa - Natal
Crassulaceae		
Crassula alcicornis Schonl.		South Africa - Cape Province
Crassula subulata Hermann var. *hispida* (Schonl. & E.G. Baker) Toelken		South Africa - Cape Province
Echeveria laui Moran & Meyran		Mexico - Oaxaca
Sedum pinetorum Brandegee	Pine City stonecrop	United States - California
Sedum polystriatum R.T. Clausen		Turkey
Tacitus bellus Moran & Meyran		Mexico - Chihuahua
Cruciferae		

Table 16.5 Extinct higher plant taxa*

MAJOR GROUP (DIVISION)
FAMILY

TAXON	COMMON NAME	HISTORIC RANGE
Ballantinia antipoda (F. Muell.) E. Shaw		Australia - Tasmania, Victoria
Caulanthus lemmonii		United States - Arizona
Diplotaxis siettiana Maire		Spain
Hutchinsia tasmanica Hook.		Australia - Tasmania
Isatis arnoldiana N. Busch.		former Union of Soviet Socialist Republics
Lepidium drummondii Thell.		Australia - Western Australia
Lepidium merrallii F. Muell.		Australia - Western Australia
Lepidium obtusatum Kirk		NEW ZEALAND - North Island
Lepidium peregrinum Thell.		Australia - New South Wales
Menkea draboides (Hook.f.) Benth.		Australia - Western Australia
Phlegmatospermum drummondii (Benth.) O. Schultz		Australia - Western Australia
Phlegmatospermum richardsii (F. Muell.) E. Shaw		Australia - South Australia, Western Australia
Rorippa coloradensis Stuckey	Colorado watercress	United States - Colorado
Stroganowia sagittata Karelin & Kir.		Asiatic former Union of Soviet Socialist Republics
Tropidocarpum capparideum Greene	caper-fruited tropidocarpum	United States - California
Cucurbitaceae		
Benincasa hispida (Thunb.) Cogn.		Australia - Queensland
Sicyos hillebrandii H. St. John		United States - Hawaii
Sicyos villosa Hook. f.		Ecuador - Galapagos
Dicrastylidaceae		
Dicrastylis morrisonii Munir		Australia - Western Australia
Dilleniaceae		
Hibbertia sargentii S. Moore		Australia - Western Australia
Epacridaceae		
Andersonia bifida L. Watson		Australia - Western Australia
Andersonia longifolia (Benth.) L. Watson		Australia - Western Australia
Choristemon humilis Williamson		Australia - Victoria
Coleanthera coelophylla (DC.) Benth.		Australia - Western Australia
Coleanthera virgata Stschegl.		Australia - Western Australia
Leucopogon cryptanthus Benth.		Australia - Western Australia
Leucopogon pogonocalyx Benth.		Australia - Western Australia
Ericaceae		
Arctostaphylos uva-ursi (L.) Sprengel var. *franciscana* (Eastw.) Roof		United States - California
Arctostaphylos uva-ursi (L.) Sprengel var. *leobreweri* Roof		United States - California
Erica acockii Compton		South Africa - Cape Province
Erica bolusiae Salter		South Africa - Cape Province
Erica jasminiflora Salisb.		South Africa - Cape Province
Erica pyramidalis Solander		South Africa - Cape Province
Erica turgida Salisb.		South Africa - Cape Province
Erica verticillata Bergius		South Africa - Cape Province
Rhododendron mucronulatum Turcz. var. *albiflora* Nakai		Republic of Korea
Erythroxylaceae		
Erythroxylum echinodendron Ekman		Cuba
Euphorbiaceae		
Acalypha rubra Roxb.		St Helena
Amperea protensa Nees		Australia - Western Australia
Beyeria cygnorum (Muell. Arg.) Benth.		Australia - Western Australia
Beyeria lepidopetala F. Muell.		Australia - Western Australia
Bonania myrcifolia (Griseb.) Benth. & Hook.		Cuba
Chamaesyce celastroides (Boiss.) Croizat & Degener var. *tomentella*	'akoko; koko; 'ekoko; kokomalei	United States - Hawaii
Claoxylon grandifolium (Poiret) Muell. Arg.		Mauritius; France - Reunion
Cnidoscolus fragrans (H.B.K.) Pohl		Cuba
Croton magneticus Airy Shaw		Australia - Queensland
Euphorbia carissoides Bailey		Australia - Queensland
Euphorbia daphnoides Balf. f.		Mauritius - Rodrigues
Pseudanthus nematophorus F. Muell.		Australia - Western Australia
Fagaceae		
Quercus boytoni Beadle	Boyton's sand post oak	United States - Texas
Frankeniaceae		
Frankenia conferta Diels		Australia - Western Australia
Frankenia decurrens Summerh.		Australia - Western Australia
Frankenia parvula Turcz.		Australia - Western Australia
Gesneriaceae		
Cyrtandra cyaneoides Rock		United States - Hawaii

Table 16.5 Extinct higher plant taxa*

MAJOR GROUP (DIVISION) / FAMILY / TAXON	COMMON NAME	HISTORIC RANGE
Cyrtandra gracilis Hillebrand		United States - Hawaii
Cyrtandra honolulensis Wawra		United States - Hawaii
Cyrtandra kohalae Rock		United States - Hawaii
Cyrtandra olona C. Forbes		United States - Hawaii
Cyrtandra pickeringii A. Gray var. *pickeringii*		United States - Hawaii
Cyrtandra waiolani Wawra var. *capitata* Hillebrand		United States - Hawaii
Cyrtandra waiolani Wawra var. *waiolani*	ha'iẇale; kanawao ke'oke'o	United States - Hawaii
Goodeniaceae		
Dampiera helmsii Krause		Australia - Western Australia
Dampiera humilis (F. Muell.) E. Pritzel		Australia - Western Australia
Dampiera rupicola S. Moore		Australia - Western Australia
Goodenia clementii Krause		Australia - Western Australia
Scaevola attenuata R. Br.		Australia - Western Australia
Scaevola macrophylla (Vriese) Benth.		Australia - Western Australia
Scaevola oldfieldii F. Muell.		Australia - Western Australia
Verreauxia verreauxii (Vriese) Carolin		Australia - Western Australia
Grossulariaceae		
Ribes kolymense (Trautv.) Komarov ex Pojark		former Union of Soviet Socialist Republics
Haloragaceae		
Gonocarpus intricatus (Benth.) Orch.		Australia - Western Australia
Haloragis stricta R. Br.		Australia - New South Wales, Queensland
Haloragis tenuifolia Benth.		Australia - Western Australia
Haloragodendron lucasii (Maiden & E. Betch) Orch.		Australia - New South Wales
Meziella trifida (Nees) Schindler		Australia - Western Australia
Hydrophyllaceae		
Phacelia amabilis Constance	Saline Valley phacelia	United States - California
Phacelia cinerea Eastw.	ashy phacelia	United States - California
Phacelia nevadensis J. Howell	Nevada phacelia	United States - Nevada
Labiatae		
Haplostachys bryanii Sherff var. *bryanii*		United States - Hawaii
Haplostachys linearifolia (Drake) Sherff var. *linearifolia*		United States - Hawaii
Haplostachys munroi C. Forbes		United States - Hawaii
Haplostachys truncata (A. Gray) Hillebrand		United States - Hawaii
Hemigenia exilis S. Moore		Australia - Western Australia
Hemigenia obtusa Benth.		Australia - Western Australia
Hemigenia pimelifolia F. Muell.		Australia - Western Australia
Hemigenia podalyrina F. Muell.		Australia - Western Australia
Hemigenia ramosissima Benth.		Australia - Western Australia
Hemigenia tysoni F. Muell.		Australia - Western Australia
Hemigenia tysonii F. Muell.		Australia - Western Australia
Microcorys pimeloides F. Muell.		Australia - Western Australia
Monardella leucocephala A. Gray	Merced monardella	United States - California
Monardella pringlei A. Gray	Pringle monardella	United States - California
Phyllostegia brevidens A. Gray var. *brevidens*		United States - Hawaii
Phyllostegia hillebrandii Mann ex Hillebrand		United States - Hawaii
Phyllostegia immunata (Sherff) St. John		United States - Hawaii
Phyllostegia knudsenii Hillebrand		United States - Hawaii
Phyllostegia rockii Sherff		United States - Hawaii
Phyllostegia variabilis Bitter		United States - Hawaii
Phyllostegia wawrana Sherff		United States - Hawaii
Prostanthera staurophylla F. Muell.		Australia - New South Wales
Pycnanthemum monotrichum Fern.	mountain mint	United States - Virginia
Stenogyne cinerea Hillebrand		United States - Hawaii
Stenogyne haliakalae Wawra		United States - Hawaii
Stenogyne oxygona Degener & Sherff		United States - Hawaii
Stenogyne viridis Hillebrand		United States - Hawaii
Teucrium leucophyllum Montbret & Aucher ex Bentham		Turkey
Thymus oehmianus Ronn. & Soska		Yugoslavia
Lauraceae		
Cassytha pedicellosa J.Z. Webb		Australia - Tasmania
Leguminosae		
Acacia forrestiana E. Pritzel		Australia - Western Australia
Acacia murruboensis Maiden & Blakely		Australia - New South Wales
Acacia prismifolia E. Pritzel		Australia - Western Australia
Acacia vassalii Maslin		Australia - Western Australia
Aspalathus variegata Ecklon & Zeyher		South Africa - Cape Province
Astragalus pseudocylindraceus Bornm.		Turkey
Astragalus robbinsii (Oakes) A. Gray var. *robbinsii*		United States - Vermont

Table 16.5 Extinct higher plant taxa*

MAJOR GROUP (DIVISION) / FAMILY / TAXON	COMMON NAME	HISTORIC RANGE
Chorizema varium Benth.		Australia - Western Australia
Crotalaria urbaniana Senn		Cuba
Gastrolobium crispifolium Domin		Australia - Western Australia
Genista melia Boiss.		Greece
Jacksonia hemisericea D. Herbert		Australia - Western Australia
Lathyrus dominianus Litv.		former Union of Soviet Socialist Republics
Lupinus sublanatus Eastw.	Santa Catalina Island desert-thorn	United States - California
Mirbelia densiflora C. Gardner		Australia - Western Australia
Onobrychis aliacmonia Reich. f.		Greece
Orbexilum macrophyllum Rydb.		United States - North Carolina
Oxylobium acutum (Benth.) Benth.		Australia - Western Australia
Phyllota gracilis Turcz.		Australia - Western Australia
Psoralea macrophylla Rowlee ex Small	bigleaf scurpea	United States - North Carolina
Psoralea stipulata Torrey ex A. Gray	scurf-pea	United States - Indiana, Kentucky
Pultenaea pauciflora M. Scott		Australia - Western Australia
Sophora toromiro (Philippi) Skottsb.	toromiro	Chile - Easter Island
Streblorrhiza speciosa Endl.		Australia - Norfolk Island
Taverniera sericophylla Balf. f.		Democratic Yemen - Socotra
Tephrosia kassasi Boulos		Egypt
Tetragonolobus wiedemannii Boiss.		Greece
Trifolium amoenum E. Greene	showy indian clover	United States - California
Vicia dennesiana H.C. Watson		Portugal - Azores
Lentibulariaceae		
Utricularia mairii Cheeseman		New Zealand - North Island
Loasaceae		
Mentzelia nitens Greene var. *leptocaulis* J. Darl.		United States - Arizona
Loganiaceae		
Mitrasacme palustris W. Fitzg.		Australia - Western Australia
Loranthaceae		
Dendrophthora terminalis Kuijt		Costa Rica
Psittacanthus nudus (A. Molina) Kuijt & Feuer		Honduras
Trilepidea adamsii (Cheeseman) Tieghem		New Zealand - North Island
Malvaceae		
Abutilon mauritianum (Jacq.) Medik.		Mauritius
Anisodontea alexandri (Baker f.) Bates		South Africa - Cape Province
Hibiscadelphus bombycinus C. Forbes		United States - Hawaii
Hibiscadelphus crucibracteatus Hobdy		United States - Hawaii
Hibiscadelphus wilderianus Rock		United States - Hawaii
Hibiscus nelsonii Rose & Standley		Mexico
Kokia lanceolata Lewton		United States - Hawaii
Malacothamnus abbottii (Eastw.) Kearney	Abbott's bush-mallow	United States - California
Malacothamnus mendocinensis (Eastw.) Kearney	Mendocino bush-mallow	United States - Arkansas, California
Sida pritzellii C. Gardner		Australia - Western Australia
Sidalcea keckii Wiggins	Keck sidalcea; Keck checker-mallow	United States - California
Sphaeralcea procera C.L. Porter	Luna County globemallow	United States - New Mexico
Menispermaceae		
Hyperbaena obovata Urban		Cuba
Menyanthaceae		
Nymphoides stygia (J. Black) H. Eichler		Australia - South Australia
Myoporaceae		
Eremophila adenotricha F. Muell.		Australia - Western Australia
Eremophila scaberula Fitzg.		Australia - Western Australia
Myrsinaceae		
Badula ovalifolia A.DC.		France - Reunion
Myrsine mezii Hosaka		United States - Hawaii
Myrtaceae		
Calothamnus blepharantherus F. Muell.		Australia - Western Australia
Hypocalymma longifolium F. Muell.		Australia - Western Australia
Melaleuca arenaria C. Gardner		Australia - Western Australia
Melaleuca arenicola S. Moore		Australia - Western Australia
Melaleuca graminea S. Moore		Australia - Western Australia
Monimiastrum fasciculatum Gueho & A.J. Scott		Mauritius
Syzygium balfourii (Baker) Gueho & A.J. Scott		Mauritius - Rodrigues
Verticordia carinata Turcz.		Australia - Western Australia
Nyctaginaceae		
Pisonia floridana Britton	rock dey devil's-claws	United States - Florida
Ochnaceae		
Ouratea alternifolia (A. Rich.) M. Gomez		Cuba
Oleaceae		
Hesperelaea palmeri A. Gray		Mexico
Onagraceae		

Table 16.5 Extinct higher plant taxa*

MAJOR GROUP (DIVISION) FAMILY TAXON	COMMON NAME	HISTORIC RANGE
Clarkia mosquinii E. Small ssp. *xerophila* E. Small		United States - California
Lopezia conjugens Brandegee		Mexico
Lopezia sinaloensis Munz		Mexico
Oenothera kleinii W.L. Wagner & S.W. Mill	Klein's evening-primrose; Wolf Creek evening-primrose	United States - Colorado
Papaveraceae		
Eschscholzia rhombipetala E. Greene	diamond-petaled; California poppy	United States - California
Penaeaceae		
Stylapterus micranthus R. Dahlgren		South Africa - Cape Province
Piperaceae		
Peperomia degeneri Yuncker		United States - Hawaii
Peperomia hirta Balf. f.		Mauritius - Rodrigues
Peperomia rodriguezi Balf. f.		Mauritius - Rodrigues
Peperomia rossii Rendle		Australia - Christmas Island
Plumbaginaceae		
Armeria arcuata Welw. ex Boiss. & Reuter		Portugal
Polygalaceae		
Comesperma lanceolatum Benth.		Australia - Western Australia
Comesperma rhadinocarpum F. Muell.		Australia - Western Australia
Polygonaceae		
Eriogonum truncatum Torrey & A. Gray	Contra Costa eriogonum; Mt Diablo buckwheat	United States - California
Portulacaceae		
Calandrinia composita Nees		Australia - Western Australia
Calandrinia dielsii Poelln.		Australia - Western Australia
Calandrinia feltonii Skottsb.		Falkland Islands
Calandrinia sphaerophylla J. Black		Australia - South Australia
Primulaceae		
Lysimachia forbesii Rock		United States - Hawaii
Lysimachia minoricensis J.D. Rodriguez		Spain - Balearic Islands
Proteaceae		
Grevillea batrachioides McGillivray		Australia - Western Australia
Grevillea divaricata R. Br.		Australia - New South Wales
Grevillea flexuosa (Lindley) Meissner		Australia - Western Australia
Grevillea scabra Meissner		Australia - Western Australia
Hakea crassinervia Meissner		Australia - Western Australia
Hakea pulvinifera L. Johnson		Australia - New South Wales
Hakea tamminensis C. Gardner		Australia - Western Australia
Isopogon uncinatus R. Br.		Australia - Western Australia
Leucadendron comosum (Thunb.) R. Br. ssp. *homoeophyllum* (Meisn.) I. Williams		South Africa - Cape Province
Leucadendron spirale (Salisb. ex Knight) I. Williams		South Africa - Cape Province
Mimetes stokoei Phillips & Hutch.		South Africa - Cape Province
Persoonia leucopogon S. Moore		Australia - Western Australia
Sorocephalus tenuifolius R. Br.		South Africa - Cape Province
Triunia robusta (C. White) D. Foreman		Australia - Queensland
Pyrolaceae		
Pyrola oxypetala Austin	sharp-petal wintergreen	United States - New York
Rhamnaceae		
Cryptandra tubulosa Fenzl.		Australia - Western Australia
Cryptandra uncinata Grun.		Australia - South Australia
Spyridium kalganense Diels		Australia - Western Australia
Spyridium microcephalum (Turcz.) Benth.		Australia - Western Australia
Trymalium albicans (Steudel) Reisseck		Australia - Western Australia
Trymalium urceolare (F. Muell.) Diels		Australia - Western Australia
Rosaceae		
Potentilla multijuga Lehm.	Ballona cinquefoil	United States - California
Rubiaceae		
Danais corymbosa Balf. f.		Mauritius - Rodrigues
Gaertnera calycina Bojer		Mauritius
Gaertnera crassiflora Bojer		Mauritius
Gaertnera longifolia Bojer var. *pubescens* Verdc.		Mauritius
Gaertnera quadriseta A.DC.		Mauritius
Hedyotis foliosa (Hillebrand) Fosb.		United States - Hawaii
Oldenlandia adscensionis (DC.) Cronk		Ascension Island
Oldenlandia polyclada (F.Muell.) F. Muell.		Australia - Queensland
Oldenlandia sieberi Baker var. *sieberi*		Mauritius
Opercularia hirsuta F. Muell ex Benth.		Australia - Western Australia
Opercularia ocolytantha Diels.		Australia - Western Australia
Ophiorrhiza brunonis Wight & Arn.		India - Karnataka State, Kerala State, Tamil Nadu State
Ophiorrhiza caudata C. Fischer		India - Kerala State

Table 16.5 Extinct higher plant taxa*

MAJOR GROUP (DIVISION) / FAMILY / TAXON	COMMON NAME	HISTORIC RANGE
Ophiorrhiza radicans Gardn.		India - Kerala State; Sri Lanka
Phyllacanthus grisebachianus Hook. f.		Cuba
Psychotria banaona Urban		Cuba
Pyrostria ferruginea Verdc.		Mauritius
Rondeletia odorata Jacq. var. *breviflora* Hook.		Panama
Wendlandia angustifolia Wight		India - Tamil Nadu State
Rutaceae		
Acmadenia candida I. Williams		South Africa - Cape Province
Agathosma orbicularis Bartl. & Wendl. f.		South Africa - Cape Province
Eriostemon falcatus P.G. Wilson		Australia - Western Australia
Galipea ossana DC.		Cuba
Kodalyodendron cubensis Borh. & Acuna		Cuba
Melicope adscendens (St. John & Hume) T. Hartley & B. Stone		United States - Hawaii
Melicope ballouii (Rock) T. Hartley & B. Stone		United States - Hawaii
Melicope degeneri (B. Stone) T. Hartley & B. Stone		United States - Hawaii
Melicope lydgatei (Hillebrand) T. Hartley & B. Stone		United States - Hawaii
Melicope ovalis (St. John) T. Hartley & B. Stone		United States - Hawaii
Melicope quadrangularis (St. John & E. Hume) T. Hartley & B. Stone		United States - Hawaii
Melicope reflexa (St. John) T. Hartley & B. Stone		United States - Hawaii
Melicope wailauensis (St. John) T. Hartley & B. Stone		United States - Hawaii
Pelea fatuhivensis F. Brown		France - French Polynesia - Marquesas Is
Pelea obovata H. St. John		United States - Hawaii
Phebalium daviesii Hook. f.		Australia - Tasmania
Phebalium lachnaeoides Cunn.		Australia - New South Wales
Zanthoxylum leonis Alain		Cuba
Zieria adenophora Blakely		Australia - New South Wales
Santalaceae		
Leptomeria dielsiana Pilger		Australia - Western Australia
Santalum fernandezianum F. Philippi		Chile - Juan Fernandez
Sapindaceae		
Euchorium cubense Ekman & Radlk.		Cuba
Saxifragaceae		
Astilbe crenatiloba (Britton) Small	crenate-lobed false goat's-beard	United States - North Carolina, Tennessee
Mitella prostrata Michaux		Canada
Saxifraga lactea Turcz.		former Union of Soviet Socialist Republics
Saxifraga oppositifolia L. ssp. *amphibia* (Sunderm.) Braun-Blanquet		Germany; Switzerland
Scrophulariaceae		
Agalinis stenophylla Pennell	narrow-leaved false foxglove	United States - Florida
Agalinis strictifolia Pennell		United States - Louisiana
Castilleja cruenta Standley	indian paintbrush	United States - Arizona
Castilleja leschkeana J. Howell	Point Reyes indian paintbrush	United States - California
Euphrasia arguta R. Br.		Australia - New South Wales
Euphrasia collina R.Br. ssp. *muelleri* (Wettst.) Barker		Australia - New South Wales, South Australia, Victoria
Limosella pubiflora Pennell	mudwort	United States - Arizona
Micranthemum micranthemoides (Nutt.) Wettst.	Nuttall's micranthemum	United States - Delaware, District of Columbia, Maryland, New Jersey, New York, Pennsylvania, Virginia
Mimulus brandegei Pennell	Santa Cruz Island monkey-flower	United States - California
Mimulus clementii Domin		Australia - Western Australia
Mimulus traskiae A.L. Grant	Santa Catalina monkey-flower	United States - California
Mimulus whipplei A.L. Grant	Whipple's monkey-flower	United States - California
Orthocarpus pachystachyus A. Gray	shasta owl-clover	United States - California
Penstemon leptanthus Pennell	Sevier Plateau beardtongue	United States - Utah
Penstemon pulchellus Lindl.	beautiful beardtongue	United States - New Mexico
Seymeria havardii (Pennell) Stand	Eagle Pass seymeria	United States - Texas
Verbascum calycosum Hausskn. & Murb.		Turkey
Veronica euxina Turrill		Bulgaria
Solanaceae		
Lycium hassei Greene		United States - California
Mellissia begonifolia (Roxb.) Hook. f.		St Helena
Solanum bahamense L. var. *rugelii* D'Arcy		United States - Florida
Solanum bauerianum Endl.		Australia - NSW Lord Howe Island Australia - Norfolk Island

MAJOR GROUP (DIVISION) FAMILY TAXON	COMMON NAME	HISTORIC RANGE
Solanum cajamarcense Ochoa		Peru
Solanum nava Webb & Berthel.		Spain - Canary Islands
Sterculiaceae		
Astiria rosea Lindley		Mauritius
Sterculia khasiana Deb.		India - Meghalaya State
Trochetia parviflora Bojer ex Baker		Mauritius
Stylidiaceae		
Stylidium merrallii (F.Muell.) E. Pritzel		Australia - Western Australia
Stylidium neglectum Mildbr.		Australia - Western Australia
Stylidium pseudocaespitosum Mildbr.		Australia - Western Australia
Styracaceae		
Styrax portoricensis Krug & Urban		Puerto Rico
Theaceae		
Franklinia alatamaha Marshall	Franklin tree	United States - Georgia
Tremandraceae		
Tetratheca deltoidea J. Thompson		Australia - Western Australia
Tetratheca elliptica J. Thompson		Australia - Western Australia
Tetratheca fasciculata J. Thompson		Australia - Western Australia
Tetratheca gunnii Hook. f.		Australia - Tasmania
Umbelliferae		
Geocaryum bornmuelleri (Wolff) Engstr.		Greece
Geocaryum divaricatum (Boiss. & Orph.) Engstr.		Greece
Platysace dissecta (Benth.) Norman		Australia - Western Australia
Platysace eatoniae (F. Muell.) Norman		Australia - Western Australia
Trachymene croniniana F. Muell.		Australia - Western Australia
Xanthosia singuliflora F. Muell.		Australia - Western Australia
Zizia latifolia Small	bristol golden alexanders	United States - Florida
Urticaceae		
Pilea thouarsiana Wedd.		Mauritius
Pilea trilobata (Poiret) Wedd.		Mauritius
Valerianaceae		
Valeriana pratensis (Benth.) Steud.		Mexico
Violaceae		
Isodendrion pyrifolium A. Gray	wahine noho kula	United States - Hawaii
Viola cryana Gillot		France
Zygophyllaceae		
Fagonia taeckholmiana Hadidi		Egypt
Monocots		
Amaryllidaceae		
Caliphruria tenera Baker		Colombia
Eucharis lehmannii Regel		Colombia
Eucrosia mirabilis (Baker) Pax		Ecuador
Gethyllis esterhuyseniae		South Africa - Cape Province
Gethyllis latifolia Masson ex Baker		South Africa - Cape Province
Habranthus caeruleus (Griseb.) Traub		Argentina
Mathieua galanthoides Klotzsch		Peru
Plagiolirion horsmannii Baker		Colombia
Araceae		
Anthurium leuconeurum Lemaire		Mexico
Philodendron clementis C.Wright ex Griseb.		Cuba
Burmanniaceae		
Thismia americana N. Pfeiffer	thismia	United States - Illinois
Centrolepidaceae		
Centrolepis caespitosa D. Cooke		Australia - Western Australia
Commelinaceae		
Sauvallea blainii C. Wright		Cuba
Cyperaceae		
Bulbostylis neglecta (Hemsley) C.B. Clarke		St Helena
Carex aboriginum M.E. Jones	Indian Valley sedge	United States - Idaho
Carex paupera Nelmes		Australia - Victoria
Carex repanda C.B. Clarke		India - Meghalaya State
Cladium drummondii C.B. Clarke		Australia - Western Australia
Eleocharis bermudiana Britton		Bermuda
Fimbristylis compacta Turrill		Australia - Northern Territory
Schoenus acuminatus R. Br.		Australia - Western Australia
Schoenus natans (F. Muell.) Benth.		Australia - Western Australia
Tetraria australiensis C.B. Clarke		Australia - Western Australia
Dioscoreaceae		
Dioscorea pentaphylla L.		Australia - Queensland
Rajania prestoniensis Knuth		Cuba
Eriocaulaceae		
Eriocaulon echinospermoideum Ruhl.		Cuba

Table 16.5 Extinct higher plant taxa*

MAJOR GROUP (DIVISION) / FAMILY	TAXON	COMMON NAME	HISTORIC RANGE
	Eriocaulon johnstonii Ruhl.		Mauritius
	Eriocaulon minutessimum Ruhl.		Cuba
	Lachnocaulon cubense Ruhl.		Cuba
Gramineae			
	Agrostis adamsonii Vick.		Australia - Victoria
	Agrostis limitanea J. Black		Australia - South Australia
	Bromus brachystachys Hornung		Germany
	Bromus bromoideus (Lej.) Crepin		Belgium
	Bromus grossus Desf. ex DC.		Belgium; Luxembourg
	Bromus interruptus (Hackel) Druce		United Kingdom
	Cenchrus agrimonioides Trin. var. *laysanensis* F. Brown	kamanomano; kumanomano	United States - Hawaii
	Deyeuxia drummondii (Steudel) Vick.		Australia - Western Australia
	Deyeuxia lawrencei Vick.		Australia - Tasmania
	Digitaria pittieri (Hackel) Henrard		Costa Rica
	Dissanthelium californicum (Nutt.) Benth.	California dissanthelium	United States - California; Mexico
	Eragrostis deflexa Hitchc.	Pacific lovegrass	United States - Hawaii
	Eragrostis fosbergii Whitney	Fosberg's lovegrass	United States - Hawaii
	Eragrostis hosakai Degener		United States - Hawaii
	Eragrostis mauiensis Hitchc.		United States - Hawaii
	Eragrostis rottleri Stapf		India - Tamil Nadu State
	Eriochrysis rangacharii Fischer		India - Tamil Nadu State
	Festuca benthamiana Vick.		Australia - South Australia
	Glyceria drumondii (Steudel) C.E. Hubb.		Australia - Western Australia
	Heterachne baileyi C.E.Hubb.		Australia - Queensland
	Homopholis belsonii C.E.Hubb.		Australia - New South Wales, Queensland
	Hubbardia heptaneuron Bor		India - Karnataka State
	Paspalum amphicarpum Ekman		Cuba
	Paspalum jimenezii Chase		Costa Rica
	Plectrachne bromoides (F. Muell.) C.E. Hubb.		Australia - Western Australia
	Poa manii Munroe ex Hillebrand	Mann's bluegrass	United States - Hawaii
	Poa mannii Munro		United States - Hawaii
	Streptochaeta angustifolia Soderstrom		Brazil
	Sucrea sampaiana (A. Hitch.) Soderstrom		Brazil
	Trisetum burnoufii Req. ex Parl.		France - Corsica
	Zea mays L. ssp. *mexicana* (Schrad.) Wilkes raza durango		Mexico
Hydatellaceae			
	Hydatella australis Diels		Australia - Western Australia
	Hydatella leptogyne Diels		Australia - Western Australia
Hydrocharitaceae			
	Elodea linearis H. St. John	Nashville waterweed	United States - Tennessee
	Elodea schweinitzii (Planchon) Casper	Schweinitz's waterweed	United States - Pennsylvania
Iridaceae			
	Gladiolus alatus L. var. *algoensis* Herb.		South Africa - Cape Province
	Hesperantha saldanhae P. Goldblatt		South Africa - Cape Province
	Iris antilibanotica Dinsm.		Syria
	Iris damascena Mont.		Syria
	Iris westii Dinsm.		Lebanon
	Moraea incurva Lewis		South Africa - Cape Province
	Romulea papyracea W. Dod		South Africa - Cape Province
	Romulea sulphurea Beguinot		South Africa - Cape Province
	Sisyrinchium farwellii Bickn.	Farwell's blue-eyed-grass	United States - Michigan
	Sisyrinchium hastile Bickn.	spear-like blue-eyed-grass	United States - Michigan
Juncaceae			
	Juncus griscomii	Griscom's rush	United States - Virginia
	Juncus oronensis Fern.	Maine rush	United States - Maine
	Juncus pervetus Fern.	Barnstable bog rush; old veteran rush	United States - Massachusetts
Liliaceae			
	Allium rouyi Gaut.		Spain
	Calochortus indecorus Ownbey & M. Peck	Sexton Mt mariposa-lily	United States - Oregon
	Calochortus monanthus Ownbey	Shasta River mariposa; single-flowered mariposa lily	United States - California
	Dipcadi concanense (Dalz.) Baker		India
	Dipcadi reidii Deb & Dasgupta		India
	Ipheion tweedianum (Griseb.) Traub		Argentina
	Lachenalia mathewsii Barker		South Africa - Cape Province
	Smilax leptanthera Pennell	catbrier	United States - Georgia
	Tulipa sprengeri Baker		Turkey
	Urginea duthiae Adamson		South Africa - Cape Province
	Urginea ecklonii Baker		South Africa - Cape Province

Table 16.5 Extinct higher plant taxa*

MAJOR GROUP (DIVISION) / FAMILY / TAXON	COMMON NAME	HISTORIC RANGE
Urginea polyphylla Hook. f.		India
Orchidaceae		
Acrolophia ustulata Schlecther & Bolus		South Africa - Cape Province
Caladenia atkinsonii Rodway		Australia - Tasmania
Caladenia pumila R. Rogers		Australia - Victoria
Calanthe whiteana King & Pantl.		India - Sikkim State
Corycium vestitum Sweet		South Africa - Cape Province
Diuris fastidiosa R. Rogers		Australia - Victoria
Paphiopedilum delenatii Guillaumin		Vietnam
Pleione lagenaria Lindley		India - Meghalaya State
Prasophyllum colemaniae R. Rogers		Australia - Victoria
Prasophyllum subbisectum Nicholls		Australia - Victoria
Satyrium guthriei Bolus		South Africa - Cape Province
Triphora latifolia G. Luer	nodding cape	United States - Florida
Zeuxine boninensis Tuy		Japan - Ogasawara-Shoto
Palmae		
Acrocomia subinermis Leon ex L.H. Bailey		Cuba
Corypha taliera Roxb.		India
Paschalococos disperta Dransfield		Chile - Easter Island
Pritchardiopsis jennencyi Becc.		France - New Caledonia
Pandanaceae		
Pandanus barklyi Balf. F. var. *macrocarpus* Vaughan & Wiehe		Mauritius
Pandanus conglomeratus Balf. f.		Mauritius
Pandanus iceryi Horne ex Balf. f.		Mauritius
Pandanus incertus Vaughan & Wiehe		Mauritius
Pandanus macrostigma Martelli		Mauritius
Pandanus obsoletus Vaughan & Wiehe		Mauritius
Pandanus spathulatus Martelle		Mauritius
Restionaceae		
Elegia extensa Pillans		South Africa - Cape Province
Elegia fastigiata Mast.		South Africa - Cape Province
Leptocarpus ramosissimus Pillans		South Africa - Cape Province
Lepyrodia heleocharoides Gilg		Australia - Western Australia
Restio chaunocoleus F. Muell		Australia - Western Australia
Tecophilaeaceae		
Tecophilaea cyanocrocus Leybold		Chile
Zingiberaceae		
Hedychium marginatum C.B. Clarke		India - Nagaland State

Notes: This list represents information available to WCMC in computerised form as of March 1992. It is intended to include species that are extinct (or presumed extinct) in the wild, whether or not they are in cultivation. Several of these plants, such as *Franklinia alatamaha*, *Paphiopedilum delenatii* and *Tecophilaea cyanocrocus*, are, in fact, well known in the horticultural trade. Others, such as *Encephalartos woodii*, are known only from relatively few specimens, mostly held in botanic gardens. A few others have become extinct in the wild but have been reintroduced from cultivated material grown in botanic gardens.

The information available is strongly biased geographically: many other species of higher plants have undoubtedly become extinct but lack of country-based data prevents their inclusion here. Some of the species in this list are almost certainly still extant in remote, isolated areas; publication of this list should stimulate searching for them.

* Includes some taxa below species level.

Notes for Table 16.6, overleaf: (1) Two amphibians (USA and Israel), one coral (Panama) and one mammal (Caribbean) are not included in this table. (2) The above species may have lived in more than one country therefore total numbers do not necessarily agree with other tables. * indicates islands which are not on the standard country list. They have been included separately because of the importance of islands when considering extinctions.

Table 16.6 Known animal extinctions since c. 1600 by country

	MOLLUSCS	INSECTS	FISHES	REPTILES	BIRDS	MAMMALS	TOTAL
ASIA							
India					3		3
Indonesia					2		2
Nansei–shoto (Japan)*					2		2
Nepal					1		1
Ogasawara–shoto (Japan)*					3		3
Philippines					1	3	4
Taiwan		1					1
Thailand						1	1
'USSR'					1		1
Bering Straits ('USSR')*					1	1	2
EUROPE							
Austria	1						1
Canary Islands (Spain)*					1		1
Corsica (France)*						1	1
Faeroe Islands					1		1
Germany		1				1	2
Iceland					1		1
Sardinia (Italy)*						1	1
United Kingdom					1		1
Yugoslavia	1						1
NORTH & CENTRAL AMERICA							
Bahamas					1		1
Barbados				1		1	2
Canada			2		2		4
Cayman Is					1	3	4
Cuba					1	4	5
Dominican Rep						6	6
Greenland					1		1
Guadalupe (Mexico)*					2		2
Guadeloupe	1			1			2
Guatemala					1		1
Haiti						6	6
Jamaica				2	1		3
Little Swan Island (Honduras)*						1	1
Martinique				3		1	4
Mexico			12		1	2	15
Navassa Island (USA)*				1			1
Saint Lucia						1	1
Saint Vincent and the Grenadines						1	1
United States	38	9	17		4		68
Virgin Islands (US)				1			1
SOUTH AMERICA							
Brazil					1		1
Colombia					1		1
Galapagos (Ecuador)*	1					4	5
Uruguay					1		1
OCEANIA							
American Samoa	2						2
Australia				1		17	18
Bougainville (PNG)*					1		1
Chatham Island (NZ)*		1			3		4
Christmas Island (Australia)*						2	2
Cook Islands	14				1		15
Fiji		1					1
French Polynesia	33				5		38
Guam	1				1	1	3
Hawaii (USA)*	29	42			15		86
Kangaroo Island (Australia)*					1		1
Lord Howe Island (Australia)*		1			2		3
Micronesia, Federated States of					3		3
New Caledonia	2				3		5
New Zealand		1	1	1	10	1	14
Norfolk Island (Australia)*					1		1
Palau						1	1
Phillip Island (Australia)*					1		1
Solomon Islands					1		1
Stephens Island (NZ)*		1			1		2
Tasmania (Australia)*						1	1
Wake Island (USA)*					1		1
Western Samoa					1		1
ANTARCTICA							
Falkland Islands (Malvinas) & dependencies						1	1
AFRICA							
Algeria						1	1
Ascension Island (UK)*					1		1
Cape Verde				1			1
Madagascar					3		3
Madeira (Portugal)*	14						14
Mauritius	23	1		5	11	1	41
Réunion	1			2	11	1	15
Rodrigues (Mauritius)*	6			4	11		21
Saint Helena	22	2					24
Seychelles	2				1		3
South Africa						2	2
Tristan da Cunha (UK)*					1		1

Figure 16.6 Known animal extinctions since c. 1600: Molluscs

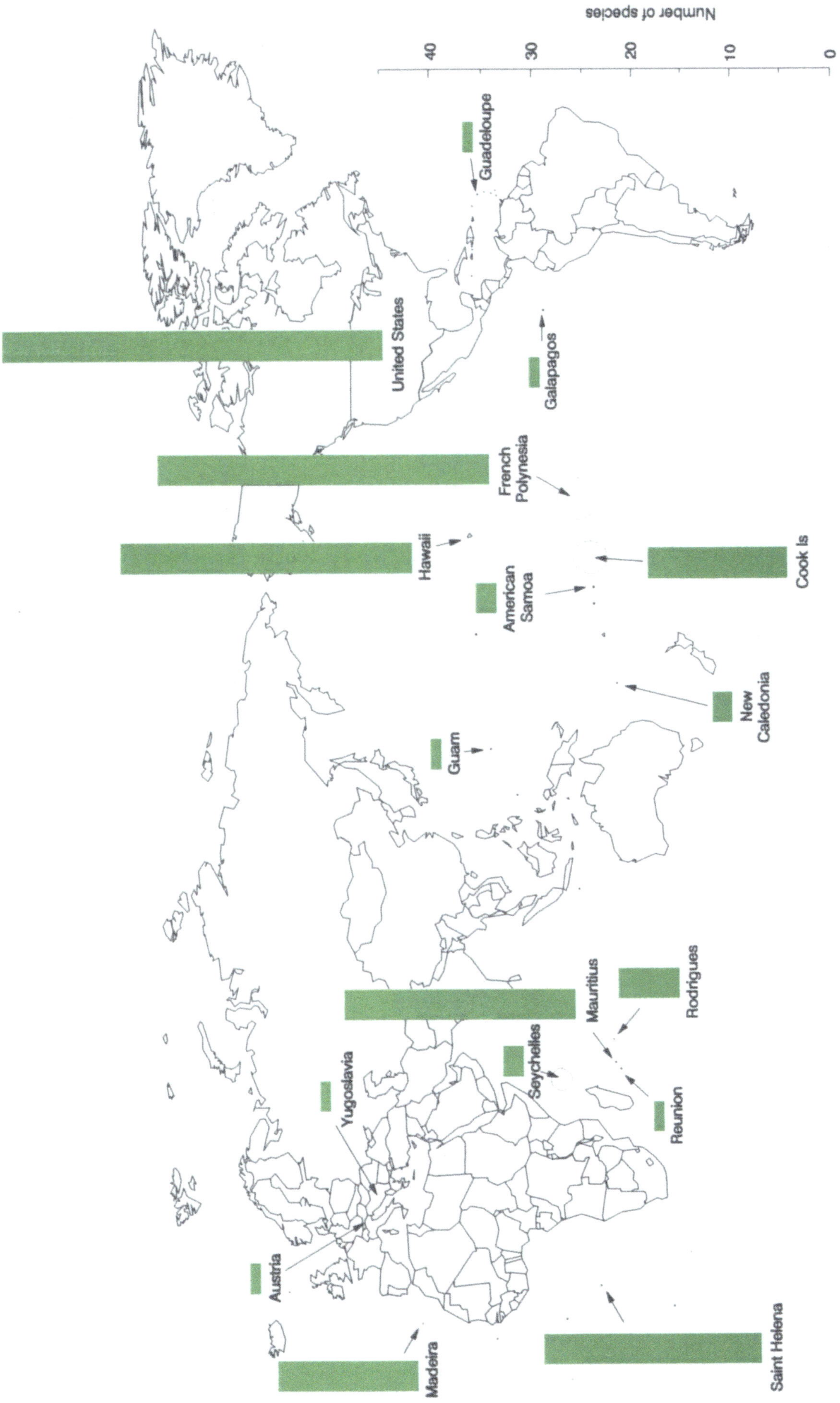

Figure 16.7 Known animal extinctions since c. 1600: Arthropods

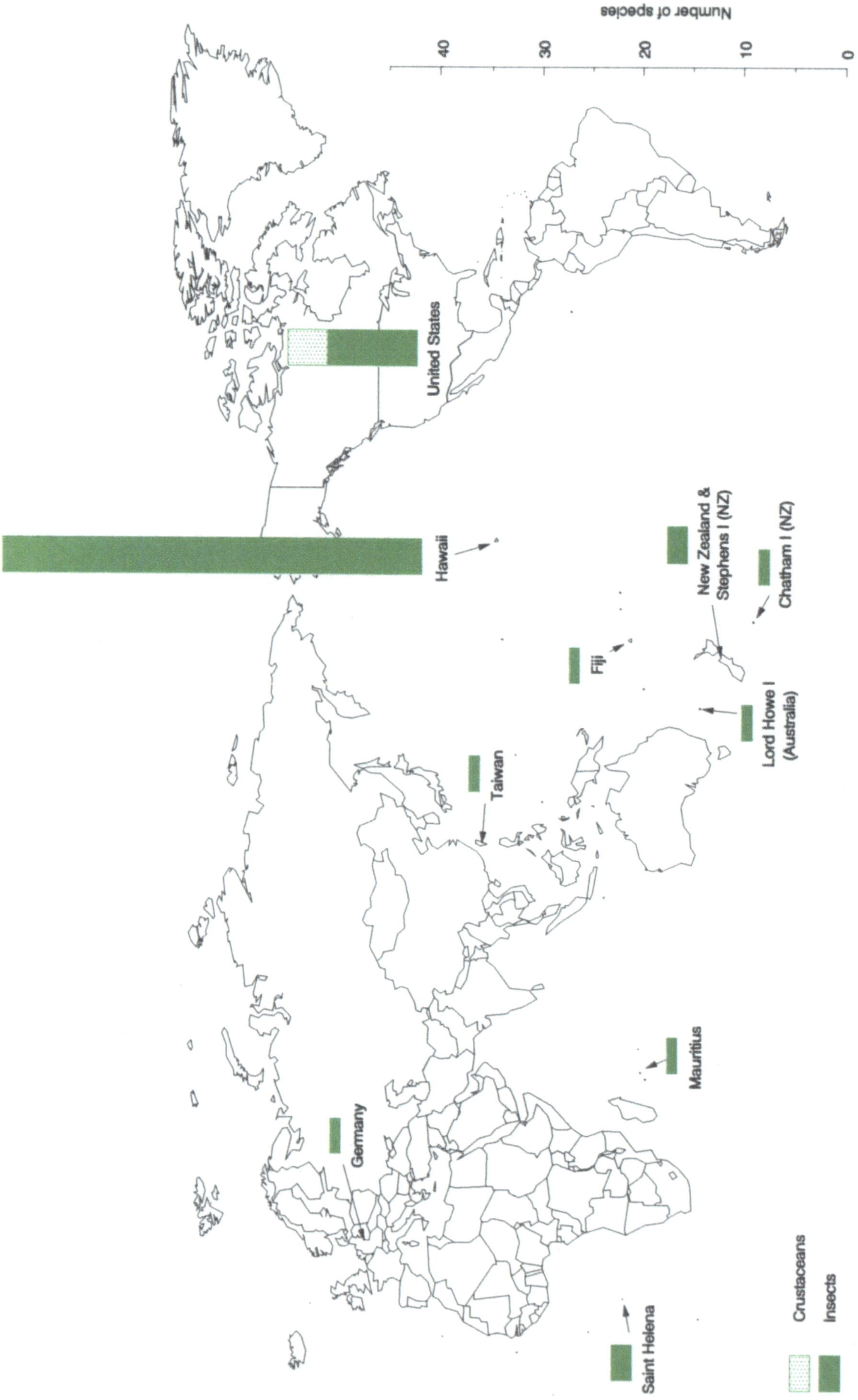

Figure 16.8 Known animal extinctions since c. 1600: Fishes, reptiles and amphibians

Figure 16.9 Known animal extinctions since c. 1600: Birds

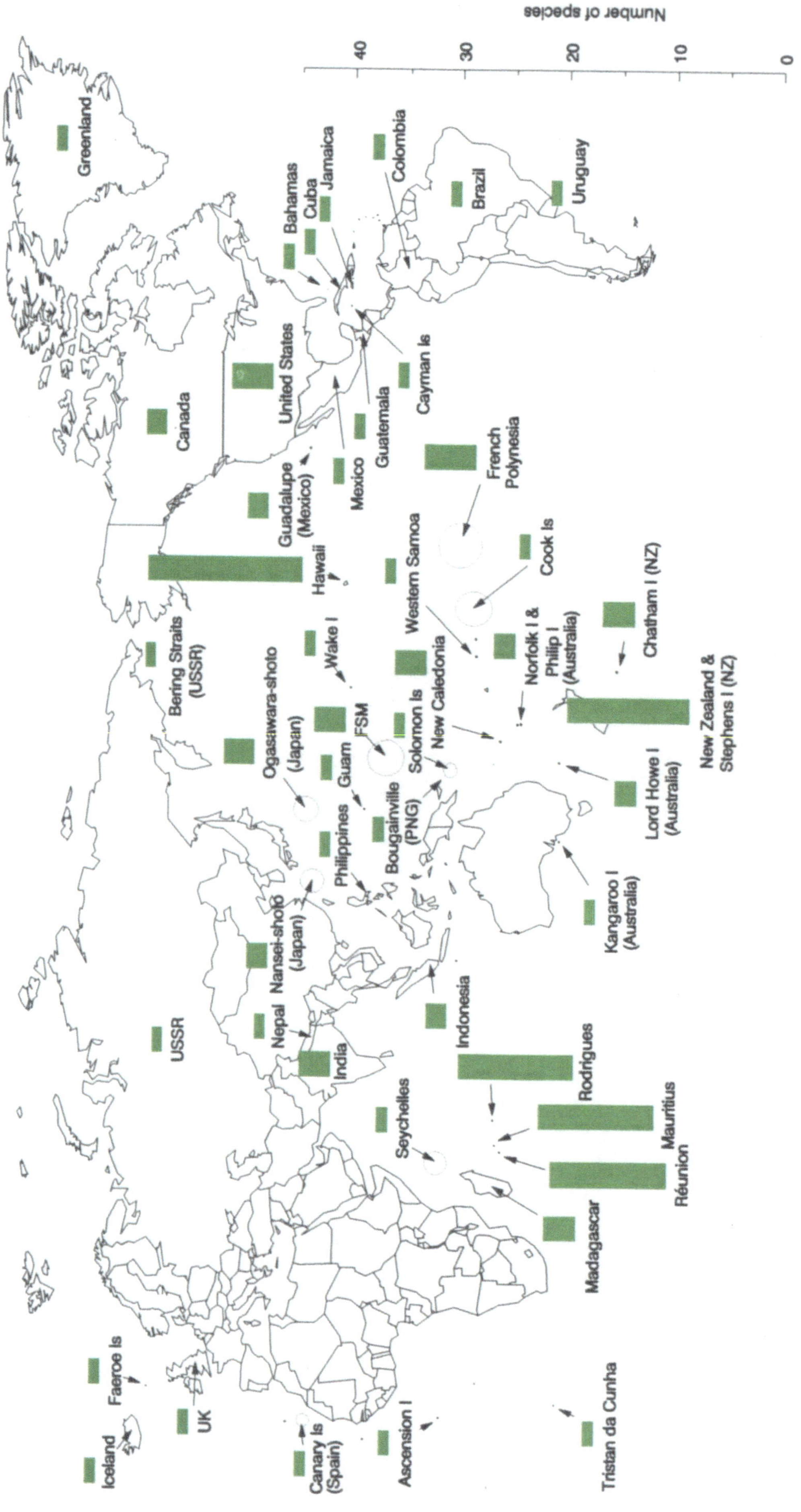

Figure 16.10 Known animal extinctions since c. 1600: Mammals

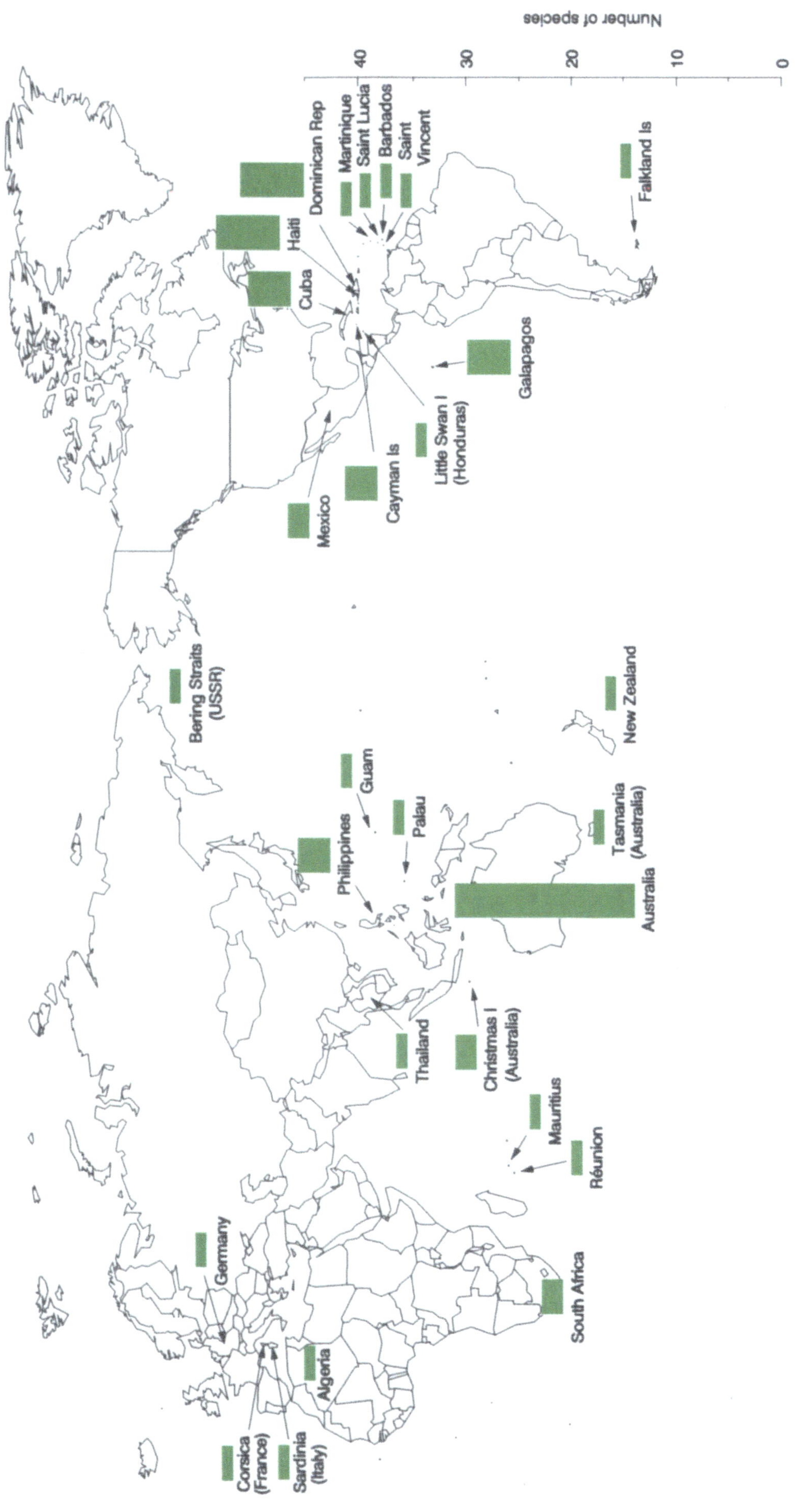

Table 16.7 Animal species surviving only in captivity

SPECIES	ENGLISH NAME	NOTES
MOLLUSCS		
Order STYLOMMATOPHORA		
Family Partulidae		
Partula spp.	Viviparous Tree Snails	French Polynesia. Exterminated in wild after introduction of *Euglandina rosea* in 1977. Various captive colonies around the world. Population status information as at March 1991 *.
Partula affinis		Tahiti. Functionally extinct, only 1 left alive.
Partula aurantia		Moorea. Functionally extinct, only 1 left alive.
Partula clara		Tahiti. Critical.
Partula hyalina		Tahiti. Increasing.
Partula mirabilis		Moorea. Critical.
Partula mooreana		Moorea. Seriously declining.
Partula nodosa		Tahiti. Increasing.
Partula otaheitana		Tahiti. Increasing but low numbers.
Partula suturalis		Moorea. Declining/stable.
Partula taeniata		Moorea. Increasing, good numbers.
Partula tohiveana		Moorea. Increasing but all from 4 individuals.
FISHES		
Order CYPRINODONTIFORMES		
Family Cyprinodontidae **		
Cyprinodon alvarezi		Mexico. Last specimens removed from wild February 1992.
Megupsilon aporus		Mexico. Last specimens removed from wild February 1992; a number of captive populations exist.
Family Poeciliidae		
Xiphophorus couchianus	Monterrey Platyfish	Mexico. Extinct in the wild in 1960s; three captive populations.
Family Goodeidae		
Skiffia francesae	Golden Sawfin	Mexico; widespread in captivity.
BIRDS		
Order CICONIIFOMES		
Family Ciconiidae		
Gymnogyps californianus	Californian Condor	USA. Last individual taken from wild 1987. 52 in captivity at end of 1991.
Family Columbidae		
Zenaida graysoni	Socorro Dove	Socorro I (Mexico). Extinct post-1958. Large captive populations.
MAMMALS		
Order PERISSODACTYLA		
Family Equidae		
Equus ferus	Wild Horse	China, Mongolia. Some disagreement on taxonomic status. *E. ferus gmelini*, the Tarpan, exterminated late 19th century. *E. f. przewalskii*, Przewalski's Horse survives in zoos, last seen in wild in 1968.
Order ARTIODACTYLA		
Family Bovidae		
Bos taurus	Domestic cattle	Europe, North Africa and the Near East. The Aurochs *B. t. primigenius*, the wild ancestor was exterminated in 1627.

Notes: * Reference: Partula '91, Proceedings of the Partula Propagation Group Meeting, 16 May 1991. Compiled by S. Tonge, JWPT. ** Note two further *Cyprinodon* species, Charco Azul and Charco Palma will probably also soon be extirpated in the wild (P. Loiselle, pers. comm.).

Table 16.8 Animal species extirpated in wild and reintroduced

BIRDS		
Order GRUIFORMES		
Family Rallidae		
Rallus owstoni	Guam Rail	Guam (USA). Extinct in wild 1985. Reintroduced 1990/91.
MAMMALS		
Order CARNIVORA		
Family Mustelidae		
Mustela nigripes	Black-footed Ferret	USA. Last specimen taken from wild in 1987. Reintroduced 1990/91.
Family Canidae		
Canis rufus	Red Wolf	USA. Extinct in wild 1980, reintroduced late 1980s.
Order ARTIODACTYLA		
Family Bovidae		
Bison bonasus	Wisent	Europe. Exterminated in wild by 1927. Reintroduced to several locations.
Oryx leucoryx	Arabian Oryx	Middle East. Last recorded in the wild in 1972. Reintroduced in Oman in 1982.
Family Cervidae		
Elaphurus davidianus	Père David's Deer	Discovered in captivity in 1861. Now exists in zoos worldwide. Reintroduced to China.

17. THREATENED SPECIES

A threatened species is one thought to be at significant risk of extinction in the foreseeable future, because of stochastic or deterministic factors affecting its populations, or by virtue of inherent rarity. This convenient working definition is deceptively simple; deciding what level of risk is significant, and what part of the future is foreseeable, is problematic.

WHAT IS A THREATENED SPECIES?

The growth in public awareness of the problem of depletion and possible extinction of species is largely attributable to the development of the *Red Data Book* (RDB) concept by Sir Peter Scott during the 1960s. This involves an attempt to categorise species at risk according to the severity of the threats facing them and the estimated imminence of their extinction. The RDBs were compiled on a global basis by IUCN, so far as available information allowed, but the concept was soon adopted at a national or sub-national level in several countries. Attention also spread from the terrestrial vertebrates, which were the principal focus of early RDBs, to invertebrates and plants.

As the volume of information has increased, the traditional Red Data Book approach, which included publication of a range of data on each threatened species, has been to some extent replaced by a direct listing of globally-threatened species recognised by IUCN. The IUCN Red List of Threatened Animals (IUCN, 1990, latest edition) is the only accepted worldwide attempt to list threatened animal species individually, and has provided the basis for the discussion below.

The animals Red List has been compiled every two years since 1986 by the World Conservation Monitoring Centre, in collaboration with the IUCN Species Survival Commission network of Specialist Groups. The Red List is based on information provided by numerous scientists, naturalists and conservationists working in the field, much of it collated by the IUCN SSC Specialist Groups. The categorisation of threatened bird species is undertaken by the International Council for Bird Preservation (ICBP).

Each species covered in the Red List is assigned a threat category determined by a review of the factors affecting it and the extent of the effect that these are having throughout its range. Key factors examined include changes in distribution or numbers, degree and type of threat, and population biology. IUCN Red List categories are applied to species on an international or global scale, and should not be confused with the national threat categories assigned to species by countries which have prepared Red Lists or Red Data Books dealing with the status of species within their own borders.

It is important to note that although the IUCN Red List is a comprehensive global compendium of animal species *known* to be threatened, many more species than those listed will in fact be threatened. Those not listed fall into two categories: first, and probably the largest number of species, are those not yet described by science; and second, the status of many described species has not been reviewed. Birds have been comprehensively reviewed by ICBP; only 50% of mammal species, and probably less than 20% of reptiles, 10% of amphibians and 5% of fish are estimated to have been reviewed.

IUCN threat categories

The main IUCN threat categories currently used, together with their definitions (as used in the Red Lists) are:

Extinct (Ex)
Species not definitely located in the wild during the past 50 years. On a few occasions, the category Ex? has been assigned, denoting that it is virtually certain that the taxon has recently become extinct.

Endangered (E)
Taxa in danger of extinction and whose survival is unlikely if the causal factors continue to operate. Included are taxa whose numbers have been reduced to a critical level or whose habitats have been so drastically reduced that they are deemed to be in immediate danger of extinction. Also included are taxa that may now be extinct although they have been seen in the wild in the past 50 years.

Vulnerable (V)
Taxa believed likely to move into the Endangered category in the near future if the causal factors continue operating. Included are taxa of which most or all the populations are decreasing because of over-exploitation, extensive destruction of habitat or other environmental disturbance; taxa with populations that have been seriously depleted and whose ultimate security has not been assured; and taxa with populations that are still abundant but are under threat from severe adverse factors throughout their range.

Rare (R)
Taxa with small world populations that are not at present Endangered or Vulnerable but are at risk. These taxa are usually localised within restricted geographical areas or habitats or are thinly scattered over a more extensive range.

Indeterminate (I)
Taxa *known* to be Endangered, Vulnerable or Rare but where there is not enough information to say which of the three categories is appropriate.

Insufficiently Known (K)
Taxa that are *suspected* but not definitely known to belong to any of the above categories, because of lack of information.

The general term *threatened* is used to refer to a species considered to belong to any one of the above categories. The same definitions have been applied to plants, although they have often been interpreted in a significantly different manner, mainly because of biological differences between animals and plants, and intermediate categories (e.g. Ex/E or E/R) are also employed.

The definition and application of such status categories has been a matter of some discussion, principally because they provide such an important tool in assessing needs and mobilising resources for conservation at the international, national or sub-national level. In the opinion of many scientists, the existing IUCN threat category definitions are excessively subjective, and as a result categorisations made by different authorities can vary and may not accurately reflect real extinction risks. Mace and Lande (1991) have recently proposed a new system based on quantitative (and therefore theoretically objective) Population Viability Analysis techniques.

The threats

Most of the causal factors currently threatening species are anthropogenic in nature, i.e. induced or influenced by man. These factors include:

- Habitat loss or modification, often associated with habitat fragmentation. Causes include pastoral development, cultivation and settlement, forestry operations and plantations, fire, and pollution
- Over-exploitation for commercial or subsistence reasons, including meat, fur, hides, collection of live animals for the pet trade and plants for the horticultural trade
- Accidental or deliberate introduction of exotic species, which may compete with, prey on or hybridise with native species
- Disturbance, persecution and uprooting, including deliberate eradication of species considered to be pests
- Incidental take, particularly the drowning of aquatic reptiles and mammals in fishing nets
- Disease, both exotic and endemic, exacerbated by the presence of large numbers of domestic livestock or introduced plant species
- Limited distribution, which may compound the effects of other factors.

In the majority of cases individual species are faced by several of these threats operating simultaneously, and it is often difficult or impossible to identify with confidence the primary cause of decline.

Some understanding of the relative importance of different threat types, as measured by frequency of occurrence, can be gained from an examination of threats facing the mammals (excluding Cetacea) of Australasia and the Americas (comprehensively reviewed by Thornback and Jenkins, 1982), and those facing the birds of the world (Diamond, 1987).

Of the 119 species of mammals from these continents considered threatened, 75% (94) are threatened by more than one factor, and of these, 27 face four or more threats.

The major category of threat, which affects 76% of species, is habitat loss and modification (Fig. 17.1). This has a variety of causes, of which the most frequent is cultivation and settlement. Over-exploitation affects half the species, the most significant cause being hunting for meat. Introduced predators and competitors affect 18% of threatened species. The most serious other factor is limited distribution, which affects one quarter of species.

Fig. 17.2 compares the major threats affecting the birds of the world with those affecting the mammals of Australasia and the Americas. There is a high degree of similarity between the two groups. Habitat destruction is the single most important threat, affecting 60% of birds and 76% of

Figure 17.1 Analysis of threats: mammals

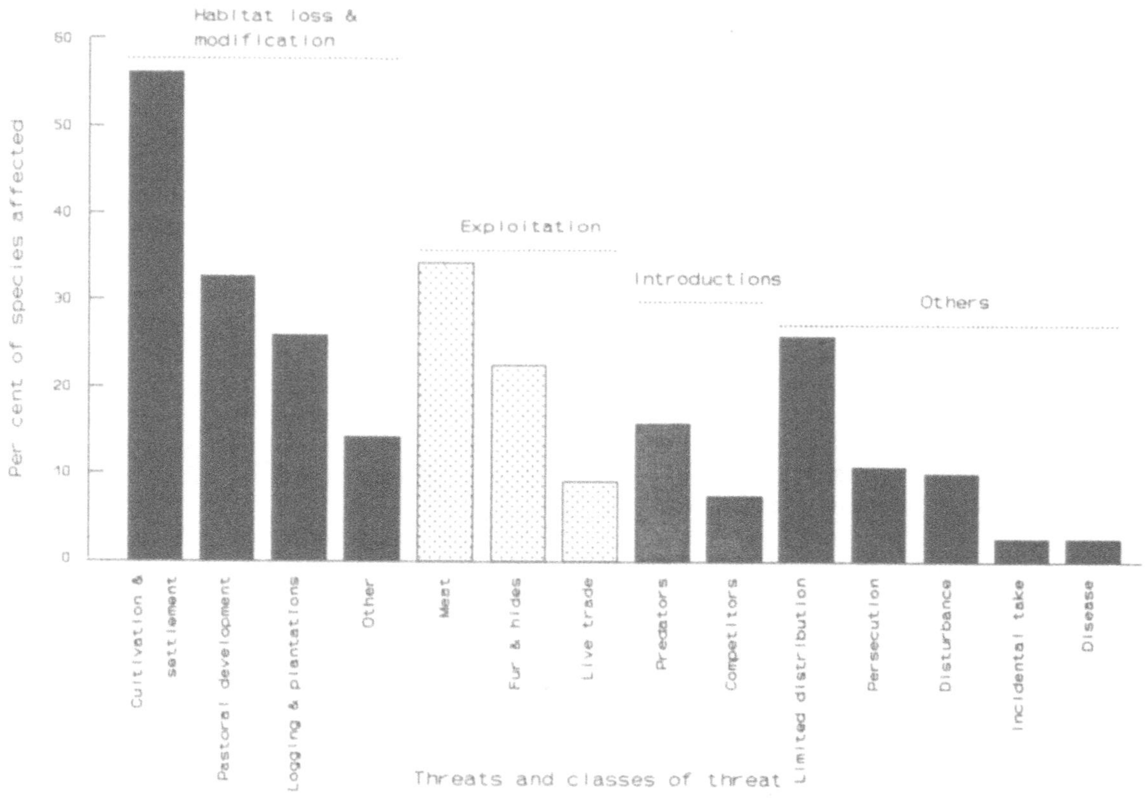

Figure 17.2 Analysis of threats: mammals and birds

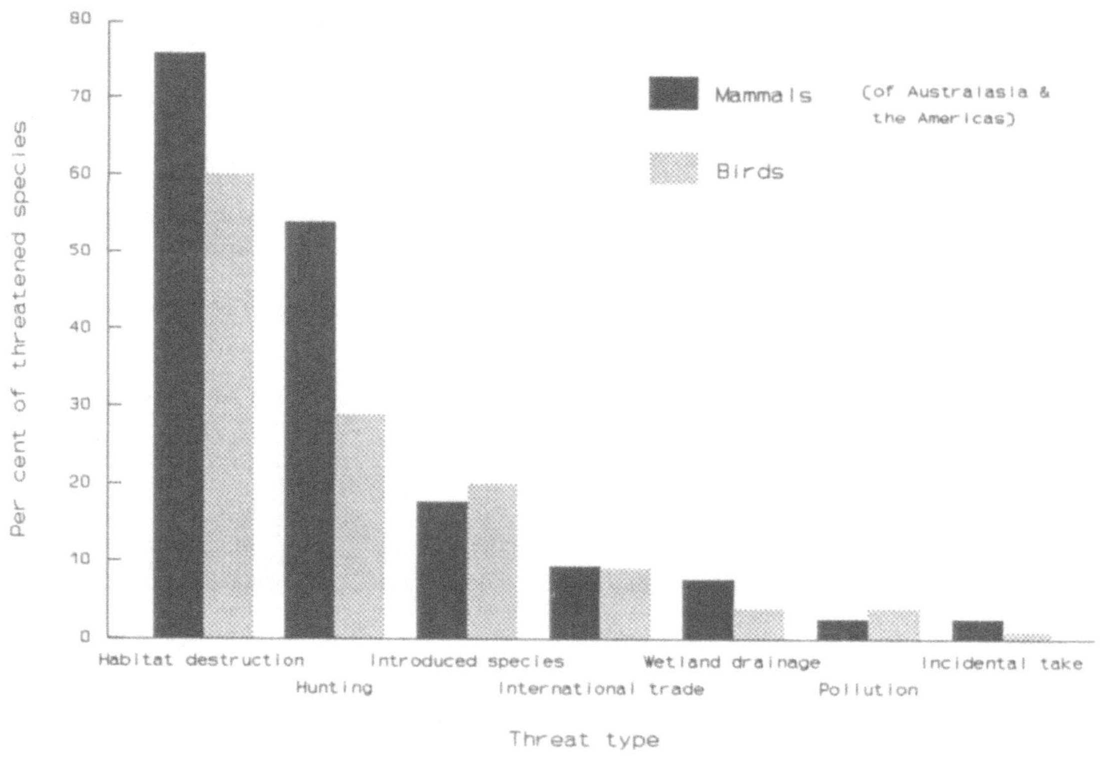

mammals. A major difference is that almost double the number of mammals as birds are threatened by hunting (54% versus 29%).

GLOBALLY THREATENED ANIMALS

Taxonomic distribution of threatened animals

The term 'threatened' in the following discussion refers to taxa assigned a relevant status category by IUCN. In all, some 4,452 animal species are listed as threatened in the 1990 Red List, or much less than 0.5% of the world's estimated total of well over 1.5 million described animal species (Tables 17.1 and 17.2). Some species are also listed in part only, i.e. one or more subspecies are included in the Red List, but only full species are considered here.

The two classes with the greatest number of threatened species are birds with 1,029 and insects with 1,083. Other major listings include 507 mammals, 169 reptiles, 57 amphibians, 713 fish, 409 molluscs, 154 corals and sponges, 139 annelid worms and 126 crustaceans. Clearly, the number of threatened species in a taxonomic group is not directly proportional to the overall number of species in that group: some groups, particularly vertebrates, have higher proportions listed as threatened than other groups.

The four major groups with the highest percentage of threatened species are mammals (11.7% threatened), birds (10.6%), fish (3.6%) and reptiles (3.5%). In comparison, although a large number (1,083) of insects is listed, this represents less than 0.15% of the world's total. This dichotomy between vertebrates and invertebrates becomes even more extreme when Endangered species, the most severely threatened category, are examined. Each of the five vertebrate groupings have a higher percentage of listed Endangered species than all of the invertebrate taxa added together (Fig. 17.3).

Considering only the mammals among vertebrates, several smaller orders have a very high proportion of threatened species (Proboscidea with two out of two species, Sirenia with four out of four species and Perissodactyla with 12 out of 16 species). Among the larger orders, Primates, Carnivora and Artiodactyla are the most threatened, with respectively 53%, 32% and 31% of their constituent species listed. Although these three orders combined only contain some 14.6% of the world's mammal species, they account for just under half of the listed threatened species and just over half of the Endangered species.

To some extent, vertebrates may be more vulnerable to extinction than invertebrates because they are typically much larger and therefore require more resources and larger ranges. On the other hand, many invertebrates have an extremely small range, which would render them liable to extinction by habitat loss. It seems reasonable to conclude that the proportion of species in a group listed as threatened reflects popular and scientific attention in addition to biological reality.

Geopolitical distribution of threatened animals

Table 17.3 shows the geopolitical distribution of threatened animal species according to the IUCN Red List (1990) together with threatened plants; Table 17.4 shows a subset of the animal data, with the countries listed in descending order according to the number of threatened species in each higher grouping. The top ten countries are listed for each taxon.

Table 17.1 IUCN Threatened Vertebrates (1990 Red List)

CLASS / ORDER	NUMBER OF SPECIES THREATENED	NUMBER OF SPECIES ENDANGERED	APPROXIMATE TOTAL OF DESCRIBED SPECIES
MAMMALS	507	140	4,327
Monotremata	1	0	3
Marsupialia	25	6	282
Xenarthra	6	1	29
Insectivora	79	3	365
Scandentia	0	0	16
Dermoptera	0	0	2
Chiroptera	45	11	977
Primates	106	47	201
Pholidota	0	0	7
Lagomorpha	9	6	65
Macroscelidia	2	0	15
Rodentia	54	15	1,793
Cetacea	21	6	77
Carnivora	76	12	235
Pinnipedia	4	2	34
Sirenia	4	0	5
Proboscidea	2	1	2
Perissodactyla	12	7	16
Hyracoidea	1	0	8
Tubulidentata	0	0	1
Artiodactyla	60	23	194
BIRDS	1,029	132	9,672
Struthioniformes	1	0	10
Tinamiformes	8	0	47
Sphenisciformes	3	0	17?
Podicipediformes	4	2	21?
Procellariformes	25	4	115?
Pelecaniformes	8	3	9?
Ciconiiformes	21	8	19?
Anseriformes	20	3	168
Falconiformes	45	6	311?
Galliformes	68	11	214
Gruiformes	51	9	196
Charadriiformes	31	4	350?
Columbiformes	49	6	313
Psittaciformes	78	16	358
Cuculiformes	11	2	143
Strigiformes	20	1	178?
Caprimulgiformes	11	0	105?
Apodiformes	39	3	103
Trogoniformes	3	0	39
Coraciiformes	20	0	152
Piciformes	14	4	355
Passeriformes	499	50	5,712
REPTILES	169	38	4,771
Testudines	78	11	
Rhynchocephalia	1	0	2?

Table 17.1 IUCN Threatened Vertebrates (1990 Red List)

CLASS / ORDER	NUMBER OF SPECIES THREATENED	NUMBER OF SPECIES ENDANGERED	APPROXIMATE TOTAL OF DESCRIBED SPECIES
REPTILES (continued)			
Sauria	43	9	2,000
Serpentes	33	7	2,500
Crocodylia	15	11	
AMPHIBIANS	57	8	4,014
Caudata	25	2	
Anura	32	6	
FISHES	713	368	20,000
LAMPREYS	3	0	
SHARKS, etc.	3	0	
BONY FISH	707	368	
TOTAL VERTEBRATES	2,475	686	42,784

Sources: World species totals for groups of animals are derived from the following sources - mammals: Corbet, G.B. and Hill, J.E. 1991. *A World List of Mammalian Species*. 3rd edn. Natural History Museum, London and Oxford Univerity Press, Oxford; birds: Sibley, C.G. and Monroe, B. L. 1990. *Distribution and Taxonomy of Birds of the World*. Yale University Press, New Haven & London; reptiles, amphibians and fishes: various sources.
Note: Table only includes groups of animals of which one or more species are listed as threatened, with the exception of mammals for which all orders are included. Species categorised as Extinct are not included, those as Extinct? are.

Table 17.2 IUCN Threatened Invertebrates (1990 Red List)

PHYLUM	CLASS	NO. OF SPECIES THREATENED	NO. OF SPECIES ENDANGERED	APPROXIMATE TOTAL OF DESCRIBED SPECIES
CILIOPHORA		1	0	?
CNIDARIA		154	0	9,000
PLATYHELMINTHES	TURBELLARIA	4	2	12,700
NEMERTEA		10	0	650
MOLLUSCA		409	85	50,000
ANNELIDA		139	2	8,700
ARTHROPODA	INSECTA	1,083	56	750,000
	MEROSTOMATA	4	0	4
	ARACHNIDA	18	1	68,000
	CRUSTACEA	126	3	42,000
ONCHYOPHORA		27	0	65
ECHINODERMATA		2	0	6,000
TOTAL INVERTEBRATES		1,977	149	947,119

Sources: various.
Notes: Table only includes groups of animals of which one or more species are listed as threatened.
Species categorised as Extinct are not included, those as Extinct? are.

Table 17.3 Country totals of threatened plants and vertebrates

	PLANTS	MAMMALS	BIRDS	REPTILES	AMPHIBIANS	FISH
ASIA	6608	497	918	146	9	124
Afghanistan	4	13	13	1	1	0
Bahrain	0	1	4	0	0	1
Bangladesh	33	15	27	14	0	0
Bhutan	15	15	10	1	0	0
British Indian Ocean Territory	0	0	0	0	0	0
Brunei	40	9	10	3	0	2
Cambodia	11	21	13	6	0	5
China	350	40	83	7	1	7
Cyprus	43	1	17	1	0	0
Hong Kong	5	1	9	2	0	0
India	1336	39	72	17	3	2
Indonesia	70	49	135	13	0	29
Iran, Islamic Rep	301	15	20	4	0	2
Iraq	1	9	17	0	0	2
Israel	3	8	15	1	1	0
Japan	41	5	31	0	1	3
Jordan	752	5	11	0	0	0
Korea, Dem People's Rep	0	5	25	0	0	0
Korea, Rep	33	6	22	0	0	0
Kuwait	1	5	7	0	0	0
Laos	3	23	18	5	0	5
Lebanon	5	4	15	1	0	0
Malaysia	522	23	35	12	0	6
Maldives	0	1	1	0	0	0
Mongolia	0	9	13	0	0	0
Myanmar		23	42	10	0	2
Nepal	33	22	20	9	0	0
Oman	2	6	8	0	0	2
Pakistan	14	15	25	6	0	0
Philippines	159	12	39	6	0	21
Qatar	0	0	3	0	0	0
Saudi Arabia	2	9	12	0	0	0
Singapore	19	4	5	1	0	1
Sri Lanka	220	7	8	3	0	12
Syria	11	4	15	1	0	0
Taiwan	95	4	16	0	0	0
Thailand	68	26	34	9	0	13
Turkey	1944	5	18	5	1	5
United Arab Emirates	0	4	7	0	0	0
Viet Nam	338	28	34	8	1	4
Yemen	134	6	9	0	0	0
'USSR'		20	38	3	0	5
EUROPE	2677 **	66	396	16	15	48
Albania	76	2	14	1	0	1
Andorra	0	0	1	0	0	0
Austria	25	2	13	0	0	2
Belgium	9	2	13	0	0	1
Bulgaria	88	3	15	1	0	3
Czechoslovakia	29	2	18	0	0	2
Denmark	7	1	16	0	0	0
Faeroe Islands	0	0	2	0	0	0
Finland	11	3	12	0	0	1
France	143	6	21	2	1	3
Germany	**	2	17	0	0	3
Greece	526	4	19	3	0	6
Hungary	21	2	16	0	0	2
Iceland	2	1	2	0	0	1
Ireland	4	0	10	0	0	1
Italy	210	3	19	2	7	3
Liechtenstein	0	0	3	0	0	0
Luxembourg	1	1	8	0	0	0
Malta	4	0	13	0	0	0
Monaco	0	0	0	0	0	0
Netherlands	7	2	13	0	0	1
Norway	13	3	8	0	0	1
Poland	16	4	16	0	0	1
Portugal	240	6	18	0	1	0
Romania	67	2	18	1	0	4
San Marino	0	0	0	0	0	0
Spain	936	6	23	5	3	2
Sweden	10	1	14	0	0	1
Switzerland	18	2	15	0	1	3
United Kingdom	24 [1]	3	22	0	0	1
Vatican City	0	0	0	0	0	0
Yugoslavia	190	3	17	1	2	5
NORTH AND CENTRAL AMERICA	5747	145	219	88	27	277
Anguilla		0	0	0	0	0
Antigua and Barbuda	1	0	2	0	0	0
Aruba	0	0	0	0	0	0
Bahamas	24	2	4	3	0	0
Barbados	1	1	1	0	0	0
Belize	36	8	4	3	0	0
Bermuda	11	0	2	0	0	0
Canada	12	5	6	0	0	15
Cayman Islands	0	0	2	2	0	0
Costa Rica	419	10	14	2	0	0

Table 17.3 Country totals of threatened plants and vertebrates (continued)

	PLANTS	MAMMALS	BIRDS	REPTILES	AMPHIBIANS	FISH
NORTH AND CENTRAL AMERICA (continued)						
Cuba	860	11	15	4	0	0
Dominica	62	0	3	0	0	0
Dominican Republic	50	1	5	4	0	0
El Salvador	26	6	2	1	0	0
Greenland (Denmark)	0	2	1	0	0	0
Grenada	4	0	2	0	0	0
Guadeloupe	14	0	1	0	0	0
Guatemala	282	10	10	4	0	0
Haiti	13	1	4	4	0	0
Honduras	43	7	11	3	0	0
Jamaica	10	5	2	3	0	0
Martinique	12	0	3	0	0	0
Mexico	883	25	35	16	4	98
Montserrat	1	0	1	0	0	0
Netherlands Antilles	0	0	3	2	0	0
Nicaragua	68	8	7	2	0	0
Panama	549	13	14	2	0	0
Puerto Rico	84	2	4	5	1	0
St Lucia	3	0	5	0	0	0
St Vincent and the Grenadines		0	3	0	0	0
St Kitts and Nevis	0	0	1	0	0	0
Trinidad and Tobago	5	1	3	0	0	0
Turks and Caicos Islands	1	0	0	1	0	0
United States	2262	27	43	25	22	164
Virgin Islands (British)	1	0	3	1	0	0
Virgin Islands (US)	10	0	3	1	0	0
SOUTH AMERICA	**2061**	**239**	**535**	**58**	**2**	**14**
Argentina	159	23	53	4	1	1
Bolivia	39	21	34	4	0	1
Brazil	318	40	123	11	0	9
Chile	284	9	18	0	0	1
Colombia	327	25	69	10	0	0
Ecuador{a}	256	21	64	8	0	0
French Guiana	47	10	5	2	0	0
Guyana	68	12	9	3	0	1
Paraguay	15	14	34	4	0	0
Peru	360	29	75	6	1	1
Suriname	68	11	6	1	0	0
Uruguay	14	5	11	2	0	0
Venezuela	106	19	34	3	0	0
OCEANIA	**2673**	**60**	**168**	**21**	**7**	**18**
American Samoa		1	1	0	0	0
Australia	2024	38	39	9	3	16
Cook Islands	0	0	1	0	0	0
Fiji	25 [2]	1	5	4	1	0
French Polynesia	65	0	20	0	0	0
Guam	12	2	4	0	0	0
Kiribati	0	0	2	0	0	0
Marshall Islands	0	0	1	0	0	0
Micronesia, Federated States of	0	5	3	1	0	0
Nauru	0	0	2	0	0	0
New Caledonia	168	1	5	0	0	0
New Zealand	232	1	26	1	3	2
Niue	0	0	0	0	0	0
North Marianas Islands	8	1	2	0	0	0
Palau	0	1	3	0	0	0
Papua New Guinea	88	5	25	1	0	0
Pitcairn Island	3	0	1	0	0	0
Solomon Islands	28	2	20	3	0	0
Tokelau	0	0	0	0	0	0
Tonga	0	0	2	1	0	0
Tuvalu	0	0	1	0	0	0
Vanuatu	8	1	3	1	0	0
Wallis and Futuna Islands		0	0	0	0	0
Western Samoa	12	1	2	0	0	0
ANTARCTICA	**4**	**0**	**0**	**0**	**0**	**0**
Antarctica	0	0	0	0	0	0
Falkland Islands (Malvinas)	4	0	0	0	0	0
French Southern Territories	0	0	0	0	0	0
AFRICA	**3308**	**688**	**453**	**89**	**8**	**49**
Algeria	145	12	15	0	0	1
Angola	19	14	12	2	0	0
Benin	3	11	1	2	0	0
Botswana	4	9	6	1	0	0
Burkina Faso	0	10	1	2	0	0
Burundi	0	4	5	1	0	0
Cameroon	74	27	17	2	1	11
Cape Verde	1	0	3	1	0	0
Central African Rep	0	12	2	2	0	0
Chad	14	18	4	2	0	0
Comoros	3	3	5	0	0	1
Congo	4	12	3	2	0	0
Cote d'Ivoire	70	18	9	1	1	0
Djibouti	3	6	3	0	0	0
Egypt	91	9	16	2	0	1

Table 17.3 Country totals of threatened plants and vertebrates (continued)

	PLANTS	MAMMALS	BIRDS	REPTILES	AMPHIBIANS	FISH
AFRICA (continued)						
Equatorial Guinea	8	15	3	2	1	0
Ethiopia	44	25	14	1	0	0
Gabon	80	17	4	2	0	0
Gambia	0	7	1	2	0	0
Ghana	34	13	8	2	0	0
Guinea	36	17	6	1	1	0
Guinea-Bissau	0	5	2	2	0	0
Kenya	144	17	18	2	0	0
Lesotho	7	2	7	0	0	0
Liberia	1	18	10	2	0	0
Libya	58	12	9	1	0	0
Madagascar	194	50	28	10	0	0
Malawi	61	10	7	1	0	0
Mali	15	16	4	2	0	0
Mauritania	3	14	5	1	0	0
Mauritius	269	3	10	6	0	0
Mayotte		0	1	0	0	0
Morocco	194	9	14	0	0	1
Mozambique, People's Rep	89	10	11	1	0	1
Namibia	17	11	7	2	0	4
Niger	1	15	1	1	0	0
Nigeria	9	25	10	2	0	0
Reunion	96	0	1	0	0	0
Rwanda	0	11	7	2	0	0
St Helena		0	1	0	0	0
Sao Tome and Principe	1 [3]	1	7	0	0	0
Senegal	32	11	5	2	0	0
Seychelles	75	1	9	2	3	0
Sierra Leone	12	13	7	2	0	0
Somalia	52	17	7	1	0	0
South Africa	1016	25	13	3	1	28
Sudan	9	17	8	1	0	0
Swaziland	25	0	5	1	0	0
Tanzania	158	30	26	3	0	0
Togo	0	9	1	2	0	0
Tunisia	26	6	14	1	0	0
Uganda	11	16	12	1	0	0
Western Sahara	0	5	5	0	0	0
Zaire	3	31	27	2	0	1
Zambia	1	10	10	2	0	0
Zimbabwe	96	9	6	1	0	0

Sources: IUCN 1990. *1990 IUCN Red List of Threatened Animals.* IUCN, Gland and Cambridge. Additional range data from WCMC Animal Database and other sources. Bird ranges from Sibley, C.G. and Monroe, B.L. 1990. *Distribution and Taxonomy of Birds of the World.* Yale University Press, New Haven and London.

Notes: Plants: numbers include many taxa below species level and also Ex/E species. Vertebrates: marine species are excluded. Extinct taxa are excluded. Only full species are accounted for. Includes K categories - i.e. all threatened species as defined by IUCN. Mammals: cetaceans are excluded. Birds: the countries within the breeding and wintering range are included (where data available). Fishes: not included are c. 252 spp. of Lake Victoria cichlids, many thought to be extinct or severley threatened. ** excludes figures for Germany (German Dem Rep = 11; Germany, Fed Rep = 15). [1] includes Gibralter (UK = 23; Gibralter = 1). [2] includes Rotuma. [3] total for Sao Tome only.

The majority of threatened mammalian species occur in mainly tropical countries, with highest numbers recorded from Madagascar (53), Indonesia (49), China (40) and Brazil (40). India, Australia, Zaire and Tanzania also have large numbers of species at risk, as do Mexico, USA, 'USSR' and most South American and Southeast Asian countries.

A regression analysis (Fig. 17.4) shows that Madagascar and Indonesia in particular have more threatened species in relation to country area than would be predicted statistically (points above the line) whereas USA, for example, has fewer.

There are approximately twice as many threatened bird species as mammals (1,029 versus 507) but they show a similar distributional pattern. The majority are concentrated in southern and Southeast Asia, USA, Mexico, and South America. The ten countries listed all have more than 40 threatened species. In comparison, Europe, Africa, Canada, the Middle East and the Arabian Peninsula have relatively few globally threatened bird or mammal species.

Figure 17.3 Per cent of known species classed as Endangered

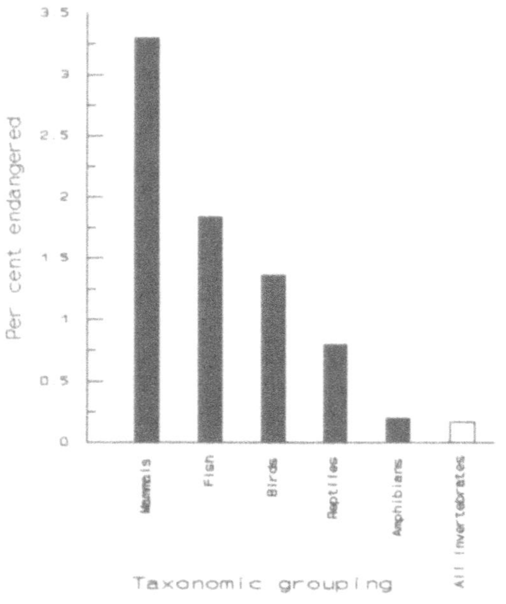

Table 17.4 Countries with greatest numbers of threatened vertebrates

MAMMALS		BIRDS		REPTILES		AMPHIBIANS		FISHES	
COUNTRY	TOTAL	COUNTRY	TOTAL	COUNTRY	TOTAL	COUNTRY	TOTAL	COUNTRY	TOTAL
Madagascar	53	Indonesia	135	USA	25	USA	22	USA	164
Indonesia	49	Brazil	123	India	17	Italy	7	Mexico	98
Brazil	40	China	83	Mexico	16	Mexico	4	Indonesia	29
China	40	India	72	Bangladesh	14	Australia	3	South Africa	28
India	39	Colombia	69	Indonesia	13	India	3	Philippines	21
Australia	38	Peru	65	Malaysia	12	New Zealand	3	Australia	16
Zaire	31	Ecuador	64	Brazil	11	Seychelles	3	Canada	15
Tanzania	30	Argentina	53	Colombia	10	Spain	3	Thailand	13
Peru	29	USA	43	Madagascar	10	Yugoslavia	2	Sri Lanka	12
Viet Nam	28	Myanmar	42	Myanmar	10			Cameroon	11

Sources: IUCN 1990. *1990 IUCN Red List of Threatened Animals*. IUCN, Gland and Cambridge; WCMC 1991. The World Conservation Monitoring Centre Animal Database. WCMC, Cambridge; Bird ranges estimated from Sibley, C.G. and Monroe, B.L. 1990. *Distribution and Taxonomy of Birds of the World*. Yale University Press, New Haven and London.

Notes: Mammals - Cetacea (whales, dolphins) are excluded; Birds - estimates include breeding and overwintering species (where data available); Reptiles - marine turtles are excluded; Fishes - the estimates do not include *c.* 252 species of cichlids in Lake Victoria. Marine species are also excluded. Extinct taxa in all groups are excluded; only full species, not subspecies, are accounted for. Numerous countries had 1 threatened amphibian species, therefore the last row could not be filled for this column.

Several factors may be involved in this distribution. Other things being equal, the number of threatened species in a country should be correlated with the total number of species present, and tropical countries generally have a higher species richness than temperate ones. The high current rate of human population increase, and consequent high rates of habitat loss and modification in tropical countries, is doubtless an important factor.

The global distribution of species richness, the non-matching and uneven geographic spread of conservation activity and field survey work, and the patchy review to which most taxonomic groups have been subjected jointly mean that the IUCN Red List gives an as yet incomplete picture of the global distribution of species which may be under threat.

Habitat distribution of threatened animals

Information on habitat requirements is not consistently available for all threatened species. A useful indication of the global situation can be derived from analysis of the threats facing, and habitat types occupied by, the mammals of Australasia and the Americas and the birds of the world.

As stated above, habitat loss or modification is the main category of threat affecting these species. The two habitat types in which the largest number of threatened mammals occur are lowland tropical rain forest (TRF) (37%) and montane TRF (19%), which together are occupied by 43% of all threatened Australian and American mammal species (Fig. 17.5). Both these habitat types are found exclusively in tropical regions, between latitudes 28°S and 28°N. Other tropical and subtropical habitats such as dry savanna, humid savanna, desert and semi-desert also possess large numbers of threatened mammals. In contrast, temperate and polar habitats such as coniferous and boreal forest, Mediterranean forest and scrub, tundra and polar ice harbour relatively few threatened species.

In general the world's threatened bird species occupy a range of habitat types remarkably similar to the threatened mammals of Australasia and the Americas, with 43% occurring in TRF. The percentages occurring in marine, freshwater, grassland and polar habitats are also very similar, but there are some notable differences (Fig. 17.6). The major disparity is that some 38% of threatened birds are found on oceanic islands. These are primarily flightless or ground-nesting species which are threatened by introduced predators, for example rats and mongooses. A direct comparison with mammals is not possible because Thornback and Jenkins did not include oceanic islands; however, there are few mammals on such islands. A higher percentage of threatened birds than mammals occurs in seasonal woodlands (20% *v.* 8.4%), while this trend is reversed in arid (1% *v.* 14%), and coastal and estuarine habitats (5% *v.* 14%).

Madagascar has the highest number of threatened mammal species (50). Most of these are forest-dwelling lemurs. Harcourt and Thornback (1990) identify habitat destruction as the main threat to lemurs, and estimate that at current rates of cutting (1.2% per year) only forests on the steepest slopes will survive the next 35 years.

AQUATIC HABITATS

These systems have received little attention in comparison with terrestrial habitats, and very little survey work has so far been undertaken in tropical areas. A recent synthesis (Moyle and Leidy, in press) demonstrated that fishes provide reliable indicators of trends in aquatic diversity.

Figure 17.4 Relationship between number of threatened species and country area

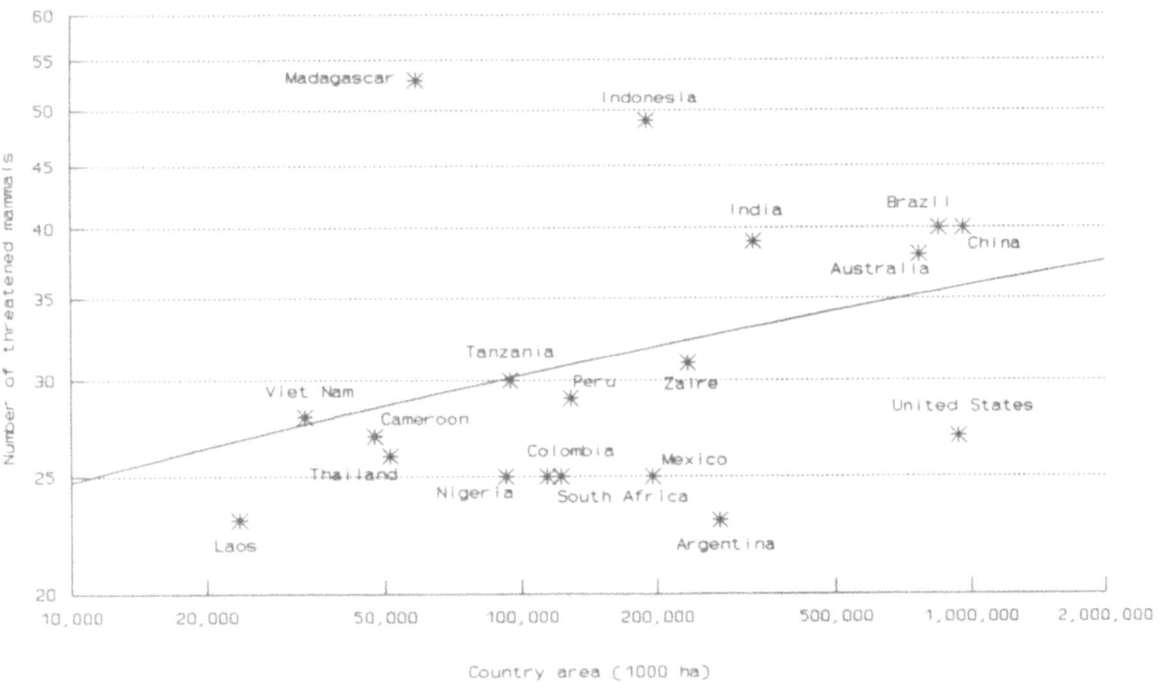

Figure 17.5 Habitat distribution of threatened mammals

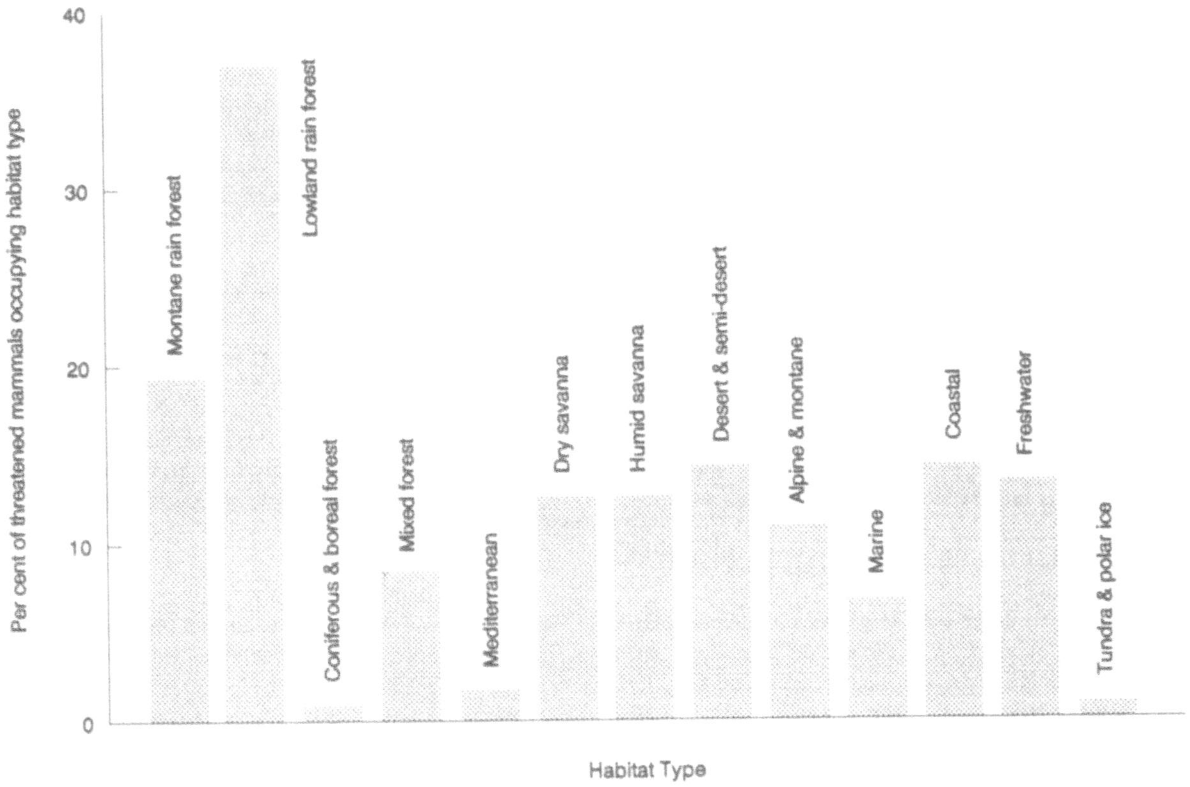

Note: Data for Australasia and the Americas, excludes cetaceans.

Figure 17.6 Habitat distribution of threatened mammals and birds

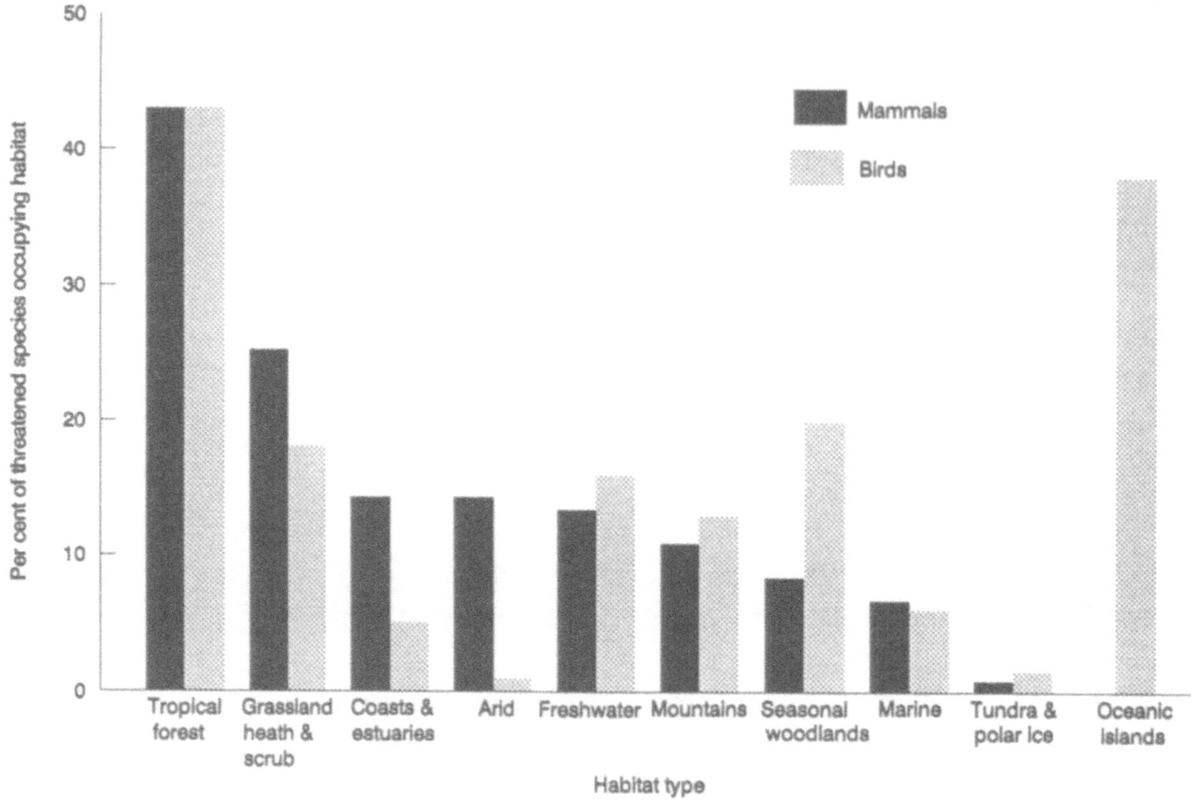

Note: Mammal data for Australasia and the Americas, excludes Cetacea; bird data are global.

Information on the fish faunas of North America, Europe, Iran, South Africa, Sri Lanka, Australia, Costa Rica, Brazil and Chile was analysed. The well-supported conclusion of this review was that at least 20% (c. 1,800 species) of the world's freshwater fish species are seriously threatened or extinct. Declines usually resulted from cumulative effects of several long-term factors. Habitat modification (competition for water, drainage, pollution), introduced species and commercial exploitation were identified as the major causes of decline. Recent fieldwork in Madagascar (Reinthal and Stiassny, 1991) corroborates these general conclusions: the native fish fauna in eastern and central Madagascar had declined severely because of introductions and habitat degradation as a result of forest clearance. These trends, coupled with inadequate knowledge of freshwater faunas and the strong representation of freshwater species in the list of known extinct species (see Chapter 16), indicate that aquatic systems require increased conservation attention.

THREATENED SPECIES ON ISLANDS: PLANTS

About one in six plant species grows on oceanic islands; one in three of all known threatened plants are island endemics. This is a measure of the diversity and fragility of island ecosystems and their importance in plant conservation.

Damage to most island floras occurred in the era of European exploration and colonisation, when oceanic islands became strategically important to the maritime powers. Most island floras evolved in the absence of large grazing animals and few endemic plants had defences against grazing animals.

On St Helena, goats were introduced in 1513 and within 75 years had formed vast herds. Botanists only reached the island in 1805-10, long after the damage had been done, and so one can only speculate on the original flora. Today 46 endemic species are known, seven of them extinct (Cronk *in litt.*, 1991), but J.D. Hooker estimated that there must have been originally over 100 endemic species (quoted in Lucas and Synge, 1978). Most of these species will never be known.

Philip Island, near the penal colony of Norfolk Island, has been affected even more severely. The island was believed to have carried a mixture of scrub and dense forest when discovered by Captain Cook in 1744. The introduction of goats, pigs and later rabbits reduced this vegetation to a near desert in which by 1964 the endemic Philip Island Glory Pea *Streblorrhiza speciosa* had become extinct and the endemic hibiscus *Hibiscus insularis* reduced to four aged bushes.

Whereas goats, sheep, pigs and even rabbits can be controlled and even eliminated, the problem of introduced plants is much more intractable. Enthusiastic gardeners often brought to islands the plants they used to grow at home, and some of these plants proved to be devastatingly invasive in the native vegetation, outcompeting the native flora. In Mauritius, for example, visitors today see rich green thickets and forests covering the hills, but few realise that virtually all this vegetation is of introduced plants. The only viable strategy for saving the Mauritian endemic flora in the short term is to make small weeded plots within the forest, a few hectares at a time. Other islands where the native flora is greatly threatened by introduced plants include Rodrigues, St Helena, Hawaii and Juan Fernández. It is noticeable that introduced plants tend to be much more destructive of island ecosystems than of continental ones.

Following the TDWG geographical classification (see Chapter 14), there are about 80 islands or island groups with significant endemic floras (here defined as more than five endemic species). For nearly half of these islands, a detailed assessment has been made of which species are threatened (Table 17.6).

Degree of threat to species varies greatly from one island or island group to another. Islands with severely affected floras include:

- Hawaii: 108 endemic taxa have already gone extinct, 15 are either Extinct or Endangered, 138 are Endangered, 37 are Vulnerable, 126 are Rare, and 9 are Indeterminate - a total of 433 threatened taxa. Hawaii has, therefore, one of the most distinctive and one of the most threatened floras in the world.
- St Helena, in the Atlantic Ocean, where all of the 46 endemic known species are threatened, 7 of them Extinct and 19 Endangered
- Bermuda, north of the Caribbean: all but one of the 15 endemic species are threatened, 3 of them Extinct and 4 Endangered
- Rodrigues, a dependency of Mauritius in the Indian Ocean: all but 2 of the 45 endemic species are threatened, 27 of them Endangered or Extinct
- Norfolk Island, east of Australia: where all but 2 of the 36 endemic species are threatened, 1 of them Extinct and 11 Endangered.

On each of these islands, the native plants are reduced to small patches of relict vegetation, and often have populations of ten individuals or fewer. It is, however, encouraging to see that on all the four islands listed above, there are active programmes to rescue the threatened plants although it may take centuries to restore the native vegetation.

Other islands have fared better. For example, the native forests on Lord Howe Island, a dependency of Australia, are still intact and are now well protected in a national park. Of the 84 endemic species in the Table, only one is Extinct and three Endangered, but 72 are Rare, meaning their world populations are low but they are not under threat. Among coral islands, the important endemic floras of Aldabra (Indian Ocean) and Henderson Island (Pacific Ocean) are intact and both are now effectively protected as nature reserves.

For some of the islands with larger floras, the flora has only been partly assessed. The true numbers of threatened species may be higher than those quoted. This is probably true, for example, of Cuba and Jamaica, with their very large endemic floras.

The islands listed in Table 17.5 all have more than 10 endemic species of plants but the conservation status of those plants is not known. The immediate priority here is for field surveys to assess the situation and provide a basis for conservation action. For more details see Table 14.1.

Table 17.5 Priority islands for surveys of endemic flora

	ESTIMATED ENDEMIC PLANT SPECIES
AFRICA	
Annobon	17
Bioko	49
Cape Verde	92
Príncipe	35
São Tome	108
CARIBBEAN	
Bahamas	112
Cayman Is	18
Dominican Republic/Haiti	1800
Trinidad-Tobago	215
Virgin Is, US and British	28 +
Most Lesser Antillean Islands	total 327
INDIAN OCEAN	
Andaman Is	144
Nicobar Is	72
Comoros	136
PACIFIC	
American Samoa	27
Coco, Isla del	15
Fiji	700
Marquesas Is	105
New Caledonia	2480
Northern Marianas	81
Society Is	?
Taiwan	892
Tonga	25
Tuamotu Is	20
Tubuai Is	140
Vanuatu	150
Western Samoa	57

Table 17.6 covers higher plants (flowering plants, ferns, gymnosperms) endemic to the island or island group concerned. The main figures are of species; the figures in brackets are of additional endemic infraspecies (subspecies and varieties). Where an endemic species is divided into several infraspecies in the database, it has been counted only at the infraspecies level; however the parallel table in Chapter 14 adds these endemic species into the endemic species totals. Thus, the total for Mauritius here is 236 (54) but above is 246 species, since the 54 infraspecies include 10 species that are wholly endemic to Mauritius.

THREATENED SPECIES ON ISLANDS: BIRDS

Islands are important for bird conservation: over 1,750 species (some 17% of the world's bird species) are confined to islands and of these, 402 (23%) are threatened (Johnson and Stattersfield, 1990) compared with only 11% of birds worldwide (Collar and Andrew, 1988). In addition, island birds have suffered the majority of bird extinctions which have occurred during historic times.

Distribution of island endemics

A high proportion of threatened island species are concentrated in a few geopolitical units: a total of 92 such units have one or more threatened species; 11 of these (Cuba, Hawaiian Islands, Indonesia, Marquesas Islands, Mauritius, New Zealand, Papua New Guinea, Philippines, São Tome, and Príncipe, Seychelles and Solomons) support

Table 17.6 Threatened endemic plant species on oceanic islands

ISLAND	Ex	Ex/E	E	E/V	V	V/R	R	I	?	nt	TOTAL
Cuba	25	0	322 (3)	0	294 (2)	0	142 (6)	49	-	-	3233
Jamaica**	0	0	76 (6)	0	137 (8)	0	122 (5)	48 (3)	-	-	827
Hawaiian Is	87 (21)	15	106 (32)	0	23 (14)	0	92 (37)	8 (1)	12	388 (152)	731 (254)
Canaries*	1	0	127	0	120	0	129	4	51	162	593
Mauritius	21 (3)	9	75 (21)	0	45 (8)	0	55 (15)	1 (3)	6	24 (4)	236 (54)
Socotra	1	0	33	0	15	0	52 (1)	27	90 (8)	49 (2)	267 (11)
Ogasawara	4	0	22	0	21	0	21	0	-	-	152
Galápagos Is	2	0	8 (4)	0	11	0	54 (42)	6 (3)	1 (2)	66 (23)	148 (74)
Juan Fernández	1	0	54	0	38 (1)	0	0	7	0	23 (1)	123 (2)
Réunion	1	1 (1)	12 (1)	3	11 (1)	8	14 (1)	5 (1)	32 (5)	33 (18)	120 (28)
Madeira	0	0	17 (1)	0	29 (1)	0	34 (5)	0	16 (2)	22 (1)	118 (10)
Marquesas Is	1	0	18	0	13	0	7	22	40	4	105
Cyprus	0	0	7 (2)	0	9 (3)	0	28 (2)	0	2 (5)	44 (14)	90 (26)
Lord Howe I	1	0	3 (1)	0	1	0	72 (4)	1	0	6 (1)	84 (6)
Balearic Is	1	0	8 (1)	0	11 (3)	0	23 (18)	0	0	28 (8)	70 (30)
Seychelles	0	2 (1)	17 (1)	7	19 (1)	11	7 (1)	0	1	1	63 (4)
Azores	1	0	1	0	4 (1)	0	17 (1)	7	6 (4)	13	49 (6)
St Helena	7	0	19	0	1	0	18 (2)	1	0	0	46 (2)
Corsica	1	0	0 (1)	0	4 (2)	0	6 (1)	1	19 (38)	16	46 (42)
Rodrigues	8	2	18 (2)	0	7 (1)	0	9 (1)	0	0	2	46 (4)
Chatham Is*	0	0	6	0	4	0	6	0	25	0	41
Sicily	0	0	4 (2)	0	5 (1)	0	15 (1)	3	3	11 (3)	41 (7)
Tristan	0	0	0	0	0	0	6	10 (1)	0 (1)	24 (4)	40 (6)
Norfolk I	1	0	11 (2)	0	2 (1)	0	18 (4)	0	0	2	36 (7)
Sardegna	0	0	5	0	4	0	9	0	0	8	26
Christmas I	1	1	0	1	2	1	9	0	1 (1)	1	17 (1)
Bermuda*	3	0	4	0	1	0	6	0	0	1	15
Guam	0	1	3	0	3	0	2	2	1	2	14
Falkland Is	1	0	1	0	0	0	3	0	0	7 (1)	12 (1)
Ascension	1	3	2	0	0	0	4 (1)	0	0	0	10 (1)
Kazan Retto	0	0	0	0	1	0	4	0	4	0	9
Malta	0	0	0	0	1	0	2	1	0	1	5
Antipodean Is	0	0	0	0	0	0	4	0	0	0	4
Selvagens	0	0	2	0	1	0	1	0 (4)	0	0	4 (4)

Notes: * Totals include species and infra-specific taxa together. ** Only flowering plants covered. Ex: Extinct; E: Endangered; V: Vulnerable; R: Rare; I: Indeterminate; ?: Insufficiently Known; nt: neither rare nor threatened. For definitions see beginning of chapter. Combination categories (Ex/E, E/V and V/R) are used for taxa that are in either one or the other of the two categories concerned.

Sources: In all cases the figures are derived from counting entries on threatened plant lists, in most cases from the WCMC Plants database, which have been updated by numerous specialists during 1991 as part of the process of preparing the Biodiversity Status Report. A handout giving the source of the list for each island or island group is available from the Plants Group of WCMC on request.

over half the threatened species restricted to islands. Over 90% of threatened species restricted to islands are endemic to their geopolitical units, with a few island groups having particularly large numbers of threatened endemics (e.g. Indonesia 91, and the Philippines 34). Some 25 islands support a single threatened endemic only.

After Indonesia and the Philippines, the islands of the Pacific Ocean support the largest number of threatened species (110). Although when compared to the Atlantic islands this constitutes a much lower portion of the endemics occurring in the region (38% and 50%, respectively) it nonetheless accounts for 27% of threatened species restricted to islands.

Degree of threat

Of the 402 species restricted to islands, the greatest number of those considered Endangered or Vulnerable occur within the Pacific region: 31 of the 66 Endangered species and 29 of the 71 Vulnerable species. These include a wide range of species, such as the severely endangered Barred-wing Rail *Nesoclopeus poeciloptera*, known only from Fiji, and the New Caledonian endemic Kagu *Rhynochetos jubatus*, belonging to a monotypic family and therefore regarded as a high priority for conservation action.

Habitat requirements

The majority of threatened island birds are forest species. Rain forest supports 200 (50%) of the threatened species. Lowland and montane forests contribute almost equally, being used by 101 and 112 species, respectively (42 species use both types; 29 rain forest species could not be assigned to the lowland/montane division). The other major forest-type, seasonal/temperate forest, supports 113 species. In total, forests of all categories support 310 species, accounting for 77% of threatened island endemics.

Threats

The most important factor threatening island species is habitat destruction, affecting over 50% of threatened island species. Given the number of extinctions attributable to introductions, it is of interest that introduced species now appear to be a major threat to only 20% of threatened island endemics, a much smaller proportion than might be expected and a considerably smaller proportion than the 41% of island species which are at risk simply by having a limited range. Other factors (hunting, trade, human disturbance, natural causes and fisheries) each affect less than 10% of threatened island birds. For some 60 threatened island endemic birds, further field research is needed to identify the cause of decline.

References

Collar, N.J. and Andrew, P. 1988. *Birds to Watch: the ICBP world checklist of threatened birds*. ICBP Technical Publication No.8, ICBP Cambridge, UK.

Corbet, G.B. and Hill, J.E. 1991. *A World List of Mammalian Species*. Third edition. Oxford University Press, UK.

Diamond, A.W. 1987. *Save the Birds*. ICBP, Girton, Cambridge.

Harcourt, C. and Thornback, J. 1990. *Lemurs of Madagascar and the Comoros. The IUCN Red Data Book*. IUCN, Gland, Switzerland and Cambridge, UK.

IUCN, 1990. *1990 IUCN Red List of Threatened Animals*. IUCN, Gland and Cambridge.

Johnson, T.H. and Stattersfield, A.J. 1990. Global review of island endemic birds. *Ibis* 132:167-180.

Lucas, G. and Synge, H. 1978. *The IUCN Plant Red Data Book*. IUCN, Switzerland.

Mace, G.M. and Lande, R. 1991. Assessing extinction threats: toward a re-evaluation of IUCN threatened species categories. *Conservation Biology* 5(2):148-157.

Moyle, P.B. and Leidy, R.A. (in press). Loss of biodiversity in aquatic ecosystems: evidence from fish faunas. In: Feidler, P.L. and Jain, S.K. (Eds), *Conservation Biology: the theory and practice of nature conservation, preservation, and management*. Chapman and Hall, New York.

Reinthal, P.N. and Stiassny, M.L.J. 1991. The freshwater fishes of Madagascar: a study of an endangered fauna with recommendations for a conservation strategy. *Conservation Biology* 5(2):231-243.

Sibley, C.G. and Monroe, B.L. 1990. *Distributionand Taxonomy of Birds of the World* Yale University Press, New Haven and London.

Thornback, J. and Jenkins, M. 1982. *The IUCN Mammal Red Data Book Part 1*. IUCN, Gland, Switzerland and Cambridge, UK.

The section on threatened plants on oceanic islands was prepared by Hugh Synge.

18. GLOBAL HABITAT CLASSIFICATION

The world encompasses an enormous range of terrestrial and aquatic environments, from polar ice-caps to forests, and coral reefs to deep ocean trenches. The classification of this immense range of variation into a manageable system is a major problem in biology and underpins much of the sciences of ecology and biogeography. It has not merely theoretical interest, but is of fundamental importance in the management and conservation of the biosphere.

Within ecology, a wide variety of terms has been coined - community, habitat, ecosystem, biome - intended to help in such a classification. Some of these can be seen as forming a loose and ill-defined hierarchy analogous in some ways with the taxonomic system developed for classifying organisms, discussed fully in Chapters 2 and 3. However, the classification of the natural environment is far more problematic than the classification of organisms and none of the above terms has a rigid, satisfactory and universally accepted definition. Indeed there are good theoretical grounds for questioning the basis of such a classification. This is because these systems are ultimately based on an assumption that the natural environment can be divided into a series of discrete, discontinuous units rather than representing different parts of a highly variable natural continuum, whereas in reality the latter is undoubtedly a more accurate description of the world.

In general, attempts to classify ecological units are based on identification of the species which occur in them along with a description of the physical characteristics of the area. Most terrestrial ecosystems, for example, are generally identified on the basis of plant communities, that is areas with similar plant species composition and structure. The basic principle underlying this is that different species may habitually be closely associated with each other over a wide geographical range. The extent to which this is true is still controversial - it can reasonably be argued that the distribution of plant species is generally dependant on the physical environment and historical accident rather than on the occurrence or otherwise of other plant species, although within a particular geographical region, species with similar ecological requirements may, of course, be expected to have similar distributions. Even if the concept of a community is accepted, it is widely acknowledged that the more rigidly a community is defined the more site-specific it becomes and hence the more limited its use in analysis and planning.

At the other extreme, very general habitat classifications ('forests', 'grasslands', 'wetlands') are based on the physical characteristics and appearance of an area, independent of species composition. They cover such a wide range of possible conditions that they have little heuristic use: the term 'forest' applies both to highly diverse lowland tropical rainforest and coniferous monoculture, two systems which may have no, or virtually no, species in common. Furthermore these general terms are virtually impossible to define and delimit in a universally applicable way. Thus, for example, the density of tree cover necessary before an area can be called a woodland is undefinable and any limit used will always be arbitrary. Similarly, it is impossible to determine for how long and how intensely an area must be flooded before it can be classified as a wetland rather than a terrestrial ecosystem. This naturally makes any mapping of habitats a problematic task.

ECOSYSTEM MAPPING

Most global habitat classification systems have attempted to steer a middle course between the complexities of community ecology and the oversimplified terms discussed above, although they too have the same problems of definition and delimitation. Generally these systems will use a more or less elaborate combination of a general definition of habitat type with a climatic descriptor (e.g. 'tropical moist forest', 'temperate grassland', 'warm deserts and semi-deserts'). Some systems also incorporate global biogeography to take into account the floristic and faunistic differences between regions of the world which may have very similar climate and physical characteristics.

Further, ecosystem mapping may either take into account man's activities to attempt to produce a realistic, contemporary map of land-cover types, or may create a potential vegetation map from an analysis of climatic or other environmental variables. The potential vegetation maps produced from this approach are independent of actual disturbances on the landscape.

Four of the major global classification systems are presented here. The Classification of Biogeographical Biomes of the World map (Plate 1), provides a modest classification based largely on geography and potential vegetation. The Ecoregions of the Continents map (Plate 2) and the Major World Ecosystems map (Plate 3) are produced from a combination of potential vegetation and actual land-use. The Holdridge Life Zone Classification map (Plate 5) depicts potential vegetation using the life zone classification system developed by Holdridge (1967).

The map in Plate 1 depicts the terrestrial biogeographic realms of the world and was produced for IUCN (Udvardy, 1975). This map provides a generalised framework to represent the distribution of biogeographical regions, biotic provinces, and biomes. The approach used to produce this map utilised vegetation and forest maps to produce the map categories. Over the past decade, this map has served IUCN and UNESCO as a primary global biogeographical guideline for conservation planning purposes. The distribution of protected areas throughout the globe within these biogeographical provinces is presented in Part 3.

The Ecoregions of the Continents Map in Plate 2 shows the distribution of ecosystems at the regional scale across the globe based upon existing climatic and vegetation data (Bailey, 1989a, b). The three levels of hierarchy used for representing ecosystems on this map are domains, divisions, and provinces. These categories are obtained by defining aggregates of ecosystems into larger biome categories. This map therefore represents a generalised depiction of ecosystem distribution across the globe. Table 18.1 presents the area contained in each region and its percentage of the global land area.

Table 18.1 Ecoregions of the continents (Bailey)

ECOREGION DOMAINS, DIVISIONS, AND PROVINCES	km²	Percent
100 POLAR DOMAIN	38,038,000	26.00%
110 Icecap Division	12,823,000	8.77%
M110 Icecap Regime Mountains	1,346,000	0.92%
120 Tundra Division	4,123,000	2.82%
121 Polar deserts	283,000	0.19%
122 Arctic tundras	1,231,000	0.84%
123 Oceanic moss-and-grass tundra	184,000	0.13%
124 Continental moss-and-lichen (typical) tundra	1,981,000	1.35%
125 Continental bush-and-shrub tundra	445,000	0.30%
M120 Tundra Regime Mountains	1,675,000	1.14%
M120 Tundra regime mountains (Antarctica)	60,000	0.04%
M121 Tundra-polar desert	795,000	0.54%
M122 Polar desert	820,000	0.56%
130 Subarctic Division	12,259,000	8.38%
131 Continental dark evergreen needleleaf open forest	2,285,000	1.56%
132 Continental light deciduous needleleaf open forest	1,286,000	0.88%
133 Eastern oceanic tayga	918,000	0.63%
134 Moderate continental dark evergreen needleleaf tayga	2,692,000	1.84%
135 Continental dark evergreen needleleaf tayga	1,880,000	1.29%
136 Continental and extreme continental light deciduous tayga	2,237,000	1.53%
137 Moderate continental small-leafed forest	251,000	0.17%
138 Continental mixed coniferous and small-leafed forest	710,000	0.49%
M130 Subarctic Regime Mountains	5,812,000	3.97%
M131 Open woodland-tundra	1,750,000	1.20%
M132 Open woodland-creeping trees-tundra	1,806,000	1.23%
M133 Forest-tundra of moderately and continental climate	686,000	0.47%
M134 Forest-creeping trees-tundra of extreme continental climate	1,203,000	0.82%
M135 Oceanic forest-tundra	367,000	0.25%
200 HUMID TEMPERATE DOMAIN	22,455,000	15.35%
210 Warm Continental Division	2,187,000	1.49%
211 Eastern oceanic mixed monsoon forest	65,000	0.04%
212 Moderate continental mixed forests	2,122,000	1.45%
M210 Warm Continental Regime Mountains	1,135,000	0.78%
M211 Oceanic forest-tundra	67,000	0.05%
M212 Oceanic forest-creeping trees	331,000	0.23%
M213 Forest-tundra of moderately continental and continental climate	736,000	0.50%
220 Hot Continental Division	1,670,000	1.14%
221 Permanently humid eastern oceanic broadleaf forests	788,000	0.54%
222 Moderately humid broadleaf forest in moderately continental climate	882,000	0.60%
M220 Hot Continental Regime Mountains	485,000	0.33%
M221 Forest-alpine meadows	485,000	0.33%
230 Subtropical Division	3,568,000	2.44%
231 Oceanic mixed constantly humid forests	3,568,000	2.44%
M230 Subtropical Regime Mountains	1,543,000	1.05%
M231 Forest-meadow of eastern oceanic (monsoon climate)	1,264,000	0.86%
M232 Oceanic constantly humid forest-alpine meadows	278,000	0.19%
240 Marine Division	1,347,000	0.92%
241 Oceanic meadow	92,000	0.06%
242 Western oceanic coniferous and mixed forests	210,000	0.14%
243 Permanently humid western oceanic broadleaf forests	951,000	0.65%
244 Western oceanic tayga	95,000	0.07%
M240 Marine Regime Mountains	2,194,000	1.50%
M241 Oceanic meadow-tundra	21,000	0.01%
M242 Oceanic forest-tundra	1,068,000	0.73%
M243 Forest-alpine meadows	1,105,000	0.76%
250 Prairie Division	4,419,000	3.02%
251 Temperate prairies (humid steppes and wooded steppes) of eastern parts of continents	752,000	0.51%
252 Broadleaf-wooded steppes and meadow steppes of moderately continental climate	1,172,000	0.80%
253 Small-leafed and coniferous wooded steppes of continental climate	787,000	0.54%
254 Open woodland, savannas, and shrub of eastern parts of continents	925,000	0.63%
255 Subtropical prairies (humid steppes and wooded steppes) of eastern	783,000	0.54%
parts of continents	0	
M250 Prairie Regime Mountains	1,256,000	0.86%
M251 Continental steppe-forest-tundra and steppe-forest-meadow	690,000	0.47%
M252 Forest-alpine meadows	566,000	0.39%
260 Mediterranean Division	1,090,000	0.75%
261 Western oceanic mixed sclerophyll forests and shrub	927,000	0.63%
262 Dry steppes and shrub of moderate continental climate	163,000	0.11%
M260 Mediterranean Regime Mountains	1,561,000	1.07%
M261 Forest-alpine meadows of western oceanic (mediterranean) climate	567,000	0.39%
M262 Shrub-forest-meadow of mediterranean climate	995,000	0.68%
300 DRY DOMAIN	46,806,000	32.00%
310 Tropical/subtropical Steppe Division	9,838,000	6.73%
311 Steppes and shrub of moderate continental climate	364,000	0.25%
312 Dry steppes, open woodland, and shrub of continental climate	846,000	0.58%
313 Shrub and semi-shrub semi-deserts of continental climate	1,392,000	0.95%
314 Desert-like savannas, open woodland, and shrub	5,807,000	3.97%
315 Dry steppes and shrub of moderate continental climate	1,429,000	0.98%
M310 Tropical/subtropical Steppe Regime Mountains	4,555,000	3.11%
M312 Forest-meadow-steppe of continental climate	670,000	0.46%
M313 Open woodland-steppe of continental climate	2,714,000	1.86%
M314 Open woodland-shrub-desert	770,000	0.53%
M315 Open woodland-steppe	400,000	0.27%

Table 18.1 Ecoregions of the continents (Bailey)

ECOREGION DOMAINS, DIVISIONS, AND PROVINCES	km²	Percent
300 DRY DOMAIN (continued)		
320 Tropical/subtropical Desert Division	17,267,000	11.80%
321 Shrub and semi-shrub semi-deserts and deserts of continental climate	1,321,000	0.90%
322 Semi-deserts and deserts	665,000	0.45%
323 Inner continental shrub semi-desert	3,674,000	2.51%
324 Inner continental deserts of continental climate	7,921,000	5.42%
325 Western oceanic semi-deserts and deserts with high relative humidity	958,000	0.65%
326 Inner continental semi-deserts and deserts of extreme continental climate	2,727,000	1.86%
M320 Tropical/subtropical Desert Regime Mountains	3,199,000	2.19%
M321 Desert-steppe and desert-steppe-desert of continental climate	1,193,000	0.82%
M322 Extreme continental desert	899,000	0.61%
M323 Desert-steppe	471,000	0.32%
M324 Desert	636,000	0.44%
330 Temperate Steppe Division	4,780,000	3.27%
331 Dry steppes of continental climate	1,790,000	1.22%
332 Steppes of moderately continental climate	1,581,000	1.08%
333 Dry steppes of extreme continental climate	1,409,000	0.96%
M330 Temperate Steppe Regime Mountains	1,066,000	0.73%
M331 Forest-alpine meadows	893,000	0.61%
M332 Continental open woodland-steppe	173,000	0.12%
340 Temperate Desert Division	5,488,000	3.75%
341 Semi-deserts and deserts of continental climate	922,000	0.63%
342 Semi-deserts of continental climate	1,213,000	0.83%
343 Deserts of continental climate	1,647,000	1.13%
344 Semi-deserts of extreme continental climate	399,000	0.27%
345 Deserts of extreme continental climate	1,306,000	0.89%
M340 Temperate Desert Regime Mountains	613,000	0.42%
M341 Extreme continental desert-steppe	613,000	0.42%
400 HUMID TROPICAL DOMAIN	**38,973,000**	**26.64%**
410 Savanna Division	20,641,000	14.11%
411 Seasonally humid mixed (deciduous and evergreen) forests	1,346,000	0.92%
412 Savannas, open woodland and shrub with seasonal moisture supply	2,496,000	1.71%
413 Seasonally humid, predominantly deciduous forests	4,951,000	3.38%
414 Humid tall-grass savannas and savanna forests	3,699,000	2.53%
415 Moderately humid grassy savannas	4,771,000	3.26%
416 Dry savannas and open woodland	3,379,000	2.31%
M410 Savanna Regime Division	4,488,000	3.07%
M411 Forest-steppe and forest-meadow of seasonally humid type	1,102,000	0.75%
M412 Forest-meadow, seasonally humid	1,220,000	0.83%
M413 Forest-steppe, inner continental and leeward slopes	2,167,000	1.48%
420 Rainforest Division	10,403,000	7.11%
421 Eastern oceanic constantly humid forests	1,843,000	1.26%
422 Mixed forests with short dry season	2,893,000	1.98%
423 Constantly humid evergreen forests	4,280,000	2.93%
424 Humid forests with short dry season	1,387,000	0.95%
M420 Rainforest Regime Mountains	3,440,000	2.35%
M421 Forest-meadow of constantly humid eastern oceanic type	728,000	0.50%
M422 Forest-paramo and forest-meadow of constantly humid oceanic (and windward-slope) type	1,013,000	0.69%
M423 Forest-paramo and forest-meadow	1,700,000	1.16%

Source: Bailey, R.G. 1989. *Ecoregions of the Continents.* U.S. Department of Agriculture, Forest Service, Washington, D.C.

The map in Plate 3 (Olson *et al.*, 1983) is a global ecology map and represents the primary world ecosystem types as of 1980. The regions on this map represent large areas within which local ecosystems are present more or less in a predictable fashion. Table 18.2 shows a country-by-country breakdown of major ecosystem types along with an estimate of the mass of carbon per unit area in live vegetation in each country.

The Holdridge Life Zone Classification system, represented in Plate 4, is a predictive scheme for identifying undisturbed vegetation based generally upon the effects of temperature, rainfall and evapotranspiration (Holdridge, 1967). This system was used to produce the map in Plate 5 that depicts the generalised distribution of eco-climatic zones across the globe. Plate 5 therefore represents the potential distribution of ecosystems in the world based on a consideration of current climatic patterns. The Holdridge Life Zone Classification System is described in further detail in the Chapter 19.

ESTIMATING RATES OF CHANGE OF ECOSYSTEMS

Given the difficulties of ecosystem definition and delimitation outlined above, it is, unsurprisingly, extremely difficult to measure existing areas of any given ecosystem or habitat, and even more problematic to estimate rates of loss. In large part this is because habitat alteration covers a wide spectrum of change, from short-term, slight and reversible disturbance to complete, and effectively irreversible, destruction. Just as it is impossible to define rigidly the limits of any given ecosystem or habitat, so it is impossible to determine how much a given area of ecosystem or habitat has to change before it can be considered destroyed or converted. This problem is compounded by the fact that the natural environment is not static but rather dynamic, sometimes highly so, on a time scale ranging from hours to millions of years. It is thus difficult even to define an undisturbed ecosystem or habitat as a standard against which to measure degree of disturbance.

Table 18.2 Estimates of vegetation type and percent cover

	OTHER COASTAL AQUATIC	MAJOR WETLANDS	DESERT & SEMI-DESERT	POLAR AND ALPINE	GRASS AND SHRUB	CROP & SETTLE-MENTS	INTER-RUPTED WOODS	MAJOR FORESTS	CARBON (Kg/m²)
WORLD	**4%**	**2%**	**13%**	**12%**	**20%**	**11%**	**17%**	**22%**	**3.1**
ASIA	**4%**	**1%**	**16%**	**9%**	**24%**	**17%**	**10%**	**18%**	**2.6**
Afghanistan			11%	16%	62%		7%	4%	1.2
Bangladesh	7%				5%	42%	25%	20%	4.6
Bhutan				25%		25%	19%	31%	3.7 *
Brunei		67%				33%			2.3 **
Cambodia	7%	4%				19%	5%	65%	6.5
China	1%	1%	14%	22%	21%	17%	5%	18%	2.4
Cyprus						100%			0.8 **
India	3%	0%	2%	2%	12%	44%	23%	14%	2.7
Indonesia	24%	9%			4%	9%	14%	40%	5.4
Iran, Islamic Rep	1%		30%		41%	7%	17%	4%	1.4
Iraq			30%		34%	33%	2%		1.0
Israel					40%	30%	30%		1.7 *
Japan	21%			1%	5%	18%	27%	30%	4.2
Jordan			44%		49%	7%			0.6
Korea, Dem People's Rep	12%				18%	14%	16%	39%	3.8
Korea, Rep	10%				24%	20%	24%	22%	3.4
Kuwait	9%		91%						0.3 *
Laos						7%	7%	86%	9.1
Lebanon						25%	75%		2.5 **
Malaysia	8%	3%			2%	10%	33%	45%	6.7
Mongolia			32%	5%	52%	1%	3%	7%	1.4
Myanmar	5%			1%	2%	16%	23%	52%	6.7
Nepal				23%			35%	42%	4.4
Oman	13%		46%		34%	7%			0.5
Pakistan	1%	2%	21%	7%	46%	13%	9%	0%	1.1
Philippines	32%				2%	22%	13%	31%	4.7
Qatar					100%				0.9 **
Saudi Arabia	1%		62%		33%	2%	1%	1%	0.5
Sri Lanka	32%				42%	3%	13%	10%	1.7
Syrian Arab Rep			18%		41%	36%	3%	1%	1.0
Taiwan	29%					12%	6%	53%	5.5 *
Thailand	7%					41%	7%	45%	5.5
Turkey	3%		9%		37%	18%	25%	7%	1.9
United Arab Emirates	3%		76%		21%				0.2
Viet Nam	17%	1%			4%	30%		49%	6.0
Yemen	7%		11%		74%	3%	6%		1.6
USSR (former)	**3%**	**2%**	**5%**	**26%**	**10%**	**8%**	**21%**	**26%**	**3.2**
EUROPE	**6%**	**0%**		**9%**	**4%**	**35%**	**22%**	**23%**	**3.0**
Albania							73%	27%	4.5 *
Austria				6%		36%	36%	22%	4.1
Belgium						40%	60%		2.1 *
Bulgaria	2%					59%	25%	14%	2.6
Czechoslovakia						29%	35%	36%	5.6
Denmark	31%					50%	19%		1.0 *
Finland	1%			12%		3%	12%	72%	5.3
France	5%	0%			2%	49%	28%	16%	2.9
Germany	3%					50%	24%	23%	3.8
Greece	26%	3%				40%	27%	3%	1.4
Hungary						82%	16%	2%	1.5
Iceland	9%			78%	13%				0.5
Ireland	3%					97%			0.8
Italy	19%			1%		35%	39%	6%	2.2
Luxembourg						100%			0.8 **
Netherlands	11%					84%	5%		0.8 *
Norway	6%			67%	1%	3%	5%	19%	2.2
Poland	2%				4%	45%	38%	10%	2.8
Portugal	3%				12%	18%	45%	21%	3.6
Romania	3%	4%				54%	6%	34%	3.3
Spain	5%	0%		2%	16%	30%	25%	22%	3.2
Sweden	5%			22%		5%	13%	55%	4.4
Switzerland				24%		29%	29%	19%	3.6
United Kingdom	22%				27%	50%	1%		0.9
Yugoslavia	1%					56%	29%	14%	2.8
NORTH & CENTRAL AMERICA	**5%**	**2%**	**3%**	**33%**	**9%**	**10%**	**17%**	**21%**	**3.1**
Bahamas	40%				40%	20%			0.5 *
Belize							17%	83%	10.8 *
Canada	5%	3%		44%	1%	3%	17%	27%	3.5
Costa Rica	33%					17%	17%	33%	3.5 *
Cuba	17%	5%			29%	29%	10%	12%	2.2
Dominican Rep					53%	26%	21%		0.7 *
El Salvador	20%					50%		30%	2.2 *
Greenland	1%			99%					0.5
Guatemala	12%					24%		65%	6.7
Haiti	21%				21%	57%			0.7 *
Honduras						14%	22%	65%	7.9
Jamaica	67%						33%		1.0 *
Mexico	6%	0%	15%	0%	28%	13%	11%	27%	3.1
Nicaragua	2%					2%	24%	72%	8.6
Panama	17%				10%	27%	10%	37%	5.2
United States	5%	1%	5%	15%	16%	18%	24%	18%	3.1

Table 18.2 Estimates of vegetation type and percent cover

	OTHER COASTAL AQUATIC	MAJOR WETLANDS	DESERT & SEMI-DESERT	POLAR AND ALPINE	GRASS AND SHRUB	CROP & SETTLE-MENTS	INTER-RUPTED WOODS	MAJOR FORESTS	CARBON (Kg/m²)
SOUTH AMERICA	**2%**	**3%**	**5%**	**2%**	**32%**	**8%**	**14%**	**33%**	**4.8**
Argentina	3%	6%	26%	2%	23%	19%	17%	4%	1.5
Bolivia	0%	4%	1%	4%	50%	2%	27%	13%	3.3
Brazil	2%	2%			36%	6%	12%	42%	5.8
Chile	5%		21%	7%	20%	10%	7%	28%	2.4
Colombia	1%	3%		1%	29%	3%	13%	49%	6.8
Ecuador	5%			2%	16%	7%	16%	53%	6.1
French Guiana	4%	11%				7%		78%	9.6
Guyana	5%	4%			23%	4%	1%	63%	7.9
Paraguay		4%				10%	43%	43%	6.0
Peru	3%		5%	8%	18%	6%	19%	41%	5.5
Suriname		8%			33%	6%		53%	6.9
Uruguay	5%	3%			66%	26%			0.9
Venezuela	3%	5%	1%		49%	7%	9%	27%	4.2
OCEANIA	**4%**	**1%**	**18%**	**0%**	**18%**	**5%**	**38%**	**16%**	**2.8**
Australia	2%		20%	0%	19%	5%	40%	13%	2.6
Fiji	67%						11%	22%	2.7 *
New Caledonia	56%				11%		33%		1.3 *
New Zealand	21%			2%	25%	15%	21%	16%	2.7
Papua New Guinea	17%	12%	1%		2%		11%	57%	6.9
AFRICA	**2%**	**2%**	**30%**	**0%**	**28%**	**7%**	**14%**	**17%**	**2.7**
Algeria	0%		81%		4%	3%	8%	3%	0.7
Angola	0%	0%	1%		30%	4%	18%	46%	4.6
Benin	6%				76%	6%	12%		2.3
Botswana		4%	66%		5%		22%	3%	1.1
Burkina Faso					93%	3%		4%	2.5
Burundi					27%		36%	36%	3.0 *
Cameroon		4%			31%	7%	21%	37%	5.8
Central African Rep					59%	5%		36%	5.0
Chad	1%	0%	49%		40%	1%	7%	1%	1.3
Congo		7%			15%	6%	17%	54%	7.6
Cote d'Ivoire	3%				37%	8%	12%	39%	5.7
Djibouti	11%		78%				11%		0.5 *
Egypt	3%		76%		8%	13%			0.5
Equatorial Guinea								100%	12.0 *
Ethiopia	0%	1%	2%		41%	4%	32%	20%	2.6
Gabon					7%	1%	9%	83%	10.4
Gambia					25%	50%		25%	2.8 *
Ghana	5%		1%		55%	2%	12%	25%	4.5
Guinea	1%	3%			77%	3%	1%	15%	2.8
Guinea-Bissau		25%			42%			33%	3.7 *
Kenya	3%		4%		35%	11%	42%	5%	2.5
Lesotho					75%		25%		1.4 *
Liberia					11%	18%	18%	53%	7.3
Libya	1%		86%		9%	2%	2%		0.4
Madagascar	11%				45%	25%	15%	5%	1.8
Malawi	17%	2%			2%	29%	2%	46%	3.2
Mali		6%	40%		28%	4%	21%	1%	1.6
Mauritania			69%		16%	2%	13%		0.8
Morocco	8%		23%	1%	14%	16%	34%	4%	1.8
Mozambique, People's Rep	6%	6%			18%	16%	17%	38%	4.0
Namibia	0%	1%	44%		33%		21%	1%	1.2
Niger			59%		26%		15%	0%	1.0
Nigeria	2%	2%			46%	24%	14%	12%	2.6
Rwanda	13%				13%		63%	13%	2.7 *
Senegal	2%				58%	38%		2%	2.2
Sierra Leone		12%			41%	12%		35%	4.7 *
Somalia	11%		38%		36%	1%	14%		1.4
South Africa	4%	0%	15%		29%	13%	37%	2%	1.8
Sudan	1%	2%	24%		52%	10%	8%	2%	1.5
Swaziland						25%	75%		3.0 **
Tanzania	7%				31%	14%	17%	30%	3.4
Togo	10%				40%	10%	30%	10%	2.8
Tunisia	13%		23%		32%	15%	15%	1%	1.1
Uganda	15%				48%	15%	14%	9%	2.5
Western Sahara	2%		98%						0.3
Zaire	1%	8%			19%	3%	6%	63%	7.6
Zambia		1%			2%	7%	17%	73%	5.0
Zimbabwe					14%	18%	38%	31%	3.4

Source: Olson, J.S., Watts, J.A. and Allison, L.J. 1983. *Carbon in Live Vegetation of Major World Ecosystems.* Oak Ridge National Laboratory, Oak Ridge. Olson, J.S., Watts, J.A. and Allison, L.J. 1985. *Major World Ecosystem Complexes Ranked by Carbon in Live Vegetation: a database.* Oak Ridge National Laboratory, Oak Ridge.
Notes: For smaller countries the precision is limited by the 0.5 degree resolution of the Olson dataset. * entire country includes less than 20 ½-degree cells; ** less than 5 cells

These problems notwithstanding, it is indisputable that man is having an increasing effect on the natural environment and that this effect extends to all the ecosystems considered here. Some, very general, impression of this can be gained from Table 18.3 which gives an estimate of the increase in area of cropland between the years 1700 and 1980. It also shows the decrease in forests and woodlands and grasslands and pastures over that time. It is clear from these estimates

Plate 1 A Classification of the Biogeographical Biomes of the World (after Udvardy 1975)

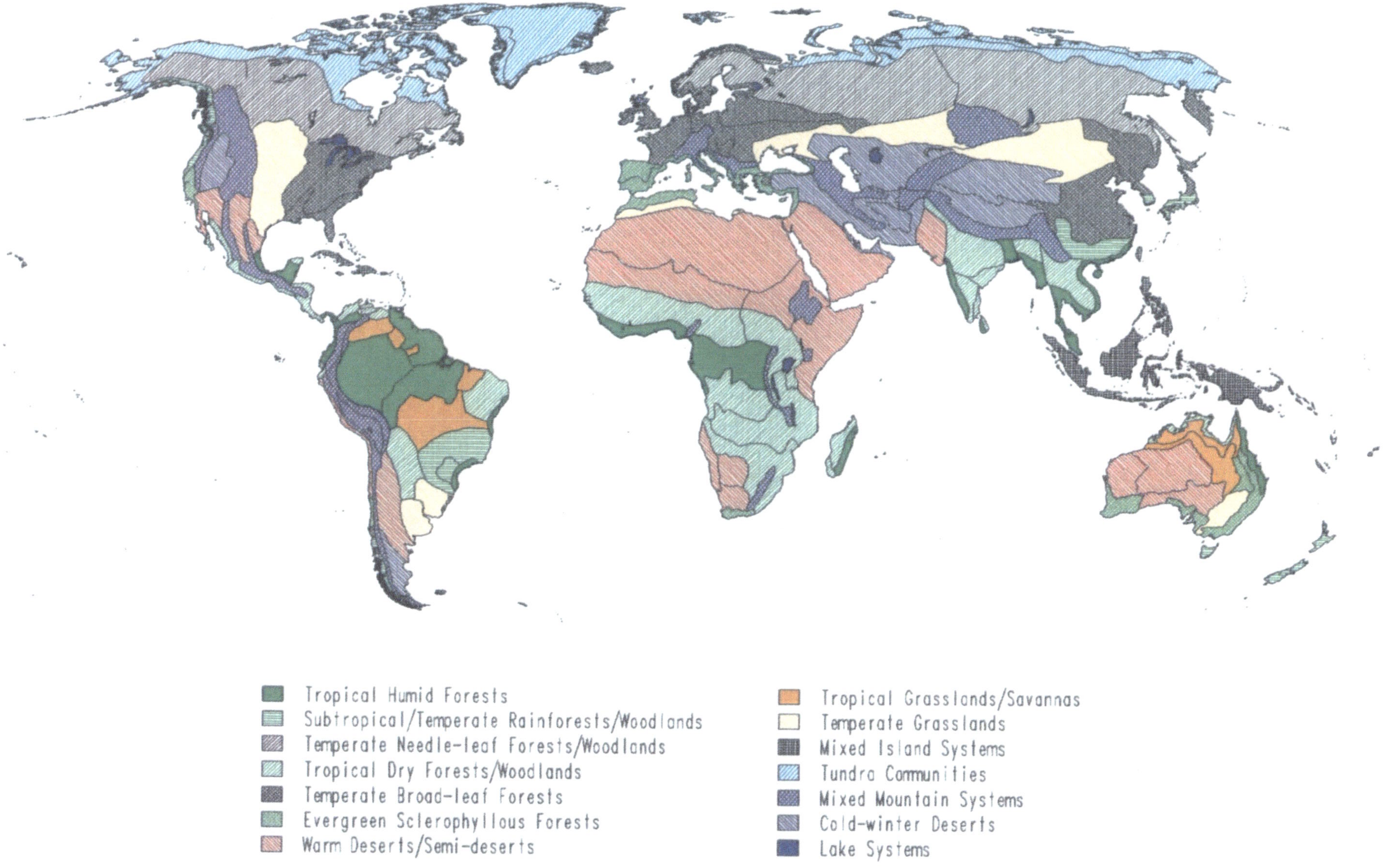

Plate 2 Ecoregions of the Continents (Bailey)

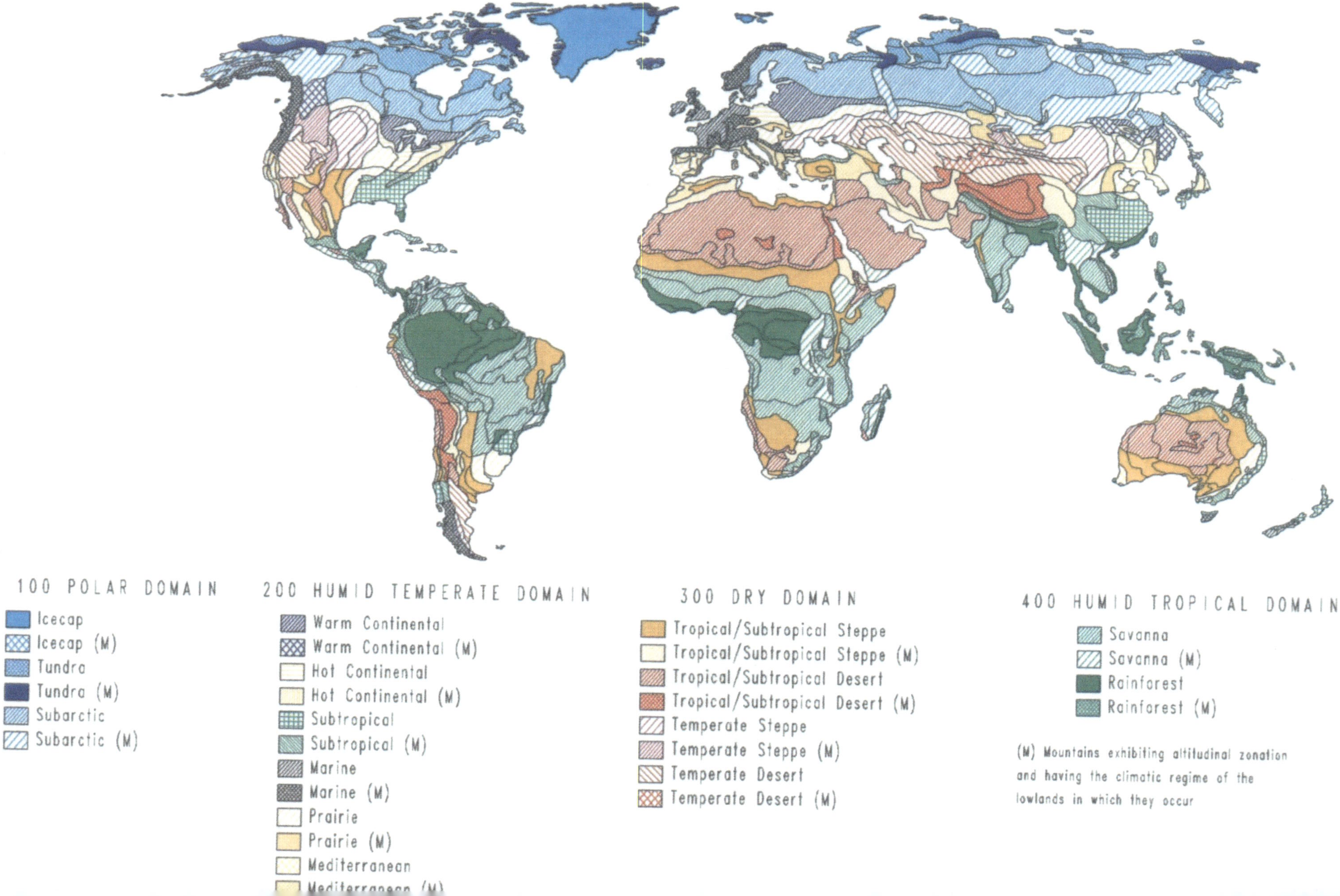

Plate 3 Major World Ecosystems (Olson)

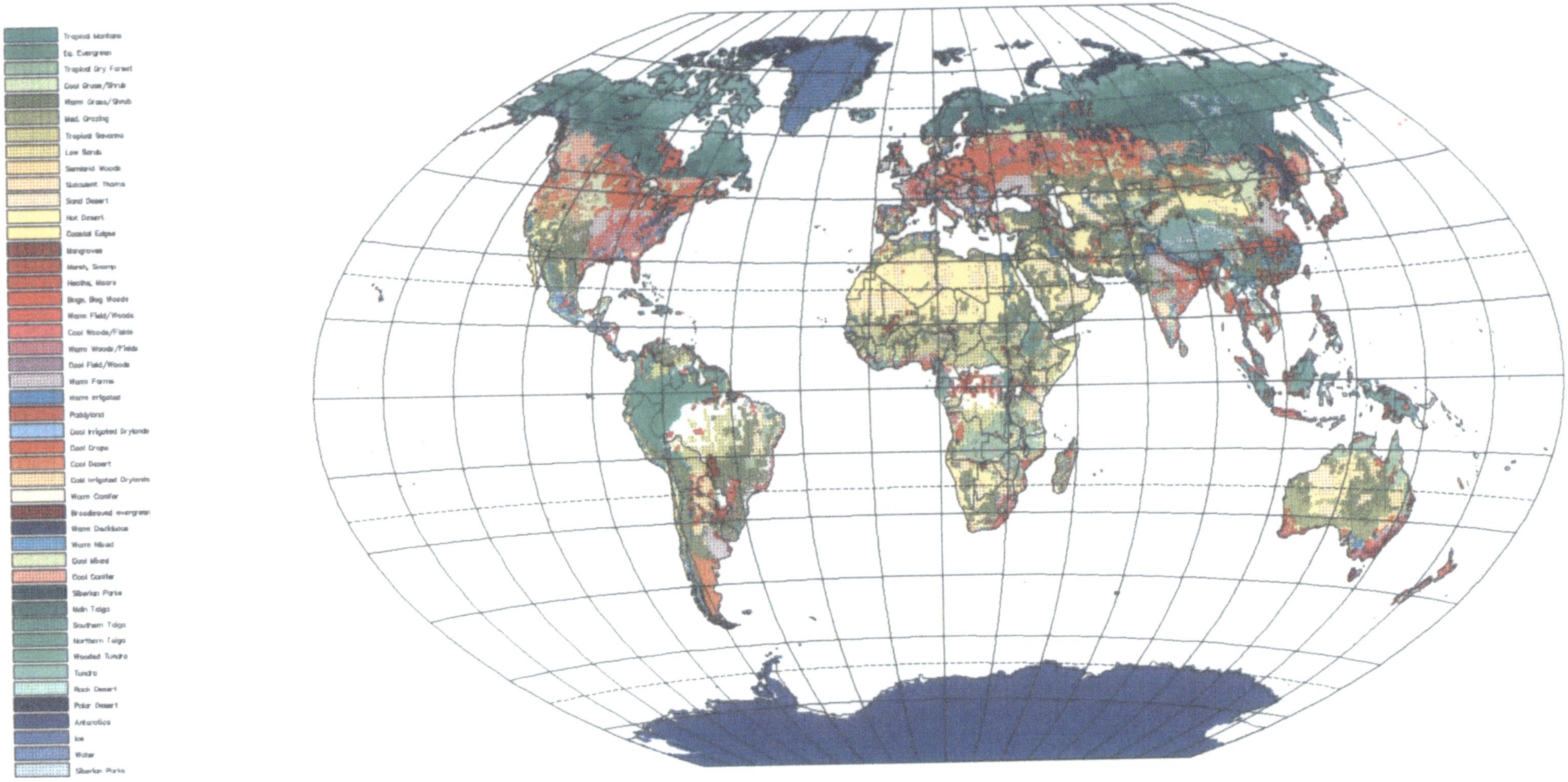

Plate 4 Holdridge Life Zone Classification scheme

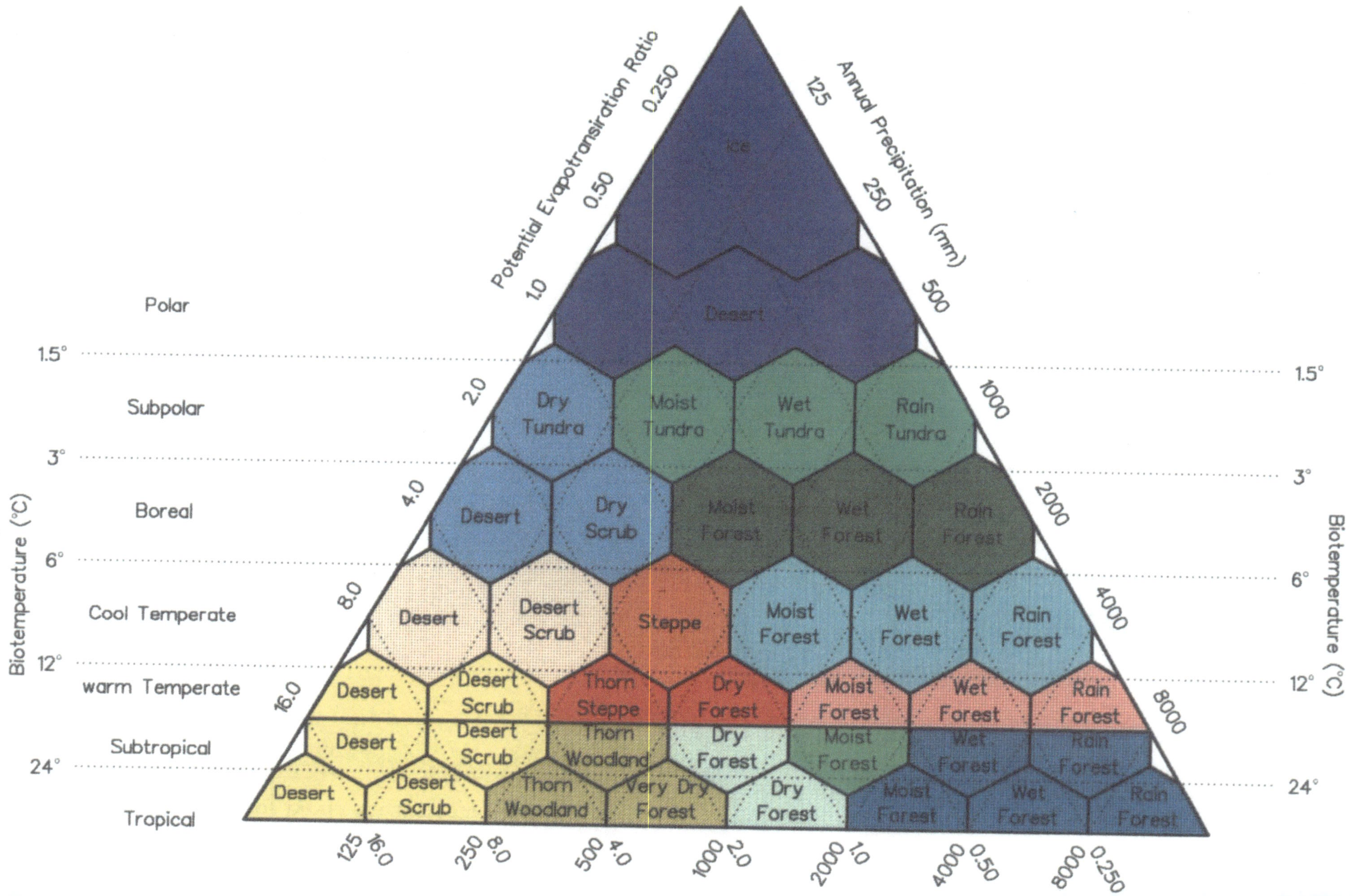

Plate 5 Holdridge Life Zone Classification for current climate

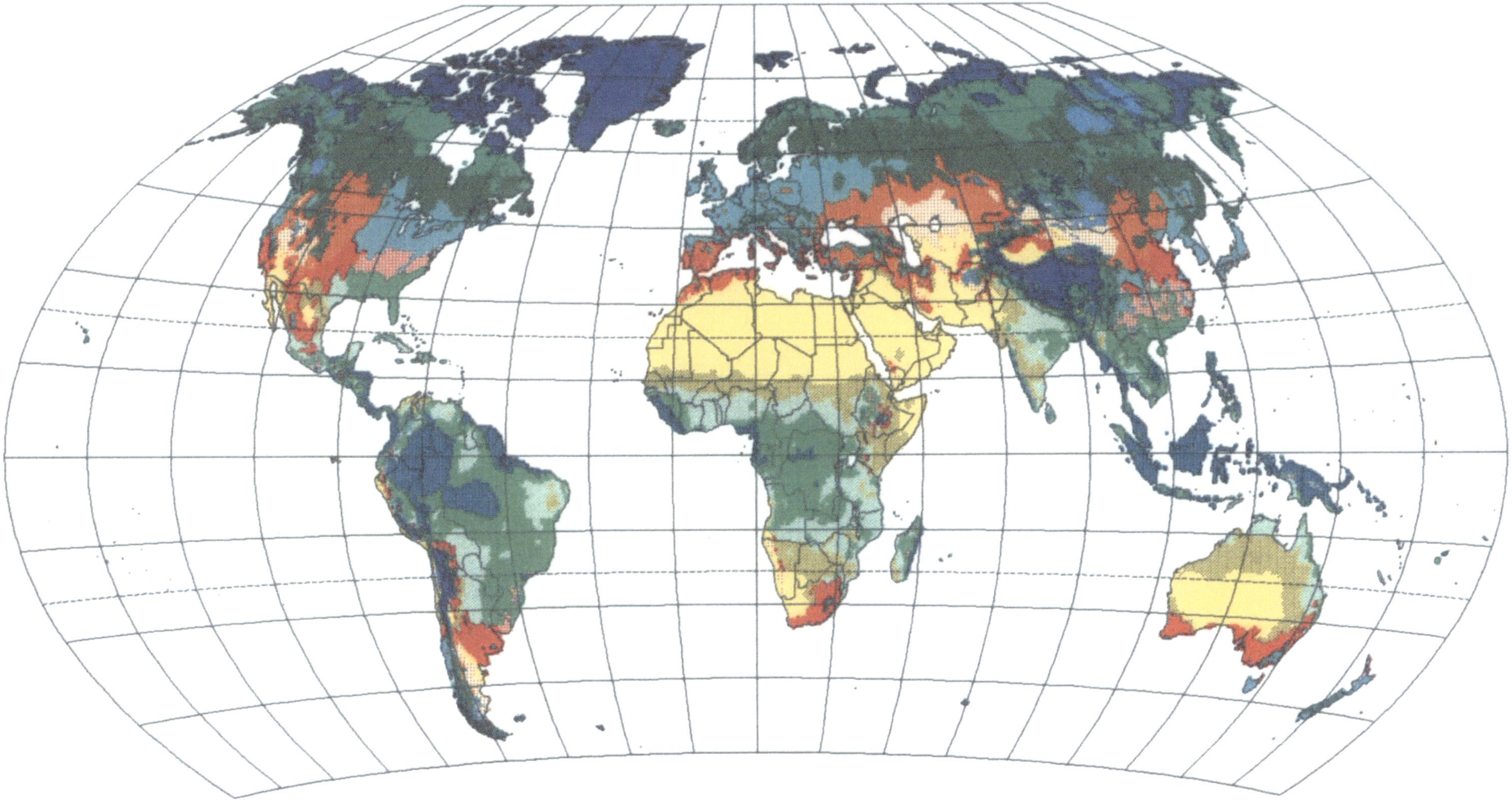

Plate 6 Holdridge Life Zone Classification predicting vegetation patterns with a doubling of carbon dioxide (NASA-GISS global climate model)

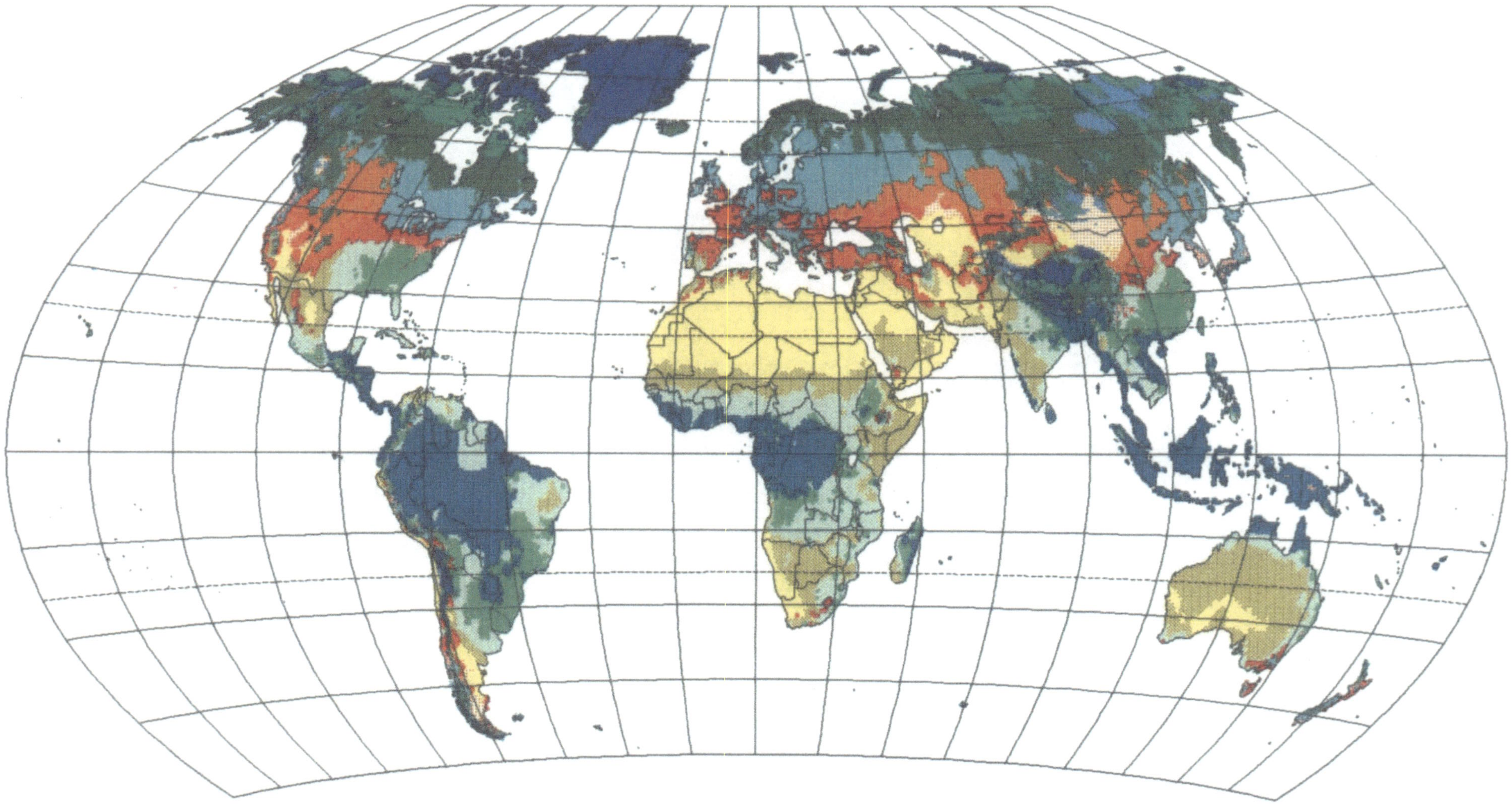

Plate 7 Existing life zones that could change to a new zone with a doubling of carbon dioxide (difference between Plates 5 and 6)

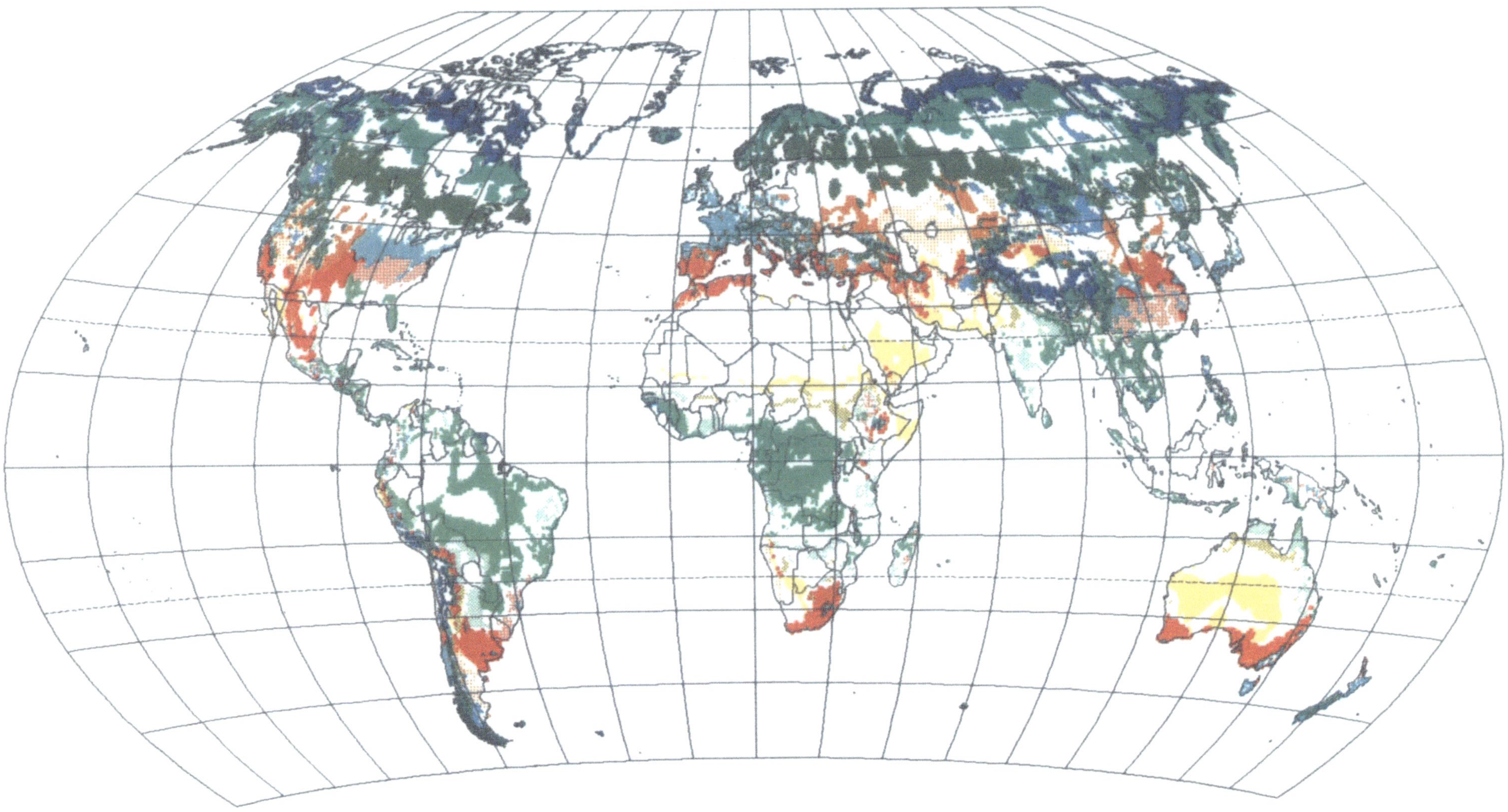

Plate 8 Existing life zones for which four different climate models predict a change to the same new zone

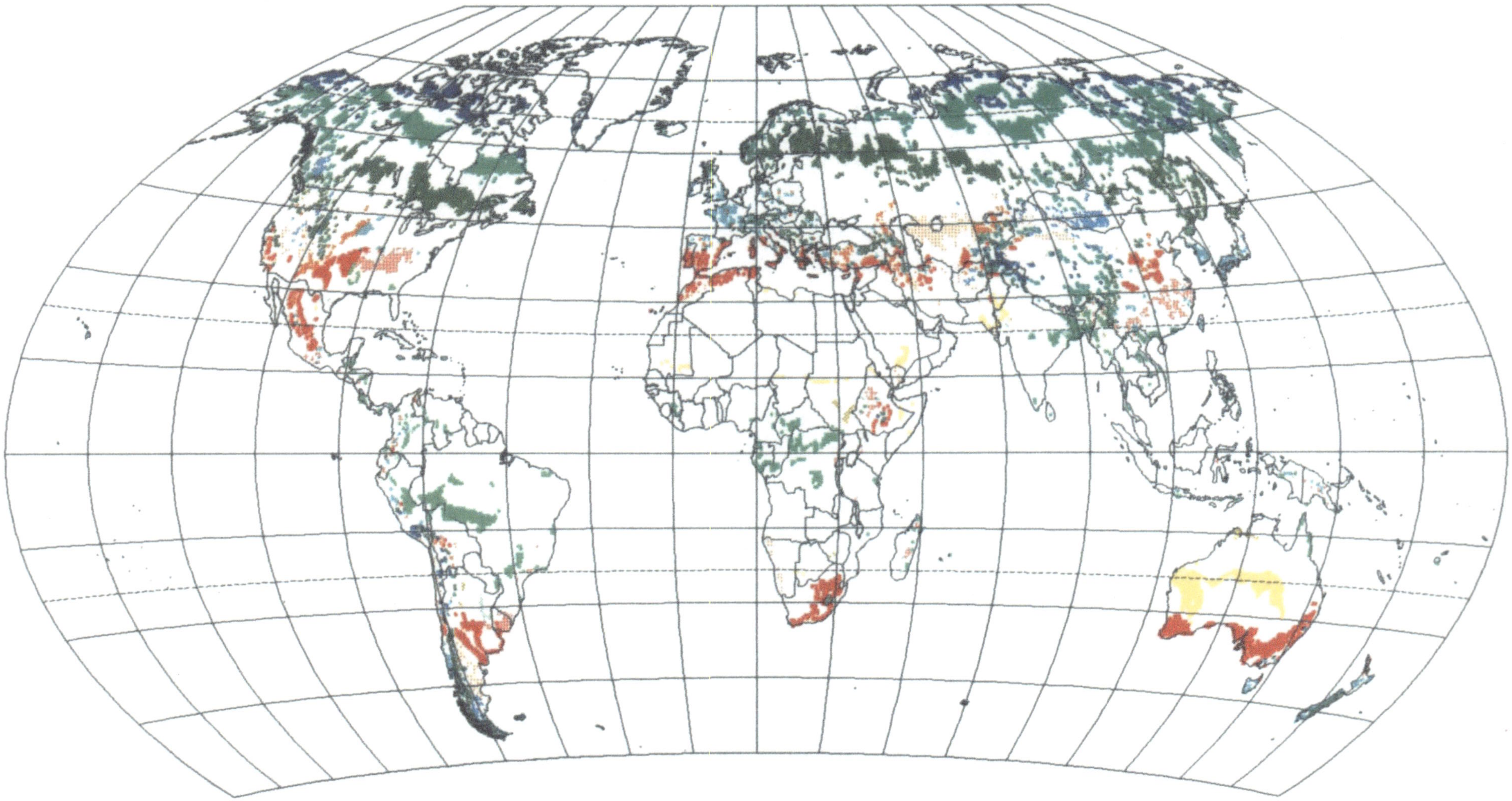

that it is the forests and woodlands that have suffered the most from conversion to croplands. Overall figures such as these may mask other changes which are deleterious to biological diversity. In Europe, for example, forest area has actually increased during the twentieth century but this is the result of large-scale planting of species-poor coniferous monoculture; the area of species-rich natural and semi-natural woodland has continued to decrease. Similarly, the area of grassland in Europe has remained static or nearly so over this period, but there has been wholesale conversion from low nutrient-input, species-rich grassland, to high input, intensively cultivated, species-poor pasture. It is extremely difficult to map these changes and to measure their effect.

Table 18.3 Global land use 1700-1980

VEGETATION TYPES	AREA (10^4 km^2)					CHANGE PERCENTAGE	CHANGE AREA (mkm²)
	1700	1850	1920	1950	1980		1700-1980
Forests and woodlands	6215	5965	5678	5389	5053	-18.7%	11.62mkm²
Grasslands and pasture	6860	6837	6748	6780	6788	-1.0%	0.72mkm²
Croplands	265	537	913	1170	1501	+466.4%	12.36mkm²

Source: Richards, J.F. 1990. Land transformation. In: Turner, B.L. (Ed), *The Earth as Transformed by Human Action*.

Chapters 20 to 24 discuss five major habitat types (tropical rain forests, grasslands, wetlands, coral reefs and mangroves) and assess the impact of man on each of these. For only the first of these are adequate data available to enable estimates for global rates of loss or conversion. These are discussed at length, with individual case studies adding detail. For the other systems considered, examples of threats and changes are given, as well as indications of their often considerable value to man.

References

Bailey, R.G. 1989a. *Ecoregions of the Continents*. U.S. Department of Agriculture, Forest Service, Washington D.C.

Bailey, R.G. 1989b. Explanatory supplement to ecoregions map of the continents. *Environmental Conservation* 16(4):307-309.

Holdridge, L.R. 1967. *Life Zone Ecology*. Tropical Science Center, San José. 206pp.

Olson, J.S., Watts, J.A. and Allison, L.J. 1983. *Carbon in Live Vegetation of Major World Ecosystems*. Oak Ridge National Laboratory, Oak Ridge.

Olson, J.S., Watts, J.A. and Allison, L.J. 1985. *Major World Ecosystem Complexes Ranked by Carbon in Live Vegetation: a database*. Oak Ridge National Laboratory, Oak Ridge.

Richards, J.F. 1990. Land transformation. In: Turner, B.L. (Ed), *The Earth as Transformed by Human Action*. Pp.163-178.

Udvardy, M.D.F. 1975. *A Classification of the Biogeographical Provinces of the World*. IUCN Occasional Paper No. 18. IUCN, Gland.

19. BIODIVERSITY AND GLOBAL CLIMATE CHANGE

The currently increasing levels of the so-called 'greenhouse' gasses (e.g. carbon dioxide, methane, chlorofluorocarbons), in the atmosphere could have large impacts on global biochemical cycles and the climate system. This increase results primarily from human industrial and agricultural activities. There is currently a growing scientific consensus that by the year 2050 global temperatures will have risen significantly (Houghton *et al.*, 1990), with many studies predicting warming of a magnitude not observed during human history. Such climatic change could lead to large impacts on individual organisms, communities, natural ecosystems and global biochemical cycles and have potentially grave impacts on biodiversity.

Large-scale patterns in the physiognomy and potential species occurrence in different vegetation types are primarily determined by climate. Climatic parameters, such as temperature and precipitation, determine the major boundaries between latitudinal zones (e.g. boreal, temperate and tropical), and vegetation types (e.g. deserts, steppes and forests). Temperature and precipitation, and their annual variation, control the potential appearance of vegetation, such as the distribution of deciduous or evergreen tree-species, or short and long grass prairie-species. The combined effects of climate, soil characteristics, vegetation history, large-scale disturbances and anthropogenic influences determine the actual vegetation both regionally and locally.

This close correlation between climate and the physiognomy of vegetation has for some time been recognised by environmental scientists and has led to the use of vegetation to create climate maps and vice versa (e.g. Köppen, 1936; Holdridge, 1967). The Holdridge Life Zone Classification (Holdridge, 1967; Plate 4) is often used for studies of the impact of climate change. The ecoclimatic zones of this model provide reasonable agreement with potential natural vegetation patterns at a global scale. The Life Zone Classification is based on the following annual climatic variables: biotemperature (mean positive temperatures), total annual precipitation and evaporation (defined as a function of biotemperature). The Life Zones are delimited by hexagons derived from a triangular graph of these three variables. Maps of the Life Zone Classification can be created for current climatic conditions (Plate 5) and for potential conditions determined by climate change. To obtain the best possible agreement with existing vegetation patterns, the Life Zones have been aggregated into biomes (large-scale vegetation assemblages).

MODELLING GLOBAL CLIMATE CHANGE

Global climate models can simulate the dynamics of the atmosphere under different conditions. Such models can be used, for example, to determine the potential climatic change equivalent to a doubling of atmospheric carbon dioxide. Detailed descriptions of these models and their results can be found in Houghton *et al.* (1990). The results used here are from the models of the Geophysical Fluid Dynamics Laboratory (GFDL), Goddard Institute for Space Studies (GISS), Oregon State University (OSU) and the United Kingdom Meteorological Office (UKMO).

Although there are differences in the magnitude of change, all models show similar patterns for a changed global climate, particularly with respect to increased temperatures. Greatest temperature increases occur during the winter season in polar regions, and could exceed 15°C. The pattern is less pronounced during the summer season, when the overall temperature increase is less. The different simulations agree less well in terms of precipitation patterns. In general, the models predict a global increase in precipitation, but there are large differences in the predicted seasonal and regional patterns. Besides, many regions that experience an increased precipitation could exhibit no change or even a negative change in moisture availability because of alterations in the balance between temperature, precipitation and evapotranspiration. The models generally predict a relatively modest rise in sea-level, unlikely to exceed one metre over the present century. This rise would be largely a result of thermal expansion of the oceans and melting of minor ice-bodies rather than any major change to the polar ice-caps.

EFFECTS OF CLIMATE CHANGE ON VEGETATION ZONES AND BIODIVERSITY

Aggregated Life Zone Classifications have been generated using the simulated climate-change scenarios (Plate 6: the GISS model). Comparison of this map and Plate 5, for current conditions, clearly displays the potential changes in global vegetation patterns. Large changes in the current extent and location of global vegetation zones are projected and the different scenarios all show a similar pattern of change (Plates 7 and 8). The changes are not consistent across the globe but depend on the non-linear change in both temperature and precipitation. Shifts of biomes are most apparent in the mid and high latitude regions, with only slight changes in the tropics. The boreal and polar biomes show the largest polewards shift, with a decrease in the extent of tundra and forested tundras. These biomes currently form a continuous circumpolar band but under a warmer climate only scattered patches remain. In comparison, the current extent of tropical forests is rather stable, with the total potential area of forest increasing. Any actual increase in the tropical forests, however, will be significantly constrained by human land-use and therefore cannot be expected to evolve to the potential mapped extent.

The maps presented in this assessment give a general indication of expected changes in the distribution of ecoclimatic zones on a global scale. The specific impacts of these changes on global biodiversity are difficult to assess definitively at present, because little is known of the physiological tolerance and potential migration capability of numerous species. However, a preliminary illustration of the potential threats these climatic changes could inflict on biodiversity protection can be given by assessing the impacts of ecoclimatic changes on a global distribution of a selection of large (>1,000ha) nature reserves.

Biodiversity protection in a changing environment will be influenced by both the magnitude and speed of environmental change and also the ability of species to respond to this change. In these terms, the effects of climate change must be viewed in the context of natural ecosystem fragmentation which may inhibit the migration of species to more suitable habitats under future climates. When the correlative ecoclimatic mapping presented above is overlaid onto a global distribution of existing nature reserves, numerous sites are shown to experience shifts in ecoclimatic types. The climatic conditions normally associated with the vegetation structure of many of these sites would be expected to shift beyond the stationary boundaries of the established reserves. Table 19.1 gives the percentage of a selection of 2,618 nature reserves that are in areas affected by large shifts in ecoclimatic zones under different climate scenarios, marked as either 'stable' or 'endangered'. It should be noted that the inherent robustness of individual reserves, for example those with a wide altitude range, has not been considered in these data. The table also indicates a CHANGE scenario which includes those reserves where all climate change scenarios agree that there will be a change from one life zone to another (cf. Plate 7). The SIMILARITY scenario further demands that all four models predict similar new life zones for the reserves concerned (cf. Plate 8).

Table 19.1 Predicted life zone changes in selected reserves

CLIMATE-SCENARIO	STABLE	ENDANGERED	PERCENT ENDANGERED
GFDL	1295	1323	50.5
GISS	1442	1176	44.9
OSU	1522	1096	41.9
UKMO	1097	1521	58.1
CHANGE	1754	864	33.0
SIMILARITY	2173	445	17.0

This translocation of ecoclimatic ranges could act to fragment habitats further as species individually respond to climatic changes. Local extinctions could occur through either direct physiological responses to climatic conditions or through changes in interspecific competition owing to alterations in the composition and population of different species groups within reserves (Peters and Darling, 1985; Hunter *et al.*, 1988). Changes in the future composition of protected habitats may also have significant impacts beyond the regional scale. Migratory species which exploit different biomes seasonally or at different stages of their life histories could be significantly affected by the climatic disruption of reserve sites which link migration corridors or flyways.

This 'biodiversity' assessment depicts shifts in climatic zones and links them with large nature reserves. The shifts were interpreted as having an ecologically significant impact on many reserve sites (Table 19.1). This assessment should, however, not be used as a direct evaluation of the potential loss or gain in biodiversity. However, the percentage of impacted reserves indicates that the current system may not provide the environmental requirements of many species and ecosystems in the near future and will thus be less capable of safeguarding biodiversity.

References

Holdridge, L.R. 1967. *Life Zone Ecology*. Tropical Science Center, San José. 206pp.

Houghton, J.T., Jenkins, G.J. and Ephraums, J.J. (Eds) 1990. *Climate change: the IPCC scientific assessment*. Cambridge University Press, Cambridge. 365pp.

Hunter, M.L., Jacobson, G.L., Jr. and Webb, T. III 1988. Paleoecology and the coarse filter approach to maintaining biodiversity. *Conservation Biology* 2:375-385.

Köppen, W., 1936. Das geographische System der Klimate. In: Köppen, W. and Geiger, R. (Eds) *Handbuch der Klimatologie*. Berlin. 46pp.

Peters, R.L. II and Darling, J.D.S. 1985. The greenhouse effect and nature reserves. *BioScience* 35:707-717.

Contributed by Rik Leemans, Global Change Department, National Institute of Public Health and Environmental Protection, the Netherlands, and P.N. Halpin, Department of Environmental Sciences, University of Virginia, Charlottesville, Virginia, USA.

20. TROPICAL MOIST FORESTS

WHAT ARE TROPICAL MOIST FORESTS?

The terms 'rain forest' and 'tropical moist forest' are often used as synonyms; although neither has a standard definition, the latter is more inclusive than the former. Schimper first used the term *rain forest* in 1903 (Schimper, 1903) and defined it as a forest that is "evergreen, hygrophilous in character, at least 30m high, rich in thick-stemmed lianas and in woody as well as herbaceous epiphytes". Sixty or so years later, Baur (1964) extended this definition somewhat to "a closed community of essentially but not exclusively broadleaved evergreen hygrophilous trees, usually with two or more layers of trees and shrubs with dependent synusiae of life forms such as vines and epiphytes. It includes the characteristic vegetation of the humid tropics, even when this has a somewhat seasonal climatic regime, as well as those of moist elevated areas of the tropics".

In the following discussion of existing areas of tropical forest and rates of change in cover, many of the data have come from FAO and refer to '*closed broadleaved forests*' which are again defined differently, and in particular include dry, deciduous forests. The most comprehensive atlas of tropical forests, two volumes of which have been used in compiling Tables 20.9 and 20.10 (Collins *et al.*, 1991; Sayer *et al.*, 1992), includes mangroves and montane forests in the estimates of tropical moist forest, as well as monsoon forests in Asia; they do not include riverine forests or dry deciduous forests. The third volume in this series, on Latin America, is still in preparation, hence the lack of comparable data for that region in the cited tables. Maps from this series, the most consistent and current available, are reproduced in simplified form in Figs 20.10-12. Many of the difficulties encountered in compiling standard statistics of forest area arise from the use of different or inconsistent definitions of vegetation type.

The significance of tropical forests

The forests are home to millions of people, providing them with shelter, food, clothing, fuel, medicines, building materials and a variety of other resources. They are also the origin of many of these same resources for countless people who do not actually live in the forest.

The commodity that is generally considered to be of the greatest economic value is timber. However, the commercial value of other products such as fruits, nuts, rattans, medicinal plants and rubber - which can be cash crops or for local use - are frequently not taken into account (Peters *et al.*, 1989). Southeast Asia, in particular, has a long history of successful export of non-timber forest products such as rattans, resins and gums (Reitbergen, 1992). Latin America's main non-timber forest exports have been rubber and brazil nuts. In general, non-timber commodities appear to be less significant among forest products of Africa.

Numerous species important to pharmaceutical companies are derived from the rain forests at present and it is predicted that many more will be found if time and money is invested in the search for them. Local people use the forest products to a considerable extent in treating their own ailments and these can form a starting point for investigation by others.

Apart from producing many resources of subsistence and commercial importance, the forests play a key role in regulating water flow, conditioning local climate and protecting against soil erosion. Although the role they play in influencing local rainfall is not well-understood, it is clear that this can be significant. In several places where forests have been destroyed there has been a reduction in rainfall. For instance in Banjul, the capital city of The Gambia, in 1965 when there was still good forest cover, annual precipitation was 1,240mm. Between 1982 and 1988, when the forest had all but disappeared, the mean level was almost halved to 650mm (Jones, 1992). There is reasonable evidence that reduction in rainfall can be a consequence of forest clearance. Preservation of the tropical rain forests is also vital for conserving biodiversity. Although they cover only 6-7% of the earth's surface, these forests probably contain more than 50%, and possibly as much as 90%, of all species of plants and animals.

Factors leading to tropical forest degradation

The timber trade is widely considered to be responsible for much of the destruction of the rain forests, partly directly, but mainly indirectly, by opening up formerly pristine areas to invasion by shifting cultivators. Mining and oil companies have the same effect, leaving roads into the forest and attracting settlers to an area. All three industries cause direct damage as well. Logging can severely degrade an area if not done selectively and with care, as is likely to occur when companies have no stake in the long-term sustainability of supplies. Pollution from mining or oil drilling gives rise to further problems. In some cases, vast quantities of fuel are needed to power a mining programme or other industry and this can be responsible for further devastation. For instance, iron-ore smelters in the Brazilian Grand Carajas Programme will consume 2,300km^2 of forest as charcoal each year.

Though not threatening the forest directly, invasion by commercial companies also displaces indigenous peoples, and this is a major cause of concern in South America and Southeast Asia. The building of dams has also resulted in large areas of forest being lost through flooding and, more importantly, this has often caused major ecological problems in nearby areas as well as encouraging road development and settlement.

It is considered by many that the most important agent of tropical forest destruction is the shifting cultivator. Poverty, population growth and unequal land ownership are the fundamental causes of this form of land conversion. In many cases, governments encourage peasants from high population areas to move into less developed, usually forested, areas (see case studies on forest destruction in Rondonia and transmigration in Indonesia). Some forest destruction, particularly in Brazil and Central America, has occurred as a result of the tax incentives offered to those who cleared forest for cattle ranches.

THE GLOBAL AREA OF TROPICAL MOIST FOREST

There are almost as many estimates of the present extent of tropical forests and rates of deforestation as there are reports about the subject. The problems associated with trying to obtain accurate figures for how much forest cover exists today are greatly multiplied when considering the cover that existed 20 or 50 years ago.

There are two major problems to be overcome before any attempt can be made at calculating either forest area (past or present) or rate of deforestation, and these are essentially problems of definition. The term 'forest' has to be defined and applied consistently throughout the study; this is frequently not done and such variation has given rise to great differences in the estimates of forest cover in some countries. The second problem is that some authors have taken 'deforestation' to mean the complete destruction of a forest while others have included areas that have been degraded (by logging in particular), and have thereby estimated much higher rates of deforestation. Increased precision over the definition and application of these terms is desirable, and clarification as to which are being included in the estimate of deforestation. It is generally very difficult to quantify degradation of an area and yet this type of disturbance can have a significant and protracted effect on the species composition (both flora and fauna), biomass and structure of a forest.

One of the first comprehensive estimates of how much tropical forest existed was made by Sommer (1976) who noted that "a global appraisal of tropical moist forests undertaken at this time can only base its research on the material available - a mass of incomplete data and a number of assumptions. It will yield rather rough results."

Sommer defined what categories of tropical forest he was including in his estimates and which countries he was covering within his regional reviews. Estimates of climax areas of tropical moist forests were taken from vegetation maps and these sources and their problems are all reported.

Figures for present areas of moist tropical forest were calculated from information at FAO headquarters. The main sources of data were land-use and vegetation maps, project reports and country statistics, occasionally supplemented by oral reports. For each country under consideration some detail on the problems encountered are presented, and Sommer makes it quite clear that he is not providing definitive figures from reliable data. He concluded that at the time of his research there were 9,350,000km^2 of tropical moist forest in the world (Table 20.1).

Sommer gives a list of 13 countries for which he was able to find figures in "various reports" for the area of forest lost per year. The character of the clearing was in most cases not reported. From this figure, of 21,600km^2, he extrapolated to all countries with tropical moist forest and obtained an estimated deforestation rate of 110,000km^2 per year.

Sommer concludes his report by noting that "an accurate appraisal of the climax and actual areas of the moist tropical forest at the global level, based on the information available at FAO headquarters, is not yet possible".

In 1980, Myers wrote a report for the National Academy of Sciences on Conversion of Tropical Moist Forests. He too noted that "sound information, especially authoritative statistical information, is not easily obtained". The data sources used by Myers were mostly published reports, generally limited to those produced in the 1970s, combined with correspondence, discussions and visits to three Southeast Asian and three South American countries.

He noted that there is no standard and objective classification of 'tropical moist forest' (TMF) but, after consulting numerous sources, he uses the definition "evergreen or partly evergreen forests, in areas receiving not less than 100mm of precipitation, in any month for two out of three years, with mean annual temperatures of 24+°C and essentially frost-free; in these forests some trees may be deciduous; the forests usually occur at

Table 20.1 Areas of tropical moist forest estimated by Sommer (1976)

REGION	PRESENT MOIST FOREST AREA (thousand km^2)	PER CENT OF WORLD'S MOIST FOREST	PER CENT OF TOTAL LAND AREA
Africa			
East Africa	70	0.7	10.6
Central Africa	1,490	15.0	65.9
West Africa	190	2.0	19.1
Total	1,750	18.7	36.2
Tropical America			
Latin America	4,720	50.5	53.5
Central America and Caribbean	340	3.6	31.9
Total	5,060	54.1	51.2
Asia			
Pacific	360	3.8	12.8
Southeast Asia	1,870	20.0	67.4
South Asia	310	3.3	24.4
Total	2,540	27.2	37.2
TOTAL HUMID TROPICS	9,350	100.0	42.8

Source: Sommer, A. 1976. Attempt at an assessment of the world's tropical moist forests. *Unasylva* 28(112-113):5-24.

altitudes below 1,300m (though often in Amazonia up to 1,800m and generally in Southeast Asia up to only 750m); and in mature examples of these forests, there are several more or less distinctive strata". However, as several other authors have noted (e.g. Lugo and Brown, 1982; Holdgate, 1982; Sayer and Whitmore, 1991), he does not subsequently use this definition when giving forest areas for some of the countries discussed.

In his chapter on the role of forest farmers, Myers (1980, p.25) states that "it is not unrealistic to suppose that forest farmers are converting at least 100,000km^2 of primary forest to permanent cultivation each year". He then continues "when considered in conjunction with other factors - timber harvesting, planned agriculture, cattle raising, etc. - it becomes possible to credit that something approaching 200,000km^2 of TMF, and possibly even more, are being converted each year". He noted earlier that conversion can range from marginal modification to fundamental transformation, and it is not clear how drastic a conversion the "other factors" are causing.

Myers' report continues with regional reviews for Southern and Southeast Asia including Melanesia (14 countries and Melanesia); Tropical Latin America (18 countries and the Caribbean); and Tropical Africa (13 countries). The degree of detail given for each country is very variable and in most instances the causes of deforestation are concentrated on, rather than the rate. No comprehensive summary of deforestation rates or forest area in each country is presented and, in many cases, this information cannot be extracted from the text. However, the report does identify areas particularly at risk from deforestation and its intention was probably never to estimate forest areas but to document the different forms and degrees of conversion of tropical moist forest that were taking place.

The most comprehensive reports on forest cover and rates of deforestation are those produced by FAO/UNEP in 1981 and updated in 1988. The 1981 study was carried out on a national basis, and this information was then used to compile a regional synthesis for each of the three tropical areas (Tropical Africa, Tropical Asia and Tropical America) from which a global assessment of forest cover and rate of deforestation was derived. An FAO Forestry paper (Lanly, 1982) was subsequently produced which collated all the findings of the Tropical Forest Resources Assessment Project in an overall synthesis for the tropical world.

Forest extent was estimated for each of the 76 countries (23 in tropical America, 37 in tropical Africa and 16 in Asia) covered by the project using the best available sources of information. The whole project occupied a total of almost seven man years. Data collection involved visits to research institutes in Europe, particularly those involved in the study and mapping of vegetation, visits to national forestry, land-use and survey departments and to regional FAO offices, considerable correspondence with the forestry services, use of satellite imagery for 18 countries and interpretation of satellite imagery for the 13 countries where no other information was available. In three cases the assessments were carried out by the countries themselves. Rates of forest removal were taken either from the degree of shifting cultivation or land-use statistics. These rates were used to extrapolate forest cover to a common baseline of December 1980. Estimates for deforestation rates for 1981-1985 were also made but these projections can be only broadly indicative of trends and future conditions and, as the report states, they have to be viewed with caution.

As a result of the differences in information quality, estimates of forest cover and deforestation rates are judged to be very reliable for only 15 of the 76 countries surveyed. In terms of total area of closed tropical forests, however, this represents 40% of forest extent, largely because Brazil is in the very reliable group. A further 38 countries (covering another 40% of closed forest) have very good baseline data on forest cover while in the remaining 23 countries, both baseline data and deforestation rates are considered to be of medium to poor quality.

The classification of vegetation types used by FAO/UNEP has placed particular emphasis on forest management. In the update of these reports (FAO, 1988), which includes a further 53 countries, a clear distinction has been made between the open and closed forests in the tabulated data provided. The closed broadleaved forests, which are the ones that are usually equated with the term tropical moist forest, are defined as "generally, but not always multi-storeyed, and may be evergreen, semi-deciduous or deciduous and occur in wet, moist or dry zones. They cover with their various storeys and undergrowth, a high proportion of the ground; and do not contain a continuous dense grass layer" (FAO, 1988).

The open forests, in contrast, "correspond to mixed forest-grassland formations, with a continuous dense grass layer in which the crowns of the trees cover more than 10% of the ground. They thus include e.g. the various forms of cerrado and chaco in America; and tree and wooded savannas and woodlands in Africa" (FAO, 1988).

In FAO's 1988 summary report, information on areas of bamboo and coniferous formations is also supplied, and each forest type is broken down into operable forests (which may be unmanaged - either virgin or logged - or managed) or inoperable forests (for either physical or legal reasons). Information on deforestation is similarly presented within these classes of forest type and management status.

The 1981 study (Lanly, 1982) concluded that in 1980 there were approximately 12 million km^2 of closed forest, of which 97% was closed broadleaved, and 7,350,000 km^2 of open tree formations remaining in the tropical world (Table 20.2).

The 1988 updated report indicated that about 75,000 km^2 of closed forest (Table 20.3) and of open formations (Table 20.4) were being cleared each year between 1981 and 1985 to allow the land to be used for other purposes. This is an average reduction rate of 0.62% for the closed forest and 0.52% for the open formations. In addition to this deforestation, considerable degradation was occurring, particularly in the open forest, caused by overgrazing, fuelwood gathering and repeated burning. This, however, is more gradual and difficult to quantify.

Table 20.2 FAO/UNEP (1981) estimates of the area of tropical forest remaining in 1980

	CLOSED FOREST[1] (km²)	OPEN FOREST (km²)
Tropical America	6,786,550	2,169,970
Tropical Africa	2,166,340	4,864,450
Tropical Asia	3,055,100	309,480
WORLD	12,007,990	7,343,900

Source: Lanly, J.-P. 1982. Tropical Forest Resources. FAO Forestry Paper 30, FAO, Rome, Italy.
Note: [1] Includes closed broadleaved, coniferous and bamboo forest.

Table 20.3 FAO (1988) estimates of annual areas deforested and deforestation rates in closed forests[1]

	TOTAL AREAS DEFORESTED (km²)		DEFORESTATION RATES (per cent)	
	1976-1980	1981-1985	1976-1980	1981-1985
Americas	41,190	43,390	0.60	0.63
Africa	13,330	13,310	0.61	0.61
Asia	18,150	18,260	0.59	0.60
TOTAL	72,670	74,960	0.60	0.62

Source: FAO 1988. *An Interim Report on the State of the Forest Resources in the Developing Countries*. FAO, Rome, Italy.
Note: [1] Includes closed broadleaved, bamboo and coniferous forests.

Table 20.4 FAO (1988) estimates of annual areas deforested and deforestation rates in open forests

	ANNUAL AREA DEFORESTED (km²) 1981-1985	DEFORESTATION RATES (per cent) 1981-1985
Americas	12,720	0.59
Africa	23,450	0.48
Asia	1,900	0.61
TOTAL	38,070	0.52

Source: FAO 1988. *An Interim Report on the State of the Forest Resources in the Developing Countries*. FAO, Rome, Italy.

Another source of figures on extent of forests in the tropics is the land-use tables in the FAO Production Yearbooks. These give data over time for each country on the extent of arable land, permanent crops, permanent pasture, forest and woodland, and other land. The area of forest and woodland "refers to land under natural or planted stands of trees, whether productive or not, and includes land from which forests have been cleared but that will be reforested in the foreseeable future." This is not, therefore, a particularly meaningful figure, and especially not when data on moist forests alone are required. However, the increase in the area under crops over time can, in some countries, be an indication of how much forest is being converted to agricultural land so some authors (e.g. Mather, 1990) have been able to use these data in their reports on deforestation. FAO is in the process of updating its information on each country and region to produce a second detailed report on forest resources with 1990 as the reference year. This should be published in late 1992. There have, however, already been two interim reports (FAO, 1990, 1991) and an evaluation of the first of these (Lanly *et al.*, 1991).

In both interim reports, forests are defined as "ecological systems with a minimum of 10% crown cover of trees and/or bamboos, generally associated with wild flora and fauna and natural soil conditions and not subject to agricultural practices". Deforestation refers to "change of land use or depletion of crown cover to less than 10%". The first interim report gives provisional estimates of forest cover and deforestation for 62 countries, mostly in the moist tropical zone (Table 20.5). It is stressed that parallel developments cannot be assumed to have taken place in the forests of the dry and mountainous zones.

By the time the second interim report was published, data on a further 25 countries had been included and existing

Table 20.5 Preliminary FAO (1990) estimates of 1990 forest area and deforestation for 62 countries in the tropics

CONTINENT	NO. OF COUNTRIES STUDIED	TOTAL LAND AREA (km²)	FOREST AREA 1980 (km²)	FOREST AREA 1990 (km²)	ANNUAL DEFOREST. 1981-90 (km²)	DEFOREST. RATE 1981-90 (%)
Africa	15	6,098,000	2,897,000	2,418,000	48,000	1.7
Latin America	32	12,636,000	8,259,000	7,530,000	73,000	0.9
Asia	15	8,911,000	3,345,000	2,875,000	47,000	1.4
TOTAL	62	27,645,000	14,501,000	12,823,000	168,000	1.2

Source: FAO 1990. *Interim Report on Forest Resources Assessment 1990 Project.* Committee on Forestry Tenth Session. FAO, Rome, Italy.

Table 20.6 Preliminary FAO (1991) estimates of forest area and deforestation for 87 countries in the tropics

CONTINENT	NO. OF COUNTRIES STUDIED	TOTAL LAND AREA (km²)	FOREST AREA 1980 (km²)	FOREST AREA 1990 (km²)	ANNUAL DEFOREST. 1981-90 (km²)	DEFOREST. RATE 1981-90 (%)
Latin America	32	16,756,000	9,229,000	8,399,000	84,000	0.9
Central America & Mexico	7	2,453,000	770,000	635,000	14,000	1.8
Caribbean Sub-region	18	695,000	488,000	471,000	2,000	0.4
Tropical South America	7	13,608,000	7,971,000	7,293,000	68,000	0.8
Asia	15	8,966,000	3,108,000	2,748,000	35,000	1.2
South Asia	6	4,456,000	706,000	662,000	4,000	0.6
Continental SE Asia	5	1,929,000	832,000	697,000	13,000	1.6
Insular SE Asia	4	2,581,000	1,570,000	1,389,000	18,000	1.2
Africa	40	22,433,000	6,504,000	6,001,000	51,000	0.8
West Sahelian Africa	8	5,280,000	419,000	380,000	4,000	0.9
East Sahelian Africa	6	4,896,000	923,000	853,000	7,000	0.8
West Africa	8	2,032,000	552,000	434,000	12,000	2.1
Central Africa	7	4,064,000	2,301,000	2,154,000	15,000	0.6
Tropical Southern Africa	10	5,579,000	2,177,000	2,063,000	11,000	0.5
Insular Africa	1	582,000	132,000	117,000	2,000	1.2
TOTAL	87	48,155,000	18,841,000	17,148,000	170,000	0.9

Source: FAO 1991. *Second Interim Report on the State of Tropical Forests by Forest Resources Assessment 1990 Project.* Tenth World Forestry Congress, September 1991, Paris, France.

Table 20.7 A comparison of forest area and deforestation rate as estimated in FAO's tropical forests resource assessment projects for 1980 and 1990

PROJECT	FOREST AREA (km²)	ANNUAL AREA DEFORESTED (km²)	DEFORESTATION RATE (%)
Reference year 1980	19,350,000	113,000 (for 1981-85)	0.6
Reference year 1990	18,820,000	169,000 (for 1981-90)	0.9

information had been updated. Although deforestation rates for Latin America and Asia remained similar to those presented in the 1990 report, the rate of change given in this report is considerably lower for Africa (Table 20.6), presumably because of the inclusion of vast areas of open forest. It must be noted that in both these interim reports, no distinction has been made between open and closed forests whereas, in the discussions above, figures from the FAO/UNEP (1981) report are for closed forests only.

When a comparison is made between the FAO assessments for 1990 and 1980 using the data from the 76 countries common to the two reports, it is apparent that the annual rate of deforestation has risen considerably (Table 20.7).

Table 20.8 Deforestation estimates for closed tropical forests[a], for selected countries

COUNTRY	FAO ESTIMATES 1981-85[1 b] ANNUAL AREA LOST (km²)	FAO ESTIMATES 1981-85[1 b] ANNUAL RATE OF LOSS (%)	RECENT NON-FAO ESTIMATES ANNUAL AREA LOST (km²)	RECENT NON-FAO ESTIMATES ANNUAL RATE OF LOSS (%)	RECENT NON-FAO ESTIMATES PERIOD OF RECENT ESTIMATES
Brazil	14,800	0.4	80,000 [2c]	2.2	1987
			35,000 [3]		1987
Cameroon[d]	800	0.4	1,000 [4]	0.6	1976-86
			1,000 [5]		*c.* 1987
Costa Rica	650	4.0	1,240 [6]	7.6	1977-83
Colombia	8,200	0.4	6,000 [7]		1960-84[e]
India[d]	1,470	0.3	15,000 [8]	4.1	1975-82
Gabon	150	0.1	1,500 [9]		*c.* 1987
Indonesia	6,000	0.5	9,000 [10]	0.8	1979-84
			10,000 [11]		*c.* 1985
Myanmar	1,050	0.3	6,770 [12]	2.1	1975-81
Peru	2,700	0.4	3,000 [13]		1988
Philippines	920	1.0	1,430 [14]	1.5	1981-88
			1,300 [15]		1988
Thailand[f]	3,790	2.4	3,970 [16]	2.5	1978-85
Malaysia	2,550	1.2	3,100 [17]		*c.* 1985
Viet Nam	650	0.7	1,730 [18]	2.0	1976-81

Sources: [1] FAO 1988. [2] Setzer, A.W. *et al.* 1988. [3] Fearnside, P.M. 1990. [4] Joint Interagency Planning and Review Mission (JIM) 1988. [5] FAO/UNDP 1988. [6] Sader and Joyce 1988. [7] Plan de Acción Forestal de Colombia, perfil de proyectos (undated). [8] Vohra, B.B. 1987. [9] IUCN 1989. [10] The World Bank 1988. [11] USAID 1987. [12] Kyaw, U.S. 1987. [13] Saavedra, C. and Suarez de Freitas, G. 1989. [14] Philippines Forest Management Bureau 1988. [15] Forest Management Bureau 1988. [16] Royal Forestry Department of Thailand 1986. [17] Thang, H.C. 1987. [18] Vo Quy 1988.

Notes: [a] Closed forests are forests in which trees cover a high proportion of the ground and in which grass does not form a continuous layer on the forest floor. Open forests are forests in which trees are interspersed with grazing lands. [b] Unless otherwise noted, annual deforestation rate is calculated from FAO 1981 estimates. [c] For Legal Amazon only. Brazil also has a small amount of closed coastal forest remaining. [d] Annual deforestation rate is calculated from data found in sources 4 and 8, above. [e] This long period over which deforestation rate has been estimated may be the cause of the comparatively low figure given. [f] Represents total forests, open and closed.

FAO (1990) and Lanly *et al.* (1991) give three possible reasons for this increase in the rate of deforestation between 1980 and 1990. These are:

- an actual increase of rate of deforestation
- an underestimation of the rate of deforestation in the 1980 assessment
- an overestimation of the rate of deforestation in the 1990 assessment.

At this stage, FAO has not been able to assess the relative contribution of these various components. It is hoped, however, that the final results, based on uniform remote sensing observations made specifically for the 1990 project, will provide a more accurate estimate of forest cover which can then be used in calculations of deforestation rates.

Reports other than those by FAO have also indicated that there has indeed been an increase in the rate of deforestation over the past decade. WRI (1990) reported a considerably higher rate of deforestation for the eight countries for which they obtained figures compared with the estimates of FAO (1988). Their data, presented in Table 20.8, are for closed forests (except in Thailand) and, if accurate, suggest that around 204,000km² of this forest type are being lost annually. Sayer and Whitmore (1991) give a table comparing annual rates of deforestation as estimated by FAO (1988) for 1981-1985 and those given by Myers (1989). These, together with other rates of deforestation from more recent sources, have been incorporated in Table 20.8. In all countries except Colombia, FAO gives lower estimates than the other sources report. IUCN and WCMC are in the process of compiling a series of atlases showing the extent of forest present today. The first volume, on Asia and the Pacific has been published, the second, on Africa is in press, while the third, on Latin America and the Caribbean, is in preparation. Data from the first two volumes have been used in compiling Tables 20.9 and 20.10 and simplified maps derived from the series are

Table 20.9 Tropical moist forests: original versus remaining extent

	APPROXIMATE ORIGINAL EXTENT OF CLOSED CANOPY TROPICAL MOIST FORESTS (KM²)	REMAINING EXTENT OF MOIST FORESTS (KM²)			% MOIST FOREST REMAINING	
		From atlas maps; rain & monsoon forests	Publication date of maps	FAO (1988) data for 1980, closed broadleaved plus coniferous forests	From map data	From FAO (1988) data
ASIA						
Bangladesh	130,000	9,730	1981–6	9,270	7.0	7.0
Brunei	5,000	4,692	1988	3,230	94.0	65.0
Cambodia	160,000	113,250	1971	71,680	71.0	45.0
China/Taiwan	340,000	25,860	1979	125,860	8.0	–
India	910,000	228,330	1986	504,010	25.0	55.0
Indonesia	1,700,000	1,179,140	1985–9	1,138,950	69.0	67.0
Laos	225,000	124,600	1987	78,100	55.0	35.0
Malaysia	320,000	200,450	–	209,960	63.0	66.0
Peninsular	(130,000)	(69,780)	1986	–	54.0	–
Sabah	(70,000)	(36,000)	1984	–	51.0	–
Sarawak	(120,000)	(94,670)	1979	–	79.0	–
Myanmar	600,000	311,850	1987	313,090	52.0	52.0
Philippines	295,000	66,020	1988	95,100	22.0	32.0
Singapore	500	20	(1980s)	–	4.0	–
Sri Lanka	26,000	12,260	1988	16,590	47.0	64.0
Thailand	250,000	106,900	1985	83,350	43.0	33.0
Viet Nam	280,000	56,680	1987	75,700	20.0	27.0
OCEANIA						
Australia	11,000	10,516	1988	10,516	96.0	–
Fiji	18,000	6,970	(1980s)	8,110	39.0	45.0
Papua New Guinea	450,000	366,750	1975	342,300	82.0	76.0
Solomon Islands	27,500	25,590	(1980s)	24,230	90.0	90.0
AFRICA						
Angola	218,200	–	–	29,000	–	13.3
Benin	16,800	424	1989–90 and 79	470	2.5	2.8
Burundi	10,600	413	1984	150	3.9	1.4
Cameroon	376,900	155,330	1985	179,200	41.2	47.5
Central African Rep	324,500	52,236	1985	35,900	16.1	11.1
Comoros	2,230	–	–	160	–	7.1
Congo	342,000	–	–	213,400	–	62.4
Cote d'Ivoire	229,400	27,464	1989–90	44,580	12.0	19.4
Djibouti	300	–	–	10	–	3.0
Equatorial Guinea	26,000	17,004	1960	12,950	65.4	49.8
Ethiopia	249,300	–	–	27,500	–	11.0
Gabon	258,000	227,500	–	205,000	88.2	79.5
Gambia	4,100	497	1985	650	12.1	15.6
Ghana	145,000	15,842	1989–90	17,180	10.9	11.8
Guinea	185,800	7,655	1989	20,500	4.1	11.0
Guinea–Bissau	36,100	–	–	6,660	–	18.4
Kenya	81,200	–	–	6,900	–	8.5
Liberia	96,000	41,238	1989–90	20,000	43.0	20.8
Madagascar	275,086	41,715	1985	103,000	15.2	37.4
Malawi	10,700	320	–	1860	3.0	17.4
Mauritius	1,850	–	–	30	–	1.6
Mozambique	246,900	–	–	9,350	–	3.9
Nigeria	421,000	38,620	1989–90	59,500	9.2	14.1
Reunion	2,500	–	–	820	–	32.8
Rwanda	9,400	1,554	(nd)	1,010	16.5	10.7
Sao Tomé and Principe	960	299	1985	560	31.1	58.3
Senegal	27,700	2,045	1985	2,200	7.4	7.9
Seychelles	270	–	–	30	–	11.1
Sierra Leone	71,700	5,064	1989–90	7,400	7.1	10.3
Somalia	21,200	–	–	14,800	–	69.8
Sudan	27,000	–	–	6,400	–	23.7
Tanzania	176,200	–	–	14,400	–	8.2
Togo	18,000	1,360	1989–90	3,040	7.6	16.9
Uganda	103,400	7,400	–	7,500	7.2	7.3
Zaire	1,784,000	1,190,737	1990	1,056,500	66.7	59.2
Zimbabwe	7,700	80	–	2,000	1.0	26.0

Notes: The data for Asian countries in column 1 are adapted from IUCN 1986. *Review of the Protected Areas System in the Indo–Malayan Realm.* IUCN, Gland. In the absence of comparable data for Australia and southern China, the totals in column 4 have been calculated using the map figures in column 2. The FAO data for India are not directly comparable with the map data, as the former includes India's extensive thorn forests. [The remaining area figure for Cambodia in column 3 is now known to be 71,500km² according to FAO 1988. *An Interim Report on the State of Forest Resources in the Developing Countries.* FAO, Rome.] Data in column 1 are mostly from MacKinnon, J. and MacKinnon, K. 1986. *Review of the Protected Area System in the Afrotropical Realm.* IUCN, Gland; (except for Gabon and Liberia which were too high). Where this source gave no estimates for forest cover of small islands it was assumed that they were once totally forested. The figures for Equatorial Guinea and Zaire in column 2 include 7,945km² and 86,547km² of degraded lowland rain forest respectively. The figure for Gabon in column 2 is from IUCN 1990. *La Conservation des Ecosystèmes Forestiers d'Afrique Centrale.* UICN, Gland. The figure in column 2 for Madagascar has been calculated by adding the figure from Green, G.M. and Sussman, R.W. 1990. Deforestation history of the eastern rain forests of Madagascar. *Science* 248:212– 215; for eastern rain forest to that calculated for mangroves from Map 26.1 in Africa plus an estimated 400km² for forest remaining in the Sambirano region. The figure for Malawi in column 2 is from Dowsett–Lemaire, F. 1989. The flora and phytogeography of the evergreen forests of Malawi. I: afromontane and mid–altitude forests. *Bulletin du Jardins Botanique National de Belgique* 59:3–131; and Dowsett–Lemaire, F. 1990. The flora and phytogeography of the evergreen forests of Malawi. II: lowland forests. *Bulletin du Jardins Botanique National de Belgique* 60:9– 71. The figure in column 2 for Uganda is from Howard, P.C. 1991. *Nature Conservation in Uganda's Tropical Forest Reserves.* IUCN, Gland. The figure for Zimbabwe in column 2 was supplied by T. Muller, in litt. The digital dataset in column 3 for Zaire was completed in 1990 but is based on 1988 data. – no data available. (nd) = no date.

Table 20.10 Tropical Moist Forests (Protected Area Coverage)

	LAND AREA (000km²)	APPROX. ORIGINAL EXTENT OF CLOSED CANOPY TMF (000km²)	REMAINING AREA OF TMF (km²)	TOTAL AREA OF PROTECTED AREAS WITH TMF			EXISTING TROPICAL MOIST FOREST PROT. AREAS AS PERCENTAGE OF:			EXISTING AND PROPOSED TMF PROT. AREAS AS PERCENTAGE OF:		
				Existing (km²)	Proposed (km²)	Totals (km²)	Land area	Original	Remaining	Land area	Original	Remaining
ASIA												
Bangladesh	134.0	130.0	9,730.0	744.0	–	744.0	0.60	0.50	7.60	0.60	0.50	7.60
Brunei	5.8	5.0	4,692.0	1,078.0	104.0	1,182.0	18.50	21.50	22.90	20.30	23.60	25.10
Cambodia	177.0	160.0	71,500.0	20,351.0	4,675.0	25,026.0	11.50	12.70	18.00	14.10	15.60	22.10
China/Taiwan	9,363.0	340.0	25,860.0	3,865.0	290.0	4,155.0	0.04	1.10	14.90	0.04	1.20	16.10
India	2,973.0	910.0	228,330.0	22,658.0	18,892.0	41,550.0	0.80	2.40	9.90	1.30	4.50	18.10
Indonesia	1,812.0	1,700.0	1,179,140.0	137,875.0	128,108.0	265,983.0	7.60	8.10	11.70	14.60	15.60	22.50
Lao People's Dem Rep	231.0	225.0	124,600.0	–	47,211.0	47,211.0	0.00	0.00	0.00	20.40	21.00	37.90
Malaysia	329.0	320.0	200,450.0	13,263.0	14,388.0	27,651.0	4.00	4.10	6.60	8.40	8.60	13.80
Peninsular	(132.0)	(130.0)	(69,780.0)	(6,181.0)	(6,519.0)	(12,700.0)	4.70	4.80	8.90	9.60	9.80	18.20
Sabah and Sarawak	(198.0)	(190.0)	(130,670.0)	(7,082.0)	(7,869.0)	(14,951.0)	3.60	3.70	5.40	7.60	7.90	11.40
Myanmar	658.0	600.0	311,850.0	5,641.0	7,399.0	13,040.0	0.90	0.90	1.80	2.00	2.20	4.20
Philippines	298.0	295.0	66,020.0	1,775.0	620.0	2,395.0	0.60	0.60	2.60	0.80	0.80	3.60
Singapore	0.6	0.5	0.1	0.7	–	0.7	0.10	0.10	70.00	0.10	0.10	70.00
Sri Lanka	65.0	26.0	12,260.0	6,309.0	–	6,309.0	9.70	24.30	51.50	9.70	24.30	51.50
Thailand	512.0	250.0	106,900.0	44,790.0	11,855.0	56,645.0	8.70	17.90	41.90	11.10	22.70	53.00
Viet Nam	325.0	280.0	56,680.0	6,252.0	–	6,252.0	1.90	2.20	11.00	1.90	2.20	11.00
OCEANIA												
Australia	7,618.0	11.0	10,516.0	7,605.0	–	7,605.0	0.09	69.10	72.30	0.09	69.10	72.30
Papua New Guinea	452.0	450.0	366,750.0	9,164.0	–	9,164.0	2.00	2.00	2.50	2.00	2.00	2.50
AFRICA												
Angola	1,247.0	218,200.0	29,000.0	*	–	–	–	–	–	–	–	–
Benin	111.0	168,000.0	424.0	none	none	–	–	–	–	–	–	–
Burundi	26.0	10,600.0	413.0	379.0	none	379.0	1.50	3.60	91.80	1.50	3.60	91.80
Cameroon	465.0	376,900.0	155,330.0	11,266.0	none	11,266.0	2.40	3.00	7.30	2.40	3.00	7.30
Central African Rep	623.0	324,500.0	52,236.0	4,359.0	10,500.0	14,859.0	0.70	1.30	8.30	2.40	4.60	28.40
Comoros	2.0	2,230.0	160.0	*	–	–	–	–	–	–	–	–
Congo	342.0	342,000.0	213,400.0	12,148.0	none	12,148.0	3.60	3.60	5.70	3.60	3.60	5.70
Cote d'Ivoire	318.0	229,400.0	27,464.0	7,095.0	none	7,095.0	2.20	3.10	25.80	2.20	3.10	25.80
Djibouti	23.0	300.0	10.0	*	–	–	–	–	–	–	–	–
Equatorial Guinea	28.0	26,000.0	17,004.0	3,150.0	none	3,150.0	11.30	12.10	18.50	11.30	12.10	18.50
Ethiopia	1,101.0	249,300.0	27,500.0	11,574.0	–	–	–	–	–	–	–	–
Gabon	258.0	258,000.0	227,500.0	17,900.0	none	17,900.0	6.90	6.90	7.90	6.90	6.90	7.90
Gambia	10.0	4,100.0	497.0	100.0	none	none	1.00	2.40	20.00	1.00	2.40	20.00
Ghana	230.0	145,000.0	15,842.0	946.0	379.0	1,325.0	0.40	0.70	6.00	0.60	0.90	8.40
Guinea	246.0	185,800.0	7,655.0	140.0	none	140.0	0.06	0.08	1.80	0.06	0.08	1.80
Guinea–Bissau	28.0	36,100.0	6,660.0	none	none	–	–	–	–	–	–	–
Kenya	570.0	81,200.0	6,900.0	13,148.0	–	–	–	–	–	–	–	–
Liberia	96.0	96,000.0	41,238.0	15,666.0	687.0	16,353.0	16.30	16.30	38.00	17.00	17.00	40.00
Madagascar	582.0	275,086.0	41,715.0	5,788.0	none	5,788.0	1.00	2.10	13.90	1.00	2.10	13.90
Malawi	94.0	10,700.0	320.0	*	–	–	–	–	–	–	–	–
Mauritius	2.0	1,850.0	30.0	*	–	–	–	–	–	–	–	–
Mozambique	782.0	246,900.0	9,350.0	*	–	–	–	–	–	–	–	–
Nigeria	911.0	421,000.0	38,620.0	2,158.0	4,060.0	6,218.0	0.24	0.50	5.60	0.68	1.58	16.10
Reunion	2.5	2,500.0	820.0	*	–	–	–	–	–	–	–	–
Rwanda	25.0	9,400.0	1,554.0	150.0	none	150.0	0.60	1.60	9.70	0.60	1.60	9.70
Sao Tomé and Principe	1.0	960.0	299.0	none	none	–	–	–	–	–	–	–
Senegal	193.0	27,700.0	2,045.0	846.0	1.0	847.0	0.40	3.10	41.40	0.40	3.10	41.40
Seychelles	0.3	270.0	30.0	*	–	–	–	–	–	–	–	–
Sierra Leone	72.0	71,700.0	5,064.0	12.0	992.0	1,004.0	0.02	0.02	0.24	1.40	1.40	19.80
Somalia	627.0	21,200.0	14,800.0	*	–	–	–	–	–	–	–	–
Sudan	2,376.0	27,000.0	6,400.0	*	–	–	–	–	–	–	–	–
Tanzania	886.0	176,200.0	14,400.0	77,008.0	–	–	–	–	–	–	–	–
Togo	54.0	18,000.0	1,360.0	none	none	–	–	–	–	–	–	–
Uganda	200.0	103,400.0	7,400.0	4,483.0	310.0	4,793.0	2.20	4.30	60.10	2.40	4.60	64.80
Zaire	2,267.0	1,784,000.0	1,190,737.0	63,130.0	none	63,130.0	2.80	3.50	5.30	2.80	3.50	5.30
Zimbabwe	387.0	7,700.0	80.0	*	–	–	–	–	–	–	–	–

Notes for ASIA section of the Table 20.10 Tropical Moist Forests (Protected Area Coverage).
Figures given in column 2 are derived from maps in chapters 12-29 of Collins, N.M., Sayer, J.A. and Whitmore, T. 1991. *The Conservation Atlas of Tropical Forests: Asia and the Pacific.* IUCN, Gland; the data varies in age.
The data in columns 3-6 for India refers only to the Western Ghats, north-east India and the Andaman and Nicobar Islands. There are no tropical rainforests outside of these regions, but monsoon forest is extensive.
The totals given in columns 4,5 and 6 are for protected areas of greater than 50km² in extent which contain at least some tropical moist forest.
The totals in column 6 are derived from those protected areas which contain tropical forest mapped in chapters 12-29 of Collins, N.M., Sayer, J.A. and Whitmore, T. 1991. *The Conservation Atlas of Tropical Forests: Asia and the Pacific.* IUCN, Gland.
The data for Australia refers only to tropical rain forest and other moist forest types are not included. Protected areas data refers only to national parks.
The remaining area figure for Cambodia in column 3 is now known to be 71,500 km² according to FAO 1988. *An Interim Report on the State of Forest Resources in the Developing Countries.* FAO, Rome. 475pp.
Notes for AFRICA section of the Table 20.10 Tropical Moist Forests (Protected Area Coverage).
Data in column 2 is taken from White, F. 1983. *The Vegetation of Africa: a descriptive memoir to accompany the UNESCO/AETFAT/ UNSO vegetation map of Africa.* Unesco, Paris; as in MacKinnon and MacKinnon, 1986, except for Gabon and Liberia where these authors indicate that both countries were originally completely forested but the figures they give are country areas rather than land areas. "Original" cover includes mosaics.
The figure in column 2 for Madagascar has been calculated by adding the figure from Green, G.M. and Sussman, R.W. 1990. Deforestation history of the eastern rain forests of Madagascar. *Science* 248:212-215; for eastern rain forest to that calculated for mangroves from Map 26.1 in Collins, N.M., Sayer, J.A. and Whitmore, T. 1991. *The Conservation Atlas of Tropical Forests: Asia and the Pacific.* IUCN, Gland; plus an estimated 400 km² for forest remaining in the Sambirano region.
The data given in column three are derived from maps in chapters 11-32 of Collins, Sayer and Whitmore (1991), unless stated otherwise.
The data in column 3 for Angola, Comoros, Congo, Djibouti, Ethiopia, Guinea-Bissau, Kenya, Mauritius, Mozambique, Reunion, Seychelles, Somalia, Sudan, and Tanzania is from FAO 1988. *An Interim Report on the State of Forest Resources in Developing Countries.* FAO, Rome. 18pp.
The figure for Equatorial Guinea in column 3 includes 7,945 km² of degraded lowland rain forest.
The figure for Gabon in column 3 is from IUCN 1990. *La Conservation des Ecosystèmes Forestiers d'Afrique Centrale.* UICN, Gland. 124pp.
The figure for Malawi in column 3 is from Dowsett-Lemaire, F. 1989. The flora and phytogeography of the evergreen forests of Malawi. I: afromontane and mid-altitude forests. *Bulletin du Jardins Botanique National de Belgique* 59:3-131; and Dowsett-Lemaire, F. 1990. The flora and phytogeography of the evergreen forests of Malawi. II: lowland forests. *Bulletin du Jardins Botanique National de Belgique* 60:9-71.
The figure for Uganda in column 3 is from Howard, P.C. 1991. *Nature Conservation in Uganda's Tropical Forest Reserves.* IUCN, Gland. 313pp.
The figure for Zimbabwe in column 3 supplied by T. Muller, *in litt.*
Data in columns 3-6 does not include forest reserves. Totals are for protected areas which contain at least some tropical moist forest as determined on the maps in Collins, Sayer and Whitmore (1991), (except in Kenya, Ethiopia and Tanzania; see last note); it is not possible to take account of fragmentation of forest within each protected area. In many cases the forest coverage will be over-optimistic.
* No data
Percentage of forest protected cannot be realistically calculated for Kenya, Ethiopia and Tanzania. This is because although there are protected areas with forest within their boundaries, the forests are often fragmented and small in size and only cover a fraction of the size of the actual protected area.

shown in Figs 20.10-12. 'Forest' in these atlases includes mangroves and montane forests as well as lowland rain forests and swamp forests. In Asia, the monsoon forests are also included in the statistics given, while in Africa dry forests are excluded, as are riverine forests in both regions.

From Table 20.9 it can be seen that, in general, considerably less of the original forest extent remains in Africa than it does in Asia. For instance, nine of the 18 Asian countries listed still have more than 50% of their estimated original forest remaining while only four of the 36 African countries have this much of their original forest left. There is also, overall, considerably more moist forest remaining in Asia than there is in Africa. Fig 20.13 represents country area, forest area and annual loss in select tropical countries (derived from FAO sources cited above).

Conclusions

It is apparent that the true extent of the remaining moist tropical forests is still unknown. However, in the past decade there have been marked advances in the capacity of satellites to achieve detailed images of vegetation cover and in the capacity of image interpreters to distinguish the different forest types (Myers, 1988). As a result, it should soon be possible to obtain a more accurate assessment of forest area. Similarly, rates of deforestation remain guesses in many instances but whatever the figures, it is generally agreed that they are high and are increasing.

FACTORS INVOLVED IN CHANGES IN FOREST COVER

Human occupation of forests dates back to 25,000-40,000 years ago in Southeast Asia and the Pacific, 10,000 years in the Amazon and perhaps 3,000 years in Africa (Poore and Sayer, 1991). However, for most of man's history, his effect on tropical forests has been limited. Until comparatively recently, populations densities were low and there was little if any harvesting of trees for timber or extensive clearing for agriculture. Changes were brought about when people moved from China into Southeast Asia, and from Europe to Africa and South America. Table 20.11, compiled by Williams (1990), gives some indication of the areas of forest cleared in the tropics through time, with a high and low estimate for each region. As he points out though, there is much guesswork involved and the figures should not be taken as definitive. It is also not entirely clear what formations are regarded as 'forest' in this calculation, but they almost certainly include dry forests and possibly woodland where the canopy is not a closed one.

From around 1600 the tropical forests were altered radically by the introduction of new crops and new methods of exploitation (Williams, 1990). Forests were cleared to make way for cash crops such as rubber in Malaysia and Indonesia, coffee in Brazil, tea in India and China, sugar in the Caribbean, tobacco and palm oil in Asia. In addition to

Table 20.11 Estimated area of forest cleared in historical time (km² x 1,000)

REGION		PRE-1650	1650-1749	1750-1849	1850-1978
Central America	H	18	30	40	200
	L	12	30	40	200
Latin America	H	18	100	170	637
	L	12	100	170	637
Asia	H	974	216	596	1220
	L	640	176	606	1220
Africa	H	226	80	-16	469
	L	96	24	42	469

Source: Williams, M. 1990. Forests. In: Turner, B.L., Clark, W.C., Kates, R.W., Richards, J.F., Mathews, J.T. and Meyer, W.B. (Eds), *The Earth as Transformed by Human Action. Global and regional changes in the biosphere over the past 300 years.* Cambridge University Press, Cambridge.
Notes: H = high estimate, L = low estimate. Data for areas outside the tropics in the original table are omitted here.

crops, domestic animals were introduced to the New World and their grazing and browsing also affected the forests, frequently ensuring that regeneration did not occur after clearing had taken place. It is, however, within the last 50 years or so that deforestation has really accelerated, and the causes of this are broadly:

- to provide more land and wood (for fuel and building materials) for a largely subsistence population, particularly in countries with the greatest population growth
- to provide hard currency and vital export earnings from the sale of timber, from cash crops grown on deforested land and from the exploitation of minerals
- the building of dams, roads, cities, etc. to meet the needs of a growing urban population. These causes are dealt with in more detail below.

Shifting cultivation

Shifting cultivation has been identified as the principal cause of forest loss in all three tropical regions, accounting for 70% of the deforestation in Africa, 50% in Asia and 35% in the Americas (Lanly, 1982).

Traditional shifting cultivators, however, who incorporate forest fallow in their rotation, do not deforest. This method involves clearing the forest and usually burning the wood so that the nutrients within it are returned to the soil. Crops will be planted for two or three years until the soil becomes less fertile or weeds encroach on the area. The farmer then moves on and repeats the process in another area, and only returns to the initial patch after it has been left to regenerate (left fallow) for at least 10 years and often more. With this long fallow period to maintain soil fertility and in areas of low population density (generally five or less per km², Myers, 1980), such systems are viable and the use of forest land sustainable (Chin, 1987).

However, below a certain minimum fallow period (depending on climatic factors, soil type, etc.) the forest fallow develops a secondary thicket, yields drop and erosion and soil degradation takes place. Forest removal becomes permanent unless active attempts are made to reforest. This permanent removal occurs where the numbers of shifting cultivators have increased greatly; in Madagascar, for instance, the population has risen from around 5 million in 1960 to nearly 12 million at present and *tavy* or shifting cultivation is the main cause of forest loss in the moist forests in the east of the country (see case study). In other cases, not only are there too many people but these people are immigrant farmers unused to local conditions who frequently deplete the soil even more rapidly by using inappropriate farming methods, and consequently clear ever larger areas of forest. For example, in the south-west of Côte d'Ivoire, smallholders successfully cultivate food crops as well as coffee and cacao in a rain forest environment. Their methods involve selective felling, thereby retaining some forest trees, light burning, no tillage and maintaining some fallow periods. In contrast, the immigrant farmers, who have moved into the area from savanna lands, clear-fell the forest, burn it heavily, and employ soil tillage (Reitbergen, 1992).

Transmigration

Social and political factors often underlie forest loss. A classic example is the transmigration programmes that have occurred in Indonesia and Brazil (see case studies). A team from the International Institute for Environment and Development with representatives from three Indonesian Ministries came to the conclusion that transmigration was "the single sectoral activity with the greatest potential to advance forest destruction - often to no constructive result. The programme ... does not support the sustainable development of Indonesia's forest lands or, for that matter, the settlements themselves" (Colchester, 1987).

In Brazil, deforestation by shifting cultivators, encouraged to settle in the forests by the government, has been compounded by large landowners moving in after them and buying up the cleared land to convert to cattle ranches. The immigrants then move on to clear more forest. Cattle ranching in Latin America is estimated to have occupied some 20,000km² of forest per year in the late 1970s (Myers, 1990) and was profitable only because of the some 20,000km^2 of forest per year in the late 1970s (Myers, 1990) and was profitable only because of the subsidies offered by the Brazilian government through tax concessions and other such incentives. Once patches of the forest have been felled, leaving piles of dry wood in an area, the area becomes particularly susceptible to fires. These have exacerbated the forest loss, particularly in the Amazon where in 1987 some 50,000km^2 of forest were burned in the two states of Rondonia and Acre alone (Myers, 1990). Indeed, fires, either natural or deliberately set, have caused forest destruction and prevented its regeneration over huge areas, especially in Africa, for many thousands of years.

Population pressure

The world's population has risen from around three billion people in 1960 to over five billion in 1990 and is estimated to reach over eight billion by 2020. Although an inverse relation appears to exist between population and forest cover, the relationship is not precise (Mather, 1990). In some countries, such as Kenya, the increase in numbers of people has caused deforestation, but in many instances it is the unequal distribution of land ownership rather than total numbers that is the root cause of the problem. In countries such as the Philippines, Brazil and Costa Rica (see case study), most of the land is owned by a very small section of the community. Most of the population has to derive a living from very little land while elsewhere large areas are underused or wrongly used. The pressure on the forests in these instances stems from the inequality of the social policies, compounded by giving priority to crops other than those which ensure that the local population is fed (Westoby, 1989). Indeed, Repetto (1990) suggests that "government policies that encourage exploitation - in particular excessive logging and clearing for ranches and farms - are largely to blame for the accelerating destruction of tropical forests."

Logging

Initially, most exploitation for tropical timber was for highly valued species such as mahogany, teak and cedar. In addition, until the 1940s the forests had been harvested using axes, handsaws and animal power so that only forests near rivers could be intensively exploited because of the problem of extracting the logs. Under these circumstances, the impact of logging forests was relatively minor. However, the last few decades have seen a huge increase in export of tropical timbers which owes much to the rising affluence of the developed countries and the consequent rise in the demand for hardwoods (Westoby, 1989). Advances in technology, including the advent of chainsaws, tractors and trucks and the buildings of roads and railways to reach the most inaccessible areas have greatly facilitated the exploitation of the forests. For example, the inland forests of Gabon have, until recently, been protected because of their inaccessibility but, with the building of the Trans-Gabonese railway, the extraction and export of logs is expected to rise considerably. Overall, exports of tropical timbers to industrialised nations has risen sixteenfold since 1950 (Poore and Sayer, 1991). Information for Bangladesh (Table 20.12) gives some indication of the increase in extraction of timber between 1977 and 1984 as well as bamboo (used for building houses) and firewood.

Inefficient logging practices contribute to the destruction of the forest. For example, selection of mature trees of the most valuable species may involve extraction of less than 10% of the timber in an area, yet it can typically result in the destruction of at least half of the remaining stock, including immature trees of the valued species as well as harvestable stocks of somewhat less desirable trees (Repetto, 1990). Repeated logging of partially harvested areas to extract more timber before stands have recovered can inflict heavy damage on the remaining trees and make regeneration impossible (Repetto, 1990). In addition, governments increasingly encourage local processing of the timber, and in many cases the outdated machinery employed in the local industry consumes considerably more timber than efficient mills would expend to produce the same output. This is the situation in Côte d'Ivoire where 30% more logs are consumed by the local mills than would be the case in an efficient operation (Repetto, 1990). In Cameroon, the average conversion of sawn timber processed locally for export is about 30% but can be as low as 20% (Gartlan, 1989).

Roads and railways

One of the major indirect effects of logging is that the roads built to enter an area are subsequently used by agriculturalists to penetrate the forest, causing yet more clearance. Indeed, in many instances the building of roads and railways paves the way for further deforestation. For

Table 20.12 Output from forests in Bangladesh

ITEM	OUTPUT (in thousand tonne)			% CHANGE
	1977-78	1980-81	1983-84	1977-84
Timber	424	597	718	+69
Firewood	507	931	1279	+152
Bamboo	805	1449	1732	+115

Source: Ahmad, M. 1987. Bangladesh: how forest exploitation is leading to disaster. In: *Forest Resource Crisis in the Third World*. Sahabat Alam Malaysia, Penang.

example, in Côte d'Ivoire it has been estimated that one hectare of forest is removed by 'follow-on' cultivators for every 5m^3 of timber removed by loggers (Myers, 1983). A classic example of road construction attracting people to an area and thereby accelerating deforestation is that of BR 364 in Rondonia, Brazil (see case study).

Large-scale commercial clearance

Conversion of forests to pasture is the foremost cause of deforestation in the countries of Central America. In Brazil, it has been estimated that 72% of the clearance in 1980 was for pastureland (Browder, 1988). Forest continues to be cleared on a large scale for plantations of oil palm, rubber, sugar cane, tea, coffee, and cacao. In Peninsular Malaysia, most clearance has been for agriculture, principally for oil palm and rubber (Brookfield *et al.*, 1990). The total area under agriculture was 21% in 1966. This had risen to 39% in 1982 and government conversion plans intend that around 45% will be under agriculture by the mid-1990s (Brookfield *et al.*, 1990).

Fuel and charcoal

Domestic collections of firewood and charcoal are not considered an important agent in the deforestation of the moist forests, as most firewood is collected from savanna woodlands, scrub and farmlands (Eckholm *et al.*, 1984; Myers, 1980). However, the demand for fuel is rising rapidly as populations increase and wood, both dead and living, will inevitably be harvested in greater quantities from the moist forests. In contrast, industrial consumers are a cause of much outright deforestation (Eckholm *et al.* 1984). For instance in Brazil, in addition to the forest cleared to grow sugar, huge areas have been cut for the fuel needed to process the cane (Williams, 1990).

Other factors

In itself, mining is a comparatively minor cause of deforestation but the associated activities such as road building and the discharge of chemicals and silt into rivers can cause considerable damage. The same is true for oil exploration, where it is the roads, and the pipelines and spillage of oil and chemicals that are the principal causes of the deforestation. War is an exceptional cause of deforestation but, in the case of Viet Nam, bombing and the extensive use of herbicides have destroyed large areas of forest (see case study).

Summary

Overall, the causes of deforestation are many and varied, and the impact of each differs between countries and even between areas. There is widespread agreement that most governments seriously underestimate the economic value of their forests both as productive sources of commodities and for the services they provide. The cost involved in transforming the capital of the natural forest into other forms of capital is not generally recognised. However, any economic judgement of 'the value' of a forest must take account of the long-term benefits of conservation, but cannot ignore the short-term financial costs to tropical countries which are already under severe economic constraints. Deforestation will be reversed only when the natural forest is seen to be more economically valuable than alternative uses for the land. Well-intentioned but misguided policies by the developed countries, such as restrictions or outright bans on the import of tropical hardwoods, may not be in the best interests of forest conservation if such actions reduce the immediate, direct value of the forests to the exporting countries. Each country with tropical moist forest within its sovereign control will have to develop, with the support of the richer countries, its own particular repertoire of methods to reverse the loss of its forests based on a comprehensive understanding of the causal agents of that loss.

MANAGEMENT PRACTICES IN TROPICAL FORESTS

There are a variety of management objectives for tropical forests and these fall broadly into three categories: 1) for the supply of products either timber or non-timber; 2) for the protection of critical soils and water catchment; and 3) for the conservation of biological diversity (Poore and Sayer, 1991). The management practices applied to the areas will obviously depend on the use of the forest but, in a recent IUCN publication (Poore and Sayer, 1991) six key principles for the management of tropical moist forest land were listed.

- Ecological constraints must be considered at the outset of the development of tropical forest land.
- The allocation of tropical forest land to other uses should be decided only after thorough economic, social and ecological evaluation, including consultation with local communities
- Tropical forest should be converted to uses other than natural forest only if it can be demonstrated that this will produce sustainable benefits in a more desirable form than the original tropical forest itself can provide
- Wherever possible, areas of tropical forest which are already degraded should be selected for uses other than natural forest rather than clearing pristine forest
- Special efforts should be made to manage carefully those large areas of tropical forest which are essential for benefits such as the maintenance of watersheds and biological diversity
- The people who live in and around tropical forests should have a major say in their management.

In essence then, the management of a tropical forest involves the sustainable and continued harvest of all products, including to satisfy the needs of the local people, the maintenance of essential ecological processes, and the conservation of biological diversity.

Management for biological diversity

The principal means of managing tropical forests for biological diversity is through conservation areas. However, only about 4% of the world's remaining tropical forests are legally protected, and in many cases these areas have no management plans and no effective protection on the ground. It is also unrealistic to anticipate expanding the network of forest protected areas to cover all species and ecological processes (Poore and Sayer, 1991). It is now well appreciated that effective management must provide real incentives, based upon an equable sharing of benefits, for local people to participate in the conservation process. Such benefits may be consumptive in the form of meat, food, building materials, medicines and other forest products, or non-consumptive, particularly in the form of tourism.

Tourism is of increasing importance in the management of forest protected areas, particularly for the income and local employment it can provide. For example, tourists visiting the Mountain Gorillas *Gorilla gorilla berengei* in the Volcanoes National Park paid over US$800,000 in park fees in 1989. With the distribution of these benefits amongst the local community, together with increased educational outreach, the proportion of local farmers who would like to see the park degazetted has declined from 50% ten years ago to only 20% (Harcourt *et al.*, 1986). Incentives to local communities in the form of increased access to credit, capital grants, and support for improved agro-forestry schemes are now increasingly used to encourage villages to respect protected area boundaries.

Estimates of the extent of protected areas within tropical moist forest are presented in Table 20.10. Very few countries have more than 10% of their forest area under protection, and only Burundi, Singapore, Australia and Sri Lanka have over half their remaining forest under direct conservation management.

Management for maintenance of service values

Most natural forests protect soil fertility, prevent soil erosion, regulate water run-off and have a moderating effect on climate. The purpose of protection forests is to maintain these values. The services can also be provided to a lesser extent by degraded forests or even artificial plantations, but the maintenance of these ecological services is often combined with preserving biodiversity values. Maintaining natural protection forest also allows harvesting of non-timber products, giving these areas both an immediate and an indirect economic value.

Management for timber

Forests can be managed for timber at a number of different levels of intensity:

- the lowest level is the demarcation of a remote area which may be economically unattractive until shortages of supply drive up market prices
- an area may be selectively logged, protected from encroachment to allow regeneration, and then periodically relogged
- selected trees may be extracted at a pre-set felling intensity that ensures maximum economic return but causes minimum damage to remaining trees that can be harvested in the future
- minimum intervention harvesting may be followed by various treatments, such as weeding or poisoning of unwanted tree species
- logging may be followed by enrichment planting of saplings of economically desirable species.

In general, the more intense the management for timber production, the less diverse the forest becomes.

At present, most of the supply of tropical hardwood timbers comes from the first cut of previously unlogged forests. Some timber does, however, come from forests that are being relogged, mostly without a management plan, with a small amount from areas that are being converted to alternative use, mainly in an unplanned fashion. Comparatively little comes from plantations, secondary regrowth or agroforestry.

As the supply from the first cut of pristine forest and from land destined for conversion to agriculture declines, the immediate market reaction to the resultant shortages and increase in price is to shift operations away from the countries where supply has dropped to those which have largely unused forest resources. For this reason, it is predicted that there will be a movement away from Southeast Asia, the major supplier at present, to South America (Grainger, 1987). In all probability, these new producer countries will then unsustainably mine their forests in the same way as their predecessors have done.

A recent study undertaken by the International Tropical Timber Organization (ITTO) (Poore *et al.*, 1989) found that virtually nowhere was it possible to demonstrate conclusively that any natural tropical forest had been successfully managed for the sustainable production of timber (see, for instance, case study on Congo). In Latin America and the Caribbean, the total area being sustainably managed at the operational level was limited to 75,000ha in Trinidad and Tobago. In Australia, an area of some 160,000ha, the whole estate of tropical forest in Queensland scheduled for logging was under sustainable production of timber, although all this area has now been taken out of logging following nomination as a World Heritage Site.

Of the total forested area of Asia, the only region practising sustainable yield management was in parts of Peninsular Malaysia where the 'selective management system' is in operation. However, even here the system has only recently come into full use and there is no certainty of its success after the first cutting cycle, although it is intended to extend the system to the total production forest estate of the country. In Africa, the report concluded that there were no sustained yield management systems currently being practised throughout the continent. With the partial exception of Ghana, forest management systems have been progressively abandoned. A selection system similar to Peninsular Malaysia has been running for eight years in the Côte d'Ivoire, and preliminary results are sufficiently encouraging for it to be extended to some 10,000ha of Yapo Forest.

The report concludes that of an estimated total area of some 828 million ha of productive tropical forest remaining in 1985, the total under sustained yield management amounted, at the very most, to about one million ha. This is the reality of the sophistication of tropical forest management throughout the world. Urgent action is required not only to ensure proper management of previously unlogged forests but also to assess the status of logged forests and degraded forest lands to bring these under sustainable protection. In view of the scale of the task, the lack of an operational definition of sustainability, and the need to develop mechanisms to monitor forest management, this is an ambitious target.

CASE STUDIES

Deforestation and degradation of forests in Sumatra

Figures for the original and present extent of rain forest in Sumatra are not easily obtained, as sources such as FAO (1988) give figures for Indonesia rather than for the separate islands. However, a series of maps showing the rapid disappearance of pristine forests in the country has recently been published (Collins *et al.*, 1991. See Fig. 20.1). In this same publication it is estimated that around 49% of the country's land area is still forested but this figure includes logged as well as the untouched forests depicted in Fig. 20.1.

Figure 20.1 Pristine forests in Sumatra

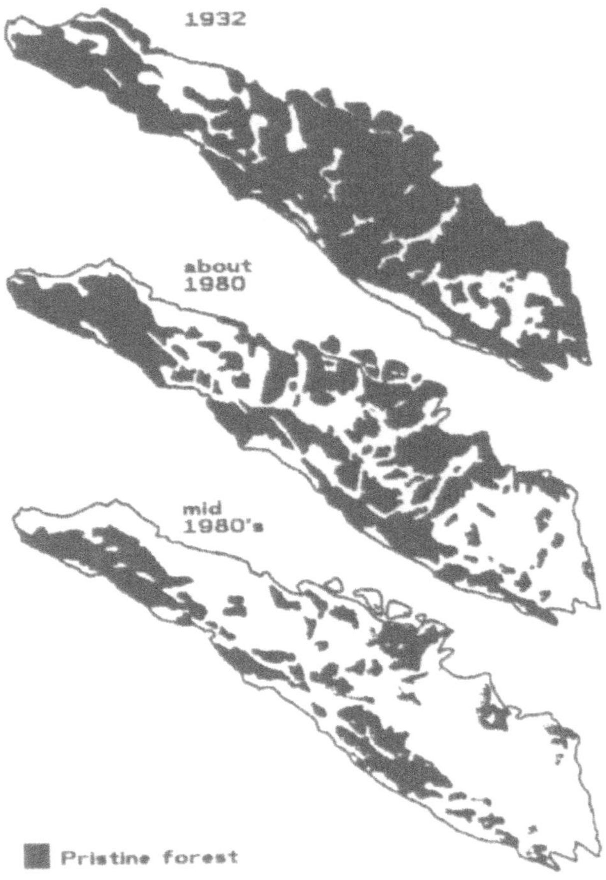

Source: Collins, N.M., Sayer, J.A. and Whitmore, T.C. (Eds) 1991. *The Conservation Atlas of Tropical Forests: Asia and the Pacific.* Macmillan Press, London, UK in collaboration with IUCN, Gland, Switzerland.

The causes of the deforestation and degradation are varied, but clearance for agricultural land is probably the primary cause. Population density is high (59 people per km² in 1980) and large areas have been cleared both for subsistence agriculture and industrial plantations (Whitten *et al.*, 1984). In addition, relatively large areas of the shallower peat swamp forests along the Malacca Straits have been drained to provide farmland for settlers who were moved there in the course of Indonesia's large transmigration scheme. There has also been considerable logging in the country. For instance, on the flat lowlands of southern Sumatra great stands of the commercially important Ironwood *Eusideroxylon zwageri*, which produces an exceptionally durable timber, have been almost entirely destroyed while, in recent years, there has been heavy logging in the lowlands east of the mountain spine. It is probable that Sumatra is losing its natural vegetation faster than any other part of Indonesia.

Deforestation in Viet Nam

An FAO (1987) report estimates that closed forest cover was 61,650km² in 1980, 48,620 in 1985 and projected that there would be only 34,060km² in 1990. In contrast, the Ministry of Forestry (1989, unpublished) using 1987 Landsat imagery, indicated that 79,054km² of closed broadleaved forest remained in 1987. Interpretation of what is believed to be the same 1987 data set by WCMC gives a total of 56,680km² of closed forest (MacKinnon and Cox, 1991). The variation in the statistics is, no doubt, due to differing interpretations of what constitutes a closed canopy forest.

The population of Viet Nam was originally centred on the Red River Delta in the north but moved south during historical times, clearing and cultivating the coastal plains and valleys and reaching the Mekong Delta a few centuries ago. These areas were consequently the first to be cleared of forest. By 1943, about 45% of the country was still forested. During the French colonial administration, which ended in 1954, extensive areas in the south were further cleared for industrial plantations, mostly banana, coffee and rubber.

From 1945-1975, there was almost uninterrupted warfare in the country. It was estimated that during the war between the North and South, 22,000km² of forest and farmland were destroyed by intensive bombing, spraying of herbicide and mechanical clearing of forest. In total some 23,000-55,000km² of forest were damaged by the bombardment (Myers, 1980). In addition, large areas of forest were cleared for agricultural land to feed the people (the population doubled between 1945 and 1985). The area cleared was larger than would normally have been needed because of the necessity to make up for the food lost when irrigation systems were bombed and crops were killed by herbicides.

Since 1975, the need to rebuild after the war and the still increasing growth in population has caused continuing loss of forest. In 1981, FAO/UNEP estimated annual deforestation of dense broadleaved forest during 1981-1985 at 600km² but FAO's report of 1987 indicated the much greater figure of 3,110km² of forest lost annually.

Figure 20.2 Viet Nam's vanishing forest cover 1943-1982

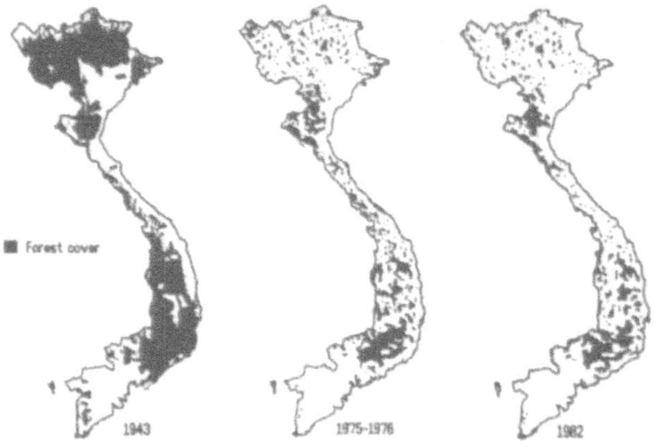

Source: Anon. 1985. Viet Nam: National Conservation Strategy. Prepared by the Committee for Rational Utilization of Natural Resources and Environmental Protection (Programme 52-02) with assistance from IUCN. WWF-India, New Delhi.

Figure 20.3 Deforestation in Madagascar's eastern rain forests

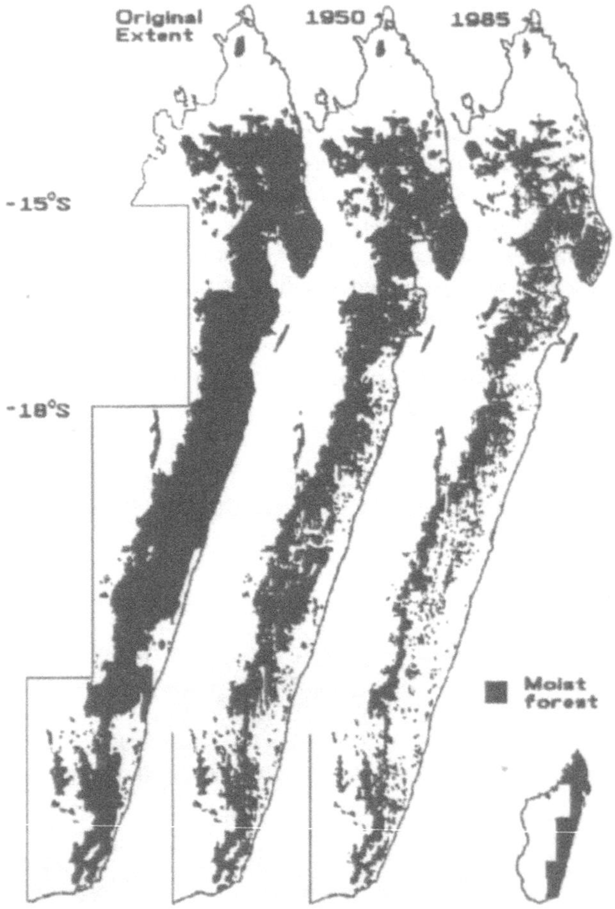

Source: Green, G.M. and Sussman, R.W. 1990. Deforestation history of the eastern rain forests of Madagascar from satellite images. *Science* 248:212-215

Eastern rain forests of Madagascar

Estimates of forest extent and rate of deforestation in the moist forests of Madagascar vary enormously. Myers (1980), using reports from a variety of authors, estimated that there were only 26,000km^2 of eastern rain forest remaining in the country and that it was being deforested at a rate of 2,000-3,000km^2 per year. In contrast, FAO/UNEP (1981) reported over 69,000km^2 remaining in 1980 with an annual deforestation rate for all the closed broadleaved forest of 400km^2 between 1976 and 1980, which it was predicted would decline during the period 1980-1985. In its 1988 report, FAO estimated an annual deforestation rate for all closed broadleaved forests of 150km^2.

A recent report, based on analysis of vegetation maps that were made from aerial photographs in 1950 and on satellite image data from 1984-1985, provides more accurate estimates of remaining forest cover and deforestation rates (Green and Sussman, 1990). The authors estimate that there were originally 11.2 million ha of eastern rain forest, that 7.6 million ha remained in 1950, and that these had been reduced to only 3.8 million ha by 1985 (Fig. 20.3). The deforestation rate between 1950 and 1985 was, therefore, 111,000ha per year.

The main cause of the deforestation in the eastern rain forests is slash-and-burn (or *tavy)* agriculture, and cutting for fuelwood to sustain Madagascar's growing population. The country supported 5.4 million people in 1960 and this had risen to 12 million by 1990. The population is still mostly rural and survives by subsistence agriculture. To obtain more land, forest areas are clear cut, the vegetation is dried and then fired some months later. Dry land rice is most commonly planted, but maize, manioc and other crops are also grown. They are cultivated for a year or two; then the land is left fallow to regain its fertility and the process is repeated elsewhere. Tavy has been practised for centuries but the increase in population has put greater pressure on the land, and it is now often left fallow for only three or four years. As a result, the soil productivity progressively deteriorates, and the area becomes unproductive grassland or, on steep slopes, erodes away to bare earth. Predictably, deforestation has been most rapid in areas of high population density and low topographic relief. If cutting continues at the present rate, Green and Sussman (1990) estimate that only the forests on the steepest slopes will survive the next 35 years.

Forest Loss in Costa Rica

Agricultural growth in Central America, as in other developing countries, is driven by an expansion of pasture and cropland rather than through intensified agriculture on existing cleared land. The area under forest shrinks as a result. The rate of transformation from forest to pasture has been increasing since 1950. In Costa Rica, 67% of the country was covered in primary forest in 1940, but only 17% remained under primary forest by 1983 and this was mostly in the mountainous areas of the country (Fig. 20.4). Deforestation has been greater in the dry western area than in the mountains as the former is comparatively easy to clear and maintain as pastureland by burning.

Figure 20.4 Loss of primary forest in Costa Rica 1940-1983

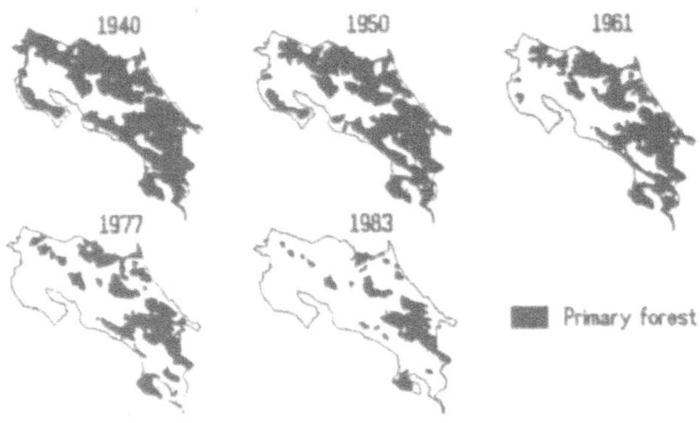

Source: After Sader, S.A. and Joyce, A.T. 1988. Deforestation rates and trends in Costa Rica 1940-1983. *Biotropica* 20(1):14.

In 1960, only 19% of the country was under permanent pasture, while by 1980 this area had risen to 31% (FAO in Leonard, 1987). Beef production takes up the majority of the converted land, with 15,580km² being devoted to cattle in 1980. This is in spite of the fact that the beef industry in Central America is very inefficient, with levels of productivity per hectare of land being considerably lower than, for example, in the USA. As in the rest of Central America, big ranchers in Costa Rica own most of the land: landowners, with 60% of all farmers occupying only 4% of the land.

Deforestation in Central America

Throughout Central America, the single most important ecological change that is taking place as a result of the current demographic pressures and economic trends is the rapid and continuing conversion of forests to other land uses (Leonard, 1987. Table 20.13). Almost all of Central America was originally forested but it is estimated that now only 40% of the seven countries are still forested (Fig. 20.5). Two-thirds of all the forest clearing has occurred since 1950 and the rates of forest clearance have increased in every decade since that date (Parson, 1976).

There are obviously some positive results arising from this deforestation: for instance, the cattle ranching and farming, both occurring on cleared forest land, are major generators of employment, national income and export revenue in the region. However, the economic contribution is predominantly indirect, that is from the land cleared of the timber rather than from the timber itself.

Despite the very rapid consumption of forests in recent years, the timber industry is not a major force in most of Central America. This is because much of the timber cut is not harvested for commercial purposes; instead it is burned in place or felled and not used. Only in Honduras does commercial harvesting of timber contribute significantly to the economy of the country. Even here it has been estimated that forests with a commercial timber value of $320 million are wasted each year (Leonard, 1987).

Figure 20.5 Deforestation in Central America 1950-1985

However, logging tracks do, here as elsewhere, open up the forest to subsequent colonisation. Road building, which has been a major goal of most governments in the region since the 1960s, has the same effect (Leonard, 1987). Demand for fuelwood is not, overall, a major force of deforestation though in arid highland areas of Guatemala, Honduras and El Salvador it does have an impact.

Deforestation in Peninsular Malaysia

In 1966, dryland forest on Peninsular Malaysia occupied 68% of the land area. It had declined to 54% by 1982 and is now less than 50% (Brookfield and Byron, 1990) (Fig. 20.6). Swamp forest diminished from 14% to 10% of the Peninsula in the same time, and now occupies about 8% of the land. The major cause of this forest depletion has been the conversion of land from forest to agriculture. Clearing the forests for large-scale farming of cash crops began on the west coast where rubber, coconut and then oil palm plantations were developed. More recently, conversion of forest has been undertaken by various federal and state governments for land development schemes to provide agricultural land and employment for landless families moved from other parts of the country. Development plans for the country encourage the further conversion of forest. For instance, the Fourth Malaysia Plan (1981-1985) suggested that another 6,075km² were to be cleared for rubber plantations, 8,470km² for oil palm and 1,500km² for settlements (Whitten, 1991).

Table 20.13 Status of lowland and tropical montane forests in Central America

COUNTRY	REMAINING (1983) PRIMARY FOREST (km²)	CURRENT ANNUAL RATE OF FOREST LOSS (km²)	% OF 1983 COVER LOST ANNUALLY
Nicaragua	27,000	1,000	3.7
Guatemala	25,700	600	2.3
Panama	21,500	500	2.3
Honduras	19,300	700	3.6
Costa Rica	15,400	600	3.9
Belize	9,750	32	0.3
El Salvador	0	0	-
Central America	118,650	3,432	2.9

Source: Nations, J.D. and Komer, D.I. 1983. Central America's Tropical Rainforests: positive steps for survival. *Ambio* 12(5):232-238.

Figure 20.6 Forested areas of the Malay Peninsula at various dates

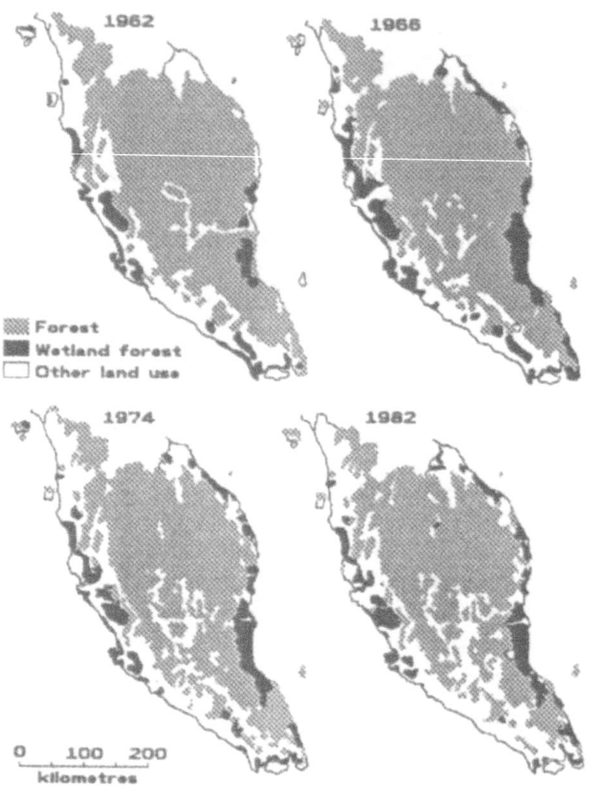

Figure 20.7 Remaining forest, after completion of present conversion plans

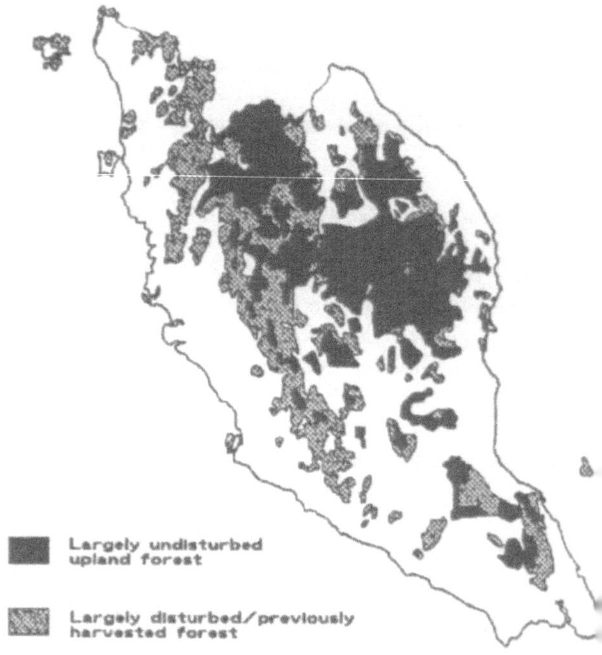

Other, though comparatively minor, causes of forest loss are construction of dams for irrigation, hydroelectric schemes and mining, particularly for alluvial tin (Collins *et al*, 1991). Logging, although rarely causing total deforestation, does result in significant ecological damage in some parts of the Peninsula (Fig. 20.7 shows remaining areas of logged and unlogged forest). The logging roads open up areas for cultivation and settlement. Shifting cultivation by indigenous people is not an important cause of deforestation in the region.

Resettlement in Rondonia, Brazil

There have been various schemes in Brazil, backed by the government and by outside agencies, to move people from over-populated areas to the Amazon basin. An extensive programme of road building has opened up the forest to these settlement schemes and to other landless people moving into the region of their own accord. For instance, small farmers were expelled from the central-south of Brazil by conversion of coffee and other labour intensive crops to mechanised soybeans and wheat, and these people migrated to the Amazonian frontier rather than to urban slums (Fearnside, 1986).

Rondonia has one of the highest rates of deforestation in Brazil's Amazonia. In 1960 the state was uninhabited except for some Amerindians and a few rubber gatherers but by the late 1970s this region had as many as 5,000 people moving in every month. Many of the immigrants were landless people from the south of Brazil. In 1981, the World Bank agreed to finance further development of the area and a major road, BR-364, was paved which increased movement into the State. The result has been progressive deforestation which shows up clearly on satellite images of the region (Fig. 20.8). The soils under the forest are generally so poor that the settlers have to clear another area within a year or two, or else sell their land to cattle ranchers. Ranching used to be profitable, in spite of the

Figure 20.8 Deforestation in Rondonia, Brazil

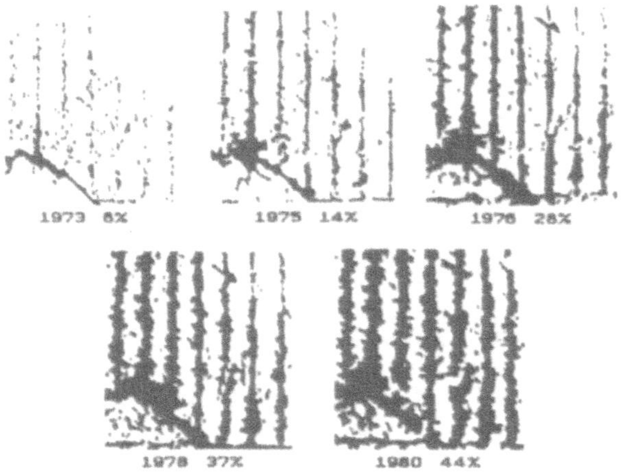

poor yields of either milk or meat, as a result of tax incentives provided by the government, but these incentives are no longer available.

This satellite imagery traces the progressive clearance (shown in black) of forest for farmland in an area of Rondonia, south-west Amazonia. The vertical lines are roads set 5km apart while the thicker black, curved line is BR-364.

Transmigration as cause of deforestation in Indonesia

In Indonesia, nearly three million people have been moved from the crowded and environmentally degraded islands of Lombak, Bali, Java and Madura to new settlements on less populated islands (Whitten, 1991). This has been the world's largest programme of voluntary assisted migration but, in addition to the sponsored migrants, it is estimated that two or three times as many unassisted people have moved to the less populated islands.

The land to which the transmigrants have been moved is, in many cases, entirely inappropriate. Some of the sites are intrinsically unsuitable for agriculture; others were inadequately prepared or inappropriately managed. As a result, loss of forests throughout the region is one of the major environmental impacts of the settlement programmes. In addition, in many cases, the removal of people from degraded land has not improved the environmental conditions they left behind. For example, in Java the population has been growing faster than the rate of transmigration and there does not appear to be any improvement in the state of the critically eroded land on this island.

The budget for this programme has been cut recently and it has been agreed that no new areas are to be cleared. Instead, development of the already existing sites will take place, roads will be built to improve communications, tree crops will be encouraged, produce will be marketed more effectively, and other improvements will be made.

Fires in Borneo

Until recently, fires were not considered to be a major factor in the fate of tropical forests. However, in 1982-1983, major fires occurred in Borneo during a drought and very large areas of forested land were burnt. Beginning late in 1982 and peaking in early 1983, numerous fires broke out in coastal and inland areas of East Kalimantan. In Sabah, an overlapping series of outbreaks occurred from early through mid-1983. Minor outbreaks also occurred in other parts of Borneo and the southern Peninsula.

The lowlands of East Kalimantan are one of the driest areas of the island of Borneo, and during the intense El Niño southern oscillation of 1982-1983 rainfall was reduced by more than 60% compared with long-term monthly averages. From February to May 1983, instead of receiving more than 135mm rain per month as normal, almost none fell (Malingreau *et al.*, 1985). Drought stress led to the shedding of leaves by evergreen species and to the accumulation of dry litter on the forest floor. The extensive fires that took place, particularly from August to October 1982 and March to May 1983, were mainly triggered by the agricultural practices used in the area, which included dry season burning as a land clearing method (Malingreau *et al.*, 1985). Accelerated settlement programmes and spontaneous migration have meant that large tracts of land in East Kalimantan are being deforested for agricultural uses, especially along the coast and main rivers, and burning in these areas almost certainly caused the major fires in that region.

It has been estimated that in East Kalimantan alone 35,000km^2 of land have been damaged by the fires. This includes 8,000km^2 of primary lowland forest, 5,500km^2 of peat swamp forest, 12,000km^2 of selectively logged forest and 7,500km^2 of shifting cultivation land (Leighton and Wirawan, 1986). It is thought that around 10,000km^2 of vegetation were damaged in Sabah (Malingreau *et al.*, 1985). The data suggest that selectively logged forest suffered greater damage than the primary forest, as the debris resulting from the logging provided further dry material to fuel the fires.

In conclusion, it is evident that increasing populations, with the resulting increase in slash-and-burn agriculture, combined with the extreme climatic conditions of 1982-1983 were the cause of considerable deforestation in Borneo. It appears that fires may well become a more common agent of deforestation in the future. Indeed, at the time of writing, considerable areas of forest on the island are once more ablaze.

Figure 20.9 Forest fires in Borneo, 1982-1983

Forest management in Congo

Congo's national forest estate has been divided into forest management units, each of sufficient size to support an independent forest industry. The industries are required to conduct an inventory of their unit and propose a management plan for ministerial approval. These plans should provide for selection felling on a 25 year cycle with a minimum diameter limit of 60cm. Extraction is subject to three year exploitation permits, which prescribe the maximum area to be logged and the minimum volume of timber to be produced. This system could have provided a sound basis for a sustainable forest industry but a variety of factors have prevented it from ever being put into practice properly.

Understaffing of the forestry service has meant that the units are not properly supervised. More important, all forest land is state property so even if an area is under management for timber, all citizens have rights to use the area. These customary rights not only allow subsistence hunting and collecting of non-timber products (neither of which harm the forest to any great extent), they also allow local people to practise shifting cultivation in the area. This has happened in the more densely populated south of the country and potential timber yields have been significantly reduced as a result. It appears that sustainability in the south can be achieved only in intensively managed plantations taken out of state ownership. In contrast, in the comparatively inaccessible, sparsely populated north, the forests remain undisturbed after logging and regenerate well. Here selective logging is practised and a near natural forest is maintained.

References

Ahmad, M. 1987. Bangladesh: how forest exploitation is leading to disaster. In: *Forest Resource Crisis in the Third World*. Sahabat Alam Malaysia, Penang.

Anon. 1985. *Viet Nam: National Conservation Strategy*. Prepared by the Committee for Rational Utilization of Natural Resources and Environmental Protection (Programme 52-02) with assistance from IUCN. WWF-India, New Delhi. 77pp.

Baur, G. 1964. *The Ecological Basis of Rainforest Management*. Forestry Commission of New South Wales, Sydney, Australia.

Brookfield, H. and Byron, Y. 1990. Deforestation and timber extraction in Borneo and the Malaya Peninsula. The record since 1965. *Global Environmental Change* 1(1):42-56.

Brookfield, H., Lian, F.J., Kwai-Sim, L. and Potter, L. 1990. Borneo and the Malay Peninsula. In: Turner, B.L., Clark, W.C., Kates, R.W., Richards, J.F., Mathews, J.T. and Meyer, W.B. (Eds), *The Earth as Transformed by Human Action. Global and regional changes in the biosphere over the past 300 years*. Cambridge University Press, Cambridge. Pp.495-512.

Browder, J.O. 1988. Public policy and deforestation in the Brazilian Amazon. In: Repetto, R. and Gillis, M. (Eds), *Public Policies and the Misuse of Forest Resources*. Cambridge University Press, Cambridge. Pp.247-297.

Chin, S.C. 1987. Do shifting cultivators deforest? In: *Forest Resource Crisis in the Third World*. Sahabat Alam Malaysia, Penang.

Colchester, M. 1987. The Indonesian Transmigration Programme: migrants to disaster. In: *Forest Resource Crisis in the Third World*. Sahabat Alam Malaysia, Penang.

Collins, N.M., Sayer, J.A. and Whitmore, T.C. (Eds) 1991. *The Conservation Atlas of Tropical Forests: Asia and the Pacific*. Macmillan Press, London, UK in collaboration with IUCN, Gland, Switzerland.

Eckholm, E., Foley, G., Barnard, G. and Timberlake, L. 1984. *Fuelwood: the energy crisis that won't go away*. IIED, London and Washington DC.

FAO 1987. *Special Study on Forest and Utilisation of Forest Resources in the Developing Region. Asia-Pacific Region. Assessment of Forest Resources in Six Countries*. FAO Field document 17. 104pp.

FAO 1988. *An Interim Report on the State of the Forest Resources in the Developing Countries*. FAO, Rome, Italy.

FAO 1990. *Interim Report on Forest Resources Assessment 1990 Project*. Committee on Forestry Tenth Session. FAO, Rome, Italy.

FAO 1991. *Second Interim Report on the State of Tropical Forests by Forest Resources Assessment 1990 Project*. Tenth World Forestry Congress, September 1991, Paris, France.

FAO/UNDP 1988 *Cameroun Tropical Forestry Action Plan. Rapport du Mission*, 2 vols. FAO/UNDP, Rome.

FAO/UNEP 1981. Tropical Forest Resources Assessment Project (in the framework of GEMS). FAO, Rome, Italy.

Fearnside, P.M. 1986. Spatial concentration of deforestation in the Brazilian Amazon. *Ambio* 15:74-81.

Fearnside, P.M. 1990. Deforestation in the Brazilian Amazonia. In: Woodwell, G.M. (Ed.), *The Earth in Transition: patterns and processes of biotic impoverishment*. Cambridge University Press, New York.

Forest Management Bureau 1988. *Natural Forest Resources of the Philippines*. Philippine-German Forest Resources Inventory Project.

Gartlan, S. 1989. *La Conservation des Ecosystèmes forestiers du Cameroun*. IUCN, Gland, Suisse et Cambridge, Royaume-Uni.

Grainger, A. 1987. *Tropform: a model of future tropical timber hardwood supplies*. Proceedings of the Symposium in Forest Sector and Trade Models. University of Washington, Seattle, USA.

Green, G.M. and Sussman, R.W. 1990. Deforestation history of the eastern rain forests of Madagascar from satellite images. *Science* 248:212-215.

Harcourt, A.H., Pennington, H. and Weber, A.W. 1986. Public attitudes to wildlife and conservation in the Third World. *Oryx* 20(3):152-154.

Holdgate, M.W. 1982. Terrestrial Biota. In: Holdgate, M.W., Kassas, M. and White G.F. (Eds), *The World Environment*. UNEP.

IUCN 1989. *La Conservation des Ecosystèmes forestiers du Gabon.* IUCN Gland, Switzerland and Cambridge, UK.

Joint Interagency Planning and Review Mission (JIM) for the Forestry Sector 1988. *Cameroon Tropical Forestry Action Plan.* JIM, Rome.

Jones, S. 1992. The Gambia and Senegal. In: Sayer, J.A., Harcourt, C.S. and Collins, N.M. (Eds), *The Conservation Atlas of Tropical Forests: Africa.* Macmillian Press, London, UK in collaboration with IUCN, Gland Switzerland.

Kyaw, U.S. 1987. National report: Burma. In: *Proceeding of Ad Hoc FAO/ECE/FINNIDA Meeting of Experts on Forest Resource Assessment, Kotka, Finland, 26-30 October 1987* (Finnish International Development Agency, Helsinki, 1987).

Lanly, J.-P. 1982. Tropical Forest Resources. FAO Forestry Paper 30, FAO, Rome, Italy.

Lanly, J.-P., Singh, K.D. and Janz, K. 1991. FAO's 1990 reassessment of tropical forest cover. *Nature and Resources* 27(2):21-26.

Leighton, M. and Wirawan, N. 1986. Catastrophic drought and fire in Borneo tropical rain forest associated with the 1982-83 El Niño southern oscillation event. In: Prance, G.T. (Ed.), *Tropical Rain Forests and the World Atmosphere.* Westview, Boulder. Pp.75-102.

Leonard, H.J. 1987. *Natural Resources and Economic Development in Central America: a regional environmental profile.* International Institute for Environment and Development. Transaction Books, Oxford.

Lugo, A.E. and Brown, S. 1982. Conversion of tropical moist forests: a critique. *Interciencia* 7(2):89-93.

MacKinnon, J. and Cox, R. 1991. Vietnam. In: Collins, N.M., Sayer, J.A. and Whitmore, T.C. (Eds), *The Conservation Atlas of Tropical Forests Asia and the Pacific.* Pp.232-239.

Malingreau, J.P., Stephens, G. and Fellows, L. 1985. Remote sensing of forest fires: Kalimantan and North Borneo in 1982-83. *Ambio* 14(6):314-321.

Mather, A.S. 1990. *Global Forest Resources.* Belhaven Press, London.

Myers, N. 1980. *Conversion of Tropical Moist Forests.* National Academy of Sciences, Washington, DC.

Myers, N. 1983. Conversion rates in tropical moist forests. In: Golley, F.B. (Ed.), *Tropical Rain Forest Ecosystems: structure and function.* Elsevier, Amsterdam. Pp.289-300.

Myers, N. 1988. Tropical deforestation and remote sensing. *Forest Ecology and Management* 23:215-225.

Myers, N. 1989. *Deforestation Rates in Tropical Forests and their Climatic Implications.* Friends of the Earth, London.

Myers, N. 1990. The world's forests and human populations: the environmental interconnections. *Population and Development Review* No.16.

Nations, J.D. and Komer, D.I. 1983. Central America's Tropical Rainforests: positive steps for survival. *Ambio* 12(5):232-238.

Parson, J.J. 1976. Forest to Pasture: development or destruction? *Revista de Biologia Tropical* 24:121-138.

Peters, C.M., Gentry, A.H. and Mendelsohn, R.O. 1989. Valuation of an Amazonian rainforest. *Nature* 339:29.

Philippines Forest Management Bureau, Department of Environment and Natural Resources 1988. *1987 Philippine Forestry Statistics,* Philippines.

Poore, D., Burgess, P., Palmer, J., Rietbergen, S. and Synott, T. 1989. *No Timber Without Trees.* Earthscan, London.

Poore, D. and Sayer, J. 1991. *The Management of Tropical Moist Forest Lands: ecological guidelines,* 2nd edn. IUCN, Gland, Switzerland and Cambridge, UK. 76pp.

Reitbergen, S. 1992. Forest management. In: Sayer, J.A., Harcourt, C.S. and Collins, N.M. (Eds), *The Conservation Atlas of Tropical Forests: Africa.* Macmillian Press, London, UK in collaboration with IUCN, Gland, Switzerland.

Repetto, R. 1990. Deforestation in the tropics. *Scientific American* 262(4):18-24.

Royal Forestry Department of Thailand 1986. *Forestry Statistics of Thailand,* 1986. Planning Division. Center for Agricultural Statistics, Office of Agricultural Economics, Ministry of Agriculture and Cooperatives, Bangkok, Thailand. Table 2, p.8.

Saavedra C. and Suarez de Freitas, G. 1989. Manu - two decades later. *WWF Reports,* June-July 1989.

Sader, S.A. and Joyce, A.T. 1988. Deforestation rates and trends in Costa Rica 1940-1983. *Biotropica* 20(1):14.

Sayer, J.A., Harcourt, C.S. and Collins, N.M (Eds) 1992. *The Conservation Atlas of Tropical Forests: Africa.* Macmillan Press, London, UK in collaboration with IUCN, Gland, Switzerland.

Sayer, J.A. and Whitmore, T.C. 1991. Tropical moist forests: destruction and species extinction. *Biological Conservation* 55:199-213.

Schimper, A.F.W. 1903. *Plant Geography upon a Physiological Basis.* University Press, Oxford, UK.

Setzer, A.W. *et al.* 1988. Relatorio de atividades do projeto IBDF-INE 'SEQE' - ano 1987. National Space Research Institute of Brazil (INPE), São José dos Campos, São Paulo, Brazil.

Sommer, A. 1976. Attempt at an assessment of the world's tropical moist forests. *Unasylva* 28(112-113):5-24

Thang, H.C. 1987. *Forest Resources Assessment in Malaysia.* Forest Management Unit, Forestry Department, Kuala Lumpur.

USAID 1987. Natural Resources and Environmental Management in Indonesia. Jakarta.

Vo Quy 1988. Vietnam's ecological situation today. *ESCAP Environment News* 6(4):4-5.

Vohra, B.B. 1987. Confusion on the Forestry Front. Advisory Board on Energy, New Delhi. Based on reconciliation of figures in Natural Remote Sensing Agency, *Mapping of Forest Cover in India from Satellite Imagery 1972-75 and 1980-82.* Government of India, Hyderbad, 1983.

Westoby, J. 1989. *Introduction to World Forestry.* Basil Blackwell, Oxford.

Whitten, A.J. 1991. Agricultural settlement schemes. In: Collins, N.M., Sayer, J.A. and Whitmore, T.C. (Eds), *The Conservation Atlas of Tropical Forests: Asia and the Pacific.* Macmillan Press, London, UK in collaboration with IUCN, Gland, Switzerland.

Whitten, A.J., Damanik, S.J., Anwar, J. and Hisyam, N. 1984. *The Ecology of Sumatra.* Gadjah Mada University Press, Yogyakarta, Indonesia.

Williams, M. 1990. Forests. In: Turner, B.L., Clark, W.C., Kates, R.W., Richards, J.F., Mathews, J.T. and Meyer, W.B. (Eds), *The Earth as Transformed by Human Action. Global and regional changes in the biosphere over the past 300 years.* Cambridge University Press, Cambridge. Pp.179-201.

World Bank, Asia Regional Office 1988. *Indonesia: forests, land and Water: issues in sustainable development,* August 1988. P.2.

WRI 1990. *World Resources 1990-1991.* Oxford University Press, Oxford.

Chapter written by Caroline Harcourt. We thank Biotropica, H. Brookfield and Y. Byron, P.M. Fearnside, G. Stephens, and R.W. Sussman for giving permission to reproduce maps originally published by them.

Figure 20.10 Tropical Moist Forest: Asia

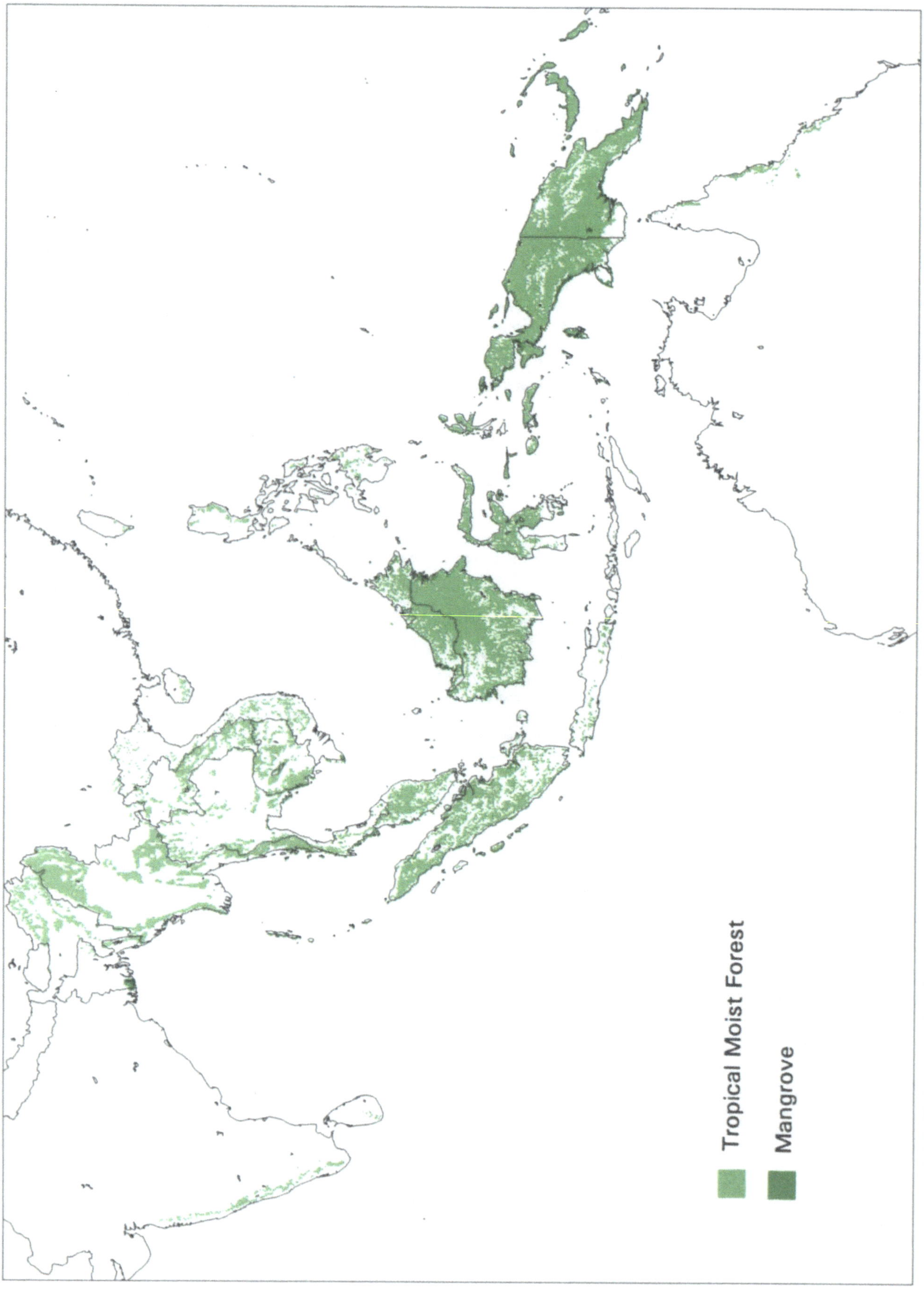

Figure 20.11 Tropical Moist Forest: Americas

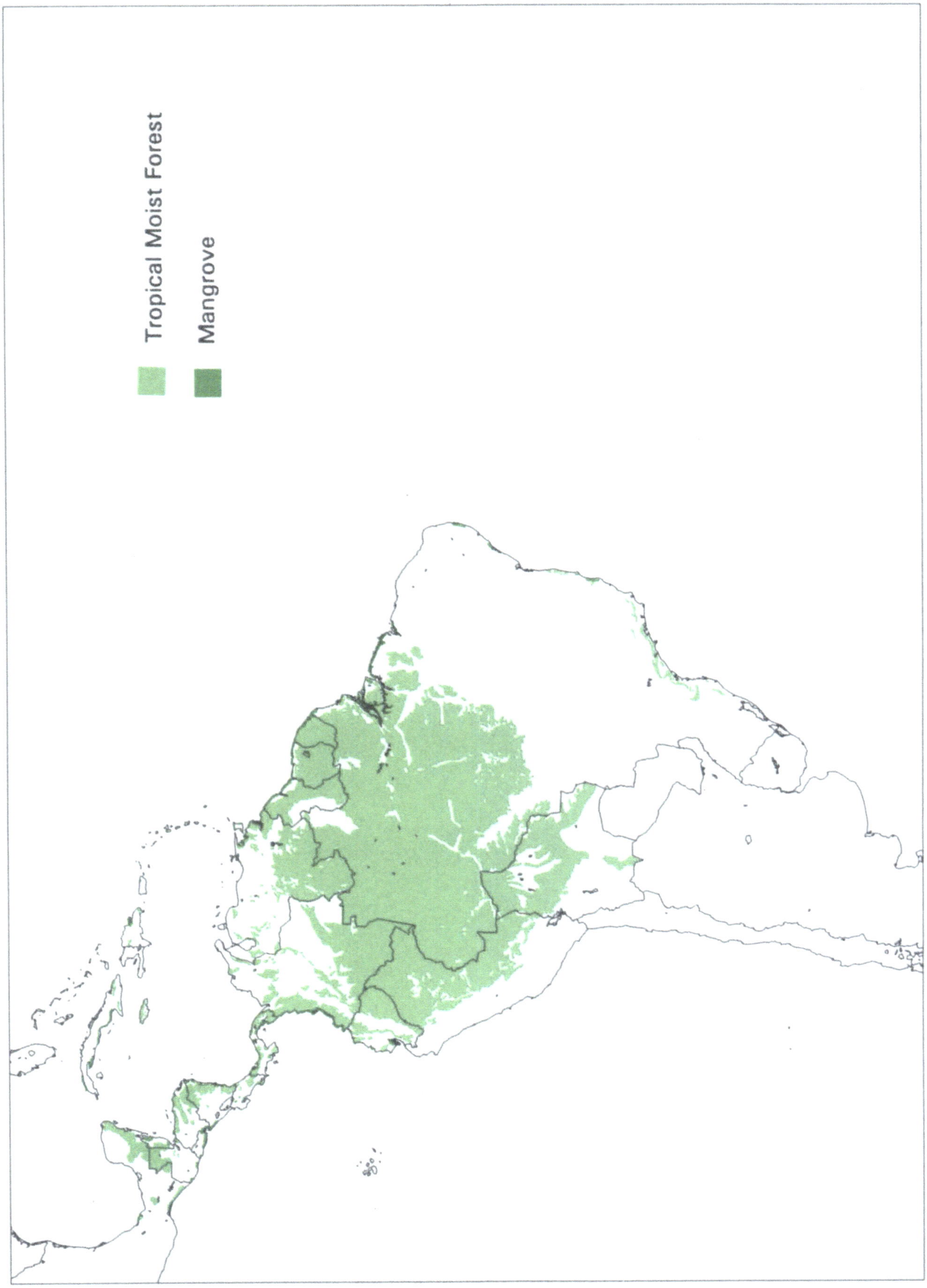

Figure 20.12 Tropical Moist Forest: Africa

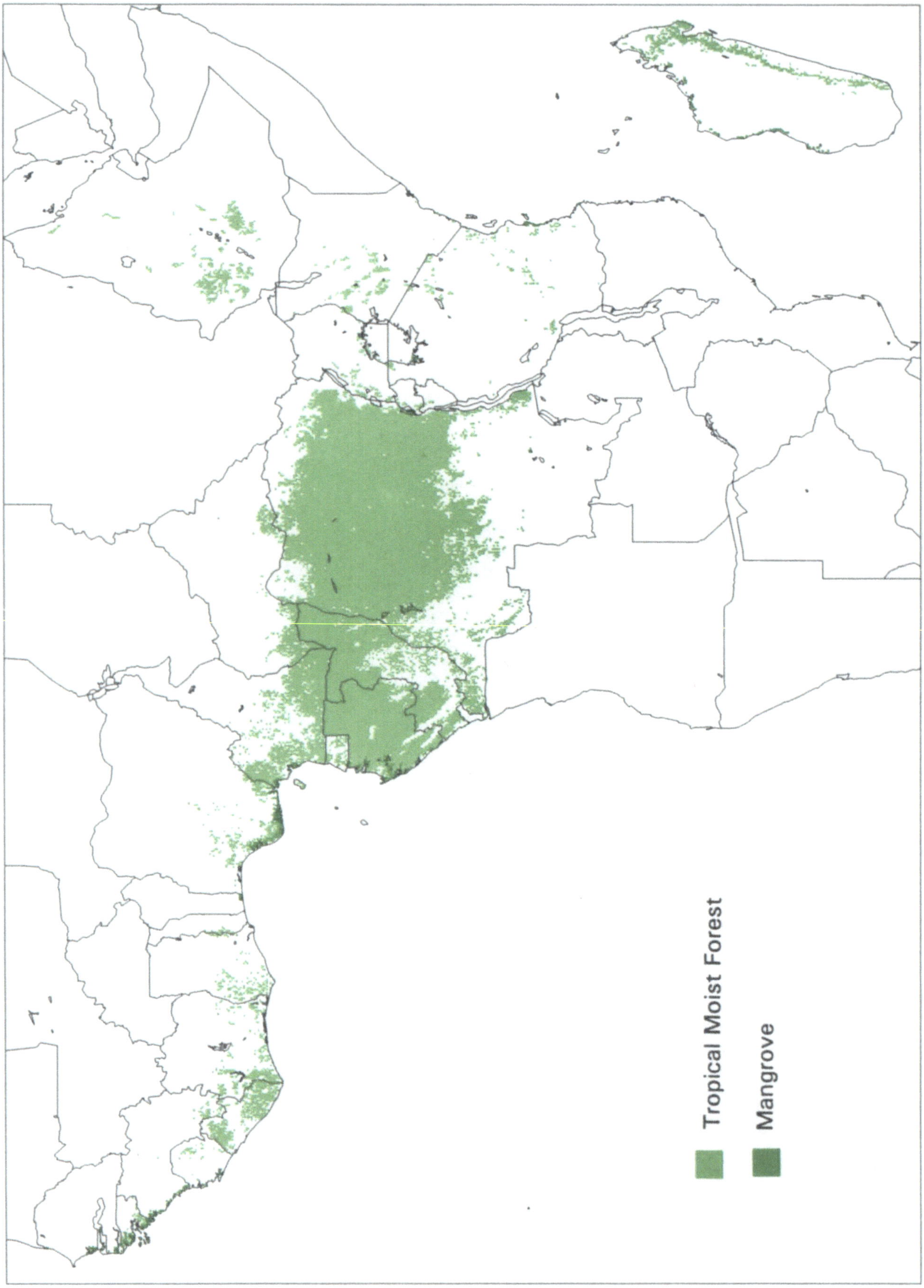

Figure 20.13 Country area, closed forest and annual loss

21. GRASSLANDS

Grassland can be described as a type of vegetation that is subjected to periodic drought, is dominated by grass and grass-like species, and grows where there are fewer than 10-15 trees per hectare. This definition is somewhat arbitrary and is one of several that may be used in discussion of grasslands and the area they cover. Different vernacular terms are used depending on the part of the world under consideration; thus grasslands may be called steppes in Eurasia, prairies in North America, llanos, cerrados or pampas in South America, savannas in Africa and rangelands in Australia.

Although virtually all the world's grasslands have been affected to some extent by man and his domestic stock, natural grasslands appear to have developed in two kinds of area:

- areas where the growth of trees is prevented by edaphic or climatic factors (these are limited in extent, being confined to areas with nutrient-poor soil and / or low rainfall)
- areas where, over a very long period, browsing by wild herbivores has prevented the establishment and growth of trees, and where wild species are still present and outnumber domestic livestock.

In both these types of natural grassland the dominant ecological factors have prevailed for sufficiently long for plants and animals to have adapted and established a natural balance. Two of the essential qualities of such natural grasslands are that the vegetation is unsown and that the balance between plant species has not been significantly affected by human activity. From these natural grasslands there is then a complete spectrum of degrees of modification by man, finishing with the entirely sown and intensively managed short-term rye-grass ley of western Europe, which has almost no significance for the maintenance of biodiversity. Semi-natural grasslands which are unsown but strongly modified by the grazing of domestic livestock are of much greater importance. A large proportion of the world's grassland species are able to use such habitats, and many species are, indeed, dependent on them.

Many natural and semi-natural grasslands have high levels of floristic diversity, at some scales and in some areas approaching that of tropical forests. Animal species richness appears to be generally low, although only vertebrates are well recorded. For example, those birds that are considered to be primarily adapted to grasslands and dependent on them number around 477 species worldwide; this is less than 5% of the world's bird species. One characteristic of grassland birds is a tendency for rapid and apparently erratic dispersal, which enables them to exploit sparsely distributed food resources in an environment in which the climate is unpredictable. Many species habitually move over very large areas. These factors mean that it is difficult to conserve grassland birds through protection and management of important wildlife areas (Grimmett and Jones, 1989). Many more species would by now have become globally threatened were it not for their ability to use mixed farmland.

Similarly, a total of 245 of the world's mammal species are considered to be primarily adapted to grassland conditions. This represents about 6% of the world's described mammal species. They can be broadly divided into large and small mammals, with 77 of the former and 168 of the latter. The overwhelming majority of the large mammals are grazers and/or browsers, and only 19 are predators or scavengers. In general, mammalian predators are adapted to a range of habitats rather than being restricted to grassland. The small mammals of grasslands are mostly seed-eaters or omnivores.

THE WORLD AREA OF GRASSLAND

It has been estimated that grasslands covered approximately 40% of the earth's surface prior to the impact of man and his domesticated animals (Clements and Shelford, 1939). Estimates of the area of grassland present today are generally much lower than this but are very variable. One of the highest estimates suggests that grasslands occupy 27% of the world's natural vegetation cover (Knystautas, 1987). Data on savanna and temperate grasslands from other sources are incorporated here into Table 21.1 and the percentage of the world's land area occupied by these habitats has been calculated. These result in estimates of the area of savanna and temperate grasslands ranging from 16.1% to 23.7% of the world's land area (or 17.9% to 26.5% if Antarctica is excluded).

The land-use statistics produced by FAO (e.g. FAO, 1987) on a country basis include a category for pasture, but this is not clearly defined and it is certainly not restricted to long established, semi-natural or biologically significant habitats. A great deal of the pasture is either newly created from woodland or arable, and it may be managed in rotation with other farm crops. Other areas regarded as permanent pasture by FAO may have had long continuity of grazing but their native flora and fauna may have been completely lost because of agricultural intensification.

Whatever the actual area of grassland present today, two facts are clear: there used to be considerably more natural grassland in the world, and its area is continuing to diminish.

ORIGINS AND FLORAL DIVERSITY OF GRASSLANDS

The best examples of grasslands that formed where soil and climate favoured the production of grass and herbaceous species, rather than trees, are those in northern South America and in South Africa.

In other parts of the world, the impact of large herbivores was more significant. The main areas where grassland formation was influenced by large herbivores were the savanna zones of Africa, the steppes of Asia and Eastern Europe, and the prairies of North America. In Africa the large wild herbivore community was dominated by ungulates such as antelopes and zebras; in Eurasia by gazelles, goats, camels, bison and wild horses; and in North America by deer and the North American bison. The effects

Table 21.1 Estimates of the area* of the world's grasslands

	WHITTAKER AND LIKENS (1975)	ATLAY, KETNER AND DUVIGNEAUD (1979)	OLSON, WATTS AND ALLISON (1983)
Savanna	15.0	22.5	24.6
Temperate grassland	9.0	12.5	6.7
Total grassland	24.0	35.0	31.3
Grassland as % of world land area	16.1%	23.7%	20.7%
Grassland as % of world land area (excluding Antarctica)	17.9%	26.5%	23.1%

Note: * in million km².

of these larger species were supplemented by vast numbers of small mammals such as marmots, pikas, ground squirrels, gerbils and voles. In the African savanna and Australian rangelands, termites are extremely important: they may consume up to one-third of the total annual production of dead wood, leaves and grass, and their biomass may reach as high as 22g/m^2, more than twice that of the greatest densities of vertebrates on Earth, found in the migrating herds of ungulates on the Serengeti plains, Tanzania.

South America

Long-established, near-natural savannas occur in tropical South America in regions where the climate is in no way inimical to the growth of trees. These tropical savannas occur over huge areas as a mosaic of grassland and forest, sometimes as extensive grassy plains with scattered trees, sometimes as grassland with strips of woodland and sometimes as islands of grassland in vast tracts of forest. Fig. 21.1 from Sarmiento (1983) shows the distribution of these habitats. To the south of the area shown, the grasslands are all secondary and less biologically diverse.

Figure 21.1 South America: major tropical savanna regions

Source: Sarmiento, G. 1983. In: Boulière, F. (Ed.), *Tropical Savannas. Ecosystems of the World*, 13. Elsevier, Amsterdam.

The main areas of savanna are listed in Table 21.2, along with their area and floristic richness. The numbers of species refer to entire regions, so they include species which are not primarily adapted to grasslands. It is difficult to make comparisons between the regions but two areas stand out as being both large in extent and rich in plant species other than trees and shrubs. These are the Colombian-Venezuelan *llanos*, to the west and north of the Orinoco river, and the very extensive central Brazilian *cerrados*. Both are rich in plant species and communities, the difference between them being related mainly to soils and drainage (Huber, 1987).

Sarmiento gives a few figures for floristic richness of the cerrados on a smaller, repeatable scale but unfortunately comparable data from other areas have not been located. He states that more than 300 plant species are recorded per hectare of protected cerrados near Brasilia. The sampling was done in 20 x 20 metre plots, with the number of species per plot varying from 52 to 117.

It is interesting to note that although the numbers of species in these neotropical savanna formations (Table 21.2) are generally high, they are not as rich in grasses and herbs as some of the temperate South American grasslands. The *pampas* of Argentina and Uruguay, which is of secondary origin, has over 400 species of grasses (Cabrera, 1970).

The pampas has generally been grazed by cattle throughout historic times. This has encouraged colonisation by species which would not otherwise be able to compete with more aggressive species. The pampas is, therefore, more species-rich but less natural than the cerrados.

Fire plays a more important role than herbivores in maintaining these South American savannas. Fire is a naturally occurring phenomenon but man has increased the frequency of burning in recent centuries and this has had a significant influence on plant communities. Nutrient deficiency, usually related to aluminium toxicity and water availability (Folster and Huber, 1984), is also believed to be of critical importance in maintaining the openness of the vegetation. The largest of the secondary grasslands are the pampas and the *campos*. The latter are open rolling plains on the central plateau of interior Brazil. The campos tend to merge into the cerrados, with a gradual increase in the proportion of trees. All these areas have been extensively modified by frequent burning and agricultural development.

Table 21.2 Floristic richness of various Neotropical savanna formations

FORMATION	AREA (km²)	NO. OF TREES AND SHRUBS	NO. OF SUBSHRUBS HALF-SHRUBS HERBS VINES, etc	NO. OF GRASS SPECIES	TOTAL NO. OF SPECIES
Cerrado in north-western São Paulo	50	45	175	17	237
Cerrado in western Minas Gerais	15,000	c. 200	c. 330	73	c. 600
Whole cerrado region	2,000,000	429 (774)[1]	181	108	718 (1063)[1]
Rio Branco savannas	40,000	40	87	9	136
Rupununi savannas	12,000	c. 50	291	90	431
Northern Suriname savannas	c. 3,000	15	213	44	272 (445)[2]
Central Venezuelan llanos	3	69 (16)[3]	175	44	288
Venezuelan llanos	250,000	43	312	200	555
Colombian llanos	150,000	44	174	88	306

Notes: [1] Total flora including other plant formations; [2] Total flora including bushes; [3] Number of savanna trees excluding groves. See Sarmiento, 1983, for data sources.

Africa

The parts of Africa covered by the various forms of vegetation described as savanna are shown in Fig. 21.2, reproduced from Menaut (1983). Everything which is not forest, desert or montane vegetation is regarded as a form of savanna, though clearly not all of this is grassland. Much of the land is cultivated and probably an even bigger area is savanna woodland. Nevertheless, a high proportion of Africa supports dry, semi-natural vegetation in vast unbroken tracts. The considerable age of the habitat and its great geographical continuity are vital factors in explaining the biodiversity of African grasslands.

Table 21.3 Africa: areal richness zones

REGION	RICHNESS
Guineo-Congolese region, peripheral domain	
Northern district	1,440
Southern district	1,680
Sudano-Zambezian region	
Sahelian and Sudanian domains	1,060
Zambezian domain	2,590
Eastern transition zone	
Sahelian type	1,270
Sudano-Zambezian type	2,330
Kalahari domain	1,020
Madagascar	5,410

Source: Menaut, J.-C., 1983. The vegetation of African savannas. In: Boulière, F. (Ed.), *Tropical Savannas. Ecosystems of the World*, 13. Elsevier, Amsterdam.

The best information available on floristic richness of dry tropical Africa was assembled by Lebrun and summarised by Menaut (1983). Lebrun pointed out that to compare the species richness of different areas it was essential that the comparisons were made between units of similar size. He chose a unit of 10,000km² as a standard and called the average number of species in that area the areal richness of the region. Table 21.3 shows the average areal richness for the major plant-geographic (phytogeographic or chorological) zones of Africa and Fig. 21.3 shows their location. The data in Table 21.3 should not be used in a detailed comparison of the diversity of grasslands because a high proportion of the plant species listed are associated with forest, wetland or other habitats. Nevertheless, the zones show a high degree of correlation with the main savanna zones, so some broad comparisons are possible. Menaut points out that in Africa the average areal richness for savanna (c. 1,750 species) is not far below that of rain forest (c. 2,020 species), contrary to the situation in the Americas. Indeed, the heart of the savanna zone, known as the Somali-Masai Region, contains 2,500 plant species, of which 50% are ecological endemics (Stuart and Adams, 1990). Not all of these can be regarded as primarily adapted to grasslands, but a sufficiently large number are for it to be regarded as the world's richest grassland zone. Biodiversity in the region is enhanced by the fact that the savanna merges gradually into other large habitat formations, notably forest and semi-desert, rather than being confined by mountains, the sea or intensive agriculture.

The presence of large mammals is important in the ecology of African grasslands; the world's greatest concentration of large mammals is found on the savanna of northern Tanzania. However, as in South America, fire has also been a major influence in the evolution of the flora and fauna. Natural fires, caused by lightning, may have affected huge areas, limiting the build-up of dry organic matter and favouring the survival of some species at the expense of others. In addition, at least in East Africa, man has probably been burning grasslands for a minimum of 50,000 years, and very likely much longer. Therefore, here as in South America, man has greatly affected the apparently 'natural' grasslands.

Australia

The grasslands of Australia have been described by Moore (1970) using a very broad definition of grasslands which includes all regions where grasses supply a substantial proportion of the food for stock. This includes a wide range

Figure 21.2 Africa - main savanna vegetation types

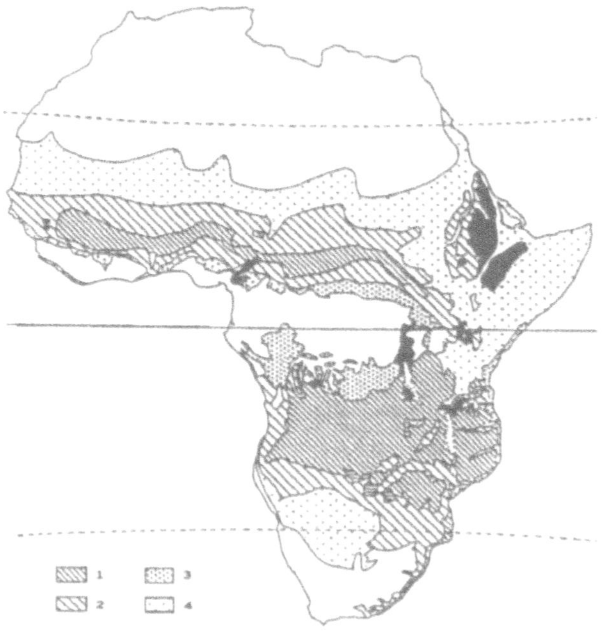

Source: Menaut, J.-C., 1983. The vegetation of African savannas. In: Boulière, F. (Ed.), *Tropical Savannas. Ecosystems of the World*, 13. Elsevier, Amsterdam.
Notes: 1 woodland; 2 tree/shrub savanna; 3 forest/savanna mosaic; 4 tree/shrub 'steppes'.

Figure 21.3 Africa: areal richness zones

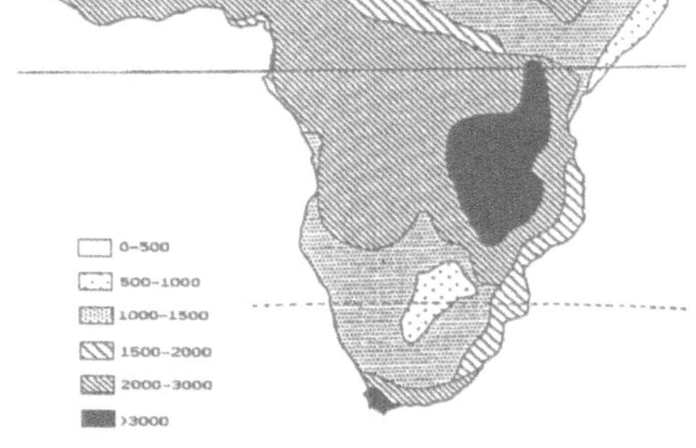

Source: Menaut, J.-C., 1983. The vegetation of African savannas. In: Boulière, F. (Ed.), *Tropical Savannas. Ecosystems of the World*, 13. Elsevier, Amsterdam.
Notes: Areal richness measured by number of species per 10,000km^2.

of semi-desert, scrub and wooded savanna. Groves (1981) considers that the only natural grasslands in Australia are those dominated by hummock grasses, i.e. species of *Triodia* and *Plectrachne*. These are very similar to the bunch grasses of North America. Hummock grassland is distributed over a very large area of south, west and northern Australia, in arid and semi-arid lands, but a high proportion of this vegetation has been subject to agricultural improvement. Relatively small areas are free of introduced species, legumes being the most common of these. Legumes raise soil nutrient levels and fertility, thus changing the ecological balance and making conditions more suitable for weedy species. Around the great zone of natural hummock grasslands are other natural and semi-natural communities, most of which contain a significant proportion of grasses. The tropical zone of northern Queensland is strongly influenced by summer rains, whereas temperate and less natural grasslands occur in a zone from north of Adelaide to northern New South Wales.

One of the main reasons for the difference in the vegetation of the arid zone of Australia and, for instance, East Africa was the limited number of indigenous grazing mammals in Australia. The larger marsupials (wallabies and kangaroos) are primarily browsers rather than grazers, and would have had little impact on the arid hummock grassland. The influence of herbivores was therefore of much less significance than in other parts of the world. However, traditional burning patterns used by the aborigines (who arrived around 40,000 years ago), together with their hunting practices, must have influenced plant communities by favouring species adapted to fire.

The arrival in Australia of settlers from Europe and their domestic animals, particularly sheep, led to the establishment of very extensive rangelands in which grasses were able to provide most of the fodder for stock. Many of the indigenous species were able to adapt to these new conditions but numerous plant and animal species were introduced from other parts of the world also. The most significant of these was the rabbit *Oryctolagus cuniculus*, whose numbers exploded in the absence of other competing herbivores. Australia's grassland species are therefore a complex mixture of desert- and fire-adapted species, species secondarily adapted to grassland and introductions.

Asia

The natural Asiatic steppe extended from Manchuria westwards as far as the land now occupied by Bulgaria and Hungary, occupying the broad zone between the *taiga* (coniferous boreal forest) and the deserts or mountains to the south. The continental climate of this vast area, with hot, dry summers and very cold winters, is inimical to the growth of trees. A large proportion of the area has not supported forest since a more favourable climate prevailed in one of the earlier interglacials. The dominant herbivores before the influence of man became widespread were horses, wild sheep and gazelles, along with a wide range of smaller mammals.

A large proportion of the Indian sub-continent supports either tropical savanna, savanna woodland or dry forest. The total range of grassland types is very broad - from semi-desert, to seasonally inundated areas, to montane habitats. In the case of India, there is no doubt that a very large area formerly supported dry woodland of various sorts, but there has been an enormous amount of clearance, followed by fire and grazing. It appears likely that there are no surviving primary grasslands in the country (although there is continuing dispute over the origin of hill grasslands in the south-west). In addition, there are only a few long-established stable communities in which the balance of species clearly reflects edaphic factors and traditional management. Indian grasslands are apparently more seral in character, always being in a phase of recovery from clearance, fire, overgrazing, erosion or abandonment. This exerts a powerful influence on the flora and fauna present. The vegetation is relatively poorly endowed with perennial herbaceous plants and floristic diversity is not particularly high. Mammals are not well-represented but a large range of bird species are able to use the grasslands and many of them are dependent on it.

Europe

The Asiatic steppes were extended at an early stage by forest clearance, initially to increase the grazing for sheep, goats and horses. Islands of forest on better soils or where rainfall was higher, were reduced or eliminated. The clearance continued westwards into Europe, into regions where the climate and soils were far more suited to the growth of deciduous forest. Thus wholly new types of grassland were created, capable of far greater productivity than the natural steppes. Man's activities enabled numerous grassland plants, particularly shorter grasses and herbs, to extend their range, accompanied by characteristic animals.

The extension of grazing enabled semi-natural grasslands to develop westwards, as far as central Spain, the Atlantic regions of France and the British Isles. While many species expanded their range from the steppes, others were no doubt lost altogether. It is very doubtful if any of the European steppes can be regarded as primary (Polunin and Walters, 1985) but some areas of secondary steppe may be very similar to the original habitats. The most natural grasslands in Europe are the Hungarian *pusztas*, traditionally managed, low-lying grasslands in the floodplain of the River Danube.

Man started grazing domestic stock on the mountain ranges of Europe at a very early time and permanent settlements were established high in the valleys. Extensive grazing by sheep and goats occurred over the high mountains in the summer but the stock was returned to pastures near the farms for the winter. Much of the land on the lower slopes was cut for hay to provide fodder for the long winters. The pattern of management was so consistent that many different plants were able to adapt to these conditions and the meadows became very species-rich. The flora of alpine hay meadows are a mixture of steppe and montane species. They form a balanced semi-natural community reflecting traditional patterns of land-use and are highly valued both aesthetically and scientifically. It must be remembered, however, that their origins are not natural.

The British Isles is a region lacking natural grasslands (according to the definition above). Here too, however, the pattern of pastoral land-use is sufficiently long-established and consistent for man-made grasslands to appear natural and they vary in composition in relation to soils, aspect and drainage. As a result, these habitats have assumed great value as resources for a variety of wildlife. For example, at least a third of Britain's 1,500 species of flowering plant are associated with grassland, and about 400 of them are most frequent in this habitat (Duffey *et al.*, 1974). In southern England, agriculturally unimproved, semi-natural grasslands may have up to 40 higher plant species per m^2. Even this level of species richness is exceeded by natural grasslands in the heart of the steppe region, which in parts of the former USSR and Mongolia may support up to 80 species per m^2 (Knystautas, 1987).

North America

In the central parts of North America, in the rain shadow of the Rockies, the dry climate naturally favours open habitats rather than woodland. Vast herds of bison also helped maintain species-rich grassland on a grand scale.

Islands

The absence of large wild grazing mammals from oceanic islands ensures that they do not generally support natural grasslands. Even where man has created grasslands on islands, such as on Madagascar or Sri Lanka, they seldom support diverse stable communities. The average areal richness of Madagascar is higher than all the other chorological territories in Africa, but this is because of the very large number of forest species. Savannas have been created in the drier parts of the island but they are species-poor, with a high proportion of introduced plants (Menaut, 1983).

An exception to this rule is New Zealand, which has some long established 'grassland' habitats. These are however high altitude, high rainfall, tussocky communities, which are very different from the rest of the world's grasslands. They support distinctive native species such as the Takahe *Porphyrio mantelli*, a flightless bird of the family Rallidae. Those grasslands which have been created in New Zealand for grazing stock are very largely composed of introduced plant species, incidentally supporting introduced wild animals such as the Red Deer *Cervus elaphus*, and they should be regarded as artificial.

The only island lacking large indigenous grazing herbivores but supporting species-rich grassland is Cuba. Floristically, the Cuban savannas compare favourably with their equivalents in South America. This may be because of climatic stability and greater ecological diversification on several different types of parent material, including some unusual substrates such as silicious rocks and serpentine (Sarmiento, 1983). It seems very surprising, however, that such diversity of grassland species could evolve in the absence of large herbivores.

THE 20TH CENTURY IMPACT ON GRASSLANDS

Until this century, the distribution of grassland species around the world had been determined by an integrated complex of various factors, including: climate, geographic and ecological isolation, the impact of large herbivores, traditional land-use practices, domestication of grazing animals and forest clearance. The richest grassland regions of the world, in descending order of importance for indigenous plants and animals, were:African savanna; Eurasian steppe; South American savanna; North American prairies; Indian savanna; and Australian grasslands.

On one hand, the original extent of these natural grassy areas has been extended by man's activities so that species-rich, semi-natural grassland now occurs in a discontinuous manner over a very much larger area. It is found throughout much of the region once occupied by the world's temperate forests and also reaches well into the tropical forest zone. On the other hand, domestic stock has often overgrazed natural grasslands, causing massive impoverishment of the ecosystem, and large areas have been converted to agricultural land.

Africa

In Africa, native people have burnt the savanna for thousands of years to improve grazing for their stock and facilitate the hunting of wild game. The frequency of these fires may have increased markedly over the last thousand years or so as population increased. Towards the end of the 19th century, when settlement became more firmly established, the area of savanna was greatly enlarged by forest clearance, burning, and massive increases in the number of cattle. Much of this was achieved by white settlers with imported European stock. Their farming activities were, however, frequently upset by the unpredictability of the climate and by parasites and diseases which became increasingly serious as livestock densities rose. Many of the imported breeds proved incapable of tolerating the indigenous diseases of Africa, the most serious of which were rinderpest, trypanosomiasis and foot-and-mouth disease.

Around the turn of the century, rinderpest spread from the north of Africa to the far south in a period of only seven years, killing 90-95% of domestic cattle as well as many wild ungulates (Rogers and Randolph, 1988). The result was widespread human starvation and the abandonment of vast areas of grazing land. The subsequent regeneration of scrub and woodland appears to have allowed unprecedented spread of tsetse flies (vectors for the parasite causing sleeping sickness in humans and trypanosomiasis in cattle). Colonial governments tried a variety of methods in an attempt to eradicate the tsetse fly, one of which was the removal of the woodland and cover which the flies require. This allowed the indigenous herbivores and, consequently, their predators to extend their ranges. These enlarged distributions have generally been maintained, assisted in recent decades by control of poaching and the establishment of national parks and other protected areas. The grassland habitat in these areas may have every appearance of naturalness but frequently it is not very old and its extent has been greatly influenced by man.

There have been large increases in the area used for growing cotton in some of the semi-arid parts of Africa, particularly Senegal, Mali and Mauritania. Persistent insecticides such as dieldrin have been used on these crops, with little regard for non-target species. The crops provide very little food for birds or mammals and the water control schemes required for their irrigation intercepts water which formerly flooded river valleys, where it provided suitable feeding areas for many species, including migrant birds (Goriup and Schulz, 1991). The loss of grassland presents a particularly serious problem for European breeding birds which winter in West Africa because the habitat is confined on its southern edge by forest, farmland or the Atlantic Ocean. In East Africa, on the other hand, there are few barriers to prevent wintering birds from moving further south to find suitable habitat.

Europe

A useful review of the surviving area of grassland in Europe has been carried out by van Dijk (1991), and a summary of this data is presented in Table 21.4. The figures quoted for permanent pasture generally correspond well with the latest FAO data. The majority of the discrepancy in total area between the two datasets is attributable to the UK although there are also major differences in the figures for Greece and Spain. In the UK the difference arises because upland areas in Scotland and Northern Ireland were included in the FAO data but not in van Dijk's data. Wherever possible van Dijk used national surveys (of various dates) but for other countries he has drawn the information from Lee (1990) or Grimmett and Jones (1989).

Figure 21.4 France: land-use 1970-1985

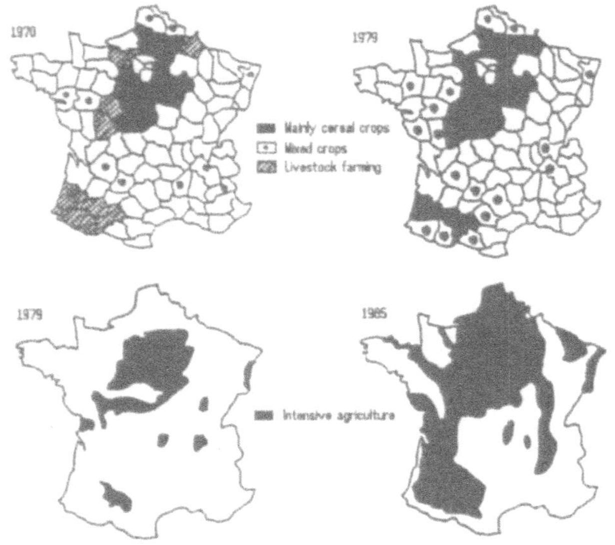

Source: Lecomte, P. and Voisin, S. 1991. Dry grassland birds in France: status, distribution and conservation measures. In: Goriup, P.D., Batten, L. and Norton, J. (Eds), *The Conservation of Lowland Dry Grassland Birds in Europe*. Proceedings of an International Seminar held at the University of Reading 20-22 March 1991.

The data clearly reveal that only a small proportion of permanent grassland can still be regarded as dry semi-natural, and therefore of biological interest. The review by van Dijk gives a lot of information on habitat loss in different countries and it is apparent from this that a great deal of the loss has taken place since the 1960s. More information on these trends is provided in the case study on grasslands in Poland.

Table 21.4 Areas of grassland and dry semi-natural grassland in Europe (1,000ha)

COUNTRY	FAO PERM (A)	VAN DIJK (B)	DRY SN (C)	PER CENT (C/B)
Belgium	688	632	0.5 +	<1
Czechoslovakia	1,646	1,600	?	<10
Denmark	217	214	?	?
France	11,740	12,000	250	2
Germany	5,707	5,700	100	2
Great Britain	11,560	4,800	*200	4
Greece	5,255	1,789	?	?
Hungary	1,210	1,350	200	15
Ireland	4,688	5,800	700	12
Italy	4,907	5,000	200 +	4 +
Netherlands	-	1,100	10	1
Norway	102	-	?	10-20
Poland	4,040	4,040	?	?
Portugal	531	761	?	?
Romania	4,410	4,400	?	?
Spain	10,210	6,645	1,452	22
Sweden	562	480	?	?
Yugoslavia	6,347	6,400	?	?
TOTAL	73,820	62,711	3,593 +	

Notes: The FAO data (A) are for permanent pasture for 1988. The second column of figures (B) have been extracted from Dijk, G. van 1991. The status of semi-natural grasslands in Europe. In: Goriup, P.D., Batten, L.A. and Norton, J.A. (Eds), *The Conservation of Lowland Dry Grassland Birds in Europe*. The third column of figures (C) are for dry semi-natural grassland, all derived from van Dijk 1991, who used various sources. The data are not all strictly comparable. This category of grassland omits wet or seasonally flooded grassland, upland grassland, acid grassland and communities dominated by ericaceous species. Where separate figures were given for calcareous and neutral grassland, these have been combined.
* includes only lowland grassland in England and Wales.

Loss of semi-natural grassland has occurred because of the enormous changes in agricultural methods. The use of inorganic fertilizers and modern pesticides has vastly increased productivity, at the expense of indigenous plants and animals. The area under intensive cultivation in Europe has increased dramatically since the 1950s. The trends have been clearly illustrated for France by Lecomte and Voisin (1991), from which Fig. 21.4 has been derived, showing increase in the area of cereal crops and intensive agriculture between 1970 and 1985.

In arid parts of Spain, Italy, the Balkans and Turkey, the prevailing pattern of agriculture for most of the last century or two has been to rotate grazing, arable cropping and fallow. This has traditionally been done in small units of land, creating a diverse landscape and normally one with a good scatter of scrub and intermittently managed vegetation. This pattern of land-use, which has been described by Goriup (1988) as pseudosteppe, provides the ecological conditions required by a very large number of plant and animal species, many of which were originally part of the steppe community. Birds, reptiles and small mammals are generally mobile enough to accommodate this gently shifting pattern of agriculture, so, even though it may be far removed from natural or even semi-natural grassland, pseudosteppe has become a very valuable wildlife resource.The pseudosteppe style of land-use is now under threat in Europe as people become increasingly dissatisfied with this hard and relatively unrewarding way of life. The trends are most apparent in the Mediterranean region but similar pressures are having adverse effects on wildlife in the western parts of the British Isles.

The Great Bustard *Otis tarda* is an example of a bird species formerly associated with extensive dry grasslands but which was able to adapt and flourish in open countryside maintained as a mosaic of grassland, arable and fallow. Great Bustards will often display on grassland but select arable land for their nest sites. Increasing dependence on this kind of low intensity farming, places the birds at great risk when agricultural intensification occurs, particularly if there is increased application of pesticides and fertilizers (Kollar, 1991). The Great Bustard has undergone significant population decline through much of its breeding range as a result. Other species particularly at risk from this threat include the Little Bustard *Tetrax tetrax*, Button Quail *Turnix sylvatica*, Sociable Plover *Chettusia gregaria* and Demoiselle Crane *Anthropoides virgo*. In the British Isles, the Corncrake *Crex crex* and the Chough *Pyrrhocorax graculus* are very much at risk from similar kinds of social and agricultural change (Goriup *et al.*, 1991). The Large Blue butterfly *Maculinea arion* illustrates similar trends for the invertebrates.

North America

The prairie zone of USA and Canada is inherently very fertile and arable farming has been developed throughout the region. A relatively small area managed to escape ploughing and cropping through various accidents of history and ownership. Much of the surviving prairie grassland is therefore secondary in origin. The extent of ploughing and agricultural improvement was greater in the western half of the prairie zone, which is known as long-grass prairie, than in the eastern short-grass prairie (Knopf, 1988). The rainfall is higher in the former region, hence there is increased agricultural production there.

A review of the survival of all the major vegetation types in USA was carried out by Crumpacker *et al.* (1989). This

Table 21.5 Area of grassland habitats in USA (km^2)

PNV TYPE*	AREA OF PNV IN USA	AREA NAT OR S/N (1967)	% NATURAL OR S/N IN USA (1967)
Mesquite savanna	23,041	-	
Mesquite-buffalo grass	70,406	51,236	72.9
Northern cordgrass prairie	3,779	2,562	67.8
Fayette prairie	7,696	4,302	55.9
Fescue oatgrass	3,565	2,110	59.2
Bluestem-Sacahuista prairie	41,457	9,908	23.9
Blackland prairie	48,461	15,216	31.4
Sea oats prairie	1,564	-	
Grama-tobosa prairie	15,196	14,831	97.6
Southern cordgrass prairie	22,292	13,108	58.8
Palmetto prairie	11,273	8,251	73.2
Bluestem-Grama prairie	150,771	53,071	35.2
Wheatgrass-Grama-Buffalo grass	2,639	2,435	92.3
California steppe	51,973	15,903	30.6
Fescue-wheatgrass	20,985	5,561	26.5
Bluestem prairie	272,567	40,612	14.9
Aleutian meadows	12,730	-	
Wheatgrass-bluegrass	36,845	25,238	68.5
Grama-buffalo grass	309,170	170,352	55.1
Wheatgrass-needlegrass shrubsteppe	27,258	25,867	94.9
Wheatgrass-needlegrass	253,707	161,414	63.8
Grama-needlegrass-wheatgrass	205,196	156,563	76.3
TOTAL	1,592,571	778,540	

Source: Crumpacker, D.W., Hodge, S.W., Friedly, D. and Gregg, W.P. 1989. A preliminary assessment of the status of major terrestrial and wetland ecosystems on Federal and Indian Lands in the United States. *Conservation Biology* 2(1):103-115.
Note: * PNV = Potential Natural Vegetation, NAT = natural, S/N = semi-natural.

is based on the concept of Potential Natural Vegetation (PNV) which is defined as the vegetation that would, if man were removed from the scene, exist in a region at the end of the sequence of plant succession. Crumpacker *et al.* give the area of each type considered to be still present in a natural or semi-natural state in 1967 and calculate this as a percentage of the potential natural vegetation. The vegetation types which broadly equate to grassland and the area of each said to have been present in 1967 are set out in Table 21.5. The figures are based on small-scale maps, however, and are subject to the usual problem of amalgamating fragmented areas of grassland into a single type. The actual areas of grasslands are therefore likely to be inflated.

The total area of grassland PNV types would occupy about 17% of USA. This compares with a figure of 25.8% for the proportion of USA occupied by permanent pasture according to FAO data. The data in Crumpacker *et al.* show that the grassland area which was considered to be natural or semi-natural in 1967 occupied about 8% of USA. As explained above, this is probably an inflated figure and it is certainly likely that there has been further significant loss since 1967. Table 21.5 shows that the ecosystems which have been reduced by the greatest extent are the bluestem prairie and the bluestem-Sacahuista prairie. A considerable area of the once very extensive bluestem-Grama prairie has also disappeared. The bluestem grasses belong to the widespread genus *Andropogon*, which constituted the dominant species over much of the long-grass prairie. In some years, after fire they can grow to a height of up to 2m, illustrating the potential productivity of this habitat and hence the probable reason for its disappearance.

Today, attempts are being made to recreate some of the original prairie grasslands of the USA and Canada. However, these newly-created prairie grasslands lack many of the original indigenous species and the densities of plants are thought to be much lower than they were in the natural prairies. Nevertheless, the dominant grasses mirror the original communities and the habitats can have a very natural appearance. More information on the fate of the prairies is given in the case studies on the Black-footed Ferret and Canada.

Asia

The eastern part of the Asian steppes, particularly Mongolia, still supports the most extensive and natural area of the world's great grasslands. Land-use practices have been extraordinarily stable over the centuries, with low intensity grazing being carried out by semi-nomadic tribesmen. Agricultural practices have developed very little and pesticides and fertilizers are largely irrelevant in this shared system of grazing. The land is all in state-ownership and people hold rights to drive their stock to wherever the grazing is best. Over a vast area approaching the size of western Europe, sparse grazing continues without fences to confine stock. The only barriers are forests and mountain ranges. Horses and sheep are the main grazers, with goats and camels in the drier regions. Gazelles are still reasonably plentiful and are highly prized as a resource for hunting. Changing social and political attitudes are now likely to change traditional land-use practices.

The state of the grasslands is very much less satisfactory in the former USSR than it is in Mongolia. A very high proportion of the original steppe has been destroyed, particularly in Kazakhstan and Uzbekistan. Very extensive irrigation projects have been carried out, permitting agricultural improvement on a huge scale. Before the irrigation works were begun, large ploughs destroyed the burrows of gerbils and other mammals. The main purpose of this was to eliminate the gerbils before people had to work on the land, because they carry a parasite which causes leishmaniasis in man. Most of the irrigated land is used for cotton crops, which are heavily treated with insecticides and provide a very sterile environment for wildlife.

The pressures on Indian grasslands have always been high but they have been accelerating this century with the growth in population (Majumdar and Brahmachari, 1988). Food production has not increased in all areas, instead there has been a major growth in production of cash crops. More information on this is given in the case study on the Lesser Florican.

Australia

In Australia the main problems in conserving grassland habitats have arisen from alterations to traditional patterns of burning. Aboriginal peoples had burned the vegetation in a rotational system, timing the burns carefully in relation to season and weather. This kept huge areas of the Australian hinterland in a broadly open condition and increased productivity for grazing animals. The fires were sufficiently frequent to ensure that woody material did not accumulate but well spaced enough to ensure that native plants and animals could recover from the fire and take advantage of the better growing conditions. Through the last hundred years or so, as more of Australia has become settled, the traditional burning patterns have been disrupted. Large areas are burned annually, and very few of the native plants and animals can cope with such a regime; pastures are becoming increasingly dominated by introduced European plants. Conversely, huge areas have been burnt very much less frequently, and this has had two consequences: firstly, species of grassland and other open habitats decline because the vegetation becomes too thick, woody and tall, and, secondly, when fires do happen they burn at a higher temperature and are more destructive. In some national parks and protected areas, efforts are now being made to return to the traditional burning patterns (Boekel, 1990), with the direct help and involvement of the indigenous people.

CASE STUDIES

The following case studies illustrate the types of threat facing declining species of grassland-adapted fauna, and those affecting the habitat as a whole.

The Lesser Florican as an indicator of grassland loss in India

The Lesser Florican *Sypheotides indica* is the smallest and formerly most widespread of the three bustard species endemic to the Indian subcontinent. The majority of the birds both breed and winter in dry grassland, though there is some migration to areas where rainfall has increased food availability. Until the 1980s, conservationists had been concentrating most of their concern on the Great Indian Bustard *Ardeotis nigriceps*. The Lesser Florican was thought to be common and widespread. It was a popular bird for the table and large numbers were shot by sportsmen and caught in nets. Indeed, until 1980 the species was still officially sanctioned as legal prey for hunters.

Field surveys were made in four separate areas of India which had been reported as strongholds of Lesser Floricans within recent decades (Goriup and Karpowicz, 1985). One of these, the Tungabadhra Wildlife Sanctuary, near Bellary in Karnataka, southern India, no longer supported any grassland as a large reservoir had been constructed and all the adjacent land in the valley had been developed for agriculture. In the Jaipur area of northern India, grassland was still present in some quantity but was overgrazed and generally unsuitable for Lesser Florican. The birds there were reported to have become very much rarer in the last 20 years. The Deccan plateau of central India had been regarded as the core of the Lesser Florican's range but the great majority of the semi-natural grassland had been converted to rice paddies. When the paddies are prepared for sowing the only remaining grass is on the embankments (*nallas*), and the villagers use these for trapping game-birds, including Lesser Floricans. Not surprisingly, the birds have become very much less frequent and the prospects for their survival in this area are very poor (Goriup and Karpowicz, 1985).

Goriup and Karpowicz concentrated the majority of their fieldwork in the Jamnagar district of the Kathiawar Peninsula in north-west India. Here the remnants of the once extensive grassland are under the control of the Gujerat Forest Department, and they occur as discrete patches known as *vidis*. The policy for these areas is that they should be cut for hay and kept free from grazing stock throughout the year. It was found, however, that the habitat had deteriorated in many of the vidis, with scrub invasion in some and cattle or buffalo grazing in others. Out of 50 vidis visited, seven were found to support Lesser Floricans, and there was a total of only 22 individuals within them. A more extensive survey of suitable areas in the region failed to produce any more birds. It was apparent that Lesser Floricans have not managed to adapt to new habitats such as fields of groundnuts. Historical data on the land-use of the Jamnagar district showed that, although the area producing food crops had fallen from 34% to 18% between 1906 and 1981, the total cultivated area had risen from 45% to 70% because of the increase in the district of cash crops, principally groundnuts, cotton and sugar cane. The agricultural pressures on the region are clearly very intense and the prospects for Lesser Floricans and other grassland species are not encouraging.

Grasslands in Poland

The extent and condition of grasslands in Poland have been effectively reviewed by IUCN (1991), much of the information having been drawn from Denisiuk (1990).

Poland occupies 312,000km^2, of which 13% (or 40,400km^2) is grassland. This area has been reduced markedly in recent years. The great majority of the semi-natural grassland is in the major river valleys and would be classified as either damp or wet. True steppe grasslands are now very rare in Poland and are confined to steep south-facing slopes. Dry species-rich, semi-natural grassland has become rare because the level of fertilizer application in the country as a whole is very high and well above the threshold for maintaining floristic diversity. Quite a high proportion of the dry grasslands is in fact former peat bogs which have been drained and grazed. Some of these drained sites have been brought under cultivation but the current estimate is that 82% of their area is now maintained as meadows, and these are classed as dry grassland. Management of this land can be particularly difficult because as the peat dries it oxidises and shrinks, thereby lowering the land surface, which then requires draining with new deep ditches. Other areas become too dry, with the result that yields fall and the grassland becomes uneconomic. This is one reason why there are about 10,000km^2 of abandoned farmland in Poland.

An important characteristic of rural Poland is that it is held in the form of very small farms, averaging only 5ha. Large farms were not formed in the same way as in other eastern European countries because of resistance to collectivisation after the Second World War. This results in a large area of little-used boundary land, which is often of value for wildlife, sometimes including grassland species. This land and the abandoned farmland is, however, much more likely to benefit the more adaptable species which require cover and woodland edge.

The pressures on grassland flora and fauna have been considerable in recent years. Agricultural intensification, particularly increasing fertilizer use, is affecting all parts of the country and more of the dry grassland has been brought under cultivation. The area of wet grassland has been reduced by drainage from 36% to 23% of the area of all grasslands between 1973 and 1988.

The economic pressures on Poland's farmers are forcing them either to improve their agricultural methods, through such measures as drainage, fertilizer use, irrigation and switching from hay to silage, or to abandon the land altogether. Further areas of grasslands have been lost to afforestation, and it seems likely that this trend will increase. For the foreseeable future, the best way of maintaining the flora and fauna of Poland's grasslands would appear to be through the establishment of national parks and nature reserves, rather than through changing the direction of agricultural development.

The Meadow Viper

The Meadow Viper *Vipera rakosiensis* is the smallest and least venomous of the European vipers. It is also the rarest, having long been restricted to a specific lowland grassland habitat in central Europe. It is found in both wet and dry grasslands but is particularly associated with the interface between the two. Sites providing large tussocks and ant-hills are favoured, especially if there is varied topography offering a range of soil moisture conditions and temperatures. The past distribution of the Meadow Viper is not well known but it is clear that the subspecies has undergone a very severe decline. The situation in each country is as follows:

Hungary. Still present in the Great Plain *pusztas* between the rivers Danube and Titza, south of Budapest. Elsewhere the habitat is severely fragmented and under pressure from agricultural improvement. Several sites have been lost as a result of grazing and grubbing by geese and pigs. One 12ha meadow has been protected in the Little Plain *Hansag*.

Romania. Recently became extinct following ploughing and agricultural improvement of the *Stipa* (feather grass) meadows with which it had long been associated.

Austria. It was formerly common in the sandy basins of Vienna and Neusiedler but agricultural improvement has destroyed almost all of its habitat. About 17ha of meadow are now protected and suitably managed but the Meadow Viper is generally thought to be extinct in Austria.

The causes of the decline of the Meadow Viper are very clear:

- Killing for bounty (when the species was more common), especially in Austria
- Land drainage and subsequent use for vineyards
- Arable farming
- Forestry
- Application of fertilizer and pesticides
- Increase in the frequency of mowing
- Rearing of pheasants, which are predators of young snakes
- Collecting, for museums and private collections.

The decline of the Meadow Viper provides an illustration of the effects of the pressures on central European meadows which were formerly managed in a casual or inefficient way. It is an example of a species which has been unable to adapt to the reduction of structural complexity and biodiversity in its grassland habitat.

The Canadian prairies

Natural grasslands were concentrated in the southern parts of the three prairie provinces, Alberta, Saskatchewan and Manitoba. Their characteristics and development have been reviewed by Mondor and Kun (1982). The Canadian prairies occupied the northern part of the north temperate zone and are situated to the south of a large zone of aspen parkland. The area of woodland was increased by farmers who planted trees for shelter but many of these farms have since been abandoned, thus allowing woodland to spread. There are still huge open plains, however, supporting the typical long-grass and short-grass prairie habitats.

At the time of settlement in the 19th century the extent of open grassland in the prairie zone of Canada was probably in the range of 360,000-400,000km^2. This had been reduced to about 80,000km^2 by 1982 and was reported to be undergoing conversion to arable at a rate of approximately 500km^2 per year (Mondor and Kun, 1982). The FAO figure for permanent pasture in the whole of Canada in 1985 was

325,000km^2, so obviously a high proportion of this is secondary and most of it will be sown or agriculturally improved pasture of little biological interest.

Cattle ranching began in the Canadian prairies in the 1870s and increased so rapidly that there was acute shortage of land by the 1890s. Mismanagement, overgrazing and hard winters forced most stockmen out of business by the early years of the 20th century and cereal production became the dominant land-use. Today open-range cattle ranching survives only in south-eastern Alberta and in an adjacent area in Saskatchewan. Elsewhere cattle grazing takes place as part of an arable rotation system and recent decades have seen very large increases in stock numbers.

The millions of bison which roamed the plains were reduced to a low ebb of approximately 1,100 by 1889. Other mammals were also reduced to very low levels, primarily through hunting for food, notably the Elk *Cervus elaphus* and Pronghorn Antelope *Antilocarpa americana*. The latter was estimated to have numbered 50 million animals, mostly on the open prairies, but by 1915 it had been reduced to only a few herds in south-eastern Alberta and south-western Saskatchewan. On the other hand, increased grazing and shorter grass benefited small mammals such as Pocket Gophers *Geomys* spp., Richardson's Ground Squirrel *Spermophilus richardsoni* and the Black-tailed Prairie Dog *Cynomys ludovicianus*. This resulted in a dramatic increase in numbers of coyotes *Canis latrans* (their predators), which became a major pest. Poisoning and trapping were undertaken on a large scale, and the Black-tailed Prairie Dog is now confined to only a few small colonies in south-western Saskatchewan.

The losses of wildlife were on such a scale that conservation efforts began at an early date. Legislation to control hunting was passed between 1905 and 1915, and the Buffalo National Park was established in 1908, a 440km^2 refuge of long-grass prairie where a herd of bison still survived. By 1922 this herd had increased to over 6,000 individuals, more than the park could support. Despite vigorous efforts to find other areas to receive surplus animals, none were found and 2,000 buffaloes were slaughtered in 1923. Buffalo National Park did not prove successful for the conservation of Pronghorn Antelope and other grassland sanctuaries were established as national parks primarily for this purpose, namely Nemiskam (21km^2), Wawaskesy (154km^2) and Menissawok (44km^2). These measures achieved their objective and allowed the Pronghorn to multiply to such an extent that it was no longer threatened in any way. Surprisingly, the three national parks were considered redundant and they were decommissioned between 1930 and 1947. Since then there have been extraordinarily protracted negotiations between the Federal and Provincial governments to establish other protected areas. A formal agreement to establish a grasslands national park was signed between Saskatchewan and the Federal government in 1981. In 1988, Saskatchewan transferred to the Federal government all the rights over a core area of 187km^2, but the balance of the proposed park, an area of 719km^2, may not be transferred until the year 2021. If all the intentions are honoured this will form a magnificent example of the prairie habitat, but prospects for survival of semi-natural grasslands outside this area are very poor.

The Black-footed Ferret - a species whose decline was not linked to habitat loss

The Black-footed Ferret *Mustela nigripes* formerly occupied a very large area within the central prairie zone of North America, from Alberta to Arizona. It was primarily associated with colonies of prairie dogs *Cynomys* spp. but was occasionally found in the burrows of ground squirrels. Ferrets were hunted by native Americans for their pelts but the main reason for the dramatic population decline in the last 100 years or so is considered to have been poisoning of prairie dog colonies (Schreiber *et al.*, 1989). This resulted in very high mortality of ferrets, presumably as a direct effect of poisoning and through loss of their main prey. It appears, however, that diseases such as canine distemper have also had a very significant impact on the reduced populations. In 1920, numbers were estimated at over 500,000 (Clark, 1987) but the catastrophic decline continued. By 1937 the species was extinct in Canada and by 1950 it was feared to be so in USA as well. This was despite the fact that there was no shortage of suitable habitat - as much as 400,000km^2, it was thought in 1970.

In 1964 a Black-footed Ferret population was discovered in prairie dog colonies in South Dakota but by 1974 this population had disappeared. In 1981 another population was found near Meeteetsee, in Wyoming, confined to a total area of about 30km^2, but scattered over 130km^2. Numbers had probably fluctuated around 100 or so individuals for about 50 years. There was a peak of 129 animals in 1984 but canine distemper reduced this to only 12 in 1985. It was considered that captive breeding provided the only hope for the species, so 24 individuals were taken into captivity between 1985 and 1987. The wild population did not survive. Initially disease caused further losses in captivity but this problem has been overcome by strict isolation and quarantine. Captive breeding has now raised numbers substantially. Plans for reintroduction to suitable areas are being drawn up. A number of large prairie sites are being managed appropriately and, outside these, farmers are being paid to protect their prairie dog colonies. The Black-footed Ferret is clearly a highly specialised grassland species unable to adapt to new conditions or to switch to other prey. It stands little chance of survival without a fully researched and properly resourced conservation strategy.

The Steppe Marmot - adapting to the changing conditions

The Steppe Marmot *Marmota bobac* was found in the short grassy steppes in eastern Europe, from Hungary to the Urals. The former very extensive range of this burrowing, hibernating herbivore was steadily reduced by cultivation and hunting until it reached a low point in the 1940s and 1950s (Bibikov, 1991). Only a few thousand survived, in a huge area between the Ukraine and the Urals. They were restricted to a handful of rather unsuitable pastures in valleys or areas where dissected relief and other factors prevented ploughing. In recent decades the Steppe Marmot has made a spectacular recovery, partly through reintroduction, protection and conservation measures but also as a result of adaptation by the species. Following rural depopulation, the marmots began to colonise abandoned farmsteads and villages and are even found in unused parts of occupied villages. There they find a variety of food plants and good conditions for burrowing. The population in USSR was thought in 1991 to be around 250,000 individuals, representing at least a ten-fold increase since the 1940s. A high proportion of these are now using farmland, including cultivated ground, and they have clearly adapted to using a wider variety of food items. Until the middle of this century the marmot would have been regarded as one of the species most typical of steppe grasslands. However, when pressure on the habitat became acute, the marmot began to demonstrate an inherent capacity for adaptation and this has resulted in a transformation of its prospects.

Native grassland and grassy woodlands in Victoria, Australia

Today there is little native grassland in the state of Victoria but before European settlement no less than 34% of the area had supported either grassland or grassy woodland. The topic has been reviewed in a convincing Conservation Strategy by Baker-Gabb and Lunt (1990).

Baker-Gabb and Lunt do not distinguish between grassland and grassy woodland, but produce abundant evidence that together they constitute the most threatened ecosystem in Victoria. For instance, at least 125 of Victoria's 866 rare and threatened plant species occur in these habitats, they include 28 composites, 14 legumes and nine orchids. No less than 31% of the endangered plant species are confined to these habitats, mostly in only one area of the state, while eight plant species that used to occur are now extinct in the state. An even greater number (26) of vertebrate species have become extinct. Indeed, of the 152 species of extinct, endangered/vulnerableand threatened species of vertebrates in Victoria, no fewer than 61 (40%) are associated with grasslands and grassy woodlands.

The rich soils of the native grasslands of Victoria were very attractive to early settlers anxious to make a living from agriculture. Ploughing, re-seeding and overstocking with sheep and cattle had disastrous effects on the native fauna and flora. The habitat loss was too complete for significant areas to be incorporated into the major national parks which have been established since the early 1970s. Today the majority of conservation areas which do contain grassland are small, isolated and surrounded by agricultural or urban land. Thus the network of national parks has done relatively little to conserve grasslands, since only 0.3% of the original area has received protection and little of the original diversity is represented.

References

Atlay, G.L., Ketner, P. and Duvigneaud, P. 1979. Terrestrial primary production. In: Bolin, B. (Ed.), *The Global Carbon Cycle*. Wiley, Chichester.

Baker-Gabb, D.J. and Lunt, I.D. 1990. *Conservation Program for Native Grasslands and Grassy Woodlands in Victoria*. Department of Conservation and Environment. Melbourne.

Bibikov, D. 1991. The steppe marmot - its past and future. *Oryx* 25:45-49.

Boekel, C. 1990. Traditional aboriginal land management practices in Australian national parks. *Parks* 1(1): 11-15.

Cabrera, A. 1970. *Flora de la Provincia de Buenos Aires: gramineas*. Instituto Nacional de Tecnología Agropecuara, Buenos Aires. 624pp.

Clark, T.W. 1987. Black-footed ferret recovery: a progress report. *Conservation Biology* 1(1):8-10.

Clements, F.E. and Shelford, V.E. 1939. *Bioecology*. Wiley, New York.

Crumpacker, D.W., Hodge, S.W., Friedly, D. and Gregg, W.P. 1989. A preliminary assessment of the status of major terrestrial and wetland ecosystems on Federal and Indian Lands in the United States. *Conservation Biology* 2(1):103-115.

Denisiuk, Z. 1990. Lowland grasslands in Poland - their natural resources, management and protection. In: *The Lowland Grasslands of Eastern Europe*. IUCN, Gland.

Dijk, G. van 1991. The status of semi-natural grasslands in Europe. In: Goriup, P.D., Batten, L.A. and Norton, J.A. (Eds), *The Conservation of Lowland Dry Grassland Birds in Europe*, Proceedings of an International Seminar held at the University of Reading 20-22 March 1991.

Duffey, E., Morris, M.G., Sheail, J. and Wells, T.C.E. 1974. *Grassland Ecology and Wildlife Management*. Chapman and Hall, London.

Folster, H. and Huber, O. 1984. *Interrelaciones Suelos-vegetación en el Area de Galipero, Territorio Federal Amazonas, Venezuela (Series Informes Tecnicos* DGSIIA/IT/144), Ministerio del Ambiente y de los Recursos Naturales Renovables.

Food and Agriculture Organization 1987. *World Crop and Livestock Statistics 1948-1995; Area, Yield and Production of Crops; Production of Livestock Products*. FAO, Rome.

Goriup, P.D. 1988. The avifauna and conservation of steppic habitats in western Europe, North Africa and the Middle East. In: Goriup, P.D. (Ed.), *Ecology and Conservation of Grassland Birds*. ICBP Technical Publication No.7. ICBP, Cambridge.

Goriup, P.D., Batten, L. and Norton, J. (Eds) 1991. *The Conservation of Lowland Dry Grassland Birds in Europe*. Proceedings of an International Seminar held at the University of Reading 20-22 March 1991.

Goriup, P.D. and Karpowicz, Z. 1985. A review of the past and recent status of the lesser florican. *Bustard Studies* 3:163-182.

Goriup, P.D. and Schulz, H. 1991. *Conservation Management of the White Stork - an international need and opportunity*. ICBP Technical Publication No.12. ICBP, Cambridge.

Grimmett, R.F.A. and Jones, T.A. 1989. *Important Bird Areas in Europe*. ICBP Technical Publication No.9. ICBP, Cambridge.

Groves, R.H. 1981. *Australian Vegetation*. Cambridge University Press, Cambridge.

Huber, O. 1987. Neotropical savannas: their flora and vegetation. *Tree* 2(3).

IUCN 1991. *The Lowland Grasslands of Central and Eastern Europe*. Cambridge.

Kollar, H.P. 1991. Status of lowland dry grasslands and great bustards in Austria. In: Goriup, P.D., Batten, L. and Norton, J. (Eds), *The Conservation of Lowland Dry Grassland Birds in Europe*. Proceedings of an International Seminar held at the University of Reading 20-22 March 1991.

Knopf, F.L. 1988. Conservation of steppe birds in North America. In: Goriup, P.D. (Ed.), *Ecology and Conservation of Grassland Birds*. ICBP Technical Publication No.7. ICBP, Cambridge.

Knystautas, A. 1987. *The Natural History of the USSR*. Century, London.

Lecomte, P. and Voisin, S. 1991. Dry grassland birds in France: status, distribution and conservation measures. In: Goriup, P.D., Batten, L. and Norton, J. (Eds), *The Conservation of Lowland Dry Grassland Birds in Europe*. Proceedings of an International Seminar held at the University of Reading 20-22 March 1991.

Lee, J. 1990. Land use trends and factors influencing change in future land use in EC-12. European Agrarian Youth Congress, Groningen.

Majumdar, N. and Brahmachari, G.K. 1988. Major grassland types and their bird communities: a conservation perspective. In: Goriup, P.D. (Ed.), *Ecology and Conservation of Grassland Birds*. ICBP Technical Publication No.7. ICBP, Cambridge.

Menaut, J-C, 1983. The vegetation of African savannas. In: Boulière, F. (Ed.), *Tropical Savannas. Ecosystems of the World*, 13. Elsevier, Amsterdam.

Moore, R.M. 1970. *Australian Grasslands*. Alexander Bros., Melbourne.

Mondor, C. and Kun, S. 1982. The long struggle to protect Canada's vanishing prairie. *Ambio* 2:286-291.

Olson, J.S., Watts, J.A. and Allison, L.J. 1983. Carbon in Live Vegetation of Major World Ecosystems. Oak Ridge National Laboratory, for US Department of Energy, Washington.

Polunin, O. and Walters, M. 1985. *A Guide to the Vegetation of Britain and Europe*. Oxford University Press, Oxford. 238pp.

Rogers, D.L. and Randolph, S.E. 1988. Tsetse flies in Africa, bane or boon? *Conservation Biology* 2(1):57-65.

Sarmiento, G. 1983. In: Boulière, F. (Ed.), *Tropical Savannas. Ecosystems of the World*, 13. Elsevier, Amsterdam. Pp.245-288.

Schreiber, A., Wirth, R., Riffel, M. and Van Rompaey, H. 1989. Weasels, Civets, Mongooses and their Relatives - An Action Plan for the Conservation of Mustelids and Viverrids. IUCN, Gland.

Stuart, S.N. and Adams, R.J. 1990. Biodiversity in Sub-Saharan Africa and its Islands. *Occasional Papers of the IUCN Species Survival Commission* No.6. Oxford.

Whittaker, R.H. and Likens, G.E. 1975. The biosphere and man. In: Leith, H. and Whittaker, R.H. (Eds), *Primary Productivity of the Biosphere*. Springer-Verlag, Berlin.

Abridged from a consultancy report written by Richard J. Hornby of the Nature Conservation Bureau (UK).

22. WETLANDS

The term 'wetlands' groups together a wide range of inland, coastal and marine habitats which share a number of common features. The Ramsar Convention defines wetlands as "*areas of marsh, fen, peatland or water whether natural or artificial, permanent or temporary, with water that is static or flowing, fresh, brackish or salt, including areas of marine water the depth of which at low tide does not exceed six metres*".

However, in spite of the apparent clarity of this definition, the classification of wetlands is fraught with problems. There are an enormous variety of wetland types, even the broadest grouping of habitat types according to their basic biological and physical characteristics gives 30 categories of natural wetlands and nine man-made ones (Dugan, 1990). In addition, wetlands are highly dynamic, changing with the seasons and over longer periods of time and it is frequently difficult to define their boundaries with precision. As a result, estimates of area of wetland vary considerably and it is not always clear what particular kinds of habitat are being discussed.

GLOBAL EXTENT AND DISTRIBUTION OF WETLANDS

Two recent papers (Aselmann and Crutzen, 1989; Matthews and Fung, 1987) give estimates of the global distribution of wetlands, but both are concerned primarily with methane production and they do not, therefore, include any salty areas as these do not emit methane to any great extent. Although very different methods were used to calculate the area of natural freshwater wetlands, the two estimates are very similar. Matthews and Fung (1987) combined three independent data sources: the first was a global vegetation database classified with the UNESCO system; the second was a global database digitised from FAO soil maps and the third was a global inundation data base compiled from Operational Navigation Charts. They concluded that 5.3 million km^2 of wetland remain. Aselmann and Crutzen (1989), using information from Gore (1983) and a variety of map sources, estimated that 5.7 million km^2 of freshwater wetlands existed. A comparison of the distribution of these wetlands according to the two reports is given in Table 22.1.

Both authors also show the distribution of the different types of wetland along 10° latitude belts, but the categories used in the two papers are very different. Fig. 22.1 shows this distribution of types according to Aselmann and Crutzen (1989) and Table 22.2 indicates their estimates of wetlands in various countries or regions. Their definitions of each vegetation type are given below.

Bogs
Peat-producing wetlands in moist climates where organic matter has accumulated over long periods. Water and nutrient input is entirely through precipitation. They are acid and nutrient deficient. Sphagnum moss typically dominates the vegetation.

Fens
Peat-producing wetlands which are influenced by soil nutrients flowing through the system. Grasses and sedges, with mosses, are the dominant vegetation. These are generally more prolific than bogs.

Swamps
Forested freshwater wetlands on waterlogged or inundated soils where little or no peat accumulation occurs.

Marshes
Herbaceous mires with vegetation commonly dominated by grasses, sedges or reeds. They may be either permanent or seasonal. Salt marshes have been excluded.

Floodplains
Periodically flooded areas along rivers or lakes. They show considerable variation in vegetation cover.

Shallow lakes
Open water bodies a few metres in depth. Regional extent and distribution of wetlands

Table 22.1 Comparison of two estimates of global wetland area* along 10° latitude belts

	LATITUDES												
	NORTH												SOUTH
SOURCE	80-70	70-60	60-50	50-40	40-30	30-20	20-10	10-0	0-10	10-20	20-30	30-40	40-50
(A)	122	1355	1235	319	128	94	276	431	484	360	333	132	3
(B)	130	1481	1445	276	156	49	85	488	1062	393	85	29	10
Difference (A) - (B)	8	126	210	-43	28	-45	-191	57	578	33	-248	-103	7

Source: (A) = Matthews, E. and Fung, I. 1987. Methane emission from natural wetlands: global distribution, area and environmental characteristics of sources. *Global Biogeochemical Cycles* 1(1):61-68.; (B) = Aselmann, I. and Crutzen, P.J. 1989. Global distribution of natural freshwater wetlands and rice paddies, their net primary productivity, seasonality and possible methane emissions. *Journal of Atmospheric Chemistry* 8:307-358.
Note: * In 1,000km^2.

Figure 22.1 Latitudinal distribution of natural wetlands

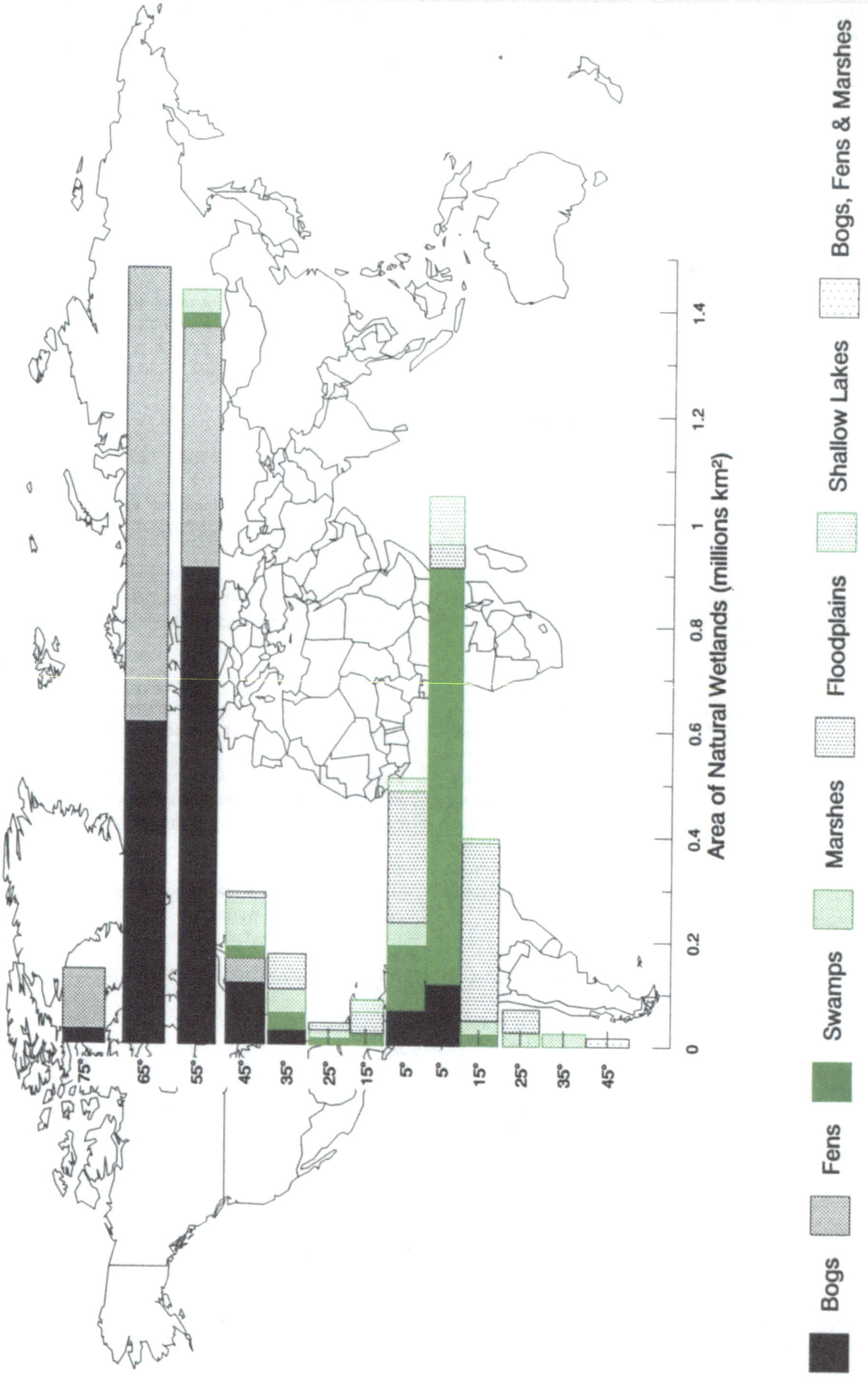

Table 22.2 Global freshwater wetland areas*

REGION	BOGS	FENS	SWAMPS	MARSHES	FLOODPLAINS	LAKES	TOTAL
'USSR'	917	531	25	39	-	-	1,512
Europe	54	93	1	4	1	1	154
Near East	-	-	-	8	-	-	8
Far East	-	-	11	-	-	-	11
China	11	-	3	18	-	-	32
Southeast Asia	197	-	44	-	-	-	241
Aust/NZ	2	3	1	-	9	-	15
Africa	-	-	85	57	174	39	355
Alaska	?	250 - 400	?	?	?	?	(325)
Canada	673	531	14	44	-	6	1,268
USA**	13	-	80	40	95	-	228
C America	-	-	15	2	1	-	18
S America	-	-	851	62	543	68	1,524
TOTAL	1,867	1,483	1,130	274	823	114	5,691

Source: Aselmann, I. and Crutzen, P.J. 1989. Global distribution of natural freshwater wetlands and rice paddies, their net primary productivity, seasonality and possible methane emissions. *Journal of Atmospheric Chemistry* 8:307-358.
Notes: * In 1,000km; ** excluding Alaska.

REGIONAL EXTENT AND DISTRIBUTION OF WETLANDS

A recent survey of the world's wetlands (Finlayson and Moser, 1991) includes maps of the distribution of selected major wetlands; some of these maps have been incorporated into Fig. 22.2. The important wetlands of Asia are shown in Fig. 22.3 (data from Scott and Poole, 1989). In many cases where regional estimates of wetlands areas have been calculated, they are very different from those given by Aselmann and Crutzen (1989). This is because of the inclusion of salty areas and, no doubt, because different definitions of wetlands and different methods of estimation were used in each set of calculations.

Europe and the Mediterranean Basin

Europe and the Mediterranean are so densely populated and have had such long histories of civilization and industrialisation that there are only a few entirely natural wetlands left in this area. Human interference has been less severe in parts of Iceland and the northern European taiga and tundra, but in most other regions the wetlands have either gone or are threatened. For instance, by the end of the 1970s, 10% of France's wetter areas and 60% of those of the UK and the Netherlands had been drained (Finlayson and Moser, 1991). There has, however, been extensive creation of artificial wetlands such as reservoirs, fishponds and gravel pits. In Tunisia, for example, 224km^2 of open water have been created while, since 1881, 190km^2 of natural wetlands have been lost (Finlayson and Moser, 1991). However, most countries have now joined the Ramsar Convention and the rate of destruction of the wetlands may at least be slowing down. Data for all the European and Mediterranean Ramsar sites reveal that only 58 of the 318 wetlands are definitely not threatened in some way.

North America

Canada is estimated to hold 24 % of all the world's wetlands, occupying over 1.27 million km^2 (Finlayson and Moser, 1991). The original wetland area of the conterminous USA (excluding Alaska and Hawaii) may have been around 890,000km^2, of which only 47 % or thereabouts remain (Dahl, 1990). There were a further 690,000km^2 in Alaska and Hawaii, with only a very small percentage in the latter state. Overall, it is estimated that around 1.11 million km^2 of wetlands remain in the whole of the USA (Dahl, 1990). Fig. 22.5 shows the distribution of wetland in the USA about 200 years ago and in the 1980s. The percentage loss of wetland in each state is also shown in Fig. 22.5.

Latin America and the Caribbean

Many of the wetlands in South America are in an almost pristine state. In contrast, most of the habitats, including the wetlands, in the Caribbean have been intensively exploited. The state of the natural habitat in Central America and Mexico is intermediate, with a fairly large area of wetlands remaining.

The wetlands of South America can be subdivided into three major systems: those of the Pacific lowlands, those of the Andean chain and those of the Atlantic-Caribbean lowlands to the north and east of the Andes. The Chilean Fjordland in the Pacific lowlands includes around 55,000km^2 of wetlands (Finlayson and Moser, 1991). The largest wetland in the Andes region is the freshwater lake, Lake Tota, in Colombia. Also in this area is the Chilean Lake District covering some 3,000km^2 of wetland. In the lowlands of the Atlantic-Caribbean region, the delta of the River Orinoco covers an area of around 30,000km^2, while that of the Amazon River covers about 35,000km^2. Also in this area is the Pantanal, covering some 200,000km^2, which is one of the largest floodplains in the world.

Many of the Caribbean islands have important wetlands, mostly coastal lagoons, mangrove swamps and inter-tidal mudflats, but there are also some freshwater lakes in old volcanic craters. Many of the flora and fauna found on the islands are endemic.

The Usumacinta Delta is the most extensive wetland on the Gulf coast of Mexico, covering around 10,000km^2.

Figure 22.2 General distribution of world wetland areas

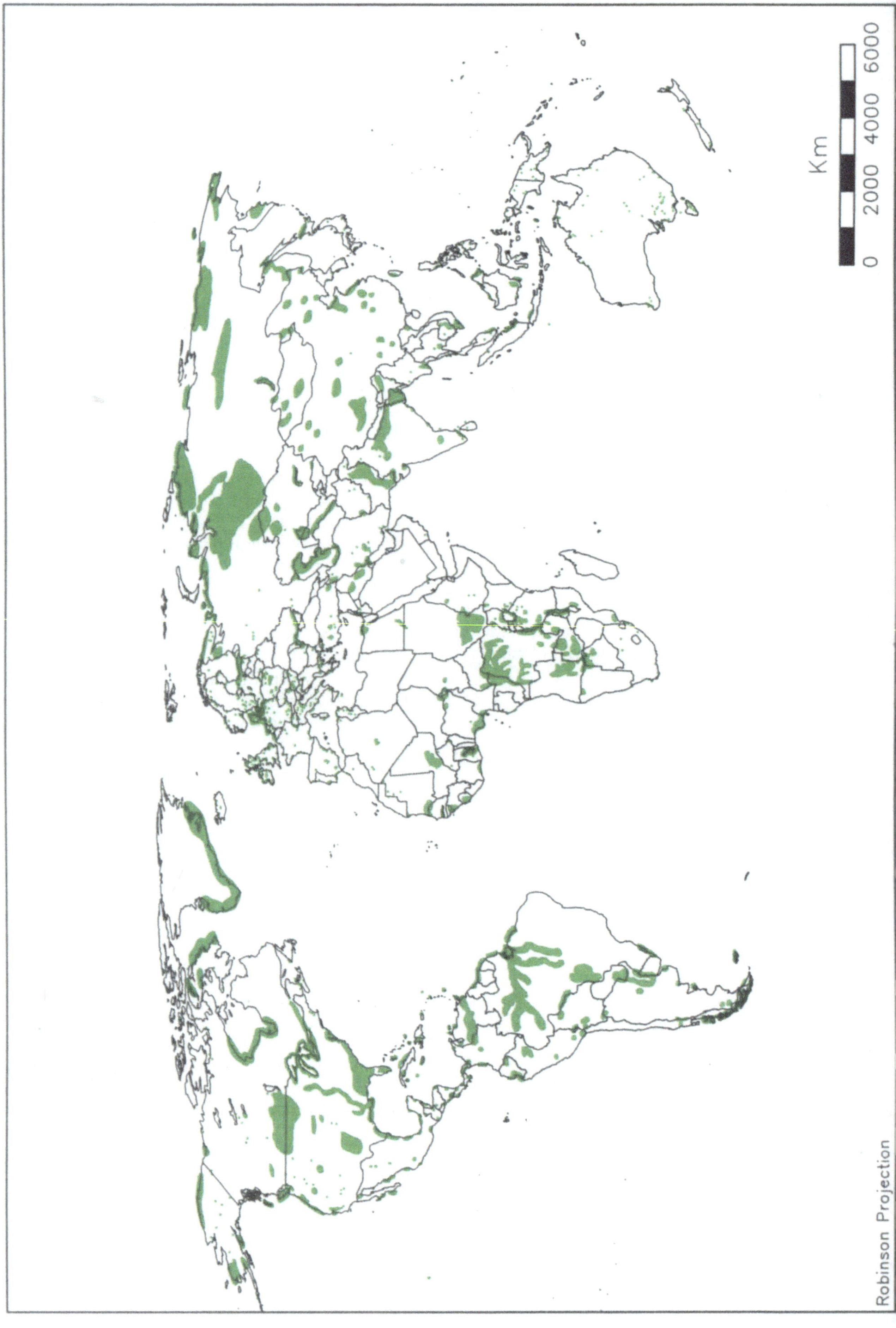

Table 22.3 Wetlands described in *A Directory of Asian Wetlands*

COUNTRY	NUMBER OF SITES	AREA OF SITES (km^2)
Bangladesh	12	67,700
Bhutan	5	85
Brunei	3	1,380
China	192	163,000
Hong Kong	3	119
India	93	54,700
Indonesia	137	87,800
Japan	85	4,750
Cambodia	4	36,500
Korea, Dem People's Rep	15	3,220
Korea, Rep	21	1,070
Laos	4	2,220
Malaysia	37	31,200
Mongolia	30	15,500
Myanmar	18	54,900
Nepal	17	356
Pakistan	48	8,580
Papua New Guinea	33	101,000
Philippines	63	14,100
Singapore	7	2
Sri Lanka	41	2,740
Taiwan	12	84
Thailand	42	25,100
Viet Nam	25	58,100
TOTAL	947	734,200

Source: Scott, D.A. and Poole, C.M. 1989. *A Status Overview of Asian Wetlands*. No.53. AWB, Kuala Lumpur, Malaysia.

Africa

Wetlands cover one per cent of Africa's total surface area (at least 345,000km^2). In Equatorial Africa, the three largest wetland systems are: the Zaire swamps (covering 80,000km^2), the Sudd in the Upper Nile (over 50,000km^2) and the wetlands of the Lake Victoria Basin (about 50,000km^2). The floodplains of the Niger and Zambezi Rivers, the Chad Basin (around 20,000km^2) and the Okavango Delta (16,000km^2) are also major wetland areas. There are also a further 12,000km^2 of wetland in southern Africa.

Asia and the Middle East

It has been estimated that there are some 830,000km^2 of peat bogs and swamps in the USSR and about 900,000km^2 of marshy ground subject to seasonal flooding (Finlayson and Moser, 1991). In the Middle East, the most extensive wetlands occur in Iraq, where the Tigris and Euphrates Rivers create a vast complex of shallow lakes and marshes covering about 15,000km^2. It is estimated that there are around 1.2 million km^2 of wetlands, excluding permanent rice paddies, in the region covered by the Directory of Asian Wetlands (Scott and Poole, 1989). This Directory gives information on 947 of the most important wetlands, covering 734,000km^2, and their distribution by country is shown in Table 22.3.

Australasia and Oceania

The major wetland types in this region are seagrass meadows, mangrove swamps, coastal salt marshes and flats, monsoonal freshwater floodplains, southern and inland swamps, lakes, river and creek channels and bogs (Finlayson and Moser, 1991).

The seagrass meadows off the coast of Australia are some of the largest in the world. Mangroves are another important wetland habitat in the region. They cover 12,000km^2 in Australia, 9,250km^2 in Papua New Guinea and about 28,500km^2 in Irian Jaya, but only 640km^2 in the Oceanic islands and a small area in New Zealand (Finlayson and Moser, 1991). Salt marshes occupy about 9,200km^2 in Australia and are also found in New Zealand. Extensive herb, woodland and forested freshwater floodplains occur in Australia and there are also numerous swamps along the rivers, both here and in New Zealand. For instance, at the confluence of the Lachlan and Murrumbidgee Rivers there are nearly 1,500km^2 of wetlands and another 400km^2 of swamp along the Macquarie River in New South Wales (Finlayson and Moser, 1991). Lakes, both saline and freshwater, permanent and temporary, are found throughout the region.

VALUES AND THREATS

Wetlands serve a wide variety of functions, including flood control, water purification, shoreline stabilisation and the control of erosion. They also support vast numbers of fish and other wildlife and numerous people depend on them for their livelihood. Table 22.4 lists these values and indicates which types of wetland provide them. This topic is developed further in Part 2.

The services provided by wetlands have tended to be taken for granted and, as a result, maintenance of natural wetlands has received low priority. Indeed, in many cases the drainage of wetlands has been seen as an advantage, with the benefits far outweighing the costs, whereas, in fact, the opposite often turns out to be nearer the truth. Table 22.5 lists the general causes of wetland loss, in particular habitat types. Major threats to wetlands in the Neotropics are summarised in Table 22.6.

A listing of general threats to wetlands in Asia is given in Table 22.7, with a more detailed country-specific analysis in Table 22.8.

In Asia, there are regional differences in the frequency of occurrence of particular types of threat. In East Asia, human settlement and encroachment occur in 20% of the 212 sites on which data have been gathered, while reclamation for urban and industrial development and pollution are both reported in 18% of the sites. In Southeast Asia, the most common threats are hunting and the disturbance associated with it (occurs at 42% of the 331 sites for which information is known), disturbance from human settlement and encroachment (34%), commercial logging and forestry (30%), wood-cutting for domestic use and drainage for agriculture (both 27%). In South Asia, hunting and its associated disturbance is, again, the most common threat, occurring at 39% of the 191 sites; fishing is also a threat in many areas (32%). The other most common threats in this region are overgrazing by domestic livestock (27%), pollution (26%) and degradation of watersheds with soil erosion and increased siltation (25%).

The most seriously threatened wetlands in Asia have been shown in Fig. 22.4; those marked in black are considered to be already too degraded to merit any special conservation effort (Scott and Poole, 1989). They are also listed in Table 22.9, where the degraded sites are marked with an asterisk.

LOSS OF WETLANDS

The rate of wetland loss cannot be quantified in most countries, but is relatively well-documented in the USA. Fig. 22.5 illustrates the difference between the distribution of wetlands some 200 years ago and at the present day, and the lower map in Fig. 22.5 also shows percentage loss between the 1780s and the 1980s in each of the states.

Table 22.10 gives figures for the loss of wetlands in some of the states.

Table 22.4 Wetland values

	Estuaries (without mangroves)	Mangroves	Open coasts	Floodplains	Freshwater marshes	Lakes	Peatlands	Swamp forest
Functions								
Groundwater recharge				●	●	●	o	o
Groundwater discharge	o	o	o	o	●	o	o	●
Flood control	o	●		●	●	●	o	●
Shoreline stabilisation/Erosion control	o	●	o	o	●			
Sediment/toxicant retention	o	●	o	●	●	●	●	●
Nutrient retention	o	●	o	●	●		●	●
Biomass export	o	●	o	●	o	o		o
Storm protection/windbreak	o	●	o	o				
Micro-climate stabilisation		o		o	o	o		o
Water transport	o	o		o		o		
Recreation/Tourism	o	o	●	o	o	o	o	o
Products								
Forest resources		●		o				●
Wildlife resources	●	o	o	●	●	o	o	o
Fisheries	●	●	o	●	●	●		o
Forage resources	o	o		●	●			
Agricultural resources				●	o	o	o	
Water supply				o	o	●	o	o
Attributes								
Biological diversity	●	o	o	●	o	●	o	o
Uniqueness to culture/heritage	o	o	o	o	o	o	o	o

Notes: ● Common and important value of that wetland type; o less common/important.
Source: Dugan, P.J. (Ed.) 1990. *Wetland Conservation: a review of current issues and required action*. IUCN, Gland, Switzerland.

Table 22.5 The causes of wetland loss

	Estuaries	Open coasts	Floodplains	Freshwater marshes	Lakes	Peatlands	Swamp forest
Human actions							
Drainage for agriculture, forestry and mosquito control	●	●	●	●	○	●	●
Dredging and stream channelisation for navigation and flood protection	●			○			
Filling for solid waste disposal, roads, and commercial, residential and	●	●	●	●	○		
Conversion for aquaculture/mariculture	●	○	○	○	○		
Construction of dykes, dams, levees, and seawalls for flood control, water	●	●	●	●	○		
Discharges of pesticides, herbicides, nutrients from domestic sewage and	●	●	●	●	●		
Mining of wetlands for peat, coal, gravel, phosphate and other materials	○	○	○		●	●	●
Groundwater abstraction			○	●			
Sediment diversion by dams, deep channels and other structures	●	●	●	●			
Hydrological alterations by canals, roads and other structures	●	●	●	●	●		
Subsidence due to extraction of groundwater, oil, gas and other minerals	●	○	●	●			
Human actions							
Subsidence	○	○			○	○	○
Sea-level rise	●	●					●
Drought	●	●	●	●	○	○	○
Hurricane and other storms	●	●				○	○
Erosion	●	●	○			○	
Biotic effects			●	●	●		

Notes: ● common and important cause of wetland degradation and loss; ○ present, but not a major cause of loss.
Source: Dugan, P.J. (Ed.) 1990. *Wetland Conservation: a review of current issues and required action*. IUCN, Gland, Switzerland.

Table 22.6 Major threats to wetlands in Latin America and the Caribbean

THREAT	INCIDENCE (% OF SITES)
Pollution	31.0
Hunting and associated disturbance	30.5
Drainage for agriculture and ranching	19.0
Disturbance from recreation	11.5
Reclamation for urban and industrial development	10.5
Forestry activities	10.0
Fishing and associated disturbance	10.0

Source: Finlayson, M. and Moser, M. (Eds) 1991. *Wetlands*. Facts on File Limited, Oxford. Based on information from 620 wetland sites described in Scott, D.A. and Carbonell, M. (Eds) 1991. *A Directory of Neotropical Wetlands*. IUCN, Cambridge and IWRB, Slimbridge.

Table 22.7 Major threats to wetlands in Asia

THREAT	INCIDENCE (% OF SITES)
Hunting and associated disturbance	32
Human settlement/encroachment	27
Drainage for agriculture	23
Pollution	20
Fishing and associated disturbance	19
Commercial logging/forestry	17
Wood cutting for domestic use	16
Degradation of watershed/soil erosion/siltation	15
Conversion to aquaculture ponds or salt pans	11
Diversion of water supply	9
Overgrazing by domestic stock	9

Source: Scott, D.A. and Poole, C.M. 1989. *A Status Overview of Asian Wetlands*. No.53. AWB, Kuala Lumpur.
Notes: These are for 734 sites for which information on threats is known. Only 107 of these are not threatened in one way or another (Table 22.8).

Figure 22.3 Asian wetlands: distribution of sites

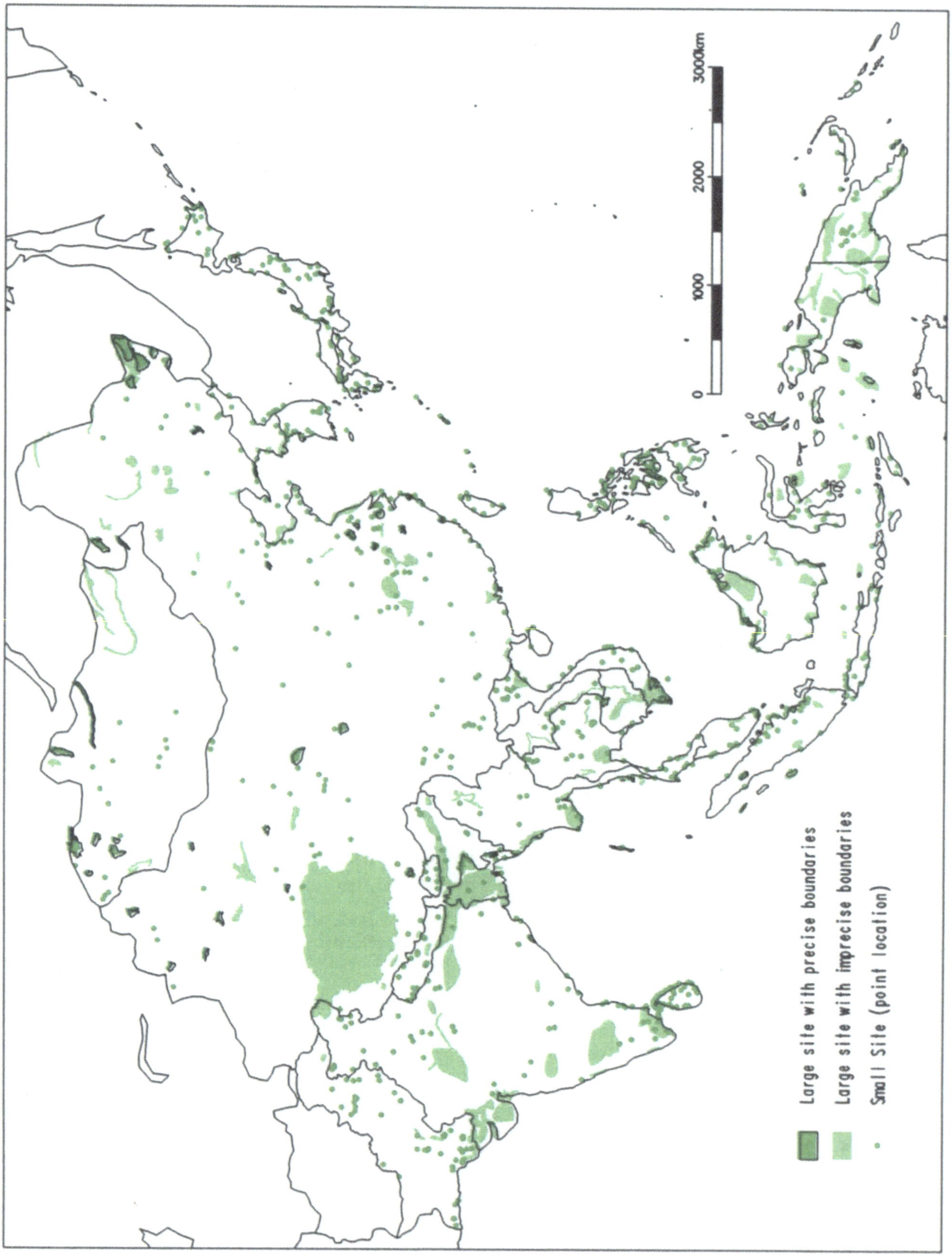

Figure 22.4 Asian wetlands: threatened sites

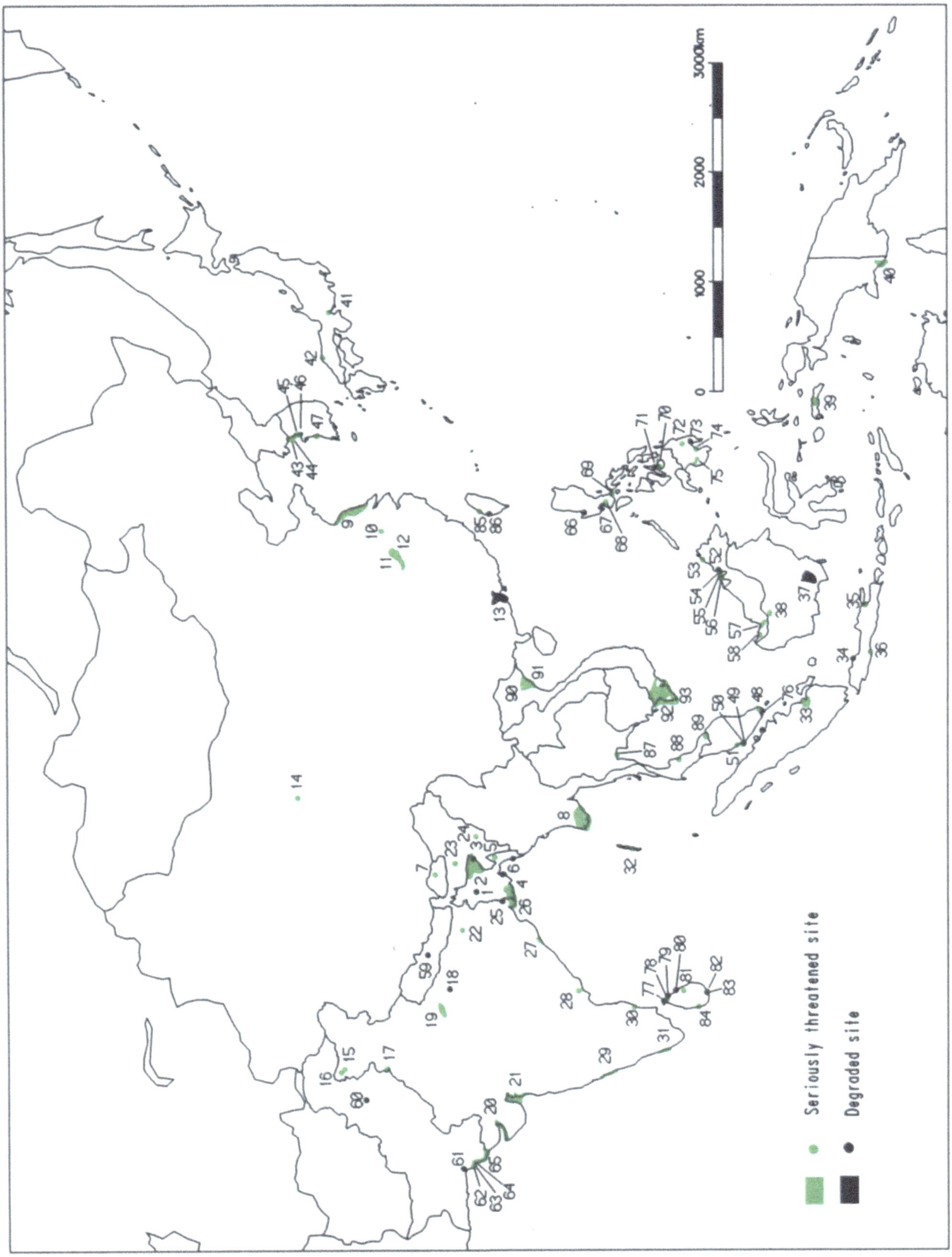

Table 22.8 Severity of threats to wetlands of international importance in Asia

	NUMBER OF SITES KNOWN	DEGREE OF THREAT				% SITES WITH MODERATE TO HIGH THREAT
		NONE	LOW	MOD	HIGH	
Bangladesh	11	1	1	5	4	82
Bhutan	5	3	-	1	1	40
Brunei	3	-	2	1	-	33
China	105	30	34	36	5	39
Hong Kong	3	1	-	2	-	67
India	88	4	44	22	18	45
Indonesia	129	1	54	66	8	57
Japan	38	8	11	17	2	50
Cambodia	3	-	1	2	-	67
Korea, DPR	5	5	-	-	-	0
Korea, Rep.	19	5	3	6	5	58
Laos	3	-	1	2	-	67
Malaysia	37	-	5	22	10	86
Mongolia	30	23	5	2	-	7
Myanmar	16	-	7	8	1	56
Nepal	14	2	7	4	1	36
Pakistan	42	1	20	15	6	50
Papua New Guinea	27	14	8	4	-	15
Philippines	49	2	13	24	10	69
Singapore	6	-	2	3	1	67
Sri Lanka	31	2	8	13	8	
Taiwan	12	1	4	5	2	58
Thailand	36	1	18	14	3	47
Viet Nam	23	3	14	4	2	26
TOTAL	734	107	262	278	87	50

Source: Scott, D.A. and Poole, C.M. 1989. *A Status Overview of Asian Wetlands*. No.53. AWB, Kuala Lumpur, Malaysia.

In New Zealand, it is estimated that over 90% of natural wetlands have been destroyed since European settlement and drainage is still occurring (Dugan, 1990). Fig. 22.6 shows the wetland areas of Waikato Basin in North Island in 1840 and in 1976. There were 1,614km^2 of wetland in 1840 and 2,62km^2 in 1976; even less remains now (Finlayson and Moser, 1986).

Dams have been the cause of considerable reduction in the areas of floodplains. Table 22.11 gives some estimates of how floodplains in Africa will be reduced by the year 2020 as a result of construction of dams upriver.

CONSERVATION OF WETLANDS

Protected Areas

Until recently, wetland conservation was largely confined to establishing protected areas, but wetlands are influenced by activities well beyond their boundaries so this often does not have the required effect. For instance, Scott and Poole (1989) give tables in which the degree of threat to fully protected sites in Asia and to those which are unprotected are estimated (Table 22.12). They note that the creation of protected areas in Asia has been successful to a certain extent, but that an analysis of threats to the sites show that the sole or principal threat is frequently from an external source outside the control of the reserve manager. While problems such as illegal settlement, overhunting, logging and overgrazing can, in theory, be controlled within a protected area, threats such as siltation from soil erosion in the watershed, pollution from urban or industrial areas, contamination with pesticides or fertilizers and disruption of water supplies due to damming and flood control projects upstream are all usually outside the jurisdiction of the authorities concerned with conserving a protected area. There are 69 protected wetlands in Asia that are moderately or highly threatened (Table 22.12). For 20 of these, the problems are solely or principally external in origin and for a further 13 this is a significant cause of the threat (Scott and Poole, 1989).

The Ramsar Convention and conservation

The Convention on Wetlands of International Importance especially as Waterfowl Habitat (or the Ramsar Convention) provides the principal intergovernmental forum for the promotion of international cooperation for wetland conservation. There are more than 60 Contracting Parties. Some of the principal obligations of the parties are:

- to designate wetlands of international importance for inclusion on a list of 'Ramsar sites' and to advise the Bureau of any change in their ecological character
- to formulate and implement planning so as to promote conservation of listed sites
- to formulate and implement planning so as to promote the wise use of wetlands

Table 22.9 The most seriously threatened wetlands in Asia

Bangladesh
1 Chalan Beel*
2 Haor Basin of Sylhet and Eastern Mymensingh
3 Dubriar Haor*
4 The Sundarbans
5 Wetlands in Pablakhali Wildlife Sanctuary
6 Chokoria Sundarbans*

Bhutan
7 Boomthang Valley

Burma (Myanmar)
8 Irrawaddy Delta

People's Republic of China
9 Yancheng Marshes
10 Shijiu Hu
11 Shengjin Hu and the lower Yangtze Lakes
12 Shengjin Hu
13 Xi Jiang (Pearl River) Delta*
14 Tuosu Hu (Kurlyk Nor) and Kuerhleiko Hu

India
15 Dal Lake
16 Wular Lake
17 Harike Lake
18 Jheels in the vicinity of Haidergarh*
19 Dahar and Sauj (Soj) Jheels
20 Southern Gulf of Kutch
21 Gulf of Khambhat
22 Khabartal
23 Dipor (Deepar) Bheel
24 Logtak Lake
25 Salt Lakes Swamp*
26 The Sunderbans
27 Chilka Lake
28 Kolleru Lake
29 Estuaries of the Karnataka coast
30 Kaliveli Tank and Yedayanthittu Estuary
31 The Cochin Backwaters
32 Wetlands in the Andaman and Nicobar Islands

Indonesia
33 Banyuasin Musi River Delta
34 Muara Cimanuk*
35 Sukolilo
36 Cilacap and Segara Anakan
37 Danau Bankau and other swamps in the Barito Basin*
38 Banau Sentarum
39 Wetlands in Manusela Proposed National Park
40 Wasur and Rawa Biru

Japan
41 Shonai-Fujimae Tidal Flats and Inner Ise Bay
42 Lake Shinji and Lake Nakaumi

Republic of Korea
43 South Kanghwa and North Yongjong Mudflats
44 Mudflats of South Yongjong and adjacent islands
45 Namyang Bay
46 Asan Bay
47 Kum, Mankyung and Tangjin Estuaries

Malaysia
48 Sedili Kecil Swamp Forest
49 Klang Islands: Pulau Ketam*
50 Kapar Forest Reserve
51 North Selangor Swamp Forest
52 Marintaman Mengalong*
53 Tempasuk Plain
54 Lawas Mangroves
55 Trusan-Sundar Mangroves
56 Limbang Mangroves
57 Maludam Swamp Forest
58 Sarawak Mangrove Forest Reserve

Nepal
59 Begnas Tal*

Pakistan
60 Khabbaki Lake*
61 Siranda Lake*
62 Hawkes Bay/Sandspit Beaches and adjacent creeks
63 Clifton Beach
64 Korangi and Gharo Creeks
65 The Outer Indus Delta

Philippines
66 Pangasinan Wetlands*
67 Manila Bay*
68 Laguna de Bay
69 Tayabas Bay including Pagbilao Bay
70 Inabanga Coast
71 Mactan, Kalawisan and Cansaga Bays*
72 Agusan Marsh
73 Lake Leonard*
74 Davao Gulf
75 Liguasan Marsh

Singapore
76 Serangoon Estuary*

Sri Lanka
77 Thandamannar Lagoon*
78 Chundikkulam Lagoon
79 Chalai Lagoon*
80 Periyakarachchi and Sinnakarachchi Lagoons*
81 Mahaweli Ganga Floodplain System
82 Maha Lewaya and Karagan Lewaya
83 Lunama Kalapuwa and Kalametiya Kalapuwa*
84 Bellanwilla-Attidiya Marshes

Taiwan, R.O.C.
85 Tatu Estuary
86 Tungshih (Ton-Shou) Mangroves*

Thailand
87 Gulf of Thailand
88 Pak Phanang Estuary
89 Pa Phru

Socialist Republic of Viet Nam
90 Red River Delta
91 Red River Estuary
92 Mekong Delta
93 Nam Can Mangrove Forest

Source: Scott, D.A. and Poole, C.M. 1989. *A Status Overview of Asian Wetlands*. No.53. AWB, Kuala Lumpur, Malaysia.
Note: * Sites marked with an asterisk are considered to be already too degraded to merit any special conservation effort.

- to make national wetland inventories
- to make environmental impact assessments before transformations of wetlands
- to establish nature reserves on wetlands and provide adequately for their wardening
- to train personnel to manage, research and warden wetlands.

Each country's Wetlands Conservation Programme will, obviously, be determined by the characteristics of its natural resources, the problems they face and the status of the current conservation efforts, as well as by the financial and human resources available (Dugan, 1990).

The quantity and quality of information on wetlands needs to be increased and improved and it is essential that their value is understood and that the benefits of them are seen. Only when this happens will the rate of wetland loss be substantially reduced.

Figure 22.5 Loss of wetlands in selected states of the USA

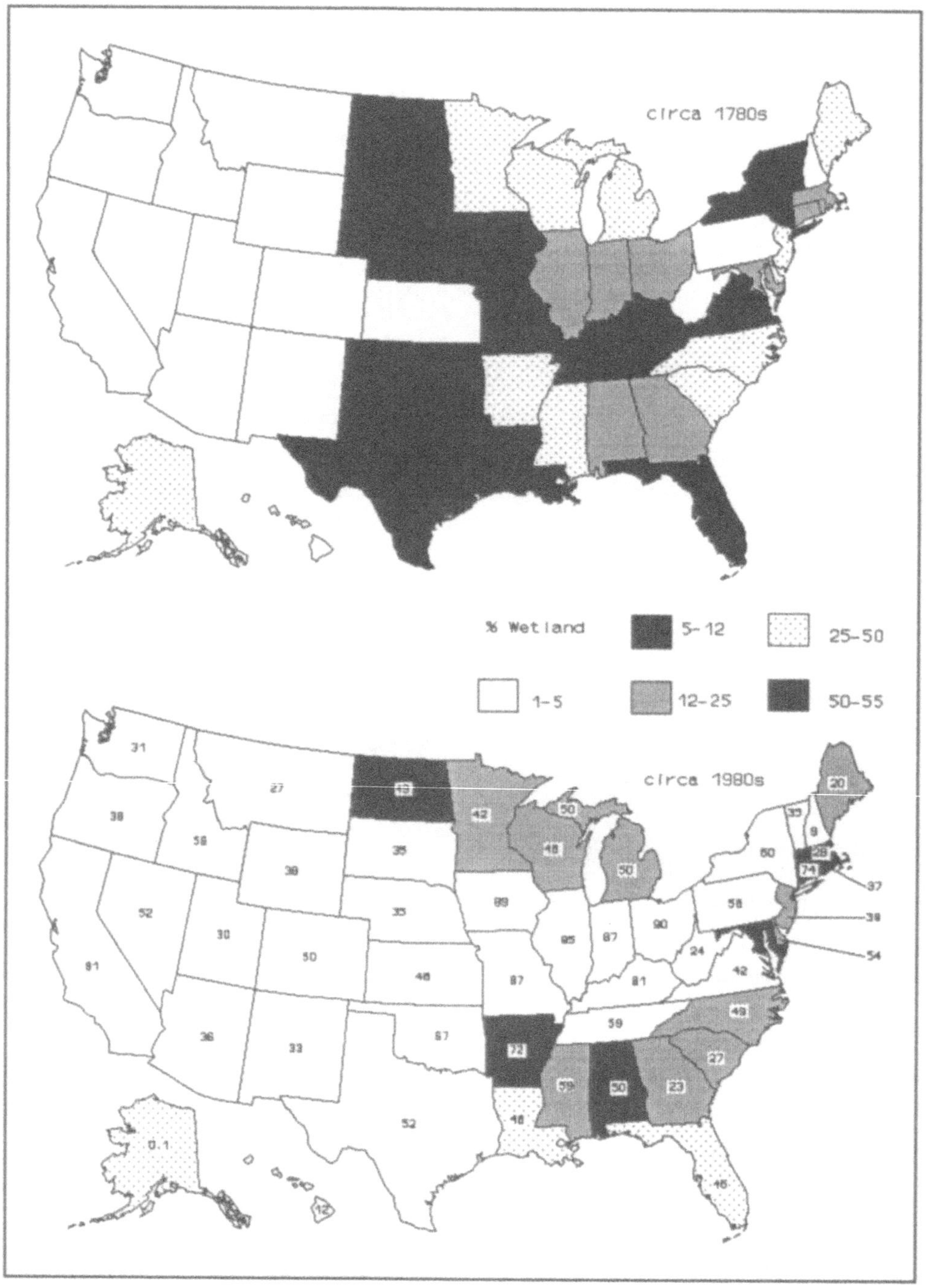

Source: Dahl, T.E. 1990. *Wetlands Losses in the United States 1780s to 1980s*. US Department of the Interior, Fish and Wildlife Service, Washington, DC.

Figure 22.6 Wetland loss in New Zealand, 1840 to 1976

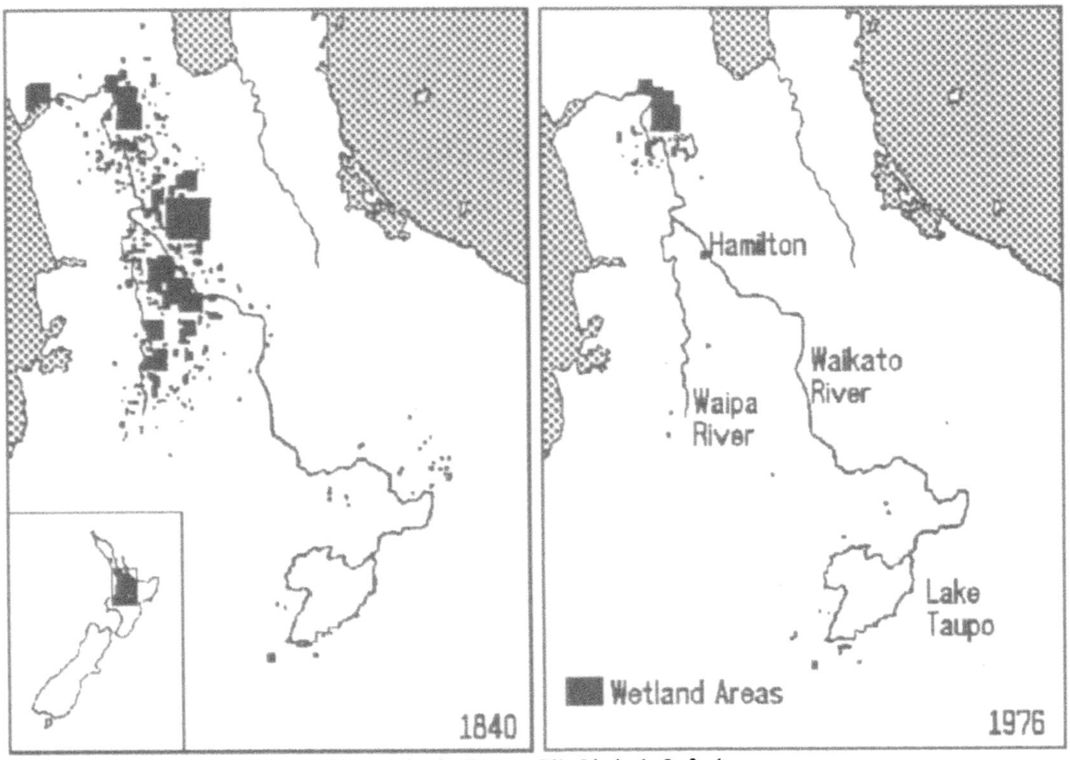

Source: Finlayson, M. and Moser, M. (Eds) 1991. *Wetlands*. Facts on File Limited, Oxford.

Table 22.10 Loss of wetlands in selected states of the USA

STATE	ESTIMATES OF WETLAND PRESENT IN 1780s (km^2)	ESTIMATES OF WETLAND PRESENT IN 1980s (km^2)	% LOST
California	20,000	1,837	91
Ohio	20,000	1,954	90
Iowa	16,200	1,707	89
Indiana	22,700	3,038	87
Illinois	33,000	5,077	85
South Carolina	25,960	18,855	27
West Virginia	540	413	24
Georgia	27,694	21,442	23
Maine	26,140	21,041	20
New Hampshire	890	809	9
TOTAL*	895,000	422,397	53
Alaska	688,790	687,980	0.1

Source: Dahl, T.E. 1990. *Wetlands Losses in the United States 1780s to 1980s*. US Department of the Interior, Fish and Wildlife Service, Washington, DC.
Note: * for all 48 conterminous states.

Table 22.11 Projected reduction in floodplain area as a result of dams

FLOODPLAIN	AREA IN 1960 (km^2)	AREA IN 2020 (km^2)
Senegal Delta	3,000	300
Senegal Valley	5,500	550
Niger Delta	30,000	27,000
Niger Valley	3,000	1,500
Sokoto and Rima	1,000	500
Hadejia Komadugu	3,800	380
Logone	11,000	6,600

Source: Dugan, 1990; modified after Drijver, C.A. and Rodenburg, W.F. 1988. *Water Management at a Cross Roads: the case of the Sahelian wetlands*. Paper presented at the International Symposium on Hydrology of Wetlands in Semi-arid and Arid regions. Seville, Spain.

Table 22.12 Degree of threat to protected and unprotected Asian wetlands

Fully protected sites

REGION	NO. OF SITES	DEGREE OF THREAT				% MOD & HIGH
		NONE	LOW	MODERATE	HIGH	
South Asia	70	6	39	20	5	36
Southeast Asia	64	2	33	27	2	45
East Asia	57	27	15	14	1	26
TOTAL	191	35	87	61	8	36

Unprotected sites

REGION	NO. OF SITES	DEGREE OF THREAT				% MOD & HIGH
		NONE	LOW	MODERATE	HIGH	
South Asia	107	4	46	33	24	53
Southeast Asia	176	18	65	77	16	53
East Asia	104	41	31	25	7	31
TOTAL	387	63	142	135	47	47

Source: Scott, D.A. and Poole, C.M. 1989. *A Status Overview of Asian Wetlands*. No.53. AWB, Kuala Lumpur, Malaysia.

References

Aselmann, I. and Crutzen, P.J. 1989. Global distribution of natural freshwater wetlands and rice paddies, their net primary productivity, seasonality and possible methane emissions. *Journal of Atmospheric Chemistry* 8:307-358.

Dahl, T.E. 1990. *Wetlands Losses in the United States 1780s to 1980s*. US Department of the Interior, Fish and Wildlife Service, Washington, DC.

Drijver, C.A. and Rodenburg, W.F. 1988. *Water Management at a Cross Roads: the case of the Sahelian wetlands*. Paper presented at the International Symposium on Hydrology of Wetlands in Semi-arid and Arid regions. Seville, Spain.

Dugan, P.J. (Ed.) 1990. *Wetland Conservation: a review of current issues and required action*. IUCN, Gland, Switzerland.

Finlayson, M. and Moser, M. (Eds) 1991. *Wetlands*. Facts on File Limited, Oxford.

Gore, A.J.P. 1983. Introduction. In: Gore, A.J.P. (Ed.), *Ecosystems of the World (4A). Mires: swamp, bog, fen and moor. Vol. 1.* Elsevier, Amsterdam. Pp.1-34.

Matthews, E. and Fung, I. 1987. Methane emission from natural wetlands: global distribution, area and environmental characteristics of sources. *Global Biogeochemical Cycles* 1(1):61-86.

Scott, D.A. and Carbonell, M. (Eds) 1986. *A Directory of Neotropical Wetlands*. IUCN, Cambridge and IWRB, Slimbridge.

Scott, D.A. and Poole, C.M. 1989. *A Status Overview of Asian Wetlands*. No.53. AWB, Kuala Lumpur, Malaysia.

Chapter contributed by Caroline Harcourt. World map of distribution of wetland areas based, by kind permission, on continent maps in Finlayson and Moser (1991).

23. CORAL REEFS

OCCURRENCE OF REEFS

Coral reefs are tropical shallow water ecosystems largely restricted to the seas between the latitudes of 30°N and 30°S. The exact extent of coral reefs in the world is unknown and is very difficult to estimate. Smith (1978) has calculated that there are 600,000km^2 of reefs to a depth of 30m. About 60% of this total occurs in the Indian Ocean region; approximately 14% in the Caribbean, 13% in the South Pacific (including eastern Australia) and 12% in the North Pacific (including the Galápagos and west coast of North America). The remaining 1% is divided between the South Atlantic and the Eastern Pacific. The map in Fig. 23.1 shows the general location of the coral reef systems of the world.

Coral reefs are one of the most productive and diverse of all natural ecosystems; they are the marine equivalents of the rain forests (Bourlière and Harmelin-Vivien, 1989). Their richness stems from the steady availability of a wide and diverse array of food resources and the extreme heterogeneity of the environment, with the corals forming a complex tri-dimensional structure providing a vast array of habitats for a great variety of organisms. Data on generic richness of corals and reef fish diversity at a number of representative sites are shown in Table 12.5, with notes on the fishes of coral reefs (Chapter 12).

The true reef-building coral polyps (stony or hermatypic corals) are ones that collectively deposit calcium carbonate to build colonies. The term 'reef' is used here for a population of stony corals which continues to build on products of its own making (Stoddart, 1969). However, not all reefs are constructed predominantly of coral. For instance, several genera of red algae grow as heavily calcified encrustations which bind the reef framework together, forming structures such as algal ridges. Alternatively, populations of ahermatypic and non-symbiotic corals exist which do not build reefs, while other populations do not build on themselves. These have been termed coral assemblages or communities.

Reefs fall into two basic categories: shelf reefs, which form on the continental shelf of large land masses; and oceanic reefs, which develop in deeper waters often in association with oceanic islands. Within these two categories are a number of different reef types: *fringing reefs* which grow close to the shore; *patch reefs* which form on irregularities on shallow parts of the sea bed; *bank reefs* which occur in deeper waters, both on the continental shelf and in oceanic waters; *barrier reefs* which develop along the edge of a continental shelf or through land subsidence in deeper waters and are separated from the mainland or island by a relatively deep, wide lagoon; and *atolls*, which are roughly circular reefs around a central lagoon and are typically found in oceanic waters, probably corresponding to the fringing reefs of long submerged islands.

VALUES AND THREATS

The World Conservation Strategy (IUCN/UNEP/WWF, 1980) identifies coral reefs as one of the "essential ecological processes and life-support systems" necessary for food production, health and other aspects of human survival and sustainable development. Reefs protect the coastline against waves, prevent erosion and contribute to the formation of sandy beaches and sheltered harbours. They also provide nutrients and breeding grounds for many commercial and subsistence fish species, as well as a habitat for numerous molluscs and crustaceans that are also caught for food. The tourist industry is another important source of income to many countries and much of this is related to the presence and condition of reefs.

Damage to coral reefs can be caused through natural events, including storms and hurricanes, climate changes, disease and predators of coral. Humans also have an impact through pollution (sewage, pesticides, fertilizers, industrial waste, etc.), sedimentation (often following land clearance and subsequent erosion inland), and over-exploitation of reef resources (intensive recreational use, coral mining).

Table 23.1 below lists in summary form: the reef resources found in tropical countries, their use, disturbances to them and what legislation or management occurs. The information has been extracted from UNEP/IUCN (1988a,b,c). Fig. 23.1 is based largely on this same source.

References

Bourlière, F. and Harmelin-Vivien, M.L. 1989. Species diversity in tropical vertebrates: an ecosystem perspective. In: Harmelin-Vivien, M.L. and Bourlière, F. (Eds), *Vertebrates in Complex Tropical Systems*. Springer-Verlag, New York.

IUCN/UNEP/WWF 1980. *World Conservation Strategy: living resource conservation for sustainable development.* IUCN/UNEP/WWF, Gland, Switzerland.

Smith, S.V. 1978. Coral reef area and contributions of reefs to processes and resources of the world's oceans. *Nature* 273:225.

Stoddart, D.R. 1969. Ecology and morphology of recent coral reefs. *Biological Review Cambridge Philosophical Society* 44:433-498.

UNEP/IUCN (1988a). *Coral Reefs of the World. Vol. 1: Atlantic and Eastern Pacific.* UNEP Regional Seas Directories and Bibliographies. IUCN, Gland, Switzerland and Cambridge, UK/UNEP, Nairobi, Kenya. 373pp., 38 maps.

UNEP/IUCN (1988b). *Coral Reefs of the World. Vol. 2: Indian Ocean, Red Sea and Gulf.* UNEP Regional Seas Directories and Bibliographies. IUCN, Gland, Switzerland and Cambridge, UK/UNEP, Nairobi, Kenya. 389pp., 36 maps.

UNEP/IUCN (1988c). *Coral Reefs of the World. Vol. 3: Central and Western Pacific.* UNEP Regional Seas Directories and Bibliographies. IUCN, Gland, Switzerland and Cambridge, UK/UNEP, Nairobi, Kenya. 329pp., 30 maps.

Figure 23.1 Coral reef systems of the world

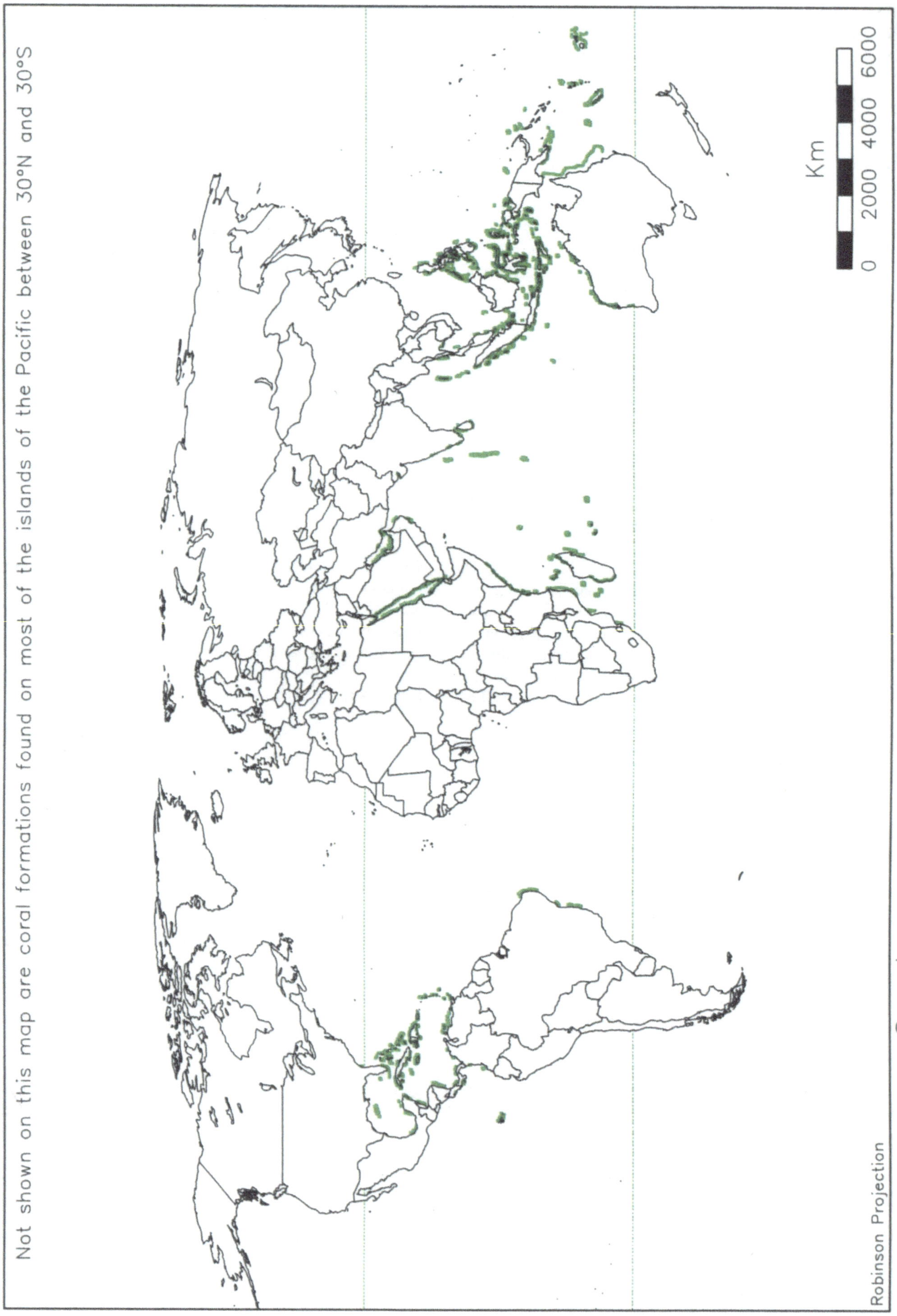

Table 23.1 Coral reefs: distribution, resources and conservation

COUNTRY	DESCRIPTION OF REEFS	REEF RESOURCES	DISTURBANCES	LEGISLATION/MANAGEMENT
ASIA				
Bahrain	The largest reefs are Fasht Adhm off the north-east coast and Fasht al Jarim in the north. Coral fringes occur around the north and east coast of Bahrain and off Al Muharraq and Sitrah Islands.	Fishing and prawn collecting have traditionally played a major role in the economy. There is an important artisanal fishery. Recreational use of the reefs is small.	There is a potential threat from oil and chemical pollution. Infilling and land reclamation have caused problems in some areas.	The southern part of the main island and most subsidiary islands are out of bounds except for fishermen and people with special permission; this affords considerable protection.
China	The mainland coast has only patchy coral growth. Coral communities are restricted to offshore islands.	Little information on the economic importance. Turtles are heavily exploited.	Cyclones may cause some damage. Siltation and over-fishing is causing damage to the Hainan reefs. Mining of corals for the construction industry, collection of live corals for aquaria and dredging also cause problems here.	A marine Environmental Protection Law exists which covers a wide range of issues.
Hong Kong	There are no true reefs but coral communities grow in the narrow coastal fringe on the eastern and south-eastern shores of the numerous islands in the region and in oceanic waters.	Fisheries are important but not directly related to the reefs. Diving is increasingly popular.	Pollution and land reclamation are major threats to the coral communities of the north-east. Over-fishing, coral collecting for lime, shell collecting and increased diving also threaten the reefs.	No coastal marine community is given specific protection. Explosives and poison are prohibited.
India	Coral reefs are present on only a few widely scattered parts of the mainland coast: the Gulf of Kutch in the north-west and off the southern coast and also around a series of small islands opposite Sri Lanka. They are mainly fringing reefs.	Corals and shells are commercial exports. There is a spiny lobster fishery along the south-east coast.	Heavy oil, sewage and industrial pollution affects the Bombay area. Coral mining has caused major damage in some regions. Siltation, as a result of deforestation, is affecting the corals round the Andaman Islands. Dredging of sand for the cement industry has had a major effect in the Gulf of Kutch.	A number of marine species are protected. A Coral Reef Committee has been set up to look into ways to preserve the reefs. There is legislation to control oil spills. Some reefs are in protected areas.
Indonesia	Almost 14,000 islands are included. The most prolific reef development is toward the eastern end of the archipelago. Fringing, patch and barrier reefs are found; there are few atolls.	Coral reefs are extremely important for both subsistence and commercial fisheries. There is a significant trade in aquarium fish. Turtles, molluscs, crustaceans and echinoderms are also collected. Coral reefs are a traditional source of limestone. Reef-related tourism is developing.	The 1983 El Niño caused coral mortality. The main causes of damage are fish blasting and coral mining. Coral is over-collected as are several other invertebrates, some fish species, turtles and dugongs. Siltation and pollution affect some areas.	Several protected areas include coral reefs. A number of traditional practices control collection of marine resources. The use of explosives and poisons is forbidden. Harvest of sponges and pearls is prohibited.
Iran	Substantial coral reefs surround some of the offshore islands along the easternmost stretch of the Gulf coast. Reefs also occur around the bays of Chah Bahar and Pozm in the Gulf of Oman.	Apparently not of major importance (large artisanal fishery based on pelagic not reef species).	Reportedly minor oil pollution, but no recent data.	There is only one protected area which includes reefs. Levels of marine discharged are regulated.

Table 23.1 Coral reefs: distribution, resources and conservation

COUNTRY	DESCRIPTION OF REEFS	REEF RESOURCES	DISTURBANCES	LEGISLATION/MANAGEMENT
ASIA (continued)				
Israel	Most of the coastline has either fringing reefs or large offshore coral knolls	Tourism is of major importance.	Rapid development for tourism at Eilat threatens the reefs. Pollution from crude oil and phosphates from industrial facilities is a problem.	The Nature Reserves Authority actively promotes reef conservation. There is a coral reserve at Eilat.
Japan	Reefs of the Okinawa Prefecture cover a total of about 80,000ha. Other coral assemblages further north cover another 6000ha or thereabouts.	The reefs are an important source of food for the Japanese. Edible seaweed, giant clams, various gastropods, squid, octopus, sea urchin and many reef fish are important. Snorkelling and diving are popular.	Coastal construction, land reclamation and intensive agriculture have caused increasing sedimentation. Over-collecting of reef resources is considered to cause significant disturbance.	There are numerous marine parks containing reefs. There are size limits, closed areas and closed seasons for some marine organisms. Net mesh size is controlled.
Jordon	The coast has a discontinuous series of fringing reefs over a length of l3km.	The coast is of recreational value, mostly for local people.	Pollution from chemical and thermal effluents, sewage and oil damage some reefs. Anchors and chains, dynamiting by fisherman and infilling during expansion of the port destroy the corals.	Fishing is controlled, coral collecting, spearfishing and dynamiting are prohibited. Commercial ships are fined for any pollution they cause. One Marine Reserve exists.
Kuwait	Isolated corals exist on rocky outcrops on the southern mainland, but reefs are largely restricted to the offshore coral cays. There are small patch and platform reefs and a remnant of fringing reef. Total reef area does not exceed $4km^2$.	Limited amount of reef fishing. The reefs are popular recreational areas.	Sewage, industrial effluents and oil pollute parts of the coast.	There are no marine protected areas. Some fishes have minimum size limits and net mesh size is controlled.
Malaysia	Typically, shallow fringing reefs, but also isolated coral patches, occur on the east coast of Peninsular Malaysia, mostly around offshore islands. West coast islands have fewer reefs. Islands off the west coast of Sabah have fringing reefs. Sarawak has some offshore coral communities.	Important for both commercial and subsistence fishing. Reef molluscs are collected for their shells and used as food. Tourism is a fast growing industry.	Coral is removed for building, broken off when fish are blasted and by trawlers and it suffers from siltation and pollution from sewage.	A Coral Reef Committee exists. Three marine parks have been established in Sabah. Provisions for their establishment and management in Peninsular Malaysia and Sarawak are incorporated into the Fisheries Act. Exports of coral from Sabah is prohibited.
Maldives	The 1,300 or so islands are all atolls and associated coralline structures.	Fishing has traditionally been a main activity of the islanders. Reef fish are exported for the aquarium trade. Shells and corals are collected for the souvenir trade, black coral is exported. Coral rock is used as building material. Tourism is important.	Greatest threats come from the developing tourist industry and the introduction of mechanised fishing. Coral mining has caused significant damage. Speed boats cause anchor damage and siltation. Pollution from sewage and industrial waste is increasing.	A minimum size for collection of turtles and spiny lobsters exists but is rarely enforced. There is a quota on the number of live fish exported for the aquarium trade. Concrete blocks, rather than coral, are now meant to be used for building. There are no marine protected areas.

Table 23.1 Coral reefs: distribution, resources and conservation

COUNTRY	DESCRIPTION OF REEFS	REEF RESOURCES	DISTURBANCES	LEGISLATION/MANAGEMENT
ASIA (continued)				
Myanmar	The main reef areas lie in the Mergui Archipelago but few data on them are available. There are no coral reefs known along the mainland coast.	The people of southern Myanmar depend partly on coastal and coral reef fisheries. There is little marine-related tourism.	The reefs are probably virtually undisturbed.	Although there are two coastal protected areas, neither specifically protects the marine environment.
Oman	Major coral growth is restricted to four areas: the Musandam Peninsula projecting into the Gulf; adjacent to the Musqat area in the Gulf of Oman; west of Jazirat Masirah; and around the islands of Zufar and Kuria Muria in the Arabian Sea.	About 10% of Oman's population fish. There is a commercial spiny lobster fishery and a local fishery based on abalone. Recreational use of reefs is increasing.	There are only a few reports of disturbance by humans. Spiny lobsters are over-exploited.	Marine pollution is monitored and an oil spill contingency plan exists. There are restrictions on potentially destructive fishing gear and on fishing in certain areas. Spearfishing is prohibited.
Philippines	Virtually the whole coastline (of over 7,000 islands) is dotted with coral reefs or coral communities. They are estimated to cover 27,000km², with the largest concentration in the south-west of the country.	The coral reefs are extremely important to the country's economy. Fish yields are high and the aquarium fish trade is substantial. There is a large trade in ornamental shells. Stony coral is collected for building. Tourism is developing rapidly.	Typhoons cause some damage. Coral and shell collection causes localised damage at least. Dynamiting for fish is a major problem. Siltation as a result of mining has had some impact.	One marine park and some reserves have been established. Destructive fishing methods are prohibited, neither turtles nor their eggs may be collected. Enforcement of legislation is poor.
Qatar	There is extensive coral growth on the northern and eastern coasts.	Little information is available. There are some local fisheries but reefs may not be important.	Some pollution from oil and sewage.	A law which covers protection and use of marine resources exists.
Saudi Arabia	There are extensive fringing reefs along the Red Sea coast. There are hundreds of patch reefs off the Gulf coast.	Important artisanal fisheries exist. Black coral is collected. Local recreational use of the reefs is increasing.	The Red Sea coastline is virtually pristine although some areas are being altered by urban and industrial development. Industrial plants, oil and petrochemical industries are a threat on the Gulf coast. Reef invertebrates are over-exploited in some areas.	There are minimum size limits for some fish but little other regulation of fishing. Long-term monitoring of the coastal and marine habitats occurs.
Singapore	Reefs occur mainly round the islands south of Singapore, these are fringing reefs. On the mainland only small coral colonies are found.	The reefs are not important for fisheries. Tourism is increasing, with reef-related recreational activities being promoted on the southern islands.	Urban and industrial development with its accompanying pollution, sedimentation and physical damage are the main threats to the reefs. Shells have been over-collected. Coral is collected for export.	There are no protected areas for reefs.
Sri Lanka	The eastern coast has the most extensive coral areas.	Collection and export of coral reef fish is a substantial business. Reef-based tourism was important in several areas but the political situation has altered this.	Pollution, quarrying for lime, use of explosives for fishing and the collection of corals, shells and fish for the tourist or aquarium trade all affect the reefs. Spiny lobster numbers have declined.	Collection of spiny lobsters is regulated. There is one marine protected area but regulations are poorly enforced.

Table 23.1 Coral reefs: distribution, resources and conservation

COUNTRY	DESCRIPTION OF REEFS	REEF RESOURCES	DISTURBANCES	LEGISLATION/MANAGEMENT
ASIA (continued)				
Taiwan	Corals are present in all the waters around Taiwan except on the west coast. The main reef area is in the south. Fringing reefs occur around offshore islands.	Reefs are important for tourism and fisheries.	Heavy tourism, siltation from unrestricted dredging and construction, explosive fishing, collecting of coral and fish for the aquarium trade and pollution threaten the reef system.	One national park includes coral reefs and there are several coastal conservation zones which include reefs.
Thailand	The best developed reefs are on the west coast in the Andaman Sea. They occur round offshore islands both here and on the east coast. There are few reefs on the mainland coast.	The reefs are exploited for the aquarium fish trade and for ornamental shells. Reef-related tourism is important. Reef-based fisheries are probably important.	Over-fishing, especially the use of dynamite and bottom trawls, threatens the reefs. Sedimentation from tin mining, mostly along the west coast, causes problems.	Several national parks include coral reefs. Coral collection and the use of dynamite or poison is prohibited. Turtles and dugongs are protected. Enforcement of legislation is limited.
United Arab Emirates	Along the Gulf coast, corals occur as patch reefs and submerged banks over broad areas.	Limited recreational diving occurs.	No information is available.	Few data exist. There are no marine protected areas.
Yemen (North)	About 25% of the shore is fringed with shallow reefs or coral communities. They are also found round the northern offshore islands.	Fish and crustaceans, including lobsters, are commercially important. Reef fisheries are important locally.	No major impacts have been noted.	A permit/licensing system exists for fisheries; explosives and harmful substances are banned.
NORTH AND CENTRAL AMERICA				
Anguilla	The 17km stretch of reef along the south-east coast is one of the most important in the Eastern Caribbean; others occur along the north.	Tourism (which is under-developed) and fishing occur.	Relatively few disturbances have been recorded. There is some hurricane damage and anchors destroy coral in places.	There is no protection of marine habitats. A Resources Development Project was set up in 1980.
Antigua and Barbuda	Estimated 25.45km² of reef coverage for both islands, most is fringing reef.	Important conch and lobster resources on both islands. Limited tourism.	Comparatively undisturbed, some over-harvesting of conch and lobster and some damage from boats occur.	Fisheries and turtle harvesting are controlled. A comprehensive Marine Parks Act has been in effect since 1972.
Bahamas	1,832km² of the Great Bahama Bank and 324km² of the Little Bahama Bank are reef covered. The reefs fringe most of the windward northern and eastern coasts and the bank edges.	Tourism is very important. There is some fishing, particularly of spiny lobster.	Reefs were deteriorating as early as the 1950s due to pollution and over-fishing. Coral collecting, dynamiting for fish and use of chlorine bleach to catch lobsters are other problems.	Spearfishing, use of dynamite and bleach, collection of lobsters, turtles and fish are controlled or prohibited. Export and collection of marine products are restricted to Bahamanians. Several marine parks exist.
Barbados	Fringing reefs, generally poorly developed, are found round the west side of the island.	A small but growing seasonal fishing industry and tourism occur. Sea urchins, conch, black coral and spiny lobster are collected.	Impact from the very dense human population and rapid coastal development. Pollution from sewage, thermal effluent and fertilizers. Increased sea level and winter storms are eroding the coastline.	The Marine Areas Act of 1976 is the legal basis for the establishment of marine reserves. A Coastal Conservation Project Unit is maintained to monitor coastal change and advise on development.

Table 23.1 Coral reefs: distribution, resources and conservation

COUNTRY	DESCRIPTION OF REEFS	REEF RESOURCES	DISTURBANCES	LEGISLATION/MANAGEMENT
NORTH AND CENTRAL AMERICA (continued)				
Belize	There is an almost continuous barrier reef stretching for 257km. It is unique in the Western Hemisphere on account of its size, its array of reef types, the luxuriance of corals and their pristine condition.	Economic potential has yet to be recognised. Spiny lobster in particular, but also conch, grouper and snapper are the main catches. There is a small tourist industry centred on diving and visits to the reefs.	Hurricanes cause heavy damage. Pollution from fertilizers and herbicides and siltation from soil erosion may increase as agriculture expands. The reefs are threatened by development of the tourist industry. Hunting of turtles and manatees are reducing their populations. Some over-fishing occurs.	The Fisheries Regulations sets size limits for lobsters, conch, bonefish and turtles and a quota and closed season for lobsters. Some of the reefs are in protected areas.
Bermuda	Total reef area is estimated to be about 190km², of which 101km² are offshore reef, 70km² are patch reef and 17km² are fringing reef.	Fish traps are used extensively; there is also active line fishing principally by charter fishing vessels. Lobsters are trapped. The tourist industry is the basis of the island's economy.	Grounding and subsequent salvaging of vessels has caused considerable damage. The very high human population is a potential threat. Blasting for geological research and for widening the eastern channel has caused some destruction.	Regulations controlling both commercial and recreational fishing exist. An oil spill contingency plan has been developed. Coral Reef Reserves protect reef resources.
Cayman Islands	Shallow water fringing reefs encircle most of the three islands of Grand Cayman, Little Cayman and Cayman Brac.	All islands are popular with divers and the tourist industry is developing. Fish and lobsters are caught.	Hurricanes frequently damage the reefs. The tourist industry is causing environmental degradation, particularly on Grand Cayman. Spiny lobsters, conch and fish are over-exploited. Tar derived from Arabian crude oil causes pollution.	Conch and lobster fishing is restricted; turtle collection is controlled. Corals and shells may be taken only with permits. A marine parks scheme has been developed.
Costa Rica	There are three main reef zones on the Atlantic coast with an estimated 10km² of living reef. Coral development is poor along the north of the Pacific coast, but richer in the south. Even here, the reef communities are generally small.	Fisheries and tourism are important.	Several reefs on the Atlantic coast are stressed by human pressure, collecting, siltation and urban pollution. Reef fishes and lobsters are declining.	There is no legislation to prevent the collecting of marine resources, only commercial fishing is restricted. One reef is protected on the Atlantic coast and four are partially protected on the Pacific coast.
Cuba	Reef length is estimated at 2150km on the north coast and 1816km on the south coast. Those on the north form an almost continuous stretch of reef.	Fishing is important, lobsters particularly so. Tourism is developing.	Human activities are having an increasing impact on the marine environment. Effluent from industrial complexes and oil pollution is affecting the reefs. Conch stocks have been over-fished in some areas.	No information is available on efforts being made to protect the reefs. Legislation was introduced in 1977 to control over-exploitation of the conch. Collecting of turtles is controlled.
Dominica	Coral reefs are limited, they occur mainly on the west coast and on the northern side of promontories elsewhere.	Fishing is important for the local people. Some coral is collected. The tourist industry is fairly limited.	The islands were devastated by a hurricane in 1979 but its impact on the reefs is unknown.	One proposed national park contains some reefs. There is legislation to protect turtles, including minimum size limits, closed seasons and total protection for eggs and nesting turtles.

Table 23.1 Coral reefs: distribution, resources and conservation

COUNTRY	DESCRIPTION OF REEFS	REEF RESOURCES	DISTURBANCES	LEGISLATION/MANAGEMENT
NORTH AND CENTRAL AMERICA (continued)				
Dominican Republic	About 166km of coast is bordered by reefs. Patch, fringing and barrier reefs are found.	The reefs are very important for local fisheries. The tourist industry is developing.	Hurricanes can cause considerable damage. Poor coastal management has led to erosion and subsequent damage to reefs through high sediment run-off. Dredging for coralline sands and for expansion of the harbour causes problems. Turtles, lobster and conch stocks have declined. Corals are collected.	Legislation exists to control coral collecting, the exploitation of turtles and spearfishing for lobsters, but it does not appear to be enforced. Two national parks contain coral reefs.
Grenada	Living reefs are found patchily distributed around all except the west coast of Grenada. Total reef cover is estimated to be 250km². Carriacou has a large bank barrier reef complex on its windward side.	Tourism is under-developed, it is concentrated on part of the southern coast of Grenada. Reef fish, conch and lobster are caught.	There is some storm damage throughout the area. Run-off of pesticides and herbicides has an impact. Pollution, removal of sand for construction and poorly planned coastal development are potential threats.	A Beach Protection Ordinance exists. There is a closed season for turtle harvesting. There are no marine protected areas.
Guadeloupe	True reefs are absent on the leeward side of Basse Terre. The best developed coral regions are on the sheltered areas to the north and east of the island.	Tourism is becoming increasingly important. Subsistence fishing, using traps, is heavy.	Hurricanes periodically damage the reefs. Urban development, industrial and agricultural pollution, fish traps and collection of corals are all threats in the area. Subsistence fishing is over-exploiting the resources. Turtles have declined because of trade in their products, mostly as tourist souvenirs.	Spiny lobsters under 14cm, lobsters with eggs, leatherback turtles, turtle eggs and corals are protected. There are closed seasons and minimum size limits for hawksbill and green turtles.
Haiti	The north coast is bordered by a barrier reef and there are seven other major areas of coral reef distributed around the country and offshore islands.	Neither tourism nor fishing is well developed. The conch is an important resource.	Little information is available. Reefs near the capital are being affected by pollution and sedimentation as a result of deforestation. The coral trade is damaging the reefs in some areas; the conch is over-harvested.	Coral reefs are not included in any protected areas. The export of corals is prohibited, but this is not enforced.
Honduras	No information is available on the reefs of mainland Honduras. The Bay Islands have well-developed reefs.	Diving tourism is of major importance to the economy of Roatan, the largest of the Bay Islands. The small patch reefs are important sanctuaries for commercial fish and are not over-exploited. In contrast, the fringing reef across the bay is heavily exploited.	Disturbances from tourists are presently minimal, but could increase. Sedimentation from run-off due to vegetation clearance is the greatest threat.	Most dive resorts prohibit or discourage souvenir collecting. The Bay Islands have been gazetted as a national park.
Jamaica	Spectacular and diverse reefs fringe most of the north coast. Reef development is more diverse but not continuous on the south coast.	The developing economy of Jamaica depends largely on tourism and fisheries. Coral reefs play a vital role in both activities.	Hurricanes inflict considerable damage. Human impact is increasing. Reefs are over-fished, corals and shells are collected. Pollution and sedimentation are a problem.	Black coral, all marine turtles, iguanas, crocodiles and Caribbean manatee are protected. Construction, drainage, etc. near the coast require a licence. Several marine parks exist.

Table 23.1 Coral reefs: distribution, resources and conservation

COUNTRY	DESCRIPTION OF REEFS	REEF RESOURCES	DISTURBANCES	LEGISLATION/MANAGEMENT
NORTH AND CENTRAL AMERICA (continued)				
Martinique	The south coast has the richest coral formations. There is an extensive bank barrier reef system along the east of the island.	Little published information is available. Both fisheries and tourism are important. Demand for reef fish is higher than yield.	Some hurricane damage occurs. Conch, lobster, turtles and possibly large fish are being over-exploited. Certain reefs are seriously threatened by siltation.	Little legislation directly related to the marine environment exists. There is a closed season and minimum size for lobster catching. Collection of turtle eggs is prohibited but not enforced.
Mexico	There are true reefs off the coast of Veracruz. The Campeche Bank is another important reef area. The reefs off the Yucatan Peninsula are predominantly barrier-cum-fringing reefs.	Tourism and fisheries are important on the Atlantic coast. Those on the Pacific coast are less important economically.	Intense fishing and oil spills threaten the corals of the Gulf of Mexico. Tourism, especially indirectly through development of the coast, is also a problem.	Legislation exists to control marine pollution and various forms of fishing, including lobster catching. A number of reefs are protected. Reef management strategies exist.
Montserrat	Small scattered patches of reef are present along all but the windward coast of the island.	Development of local fisheries is limited but they provide an important food resource. Dive and snorkelling sites are an asset to the tourist industry.	Little information exists. Sand mining for construction purposes occurs at several beaches and has altered their profile. Turtle populations are declining.	No marine protected areas exist. There are closed seasons and minimum size limits for turtle catching. Artificial reef studies to enhance lobster stocks have been carried out.
Netherlands Antilles	Curaçao is surrounded by fringing reefs 50-100m wide. Both other islands in the Leeward group have some corals around them. No major reefs are found in the Windward group of islands	Fishing by locals is important. Black coral is collected. Diving is an essential part of the increasing tourist industry.	Reefs of the Leeward group have been affected to various extents by over-fishing, collecting and industrial pollution. Little information exists for the Windward group, but decline of the spiny lobster is reported around St Maarten.	Several marine parks exist. There are ordinances on some islands controlling spearfishing, the catching of lobsters and turtles and the collection or destruction of coelenterates or crustose algae.
Panama	The Caribbean coast has about 250km of fringing reefs, which are often exposed at low tide. Fewer reefs are present on the Pacific coast, they are mostly subtidal.	Little information is available. Reefs are important for local fishermen, shrimp fishing occurs on the Pacific side. Tourism is not yet developed.	The abnormal El Niño in 1983 caused extensive damage. Industrial development threatens specific sites but generally few data exist on effects of humans.	Regulations exist to control turtle collection. The Smithsonian Tropical Research Institute maintains Galeta Reef as a private biological reserve. There are two state-owned marine parks.
Puerto Rico	Corals grow round much of the island but there is only localised reef formation. Small reefs are abundant on the wide insular shelf on the south coast; greatest development and diversity is in the south-west. There are few reefs on the north coast.	Most fishing is artisanal but it is a major activity. Most important areas for tourism are the north-east and south-west. Increasing numbers of condominiums and marinas have been built in formerly rural coastal areas.	Hurricanes cause extensive damage. Areas being developed for tourism have suffered considerably. Most damaging is siltation, mainly caused by clearing of upland vegetation. Pollution from urban and industrial development and oil spills is a problem.	Regulations for control of soil erosion have been drafted. Management plans have been prepared for reef fish and spiny lobster. An oil spill contingency plan exists. There is a coastal management programme.
St Kitts-Nevis	Bank barrier reefs with associated fringe or bench reefs occur along much of the coast of both islands.	Conch and lobster fishing are important. The tourist industry is expanding rapidly.	Inshore reefs are being over-fished, especially for conch and lobster. Tourism will probably lead to deterioration of the environment. There is a decline in turtle numbers.	Lobster numbers are monitored. There are closed seasons and minimum size limits for turtle catching. No marine protected areas exist.

Table 23.1 Coral reefs: distribution, resources and conservation

COUNTRY	DESCRIPTION OF REEFS	REEF RESOURCES	DISTURBANCES	LEGISLATION/MANAGEMENT
NORTH AND CENTRAL AMERICA (continued)				
St Lucia	Reefs are found on almost all coasts, but they are generally small and not well developed.	Over 30% of St Lucia's revenue is derived from the marine sector. Sand mining, fishing, tourism and transport are all important. There is little commercial fishing. Lobster, conch and coral are collected.	Hurricanes damage reefs. Sedimentation associated with construction and dredging is an increasing problem. Coral collection, fishing with explosives, anchor damage and recreational divers are also problems.	Beaches are managed by a commission. Legislation protects turtles, controls conch collecting and prohibits coral collection and trade in reef fish without a permit. There is extensive work on coral reef management. A number of marine reserves exist.
St Martin and St Barthélémy	The reefs on St Martin extend more or less continuously for 25km round the island. On St Barts small reefs occur across the entrance of most bays on the north-east and south-east coasts.	There is some commercial fishing, line and pot fishing are major activities. Tourism is becoming increasingly important.	Recreational pressure, pollution, sand extraction and coral mining are all problems in the area.	There is a marine reserve off St Martin. St Barts has no protected areas.
St Vincent	The exposed north and east coast lack reefs, but on the southern, south-eastern and western coasts there are several small fringing reefs.	Tourism is an important part of the economy in the south-east. Fisheries have not been fully developed.	Coral mining occurs and the south coast is subject to pollution. Corals are sold as souvenirs. Conch has been over-exploited. Some dynamiting of fish occurs.	Little information exists. There is a closed season for turtles. There are no marine protected areas.
Trinidad and Tobago	On Trinidad, the north coast has the greatest coral development, the richest community is in the extreme north-east. Tobago has richer, more diverse, but not extensive, coral communities	Diving and reef-related activities are being promoted, particularly on Tobago, where Buccoo Reef is a popular resort.	Siltation is a potential threat. Buccoo Reef could be damaged by its intensive use as a recreational area. Corals are collected for construction work.	Buccoo Reef is a protected marine area. Reef studies have provided baseline data for management plans. Legislation covers the exploitation of fish, crustaceans and turtles and the control of oil pollution.
Turks and Caicos	The south sides of the Caicos Bank are fringed with patchy boulder coral heads. Barrier and fringing reefs are found along the northern shores of the Caicos Islands. Patch and fringing reefs are found around most of the islands in the Turks group.	Pristine reefs and beaches are accessible from all islands. Tourism is developing rapidly. The fishing industry is small though potentially rich. Spiny lobster are a source of foreign revenue.	Rapid development, including tourism, threatens the reefs. Pollution is minimal as there is little industry. Poaching, of fish and lobsters, is a problem.	A licensing system limits the taking of marine products. There are closed seasons for lobsters and minimum sizes for these, hawksbill and green turtles and conch. Conch are cultured commercially. A foundation exists concerned with management, sustainable use of the reefs and education.
USA	The reefs south of Florida are the only coral assemblages of significance in the USA. There are over 6,000 patch reefs in the area.	The reefs within Monroe County, directly or indirectly, generate as much as $50 million annually from, for instance, fishing, diving, education and research.	Hurricanes damage the reefs. Dredging, causing heavy siltation, is a major problem. Heavy metals and raw sewage pollute the reefs. Some corals and spiny lobsters are over-exploited.	Fishery Management Plans for corals, reefs, spiny lobster and some fish exist. Collection of certain corals is controlled. There are several marine sanctuaries. Environmental monitoring and research is carried out.

Table 23.1 Coral reefs: distribution, resources and conservation

COUNTRY	DESCRIPTION OF REEFS	REEF RESOURCES	DISTURBANCES	LEGISLATION/MANAGEMENT
NORTH AND CENTRAL AMERICA (continued)				
Hawaii and Central Pacific U.S. Dependencies	All islands north-west of Gardner Pinnacles are atolls, coral islands or reefs and shoals of limestone construction. Subtidal reefs are found off Hawaii. Fringing reefs occur round most of the other large islands.	Subsistence and commercial fishing occurs. Many recreational activities take place in Hawaii. Algae, black coral, lobsters, crabs, octopus, etc. are also harvested commercially.	Storm damage is greater than that caused by man. Pollution from sewage, power plant effluents and dredging cause localised problems, though they can be serious. Over-fishing does occur.	There are numerous protected areas that include reefs. Dredging, filling and construction are regulated. Hawaii has a Coastal Zone Management Programme.
Virgin Islands (British)	Most of the 40 islands have reefs round them, but few detailed data are available. One of the largest islands, Anegada, has a continuous fringing reef.	Recreational boating is important. Diving is a particularly popular tourist activity. Some fishing, especially using traps, occurs.	Hurricanes cause some damage. Sand extraction has lead to erosion and disturbed a major turtle nesting site. Pollution and physical habitat degradation are causing increasing environmental degradation. Trade in black coral is unregulated and may be unsustainable. Lobster and conch stocks are depleted.	Protected areas can be formed under the Fisheries Ordinance. Fishing is licensed, lobster catching is controlled and there is a minimum mesh size for fishing nets. The impact on heavily used anchorage sites is being monitored.
Virgin Islands (US)	St Croix has the most extensive reefs with miles of bank barrier reefs extending round half the island. Fringing and patch reefs also occur. Fringing reefs on St John are poorly developed.	Traditionally, fishing is important. Spiny lobsters, black coral and queen conch are major resources. Tourism is now the leading economic activity on the islands, most visitors use the sea and reefs.	Hurricanes cause damage. Visitors cause some stress, as do boat groundings, coral collection and sewage pollution. Dredging for sand causes problems in some areas.	A coastal zone management plan exists and several organisations are involved with reef management. Size limits and quotas for lobsters exist. Harvesting of corals is prohibited, turtles are protected. Reefs occur in several marine parks.
SOUTH AMERICA				
Brazil	Some 3,000km of coast has reefs, though not all are true coral reefs. They are interesting on account of their high proportion of endemic species.	Few data are available on economic importance. Some tourism occurs and coastal fisheries are present. Manatees and turtles are heavily exploited.	Deforestation causes a high sediment load in the rivers which stresses the corals. Coral is extracted for lime, decorative purposes and construction work in some areas. Tourist and industrial development threaten the coastline.	Turtle collection is controlled but information on other legislation is unavailable. Some reefs are in protected areas.
Chile	No coral is found off mainland Chile, but Easter Island, 3568km west, has significant coral communities though no true reefs.	The reef fish fauna is heavily exploited by the islanders. The marine environment is not used to any great extent by tourists.	Little information is available. There may be local effects of coral and shell collecting.	Lobsters are protected but the legislation is not enforced. None of the marine environment is protected or managed.
Colombia	There is extensive coral growth around offshore islands. Along most of the Pacific and Caribbean coastline, conditions are suboptimal for coral growth.	Little information available, but reefs off the Caribbean coast are undoubtably important for fisheries and tourism. Data on the economic value of the Pacific reefs are unavailable.	El Niño of 1983 caused some mortality. Construction work, sewage pollution, siltation, illegal fishing, anchor damage and boat traffic cause problems.	Three national parks protect some of the most important reef areas. There is a size limit for the taking of spiny lobsters.

Table 23.1 Coral reefs: distribution, resources and conservation

COUNTRY	DESCRIPTION OF REEFS	REEF RESOURCES	DISTURBANCES	LEGISLATION/MANAGEMENT
SOUTH AMERICA (continued)				
Ecuador	Small coral formations are present along the mainland coast, but the only detailed information is for the Galápagos where most islands have some coral communities around their shores.	Many of the residents live mainly by fishing. Tourism, though not particularly reef-related, is very important. The marine environment around these islands is varied and scientifically noteworthy.	El Niño's warming of the seas caused some damage. There is increasing pressure from human activities. Black coral may be over-exploited. Most of the reefs are relatively pristine.	All the Galápagos islands are included in a national park and are well protected.
Venezuela	Comparatively few areas are optimal for reef growth, the best are around offshore islands.	The reefs are a major fisheries resource. Tourism is limited but developing.	Few data exist except for the Parque Nacional de Morrocoy where sedimentation is a particular problem. Pollution from urban and industrial waste also affects the reefs.	Turtle capture is prohibited; no information on other legislation. Reef habitats are protected in at least three national parks.
OCEANIA				
American Samoa	Many of the islands have fringing reefs; they are mostly relatively narrow.	The Samoan people have historically relied on reef organisms for a substantial part of their diet. Shoreline recreational fishing is important. In some areas, tourism is significant.	Pollution, siltation of coral, fish dynamiting and poisoning are threats. Dredging and blasting are destroying the coral. Turtle numbers have declined.	Two protected areas include reefs. The American Samoa Coastal Management Programme contains several policies governing the use of the coastal zone. Pollution control exists.
Australia, Eastern	The Great Barrier Reef includes some 2,500 major reefs and almost as many islands. Other reefs and atolls also occur.	The Great Barrier Reef has major economic importance for commercial fisheries and the tourist industry.	The reefs are relatively undisturbed although some areas may be damaged by tourists.	The Great Barrier Reef Marine Park protects most of the barrier reef. Other protected areas exist.
Australia, Western	Open ocean atolls, fringing and veneer reefs, continental shelf atolls, platform reefs and an extensive barrier/fringing reef tract all occur.	Some, but not many, of the reefs are affected by human activities. There is limited tourism.	Most are in pristine condition.	Some protected areas include reefs. A Marine Parks and Reserves Selection Working Group exists.
Cook Islands	Fringing and lagoon reefs are common. A barrier reef and atolls are also found.	Islanders are very dependent on reef resources for their protein. The tourist industry is increasing.	Hurricanes periodically cause damage. Most problems from human activities occur on Rarotonga, the most developed island. Pollution, siltation, construction work and port improvement threaten the reefs. Elsewhere, reefs are in good condition.	Traditional legislation is important; this enables access to particular areas to be restricted. National parks protecting corals and fishing reserves have been set up.
Fiji	Reefs are found associated with all the island groups. Many of the reef systems are extensive, and complex and include fringing, barrier and platform reefs. There are 844 islands and islets in the area. The Great Sea Reef here is one of the world's major barrier reefs.	Subsistence fishing occurs. *Trochus*, mother of pearl, bêche-de-mer and giant clams are commercially important. Lobsters and turtles are exploited. Tourism is a major foreign exchange earner.	Hurricanes cause periodic damage. Oil pollution, sand dredging, over-fishing, and coral collecting threaten the reefs. Sedimentation, from logging and coastal development, is most significant. Giant clams are very depleted.	The Fisheries Act prohibits the use of dynamite and poisons, provides minimum size limits for harvesting a number of species, including turtles, and protects others completely. There are no marine protected areas.

Table 23.1 Coral reefs: distribution, resources and conservation

COUNTRY	DESCRIPTION OF REEFS	REEF RESOURCES	DISTURBANCES	LEGISLATION/MANAGEMENT
OCEANIA (continued)				
French Polynesia	Most (84 of 130) islands are atolls	Fishing is an important activity, most of the catch is consumed locally. Pearls are cultured and are a significant export. Tourism is increasing.	Hurricanes cause some destruction. The reefs are being affected by dredging, coastline alteration and filling and discharge of sewage. There is considerable pollution on Tahiti. Some damage is caused by tourists.	Spearfishing is controlled. Exploitation quotas are set for pearls. Some marine reserves have been established.
Guam	The majority of the reefs are fringing, but there are two barrier reef lagoons. The reefs occur around most of the coast.	Coral, shells and algae are harvested. Several species of gastropod and bivalve are taken for food. Diving is popular on a number of the reefs.	Thermal effluent from power plants is having an adverse effect on some reefs. Siltation is a problem.	Coastal resources are subject to planning controls and legislation. Fishing regulations prohibit poisons and explosives and limit exploitation of live coral, coconut crabs and spiny lobsters. Pollution controls exist. Two protected areas include reefs.
Kiribati	All the islands in this republic, except Banaba (a raised reef) are low islands related to atolls.	Reefs are very heavily fished, although virtually all on a subsistence basis. Tourism is at a low level.	Some species, including turtles, giant clams and coconut crabs, have been over-exploited.	Fishing without a licence, or with explosives, gases and poisons is prohibited. Islands Councils restrict fishing on many of the atolls.
Marshall Islands	There are 29 coral atolls and five or so coral islands in this group.	Subsistence fishing is important. Tourism is not high.	Tropical storms affect the reefs. Many of the islands have been modified by military activity, including bombing and dredging.	There is no environmental legislation nor are there any protected areas.
Micronesia, Federated States	Most of the islands in this group are coral atolls. Low coral islets also occur.	Fisheries and tourism are both important.	Sewage pollution is a problem. Dredge and landfilling activities cause some trouble. Dynamiting for fish is a widespread problem. Giant clams have been over-exploited.	Traditional customs are important for reef management. With the termination of the Trust Territory, there are now few legislative controls in FSM.
Nauru	No true reefs exist but a wide intertidal platform surrounds the island and this has numerous emergent coral pinnacles.	Subsistence fishing occurs.	Some slight oil pollution has occurred but there is no other recorded damage.	There are no marine protected areas. A Marine Resources Act provides for the exploitation, conservation and management of fish and aquatic resources.
New Caledonia	Grande Terre has an almost continuous barrier reef around it, more than 1,600km in length. Most of the smaller islands are coral atolls or have extensive reefs around them.	Fish, crustaceans and shellfish are exploited in certain parts of Grande Terre's lagoon. Commercial fisheries occur and *Trochus* is exported in large quantities. There are small-scale coral and aquarium fish trades. Tourism is a minor activity.	Most of the reefs are in good condition. Industrial and domestic pollution is a problem only around Noumea where most of the population is concentrated.	Collection of turtles and *Trochus* is controlled. Fishing is regulated, dynamiting is prohibited. Coral collectors require a permit in some areas. There are some marine protected areas.

Table 23.1 Coral reefs: distribution, resources and conservation

COUNTRY	DESCRIPTION OF REEFS	REEF RESOURCES	DISTURBANCES	LEGISLATION/MANAGEMENT
OCEANIA (continued)				
New Zealand	The only reef-forming corals occur in the Kermadec Islands.	A few divers visit the islands. Fishing is limited but may increase as areas closer to the mainland are over-exploited.	Little information. The spotted black grouper may be over-fished.	The islands are a nature reserve.
Niue	The island is a raised atoll of coralline limestone. It has no true reef.	Fishing is one of the main activities of the 3000 or so inhabitants. Tourism is at a very low level.	Very little disturbance occurs. There is limited shell collecting.	Fishing is controlled by each village and customary restrictions ensure resources are not over-exploited. There are no marine protected areas.
North Marianas Islands	Barrier reefs and well developed fringing reefs are found on three of the islands (Rota, Tinian and Saipan). A few corals are found round some of the other islands.	Fishing and tourism are the main sources of income on Saipan where most of the population lives.	Dynamiting and poisoning of fish occurs. The rapidly growing tourist industry threatens the area. Many of the reefs around Saipan are not growing, the reason is unknown.	Coral reef areas on Saipan have been protected under the Coastal Resources Management Program but there is no enforcement. Public information and education are provided by the same programme.
Palau	A complex of fringing and patch reefs surround the numerous islands and islets. They are considered the richest reefs in the Pacific. Barrier reefs are also present.	Mollusc fisheries are important. There is some commercial fishing. A diver-orientated tourist industry is developing.	Few disturbances at present.	Under traditional custom, each village controls the adjacent reefs and fishing grounds. There are provisions for the control of fishing with explosives and poisons. Collecting of turtles and sponges is controlled.
Papua New Guinea	There are estimated to be 170,000km² of coralline shelf in depths of less than 20m and 40,000km² of reef and associated shallow water in depths of 30m or less. Milne Bay Province has the greatest concentration.	Reefs have been traditionally exploited for food, and shells. Coral reef fisheries occur. Crayfish are an important catch. Tourism is poorly developed.	The reefs are virtually pristine. Corals in Port Moresby Harbour are subject to pollution and physical degradation. Giant clams and turtles are probably over-exploited. Fishing with explosives occurs on the north coast of PNG and in the northern islands.	There are several protected areas which include coral reefs, many of them are managed. Harvesting of clams, *Trochus*, etc. is controlled. Shells can be marketed only by nationally owned companies. An oil spill contingency plan exists.
Pitcairn Islands	Oeno and Ducie are coral atolls; Henderson Island has fringing reefs, mainly around its northern end.	Fishing is relatively unimportant and is not reef-related.	There is little threat to the reefs.	There appears to be little legislation but access to Henderson Island requires a permit.
Solomon Islands	Several of the islands are atolls, others have fringing reefs, and most are poorly developed. Many reef flats are devoid of living corals.	Subsistence fisheries are important. Coral sand and rock are extracted and shells and coral are collected. There is little tourism but reef-related recreational activities are being developed.	Many coral reefs are dead but the reasons for this are unknown. There is little evidence of human-induced damage. There is some pollution from light industry. Turtles have declined due to over-exploitation.	There are size limits for the collection of *Trochus* and for the sale of turtles. A giant clam hatchery has been established. There are no marine protected areas.

Table 23.1 Coral reefs: distribution, resources and conservation

COUNTRY	DESCRIPTION OF REEFS	REEF RESOURCES	DISTURBANCES	LEGISLATION/MANAGEMENT
OCEANIA (continued)				
Tokelau	Three low reef-bound atolls make up the territory.	Subsistence fisheries occur and are an important source of food. There is no tourism.	The abnormal El Niño may have caused the extensive coral mortality found in 1983. Hurricane damage occurs. Giant clams have been depleted.	There is no protected area system. Traditional practices have conserved resources.
Tonga	Most of the 171 islands are elevated coral reefs.	Fisheries are important, but demand exceeds supply. Giant clams and black coral are harvested. Tourism is low but growing.	Damage is caused by cyclones. Storms, pollution, causeway construction and destructive fishing techniques have stressed the corals. Some reefs have been over-fished and black coral is over-collected.	There are a number of marine and coastal protected areas. Some marine taxa are protected but legislation concerning turtles is poorly enforced. Dynamiting and poisoning of fish is prohibited.
Tuvalu	There are five atolls and four reef islands in the region.	Subsistence fisheries are important. Shellfish are also collected extensively.	Some hurricane damage occurs. Giant clams and turtles are declining.	Fishing with explosives and poisons is prohibited. Sand and coral removal, pollution and waste disposal is regulated.
Vanuatu	Reefs are generally more extensive in the western than in the eastern part of the island chain. The best fringing reefs are around Anatom.	Subsistence fisheries are very important. *Trochus* is collected for food. Turtles and their eggs are exploited. Tourism is developing rapidly.	Hurricanes have damaged some reefs. Some corals have been heavily collected. Increased urban growth could lead to pollution.	There is a minimum size limit for a number of species including spiny and slipper lobsters and coconut crabs. Marine mammals are protected. Permits are needed for export of coral, aquarium fish, crustaceans, etc. Four small reserves include coral.
Wallis and Futuna	Wallis is composed of Uvea, which is surrounded by a barrier reef, and about 22 reef islets. Futuna is surrounded by reef flat. Alofi has a small patch of fringing reef.	Fish provide the major protein supply to the islanders. *Trochus* is exported.	The reefs are mostly undisturbed. Over-fishing occurs around Futuna.	There are no reserves and no conservation legislation.
Western Samoa	The main islands are Savai'i, which has widely distributed fringing reefs, and Upolu, which has extensive barrier and fringing reefs.	Fisheries provide a major source of protein. Tourism is being developed, but is limited at present.	Dynamite fishing and poisoning are problems. Some species, including giant clams, palolo worms and edible seaweed, have been over-exploited. Siltation is a problem.	Dynamiting and some poisons are prohibited. One marine reserve contains coral reefs.
AFRICA				
Comoros	Fringing reefs occur around the three northern islands. Mayotte has a substantial barrier reef.	Reef fishing occurs and is particularly important on the three western islands. There is only very limited tourism.	Silt from the heavily eroded land has had a marked impact on the reefs. Mining of reef rock and extraction of beach sand for building are problems. Lobsters are over-fished.	The quantities of fish or marine invertebrates taken are not regulated. Fishing with poison and dynamite is prohibited but not enforced.

Table 23.1 Coral reefs: distribution, resources and conservation

COUNTRY	DESCRIPTION OF REEFS	REEF RESOURCES	DISTURBANCES	LEGISLATION/MANAGEMENT
AFRICA (continued)				
Djibouti	Generally shallow reefs are distributed round the Golfe de Tadjoura and outlying islands.	Tourism is being developed.	The reefs by the port of Djibouti are becoming silted. Over-fishing is serious. Collection of corals and shells, dredging and boat traffic are problems.	Spearfishing has been prohibited. A number of marine protected areas have been set up.
Egypt	There are extensive reefs along most of Egypt's coast; the fringing reefs are less well developed in the Gulf of Suez.	The Sinai Peninsula is being developed for tourism; several diving centres are established there. Fishing is particularly productive in the Gulf of Suez.	Sea urchin numbers have increased, probably because of over-fishing of their predators, and have caused considerable damage to corals. Spearfishing and souvenir collecting are problems. Oil pollution is severe.	Efforts to manage and protect reefs have been largely confined to the Sinai Peninsula. The taking of fish and other marine creatures by nets or dynamite is forbidden in some areas around the Peninsula.
Ethiopia	Little information is available. Shallow fringing reefs probably occur along the mainland coast. Most of the inner islands of the Dahlak Archipelago have poorly developed reefs, those around the outer islands are probably better developed.	Limited commercial fishery occurs around the islands of the Dahlak Archipelago. A tourist industry is being developed on these islands.	No information exists for the mainland. Recreational use of the reefs of the Dahlak Archipelago cause some damage. Coral and shell collecting occurs and turtles are exploited.	One Proclamation has the prevention of marine pollution as its main purpose.
Kenya	Fringing and patch reefs occur along most of the coast, lying 0.5-2km offshore.	There is some reef fishing. Reef-related recreational activities are important, with tourism generating considerable income.	Sedimentation as a result of erosion inland is an increasing problem. Pollution is a potential problem, especially around Mombasa. Collection of corals and shells is a major problem.	A large proportion of the reefs are included in protected areas.
Madagascar	Main reef formations are on the north-west and south-west coast, they cover more than 1,000km. Fringing reefs occur in the north and there are also barrier reefs in the south. There are rudimentary but extensive fringing reefs on the east coast.	There is a major artisanal fishery. Ornamental shells and corals are collected for export. Several species of spiny lobster are exploited. Tourism is becoming important in Nosy Bé.	Cyclones cause occasional damage. Over-fishing is becoming a problem, especially around Toliara. Sedimentation as a result of massive deforestation may have an impact.	There is a closed season for lobsters, but this is not enforced. There is one marine national park.
Mauritius	There are 150km (300km²) of almost continuous fringing reef round the island of Mauritius. Agalega is surrounded by some 100km² of fringing reef. Rodrigues has a wide expanse of unbroken reef platform extending for 90km around it. About 190km² of reef occur round the Cargados Carajos Shoals.	There are numerous local fishermen on the main island. Reef fish are exported for aquaria. Tourism is becoming increasingly important; there are a number of diving and snorkelling centres. Fisheries are important around Cargados Carajos.	Occasional damage caused by cyclones. Coral destroying urchins have increased, possibly because of over-collection of their predators. Coral and coral sand is taken for building and for lime. Shells and corals are collected for tourists. Over-fishing is a problem. Pollution and sedimentation occurs.	The use of poisons and explosives and spearfishing is prohibited; live fish, corals and shells cannot be exported without a permit, turtles and marine mammals cannot be collected. Some nature reserves include coral reefs.

Table 23.1 Coral reefs: distribution, resources and conservation

COUNTRY	DESCRIPTION OF REEFS	REEF RESOURCES	DISTURBANCES	LEGISLATION/MANAGEMENT
AFRICA (continued)				
Mozambique, People's Rep	Abundant fringing reefs are found along the northern coastline. South of Mocambo Bay, reefs are found only on offshore islands.	Local fisheries are important.	Silt may be a problem. Pollution from fertilizers, pesticides, herbicides and industrial effluents is a problem. Spearfishing and collection of curios are threats. Turtles are over-exploited.	At least two protected areas contain reefs. Turtles are protected but this is not enforced.
Réunion	Little reef development, only 10-12km of discontinuous fringing reef on the south-west coast. Réunion's five island dependencies are coral atolls.	The fishing industry is limited. Tourism is an important element of the island's economy.	Chemical pollution, siltation and over-collecting of reef resources are problems. Spearfishing and dynamiting cause damage.	Fishing is controlled, size limits and closed seasons are imposed for some species, dynamiting and poisoning is prohibited. Turtles are protected. Much legislation is ineffective.
Seychelles	The coral reefs of the Seychelles are among the most extensive in the world and are scattered over a vast area of the Western Indian Ocean. In the granitic islands, there are scattered fringing reefs and patch reefs in many areas.	Tourism is very important. Fisheries are developing, marine resources are important to the islanders.	Landfill and dredging may be a threat. Corals and shells are collected extensively.	Fishing is regulated, use of dynamite is prohibited and spearfishing is illegal. All turtles are protected. There are at least four marine national parks which include reefs.
Somalia	There is an interrupted barrier reef along the south coast of the country from Cadale to the Kenyan border.	Artisanal fishing occurs along the entire coast. Spiny lobsters are one of the most important species caught. There is considerable potential for tourism.	Pollution is not a major problem. Sedimentation as a result of soil erosion could be a problem.	There are no marine protected areas and no legislation concerning their establishment or management.
South Africa	Coral communities, not true reefs, occur off the Maputaland coast in the north-east.	Shells are collected along the east coast. There is much recreational fishing. Shellfish provide significant quantities of protein for the Tonga people in the St Lucia region.	No problems are reported, though tourism could become a threat.	All the Maputaland corals fall within proclaimed marine reserves. Collection of live molluscs requires a permit.
Sudan	Much of the 750km coastline is bordered by fringing reefs. Barrier reefs, 1-14km wide, parallel these. These are some of the richest reefs in the Red Sea.	Coral reefs are the main tourist attraction and they are used by local people. Shrimps, sponges and lobsters are collected; fishing is not particularly important.	Threats are largely limited to the Port Sudan area where tourists are damaging the reefs. Pollution is becoming an increasing threat.	A Sudanese Marine Conservation Committee exists which is concerned with control and management of the marine environment. There are no marine protected areas.
Tanzania	About 600km of the country's shelf are covered with coral reefs. Fringing and patch reefs predominate. Many islands, including Zanzibar, are surrounded by fringing reefs.	The reefs are widely used for fishing. Several of the reefs, particularly around Dar es Salaam, are important for tourism and tourism is being developed in Zanzibar.	Silting of reefs as a result of inland deforestation is an increasing problem. Dynamiting is a serious threat. The marine curio trade has had a significant impact. Green turtles are over-exploited.	Turtles are protected but regulations are not enforced. Dynamite fishing is prohibited but only Zanzibar is effective in enforcing this.

Source: data extracted from UNEP/IUCN 1988a,b,c.

24. MANGROVES

THE MANGROVE HABITAT

Mangroves are the characteristic littoral plant formations of sheltered, low-lying tropical and subtropical coasts.

The species found in these habitats are a diverse collection of trees and shrubs that have adapted to salty, inundated environments. The mangrove species can be divided into two groups: the exclusive species found only in mangrove habitats and the non-exclusive species which may be important in the mangrove community but are not restricted to it. Saenger *et al.* (1983) list 60 species in the former group, in 22 genera, and give a list of 23 species, in 16 genera, which are some of the important non-exclusive species. The appearance of mangroves is far from uniform; they vary from closed forests 40-50m high in parts of South America to stunted shrubs less than 1m high, which can be in discrete and widely separated clumps (Finlayson and Moser, 1991).

In addition to the mangroves themselves, a wide variety of organisms are associated with the mangrove system, and the habitat is critical to many. Such organisms include a number of epiphytes, parasites and climbers among the flora, and large numbers of crustaceans, molluscs, fishes and birds among the fauna. Table 24.1 provides indicative information on species richness in these groups.

Current information on the area of mangrove habitat in each country where it occurs, together with the number of protected areas known to include mangroves, is given in Table 24.2.

Table 24.1 Species richness in groups associated with mangroves

TAXONOMIC GROUP	ASIA	CARIBBEAN W. and ATLANTIC
Bacteria	10	-
Fungi	25	-
Algae	65	105
HIGHER PLANTS		
Bryophytes/Ferns	35	2
Monocotyledons	73	20
Dicotyledons	110	28
ANIMALS		
Protozoa	18	3
Sponges/Bryozoa	5	36
Coelenterata/Ctenophora	3	42
Non-polychaete worms	13	13
Polychaetes	11	33
Crustaceans	229	87
Insects/Arachnids	500	-
Molluscs	211	124
Echinoderms	1	29
Ascidians	0	30
Fish	283	212
Reptiles	22	3
Amphibians	2	2
Birds	177	138
Mammals	36	5

Source: Abridged from Saenger, P., Hegerl, E.J. and Davie, J.D.S. (Eds) 1983. *Global Status of Mangrove Ecosystems*. Commission on Ecology Papers Number 3. IUCN, Gland, Switzerland.

Note: Estimates refer to two regions only; they are not fully comprehensive but provide general indications of relative species richness; missing data shown by -.

VALUE OF MANGROVES

The uses and values of mangroves to humans are many and varied. The wood from mangrove trees is used by local people for building materials for houses, fence-poles, material for fish traps, and so forth, and is also harvested on a large scale by international companies, particularly for pulp and particle board. The mangroves are also a significant source of fuel, both firewood and charcoal. The most important species for this purpose are those belonging to the genus *Rhizophora*, as this wood is heavy and clean burning. Another potential source of fuel in the mangrove habitat is the Nypa palm (*Nypa fruticans*), which produces a sugar that can be converted into alcohol and be used as a transport fuel (Saenger *et al.*, 1983).

Other products from the mangrove habitat include shellfish, crustaceans and fish. These are harvested both on a subsistence basis and commercially. For instance, in the mangroves of Sierpe, Costa Rica, the five million shellfish (*Anadara* sp.) harvested annually are worth US$85,000 to the local communities which collect them (Lahmann, 1989). Most of the larger commercial penaeid shrimps are mangrove-dependent, and fisheries for shrimps and prawns are major sources of export earnings in many tropical countries. In addition to providing habitat for adult fish, the mangroves are essential spawning and nursery areas for many species of marine fish. In the Gulf of Mexico, 90% of the fish harvest, worth US$700 million per year, consists of species which are dependent on mangroves and other coastal wetlands at some stage in their life cycle (Dugan, 1990). Mangrove communities are amongst the most productive ecosystems in the world.

Domestic animals feed on mangrove foliage in many countries. In Pakistan, camels are herded down to the coast to feed on the mangroves in the dry season; water buffalo graze them in parts of Asia and Australia, while in Africa goats and cattle feed on the foliage (Dugan, 1990).

Mangroves stabilise shorelines and decrease coastal erosion by reducing the energy of waves and currents and by holding the bottom sediment in place with plant roots. They also act as windbreaks and protection from coastal storms, forming a cost-free, self-repairing barrier.

Mangrove habitats can be used for tourism, education and scientific study. For instance, the Bengal Tiger *Panthera tigris* population in Sundarbans mangrove is a focus for tourism in Bangladesh, and thousands of visitors go to Trinidad's Caroni Swamp mangrove area every year to view the great numbers of Scarlet Ibis *Eudocimus ruber* and other rare or endangered birds found there.

THREATS TO MANGROVE HABITATS

Vast areas of mangroves are being destroyed either directly or as a secondary result of other activities. There are numerous reasons for the destruction and frequently short-

Table 24.2 Mangroves

COUNTRY	TOTAL AREA OF MANGROVES (in hectares)	PROTECTED AREAS WITH MANGROVES
ASIA		
Bahrain	40	1
Bangladesh	410,000	5
Brunei	7,000	3
Cambodia	10,000	0
China	67,000	24
Hong Kong	*	9
India	356,000	(34)
mainland	(306,000)	(9)
Andaman Is	(50,000)	25
Indonesia	4,251,011	152
Irian Jaya	2,934,000	19
Java and Bali	51,885	30
Kalimantan	383,450	28
Lesser Sunda Is	5,508	16
Moluccas	100,000	12
Sulawesi	99,833	18
Sumatra	657,335	29
Japan (Ryuku Is)	400	4
Iran	23,717	3
Malaysia	630,000	99
Peninsula	105,000	48
Sabah	350,000	35
Sarawak	175,000	16
Maldives	*	0
Myanmar	517,000	6
Oman	*	-
Pakistan	249,500	2
Philippines	400,000	59
Qatar	*	0
Saudi Arabia	*	-
Singapore	1,800	2
Sri Lanka	120,000	9
Taiwan	174	4
Thailand	268,693	17
Viet Nam (South)	370,000	2
UAE	*	6
Yemen	*	5
NORTH AND CENTRAL AMERICA		
Anguilla	*	0
Antigua and Barbuda	600 & 900	0
Aruba	100	2
Bahamas	233,200	10
Barbados	12	1
Belize	78,317	13
Bermuda	17	8
Cayman Islands	11,655	11
Costa Rica	35,000	11
Cuba	626,000	(23)
North coast	131,000	-
N coast Is and arch.	114,000	-
South coast	318,000	-
S coast Is	38,000	-
S Is & I de Pinos	25,000	-
Dominica	*	1
Dominican Rep.	23,500	6
El Salvador	45,000	6
Grenada	*	11
Guadeloupe	5,700	2
Guatemala	16,000	6
Haiti	18,000	0
Honduras	117,000	10
Jamaica	20,200	12
NORTH AND CENTRAL AMERICA (continued)		
Martinique	2,200	2
Mexico	1,420,200	(19)
Montserrat	7	0
Netherlands Antilles (total)	2,200	5
Bonaire	1,000	2
Curaçao	300	1
St Martin and Barthélémy	100	0
Nicaragua	60,000	9
Panama	297,532	23
Puerto Rico	6,497	13
St Lucia	179	5
St Kitts and Nevis	20	-
St Vincent	*	2
Trindad and Tobago	9,000	6
Turks and Caicos Is	*	5
USA	280,594	-
Alabama	25	-
California	150	-
Florida	274,857	(47)
Louisiana	2,956	(2
Mississippi	250	-
Texas	2506	(3)
Hawaii	*	0
Virgin Islands (British)	*	10
Virgin Islands (United States)	310	4
SOUTH AMERICA		
Brazil	250,000	(32)
Colombia	501,300	12
Ecuador	182,108	6
Galápagos Is	(200)	4
French Guiana	5,500	3
Guyana	80,000	-
Peru	6,346	3
Suriname	115,000	6
Venezuela	673,569	13
OCEANIA		
American Samoa	*	0
Australia	1,161,700	218
Capital Territory	*	2
New South Wales	10,673	33
Northern Territory	*	20
Queensland	*	113
South	20,100	21
Victoria	*	18
Western	*	11
Fiji	19,700	3
French Polynesia (Society Is)	*	0
Guam	*	1
Kiribati (Gilbert Is)	*	0
Marshall Is	*	0
Micronesia, Fed. States	*	0
Nauru	2	0
New Caledonia	20,000	2
New Zealand	19,800	(15)
Palau	4,708	1
Papua New Guinea	200,000	31
Solomon Islands	64,200	0
Tonga	1,000	2
Tuvalu	47	0
Vanuatu	*	0
Western Samoa	<1,000	2

Table 24.2 Mangroves (continued)

COUNTRY	TOTAL AREA OF MANGROVES (in hectares)	PROTECTED AREAS WITH MANGROVES	COUNTRY	TOTAL AREA OF MANGROVES (in hectares)	PROTECTED AREAS WITH MANGROVES
AFRICA			AFRICA (continued)		
Angola	110,000	2	East coast	4,815	-
Benin	3,000	1	West coast	320,745	-
Cameroon	306,000	1	Mauritania	*	3
Comoros	*	0	Mauritius	7	6
Congo	(2,000)	1	Mozambique	85,000	8
Côte d'Ivoire	(2,000)	2	Nigeria	3,238,000	1
Djibouti	*	0	Réunion	*	2
Egypt	*	2	Seychelles	*	2
Equatorial Guinea	20,000	3	Senegal	169,000	3
Ethiopia	*	1	Sierra Leone	250,000	6
Gabon	250,000	2	Somalia	10,000	6
Gambia	66,000	3	South Africa	673	11
Ghana	(2,000)	3	Sudan	*	1
Guinea	223,000	1	Tanzania	133,540	2
Guinea-Bissau	236,000	3	Mainland	115,476	-
Kenya	45,000	14	Zanzibar Is	18,064	-
Liberia	(20,000)	2	Togo	*	0
Madagascar	325,560	4	Zaire	53,000	1

Source: WCMC Protected Areas Data Unit. 1991. *Draft list of Protected Areas with mangrove habitats*. Much information in this list was provided by S.C. Snedaker.
Notes: * Country notes are provided in the *Draft list*; in many cases these notes indicate that no numerical estimate was obtainable; in others, additional island-specific estimates are given. (-) No information. ( ) Figure from list of protected areas, considered an estimate.

term exploitation for economic gain takes precedence over long-term benefits which have both economic and natural value.

In many areas, the demand for fuelwood from mangroves is well above a sustainable level and it is increasing as the human population increases. In addition, the commercial use of the wood, for pulp in particular, results in some areas being more or less clear-felled. Natural regeneration frequently does not occur and often the area is converted to other forms of land-use such as agriculture or aquaculture. In many instances where attempts have been made to convert mangroves to agricultural land, the soil becomes extremely acid because of the oxidation of the pyrite sulphur which is commonly found in large quantities in the mangrove soils. This, combined with the high concentration of soluble salts in the soil, leads either to crop loss or to a considerable reduction in productivity (Saenger *et al.*, 1983).

The conversion of mangroves areas to aquaculture gives rise to a number of problems. Much of the mangrove flora and fauna in the areas surrounding the ponds is destroyed because of major changes in drainage conditions, nutrient availability and frequency of tidal inundation, as well as being adversely affected by run-off from ponds and channels. The acid sulphate soils that have an adverse effect on crops also inhibit algae growth, which the fish feed on, and may kill the prawns or fish directly by poisoning them. Conversion to aquaculture ponds is a particular threat in the Asian region, although in the Indo-Pacific area it was estimated, in 1977, that 1.2 million ha of mangrove forest had already been converted to aquaculture ponds (Saenger *et al.*, 1983). The building of ponds for extraction of salt water can, especially in arid and semi-arid areas, cause extensive damage to mangroves. The land has to be cleared of all trees and shrubs, levelled and dyked; a canal system has to be built and the soil surface compacted, so that even if the ponds are later abandoned the chemical and physical properties of the soil have been so changed that recolonisation by mangroves is impossible.

Mangroves are also converted for urban and industrial development, commonly for housing, tourist facilities, airports and small ports. Many of the mangroves that are not directly destroyed by these developments are affected by loss of freshwater and by pollution from numerous different sources. Rubbish and solid wastes are often deliberately dumped in mangrove habitats. Mining within the mangrove system completely destroys the habitat, while mining in adjacent areas causes variable adverse effects, foremost among these being excessive silt deposition in the mangrove system causing tree loss or reduced productivity. Chemical wastes from mines are also frequently carried into coastal areas where mangroves occur, with similar effects. Drilling for oil occurs in some mangroves and both spillage and the associated pipelines and roads which alter the drainage of the area can be very destructive to the ecosystem.

Another threat to mangroves is a diversion or alteration of the freshwater flow into them. In arid, semi-arid or seasonally dry regions the mangroves are particularly dependent on periodic inputs of freshwater, but in these regions there is a high demand for freshwater and its flow into the oceans is regarded as wasteful. Consequently, rivers are often dammed or diverted so that their waters can be used on land. Changes in land-use upstream, such as the logging of a forest, can also affect the freshwater flow into the mangroves. The reduction in freshwater results in the gradual replacement of mangrove species with more salt-tolerant and possibly less useful species. Mammals within the mangrove system are affected by the lack of freshwater, while fishery resources may be depleted by the higher salinity and the reduced nutrients.

Much of the conversion of mangroves has occurred because this habitat has, traditionally, been regarded as unproductive wasteland. In many cases, government policies have contributed to the destruction of the mangroves and it is only as adverse effects of their disappearance are noted that these policies are changing.

References

Dugan, P.J. (Ed.) 1990. *Wetland Conservation: a review of current issues and required action*. IUCN, Gland, Switzerland.

Finlayson, M. and Moser, M. (Eds) 1991. *Wetlands*. Facts on File Ltd, Oxford.

Lahmann, E. 1989. Formulación de un proyecto de conservación de los recursos naturales para la zona de manglares de estero real, Nicaragua. Mimeographed report, IUCN, San José, Costa Rica.

Saenger, P., Hegerl, E.J. and Davie, J.D.S. (Eds) 1983. *Global Status of Mangrove Ecosystems*. Commission on Ecology Papers Number 3. IUCN, Gland, Switzerland.

PART 2

USES AND VALUES OF BIODIVERSITY

Part 1 of this book outlined the nature of biological diversity, the elements of which it is comprised, and some ways in which diversity is measured. These themes were illustrated by discussion and data on selected groups of organisms and habitats.

The intention in Part 2 is to introduce some of the ways in which humans use and benefit from components of biodiversity, and to discuss aspects of the problems involved in attempting to assign appropriate economic values to goods and services provided by them.

Part 2 includes three chapters. Chapters 25 and 26 provide an introduction to human uses of plants and animals respectively. The aim is to outline some of the principal uses and selectively to present further detail, where possible by means of data tables, and usually where the subject is of particular interest or is unfamiliar to many. No attempt has been made to document comprehensively the entire range of uses to which natural resources are put, nor to catalogue all the species involved.

It is an ecological imperative that humans depend on plants and, to a lesser extent, on other animals for the basic requirements of existence, so we have not stressed this point. Nor have we detailed the ecological functions at the habitat and landscape level that collectively provide benefits in the form of services, largely because the role of *diversity* in these functions is poorly-understood and difficult to quantify, although an area of active international research. Some service functions, such as carbon-fixing and watershed protection, could probably be performed as well by plantation forest monocultures as by native multi-species forests.

Chapter 27 draws upon a growing literature on the application of economic theory to biodiversity, in particular on the ways in which values can be attached to natural resources or habitats. Whilst the discipline of the economist provides an interesting perspective on biodiversity conservation, giving particular insight into the general forces which drive habitat conversions, the analytic methods used can, again, be difficult to apply to the concept of biological *diversity*.

The values that can be assigned to species or habitats may be sufficiently high to suggest that cost/benefit analyses must often in the past have greatly undervalued their worth in comparison to developments that impact upon them. On the other hand, whilst the obvious value of keeping open future options provides a very powerful general argument for conserving biological diversity, the economic values to be derived are extremely difficult to quantify and can even be negative; many would regard the economist's viewpoint as here subsidiary to aesthetic and moral arguments.

25. PLANT USE

Plant species provide an extremely wide range of useful products relied on by people in all countries of the world. A mixture of direct harvesting from the wild and cultivation ranging from basic subsistence farming to sophisticated agricultural systems supplies food, medicines and a wealth of raw materials. Plant biodiversity as a global resource remains poorly understood, inadequately documented and often wasted, but still retains immense potential for further development of natural products.

Indigenous people in developing countries retain a basic reliance on wild and traditionally cultivated plant species that directly supply a wide range of their needs and often display a remarkable knowledge of these local, often undocumented, plant resources.

As well as the more obvious plant products such as food, medicines, ornamental plants and timber, plants provide a wide variety of resources used in industry and commerce. To mention but a few, plant extracts are used in the manufacture of glue, soaps, cosmetics, dyes, plastics, lubricants and polishes. Plants provide an important source of renewable energy, with Brazil, for example, obtaining 28% of its energy needs from sugarcane biomass resources. This represents a major saving on petrol imports amounting to US$8.9 billion from 1976 to 1985. An economic analysis of various aspects of the value of plant diversity is given in Chapters 27. This chapter provides a more general outline of the importance of plants in five major categories: food plants; timber; rattans; medicinal plants; ornamentals.

FOOD PLANTS

One of the most fundamental values of plant biodiversity is in supplying the world's food. Originally plants were consumed directly from the wild and gathering of wild produce continues throughout the world today. Through the processes of domestication wild plants became reservoirs of new crop species and they are now an invaluable source of genes needed to improve the world's crops.

Of the estimated 250,000 species of flowering plants, only about 3,000 have been regarded as a food source, although most have probably been sampled at one time or another. Others will have provided forage and browse for animals in turn hunted or farmed by people. Around 200 plant species have been domesticated for food, and of these about 15-20 are crops of major economic importance.

Relatively few botanical families account for the world's main domesticated plants. Gramineae and Leguminosae are the most important, followed by the Cruciferae, Rosaceae, Umbelliferae, Solanaceae and Labiatae. Other significant families are the Chenopodiaceae, Araceae, Cucurbitaceae and Compositae. Table 25.1 lists the nutritionally important plants of the world and reflects the predominance of these families. The species included are not all crops of major economic importance but are the plants that account for the bulk of food production.

Although relatively few plants contribute to food production globally, at a local level plant resources provide a varied source of nutritional needs. In one region of Peru, fruits of 193 species are regularly consumed; of these, 120 species are exclusively wild-collected and a further 19 originate from both wild and cultivated sources. Locally consumed species such as these hold considerable potential as food plants for wider use.

The history of food crops

The evolution of crop plants began between 5,000 and 10,000 years ago. It is now generally thought that agriculture originated more or less simultaneously in various parts of the world. The Fertile Crescent of the Near East, centred on the area which is now Iraq, is well known as the source of domestication of wheat and barley together with certain pulses such as the lentil. Early agricultural development based on the domestication of millets, also took place in the loess regions north of the Huang He (Yellow River) in China and in southern Mexico where squashes, beans, peppers and maize were domesticated. Agriculture is also thought to have developed independently in the South American Andes.

From these early centres of agriculture the spread of domesticated plants took place and, following conscious and unconscious selection pressures, individual crops became increasingly diverse.

In traditional agro-ecosystems newly domesticated plant types and primitive cultivars diverged from their wild ancestors. Nevertheless, occasional crosses continued to occur between the early crops and their wild relatives and allowed, for example, the incorporation of disease and pest resistance genes harboured by wild parent plants. The introduction of genes from wild and weedy relatives increased the availability of crop genetic diversity for further selection and improvement by farmers, and increased the potential of crops to respond to changing environmental conditions. Many cultivated species may not have survived in domestication without the interchange of genes between wild and crop populations (Oldfield, 1984).

Human migration and trade also contributed significantly to the evolution of crop plants. When a species is introduced to a new environment it often changes relatively quickly, adapting to new ecological conditions. Furthermore, when crops are taken from their source areas they may encounter different wild relatives and cross with them. This has occurred across continents, for example, with different species of rice. In West Africa, an indigenous cultivated rice *Oryza glaberrima* has hybridised with the introduced common Asian rice *O. sativa*, enriching the rice gene pool in the region. In this way, crops may develop secondary centres of variation with greater genetic diversity than their original centres of origin.

Processes of natural selection in response to new ecological conditions have increased crop diversity and this in turn has been enhanced by farmers selecting for particular characteristics over thousands of years of cultivation. The range of crops cultivated also increased through time as additional species were brought into cultivation, and in some cases weeds of the primary crops became important food plants in their own right.

Table 25.1 Food crops of the world

FAMILY	SPECIES	FOOD	ORIGIN	CONSERVATION
Anacardiaceae	*Mangifera indica*	Mango	NE India, the majority of fruit-bearing trees are more or less wild.	Wild species of mango are threatened in Southeast Asia as a result of deforestation and replacement by commercial species. WWF is funding conservation of wild fruit trees in Peninsular Malaysia.
	Pistacia vera	Pistachio	Native to the Near East and western Asia, cultivated in the Mediterranean and western Asia for 3000-4000 years.	Many wild populations have been destroyed by forest clearance, over-cutting for charcoal and grazing.
Araceae	*Colocasia esculenta*	Taro	India	Collection, preservation and research are needed for aroid cultivars. More than 1,000 cultivars of *Colocasia* exist as a result of efforts by subsistence farmers.
	Xanthostoma sagittifolium	Yautia	A tropical American plant developed by Amerindian people.	
Aquifoliaceae	*Ilex paraguariensis*	Mate	Native to S. Brazil, Paraguay and N. Argentina, cultivated throughout its natural range. Leaves are also still collected from wild plants.	
Betulaceae	*Corylus avellana*	Hazel	Europe and SW Asia. Domesticated in the 17th century.	
	Corylus maxima	Filbert	SE Europe and western Asia.	
Bromeliaceae	*Ananas comosus*	Pineapple	Thought to be a lowland South American domesticate.	Species of wild pineapple are native to botanically under-explored parts of lowland South America. They are now being used in breeding programmes. Collection and conservation of clones from the upper Amazon and Upper Orinoco is considered desirable.
Camelliaceae	*Camellia sinensis*	Tea	Probably the lower Tibetan mountains or Central Asia	Truly wild teas probably no longer exist. In cultivation a substantial loss of genetic variability has been anticipated which needs to be countered by deliberate conservation measures.
Caricaceae	*Carica papaya*	Papaya	Lowlands of eastern Central America	
Chenopodiaceae	*Beta vulgaris*	Sugar Beet	Europe, developed as a crop for sugar in the 18th century.	Wild populations of the related *Beta maritima* are threatened in parts of the Mediterranean.
	Chenopodium quinoa	Quinoa	A native American crop of the high central Andes developed by Indian agriculturists in pre-Colombian times.	
	Spinacia oleracea	Spinach	Native to SW Asia	
Compositae	*Carthamus tinctorius*	Safflowerseed	The cultivated species had its origins in the Near East	One of the wild relatives of safflower, the Moroccan endemic *Carthamus rhiphaeus* is considered Rare by IUCN.
	Cynara scolymus	Artichoke	Native to the Mediterranean area and Canary Islands, domesticated several thousand years ago.	

Table 25.1 Food crops of the world (continued)

FAMILY	SPECIES	FOOD	ORIGIN	CONSERVATION
Compositae (continued)	*Helianthus annuus*	Sunflowerseed	Domesticated in central USA probably before the arrival of maize, beans and squash	Some of the American varieties have been preserved. A large genetic reservoir exists among the weed and wild sunflowers. Wild gene pools are disappearing owing to habitat loss.
	Lactuca sativa	Lettuce	Mediterranean	
Convolvulaceae	*Ipomoea batatas*	Sweet Potato	Central and South America	The conservation of variability is a major concern in breeding for subsistence agriculture.
Cruciferae	*Brassica oleracea/B. rapa*	Cabbage	The wild cabbage is native to Europe; development of cultivars took place in the Mediterranean region.	IBPGR has designated the collection of wild forms of *B.oleracea* as a conservation priority. Several related Mediterranean taxa are threatened in the wild.
	Brassica juncea	Mustardseed	The primary centre of origin is believed to be Central Asia - Himalayas.	Large collections serve as substantial gene pools and wild material is widely distributed.
	Brassica napus, B. rapa	Rapeseed	*B. napus* probably does not exist in the wild.	
Cucurbitaceae	*Citrullus lanatus*	Melonseed	Native to S Africa, chiefly in the Kalahari Desert.	
	Cucumis melo	Melon/Water-melon	Africa, wild forms found in eastern tropical Africa.	
	Cucumis sativus	Cucumber	Native to India, probably cultivated for over 3,000 years.	
	Cucurbita maxima, C. moschata, C. pepo	Pumpkin, Squash, Gourd	Domesticated in the Americas at least 10,000 years ago.	Many of the wild *Cucurbita* species have restricted ranges.
Dioscoreaceae	*Dioscorea* spp.	Yam	Domestication of yams in Asia, Africa and tropical America took place separately with different species involved.	Serious genetic erosion has occurred among cultivated yams and there is an urgent need to collect and conserve genetic diversity. There is little information on the status of wild relatives of yams.
Euphorbiaceae	*Manihot esculenta*	Cassava	A cultigen, unknown in the wild state.	The virtually unexplored wild relatives are an important genetic resource for crop improvement. Centre of diversity of wild relatives are in east-central Brazil, NE Brazil and SW Mexico.
Gramineae	*Avena sativa*	Oats	Generally regarded as a secondary crop, evolved in W and N Europe from weed oat components of wheat and barley crops.	The potential of wild populations in breeding programmes remains to be determined.
	Echinochloa frumentacea	Japanese Barnyard Millet	Different strains are thought to have at least partially different origins.	
	Eleusine coracana	Finger Millet	Central Africa. Taken to India probably over 3,000 years ago where a second centre of diversity became established.	This species is still capable of genetic exchange with related wild forms living in the same area.
	Digitaria exilis	Fonio	West Africa, thought to be a cultigen.	
	Hordeum vulgare	Barley	One of the first crops domesticated in the Near East.	Concern about genetic erosion e.g. in Ethiopia, where cultivars are valuable for genetic resistance to disease and improved nutritional quality.

Table 25.1 Food crops of the world (continued)

FAMILY	SPECIES	FOOD	ORIGIN	CONSERVATION
Gramineae (continued)	*Oryza glaberrima, O. sativa*	Rice	The origin of Asian rice *O. sativa* is uncertain. The African *O. glaberrima* probably originated 3,500 years ago. Its primary centre of diversity is the swampy area of the Upper Niger.	As rice cultivation has become more intensive, many wild populations have disappeared. The International Rice Research Centre in the Philippines coordinates the collection of indigenous varieties. Little effort has been made to conserve *O. glaberrima* and its wild relatives, however.
	Panicum miliaceum	Common Millet	A millet of ancient cultivation which is not known in its wild state.	
	Pennisetum americanum	Bulrush Millet	Probably in western tropical Africa where the greatest number of cultivated and related wild forms occur. A second centre of diversity became established in India.	This species is still capable of genetic exchange with related wild forms living in the same area.
	Saccarhum officinarum	Sugarcane	New Guinea	Valuable germplasm of wild sugarcane and related species has been lost as a result of habitat destruction in Malaysia, Indonesia and Papua New Guinea.
	Secale cereale	Rye	SW Asia, arising as a weed of wheat and barley	
	Setaria italica	Foxtail Millet	Origin unknown in the wild state, the crop is thought to have arisen from the common Old World weed *S. viridis*.	
	Sorghum bicolor	Sorghum	Developed primarily from the wild *S. arundinaceum* in Africa.	
	Triticum aestivum, T. turgidum	Wheat	Mediterranean and Near East	A number of wild relatives are restricted to small areas. There is a need for further *ex situ* conservation.
	Zea mays	Maize	Maize was domesticated in prehistoric times in Mexico and Central America.	A wild species *Z. perennis* was presumed extinct in the wild until its rediscovery in 1977. A new species was also discovered, *Z. diploperennis*, and is now protected in the Sierra de Manantlan Biosphere Reserve, Mexico.
Grossulariaceae	*Ribes nigrum, R. rubrum*	Currants	Black and red currants are native to northern Europe and northern Asia, with the black currant extending to the Himalayas. Domesticated in northern Europe within the past 500 years.	
Illiciaceae	*Illicium verum*	Star Anise	China, Viet Nam	
Juglandaceae	*Juglans regia*	Walnut	Native from SE China to Europe	
Lauraceae	*Persea americana*	Avocado	The crop originated in Central America and has been cultivated for several thousand years.	Primitive wild relatives are restricted to small areas in Central America. The endangered caoba tree from Ecuador *Caryodaphnopsis (Persea) theobromifolia* is a wild relative resistant to blight.

Table 25.1 Food crops of the world (continued)

FAMILY	SPECIES	FOOD	ORIGIN	CONSERVATION
Leguminosae	*Arachis hypogaea*	Groundnut	A cultigen domesticated thousands of years ago in South America.	Much unexplored genetic variability in wild relatives of potential importance in breeding programmes. The protection of perennial *Arachis* species in Latin America is considered a conservation priority.
	Cajanus cajan	Pigeonpea	The centre of origin is assumed to be India	
	Cicer arietinum	Chickpea	Western Asia	Many of the wild relatives of chickpea are threatened or rare.
	Glycine max	Soybean	A cultigen not known in the wild, soybean is thought to have arisen as a domesticate in the eastern half of northern China.	Soybean cultivars grown in the USA show a high degree of genetic uniformity. The germplasm base in Asian countries is being destroyed partly through the introduction of modern cultivars. Conservation of traditional land races is urgently needed.
	Lablab purpureus	Lablab bean	Thought to be of Asian origin, now widespread in the tropics	
	Lens culinaris	Lentil	The wild progenitor of the cultivated lentil is *Lens orientalis*, a Near Eastern species.	
	Lupinus mutabilis	Lupin	A very variable cultigen of the high Andes.	
	Phaseolus lunatus	Lima bean	It is thought that separate domestications occurred in Central and South America from conspecific geographic races.	Most wild relatives are widespread but populations of several taxa are being lost to overgrazing in south-west USA and northern Mexico
	Phaseolus vulgaris	Haricot bean	It is thought that separate domestications occurred in Central and South America from conspecific geographic races.	Most wild relatives are widespread but some forms in Mexico are worthy of conservation attention.
	Pisum sativum	Pea	The wild progenitor is unknown and the early history of the pea crop is unclear. Probable centres of origin are Ethiopia, the Mediterranean and Central Asia.	Breeding relies on a fairly narrow genetic resource base and efforts to conserve genetic variability of the cultivated crop have been fairly limited.
	Vicia faba	Broad bean	Near East	
	Vigna unguiculata	Cowpea	The common cultivated subspecies is thought to be derived from wild plants in Ethiopia several thousand years ago.	
Lecythidaceae	*Bertholletia excelsa*	Brazil nut	Tropical South America. Nuts are still collected from wild trees as experimental plantations have mainly failed.	The species is threatened in the wild because of logging for its valuable timber. Commercial collection of wild nuts is a sustainable form of forest exploitation and is being promoted in extractive reserves.
Liliaceae	*Allium cepa: Allium fistulosum*	Onion	Central Asia	
	Allium sativum	Garlic	Known only in cultivation. *A. longicuspis*, a species endemic to central Asia, may be its wild ancestor.	
Malvaceae	*Gossypium barbadense, G. hirsutum*	Cottonseeds	South America	
Moraceae	*Ficus carica*	Fig	Southern Arabia	

Table 25.1 Food crops of the world (continued)

FAMILY	SPECIES	FOOD	ORIGIN	CONSERVATION
Musaceae	*Musa acuminata; M. x paradisiaca*	Banana and Plantain	Wild bananas occur in SE Asia and the Pacific. The primary centre for *M. acuminata* was the Malay peninsula.	The genetic base of banana breeding is narrow. Forest clearance is threatening the variability of wild bananas *M. acuminata* and other *Musa* spp. Protection of wild species in Asia is an IBPGR conservation priority.
Myrtaceae	*Pimenta dioica*	Pimento	West Indies and Central America	
Oleaceae	*Olea europaea*	Olive	Originated as a hybrid in the eastern Mediterranean	Olive production is in decline and the loss of traditionally managed olive groves has serious consequences for wildlife in the Mediterranean region. In Algeria and Niger the wild olive relative *Olea laperrinei* is threatened partly by over-cutting for cattle fodder.
Palmae	*Cocos nucifera*	Coconut	The origin of the coconut is obscure. Wild types predominate on the African and Indian coasts of the Indian Ocean, and scattered in Southeast Asia and the Pacific	The tendency to plant uniform, improved hybrids is reducing genetic variation particularly in domesticated types.
	Phoenix dactylifera	Date	A food plant of ancient cultivation in North Africa and the Middle East.	One wild relative is restricted to Crete where it is Vulnerable.
	Elaeis guineensis	Oil Palm	West Africa, originally a species of the transition zone between savanna and rain forest.	In West Africa oil palm groves are being thinned to make way for other food crops. Conservation of the entire genepool in Africa and parts of Latin America is considered a priority by IBPGR.
Pedaliaceae	*Sesamum orientale*	Sesameseed	Possibly Ethiopia or peninsular India	
Piperaceae	*Piper nigrum*	Pepper	Wild pepper plants grow in the Western Ghats of Malabar, southwestern India and this is presumed to be the crop's centre of origin.	
Rosaceae	*Fragaria x ananassa*	Strawberry	A hybrid between two American species, *F. chiloensis* and *F. virginiana*. Both species were harvested from the wild and also planted by Indians before European settlement. Crossing took place in Europe in the 18th century.	
	Malus pumila	Apple	Central Asia and Himalayan region	Conservation of wild relatives of *Malus* in Europe and Asia is an IBPGR priority. The Chatkal Mts Biosphere Reserve, USSR, conserves apples and various other fruit trees.
	Prunus amygdalus	Almond	Central to western Asia	A reserve for the conservation of almond and other important fruit trees has been created in the Kopet Mountains (USSR)
	Prunus armeniaca	Apricot	Western China	Wild apricots are protected in the Kopet Mountains Reserve
	Prunus avium	Cherry	Western Asia	
	Prunus communis	Pear	Central Asia and the Himalayas	Protection of wild species in Europe and Asia is considered a conservation priority by IBPGR
	Prunus domestica	Plum	Europe	
	Prunus persica	Peach	Western China	

Table 25.1 Food crops of the world (continued)

FAMILY	SPECIES	FOOD	ORIGIN	CONSERVATION
Rubiaceae	*Coffea arabica*	Coffee	Ethiopia	Coffee grows wild in the threatened forests of the Ethiopian massif. Much of the forest habitat in Ethiopia has been destroyed. Habitats of wild coffee are also threatened in Kenya. Protection of *C. arabica* in the wild is a conservation priority.
Rutaceae	*Citrus aurantiifolia*	Lime	Cultivated hybrid with obscure origins.	Protection of wild *Citrus* species in Asia is a conservation priority.
	Citrus grandis	Pomelo	Thailand. The origins of cultivated citrus fruits are obscure.	Protection of wild *Citrus* species in Asia is a conservation priority.
	Citrus limon	Lemon	Cultivated hybrid with obscure origins.	Protection of wild *Citrus* species in Asia is a conservation priority.
	Citrus x paradisi	Grapefruit	The origins of cultivated citrus fruits are obscure. The grapefruit is thought to be a cross between the pomelo and the sweet orange.	Protection of wild *Citrus* species in Asia is a conservation priority.
	Citrus reticulata	Tangerine	Southeast Asia	
	Citrus sinensis	Orange	A hybrid, probably originating in China.	
Sapotaceae	*Vitellaria paradoxa*	Karite nut, Sheanut	West Africa, grown in plantations in Ghana and Nigeria.	
Solanaceae	*Capsicum annuum*	Chili Pepper, Sweet Pepper	Domestication first occurred in Middle America	Wild peppers are still collected and sold locally. A large number of yet unexploited varieties exist in the Tropics. More collection for seed banks is needed.
	Lycopersicon esculentum	Tomato	The genus is native to South America. Mexico was probably the centre of domestication.	The wild relatives of the tomato have limited ranges. The crop's wild gene pools are prone to erosion by habitat destruction.
	Solanum melongena	Eggplant	India	
	Solanum tuberosum	Potato	The area of domestication is assumed to be the high plateau of Bolivia-Peru.	There are over 150 wild species of potato, many of which have limited natural distributions; 3,000-5,000 varieties are recognised by farmers in the Andes. Conservation of genetically valuable local varieties is being carried out at the International Potato Centre in Peru.
Sterculiaceae	*Theobroma cacao*	Cocoa	Centre of origin is the eastern slopes of the Andes and the centre of cultivation is Central America.	Cultivated varieties suffer from a lack of genetic variation. Forests harbouring genetic diversity in the wild are being rapidly destroyed.
Umbelliferae	*Daucus carota*	Carrot	The species is widespread in Europe and Asia. The primary centre of origin for cultivated forms is thought to be Afghanistan.	
Vitaceae	*Vitis vinifera*	Grape	10,000 Old World cultivars are thought to be derived from this single wild species which still occurs in Middle Asia.	Wild relatives are suffering genetic erosion in the USA.
Zingiberaceae	*Elettaria cardamomum*	Cardamom	Native to India	Collection from the wild contributes to the commercial trade.

Source: Compiled from multiple sources. Species list based on Prescott-Allen, C. and Prescott-Allen, R. 1990. How many plants feed the world? *Conservation Biology* 4(4):365-374. Historic, cultivation and nomenclatural data from Simmonds, N.W. (Ed.) 1976. *Evolution of Crop Plants*. Longman Scientific and Technical. Conservation data from multiple sources.

Most of the world's major food crops were domesticated and widely dispersed by 2,000 years ago. Diversification continued during colonial periods from the spread of Roman civilisation through to European settlement of the tropics. Colonial expansion undoubtedly contributed to the loss of genetic diversity of cultivated plants in the New World as a result of devastation of farming communities by invasion and disease. It has been suggested, however, that more varieties arose as a result of crop interchange between continents and islands during colonial expansion than were lost through cultural disintegration (Plucknett *et al.*, 1987). Maize and cassava were, for example, introduced to Africa by the Portuguese in the 16th century and diversified as they were grown under a new range of ecological and cultural conditions.

The history of food crops is complex and the exact origins of some cultivated food plants are obscure. Nevertheless, the geographical origins of major crops can be traced back through time.

From the earliest stages of agriculture, regions of diversity developed which remain important centres of crop biodiversity today. In addition, there are a number of minor centres of origin where a few crops can be related to their initial domestication in particular localities.

The Russian botanist N.I. Vavilov first described and mapped centres of diversity for individual crops which he believed represented their centres of origin. It became apparent that the centres of diversity of different crops coincided to give remarkable concentrations of crop plant variation. Vavilov (1951) ultimately recognised eight such centres; later authors have modified the centres and identified new ones (Fig. 25.1).

In general the concept of 'Vavilov Centres' where a centre of current diversity is taken to indicate the centre of origin of crops is now considered an oversimplification. Nevertheless, geographical concentrations of crop variation are real and these areas are of immense conservation importance. The reasons for the diversity are: the great age of cultivation in such centres, the wide range of ecological conditions and farming practices found within them, and the processes of natural selection caused by the presence of many different pathotypes of pests and diseases and by the variable ecological conditions. Some features of one of Vavilov's Centres of Diversity, the Ethiopian centre, are described in the case study below.

Crop genetic resources

Genetic resources can be defined as the genetically transmitted characteristics of organisms which are of actual or potential value to people. Such characteristics may include rapid growth, high yields, disease- and pest-resistance and environmental adaptation. The genetic resources of crop plants represent the total genetic diversity of cultivated species and their wild relatives, much of which is of immense value in crop breeding programmes. Many of the species from which crop plants have been selected continue to survive in the wild today. These, together with closely related species, comprise the wild relatives of crops. They continue to evolve under natural conditions and provide a largely untapped reservoir of genetic diversity. Gene flow between cultivated crops and wild relatives continues to occur today, and is encouraged in areas where traditional forms of agriculture are still practised. In Mexico, for example, some traditional farmers still utilise teosinte, the closest wild or weedy relative of maize, to increase corn yields. The weedy plants are allowed to remain within or near cultivated maize populations so that natural crosses may occur and produce fertile hybrid stock that can be selected for desirable characteristics.

The genetic resources of wild crop relatives can be classified according to the ease with which the species can be crossed with the cultivated crop. The primary gene pool consists of relatives that are interfertile and hybridise readily with the cultivated crop. Wild forms of cocoa *Theobroma cacao* occurring in the Amazon forests can, for example, be crossed readily with cultivated cocoa and constitute the 'primary gene pool' for the crop. The same applies to certain wild relatives of maize *Zea mays*.

The 'secondary gene pool' consists of species which can be crossed using conventional breeding methods but crossing is difficult and only a small proportion of first generation progeny may be fertile. The secondary gene pool of maize includes, for example, wild relatives *Zea perennis* and *Tripsacum* species. The 'tertiary gene pool' of a crop plant consists of species that are more distantly related. The genetic diversity in tertiary gene pools can only be utilised by experimental techniques in plant breeding such as using another species as a bridge.

The most common use of wild genetic resources in crop breeding programmes has been in the introduction of resistance to pests and diseases. Wild tomato species, *Lycopersicon pimpinellifolium* and *L. peruvianum* have, for example, been used in breeding programmes to confer resistance to various forms of bacterial wilt. Genes from wild relatives of the tomato have also conferred resistance to a range of viruses, moulds, and other pests. Likewise, wild potato relatives have been crossed with cultivars for about a hundred years, the wild species yielding genes for resistance to viruses, bacterial wilt, nematodes, aphids and a range of other potato disorders.

In addition to wild crop relatives a second important storehouse of genetic crop diversity is the range of variation shown by 'land races'. These are races or populations of crops that have become adapted under natural and artificial selection processes to the local conditions under which they are cultivated. Land races have not been deliberately bred but have been developed over centuries of traditional agriculture. They are now being explored as a source of genetic material for crop improvement programmes. Recent work in the Himalayan foothills of north-east India has, for example, revealed a large number of primitive rice cultivars with resistance to major pests and diseases including bacterial blight, tungro virus, gall midge and stem borer.

Genetic erosion

The evolution of food crops under centuries of domestication has increased variation as seen in the main regional centres of crop diversity. But the development of

Figure 25. 1 Regions of diversity of crop plants

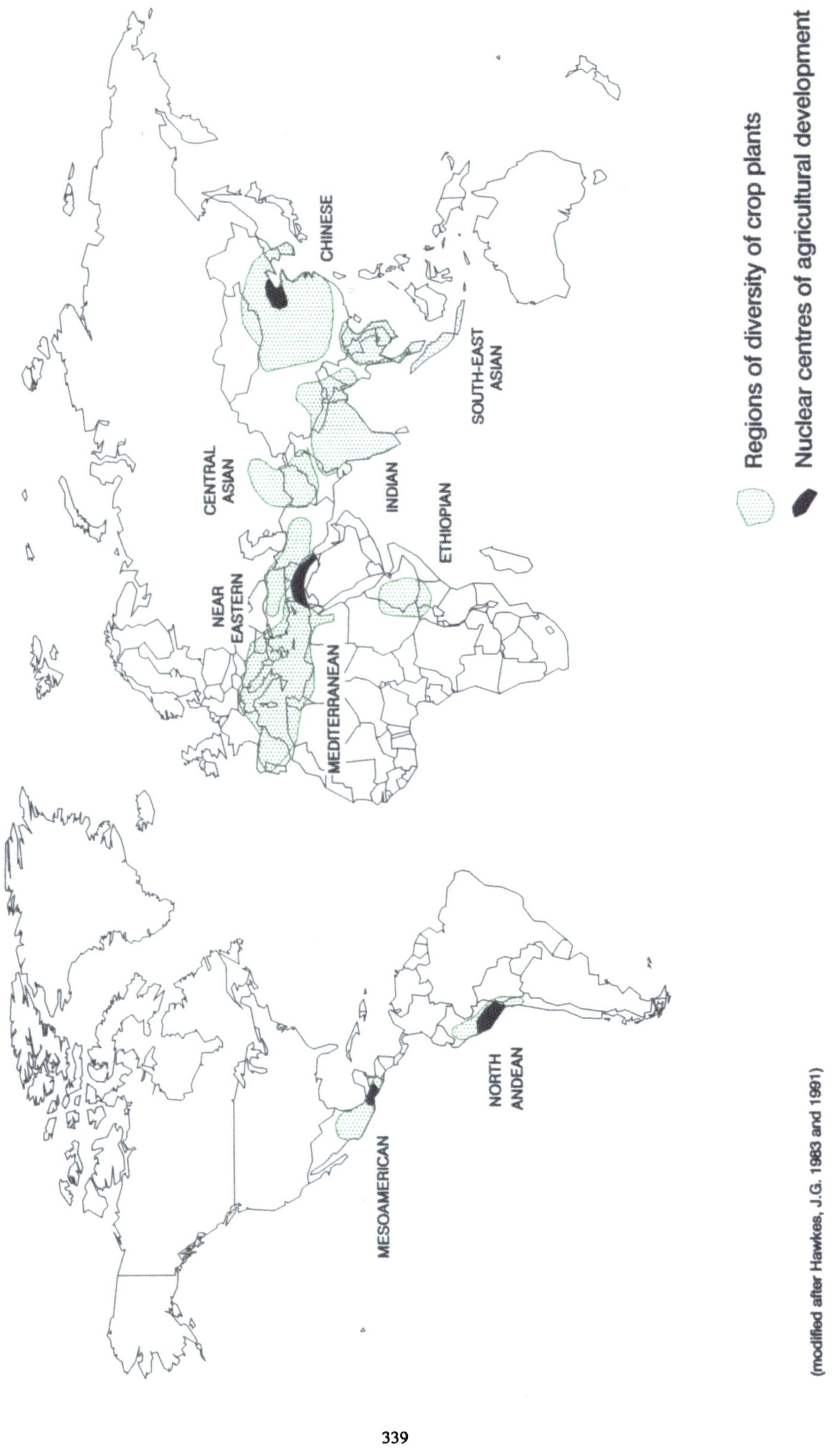

high-yielding modern cultivars for intensive agriculture is now rapidly reversing this trend, leading to a dangerous reliance on genetically uniform crops.

Genetic erosion, or the loss of genetic diversity, of the world's food plants is an issue of serious concern with implications for the long-term maintenance of global food supplies. At a time when more genetic diversity is needed in crop breeding programmes to increase food production this diversity is rapidly disappearing or has already been lost. Economic aspects of this process are discussed in Chapter 27.

Various factors contribute to genetic erosion. The worldwide threats to wild species through habitat destruction and modification have an obvious impact on wild crop relatives. Habitat destruction is having a direct effect on, for example, wild forms of cocoa: large parts of the centre of genetic diversity of *Theobroma cacao* in Colombia, Ecuador and Peru have been destroyed as a result of petroleum exploration and exploitation, and by agricultural expansion. Similarly, around 90% of the Ethiopian highland forests, which harbour wild coffee *Coffea arabica*, have been destroyed.

Of equal concern is the loss of old land races through replacement by modern highly-bred crop cultivars. This may lead to improvements in yield but also results in increased reliance on agrochemicals and all the problems associated with monoculture cultivation.

The extent of genetic erosion differs for various crops. In general, the wild relatives of cereals are widespread, weedy and thrive in disturbed ground. There is some evidence of genetic erosion of wild relatives, however, and conservation attention is a priority for those of rice, wheat and maize. The loss of local land races for these major cereals has been a particularly serious problem in various parts of the world. The introduction of new high-yielding varieties of wheat has, for example, caused severe genetic erosion in Turkey, Iraq, Afghanistan, Pakistan and India. In Greece, 95% of the native varieties of wheat have been lost in 40 years (Davies, 1991).

The wild relatives of root crops are also suffering loss of genetic diversity. More than half the wild species in the genera *Solanum* (potato) and *Manihot* (cassava) are narrowly endemic in South, Central and Middle America (FAO, 1984). Conservation of centres of potato diversity is an urgent concern, as some species have already become extinct. Diversity of natural populations of *Manihot* species is declining owing to conversion of their habitats to pasture and elimination of the plants, which are poisonous to grazing animals.

All the wild relatives of the tomato *Lycopersicon esculentum* have limited natural distributions. Clearance of habitats for agriculture, housing and industry has led to the loss of wild populations of the tomato species *Lycopersicon hirsutum* and *L. peruvianum*. As yet no *in situ* conservation areas have been established for wild tomato plants. Loss of genetic resources from the wild can also be seen with *Brassica oleracea*. This species, native to southern England, western France and northern Spain, provides a number of cultivars including cabbages, cauliflowers and Brussels sprouts. The populations related to *B. oleracea* form a group of about 12 perennial species (often considered to be subspecies of *B. oleracea*) most of which are endemic to the western Mediterranean. These species are composed of small populations often reduced to a few specimens and isolated geographically (Valdes, 1991). Several of the species are considered by IUCN to be Rare, Endangered or threatened. Other wild taxa in the group may not be endangered as species although many of their populations are. In many cases, wild *Brassica* populations in the Mediterranean are protected by the inaccessible nature of their rocky habitats but elsewhere they are threatened by competition from maquis or garrigue scrub or by habitat destruction (Olivier, 1991).

Examples of genetic erosion in wild populations and traditional cultivars can be found in all groups of food crops throughout the world. Summary information is given in Table 25.1. Lack of knowledge of intra-specific genetic variation remains a problem in detecting the degree of threat to plant gene pools, but certain priorities have been established for international conservation action. The conservation of plant genetic resources is discussed in Part 3.

Case study: the Ethiopian centre of crop genetic diversity

Ethiopia represents one of the world's eight major centres of crop plant diversity. It is the probable area of domestication for many crops and for others, where no wild relatives are known within the country, it is a secondary centre of diversity. The rich variation within crop plants results from the highly dissected topography of the country allowing crops to evolve in isolation under primitive agricultural conditions, together with the ancient and very diverse cultural history of the country. The geographical position of Ethiopia, at the crossroads between the Near East and Indian centres of diversity, also accounts for the genetic richness of the country's crop plants.

Below are described 12 widespread crops which are believed to have their centres of diversity within the region along with three other Ethiopian crops - chat, ensete and noog - which have originated and evolved within the country (Engels *et al.*, 1991).

In addition to the cultivated crop species, there are many wild plants used for food in Ethiopia, particularly in times of food shortage. Some have considerable potential as new crop plants. One such example is the yeheb nut *Cordeauxia edulis*. This species is endemic to eastern Ethiopia and part of Somalia. It has long been valued for its highly nutritious nuts but is now Endangered in the wild because of over-exploitation and overgrazing.

Many other plant species are of local importance. Medicinal plants, for example, remain important sources of drugs for nearly 80% of the Ethiopian population. Most of Ethiopia's major medicinal plants are not in cultivation.

Plant species have a wide range of other uses in Ethiopia. Fuelwood, for example, provides for over 90% of the country's total energy consumption. Timber, resins, gums, cosmetics, perfumes, dyes, inks, fibres, and forage are all derived from wild plants. Ethiopia's rich plant biodiversity is clearly of immense importance. The country's flora and vegetation types are, however, still incompletely known and information on the conservation status of individual species is sparse. The almost complete deforestation of Ethiopian highlands, changes in land-use and agricultural practices are undoubtedly having serious consequences for both wild and cultivated plant diversity.

Concern about the loss of plant genetic resources led to the establishment of the Plant Genetic Resources Centre (PGRC/E) in 1976. The Centre is involved in exploration, collection and preservation of crop germplasm, together with the provision and exchange of germplasm for crop breeding programmes. Collection of genetic resources includes the collection of land races from drought-prone areas for storage at seed reserve centres and redistribution to farmers when required. This is an insurance measure to prevent major losses of crop genetic diversity by consumption of seed in times of famine or the replacement of traditional varieties by imported seeds distributed through relief agencies. The involvement of farmers in the conservation of germplasm is being considered. Already measures are under way to conserve semi-cultivated coffee on peasant farms and 'backyard' coffee in cooperatives, as part of the national coffee conservation programme.

Coffee (*Coffea arabica*; Rubiaceae) *Coffea arabica* accounts for over 80% of the world's coffee production. Almost the entire diversity of this crop originated in Ethiopia, mainly in the south-western rain forest area. *C. arabica* still occurs as a wild plant in these moist montane forests and as a semi-wild or cultivated crop in the same areas. In drier parts of Ethiopia it is grown as an irrigated crop and elsewhere as a garden plant often mixed with fruit trees and herbs. There is extremely high genetic diversity within Ethiopian coffee but this rich diversity is under considerable threat. Deforestation, replacement of primitive coffee populations by other crops and changing patterns of land-use are leading to severe problems of genetic erosion.

Barley (*Hordeum vulgare*; Gramineae) Ethiopia's third most important cereal crop, barley, was introduced from the Near East in ancient times. Ethiopia is a secondary centre of diversity for the species and the crop has developed many important and unique characteristics within the country. Extensively grown land races can still be found, but genetic erosion is resulting from replacement by other cereals.

Sorghum (*Sorghum bicolor*; Gramineae) Ethiopia and Sudan are assumed to be the primary centre of origin and diversity of sorghum. This cereal is probably the most diverse of all Ethiopia's crops. It is grown in a wide range of ecological conditions throughout the country. Disease-, pest- and drought-resistance have all been reported. Use of improved local land races and imported varieties together with replacement by maize and other crops are causing the loss of genetic diversity.

Wheat (*Triticum* spp.; Gramineae) Ethiopia represents a secondary centre of diversity for wheat. Durum wheat is the main type grown. It exhibits high phenotypic diversity within the country and agronomically important genes have been located in Ethiopian germplasm. Genetic erosion is occurring because of replacement by other crops.

Teff (*Eragrostis tef*; Gramineae) Teff is the most widely grown crop in Ethiopia. It is used mainly for making a pancake-like bread called 'injera' and also to make porridge and alcoholic drinks. The straw is used as a cattle feed and in house construction. Elsewhere teff is cultivated only in North and South Yemen although *E. tef* has a wide distribution in Africa. Ethiopia is the centre of origin for teff and domestication is thought to have first taken place in the northern highlands. Over 50 *Eragrostis* spp. occur in Ethiopia, of which 14 are endemic. At least 35 land races of teff are known in the country. Genetic erosion is not a problem for the crop which is still expanding its acreage. Improved varieties are not being introduced.

Niger seed, noog (*Guizotia abyssinica*; Compositae) Noog is the most important oil crop in Ethiopia, and the area under cultivation is expanding. It is thought that the crop originated in the highlands of Ethiopia and that it was one of the earliest crops to be domesticated within the country. It probably originated from the wild species *Guizotia scabra*, which now frequently occurs as a weed of noog fields. Genetic erosion is not currently a problem faced by noog, but improvement of agricultural practices may lead to genetic erosion of *G. scabra*.

Linseed (*Linum usitatissimum*; Linaceae) Linseed is the second most important oil crop in Ethiopia. It was introduced in ancient times from Asia, and Vavilov considered Ethiopia to be a centre of flax diversity. This diversity is seriously threatened by genetic erosion.

Sesame (*Sesamum indicum*; Rubiceae) Ethiopia is likely to be the centre of origin of sesame, but this remains uncertain. Economically, sesame is the third most important oil crop in Ethiopia. The crop exhibits considerable diversity but is facing critical genetic erosion.

Castor bean (*Ricinus communis*; Euphorbiaceae) Castor bean is not cultivated as a commercial crop in Ethiopia. It is widely distributed throughout the country as a wild plant or weed and is used as a medicinal plant or source of oil for lighting. Phenotypic diversity is enormous, and this has led to suggestions that the cultivated castor bean might be of Ethiopian origin. There is no threat of genetic erosion at present.

Pea (*Pisum sativum*; Leguminosae) A unique subspecies occurs in Ethiopia. It has been suggested that Ethiopia is one of four possible centres of diversity of the pea. Phenotypic diversity is rather limited and the degree of genetic erosion is expected to be low.

Chickpea (*Cicer arietinum*; Leguminosae) Chickpea is an ancient crop in Ethiopia, and the country is a centre of diversity for the cultivated plant. Phenotypic variation is considerable, and initial testing has shown some disease resistance and drought tolerance. Genetic erosion is not a significant threat.

Lentil (*Lens culinaris*; Leguminosae) The lentil was an early introduction into Ethiopia from west Asia. The crop shows a high degree of diversity. Genetic erosion is expected as the acreage of lentils is declining.

Ensete (*Ensete ventricosum*; Musaceae) Ensete is a crop species unique to Ethiopia. Both wild and cultivated forms occur throughout the country wherever there is sufficient moisture. The pseudocorm is processed to form a staple food; other parts of the plant are used as fodder, fuel, packing material, to wrap bread during cooking and to make ropes. The crop shows considerable variation and over 70 named varieties have been described. Bacterial wilt and drought are contributing to genetic erosion.

Chat (*Catha edulis*; Celastraceae) The leaves of this evergreen shrub are used as a stimulant. The plant was first domesticated in Ethiopia. Cultivation is now expanding and is leading to the replacement of coffee in the eastern part of the country. No genetic erosion is currently taking place.

TIMBER

Wood is one of the basic commodities utilised worldwide that is still predominantly harvested from the wild. It provides the primary source of fuel in many developing nations, shelter in traditional home-building and sophisticated construction, and the basis for the international pulp and paper industry. Wood is one of the most important commodities in international trade and accounts for a particularly significant proportion of the export earnings of developing tropical countries. In 1989 the total worldwide value of wood exports was around US$6 billion.

Table 25.2 provides figures for wood production and trade for 1989. The bulk of the wood in world trade comes from temperate sources, with the major exporters being USA, the former USSR, and Canada for logs and sawnwood and USA, the former USSR and Finland for plywood. The main tropical source countries are Malaysia, Papua New Guinea and Gabon for logs, and Malaysia and Indonesia for sawnwood and plywood.

Within developing countries there is a trend towards value-added processing in the timber industry. The export of timber in log form is increasingly being restricted partly to retain wood within the country for further processing and partly as a conservation measure. Nevertheless, logs still account for a significant proportion of world trade.

Overall, developing countries still retain a relatively small proportion of the financial value of their timber resources despite increasing industrialisation of the forestry sector. The trade imbalance remains heavily in favour of the developed nations. It has been shown, for example, in a recent study that 65-90% of the growth in value of tropical forest products occurs in consumer countries, made up of operating costs, tax revenue and profits (Oxford Forestry Institute, 1991).

In general it is difficult to assess the extent to which timber for domestic consumption or international trade is derived from plantations. Industrial timber plantations mainly consist of conifers which lend themselves well to cultivation as pure crops. Relatively few hardwoods have been cultivated as plantation timbers. The majority of hardwoods in international trade are derived from natural forests which are subject to varying degrees of management. In tropical regions relatively few examples of successful forest management for sustainable timber production are known.

Detailed information on levels of production and trade in individual timber species is scarcely assembled at an international level. The conventional division of timber products into hardwoods (non-coniferous) and softwoods (coniferous), for trade purposes, disguises the great diversity of wood as a natural product. Timber species richness is particularly high in tropical regions. Ghana, for example, has 674 tree species reaching timber size and timber from about 60 of these has been exported in the past 20 years. Peninsular Malaysia has at least 3,000 tree species of which over 400 have been traded on international markets. Developing countries are attempting to diversify their timber exports by promoting lesser known species but consumer demand remains conservative in importing countries. Potentially valuable timber resources are under threat in many parts of the world through inadequate management, habitat loss and over-harvest (Table 25.3).

Ghana Timber Species Case Study

Timber is Ghana's third most important export commodity after cocoa and minerals. Ghana's share of the world's tropical timber trade is about 1% and it accounts for about 3% of West European imports of tropical hardwoods. Europe takes over 90% of Ghana's timber exports. Logs are predominantly exported to Germany and the UK; Germany and Ireland are leading sawn timber importers and sliced veneer goes mainly to Germany.

The timber export trade began a century ago, concentrating on species of *Entandrophragma* and *Khaya*, the so-called African Mahoganies. In total around 674 tree species reach timber size in Ghanaian forests and timber of about 60 of these has been exported over the past 20 years. Commercial exploitation over the past century, together with the reduction of natural forest from 8 million ha to below 2 million ha, has placed considerable pressure on the commercial timber species.

The report of the Fifth Session of the FAO Panel of Experts on Forest Gene Resources drew attention to the fact that in Ghana, "some of the most valuable commercial species *Pericopsis elata*, *Gossweilerodendron balsamiferum*, *Lovoa trichilioides*, *Entandrophragma utile*, *Nauclea diderrichii*, *Terminalia ivorensis*, *T. superba*, *Antiaris africana*, *Triplochiton scleroxylon* and *Hallea ledermannii*, are threatened with extinction in their areas of natural distribution because of massive exploitation."

More recently a full inventory of Ghana's timber resources has been carried out in a project funded by the UK's Overseas Development Administration. Information from this study suggests that immediate, serious problems of over-logging apply to timbers of the Meliaceae, especially *Khaya ivorensis*, and also to *Pericopsis elata*. Taxa such as *Terminalia* and *Triplochiton scleroxylon* are, in fact, regenerating well in disturbed forest and are relatively fast

Table 25.2 Wood production and trade, 1989

	ROUNDWOOD PRODUCTION[1]			TIMBER PRODUCTION		PAPER +	NET
	TOTAL m3	FUEL AND CHARCOAL m3	INDUSTRIAL ROUNDWOOD m3	SAWNWOOD & SLEEPERS m3	WOOD-BASED PANELS m3	PAPERWOOD PRODUCTION metric tons	TRADE[2] ROUNDWOOD m3
ASIA	1.07E+09	793913140	273434972	106701500	27453931	5122400	110238000
Afghanistan	6104000	4609000	1495000	400000	1400		
Bahrain							36000000
Bangladesh	30144992	29272000	873000	79000	8000	96000	
Bhutan	3224000	2946000	278000	5000			-7000
Brunei	294000	79000	215000	90000			
China	274589952	177610016	96980000	24958000	3650000	15336000	13382679000
Cyprus	78300	22300	56000	57300	22000		
Hong Kong	187000	187000		248000	12000	40000	699637000
India	269450752	245126992	24324000	17460000	441700	1940000	902324000
Indonesia	175730496	136079008	39651488	10390500	8838427	974000	-1131000
Iran, Islamic Rep	6829000	2453000	4376000	163000	54200	78000	117300000
Iraq	149000	99000	50000	8000	3000	28000	1200000
Israel	118000	11000	107000		148000	180000	212500000
Japan	31935904	571000	31364896	30542000	8993000	26809000	51809890000
Jordan	9000	5000	4000			10000	16900000
Kampuchea, Dem	5803000	5236000	567000	43000	2300		
Korea, Dem People's Rep	4761000	4161000	600000	280000		80000	70600000
Korea, Rep	6803000	4491000	2312000	4014000	1453000	4018000	7384300000
Kuwait							60493000
Laos	3972000	3660000	312000	16000	10000		-34000
Lebanon	503000	482000	21000	27000	46300	37000	15500000
Malaysia	50536688	8258000	42278688	8275000	1630000	70000	282530000
Mongolia	2390000	1350000	1040000	470000	3500		
Myanmar	22287008	17407008	4880000	463700	15200	8000	-360000
Nepal	17804000	17244000	560000	220000		2000	
Oman							22500000
Pakistan	24408992	23226000	1183000	751000	93500	151000	34700000
Philippines	38503008	33075008	5428000	950000	425000	334000	397895000
Qatar							35700000
Saudi Arabia							221300000
Singapore				206000	489000	10000	170688000
Sri Lanka	8988400	8302000	686400	20000	9604	28000	-33000
Syrian Arab Rep	48300	14800	33500	9000	26700	19000	22396000
Thailand	38734000	34115008	4619000	1279000	257000	520000	1410029000
Turkey	15449000	9721000	5728000	4923000	781000	400000	467570000
Viet Nam	27188000	23776000	3412000	354000	40100	56000	41100000
Yemen, People's Dem Rep	324000	324000					5199000
USSR*	382099968	80700000	301399808	100000000	14635000	10654000	136949000
EUROPE	368250792	53890800	314360008	86292500	38232200	66749000	-29257000
Albania	2330000	1608000	722000	200000	12000	24000	
Austria	16086000	1413000	14673000	7054000	1538000	2754000	4830078000
Belgium and Luxembourg	4757000	572000	4185000	1114000	2247000	1237000	4127038000
Bulgaria	4455000	1810000	2645000	1283000	496000	438000	134661000
Czechoslovakia	18552000	1532000	17020000	4993000	1441000	1312000	50552000
Denmark	2118000	467000	1651000	861000	331200	326000	382296000
Finland	46262000	2984000	43278000	7763000	1482000	8752000	6783253000
France	43726992	10436000	33290992	10559000	3018000	6754000	1756597000
German Dem Rep	10897000	626000	10271000	2555000	1179000	1351000	539809000
Germany, Fed Rep	35332000	3656000	31676000	11405000	8529000	11259000	3903311000
Greece	3289000	2320000	969000	355200	398000	282000	244697000
Hungary	6609000	2949000	3660000	1256500	449000	504000	1215830000
Iceland							1200000
Ireland	1527000	50000	1477000	300000	236000	34000	7688000
Italy	8846000	4177000	4669000	1998000	4342000	5555000	6458978000
Malta							400000
Netherlands	1331000	116000	1215000	465000	94000	2570000	1327486000
Norway	11039000	936000	10103000	2491800	611000	1789000	1667565000
Poland	22348992	3123000	19226000	4963000	1845000	1406000	247260000
Portugal	10341800	597800	9744000	1650000	1025000	740000	819269000
Romania	19306992	2790000	16517000	2851000	1525000	819000	56884000
Spain	17182000	2384000	14798000	2724000	2295000	3446000	1940367000
Sweden	55704016	4424000	51280016	11487000	1307000	8362000	8207344000
Switzerland	4562000	850000	3712000	1282000	918000	1259000	961168000
United Kingdom	6462000	209000	6253000	2191000	1676000	4475000	901083000
Yugoslavia	15186000	3861000	11325000	4491000	1238000	1302000	1072728000
NORTH & CENTRAL AMERICA	771385180	172135000	599250156	166232942	40996200	89698000	-33376000
Bahamas	115000		115000	1400			300000
Barbados							500000
Belize	187600	126000	61600	14300			-8000
Canada	176976016	6834000	170142016	59224992	6913000	16555000	4263107000
Costa Rica	4037000	2886000	1151000	515400	57500	18000	-4000
Cuba	3122000	2511000	611000	130100	149000	168000	5700000
Dominica							700000
Dominican Rep	982300	976000	6300	0		10000	40500000
El Salvador	4440000	4320000	120000	54000		17000	100000
Guadeloupe	17000	15000	2000	1000			900000
Guatemala	7604000	7490000	114000	83000	6000	17000	-12000
Haiti	5727000	5488000	239000	13800			
Honduras	6056000	5172000	884000	441000	10000		-21000
Jamaica	218000	13000	205000	40000	0	4000	200000
Martinique	11900	10000	1900	950			
Mexico	22628992	15204000	7425000	2410000	645300	3375000	19194000
Netherlands Antilles							300000
Nicaragua	3972000	3092000	880000	222000	3400		200000
Panama	2047300	1708000	339300	18000	12000	20000	1700000
Trinidad and Tobago	75200	22000	53200	23000			1100000
United States	533167872	116268000	416899840	103040000	33200000	69514000	3442768000

Table 25.2 Wood production and trade, 1989 (continued)

	ROUNDWOOD PRODUCTION TOTAL m3	ROUNDWOOD PRODUCTION FUEL AND CHARCOAL m3	ROUNDWOOD PRODUCTION INDUSTRIAL ROUNDWOOD m3	TIMBER PRODUCTION SAWNWOOD & SLEEPERS m3	TIMBER PRODUCTION WOOD-BASED PANELS m3	PAPER + PAPERWOOD PRODUCTION metric tons	NET TRADE ROUNDWOOD m3
SOUTH AMERICA	**335574900**	**234266700**	**101308208**	**26624292**	**4112500**	**7572000**	**-4763000**
Argentina	10819000	4332000	6487000	1446000	354000	917000	3299000
Bolivia	1556500	1301000	255500	95000	3900	2000	
Brazil	255455008	182806000	72649008	18178992	2892000	4806000	25854000
Chile	16864000	6540000	10324000	2713000	282000	445000	-4679000
Colombia	18478992	15806000	2673000	721200	113000	501000	
Ecuador	9728000	6642100	3085900	1491600	145400	35000	
French Guiana	253600	65600	188000	19000	0		-5000
Guyana	228000	19000	209000	57000	0		-22000
Paraguay	8394000	5288000	3106000	906000	106500	11000	
Peru	8785800	7669000	1116800	541500	36000	260000	100000
Suriname	235000	20000	215000	73000	9700		-11000
Uruguay	3295000	3038000	257000	57000	10000	70000	400000
Venezuela	1482000	740000	742000	325000	160000	524000	11800000
OCEANIA	**39795692**	**8738000**	**31057692**	**6004573**	**1774100**	**2605000**	**-11923000**
Australia	20040992	2886000	17154992	3612000	1081000	1870000	-5814000
Fiji	306800	37000	269800	93600	16100		-148000
French Polynesia							1300000
New Caledonia	12100		12100	5400			400000
New Zealand	10557000	50000	10507000	2131000	658000	735000	2830000
Papua New Guinea	8231000	5533000	2698000	117000	19000		
Solomon Islands	449000	138000	311000	16000	0		
Tonga	4600		4600	1573			
Vanuatu	63200	24000	39200	7000			
Western Samoa	131000	70000	61000	21000	0		
AFRICA	**498508268**	**441864308**	**56644000**	**8829600**	**1904423**	**2460000**	**-5292000**
Algeria	2131000	1874000	257000	12800	49500	120000	210000000
Angola	5402000	4335000	1067000	5000	2000	15000	
Benin	5000000	4738000	262000	11000			
Botswana	1321000	1239000	82000				
Burkina Faso	8526400	8141000	385400	750			
Burundi	4083000	4034000	49000	3000			
Cameroon	12850000	10142000	2708000	653000	80000	5000	
Central African Rep	3455000	3055000	400000	52000	3900		
Chad	3936000	3380000	556000	1000			
Congo	3300000	1776000	1524000	46000	54400		
Côte d'Ivoire	13243000	9830000	3413000	775000	266000		-550000
Djibouti	0	0	0				
Egypt	2266000	2161000	105000		80000	160000	200100000
Equatorial Guinea	607000	447000	160000	51000	10000		
Ethiopia	39640000	37884000	1756000	34000	15300	10000	
Gabon	3700000	2478000	1222000	126000	228000		-913000
Gambia, The	921600	901000	20600	1000			
Ghana	17168992	16068000	1101000	537000	53000		-201000
Guinea	4669000	4022000	647000	90000	0		-8000
Guinea-Bissau	567000	422000	145000	15700			
Kenya	35650000	33884000	1766000	185000	52200	108000	
Lesotho	596000	596000					32700000
Liberia	5960000	4800000	1160000	411000	5000		-701000
Libya	642000	536000	106000	31000		6000	29800000
Madagascar	7856000	7049000	807000	234000	5000	6000	-2000
Malawi	7621000	7275000	346000	31000	6200		
Mali	5515800	5163000	352800	12800			
Mauritania	12000	7000	5000				
Mauritius	31000	16500	14500	4600	0		100000
Morocco	2110600	1363600	747000	83000	147000	109000	600500000
Mozambique, People's Rep	16027000	15022000	1005000	35750	2800	2000	-1000
Niger	4418000	4146000	272000				
Nigeria	108298000	100430000	7868000	2712000	233000	73000	984000
Réunion	33300	31000	2300	2200			1600000
Rwanda	5842000	5602000	240000	13000	1723		
Sao Tome and Principe	9000		9000	5400			
Senegal	4391000	3786000	605000	11000			25300000
Sierra Leone	3014000	2874000	140000	12000			
Somalia	6986000	6896000	90000	14000	0		300000
South Africa	19360992	7078000	12283000	1873000	398000	1636000	26257000
Sudan	22198992	20112000	2087000	12500	1500	10000	
Swaziland	2223400	560000	1663400	136000	8000		71830000
Tanzania	33102992	31114000	1989000	156000	14700	28000	
Togo	866000	683000	183000	5000			200000
Tunisia	3177500	3015000	162500	20000	97000	82000	15800000
Uganda	14365000	12507000	1858000	28100	3300	2000	
Zaire	35348000	32557008	2791000	121000	52500	2000	-117000
Zambia	12204000	11565000	639000	76000	8200	4000	
Zimbabwe	7861700	6269200	1592500	190000	26200	82000	-2000

Source: FAO Yearbook 1989, Forest products.
Note: [1] Roundwood refers to all wood in the rough destined for either fuel or industrial uses. It includes sawlogs, veneer logs and pulpwood. [2] Net trade is the balance of imports minus exports. * Former USSR.

growing. A quantitative measure of the status of various timbers was provided by an estimation of their resource life undertaken as part of the inventory project. Estimates based on the results of the project for resource data combined with information on growth and extraction rates suggest that the resource life of *Pericopsis elata* is already zero.

The various assessments of frequency, resource life and conservation status of a number of Ghana's commercial timbers are summarised in Table 25.4. This table also

Table 25.3 Endangered tree species and provenances

FAMILY	SPECIES	DISTRIBUTION	STATUS	THREATS
Anacardiaceae	*Astronium urundeuva*	Brazil, Argentina, Paraguay	Endangered	Exploitation for timber, tannin and medicinal purposes
	Schinopsis brasiliensis	North-eastern Brazil	Suffering a slow decline	
Apocynaceae	*Aspidosperma polyneuron*	Brazil, Argentina, Paraguay and Peru		Intensive exploitation and habitat conversion
Aquifoliaceae	*Ilex paraguaiensis*	South America	Over-exploited	Leaves used to produce mate, a tonic and stimulant drink
Araliaceae	*Didymopanax morototoni*	Central and South America and Caribbean islands	Abundant but in need of conservation attention	Heavy utilisation
Araucariaceae	*Araucaria angustifolia*	Brazil and Argentina	Endangered in parts of its range	Excessive exploitation of wild stands
	Araucaria cunninghamii	Irian Jaya and Papua New Guinea	Endangered in parts of its range	Habitat destruction, logging, low natural regeneration
	Araucaria hunsteinii	Papua New Guinea	Endangered in parts of its range	Shifting agriculture, fire, over-exploitation
Betulaceae	*Alnus acuminata*	Mexico, Central America and the Andes	In danger of genetic impoverishment	Substitution with introduced fast-growing species
Bignoniaceae	*Tabebuia impetiginosa*	Brazil	Suffering a slow decline	
	Zeyhera tuberculosa	South-eastern Brazil	Threatened	Agriculture, livestock and charcoal production
Bombacaceae	*Bombacopsis quinata*	Tropical America	Severely threatened at the provenance level	Excessive felling and forest clearance
Boraginaceae	*Cordia milleni*	Tropical Africa	Rare in Kenya	Forest clearance and utilisation of species for timber and medicinal purposes
Cercidiphyllaceae	*Cercidiphyllum japonicum*	Japan and China	Provenances are endangered in China	
Chenopodiaceae	*Atriplex repanda*	Chile	Endangered in parts of its range	Over-utilisation for livestock
Compositae	*Brachylaena huillensis*	Central Africa	Endangered	Habitat clearance and excessive felling
Cupressaceae	*Cupressus atlantica*	Morocco	Endangered	Exploitation and increasing human pressure
	Cupressus dupreziana	Algeria	Endangered	Grazing and exploitation for firewood
	Juniperus bermudiana	Bermuda		Approx. 90% of the trees died between 1944 and 1950 as a result of severe insect infestation
	Juniperus procera	Arabia and Tropical Africa	Outlying populations endangered	Fire, browsing pressure particularly from buffalo and elephant, logging and plantation development

Table 25.3 Endangered tree species and provenances (continued)

FAMILY	SPECIES	DISTRIBUTION	STATUS	THREATS
Ebenaceae	*Diospyros hemiteles*	Mauritius	Endangered; one individual remains in the wild	Cultivation, illegal wood cutting, monkey and deer damage, invasive plant species
Euphorbiaceae	*Joannesia principes*	Brazil	Threatened	Forest clearance and commercial exploitation
Fagaceae	*Fagus longipetiolata*	China	Endangered in parts of its range	
Hamamelidaceae	*Liquidambar styraciflua*	Southern USA and Central America	Endangered in parts of its range	Land clearance for agriculture and grazing pressure
Irvingiaceae	*Irvingia gabonensis*	Tropical Africa	In danger of genetic impoverishment	Logging operations and settlement
Lauraceae	*Aniba duckei*	Amazon region		Excessive exploitation
	Ocotea porosa	Brazil	Threatened	Clearance for agriculture, livestock and plantation development, timber exploitation
Lecythidaceae	*Bertholetia excelsa*	Bolivia, Brazil, Colombia, Peru and Venezuela		Habitat destruction and over-exploitation
Leguminosae	*Acacia albida*	Sub-Saharan Africa and parts of the Middle East	Some stands threatened in Israel	Use as fuelwood, fodder and browse and changing land-use patterns
	Acacia caven	Bolivia, Argentina, Uruguay and Chile	Endangered in parts of its range	Use as fuelwood and grazing pressures
	Acacia tortilis ssp. raddiana	North Africa, Egypt, Israel, Jordan, Saudi Arabia	Endangered in parts of its range	Over-grazing and human pressure
	Acacia tortilis ssp. tortilis	Somalia, Ethiopia, Sudan, Egypt, Israel, Arabia	Endangered in parts of its range	Over-grazing and human pressure
	Anadenanthera macrocarpa	Brazil, Argentina, Peru, Bolivia, Paraguay	Suffering a slow decline	Utilisation of wood and bark
	Caesalpinia dalei	Kenya	Endangered	Intensive agriculture
	Dalbergia nigra	Brazil	Endangered	Intensive logging
	Dipterix alata	Brazil	Threatened	Forest destruction and exploitation for wood and medicinal purposes
	Gigasiphon macrosiphon	Kenya and Tanzania	Rare or Endangered	Long-term climatic changes; forest clearance for settlement and cultivation, mineral exploitation
	Gossweilero-dendron balsamiferum	Nigeria to Zaire	Likely to be endangered in parts of its geographic range	Heavy exploitation, plantation development
	Machaerium villosum	Brazil	Threatened	Timber exploitation, forest clearance for pasture and plantations

Table 25.3 Endangered tree species and provenances (continued)

FAMILY	SPECIES	DISTRIBUTION	STATUS	THREATS
	Mimosa caesalpiniaefolia	Brazil	Suffering a slow decline	
Leguminosae (continued)	*Mimosa verrucosa*	Brazil	Suffering a slow decline	Wood exploitation
	Pericopsis elata	West Africa to Zaire	Endangered in parts of its range and subject to genetic impoverishment throughout	Excessive exploitation for the world timber market and poor natural regeneration
	Piptadenia peregrina	Southern Brazil	Threatened	Clearance for agriculture and cattle rearing
	Plathymenia foliosa	Brazil	Suffering a slow decline	Selective exploitation
	Prosopis cineraria	Arabia to India	Endangered in parts of its range	Increasing human pressure and changing land-use patterns
	Pterogyne nitens	Argentina, Brazil and Paraguay	Becoming rare	Timber exploitation
	Stuhlmannia moavi	Tanzania	Endangered	Settlement and forest clearance
Meliaceae	*Cedrela fissilis*	Costa Rica to Argentina	The best phenotypes of many provenances have mostly disappeared	Over-exploitation
	Cedrela odorata	Mexico to Argentina and Caribbean Is	Endangered in parts of its range	Over-exploitation and clearance of lowland forest
	Entandrophragma angolense	Tropical Africa	Threatened in parts of W. Africa, severe genetic erosion noted in Nigeria	Commercial exploitation
	Khaya senegalensis	Tropical Africa	Populations of best trees are in danger of genetic erosion	Exploitation for timber
	Lovoa swynnertonii	Tropical Africa	Rare	Forest clearance, excessive exploitation, poor natural regeneration
Moraceae	*Milicia excelsa*	Tropical and sub-tropical Africa	Widespread but threatened in some areas	Extensive logging
Myrtàceae	*Eucalyptus deglupta*	Philippines, Irian Jaya and PNG	Endangered in parts of its range	Limited regeneration, logging and clearance for agriculture
	Eucalyptus globulus ssp. globulus	Southern Australia and Tasmania	Threatened	Development of pine plantations
Pinaceae	*Abies guatemalensis*	Mexico to El Salvador	Extremely rare and threatened with extinction	Illegal felling of small Christmas trees
	Abies nebrodensis	Sicily	Endangered; about 20 wild trees remain	Exploitation and increasing human pressure
	Abies numidica	Algeria	In danger of slow decline	Grazing, local use of wood
	Cedrus libani	Lebanon and Turkey	Threatened in Lebanon	Exploitation and grazing

Table 25.3 Endangered tree species and provenances (continued)

FAMILY	SPECIES	DISTRIBUTION	STATUS	THREATS
	Pinus armandii var. amamiana	Japan	Endangered	Lumbering
Pinaceae (continued)	*Pinus eldarica*	USSR, Afghanistan and Pakistan	Endangered in the USSR	A relic species
	Pinus koraiensis	Japan and the Korean Peninsula	In danger of depletion in parts of its range	Logging
	Pinus patula ssp tecunumanii	Central America	Under threat throughout its entire range	Clearance for agriculture and attacks by Bark beetle
	Pinus pentaphylla	Japan and island of Ullung-do (Korea)	In danger of depletion in parts of its range	Logging operations
	Pinus pseudostrobus	Central America	Some provenances are endangered	Selective logging
	Pinus radiata	California and Mexico	5 populations are known; 1 is endangered and the genetic integrity of 2 others is endangered	Grazing, urbanisation and contamination by cultivated stock
	Pseudotsuga gaussenii	Eastern China	Endangered	
	Pseudotsuga sinensis	China	Naturally rare	
Platanaceae	*Platanus orientalis*	E Mediterranean to the Himalayas	Endangered in parts of its range	Agricultural expansion and modification of the water table through irrigation
Rutaceae	*Balfourodendron riedelianum*	Brazil, Paraguay and Argentina	Becoming scarce	Habitat destruction and exploitation
	Esenbeckia leiocarpa	Brazil and Zaire	Threatened in Brazil	Forest clearance and commercial felling
	Vepris glandulosa	Kenya	Endangered	Settlement and forest clearance
Salicaceae	*Populus ilicifolia*	Kenya	Endangered	Habitat clearance
Simaroubaceae	*Gymnostemon zaizou*	Cote d'Ivoire	Restricted distribution	
Sterculiaceae	*Nesogordonia papaverifera*	West Africa	Endangered in parts of its range and subject to genetic impoverishment in outlying populations	Logging
Taxodiaceae	*Glyptostrobus lineatus*	Widely cultivated in parts of China, not known in the wild	Natural populations are extinct	
	Taiwania cryptomerioides	Taiwan and possibly Myanmar	Endangerd in certain areas	Large scale clear-cuttings
	Taiwania flousiana	Myanmar, Tibet and Yunnan	Endangered	
Ulmaceae	*Ulmus wallichiana*	Afghanistan to Nepal	Endangered	Use as fodder
Verbenaceae	*Tectona hamiltoniana*	Myanmar	Likely to be endangered	Local use for fuel and construction and forest fires
	Tectona philippinensis	Philippines	Likely to be endangered	Naturally rare and sought after for general construction

Source: FAO 1986. *Databook on Endangered Tree and Shrub Species and Provenances.* FAO Forestry Paper 77. FAO, Rome.

Table 25.4 Commercial timber species of conservation concern in Ghana

FAMILY	SPECIES	TRADE NAME	NO. OF TREES PER km^2	EXPORT OF LUMBER IN 1989	CONSERVATION AND LEGAL STATUS
Combretaceae	*Terminalia ivorensis*	Emire	6	4697	Priority for *in situ* and *ex situ* conservation. Threatened by over-exploitation (FAO, 1984).
	Terminalia superba	Ofram	45	32	Priority for *in situ* and *ex situ* conservation. Threatened by over-exploitation (FAO, 1984).
Leguminosae	*Guibourtia ehie*	Anokye-hyedua	3	0	ERL 18 years. Log export ban.
	Pericopsis elata	Kokrodua (Afrormosia)	2	2204	Threatened by over-exploitation (FAO, 1984). Vulnerable. ERL 0 years. Log export ban.
Meliaceae	*Entandrophragma angolense*	Edinam	13	5155	Priority for *in situ* conservation. ERL 18 years. Log export ban.
	Entandrophragma candollei	Penkwa-akoa (Candollei,Omu)	2	807	Log export ban.
	Entandrophragma cylindricum	Penkwa (Sapele)	12	6749	Priority for *in situ* conservation. ERL 25 years. Log export ban
	Entandrophragma utile	Efoobrodedwo (Utile)	5	7063	Priority for *in situ* conservation. Threatened by over-exploitation (FAO,1984). ERL 20 years. Log export ban.
	Guarea cedrata	Kwabohoro (Guarea)	4	531	Priority for *in situ* conservation. ERL Guarea spp. 82 years.
	Guarea thompsonii	Kwadwuma (Black Guarea)	1	0	Priority for *in situ* conservation. ERL Guarea spp. 82 years.
	Khaya anthotheca/ grandifoliola	Krumben/Kruba (Ahafo)	3		Priority for *in situ* and *ex situ* conservation. ERL *Khaya* spp. 20 years. Log export ban.
	Khaya ivorensis	Dubini (Mahogany)	18	10463	Suffering from over-logging in Ghana. Priority for *in situ* and *ex situ* conservation.
	Lovoa trichilioides	Dubinibiri (Walnut)	4	854	Priority for *in situ* and *ex situ* conservation. Threatened by over-exploitation (FAO,1984) Log export ban.
	Turraeanthus africanus	Apapaye (Avodire)	8	55	Priority for *in situ* conservation Log export ban.
Moraceae	*Milicia excelsa*	Odum		37747	Priority for *in situ* conservation. Vulnerable. ERL 10 years. Log export ban.
	Milicia regia			combined with above	Priority for *in situ* conservation. Vulnerable. Log export ban.
Rubiaceae	*Hallea ledermannii/ H. stipulosa*	Subaha (Abura)	1	95	*H. ledermannii* threatened by over-exploitation (FAO, 1984).
	Nauclea didderichii	Kusia (Opepe)	5	674	Threatened by over-exploitation (FAO, 1984).
Sapotaceae	*Aningeria robusta*	Samfena (Aniegre, Asanfona)	14		Log export ban.
	Tieghemella heckelii	Baku (Makore)	4	5668	Log export ban.
Sterculiaceae	*Mansonia altissima*	Oprono (Mansonia)	4	778	Priority for *in situ* conservation. Log export ban.
	Heritiera utilis	Nyankom (Niangon)	6	1624	Log export ban.
	Nesogordonia papaverifera	Danta	11	428	Log export ban.
	Triplochiton scleroxylon	Wawa	152	64818	Priority for *in situ* and *ex situ* conservation. Threatened by over-exploitation (FAO,1984).

Source: Compiled from multiple sources.

shows the species that are subject to a Ghanaian log export ban. In addition to the log export ban, the Ghanaian Government introduced Forest Improvement Levies in November 1990. The highest rate of 50% is charged on *Nauclea diderrichii*, 40% on *Guarea cedrata* and three other species, and 10% on *Triplochiton scleroxylon* and one other species. Levies on green/air-dried sawnwood were: 50% on *Pericopsis elata, Entandrophragma utile, Guibortia ehie* and *Tieghemella heckelii* and 8.5% on *Milicia excelsa* and *M. regia*. The Government plans to follow up the levies by introducing a ban on exports of green sawn timber in January 1994.

In addition to the species that have suffered genetic erosion because of heavy exploitation, there are many rare timber species in Ghana that are not currently recorded in the timber export trade. A recent field guide to the forest trees (Hawthorne, 1990), for example, notes 27 timber species as being uncommon, rare, or very rare. Some of these are of more immediate conservation concern than the major commercial species. Overall the most serious threat to tree species in Ghana comes from fire damage, which has severely undermined the regeneration of trees even in the most productive moist semi-deciduous forest zone.

RATTANS

After timber, rattans (lianoid palms) provide the second most important source of export earnings from tropical forests. Most of the 600 or so species are native to South and Southeast Asia. Countries with major rattan industries include the Philippines, China, Indonesia, India, Sri Lanka and Thailand, and these provide full-time employment for at least half a million people. For the international market, rattans are mainly used in the production of cane furniture. Local uses include the production of mats, baskets, fish traps, dyes and medicines.

The rattan industry relies almost entirely on wild stocks. About 90% of the world's raw material supply is extracted from the wild and the remaining 10% from plantations in Central and South Kalimantan. Exploitation combined with habitat destruction has led to the decline of major commercial rattan species and species that are valuable in local use and local markets. Table 25.5 lists the major commercial rattan species with notes on their conservation status.

Indonesia is the world's main producer of rattans, supplying about 90% of the total raw material utilised. Export of raw rattan from the country has been banned since 1979. There has been relatively little downstream processing of rattans into finished products within Indonesia and, in an attempt to boost local value-added production, a ban on export of non-finished products was introduced in 1989. This has led to concern about increasing commercial pressure on wild stocks elsewhere.

The centre of diversity for rattans is the Malay Peninsula. A total of 104 species occur within this area, of which about 38% are endemic. Of these Malay Peninsula species, only two are considered to be not threatened and 98 are categorised as Vulnerable or Endangered (Kiew and Dransfield, 1987). Research has begun on the taxonomy and silviculture of Malaysian rattans as a prelude to bringing these into cultivation. In the meantime, it is uncertain how many of the 104 species in Peninsular Malaysia occur within the State's existing national park (Taman Negara). Illegal removal of commercial species remains a threat within the protected area. *Ex situ* conservation of rattan species in seed banks is not currently a viable proposition because rattans have recalcitrant seeds. The most attractive form of genetic conservation for rattans in Peninsular Malaysia will probably be through their cultivation in logged-over hill dipterocarp forests.

MEDICINAL PLANTS

Around 119 pure chemical substances extracted from some 90 species of higher plants are used in medicines throughout the world. At a local level an extremely wide range of plant species is used medicinally. The World Health Organization has listed over 21,000 plant names (including synonyms) that have reported medical uses around the world. Very few of these medicinal plants have been subject to scientific scrutiny. In all about 5,000 higher plant species have been thoroughly investigated as potential sources of new drugs. Most of these are temperate species and the biochemical potential of tropical plants has been largely overlooked. Nevertheless around 80% of people in developing countries rely on traditional medicines. Table 25.6 shows some of the most important plant species whose derivatives are used in orthodox medicine along with an indication of whether analogous uses have been reported in traditional medicine.

Medicinal plant species are still to a large extent harvested from the wild and relatively few are cultivated as crop plants. For example in Germany two-thirds of the species used are still wild collected and cultivation of major medicinal plants such as *Gentiana lutea, Valeriana mexicana, Echinacea* and *Arnica* has only begun in the past 20 years. Plant breeding has only taken place with the commercially most important plants such as *Papaver somniferum, Papaver bracteatum, Cinchone* sp., *Chamomilla recutita* and *Mentha piperita* (Schumacher, 1991). In many cases, biochemicals extracted from plants have been used as blueprints for the synthesis of drugs and the natural source material is no longer required. Nevertheless, the USA annually imports over US$20 million worth of rain forest plants for medicinal purposes. Important drugs include tubocuranin, derived from plant-based curare and used as a muscle relaxant during surgery, and curianol, a Guyanese fish poison used in heart operations. Economic aspects of the production of pharmaceuticals from plants are discussed in Chapter 27.

The US National Cancer Institute has identified over 1,400 tropical forest plants with the potential to fight cancer. One such plant is the Rosy Periwinkle *Catharanthus roseus* native to Madagascar. Used for generations by tribal healers, this species is now used in the production of drugs effective against Hodgkins disease and other forms of cancer. The Rosy Periwinkle yields vinca alkaloids, which are complex molecules difficult to synthesise chemically. It remains cheaper to collect leaves of living plants for extraction of the valuable medicinal products. *Catharanthus roseus* is now a widespread weedy species in the tropics and is commonly cultivated. All other species of the genus are

Table 25.5 Main commercial species of Rattan (Palmae: *Calamus*)

SPECIES	RANGE	STATUS AND THREATS
Calamus caesius Blume	Malay Peninsula, Borneo, Sumatra, Philippines (Palawan); Thailand (possibly introduced)	Domesticated in Kalimantan. Supply of wild stocks threatened by over-exploitation
Calamus diepenhorstii Miq.	Malay Peninsula, Singapore, Sumatra, Borneo (Sabah), Philippines (Palawan)	
Calamus manan Miq	Malay Peninsula, Borneo, Sumatra, south Thailand	Threatened: viable populations largely limited to a few inaccessible areas as a result of excessive and premature exploitation
Calamus maximus Merr.	Philippines (Basilan, Luzon, Mindanao, Mindoro)	Any accessible populations have been exploited; but the species clusters and so is not as vulnerable to over-exploitation as is (say) *C. manan*. However, there has been extensive habitat destruction by logging, shifting cultivation and spontaneous settlement
Calamus mindorensis Becc.	Philippines (Luzon, Mindoro)	*
Calamus optimus Becc.	Borneo	Endangered: a rare and much sought after species; so much so that it is very difficult to find mature long canes even in Mulu National Park (Sarawak)
Calamus ornatus Bl.	Malay Peninsula, Borneo, Sumatra, Sulawesi, south Thailand, Philippines (Luzon, Mindanao, Mindoro, Negros, Palawan, Polilo)	*
Calamus peregrinus Furtado	Malay Peninsula, Thailand	*
Calamus rudentum Lour.	Thailand, Laos, Cambodia, Viet Nam	Extensive habitat destruction
Calamus scipionum Lour.	Malay Peninsula, Singapore, Borneo, Sumatra, Philippines (Palawan)	*
Calamus subinermis H. Wendl.	Sabah	Present stocks are limited and their exploitation requires strict control
Calamus trachycoleus Becc.	Kalimantan	Domesticated in Kalimantan
Calamus tumidus Furtado	Malay Peninsula, Sumatra	Largest known populations threatened by agriculture
Calamus zollingeri Becc.	Sulawesi	*

Source: Dransfield, J. 1979a. *A Manual of the Rattans of the Malay Peninsula.* Malayan Forest Records 29. Forest Department. Ministry of Primary Industries, Malaysia. Dransfield, J. 1979b. *Report of Consultancy on Rattan Development carried out in Thailand, Philippines, Indonesia and Malaysia*. 14 March-8 May 1979. For FAO Regional Office for Asia and the Far East, Bangkok. Dransfield, J. 1981. The biology of Asiatic rattans in relation to the rattan trade and conservation. In: Synge, H. (Ed.), *The Biological Aspects of Rare Plant Conservation*. Wiley, Chichester. Dransfield pers. comm. 18 February 1981.
Notes: * No information.

endemic to Madagascar, where several are used medicinally. One species which has not been tested phytochemically is close to extinction.

Regions that are known to have important concentrations of major medicinal plants include Mexico and Central America, the west-central region of South America (Colombia, Ecuador and Peru), the Indian subcontinent, west Asia and parts of north-eastern Africa. Over-exploitation of medicinal plants extracted from the wild is leading to problems of genetic erosion in some of these regions. In India, for example, where 2,500 plant species are used by traditional healers, species of *Aconitum*, *Dioscorea* and *Ephedra* are some of the medicinal plants under threat in the wild.

Dioscorea deltoidea, a species that grows in the Himalayan foothills of northern India, is a major source of diosgenin used in the manufacture of contraceptive pills. Over-collection has led to the decline of this species in the wild and it is now subject to international trade controls. The remaining small specimens of the plant yield less than 15% of the diosgenin found in the large, old tubers which have mostly been removed from the wild.

Also threatened in India is *Rauvolfia serpentina*, a forest shrub known as serpentine root. This has been used in traditional medicine for 4,000 years to treat snakebite, nervous disorders, dysentery, cholera and fever. An extract from the plant, reserpine, became the principal source of materials for modern tranquilisers following research around 50 years ago. In Thailand, *Rauvolfia serpentina* is collected both for use in local medicine and for sale, via middlemen, to national and international pharmaceutical companies.

Table 25.6 Principal plant species with constituent compounds used as drugs

PLANT NAME	COMPOUND NAME	THERAPEUTIC CATEGORY IN MEDICAL SCIENCE	PLANT USES IN TRADITIONAL MEDICINE	CORRELATION BETWEEN TWO USES	COUNTRY OF PRODUCTION OR CULTIVATION
Ammi spp.	Xanthotoxin	Pigmenting agent	Leukoderma; Vitiligo	Yes	Cultivated in Asia and the Mediterranean region
	Khellin	Bronchodilator	Asthma	Yes	
Atropa belladonna	Atropine	Anticholinergic	Dilate pupil of eye	Yes	Central and Southern Europe, cultivated in USA, UK, Eastern India, Europe, China
Berberis vulgaris	Berberine	Antibacterial	Gastric ailments	Yes	Europe, Asia
Carica papaya	Chymopapain	Proteolytic; mucolytic	Digestant	Yes	Cultivated in Sri Lanka, Zaire, Uganda, Mozambique, Tanzania, South Africa, India
	Papain	Proteolytic; mucolytic	Digestant	Yes	
Cassia spp.	Danthron	Laxative	Laxative	Yes	*C. acutifolia* cultivated in India *C. senna* cultivated in Egypt
Catharanthus roseus	Vinblastine	Antitumor agent	Not used	No	Pantropical, cultivated in US, India and other countries
	Vincristine	Antitumor agent	Not used	No	
Cephaelis ipecacuanha	Emetine	Amebicide; emetic	Amebicide; emetic	Yes	Brazil, much collected in Mato Grosso
Cinchona ledgeriana	Quinidine	Antiarrhythmic	Malaria	No	Cultivated in Indonesia, Zaire, Tanzania, Burundi, India, Kenya, Guatemala, Peru, Ecuador, Bolivia, Rwanda, Sri Lanka, Colombia, Costa Rica
	Quinine	Antimalarial; antipyretic	Malaria	Yes	
Datura matel	Scopolamine	Sedative	Sedative	Yes	Cultivated in Asia
Digitalis spp.	Acetyldigitoxin	Cardiotonic	Not used	Indirect	*D. lanata* Cultivated in southern Europe and Asia, *D. purpurea* cultivated in India and temperate zones
	Deslanoside	Cardiotonic	Not used	Indirect	
	Digitoxin	Cardiotonic	Cardiotonic	Yes	
	Lanatosides	Cardiotonic	Not used	Indirect	
	Digitalin	Cardiotonic	Cardiotonic	Yes	
	Gitalin	Cardiotonic	Cardiotonic	Yes	
Ephedra sinica	Ephedrine	Sympathomimetic	Chronic bronchitis	Yes	China
	Pseudoephedrine*	Bronchodilator	Chronic bronchitis	Yes	
Glycyrrhiza glabra	Glycyrrhizin (Glycyrrhetic acid)	Sweetener	Sweetener	Yes	Cultivated in Spain, Turkey, Iraq, China, Mongolia, USSR, South Africa, USA, France, Italy, Iran, Afghanistan, Syria, Lebanon, Israel, UK, China
Hyoscyamus niger	Hyoscyamine	Anticholinergic	Sedative	Yes	Cultivated in temperate zones
Papaver somniferum	Codeine	Analgesic; antitussive	Analgesic; sedative	Yes	Cultivated in Turkey, India, Burma, Thailand
	Morphine	Analgesic	Analgesic; sedative	Yes	
	Noscapine (narcotine)	Antitussive	Analgesic; sedative	Yes	
	Papaverine*	Smooth muscle relaxant	Sedative; analgesic	No	
Pausinystalia yohimbe	Yohimbine	Adrenergic blocker; aphrodisiac	Aphrodisiac	Yes	Cameroon, Nigeria, Rwanda
Physostigma venenosum	Physostigmine (eserine)	Anticholinesterase	Ordeal poison	Indirect	Sierra Leone, Cameroon, introduced to India and Brazil
Pilocarpus jaborandi	Pilocarpine	Parasympathomimetic	Poison	Indirect	Tropical America especially Brazil
Rauvolfia spp.	Ajmalicine	Circulatory stimulant	Tranquilizer	Indirect	e.g. *R. serpintina*; Thailand, Zaire, India, Bangledash, Sri Lanka, Burma, Malaysia, Indonesia, Nepal
	Rescinnamine	Antihypertensive; tranquilizer	Tranquilizer	Yes	
	Reserpine	Antihypertensive; tranquilizer	Tranquilizer	Yes	
Silybum marianum	Silymarin	Antihepatotoxic	Liver disorders	Yes	Mediterranean region
Urginea maritima	Scillaren A	Cardiotonic	Cardiotonic	Yes	Mediterranean region, Egypt, Turkey
Valeriana officinalis	Valepotriates	Sedative	Sedative	Yes	Cultivated in Eastern Europe, Netherlands, Japan

Note: * Also now synthesised commercially. **Source:** Farnsworth, N.R. 1988. Screening plants for new medicines. In: Wilson, E.O. (Ed.), *Biodiversity*. National Academy Press, Washington. Pp.83-97; Farnsworth, N.R. and Soejarto, D.D. 1991. Global importance of medicinal plants. In: Akerele, O., Heywood, V. and Synge, H. (Eds), *The Conservation of Medicinal Plants*. Proceedings of an International Consultation 21-27 March 1988. Chiang Mai, Thailand. Cambridge University Press, Cambridge, UK.

In many parts of the tropical world there is a serious lack of knowledge about the genetic resources and conservation status of the medicinal plants on which most people rely. In Malaysia, for example, the genetic resources of medicinal plants have scarcely been evaluated. At present although some medicinal plants are cultivated on a small scale, most herb traders and local medicine men rely on wild resources. There is some concern that collectors do not know the status of individual species and may contribute to the loss of populations of threatened species. Clearance of Malaysian forests is likely to lead to the loss of medicinal plant species in families such as Apocynaceae, Annonaceae, Rutaceae, Dioscoreaceae, Leguminosae, Lauraceae and Zingiberaceae. Unlike the fruit tree resources, which have survived under semi-wild cultivation in village orchards, medicinal plant resources will be lost from forest areas.

In Thailand most of the plants used in rural medicines are collected from forests. Many of these have commercial value as raw materials used by over 1,000 traditional drug manufacturing companies, modern drug companies and in export as crude extracts. Thailand is expanding its domestic production of herbal medicines but increasingly depends on imported raw materials as local resources are lost.

The use of plant resources in Indonesia for the production of 'jamu' (herbal medicine) has generally been sustainable for home consumption. Increasing commercialisation is, however, putting pressure on wild populations of medicinal plants. Modern jamu industries and pharmaceutical companies are using large quantities of plant materials and this is leading to genetic erosion of species that are not in cultivation. Species that are being affected include *Curcuma* spp., which were previously abundant in East Java but now have to be imported from other islands.

Another species, *Parkia roxburghii*, has also become rare in parts of East Java, owing to excessive harvesting of the fruits. Increasing international demand for *Curcuma* spp., and others such as *Voacanga gradifolia, Orthosiphon aristasus* and *Rauvolfia*, is leading to the loss of economically valuable plant genetic resources that could provide for a sustainable source of foreign exchange if brought into cultivation.

One of the few medicinal plants which has been developed as a major crop species is quinine *Cinchona* spp. The main use of quinine, extracted from bark of the *Cinchona* trees, is in anti-malarial drugs. Initially the whole world supply came from wild trees in the Andes. This led to concern about the possible extinction of the species and as a result, plantations were developed in the middle of the last century, for example in India and Indonesia. Synthetic alternatives are available but quinine remains an important drug with new applications being found. The genetic base of the crop is very narrow and conservation of wild stands of *Cinchona* is important for future breeding work.

ORNAMENTAL PLANTS

The discovery, domestication and cultivation of ornamental plants have a long history, comparable to that of food crops. Lilies, for example, have been cultivated in China for both medicinal and decorative purposes for around two thousand years. In Roman times, roses, lilies, violets, anemones, narcissi and lavender were grown as garden plants in Europe. Today, the diversity of decorative plant species established in cultivation far surpasses the variety of plants commonly grown for food around the world. In the UK alone, an estimated 3,000 species are in general cultivation in addition to the wide range of cultivars and hybrids. At least five times as many species have been introduced at various times in the past. Novelty and variety remain important factors in the horticultural market.

Ornamental plants are an important commodity in international trade, with an expanding international market. Total world imports of cut flowers, cut foliage and plants amounted to US$2,488 million in 1985. The value of world trade in cut flowers and live plants for 1981 to 1985 is shown in Table 25.7. It should be stressed that virtually all this value is in artificially-propagated stock.

Despite the economic importance of ornamental plants, the conservation of these genetic resources is usually given a low priority both nationally and internationally when compared to food, fruit and forage crops (Chin, 1989). Wild species of horticultural value are under threat around the world, both through the processes of habitat destruction and through direct exploitation for local use and international trade.

Although sophisticated propagation techniques have been developed for ornamental plants, significant quantities of plants in some groups continue to be dug from the wild for the world market. This is apparent for example with bulbs, orchids, cacti and other succulent plants, cycads and insectivorous plant species. Concern about the level of exploitation of some of these plants has led to their listing on the Appendices of CITES. As a result, data on levels of international trade in both wild-collected and artificially propagated plants have been recorded. Summary figures for cactus and orchid trade for 1989 are given in Table 25.8.

Orchids

Over 5,000 orchid species have been recorded in CITES trade statistics during the period 1983-1989, with the average annual number of plants in international trade being nearly five million. This figure excludes orchids in flasks and cut flowers. Around 80% of the orchids in trade are reported to be artificially propagated and most trade is in artificially propagated hybrids. There is still, however, significant international demand for species orchids and a large part of this trade is satisfied through the collection and export of wild plants. The major source country for orchids in international trade is Thailand. Orchids are propagated in commercial nurseries within the country but, at the same time, there is a huge trade in both native and imported wild orchids.

The most heavily traded orchid genus exported from Thailand is *Dendrobium*. The numbers of native species exported together with the levels of plants reported to be propagated are shown in Table 25.9.

Table 25.7 Value of world trade in flowers and plants, 1981–1985, (US$ millions)

CUT FLOWERS					
IMPORTS	ANNUAL MEAN	%	EXPORTS	ANNUAL MEAN	%
	1238.79			1101.79	
EEC			Netherlands	701.51	63.67
Germany, Fed Rep	535.54	43.23	Colombia	121.68	11.04
France	85.55	6.91	Israel	73.31	6.65
United Kingdom	70.27	5.67	Italy	73.35	6.66
Netherlands	53.82	4.34	Spain	17.96	1.63
Italy	25.49	2.06	Thailand	16.65	1.51
Belgium–Luxemburg	29.33	2.37	France	14.03	1.27
Denmark	11.28	0.91	Kenya	7.60	0.69
Ireland	3.48	0.28	Taiwan Province (China)	6.28	0.57
Greece	0.41	0.03	Germany, Fed Rep	5.59	0.51
United States	206.90	16.70	United States	9.92	0.90
Canada	24.17	1.95	South Africa	6.17	0.56
Japan	18.08	1.46	Singapore	6.83	0.62
Switzerland	63.14	5.10	United Kingdom	4.36	0.40
Austria	37.81	3.05	Peru	3.18	0.29
Sweden	30.19	2.44	New Zealand	3.03	0.28
Norway	12.05	0.97	Mexico	3.49	0.32
Singapore	5.07	0.41	Costa Rica	1.29	0.12
Finland	6.38	0.52	Brazil	2.31	0.21
Hong Kong	5.21	0.42	Ethiopia	0.86	0.08
Saudi Arabia	3.58	0.29	Morocco	1.11	0.10
Australia	2.65	0.21	Malaysia	1.34	0.12
Kuwait	1.16	0.09	Mauritius	0.70	0.06
United Arab Emirates	1.01	0.08			
Spain	0.56	0.05			
LIVE PLANTS					
IMPORTS	**915.76**			**882.15**	
EEC			Netherlands	389.70	44.18
Germany, Fed Rep	228.64	24.97	Denmark	123.77	14.03
France	116.20	12.69	Belgium–Luxemburg	92.62	10.50
United Kingdom	80.00	8.74	Germany, Fed Rep	51.86	5.88
Italy	57.14	6.24	France	37.30	4.23
Netherlands	50.99	5.57	United States	33.64	3.81
Belgium–Luxemburg	41.72	4.56	Italy	26.94	3.05
Denmark	13.94	1.52	Canada	18.16	2.06
Greece	4.36	0.48	Spain	16.08	1.82
Ireland	3.99	0.44	Guatemala	10.00	1.13
United States	37.67	4.11	Costa Rica	8.31	0.94
Canada	43.69	4.77	Japan	6.20	0.70
Japan	7.15	0.78	Israel	5.32	0.60
Sweden	60.03	6.56	United Kingdom	4.39	0.50
Switzerland	41.91	4.58	Côte d'Ivoire	4.15	0.47
Austria	20.99	2.29	New Zealand	2.49	0.28
USSR	21.91	2.39	Honduras	3.36	0.38
Spain	16.52	1.80	Hungary	1.89	0.21
Finland	14.07	1.54	Brazil	1.77	0.20
Norway	12.24	1.34	Singapore	2.08	0.24
Saudi Arabia	7.65	0.84	Malaysia	0.73	0.08
Colombia	4.20	0.46	Egypt	1.41	0.16
Hong Kong	2.77	0.30	Jamaica	1.05	0.12
Algeria	2.23	0.24	Thailand	0.78	0.09
Singapore	1.82	0.20	Colombia	1.56	0.18
			Turkey	0.79	0.09

Source: International Trade Centre UNCTAD/GATT. 1987. *Floricultural products: a study of major markets*. Genera.

The impact of orchid collection within Thailand has been highly detrimental. The conservation status of native species is scarcely known, but it is apparent that some species have been virtually eradicated even within national parks where collection is banned. Increasing prices reflect the increased scarcity of desirable species.

Orchid collection, together with habitat destruction, has led to the decline of wild orchid species in many other countries. In Japan, for example, more than 70 orchid taxa are included in the Japanese Plant Red Data List, of which 50 are threatened by over-collection. The genus *Calanthe* has been particularly popular with collectors and is under great pressure in the wild. Other genera which have been seriously over-collected include the Asian slipper orchids in the genus *Paphiopedilum* and the Latin American slipper orchids in the genus *Phragmipedium*. Both these genera are now included in Appendix I of CITES which effectively bans commercial international trade in wild-collected specimens.

Table 25.8 Cactus and orchid trade data for 1989

	CACTI		ORCHIDS	
	IMPORTS[1]	EXPORTS[1]	IMPORTS[1]	EXPORTS[1]
WORLD	6513647	6513647	8313088	8313088
ASIA	209716	3749060	6017522	7133797
Brunei	0	0	2388	0
China	200	0	0	165505
Cyprus	17801	0	356	0
Hong Kong	45313	0	114841	0
India	3	0	0	8423
Indonesia	6	0	4888	0
Israel	0	0	1660	0
Japan	0	1708096	5509995	0
Korea, Dem People's Rep	0	0	1100	0
Korea, Rep	0	2040964	381027	0
Macau	0	0	21	0
Malaysia	9000	0	0	22117
Nepal	1105	0	0	16
Pakistan	3	0	0	0
Philippines	0	0	0	50971
Saudi Arabia	1371	0	40	0
Singapore	56204	0	0	27275
Sri Lanka	0	0	0	20283
Syria	0	0	5	0
Taiwan	71464	0	0	1594732
Thailand	4	0	0	5244450
Turkey	330	0	52	0
United Arab Emirates	6912	0	1149	0
Viet Nam	0	0	0	25
USSR*	0	0	0	1499
EUROPE	3038304	100906	1671781	984041
Austria	195652	0	67590	0
Belgium	0	6494	8077	0
Czechoslovakia	0	433	2427	0
Denmark	6533	0	1753	0
Finland	44136	0	4942	0
France	0	70991	9600	0
German Dem Rep	2	0	2519	0
Germany, Fed Rep	189264	0	1138720	0
Hungary	0	0	4732	0
Iceland	0	0	14	0
Ireland	0	0	25	0
Italy	28138	0	14841	0
Luxembourg	0	0	456	0
Malta	258	0	388	0
Monaco	0	3	0	0
Netherlands	1991664	0	305494	0
Norway	12956	0	131	0
Poland	0	0	0	24450
Portugal	0	0	110	0
Spain	0	22985	154	0
Sweden	165480		14431	0
Switzerland	382916	0	95377	0
United Kingdom	18153	0	0	959591
Yugoslavia	3152	0	0	0
NORTH & CENTRAL AMERICA	3204234	1462004	408654	40126
Antigua	0	0	81	0
Aruba	0	8	203	0
Bahamas	1054	0	1683	0
Barbados	1594	0	259	0
Belize	0	6	0	4387
Bermuda	0	0	4098	0
Canada	0	1032492	118474	0
Cayman Islands	0	0	978	0
Costa Rica	0	0	3499	0
Dominica	0	0	0	180
Dominican Rep	0	347857	692	0
El Salvador	0	0	197	0
Guadeloupe	0	0	45	0
Guatemala	0	4	0	69
Haiti	0	3739	734	0
Honduras	0	1	0	13205
Jamaica	0	1	0	1419
Martinique	0	0	576	0
Mexico	0	77896	0	20716
Montserrat	0	0	760	0
Netherlands Antilles	0	0	1	0
Nicaragua	0	0	0	150
Panama	300	0	15982	0
Puerto Rico	0	0	221	0
St Lucia	300	0	318	0
Trinidad and Tobago	1495	0	11494	0
United States	3199272	0	248359	0
Virgin Islands (British)	219	0	0	0
SOUTH AMERICA	17	1127863	5203	124525
Argentina	0	396	9	0
Bolivia	0	0	20	0
Brazil	0	1127181	0	93426
Chile	0	59	68	0
Colombia	0	17	0	8965

Table 25.8 Cactus and orchid trade data for 1989 (continued)

	CACTI		ORCHIDS	
	IMPORTS[1]	EXPORTS[1]	IMPORTS[1]	EXPORTS[1]
SOUTH AMERICA (continued)				
Ecuador	0	0	0	795
Guyana	0	0	5030	0
Paraguay	0	7	0	872
Peru	0	186	0	18078
Suriname	0	0	76	0
Uruguay	17	0	0	0
Venezuela	0	17	0	2389
OCEANIA	**60327**	**0**	**138112**	**11023**
Australia	56632	0	71134	0
Fiji	0	0	23837	0
French Polynesia	0	0	41286	0
New Caledonia	0	0	1811	0
New Zealand	3695	0	0	8149
Palau	0	0	44	0
Papua New Guinea	0	0	0	2668
Vanuatu	0	0	0	206
AFRICA	**359**	**73807**	**56273**	**18040**
Botswana	0	0	21	0
Cameroon	76	0	46	0
Comoros	0	0	0	50
Côte d'Ivoire	0	0	0	526
Ethiopia	0	0	0	66
Gabon	0	0	1017	0
Ghana	0	0	0	2
Kenya	0	0	0	4819
Liberia	0	0	0	6
Madagascar	0	3	0	12459
Mauritius	0	0	1271	0
Morocco	0	51607	41	0
Mozambique	0	0	50	0
Nigeria	0	0	0	9
Reunion	60	0	4718	0
South Africa	0	22197	49072	0
Togo	2	0	0	0
Tunisia	0	0	8	0
Zaire	0	0	14	0
Zambia	0	0	15	0
Zimbabwe	221	0	0	103
OTHER	**690**	**7**	**15543**	**37**
Country Unknown	690	0	15527	0
Other	0	7	16	37

Notes: [1] Figures are net. * Former USSR.
Sources: Annual reports of Parties to CITES compiled by WCMC.

Table 25.9 *Dendrobium* Orchids from Thailand

SPECIES	AVERAGE NO. IN ANNUAL TRADE 1983-1989	% ARTIFICIALLY PROPAGATED
D. aphyllum #	740	11
D. bellatulum #	1526	19
D. chrysotoxum #	5110	25
*D. densiflorum **	777	30
D. draconis	985	13
D. farmeri #	2396	18
D. fimbriatum	464	16
D. harveyanum	1022	14
D. nobile	812	90
D. parishii	1960	45
D. scabrilingue #	1779	11
D. senile #	1160	30
D. thrysiflorum #	2959	27
D. unicum #	1175	4

Notes: # Known to be sold as wild-collected plants in Thailand.
* Doubtfully native to Thailand.
Source: Oldfield, S. 1991. Review of significant trade in species of plants included in Appendix II of CITES, 1983-1989. Report prepared for the 8th meeting of the Conference of Parties. Unpublished Report.

Cacti and other succulents

The average annual international trade in cacti as recorded in CITES statistics is close to 14 million. This is probably an underestimate of the real trade, because a single nursery in the Netherlands (which exports most of its production) produces over 18 million cacti annually and commercial cactus production in the USA has been estimated at 10-50 million per year. The bulk of cacti in international trade are propagated but collection continues to put pressure on certain desirable species which are close to extinction in the wild. Mexico, one of the main centres of diversity of the cactus family, exports around 50,000 cacti annually according to CITES figures. A high proportion of these are wild-collected and exported illegally.

A wide range of other succulent plants, including species of *Aloe*, *Euphorbia* and *Pachypodium*, are also traded internationally. One of the main source countries is Madagascar which has exported around 135,000 CITES-listed succulents annually, all of which are wild-collected. This trade poses a severe threat to Madagascar's unique plants and adds to the pressures of habitat destruction.

Table 25.10 Trade and conservation status of Turkish bulbous species

GENUS/SPECIES	CONSERVATION CATEGORY	EXPORT FIGURES (1987)	EXPORT CONTROLS	CULTIVATION IN TURKEY (WHERE KNOWN)
Allium roseum	V		B	
Anemone blanda	V	7,500,000	Q	
Arum spp.	V		Q	
Crocus spp.	V		B	
Cyclamen spp.	V	995,000	Q	
Cyclamen cilicium	V			
C. graceum	V			
C. hederifolium	V			
C. mirabile	V			
C. persicum	V			
C. repandum	I			wild-transplanted
Dracunculus spp.			Q	
Eranthis hyemalis	V	10,000,000	Q	artificial propagation
Fritillaria imperialis	E	275,000	B	artificial propagation
F. persica	E	275,000	B	wild-transplanted
Galanthus spp.	V	30,000,000 (*G. elwesii* and *G. ikariae*)	Q	
Hyacinthus orientalis orientalis	V		B	
Leucojum aestivum	V	8,500,000	Q	artificial propagation
Lilium candidum	E	1,335		wild-transplanted
L. martagon	E			artificial propagation
Muscari spp.			B/Q	
Narcissus spp.			Q	
N. serotinus	R			
Pancratium maritium	V		B	
Scilla spp.	R	100,000		
Sternbergia spp.		450,000 (*S. lutea* (V) and *S. clusiana*)		wild transplanted
Tulipa spp.	R			
T. praecox	E			
T. humilis			B	
Urginea maritima		37,000	Q	

Sources: Ekim, T., Koyuncu, M., Erik, S. and Darslan, R. 1989. *List of Rare, Threatened and Endemic Plants in Turkey, prepared according to IUCN Red Data Book categories.* Turkish Association for Conservation of Nature and Natural Resources, Ankara. Series No. 18. McGough, H.N., Mathew, B.F., Peter, H., Read, M., Wertel N. and Wijnands, O. 1989. A report on the status and cultivation of *Cyclamen* species and other geophytes in Turkey. Paper prepared for the Scientific Working Group of the EC CITES Committee.
Notes: E Endangered; V Vulnerable; R Rare; I Indeterminate; B Ban on export; Q Quota system for exports

Bulbs

Information on levels of international trade in wild bulbs is less readily available because most genera are not covered by CITES. Commercial cultivation of most bulbous genera is well-established but collection from the wild takes place routinely for certain so-called minor bulbs such as snowdrops (*Galanthus*) and *Cyclamen*. In general, it is difficult to assess the impact of collection on wild bulb populations but genetic erosion is a serious problem for species of horticulturally popular genera.

The daffodil genus *Narcissus* has around 40 species, with its centre of diversity in Spain and Portugal. Hundreds of daffodil cultivars and hybrids have been developed and daffodils are an important horticultural crop in various countries. The UK is the major exporter of daffodil bulbs, with five varieties dominating commercial production. In 1987 the UK exported nearly 87 million *Narcissus* bulbs with a value of over £4 million. At present there is very limited UK production of small *Narcissus* species which are becoming increasingly popular. One of the main sources of these species is Portugal where the bulbs are dug from the wild.

Ten taxa of *Narcissus* are considered to be threatened in Portugal and to be in need of protection. Several of these, including *N. asturiensis* and *N. cyclamineus*, are exported to the Netherlands for re-export around the world.

The main source country for wild-collected bulbs in international trade is Turkey. The country has a very rich bulbous plant flora and is the origin of many of the attractive bulbs in cultivation. Commercial exports of bulbs from Turkey are subject to licensing by the Turkish Government and official statistics are based on the quantities licensed. Turkish exports of the main commercial genera are given in Table 25.10 above.

References

Chin, H.F. 1989. Ornamental germplasm in Malaysia. In: Zakri, A.H. (Ed.), *Genetic Resources of Under-utilised Plants in Malaysia.* Malaysian National Committee on Plant Genetic Resources.

Davies, J.C. 1991. Global support and coordination: conserving germplasm of world crop species and their relatives. In: Hawkes, J.G. (Ed.), *Genetic Conservation of World Crop Plants.* Published for the Linnaean Society of London by Academic Press.

Dransfield, J. 1979a. *A Manual of the Rattans of the Malay Peninsula.* Malayan Forest Records 29. Forest Department. Ministry of Primary Industries, Malaysia. 270pp.

Dransfield, J. 1979b. *Report of Consultancy on Rattan Development carried out in Thailand, Philippines, Indonesia and Malaysia.* 14 March-8 May 1979. For FAO Regional Office for Asia and the Far East, Bangkok. 44pp.

Dransfield, J. 1981. The biology of Asiatic rattans in relation to the rattan trade and conservation. In: Synge, H. (Ed.), *The Biological Aspects of Rare Plant Conservation.* Wiley, Chichester. Pp.179-186.

Ekim, T., Koyuncu, M., Erik, S. and Darslan, R. 1989. *List of Rare, Threatened and Endemic Plants in Turkey, prepared according to IUCN Red Data Book categories.* Turkish Association for Conservation of Nature and Natural Resources, Ankara. Series No.18.

Engels, J.M.M., Hawkes, J.G. and Melaku Worede (Eds) 1991. *Plant Genetic Resources of Ethiopia.* Cambridge University Press, UK.

FAO 1984. *In Situ Conservation of Wild Plant Genetic Resources: a status review and action plan.* Forest Resources Division, FAO, Rome.

Hawkes, J.G. 1983. *The Diversity of Crop Plants.* Harvard University Press.

Hawkes, J.G. (Ed.) 1991. *Genetic Conservation of World Crop Plants.* Published for the Linnaean Society of London by Academic Press.

Hawthorne, W. 1990. *Field Guide to the Forest Trees of Ghana.* Natural Resources Institute, for the Overseas Development Administration, London. Ghana Forestry Series 1.

Hoyt, E. 1988. *Conserving the Wild Relatives of Crops.* IBPGR, IUCN and WWF.

Kiew, R. and Dransfield, J. 1987. The conservation of palms in Malaysia. *Malayan Naturalist* 41:24-31.

Lundborg, G. 1989. African plant genetic resources. *Development and Cooperation* 4:24-26.

McGough, H.N., Mathew, B.F., Peter, H., Read, M., Wertel N. and Wijnands, O. 1989. A report on the status and cultivation of *Cyclamen* species and other geophytes in Turkey. Paper prepared for the Scientific Working Group of the EC CITES Committee.

Oldfield, M.L. 1984. *The Value of Conserving Genetic Resources.* USDI National Park Service, Washington DC.

Olivier, L. 1991. The role of the national botanic conservancies and protected natural areas. In: *The Conservation of Wild Progenitors of Cultivated Plants.* Environmental Encounters Series, No.8. Council of Europe, Strasbourg.

Oxford Forestry Institute 1991. Pre-project report on incentives in producer and consumer countries to promote sustainable development of tropical forests.

Plucknett, D.L. 1987. *Gene Banks and the World's Food.* Princeton University Press, Princeton, New Jersey.

Prescott-Allen, C. and Prescott-Allen, R. 1990. How many plants feed the world? *Conservation Biology* 4(4):365-374.

Schumacher, H.M. 1991. Biotechnology in the production and conservation of medicinal plants. In: Akerele, O., Heywood, V. and Synge, H. (Eds), *The Conservation of Medicinal Plants.* Cambridge University Press, Cambridge. Pp.179-198.

Simmonds, N.W. (Ed.) 1976. *Evolution of Crop Plants* Longman Scientific and Technical.

Valdes, B. 1991. Phytotaxonomical studies for the investigation of species and their distribution. In: *The Conservation of Wild Progenitors of Cultivated Plants.* Environmental Encounters Series, No.8. Council of Europe, Strasbourg.

Vavilov, N.I. 1951. The origin, variation, immunity and breeding of cultivated plants. *Chronica Botanica* 13:1-364.

Chapter provided by Sara Oldfield.

26. ANIMAL USE

INTRODUCTION

Wildlife can be used in a variety of ways, involving different degrees of human intervention and modification of natural habitats, with varying effects on conservation. At one end of the spectrum lie *in situ* harvesting regimes, such as the harvesting of wild plants and animals for subsistence use by local communities. Many, perhaps most, harvests of wild species have in the past resulted in population declines but, if practised at appropriate intensity, these forms of use could be sustainable and need entail little alteration of natural ecological processes. In such cases they might be of conservation benefit if they also provided an economic incentive for conserving natural habitats. At the other end of the scale, wildlife can be brought into captivity and reared in a controlled environment (a process leading eventually to domestication) which usually has little or no conservation value. In between these two extremes there is a range of different containment or husbandry systems, varying from the intensive to the extensive. The implications for the conservation of biodiversity of these different forms of use are considered below.

Many of the products obtained from wildlife are exploited commercially. Worldwide, the commercial trade in wild plants and animals was valued by Hemley (1988) at US$5 billion. Many of the arguments used to justify the conservation of biodiversity rely on the benefits that can be obtained, both economic and otherwise, from the sustainable use of wild resources. However, it should be stressed that, whatever the management system involved, arguments for the conservation of wildlife on purely economic grounds may be insufficient to ensure its long-term preservation. If a landowner is convinced that wildlife should be conserved because of its profitability compared to other forms of land-use, then the logical extension of the argument is that if it ceases to be profitable he should remove it. It should therefore be remembered that aesthetic and moral grounds are just as valid as financial arguments as justifications for the conservation of wildlife.

FOOD: TERRESTRIAL ANIMALS

Vertebrates

Although most of the human diet is now more generally provided by domesticated animals and plants, and fisheries, other wildlife still feature as an important source of nutrition (Table 26.1). Information on wildlife consumption is sparse and typically non-quantitative, partly because of the nature of the consumption, which is generally on a subsistence basis and therefore unrecorded by the normal accounting processes. Estimated figures vary wildly and are often contradictory. However, case-studies reveal the pervasive nature of subsistence-based wildlife use in many cultures and societies. For example, in the Huallaga Central Region of Peru, new settlers and the indigenous peoples rely on wildlife for as much as 80% of their animal protein (Library of Congress, 1979). In northern Alaska in 1974, the people of the Anaktuvuk Pass each consumed an average of 755kg of meat from wild animals, some 88% of their diet. By 1984, this proportion had fallen to 70% (Klein, 1989). The Mbuti pygmies of Zaire obtain up to 60% of their calorific intake from hunting (Marks, 1989). The majority of animal protein consumed by rural communities around Kisangani (Zaire) derives from wild animals, mostly duikers, rodents, primates and other small mammals such as bush pigs, bats and pangolins (Colyn *et al.*, 1988). In Liberia 70% of the population is reported to consume some bushmeat or to sell it. Estimates of game meat consumption in a variety of other African countries are given in Table 26.2. In Nicaragua wildlife provided over 98% of the meat and fish consumed by the Miskito Indians (Nietschmann, 1973). Studies of hunting and wild meat consumption in Sarawak have estimated the total value of wild meat production of about 18,000 tonnes as having a replacement value (cost of domestic substitutes) approaching M$100 million. Even in an industrialised country such as the USA, sport hunting of large ungulates alone was estimated to yield 150,000 tonnes of meat a year with a replacement value of US$450 million (Payne, 1989). In Sweden, the shooting by sport hunters of 186,000 Elk *Alces alces* in 1983 yielded 3.0-3.4kg of meat of this single species per head of population (Bubenik, 1989). Many different species of wild animal are exploited as sources of food, providing a variety of dietary essentials such as protein, fats, and oils. The most conspicuous terrestrial source of wild animal protein is medium to large mammals. Subsistence hunters generally take more mammals than birds, and more birds than reptiles. This is reflected in the number of species taken: the bush people of Suriname take at least 27 mammal species, 24 birds, three turtles and two species of lizards (Redford and Robinson, 1991).

Amongst mammals, ungulates, primates, and large rodents all figure prominently in the bushmeat trade in Africa (see Table 26.3) and South America. Edentates (anteaters and armadillos) are also taken in the New World, while fruit bats of the genus *Pteropus* are considered a delicacy in Oceania.

Birds generally provide meat and eggs for human consumption, but even their nests may be eaten in the case of the cave swiftlets of Southeast Asia. For example, the 49 species of Cracidae constitute an important source of meat for the campeseino and indigenous Indian populations. The eggs of the Greater Rhea *Rhea americana*, Black-bellied Whistling-duck *Dendrocygna autumnalis*, and flamingos are all collected for consumption in South America (Redford and Robinson, 1991). The eggs and young of seabirds are eaten in many parts of the world, probably because the colonial nesting habit makes them particularly easy to collect in large quantities.

Among reptiles, monitor lizards are widely eaten in Africa, iguanas in South America, and sea turtles, particularly Green Turtles *Chelonia mydas*, provide meat for many littoral peoples worldwide. In many areas, reptiles are an even more important source of eggs than birds: eggs of sea turtles, freshwater turtles, Green Iguana *Iguana iguana* and the tegu *Tupinambis* spp. are all consumed by local people and exploited as a source of income.

Table 26.1 Daily per capita consumption of animal protein (g) in countries obtaining more than half of their average supply from wild animals

COUNTRY	TOTAL ANIMAL PRODUCTS	GAME MEAT	FISH AND SEAFOOD	TOTAL WILD MEAT	WILD AS % OF TOTAL
ASIA					
Bangladesh	6.7	-	3.5	3.5	52.2
Indonesia	5.3	-	3.6	3.6	67.9
Korea, DPR	12.1	-	8.0	8.0	66.1
Korea, Rep	13.1	-	9.0	9.0	68.7
Malaysia (Sabah)	22.1	-	11.3	11.3	51.1
Malaysia (Sarawak)	15.5	-	9.3	9.3	60.0
Maldives	30.6	-	28.4	28.4	92.8
Philippines	16.9	-	8.9	8.9	52.7
Thailand	13.2	-	6.8	6.8	51.5
Viet Nam	14.0	-	7.8	7.8	55.7
OCEANIA					
Papua New Guinea	18.2	2.6	8.3	10.9	59.9
AFRICA					
Benin	8.3	1.1	3.7	4.8	57.8
Congo	11.8	2.0	6.9	8.9	75.4
Ghana	15.2	1.4	10.0	11.4	75.0
Liberia	9.2	1.4	4.6	6.0	65.2
Senegal	17.1	-	9.6	9.6	56.1
Sierra Leone	10.0	0.3	7.2	7.5	75.0
Togo	6.8	0.9	3.1	4.0	58.8
Zaire	7.1	1.9	2.6	4.5	63.4

Source: Modified from Prescott-Allen, R. and Prescott-Allen, C. 1982. *What's Wildlife Worth?* Earthscan; FAO, 1977. *Provisional Food Balance Sheets: 1972-74 average.*

Table 26.2 Estimated annual game output and per capita supply in selected African countries

	OUTPUT (thousand tonnes) 1972-74	OUTPUT (thousand tonnes) 1977	SUPPLY PER CAPITA (kg) 1972-74	SUPPLY PER CAPITA (kg) 1977	CONTRIBUTION OF GAME MEAT TO PER CAPITA ANIMAL PROTEIN SUPPLY (%)
Angola	6	6	1.0	0.9	4.5
Benin	6	6	2.2	1.9	13.2
Botswana	5	6	7.6	7.5	15.8
Cameroon	4	4	0.6	0.6	2.8
Chad	3	3	0.8	0.7	3.2
Congo	5	6	4.0	4.0	16.9
Côte d'Ivoire	13	13	2.8	2.5	7.4
Ethiopia	7	7	0.2	0.2	1.0
Gambia	1	1	1.7	1.8	7.8
Ghana	26	28	2.8	2.7	9.2
Guinea	4	4	0.9	0.8	10.0
Kenya	7	7	0.6	0.5	2.6
Lesotho	3	4	3.0	2.9	13.4
Liberia	5	5	2.8	3.0	15.2
Namibia	2	2	2.6	2.6	3.9
Nigeria	87	95	1.2	1.2	13.0
Rwanda	5	6	1.3	1.3	25.0
Sudan	6	7	0.4	0.3	1.0
Tanzania	7	8	0.5	0.5	1.6
Togo	4	4	1.9	1.7	13.2
Uganda	12	14	1.1	1.2	5.0
Zaire	90	68	3.9	2.6	26.8
Zambia	17	20	3.7	3.7	13.4

Source: Modified from Prescott-Allen, R. and Prescott-Allen, C. 1982. *What's Wildlife Worth?* Earthscan; FAO data.

The only amphibians widely used as a source of food are frogs. The frogs' leg trade, mainly based on wild-caught individuals of the genus *Rana*, is economically important in Asia. The majority of the specimens caught are destined for export to supply frogs' legs to the gourmet market in Europe and North America, although some local consumption does occur.

Table 26.3 Wild animals of the bushmeat trade in Ghana*

NORTHERN GUINEA SAVANNA	SEMI-DECIDUOUS FOREST	COASTAL PLAINS SAVANNA
Warthog	Baboon	Grasscutter
Baboon	Warthog	Giant Rat
Hartebeest	Grasscutter	Royal Antelope
Bushbuck	Hartebeest	Bushbuck
Crowned Duiker	Kob	Bat
Aardvark	Bushbuck	Green Monkey
Grasscutter	Roan Antelope	Crowned Duiker
Roan Antelope	Aardvark	Black Duiker
Buffalo	Waterbuck	Red River Hog
Waterbuck	Oribi	Monitor Lizard

Source: Sale, J.B. 1981. *The Importance and Values of Wild Plants and Animals in Africa*. Part 1. IUCN, Gland, Switzerland.
Note: * Top ten species, listed in descending order of importance

Invertebrates

The molluscs and the arthropods include many species used as a food resource. Marine and freshwater molluscs are more important to human nutrition on a global basis than terrestrial species, but in certain areas the latter may figure prominently in the diet. Thus the Giant African Land Snail, of the genus *Achatina*, is eaten on a large scale in West Africa and is immensely popular with people in central Ghana and parts of Nigeria. It has a protein value nearly equivalent to beef. Non-insect arthropods used for food include land crabs, centipedes, woodlice and large spiders.

Insects are an important supplementary source of calories and protein in many regions of the world. Examples of some of the 500 or so species known to be consumed are given in Table 26.4 with an indication of the region where they are eaten.

Insects of most major orders are eaten, but the most widely used species are those, such as termites, which habitually occur in very large numbers in one place, or which periodically swarm, such as locusts, or large species such as saturniid moth larvae. The seasonal abundance of certain species makes them especially important at times of year when other food resources may be lacking.

Orthoptera (grasshoppers, crickets etc.) are a valued food for many peoples. Swarming locusts can easily be gathered by the sackful, are easily dried for storage and can be a valuable resource to help tide over hard times.

The Lepidoptera (butterflies, moths) is probably the order containing the largest number of species eaten. Especially popular in Asia and Africa are many species of saturniid moths which have large fleshy larvae. Some species are dried and sold to quite a large market and are important in the local economy. The pupae of various species of silk moths are consumed in much of Southeast Asia, partly as a by-product of the silk industry. The pupae are killed by immersing them briefly in hot water before the silk is wound off, after which they may be eaten and thus provide an important source of nutrients for many silk workers (Taylor, 1975; Vane-Wright, 1991).

Isoptera (termites) are eaten almost everywhere they occur. In some areas only the swarming reproductive termites are taken; in others, the nests are dug out and all stages are eaten (TFIN 5).

The Hymenoptera (bees, wasps and ants) is another widely utilised order. Honey produced by bees from nectar and pollen is prized, and is one of the most widely accepted insect products (Table 26.5). However, the honey can be less important than the bee brood which is collected with the honey. These immature stages in the comb are eaten by many indigenous peoples and are highly nutritious (TFIN 3). Similarly, wasp brood is eaten in some areas, and less commonly, adult wasps. Larger ants are quite common as food items. In Australia, honeypot ants are or were popular with Aboriginal peoples and leafcutter ants are commonly eaten in parts of the Americas.

Beetles (Coleoptera) are eaten both as larvae and adults. In Thailand, several species are eaten as adults and can often be bought live in markets (Watanabe and Satrawaha, 1984). The abdomens of adult rhinoceros beetles are also eaten by the Yukpa of Venezuela and Colombia (Ruddle, 1973).

Despite the widespread use of insects and other invertebrates for food, they represent an under-exploited resource. Great fecundity is a feature of insects giving potential for wild harvesting and farming. There have been several studies on large-scale insect production both as human food and for animal feed (Maitipe, 1984; TFIN 1 and 4). Many species which are agricultural pests are also used as a food resource in some part of their range, or have the potential to be utilised. Palm grubs (weevil larvae), which are a pest of coconuts and oil palms throughout the tropics but are considered a delicacy by indigenous peoples, are a case in point. Several Indian tribes in the American tropics 'farm' the larvae in that logs are prepared as laying sites and then mature larvae are collected several months later. Several hundred grammes of larvae can be collected per log (TFIN 2).

Table 26.4 Selected insects used as a human food source

ORDER FAMILY	SCIENTIFIC NAME	LIFE STAGE NORMALLY CONSUMED	AREA WHERE IT IS EATEN
Odonata (Dragonflies)			
Aeschnidae		N	Southeast Asia
Libellulidae		N	Southeast Asia
(general Odonata)		A,N	West Africa, Asia, New Guinea Southeast Asia, Madagascar
Blattaria (Cockroaches)			
Blattidae	*Blatta orientalis*	E,N,A	Southeast Asia
	Periplaneta	A	Asia, Australia
Mantodea (Mantids)			
Mantidae	*Hierodula sternosticta*	A	New Guinea
	Mantis religiosa	A	Southern Africa
Isoptera (Termites)			
Rhinotermitidae	*Coptotermes formosanus*	A	Asia
Termitidae	*Macrotermes*	A	Africa
	Termes flavicolle	A	South America
Orthoptera (Grasshoppers, crickets etc)			
Tettigoniidae	*Conocephalus angustifrons*	A	Tropical South America
Gryllidae	*Acheta*	A,N	Southeast Asia
	Gryllotalpa africana	A	Southeast Asia
Acrididae	*Aidemonaazteca*	A	Tropical South America
	Locusta	A	Asia
	Oryxa	A	Asia
	Tropidacris latreillei	A	Tropical South America
Phasmoptera (Stick and leaf insects)			
Phylliidae	*Haaniella grayi*	E	Southeast Asia
Phasmatidae	*Eurycantha horrida*	A	New Guinea
	Platycrana viridana	A	Southeast Asia
Hemiptera (Bugs)			
Belistomatidae	*Lethocerus indicus*	A	Asia, Southeast Asia
Corixidae	*Corixa femorata*	E	Central America
Pentatomidae	*Erthesina fullo*	A	Asia
Homoptera (Bugs)			
Cicadidae (cicadas)		A	New Guinea
Neuroptera (Lacewings etc)			
Corydalidae	*Corydalus*	(A),L	Tropical South America
Coleoptera (Beetles)			
Dytiscidae	*Cybis*	A	Southeast Asia
	Cybista hova	A	Madagascar
Scarabaeidae	*Copris*	A	Southeast Asia
	Heliocopris bucephalus	A	Asia, Southeast Asia
	Oryctes	L	Africa, Asia and western Pacific
	Podischnus agenor	A	Tropical South America
Buprestidae	*Sternocera*	A	Southeast Asia
Tenebrionidae	*Tenebrio*	A	South-west Asia
Bruchidae	*Caryobruchus*	L	Tropical South America
Curculionidae	*Anthonomus*	A	Tropical South America
	Phyncophorus	L	Tropics worldwide
Stratiomyidae	*Chrysochlorina*	L	Tropical South America
Trichoptera (Caddis flies)			
Hydropsychidae	*Leptonema*	L	Tropical South America
Lepidoptera (Butterflies and moths)			
Hepialidae	*Hepialus amoricanus*	L	Asia
Cossidae	*Xyleutes leuchomochla*	L	Australia
Hesperidae	*Acentrocneme hesperialis*	L	Central America
Megathymidae	*Aegiale hesperialis*	L	
Pieridae	*Eucheira socialis*	L	Central America
Bombycidae	*Bombyx mori*	P	Asia
Saturniidae	*Athletes semialba*	L	Central Africa
	Bunaeopsis aurantiaca	L	Central Africa
	Cirina forda	L	Central Africa
	Gonimbrasia	L	Africa
	Gynasia maja	L	Central Africa
	Lobobunaea saturnus	L	Central Africa
	Urota sinope	L	Central Africa

Table 26.4 Selected insects used as a human food source (continued)

Lepidoptera (Butterflies and moths) (continued)			
Notodontidae	*Antheua insignata*	L	Central Africa
	Elaphrodes lactea	L	Central Africa
Thaumatopoeidae	*Anaphe panda*	L	Central Africa
Noctuidae	*Laphygma frugiperda*	L	Tropical South America
	Nyodes prasinodes	L	Central Africa
Hymenoptera (Ants, wasps, bees etc)			
Formicidae	*Atta*	A	Tropical South America
	Liometopum apiculatum	A?,E,L?,P?	Central America
	Melophorus bagoti	A	Australia
	Oecophylla smaragdina	L,P	Southeast Asia
Vespidae	*Mischocyttarus*	L	Tropical South America
	Polistes	L	Tropical South America
	Polybia ignobilis	L	Tropical South America
	Vespula lewisi	A,L,P	Asia
Apidae	*Apis dorsata*	L,P	Southeast Asia
	Apis laboriosa	L,P	Asia
	Trigona	A?,L	Tropical South America, Australia

Source: See references; drawn from various texts on entomology.
Notes: A = adult, E = egg, L = larva, N = nymph, P = pupa.

Table 26.5 Countries with honey production at or in excess of 10,000 tonnes in 1989

ASIA		**NORTH AND CENTRAL AMERICA**	
China	177,000 *	Canada	28,100
India	50,000 F	Mexico	52,530 *
Turkey	40,000 F	United States	80,000
		SOUTH AMERICA	
USSR	230,000	Argentina	38,000 *
		Brazil	16,000 F
EUROPE		**OCEANIA**	
Czechoslovakia	10,000 F	Australia	22,619
France	26,000 F		
Germany	23,300 F	**AFRICA**	
Greece	11,650 F		
Hungary	16,000 F	Angola	15,000 F
Poland	15,000 F	Egypt	13,000 F
Romania	17,000 F	Ethiopia	22,600 F
Spain	21,000 F	Kenya	16,000 F
		Tanzania	14,000 F

Source: FAO, 1990a. *FAO Yearbook: Production 1989*. Food and Agriculture Organization of the United Nations, Rome.
Notes: F = FAO estimate; * = unofficial figure.

Nutritionally, as demonstrated in Table 26.6, insects compare well with other animal products; furthermore, the efficiency of food conversion to biomass is also favourable (Table 26.7).

Trade in meat products

In all its nutritional forms, wildlife has great economic value to local people. Thus Ajayi (1971) calculated that the annual replacement value of wild animal protein (inclusive of bushmeat, wildfowl and fish) used in Nigeria totalled some £30 million which at the time was approximately equal to 4% of Nigeria's gross domestic product.

Many species are also traded in local markets, providing direct revenue for hunters and traders alike. Around Kisangani (Zaire), the village hunters take the higher value carcasses, especially duikers, to the towns for sale, while the rodents and other smaller animals are mostly consumed by the hunters and their families (Colyn *et al.*, 1988). The advent of efficient transportation and storage has allowed the development of commercial food industries based on international trade in wildlife meat. Green Turtles *Chelonia mydas* have long been exported from the Caribbean and Indian Ocean to Europe and elsewhere for the gourmet trade but this is now prohibited by the Convention on International Trade in Endangered Species (CITES). Most species traded for meat are relatively common and therefore not listed in the CITES statistics. However Customs statistics often contain a category for "game meat", and an analysis of this produced an estimate of an annual average trade of some 32,000 tonnes (Table 26.8). The main exporting countries were Argentina, the UK and several Eastern European countries, particularly Hungary and

Table 26.6 Nutritional values of selected vertebrate and invertebrate products

	LIFESTAGE	PROTEIN (%)	FAT (%)	CARBO-HYDRATE (%)	WATER (% weight)	ASH (minerals) (%)	CALORIES /100g
Vertebrate products							
Beef		17.4-19.4	15.8-25.1	0.0	56.7-63.9	0.8-0.9	225-301
Lamb		15.4-16.8	19.4-27.1	0.0	56.3-62.5	1.2-1.3	247-310
Pork		14.6-16.7	22.7-31.4	0.0	52.6-59.5	1.0-1.2	276-346
Chicken		20.6-23.4	1.9-4.7	0.0	73.7	1.0	117-130
Fish							
Lake trout		18.3	10.0	0.0	70.6	1.1	168
Halibut		20.9	1.2	0.0	76.5	1.4	100
Milk		3.5	3.7	4.9	87.2	0.7	66
Eggs		12.9	11.5	0.9	73.7	1.0	163
Invertebrates							
Isoptera							
(Living, species not known)	A	23.2	28.3		44.5		347
(Fried, species not known)	A	36.0	44.4		6.0	6.4	561
Orthoptera							
(Living: mixture of species)	A	15.3-46.1	2.4-9.6	6.8-7.5	10.5-70.6	0.8-5.0	
(Sun-dried: mixture of	A	49.7-75	10.1-18.4	6.4-16.1	5,0	3.7-18.9	
Coleoptera							
Scarabaeidae							
Lachnosterna sp.	L	11.1	3.1	2.3	79.9	2.0	
	A	20.1	4.9	0.3	69.4	1.6	
Curculionidae							
Polycleis equestris	A	30.3	2.2		51.8		
Diptera							
Muscidae							
Musca domestica	P	63.1	15.5		3.9	5.3	
Lepidoptera							
Bombycidae							
Bombyx mori	P	23.1	14.2		60.7	1.5	207
Saturnidae							
Athletes semialba	L	15.5	4.5	1.2	78.0	0.8	504*
Bunaea alcinoe	L	5.9	0.9	1.7	91.0	0.4	443*
Gonimbrasia richelmanni	L	15.9	2.1	0.7	80.0	1.3	447*
Gynanisa maja	L	10.1	3.3	1.0	84.5	1.1	495*
Imbrasia rubra	L	11.8	2.1	0.2	83.0	1.3	445*
Notodontidae							
Drapedites uniformis	L	10.8	4.0	3.1	79.5	1.1	452*
Elaphrodes lactea	L	16.3	5.9	1.8	72.0	1.2	461*
Hymenoptera							
Apidae							
Apis mellifera	L	15.4	3.7	0.4	77.0	3.0	
	P	18.2	2.4	0.8	70.2	2.2	
Araneae							
Theraphosidae							
Melopoeus albostriatus	A	63.4	9.8				

Source: Adapted from Malaisse and Parent, 1980. Les chenilles comestibles du Shaba meridional (Zaire). *Naturalistes Belges* 61(1):2-24 and Taylor, R.H. 1975. *Butterflies in my Stomach*. Woodbridge Press Publishing Co., Santa Barbara, California (see also table in The Food Insects Newsletter, IV (1), on fatty acids). **Notes:** * Dry weight.

Table 26.7 Efficiency of food conversion for selected animals

	Efficiency %*
Chicken (broilers)	38-40
Turkeys	21
Sheep and lambs	5.3
Beef cattle and calves	c 10
Pigs	17-20
Fish and shellfish	10-20
Blattaria (Cockroaches)	
Blatella germanica	38-40
Orthoptera (Grasshoppers etc.)	
Gryllus domesticus	12
Melanoplus bilituratus	8-11
Schistocerca gregaria	8-9
Phasmoptera (Stick Insects etc.)	
Carausius morosus	8
Hemiptera (True Bugs)	
Cimex lectularius	21-41
Phonoctonus nigrofasciatus	33-53
Rhodnius prolixus	19-24
Stalia major	41-54
Coleoptera (Beetles)	
Lasioderma serricorne	21.4
Paropsis atomaria	20
Tribolium confusum	40
Diptera (Flies)	
Cochliomyia hominivorax	35.7
Lepidoptera (Moths etc.)	
Aglais urticae	14-17
Agrotis orthogonia	16-37
Bombyx mori	19-31
Chilo suppressalis	13
Dendrolimus pini	13
Galleria mellonella	20-30
Hepialus humuli	2
Hyphantria cunea	12
Malacosoma neustria	14-22
Mamestra brassicae	3-14
Phalera bucephala	28-34
Pieris brassicae	13-16
Prodenia eridania	8-38
Protoparce secta	12-36
Smerinthus populi	32
Tineola bisselliella	16-22

Source: Compiled from data in Taylor, R.H. 1975. *Butterflies in my Stomach*. Woodbridge Press Publishing Co., Santa Barbara, California.
Note: * % of food consumed converted into animal tissue (optimum).

Table 26.8 Estimated minimum net exports of game meat (tonnes)

NET EXPORTERS	1980	1981	1982	1983	1984	1985
Algeria	2	--	--	--	--	--
Argentina	12,098	10,468	9,304	9,599	8,986	11,627
Australia	382	1,860	1,403	337	146	103
Austria	1,015	999	873	953	192	718
Brazil	--	--	10	22	20	31
Bulgaria	108	58	31	150	194	174
Canada	24	--	--	--	--	--
Chile	--	50	49	189	137	147
China	1,444	876	853	1,719	2,110	1,528
Czechoslovakia	977	918	1,180	1,184	1,488	1,316
Finland	24	53	21	31	95	59
Germany, DR	--	--	--	--	--	32
Greece	--	--	--	--	30	17
Greenland	39	51	51	41	56	38
Hungary	3,086	2,534	2,369	2,934	2,763	3,097
Iceland	--	--	--	--	--	16
Ireland	26	26	52	70	54	63
Israel	--	8	--	--	--	--
Mongolia	121	133	138	167	--	166
Morocco	--	--	--	--	14	15
Netherlands	*	627	746	710	350	313
New Zealand	1,017	1,641	1,197	929	831	1,291
Norway		*	*	*	6	*
Poland	1,729	2,217	1,858	3,449	3,119	2,850
Romania	678	429	523	291	293	89
South Africa	3,479	2,197	1,546	560	835	1,033
Spain	1,296	1,560	1,620	1,574	1,758	1,632
Sweden	270	*	*	*	*	*
Tunisia	--	26	--	28	34	10
Turkey	33	--	--	--	--	--
UK	5,041	4,761	3,857	4,061	4,788	4,695
Uruguay	358	237	181	218	280	495
USA	37	37	--	*	*	*
USSR	883	1,133	1,321	578	1,075	1,433
Yugoslavia	1,081	1,144	918	850	1,040	1,278
Country unknown	44	*	*	15	43	56
Total tonnes	35,292	34,043	30,101	30,659	30,737	34,322

Source: Luxmoore, R.A. 1989. International trade. In: Hudson, R.J., Drew, K.R. and Baskin, L.M. (Eds), *Wildlife Production Systems: economic utilisation of wild ungulates*. Cambridge University Press, Cambridge, UK.
Note: * Net importers in this year.

Poland. Exports from Argentina comprise largely Cape Hare *Lepus capensis*, and from the UK, Red Deer *Cervus elaphus* and a variety of game birds. Because of veterinary health controls, international trade in fresh meat is very much more closely controlled than most animal products. This effectively prohibits exports of game meat from much of Africa to Europe, which constitutes the main market, and explains why South Africa is virtually the only exporter listed in Table 26.8.

FOOD: FISHERIES

Global fish production exceeds that of cattle, sheep, poultry or eggs, and is the largest source of either wild or domestic animal protein for the world's expanding human population (Norse, 1992). It is particularly important in the developing countries, as is evidenced by the large contributions fish and seafood make to the totals in Table 26.1.

The fisheries industry is a large and expanding one. Annual world landings of aquatic resources have increased more than four-fold in the last 40 years, from 21.9 million tonnes per year between 1948 and 1952 to 99.5 million tonnes in 1989 (FAO, 1990b, 1991a). The majority of these landings originated from marine fisheries and were destined for human consumption. Marine landings comprised 85.8 million tonnes (86.2% of total landings in 1989) while inland fisheries (aquaculture and capture fisheries) accounted for the remaining 13.8 million tonnes, or 13.8% (FAO, 1991a). Almost 70% (69.2 million tonnes) of total landings were used for human consumption, while the remainder were used for animal feed, fertilizer etc. The vast majority of the catch (92.1%) comprises fishes (marine, diadromous and freshwater), with molluscs, crustacea and other animals being relatively unimportant in terms of global landings (Fig. 26.1). However, many of these groups command high prices per kg and have a disproportionately high economic value (FAO, 1991b).

Distribution of marine fisheries

The location of the world's marine fisheries is governed principally by the distribution of the floating plants on which they depend for food. Phytoplankton production is principally dependent on adequate supplies of nutrients, and is largest in areas of upwelling.

Climatic fluctuations can greatly alter the pattern of ocean

Figure 26.1 World nominal catches in 1989 by groups of species

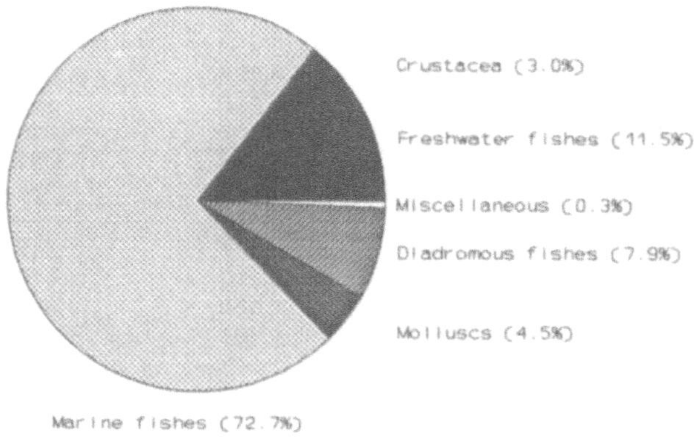

Source: FAO 1991a. FAO Fishery Statistics Yearbook: catches and landings 1989. Vol. 68. FAO, Rome.
Note: Total world nominal catch = 99,534,584 tonnes.

circulation, and hence fisheries production, around the world. Perhaps the most famous of these events is the disruption in some years of the circulation pattern off the coast of Peru, a phenomenon known as 'El Niño', which intermittently leads to the near total collapse of the coastal fisheries.

The relative importance of the catches in the different fishing areas reflect the differences in production. The four major fishing areas (as defined for statistical purposes by FAO) in descending order of annual tonnage of landings are the North-west Pacific, the South-east Pacific, the North-east Atlantic and the Western Central Pacific (Fig. 26.2, Table 26.9).

Composition of marine fisheries

Although there are approximately 22,000 species of fish, of which more than 13,000 are marine (Nelson, 1984), only a very small fraction are of major commercial importance. FAO statistics (FAO, 1991a) break down aquatic animals and plants into 980 "species items" (species, genera, or families) which are then further categorised into 51 groups of species. Of these, only 17 contributed more than 1% (= one million tonnes) towards total recorded world landings, which approached 100 million tonnes in 1989 (Fig. 26.3). The most important groups were the herrings, sardines and anchovies, of which 24.5 million tonnes were landed in 1989, followed by the cods, hakes and haddocks, of which 12.8 million tonnes were landed.

The fisheries industry is based on a remarkably small number of species. Over one million tonnes each of 12 individual fish species (10 marine and two freshwater, see Table 26.10) were caught in 1989: together these comprised 34.7 million tonnes, or 34.9% of the total world catch. The single largest species fishery was the Alaska Pollock *Theragra chalcogramma* of which 6.3 million tonnes were landed, while over five million tonnes of both the Anchoveta *Engraulis ringens* and Japanese Pilchard *Sardinops melanostictus* were also caught. Of the six largest fisheries, five are located in the Pacific (three in the South-east Pacific and two in the North Pacific) while one is from the North Atlantic.

Recent trends in marine fisheries

Reported world landings have generally increased over the past 25 years. During the 1960s (Fig. 26.4), total landings increased steadily as new stocks were discovered, while improved fishing technology and an expansion of fishing effort enabled fuller exploitation of existing stocks of both pelagic (surface water or open sea) and demersal (deep water or bottom-dwelling) species. Long-range fleets increased in size during this period, concentrating their efforts in the richest ocean areas, and were largely responsible for the rapid increase in world catches.

In the 1970s, following the collapse of the Peruvian anchovy fishery there was very little increase in the total catch. Landings of most demersal fish stocks remained relatively constant, implying that they were close to full exploitation and, whilst landings of pelagic fish stocks changed from one species to another in certain areas, there was no appreciable change in total pelagic landings (FAO, 1990b). Long-range fleets continued to expand in importance.

The 1980s once again saw a period of continuous growth (averaging 3.8% a year) in world landings. Because most demersal stocks were (and still are) fully fished, shoaling pelagic species provided most of the increase in fish production. In fact, just three pelagic species (Peruvian Anchovy *Engraulis ringens*, South American Sardine *Sardinops sagax*, and Japanese Sardine *Sardinops melanostictus*) and one semi-demersal species (Alaska Pollock *Theragra chalcogramma*) accounted for 50% of the increase in world landings during the 1980s (FAO, 1990b). Most of this increase appears to have been because of favourable climatic effects on stock sizes rather than new fishery developments or improved management practices (FAO, 1990b).

A concurrent change in the fishing industry in the 1980s was the increase in levels of national and international controls designed to ensure the conservation of fish stocks. This reduced the importance of long-range fishing in many areas and allowed the development of short- or medium-range fishing fleets, (FAO, 1990b). Thus in the early 1970s long-range catches formed 79% of the North-eastern Pacific catch, but had declined to only 8% in 1988, having been replaced by local fleets and joint fishing ventures (FAO, 1990b).

The regional trend in marine landings over the period 1983-1989 was upwards in all but three of the FAO designated fishing areas: the Mediterranean and Black Sea, where output was relatively stable; and the North-east Atlantic and Western Central Atlantic where output declined slightly (Fig. 26.2, Table 26.9). The largest increases occurred in the North-east Pacific and the South-east Pacific.

Although annual world landings of aquatic resources have grown steadily since the 1950s (Fig. 26.4) fishery resources around the world are now thought to be close to their maximum catch limits, and many show signs of biological degradation (FAO, 1990b). Total world marine catch in 1989 was 85.8 million tonnes (FAO, 1990b) and it has been

Figure 26.2 Catches in FAO fishery areas, 1984-1989

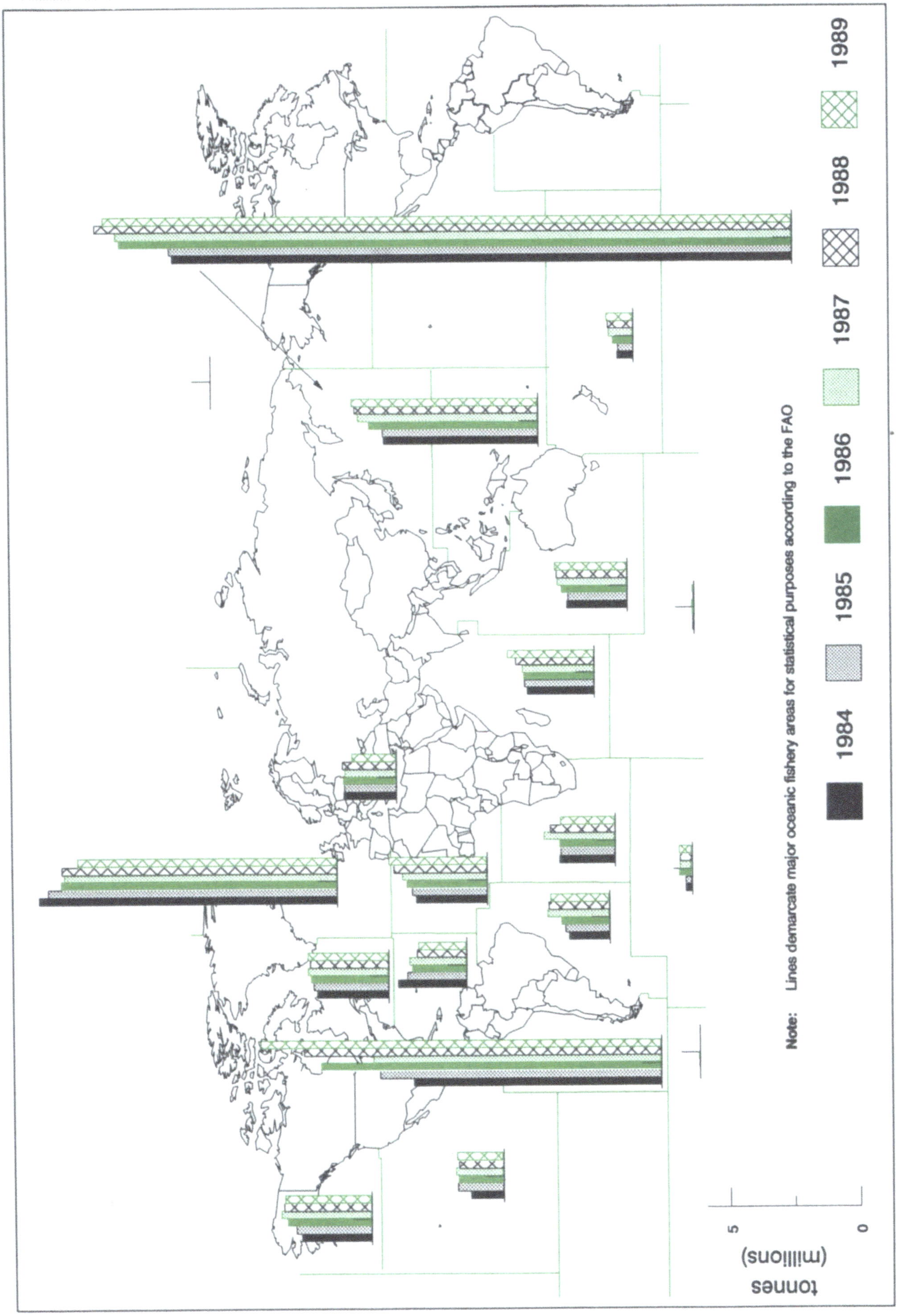

Table 26.9 World nominal catches of fish, crustaceans and molluscs, 1984-1989

FAO FISHING AREA Region	No.	YEAR 1984	1985	1986	1987	1988	1989
INLAND							
Africa	1	1535000	1542900	1667200	1746400	1842800	1871400
America, North	2	433900	443300	484500	572900	535100	528800
America, South	3	338500	328900	362100	386200	356400	321900
Asia	4	6405800	7025400	7840400	8564000	9175900	9535700
Europe	5	413800	430600	459400	448700	474800	476000
Oceania	6	20000	20400	20500	22300	23400	23100
USSR	7	881500	905600	926900	988400	995600	1019700
MARINE							
Arctic Sea	18	0	0	0	0	0	0
Atlantic, Northwest	21	2734300	2870000	2961900	3079800	3020900	3079300
Atlantic, Northeast	27	11460600	11118700	10589100	10457400	10567500	9931000
Atlantic, Western Central	31	2598800	2246400	2044800	2144000	1874500	1791900
Atlantic, Eastern Central	34	2692000	2844800	3027100	3194100	3533500	3702300
Mediterranean & Black Sea	37	2016000	1979000	2012400	1946500	2071300	1673000
Atlantic, Southwest	41	1567200	1700700	1846500	2370300	2329900	2254200
Atlantic, Southeast	47	2143700	2104100	2125100	2728800	2499400	2095000
Atlantic, Antarctic	48	225200	228200	462000	434400	443100	465200
Indian Ocean, Western	51	2542900	2654200	2661600	2719600	2985300	3290900
Indian Ocean, Eastern	57	2328000	2271300	2513900	2656600	2720400	2758200
Indian Ocean, Antarctic	58	35600	31300	37200	39100	14900	31400
Pacific, Northwest	61	23717100	23841700	25709600	25848300	26658000	26310500
Pacific, Northeast	67	2689200	2882100	3204900	3447400	3338600	3290700
Pacific, Western Central	71	5867500	5903500	6416900	6829000	6990800	7076800
Pacific, Eastern Central	77	1213600	1700700	1642100	1753000	1655400	1705200
Pacific, Southwest	81	612800	577500	755100	907900	969000	990600
Pacific, Southeast	87	9465800	10741600	13000500	10988000	13665800	15310600
Pacific, Antarctic	88	800	4700	3900	400	0	1100
Total		73911100	75700500	81014600	81544600	85338300	85757900

Source: FAO 1991a. *FAO Yearbook, fishery statistics (catches and landings).* Vol. 68, 1989. FAO, Rome.

Figure 26.3 FAO species groups contributing over 1% to world catches in 1989

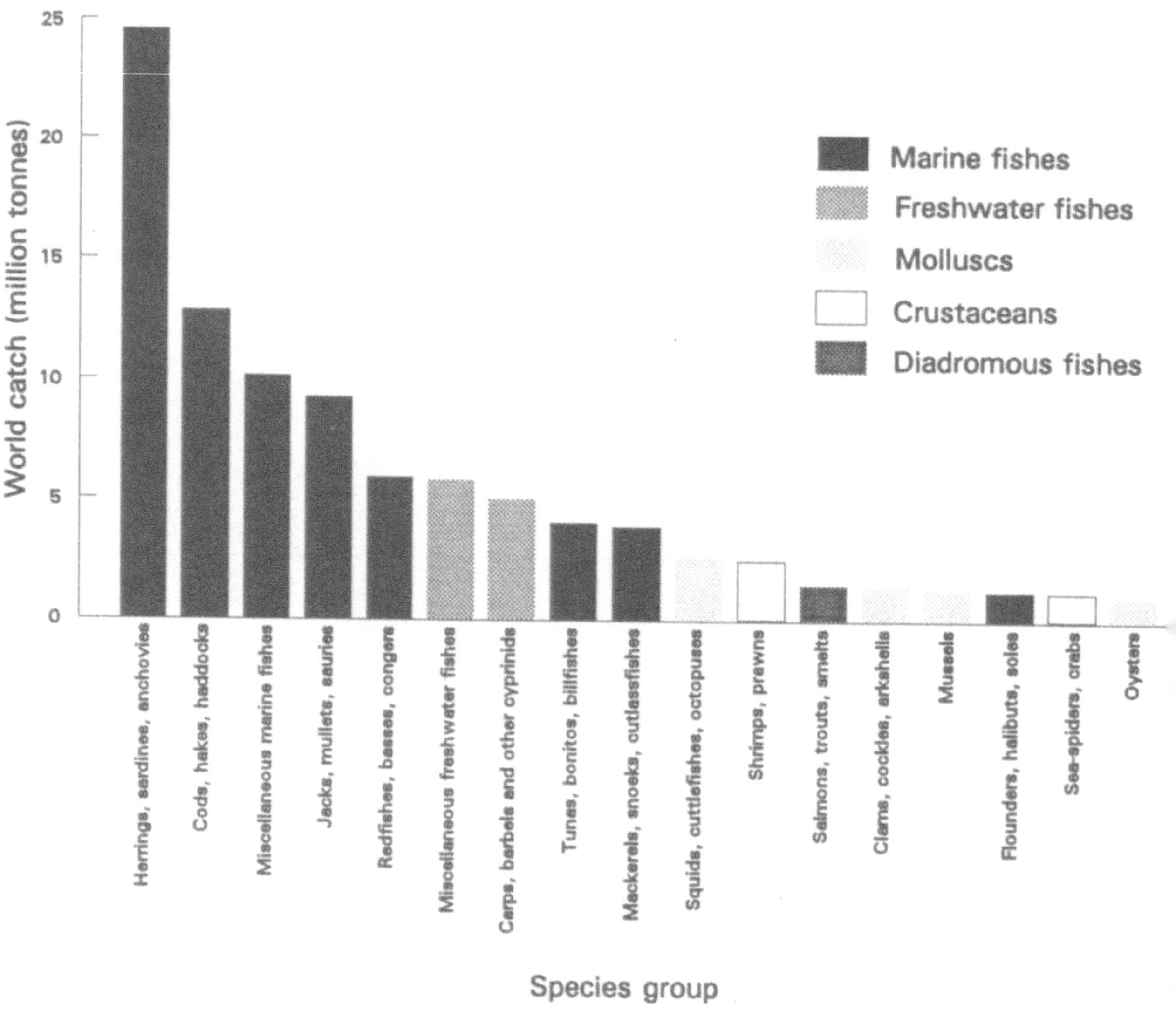

Table 26.10 World total and area nominal catches for the principal fishery species*

SPECIES		AREA	AREA CATCH (tonnes)	WORLD CATCH (tonnes)
Alaska Pollock	*Theragra chalcogramma*	Pacific, Northwest	4,741,659	6,259,058
		Pacific, Northeast	1,517,399	
Anchoveta	*Engraulis ringens*	Pacific, Southeast	5,407,527	5,407,527
Japanese Pilchard	*Sardinops melanostictus*	Pacific, Northwest	5,111,525	5,111,583
		Pacific, Northeast	58	
South American Pilchard	*Sardinops sagax*	Pacific, Southeast	4,196,169	4,196,169
Chilean Jack Mackerel	*Trachurus murphyi*	Pacific, Southeast	3,654,628	3,654,628
Atlantic Cod	*Gadus morhua*	Atlantic, Northwest	630,170	1,782,582
		Atlantic, Northeast	1,152,412	
Chub Mackerel	*Scomber japonicus*	Atlantic, Northeast	6,889	1,671,070
		Atlantic, Western Central	607	
		Atlantic, Eastern Central	313,727	
		Mediterranean and Black Sea	28,095	
		Atlantic, Southwest	13,150	
		Atlantic, Southeast	30,200	
		Indian Ocean, Western	98	
		Pacific, Northwest	986,333	
		Pacific, Northeast	1,470	
		Pacific, Western Central	697	
		Pacific, Eastern Central	43,724	
		Pacific, Southwest	973	
		Pacific, Southeast	245,107	
Atlantic Herring	*Clupea harengus*	Atlantic, Northwest	275,110	1,612,186
		Atlantic, Northeast	1,337,076	
European Pilchard	*Sardina pilchardus*	Atlantic, Northeast	172,842	1,400,030
		Atlantic, Eastern Central	973,643	
		Mediterranean and Black Sea	253,545	
Silver Carp	*Hypophthalmichthys molitrix*	America, North (inland)	1,509	1,359,724
		Asia (inland)	1,339,556	
		Europe (inland)	18,650	
		Atlantic, Northeast	9	
Skipjack Tuna	*Katsuwonus pelamis*	Atlantic, Northwest	41	1,180,121
		Atlantic, Northeast	5,931	
		Atlantic, Western Central	4,203	
		Atlantic, Eastern Central	84,851	
		Atlantic, Southwest	23,053	
		Atlantic, Southeast	585	
		Indian Ocean, Western	217,323	
		Indian Ocean, Eastern	14,074	
		Pacific, Northwest	120,135	
		Pacific, Northeast	682	
		Pacific, Western Central	589,247	
		Pacific, Eastern Central	78,639	
		Pacific, Southwest	10,461	
		Pacific, Southeast	30,896	
Common Carp	*Cyprinus carpio*	Africa (inland)	812	1,085,341
		America, North (inland)	25,050	
		America, South (inland)	688	
		Asia (inland)	679,315	
		Europe (inland)	123,885	
		USSR (inland)	255,706	
		Atlantic, Northeast	70	
		Mediterranean and Black Sea	15	

Source: FAO 1991a. *FAO Yearbook Fishery Statistics: catches and landings 1989*. Vol. 68. FAO, Rome.
Note: * Those with catches of over one million tonnes; data for 1989.

estimated that there is now little scope for increased catches of any of the traditionally fished marine species. There is an increasing need for conservation measures to protect and manage fish stocks in order to sustain current levels of take and rehabilitate degraded fisheries (FAO, 1990b).

The most important step to facilitate the sustainable exploitation of fish stocks has been the establishment by coastal states of jurisdiction up to 200 miles from their shores; 99% of the marine fisheries catch is currently taken within this limit (FAO, 1990b). Most countries are now declaring or have declared 200-mile fishing exclusion zones around their coasts, providing increased potential for rational and sustained use of resources, (FAO, 1981). Further discussion of fisheries management practices, with particular reference to international agreements, is provided in Part 3.

Trends in fish stocks

All species of fish are subject to population and recruitment fluctuations which vary according to the species' biology, migratory habits, food resource availability, natural hydrographic factors, fishing practices and management. Brief details of two stocks are given below in order to illustrate the nature of such fluctuations.

Atlantic Herring (*Clupea harengus harengus*)

The Atlantic Herring provides an example of an over-exploited fish population that has recovered under sound management. It is widely distributed on both sides of the North Atlantic in many reproductively independent groups (races). Following a long period of overfishing and the failure of management controls, most stocks declined to very low levels in the 1970s. At that point, bans and subsequent catch quotas were introduced which allowed stocks to recover (Fig. 26.5).

The North Sea Herring reached a minimum of 75,000 tonnes in 1975, but a ban on fishing from 1977 to 1981 has allowed the population to build up to 1.4 million tonnes which, although lower than the post-War size, is about the same level as in the 1950s and 1960s prior to the collapse (Corten and van de Kamp, 1991).

The Norwegian spring-spawning stock, once the largest herring stock in the world with a spawning biomass of 10 million tonnes in 1957, collapsed to virtually zero in the 1970s as a result of excess fishing pressure and subsequent poor recruitment. Landings decreased to only 10,000-20,000 tonnes between 1972 and 1983, but management measures (e.g. fishing quotas, minimum mesh sizes etc.) permitted the stock to recover. Landings in 1988 were recorded as 125,000 tonnes and the spawning stock biomass estimated at 1.3 million tonnes (ACFM, 1991).

Certain races have never recovered from the earlier overfishing: the spring-spawning stock of Icelandic Herring is now effectively extinct, but Jakobsson (1985) believes that the failure of this stock to recover may have been associated with a sharp decline in the level of primary production in the area.

Western Atlantic Bluefin Tuna (*Thunnus thynnus*)

The Bluefin Tuna is another species which has suffered a catastrophic decline from over-exploitation; however, management practices have not yet produced a recovery in adult numbers.

The Bluefin Tuna is found on both sides of the Atlantic and both the Eastern and Western Pacific. In the Western Atlantic it ranges from Labrador to Brazil (ICCAT, 1990). The breeding population in the western Atlantic has been on the decline for two decades. The population of 'giant', adult fish (age 10+ years) is estimated by ICCAT (International Commission for the Conservation of Atlantic Tunas) to have declined by nearly 95% since 1970 (Fig. 26.6).

Because of the extremely high value of tunas (up to US$30,000 a fish) there has been a marked reluctance to curtail catches even when the need to do so was evident. A total allowable catch quota was set by ICCAT in 1982 but this was doubled in 1983 and has remained the same ever since. Intense publicity was directed at the species in 1992 as a result of a proposal to include it in Appendix I of CITES. The proposal was eventually withdrawn under political pressure but a reduction in catch quotas (10% reduction on the 1991 quota in 1992-1993 and a 25% reduction in 1994-1995) was agreed, entering into effect in May 1992 (ICCAT, 1991). This was expected to allow a very slow recovery in population size but obviously not as swift as would occur if fishing were to be halted altogether. The amount of "illegal" catch (i.e. catch exceeding the ICCAT quota) causes these quotas to be exceeded and further slows recruitment.

Long-lived species, such as the Bluefin, have relative stability in the numbers of young which survive each year and, because of their longevity (20 years or more), they have a steady but slow recruitment each year. In contrast, short-lived species, such as the Atlantic Herring, have a highly variable recruitment but are capable of recovering rapidly. These characteristics partially explain the differences in the success of the management programmes for the two species but it is probable that economic factors played a greater role. As Beverton (1991, *in litt.*) has pointed out, "high prices lead to depletion, even when abundance is low".

Future development in marine fisheries

Most fishery stocks currently used are believed to be fully- or even over-exploited, but demand for fishery products is predicted to rise by the end of this century (Norse, 1992). New stocks, species, or techniques will therefore be needed. Two further options available to increase fisheries production are the exploitation of high seas resources, and mariculture.

High seas resources - those over 200 miles from shore, beyond national jurisdiction - are increasingly under pressure from long-range fleets, which in many cases are exploiting them as a direct result of being banned from traditional fishing grounds in newly-formed fishery exclusion zones. More than 400 fishery species are

Figure 26.4 Trends in fisheries catches, 1963-1989

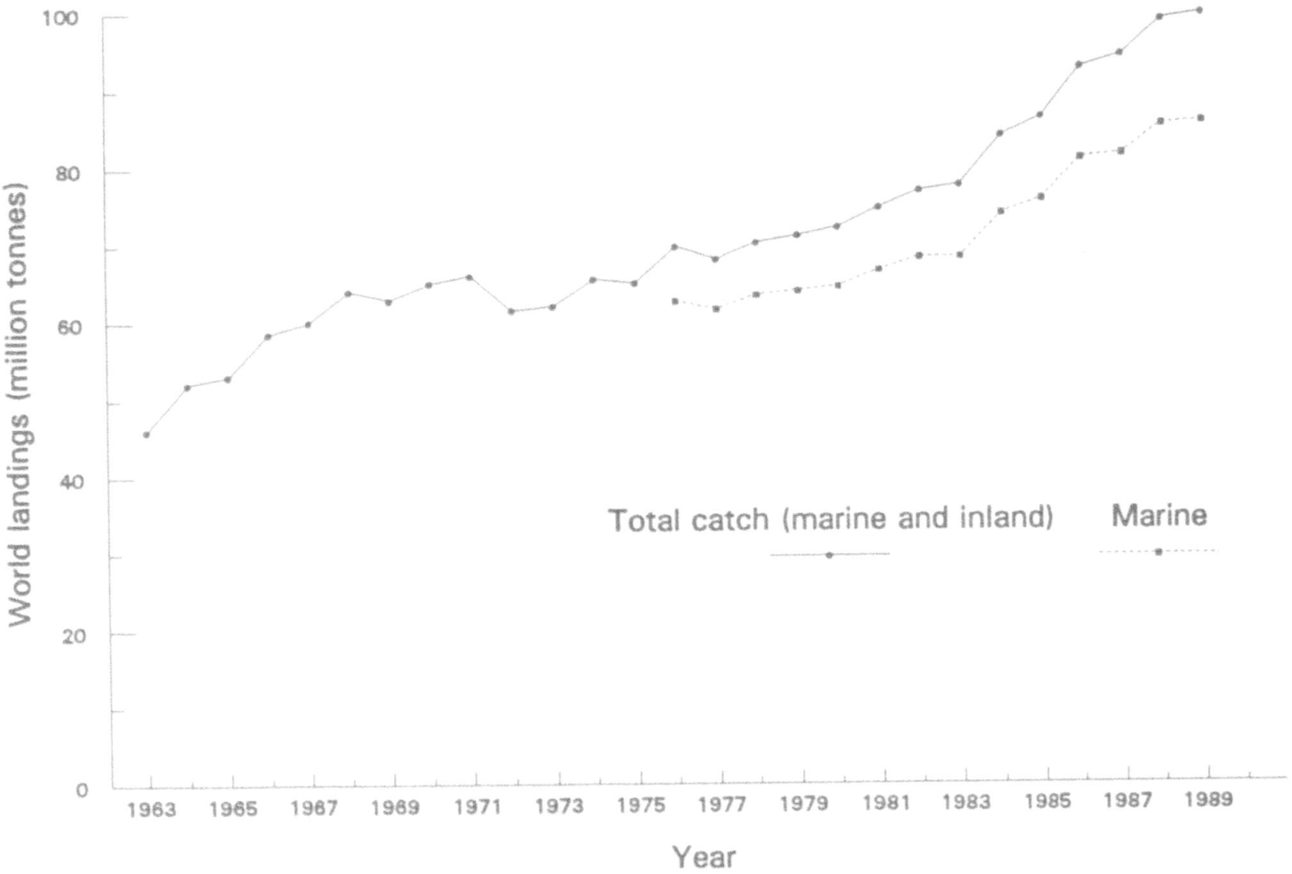

considered to be high seas or oceanic: these include 50 species of cephalopods, 40 species of sharks, 60 species of marine mammals and 230 species of bony fish (FAO, 1990b). Most stocks of these species are dispersed and difficult to harvest or study, and occur at much lower densities than those in upwellings and coastal zones. The main technique used to harvest them is pelagic drift-netting, in which monofilament gillnets are set hanging from the surface to a depth of 10m and left to drift overnight. This practice causes large-scale accidental mortality of non-target species (Norse, 1992). However, because of the concern about the level of incidental catch, restrictions have been progressively introduced to curb drift-netting.

The group with the greatest potential for fishery development are the oceanic squids. Although the main species are already fully fished or overfished, new species and areas have development potential. In the Antarctic, krill fisheries could be expanded, though the economic viability of such a project is very doubtful at the moment and there are serious problems of ecosystem management which have yet to be solved (FAO, 1990b). Most high sea resources under international management suffer from excessive effort and depletion, and practices for responsible fishing need to be agreed by participating nations (FAO, 1990b).

Mariculture is expected to contribute increasingly to world fishery production. Both intensive and extensive mariculture production have grown considerably more than capture fisheries in the past few years. In 1988, 14.6 million tonnes, or 14.8% of the total world catch, was obtained from aquaculture activities, and by the year 2000 this proportion may increase to 33% (Norse, 1992). Freshwater culture of carps and other cyprinid fishes was by far the largest component of this production, but marine species such as mussels, oysters, salmon, shrimps and prawns were also highly significant.

Inland fisheries

Inland fisheries (aquaculture and capture fisheries) grew steadily by 32% over the five-year period 1984-1988 and, in 1989, contributed 13.8% or 13.8 million tonnes of world landings of aquatic resources (FAO, 1991a). Asia is the major inland fisheries producer: in 1989 the continent harvested approximately 9.5 million tonnes, 69.2% of the world total inland catch; 98% of the Asian catch consisted of carp and tilapia species (FAO, 1990), much of which was produced in fish farms. Africa and the "USSR" also produced significant quantities of freshwater fish, mainly wild-caught, but in other continents production was negligible in terms of volume compared to that from marine fisheries (Fig. 26.7).

Despite their small volume relative to marine fisheries, inland fisheries are frequently of particular subsistence value to local communities. Aquaculture grew rapidly in the five years 1984-1988. Global aquaculture production from inland waters increased by 42%, with spectacular increases in Oceania (261%) and Africa (99%). In contrast, inland capture fisheries remained relatively stable or declined over the same period, especially in the industrialised countries. The general trend in the industry is thus for aquaculture to substitute increasingly for production from capture fisheries (FAO, 1990).

Figure 26.5 Trends in herring stocks in the past 20 years

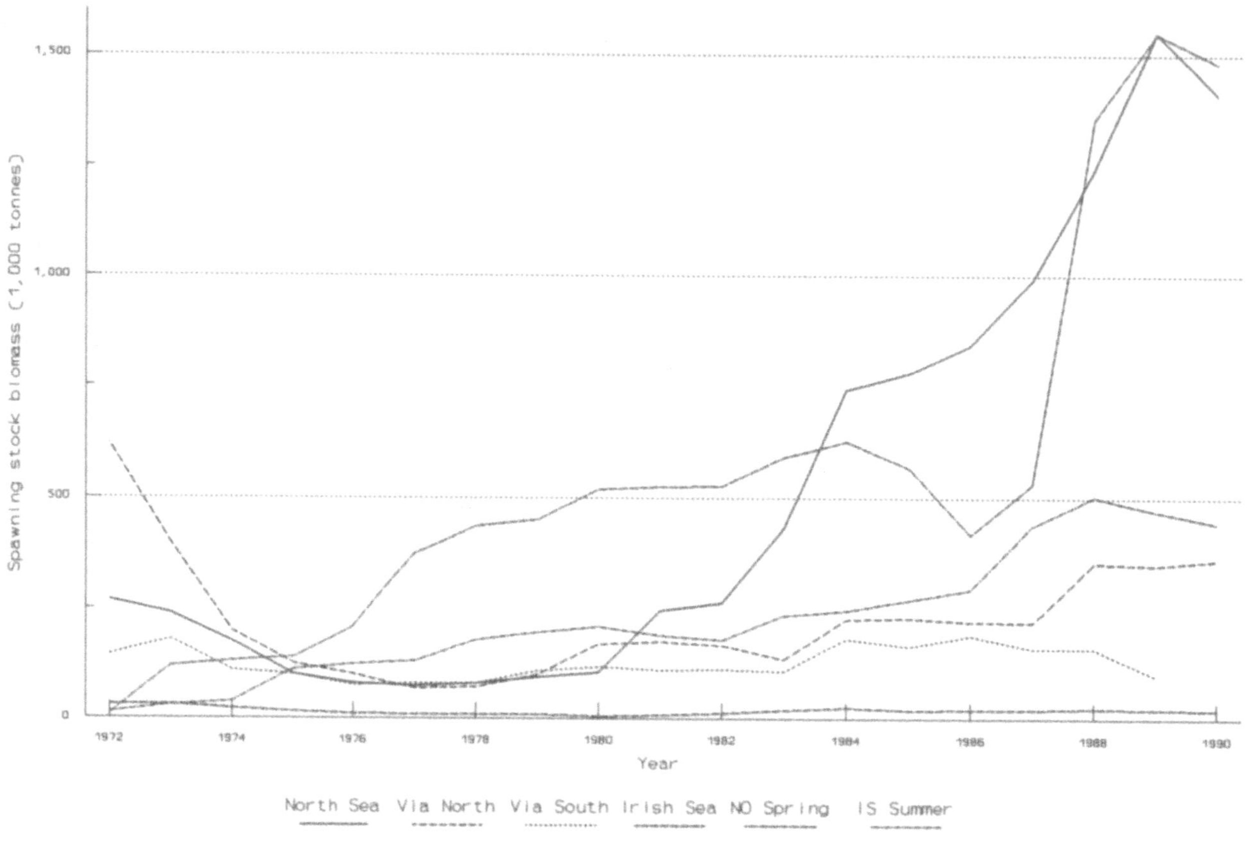

Figure 26.6 Trends in Bluefin Tuna populations of age 10+ in the last 20 years

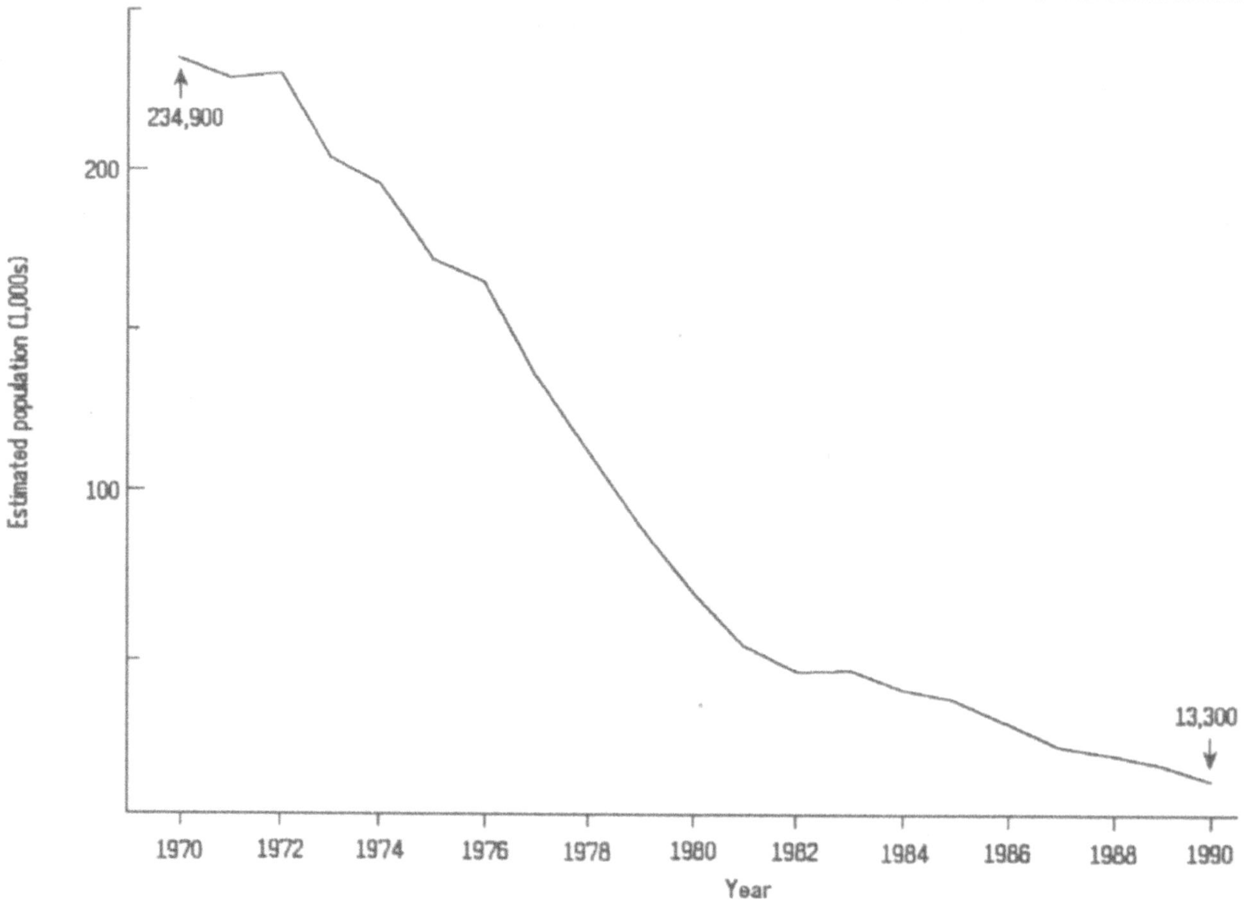

Figure 26.7 Catches in inland waters, 1984-1989

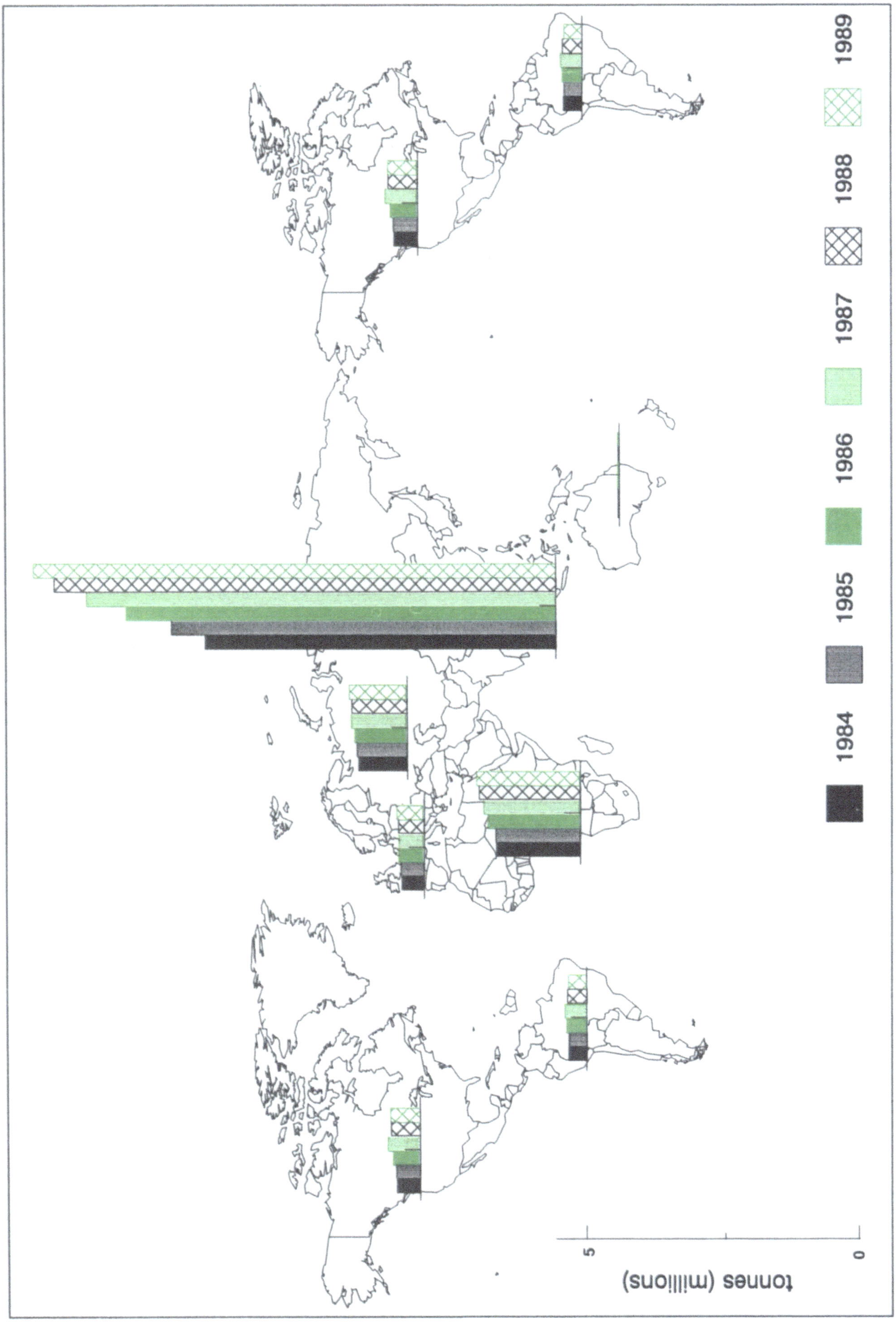

NON-FOOD USES

Utilitarian uses

In addition to its nutritional value wildlife may provide important utilitarian products for both domestic and commercial markets. Fur, hides, scales, bones, and feathers may be used to make a variety of clothing and utensils, while fat may be rendered for oil. Glue and household implements such as needles and hooks can be made from bones, scales and fins. The sale of these products on a small or large scale can generate significant income. It is worth noting that the exploitation of wild species does not necessarily entail killing individual animals. For example, the great seabird colonies of Peru are a source of guano, used as fertilizer, while for centuries in Iceland, the down of Eider Ducks *Somateria molissima* has been collected from their nests. In Peru Vicuña *Vicugna vicugna* are periodically rounded up and shorn of their extremely fine hair, and the fibre from the Musk Ox *Ovibos moschatus* can be collected from the ground during the annual moult.

Ornamental uses

Other wildlife products are valued for their ornamental, decorative or ceremonial purposes. Elephant ivory, tortoiseshell (derived from the Hawksbill Turtle *Eretmochelys imbricata*) and furs have been much prized and in international commerce for many centuries. Both are now prohibited from international trade by their inclusion in Appendix I of CITES, but ivory was formerly much in demand for fabrication into billiard balls, piano keys and a variety of jewellery and artefacts. Fig. 26.8 shows the changes in the amount of ivory exported between 1979 and 1988.

The production of reptile leather has risen in importance since the start of the present century, and demand is continuing for manufacture of shoes and fancy goods.

Since sea turtles have been included in Appendix I, reptile skin trade is now principally confined to three groups: the crocodilians, lizards and snakes. Crocodilians have the highest value skins, particularly those described by the industry as 'classic skins', members of the genera *Crocodylus* and *Alligator*. The Latin American Spectacled Caiman *Caiman crocodilus* is now much more numerous in trade, as it is in the wild, but has a cheaper skin and is therefore well suited to the mass market.

Amongst the lizards, the tegus (genus *Tupinambis*) and the monitors (genus *Varanus*) are large enough to produce useful skins. Both are traded in huge volumes, up to nearly three million skins a year for *Tupinambis*, almost all of which come from Argentina. The monitor lizards can be divided into the African species, *V. niloticus* and *V. exanthematicus*, which are mostly imported to Europe, the Southeast Asian species, *V. salvator*, which is traded to Japan and Europe, and the Indian species, *V. bengalensis* and *V. flavescens*, which almost all go to Japan. The last two are included in Appendix I but are imported by Japan under the terms of a "reservation" which enables it to continue importing them as if they were not covered by CITES. For this reason the statistics are incomplete.

The most valuable of the snakes are the large boids, particularly the pythons, *Python reticulatus*, *P. molurus*, and *P. curtus* from Southeast Asia and the African Rock Python *P. sebae*. Latin American species include the anacondas *Eunectes* spp. and the Boa Constrictor. Trade in the skins of boids has remained at around three-quarters of a million, the majority being from *P. reticulatus*. In order to serve a mass fashion market, the industry has recently been shifting away from boids towards some smaller and cheaper species from Asia such as the Rat Snake *Ptyas mucosus* and the Dog-faced Water Snake *Cerberus rhynchops*. Both were formerly included in CITES Appendix III, and were therefore incompletely recorded in the statistics, but have recently been transferred to Appendix II. This means that the trends shown in Table 26.11 are misleading for such species (marked with *) but otherwise they give an overall impression of the volume and composition of the trade. Sea snakes, especially of the genus *Lapemis* and the brackish water genus *Homalopsis*, are also traded in large numbers, particularly from the Philippines, but are not included in CITES and therefore do not appear at all in the statistics.

Most of the trade in furs derives from farms but cat (felid) species are not farmed and the entire trade derives from the wild. Trade in cat skins is summarised in Table 26.12. Europe and Japan provide the main markets but North American countries are net exporters of cat skins in most years, the two species exported being the Lynx *Felis lynx canadensis* and Bobcat *F. rufa*. Exports of both have gradually declined since the early 1980s owing, principally, to the decreasing popularity of furs as fashion items. This shift in demand has largely been responsible for the very marked decline in trade in cat skins from all sources.

Latin America was the main source of skins in the early 1980s, especially Paraguay, Bolivia and Argentina, but this trade declined sharply in 1985 as a result of import restrictions brought in by the EEC. The species in trade were Ocelot *Felis pardalis*, Little Spotted Cat *F. tigrina*, Margay *F. wiedii* and Geoffroy's Cat *F. geoffroyi*. The first three of these were transferred to CITES Appendix I in 1989 and trade in the fourth was virtually confined to old, stockpiled skins.

Largely as a result of these legal restrictions on the supply of the South American species, the trade has shifted in the late 1980s to China which has been the single largest source of skins, almost all of the one species, the Leopard Cat *F. bengalensis*.

The trade in cat skins has therefore reflected major changes in fashion, coupled with alterations in the legal control under CITES. This has affected not only the overall volume of trade but, possibly more importantly in biological terms, the sources and species in trade.

Bird feathers are also used as items of adornment in many parts of the world, often being incorporated into traditional dress to indicate status or hierarchy. In Latin America feathers of birds such as the Quetzal *Pharomachrus mocinno*, the Roseate Spoonbill *Ajaia ajaia* and macaws *Ara* spp. are prized for their decorative qualities. These feathers have commanded great value throughout many generations;

Figure 26.8 Ivory exports from Africa, 1979-1988

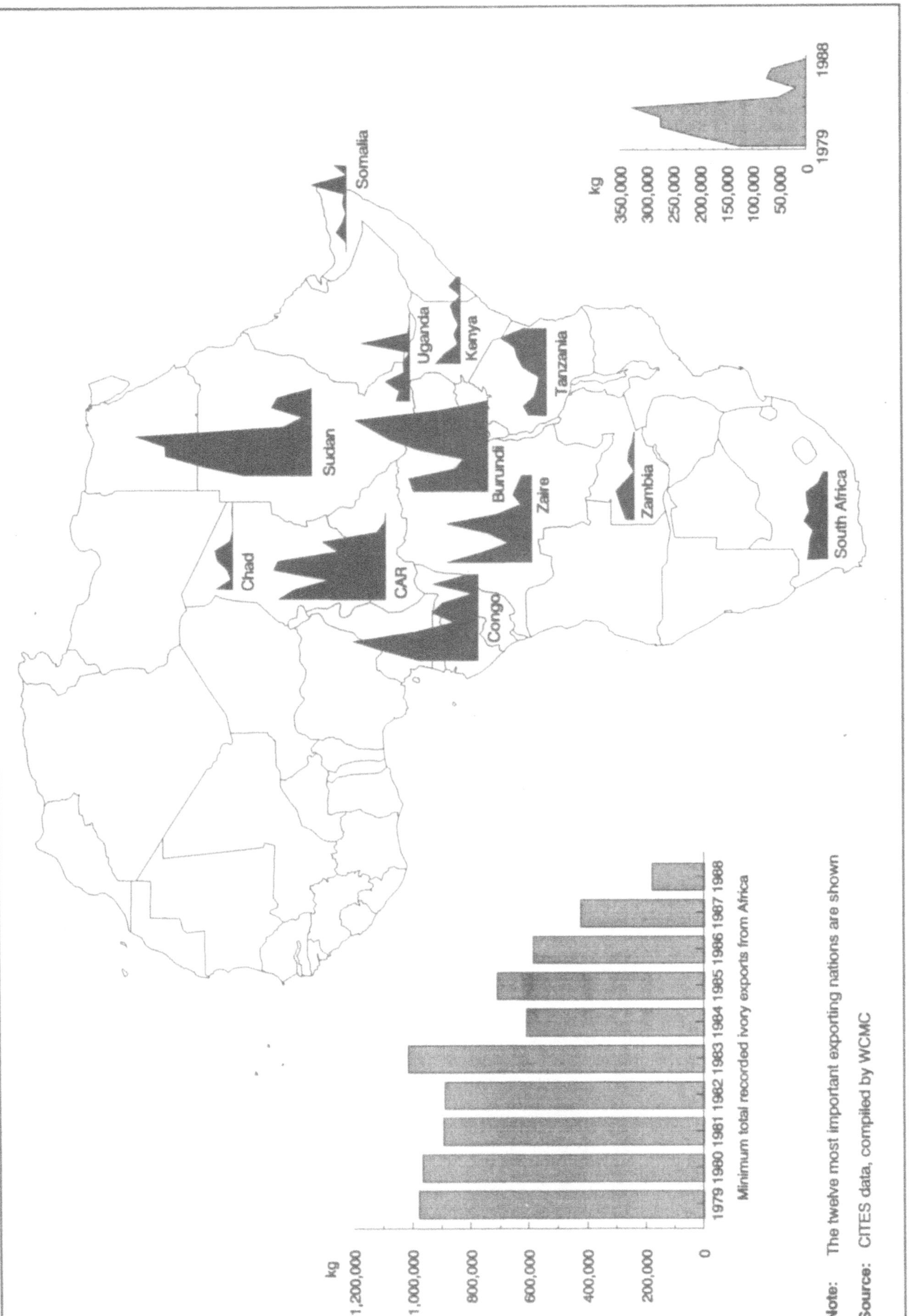

Note: The twelve most important exporting nations are shown

Source: CITES data, compiled by WCMC

Table 26.11 Approximate total net trade in reptile skins 1983–1989

TAXON	1983	1984	1985	1986	1987	1988	1989	AVERAGE
Chelonia mydas	5,716	1,000	0	0	0	0	0	5,575
CROCODYLIA spp.	0	0	0	0	0	0	2	2
Alligator mississippiensis	17,826	13,057	13,228	24,023	33,080	38,720	63,860	32,290
Caiman crocodilus	1,523,421	1,502,191	1,738,423	862,059	639,780	882,305	238,409	1,155,371
Crocodylidae spp.	0	0	17	0	0	0	0	17
Crocodylus acutus	800	0	0	0	0	0	59	173
Crocodylus cataphractus	9,911	2,030	0	11	149	1,193	570	2,348
Crocodylus johnsoni	0	157	0	0	826	0	614	167
Crocodylus niloticus	33,474	7,027	9,378	18,753	24,156	29,524	35,415	23,376
Crocodylus novaeguineae novaeguineae	30,995	30,061	56,924	43,972	39,340	38,171	23,768	35,430
Crocodylus porosus	5,495	5,839	9,160	7,340	8,431	10,259	6,008	7,409
Crocodylus siamensis	0	0	351	605	2,132	2,050	0	1,219
Crocodylus spp.	0	3	21	0	5	0	0	4
Melanosuchus niger	0	452	0	0	0	0	0	452
Osteolaemus tetraspis	0	20	38	41	0	28	224	63
Dracaena guianensis	51,424	71,541	0	26,639	0	0	0	35,667
Iguana iguana	0	0	0	0	18,755	179	0	9,294
Tupinambis rufescens	0	0	4,300	0	147,687	1,047,429	1,196,104	223,778
Tupinambis spp.	280,195	0	0	194,190	96,913	96,657	86,956	119,842
Tupinambis teguixin	1,476,779	1,858,403	1,480,744	1,393,044	1,466,912	873,625	817,675	1,440,234
Tupinambis teguixin nigropunctatus	1,254,300	799,076	166,842	33,894	7,001	0	0	990,163
Uromastyx spp.	0	0	0	40,000	0	0	0	40,000
Varanus bengalensis	0	474,491	296,684	24,040	0	0	0	76,687
Varanus exanthematicus	28,045	14,315	144,460	44,230	4,297	76,461	7,212	40,941
Varanus flavescens	0	56,274	196,316	0	0	0	0	69,055
Varanus niloticus	280,617	354,700	444,295	302,747	712,997	722,532	651,753	507,848
Varanus salvator	1,030,707	1,222,605	1,218,145	1,215,784	1,880,726	1,614,836	1,540,980	1,436,327
Varanus salvator cumingi	0	0	0	15,996	6,951	29,167	0	13,973
Varanus spp.	320	0	0	1,734	16,991	1,374	0	9,310
Atretium schistosum	0	0	0	0	0	28,290	0	28,290
Boa constrictor	143,809	32,517	20,889	25,591	4,919	1,443	2,125	37,785
Boa constrictor constrictor	17,004	15,111	0	0	688	0	0	16,211
Boa constrictor occidentalis	0	0	0	0	901	0	0	901
Boidae spp.	0	0	0	58	530	0	0	59
Cerberus rhynchops *	0	19,250	204,224	821,964	700,072	771,428	38,604	295,753
Eryx muelleri	0	0	0	1	0	0	0	1
Eunectes murinus	9,842	25,331	10,643	6,829	10,759	15,505	2,081	12,021
Eunectes notaeus	17,883	44,376	22,483	19,110	7,160	19,839	0	21,348
Morelia spilota	0	0	0	0	0	13	0	13
Morelia spilota spilota	0	0	0	0	0	3,000	0	3,000
Naja naja *	0	2,463	8,966	266,600	62,510	77,009	30,593	24,440
Ophiophagus hannah *	0	0	662	1,455	320	0	0	651
Ptyas mucosus *	0	712,671	819,055	2,658,195	1,957,577	1,615,906	1,061,874	1,531,584
Python curtus	43,929	42,204	0	83,922	77,293	172,203	29,923	71,244
Python molurus	155	0	0	0	0	0	0	155
Python molurus bivittatus	117,475	156,486	211,414	45,843	70,844	55,132	20,054	112,559
Python regius	0	0	17	136	0	152	88	78
Python reticulatus	478,901	591,168	539,265	569,084	736,847	767,272	456,419	593,684
Python sebae	1,047	782	2,494	19,735	15,569	64,594	9,666	16,686
Python spp.	215	945	38	3,508	57	0	1,205	1,587
Sanzinia madagascariensis	0	0	1	0	0	0	0	1
Vipera russellii	0	0	0	227,163	22,303	42,189	17,485	112,551

Note: * indicates species which were moved from Appendix III of CITES to Appendix II within the period 1983–89 and have thus been incompletely recorded in statistics.
Source: Annual reports of Parties to CITES, compiled by WCMC.

indeed the Incas collected tribute in exotic bird feathers from their Amazonian subjects. The trade in feathers to supply Western fashions, which was at its height in the early 1900s, has largely disappeared, but was once of great commercial significance. Between 1899 and 1920 over 15,000kg of egret and heron feathers, representing plumes from an estimated 15-20 million birds, were exported from South America (Redford and Robinson, 1991). The Ostrich farming industry, established in South Africa but having now spread to the USA and other African countries, was originally almost exclusively for the feather trade, but skins and meat are now more economically important.

The butterfly trade is largely based on ornamental use with some research and education value. Butterflies of the family Papilionidae feature prominently in trade. This family contains the spectacular 'birdwings' which include the world's largest species of butterfly, Queen Alexandra's Birdwing *Ornithoptera alexandrae* with a wingspan of up to 250mm. The large butterflies are generally traded as individual items and some are ranched in order to produce perfect intact specimens. Many of the most spectacular and endangered species have various levels of protection under CITES as well as under local legislation. There is also a major trade in less spectacular tropical species for incorporation in ornaments and souvenirs. This latter trade centres on Taiwan, where estimates suggest between 15 million and 500 million butterflies are traded annually (Pyle, 1981). There has been little monitoring to assess whether this trade is sustainable. Another more recent trade involves live butterflies which are transported (most often now as pupae) to provide exhibits at a variety of locations,

Table 26.12 Net imports and exports of cat skins, 1980–1989

	NET IMPORTS										NET EXPORTS									
	1980	1981	1982	1983	1984	1985	1986	1987	1988	1989	1980	1981	1982	1983	1984	1985	1986	1987	1988	1989
WORLD	**293809**	**232322**	**173887**	**320772**	**171573**	**115214**	**143638**	**176375**	**125392**	**80257**	**289773**	**226792**	**169821**	**318377**	**163853**	**109497**	**135396**	**157701**	**136826**	**68790**
ASIA	**1064**	**8382**	**12631**	**4085**	**8787**	**5490**	**15901**	**41270**	**39510**	**54398**	**20463**	**10181**	**13160**	**3628**	**60332**	**58557**	**72768**	**58946**	**89656**	**48223**
Afghanistan	0	0	0	0	0	0	0	0	0	0	0	0	0	0	0	1	0	0	0	5
Bahrain	0	0	0	0	0	1	1	0	0	0	0	0	0	0	0	0	0	0	0	0
Bangladesh	0	0	0	0	0	0	0	0	0	0	1	0	0	0	1	1	1	1	0	0
China	0	0	0	1492	0	0	0	0	0	0	4221	7844	8231	0	43172	57492	68273	58939	89650	48111
Cyprus	0	0	22	2	0	0	1	0	7	2	0	2	0	0	0	0	0	1	0	0
Hong Kong	540	271	948	927	8621	3419	996	2069	1772	1960	0	0	0	0	0	0	0	0	0	0
India	0	0	0	0	3	3	0	5	0	2	195	8	121	1	0	0	1	0	0	0
Indonesia	0	0	0	0	2	0	0	0	0	0	0	1	0	0	0	0	0	0	0	0
Iran, Islamic Rep	1	0	0	0	0	0	0	0	0	0	0	0	0	0	0	0	0	0	0	0
Israel	175	78	111	43	127	1	0	399	0	3	0	0	0	0	0	0	4490	0	0	0
Japan	348	8020	11486	1499	0	1907	14290	38390	34696	52256	0	0	0	0	16778	0	0	0	0	0
Jordan	0	0	0	0	0	0	0	0	1	0	0	0	0	0	0	0	0	0	0	0
Korea,Dem People's Rep	0	0	0	0	0	0	0	0	0	0	0	0	0	950	0	0	0	0	0	0
Korea, Rep	0	0	0	0	0	0	339	84	2861	75	0	500	4435	0	0	1059	0	0	0	0
Kuwait	0	0	0	0	1	0	0	0	1	0	4	0	0	0	0	0	0	0	0	0
Lebanon	0	12	51	0	2	0	31	20	14	0	0	0	0	0	0	1	0	0	0	0
Malaysia	0	0	0	0	0	0	0	0	0	0	0	0	0	0	0	0	0	0	1	0
Mongolia	0	0	0	0	0	0	0	0	0	0	0	0	9	2614	369	0	0	0	0	100
Nepal	0	0	0	0	0	0	0	0	0	0	1	0	0	1	1	2	1	2	2	1
Pakistan	0	0	0	2	1	1	0	22	2	3	16037	0	0	0	0	0	0	0	0	0
Philippines	0	0	0	1	0	1	3	0	0	0	0	0	0	0	0	0	0	0	0	3
Qatar	0	0	0	1	0	0	0	0	0	0	0	0	0	0	0	0	0	0	0	0
Saudi Arabia	0	1	0	0	0	1	1	0	13	0	0	0	0	4	1	0	0	0	0	1
Singapore	0	0	0	0	0	1	0	2	1	2	2	1	0	0	0	0	1	0	0	0
Sri Lanka	0	0	0	1	0	0	0	0	0	0	0	0	1	0	5	1	0	0	2	0
Syria	0	0	0	0	0	0	0	0	0	0	0	0	363	58	0	0	0	0	0	0
Taiwan	0	0	3	0	10	5	17	10	15	16	0	1819	0	0	0	0	0	0	0	0
Thailand	0	0	0	0	0	0	0	0	0	3	0	1	0	0	5	0	0	2	0	0
Turkey	0	0	10	117	18	150	222	266	127	74	0	0	0	0	0	0	0	0	0	0
United Arab Emirates	0	0	0	0	2	0	0	3	0	2	2	3	0	0	0	0	1	0	0	0
Viet Nam	0	0	0	0	0	0	0	0	0	0	0	2	0	0	0	0	0	0	1	2
Yemen	0	0	0	0	0	0	0	0	0	0	0	0	0	0	0	0	0	1	0	0
USSR (former)	0	0	0	0	0	0	0	0	11434	0	4036	5530	4066	2395	7720	5717	8242	18674	0	11467
EUROPE	**292178**	**223797**	**160549**	**316595**	**162677**	**108627**	**127507**	**126533**	**84844**	**25373**	**23322**	**20976**	**47404**	**197130**	**1765**	**1970**	**3680**	**4586**	**2179**	**720**
Albania	0	0	0	0	0	0	0	0	0	1	0	84	0	166	76	0	0	0	0	0
Andorra	0	0	0	0	0	0	22	22	127	104	0	0	0	0	0	0	0	0	0	0
Austria	5341	6037	1543	5362	3390	2175	602	1800	1157	286	0	0	0	0	0	0	0	0	0	0
Belgium	37547	11183	2268	1958	1080	179	45	2897	329	344	0	0	0	0	0	0	0	0	0	0
Bulgaria	0	89	0	0	0	0	0	0	0	0	0	0	0	199	150	0	0	0	0	1
Czechoslovakia	0	0	0	0	0	0	0	0	0	0	0	0	0	1	0	0	0	0	1	0
Denmark	8185	5686	0	4026	0	0	6929	3178	957	0	0	0	1264	0	171	1911	0	0	0	254
Finland	163	0	1780	0	4977	2145	5477	1798	0	0	0	103	0	182	0	0	0	0	348	42
France	9544	889	0	0	45476	3753	7665	2292	3441	360	0	0	37024	195453	0	0	0	0	0	0
German Dem Rep	0	0	0	0	0	382	0	0	0	0	2877	5338	2398	631	455	0	2743	4586	521	423
Germany, Fed Rep	164018	166890	133313	274741	80063	66333	78076	52248	21637	2851	0	0	0	0	0	0	0	0	0	0
Greece	17242	668	5896	1748	211	1768	1885	2892	1448	2065	0	0	0	0	0	0	0	0	0	0
Hungary	0	0	0	0	0	0	22	0	5	1	0	0	0	0	0	0	0	0	0	0
Iceland	0	0	0	0	1	0	0	0	1	0	0	0	0	0	0	0	0	0	0	0
Ireland	4	4	1	2	1	1	1	2	1	0	0	0	0	0	0	0	0	0	0	0
Italy	29908	22619	11863	15490	5790	13984	8873	24130	7978	10290	0	0	0	0	0	0	0	0	0	0

Table 26.12 Net imports and exports of cat skins, 1980–1989 (continued)

	NET IMPORTS										NET EXPORTS									
	1980	1981	1982	1983	1984	1985	1986	1987	1988	1989	1980	1981	1982	1983	1984	1985	1986	1987	1988	1989
EUROPE (continued)																				
Liechenstein	768	0	0	0	0	0	0	0	94	0	0	0	0	0	0	0	0	0	0	0
Luxembourg	61	140	66	25	0	0	0	0	0	0	0	0	0	0	0	0	0	0	0	0
Malta	7	0	0	0	440	0	20	0	0	0	0	0	0	0	0	46	0	0	0	0
Monaco	0	0	0	33	0	0	0	0	0	0	0	0	0	0	0	0	0	0	0	0
Netherlands	145	13	34	682	9	2	12	13	3	74	0	0	0	0	0	0	0	0	0	0
Norway	28	157	0	102	12	13	20	14	122	19	0	0	83	0	0	0	0	0	0	0
Poland	0	0	0	1	0	0	0	0	30299	0	0	0	0	0	0	0	0	0	0	0
Portugal	477	0	0	0	2	5	4	5	3	7	0	0	0	0	0	0	0	0	0	0
Romania	0	0	0	0	0	0	0	0	0	0	0	0	1	0	0	0	0	0	0	0
Spain	6402	8855	1613	4400	4618	1930	2260	26486	6956	5916	0	0	0	0	0	0	0	0	0	0
Sweden	0	567	0	27	0	0	0	1229	16	28	822	0	487	0	325	13	26	0	0	0
Switzerland	12338	0	2172	5646	1816	5332	0	6479	10270	2350	0	14941	0	0	0	0	911	0	0	0
United Kingdom	0	0	0	2352	14791	10625	15594	1046	0	568	19623	510	6144	0	0	0	0	0	1309	0
Yugoslavia	0	0	0	0	0	0	0	2	0	109	0	0	3	498	588	0	0	0	0	0
NORTH & CENTRAL AMERICA	**154**	**86**	**507**	**5**	**11**	**3**	**10**	**6595**	**740**	**34**	**84862**	**66931**	**37782**	**78693**	**53865**	**43429**	**50795**	**41941**	**28763**	**17714**
Antigua	0	0	0	0	0	0	0	0	0	0	0	0	0	0	0	0	1	0	0	0
Belize	0	0	0	0	0	0	0	0	0	0	341	3	3	2	1	3	1	0	0	2
Canada	0	0	0	0	0	0	0	6343	0	0	26697	14609	31049	18570	15692	2499	2993	0	9160	934
Costa Rica	0	0	6	0	0	0	1	0	0	0	1	0	0	2	1	0	0	1	1	2
El Salvador	0	0	0	0	0	0	0	0	0	0	0	1	1	0	0	1	0	2	2	2
Guadeloupe	0	0	500	0	0	0	0	0	0	0	0	0	0	0	0	0	0	0	0	0
Guatemala	0	0	0	0	0	0	0	0	0	0	1	0	1	0	2	5	2	4	1	3
Honduras	0	0	0	0	0	0	0	0	0	0	1	11	4	3	3	1	2	1	3	3
Jamaica	0	0	1	0	0	0	0	0	0	0	0	0	0	0	8	0	0	1	0	2
Mexico	154	86	0	0	0	0	0	0	0	0	0	0	20	25	26	36	29	67	56	49
Netherlands Antilles	0	0	0	1	0	0	0	0	0	0	0	0	0	0	0	0	0	0	0	0
Nicaragua	0	0	0	0	0	0	0	0	0	0	2	4	7	1	1	2	2	2	0	4
Panama	0	0	0	0	3	3	9	252	740	34	4011	91	0	1	0	0	0	0	0	0
Puerto Rico	0	0	0	0	0	0	0	0	0	0	0	0	0	0	1	0	0	0	0	0
St Lucia	0	0	0	3	8	0	0	0	0	0	0	0	0	0	0	0	0	0	0	0
Trinidad and Tobago	0	0	0	0	0	0	0	0	0	0	0	0	0	0	0	0	0	0	0	1
United States	0	0	0	0	0	0	0	0	0	0	53808	52212	6697	60089	38130	40882	47765	41863	19540	16712
Virgin Islands (British)	0	0	0	1	0	0	0	0	0	0	0	0	0	0	0	0	0	0	0	0
SOUTH AMERICA	**27**	**0**	**118**	**1**	**7**	**1005**	**10**	**118**	**4**	**176**	**152034**	**127235**	**70513**	**37719**	**45660**	**3354**	**6147**	**50439**	**15032**	**1176**
Argentina	0	0	117	0	0	0	7	0	0	0	9857	1086	0	7	12225	0	0	50418	15015	1163
Bolivia	0	0	0	0	0	0	0	0	0	0	12	8	3	3315	30828	3320	6124	1	1	2
Brazil	27	0	0	0	3	5	3	0	2	0	0	0	0	0	0	0	0	5	0	0
Chile	0	0	0	0	0	0	0	0	0	0	0	0	0	0	0	0	0	0	1	2
Colombia	0	0	0	0	0	0	0	0	0	0	8	16	19	12	3	8	1	2	5	4
Ecuador	0	0	0	0	0	0	0	0	0	175	4	11	24	2	2	8	11	9	1	0
Guyana	0	0	0	0	0	0	0	0	0	0	1	1	0	2	2	0	3	2	3	1
Paraguay	0	0	0	0	0	0	0	0	0	0	139641	126092	70459	34375	2600	2	1	0	4	0
Peru	0	0	0	0	0	0	0	1	0	1	2511	20	8	6	0	7	3	0	0	0
Suriname	0	0	0	0	0	0	0	0	2	0	0	0	0	0	0	0	1	0	0	0
Uruguay	0	0	0	0	2	1000	0	117	0	0	0	0	0	0	0	0	0	0	0	0
Venezuela	0	0	1	1	2	0	0	0	0	0	0	1	0	0	0	9	3	2	2	4
OCEANIA	**21**	**17**	**24**	**79**	**12**	**4**	**17**	**24**	**14**	**13**	**0**	**0**	**0**	**0**	**0**	**0**	**0**	**0**	**2**	**0**
Australia	21	17	22	77	8	3	14	19	13	9	0	0	0	0	0	0	0	0	0	0
New Zealand	0	0	2	2	4	1	3	5	1	4	0	0	0	0	0	0	0	0	0	0
Papua New Guinea	0	0	0	0	0	0	0	0	0	0	0	0	0	0	0	0	0	0	2	0
ANTARCTICA	**0**	**0**	**0**	**1**	**0**	**0**	**0**	**0**	**0**	**0**	**0**	**0**	**0**	**0**	**0**	**0**	**0**	**0**	**0**	**0**
Heard and Macdonald Is.	0	0	0	1	0	0	0	0	0	0	0	0	0	0	0	0	0	0	0	0

Table 26.12 Net imports and exports of cat skins, 1980–1989 (continued)

	NET IMPORTS										NET EXPORTS									
	1980	1981	1982	1983	1984	1985	1986	1987	1988	1989	1980	1981	1982	1983	1984	1985	1986	1987	1988	1989
AFRICA	**365**	**40**	**2**	**6**	**79**	**75**	**193**	**526**	**218**	**158**	**9035**	**772**	**962**	**1043**	**2127**	**1179**	**1887**	**1789**	**1194**	**957**
Benin	0	0	0	0	0	0	0	0	0	0	0	0	0	0	0	0	3	0	0	0
Botswana	0	0	0	0	0	0	0	38	193	0	71	192	23	148	353	270	14	0	0	186
Burundi	0	0	0	0	0	0	0	0	0	0	0	1	0	0	0	0	2	0	0	0
Cameroon	0	0	0	0	13	0	0	14	11	8	0	1	0	1	0	22	3	0	0	0
Central African Rep	0	0	0	0	0	0	0	0	0	0	1	4	9	3	5	17	37	40	43	16
Chad	0	0	1	0	1	0	0	0	0	0	0	0	0	0	0	0	0	0	0	0
Congo	0	0	0	0	0	8	26	0	0	0	0	0	0	0	7	0	0	1	1	1
Côte d'Ivoire	0	0	0	0	5	1	0	0	0	0	0	0	0	0	0	0	0	0	0	1
Djibouti	0	0	0	0	0	0	0	0	0	0	0	0	0	1	0	0	0	0	0	0
Egypt	0	0	0	2	0	0	0	2	1	0	0	1	0	0	0	0	0	0	0	0
Ethiopia	0	0	1	0	0	0	0	0	0	0	294	3	0	9	6	10	1	13	7	2
Gabon	0	0	0	0	59	58	142	34	10	9	0	0	0	1	0	0	0	0	0	0
Gambia	0	0	0	0	0	0	0	0	0	0	0	0	0	0	0	1	0	0	0	0
Ghana	0	0	0	0	0	0	0	0	0	0	1	0	1	0	0	0	0	2	0	0
Guinea	0	0	0	0	0	0	0	0	0	0	0	0	4	0	0	9	45	118	1	9
Guinea Bissau	0	0	0	0	0	0	0	0	0	0	0	0	0	0	0	0	0	0	59	2
Kenya	0	0	0	0	0	0	0	0	0	0	4	1	4	1	4	15	19	195	5	13
Lesotho	0	0	0	0	1	5	0	0	0	0	0	0	0	0	0	0	1	0	0	0
Liberia	0	0	0	0	0	0	0	0	0	0	0	5	0	4	1	6	2	0	1	0
Libya	0	0	0	0	0	0	0	1	0	0	0	0	0	0	0	0	0	0	0	0
Malawi	0	1	0	0	0	0	0	0	0	0	1	0	11	3	3	7	2	5	7	4
Mali	0	0	0	4	0	0	0	0	0	0	710	0	0	0	0	0	500	0	0	1
Mauritius	0	0	0	0	0	1	0	0	0	0	0	0	0	0	0	0	0	0	0	0
Morocco	0	0	0	0	0	0	0	0	0	3	0	0	0	0	0	0	0	0	0	0
Mozambique	0	2	0	0	0	0	0	0	1	0	0	0	1	1	1	0	3	1	0	0
Namibia	0	0	0	0	0	0	0	0	0	0	6489	9	23	17	81	116	98	67	84	44
Niger	0	0	0	0	0	0	0	0	0	0	0	0	0	2	0	0	0	0	0	0
Nigeria	0	0	0	0	0	0	0	0	0	0	4	6	5	14	9	30	42	8	14	25
Reunion	0	0	0	0	0	0	0	0	0	1	0	0	0	0	0	0	0	0	0	0
Rwanda	0	0	0	0	0	0	0	0	0	0	0	0	0	0	0	1	0	0	1	12
Senegal	0	0	0	0	0	0	2	0	1	3	0	2	0	0	0	0	0	0	0	0
Seychelles	0	0	0	0	0	0	0	0	1	0	1	0	0	0	0	0	0	0	0	0
Sierra Leone	0	0	0	0	0	0	0	0	0	0	0	0	0	1	0	0	0	0	0	0
Somalia	0	0	0	0	0	0	0	0	0	0	0	2	0	0	0	0	0	0	0	0
South Africa	0	0	0	0	0	0	0	0	0	0	1409	489	607	327	639	396	654	692	634	129
Sudan	0	0	0	0	0	0	0	0	0	0	1	11	4	4	2	5	4	2	0	1
Swaziland	0	5	0	0	0	2	7	1	0	9	0	0	0	0	0	0	0	0	0	0
Tanzania	0	0	0	0	0	0	0	0	0	0	13	13	194	135	151	124	342	210	70	69
Togo	0	0	0	0	0	0	0	0	0	0	0	0	0	0	0	0	0	0	0	3
Tunisia	365	32	0	0	0	0	16	436	0	125	0	0	0	0	0	0	0	0	70	0
Uganda	0	0	0	0	0	0	0	0	0	0	1	2	0	0	0	1	2	0	0	0
Zaire	0	0	0	0	0	0	0	0	0	0	0	1	3	0	4	4	3	12	11	3
Zambia	0	0	0	0	0	0	0	0	0	0	11	20	27	141	129	37	41	67	38	18
Zimbabwe	0	0	0	0	0	0	0	0	0	0	24	9	46	230	732	108	69	356	148	418
OTHER	**0**	**0**	**56**	**0**	**0**	**10**	**0**	**1309**	**62**	**105**	**57**	**697**	**0**	**164**	**104**	**1008**	**119**	**0**	**0**	**0**
Other	0	0	0	0	0	10	0	0	6	0	0	5	0	0	0	0	0	0	0	0
Country Unknown	0	0	56	0	0	0	0	1309	56	105	57	692	0	164	104	1008	119	0	0	0

Source: Annual reports of Parties to CITES, compiled by WCMC

Source: Annual reports of Parties to CITES, compiled by WCMC

such as zoos and butterfly houses in various parts of the world. The livestock for display is mainly captive bred at butterfly farms in various tropical locations.

Among invertebrate-based products, the silk industry produces a luxury commodity on a massive scale and is of major importance to many countries with rural-based economies. Most world trade uses the product of one species, the Mulberry Silk Moth *Bombyx mori* (Bombycidae), a domesticated species. The silk produced by this species is the finest quality of those available. Other species which contribute to the world trade are mostly from the family Saturniidae. These are all known as 'wild' silkworms, even though the eri (*Attacus ricini*), tussah (*Antheraea* spp.) and muga (*A. assama*) silkworms are partially domesticated. Sericulture (silkworm husbandry) is an important source of revenue for many people. Table 26.13 shows the main countries producing raw mulberry silk and silk waste and their estimated 1989 production. The process of silk production is very efficient in that little is wasted. The high-protein byproducts of sericulture (mulberry leaves and silkworm rearing litter) make good animal feeds; the moth pupae killed after spinning their cocoons can be used as human or animal food and are also used in soap and cosmetics; and the mulberry trees which are cultivated to feed the silk worm can be used for timber and will often grow on areas of land unsuitable for other agriculture (Greenhalgh, 1986).

Other important insect products include lac, from *Laccifer lacca* and cochineal, from *Dactylopius coccus*, both of which are Homopteran scale insects. Lac is refined to produce shellac, widely used as a base for polishes and other wax products, with an estimated US$ 9 million used in the USA in 1981 (Lindberg, 1988). Cochineal, a brilliant red colouring agent, formerly widely used in the food and cloth industries, is commercially obtained from a semi-domesticated form of the insect. The cochineal trade was very much reduced by the invention of synthetic dyes and today only Peru and the Canary Islands still produce the dye for export.

Medicinal and biomedical uses

Animal products are widely used in medicines by traditional societies, and even urbanised societies may retain their faith in traditional animal-based remedies, particularly in Asia. For example, Levy-Luxereau (1972) described 181 animal-based remedies used by the Hausa tribe in Niger, most of which were derived from wild species. In local markets in Brazil, dried lizards of several species, the genitalia of dolphins, fox fur, and many other pieces of wild animals are sold for medicinal and magical purposes (Redford and Robinson, 1991). Medicinal wildlife products are frequently traded internationally particularly to satisfy the demand for traditional oriental medicine, a trade which can be extremely lucrative. International trade in the medicinal products of deer, especially antlers, tendons and musk was calculated to be worth some US$30 million a year, almost as much as the international trade in meat (Luxmoore, 1989). Some products are so valuable that the trade can continue even when the species become extremely rare in the wild and, in such cases, it may pose a severe threat to their survival. Trade in musk (from musk deer *Moschus* spp.), bears' gall bladders, tiger bones and, most notably, rhino horn have all been blamed for the decline in populations of several species.

The use of leeches for medical purposes probably began in India but was first described in writing by the Greeks in the 2nd century BC (Conniff, 1987). Leeching became very fashionable in Europe during the mid-19th century to the point where the European Medicinal Leech *Hirudo medicinalis* is now threatened in the wild and included in CITES Appendix II. Present day usage of the live animal is now mostly restricted to micro-surgery where the sucking action and substances produced by leeches during feeding help survival of accidentally severed parts, such as fingers and ears, after re-attachment. Leech saliva contains anticoagulants, anaesthetics, vasodilatory agents and a spreading factor (which allows the other agents to spread far beyond the edges of the incision) all of which have potential uses in a range of research and medical fields. For instance, *Hirudo medicinalis* is used as a source for hirudin, an anticoagulant which can help prevent blood clots from forming. The Giant Amazon Leech *Haementeria ghilianii* uses a different chemical, hementin, for a similar purpose. However this factor is not only capable of preventing clotting, it can dissolve already formed clots. The spreading agent is also produced commercially as are several other leech-derived products. Leeches of various species are farmed commercially in the UK to supply these different outlets.

Horseshoe Crabs (*Limulus polyphemus*) are extremely ancient in evolutionary terms and possess several unique features. They have been used in fundamental research into vision and the clotting property of Horseshoe Crab blood is exploited in human blood testing. Gram-negative bacteria are responsible for a wide range of serious diseases in man, such as spinal meningitis and gonorrhoea, and are sometimes responsible for contamination of manufactured drugs. The blood of Horseshoe Crabs clots rapidly as soon as it comes into contact with gram-negative bacteria or their endotoxins. Refined and freeze-dried samples of Horseshoe Crab blood can therefore be used in a very accurate assay for the presence of these endotoxins, allowing rapid diagnosis of disease and routine checking of purity of drug samples. These tests have largely replaced less accurate assays carried out on rabbits. Although blood is obtained without killing the animals, which are collected, bled and released, there has been some concern over the long-term impact of this practice on Horseshoe Crab populations.

Substances with great potential uses in medicine have also been isolated from snake venoms; these include coagulating enzymes, anticoagulants, neurotoxins and cytotoxins. The first two have been used in the study and control of bleeding disorders in man and the development of fast and accurate assay methods to test prothrombin in human blood, so helping prevent thrombosis, or blood clot formation. Neurotoxins may be useful as anaesthetics, and cytotoxins could prove useful in cancer treatments.

An important use of live animals, particularly primates, is as experimental animals in the biomedical trade. The total volume of trade in 1989 was about 42,000 (Table 26.14), the majority again being imported to Europe and the USA.

Table 26.13 Silk production: FAO estimates for 1989 in tonnes

ASIA		**EUROPE**	
Afghanistan	60	Bulgaria	160
Cambodia	14	Greece	11
China	42,044	Italy	20
India	10,500	Poland	3
Iran, Islamic Rep	850	Romania	185
Japan	7,000	Spain	15
Korea, Dem People's Rep	3,000	Yugoslavia	45
Korea, Rep	1,400	**SOUTH AMERICA**	
Lebanon	6		
Thailand	1,250	Brazil	1,900
Turkey	250		
Viet Nam	450	**AFRICA**	
USSR	4,400	Egypt	11
		Madagascar	15

Source: FAO, 1990a,b. *FAO Yearbook: Production 1989*. Food and Agriculture Organisation of the United Nations, Rome.
Note: All figures are estimates.

The main exporting countries are Indonesia and the Philippines, most of the exports comprising the single species, the Cynomologus Macaque *Macaca fascicularis*. The same species occurs in large feral populations in Mauritius, and this small island constituted the world's third largest exporter of live primates. In the past, South America has been a major source of primates but this has now declined owing to trade restrictions and any residual need is filled by captive breeding. The species most commonly exported were Common Marmosets *Callithrix jacchus* and Squirrel Monkeys *Saimiri sciureus*, followed by marmosets of the genus *Saguinus*, and Night Monkeys *Aotus trivirgatus* (Redford and Robinson, 1991).

Chimpanzees *Pan troglodytes* are also widely used in biomedical research, and in 1990 approximately 1,300 were held by biomedical facilities in USA alone. International trade is now banned by the inclusion of the species in CITES Appendix II but illegal capture of Chimpanzees from the wild and export from Africa continue, driven by the high market value of infants - around US$25,000. This trade is blamed for the continuing decline of wild populations (CCCC, 1990).

Increasingly, primates are being bred in captivity for biomedical research. This is preferable as it does not risk depleting wild populations and it produces a genetically uniform and disease-free stock. Some species are only bred in very small numbers on an experimental scale, but others, such as *M. fascicularis* and *C. jacchus*, are bred commercially.

Working animals

A number of wild animal species are trained to assist in various human activities. For example, the Indian Elephant *Elephas maximus* has been used for centuries as a draught animal in forest industries, warfare, and for ceremonial purposes. There are currently approximately 16,000 tame elephants, the majority employed in the timber industry in India, Myanmar and Thailand. Although the species breeds well in captivity most working elephants are captured from the wild and subsequently tamed. Both otters *Lutra* spp. and cormorants *Phalacrocorax* spp. are used for fishing in China and Southeast Asia. After training, the animals are fitted with restraints to prevent them damaging or swallowing the fish they catch (leather straps over the canines in the case of otters, neck rings on the cormorants), and generally kept tethered to the fisherman's boat during fishing. Pig-tailed Macaques *Macaca nemestrina* are often captured in Southeast Asia and trained to climb trees and throw down ripe fruit and coconuts.

Trained wild animals are also widely used in sport hunting. Falconry, the sport of using falcons, hawks and sometimes eagles to capture and kill wild game, relies on the training of both young and old birds taken from the wild. Practised worldwide, it is still a popular pastime in India, Pakistan and Saudi Arabia. The Cheetah *Acinonyx jubatus* was once used by wealthy Indians as a trained courser, and might well have become domesticated if it had not been for its reluctance to breed in captivity.

Mutually profitable associations can even arise between men and totally untrained wild animals. Thus some Amazon Indians have developed close fishing partnerships with individual Amazon River Dolphins *Inia geoffrensis*. The man calls his dolphin by whistling and it feeds on the opposite side of the river from his canoe. The dolphin's activities drive the fish towards the man, and vice versa, resulting in more successful fishing for both. The association between honey guides *Indicator* spp. and men, in which the bird leads the hunter to wild bees' nests by calling insistently and fluttering its wings in return for a share of the spoils of honey, wax and grubs, is another example (Barton, 1986).

Pollination

A vital operation carried out by insects, mostly independently of man, is pollination. Many of man's most important crops rely on pollination by insects, and of these insects bees are by far the most important. The bees fall into two categories, wild bees and domesticated honeybees. Honeybees of the genus *Apis* have been cultivated by man for many centuries in various regions of

Table 26.14 Live reptile and primate trade in 1989

	LIVE REPTILES		LIVE PRIMATES	
	IMPORTS[1]	EXPORTS[1]	IMPORTS[1]	EXPORTS[1]
WORLD	**438875**	**386754**	**40619**	**42249**
ASIA	**32911**	**57297**	**4815**	**26918**
Bangladesh	0	0	8	0
Brunei	0	0	4	0
China	17	0	0	1292
Cyprus	43	0	0	0
Hong Kong	135	0	0	24
India	2	0	29	0
Indonesia	0	7914	0	16501
Iran	0	0	180	0
Israel	0	30	69	0
Japan	30623	0	4184	0
Jordan	0	6	0	0
Korea, Rep	2	0	2	0
Laos	0	150	0	44
Lebanon	0	0	1	0
Malaysia	0	19708	13	0
Myanmar	0	0	0	56
Pakistan	0	1	0	0
Philippines	0	19059	0	8963
Saudi Arabia	10	0	53	0
Singapore	1965	0	0	32
Sri Lanka	2	0	0	2
Taiwan	0	4	223	0
Thailand	0	2413	28	0
Turkey	0	7504	0	0
United Arab Emirates	112	0	21	0
Viet Nam	0	501	0	4
Yemen	0	7	0	0
USSR (former)	0	52121	1630	0
EUROPE	**131561**	**224**	**14626**	**3**
Austria	32252	0	135	0
Belgium	1270	0	1143	0
Czechoslovakia	0	146	35	0
Denmark	29	0	60	0
Finland	2	0	0	1
France	31522	0	2333	0
German Dem Rep	225	0	13	0
Germany, Fed Rep	23745	0	190	0
Greece	0	29	3	0
Hungary	591	0	85	0
Ireland	2	0	0	1
Italy	3692	0	1169	0
Malta	10	0	0	0
Monaco	254	0	3	0
Netherlands	16912	0	2264	0
Norway	5	0	0	1
Poland	0	40	1	0
Portugal	0	0	60	0
Romania	9	0	18	0
Spain	1706	0	76	0
Sweden	97	0	698	0
Switzerland	6912	0	111	0
United Kingdom	12326	0	4183	0
Yugoslavia	0	9	2046	0
NORTH & CENTRAL AMERICA	**272038**	**128416**	**20914**	**1515**
Antigua	0	180	0	0
Bahamas	2	0	0	0
Barbados	0	0	0	986
Belize	0	4	0	0
Canada	5286	0	1350	0
Cayman Islands	0	19	0	0
Costa Rica	0	48	0	0
Dominican Rep	0	11	20	0
El Salvador	0	8801	0	0
Haiti	0	5635	0	0
Honduras	0	112625	0	528
Mexico	0	12	221	0
Netherlands Antilles	0	75	0	0
Nicaragua	0	1000	0	0
Panama	0	2	0	0
Trinidad and Tobago	0	4	0	1
United States	266750	0	19323	0
SOUTH AMERICA	**120**	**57427**	**110**	**3501**
Argentina	0	2183	0	91
Bolivia	0	0	0	5
Brazil	105	0	0	231
Chile	0	903	47	0
Colombia	0	7985	0	1
Ecuador	15	0	0	2
French Guiana	0	0	63	0
Guyana	0	9842	0	2822
Peru	0	9765	0	342
Suriname	0	26749	0	3
Venezuela	0	0	0	4

Table 26.14 Live reptile and primate trade in 1989 (continued)

	LIVE REPTILES		LIVE PRIMATES	
	IMPORTS[1]	EXPORTS[1]	IMPORTS[1]	EXPORTS[1]
OCEANIA	**272**	**3142**	**2**	**272**
Australia	272	0	0	272
Fiji	0	0	2	0
New Zealand	0	1	0	0
Papua New Guinea	0	1	0	0
Vanuatu	0	16	0	0
Solomon Islands	0	3124	0	0
AFRICA	**1813**	**140157**	**17**	**9613**
Angola	0	0	3	0
Benin	0	0	0	1
Botswana	0	671	1	0
Burkina Faso	0	0	2	0
Burundi	0	4326	0	0
Cameroon	0	0	0	80
Comoros	0	428	0	0
Congo	0	2	0	4
Côte d'Ivoire	0	12	0	0
Egypt	0	1060	0	0
Ethiopia	0	0	0	492
Gabon	0	0	0	2
Ghana	0	25400	0	249
Guinea	0	6	0	2
Kenya	0	53	0	2176
Lesotho	0	1462	0	0
Liberia	0	150	0	0
Libya	12	0	2	0
Madagascar	0	24901	0	4
Mali	0	10	0	0
Mauritius	388	0	0	3215
Morocco	0	15	4	0
Mozambique	0	368	0	0
Namibia	0	1028	1	0
Niger	0	0	0	2
Nigeria	0	2	0	0
Reunion	19	0	0	0
Senegal	0	0	0	607
Seychelles	0	176	0	0
South Africa	1368	0	0	13
Sudan	0	2	0	0
Swaziland	0	0	2	0
Tanzania	0	4416	0	2387
Togo	0	74975	0	363
Tunisia	26	0	0	3
Uganda	0	0	1	0
Western Sahara	0	126	0	0
Zaire	0	226	0	12
Zambia	0	195	1	0
Zimbabwe	0	147	0	1
OTHER	**160**	**91**	**135**	**427**

Notes: [1] figures are net.
Sources: Annual reports of Parties to CITES compiled by WCMC.

the Old World where they occur naturally. In the Americas where there are no native *Apis*, 'stingless' bees of the family Meliponidae have traditionally been kept, but the more recently introduced *A. mellifera* from Europe has become the main species in the beekeeping industry. A main reason for the domestication of bees has always been the production of honey and beeswax (Table 26.5). However, the importance of these insects for pollination has not been missed and apiculture has usually gone hand-in-hand with agricultural production of crops requiring bee pollination (see Table 26.15).

Wild bees of hundreds of species are also important pollinators and are more effective than the honeybees for certain crops and in colder climates.

Sport hunting

In many societies animals are hunted for pleasure and, in affluent societies, private individuals may pay large sums of money for the privilege. In many cases the offtakes are controlled by a system of hunting licences or permits sold by the government, which can raise significant revenues for central or local treasuries. For example, deer, gamebird and wildfowl harvests are regulated in this manner in North America. Owners of private land, including in some African countries, are able to sell the right to shoot animals on their land to visitors, considerably enhancing the economic value of wildlife.

Recreation, tourism, aesthetic value

Wildlife also has enormous recreational and aesthetic value. Many people derive pleasure from wildlife either by observing them in the course of their daily lives, by making special excursions to view them, by watching them on film or television, or simply by knowing that they continue to exist, without necessarily wishing to see for themselves. This kind of non-consumptive use is very difficult to evaluate but it is possibly the single greatest economic value of wildlife. One of the easier techniques is to quantify what tourists are prepared to pay to observe wild animals in their natural habitat. This is of considerable interest in developing countries because overseas tourists bring in much-needed foreign exchange. It has been estimated that visitors pay almost US$200 per person to spend an hour with wild but habituated Mountain Gorilla *Gorilla gorilla beringei* groups in Rwanda. These visits generate nearly

Table 26.15 Selected crops of commercial importance in the EC for which there is agreement of the importance of insect/bee pollination from several sources

FAMILY	GENUS AND SPECIES	COMMON NAME	EC PRODUCTION (x1000 tons)	REPORTED VISITORS	NEED FOR INSECT POLLINATION D	P
Compositae	*Helianthus annuus*	Sunflower	3,908	H,B,S	1.0	0.9
Cruciferae	*Brassica campestris*	Turnip rape	?	H,B,S,	-	-
Cucurbitaceae	*Citrullus lanatus*	Water melon	1,838	H,S	0.7	0.9
	Cucumis sativus	Cucumber/gherkin	1,372	H,S	0.9	0.9
	Cucumis melo	Melon	1,654	H	0.8	0.9
Lauraceae	*Persea americana*	Avocado	29	H,S	1.0	0.9
Leguminosae	*Medicago sativa*	Lucerne/alfalfa	?	H,B,S	1.0	0.6
	Phaseolus multiflorus	Runner bean	?	H,B	-	-
	Trifolium pratense	Red Clover	?	H,B,S	-	-
Rosaceae	*Prunus amygdalus*	Almond	347	H	1.0	1.0
	Prunus sp.	Cherry	546	H,B	0.9	0.9
	Prunus sp.	Pear	2,631	H,B	0.5	0.9
	Pyrgus malus	Apple	9,321	H,B,S	1.0	0.9
Rubiaceae	*Coffea* sp.	Coffee	?	H,S	-	-

Source: Compiled from several tables in : Corbet, S.A., Williams, I.H., and Osborne, J.L. 1991. *Bees and the Pollination of Crops and Wild Flowers: changes in the European Community.* Review commissioned by Scientific and Technical Options Assessment, European Parliament. These tables are themselves drawn from a number of studies, some carried out in the USA.
Notes: ? Statistics unknown. H = honeybee. B = bumblebee. S = solitary bee. D = proportion of yield attributable to insect pollination. P = proportion of effective insect pollinators that are bees.

US$1 million per year in direct park revenues (Vedder and Weber, 1990). In Kenya, the African Elephant was estimated to be worth some US$25 million a year to the tourist industry (Brown, 1989). The enjoyment of wildlife and wild habitats is one of the tourist industry's most rapidly expanding sectors. Thus it is primarily the wildlife which draws people to places such as East and southern Africa, the Galápagos, the Valdés Peninsula and the Brazilian Pantanal.

Social and cultural significance

Wildlife influences the philosophy, language, art, religion, and social structure itself of many societies. In African cultures wild animals figure prominently in animist beliefs, mythology, and works of art such as carvings and paintings. Economic dependence on wild resources produces a close relationship between the ecological factors governing wildlife and social organisation in some tribes. For example, seasonal alterations in hunting technique (individual hunting in the wet season, cooperative netting of game in the dry season) in the Babinga people of the Central African Republic affect the location of encampments, seasonally influencing group composition, social rapport and material exchange. In many cultures a man's social worth is measured by his prowess as a hunter.

Companion animals

Wild animals have been captured and kept in captivity for a variety of reasons for centuries, including as pets or companion animals, for entertainment, and for private and public display in zoos and menageries. In South America, a survey of four Kayapo villages revealed that at least 31 species of animal were kept as pets, including five species of turtle, 16 species of parrot and macaw, a lizard and a spider (Redford and Robinson, 1991). Pet-keeping is an almost ubiquitous human activity, and there is a thriving export trade in wild animals for pets from many tropical countries to the developed countries, involving large numbers of species and significant sums of money. For example, in the USA the annual retail turnover of parrots alone, both wild and captive-bred, has been estimated at US$300 million (Hemley, 1988). The gross retail value of parrots exported from Neotropical countries from 1982 to 1986 was estimated to be some US$1.6 billion. Parrots are more valuable than the smaller species of birds used in the pet trade and, although they are traded in smaller numbers, they contibute a disproportionate amount to the total value of the trade. Bird imports to the USA from Indonesia in 1986 and 1987 were worth US$4.4 million, an average of US$79 a bird, while the larger numbers of birds imported from Senegal over the same period were only worth an average of US$1.37 each because they comprised mostly the smaller seed-eating species (Thomsen *et al.*, 1992).

Table 26.16 shows a summary of the trade in live parrots from 1980 to 1989 recorded in CITES annual reports. The overall levels of trade appear to have increased from 76,629 in 1980 to a peak of 625,799 in 1988 before declining slightly in 1989. This finding must be treated with some caution because the standard of reporting of trade has not remained constant over this period. The number of Parties to CITES has increased from 1980 to 1989 and this will almost inevitably entail an increase in the volume of trade reported. Furthermore, the ability of Parties to monitor their trade has improved as more sophisticated mechanisms and procedures have been set up. Thus, although France has been a Party to CITES since 1978, its reported volume of parrot imports has climbed from zero in 1980 to 34,643 in 1989. There may have been some increase in the volume of trade over this period but the major increase, apparent between 1984 and 1985, was the result of the adoption of new procedures to report on imports of Appendix II species. Nevertheless, data from a range of countries suggest that there has been a genuine increase in trade over the period 1982-1988. The apparent decline in the volume of trade in 1989 is probably attributable to the delay in

Table 26.16 Imports and exports of live parrots, 1980–1989

	IMPORTS										EXPORTS									
	1980	1981	1982	1983	1984	1985	1986	1987	1988	1989	1980	1981	1982	1983	1984	1985	1986	1987	1988	1989
WORLD	**76621**	**319063**	**476836**	**495616**	**523782**	**531892**	**588649**	**624583**	**624665**	**543326**	**76629**	**319067**	**476917**	**497719**	**526001**	**532514**	**588884**	**624680**	**625799**	**543739**
ASIA	**2**	**29517**	**52446**	**48374**	**71433**	**54245**	**47303**	**91309**	**76474**	**45696**	**37**	**45198**	**84000**	**113392**	**98637**	**75574**	**112384**	**179593**	**135210**	**120082**
Afghanistan	0	0	0	10	0	0	0	0	0	0	0	0	0	0	0	0	0	0	0	0
Bahrain	0	0	0	52	462	778	431	545	373	343	0	0	1	0	0	0	0	1	0	0
Bangladesh	0	0	0	0	48	0	0	0	1238	265	0	0	0	0	0	1	9829	20492	0	0
Bhutan	0	0	0	0	0	0	0	0	0	0	0	0	0	0	0	110	500	0	0	0
Brunei	0	0	0	0	0	2	152	106	173	0	0	0	0	1	0	0	0	0	0	284
China	0	0	880	0	0	117	0	1304	118	111	0	260	0	178	1163	0	489	0	0	0
Cyprus	0	0	1	1	503	704	661	396	451	1	0	0	0	0	0	0	0	0	0	0
Hong Kong	0	1961	1291	2420	1201	2459	6699	9295	8441	6354	0	0	0	0	0	0	0	0	0	0
India	2	0	0	0	0	0	0	0	0	0	0	12865	15858	27698	16075	21900	15283	15342	8781	38404
Indonesia	0	0	0	0	0	0	0	0	0	0	0	26068	32861	77445	79044	43991	58686	77761	87830	62728
Iran	0	0	0	0	0	0	1	0	0	0	0	0	0	0	0	2	0	0	0	0
Iraq, Islamic Rep	0	0	1	4	6	1	1	0	8	105	0	0	0	0	0	0	0	0	0	0
Israel	0	0	1	1	8	1969	2546	2510	2242	2066	0	0	0	0	0	0	0	0	0	0
Japan	0	20755	44807	38533	38768	22970	27877	36628	35097	12380	0	0	0	0	0	0	0	0	0	0
Jordan	0	0	0	0	773	482	374	13631	10248	271	0	0	0	0	0	0	0	0	0	0
Korea, Dem People's Rep	0	0	0	0	0	20	0	0	0	0	0	0	0	0	1	0	0	0	0	0
Korea, Rep	0	360	309	11	1081	308	0	150	363	349	0	0	0	0	0	0	149	0	0	0
Kuwait	0	649	492	1013	2958	2917	1732	713	1127	0	0	0	0	0	0	0	0	0	0	3
Laos	0	0	0	0	0	0	0	0	0	0	0	0	0	0	0	0	7	0	0	0
Lebanon	0	1	15	246	821	123	6	9	51	30	0	0	0	0	0	0	0	0	0	0
Macau	0	0	0	0	0	0	0	0	0	0	0	0	0	0	0	2	0	0	0	0
Malaysia	0	0	0	2208	5250	0	0	0	0	0	1	1468	6876	0	0	5030	14335	8255	6818	11159
Nepal	0	0	0	0	0	0	18	0	35	0	0	0	0	0	0	0	0	0	0	1
Mongolia	0	0	1	0	0	0	0	0	0	0	0	0	0	0	0	0	0	0	0	0
Myanmar	0	0	0	0	1	0	0	0	0	0	0	0	0	0	0	0	0	0	597	1362
Oman	0	0	0	1	229	1328	362	59	72	3	0	0	1	0	0	0	0	0	0	0
Pakistan	0	0	200	3	712	571	426	0	0	3338	0	0	0	0	0	0	0	10202	6394	0
Philippines	0	0	0	0	0	0	0	0	0	0	13	48	168	1287	139	1197	482	2131	661	252
Qatar	0	0	20	0	946	520	837	809	910	674	0	0	0	0	0	0	0	0	0	0
Saudi Arabia	0	0	0	314	13666	11523	1994	2654	2456	3811	1	1	7	0	0	0	0	0	0	0
Singapore	0	2757	2479	0	0	132	0	17940	12130	13672	6	0	0	522	967	0	190	0	0	0
Sri Lanka	0	4	14	0	79	406	345	84	157	41	0	0	0	25	0	0	0	0	0	0
Syria	0	0	0	0	0	110	0	0	2	2	0	0	0	0	1	0	2	0	0	0
Taiwan	0	0	0	0	0	1921	0	0	0	0	0	4488	21738	6235	1247	0	8171	26280	21541	1350
Thailand	0	2169	0	891	659	0	0	0	0	0	16	0	6490	0	0	3326	4261	8493	319	654
Turkey	0	0	2	0	1	185	3	0	23	1260	0	0	0	1	0	0	0	6	0	0
United Arab Emirates	0	861	1933	2666	3151	4699	2838	4476	759	620	0	0	0	0	0	0	0	0	0	0
Viet Nam	0	0	0	0	10	0	0	0	0	0	0	0	0	0	0	15	0	10630	2269	3885
Yemen	0	0	0	0	100	0	0	0	0	0	0	0	0	0	0	0	0	0	0	0
USSR (former)	8	4	81	2103	2219	622	235	98	1137	437	0	0	0	0	0	0	0	0	0	0
EUROPE	**1836**	**43558**	**123475**	**131198**	**124412**	**152364**	**217851**	**232099**	**222952**	**242326**	**46635**	**35735**	**2257**	**8467**	**35042**	**8426**	**4417**	**4119**	**7660**	**9266**
Albania	0	0	0	0	0	28	0	0	0	0	0	0	0	0	0	0	0	0	0	0
Andorra	0	0	0	0	0	350	0	0	0	50	0	0	0	0	0	0	0	0	0	0
Austria	0	216	1328	400	1325	2230	3174	3889	4472	4637	0	0	0	0	0	0	0	0	0	0
Belgium	0	0	1645	0	2444	7630	7231	9329	0	12571	957	4285	0	1125	0	0	0	0	1801	0
Bulgaria	0	0	0	8	11	1	13	40	3	0	0	0	0	0	0	0	0	0	0	1
Czechoslovakia	0	7	0	0	0	0	0	0	0	0	35	0	1526	34	473	110	1217	1545	1616	567
Denmark	0	1510	1309	3996	4194	6224	7984	2294	5205	2232	517	0	0	0	0	0	0	0	0	0
Finland	0	0	9	1	0	0	0	0	4	1637	0	0	0	0	30	41	2	4	0	0
France	0	1061	2866	4667	6316	16498	24198	30537	30123	34643	10	0	0	0	0	0	0	0	0	0
German Dem Rep	0	228	0	0	0	0	0	0	0	0	0	0	651	7306	11042	6156	3198	2369	2102	5057

Table 26.16 Imports and exports of live parrots, 1980–1989 (continued)

	IMPORTS										EXPORTS									
	1980	1981	1982	1983	1984	1985	1986	1987	1988	1989	1980	1981	1982	1983	1984	1985	1986	1987	1988	1989
EUROPE (continued)																				
Germany, Fed Rep	1764	33844	56030	54546	52265	54375	63854	51633	54906	64722	0	0	0	0	0	0	0	0	0	0
Gibraltar	0	0	0	0	22	0	47	0	0	0	0	0	0	0	0	0	0	0	1	0
Greece	0	0	9	5	2	1311	955	2567	4064	1823	0	1	0	0	0	0	0	0	0	0
Greenland	0	0	0	0	0	30	3	3	4	1	0	0	0	0	0	0	0	0	0	0
Hungary	0	21	0	27	7	757	1521	0	51	188	1	0	1	0	0	0	0	1	0	0
Iceland	0	0	0	0	1	0	0	0	0	124	0	0	0	0	0	0	0	0	5	0
Ireland	0	4	0	0	0	0	50	0	17	0	0	0	78	0	0	0	0	0	0	2
Italy	0	1572	16846	25863	12774	7949	17925	19704	29723	24563	1	0	0	0	0	0	0	0	0	0
Liechenstein	0	0	0	0	0	0	7	0	0	0	0	0	0	0	0	0	0	200	0	0
Luxembourg	0	0	0	0	1	0	1	0	2	0	0	0	0	0	0	0	0	0	0	0
Malta	0	0	60	0	80	801	423	529	1131	549	50	0	0	0	0	2	0	0	0	0
Monaco	0	0	0	0	0	0	125	1	0	1	0	0	0	0	0	0	0	0	1	0
Netherlands	0	0	8713	371	0	0	14346	16267	0	0	19594	27391	0	0	23497	2106	0	0	2134	3639
Norway	0	0	0	0	17	0	64	13	12	42	775	216	1	2	0	11	0	0	0	0
Poland	0	0	0	5	47	51	154	6	0	285	0	0	0	0	0	0	0	0	0	0
Portugal	0	127	4703	388	983	1198	4934	15980	9019	6006	4	0	0	0	0	0	0	0	0	0
Romania	0	0	0	1	2	5	0	24	16	137	0	0	0	0	0	0	0	0	0	0
Spain	0	1502	4902	5658	14806	13288	16851	31917	30937	38539	2	0	0	0	0	0	0	0	0	0
Sweden	0	0	5311	7484	11403	14638	13962	7343	8811	4685	1	3842	0	0	0	0	0	0	0	0
Switzerland	72	446	3053	3395	2149	2697	3655	2016	3402	4773	0	0	0	0	0	0	0	0	0	0
United Kingdom	0	3015	16691	24371	15556	22301	36370	37985	40771	39044	24688	0	0	0	0	0	0	0	0	0
Yugoslavia	0	5	0	12	7	2	4	22	279	1074	0	0	0	0	0	0	0	0	0	0
NORTH & CENTRAL AMERICA	**74782**	**239035**	**292488**	**312631**	**320774**	**319562**	**316501**	**275316**	**281203**	**213981**	**24888**	**10044**	**14714**	**10420**	**18963**	**29848**	**22151**	**21384**	**29836**	**22680**
Anguilla	0	0	0	0	0	1	0	0	0	0	0	0	0	0	0	0	0	0	0	0
Antigua	0	0	0	0	0	0	405	1	3	0	0	0	0	0	4	0	0	0	0	298
Aruba	0	0	0	0	0	229	181	17	16	1	0	0	0	0	0	0	0	0	0	0
Bahamas	0	0	2	7	212	378	222	103	283	7	0	0	0	0	0	0	0	0	0	0
Barbados	0	0	0	0	10	0	113	19	58	0	0	0	2	0	0	0	0	0	0	122
Belize	0	0	0	0	0	0	0	0	0	0	0	5	13	29	7	15	42	24	10	5
Bermuda	0	0	0	0	0	0	0	7	292	116	0	0	0	1	0	0	2	0	0	0
Canada	74782	62948	31688	23434	10002	10744	9410	7525	5187	6405	0	0	0	0	0	0	0	0	0	0
Cayman Islands	0	16	1	0	16	97	96	57	115	57	2	0	0	0	0	0	0	0	0	0
Costa Rica	0	4	0	0	42	0	14	0	0	0	0	0	30	43	0	302	0	771	4632	1200
Cuba	0	48	9	11	13	16	2	7	0	11	0	0	0	0	0	0	0	0	181	0
Dominica	0	0	0	0	0	0	0	0	0	0	0	0	0	2	0	2	0	1	0	0
Dominican Rep	0	0	0	4	155	253	71	276	455	240	0	602	433	0	0	0	0	0	0	0
El Salvador	0	0	0	0	0	0	0	13	81	42	0	124	874	549	1248	2049	19	0	0	0
Grenada	0	0	0	0	0	0	0	1	3	2	0	0	0	0	0	0	0	0	0	327
Guatemala	0	0	0	0	0	0	0	37	116	67	1	544	86	68	3139	10083	3628	0	0	0
Haiti	0	0	0	0	0	0	0	17	0	9	1	2	6	19	7	0	1	0	3	0
Honduras	0	0	0	0	0	0	0	0	0	0	2	3160	6496	9251	14456	17144	15774	19386	24515	18935
Jamaica	0	0	0	0	0	0	0	28	0	2	0	0	2	84	5	2	296	0	3	0
Martinique	0	1	0	0	0	0	1	0	0	0	0	0	0	0	0	0	0	0	0	0
Mexico	0	0	0	0	326	94	142	891	3111	6633	6	5278	6534	132	0	0	0	0	0	0
Netherlands Antilles	0	1	5	4	2345	1353	798	1352	835	190	10	0	0	0	0	0	0	0	0	0
Nicaragua	0	0	0	0	0	0	0	0	0	0	0	49	56	108	96	251	2389	1202	479	1743
Panama	0	0	0	0	198	282	230	323	120	67	0	280	182	132	0	0	0	0	0	0
Puerto Rico	0	0	2	0	0	4	0	4	3	0	0	0	0	0	1	0	0	0	0	2
St Lucia	0	0	0	0	0	0	0	0	2	0	0	0	0	0	0	0	0	0	0	48
Trinidad and Tobago	0	195	401	132	1265	603	15	71	102	80	0	0	0	0	0	0	0	0	0	0
United States	0	175822	260380	289039	306190	305508	304801	264562	277421	200052	24866	0	0	0	0	0	0	0	0	0
Virgin Islands (British)	0	0	0	0	0	0	0	5	0	0	0	0	0	0	0	0	0	0	1	0
Virgin Islands (US)	0	0	0	0	0	0	0	0	0	0	0	0	0	2	0	0	0	0	12	0

Table 26.16 Imports and exports of live parrots, 1980–1989 (continued)

	IMPORTS										EXPORTS									
	1980	1981	1982	1983	1984	1985	1986	1987	1988	1989	1980	1981	1982	1983	1984	1985	1986	1987	1988	1989
SOUTH AMERICA	**1**	**479**	**1314**	**1187**	**332**	**189**	**4**	**77**	**82**	**185**	**118**	**175109**	**236455**	**245875**	**255244**	**261973**	**270713**	**213808**	**274577**	**232056**
Argentina	0	0	0	0	0	0	0	0	0	0	101	73007	88801	114018	108766	179113	177971	151025	179762	171531
Bolivia	0	0	0	0	0	0	0	0	0	0	0	39872	56340	48771	11584	115	18	3	4	11
Brazil	1	0	26	28	53	57	0	0	58	95	0	14	0	0	0	0	39	27	0	0
Chile	0	0	0	945	162	0	0	0	0	0	1	1	16	0	0	502	815	716	826	296
Colombia	0	0	0	0	0	0	0	0	0	0	0	209	256	219	150	21	26	33	21	7
Ecuador	0	0	0	0	0	0	0	0	0	0	0	5395	3648	398	3936	201	62	8	10	8
French Guiana	0	0	0	0	0	13	4	0	0	0	0	0	0	0	0	0	0	0	0	0
Guyana	0	0	0	0	0	0	0	0	0	0	10	28941	26555	27588	38176	27259	30322	9322	26935	13926
Paraguay	0	0	0	0	0	0	0	76	0	0	0	1082	10	7	24	5	42	0	3	37
Peru	0	0	0	0	0	0	0	0	0	0	4	15562	39303	19462	51671	33785	16735	9639	16760	16158
Suriname	0	0	0	0	0	0	0	0	0	0	0	2106	1982	1896	1727	2517	6825	8155	9719	8249
Uruguay	0	0	0	0	0	0	0	0	0	0	0	8920	19544	33516	39210	18455	37850	34880	40537	21833
Venezuela	0	479	1288	214	117	119	0	1	24	90	2	0	0	0	0	0	8	0	0	0
OCEANIA	**0**	**0**	**0**	**654**	**281**	**327**	**110**	**195**	**288**	**254**	**29**	**281**	**437**	**16**	**432**	**377**	**102**	**1568**	**1148**	**48**
Australia	0	0	0	249	0	0	0	0	0	250	23	168	252	0	357	377	94	1544	1129	0
Fiji	0	0	0	0	0	0	0	0	133	4	0	0	0	2	0	0	0	0	0	0
Guam	0	0	0	0	0	0	0	0	0	0	0	0	0	0	1	0	0	0	0	0
New Caledonia	0	0	0	0	0	0	0	0	0	0	6	0	0	0	0	0	0	0	15	8
New Zealand	0	0	0	405	281	327	110	195	155	0	0	111	182	0	0	0	0	0	0	40
Papua New Guinea	0	0	0	0	0	0	0	0	0	0	0	2	3	12	41	0	8	22	4	0
Solomon Islands	0	0	0	0	0	0	0	0	0	0	0	0	0	2	33	0	0	2	0	0
AFRICA	**0**	**234**	**3567**	**1572**	**6550**	**5205**	**6880**	**25573**	**31034**	**15993**	**4917**	**52700**	**139037**	**117327**	**117150**	**155525**	**178495**	**201214**	**177367**	**159604**
Algeria	0	3	0	0	8	0	4	191	4	0	0	0	0	0	0	0	0	0	0	3
Angola	0	0	0	0	0	0	0	0	0	0	0	9	14	8	14	3	5	5	6	1
Benin	0	0	0	0	16	2	0	0	0	0	0	1	0	4	0	0	1	8	0	7
Botswana	0	0	1	0	0	120	1083	12	1	4	0	0	0	0	0	0	0	0	0	0
Burkina Faso	0	0	0	0	0	0	0	3	0	0	0	1	0	3	4	1	0	0	9	0
Burundi	0	0	0	0	0	0	0	0	0	0	0	0	0	0	0	0	34	147	1	73
Cameroon	0	0	0	0	0	0	0	0	0	0	0	5760	4955	10341	12145	11198	10318	14444	14191	13085
Cape Verde	0	0	0	0	0	0	0	0	0	4	0	0	0	0	0	0	1	0	0	0
Central African Rep	0	0	0	0	0	0	0	0	0	0	0	0	0	237	56	788	1085	417	465	200
Chad	0	0	0	0	0	2	0	6	0	0	0	0	0	0	0	0	0	0	0	1
Comoros	0	0	0	0	1	0	0	0	1	0	0	0	0	0	0	0	0	0	0	0
Congo	0	0	0	0	0	0	0	0	0	0	0	0	0	30	92	38	22	32	36	44
Côte d'Ivoire	0	0	0	0	18	0	0	0	0	0	0	0	102	0	0	338	1701	2336	1053	10023
Djibouti	0	0	0	0	0	12	0	0	0	0	0	0	0	0	0	0	0	0	0	0
Egypt	0	198	492	2	1588	3139	1603	906	1412	1218	0	0	0	0	0	0	0	0	0	0
Equatorial Guinea	0	0	0	0	0	0	0	0	0	0	0	0	1	0	0	0	0	373	50	0
Ethiopia	0	0	0	0	0	0	0	0	3	0	0	0	0	0	0	0	0	0	0	0
Gabon	0	0	0	0	0	0	0	0	2	0	0	0	2	0	0	280	3	5	0	13
Gambia	0	0	0	0	2	0	0	1	0	0	0	0	1	1	0	1	0	0	0	0
Ghana	0	0	0	0	0	0	0	0	0	0	4871	950	5836	10684	6598	11266	6561	1634	4141	1434
Guinea	0	0	0	0	0	0	0	0	0	0	0	2	0	4	0	159	13094	8891	29878	8896
Guinea Bissau	0	0	0	0	0	0	0	0	0	0	0	0	1	0	4	1	2	0	1	1
Kenya	0	0	0	0	0	0	0	0	0	0	2	8	2	6	295	21	24	13	7	11
Lesotho	0	0	0	200	0	0	0	4	0	0	0	0	0	0	0	0	0	0	0	0
Liberia	0	0	0	0	0	0	0	0	0	0	0	3102	8905	3895	6373	5275	9735	10046	7693	4461
Libya	0	9	0	3	16	19	20	29	8	11	0	0	0	0	0	0	0	0	0	0
Madagascar	0	0	0	0	0	0	0	0	0	0	0	666	2602	3168	5527	11660	9433	7475	12657	5082

Table 26.16 Imports and exports of live parrots, 1980–1989 (continued)

	IMPORTS										EXPORTS									
	1980	1981	1982	1983	1984	1985	1986	1987	1988	1989	1980	1981	1982	1983	1984	1985	1986	1987	1988	1989
AFRICA (continued)																				
Mali	0	0	0	0	0	0	0	0	0	0	0	7832	13672	9211	4176	5193	3123	5717	996	2547
Mauritius	0	20	0	0	0	290	1060	3	304	1126	0	0	0	0	0	0	0	0	0	0
Morocco	0	0	3	0	0	20	40	4	26	63	0	0	0	0	3	0	0	0	0	0
Mozambique	0	0	0	0	0	0	2	8	0	4	0	0	0	2	0	0	0	0	0	0
Namibia	0	0	0	1	19	23	46	93	58	47	0	0	0	0	0	0	0	0	0	0
Niger	0	0	40	2	0	100	0	0	0	0	0	0	0	0	0	0	6	8	0	3
Nigeria	0	0	0	0	0	0	0	0	0	0	4	15	66	99	51	26	39	15	20	14
Reunion	0	1	0	29	549	607	631	341	743	1040	0	0	0	0	0	0	0	0	0	0
Rwanda	0	0	0	0	0	18	0	0	0	0	0	0	2	0	1	0	3	2	4	3
Senegal	0	0	0	0	0	0	0	0	0	0	28	7665	25270	20975	26674	19874	34663	36654	33729	60114
Seychelles	0	0	0	0	0	0	0	0	0	0	0	0	0	0	0	0	0	0	0	1
Sierra Leone	0	0	0	0	0	0	0	0	0	0	0	0	0	2	0	1	0	4	1	1
Somalia	0	0	0	0	0	1	0	20	0	0	0	0	0	0	0	0	0	0	0	2
South Africa	0	0	131	1328	4088	0	72	20992	26165	11038	4	8	0	0	0	10550	0	0	0	0
Sudan	0	0	0	1	239	84	53	3	5	38	0	0	1	0	0	0	0	0	0	0
Swaziland	0	0	2900	5	0	760	2264	2950	1755	1398	0	0	0	0	2	0	0	0	0	0
Tanzania	0	0	0	0	0	0	0	0	0	0	2	22233	74151	51607	44889	70418	84228	104028	65091	45410
Togo	0	0	0	0	0	0	0	0	0	0	0	3442	2535	2069	7962	7917	3949	7935	5985	5211
Tunisia	0	3	0	1	4	8	2	7	10	2	0	0	1	0	0	0	0	0	0	0
Uganda	0	0	0	0	2	0	0	0	0	0	0	0	0	1	0	4	0	6	3	5
Zaire	0	0	0	0	0	0	0	0	0	0	5	1005	917	4740	2256	177	81	197	348	1963
Zambia	0	0	0	0	0	0	0	0	0	0	0	0	0	0	5	4	0	600	1002	605
Zimbabwe	0	0	0	0	0	0	0	0	537	0	1	1	1	240	23	332	384	222	0	390
OTHER	**0**	**6240**	**3546**	**0**	**0**	**0**	**0**	**14**	**4632**	**24891**	**5**	**0**	**17**	**2222**	**533**	**791**	**622**	**2994**	**1**	**3**
Country Unknown	0	6240	3546	0	0	0	0	0	4616	24891	5	0	0	2217	532	790	622	2994	0	0
Other	0	0	0	0	0	0	0	14	16	0	0	0	17	5	1	1	0	0	1	3

Source: Annual reports of Parties to CITES, compiled by WCMC.

submitting annual reports. Table 26.16 shows that Europe and North America are the largest importers of live parrots and that the main exporters are in South America, Africa and Asia. The principal individual countries are Argentina, Indonesia, Tanzania and Senegal.

Reptiles are also used in the pet trade, the total volume of specimens recorded in CITES annual reports approaching half a million specimens in 1989 (Table 26.14). Over half of the total was imported by the USA and the majority of the remainder by countries in Europe. The main source countries were in Africa and Latin America.

DOMESTIC LIVESTOCK

The development of settled agriculture and animal husbandry has enabled human societies to live at high population densities which is a prerequisite for cultural development of the kind based on extensive division of labour. Consequently, hunter-gatherers everywhere have been displaced by agriculturists and there has been an adaptive radiation of domestic mammalian livestock into a greater range of habitat types than those occupied by any wild mammal. When ecological conditions have worsened, farming systems have been adapted, usually by adopting seasonal patterns of exploitation (pastoralism).

Domestication and breeds

In the process of domestication small numbers of wild animals were enfolded into human societies, which assumed responsibility for them and exerted control over their breeding. Almost all major domestication took place in western Asia and the Near East, from about 10,000 BC.

A domestic animal can be defined as one "that has been bred in captivity for purposes of economic profit to a human community that maintains complete mastery over its breeding, organisation of territory, and food supply" (Clutton-Brock, 1987). These truly *domestic* (or 'man-made') animals may differ radically from their wild ancestors in respect of a variety of features. Animals generally regarded as domestic are listed in Table 26.17.

A second group includes *domesticated* (or 'exploited captive') animals, which are tamed or conditioned individuals from populations whose breeding is not wholly under human control and which, as a result of continuing natural selection, tend to retain features of value in demanding environments. Principal domesticated vertebrates are listed in Table 26.18.

These categories are somewhat arbitrary in so far as domestic animals (in the broad sense) represent a wide spectrum of conditions and particular cases, but the distinction can be useful. There is a clear difference between traditional European farm livestock on the one hand, and reindeer, yak and dromedary on the other. In the latter group, whilst there may be some artificial selection for particular traits (for pack or racing dromedaries, for example), artificial selection has been limited because the animals' continuing close adaption to particularly harsh environments is the feature of special value to humans. However, several intermediate groups could, with emphasis on different aspects of breeding or husbandry, be regarded as either truly domestic or 'exploited captive'. Semi-domesticated animals are thus more difficult to define as they range from those such as the Silver Fox, which has been bred for many generations to produce distinctive pelt colours, to species, such as the musk deer, which are only kept on a very small, and at present experimental, scale. Pets make up a further group of animals that is extremely important in many if not most societies and pet keeping could have been a first stage in domestication (Serpell, 1989).

It is important to distinguish animals of the above kinds from tamed animals; as Clutton-Brock (1987) points out, any young mammal, taken from its mother, can be tamed but this is not necessarily a permanent state. Whether or not the tameness persists into adult life depends on the species, and any offspring would have to be tamed in their turn.

Domestic species of ungulate (hoofed mammals) tend to be large (over 50kg) non-selective feeders whose native habitat is open terrain or mountains (Tennessen and Hudson, 1981). The smaller-bodied species from habitats like forests and swamps (including most deer and antelopes) include several semi-domesticated species but no fully domestic ones, apart from the pig. Generally, domestic ungulates are non-territorial, living in groups of 15-100. It is not clear why other ungulate species, such as the Eland *Taurotragus oryx* and the European Bison *Bison bison* were never domesticated. Perhaps human societies sharing their ranges obtained enough food by hunting and there was not the population pressure that seems to have been the impetus for domestication of other ungulates.

The changes in mammals consequent on domestication are physical and behavioural, and have a genetic basis (Clutton-Brock, 1987). Body and brain sizes were reduced and proportions altered. Ears were lengthened in most species (except the horse), and the tail was lengthened in sheep, and selected to be curly in pigs and dogs. There was greater variability in the pelage. Particularly in the dog and pig, but also in the Niata cattle of Uruguay and to some extent in Jersey cattle (Darwin, 1868), the facial region and jaws were shortened. The teeth were reduced in size and became more crowded, especially in the dog. Greater docility, and changes in vocalisations, particularly in the dog, were accompanied by a retention of juvenile patterns of behaviour, particularly playfulness. Breeding seasons were lengthened, but less so in the most primitive breeds, e.g. of dogs and of sheep (Brisbin, 1977; Lincoln, 1989). While humans selected for behavioural traits and visible markers like coat colours, the animals were also subject to local pressures of natural selection.

It was the interaction of natural (environmental) selection and human (artificial) selection that led to distinct breeds. A breed can be defined as, "a group of animals that has been selected by man to possess a uniform appearance that is inheritable and distinguishes it from other groups of animals within the same species" (Clutton-Brock, 1987).

Wild animals are adapted to the physical and biological environments, and domestic animals are also subject to

Table 26.17 Domestic livestock

DOMESTIC FORM		WILD PROGENITOR		FIRST KNOWN DOMESTICATION DATE	FIRST KNOWN DOMESTICATION PLACE	DISTRIBUTION OF WILD PROGENITOR
MAMMALS						
LAGOMORPHA						
Rabbit	*Oryctolagus cuniculus*	European rabbit	*O. cuniculus*	36 BC	S Europe	SW Europe, possibly N Africa
RODENTIA						
Guinea pig	*Cavia porcellus*		*Cavia aperea*	1000 BC	S America	S America
CARNIVORA						
Dog	*Canis familiaris*	Wolf	*Canis lupus*	12,000 BC	Iraq	N hemisphere
Ferret	*Mustela furo*	Polecat	*Mustela putorius*	20 AD	S Europe	Europe
		Steppe polecat	*Mustela eversmanni*			USSR, China
Cat	*Felis catus*	Wild cat	*Felis silvestris*	1600 BC	Egypt	Europe, Asia, Africa
PERISSODACTYLA						
Horse	*Equus caballus*	Wild horse	*Equus ferus*	3500 BC	S Ukraine	Russia, Central Asia
Donkey	*Equus asinus*	African ass	*Equus africanus*	4000 BC	Egypt	N Africa, possibly W Asia
ARTIODACTYLA						
Pig	*Sus domesticus*	Wild boar	*Sus scrofa*	7000 BC	W Asia	Europe, Asia and N Africa
Llama	*Lama glama*	Guanaco	possibly *Lama guanicoe*	5500-4200 BC	Andean plateau	S America
Alpaca	*Lama pacos*		Lama sp.			S America
Dromedary	*Camelus dromedarius*	Dromedary	*Camelus* sp.	3000 BC	W Asia	Asia, possibly N Africa
Bactrian camel	*Camelus bactrianus*	Bactrian camel	*Camelus ferus*	3000 BC	W Asia	Russia, Central Asia
Reindeer	*Rangifer tarandus*	Reindeer	*Rangifer tarandus*	?	?	Arctic, sub-Arctic (feral: Greenland, Iceland, S Georgia)
Water buffalo	*Bubalus bubalis*	Water buffalo	*Bubalus arnee*	not known	China/Indo-China	India, S Asia, possibly W Asia
Cattle (taurine)	*Bos taurus*	Aurochs	*Bos primigenius*	6200 BC	Turkey	Europe, Asia, N Africa
Cattle (zebu)	*Bos indicus*	derived from *B. taurus*				
Yak	*Bos grunniens*	Yak	*Bos mutus*	not known	not known	Tibet, Himalayas
Mithan	*Bos frontalis*	Gaur	*Bos gaurus*	2500 BC		S and SE Asia
Bali cattle	*Bos javanicus*	Banteng	*Bos javanicus*			SE Asia including Borneo
Goat	*Capra hircus*	Wild goat	*Capra aegagrus*	7-8000 BC	W Asia	W Asia
Sheep	*Ovis aries*	Mouflon	*Ovis orientalis*	7-8000 BC	W Asia	W Asia
BIRDS						
GALLIFORMES						
Chicken	*Gallus gallus*	Red junglefowl	*Gallus gallus*	4000 BC	S and SE Asia	S and SE Asia
		Ceylon junglefowl	*Gallus lafayetii*			
		Grey junglefowl	*Gallus sonneratii*			
		Green junglefowl	*Gallus varius*			
Turkey	*Meleagris gallopavo*	Wild turkey	*Meleagris gallopavo*	1500 AD	Europe	C - N America
ANSERIFORMES						
Goose	*Anser anser*	Greylag goose	*Anser anser*	500 BC		N Europe, N Asia to NW Africa
Chinese goose	*Anser cygnoides*	Swan goose	*Anser cygnoides*	500 BC		Europe, Asia, N America, N Africa
Muscovy duck	*Cairina moschata*					Mexico to Peru and Uruguay
Mallard duck	*Anas platyrhynchos*			500 BC		Europe, Asia, N America, N Africa
COLUMBIFORMES						
Pigeon	*Columba livia*			3000 BC		Europe, N Africa, India to Japan
INSECTS						
Honey bee	*Apis mellifera* and other *Apis* spp.			2000 BC		Africa, Europe
Silk worm	*Bombyx mori*		possibly *B. mandarina*	2500 BC	Asia	
Silk worms - other semi-domesticated species	e.g. *Antheraea pernyi, A. mylitta, Attacus ricini, Anaphe* spp.					
Cochineal bug	*Dactylopius coccus*		*Dactylopius coccus*	pre 1200AD	Mexico, Central America	Americas

Source: compiled from various sources; mammal data after Clutton-Brock, J. 1987. *A Natural History of Domesticated Mammals.*

Table 26.18 Domesticated and semi-domesticated vertebrates (excluding fishes) used in wildlife farming

NAME	LATIN NAME	FARMED IN	WILD IN	PURPOSE
DEER				
Elk/Moose	*Alces alces*	Canada, USSR, Sweden	N Europe, Asia, N America	Meat, velvet, milk, draught
Axis Deer	*Cervus axis*	Australia, USA	India, Sri Lanka, Nepal	Velvet, meat, medicinal
Fallow Deer	*Cervus dama*	Eurasia, Australasia, N America	S Europe, N Africa,	Velvet, meat
Red Deer/Wapiti	*Cervus elaphus*	Eurasia, Australasia, N America	Europe, USSR, N America	Velvet, meat, medicinal, trophies
Sika Deer	*Cervus nippon*	Asia, Australasia	Asia, USSR, Japan	Meat, velvet
Rusa Deer	*Cervus timorensis*	Australia, Mauritius, Malaysia	Indonesia	Velvet, meat, medicinal, reintroduction
Sambar	*Cervus unicolor*	China, Thailand, Taiwan	SW Asia	Velvet, meat, medicinal, reintroduction
Père David's Deer	*Elaphurus davidianus*	UK, USA	China	Restocking, medicinal
Dwarf Musk Deer	*Moschus berezowskii*	China	SW China, N Viet Nam	Musk
Himalayan Musk Deer	*Moschus chrysogaster*	China, N India	E Asia	Musk
OTHER UNGULATES				
Impala	*Aepyceros melampus*	South Africa, Kenya, Zimbabwe	Southern Africa	Trophies, meat
Springbok	*Antidorcas marsupialis*	South Africa, Namibia, USA	South Africa	Trophies, meat
Blesbok	*Damaliscus dorcas*	South Africa, USA	Southern Africa	Trophies, meat
Grant's Gazelle	*Gazella granti*	Kenya, USA	E Africa	Trophy hunting
Thompson's Gazelle	*Gazella thomsoni*	Kenya, USA	E Africa	Trophy hunting
Fringe-eared Oryx	*Oryx beisa*	Kenya	Southern Africa	Trophies, meat
Gemsbok	*Oryx gazella*	South Africa, Namibia, USA	Southern Africa	Trophies, meat
African Buffalo	*Syncerus caffer*	USA, Kenya, Zimbabwe	Africa	Trophy hunting
Eland	*Taurotragus oryx*	South Africa, Kenya, Zimbabwe, USA, USSR	Africa	Trophies, meat, milk
FUR-BEARERS				
Arctic Fox (Blue phase)	*Alopex lagopus*	Europe, USA, USSR	N America, Europe, Asia	Pelt
Long-tailed Chinchilla	*Chinchilla laniger*	N America, Canada, Europe	S America	Pelt
Sable	*Martes zibellina*	USSR	USSR, China	Pelt
Polecat/Fitch	*Mustela putorius*	Europe, USSR, China, USA	Europe	Pelt
Mink	*Mustela vison*	Europe, N America, Asia	Canada	Pelt
Coypu	*Myocastor coypus*	Europe, USSR	S America	Pelt
Racoon Dog	*Nyctereutes procyonoides*	Europe, China	China	Pelt
Red/Silver Fox	*Vulpes vulpes*	N America, Europe, USSR	Americas, Europe, Asia, Africa	Pelt
OTHER MAMMALS				
Bison	*Bison bison*	N America	N America	Meat, trophies
Asian Elephant	*Elephas maximus*	India, Laos, Myanmar, Thailand	S and SE Asia	Timber extraction
Capybara	*Hydrochaeris hydrochaeris*	S America	S America	Meat, pelt

Table 26.18 Domesticated and semi-domesticated vertebrates (excluding fishes) used in wildlife farming (continued)

NAME	LATIN NAME	FARMED IN	WILD IN	PURPOSE
OTHER MAMMALS (continued)				
Musk Ox	*Ovibos moschatus*	N America	N America, USSR	Meat, wool
Wild Boar	*Sus scrofa*	Europe, USA	Eurasia	Meat
Cane/Grasscutter Rat	*Thryonomys swinderianus*	W Africa	West and Central Africa	Meat
Giant Rat	*Cricetomys gambianus*	West Africa	Caribbean	Weed control
Vicuña	*Vicugna vicugna*	Peru	S America	Wool
African Civet	*Civettictis civetta*	Ethiopia	Africa	Musk
BIRDS				
Chukar Partridge	*Alectoris chuka*	Worldwide	S Eurasia	Meat, hunting
Red-legged Partridge	*Alectoris rufa*	Worldwide	S Eurasia	Meat, hunting
Northern Bobwhite/Quail	*Colinus virginianus*	Worldwide	N America	Meat, hunting, feathers
Emu	*Dromaius novaehollandiae*	Australia	Australia	Meat, skin, feathers
Common/Grey Partridge	*Perdix perdix*	Worldwide	Eurasia	Meat, hunting, feathers
Common/Ring-necked Pheasant	*Phasianus colchicus*	Worldwide	Eurasia	Meat, hunting, feathers
Ostrich	*Struthio camelus*	Southern Africa, USA	Southern Africa	Meat, skin, feathers
CROCODILIANS				
American Alligator	*Alligator mississippiensis*	USA	USA	Skin, meat
Common Caiman	*Caiman crocodilus*	E Asia, S America	C and S America	Skin, meat
Broad-nosed Caiman	*Caiman latirostris*	Italy	S America	Skin, meat
American Crocodile	*Crocodylus acutus*	Colombia, Cuba	C and S America, Florida	Skin, meat
Australian Freshwater Crocodile	*Crocodylus johnsoni*	Australia	Australia	Skin, meat
Morelet's Crocodile	*Crocodylus moreletii*	Mexico	C America	Skin, meat
Nile Crocodile	*Crocodylus niloticus*	Africa, Brazil	Africa	Skin, meat
New Guinea Crocodile	*Crocodylus novaeguineae*	Indonesia, Papua New Guinea, Singapore	Indonesia, Papua New Guinea	Skin, meat
Estuarine Crocodile	*Crocodylus porosus*	Asia, Australia	E Asia, Australia	Skin, meat
Cuban Crocodile	*Crocodylus rhombifer*	Cuba, Viet Nam	Cuba	Skin, meat
Siamese Crocodile	*Crocodylus siamensis*	Thailand	E Asia	Skin, meat
OTHER REPTILES				
*Green Turtle	*Chelonia mydas*	Cayman Is, Réunion, Suriname		Meat, shells, oil, leather
*Hawksbill Turtle	*Eretmochelys imbricata*	Indonesia		Shells
Freshwater Turtle	*Trionychidea*	Asia	Asia	Pet trade, Meat
*Green Iguana	*Iguana iguana*	Costa Rica	C and S America	Meat, restocking
AMPHIBIANS				
Frogs (various)	*Rana* spp. (etc.)	Asia	Asia	Meat (legs)

Source: Unpublished 1984 survey of wildlife ranching operations by WCMC (WTMU).
Note: Includes principal species only, grouped for convenience according to taxonomic group or by primary purpose. *Indicates semi-domesticated species bred on a small or experimental scale

these environments but must also be adapted to their commercial environment. Animal husbandry is the art or science of reconciling domestic animals with these three environments. Animal breeding is the art or science of enhancing the positive economic response to husbandry, in such a way that this enhancement is inherited. The basic principles of animal husbandry are control over the movements of the animals, the securing of supplies of feed and water, and the management of reproduction so that young are born at the right time of year.

Domestication and genetic variation

Domestic and domesticated animals with all their breeds and types appear much more variable than their wild progenitors. Domestication represents a genetic bottleneck, meaning that the small sample of wild animals that is taken into reproductive isolation may lack much of the genetic variation of the wild population. However the great range of breeds with all their different inherited characteristics argues that either the genetic bottleneck was not particularly narrow, i.e. that the total sample of wild animals was not small, or that much genetic variation has arisen since domestication.

Probably, many small new domesticate groups were set up, most of which succumbed to the deleterious effects of inbreeding. The minority that survived were groups which thrived under these novel conditions, perhaps possessing genetic material which predisposed them to cope with the stresses of life with man (Kohane and Parsons, 1988), and these groups then gave rise to all present domestic animals.

One attribute of a surviving inbred line or lines might be a genetically determined tolerance of inbreeding (Templeton and Read, 1984). Such tolerance could vary from breed to breed, but it does seem clear that livestock in general show declines in commercial productivity at similar levels of inbreeding as those which zoo and wildlife managers or laboratory animal specialists try not to exceed (Thomas, 1990; Roberts, 1982). In terms of biochemical polymorphisms, domestic animals are at least as heterozygous as wild populations, and often more so (Table 26.19).

Table 26.19 Mean heterozygosity in selected vertebrate species

Mammals (in general)	0.041
Man	0.063
Cattle (3 breeds taurine, Belgium)	0.069 - 0.084
Pig (Belgium, Austria)	0.029 - 0.067
(feral herds in USA)	0.027 - 0.053
Wild boar (Italy, France, Austria)	0.021 - 0.031
Mouse	0.088
Domestic cat	0.066
Wild cat	0.042
White tailed deer	0.049 - 0.104
Moose	0 - 0.047
Red deer and Wapitu	0 - 0.060
Fallow deer (Britain, Italy)	0 - 0.006
Cheetah	0.013
Quail (domestic and wild populations, Japan)	0.086 - 0.106
Mustelids (8 species)	0 - 0.060

Source: compiled from multiple sources.

Many breeds have been divided into strains which, while specialised, are still closer to each other than to other breeds (Hall, 1990). Originally, strains probably developed as a simple result of herds and flocks tending to acquire breeding stock from nearby areas. In 19th century Britain, new breeds spread from their points of origin by slow, steady diffusion (Walton, 1983, 1984). This pattern of local interchange of breeding stock was described by Lush (1943) as ideal for the adaptation of the breed as a whole to its local environment.

General patterns of world livestock farming

As the human population of the world continues to grow, the production of livestock will increase. Some of this increase will come from further conversion of natural habitat to agricultural use, while the rest will come from intensification. Intensification means increased production per livestock unit, or per hectare, and its pattern will vary according to socio-economic conditions in the country concerned.

Intensification will make the agroecosystem less diverse; extensification will jeopardise natural and semi-natural environments. The thrust of most development programmes has been towards intensification; for example, provision of deep wells in the Sahel region (aimed at extending the grazing season and increasing the utilisation of a given area). Perhaps because the developed world understands intensification, which is market and science led, development programmes are usually on these lines, and it has generally been concluded that intensification is the only practicable future course (Payne, 1986).

Patterns of cattle husbandry have been discussed by Meyn (1984). Some 15% of Third World cattle are kept by pastoralists (important in arid and semi-arid areas of Africa, Middle East, parts of India and Pakistan, and central Asia). About 30% are kept on ranches, which are mainly in Latin America. The remainder are kept by smallholders. These include a great variety of crop-livestock systems throughout the tropics and sub-tropics, differentiated by altitude and climate; and livestock farming without land as in Caribbean feedlots and among Indian dairy farmers who supplement roadside grazing with purchased fodder.

For other species, though there are census figures and general qualitative descriptions of husbandry systems, there are no global estimates of how the populations are partitioned among the systems. Sheep tend to be kept in the following general ways (Howe and Turner, 1984): sedentary (sheep graze out from a home base, on crop stubble, roadsides, waste land, steep hills); transhumant (sheep move between summer and winter quarters, usually highland and lowland respectively), and nomadic (no home base, tending to move along well defined routes). Generally, goats are kept in a similar range of systems.

The kinds of husbandry under which each species thrives and yields a profit are determined by its biology. The goat and the sheep are best suited to extensive systems, that is, they can thrive on minimal husbandry. The pig has a tremendous ability to revert to the feral state (in Australia there are probably about 13.5 million feral pigs; Hone,

1990) and without fencing cannot be kept profitably in free range. Chickens are too vulnerable to predators to be even partly independent of man. Cattle need large amounts of water to drink and large amounts of fodder. They, with sheep and goats, are capable of ranging long distances in search of food. Water buffalo need wallows in hot climates.

Cattle can be kept under tightly controlled conditions but goats and sheep are less amenable to intensive rearing. Cattle and water buffalo are the most versatile species to market because they can be used for meat, hides, milk, production of dung, and work. Further, cattle in rural communities usually carry greater social prestige than other species, except horses (sometimes) and camels (occasionally). As a result of these factors, most livestock development programmes worldwide have emphasised cattle.

Animal production in developing countries

Animal production has a great deal to contribute to the short- and long-term alleviation of individual and national poverty. The special place animals hold in food supply systems arises thus:

- animals can use wastes otherwise useless and can supply traction and fertilizer
- animals provide a form of low risk savings account
- if milked, mammals can provide daily income
- animal husbandry, being a year-round necessity, can provide stable rural employment.

To be sustainable, systems should be based on locally abundant feeds and human resources. The genetic potential of the animals for production should be matched to the resources available and this, typically, is best achieved by the use of local breeds. In many developing countries, animal production is not important at present and animal products are bought in rather than produced locally. Nevertheless, its wider adoption would help improve the quality of life of many people. For instance, the humid zone covers 19% of tropical Africa but has only 5% of the domestic ruminant population; even though feed is plentiful and there are many big cities, animal production has been neglected partly for reasons of tradition (Armbruster and Peters, in press).

Provided enough resources are allocated, probably any breed of livestock can survive and produce in any country, though the substantial recurring expenditure on imported feed, veterinary care and housing may mean that unless subsidies are forthcoming the enterprise would not be profitable. Intensive industrial farming systems such as those based on Holstein-Friesian cattle, Large White-Landrace pigs, and hybrid fowl, need not directly supplant local breeds. These enterprises are not sustainable in less-developed countries, however, and many regard them as entirely inappropriate subjects of aid funding. Such aid programmes have been numerous in South America, where large numbers of pure-bred North American and European dairy cattle have been sent (120,000 to Venezuela alone in the period 1983-88). Mortality rates have been extremely high and the system is far from being sustainable. However, it is clear (Vaccaro, 1990) that an element of crossbreeding with local cattle that are adapted to the environment greatly improves survival without excessive penalties on milk yield.

One advantage of the continued existence of local breeds of livestock therefore is that they provide genetic material to enable imported breeds themselves to become locally adapted, and thus to help rescue schemes which were put in place without adequate planning.

Exploiting genetic diversity

The biological diversity represented by a multiplicity of different breeds enables productive agriculture to be carried out in a wider range of environments than would be the case if there were genetic uniformity. The local adaptations of breeds can reduce dependence on veterinary care.

Breed diversity also permits more rapid genetic progress to be made. *It is always quicker to develop livestock by importing genes from outside than by selecting within a breed.* One breed can act as a source of genetic material for another. This reservoir began to be tapped as husbandry developed and market requirements changed, leading farmers and breeders to look elsewhere for breeds that could be mated with their own stock to produce more remunerative animals. For example, in the 18th century, Merino sheep, Chinese pigs and dairy cattle from the Low Countries (Hall and Clutton-Brock, 1989) were all imported and crossed with local British types to confer on them fleece quality, pork quality, and milk yield, respectively.

Sometimes, new genetic mutations manifest themselves in flocks and herds and these can act as the foundation of a new breed. The best known in recent years has been the Booroola gene found in certain Merino sheep, which enhances ovulation rate (Bindon and Piper, 1986). Another instance led not only to the foundation of a new breed but also a new industry. In 1931 a mutation in a New Zealand Romney sheep resulted in a ram lamb with a very hairy fleece. The gene for hairiness, when homozygous, resulted in a fleece 65% by weight of hair, 35% wool. This mix turned out to be ideal for carpet manufacture, which was previously not economic in New Zealand. The new breed, the Drysdale, has attained some importance (Nicholas, 1987).

During the 19th century British livestock breeds were exported to be crossed with local types all over the world (Hall and Clutton-Brock, 1989); exports dropped sharply during the first half of the 20th century and more attention was then paid to local stocks. Today there is a great deal of pressure on tropical countries to accept North American dairy cattle; hundreds of thousands of Holsteins in particular have been exported, notably to Latin America (Vaccaro, 1990). However more and more advisors are maintaining that purebred temperate zone breeds like these are not appropriate in such areas and more attention should again be paid to local breeds (McDowell, 1985; Bondoc *et al.*, 1989; Vaccaro, 1990; Wilkins, 1991).

The crossing of breeds can be conducted according to the following systems.

Breed replacement

This took place several times on the plains of North and South America. Range cattle of Spanish descent were run as vast semi-feral herds primarily for hide production until the development of railways, refrigerated ships and cold stores led to expansion of the beef market. It then became worthwhile crossing the range cattle with British breed bulls, first Shorthorns, then Aberdeen Angus and Herefords. More recently still, arid lands have been made into ranching areas by the use of drought adapted cattle such as the Santa Gertrudis, developed by adding genes of zebu bulls (Sanders, 1980). Contemporary North American range cattle and those of the pampas of South America probably include in their genetic makeup only a tiny proportion of Spanish genes, but these genes, with those of the later imports, were the material on which a combination of natural and artificial selection has acted to produce locally-adapted animals. Only a very few cattle considered to be Texas Longhorns (the original Spanish stock of the south-west) and Florida Scrub (that of the south-east) survive and these are the subject of conservation efforts (Simmons-Christie, 1984; Olson, 1987).

Formation of a synthetic breed

This can result from crossing two or more breeds and then selecting from the crossbred stock. Examples include the Jamaica Hope dairy breed (McDowell, 1985), stabilised at 80% Jersey, 15% Sahiwal (one of the very few breeds of zebu dairy cattle), and 5% Holstein. There has also been much crossbreeding of European and North American dairy cattle (of the *taurus* group) with local zebu breeds, in South America and India notably, mainly for milk production (Cunningham, 1989). Typically, age at first calving and calving interval are reduced, and first lactation milk production is increased up to 50% in step with the increasing proportion of introduced genes. If that proportion is exceeded, calving interval tends to be longer and milk performance not much improved.

Stabilised crossbreeding

In this system, breeds are bred pure but the progeny crossed. This combines in the offspring the merits of both parents. In some such systems the offspring are superior, with respect to traits of value, than the parental mean; that is, they exhibit hybrid vigour (heterosis). The standard technique for exploiting genetic distinctiveness has been to make use of the additive or heterotic effects that can arise when distinct breeds are crossed (Hall, 1990).

Strategies for genetic improvement of livestock

Use of locally existing genotypes

Advantages of this course are that such livestock may well be adequate and able to respond sufficiently to improvements in the system. Over many generations they will have acquired the ability to perform locally appropriate and multiple functions. They will probably have resistance to local diseases. Breeding stock would be locally available and their purchase would create cash flow and contribute to the confidence of those who are particularly competent breeders. The disadvantages are that it is not as glamorous an approach as the importation of new genotypes and is perhaps less likely to attract aid funding as it does not involve heavy expenditure on imports from donor countries.

Local breeds can be improved by selection without the admixture of imported genetic material, but it is hard to predict whether the results would justify the investment. This is because heritabilities of commercial traits are difficult to estimate (see Table 26.20), the possible selection intensities are likely to be low, and the programme depends critically on the collection and analysis of records. However the scheme most likely to work is a nucleus breeding scheme, whereby participating breeders contribute their best females to a central unit and are entitled to purchase stud males from the unit. Such a scheme apparently operates in Libya, with the Libyan Barbary fat-tailed sheep. Howe and Turner (1984) reported that since 1978 20,000 ewes in small flocks had been screened and a nucleus of 2,000 ewes established and subjected to selection. Rams from this nucleus flock are distributed back to the small flocks.

Table 26.20 Heritability of various traits in animals

CATTLE	
Birth weight	
zebu, tropics	0.38
taurine, temperate	0.45
Weaning weight	
zebu, tropics	0.29
taurine, temperate	0.26
Heat tolerance (zebu x, Australia)	0.44
Tick burden (zebu x, Australia)	0.39
Worm egg count (zebu x, Australia)	0.12 - 0.25
Milk and component yield (Holsteins, USA)	0.25
Body size traits (Holsteins, USA)	0.40
Milk composition traits (Holsteins, USA)	0.55
Disease susceptibility	
Mastitis (Holsteins, USA)	0.01 - 0.07
Feet and legs (Holsteins, USA)	0.10
Sum of all diseases (Holsteins, USA)	0.02 - 0.06
Milking behaviour (Holsteins, Canada)	0.12 - 0.16
Ease of handling	0.12
Aggressiveness at feeding	0.11
Calving interval (zebu, Brazil)	0.23 - 0.86
Lifetime number of calves reared	0.03
SHEEP	
Clean fleece weight	0.45
Fibre diameter	0.12 - 0.50
Staple length	0.30 - 0.60
Embryonic mortality (Romanov, France)	0.09
Litter size (Romanov, France)	0.02
Ewe fertility	0.00 - 0.17
Lambs born (per ewe lambing)	0.04 - 0.15
OTHERS	
Mohair and cashmere (goats)	
Fibre length	0.70
Fibre diameter	0.12 - 0.40
Energetic efficiency (broiler fowl)	0.30 - 0.40
Liability to myxomatosis (rabbits)	0.35
Stature (humans)	0.51

Source: compiled from multiple sources.

Notes: The higher the heritability (range 0-1.0) the greater the response to selection in the environment.

Replacement by imported genotypes

The importation of Spanish cattle to the Americas, of Merino sheep to Australia, and the rapid contemporary spread of the Holstein-Friesian are examples of this process. However, whilst industrial farming is generally not appropriate in developing countries, the possible role of imported breeds in traditional husbandry bears examination because of the basic fact that development of a breed by the introduction of genetic material from other breeds is much more rapid than the development of a breed by selection. One result of most published studies on local breeds has been to show that they already possess the genotype enabling them to respond to improved husbandry and this could make the importation of genetic material unnecessary.

Steinbach (1986) established an experimental herd of goats on a research station in Tunisia and compared the local nondescript breed with the Boer, a breed specially developed for meat production, and the Alpine, Saanen and Poitou (European dairy breeds). He found the local breed to be the most profitable, responding very well to improved husbandry and incurring the least veterinary expense. Similarly, Nguni cattle (Scholtz, 1988) from the east coast of southern Africa perform comparably to improved breeds in controlled trials under good husbandry.

Importing exotic stock into developing countries is generally a high cost, high risk strategy, which is unlikely to solve the problems of the majority of farmers. It is advantageous for donor countries because it provides continuing profits for breeders and for veterinary products and services. Disadvantages have come to light as a result of practical experience; very many introductions, particularly of dairy cattle, have failed and others are kept going only by massive and continuing importations of replacement females.

Supplementation with imported genotypes

The introduction of Indian humped (zebu) cattle to the Americas (Sanders, 1980), mostly over the last 100 years, illustrates this process. Crossbreeding among the imports, with little if any contribution from European cattle, led to the Brahman, Indu-Brazil, Gir, Guzera and others; crossbreeding in Texas with pre-existing British type cattle led to the Santa Gertrudis which is 5/8 Shorthorn and 3/8 Brahman.

Recent examples of breeds being imported to add genetic material to local breeds include the highly prolific Meishan pig, one of the Taihu breed group of China, now being widely used in breeding programmes in Europe (Sellier and Legault, 1986), also highly prolific, and the Sahiwal dairy breed of Pakistan, imported to Australia from 1960 to confer tick resistance on Friesian cattle (Turton, 1985). The Finnish Landrace sheep has been used to develop new breeds such as the Cambridge (Owen and ap Dewi, 1988) and in crossbreeding schemes, most notably with the Dorset Horn, to produce ewes which are mated with rams like the Suffolk to produce meat lambs. None of these breeds is considered rare in its native country, but it is quite possible that rarer and less well known breeds may well exist there which themselves may possess useful genes.

The most dramatic livestock development of the last 200 years, the emergence of range cattle husbandry in North and South America, arose through progressive crossbreeding of imported British breeds with the Texas Longhorn and other Criollo breeds descended from those brought from Spain soon after Columbus. Here, a slow process of natural selection led the cattle, which with each successive generation resembled more and more their purebred British ancestors, to retain the locally adapted genes of their Criollo ancestors. This is the process of upgrading. If inseminations of native cows are only by imported bulls, the average percentage of the genotype of the progeny that is of imported type will increase from generation to generation in the progression 50%, 75%, 87.5%, 93.75%. In principle, the small percentage of native genotype remaining comprises, by natural selection, the genes adapting the animal to the local environment.

Most tropical breeds of cattle have only low milk yields (Turton, 1985) but if crossed with temperate zone dairy breeds, yields of the progeny are at least double those of the local breed. The practicalities of a crossbreeding scheme that maintains the proportion of temperate zone blood in the milking cows at 50% (which has generally been found to be sufficient for maintaining high milk yield without jeopardising local adaptation) are complicated, and it seems essential that a continued input of genetic material from the temperate zone breed is necessary. An alternative would, in principle, be to create a new breed, by mating among the first crossbred generation and then selecting, and this has been tried, but the synthetic dairy breeds thus created in tropical countries have not generally been very successful.

Replacement of local stock by nearby breeds

Sometimes local breeds may be replaced by supposedly more profitable breeds from the same or neighbouring countries. In Nigeria, for example, West African Shorthorn or Muturu cattle, a dwarf trypanotolerant breed of the coastal and central zones (adult body weight about 200kg) is under threat of replacement by other West African breeds, though as the breed is still numerous this is a long-term threat. These cattle are kept under a form of communal ownership in villages where the main interest is in crop growing. Numbers suffered greatly in the civil war of the late 1960s and have not recovered, there being little local interest in their husbandry for profit. Schemes aimed at promoting cattle raising in these areas are based on the NDama, another trypanotolerant breed, mainly from Senegal. In the central zone, tsetse fly eradication and a preference by traders for larger bodied cattle mean that the White Fulani or Bunaji, a humped apparently trypanosomiasis-sensitive breed, has been replacing the West African Shorthorn. It is also possible, though data are lacking, that the Kuri (a large bodied humpless breed with giant bulbous horns), kept in the Lake Chad area by sedentary communities, could be under pressure as a consequence of fighting in Chad, the spread of cultivation around the Lake, and perhaps by pressure from the Red Bororo cattle kept by migratory pastoralists.

Use of wild relatives of domestic stock

There are many examples where plant collecting expeditions to areas of diversity for domesticated plants and their wild relatives have brought back genetic material of great value for crossing with cultivars (e.g. to improve hardiness and disease resistance). The use of interbreeding with wild animals to improve domestic livestock is much more uncommon, possibly because of the smaller number of species and the extreme rarity of most wild relations of domestic species. However, there is some potential for this: for instance, some of the impetus for tracking down the remaining Kouprey *Bos sauvelii* is the belief that they may possess natural immunity to various diseases which could be harnessed by cross-breeding.

Rare or threatened breeds

Pursuit of higher production targets, the commercial success of particular breed promoters, and, in developed countries, changes in consumer preferences have led to livestock development activities becoming concentrated in few breeds and breed groups. The corollary of this is that more breeds are declining in importance, many have been lost and the survival of many others is in considerable doubt. Concern for rare breeds has been most marked in north temperate countries with a history of specialised livestock production, but it is becoming increasingly evident that declining breeds in less developed countries can represent genetic resources of great significance. Here, it seems likely that local varieties distinct enough to be defined as breeds had European criteria been applied may already have been lost. The lack of inventories and of status reports for local breeds in developing countries is cause for concern, as is the lack of support for local breeds in development programmes.

The most authoritative world list of animal breeds (Mason, 1988), lists a total of 3,237 extant breeds of ass, buffalo, cattle, goat, horse, pig and sheep. The number of such breeds in each country with native breeds is shown in Table 26.21. Data for certain countries are shown graphically in Fig. 26.9; the countries have been selected to illustrate general global patterns of breed richness.

Some 474 of extant breeds can be regarded as rare (Hall, in press). A further 617 have become extinct since 1892; numbers of extinct breeds in each country are given in Table 26.22, and data for selected countries are shown in Fig. 26.10.

Overall breed numbers (extinct, rare and non-rare) in each continent are shown in Table 26.23.

There is likely to be significant bias in these data, particularly with regard to extinct breeds. For example, of the 1,259 cattle breeds listed by Mason (1988), 242 are indicated as extinct, of which 200 were in Europe and the former USSR; only 20 were in Africa and two in India. Because breeds tend to be less formally structured and not well documented in developing countries, genetic variation may not be adequately represented by current breed nomenclature. However, it may be that in developed countries where human populations are high, rates of breed development have also been high, in response to commercial and aesthetic demands. Breed turnover, as measured by numbers of extinct breeds, would be expected to be high in such circumstances. Whatever the explanation may be, present data show that the great majority (83%) of known breed losses occurred in Europe and the former USSR.

Reasons to conserve breeds

Threatened breeds ought to be conserved for the following economic reasons:

- they may possess adaptations to local conditions,
- they may possess adaptations which can be exploited in other geographical areas or farming systems.

One of the great advantages of having access to a diversity of breeds is that in several, breed characteristics exist which are governed by single genes. In principle, single favourable genes could be transferred from one breed to another (Davis and Hinch, 1985). In practice, the major gene may owe at least some of its efficacy to its genetic background, and in a recipient breed the background may be different and unpredictable. Even detecting an animal that carries the Booroola gene (see above) is difficult (Haley, 1991). Genetic probes can be used to identify genotypes; it has proved possible to treat spermatozoa with such probes to identify which individual bulls carry a certain gene (coding for kappa-casein) which improves the suitability of milk for cheesemaking (Medrano and Aguilar-Cordova, 1990). In California, the Jersey breed has a far higher frequency (88%) of cows homozygous for this gene than does the Holstein.

Rare breeds in protected areas

Several protected areas provide a home for notable feral populations; Chillingham Park in northern England has been the home of the Chillingham white cattle possibly since the 13th century (Hall, 1989a,b), and the St Kilda islands off north-west Scotland are a refuge for the Soay sheep (Jewell *et al.*, 1974). In New Zealand, a reserve for feral sheep was established on Pitt Island (Rudge, 1983).

There are some countries where protection has been applied to endangered breeds through areas in national parks being set aside for them. These include Ireland (Muckross National Park: Kerry cattle; O'hUigin and Cunningham, 1990), Hungary (Hortobagy National Park: Hungarian Grey cattle, Mangalica pigs, Racka sheep; Henson, 1983), Poland (Roztocze National Park: Konik pony; Sasimowski and Slomiany, 1987), Swaziland (Mkhaya Farm: Nguni cattle; Setshwaelo, 1990).

Information requirements

The first step in organising conservation is to compile an inventory and to decide on priorities. Examples of inventories are cited by Hall (1990). Worldwide, FAO is organising a global data bank (Maijala, 1990) while the European Association for Animal Production has published a list of endangered populations in Europe, to the number of 241 (Maijala *et al.*, 1984).

Many breeds of livestock are promoted by breed societies. In the British tradition of pedigree breeding, which has been adopted in very many other countries, the breed societies each operate a register of breeding stock, known as a stud (equine), flock (sheep), or herd (cattle, goat, pig) book. Such societies are almost entirely lacking in the developing world.

Table 26.21 Numbers of extant breeds in each country with native breeds

	ASS	BUFFALO	CATTLE	GOAT	HORSE	PIG	SHEEP
ASIA							
Afghanistan		1	4	3	5	9	
Bahrain				1			1
Bangladesh		4	7	1	1		1
Bhutan		1	2	1			
Burma		1	1	1	1		1
Cambodia		2	2 (1)	1	1		
China	6	12	25 (2)	41	16	103	33
Cyprus	1		3	2 (1)			1
India	1	16 (1)	56 (3)	32	7 (1)	3	56 (1)
Indonesia		1	6	4 (1)	9	5	3
Iran	4	1	8	7	12 (1)		24
Iraq		1	5	3	1		6
Israel	1	1 (1)	3	3			3
Japan			7 (2)	3 (2)	8 (8)	2	
Jordan		1	1			1	
Korea N and S			2	1	2	2	
Laos		1		1			
Lebanon			2	2			1
Malaysia		1	2	2		2	1
Mongolia			2	1	1		10
Nepal		3	4	4	5		61
Oman				1	4		2
Pakistan		4	11	30	7 (3)		44
Philippines		2	5	1		7 (1)	1
Saudi Arabia				1	4	6	
Sri Lanka		3	4	3		1	1
Syria	2	2 (1)	5	3	1		3
Taiwan		1	1			2 (1)	
Thailand		1	2	2	1	3	
Turkey		1	13	4	9 (1)	1	23
Vietnam		1	5	1	1	10	
Yemen	2		4	8	9		14
USSR (former)							
	15	1	58 (9)	19 (4)	60 (23)	34 (2)	133 (11)
EUROPE							
Albania	1	1	4	2	3	2	8
Austria			6 (2)	1	5 (1)		4 (2)
Belgium			5	2	4	4 (1)	2 (1)
Bulgaria		1	5 (2)	4	11	5 (1)	39 (13)
Czechoslovakia			5	2	7 (1)	7	7 (2)
Denmark			5 (1)	1	3 (2)	2 (1)	1
Faeroe Islands			1		1		1
Finland			5 (2)	2 (1)	1 (1)	1	1
France	2 (1)		36 (12)	10 (6)	19 (6)	9 (7)	54 (13)
Germany			26 (9)	4 (1)	20 (10)	10 (4)	18 (6)
Greece		1	6 (4)	3	3		30 (7)
Hungary			5 (1)	1	10 (1)	5	4 (2)
Iceland			1	1 (1)	2		2
Ireland			2 (3)	1 (1)	4 (1)		5
Italy	8 (6)	1	39 (12)	27 (11)	11 (9)	7 (6)	63 (24)
Malta			1 (1)	1 (1)			1 (1)
Netherlands			10 (6)	5 (2)	5 (2)	3	14 (10)
Norway			7 (2)	2	4 (1)	1	7 (2)
Poland			5 (1)	3	21 (2)	5 (2)	23 (2)
Portugal			18 (5)	11	5	5 (1)	17
Romania		2	9 (2)	2	8	11 (1)	8
Spain	7 (3)		37 (11)	14 (1)	8 (2)	8 (4)	47 (4)
Sweden			5 (3)	1	5 (1)	2	4
Switzerland			6 (2)	12 (3)	1	2	8 (4)
UK			42 (16)	9 (3)	20 (7)	16 (8)	73 (17)
Yugoslavia		1	19 (5)	2	7 (1)	10 (2)	33 (5)
NORTH AND CENTRAL AMERICA							
Central America (gen)				1		1	1
Bahamas							1
Barbados							1
Belize						1	
Canada			9		5 (2)	4 (1)	4 (1)
Costa Rica			3				
Cuba			5		1		1
Dominican Rep.			3				

Table 26.21 Numbers of extant breeds in each country with native breeds (cont.)

	ASS	BUFFALO	CATTLE	GOAT	HORSE	PIG	SHEEP
NORTH AND CENTRAL AMERICA (continued)							
Guadeloupe							1
Guatemala			1 (1)				
Honduras						1	
Jamaica		1	4				
Mexico			3 (1)	1	2	4 (1)	4
Nicaragua			1				
Puerto Rico			1				
United States	6		40 (6)	10 (4)	35 (7)	10 (4)	32 (6)
SOUTH AMERICA							
South America (gen)				1		1	
Argentina			5		3		3
Bolivia			4		1		
Brazil	5 (1)	2	26 (4)	8	8	13	7
Chile				1	2		
Colombia			9		1	3	2
Ecuador			2		1		
Paraguay			1				
Peru			1		5		1
Uruguay					1		2 (1)
Venezuela			5		1	1	2
OCEANIA							
Oceania (gen)						1	
Australia			20	2		2	23 (2)
Fiji				1			
Guam						1	
Hawaii			2 (1)				1
New Zealand			1	5	1	2	16 (3)
N Marianas			1				
Papua New Guinea						1	
AFRICA							
Algeria	1		2	2	1		10
Angola			5				4
Benin			4		1		1
Botswana			7 (2)	1			1
Burkina Faso			2	1	4		1
Cameroon			11 (4)	1		1	2
Chad			7	4	3		3
Cote d'Ivoire			1	1			
Egypt	3	4	6	6	1		9 (1)
Ethiopia	2		17 (1)		2		11
Gambia			3	1			1
Ghana			3	1		1	2
Guinea			1	1			1
Guinea-Bissau			3 (1)	1			1
Kenya	1		14	3			5
Liberia			3	1			1
Libya	1		1	1			4
Madagascar			4 (1)				1
Malawi			1	1			1
Mali			6	2	5		8
Mauritania			1	2	1		5
Morocco	1		3	3	1		29 (1)
Mozambique			2	2			1
Namibia			6	1			1
Niger			5	3	3		5
Nigeria			11	9	5	1	6
Senegal			6	1	4		3
Seychelles						1	
Sierra Leone			1	1			1
Somalia	1		8	8	1		2
South Africa			11	4	4	2	25 (2)
Sudan	4		20	11	2		16
Tanzania	2		9 (1)	1			2
Togo			4	1	1		1
Tunisia	1		4	1	2		5
Uganda			12	3			1
Zaire			8	2			3
Zambia				1			
Zimbabwe			4				3

Source: data from Mason, 1988; analysis by S.J.G. Hall (1992. Livestock breeds and their conservation. 1. World distribution. in prep.)
Note: principal numbers are estimates of extant breeds, numbers in parentheses indicate those classed as rare by Hall (included in main figure).

Figure 26.9 Numbers of living breeds of livestock in selected countries

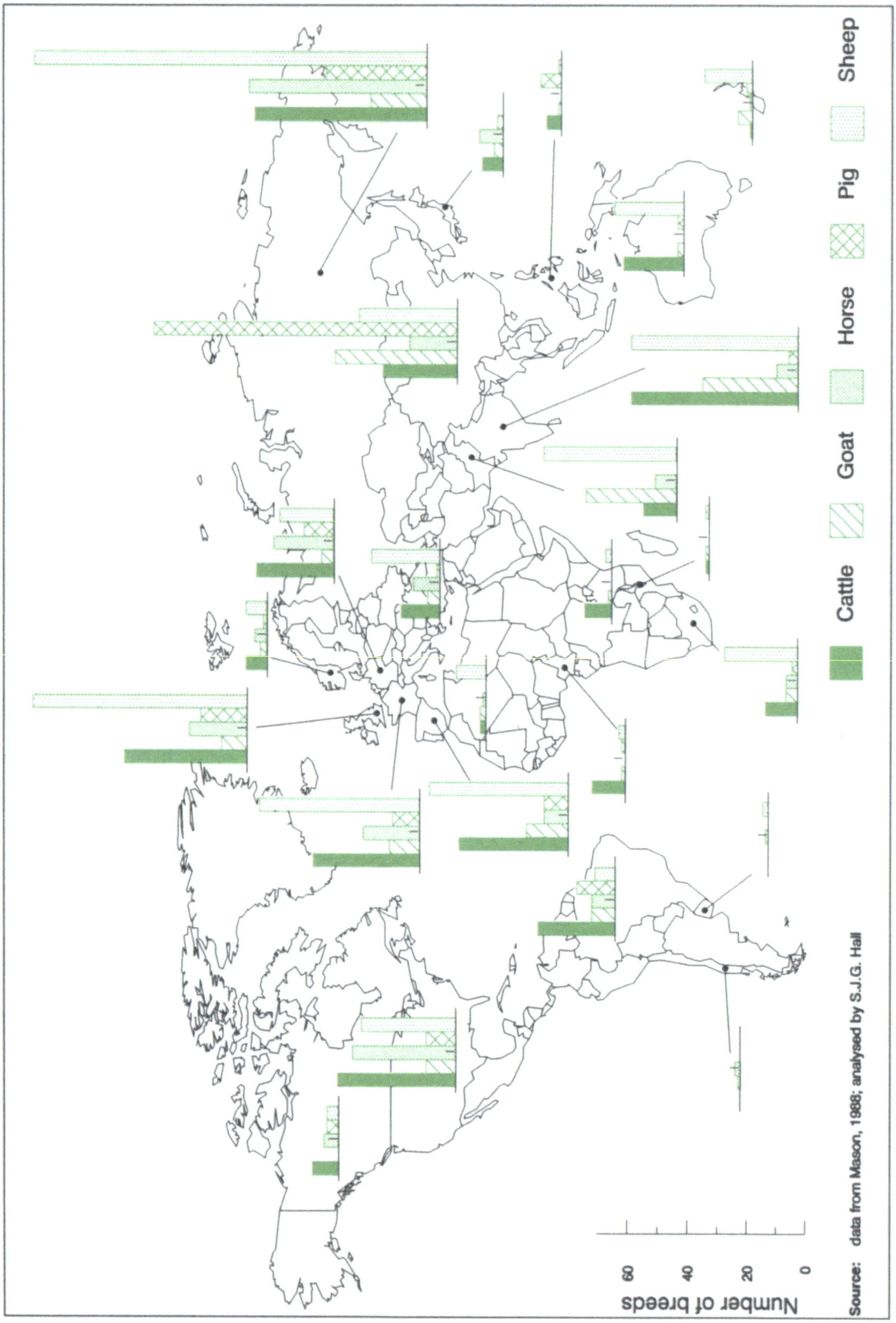

Table 26.22 All countries: numbers of breeds extinct since 1892

	ASS	CATTLE	GOAT	HORSE	PIG	SHEEP
ASIA						
China						2
Hong Kong					3	
India		2		2		
Japan				1		
Pakistan			1			
Philippines		1				
Taiwan					5	
Turkey		2				
USSR (former)						
		22	6	20	21	31
EUROPE						
Austria		16		1		9
Belgium		2				2
Bulgaria		2		2		4
Czechoslovakia		10			1	
Denmark		3				1
France		18	2	15	18	33
Germany		29	6	7	10	5
Greece		7				1
Hungary		3		1		
Ireland				1		3
Italy	4	22		5	24	15
Netherlands						1
Norway		11	5	1		
Poland		5	2		5	1
Portugal		1				
Romania		3		3		
Spain	1	12		12	8	2
Sweden		4			1	
Switzerland		2	1	4		11
United Kingdom		5	2	4	7	8
Yugoslavia		2		3	5	
NORTH AND CENTRAL AMERICA						
Canada				2		1
United States		1	1	2	17	9
SOUTH AMERICA						
Brazil		15				
Chile		1				
Uruguay		1				
Venezuela		2				
OCEANIA						
Australia		2		1		1
New Zealand					1	1
AFRICA						
Algeria		4				
Benin		1				
Cameroon		1				
Gambia		1				
Lesotho				1		
Malawi		1				
Nigeria		2				
Rwanda		1				
South Africa		4		2		1
Tanzania		4				
Zimbabwe		3				
TOTAL	5	228	26	90	126	142

Source: data from Mason, 1988; analysis by S.J.G. Hall (1992. Livestock breeds and their conservation. 1. World distribution. in prep.)
Note: Grand total worldwide = 617.

Figure 26.10 Numbers of extinct breeds of livestock in selected countries

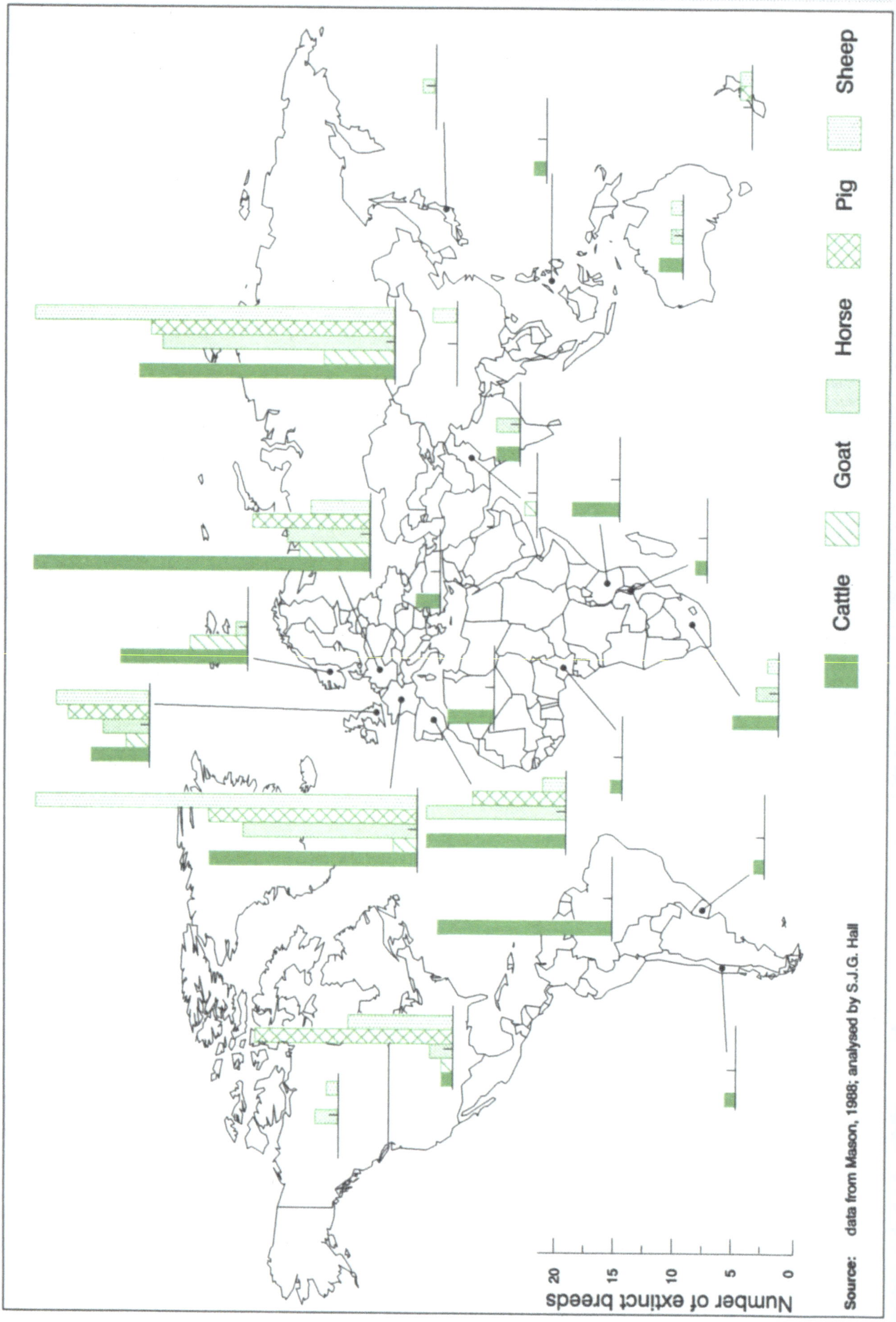

Table 26.23 Summary of world distribution of extinct, rare and commercial breeds

		ASS	BUFFALO	CATTLE	GOAT	HORSE	PIG	SHEEP	ROW TOTALS	% RARE
ASIA	Rare		2	8	4	14	2	1	31	
	Extinct			5	1	3	8	2	19	
	Commercial	17	53	180	144	71	140	219	824	
	Total	17	55	193	149	88	150	222	874	4
USSR (former)	Rare			9	4	23	2	11	49	
	Extinct			21	6	20	21	31	99	
	Commercial	15	1	49	15	37	32	122	271	
	Total	15	1	79	25	80	55	164	419	15
EUROPE	Rare	10		101	29	49	37	109	335	
	Extinct	5		154	19	58	79	97	412	
	Commercial	8	8	209	91	137	76	356	885	
	Total	23	8	464	139	244	192	562	1632	27
N AND C AMERICA	Rare			8	4	9	5	7	33	
	Extinct			1	1	4	17	10	33	
	Commercial	6	1	58	8	34	33	38	178	
	Total	6	1	67	13	47	55	55	244	16
SOUTH AMERICA	Rare	1		4				1	6	
	Extinct			19					19	
	Commercial	4	2	45	11	21	18	16	117	
	Total	5	2	68	11	21	18	17	142	5
OCEANIA	Rare			1		1	1	2	5	
	Extinct			2		1	1	5	9	
	Commercial			23	7		5	35	70	
	Total			26	7	2	7	42	84	7
AFRICA	Rare			10		1		4	15	
	Extinct			22		3		1	26	
	Commercial	15	4	168	61	33	6	131	418	
	Total	15	4	200	61	37	6	136	459	3
TOTAL		81	71	1097	405	519	483	1198	3854	

Source: data from Mason, 1988; analysis by S.J.G. Hall (1992. Livestock breeds and their conservation. 1. World distribution. in prep.)

Note: 'commercial' here indicates, for convenience, the breeds that are neither rare or extinct.

Costs and benefits of conserving breeds

A breed can be conserved (a stock maintained which continues to represent the foundation stock without too much genetic drift or inbreeding) for surprisingly small cost compared with the possible economic benefits. Either a live breeding stock can be maintained, or semen or embryos preserved, or all methods can be used. For semen and embryos, the genetic variability in a typical breed would be adequately represented by collection from 25 males, or by 25 embryos each from 25 donors (Smith, 1984). Embryo storage is not yet possible for chickens and pigs, and is not yet fully developed for equines (Guay and Poitras, 1989; Heyman and Vincent, 1988).

Live breeding stocks are much more expensive to maintain, as any farm that keeps a conservation unit of a non-commercial breed is losing the opportunity to keep a profitable breed. If it is decided to keep a live conservation population, its size is best defined by what rates of increment of inbreeding and of fixation of genes through random genetic drift are permissible. An effective population size of 30 seems appropriate (Smith, 1984). The rates should be expressed per year not per generation and recalculation yields the minimum sizes of population of each species necessary to keep annual increment of inbreeding at below 0.2%. These population sizes are surprisingly small, provided appropriate sex ratios are chosen.

Potential benefits of livestock conservation are very great (Smith, 1984). If a 1% gain in economic efficiency arises in a livestock industry through the use of a conserved breed, this benefit will exceed the cost by between 33 and 190 times. Even though the cost/benefit ratio of livestock conservation is so favourable, in absolute terms the amounts required are still large. In Europe alone, there are 241 breeds and strains that appear to justify conservation (Maijala *et al.*, 1984).

References

ACFM 1991. Report of the Herring Assessment Working Group for the area North of 62°, April, 1991. ICES, May 1991.

Ajayi, S.S. 1971. Wildlife as a source of protein in Nigeria: some priorities for development. *The Nigerian Field* 36(3):115-127.

Armbruster, T. and Peters, K.J. (in press). Traditional sheep and goat production in southern Côte d'Ivoire. 1. Reproductive performance and growth development. *Journal of Animal Breeding and Genetics*.

Barton, M.A. 1986. Hunters and gatherers. In: Messent, P.R. and Broom, D.M. (Eds), *The Encyclopaedia of Domestic Animals*. Grolier Int., UK. Pp.282-285.

Bindon, B.M. and Piper, L.R. 1986. The reproductive biology of prolific sheep breeds. *Oxford Reviews of Reproductive Biology* 8:414-451.

Bondoc, O.L., Smith, C. and Gibson, J.P. 1989. A review of breeding strategies for genetic improvement of dairy cattle in developing countries. *Animal Breeding Abstracts* 57:819-829.

Brisbin, I.L. 1977. The pariah. Its ecology and importance to the origin, development and study of pure bred dogs. *Pure-bred Dogs American Kennel Gazette* January 1977:22-29.

Brown, G. 1989. The viewing value of elephants. In: *The Ivory Trade and the Future of the African Elephant*. Ivory Trade Review Group Report prepared for the 7th meeting of the Conference of the Parties to CITES.

Bubenik, A.B. 1989. Sport hunting in continental Europe. In: Hudson, R.J., Drew, K.R. and Baskin, L.M. (Eds), *Wildlife Production Systems: economic utilisation of wild ungulates*. Cambridge University Press, Cambridge, UK. Pp.115-133.

Clutton-Brock, J. 1987. *A Natural History of Domesticated Mammals*. Cambridge University Press, Cambridge and British Museum (Natural History), London.

Colyn, M.M., Dudu, A. and Mbaelele, M.M. 1988. Data on small and medium scale game utilization in the rain forest of Zaire. In: *Wildlife Management in Sub-Saharan Africa: sustainable economic benefits and contribution towards rural development*. International Foundation for the Conservation of Game, Paris. Pp.109-141.

Committee for Conservation and Care of Chimpanzees 1990. *Status Report. The Chimpanzee Trade*. 2pp.

Conniff, R. 1987. The little suckers have made a comeback. *Discover*, 8(8):84-94.

Corbet, S.A., Williams, I.H. and Osborne, J.L. 1991. *Bees and the Pollination of Crops and Wild Flowers: changes in the European Community*. Review commissioned by Scientific and Technical Options Assessment, European Parliament.

Corten, A. and van de Kamp, G. 1991. Natural changes in pelagic fish stocks of the North Sea in the 1980s. Netherlands Institute for Fisheries Research (RIVO), ICES, 1991, Variability Symposium No.27, session 3.

Cunningham, E.P. 1989. The genetic improvement of cattle in developing countries. *Theriogenology* 31:17-28.

Darwin, C. 1868. *The Variation of Animals and Plants under Domestication* Vols 1 and 2. Popular edition, 1905. John Murray, London. Pp.xiv+566; xii+605.

Davis, G.H. and Hinch, G.N. 1985. Introduction and management of the Booroola gene in sheep flocks in New Zealand. In: Land, R.B. and Robinson, D.W. (Eds), *Genetics of Reproduction in Sheep*. Butterworths, London. Pp.139-148.

FAO 1977. *Provisional Food Balance Sheets: 1972-74 average*.

FAO 1981. *Atlas of the Living Resources of the Seas*. FAO Fisheries Department, FAO, Rome.

FAO 1990a. *FAO Yearbook: Production 1989* Food and Agriculture Organization of the United Nations, Rome.

FAO 1990b. *Review of the State of World Fishery Resources*. Marine Resources Service, FAO, Rome.

FAO 1991a. *FAO Yearbook Fishery Statistics: catches and landings 1989*. Vol. 68. FAO, Rome.

FAO 1991b. *FAO Yearbook Fishery Statistics: commodities 1989*. Vol. 69. FAO, Rome.

Greenhalgh, P. 1986. *The World Market for Silk*. Report of the Tropical Development and Research Institute. G195, ix + 117pp.

Guay, P. and Poitras, P. 1989. Preservation of equine embryos. *AgBiotech News and Information* 1:515-517.

Haley, C.S. 1991. Use of DNA fingerprints for the detection of major genes for quantitative traits in domestic species. *Animal Genetics* 22:259-277.

Hall, S.J.G and Clutton-Brock, J. 1989. *Two Hundred Years of British Farm Livestock*. British Museum (Natural History), London. 272pp.

Hall, S.J.G. (in press). Rare and traditional breeds - have they a place? *Proceedings UFAW Conference on Animal Welfare Implications of Extensive and Organic Husbandry. Cirencester.*

Hall, S.J.G. 1989a. The white herd of Chillingham. *Journal of the Royal Agricultural Society of England* 150:112-119.

Hall, S.J.G. 1989b. Running wild. Parts 1 and 2. *Ark* 16:12-15, 6-59.

Hall, S.J.G. 1990. Genetic conservation of domestic livestock. *Oxford Reviews of Reproductive Biology* 12:289-318.

Hemley, G. 1988. International wildlife trade. In: W.J. Chandler (Ed.), *Audubon Wildlife Reports*. Academic Press, New York. Pp.337-374.

Henson, E. 1983. Rare breed survival in Hungary. *Ark* 10:352-354.

Heyman, Y. and Vincent, C. 1988. (Transplantation: the embryos which came in from the cold). *Biofutur* 69:38-42.

Hone, J. 1990. How many feral pigs in Australia? *Australian Wildlife Research* 17:571-572.

Howe, R.R. and Turner, H.N. 1984. Sheep breed resources in developing countries. In: Hofmeyr, J.H. and Meyer, E.H.H. (Eds), *Proceedings of the 2nd World Congress on Sheep and Beef Cattle Breeding, Pretoria*. Pp.87-96.

ICCAT 1990. Report of the Standing Committee on Research and Statistics. International Convention for the Conservation of Atlantic Tunas.

ICCAT 1991. Recommendations for enhanced management of Western Atlantic Bluefin Tuna, Annex 7. Report of the 12th regular meeting of the Commission (provisional), 11-15 November. ICCAT, Madrid. Pp.74-76.

Jakobsson, J. 1985. Monitoring and management of the Northeast Atlantic Herring Stocks. *Canadian Journal of Fisheries and Aquatic Science* 42:207-221.

Jewell, P.A., Milner, C. and Boyd, J. Morton (Eds) 1974. *Island Survivors: the ecology of the Soay sheep of St. Kilda*. Athlone Press, London.

Klein, D.R. 1989. Northern subsistence hunting economies. In: Hudson, R.J., Drew, K.R. and Baskin, L.M. (Eds), *Wildlife Production Systems: economic utilisation of wild ungulates*. Cambridge University Press, Cambridge, UK. Pp.96-111.

Levy-Luxereau, A. 1972. *Etude Ethno-zoologique du Pays Hausa, en République du Niger*. Société d'Etudes Ethnozoologiques et Ethnobotaniques, Paris. 341pp.

Library of Congress, 1979. Draft Environmental Report on Peru. Science and Technology Division, Library of Congress, Washington.

Lincoln, G.A. 1989. Seasonal cycles in testicular activity in Mouflon, Soay sheep and domesticated breeds of sheep: breeding seasons modified by domestication. *Zoological Journal of the Linnean Society* 95:137-147.

Lindberg, C. 1988. The economic value of non-domesticated insects. Unpublished report, World Conservation Monitoring Centre, Cambridge, UK.

Lush, J.L. 1943. *Animal Breeding Plans*. Iowa State College Press, Ames, Iowa. Pp.viii+437.

Luxmoore, R.A. 1989. International trade. In: Hudson, R.J., Drew, K.R. and Baskin, L.M. (Eds), *Wildlife Production Systems: economic utilisation of wild ungulates*. Cambridge University Press, Cambridge, UK. Pp.28-49.

Maijala, K. 1990. Establishment of a world watch list for endangered livestock breeds. In: Wiener, G. (Ed.), *Animal Genetic Resources. A global programme for sustainable development*. FAO Animal Production and Health Paper 80. FAO, Rome. Pp.167-184.

Maijala, K., Cherekaev, A.V., Devillard, J.M., Reklewski, Z., Rognoni, G., Simon, D.L. and Steane, D. 1984. Conservation of animal genetic resources in Europe. Final report of EAAP working party. *Livestock Production Science* 11:3-22.

Maitipe, P. 1984. Fighting malnutrition in Tropical Asia with 'Natural Protein Capsules'. *Spirit of Enterprise - The 1984 Rolex Awards*. Pp.10-13.

Malaisse, F. and Parent, G. 1980. Les chenilles comestibles du Shaba meridional (Zaire). *Naturalistes Belges* 61(1):2-24.

Marks, S.A. 1989. Small-scale hunting economies in the tropics. In: Hudson, R.J., Drew, K.R. and Baskin, L.M. (Eds), *Wildlife Production Systems: economic utilisation of wild ungulates*. Cambridge University Press, Cambridge, UK. Pp.75-95.

Mason, I.L. 1988. *A World Dictionary of Livestock Breeds Types and Varieties*, 3rd edn. CAB International, Wallingford, UK.

McDowell, R.E. 1985. Crossbreeding in tropical areas with emphasis on milk, health and fitness. *Journal of Dairy Science* 68:2418-2435.

Medrano, J.F. and Aguilar-Cordova, E. 1990. Genotyping of bovine kappa-casein loci following DNA sequence amplification. *Bio/Technology* 8:144-146.

Meyn, K. 1984. Requirements and constraints for cattle breeding programmes in developing countries. In: Hofmeyr, J.H. and Meyer, E.H.H. (Eds), *Proceedings of the 2nd World Congress on Sheep and Beef Cattle Breeding, Pretoria*. Pp.55-66.

Nelson, J.S. 1984. *Fishes of the World*, 2nd edn. John Wiley and Son, New York.

Nicholas, 1987. *Veterinary Genetics*. Clarendon, Oxford. xvii+580pp.

Nietschmann, B. 1973. *Between Land and Water*. Seminar Press, New York.

Norse, E.A. 1992. Marine biological diversity strategy and action plan. First draft, January 6 1992. To be published by Center for Marine Conservation, Washington, DC, USA.

O'hUigin, C. and Cunningham, E.P. 1990. Conservation of the Kerry breed. *Farm and Food Research* Jan-Mar 1990:25-27.

Olson, T. 1987. Florida's native cattle - an American rare breed. *Ark* 14:89-92.

Owen, J.B. and ap Dewi, I. 1988. The Cambridge sheep - its exploitation for increased efficiency of lamb production. *Journal of Agricultural Science in Finland* 60:585-590.

Payne, C.H. 1989. Sport hunting in North America. In: Hudson, R.J., Drew, K.R. and Baskin, L.M. (Eds), *Wildlife Production Systems: economic utilisation of wild ungulates*. Cambridge University Press, Cambridge, UK. Pp.134-146.

Payne, W.J.A. 1986. Intensification in developing countries. In: Cole, D.J.A. and Brander, G.C. (Eds), *Bioindustrial Ecosystems*. Elsevier, Amsterdam. Pp.241-254.

Prescott-Allen, R. and Prescott-Allen, C. 1982. *What's Wildlife Worth?* Earthscan, London.

Pyle, R.M. 1981. Butterflies: now you see them. *International Wildlife* 11(1):4-11.

Redford, K.H. and Robinson, J.G. 1991. Subsistence and commercial uses of wildlife in Latin America. In: Robinson, J.G. and Redford, K.H. (Eds), *Neotropical Wildlife Use and Conservation*. University of Chicago Press, USA. Pp.6-23.

Roberts, R.C. 1982. Lessons to be drawn from selection experiments with laboratory animals. In: *Proceedings of the World Congress on Sheep and Beef Cattle Breeding*, Vol. I. Dunsmore Press, Palmerston North, New Zealand. Pp.253-259.

Ruddle, K. 1973. The human use of insects: examples from the Yukpa. *Biotropica* 5(2):94-101.

Rudge, M.R. 1983. A reserve for feral sheep on Pitt Island, Chatham Group, New Zealand. *New Zealand Journal of Zoology*, 10:349-63.

Sale, J.B. 1981. *The Importance and Values of Wild Plants and Animals in Africa*. Part I. IUCN, Gland, Switzerland. 41pp.

Sanders, J.O. 1980. History and development of zebu cattle in the United States. *Journal of Animal Science* 50:1188-1200.

Sasimowski, E. and Slomiany, J. 1987. Polish Koniks in the Roztocze National Park. In: Hodges, J. (Ed.), *Animal Genetic Resources. Strategies for Improved Use and Conservation*. FAO Animal Production and Health Paper 66. FAO, Rome. Pp.285-289.

Scholtz, M.M. 1988. *Proceedings of the 3rd World Congress on Sheep and Beef Cattle Breeding*, Paris. Vol. 2, pp.303-319.

Sellier, P. and Legault, C. 1986. The Chinese prolific breeds of pigs: examples of extreme genetic stocks. In: *Exploiting New Technologies in Animal Breeding*. Genetic developments. OUP, Oxford.

Serpell, J. 1989. Pet-keeping and animal domestication: a reappraisal. In: Clutton-Brock, J. (Ed.), *The Walking Larder. Patterns of domestication, pastoralism, and predation*. Unwin Hyman, London. Pp.10-21.

Setshwaelo, L.L. 1990. Live animal conservation projects in Africa. In: Wiener, G. (Ed.), *Animal Genetic Resources. A global programme for sustainable development*. FAO Animal Production and Health Paper 80. FAO, Rome. Pp.135-141.

Simmons-Christie, N. 1984. A new look at some old-fashioned cattle. *Animal Kingdom* 87:30-32.

Smith, C. 1984. Economic benefits of conserving animal genetic resources. *Animal Genetic Resources Information* 3:10-14.

Steinbach, J. 1986. Experiences with the evaluation of the production potential of local and imported goat breeds in northern Tunisia. *Animal Research and Development* 28:100-114.

Taylor, R.H. 1975. *Butterflies in my Stomach*. Woodbridge Press Publishing Co., Santa Barbara, California.

Templeton, A.R. and Read, B. 1984. Factors eliminating inbreeding depression in a captive herd of Speke's gazelle. *Zoo Biology* 3:177-179.

Tennessen, T. and Hudson, R.J. 1981. Traits relevant to the domestication of herbivores. *Applied Animal Ethology* 7:87-102.

The Food Insects Newsletter (TFIN):
1) Volume II, No.3 pp.2 and 10
2) Volume III, No.2 pp.1, 3, 4 and 6
3) Volume III, No.3 p.1
4) Volume IV, No.2 pp.3-4
5) Volume IV, No.2 p.8

Thomas, C.D. 1990. What do real population dynamics tell us about minimum viable population sizes? *Conservation Biology* 4:324-327.

Thomsen, J.B., Edwards, S.R. and Mulliken, T.A. 1992. *Perceptions, Conservation and Management of Wild Birds in Trade*. TRAFFIC International, Cambridge. 165pp.

Turton, J.D. 1985. Progress in the development and exploitation of new breeds of dairy cattle in the tropics. In: Smith, A.J. (Ed.), *Milk Production in Developing Countries*. Centre for Tropical Veterinary Medicine, Edinburgh. Pp.218-239.

Vaccaro, L. P. de 1990. Survival of European dairy breeds and their crosses with zebus in the tropics. *Animal Breeding Abstracts* 58:475-494.

Vane-Wright, R.I. 1991. Why not eat insects? *Bulletin of Entomological Research* 81:1-4.

Vedder, A. and Weber, W. 1990. Mountain Gorilla project (Volcanoes National Park) - Rwanda. In: Kniss, A. (Ed.), Living with wildlife: wildlife resource management with local participation in Africa. Unpublished report, World Bank Environment Division, Africa Region. Pp.83-90.

Walton, J.R. . 1983. The diffusion of improved sheep breeds in eighteenth- and nineteenth-century Oxfordshire. *Journal of Historical Geography* 9:175-195.

Walton, J.R. 1984. The diffusion of the Improved Shorthorn breed of cattle in Britain during the eighteenth and nineteenth centuries. *Transactions of the Institute of British Geographers* 9:22-36.

Watanabe, H. and Satrawaha, R. 1984. A list of edible insects sold at the public market in Khon Kaen, Northeast Thailand. *Tonan Ajia Kenkyu* 22(3):316-325.

Wilkins, J.V. 1991. Special regional problems of breeding resources - Latin America and Caribbean examples. In: Hickman, C.G. (Ed.), *Cattle Genetic Resources, World Animal Science, B7*. Elsevier, Amsterdam. Chapter 5, pp.91-113.

Yost, J.A. and Kelly, P.M. (1983). Shotguns, blowguns, and spears. The analysis of technological efficiency. In: Hames, R.B. and Vickers, W.T. (Eds), *Adaptive Responses of Native Amazonians*. Academic Press, New York. Pp.189-224.

Section on domestic livestock abridged from a consultancy report by Stephen J.G. Hall (Research Group in Mammalian Ecology and Reproduction, University of Cambridge), with additional tables extracted by permission from work in preparation by the same author.

27. BIODIVERSITY AND ECONOMICS

A growing literature in applied economics is demonstrating that techniques are available for obtaining concrete estimates of the value of many different facets of the environment, including the more intangible aspects of environmental quality, such as clean water, clean air and better views. These methods can be applied to biodiversity, but are subject to major limitations and problems of interpretation. One of the major difficulties is that they are based on the premise that value is determined by human willingness to pay. The range of human values can be very broad and consequently difficult to measure: many people are willing to pay for qualities that are seemingly unselfish, by placing, for example, an 'existence value' on certain natural resources that they will never personally see or experience.

The major problem, however, involved in the application of these methods to biodiversity is defining exactly what is meant by biodiversity, a notoriously intractable question. In this regard, the distinction between valuing biological resources and valuing biological *diversity* (i.e. the range of variation in biological resources, whether measured quantitatively or qualitatively) is an important one and leads to two different types of question: in the first instance, a gross estimate of the value of biological resources in a particular geographic locale is sought; in the second, attempts are made to trace the impact of changes in diversity on economic values.

VALUING THE ENVIRONMENT

The total economic value of an environmental resource may be broken down into a range of use and non-use values. The direct use of ecosystem outputs in non-consumptive, consumptive or productive activities is the impact that is most commonly measured in valuation exercises. Included as direct uses would be the harvesting of wild species for use as food, fuel, shelter or medicine. Other activities such as ecotourism involve a direct 'transaction' between people and biological resources and fall into this category of direct use values. Some direct uses of biological resources such as commercial logging, agriculture or fisheries generate products which are exchanged in the marketplace, while the products of others such as subsistence hunting and gathering go largely unmarketed. In the latter case, although these non-marketed resources have no financial value (cash price in exchange) they do have economic value as they are of importance to society.

Biological resources may also make indirect contributions to the welfare of society. Environmental functions support economic activity by recycling important elements such as carbon, oxygen and nitrogen and by acting as a buffer against excessive variations in weather, climate and other natural events outside the control of human beings. Economists are increasingly attempting to place values on these indirect use values. Since indirect use values do not enter directly into human preferences and are often widely available, their value is not often recognised and incorporated into development decisions. As natural habitat declines these ecological processes become scarcer, their economic value grows and eventually mechanisms are designed for 'marketing' these services (note the increasing trend towards user charges for water supply and waste disposal in developed countries).

In addition to direct and indirect use values, biological resources may have option and existence (non-use) values. Option values are associated with the future use of a resource and future flows of information regarding the use of resources. Risk-aversion dictates that societies should be willing to pay an additional sum above and beyond what a future use value of a biological resource is worth in order to guarantee future access. If this is the case, there is an 'extra' value that can be placed alongside the use values of the resource.

Finally, there may be non-use or existence values associated with a resource. These are benefits derived by an individual from the mere knowledge that the resource exists. For example, people who donate money to a conservation organisation with no expectation of ever visiting the habitats or hunting the species which the organisation aims to conserve must be deriving some satisfaction that is simply a result of the continued survival of the species or habitat. To sum up the different types of values that make up the total economic value of biological resources, Table 27.10 illustrates how the goods and services produced by a tropical forest fit into this 'taxonomy' of values.

A given habitat or species may have many different use and non-use values. Ancient redwoods may have ecotourism value, timber value, contribute to watershed protection and carbon storage, and have significant existence values. Because a number of values may be involved and since some techniques are better than others for measuring different types of values, any comprehensive valuation exercise of a particular habitat or even one species may involve the application of a range of valuation techniques. Care must be taken, however, to avoid simply adding the resulting values to each other to obtain a total economic value. Trade-offs between values and double-counting of benefits may occur, making simple summation of the outcomes of separate analyses of different values potentially very misleading. Despite these difficulties the techniques reviewed below provide useful methods for quantifying the benefits of environmental resources.

Changes in productivity approach

The changes in productivity approach relies on an understanding of underlying ecological relationships to derive a model indicating how changes in the supply of an environmental resource results in changes to the economic value of production. This technique can be used to investigate improvements or damage to environmental quality. For instance, soil fertility has a direct impact on agricultural productivity. Soil degradation will raise crop production costs for a certain level of output. Resulting changes in quantities and prices will cause the benefits received by consumers and producers to change. Comparing initial levels of surplus with the resultant levels provides a way of estimating the value of changes in supply of the environmental resource or quality. This technique is particularly relevant to basic resource issues in developing countries where a large proportion of economic production comes from agriculture, fisheries, forests, etc. The production function technique is a natural complement to cost-benefit analyses of projects that require estimation of the economic effects of changes in resource availability.

The Nepal Hill Forest Development Project provides a simple example of using the changes in productivity method to value improvements in environmental quality for incorporation in cost-benefit analysis. As reported by Dixon *et al.* (1988) the project involved introducing systematic hill-forest development into 38,500ha devoted to a mixture of land uses in the vicinity of Kathmandu and Pokhara. The objectives of the project were to reduce soil erosion, increase the productivity of different land uses in the watershed and provide sustainable flows of fuelwood and fodder, amongst other resources. The benefits from reductions in soil erosion were not quantified, but improvements in the physical yields of milk, fertilizer and fuelwood were calculated for the four land types: grazing land, pasture, unmanaged scrubland, and unmanaged forest. Using readily available market values the project values for milk and fertilizer production were calculated.

Three different methods were used in calculating the unit value for increased fuelwood supplies. A direct market value approach used the economic price of fuelwood (minus transport costs) in Pokhara and Kathmandu (280 rupees/m^3). As the production from the project would increase Kathmandu's fuelwood supply by 20% and because the two markets were considered small and isolated, two additional techniques were utilised. Cattle dung is the closest available substitute for fuelwood in rural areas. The economic price of this fuelwood substitute was estimated to be 65 Rs/m^3 based on the marginal loss of foodgrains that would occur if dung was diverted away from its role as a fertilizer. A final approach involved valuing fuelwood in terms of the opportunity cost of labour diverted from other employment by the need to gather fuelwood. The opportunity cost approach yielded a value of 83 Rs/m^3 for fuelwood. The correct value to use in calculating fuelwood production is the lowest value - this case that derived from indirect substitution. The total production values for the different land types were aggregated and compared with the benefits from allowing continued land and forest degradation.

Contingent Valuation Method (CVM)

The objective of any valuation exercise is to determine people's preferences for environmental quality. What this entails is discovering what people are willing to pay (WTP) for increments in environmental quality or what they are willing to accept (WTA) in compensation for forgoing such benefits. While there is little theoretical reason for suspecting that WTP and WTA should be different, empirical research has revealed that measurements of WTA often exceed those of WTP. In this discussion, WTP is used as representative of demand for environmental quality.

By undertaking surveys or administering questionnaires it is possible to elicit people's WTP (or WTA) for environmental goods or services directly. By creating a hypothetical market situation the researcher can use the respondent's replies to place values on items that are usually not marketed. The valuation is 'contingent' because the values derived from CVM depend on individual perceptions of a host of background factors that influence the market being surveyed. A poorly designed and implemented survey will produce answers that bear little resemblance to the population's true WTP. It is precisely because there is so much room for difference between consumer intentions as expressed on a questionnaire and consumer preferences as revealed in the marketplace that CVM results are often considered unreliable.

The literature has focused on overcoming the many sources of bias in CVM studies. Bias is any element in the study that consistently skews results in one direction, thereby leading survey results away from the true WTP of the population. Biases may arise from the way the sample is selected, the effect of the survey design or implementation on the responses gathered from respondents, or when the respondent misunderstands the nature of the contingent market. Resolving these difficulties involves careful design and testing of questionnaires, competent survey administration and a number of econometric tests for remaining sources of bias.

The use of CVM for valuing environmental resources is largely a North American and, to some degree, European tradition with very little work conducted in developing countries. A recent study in Brazil, however, indicates that results of CVM studies are credible even in rural areas in developing countries when respondents are well-informed about the resource in question. The study demonstrated how CVM surveys of actual and hypothetical water-use practices can provide estimates of WTP for access to clean water that vary according to household socioeconomic characteristics, and qualitative differences in water supply and delivery systems (Briscoe *et al.*, 1990).

CVM can be used to elicit values across the spectrum of total economic value. It is generally regarded as the only method for arriving at option prices and existence values. Since there are few surrogate or implicit markets for these values, indirect techniques relying on revealed preferences are often of little use. One commonly cited exception is the use of contributions to conservation organisations as a surrogate market for existence or option values. There are, however, a number of difficulties with this assertion, not the least being getting at the actual reason people make contributions. Option and existence values are discussed in greater detail later on in this chapter.

Hedonic pricing

The hedonic pricing technique relies on the observation that the value of non-marketed environmental services are frequently incorporated into the prices of other marketed goods and services. By disaggregating such market values an economist may uncover the relative contributions.of valued attributes to human welfare. Although soil fertility, scenic beauty or air quality are not directly exchanged in markets, hedonic pricing techniques enable economists explicitly to value these services that are implicit in the price of land and property, and wages.

The technique involves two stages. The first is relatively straightforward, involving an econometric estimation of the value contributed by the chosen environmental attribute to, for example, property values. The second stage involves working from this hedonic price equation back to the actual demand curve. This second stage entails overcoming both theoretical and practical obstacles. Basically, the output of the first stage identifies the price for environmental quality paid in a competitive market - not what the buyers are willing to pay. In order to identify the WTP, analysts must make simplifying assumptions about consumer preferences or gather additional data on consumer preferences.

Applications of this complex and often ponderous technique are scarce outside developed countries. The data requirements are one very large drawback to undertaking such studies in countries with a poor statistical base.

Travel cost method

The travel cost method is frequently applied to valuation problems involving ecotourism and recreational services derived from the environment. The technique requires information on the expenditures by site visitors. Aggregating the number of visitors by what it costs them to travel to and from the site provides a surrogate market indicating what people are willing to pay for access to the site. Essentially, travel costs form a variable admission price to the site. Those visitors from far away exhibit a large WTP, while those from surrounding areas reveal a low WTP. This relationship between distance and travel costs can be used to estimate the benefits that visitors gain by visiting the site. Drawbacks and potential obstacles to the techniques involve unobserved travel costs, the question of whether leisure time and travel to the site are necessarily a cost and the fact that trips are usually multi-purpose.

Applications of the travel cost methods to protected areas and other tourist and recreational sites in developing countries are rare but likely to increase as the technique is not overly demanding in terms of data or calculations.

In a recent application of the travel cost method Tobias and Mendelsohn (1991) examined the willingness of local tourists to pay to visit the Monteverde Cloud Forest Reserve in Costa Rica. The research involved gathering data from the reserve's headquarters on the frequency of ecotourist visits from each of Costa Rica's 81 cantons. Average distances from the major towns in each canton were measured and then multiplied by an estimate of the per kilometre travel costs in Costa Rica. Population density and illiteracy were included alongside this travel cost as variables that might explain the visitation rates for each canton.

These results were used to calculate the ecotourism value generated by the reserve. On average each visitor valued the experience at US$35. The present value of such trips, assuming constant flow of visitors and a real interest of 4%, came to around $2.5 million. Because only one out of five visitors to the reserve are Costa Ricans the total ecotourism value is actually much larger. Foreign visitors are likely to have far greater travel costs than local citizens, but are less likely to be travelling to Costa Rica just to see the Monteverde Cloud Forest. The authors confront this difficulty by making the safe assumption that foreigners value the experience as much as locals - leading to a present value of $12.5 million. Since the total area of the reserve is 10,000ha the value per hectare for the reserve land is estimated to be $1,250. When compared to the going price for land surrounding the reserve of $30-$100 per hectare, Tobias and Mendelsohn assert that expansion of the reserve is called for on economic grounds. Their case would be strengthened if the other direct, indirect, option and existence values were included in the calculation.

Other techniques

A number of methods for deriving the value of environmental resources exist. Although relatively inexact these methods are often the second and third best techniques that are actually used when time and money do not allow for detailed research. Estimating how much would need to be spent in order to prevent expected damage to environmental quality is one way of valuing resource degradation. Another way is to estimate the costs of replacing the environmental asset that is degraded either by the use of natural or man-made goods or services. This can be accomplished by pricing available substitutes or the cost of developing substitutes. The price of an environmental improvement or cost of degradation may also be generated by assessing the opportunity costs of the relevant action. A final method of assessing the value of resources/resource damage is to examine the cost of relocating economic activity should the resource flow be disrupted.

LOSS OF BIODIVERSITY AS AN ECONOMIC PROCESS

It is predictable that the application of economic processes to natural resources will lead to substantial losses of biological diversity. This is part of the general process whereby humans continue to modify their natural environment so that it can better satisfy their needs. To a large extent, loss of biodiversity is a concomitant of this value-maximising process.

However, it is also predictable that this process is unlikely to target an adequate amount of diversity, because persons deciding to convert their local environments do not consider the global costs of so doing. From the economic perspective, the biodiversity problem requires the regulation of local development processes for the advancement of global interests.

Conversions and loss of diversity

The economic theory of natural resources predicts that much of existing diversity will be depleted. This is because economics views the natural form of the resource as being necessarily competitive with other forms in which humans might hold these same resources: natural resources are 'natural capital' in contrast to 'man-made capital'. Humans make the choice of whether to hold the resources in their original form, or to convert them to a modified form (Solow, 1974).

From this perspective, human history has been one continuing process of the conversion of natural resources into more productive resource forms. For example, iron ore is more productive in the form of a machine than it is as a vein of sub-surface minerals. Therefore, the natural form of the capital is altered to make it more useful, resulting in the man-made form of the capital.

Similarly, a given hectare of land originally growing diverse native grasses will be converted to another plant form, such as wheat, because of the enhanced productivity of this resource. That is, there is a choice to be made not just between natural and man-made resources but also between more and less productive forms of natural resources. The biosphere can capture a limited amount of useful solar energy, and it is now unavoidably a human choice to determine which species will be used to perform this task over much of the earth's surface. Economics indicates that humans will choose to channel this energy only through those species which are most productive, eliminating the others through this competitive process.

These conversions have been driven by two important economic characteristics of resources: the relative growth rates of what humans wish to consume and the relative harvest costs. In economic terms, it is predictable that those species which exhibit relatively high growth rates and low harvest costs will displace many of the others.

Specialisation and loss of diversity

There are good reasons to believe that prevailing methods of production are biased against the maintenance of a wide range of diversity. This is attributable to the economies of scale implied by the law of economic specialisation. Biodiversity losses will not only result from the substitution of the more productive resources for the less; these losses will also occur by reason of the inertia that will develop around the more productive forms of natural resources.

The law of specialisation is one of the first laws of economics, developed by Adam Smith in the 18th century. He observed that there tended to be increasing productivity with increasing homogeneity in production methods and processes. It is far less costly, in general, to create thousands of units of an identical product than it is to make smaller numbers of differentiated products. This is why 'handcrafted' goods are more expensive than factory produced ones.

The cost differential is attributable to the application of capital goods in the process of mass production. Once capital is applied to the production of a particular good, it usually becomes much less costly to produce. However, the use of capital also implies homogeneity in the product. It is the very essence of capital that it must be fine-tuned to the production of goods of very specific characteristics.

In terms of biological products, the capital goods applied in mass production are the chemicals and machinery of intensive agriculture. These capital goods do not enhance the general productivity of the biosphere; rather, they increase its productivity by means of specialised substitution of natural resources. The diverse resources of nature are removed in favour of the specific resources for which capital goods have been developed. Cultivators and harvesters are developed to work in fields that are planted with a single crop. Chemicals are fine-tuned to eliminate all competitors of that crop.

The result of such specialisation is that an increasingly narrow spectrum of species meets all of the needs of humankind. A very small proportion of the thousands of plant species which are deemed edible produce the vast majority of the world's food. The four major carbohydrate crops (wheat, maize, rice and potatoes) feed more people than the next 26 crops combined (Witt, 1985). This also applies to animal protein sources. The tables of the Food and Agriculture Organization list only a handful of domesticated animals (sheep, goats, cattle, pigs etc.) which supply nearly all of the non-fishery animal protein for the vast majority of humans.

This concentration on a few useful species is occurring not only because these are relatively productive and manageable, but also because of the inertia resulting from specialisation. The economies realised from mass production continue to become greater as larger capital goods (larger machinery, larger farms, more chemicals) are employed, but this also implies increasingly homogeneous production. This means that the gains derived from specialisation also entail losses of diversity.

Globalisation and loss of diversity

The production of homogeneous capital goods also results in increasing economies. Producing a wide range of tractors and harvesters each tailored to a different crop is inefficient. Making a single style of machine to be employed the world over is the least costly method of producing capital. The same applies to chemicals. It will be less expensive to continue to fine-tune these to a few crop species, and mass-produce these crops, than it will be to produce chemicals adjusted to a range of different species.

Therefore, diversity losses do not occur only because of relative differences in natural productivity, and because of the inertia that develops around a given species once capital goods are applied to its production, but also because of inertia which develops around a particular type of capital good. Once a particular species has been chosen for capital-intensive production, it represents a commitment to a particular technology and mode of production. As capital spending becomes ever larger in regard to agricultural production (e.g. biotechnology investments), it becomes more important to increase the amounts of the specialised species produced in order to be able to spread the fixed costs of the investment, and to do this across both space and time. Thus, the spread of intensive agriculture across the world (including the 'Green Revolution') is predictable as a method of spreading these fixed costs across space. Specialisation and globalisation have gone hand-in-hand to generate worldwide losses of diversity in the furtherance of agricultural productivity.

Overshooting optimal diversity

In conclusion, much of the global loss of biological diversity derives from the relative advantages of particular species and particular methods of production.

From an economic perspective, the loss of some diversity is inevitable. The issue at stake is whether the decrease taking place at present is optimal in maximising benefit to mankind, or whether it is in fact excessive (in economic terms, whether the process will 'overshoot', or has already done so). There are several reasons why loss of diversity will tend to be excessive.

First, there is the possibility that current diversity is being too readily traded-off for immediate gains in productivity. At some point in time humans might decide that they would prefer to have more than four sources of carbohydrate or more than a dozen sources of protein, but by then diversity might have been reduced to such an extent that it is no longer possible. Similarly, it might be desired, if life-sustaining income levels are achieved world wide, that the world should contain more diversity to experience and enjoy; this will not be possible if diversity has been over-exploited.

Second, there is an increasing level of risk attached to a strategy of specialisation. Diversity supplies insurance against unforeseen events which specialisation does not provide. There can be an increase in the average productivity from conversions, but its variability might simultaneously increase, and because of the increasing risk of further conversions, the cost of each is not the same. There is a cost involved in converting diversity that is felt globally but not considered locally.

Third, the earth's natural ecosystems are being altered at an unprecedented rate, and at a rate far faster than our understanding of them is advancing. Our knowledge of the interactions between different parts of the biosphere is particularly inadequate.

In summary, it is possible that some aspects of resource conversion, while locally desirable, will have effects which are undesirable from a wider perspective. This asymmetry results from the element of 'globalised value' that attaches to the remaining biodiversity. From the economic perspective, the problem to be solved is how to bring this 'external value' into the equation when land-use decisions are being made. To do so, it is necessary to acquire some understanding of the nature of these values.

CURRENT USES OF DIVERSE RESOURCES

Introduction

The range of uses to which biological resources are currently put was surveyed briefly in Chapters 25 and 26. The intention here is to demonstrate how monetary value can be attached to wild resources, by reference to two kinds of use or consumption.

Diverse resources make substantial contributions to current consumption in both a relative and an absolute sense. That is, a notable portion of the world's population relies for a significant part of its sustenance upon wildlife resources, particularly in many developing countries (Prescott-Allen and Prescott-Allen, 1982). In addition, there is significant consumption of wildlife resources in those countries where the vast majority of consumption does derive from monocultures. Although the percentage value of these resources in terms of the overall economy of the countries is small, the absolute amount of value attached is very large.

Even though wild resources are being replaced by monocultures in the economic process of global conversion, they are far from valueless, even in terms of their current known use. It is important to recognise that it is the value of wild resources relative to specialised resources that will determine the extent to which wildlands will be converted. If there is no added value from converting land from a natural state, then the process of conversion will cease. One of the important problems of biodiversity conservation arises from the irreversibility of the conversion process. This means that decisions by current generations regarding the loss of biodiversity cannot in general be undone. Future generations must live with these decisions even if their values are different.

There is good reason to expect that the values of certain diverse resources will systematically increase over time. It has been argued that society's preferences shift toward natural resources and wilderness experiences, as wealth increases and natural resources become scarcer. This is evident in increased international tourism to places of natural beauty, the development of 'ecotourism' and wildlife encounters, and the willingness to pay for preservation. Some societies have long held a preference for natural products over synthetic varieties. For example, the Japanese are renowned for their dissatisfaction with man-made substitutes for wildlife products (Barbier *et al.*, 1990).

These uses of wildlife resources are ways of expressing preferences for the natural form of the habitat (and its products) over the domesticated form. They are important as means of countering the trend toward increased conversion. There are problems to be solved with regard to the use of wildlife resources, as there are with many kinds of resources, but it is important to conserve more wild resources now so that there is not a deficit of variety to meet the needs and desires of future generations.

COMMUNITY USE OF WILDLIFE RESOURCES

State provision for wildlife protection can create conflicts of interest between local communities and the protecting bodies. The establishment of protected areas for wildlife is often in direct conflict with the economic interests of local communities. Many are denied access to resources that they have traditionally exploited. Rural communities have not only lost their traditional management and use rights to local wildlife resources, but they may also bear the full costs of crop damage because of migrating wildlife. The situation has deteriorated in recent years, with rising rural populations and increased poverty. Illegal encroachment, hunting and harvesting are often the only available means of securing subsistence and income. Where local populations are not directly involved in these activities, their alienation from wildlife resources means they have little will to oppose the exploitation of these resources by others. "The breakdown of traditional common-property management regimes into virtual open-access exploitation leaves rural communities with little means to enforce sustainable management" (Barbier, 1990).

Table 27.1 Projected values from management of elephant resources in Botswana

	NET PROJECTED VALUE *		
OPTION	**5 years**	**10 years**	**15 years**
Game viewing with no consumptive uses	34.7	98.1	160.6
Game viewing with elephant cropping	91.2	198.4	288.9

Source: J. Barnes, Department of Wildlife and National Parks, Botswana
Note: * Values in million pula, discount rate 6%

Several programmes have attempted to counteract this by diverting some of the revenue generated by sustainable management of wildlife populations into the hands of the communities in which these populations are concentrated.

Community based utilisation of African Elephants

Elephant culling has been undertaken in Zimbabwe since 1965, with the objective of controlling elephant numbers. Revenues from the sale of ivory (legal and confiscated illegal ivory), skins and meat are a natural byproduct of such population control policies. These revenues fund the management programmes and compensate the local communities for elephant damage. They also support anti-poaching activities which protect the rents available from the sale of ivory.

Revenue from tourism and culling (the sale of ivory, hide and meat) generates a very persuasive argument in favour of elephant conservation. The combined value of elephants from the sale of meat products and tourism in Tanzania has been estimated in excess of US$80 million. Elephants constitute a major tourist attraction generating tourist revenues in excess of $25 million annually; were populations to recover sufficiently, sale of products could yield an additional $10 million; illegal meat hunting currently generates around $40-50 million annually. Not all the values derived from each of these activities can be realised simultaneously: sustainably managed populations cannot be subject to illegal poaching (ITC, 1989).

Comparable figures for the projected values from elephant utilisation in Botswana are presented in Table 27.1. Although game viewing alone results in significant returns, these are more than doubled if elephant culling is included.

The Nyaminyami Wildlife Management Trust, Zimbabwe

The Nyaminyami Wildlife Management Trust (NWMT) was formed by the Nyaminyami District Council in Zimbabwe. Its objective was to administer the management of wildlife resources for the benefit of the local inhabitants. The Trust established hunting and culling quotas for wildlife, pursued anti-poaching measures, and set up two Impala *Aepyceros melamprus* sites where herds could be sustainably managed. It also licensed two safari operators and metered compensation for economic losses incurred by residents as the direct result of conservation practices.

Table 27.2 Nyaminyami Wildlife Management Trust revenue*

Revenue	NWMT
Buffalo Range Safaris	148,349
Astra Wildlife	117,790
Mashonaland Hunters	6,048
Sub-total Safari Hunting	**272,187**
Cropping 1	11,554
Cropping 2	24,356
Sub-total Cropping	**35,910**
Meat and skins, Kapenta fishing	11,256
Total Revenue	**319,353**
Recurrent expenditure	
Cropping costs 1	-10,244
Cropping costs 2	-18,604
Wages and salaries	-16,378
Transport and equipment hire	-636
Vehicle maintenance/fuel/repair/insurance	-5,829
Wildlife compensation	-26,681
Kapenta licences	-2,400
Advertising, publications,printing	-4,341
Miscellaneous, bank charges	-1,469
Total Recurrent Expenditure	**-86,582**
Net Revenue	**232,771**
ZimTrust	20,093
Adjusted Net Revenue	**252,864**

Source: Adapted from Jansen, D.J. 1990. *Sustainable Wildlife Utilisation in the Zambezi Valley of Zimbabwe: economic, ecological and political tradeoffs*. Project Paper No. 10, WWF Multispecies Project, Harare.
Note: * In Zimbabwe $; in 1989 Z$2.1 = US $1.

In 1989, the Trust earned Z$319,353 in wildlife revenues; approximately 85% of this came from concession and trophy fees paid by safari hunters, and the remainder from sales of meat, skin and hides. Through additional contributions from the ZimTrust, Z$20,093 to finance recurrent expenditures and Z$191,683 for capital expenditure, NWMT ran a surplus of Z$252,865 (Table 27.2). This surplus was distributed between a reserve fund for capital expenditures (12%), levies retained by the district council (10%) and the remaining 78% to be channelled back into the communities, funding housing projects, clinics, teaching, and recreational facilities. A total of Z$198,000 was disbursed throughout the communities, constituting about 15-20% of annual household incomes. In addition, direct compensation amounting to Z$27,681 was paid for crop and animal damage, offsetting the costs associated with wildlife conservation. Meat and skins from cropping were also sold locally at a subsidised price representing an additional net gain to the local inhabitants from enlightened wildlife management.

For programmes such as this to work effectively, the revenues must be channelled back to the community. In Zimbabwe, trophy hunting fees paid by operators to the central government should have been redirected back to the community through investment in schools and clinics. However, only 57% of the nearly Z$6 million earned from wildlife over the period 1980-1987 had been returned by the end of 1987. The Zambezi valley project generated wildlife revenues of Z$2.1 million between 1981 and 1986, but by 1987 only 44% had been returned to the districts.

Community use of the Vicuña, Argentina

The Vicuña (*Vicugna vicugna*) is a wild camelid inhabiting the *puna*, a treeless pastoral zone in the central Andes of western South America. Vicuña have fine wool and are a sought-after meat delicacy. They have been hunted for centuries and the Incas are recorded as following sound management practices in harvesting them.

In 1987, Vicuña populations in the Laguna Blanca Reserve (Catamarca province) were examined to assess their potential contribution to the indigenous peasant economy. This is primarily a subsistence economy, with a small but increasing involvement in the market economy. The two main sources of income are from sheep and llama spun wool. The potential harvest of the Vicuña population was estimated using simulation techniques, calculating the maximum sustainable yield and the carrying capacity of the area (Rabinovich *et al.*, 1991). If the Vicuña population were allowed to grow from its current size of 5,000 animals to around 8,000, 15.2% of that population could be harvested each year. The monetary value of each Vicuña is estimated at US$64: $19 for the wool, $10 for the meat (assuming a 20kg animal fetches $0.50 per kg) and $35 for the hide. The estimated total income that could be derived from sustainable management of the Vicuña is US$94,464 per year. This would provide an annual household income to the peasant community of the Laguna Blanca Reserve of almost US$1,000 if equally distributed among the 95 families.

ECOTOURISM

Ecotourism, or nature tourism, is just one component of the tourism industry. A precise definition of tourism is elusive because of its complex nature, involving a combination of attractions, transport, accommodation, supporting facilities and infrastructure. It is generally defined by its spatial dimension (Pearce, 1989), and is thus often characterised by criteria such as a minimum distance of travel or travel involving at least a one-night stay away from home.

Tables 27.3 and 27.4 present data and projections on worldwide tourist arrivals and receipts from the World Tourism Organization (WTO). Note that these figures pertain to cross-border tourism; according to WTO purely domestic tourism may be worth ten times as much as the $250 billion generated by international tourism in 1990. While political and military events have strongly influenced recent tourist movement the upwards trend in arrivals and receipts during the past five years is clear. Over the 1985-1990 period tourist arrivals grew by 6% and receipts by over 16%. WTO forecasts through to 2000 envisage continued growth but at a reduced rate of 4% for arrivals and 8% for receipts.

Although Europe accounts for the bulk of the world's tourist arrivals (over 60%), Europe's share of the receipts is not as large, presumably because of short intra-European stays. The fastest growing segment of the market is in Asia. WTO predicts that Asian tourist receipts and arrivals will exceed that of the growing American market by the year 2000. Meanwhile, the African market for tourism remains a small fraction of the world market, accounting for just three per cent of world arrivals and two per cent of world receipts in 1990.

Table 27.5 presents data on arrivals in all regions from the Americas and Europe. It reveals that almost 90% of travellers from Europe go on holidays to Europe or North America. Similarly, a majority of travellers from the Americas either stay at home or go to Europe. Clearly some of the North-North travel is ecotourism (in 1986 US parks brought in foreign exchange worth $3.2 billion); however, destinations such as East Africa, Central America, and Southeast Asia, which are renowned for their wildlife, are clearly of only marginal significance in the overall tourism picture. This contrasts with the fact that the tropics are very rich in biodiversity and the temperate latitudes, including Europe and North America, much less rich.

Nonetheless, tourism revenues may be of great economic significance to local economies, particularly in developing countries that are popular destinations for nature tourism. In this regard, Swanson (1991) has calculated that tourism provides 9-13% of exports from developing nations in sub-Saharan Africa, South and East Asia and Latin America. Lindberg (1991) reports that tourism in Kenya generated $400 million; in recent years it has been Kenya's largest earner of foreign exchange. Dixon and Sherman (1990) put tourism's share of the economy in Caribbean nations at 15-30%. Tourism can obviously provide a boost to local economies, but how much of this revenue comes from ecotourism?

Defining the exact meaning of 'ecotourism' is no easier then agreeing on the coverage of the term 'tourism.' Lindberg (1991) characterises nature tourism as being distinctly different from large-scale, highly developed, 'mass' tourism. Sites that attract Lindberg's ecotourist feature natural attractions and a certain degree of solitude. Lindberg estimated that of the $55 billion in tourism revenues accruing to developing countries in 1988, 'nature tourism' brought in 4-22% of these revenues. Despite its relatively small share of the market, ecotourism, like other 'special interest' sections of the market such as cultural tourism and adventure travel, is expected to outpace the general growth of 'mass' tourism in the next decade (Dixon and Sherman, 1990).

Table 27.3 International tourist arrivals

	1985	1986	1987	1988	1989	1990	1995	2000
WORLD	322,723	330,527	356,787	381,946	414,223	429,250	515,000	637,000
EUROPE	214,263	215,396	230,752	239,347	266,946	275,500	294,000	338,000
AMERICAS	58,728	62,894	67,986	74,991	78,456	84,000	103,000	128,000
AFRICA	9,805	9,488	9,986	12,646	13,604	14,000	23,000	32,000
ASIA	39,927	42,749	48,063	54,962	55,217	55,750	95,000	140,000
east	29,408	33,128	38,372	44,703	44,387	46,500		
mid–east	7,979	6,890	6,984	7,379	7,775	6,000		
south	2,540	2,731	2,707	2,880	3,055	3,250		

Source: from World Tourism Organisation
Note: figures are given in thousands

Table 27.4 International receipts from tourism

	1985	1986	1987	1988	1989	1990	1995	2000
WORLD	116,158	139,234	170,456	196,521	209,416	249,300	343,000	527,000
EUROPE	61,181	77,024	96,341	106,746	109,007	136,300	152,000	206,000
AMERICAS	33,314	37,383	41,982	49,632	56,600	65,900	95,000	146,000
AFRICA	2,601	2,993	3,687	4,625	4,479	5,000	10,000	14,000
ASIA	19,062	21,834	28,446	35,518	39,330	42,100	86,000	161,000
east	12,851	16,118	21,281	28,394	32,405	36,500		
mid–east	4,811	4,036	5,311	5,233	4,944	3,500		
south	1,400	1,680	1,854	1,891	1,981	2,100		

Source: from World Tourism Organisation
Note: figures given in $US millions

Table 27.5 Tourist arrivals, 1988

REGION OF DESTINATION	REGION OF ORIGIN: EUROPE ARRIVALS	EUROPE % OF TOTAL	AMERICAS ARRIVALS	AMERICAS % OF TOTAL
EUROPE	**206,482,068**	**88.61**	**17,627,702**	**22.25**
ASIA		**4.69**		**5.74**
Eastern Asia	2,160,959	0.93	2,691,257	3.40
Southeast Asia	2,703,173	1.16	1,018,680	1.29
Southern Asia	1,059,245	0.45	254,362	0.32
Western Asia	5,007,702	2.15	586,395	0.74
AMERICAS		**4.12**		**70.33**
Northern America	7,300,070	3.13	36,988,699	46.69
Southern America	1,054,539	0.45	5,756,668	7.27
Caribbean	1,010,705	0.43	6,601,596	8.33
Central America	231,886	0.10	6,370,939	8.04
AFRICA		**2.26**		**0.63**
Eastern Africa	624,460	0.27	117,650	0.15
Middle Africa	38,904	0.02	5,802	0.01
North Africa	3,933,058	1.69	263,862	0.33
South Africa	259,970	0.11	61,787	0.08
Western Africa	416,808	0.18	47,750	0.06
OCEANIA		**0.31**		**1.05**
Aust. & N. Zealand	662,693	0.28	594,285	0.75
Melanesia	25,359	0.01	65,684	0.08
Micronesia	1,397	0.00	80,041	0.10
Polynesia	43,895	0.02	90,862	0.11
TOTAL	**233,016,891**		**79,224,021**	

Source: from World Tourism Organisation

More precise estimation of the size of the ecotourism market is a near impossible task which requires a far more discriminating statistical base than is currently available in the national-level figures given to the WTO. While some activities can clearly be classified as ecotourism (eg. safaris to view the Mountain Gorillas in the Parc National des Volcans in Rwanda), much tourism defies such disaggregation. Most tourism is motivated by a mixture of cultural, historical, biological, geological and personal attractions. Evidence of this is demonstrated by Table 27.6 which presents the results of a survey examining the motives of tourists travelling to five countries in Latin America. An additional problem in specifying the value of ecotourism is determining which receipts should be allocated to which type of tourism. The bulk of the receipts for tourist expenditures do not occur at tourist sites such as parks, museums and cultural festivals, but at hotels, restaurants and for travel costs.

For these reasons, attempts to demonstrate the value of ecotourism often focus on illustrating the importance of charismatic species to the conservation of particular natural sites. For example, the Parc National des Volcans in Rwanda receives approximately US$1 million a year in entrance fees, with an additional $9 million in indirect benefits to the local economy (Lindberg, 1991). In this case, the park's survival depends entirely on one species: the gorilla. If ecotourists were not enthralled by, and willing to pay for, the prospect of a face-to-face encounter with gorillas, the park's natural habitat would doubtless long since have been converted to other uses.

In other cases, tourists may be attracted by a range of species. A number of studies have put rough figures on the value of elephants and lions in Kenya. Western and Henry (1979) found that each lion in Amboseli Park in Kenya generated US$27,000 in tourism revenues, while elephant herds in the same park produced $610,000 per year. In a further study of the tourism value of lions in Amboseli, Thresher (1981) calculated that over a 15-year period a single lion would draw in $515,000 in foreign exchange. More recently, Brown and Henry (1989) used contingent valuation and travel cost methods to calculate that the value of viewing elephants in Kenya is $25 million per year. Such figures lend credence to the claim that the ecotourism value of such species is far greater than their trophy value.

Barnes *et al.* (1992) point out that management of elephants in Kenya should consider not just their ecotourism value but their total economic value. The role of uncertainty in valuing ecotourist use, the potential for large existence values and the indirect value of the elephant as a species with a large ecological role to play must also be incorporated into an estimate of their total value. Of course, conservation efforts in Kenyan national parks must also consider the total economic value of lions, zebras, acacias, and other resources. In order to make decisions that maximise the net benefits to society, the total economic value of the site and all its associated values (use and non-use) must be considered. For example, although conservation of gorillas in the Parc National des Volcans is encouraged by the incentives of ecotourism revenues, there are additional ecological and existence values that accrue to locals and the global community simply because the gorillas and their habitat are protected.

A final, cautionary note must be added when discussing the value of ecotourism and the receipts generated by the tourism industry in developing countries. Tourism may indeed have a macroeconomic multiplier effect (the indirect and positive feedback effects brought on by the direct expenditure of tourist monies); however, the extent of such a multiplier will depend on the funds remaining in the local economy. If expenditures on tourist hotels and restaurants are promptly spent on imports or repatriated by foreign companies, there will be no 'multiplying' effect. Pearce (1989) reports that small Caribbean and Pacific nations may lose half of their gross foreign exchange earnings to expenditures on tourism-related imports. The World Bank has estimated that developing countries lose 55% of gross tourism revenues in such leakage (Boo, 1990). The lower the availability of locally produced goods and services used by tourists the worse this leakage becomes. Thus, ecotourism is not a panacea that guarantees wise and effective use of biological resources. If the revenues of ecotourism do not accrue to national park systems or local communities, there will be little economic incentive for investment in the recurring costs of conservation activities.

Table 27.6 Reasons for selecting travel destinations in Latin America

REASON	RESPONDENTS	%
Natural History	167	38.3
Sightseeing	161	36.9
Visiting friends and/or relatives	132	30.3
Sun, beaches, entertainment	130	29.8
Cultural/native history	102	23.4
Business/convention	87	20.0
Archaeology	63	14.4

Source: Boo, E. 1990. *Ecotourism: the potentials and pitfalls*. World Wildlife Fund, Washington.
Note: Total number surveyed = 436

EXISTENCE VALUES

Existence values are those benefits that are completely disassociated from the use of a resource. Existence benefits occur when people are willing to pay simply for the pleasure they derive from knowing that particular species or habitats continue to exist, irrespective of any plans they may have to hunt, observe or otherwise use these biological resources.

A range of terminology is used in referring to existence values, including bequest, stewardship, vicarious and intrinsic values. If the motivation behind these expressions of value is to preserve the resource for future use, then the criterion of non-use is not met and these values are better considered as use values. An additional source of confusion and debate is the status of so-called 'intrinsic' value. Interpreted to mean the value accruing to species other than *Homo sapiens*, this value is outside the scope of economic analysis which is based solely on the expression of human preferences.

Existence values may accrue to people in both the developed and the developing world. Unfortunately, as Table 27.7 demonstrates, the results of empirical research to date comes mainly from the developed world, in particular the USA. While casual observation may lead to the expectation that existence values are a 'luxury of the rich', further empirical work is needed to deny or confirm this hypothesis. While many economists agree that people are willing to pay for the mental satisfaction of knowing species and habitats exist, the psychological nature of existence values has so far defied the emergence of serious theoretical or analytical approaches on these values. With little in the way of theory to guide empirical investigations, the insight gained from case studies of existence values is often limited to a mere examination of their size relative to use values or option prices.

A study of preservation bids (synonymous with option prices) for Bighorn Sheep and Grizzly Bears in Wyoming by Brookshire *et al.* (1983) provides an excellent illustration. Using survey questionnaires (the contingent valuation method) the authors measured the willingness of prospective hunters to pay for hypothetical future permits to hunt Bighorn Sheep and Grizzly Bears. In addition the questionnaire also identified existence values and observer preservation bids. The bids of respondents indicating that they would neither hunt nor directly observe the animals were taken to reflect existence values. The results of the study revealed a range of hunting bids from just under $10 to almost $30. Observer option bids for the two species were in the vicinity of $20. Existence bids for the Bighorn Sheep were in the $7 range while those for Grizzly Bears averaged $15. This study reveals that existence values for species may be of the same order of magnitude as option prices for such direct uses as hunting and game-watching.

Table 27.7 Empirical measures of existence values

	VALUE PER ADULT RESPONDENT IN MID-1980s (US$)
Animal Species	
Bald Eagle	11
Emerald Shiner	4
Grizzly Bear	15
Bighorn Sheep	7
Whooping Crane	1
Blue Whale	8
Bottlenose Dolphin	6
California Sea Otter	7
Northern Elephant Seal	7
Natural Amenities	
Water quality (S Platte River Basin)	4
Visibility (Grand Canyon)	22
Additional park facilities (Australia)	6

Sources: Pearce, D.W. 1990. *An Economic Approach to Saving the Tropical Forests*. LEEC Paper DP 90-06. IIED, London. Majid, I., Sinden, J.A. and Randall, A. 1983. Benefit evaluation increments to existing systems of public facilities. *Land Economics* 59:377-392.

In a survey of the willingness to pay for additional park facilities in Australia, Majid *et al.* (1983) demonstrated that the existence values for habitat are also of a comparable size to their recreational use values. The initial survey questions asked respondents how much they would pay for recreational use benefits and total benefits generated by a list of current and proposed facilities. As a measure of existence value the authors calculated the difference between the willingness to pay for recreational site visits and the total willingness to pay for each site. The results for all parks indicated that the total benefits were roughly twice as big as the use values - thus existence values were judged of equal value to recreation values.

Table 27.8 Gifts to surveyed environmental/wildlife organisations

	US$,000 1989	US$,000 1990	PERCENT CHANGE
TOTALS	208,907	273,385	31
Nature Conservancy	48,963	85,527	75
WWF and the Conservation Foundation	33,465	42,438	27
Ducks Unlimited, Inc.	25,501	29,674	16
Sierra Club	21,908	28,718	31
Natural Resources Defense Council	12,524	13,821	10
National Audubon Society	10,174	11,094	9
National Arbor Day Foundation	8,126	11,045	36
New York Zoological Society	17,073	9,531	−44
Sierra Club Legal Defense Fund	5,973	6,833	14
World Resources Institute	5,240	6,336	21
American Farmland Trust	2,716	5,195	91
International Fund for Animal Welfare	3,912	4,555	16
Resources for the Future	2,651	2,948	11
Animal Protection Institute	435	2,607	499
American Humane Association	1,992	1,903	−4
American Forestry Association	909	1,816	100
Clean Water Fund	719	1,607	124
Adirondack Council	1,178	1,542	31
American Rivers	1,728	1,502	−13
Trout Unlimited	1,180	1,309	11
Earth Island Institute	1,007	1,026	2
Rainforest Alliance	254	798	214
Soil and Water Conservation Society	388	390	1
Farm Sanctuary, Inc.	165	346	110
Alliance for Environmental Education	19	186	879
Wildlife Habitat Enhancement Council	152	182	20
Lake Michigan Federation	133	165	24
Animal Rights Network, Inc.	200	118	−41
American Cave Conservation Association	87	108	24
Peace Garden Project	135	65	−52

Source: AAFRC Trust for Philanthropy

But how are these existence and preservation bids actually expressed in the real world? One way of expressing these desires for the benefits of species and habitats is to make donations to organisations that conserve biological resources and biodiversity. Table 27.8 provides figures on such philanthropic giving in the USA over the 1989-1990 period. The data - as would be expected - reveal an upward trend in overall giving to the organisations surveyed by the AAFRC Trust for Philanthropy. Table 27.9 reveals figures for the total charitable contributions to environmental and wildlife causes in the USA in the context of total giving. While $2.3 billion is a substantial sum of money, the amount donated to the environment pales beside that donated to other philanthropic causes. While environmental giving has registered growth of 9%, 11% and, most recently, 24% per year in real terms, the fact remains that the average donation came to roughly $10 per person in the USA in 1990.

Table 27.9 Charitable contributions in the USA

Total Funds in 1990: $122.6 billion

Destination of Funds	%
Churches and synagogues	53.7
Education	10.1
Human services	9.6
Arts and culture	6.4
Public benefit	4.0
Environment	1.9
International	1.8
Undesignated	4.4
Sources of Funds	%
Individuals	83.0
Bequests	6.4
Foundations	5.8
Corporations	4.8

Source: AAFRC Trust for Philanthropy. 1990. *Giving USA*, NY.

A critical ingredient of CVM studies is the extent to which respondents are informed about the object of the studies and, correspondingly, how much information is disclosed during the survey process. Samples *et al.* (1986) investigated the effects of information disclosure on preservation bids for endangered species. Although preservation bids may be interpreted to have both use and non-use components, the authors assumed that the endangered status of the species would cause responses to reflect mainly on the value of 'saving' the species as there was little real prospect for 'using' species close to extinction.

THE VALUATION OF DIVERSE ECOSYSTEMS

Introduction

A particular cause for concern arising from the conversion of biological systems is the problem of accumulated losses of unknown ecosystem values.

Rational decisions about the conversion of one system to another should involve an assessment of relative values; typically, however, the only values that are included in that comparison are the appropriable ones. If a person or group cannot capture that value, it is unlikely to be considered important. However, many of the benefits of biological systems flow not to any one particular individual or group but to the community at large. Attributes of forests, such as oxygen production and carbon fixation, are unlikely to stop a logger from acting, even though these are very important characteristics. The total economic value, including these non-appropriable values, of biological systems must be entered into the calculation if the optimal amount of diversity is to be conserved. This is unlikely to occur at any time in the near future, simply because we do not have the capacity to do so. There is, therefore, good reason to preserve some parts of the world's diversity in general recognition of the global public goods that it provides even if these cannot be assigned a value.

THE VALUE OF TROPICAL FORESTS

The fact that tropical forests have value is not disputed: they are a source of ecological benefit and material wealth (see Chapter 20). In order to understand the consequences of decisions made about different possible uses of the forest, it is necessary to quantify and rank the values under these different uses.

The method of system valuation

The purpose of valuation techniques is to correct those prices that do not correspond to the 'true' economic values and to calculate prices for those assets that are not valued at all (Maler, 1989). The concept of *total economic value* (TEV) offers a unified approach to the valuation of tropical forests. This concept is based upon the idea that it is possible in certain cases to disaggregate the flow of goods and services from environmental resources, and then to assign monetary values to these discrete functions.

Some of the component goods from forests are traded in markets, and in these cases the market price provides an indicator of social value. But this is so only if markets are perfectly competitive and complete; then, the prices arrived at will reveal the correct marginal valuations of those goods and services exchanged. Otherwise, it is necessary to compute a *shadow price*, i.e. a price that differs from the market one but corresponds more closely to social value. However, in many cases environmental goods cannot be traded in the marketplace, hence their value is not directly revealed. In this case other methods must be employed to gauge their value and capture how that value alters with different uses.

Therefore, the object of environmental valuation is the performance of these three tasks:

- the segregation of a unitary system into discrete components
- the valuation of those components that are not traded in markets
- the correction of market values, where these differ from social values.

Table 27.10 The concept of total economic value* in a tropical forest context

USE VALUE			+	NON-USE VALUE
DIRECT + VALUE	INDIRECT + VALUE	OPTION + VALUE	{QUASI + OPTION VALUE}	EXISTENCE VALUE
Sustainable timber				
Non-timber products	Nutrient cycling	Future direct and indirect uses		Forests as objects of intrinsic value, as a bequest, as a gift to others, as a responsibility (stewardship). Includes cultural and heritage values.
Recreation	Watershed protection			
Medicine	Air quality			
Plant genetic resources	Micro-climate			
Education				
Human habitat				

Source: Pearce, D.W. 1990. *An Economic Approach to Saving the Tropical Forests.* LEEC Paper DP 90-06. IIED, London.
Notes: **Direct Value** refers to those benefits that can be observed being consumed, although their consumption might not yield a meaningful price which can be assigned to that benefit. **Indirect Value** refers to those benefits that are not observed being consumed, but that are known to be essential to the preservation and maintenance of ecosystems. **Option Value** is the value placed on securing the future consumption of goods and services yielding direct and indirect value. **Quasi-Option Value** is the value of learning about future benefits that would be precluded by development or irreversible change of the forests today. This takes account of the fact that current valuations are circumscribed by current knowledge of forest functions. **Existence Value** is that value placed on an environmental asset independent of its current or future 'usage'. This incorporates the innate value of the forest *in situ*. * Total economic value = use value + non-use value

Table 27.10 outlines the components of TEV with reference to tropical forests. Trade-offs occur between the different types of uses, direct and indirect: the supply of hardwood might diminish the amount of protection offered to watersheds by the root network of the trees; forest areas devoted to recreational facilities might displace indigenous peoples. We cannot simply add the components of TEV to obtain a measure of the ecological wealth of the forests.

Direct use values in tropical forests

Timber

Logging can be consistent with forest conservation if the forest lands are managed sustainably. The limitation to this approach to timber valuation is that market prices generally diverge from *shadow prices* (those that reflect the true opportunity cost of the good or service). In particular, if the forest is being logged without well-defined property rights (without care for its future flow of timber), then it is likely that the price of timber does not reflect the full amount of resource rent that is available. The timber price might then reflect only the social value of the labour and capital inputs into its production, not the value of the timber itself. In this case, the market price must be corrected to reflect the true social value of the resource.

The management of all environmental resources is tied closely to the problem of the valuation of those resources over time. Given that one of the main reasons to conserve today is to preserve the resource for tomorrow, it is important to value a resource not at a single moment but over a period of time. This allows the value of unused resources to come into the calculation. When future values are combined with current values, it is important to take into consideration the relative weights to ascribe to the time periods. Usually, a discount is applied to future period values, because of the uncertainties involved. The net present value calculation is very sensitive to the discount rate employed to convert the stream of future benefits into a single value.

Discounting enables the economist to represent the value of a resource or asset in terms of the flow of income deriving from that asset over a specific time span. The discount rate reflects the greater importance attached to current, compared with future, consumption and the trade-off between them.

One important study (Table 27.11) examined the comparative present values of an Indonesian forest, given a range of possible uses. It demonstrates the comparability of returns available from a range of different forest management practices for timber production. In addition, it also illustrates the method of present valuation of the production capability of a hectare of forest land.

The impact of utilisation on environmental quality depends on how the forest lands are altered as a result. SAW and PULP may threaten existing tropical forest-lands because they are typically concentrated among uniform plantings of non-indigenous species. However, where these plantations occur on previously unforested land they may still contribute to 'carbon-fixing' and have positive net environmental worth.

Table 27.11 Profitability of logging at different discounts

REGIME	DISCOUNT RATE		
	5%	6%	10%
TPI	2,705	2,409	2,177
CHR	2,690	2,593	2,553
INTD		2,746	2,203
PULP		2,926	2,562
SAW20		2,419	2,278
SAW10		2,165	2,130

Source: D.W. Pearce. 1987. *Forest policy in Indonesia*. unpublished memorandum. World Bank.
Notes: Net present value US$1,986/ha. **TPI** - selective cutting regime in which only those trees over 50cm in diameter at breast height are harvested. **CHR** - complete harvesting and regeneration; all merchantable trees are harvested then the cleared land undergoes natural or enriched regeneration. **INTD** - intensive dipterocarp management, following a plantation approach on clear-felled land. **PULP** - when plantations of fast-growing trees are harvested for wood pulp. **SAW** - refers to saw timber plantations where the trees are cut at 10/20 years respectively.

The table reveals that were an estimate of the net worth of a project to be based solely on financial profitability it would be preferable to encourage rapid-growth plantations for pulp production. The more sustainable selective cutting regime is only favoured at the lowest discount rate of 5%. Therefore, although there is no clear-cut advantage to large-scale alteration of the forest environment during utilisation, the financial incentives from logging alone might encourage this to happen. Focusing on this single use of the diverse habitat can lead to the conclusion that habitat conversion is economically optimal. This conclusion may not be the case, even when only logging is being considered. In many instances the conversion of forest lands is only financially profitable when considered in combination with government subsidies that encourage the same.

Non-timber products

This conclusion can be altered quite dramatically by the introduction of a wider range of goods and services into the analysis. Non-timber forest products are often a vital source of foreign exchange earnings and revenue. They are also essential to the rural household economy (de Beer and McDermott, 1989). In many parts of Southeast Asia, the rural population depends heavily on forest products for their daily needs. It has been estimated that in Southeast Asia at least 27 million individuals rely on the forests to satisfy their nutritional, fodder, fuelwood, and shelter requirements.

These products are seldom exchanged or sold, so care must be taken in valuing their contribution to the rural economy. It is possible for this purpose to use the value of the effort expended as a surrogate. This can be done by calculating the 'cost' of the labour inputs that are applied in gathering and harvesting the non-timber products. In Thailand it was estimated that the one million families who are forest-dwellers devote about 180 person-days a year to collecting forest-food, whilst the three million families located on the periphery of forests spend about 60 person-days harvesting food. This can be taken to indicate that these resources are valued by forest-dwellers at an amount up to half a year's salary (assuming that the hours spent gathering the fuelwood could be otherwise redirected to the labour market). In many cases the forest products represent substitutes for goods that can be bought in local markets and therefore the market values of these substitutes can be aggregated to provide a monetary equivalent of the forest-products. Of those products that are themselves marketed, such as rattan, nuts, fruit etc., values may be more easily discerned. Non-timber forest products also contribute greatly to the national economy. They generate employment, foreign exchange earnings, trading and processing revenues, consumption and import substitution opportunities.

In Thailand it was estimated that there were 200 rattan furniture manufacturers operating small-scale cottage industries which produce goods mainly for the domestic market. In 1987 the US dollar value of exports totalled US$29.1 million (Anon., 1988). Thailand also exports finished bamboo products: handicraft export values vary from US$212,413 (Ministry of Commerce) to about US$3 million (a questionably large figure). Whilst attempts to value bamboo exports produce varied results, bamboo is nevertheless a widely-used product of great importance to the national economy.

Ecotourism values

Tropical forests are also valued for their recreational benefits. In Costa Rica, Ecuador, Philippines and Thailand tourism is a vital source of foreign exchange earnings, generating more revenue than the export of timber and timber products.

The *travel cost* method is often employed to value the benefits derived from tourism. Direct costs of access, package tours, and hire of transport can all be regarded as components of the overall value attributed to recreational use. This approach has been used extensively in developed countries to value the provision of recreational goods and services. The methodology rests on the proposition that observed behaviour can be used to derive a demand function for non-marketed environmental goods and services, regarding travel costs as surrogates for variable admission or access costs. For example, European package 'explorer' holidays to the Peruvian Amazon cost about $2,300 per person for 20 days.

Medicine and plant genetics

Tropical forests provide the habitat for a great variety of species. The legal export of hides and skins, genetic materials, spices, and oils provides many developing economies with revenue. Where these products are traded we may estimate the value of forest byproducts.

A very important product of forest diversity is its plant varieties and the special information on chemical use that these represent. Plant-based pharmaceuticals are a vital source of foreign exchange earnings. In 1979 Thailand exported medical plants and spices valued at US$17 million, consuming about US$20 million domestically (Anon, 1981).

However, not all of the potential rents deriving from the sale of these products are captured. In the case of pharmaceuticals, calculation of the market value of drugs bought does not yield an estimate of the full value of the plant source because the value reflects not only the drug manufacturers' willingness to pay, but also the consumers' net gains from the use of this plant. Here, the market price is probably a poor indicator of the actual social value of the good; it needs to be adjusted upwards.

Indirect use values

Whilst some forest conversion in the tropics provides farmers and ranchers with valuable new tracts of land, much leaves only degraded soils unsuitable for sustained agricultural production. The loss of tree cover in watersheds increases flooding, erosion, soil-leaching, and downstream sedimentation. In semi-arid areas, deforestation depletes essential organic matter, exposing the soils to wind and water erosion. There is a very significant loss of ecosystem function and the benefits which these systems render, both on and off site.

The damage incurred as a result of the removal of forest cover may provide an estimate of the value of watershed protection. The loss of revenue because of declining soil fertility, decreased freshwater fish yields as the result of increased sedimentation, and reduced local rainfall can all provide a measure of the indirect use values that accrue to forest conservation.

Defensive expenditures designed to mitigate against the effects of the loss of forest cover can also provide an economic value for indirect usage. Such expenditures include the cost of building levees, windbreaks, the application of fertilizers, and increased irrigation requirements. These, however, are undertaken with the implicit assumption that the benefits from replacement exceed the costs of deforestation. For if they did not, it would not have been rational to deplete forest cover in the first place.

Where forest cover is interrupted, nutrients are released into the hydrological cycle. In general there is a net nutrient outflow which can in itself pollute local river systems and that greatly reduces the productive capacity of the cleared land.

In growing, forests fix carbon dioxide through the process of photosynthesis and give off oxygen. Once grown, there is no net exchange of carbon and oxygen, mature forests are described as being in carbon equilibrium, and in this state they release as much CO_2 as they absorb. Deforestation releases CO_2 (and other greenhouse gases such as methane) into the atmosphere, contributing to the greenhouse effect.

In valuing the carbon-fixing properties of a tropical forest, we must be careful not to double-count. Whilst preservation ensures that the damage associated with carbon release is averted, forest clearance results in a net debit. However, it would be inappropriate to ascribe both a positive value to carbon-fixing and a negative one to forest clearance in an evaluation of the net benefits deriving from conservation.

The calculation is sensitive to the method of forest clearance and the subsequent use to which the timber or forest lands are put. If the forest is clear-felled and all the timber is used to make durable wood products (housing timbers, furniture etc.), then deforestation may cause little CO_2 release because much of the carbon will remain contained in the timber products. However, clearance through a 'slash-and-burn' approach will release all carbon contained by the forest, with no offsetting gain from the productive use of the forest timber.

Non-use benefits

Most attempts to develop existence values (those not related to functional requirements) rely on the *contingent valuation approach*, which reports the 'willingness to pay' of individuals for environmental goods or services. To date there have been no studies relating directly to the existence value of tropical forests.

Cost-benefit analysis: the Korup Project, Cameroon

The following example illustrates the type of calculations that might be undertaken to elicit a value for the net benefit of a particular forest *in situ*. The aims of the Korup project are to promote conservation of the rain forest in Korup National Park in Southwest Province, Cameroon. It was undertaken on behalf of the Government of Cameroon and the World Wide Fund for Nature.

The project chose to evaluate the flow of benefits from conservation options. The net benefits deriving from sustained forest and subsistence use, tourism, genetic materials, watershed protection, soil fertility maintenance, and flood control are compared with the opportunity costs of forestry and other development options (Table 27.12).

The *opportunity costs* measure that value of timber earnings forgone by the preservation of the forest. The *direct benefits* attempt to place a value on the sustained forest use beyond the year 2020 when the forest would have disappeared had it continued to be managed under the current regime. They also give a figure for the replacement of the subsistence production of the resettled villagers; the value of tourism; the minimum expected genetic value of the forest resources, etc. The *induced benefits* value the contribution the project makes to agricultural productivity and forest activities in the locality of the forest.

The final figure is then adjusted to reflect the net positive contribution of the external funding to Cameroon, the fact that Cameroon will be able to realise only 10% of the genetic value through the operation of patents and licensing, and that some of the watershed benefits accrue to Nigeria and not to Cameroon.

Conclusion

The forest represents a wide range of values, from timber to carbon-fixing. Incorrect decisions about use will always be made if any one of these uses is considered in isolation from the others.

Valuation becomes increasingly difficult as the use becomes more removed from the marketplace. Thus, carbon-fixing

Table 27.12 Cost-benefit analysis: the Korup Project

Direct costs of conservation	-11,913
Opportunity costs	
Lost stumpage value	- 706
Lost forest use	- 2,620
	- 3,326
Direct benefits	
Sustained forest use	3,291
Replaced subsistence production	977
Tourism	1,360
Genetic value	481
Watershed protection of fisheries	3,776
Control of flood risk	1,578
Soil fertility maintenance	532
	11,995
Induced benefits	
Agricultural productivity gain	905
Induced forestry	207
Induced cash crops	3,216
	4,328
NET BENEFIT - PROJECT	1,084
Adjustments	
External trade credit	7,246
Uncaptured genetic value	- 433
Uncaptured watershed benefits	- 351
NET BENEFIT - CAMEROON	7,545

Source: Ruitenbeek, H.J. 1989. Social cost-benefit analysis of the Korup Project, Cameroon, prepared for the World Wide Fund for Nature and the Republic of Cameroon, London.
Note: NPV £,000, 8% Discount rate.

values and nutrient cycles are real sources of value, but very difficult to quantify. The Korup study demonstrates that a careful attempt to derive these values indicates they are very substantial indeed. The tropical forest resources will be depleted if their entire range of values is not fully recognised and integrated into decision-making by individuals and governments.

THE VALUE OF WETLANDS

Wetlands are areas of land that remain waterlogged for a substantial period of the year (see Chapter 22). Tropical wetlands cover 2.64 million km^2 world wide whereas wetlands in temperate and boreal regions occupy about 5.72 million km^2. They support a wide variety of plant and animal species restricted to such environments. Wetland ecosystems are among the most threatened of all environmental resources. Much of the physical loss of wetland area has been because of the conversion to industrial, agricultural and residential use. However, qualitative degradation can occur in more subtle ways: through discharge, effluent, and mechanical interference to water flows. Wetlands are acutely vulnerable to damage caused by activities located a considerable distance from the wetland site but within its drainage basin.

As with tropical forests, the functions performed by wetland systems are diverse. The structural components of wetland systems (flora and fauna) are considered as stocks, whereas the ecological functions can be regarded as flows (services that the wetlands yield over time).

An ecosystem is both a set of constituent characteristics and the sum of these components. In many cases, the value of the sum of the components is greater than the value of the individual components alone. This is because some of the functions of an ecosystem are able to continue only when some significant proportion of the components are present. Once some certain threshold is passed, the effect is to lose these synergistic values. Therefore, the task of valuing an ecosystem involves both the valuation of the components and the identification of the synergism they generate.

A study of the Petexbatun wetlands in Guatemala provides an indication of the range of values available at a single wetland site. These values include *direct use* values, from the generation of fisheries and wildlife habitat for example. Less evidently, this wetland also provides a wide range of *indirect use* values, by, for example, recharging inland groundwater supplies and providing a buffer for flood control. Finally, there are also the inappropriable values represented by a wetland as a dynamic and diverse biological system; although this is a *non-use* value, it is probably one of the most important roles of the wetland. The wide range of use and non-use values represented by this single wetland are set out in Table 27.13.

Table 27.13 Wetland values: Petexbatun, Guatemala

	DIRECT	INDIRECT	NON-USE
Components			
Forest resources	●●●		
Wildlife resources	●		
Fisheries	●●		
Forage resources	●●		
Agricultural resources	●●		
Water supply	●●●		
Functions			
Groundwater recharge/ discharge		●	
Flood and flow control		●●●	
Shoreline/ bank stabilisation		●●●	
Sediment retention		●●●	
Nutrient retention		●/●●	
External support		●●●	
Recreation/tourism		●	
Water transport		●●●	
Attributes			
Biological diversity	●●	●●	●●
Uniqueness to culture/ heritage			●

Source: Barbier, E.B. 1989. *The Economic Value of Ecosystems: 1 tropical wetlands.*
Notes: ● = low ●● = medium ●●● = high

Case study: the Hadejia-Jama'are floodplain, Nigeria

Coherent policy determining the use and exploitation of wetland resources requires that decision-makers have available to them a set of shadow prices and values which reflect the total economic value of these resources under various management regimes.

One approach to valuing the wetlands is exemplified by the case of the Hadejia-Jama'are floodplain in Nigeria. The Hadejia-Jama'are wetlands lie in an area of confused drainage between Hadejia (Kano State) and Nguru and Gashua (Borno State), where the Hadejia and Jama'are rivers flow across a fossil plain of late Quaternary sand dunes. These wetlands provide essential income and nutritional benefits for the regional inhabitants. They constitute a source of fuelwood, fishing, grazing, and agricultural opportunities. It is not only those located on the periphery of the wetlands for whom this natural resource is important. The floodplains provide dry-season grazing for semi-nomadic pasturalists and agricultural surpluses for Kano and Borno states, as well as educational and scientific benefits. They also provide a natural habitat for migratory and resident bird species.

However, the wetlands are shrinking as the result of prolonged drought coupled with upstream water developments which divert water flowing into the floodplains. The Hadejia-Jama'are wetlands comprise dry farmland and savanna, open reaches of water, swamp and seasonally-flooded grassland. Agricultural practices vary according to the terrain and comprise dryland agriculture on the better drained sands together with various forms of wetland cultivation, and seasonal grazing and fishing in the more waterlogged soils and permanently flooded stretches. The region experiences a single short wet season (May to September); consequently the growing season for rain-fed crops is short. Additionally, the region is characterised by extreme rainfall variability, producing a high variance in agricultural production. River flows are also highly seasonal, with the timing, extent and duration of flooding depending on the seasonal flood of the rivers and the height of the water table beneath the plains. Thus the area and nature of the wetlands also vary.

Direct use values

The direct uses of the floodplains encompass: fuelwood collection, grazing of floodplain pastures, floodplain agriculture and fishing, recreation, and transport.

The total cultivated area in the Hadejia-Jama'are floodplain is estimated at approximately 230,00ha, of which roughly 77,500ha are cultivated in the dry season, and 152,500ha in the wet season. The current annual net benefits from 14 agricultural crops grown in the Hadejia-Jama'are floodplain have been estimated (Table 27.14).

Fishing is concentrated in approximately 100,000ha of flooded land. Roughly 73,150 rural households in the floodplain were estimated to fish throughout the year; 12% of these households contained people whose main activity was fishing, 21% were dry season fishing households, 15% wet season fishing households and the remaining 52% comprised households that only fished at fishing festivals.

Table 27.14 Agriculture: net benefits from the Hadejia-Jama'are floodplain, Nigeria, 1989-1990

CROP	AVERAGE OUTPUT (Tonnes)	FINANCIAL PRICE (N/Kg)	ECONOMIC PRICE (N/Kg)	FINANCIAL BENEFITS (N'000)	ECONOMIC BENEFITS (N'000)	NET ECONOMIC BENEFITS (N'000)
Total [3]	281,955	26,710.50	19,092.72	563,104	366,455	54,968
Tradeable [1]						
Rice	22,335	4,770	3,144.90	106,538	70,241	10,536
Wheat	43,350	4,010	1,382.88	173,834	59,948	8,992
Soyabeans	6,000	3,310	2,137.50	19,860	12,825	1,924
Non-Tradeable [2]						
Sorghum	50,315	720	612	36,227	30,793	4,619
Maize	15,705	842	715.70	13,224	11,240	1,686
Groundnuts	3,855	4,980	4,233	19,198	16,318	2,448
Millet	50,415	842	715.70	42,449	36,082	5,412
Cow-Peas	25,035	3,310	2,813.50	82,866	70,436	10,565
Tomatoes	15,955	662.50	563.13	10,570	8,985	1,348
Onions	1,1925	662.50	563.13	7,900	6,715	1,007
Peppers	32,400	1,336	1,135.60	43,286	36,793	5,519
Sweet Potato	2,925	600	510	1,755	1,492	224
Aubergine	1,740	662.50	563.13	1,153	980	147
Pumpkins	141,500	3	2.55	4,244	3,607	541

Net economic benefits per hectare: 239
(agricultural area 230,000ha)

Source: Barbier, E.B., Adams, W.M. and Kimmage, E. 1991. *Economic Valuation of Wetland Benefits: the Hadejia-Jama'are floodplain, Nigeria.*
Notes: Values in Naire per hectare; N7.5 = US$1 [1] The economic prices of all tradeables are the c.i.f. import (border) prices converted at the official 1989 exchange rate N7.5 = US$1. [2] Non-tradeables are defined as crops whose prices exceed f.o.b. export prices but are less than the c.i.f. import prices. The economic prices of all non-tradeables are the financial prices adjusted by the standard conversion factor 0.85. [3] The total for average crop output excludes pumpkins.

Table 27.15 Fishing: net benefits from the Hadejia-Jama'are floodplain, Nigeria, 1989/90

FISHING SEASON	TOTAL CATCH (Tonnes)	FINANCIAL PRICE (N/Kg)	ECONOMIC PRICE (N/Kg)	TOTAL FINANCIAL BENEFITS (N '000)	TOTAL ECONOMIC BENEFITS (N '000)	TOTAL FISHING DAYS (DAYS)	TOTAL FINANCIAL COSTS (N '000)	TOTAL ECONOMIC COSTS (N '000)	NET ECONOMIC BENEFITS (N '000)
TOTAL	6,263			45,380	38,574	3,125,701	24,349	20,697	17,876
DSB	1,728	7.48	6.35	12,918	10,980	768,075	5,983	5,086	5,894
DSM	931	9.75	8.29	9,079	7,717	621,044	4,838	4,112	3,605
DSE	313	5.35	4.55	1,674	1,423	313,082	2,439	2,073	-650
WSB	562	7.25	6.16	4,077	3,466	375,260	2,923	2,485	981
WSM	717	6.35	5.4	4,555	3,872	478,401	3,727	3,168	704
WSE	2,012	6.5	5.53	13,077	11,116	569,839	4,439	3,773	7,342

Net economic benefits per hectare: 179
(fishing area 100,000ha)

Source: Barbier, E.B., Adams, W.M. and Kimmage, E. 1991. *Economic Valuation of Wetland Benefits: the Hadejia-Jama'are floodplain, Nigeria.*
Notes: Values in N per hectare; N7.5 = US$1. Economic values are financial values adjusted by a standard conversion factor of 0.85. Financial costs based on a rural cost of agricultural day labourer of N7.79 per day. Total fishing days calculated on the average number of fishing days per household per season (73,150 fishing households total).
DSB = Dry season beginning; DSM = Dry season middle; DSE = Dry season end; WSB = Wet season beginning; WSM = Wet season middle; WSE = Wet season end.

Table 27.16 Fuelwood: net benefits from the Hadejia-Jama'are floodplain, Nigeria, 1989/90

FUELWOOD CONSUMERS	TOTAL OUTPUT (Tonnes)	FINANCIAL PRICE (N/t)	TOTAL ECONOMIC PRICE (N/t)	TOTAL FINANCIAL BENEFITS (N '000)	TOTAL ECONOMIC BENEFITS (N '000)	TOTAL FINANCIAL COSTS (N/t)	NET ECONOMIC COSTS (N '000)	ECONOMIC BENEFITS (N'000)
TOTAL	115,100	120	102	13,812	11,740			8,265
Rural	51,600	120	102	6,192	5,263	30	26	3,947
Non-Rural	63,500	120	102	7,620	6,477	40	34	4,318

Net economic benefits per hectare: 21
(fuelwood area 400,000ha)

Source: **Barbier, E.B., Adams, W.M. and Kimmage, E. 1991.** ***Economic Valuation of Wetland Benefits: the Hadejia-Jama'are floodplain, Nigeria.***
Notes: **Values in N per hectare; N7.5 = US$1 (1989-90). Economic values are financial values adjusted by a standard conversion factor of 0.85. Financial costs are based on an initial felling/trimming cost of N18/t, cost of chopping down into bundles of N12/t and a transport cost to local urban centres of N10/t.**

Table 27.17 Net present value of benefits from the Hadeja-Jama'are floodplain, Nigeria

BASE CASE	(8%, 50 YEARS)	(8%, 30 YEARS)	(12%, 50 YEARS)	(12%, 30 YEARS)
TOTAL	1,360	1,251	922	895
Agriculture	921	848	625	607
Fishing	300	276	203	197
Fuelwood	139	127	94	91
Adjusted agriculture	838	773	574	558
Adjusted total	1,276	1,176	872	846

Source: Barbier, E.B., Adams, W.M. and Kimmage, E. 1991. *Economic Valuation of Wetland Benefits: the Hadejia-Jama'are floodplain, Nigeria.*
Notes: values in Naire per hectare; N7.5 = US$1 (1989-90)

Fuelwood production provides an important source of both rural income and domestic inputs. With the estimated 86,000 rural households in the region each consuming an average of 50kg of fuelwood per month, total annual rural fuelwood consumption is approximately 51,600 tonnes annually.

However these estimated benefits accrue over the lifetime of the wetlands. The flow of benefits over time from the continued existence of the wetlands must be converted to a single number reflecting their discounted net *present value*. The overall calculation is acutely sensitive to the discount rate employed and the time horizon considered.

Indirect use values

In addition to the fuelwood products, wetland forest reserves yield other non-timber products that are vital for the rural household economy in developing countries. In the Hadejia-Jama'are wetlands leaves are harvested from the doum palm which can be processed into mats and other household materials or sold unprocessed. Baobab leaves provide a staple food source as an ingredient in soups and stews. Mats and other doum products such as baskets and rope are sold in regional markets or exported to other localities. The leaves fetch about N20 per sack. Whilst many of these products are not directly sold, in theory a value could be attributed to their consumption, using the price of available substitutes. Livestock and grazing are also supported by the Hadejia-Jama'are wetlands. To value the contribution of these inputs to agricultural production we could assess the costs of alternative means of providing feed and shelter for livestock. Gradually a figure for each of the component parts of total economic value that are explicitly marketed or that have readily available marketed substitutes can be developed.

Wetland recreational values: USA

The value of a wetland site for recreational purposes proves to be a more difficult valuation problem. Attempts to place a value on recreational services yielded by wetlands have focused on the *travel cost approach* and *contingent valuation methods*. The travel cost approach uses the cost of travelling to the site as a surrogate for the value yielded to the recreational consumer of enjoying access to that site. One study of a wetlands system in Terrebonne Parish, Louisiana (Farber, 1988), employed windshield questionnaires to assess the costs of travel of respondents. The costs of access to the site for the different groups were aggregated to provide a single estimate of 'willingness to pay' for the site of $3,898 million (Table 27.18).

Table 27.18 Estimating willingness to pay for wetland recreation site

METHOD OF VALUATION	ANNUAL WTP ($)	PRESENT VALUE ($)*
Full Wage	3.898	72.185
0.6 Full Wage	2.733	50.611
0.3 Full Wage	1.860	34.444

Source: Farber, S. and Costanza R. 1987. *The Economic Value of Wetlands Systems.*
Notes: Amounts in million US$; * 8% interest and an annual population growth of 2.6%

Hedonic pricing provides analysts with another means of deriving values for qualitative environmental attributes. The assumption is that land values and property values have such environmental values governing location, air quality, proximity to sites of natural beauty etc., capitalised in them. Some portion of the final value of the good or service exchanged reflects these values. Many studies have focused on an array of environmental and aesthetic factors to attribute values to coastal waterfront sites, capturing part of the aggregate wetland value. Data for land sales in Virginia Beach, Virginia, covering the period 1953-1976 were used to estimate a hedonic price equation. It was found that an increased level of amenity was reflected in increased values, and that over time the annual value of the amenity was rising.

Flood control benefits can be estimated using defensive expenditures: that is the amount needed to be spent to mitigate against the effect of degrading the wetlands. One study of the Charles River Basin in metropolitan Boston, undertaken by the US Department of the Army, recommended preservation of 8,422 acres of natural storage areas in the river basin. The estimated value of flood damages averted by preserving these wetlands was approximately $80 per acre per year. The value of an acre of wetland when the flow of flood control benefits had been capitalised into a single number was estimated at $1,488 (Corps of Engineers).

Table 27.19 Values of waterfront amenity in Virginia Beach, Virginia

	ARTIFICIAL CHANNEL		NATURAL BAY	
Year	land unit 0.1214ha frontage 30.48m	land unit 0.3035ha frontage 45.72m	land unit 0.1214ha frontage 30.48m	land unit 0.3035ha frontage 45.72m
1955	182	461	466	515
1965	192	473	568	763
1975	203	557	699	1,064
1985	216	652	848	1,402
1995	231	760	1,016	1,783

Source: Adapted from Shabman, L. and Bertelsen, M.K. 1979. The use of development value estimates for coastal wetland permit decisions. *Land Economics* 55:213-222.
Note: Amounts in million US$

Conclusion

Economic development constitutes one of the major threats to the world's wetland systems. Land reclamation, diverting water for irrigation purposes, damming, forestry, and industrial development are all viewed as the necessary consequences of advancement and technical change. The purpose of valuation is to provide a coherent theoretical basis to examine the costs and benefits of wetland conversion, in order that the same principles as those governing other types of investment decisions be adhered to. Attempts to value wetlands are essentially circumscribed by knowledge of the functions that the wetlands perform. This understanding, and thus the associated values may as yet be imperfect, but are an attempt to ensure that some values are incorporated into the decision-making process where perhaps none were before.

PRESERVING FUTURE OPTIONS

Introduction

It is widely argued that a major value of the conservation of diverse resources arises from their potential contribution to future mainline production techniques. At some time in the future it might be useful to include part of the diversity that now exists in the specialised processes of production. It follows from this that some amount of variety is then desirable for the 'options' that it represents.

However, studies indicate that valuing future options is not straightforward. A large amount of theoretical literature and a small number of empirical studies have examined the role of uncertainty in valuing uses of natural resources. Option value (Weisbrod, 1964) reflects the willingness of a risk-averse society to pay a premium, on top of the use value itself, for guaranteeing access to a resource of uncertain future supply. A related concept, often called quasi-option value (Arrow and Fisher, 1974), is based on information and the irreversibility of much of the degradation of biological resources. Quasi-option value attempts to evaluate the extent to which irreversible changes in natural resources deny us the opportunity to use future information indicating new and valuable resource applications.

The intuitive appeal of the concept of option value often leads conservationists to stress the economic importance of 'preserving options' by maintaining biodiversity. However, it should be emphasised that not only has little empirical work been carried out confirming the magnitude of the range of possible option values but in theory both option and quasi-option values may be negative as well as positive.

If future demand for a resource is uncertain, a discount instead of a premium may be applied to the use value. Indeed, future demand may become more uncertain as time passes. Future discoveries may even provide substitutes for biological resources rather than indicating additional, profitable uses, leading quasi-option value to be negative. Randall (1991) suggests that by the early 1980s the theoretical debate and empirical results had led to a consensus amongst economists that the sign of option value is generally indeterminate and that there is little reason to expect option values necessarily to be of substantial magnitude.

Nevertheless, there remain persuasive arguments for believing that the potential value of diversity can be very large.

First, it is very likely that our current menu of production does not include the most useful varieties, even under existing conditions. Active screening programmes have identified species such as the Rosy Periwinkle which has yielded great benefit in the treatment of previously untreatable cancers, including Hodgkin's disease and leukaemia. Similarly, close wild relatives of tomatoes and maize, with extremely valuable characteristics, have been found in Latin America.

Moreover, because the information and technology that we have is constantly evolving, there is good reason to expect that the future usefulness of a particular plant or animal may be very different from that now recognised. The rediscovered importance of a dwarf variety of rice provides a concrete example. As the rice grains were developed for increased mass, the strength of the stalk became important (wind could cause all of the nearly mature plants to collapse making harvest difficult). A dwarf variety of rice with a short but strong stalk was rediscovered and used in crop improvement; if this rice variety had earlier been lost there would have been no way to meet the unforeseen need. Thus, the conservation of biodiversity provides insurance against new needs that arise with changing conditions.

Changing conditions arise for more fundamental reasons than technological advance. Perhaps the most important of these are climatic changes and the constant evolution and changes in distribution of pest organisms and pathogens. In essence, the movement towards monocultural production results in a high proportion of the agricultural product being vulnerable to a single pest. The consequence can be the collapse of significant portions of the crop. Many of the crop failures of recent times have been the result of this interaction between pest and monoculture. Against this unstable background, it is difficult to know which particular varieties will be most useful in the long run.

THE VALUE OF DIVERSITY IN PROVIDING INSURANCE: CROP YIELDS

There will always be a trade-off inherent in specialisation in production methods. Specialisation implies increased productivity, manifest in increased average yield, but it also implies a decreased range of productive assets. Maximum security is obtained from having the widest possible range of productive assets; this is known in economics as 'the portfolio effect'. It is the basic reason why people tend to hold their assets in a variety of different forms (e.g. stocks, bonds, gold and cash). It provides a hedge against the numerous different risks relating to any one form of asset. Holding our biological assets in the widest possible variety of forms would provide this portfolio effect but, potentially, at the cost of reduced average productivities.

The last 20 years have seen a dramatic rise in international food-grain production as a result of specialisation, involving development of high-yield crop varieties, higher inputs, intensive cultivation and more homogenised farming techniques. However, as agricultural output has risen, so has its variability. This increased variance concerns farmers, governments and policy-makers alike. The instability of agricultural output may give rise to famines, regional shortfalls in agricultural produce and at the very least often results in income instability. The costs of this variability are not insignificant.

Some theorists argue that these costs are the inevitable consequence of concentrating the genetic base of many crops through hybridisation, and that the costs of measures to reduce the effects of yield variability provide a natural measure of the costs of loss of biodiversity. The Green Revolution changed the fortunes of the developing and developed worlds, helping to alleviate the predicted famines of the early 1960s and 1970s. It may also provide a key to developing a value for biodiversity.

The Green Revolution

The 'Green Revolution' refers to the rapid increase in wheat and rice yields in developing countries, brought about by the use of improved seed varieties and the application of fertilizers and other chemical inputs. These have made high-yielding crop varieties comparatively more profitable than other traditional grains and vegetables and as a consequence the area sown with improved seed has increased dramatically.

The success of the high-yield varieties is indicated by the speed at which they have spread across the developing world. It has been estimated that between one-third and a half of the area devoted to rice in the developing countries is now sown with them. CIAT (International Centre for Tropical Agriculture, Colombia) estimated, for example, that in the mid-1980s high-yielding varieties were grown on 90% of the 3 million ha devoted to rice in Latin America. Table 27.20 indicates the area devoted to modern rice varieties in 11 Asian countries.

There can be no doubt that the Green Revolution has worked miracles in improving food production in many parts of the world. Again, CIAT (1981) estimates that yield increases are between one ton/ha on irrigated areas and 0.75 tons/ha on upland rice areas. This constitutes an annual increase of about 2.75 million tons of rice, which at an average price of $200/ton is an increase in the value of production of approximately $550 million.

Variability of world cereal production

Whilst the beneficial impact of the Green Revolution has been a greatly increased volume of food-grain production since the 1960s, one of the hidden costs has been a simultaneous increase in production variability. While world cereal production grew at an average yearly rate of 2.7% between 1960 and 1983, the coefficient of variation (a measure of variability) increased from 0.028 during the period 1960-71 to 0.034 in the period 1971-83. This increased variability appears to result from reduced diversity in the varieties and practices used in food production.

In the absence of explicit stabilisation policies, large fluctuations in agricultural output can feed through into extreme price variability. Small farmers and the very poor are particularly vulnerable to such price movements. The degree of price instability induced can be substantial in countries with a large agricultural base.

There are two major components of the increase in the variability of world cereal production:

- increased yield variances (the year-on-year variability of production from the same field increases);
- increased correlations between the yields of different crops and countries (there is less regional and global variety available to average out the effects of local variability).

The second factor is usually the more important of the two: the loss of diversity is having its greatest impact by reason of the loss of the 'insurance' role that such variety can provide on a regional basis. For example, comparing yields for pre- and post-Green Revolution India reveals that increased variances in grain yields within crops accounts for less than 10% of the increase in the variance of India's total cereal production. The factor that contributed most to variations in aggregate agricultural output were increased synchrony in output between regions. Prior to the Green Revolution, the pattern of agricultural output had been more

Table 27.20 Area devoted to modern rice varieties in 11 Asian countries

COUNTRY	YEAR	1000ha	% OF RICE AREA
Bangladesh	1981	2,325	22
India	1980	18,495	47
Indonesia	1980	5,416	60
Korea, Rep	1981	321	26
Malaysia W	1977	316	44
Myanmar	1980	1,502	29
Nepal	1981	326	26
Pakistan	1978	1,015	50
Philippines	1980	2,710	78
Sri Lanka	1980	612	71
Thailand	1979	80	09

Source: Hazell, P.B.R. 1985. The impact of the Green Revolution and the prospects for the future. *Food Reviews International* 1(1).

Table 27.21 Extent of genetic uniformity in selected crops

CROP	COUNTRY	NUMBER OF VARIETIES	SOURCE
Rice	Sri Lanka	From 2,000 varieties in 1959 to 5 major varieties today	Rhoades, 1991
		75% of varieties descended from one maternal parent	Hargrove *et al.*, 1988
Rice	India	From 30,000 varieties to 75% of production from less than 10 varieties	Rhoades, 1991
Rice	Bangladesh	62% of varieties descended from one maternal parent	Hargrove *et al.*, 1988
Rice	Indonesia	74% of varieties descended from one maternal parent	Hargrove *et al.*, 1988
Wheat	USA	50% of crop in 9 varieties	NAS, 1972
Potato	USA	75% of crop in 4 varieties	NAS, 1972
Cotton	USA	50% of crop in 3 varieties	NAS, 1972
Soybeans	USA	50% of crop in 6 varieties	NAS, 1972

diverse, with many different regions cultivating more diverse crops by more diverse methods and faring differently according to regional weather and disease outbreaks. However many of the techniques and crops have now been standardised, and thus different regions follow similar output responses. The result is that yields now have a strong tendency to move up or down together over large areas of India (Hazell, 1984). This phenomenon is not exclusive to the developing world. Maize yields in the USA exhibit a similar, though more pronounced, trend. The rate of grain yield increased dramatically in the mid-1950s, rising from about 57kg/ha/yr in the period 1930-1955 to approximately 133kg/ha/yr for 1955-1985. The variation around this rising mean has also increased: the coefficient of variation for 1950-66 is 0.06, but 0.105 for 1967-85. The other source of increased variability is the common genetic base of the different crop varieties. The existence of genetic variety within the species itself provides insurance in the same way as variety in crop and cultivation practice within a nation. With increasing genetic uniformity at the species level (Table 27.21), production over a wide area become susceptible to a single external impact, such as a particular pest or disease.

For example, in 1970 a particular form of corn leaf blight (*Helminthosporium maydis*) struck in Florida, spreading northwards throughout the corn belt, drastically reducing yields. Only certain types of hybrids were known to be susceptible to this blight. Unfortunately, a large proportion of maize growers in the USA were cultivating such varieties. Competition over yields and quality had led farmers to concentrate their dependence on a few plant varieties with a narrow genetic base. Table 27.22 demonstrates that this phenomenon has been associated with numerous large-scale crop failures.

Concentrating the genetic base: the case of rice

The widespread adoption of a relatively small number of improved rice varieties, many of which are closely related genetically, has gradually reduced the genetic diversity of the crop. It has been estimated that about 40% of the world's rice crops comprise high-yielding varieties. One variety introduced into Asia in the late 1960s was IR8. This had a high yield potential with short stiff straw that allowed it to produce heavy panicles of grain without falling over. IR8 was also insensitive to photoperiod (daylight hours) or growth duration, which meant that it could be grown at any time of the year. IR8 and other semi-dwarf rice varieties were rapidly adopted throughout Asia, with dramatic results. Average rice yields for Asia in the period 1971-1980 were 42% higher than in 1951-1960. Total production rose by 77% at a time when the land area devoted to rice cultivation rose by only 25%.

Table 27.22 Past crop failures attributed to genetic uniformity

DATE	LOCATION	CROP	CAUSE AND RESULT	SOURCE
900	Central America	Maize	Anthropologists speculate that the collapse of the Classic Mayan Civilization might have been a result of a maize virus	Rhoades, 1991
1846	Ireland	Potato	Potato blight led to famine in which 1 million died and 1.5 million emigrated from their homeland	Hoyt, 1988
late 1800s	Sri Lanka	Coffee	Fungus wiped out homogenous coffee plantations on the island	Rhoades, 1991
1940s	USA		US crops lost to insects has doubled since the 1940s	Plucknett and Smith, 1986
1943	India	Rice	Brown spot disease aggravated by typhoon destroyed crop starting the 'Great Bengal Famine.'	Hoyt, 1988
1953-54	USA	Wheat	Wheat stem rust affected most of hard wheat crop	Hoyt, 1988
1960s	USA	Wheat	Stripe rust reached epidemic proportions in Pacific Northwest	Oldfield, 1984
1970	USA	Maize	Decrease in yield of 15%, $1 billion lost*	NAS 1972, Tatum, 1971
1970	Philippines & Indonesia	Rice	HYV rice attacked by leafhoppers spreading tungro virus	Hoyt, 1988
1972	USSR	Wheat	Crop badly affected by weather	Plucknett *et al.* 1987
1974-77	Indonesia	Rice	Grassy stunt virus destroyed over 3 million tonnes of rice - from the late 1960s to the late 1970s the virus plagued South and Southeast Asian rice production	Hoyt, 1988
1984	Florida	Citrus	Bacterial disease caused 135 nurseries to destroy 18 million trees	Rhoades, 1991

Notes: * Duvick (1986) reports that although the leaf blight attacked a widespread and uniform genotype, the problem was uniformity of cytoplasm - introduced to eliminate the chore of detasseling - not the genetic material in the nucleus of the seed.

High yielding varieties have had similar successes elsewhere in the developing world. Some theorists stress that no conclusive evidence has yet been found that a common ancestry might contribute to production variability and it could be argued that certain modern varieties (such as IR64) have a very diverse parentage and should perform well under a wide range of conditions. However, it is increasingly apparent that some varieties share many genetic elements. The genetic parentage of IR8 can be traced to 1914, when a variety of rice called Cina was introduced from China into Indonesia where it attained rapid popularity because of its photoperiod insensitivity, its high yields and grain quality. In 1934 plant breeders in Indonesia crossed Cina and Latisail (an Indian variety) to form Peta. In 1962 Peta was used as the female parent in a cross with Dee-geo-woo-gen, a semi-dwarf stiff-strawed rice variety from China. IR8 was one of the progeny from this cross. In the 1970s and early 1980s further hybridisation using IR8 and Peta derivatives as parents produced many of the semi-dwarf varieties that are now grown worldwide. Components of the cytoplasm (the protoplasm of a cell excluding its nucleus) are inherited through the female parent, so varieties with Cina as their ultimate maternal ancestor probably carry similar cytoplasm. In 1983-1984, 38% of a sample of the female parents used in 106 crosses were the maternal progeny of Cina, which implies that many of the varieties selected from these crosses and released in the late 1980s will also carry similar cytoplasm.

The rice plant is most vulnerable to stress during its reproductive growth phase. Photoperiod-sensitive varieties grown traditionally in tropical Asia entered the reproductive phase during the peak rainfall period when risks were minimised. These varieties then ripened at the end of the rainy season. With the introduction of IR8 and other similar varieties with photoperiod insensitivity, farmers in many latitudes were able to plant and cultivate at any time of the year. Those in irrigated areas were also able to plant several rice crops annually instead of one, as previously. Varieties that mature a fixed period after seeding are often more vulnerable to climatic changes and natural disasters such as typhoons or droughts. This can be offset by strategic staggered planting, which would ensure staggered maturity. However, because planting is a particularly labour-intensive activity, staggered planting would require radical changes in the agricultural labour market (from seasonal labour flows to continual rolling employment). Various institutional changes would have to come about to ensure this, and until such changes occur the fixed growth duration varieties may have less 'buffering capacity' to withstand extreme climatic variation than the traditional photoperiod-sensitive varieties.

Reduced plant height is one of the more obvious characteristics of modern rice plant varieties. This improves the harvest index and the ratio of grain to straw, and allows the plant to remain standing after heavy doses of nitrogen fertilizer. It is this characteristic that has been hailed as the one most responsible for the production yield gains of the Green Revolution. Unfortunately drought during the vegetative growth stage can shorten the height of semi-dwarf varieties to significantly below optimal levels. This undermines their ability to withstand subsequent floods and weed growth. Although most modern rice varieties continue to be semi-dwarfs there may be a gradual shift towards taller varieties. In 1975 69% of new varieties were semi-dwarf with a mean height of less than 130cm whereas in 1984 54% of new varieties were semi-dwarf.

Traditional rice varieties were naturally selected over a period of centuries for their resistance to or tolerance for their local environments. Modern rice varieties are the product of less than a century of genetic experimentation, with many varieties being selected in less than a decade (IR36 was developed in five years). The experimental sites have often been radically altered by the application of pesticides and fertilizers. As a result of this experimentation, pest-resistant varieties have been developed with inbuilt responses to many common rice pests for which they have been specifically screened. However, it seems unlikely that modern varieties could have generic resistance comparable to that of traditional types. It is even feasible that pest attacks are more concentrated or widespread because extensive areas are being sown with more homogenous crops.

Drought tolerance is not generally a feature of modern rice varieties. Modern rice plants are bred to produce the maximum amount of grain under optimal conditions. In the process of genetic refinement, many of these varieties have lost the deep root system required for drought tolerance. IR52 has been developed in response to this loss in root depth, but the apparent yield potential of such varieties is significantly lower than that of their less drought-tolerant counterparts.

Similar case histories can be related for wheat and maize which emphasise that, whilst a common genetic base is not necessarily cited as the sole cause of increased yield variability, it may play an important part in causing co-movements in grain yields world wide.

SOURCES OF YIELD VARIABILITY

Tables 27.23 and 27.24 illustrate the main components of variability in world agricultural cereal production. These figures reveal that increases in mean yields account for about 70% of the increase in total cereal production and expansion in area for 20%, and also that wheat and maize contribute greatly to the change in mean total cereal production (32.65% and 35.18% respectively). Table 27.24 indicates the percentage change in the variance of world cereal production attributed to its components. The column sums show that 95.93% of the increase in the variance of world cereal production is attributable to changes in the variances and covariances of crop yields. The change in maize yield variances and covariances accounts for 17.16% of the overall increase in the variance of world cereal production. Changes in area-yield covariances exerted an important stabilising effect on world cereal production, reducing the variance of total cereal production by 42.28%. Virtually all of this reduction can be attributed to a decline in area-yield correlations, of which the strongest declines appear to be between crop yields in one country and the sown areas of different crops in different countries 28.51%.

Genetic uniformity and crop yield variability

Genetic uniformity has been cited as one of the major causes of widespread yield reductions in maize in 1970 in the USA. At that time approximately 80% of US maize was based on T cytoplasm, which is particularly susceptible to the T race of southern corn leaf blight (Tatum, 1971). The rapid spread of this fungus across the eastern part of the nation was aided by an abnormally wet summer which increased the germination and dissemination of its spores. Hybrids without the T cytoplasm were unaffected.

The direct costs of genetic uniformity and of monocultural production have been an increase in the vulnerability of crops and regions to climatic variations and to disease. The fact that common wheat and rice varieties such as Bezostaia wheat in Eastern Europe and IR36 rice in Asia have been cultivated extensively (more than 10 million ha in each case), increases the risk of crop failure in the event of an epidemic.

Technical uniformity: inputs and crop yield variability

With the international adoption of genetically engineered seed types there has been a worldwide revision in cultivation techniques. The application of chemical inputs has dramatically increased over the last three decades. Many argue that one of the main causes of agricultural output variability, and especially of grains and cereals, is varying levels of input use in response to price and interest rate movements.

Table 27.23 Components of change in world average cereal production 1960-1971 to 1971-1983

COMPONENTS OF CHANGE	WHEAT	MAIZE	RICE	BARLEY	MILLET	SORGHUM	OATS	OTHER	TOTAL
% change in mean yields	80.93	64.21	60.62	39.52	63.64	45.63	-528.08	-179.99	72.40
% change in mean areas	14.94	28.61	33.64	49.11	44.76	44.42	534.84	220.53	22.36
Change in area-yield covariances	0.19	0.09	-0.02	0.45	2.96	0.20	15.21	-1.08	00.14
Contribution of crop to change in mean production of total cereals	32.65	35.18	11.50	18.28	0.55	4.34	-0.47	-2.03	100.00

Source: Hazell, P.B.R. 1989. Changing patterns of variability in world cereal production. In: Anderson, J. and Hazell, P. (Eds), *Variability in Grain Yields, Implications for Agricultural Research and Policy in Developing Countries*.
Note: In per cent; excluding China.

Table 27.24 Components of change in the variance of world cereal production 1960-1971 to 1971-1983

	SOURCE OF CHANGE				
	CHANGE IN MEAN YIELDS	CHANGE IN MEAN AREAS	CHANGE IN YIELD VARIANCES AND COVARIANCES	CHANGE IN AREA VARIANCES AND COVARIANCES	CHANGE IN AREA-YIELD COVARIANCE
Crop variances					
Wheat	2.06	-2.38	5.27	-0.57	3.57
Maize	6.67	1.94	17.16	-6.15	-5.01
Rice	0.11	0.25	0.45	0.12	0.16
Barley	0.43	2.30	1.87	0.86	1.37
Millet	0.01	-0.01	0.04	0.01	0.06
Sorghum	0.19	0.07	0.57	-0.23	0.12
Oats	0.83	0.27	0.11	-1.25	-0.54
Other	0.14	-0.15	0.93	-0.14	0.29
Sum Crop Variances within Countries	10.44	2.28	26.40	-7.36	0.01
Intercrop Covariances within Countries	0.97	4.48	36.68	-0.94	-9.38
Intercountry Variances within Crops	0.09	1.61	11.49	-3.61	-4.40
Covariances between different crops in different countries	2.75	0.85	21.36	19.13	-28.51
Column sums	14.24	9.22	95.93	7.22	-42.28

Source: Hazell, P.B.R. 1989. Changing patterns of variability in world cereal production. In: Anderson, J. and Hazell, P. (Eds), *Variability in Grain Yields, Implications for Agricultural Research and Policy in Developing Countries.*

Most of the studies in Table 27.25 support the view that increased application of nitrogen increases variance in yield. In most cases, the change in variance with respect to nitrogen is higher than change in mean yield. According to Byerlee and Anderson (1969), with a nitrogen level of 20 kg/ha, a 1% increase in the nitrogen level would result in a 0.08% increase in the mean yield and a 0.44% increase in the variance of output. Where the nitrogen level is 40 kg/ha, a 1% increase in nitrogen level results in a 0.04% increase in the mean yield but a 0.62% increase in the variance of output. The supply of other inputs such as irrigation or the application of pesticides and herbicides may also affect the variability of crop yields. How the application of such inputs affects yields and their variance is not yet fully recognised.

CROP INSURANCE: THE RESPONSE TO INCREASED AGRICULTURAL RISK

One possible response to increased yield instability and consequent income variability is to acquire insurance. Insurance schemes typically offer a means of guaranteeing expected future income in the face of uncertainty. This is accomplished through the payment of a premium which ensures that an indemnity is received in the event of an undesirable outcome. The amount of this indemnity is usually sufficient to compensate the individual for the loss. In other words, the essence of an insurance programme is the sacrifice of some amount (the 'premium') of the average return from the activity in return for a reduction in the long-term variability of returns.

This is one of the major roles of diversity in agricultural production. Reduction in variety of species and techniques has raised average returns but also increased variability. Conversely, increases in diversity in agriculture provide insurance, by reducing variability in return for a reduced mean return. Biodiversity can be said to provide a form of natural insurance.

One means of reducing the risks associated with yield fluctuations is to diversify the portfolio of crops, moving away from monoculture. Intercropping, spatial diversification, staggered planting, and hoarding are surprisingly efficient in reducing income risks. Such practices have been employed by agriculturalists for centuries.

However, diversity is not the only means of providing crop insurance. The market itself will do so, if the risks are insurable. There is not necessarily any reason to intervene if this is the case, because farmers themselves could then choose the least expensive basis for insuring their crops, allocating their 'insurance policies' between the market and diversity.

However, this is only the case if insurance markets are able to operate effectively. In many cases they do not. This is because market insurance operates by means of the pooling of independent risks. That is, in many circumstances individuals may face uncertainty, but society as a collective of individuals faces approximate certainty. This is attributable to 'the law of large numbers'. In essence, insurance works effectively when an individual farmer does not know whether his/her crops will fail this year, even though the failure rate for crops in that region for any given year is known and relatively stable over time.

Table 27.25 Changes in mean and variance of crop yield with respect to nitrogen fertilizer

STUDY/SOURCE	CROP	NITROGEN LEVEL (Kg/ha)	MEAN	VARIANCE
Anderson, 1973	Wheat, Australia	40	0.14	0.22
		80	0.06	0.19
Smith and Umali, 1985	Rainfed rice, Philippines normal[a]	40	0.28	0.36
	Rainfed rice, Philippines gamma[a]	80	0.16	-0.08
Antle and Crissman, 1986	Rice, Philippines 1975-76*	11	0.16	0.22
	1977-79*	21	0.25	-0.35
Byerlee and Anderson, 1969	Wheat, Australia	20	0.08	0.44
		40	0.04	0.62
Ryan and Perrin, 1973	Potatoes, Peru	100	0.10	0.20
		200	0.17	0.35
Roumasset, 1974	Rice, Philippines			
	Village 1	40	0.49	0.49
		80	0.16	0.32
	Village 2	40	0.22	0.45
		80	0.03	0.06
	Village 3	40	0.19	0.37
		80	0.00	-0.01
Rosegrant and Herdt,	Rice, Philippines			
1981	Irrigated	40	0.20	0.29
		80	0.19	0.42
	Rainfed	40	0.14	0.31
		80	0.10	0.30
Smith *et al.*, 1984	Rainfed rice, Philippines			
	Wet Season	40	0.16	0.24
		80	0.10	0.48
	Dry Season	40	0.15	0.26
		80	0.10	0.48
Rosegrant and Roumasset,	Rice, Philippines			
1985	Good irrigation, dry season	40	0.20	0.03
		80	0.24	0.36
	Average irrigation, dry season	40	0.19	0.12
		80	0.21	0.54
	Average irrigation, wet season	40	0.14	0.06
		80	0.13	0.49
	Rainfed, wet season	40	0.13	0.14
		80	0.10	0.59

Source: Adapted from Roumasset, J.A., Rosegrant, M.W., Chakravorty U.N. and Anderson J.R. 1989. In: Anderson, J.R. and Hazell, P.B.R. (Eds), *Variability in Grain Yields, Implications for Agricultural Research and Policy in Developing Countries.*
Notes: * Reported elasticities are computed at mean input levels, expressed in Pesos per hectare. Figures given are estimated mean nitrogen use given prevailing prices. [a] Yield distribution is assumed to be normal or gamma as specified.

The primary assumption that drives the insurance principle is that the probability of a crop failure for any given individual is independent of that for anyone else. That is, when risks are faced by all persons uniformly, it is not possible for an insurance market to operate. This is because it does no good to 'pool' a risk if everyone will incur the loss at the same time.

It is apparent that the assumption of independence fails in the case of crop insurance in the USA. The agricultural sector is one that faces pronounced co-movements in output. It is self-evident that individual agents' probabilities of experiencing a crop failure are not independent when techniques and varieties become standardised. This is borne out heavily in the data. The government is required to subsidise the insurance companies in order that continued cover can be provided.

In short, the crop insurance market in the USA has not operated effectively, probably on account of the correlation of risks. The US experience demonstrates the difficulty in developing and administering crop insurance cover, with the private sector being unwilling to provide complete insurance. The current insurance programme dates only to the Crop Insurance Act of 1980, which allowed private insurance schemes to operate in this area, but the evidence from this period is clear. The Federal Crop Insurance Corporation (FCIC) currently subsidises the premiums paid by farmers by about 30%. The amount of government subsidy can be seen in the difference between Total Premium and Farmer Premium (Table 27.26). The total costs of the protection offered including the subsidy and administration costs are shown in Table 27.27.

During the 1980s, the US government spent $3.8 billion on crop insurance programmes for US farmers. This is very important for two reasons. First, it is indicative of the extent of crop failures occurring under specialised agriculture. Second, and more important, it is obvious that these markets were requiring substantial government

Table 27.26 Summary[1] of multiple peril crop insurance protection in USA

YEAR	PROTECTION Million $	TOTAL ACRES INSURED Thousands	TOTAL PREMIUM Million $	FARMERS PREMIUM Million $	LOSSES PAID Million $	LOSS RATIO	FARMERS BENEFIT/COST RATIO
Total '81-90	75,592	753,468	4,751	3,648	6,912	1.46	1.89
1981	5,981	58,324	377	330	407	1.08	1.23
1982	6,125	54,918	396	305	529	1.34	1.74
1983	4,370	36,542	286	222	584	2.04	2.63
1984	6,620	55,492	434	336	638	1.4	1.90
1985	7,167	63,360	440	340	683	1.55	2.01
1986	6,219	64,004	380	291	613	1.62	2.10
1987	6,079	64,794	365	277	369	1.01	1.33
1988	6,957	73,799	436	328	1,049	2.41	3.20
1989	13,563	139,365	816	610	1,189	1.46	1.95
1990	12,511	142,870	821	609	851	1.04	1.40

Source: American Association of Crop Insurers (1991).
Note: [1] Summary of all crops for all states by year.

Table 27.27 Nature and extent of all government costs

	PREMIUM SUBSIDY	EXCESS LOSSES	FCIC COSTS	MMA COSTS *	REINSURANCE COSTS	TOTAL COSTS
Total	684,583	1,731,597	562,356	178,705	560,583	3,717,824
1981	46,995	30,471	60,630	27,658	3,663	169,417
1982	91,990	132,250	69,190	46,978	23,138	363,546
1983	63,669	297,971	69,745	25,958	35,603	492,946
1984	98,296	204,314	73,632	25,235	78,887	480,364
1985	100,224	242,438	79,009	17,711	102,888	542,270
1986	88,043	233,806	85,027	10,765	97,711	515,352
1987	87,536	4,669	60,046	12,700	97,148	262,099
1988	107,830	585,678	65,077	11,700	121,545	891,830

Source: *Report of the Commission for the Improvement of the Federal Crop Insurance Program*. Washington DC.
Note: Figures in thousands US$; * direct agent costs Master Marketers.

subsidies for operation. In the period 1981-1988, the US government spent $685 million on direct subsidies in order to encourage the operation of the market.

This crop insurance programme both indicates the value of diversity and discriminates against it. If diversity can itself provide insurance against widespread crop failures, then this value would accrue to practices which maintained diversity. Although insurance through diversity would not be a policy operated through the financial markets, it could just as effectively generate this value as one that does.

being allowed to operate. It is instead being pre-empted by a government policy that is encouraging, through subsidy, the substitution of the financial market. This sort of policy discourages farmers from using natural diversity for the provision of insurance, even when it is the most effective means of doing so (Swanson, 1992).

THE VALUE OF AGRICULTURAL GENETIC DIVERSITY

One area in which the actual value of qualitative diversity has been estimated is agricultural genetic diversity. Here, the closest relatives to the small number of domesticated species are often investigated to ascertain their potential for contributing to the productivity or resilience of the domestic variety.

Yield gains in agriculture are typically broken down into a technology component (encompassing chemicals and capital machinery) and a genetic component. Gains from crop breeding arise from genetic improvements in a number of different fashions:

- the environmental conditioning of the plant (e.g. better standibility, drought resistance, etc.)
- pest and disease resistance
- suitability to changing cultivation technology (e.g. response to fertilizers)
- more productive genotypes (e.g. number or size of kernels)
- quality characteristics (e.g. changes in protein or oil content).

A considerable amount of work has been carried out in estimating the often substantial value of genetic improvements to crops. Some of the more important studies are summarised in Table 27.28.

The aggregate value of the raw genetic materials used in crop-breeding is best ascertained by reference to the industry's spending on research and development. This is because, as with so many of the facets of biodiversity, the value of genetic variety for crop breeding lies in the potential value of future finds from the existing genetic

Table 27.28 Genetic diversity and agriculture: genetic contributions of cultivars to crop yields

CROP	LOCATION	PERIOD	EFFECT ON PRODUCTION	SOURCE
All crops	USA	1980s	$1.0 billion/year	OTA, 1987, USDA est.
Maize	USA	1930-80	≈ ½ of a fourfold increase in yields	OTA, 1987
	USA	1930-80	89% of yield gain of 103 kg/ha/yr in commercials	Duvick, 1984
	USA	1930-80	71% of yield gains in single cross hybrids	Duvick, 1984
	USA	1985-89	Genetic gains to N. Dakota of $2.3 million/year	Frohberg, 1991
Rice	Asia	GR	$1.5 billion/year	Walgate, CALP
	USA	1930-80	≈ ½ of a doubling in yields	OTA, 1987
Wheat	Asia	GR	$2.0 billion/year	Walgate, CALP
	USA	1930-80	≈ ½ of a doubling in yields	OTA, 1987
	USA	1958-80	0.74% genetic gain per year - ½ of 32% yield gain	Schmidt, 1984
	UK	1947-75	50% of an 84% gain in yields	Silvey, 1978
	World	1970-83	43% of genetic gain totalling 46% (best data)	Kuhr *et al.*, 1985
			55% of genetic gain totalling 32% (all sites)	Kuhr *et al.*, 1985
Sorghum	USA	1930-80	≈ ½ of a fourfold increase in yields	OTA, 1987
		1950-80	1-2% genetic gain per year from manipulating kernel numbers, plant weight, height and leaf area	Miller and Kebede, 1984
Barley	USA	1930-80	≈ ½ of a doubling in yields	OTA, 1987
Potato	USA	1930-80	≈ ½ of a fourfold increase in yields	OTA, 1987
Soybeans	USA	1930-80	≈ ½ of a doubling in yields	OTA, 1987
	USA	1902-77	79% of 23.7 kg/ha annual yield gains	Specht and Williams, 1984
Pearl Millet	India	at present	genetic improvements worth $200 million annually	ICRISAT, 1990
Cotton	USA	1930-80	≈ ½ of a doubling in yields	OTA, 1987
		1910-80	0.75% genetic gain per year	Meredith, Jr and Bridge, 1984
Sugar cane	USA	1930-80	≈ ½ of a doubling in yields	OTA, 1987
Tomato	USA	1930-80	≈ ½ of a threefold increase in yield	OTA, 1987

Table 27.29 Genetic diversity and agriculture: specific contributions made by wild relatives of crops

CROP	FOUND IN	EFFECT ON PRODUCTION	SOURCE
Wheat	Turkey	Genetic resistance to disease valued at $50 million per year	Witt, 1985
Rice	India	Wild strain proved resistant to the grassy stunt virus	
Barley	Ethiopia	Protects California's $160 million per year crop from yellow dwarf virus	Witt, 1985
Hops		Added $15 million to British brewing industry in 1981 by improving bitterness	Witt, 1985
Beans	Mexico	The International Center for Tropical Agriculture in Colombia used genes from the Mexican bean to beat the Mexican bean weevil which destroys as much as 25% of stored beans in Africa and 15% in South America	Rhoades, 1991
Grapes	Texas	Texas rootstock (from land now covered by the Dallas-Fort Worth Airport) was used to revitalise the European wine industry in the 1860s after a louse infection	Rhoades, 1991

stock. An indication of this value is given by the returns realised from past efforts at developing the previously existing gene pool for commercial use, as well as by the amounts currently being invested in such efforts.

The top 25 agricultural biotechnology - or crop breeding - firms spent $330 million on research and development in 1988 (Hobbelink, 1991). Crop breeding has generated a large return in the past - US public and private expenditures on corn research totalled $100 million in 1984 contrasted with an estimated return of $190 million (Huffman and Evenson, 1991). These figures both indicate that there is considerable value to be had from retaining substantial variety in the plants that are most closely related to our domesticated crops. Several important examples are given in Table 27.29. These varieties represent only a fraction of existing biological diversity, but are probably some of the most valuable species to retain on account of the ease of their introduction into mass production.

The calculated value-gains from crop-breeding efforts are not, however, equivalent to the value of the raw genetic material that exists in the wild, for two main reasons. First, such gains may be achieved using raw materials from a variety of sources: existing cultivated varieties (cultivars), varieties husbanded by traditional farmers (land races), wild relatives of crops or even - with the advent of genetic engineering - completely unrelated species. Second, these gains must be apportioned amongst a number of factors which, together with these raw genetic materials, generate this increased value, including scientific effort, technology and commercial development.

THE VALUE OF BIODIVERSITY IN THE PRODUCTION OF PHARMACEUTICALS

The medicinal value of plants and their derivatives has been recognised for millennia (see Chapter 25). Estimating the importance and economic value of the biodiversity which gives rise to the possibility of more discoveries is a very recent field of interest.

The basis of much of the estimation is a very detailed survey which was carried out on those prescription drugs (in the USA) which were derived in some way from flowering plants (Farnsworth and Soejarto, 1985). The study involved determining the basic materials in all of the thousands of different drugs prescribed in the USA over the period 1959 to 1973 and then identifying those which were plant-based (see Table 25.6 for examples). This was taken to include those drugs which contained crude plant extracts, semi-purified mixtures of active principles, single active principles or active principles which had been chemically modified.

It was found that, for the period examined, the proportion of plant-based drugs was just over 25% of all prescription drugs (in a market where the 1973 value of the total prescription drugs sales was over $6.3 billion at retail prices). On this basis the value of plant-based prescription drugs was estimated to be about $1.6 billion in 1973 and the additional value of the same drugs provided directly through hospitals and clinics was probably as much again, giving a total value of about $3.2 billion.

The authors also estimated a figure for 1980 on the same basis and obtained a total of around $8.2 billion (in current prices). A later study (Principe, 1991) using a variation of this approach but including an estimate of non-prescription drugs revised the 1980 figure to $9.8 billion and calculated a 1985 value of $18 billion (all of these figures being for US sales alone).

Interestingly, the pharmaceutical industry use of plant diversity has been dependent upon a small number of species. The authors of the first study found that, of the 25% of pharmaceuticals traceable to plant-based origins, a mere 40 species of plants were at the ultimate source. Using their figure of total retail value of $8 billion gives an average value per species utilised of $200 million, though of course there is a large amount of variability.

These figures give an indication of the direct retail value of plant-based materials in medicine. The numbers are very large and can probably be trebled to give a worldwide total because the US market represents about one-third of world pharmaceutical sales.

However, it must be remembered that these values are retail market figures, and not only the value of the plant material on which the drugs are based. The price of the raw materials themselves may be of the order of only a few per cent of the final market value but their economic value to the drug industry is far more than their basic cost. Estimation of the real economic value is a conceptual problem as much as a practical one and is discussed below. The value of the underlying biodiversity which has generated these plant-based drugs and which may give rise to many others is an even more difficult issue.

With successful plant-based drugs having a very high potential value it might be expected that the pharmaceutical industry would be very active in research in this area but the industry's attitude appears to be somewhat ambivalent. New drugs are developed through two broad approaches: the screening of potentially active material for medical usefulness and/or the synthesis of specific types of compounds based on the understanding of biochemical reactions within the human body. Recently, many of the most successful modern drugs have come through the application of the techniques of biotechnology and genetic engineering, and there has been a movement away from lengthy and costly screening processes. Even more recently, however, there appears to be a resurgence of interest, on a small scale at least, in screening approaches (Findeisen, 1991) . The reasons for these shifts in emphasis will also be discussed below.

What role do plants play in pharmaceutical production? Three major ways have been identified in which plants are used within the pharmaceutical industry (Principe, 1991). These are:

- constituents isolated from plants are used *directly* as therapeutic agents
- plant constituents are used as *base materials* for the synthesis of useful drugs
- natural products are used as *models* for the synthesis of pharmacologically active compounds.

The first two of these uses represent market values of natural plants as raw materials consumed directly in the pharmaceutical industry. These are the uses which have been valued in the billions of dollars by the studies cited above. However, it has already been noted that the raw material value is usually only a very small proportion of the overall retail price of the drugs which includes factors such as store rental, employees' salaries, transport and taxes. Therefore, estimates based on retail value necessarily represent upper-bounds on the raw material values.

There is good reason to believe that the cost of the raw materials used directly in pharmaceutical manufacturing must remain low. This is because it is generally possible to synthesise chemical substances artificially if the costs of the natural material are too high. Once the method of operation is identified, the cost of chemical batch processing is generally very low, and artificial synthesis of the active ingredients usually becomes the least-cost mode of production for mass-produced substances. For example, aspirin is now produced synthetically although the original source was the bark of the willow tree.

For this reason, it cannot be expected that the *direct* use of plant variety for pharmaceutical manufacture will ever be very substantial, or that it will be possible to claim high returns for presently unpatentable natural products. For example, the Mexican government has historically been a major producer of the yam *Dioscorea*, which has been the source of the basic material used in the production of steroid drugs sold as oral contraceptives and cortisone. This market was producing nearly $83 million annually for Mexico in 1976 (Oldfield, 1984). However, as the Mexican government attempted to extract a higher return from the export of the yam by raising prices, the pharmaceutical manufacturers turned to synthetic processes and the market for *Dioscorea* collapsed (Principe, 1991). Therefore, given the ready alternative of artificial synthesis, direct use values will never be very substantial (there are exceptions to this general rule, namely: reserpine, codeine, morphine, digitoxin, and atropine (Oldfield, 1984)).

Despite advances in medical science and progress in biochemical engineering, there are many conditions and diseases for which we currently have no effective treatment. As long as untried or unknown plant species exist so do the possibilities for discovering materials which could lead to important new drugs.

A topical example of this is the development of the drug Taxol and its derivatives. Taxol is a compound obtained from the bark of the Pacific Yew *Taxus brevifolia* and has been demonstrated in clinical trials to be effective in treating certain difficult ovarian and breast cancers. Unfortunately the Pacific Yew tree is extremely slow growing and the bark from several trees would be required to provide sufficient Taxol to treat one patient. Several lines of development are being pursued, from high technology chemical synthesis techniques (which have so far had meagre success) to the planting of large numbers of yews in commercial forests. A promising approach is the isolation of a related but possibly more powerful compound from the leaves of the same yew tree, leading to the prospect of harvesting the compound without killing the tree (Potier, 1991).

This example illustrates the potential for plant products. A highly promising drug is being developed, based on the efficacy of a natural compound. The active ingredient is very difficult to synthesise but research continues on synthesis and on naturally occurring variations. Whatever form the final commercial product takes it will have been derived from the discovery of the properties of the basic natural compound. Nature has, in effect, provided the blueprint for a drug which is effective in fighting cancer, and while biochemical engineers may modify the original design these are only incremental changes.

Thus, the most important value of plants in this context lies in the *information* which they can provide; specifically, information about the possible existence (and possible loss) of natural blueprints for drug design.

What is the value of the information contained in plant and animal diversity? First, it is the value of the chance discovery, i.e. one that proceeds from mere trial and error. One straightforward attempt at such a valuation has been attempted (Farnsworth and Soejarto, 1985; Principe, 1991). Its method was to look at the success rate for those plants that have been surveyed for their pharmaceutical benefits, assuming those species to be randomly chosen. As earlier studies had estimated that 5,000 plant species had been thoroughly examined for medicinal effectiveness, and since there are 40 species in use in prescription drugs, the assumption of randomness would suggest that one in 125 randomly selected species would be developed to a successful product. Thus for every 1,000 species which becomes extinct, eight potentially useful plant-derived drugs would be lost. At the average retail value of $200 million, this would lead to pharmaceutical losses of $1.6 billion in retail value. In this case, retail value is a useful measure of the willingness-to-pay for the information which is assumed to be a prerequisite to the existence of the particular drug. Consumers demonstrate that they value the existence and discovery of this information through their willingness to purchase the drug at its shelf price. However, it should be stressed that in practice species are not chosen for medical screening at random but are pre-selected. Therefore among 1,000 plant species chosen at random, there may be expected to be fewer than eight potentially useful plant-derived drugs; nevertheless this form of valuation gives a useful approximation of what may be lost.

This valuation methodology stresses the experimental nature of pharmaceutical company research. Although this example requires the use of averages, in fact the pursuit of new drugs is much more of a lottery than even these numbers would suggest. If the company's experiments result in a major discovery, such as Taxol appears to be, a single drug can be as valuable as many other entire industries. The sales and profits of a best seller can be very high: in 1990, the top selling drug world wide (Zantac - an ulcer medicine) grossed sales of about $2.4 billion. Nine drugs earned over $500 million each in the USA alone (which probably indicates per drug earnings of about $1 billion world wide). Pharmaceutical companies must reject hundreds if not thousands of possibilities before one of these discoveries is unearthed. Nevertheless, this method of research and discovery is not haphazard, although imbued with chance: five of the top 20 most profitable companies in the world are pharmaceutical companies.

This profitability is partially attributable to the fact that significant discoveries are awarded monopoly rights for a period of 10-20 years, which generates substantial returns to the successful experiment. However, this profitability is attributable to the fact that the search is not entirely random. The companies utilise all of the information on chemistry, physiology, and other experimental evidence that is available in order to guide them.

One very important form of experimental evidence available to pharmaceutical companies is the experience of peoples living in contact with plant and animal species. These communities have had, in most cases, thousands of years of trial and error experimentation in order to build a record regarding plant usefulness. This indigenous knowledge is the directory which provides the indicator concerning which species are most useful in terms of chemical effects. With the use of this knowledge, search by pharmaceutical companies need not be random.

Recently, a return to greater interest in plant opportunities and to screening approaches seems to be occurring. In 1988, 17% of total pharmaceutical industry research and development spending in the USA went on 'Biological Screening and Pharmacological Testing' (Pharmaceutical Manufacturers Association, 1988-1990). This represents expenditures of over $1 billion dollars although the amount actually spent on investigating new plant products would only be a small fraction of this aggregate figure.

It is probably to be expected that research in this industry would follow an extensive-intensive cycle, where new useful chemical substances are first discovered through extensive exploration and then developed through intensive laboratory applications. It is only in the first phase of pharmaceutical research and development that diversity, biological and cultural, figures largely; however, from time to time this input may be crucial for progress to continue.

Finally, it is important to note that the real economic value lost from possible plant extinctions will be considerably greater than the financial losses that are identified in these studies. The market prices do not include the savings to society in health care and the pain and suffering avoided through the development of drugs. (In strict terms these effects should be considered as the marginal difference over the next best form of treatment.) An estimate of the annual *economic* benefits of plant-based drugs currently in use in the USA gave a range of $34-$300 billion (in 1984 dollars). This range is very wide because of the wide range in estimates of the 'value of a life' - i.e. actually the value of a small change in a small risk that affects a very large number of people. Whatever the precise value, the economic values involved are clearly very large and are an order of magnitude greater than the retail market values.

This survey of the value of plant-based pharmaceuticals demonstrates that there is very real and concrete value attached to the information derived from genetic variety. The difficulties in harnessing this value to conserve the diversity within which it is embedded lie in the impossibility of knowing which species have the potential to contribute economic value. Although it is probabilistically known that these species have substantial economic value in aggregate, discovering precisely which species are valuable will take years of extensive research.

There are a number of issues which will have to be resolved before the market system can develop real incentives to preserve biodiversity for pharmaceutical purposes. As it is very difficult to price values as intangible as information and options, it is necessary to focus on the creation of mechanisms that can assist in this. These include the development of patent rights and royalty payments in *natural* variety. There is a slowly growing acceptance of the potential value of biodiversity but there is no real incentive yet to halt the rapid loss of an irreplaceable resource, despite the economic value that can be attributed to it. The creation of systems that can recognise and appropriate these clear but intangible values is a necessary step.

References

Anderson, J.R. 1973. Sparse data, climatic variability, and yield uncertainty in response analysis. *American Journal of Agricultural Economics* 55:77-82.

Anderson, J.R. and Hazell, P.B. 1989. Variability in grain yields. Food Policy Statement, International Food Policy Research Institute, No. 11.

Anon. 1981. Mahidol University, Thailand.

Anon. 1988. Economic Supplement, *Bangkok Post*, December 1988:79.

Antle, J.M., and Crissman, C.C. 1986. Measuring technical efficiency in risky production during technical change. Mimeo. Department of Agricultural Economics, University of California, Davis.

Arrow, K.J. and Fisher, A.C. 1974. Environmental preservation, uncertainty, and irreversibility. *Quarterly Journal of Economics* 88:313-319.

Barbier, E.B. 1989. *The Economic Value of Ecosystems: 1 tropical wetlands*. LEEC Discussion Paper 91-02. IIED.

Barbier, E.B. 1990. *Community Based Development in Africa*. London Environmental Economics Centre/UCL.

Barbier, E.B., Adams, W.M. and Kimmage, E. 1991. *Economic Valuation of Wetland Benefits: the Hadejia-Jama'are floodplain, Nigeria*. LEEC Discussion Paper. IIED.

Barbier, E.B., Burgess, J.C., Swanson, T.M. and Pearce, D.W. 1990. *Elephants, Economics and Ivory*. Earthscan, London.

Barnes, J., Burgess, J. and Pearce, D.W. 1992. Wildlife tourism. In: *Economics for the Wilds*. Earthscan, London.

Beer, J.H. de and McDermott, M.J. 1989. *The Economic Value of Non-timber Forest Products in Southeast Asia*. Netherlands Committee for IUCN, WWF.

Boo, E. 1990. *Ecotourism: the potentials and pitfalls*. World Wildlife Fund, Washington.

Briscoe, J., Furtado de Castro, P., Griffin, C., North, J., Olsen, O. 1990. Toward equitable and sustainable rural water supplies: a contingent valuation study in Brazil. *The World Bank Economic Review* 4(2):115-134.

Brookshire, D.S., Eubanks, L.S. and Randall, A. 1983. Estimating option prices and existence values for wildlife resources. *Land Economics* 59:1-15.

Brown, G. and Henry, W. 1989. *The Economic Value of Elephants*. London Environmental Economics Centre Discussion Paper 89-12. IIED, London.

Byerlee, D.R. and Anderson, J.R. 1969. value of prediction of uncontrolled factors in response functions. *Australian Journal of Agricultural Economics* 13:28-37.

CIAT 1981. Report on the Fourth IRTP Conference in Latin America, Cali.

Dixon, J., Carpenter, L.A., Fallon, Sherman, P.B. and Manipomoke, S. 1988. *Economic Analysis of the Environmental Impacts of Development Projects*. Earthscan, London.

Dixon, J.A. and Sherman, P.B. 1990. *Economics of Protected Areas: a new look at benefits and costs*. Earthscan, London.

Duvick, D.N. 1984. Genetic contributions to yield gains of U.S. hybrid maize, 1930 to 1980. In: Fehr, W.R. (Ed.), *Genetic Contributions to Yield Gains of Five Major Crop Plants*. Crop Science Society of America, Special Publication 7, Madison. Pp.15-47.

Duvick, D.N. 1986. Plant breeding: past achievements and expectations for the future. *Economic Botany* 40:289-297.

Farber, S. 1988. The value of coastal wetlands for recreation: an application of travel cost and contingent valuation methodologies. *Journal of Environmental Management* 26:299-312.

Farber, S. and Costanza R. 1987. The Economic Value of Wetlands Systems. *Journal of Environmental Management* 24:41-51

Farnsworth, N.R. and Soejarto, D.D. 1985. Potential consequences of plant extinctions in the United States on the current and future availability of prescription drugs. *Economic Botany* 39(3).

Findeisen, C. 1991. Natural Products Research and the Potential Role of the Pharmaceutical Industry in Tropical Forest Conservation. Rainforest Alliance, New York.

Frohberg, R.C. 1991. Economic impact of plant breeding programs. *Farm Research* 48:3-8.

Hargrove, T.R., Cabanilla, V.L. and Coffman, W.R. 1988. Twenty years of rice breeding. *BioScience* 38:675-681.

Hazell, P. 1989. Changing patterns of variability in world cereal production. In: Anderson, J. and Hazell, P. (Eds), *Variability in Grain Yields, Implications for Agricultural Research and Policy in Developing Countries*.

Hazell, P.B.R. 1984. Sources of increased instability in Indian and U.S. cereal production. *American Journal of Agricultural Economics* 66.

Hazell, P.B.R. 1985. The impact of the Green Revolution and the prospects for the future. *Food Reviews International* 1(1).

Hobbelink, H. 1991. *Biotechnology and the Future of World Agriculture*. Zed Books, London.

Hoyt, E. 1988. *Conserving the Wild Relatives of Crops*. IPBGR, IUCN, WWF, Rome and Gland.

Huffman, W.E. and Evenson, R.E. 1991. Science for agriculture. Department of Economics, Iowa State University, Ames. Mimeo.

ICRISAT 1990. *ICRISAT's Contribution to Pearl Millet Production*. ICRISAT, Cereals Program, Andhra Pradesh.

International Trade Centre, 1989. Report on Development and Promotion of Wildlife Utilisation. Ministry of Lands, Natural Resources and Tourism, Government of Tanzania, Dar es Salaam.

Jansen, D.J. 1990. *Sustainable Wildlife Utilisation in the Zambezi Valley of Zimbabwe: economic, ecological and political tradeoffs*. Project Paper No.10, WWF Multispecies Project, Harare.

Kuhr, S.L., Johnson, V.A., Peterson, C.J. and Mattern, P.J. 1985. Trends in winter wheat performance as measured in international trials. *Crop Science* 25:1045-1049.

Lindberg, K. 1991. *Policies for Maximising Nature Tourism's Ecological and Economic Benefits*. World Resources Institute, Washington.

Majid, I., Sinden, J.A. and Randall, A. 1983. Benefit evaluation increments to existing systems of public facilities. *Land Economics* 59:377-392.

Maler, K.-G. 1989. Valuation of costs and benefits from resource use. Unpublished report, Stockholm School of Economics.

Meredith, W.R. Jr and Bridge, R.R. 1984. Genetic contributions to yield changes in upland cotton. In: Fehr, W.R. (Ed.), *Genetic Contributions to Yield Gains of Five Major Crop Plants*. Crop Science Society of America, Special Publication 7, Madison. Pp.75-87.

Miller, F.R. and Kebede, Y. 1984. Genetic contributions to yield gains in sorghum, 1950 to 1980. In: Fehr, W.R. (Ed.) *Genetic Contributions to Yield Gains of Five Major Crop Plants*. Crop Science Society of America, Special Publication 7, Madison. Pp.1-13.

NAS 1972. *Genetic Vulnerability of Major Farm Crops*. Committee on Genetic Vulnerability of Major Farm Crops, Agricultural Board, National Research Council, National Academy of Sciences, Washington.

Oldfield, M.L. 1984. *The Value of Conserving Genetic Resources*. Sinauer Associates, Sunderland, Mass.

OTA 1987. *Technologies to Maintain Biological Diversity*. OTA-F-330. U.S. Government Printing Office, Washington.

Pearce, D. 1989. *Tourist Development*. John Wiley, New York.

Pearce, D.W. 1990. *An Economic Approach to Saving the Tropical Forests*. LEEC Paper DP 90-06. IIED, London.

Pharmaceutical Manufacturers Association 1988-1990. Annual Survey Report.

Plucknett, D.L. and Smith, N.J.H. 1986. Sustaining agricultural yields. *BioScience* 36:40-45.

Plucknett, D.L., Smith, N.J.H., Williams, J.T. and Murthi Anishetty, N. 1987. *Gene Banks and the World's Food*. Princeton University Press, Princeton.

Potier, P. 1991. A report to the Royal Society of Chemistry. Reported in the *Independent* newspaper, London, 11 April.

Prescott-Allen, R. and Prescott-Allen, C. 1982. *What's Wildlife Worth?* Earthscan.

Principe, P.P. 1991. Valuing the biodiversity of medicinal plants. In: Akerele, O., Heywood, V. and Synge, H. (Eds), *The Conservation of Medicinal Plants. Proceedings of an International Consultation 21-27 March 1988 held at Chiang Mai, Thailand*. Cambridge University Press, Cambridge, UK. Pp.79-124.

Rabinovich, J.E., Capurro, A.F. and Pessina, L.L. 1991. Vicuña use and the bioeconomics of an Andean peasant community in Catamarca, Argentina. In: Robinson, J.G. and Redford, K.H. (Eds), *Neotropical Wildlife Use and Conservation*. University of Chicago Press.

Randall, A. 1991. Total and nonuse values. In: Braden, J.B. and Kolstad, C.D. (Eds), *Measuring the Demand for Environmental Quality*. Elsevier Science Publishers B.V., Amsterdam. Pp.303-321.

Rhoades, R.E. 1991. The world's food supply at risk. *National Geographic*. 179(4):74-103.

Rosegrant, M.W. and Herdt, R.W. 1981. Simulating the impacts of credit policy and fertiliser subsidy on Central Luzon rice farmers, Philippines. *American Journal of Agricultural Economics* 63:655-665.

Roumasset, J.A. 1974. Estimating the risk of alternate techniques: nitrogenous fertilization of rice in the Philippines. *Review of Marketing and Agricultural Economics* 42:257-294.

Roumasset, J.A., Rosegrant, M.W., Chakravorty U.N. and Anderson J.R. 1989. In: Anderson, J.R. and Hazell, P.B.R. (Eds), *Variability in Grain Yields, Implications for Agricultural Research and Policy in Developing Countries*.

Ruitenbeek, H.J. 1989. Social cost-benefit analysis of the Korup Project, Cameroon, prepared for the World Wide Fund for Nature and the Republic of Cameroon, London.

Ryan, J.G. and Perrin, R.K. 1973. *The Estimation and Use of a Generalised Response Function for Potatoes in the Sierra of Peru*. Technical Bulletin No. 214. North Carolina Agricultural Experiment Station, Raleigh.

Samples, K.C., Dixon, J.A. and Gowen, M.M. 1986. Information disclosure and endangered species valuation. *Land Economics* 62(3):306-312.

Schmidt, J.W. 1984. Genetic contributions to yield gains in wheat. In: Fehr, W.R. (Ed.) *Genetic Contributions to Yield Gains of Five Major Crop Plants*. Crop Science Society of America, Special Publication 7, Madison. Pp.89-101.

Shabman, L. and Bertelsen, M.K. 1979. The use of development value estimates for coastal wetland permit decisions. *Land Economics* 55:213-222.

Silvey, V. 1978. The contribution of new varieties to increasing cereal yield in England and Wales. *Journal of the National Institute of Agricultural Botany* 14:367-384.

Smith, H. and Umali, G. 1985. Production risk and optimal fertiliser rates: a random coefficient model. *American Journal of Agricultural Economics* 67.

Smith, H., Umali, G., Rosegrant, M.W. and Mandac, A.M. 1984. Risk and fertilizer use of rainfed rice: Bicol, Philippines. Mimeo. International Rice Research Institute, Los Banos, Philippines.

Solow, 1974. The economics of resources or the resources of economics. *American Economic Review* 64.

Specht, J.W. and Williams, J.H. 1984. Contributions of genetic technology to soybean productivity - retrospect prospect. In: Fehr, W.R. (Ed.), *Genetic Contributions to Yield Gains of Five Major Crop Plants*. Crop Science Society of America, Special Publication 7, Madison. Pp.15-73.

Swanson, T. 1991. *Wildlife Utilization as an Instrument of Natural Habitat Conservation: a survey of the literature and of the issues*. London Environmental Economics Centre Discussion Paper 91-03. IIED, London.

Swanson, T. 1992. The economics of a Biodiversity Convention. *Ambio* Paper 92-02, Centre for Social and Economic Research in the Global Environment, London.

Tatum, L.A. 1971. The southern corn leaf blight epidemic. *Science* 171:1113-1116.

Thresher, P. 1981. The economics of a lion. *Unasylva* 33(134):34-5.

Tobias, D. and Mendelsohn, R. 1991. Valuing Ecotourism in a Tropical Rain-Forest Reserve. *Ambio* 20(2):91-93.

Weisbrod, B.A. 1964. Collective-consumption services of individual-consumption services of individual-consumption goods. *Quarterly Journal of Economics* 78:471-77.

Western, D. and Henry, W. 1979. Economics and conservation in Third World national parks. *Bioscience* 29(7):414-418.

Witt, S. 1985. *Biotechnology and Genetic Diversity*. California Agricultural Lands Project, San Francisco.

Witt, S.C. 1985. *BriefBook: biotechnology and genetic diversity*. California Agricultural Lands Project, San Francisco.

Abridged from material assembled under the supervision of Timothy M. Swanson. Authors as follows: Valuing the environment, Bruce Aylward (LEEC/IIED); Loss of biodiversity as an economic process, Timothy Swanson; Current uses of diverse resources, Sarah Gammage; Ecotourism, Bruce Aylward, Shirra Freedman; Existence values, Bruce Aylward; Valuation of diverse resource systems, Sarah Gammage; Preserving future options, Bruce Aylward, Sarah Gammage; Crop insurance, Timothy Swanson; Agricultural genetic diversity, Bruce Aylward; Pharmaceuticals, David Hanrahan.

PART 3

CONSERVATION AND MANAGEMENT OF BIODIVERSITY

The first parts of this book outlined the nature and status of selected elements of biological diversity (Part 1), and then discussed the uses made of plants and animals and the economic values which can be associated with biodiversity (Part 2).

Part 3 will introduce some of the policies, systems, institutions and practices employed in the conservation and management of biodiversity. The eight chapters are grouped into four principal sections.

The first section treats two principal mechanisms for management: national legislation (Chapter 28) and protected area systems (Chapter 29). The latter also covers sites which are components of international protected area systems. These particular topics have been selected from among the many national-level approaches because of their direct impact on biodiversity management. Because almost all the elements of which biological diversity is comprised occur within national boundaries, national policies are self-evidently central to conservation.

The second section focuses on international policies and instruments which are intended either to support national approaches, or to deal with resources which lie outside national boundaries and thus demand international management. Within this section, Chapter 30 includes a tabulation of existing multilateral treaties, and outlines some of the formal procedures involved in their genesis; many deficiencies exist in the effectiveness of these treaties but a handful have come to be of considerable global significance. Chapter 31 discusses some ways in which international policy and legal initiatives have supported national efforts, or could increasingly do so, while Chapter 32 covers additional assistance which is directly financial in nature (this chapter includes an attempt to discover to what extent biodiversity is explicitly identified as a sector for aid support). Finally in this section, Chapter 33 details two examples (fisheries, Antarctica) where international measures have been designed, with varied success, to manage international resources.

Section three (Chapter 34) moves to a different viewpoint, away from policy and legal issues, and focuses on current practices in biodiversity conservation and the institutions involved in implementing them. Emphasis is placed on threatened species and genetic resources. *In situ* and *ex situ* approaches to plant and animal conservation are compared, and the need for coordinated planning at national and international levels is stressed.

The fourth and final section (Chapter 35) outlines the origin and development of the Convention on Biological Diversity. Negotiations to date have been difficult, as participating countries have a wide variety of perceptions of the role of such a convention. Some see it purely as a mechanism for ensuring the maintenance of biodiversity as part of the global heritage, while others regard it as a means of increasing the returns from genetic resources within their boundaries and ensuring a more equitable distribution between countries of the costs and benefits derived from maintaining biodiversity.

28. NATIONAL LEGISLATION

Conservation action typically is carried out within policy and legal systems established by national governments (or in a few instances, by regional or provincial governments). With the exception of Antarctica, virtually all the world's terrestrial biodiversity occurs within national boundaries and measures taken by national governments are thus of fundamental significance.

A wide range of different national policy and legal measures for the conservation of biodiversity exists which vary from country to country depending on the social, political and economic environment. Despite this variety, there are a number of common legislative techniques in use throughout the world; this chapter will describe some of the more important of these. National legislation in this area is often divided along sectoral lines, with different legislation covering the protection of flora, fauna and habitats.

THE PROTECTION OF WILD FLORA

The conservation of wild flora has generally had a rather low priority. As a result, initiatives and legislation at the national level for the specific protection of wild flora are rare and on the whole confined to the developed world.

Most European countries have now adopted legislation to protect wild plants. In the USA, endangered species of wild flora are protected under the federal Endangered Species Act and certain States have enacted additional legislation. In other parts of the world, comprehensive legislation for the conservation of wild flora exists, for example, in Israel, Canada, most Australian states and South Africa. Certain other countries protect wild flora through legislation on forests. This is commonly the situation in Africa. Experience has shown that the degree of protection afforded to wild flora through such legislation is very limited.

Four types of measures common to many countries that have enacted legislation for the protection of wild flora are described below.

Collection and possession

The earliest form of legislative protection specifically for wild flora was restriction on the collection of specimens. The first such restriction was imposed on the collection of edelweiss *Leontopodium alpinum* in the Swiss canton of Zug in 1911. Most countries which have such legislation have a differentiated system of protection, with some species being fully protected and others receiving partial protection. Full protection of wild flora is normally provided to plants which have been 'listed' under the relevant legislation. The legislation typically includes prohibitions on taking, destroying or damaging plants of listed species or any part of them. Full protection for listed wild plants is, however, normally limited to public land and exemptions to the prohibition on collection are usually granted for scientific or educational purposes.

In some countries, legislation provides for the protection of all species in certain areas, as opposed to specific plants. In Austria, for example, collection prohibitions apply to the alpine flora of several mountain regions. In the Swiss canton of Ticino there is a general prohibition on the collection of flora in marshes and peatbogs and on river banks and lake shores. In addition, collection is banned from certain areas designated because of their scientific interest. In Italy, collection of all plants growing on rocks or wetlands in certain areas is banned. In South Africa and Swaziland there is a complete ban on the collection of wild flora along public highways for a distance of about 100m on either side of the road, and several US States have prohibited the removal of plants along public highways.

Partial protection, in many countries, takes the form of a ban on mass collection or destruction of wild flora without good reason. Examples of this type of restriction exist in Luxembourg, Zimbabwe and parts of Australia. In the UK there is a general prohibition on uprooting wild flora, except by landowners, persons authorised by them or by local authorities. In other jurisdictions (parts of Italy and Switzerland) there is an additional prohibition on picking the aerial parts of plants except in limited numbers. In some areas one is permitted to pick no more than a small bunch whilst in others the root or bulb of the plant is protected but gathering of the aerial parts is allowed without limit. Some jurisdictions (e.g. Belgium, Czechoslovakia and parts of Austria) which have adopted the latter approach also stipulate that care must be taken not to damage the root when picking the flower.

Because of the difficulty of catching offenders in the act of collecting, the control of possession is a necessary complement to prohibiting collection and legislation usually restricts both activities.

One of the common problems with controls on collection is that they are often limited to public land. On private land, the owner or occupier may generally collect the flora growing on that land without restriction and other collectors need only seek the permission of the owner or occupier.

This is the situation in most common law countries. In the UK and South Africa, the general restrictions on collecting and uprooting of all species of wild flora are not applicable to landowners. In the USA, under the Endangered Species Act the collection of listed species is only prohibited on federal land. Wild flora outside federal land is not covered by this Act, unless the same species are also protected by State legislation applicable to private land, or are collected in the course of the violation of a State trespass law; in this case, under the 1988 amendment to the Endangered Species Act State offences automatically become federal offences as well.

The reason that legislative protection of wild flora rarely extends to private land is because plants are normally considered the property of the landowner and any attempt to curtail the use of this property is seen as an infringement of property rights.

Trade restrictions

Another common legislative mechanism used for protection of wild flora is legislation imposing restrictions on its trade.

The extent of restriction varies considerably from one country to another. Some national laws contain exhaustive lists of prohibited activities (e.g. banning possession, transport, exhibition, offer for sale, sale, purchase); others merely state that the sale and/or possession of protected plants without a permit is prohibited. Where the purpose of legislation is not to prohibit trade altogether but to ensure the rational utilisation of a natural resource, fairly complex permit systems have sometimes been developed.

The aim of trade controls is usually to reinforce collection bans by eliminating the economic incentives for unlawful taking of wild flora. Thus, many legal systems completely prohibit trade in fully or partially protected species.

Trade may be prohibited to prevent the exploitation of certain plants for profit whilst collection for personal use remains legal. In several Swedish counties, for instance, certain species may be freely picked but not sold. The Belgian plant protection order of 1976 contains a list of taxa in respect of which only collection for commercial purposes is prohibited. In Costa Rica there is a trade ban on all species of orchid but no restrictions on collection.

Trade restrictions are usually implemented by requiring permits for the commercial collection and sale of wild plants. They are designed to prevent over-exploitation and to ensure the rational utilisation of economically valuable plants. In France, for instance, the 1982 Plant Protection Order contains a list of species which may only be collected for commercial purposes under a permit from the Ministry of the Environment. In Italy, the commercial collection and sale of medicinal plants is also subject to the granting of a permit. Other examples are found in the legislation of most Australian states and of Zaire which provides for a licensing system for the collection of *Rauvolfia* species. Other jurisdictions are now attempting to bring under control the commercial exploitation of a large variety of wild plants and forest products, such as berries, fungi and mosses, which until recently were considered almost everywhere as a free product of nature.

As enforcement is usually difficult, the legislation tends to be complex. For instance, under the Californian Desert Native Plants Act of 1981 collection permits are issued by the local counties. Permits specify the species which may be harvested, the area from which they may be harvested and the collection methods authorised. The number of specimens that can be taken by the permit holder may also be specified. In addition to the collection permit, the permission of the landowner must be sought. Detailed information tags are issued with the permit and must be attached to the harvested specimens from the time of collection until they reach their ultimate owner. The owner must retain the tag as proof of ownership. This elaborate system is a rather expensive form of conservation which limits its use to a relatively small number of species and countries.

Destruction

Many countries have enacted prohibitions on the destruction of protected or listed species. There are, however, often serious flaws in this type of protection which limit its effectiveness. The prohibition is often expressed in such vague terms as to be very difficult to enforce, and is often limited by so many exceptions that the ban is of little practical use.

The prohibition also rarely extends to the habitat of wild flora. One example where controls do extend to include the habitat of the species is the US Endangered Species Act. Under this Act federal agencies are not only prohibited from carrying out any activity which is likely to jeopardise the existence of listed species, they are also prohibited from carrying out any action which may result in the destruction or adverse modification of their critical habitat.

Other examples of specific connections between protection for a particular species of wild flora and protection of their habitats exist. In Norway, for example, Article 9 of the Nature Protection Act of 1970 prohibits development, construction, pollution and other encroachments in areas of major importance for protected species to preserve the habitats. This provision has been applied to Mistletoe *Viscum album*, a rare plant in Norway, and an order of 1976 prohibits the felling of trees on which this plant grows.

Controlling the introduction of exotic species

The introduction of new exotic species can have drastic consequences for native flora, fauna and natural habitats, and exotic species pollution is an important threat to biodiversity in many parts of the world. Preventative action is essential and legislation controlling deliberate introductions has now been adopted in many countries. The system of control is usually regulated by quarantine laws. Typically these will allow the importation of exotic species only for limited purposes such as for zoological or botanical gardens or for research purposes and in many cases only after it is ascertained that specimens are disease-free. Commonly, the importation of such species is restricted to a limited number of entry points in a country where the customs officials have the capacity to investigate the consignment to ensure that it complies with the law. Importation of endangered exotic species is in most countries subject to additional controls under legislation implementing the obligations of the Convention on International Trade in Endangered Species of Wild Fauna and Flora (CITES) (see Chapter 31).

In certain cases concern over the inadvertent introduction of exotic species has also caused the promulgation of some rather drastic measures. In some countries there is a complete prohibition on the import of certain potentially harmful species. Many countries have also enacted strict requirements as to packaging of imports in order to prevent accidental introduction of invertebrates. The determination of some countries to keep exotic species pollution to a minimum is well illustrated by Australia, where even ship discharges are now regulated in order to prevent the spread of toxic algal blooms.

THE PROTECTION OF WILD FAUNA

The protection of wild fauna has generally been given much more attention than the protection of wild flora. Specific legislation for the protection of wild animals has existed for many centuries and the legislative mechanisms used are often very similar to those described above for wild flora. Indeed, most modern examples of species-specific legislation cover both flora and fauna.

Taking

The oldest and most common form of protection for wild fauna has been restriction on taking. Such legislation has existed in some areas for many hundreds of years (restrictions associated with medieval hunting reserves in Europe, for example). In most countries there is a differentiated system of protection, with some species being fully protected and others only partially so. Typically, this differentiated protection is implemented through the use of appendices containing lists of the species at different levels. Usually the degree of protection a species receives is proportional to the seriousness of the perceived threat to its survival.

Partial protection can vary from strict controls which in practice are little different from those applied to fully protected species, to cases where the restrictions have little practical effect. A typical example of this system can be found in India, where the principal legislation is The Wildlife Protection Act 1972. Under this Act there are five Schedules. Species listed in the first Schedule are fully protected; those in the other schedules are provided varying degrees of protection. Species listed in Schedules II, III or IV are protected from hunting except in accordance with a licence issued by the relevant government official. The Act provides for the following kinds of licences: special game hunting licences for Schedule II species, big game hunting licences for Schedule III species and wild animal trapping licences for Schedule IV species. Any Schedule II or III species which is killed, wounded or captured must be reported to an authorised government official in accordance with specified procedures. No licence is required to hunt Schedule V species. The Act prohibits the hunting of any wild animals in wildlife sanctuaries and national parks. The chief warden of such an area may, however, permit hunting with the prior approval of the state government, provided it is necessary for the better protection of wildlife in the particular sanctuary or park.

Legislation for the partial protection of wild fauna is also commonly found in the controls placed on recreational hunting and fishing. These laws typically attempt to limit the taking of species to sustainable levels. Common techniques employed to achieve these objectives are the creation of seasons which limit hunting to certain times of the year, prohibition on taking in certain areas (such as national parks and game reserves), limitations on the types of equipment which can be employed, licensing of operators and establishment of total catch to try to maintain stocks.

Many countries also have elaborate legislative controls for the commercial exploitation of wild species. Typical of this is the control exercised over the fishing industry (see Chapter 33 for some international examples). Here the controls, although different in degree, are similar to the types of control exercised over recreational hunting.

Possession and trade

Another common form of legislative protection is restriction on trade of wild species and their products. Frequently these controls are provided for in the same legislation as that which controls taking. This type of control usually operates on a permit basis and these are granted to specific persons normally on a restricted basis which enables them to trade in a specified number of animals or their products. The extent of the restriction varies considerably not only from country to country but also from species to species within a particular country. The aim of such restrictions is the same as in the case of wild flora, i.e. to restrict the economic incentives for unlawful taking of protected species. As with restrictions on the trade of wild flora, a necessary addition to this type of control is restriction on possession. Thus, most legislation which establishes restriction on the taking of wild fauna also restricts possession of such species and their products.

Controls on the import and export of wild fauna also play an important role in the protection of threatened species by reinforcing the effectiveness of the trade controls that exist in a country. Legislative efforts in this regard are influenced by the work of CITES and in quite a few cases are confined to implementation of national obligations arising from CITES. In New Zealand, for example, import and export of wildlife is principally regulated by the Trade in Endangered Species Act of 1989 which was specifically enacted to implement CITES. The Act regulates trade in endangered, threatened and exploited species identified in one of the three schedules, which are equivalent to CITES Appendices I, II and III. Any person wishing to trade in any specimen of such species must apply to the government for the appropriate permit or authorisation. With regard to obtaining the necessary permits, separate conditions apply to export, import, re-export or introduction from the sea of endangered species, threatened species and exploited species. In general a permit authorises the holder to undertake on one occasion the type of trade to which the authorisation relates. Such permits are non-transferable and remain in force for six months unless revoked or surrendered. The Act also provides for extensive powers of inspection and gives customs officers broad powers of search and seizure with respect to listed species being traded in contravention of the Act.

One of the most extensive and innovative regimes established to control the import and export of wildlife is found in the USA. The two principle pieces of legislation establishing this regime are the Endangered Species Act 1973 and the Marine Mammal Protection Act 1972. The Endangered Species Act 1973 makes it illegal for any person to import or export endangered species within the US, to take endangered species within the US or territorial seas of the US, to take endangered species upon the high seas, or to sell or offer for sale any endangered species in interstate or foreign commerce. The Act also makes it unlawful for any person subject to US jurisdiction to engage in any trade in specimens or to possess any specimens in violation of CITES. This Act comprehensively implements the obligations contained in CITES. These legal norms are also backed up by extensive administrative resources which ensure the practical implementation of the Act. An unusual feature of this Act is the extension of its requirements outside the US itself. Thus US nationals are still bound by the requirements contain in this Act even though they themselves may be outside the country.

The Marine Mammal Protection Act 1972 prohibits the taking on the high seas of marine mammals by any persons or vessel subject to US jurisdiction; the taking of such animals by any person in waters or on lands subject to US jurisdiction; and the importation of marine animals, products and parts. The Act also has the extraterritorial application of the Endangered Species Act. One of the main purposes of the Act is to control commercial exploitation in order to reduce to insignificance the incidental killing or serious injury of marine mammals as a result of commercial fishing operations. To this end, the Act provides that it is to be administered for the benefit of protected species rather than the benefit of commercial exploitation. The Act therefore represents an unusual primacy of conservation over commercial interests. The Act provides a scheme to determine the number and kind of marine animals which can be taken incidentally to commercial fishing, which in practice essentially requires commercial fishing operations to adopt modern techniques and equipment to reduce the hazard to protected species. The products of commercial fishing operations which are conducted in contravention to this scheme are banned from importation into the US. Several such bans have been implemented, the most notable in relation to control of tuna products because of the incidental killing of small cetaceans. This type of control of commercial interests, ensuring that they take account of protected species, is quite unusual but if properly implemented a very important means of protection for wild species.

LIMITATIONS OF SPECIES LEGISLATION

There are a number of common problems with species-specific legislation. The usual method for providing legal protection to species consists in laying down prohibitions or restrictions together with penalties for non-compliance. The species to which these rules apply are usually listed in an annex or schedule to the legislation. Normally, the appropriate government minister, or other authority, is empowered to amend the list of species by statutory instrument, thus making it unnecessary to go through the elaborate process of adopting a new act each time a change in the list is required. There are usually no criteria laid down for listing or delisting the species, and so this remains entirely at the discretion of the particular authority.

An analysis of wildlife protection legislation shows that in most cases the lists of protected taxa are relatively short, rarely exceeding 100 entries. Often the lists are largely dominated by spectacular species attractive to collectors or the public and do not comprehensively cover the threatened species in a particular country. The extent of coverage for plants and invertebrates is frequently very limited. For instance, a recent survey of plant legislation found that only five jurisdictions (including France, Greece and Hungary) protect a large number of species, and often the list reflects the personal bias of the people working in the relevant authority. This situation points to the need for public and accountable procedures for listing.

One example where this is the case is the US Endangered Species Act. This provides for a detailed listing procedure involving a preliminary listing, an inquiry and, if requested, public hearings. An unusual feature of this procedure is that it can be initiated by any interested person. Only those species which are determined by the Secretary of the Interior to be endangered or threatened may be listed. Plant protection legislation in individual US States usually uses the same listing criteria but procedural requirements are generally simpler.

A common problem with much species-specific legislation is the restricted definition of taking. In some cases the definition is so narrow as to limit severely the effectiveness of the legislation. However, perhaps the most important deficiency with most legislation of this type is the absence of any provisions for the maintenance of the habitat of the species. This is despite the fact that protecting critical habitats is universally recognised as a basic requirement for species preservation. Even where there is such a provision it is usually in such general terms that implementation or enforcement is difficult. Again a notable exception to this is the US Endangered Species Act.

Species-specific legislation is thus fraught with many problems and of limited efficacy in the conservation of biodiversity. Consequently it is only really effective for species primarily affected by excess exploitation, or as a last resort measure for rare and endangered species.

THE PROTECTION OF NATURAL HABITATS

The most important form of legislative measure for the conservation of biodiversity is that for the protection of ecosystems and habitats. Control over the use of land is the essential means by which such systems are managed and protected. National legislation is the most common way for these controls to be established. There are several common types of such mechanism.

Protected species habitats

The US Endangered Species Act was cited above as an example of legislation which extended to protection of habitats, in this case 'critical habitats' of threatened species, these being defined as areas which are essential to the conservation of the species concerned. These areas must be designated and their boundaries precisely described in the *Federal Register*. As of October 1987, of the 168 listed species of wild flora there were 23 species for which critical habitats had been designated.

The critical habitat concept has also been used in the 1988 Flora and Fauna Guarantee Act of Victoria, Australia. Under this Act, where a critical habitat designation is made landowners are prohibited from collecting protected flora in the critical habitat. The Act also gives the Minister power to make interim conservation orders prohibiting or regulating any activity which takes place within or could have adverse effect on the designated critical habitat. An order may also contain a positive requirement that specified works or activities be undertaken. Interim conservation orders must be complied with by all persons and may be applicable to any land. However, the designation can only be made for a period of two years. Before it expires, the Minister must take all reasonable steps, including the conclusion of management agreements, to ensure the long-term conservation of the taxa, communities or critical habitats for which they were made.

The French Nature Conservation Act of 10 July 1976 contains a general provision prohibiting the destruction, alteration or degradation of the habitat of protected species. A decree adopted in 1977 to implement the Act provides that the central government representatives (*préfets*) may make regulations to promote the conservation of the habitat of listed protected species. The establishment of these protected areas, known as *arrêtés de biotope*, is not automatic. Three conditions have to be fulfilled. There must be an individual order from the *préfet* designating a certain area where particular prohibitions apply. The order may only prohibit activities that can affect the habitat of a species. It may only apply to a protected species, that is to say to a species listed in regulations made by the Minister of the Environment. Subject to these limitations the powers of the *préfet* are quite broad, as he may prohibit or otherwise regulate activities such as vehicle traffic, farming, drainage, construction or any other action which may be detrimental to the conservation of the species habitat. No compensation is provided to landowners.

An important feature of the *arrêtés de biotope* is the flexibility and simplicity of the procedure underlying their adoption. In contrast to the establishment of nature reserves, which requires a long and protracted consultation procedure, the *arrêtés de biotope* may be adopted with a minimum of formalities. They are, therefore, increasingly used as a substitute for nature reserves, which are meeting with growing opposition from local populations and authorities.

The network of *arrêtés de biotope* began to develop after 1982-1983. Most of the areas so protected are designed to preserve the habitat of animal species, for instance heronries, and the number which exclusively concern plants is still small. Examples are a few peatlands harbouring rare and specialised flora species (e.g. *Andromeda polifolia*, *Drosera* spp.) and certain sites of botanical interest containing species such as *Gagea bohemica, Gagea lutea* and *Crambe maritima*.

Protected areas

The world's protected area network, the status of which is examined in Chapter 29, plays a vital and essential role in the conservation of habitats and ecosystems. With 169 countries in the world having recognised protected area networks, their use for the conservation of biodiversity is universal. Whereas the initial purpose of many such areas was to protect spectacular scenery and provide recreational facilities, in recent years the concept has evolved to encompass habitats of endangered species and ecosystems rich in biodiversity. Even though the legislation used to establish such areas varies technically from jurisdiction to jurisdiction the mechanisms used to control or prohibit certain activities, the essence of the concept of a protected area, are more or less universal.

In countries where there are large tracts of public lands the establishment of protected areas under public ownership is relatively straightforward in theory in that the government can if it so wishes simply manage the area as a protected area. Unfortunately the simplicity of this solution from a legal point of view belies the practical difficulties which often arise. Frequently, the change of management will also require that control of the land changes from one government department to another; this change is often problematic. In some instances it will require legislative measures to be promulgated, in others cases it will require the transfer of the property at market prices even though the 'purchaser' is another government department.

One simple and effective way to ensure that government departments preserve natural habitats on public land is the 'wilderness area' concept as used in the USA. Pursuant to the Wilderness Act of 1964 it is possible to ban the construction of all roads and tracks and other means of access within a specified area. The National Wilderness Preservation System, which is made up of these specified areas, has developed rapidly and is intended to cover some 400,000km^2 of federal land under the control of various government departments. The potential of this type of measure is obvious because threats typically escalate following increasing access to wilderness areas by road construction.

If the land requiring protection is in private hands, governments have used a variety of mechanisms to establish the necessary protection. In some instances they have simply acquired the land from the owner. This mechanism can be expensive. One way that governments have sought to ameliorate this cost is to acquire a lessor interest in the land, such as the right of drainage, where such rights are separable.

Alternatively governments can and have used their rights of expropriation to force private owners to either relinquish the land or agree to controls over the use of the land. Governments are now reluctant to use such powers especially for conservation purposes. More commonly governments will impose restraints on the use of land by private persons by, for instance, banning all forms of use which are detrimental to the ecosystems present in the area.

Such forms of control are not always constitutionally possible, as in common law countries where such a curtailment of rights is generally perceived as unlawful. In these countries, the government is generally only able to impose such controls under a voluntary management agreement with the owner. Under voluntary agreements the owner commits himself not to use the land for certain purposes. One example of this type of agreement is that found in England where, under the Wildlife and Countryside Act 1981, English Nature (formerly the NCC) can enter into agreement with the owners of Sites of Special Scientific Interest (SSSIs).

Protection of private land is also facilitated by the legal system through the use of caveats. These rights attach to the land itself and will bind future owners. Such rights exist in most common law jurisdictions. In some countries the government has pre-emptive rights over the sale of certain land should it happen to be sold by the owner. Such provisions exist in several European countries. In the USA the government frequently negotiates a pre-emptive right individually with the owner. In France the pre-emptive right is also linked with a mechanism to finance the purchase of such properties which come onto the market. Under the legislation creating the pre-emptive right the particular department is also empowered to collect a tax on the construction of buildings the proceeds of which are hypothecated to the acquisition of private land.

Land-use controls

Many countries have legislation limiting the use to which land may be put. Such land-use controls or zoning restrictions typically control activities such as construction or mining and are normally restricted to the urban environment. In a few countries zoning restrictions also extend to rural areas; however, agricultural and forestry activities are normally exempted from their provisions.

In a few countries such mechanisms are used to protect natural habitats. Examples of such mechanisms include: special protection orders for specific sites; the use of specially protected areas in local zoning plans; or the prohibition on altering of certain habitats without a permit.

A prime example of such a mechanism being used to protect natural habitats is the Danish Nature Conservation Act of 1969 (as amended). This Act establishes a strict system of permits applicable to all activities which may have an adverse effect on river beds, lakes, peatbogs, salt marshes, coastal vegetation and natural grasslands. This type of approach has also been adopted in many European countries, North America and parts of Australia.

Another important and common land-use control is restriction on felling of private forests. In most cases, the restrictions are not applicable to the government forestry department itself. One exception to this is found in the USA where, under the US Federal Forests and Rangelands Renewable Resources Planning Act of 1976 (as amended), the discretionary authority of the Forest Service is curtailed and the objective of the organisation must now include the maintenance of all plants and animal species and the promotion of the recovery of endangered species. Forest plans must be drawn up for each unit in the National Forest System using an interdisciplinary approach and including public participation. A common problem with this mechanism is that the purpose of the legislation is often not the preservation of natural forests but simply the maintenance of forest cover. This means that the replacement of native forest, rich in biodiversity, with comparatively sterile monocultures of production timber is not regulated by such controls.

Incentives

A common legislative mechanism to help conserve natural habitats is the provision of incentives or disincentives to influence the activities of land users to conserve natural habitats. Examples of such mechanisms are the EC regulation providing for the subsidy payments to farmers to maintain the natural environment on their land, and the granting of land tax credits for the preservation of wetlands or natural prairie areas, or for the conservation of river banks, in the US State of Minnesota. Another important example of an incentive, although an indirect one, is the tax exemptions granted in many countries to many conservation organisations on the basis of their charitable status. In the USA, land owned by conservation organisations or land dedicated to conservation is frequently exempt from land tax.

Many countries not only provide incentives to preserve natural habitats but also penalise environmentally harmful activities. Measures of this sort include the refusal of subsidies and the imposition of special taxes on such activities. The UK Wildlife and Countryside Act 1981 contains such a mechanism. It provides that agricultural subsidies may be refused for activities which will adversely affect the flora, fauna and physiogeographical features of national parks or in areas specially designated for that purpose (e.g. SSSIs). The US Food Security Act of 1985 is also another example of such a mechanism. The purpose of this Act is to remove up to 40 million acres (16 million hectares) of erodible land from agricultural production to, *inter alia*, reduce erosion and enhance wildlife. It seeks to achieve this by removing a number of subsidies from crops produced on highly erodible soil or altered wetland.

Indirect legislation

The types of legislative mechanisms described above are all examples of direction protection of biodiversity. In many countries there exist numerous legislative mechanisms which while not directly protecting biodiversity do nonetheless play a vital role in its conservation. Examples of this type of legislation are pollution control laws or legislation regulating development and investment in a country. Such controls can and do have an important effect on the conservation of biodiversity in a country. If properly framed, they can be powerful forces for the conservation of biodiversity; if not, such regimes can have drastic consequences for its conservation.

Chapter contributed by Sam Johnston.

29. PROTECTED AREAS

Natural ecosystems and the habitats they contain are subject to some degree of control and protection in every country in the world. Many different legal and administrative mechanisms are used by governments to manage habitats for the conservation of biodiversity. Protected area systems are central to such management. This section will provide information on protected areas which contribute to such systems, charting the growth in protected areas over the past century. It will also examine the extent to which different geographic and biogeographic regions, and biome types are covered by protected area systems, and highlight major gaps in the network.

NATIONAL PROTECTED AREA SYSTEMS

There is considerable variation between countries in the mechanisms used to create and maintain systems of protected areas. Some standard means of classification needs to be used in making international comparisons. The IUCN, through its Commission on National Parks and Protected Areas (CNPPA), has developed a system of classification for different types of protected area, based upon management objectives. This system has 10 different classes of protected areas, two of these, World Heritage Sites (X) and Biosphere Reserves (IX) being international designations.

In the analysis in this chapter the term 'protected area' is defined as an area of 1,000ha or more in IUCN Management Categories I-V, managed by the highest competent authority. These are the criteria used in compiling the *1990 United Nations List of National Parks and Protected Areas* (IUCN, 1990).

However, statistics prepared using such standard criteria omit a range of significant sites. For instance, the statistics presented here do not include: sites which are in other management categories (such as multiple-use areas (VIII)), areas under 1,000ha (such as the numerous small reserves in Europe), areas outside the IUCN Categories altogether, such as partially protected areas (e.g. hunting reserves), and areas not managed by the 'highest competent authority' but protected by private organisations (such as NGOs), superstition, isolation or military activity. All of these conserve significant amounts of biodiversity. Whilst information on such sites is available, it is not yet consistent and has.

The wise management of areas which are devoted to agriculture, through management techniques such as non site-specific legal instruments, planning control, voluntary agreements, and integrating conservation principles into land-use planning, also play an essential role in conservation of biodiversity. Indeed, in most countries, management of land-use outside the national network of protected areas will play as important a role in the conservation of biodiversity as will the network itself. In order to examine comprehensively the role that land management plays in the conservation of biodiversity, it would be necessary to survey the use of these other areas and techniques as well. Unfortunately this is not possible at this stage because of the paucity of reliable data on these important measures.

Categories and management objectives of protected areas

The following categories and criteria for protected areas are abridged from IUCN (1984).

I *Scientific Reserve/Strict Nature Reserve*: to protect nature and maintain natural processes in an undisturbed state in order to have ecologically representative examples of the natural environment available for scientific study, environmental monitoring, education, and for the maintenance of genetic resources in a dynamic and evolutionary state.

II *National Park*: to protect natural and scenic areas of national or international significance for scientific, educational and recreational use.

III *Natural Monument/Natural Landmark*: to protect and preserve nationally significant natural features because of their special interest or unique characteristics.

IV *Managed Nature Reserve/Wildlife Sanctuary*: to assure the natural conditions necessary to protect nationally significant species, groups of species, biotic communities, or physical features of the environment where these require specific human manipulation for their perpetuation.

V *Protected Landscape or Seascape*: to maintain nationally significant natural landscapes which are characteristic of the harmonious interaction of man and land while providing opportunities for public enjoyment through recreation and tourism within the normal life style and economic activity of these areas.

Other categories defined by IUCN but not analysed here are Category VI (*Resource Reserve*), Category VII (*Natural Biotic Area/Anthropological Reserve*) and Category VIII (*Multiple-Use Management Area/Managed Resource Area*). The classes and their different management objectives are given in Table 29.1.

Development

Areas that are in some sense 'protected', in that access or forms of use are controlled, have existed for many thousands of years. In India, protected areas have existed since the 4th century BC, with the establishment of Abhayaranxyas or forest reserves. In the Pacific region, the imposition of tapu (taboo) effectively created protected areas; the existing protected area on Niue, for example, consists of a tapu forest. Hunting reserves have existed in Europe for hundreds of years. The first modern examples of protected areas were established towards the end of the 19th century.

Table 29.1 Protected area objectives

	PROTECTED AREA DESIGNATION (IUCN CATEGORY NUMBER)							
CONSERVATION OBJECTIVE	Scientific Reserve I	National Park II	Natural Monument III	Managed Nature Reserve IV	Protected Landscape V	Resource Reserve VI	Natural Biotic Reserve VII	Multiple-Use Area VIII
Maintain sample ecosystem in natural state	•	•	•	•	o	o	•	
Maintain ecological diversity and environmental regulation	o	•	•	o	o	o	•	o
Conserve genetic resources	•	•	•	•	o	o	•	o
Provide education, research and environmental monitoring	•	o	•	•	o	o	o	o
Conserve watershed, flood control	o	•	o	o	o	o	o	o
Control erosion and sedimentation	o	o	o	o	o	o	o	o
Maintain indigenous use or habitation					•	o	•	o
Produce protein from wildlife				o	o	o	o	•
Produce timber, forage or extractive commodities					o	o	o	•
Provide recreation and tourism service		•	•	o	•		o	•
Protect sites and objects of cultural, historical, or archaeological heritage		o	o	o	•	o	•	o
Protect scenic beauty	o	•	o	o	•			o
Maintain open options, management flexibility, multiple-use					o	•		•
Contribute to rural development	o	•	o	o	•	o	o	•

Sources: Miller, K.R. 1980. *Planning National Parks for Ecodevelopment*, Center for Strategic Wildland Management Studies, Ann Arbor; IUCN/UNEP 1986. *Managing Protected Areas in the Tropics*. IUCN, Gland, Switzerland.
Notes: • = Primary Objectives, o = Compatible Objectives.

By the beginning of this century many countries had either already established protected areas or were contemplating doing so. The concept, however, was slow to develop to a stage where any one country had developed a comprehensive network of actively managed protected areas. It was not until the 1940s that protected areas were beginning to be established in any significant number. After World War II, the number of protected areas established continued to be low, and the rate at which land was being incorporated into the system did not increase above pre-World War II levels until the early 1960s. In 1962 the establishment of protected areas began to increase dramatically. An important stimulus for this increase may have been the first World Parks Congress held in Seattle, USA in 1962. This meeting signified the emergence of the modern protected area network with over 80% of the world's protected areas being established since then. Table 29.2 shows that the increase experienced during this period has continued unabated until the present day.

The rates of growth of protected areas on global and regional bases are illustrated in Fig. 29.1 and 29.2 respectively, showing the number of sites and the area protected. It should be noted that the creation of Greenland National Park in 1974, which covers some 97 million ha, and the creation of Great Barrier Reef Marine Park in the 1980s, which covers some 34 million ha, has a marked effect on the area protected for the relevant period.

Table 29.2. Dates of establishment of protected areas

DATES	NUMBER	AREA (km²)
Pre-1962	1,433	1,324,600
1962-1971	1,372	862,800
1972-1981	2,258	3,559,800
1982 onwards	2,140	1,684,100
Date unclear	1,288	303,600
TOTAL	8,491	7,734,900

Fig. 29.2 shows that there is considerable regional variation in the development of protected area networks. Reasons for this variation include: cultural and historical factors, the development of interest in wildlife and conservation in the region, and patterns of settlement and land-use.

Despite the regional variations the graphs do illustrate a number of global trends. For most regions, networks of protected areas are a recent phenomenon, with only Africa

Figure 29.1 World growth of the protected areas network

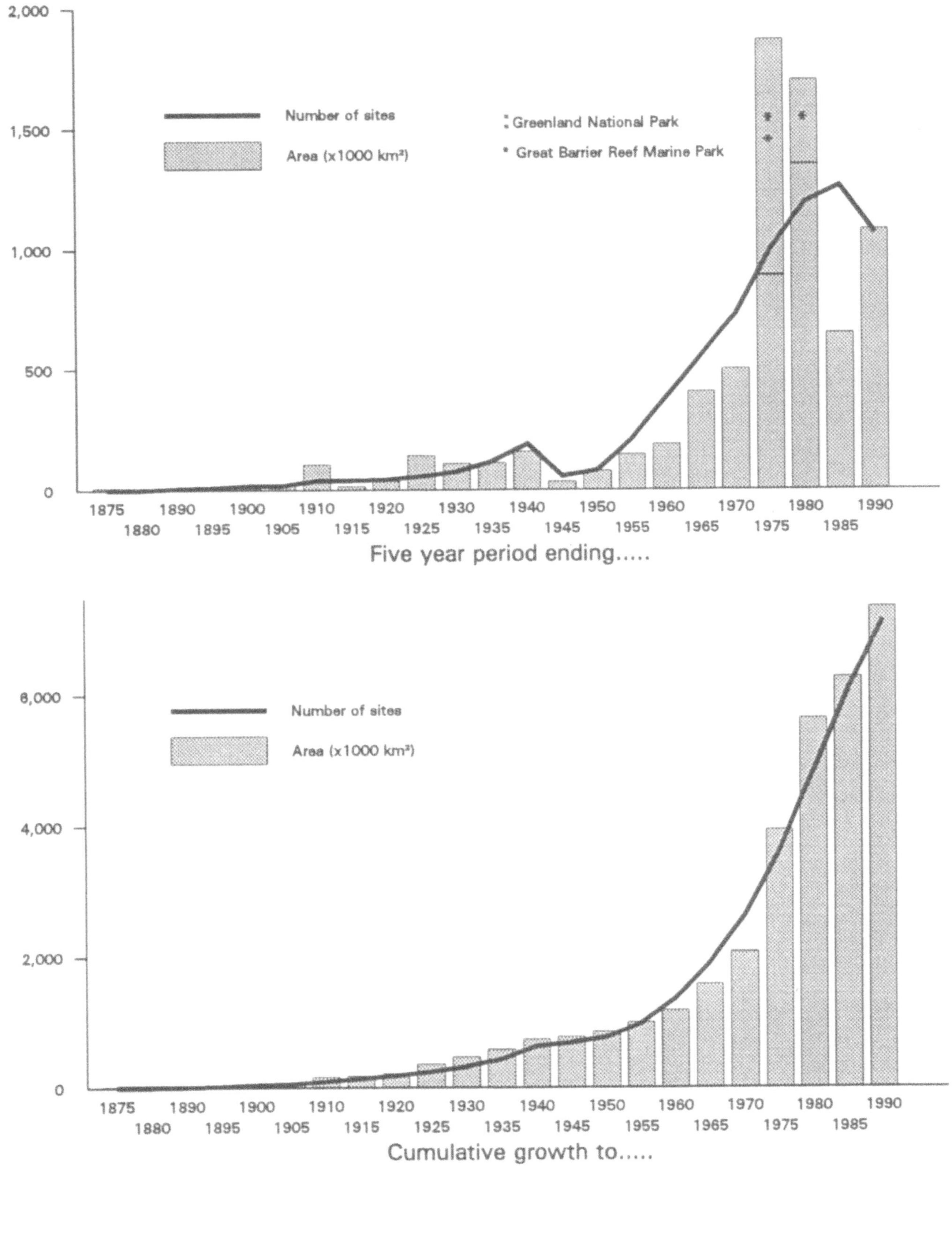

Figure 29.2 Regional growth of the protected areas network

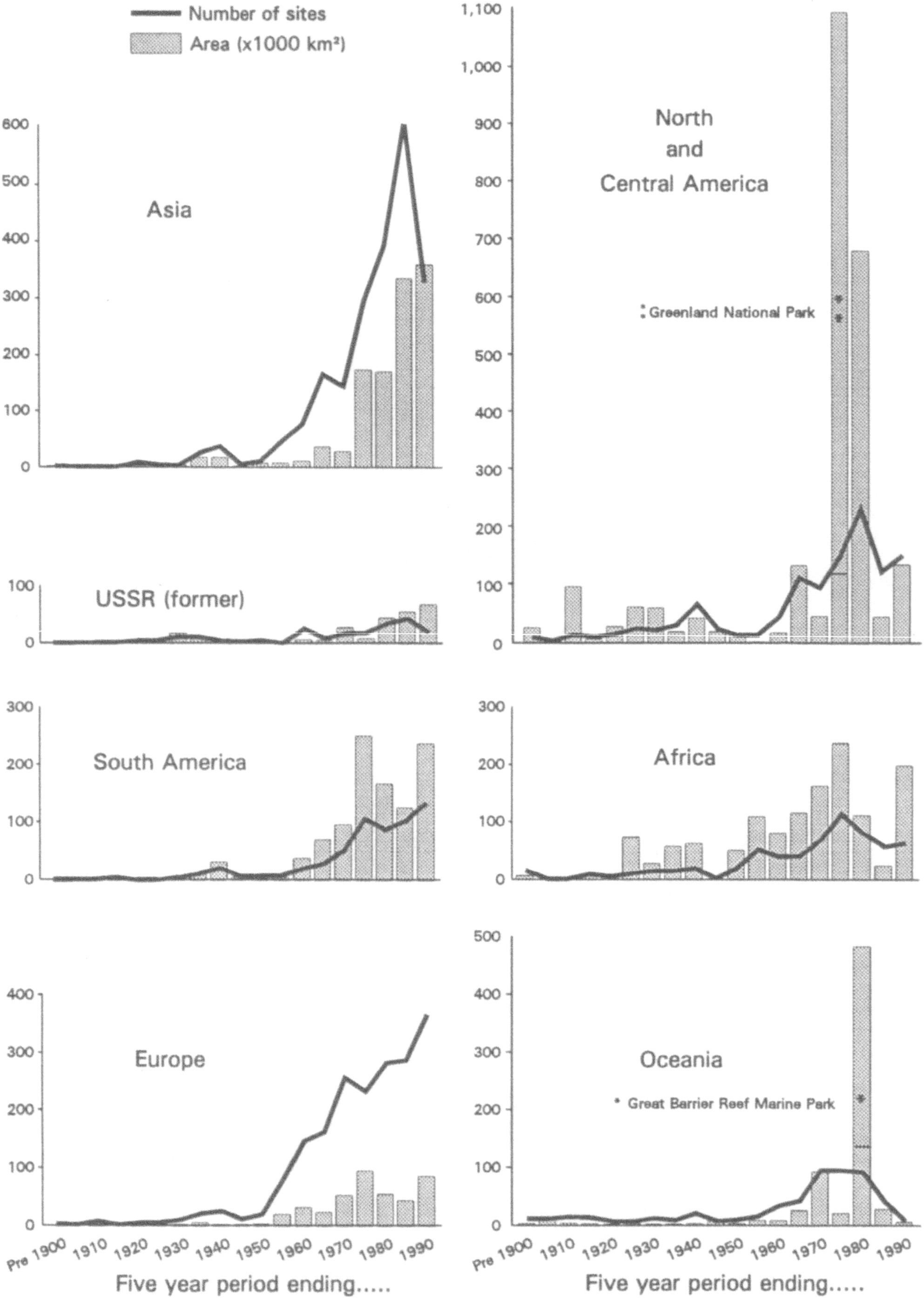

and North America having a significant history of protected areas before 1962. The rates of growth for both number of sites and area protected are still high throughout much of the world. There is a tendency for larger protected areas to be established in the developing world.

Present status of national systems

Protected areas meeting the criteria given now exist in 169 countries in the world. There are currently some 8,491 sites covering some 7,734,900km^2 or some 5.19% of the earth's land area. The largest protected area is Greenland National Park, which covers 972,000km^2. In 115 countries, 1,328 sites covering some 3,061,300km^2 have marine or coastal elements within them. Of these, 94 sites have coral reefs. The largest marine protected area is Great Barrier Reef Marine Park, which covers some 340,000km^2.

The relative proportion of each type of IUCN Category is illustrated in Fig. 29.3. Managed nature reserves/wildlife sanctuaries (Category IV) are the most prevalent type of protected area in terms of number of sites. National parks (Category II) cover more area than any other category. The extent to which each category is applied varies considerably from region to region, as a result of cultural, demographic and geographic factors. Thus, in Europe, where there is very little natural habitat and where man has extensively altered that which remains, most large protected areas are managed as protected landscapes. In Australia, where man's influence is less pervasive, the predominant protected area category is the national park.

The size distribution of protected areas is illustrated in Fig. 29.4. The most common size for a protected area on a worldwide basis is only 10-30km^2. However, the majority of the world's 7.7 million km^2 of protected area is contained in a relatively few large sites. These figures suggest that fragmentation may be a problem in providing protection to many of the world's natural habitats. Significant regional differences in size distribution can also be deduced from Fig. 29.2.

Table 29.3 presents the distribution of protected areas according to the World Bank classification of the country's economy. The classes are based on per capita income. The low income class is subdivided according to country size

Figure 29.3 Protected areas by IUCN category

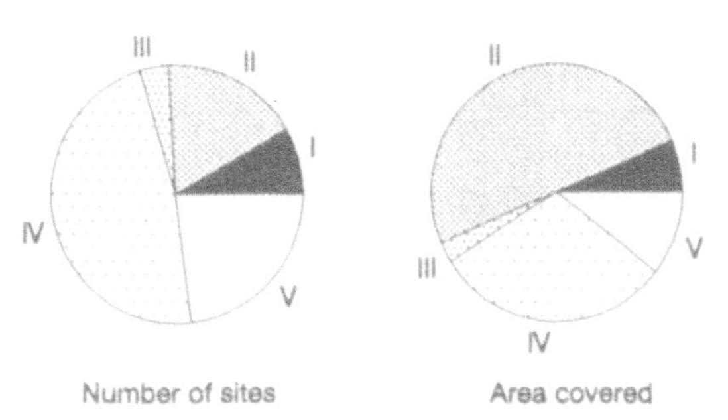

('large' includes India and China). The 'middle income (upper)' class is distorted by the former USSR, where the protected areas network has extremely low coverage and the country area is very large. The high income class is divided by membership in the Organization for Economic Cooperation and Development (OECD). The extraordinary figures for the Non-OECD group are because of Greenland National Park. Protected areas are fairly evenly spread by income groups and quite high even for very poor countries. The smaller average size of a protected areas for the large low income group and the OECD countries probably reflects the high population densities of these countries.

A major objective of the protected area system of the world is maintaining the diversity of species and ecosystems. Biogeographical analysis of protected area coverage provides information on how effectively the various natural ecosystems are being conserved.

A basic system of biogeographic analysis has been worked out for terrestrial ecosystems by Udvardy (1975). He divides the world into eight *biogeographical realms*, continent or subcontinent-sized areas, which are further divided into 193 *provinces* defined by significant differences in flora, fauna, or vegetation structure. The provinces are associated with 14 *biomes*, which are major regional ecological communities of plants and animals. It should be noted that a protected area located within a particular province may not necessarily contain vegetation typical of that province. Thus a protected area within the Congo Rainforest province may not necessarily contain tropical humid forest, and although insular Malaysia, Indonesia and the Philippines are classified as mixed island systems, they all contain extensive tropical humid forests.

Table 29.4 presents the extent to which each biome is covered by protected areas. This table shows that temperate grasslands and lake systems are poorly represented in the protected area network, and that this is an area requiring attention. The conclusions that can be drawn from the high level of aggregation at the biome level are limited. A more accurate picture of ecosystem protection can be gained from an analysis of protected area coverage at the province level. Table 29.5 lists in descending order the percentage coverage of each province. The analysis of protected area coverage at this level still suffers from the problems mentioned above, albeit in a reduced way. These data are also presented in map form in Fig. 29.5.

Fig. 29.6 illustrates the distribution of marine and coastal protected areas throughout the world. There are also 559 sites that have an altitudinal range of 1,500m or more and Fig. 29.7 illustrates the distribution of these mountainous areas.

Studies of protected area coverage at regional and national levels would provide a much better assessment of priorities, and many such studies have been undertaken. The mechanisms for assessment used in these studies vary very widely, so an assessment of coverage based on these studies has not been attempted. A range of regional studies have been published by IUCN and others.

Figure 29.4 Protected areas by size class frequency

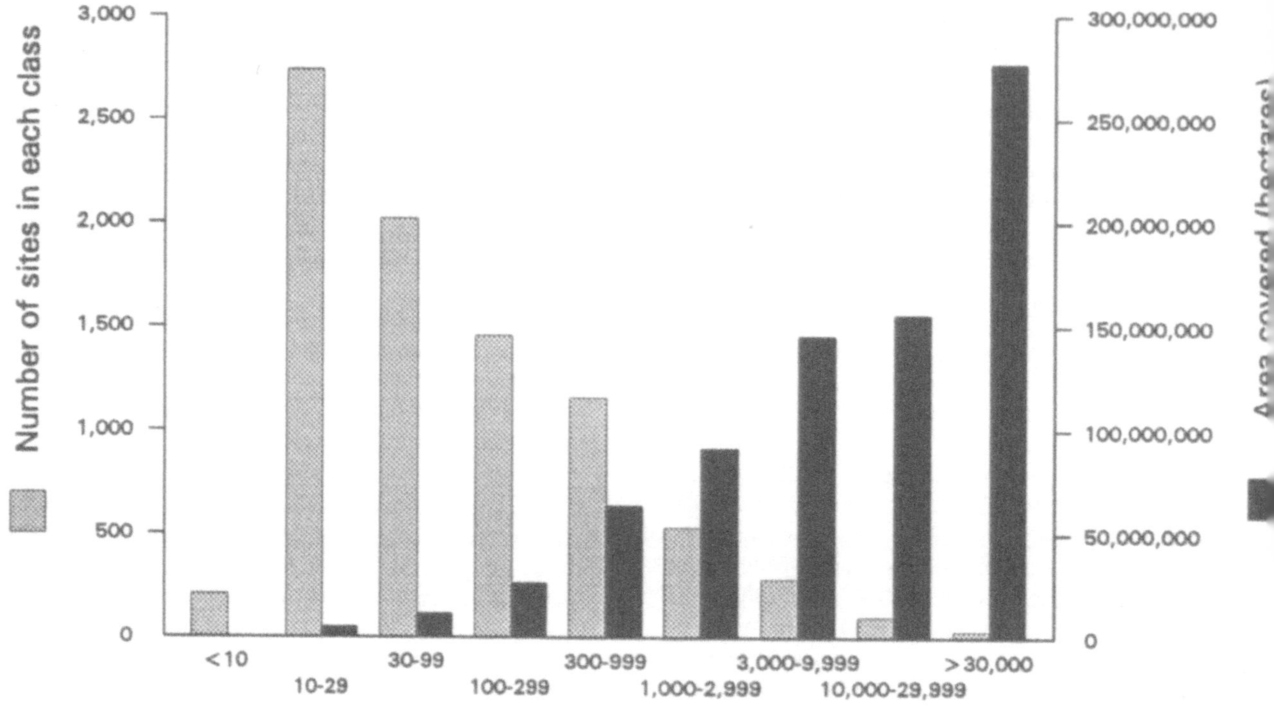

Table 29.3 Distribution of protected areas by World Bank income groups

INCOME GROUP AND SUBGROUP	Number	% of Total No.	Area (km²)	% of Total Area	Average Size (km²)	Country Area (km²)	% of country Area
Low Income (large)	758	8.9	421,300	5.5	556	12,764,000	3.3
Low Income (small)	734	8.6	1,067,300	13.8	1,454	24,636,000	4.3
Middle Income (lower)	1,051	12.4	1,338,500	17.3	1,274	23,173,000	5.8
Middle Income (upper)	1,126	13.3	1,200,400	15.5	1,066	41,404,000	2.9
High Income (OECD)	4,713	55.5	2,677,100	34.6	568	31,079,000	8.6
High Income (Non-OECD)	62	0.7	990,600	12.8	15,977	2,381,000	41.6
Income not assigned	47	0.6	39,700	0.5	845	13,677,000	0.3
TOTAL	8,491	100.0	7,734,900	100.0	911	149,114,000	100.0

Table 29.4 Distribution and coverage of protected areas by biome type

BIOME TYPE	PROTECTED AREAS NUMBER	AREA (km²)	BIOME AREA (km²)	% OF TOTAL AREA
Subtropical/temperate rainforests/woodlands	935	366,100	3,928,000	9.32
Mixed mountain systems	1,265	819,600	10,633,000	7.71
Mixed island systems	501	246,300	3,244,000	7.59 **
Tundra communities	81	1,643,400	22,017,000	7.46
Tropical humid forests	501	522,000	10,513,000	4.96
Tropical dry forests/woodlands	807	818,300	17,313,000	4.73
Evergreen sclerophyllous forests	786	177,400	3,757,000	4.72
Tropical grasslands/savannas	56	198,200	4,265,000	4.65
Warm deserts/semi–deserts	296	957,700	24,280,000	3.94
Cold–winter deserts	139	364,700	9,250,000	3.94
Temperate broad–leaf forests	1,509	357,000	11,249,000	3.17
Temperate needle–leaf forests/woodlands	440	487,000	17,026,000	2.86
Lake systems	18	6,600	518,000	1.28
Temperate grasslands	196	70,000	8,977,000	0.78
Classification unknown	961	700,800	0	NA
TOTAL	**8,491**	**7,734,900**	**146,968,000**	**5.26**

Notes: ** Protected area includes significant marine areas, inflating the % figure. Biome definitions after Udvardy, 1975.

Table 29.5 Distribution and coverage of protected areas by biogeographic province

PROVINCE	REALM	PROTECTED AREA NUMBER	PROTECTED AREA AREA (km^2)	PROVINCE AREA (km^2)	% OF TOTAL AREA
Cocos Island	Neotropical	1	24	24	100.0 **
Everglades	Neotropical	17	8,080	6,800	100.0 **
Fernando De Noronja Island	Neotropical	1	362	17	100.0 **
Galapagos Islands	Neotropical	1	7,600	7,600	100.0 **
Aleutian Islands	Nearctic	9	79,100	124,500	63.5 **
Campos Limpos	Neotropical	6	108,640	207,300	52.4
Sitkan	Nearctic	85	172,010	350,500	49.1
Arctic Desert and Icecap	Nearctic	2	982,500	2,119,500	46.4
Valdivian Forest	Neotropical	12	40,160	111,900	35.9
Lesser Antillean	Neotropical	26	2,240	6,600	34.0 **
Alaskan Tundra	Nearctic	25	309,420	958,500	32.3
Chilean Nothofagus	Neotropical	7	39,200	123,700	31.7
Colombian Montane	Neotropical	23	42,500	154,800	27.5
Queensland Coastal	Australian	78	81,690	300,200	27.2
Namib	Afrotropical	7	95,970	364,600	26.3
Panamanian	Neotropical	9	10,350	40,100	25.8
Cocos-Keeling and Christmas Islands	Indomalayan	1	87	337	25.8
Ceylonese Monsoon Forest	Indomalayan	42	7,760	34,900	22.2
Tasmanian	Australian	28	13,910	68,000	20.5
Yukon Taiga	Nearctic	27	203,110	1,019,600	19.9
Sonoran	Nearctic	38	100,540	507,800	19.8
Kalahari	Afrotropical	10	97,770	504,900	19.4
Venezuelan Dry Forest	Neotropical	41	50,640	270,300	18.7
Comores Islands and Aldabra	Afrotropical	1	350	1,920	18.2
Arctic Desert	Palaearctic	5	34,910	195,900	17.8
Seychelles and Amirantes Islands	Indomalayan	3	36	204	17.5
Central European Highlands	Palaearctic	381	62,620	369,900	16.9
Hawaiian	Oceanian	19	2,790	16,700	16.7
Insulantarctica	Antarctic	11	3,140	19,200	16.3
Southern Andean	Neotropical	51	106,940	662,900	16.1
Ryukyu Islands	Palaearctic	5	391	2,500	15.8
Southeastern Polynesian	Oceanian	12	640	4,200	15.4
Venezuelan Deciduous Forest	Neotropical	16	8,960	58,900	15.2
Scottish Highlands	Palaearctic	43	7,110	46,800	15.2
New Caledonian	Oceanian	16	2,510	16,700	15.0
British Islands	Palaearctic	103	39,560	266,600	14.8
Serro Do Mar	Neotropical	56	35,980	243,800	14.8
Northern Andean	Neotropical	18	37,380	256,500	14.6
Cape Sclerophyll	Afrotropical	51	18,670	129,700	14.4
Sulawesi (Celebes)	Indomalayan	38	25,230	196,700	12.8
Andaman and Nicobar Islands	Indomalayan	42	797	6,200	12.8
Sierra-Cascade	Nearctic	82	28,380	228,700	12.4
Guinean Highlands	Afrotropical	3	9,720	80,000	12.1
Canadian Tundra	Nearctic	20	198,210	1,733,400	11.4
Central African Highlands	Afrotropical	7	30,630	269,500	11.4
Neozealandia	Antarctic	150	29,050	266,000	10.9
Japanese Evergreen Forest	Palaearctic	481	28,370	266,900	10.6
Sumatra	Indomalayan	38	49,080	461,900	10.6
Bahamas-Bermudean	Neotropical	6	1,350	12,800	10.6
Greater Antillean	Neotropical	35	9,990	95,800	10.4
Himalayan Highlands	Palaearctic	79	82,570	860,100	9.6
Pamir-Tian-Shan Highlands	Palaearctic	30	60,910	643,200	9.5
East African Woodland/Savanna	Afrotropical	71	142,820	1,510,600	9.5
Yungas	Neotropical	18	44,100	483,100	9.1
Anatolian-Iranian Desert	Palaearctic	47	200,300	2,203,800	9.1
Icelandian	Palaearctic	23	9,170	101,600	9.0
Iberian Highlands	Palaearctic	114	28,450	316,100	9.0
Central Polynesian	Oceanian	4	362	4,200	8.7
Rocky Mountains	Nearctic	145	127,850	1,578,500	8.1
Macaronesian Islands	Palaearctic	11	1,130	14,000	8.0
Taiwan	Indomalayan	5	2,890	36,600	7.9
Oregonian	Nearctic	32	9,430	124,600	7.6
Amazonian	Neotropical	34	184,090	2,509,400	7.3
Malayan Rainforest	Indomalayan	23	12,740	179,200	7.1
Papuan	Oceanian	36	68,180	960,100	7.1
Java	Indomalayan	43	9,730	137,900	7.1
Southern Mulga/Saltbush	Australian	14	58,700	837,000	7.0
Southern Sclerophyll	Australian	67	17,080	246,700	6.9
Western Sclerophyll	Australian	125	26,600	410,800	6.5
Indochinese Rainforest	Indomalayan	60	29,090	452,500	6.4

Table 29.5 Distribution and coverage of protected areas by biogeographic province (continued)

PROVINCE	REALM	PROTECTED AREA NUMBER	PROTECTED AREA AREA (km²)	PROVINCE AREA (km²)	% OF TOTAL AREA
Thar Desert	Indomalayan	39	45,680	711,800	6.4
Cuban	Neotropical	34	6,910	109,800	6.3
Lesser Sunda Islands	Indomalayan	20	5,390	86,600	6.2
Atlantic	Palaearctic	120	44,300	715,900	6.2
Miombo Woodland/Savanna	Afrotropical	38	148,400	2,432,100	6.1
Borneo	Indomalayan	66	42,310	741,000	5.7
Eastern Sclerophyll	Australian	169	35,920	643,800	5.6
Central Desert	Australian	17	98,960	1,777,100	5.6
Takla–Makan–Gobi Desert	Palaearctic	19	120,000	2,184,600	5.5
West African Woodland/Savanna	Afrotropical	80	177,370	3,247,600	5.5
Caucaso–Iranian Highlands	Palaearctic	66	49,940	936,000	5.3
Puna	Neotropical	19	23,390	464,900	5.0
Mahanadian	Indomalayan	29	10,970	219,400	5.0
Lake Titicaca	Neotropical	1	362	7,200	5.0
Colombian Coastal	Neotropical	9	11,280	237,200	4.8
South African Woodland/Savanna	Afrotropical	104	80,350	1,694,800	4.7
Middle European Forest	Palaearctic	401	68,870	1,467,300	4.7
Malabar Rainforest	Indomalayan	43	10,140	223,600	4.5
Northern Savanna	Australian	10	26,100	580,900	4.5
Central American	Neotropical	58	13,900	310,000	4.5
Balkan Highlands	Palaearctic	103	9,830	221,200	4.4
Campechean	Neotropical	19	11,400	259,200	4.4
Szechwan Highlands	Palaearctic	52	24,660	578,600	4.3
Eastern Sahel	Afrotropical	4	48,460	1,169,700	4.1
East African Highlands	Afrotropical	11	2,680	65,500	4.1
Indus–Ganges Monsoon Forest	Indomalayan	129	55,680	1,412,200	3.9
Kamchatkan	Palaearctic	1	10,990	283,300	3.9
Congo Rain Forest	Afrotropical	24	71,900	1,921,900	3.7
Northern Coastal	Australian	14	12,890	350,400	3.7
Equadorian Dry Forest	Neotropical	4	1,840	50,300	3.7
Canadian Taiga	Nearctic	286	180,310	5,127,200	3.5
Pannonian	Palaearctic	33	3,520	102,500	3.4
Sahara	Palaearctic	17	226,970	6,960,900	3.3
Patagonian	Neotropical	25	13,200	413,100	3.2
Ethiopian Highlands	Afrotropical	7	16,060	505,400	3.2
Babacu	Neotropical	6	9,030	293,000	3.1
Madrean–Cordilleran	Nearctic	83	23,410	763,200	3.1
Mediterranean Sclerophyll	Palaearctic	227	36,590	1,194,700	3.1
Western Mulga	Australian	15	22,610	778,100	2.9
Congo Woodland/Savanna	Afrotropical	6	37,740	1,356,800	2.8
Yucatecan	Neotropical	3	1,070	40,000	2.7
Manchu–Japanese Mixed Forest	Palaearctic	167	32,730	1,252,300	2.6
Llanos	Neotropical	3	11,410	438,000	2.6
Higharctic Tundra	Palaearctic	2	22,290	859,900	2.6
Bengalian Rainforest	Indomalayan	20	4,630	179,900	2.6
Chilean Sclerophyll	Neotropical	8	1,470	57,300	2.6
Thailandian Monsoon Forest	Indomalayan	62	24,530	959,700	2.6
Uruguayan Pampas	Neotropical	12	12,890	522,200	2.5
Guinean Rain Forest	Afrotropical	23	14,960	607,000	2.5
Guyanan	Neotropical	27	24,860	1,009,100	2.5
Iranian Desert	Palaearctic	9	9,810	403,500	2.4
Oriental Deciduous Forest	Palaearctic	200	64,100	2,751,400	2.3
Lake Ladoga	Palaearctic	1	410	17,600	2.3
Malagasy Rain Forest	Afrotropical	16	4,560	200,600	2.3
Eastern Forest	Nearctic	192	49,950	2,223,000	2.2
Mascarene Islands	Afrotropical	5	100	4,500	2.2
Chinese Subtropical Forest	Palaearctic	78	18,790	863,000	2.2
South African Highlands	Afrotropical	38	4,330	199,000	2.2
Altai Highlands	Palaearctic	7	22,820	1,048,300	2.2
Austroriparian	Nearctic	98	12,200	596,900	2.0
Campos Cerrados	Neotropical	25	36,280	1,778,600	2.0
Great Lakes	Nearctic	13	5,140	254,500	2.0
Somalian	Afrotropical	27	43,280	2,166,800	2.0
Arabian Desert	Palaearctic	31	59,300	2,996,100	2.0
Philippines	Indomalayan	27	5,730	292,200	2.0
Burma Monsoon Forest	Indomalayan	29	5,800	297,200	2.0
Subarctic Birchwoods	Palaearctic	13	2,530	132,500	1.9
Eastern Grasslands and Savannas	Australian	51	10,080	527,800	1.9
Malagasy Woodland/Savanna	Afrotropical	19	6,150	324,100	1.9
South Chinese Rainforest	Indomalayan	53	3,540	189,000	1.9
Coromandel	Indomalayan	4	1,570	88,400	1.8

Table 29.5 Distribution and coverage of protected areas by biogeographic province (continued)

PROVINCE	REALM	PROTECTED AREA NUMBER	PROTECTED AREA AREA (km²)	PROVINCE AREA (km²)	% OF TOTAL AREA
Brigalow	Australian	12	3,940	231,600	1.7
Californian	Nearctic	13	8,650	526,500	1.6
Sinaloan	Neotropical	5	2,970	192,100	1.5
Boreonemoral	Palaearctic	152	18,460	1,285,300	1.4
Western Sahel	Afrotropical	9	39,510	2,814,700	1.4
Chilean Araucaria Forest	Neotropical	2	454	32,900	1.4
Gran Chaco	Neotropical	14	12,830	988,500	1.3
Karroo	Afrotropical	18	4,660	377,700	1.2
West Eurasian Taiga	Palaearctic	117	65,400	5,342,600	1.2
Great Basin	Nearctic	21	7,230	660,400	1.1
Micronesian	Oceanian	4	23	2,200	1.1
Chihuahuan	Nearctic	19	5,820	577,200	1.0
Brazilian Rain Forest	Neotropical	59	15,030	1,533,800	1.0
Mongolian–Manchurian Steppe	Palaearctic	17	22,530	2,605,100	0.9
Hindu Kush Highlands	Palaearctic	5	1,830	217,100	0.8
East Siberian Taiga	Palaearctic	11	45,750	5,536,100	0.8
Guerreran	Neotropical	6	1,290	158,400	0.8
Northern Grasslands	Australian	6	6,700	967,000	0.7
Monte	Neotropical	26	8,490	1,234,800	0.7
Argentinian Pampas	Neotropical	17	3,470	512,200	0.7
Turanian	Palaearctic	15	13,940	2,116,800	0.7
Lake Ukerewe (Victoria)	Afrotropical	1	460	69,500	0.7
Pontian Steppe	Palaearctic	25	12,550	1,945,400	0.6
Malagasy Thorn Forest	Afrotropical	2	450	70,700	0.6
Madeiran	Neotropical	4	10,460	1,671,800	0.6
Lowarctic Tundra	Palaearctic	1	13,490	2,158,100	0.6
Deccan Thorn Forest	Indomalayan	9	1,940	338,400	0.6
Grasslands	Nearctic	69	7,690	2,442,300	0.3
Lake Malawi (Nyasa)	Afrotropical	1	87	28,900	0.3
West Anatolian	Palaearctic	4	107	37,600	0.3
Aral Sea	Palaearctic	1	183	67,500	0.3
Caatinga	Neotropical	6	2,430	899,700	0.3
Ceylonese Rainforest	Indomalayan	1	80	31,100	0.2
Tamaulipan	Nearctic	3	500	210,400	0.2
Atlas Steppe	Palaearctic	6	910	421,500	0.2
East Melanesian	Oceanian	2	50	27,000	0.2
Pacific Desert	Neotropical	2	490	290,400	0.2
Burman Rainforest	Indomalayan	2	200	257,600	0.1
Brazilian Planalto	Neotropical	3	140	219,200	0.1
Tibetan	Palaearctic	3	240	1,268,100	0.0
Maudlandia	Antarctic	5	340	10,465,200	0.0
Marielandia	Antarctic	1	3	2,194,000	0.0
Arctic Archipelago	Nearctic	0	0	690,000	0.0
Ascension and St Helena Islands	Afrotropical	0	0	200	0.0
Greenland Tundra	Nearctic	0	0	498,600	0.0
Laccadives Islands	Indomalayan	0	0	32	0.0
Lake Baikal	Palaearctic	0	0	32,300	0.0 *
Lake Rudolf	Afrotropical	0	0	7,300	0.0 *
Lake Tanganyika	Afrotropical	0	0	32,800	0.0 *
Maldives and Chagos Islands	Indomalayan	0	0	36	0.0
Revilla Gigedo Island	Neotropical	0	0	200	0.0
South Trinidade Island	Neotropical	0	0	11	0.0
Classification unknown		951	692,830	0	NA
TOTAL		**8,491**	**7,734,900**	**146,970,700**	**5.3**

Source: WCMC.
Note: * Shorelines are often protected but not included in this figures. ** Protected area includes significant marine areas, inflating the % figures. Province definitions after Udvardy, 1975.

Figure 29.5 Percentage of Udvardy province protected

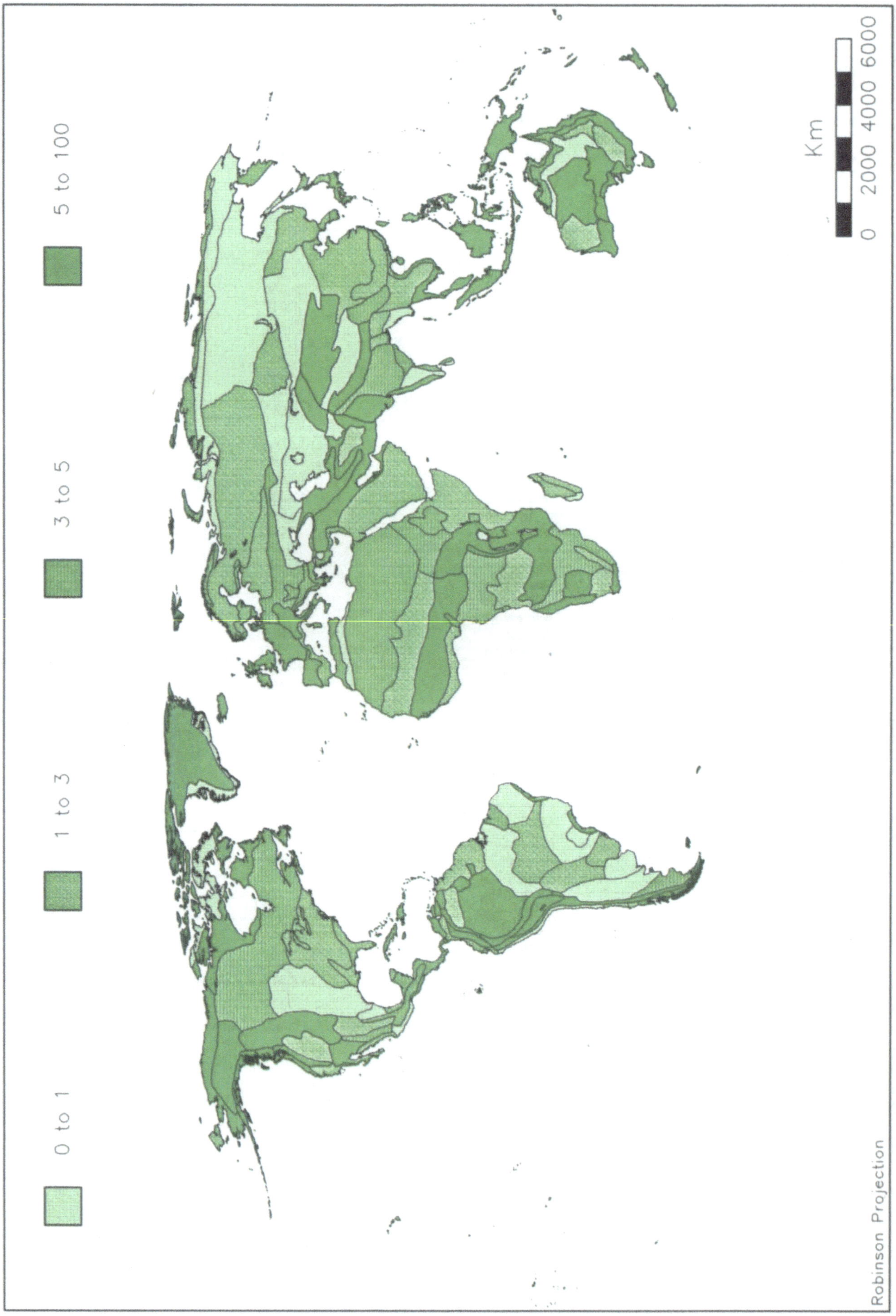

Figure 29.6 Marine and coastal sites

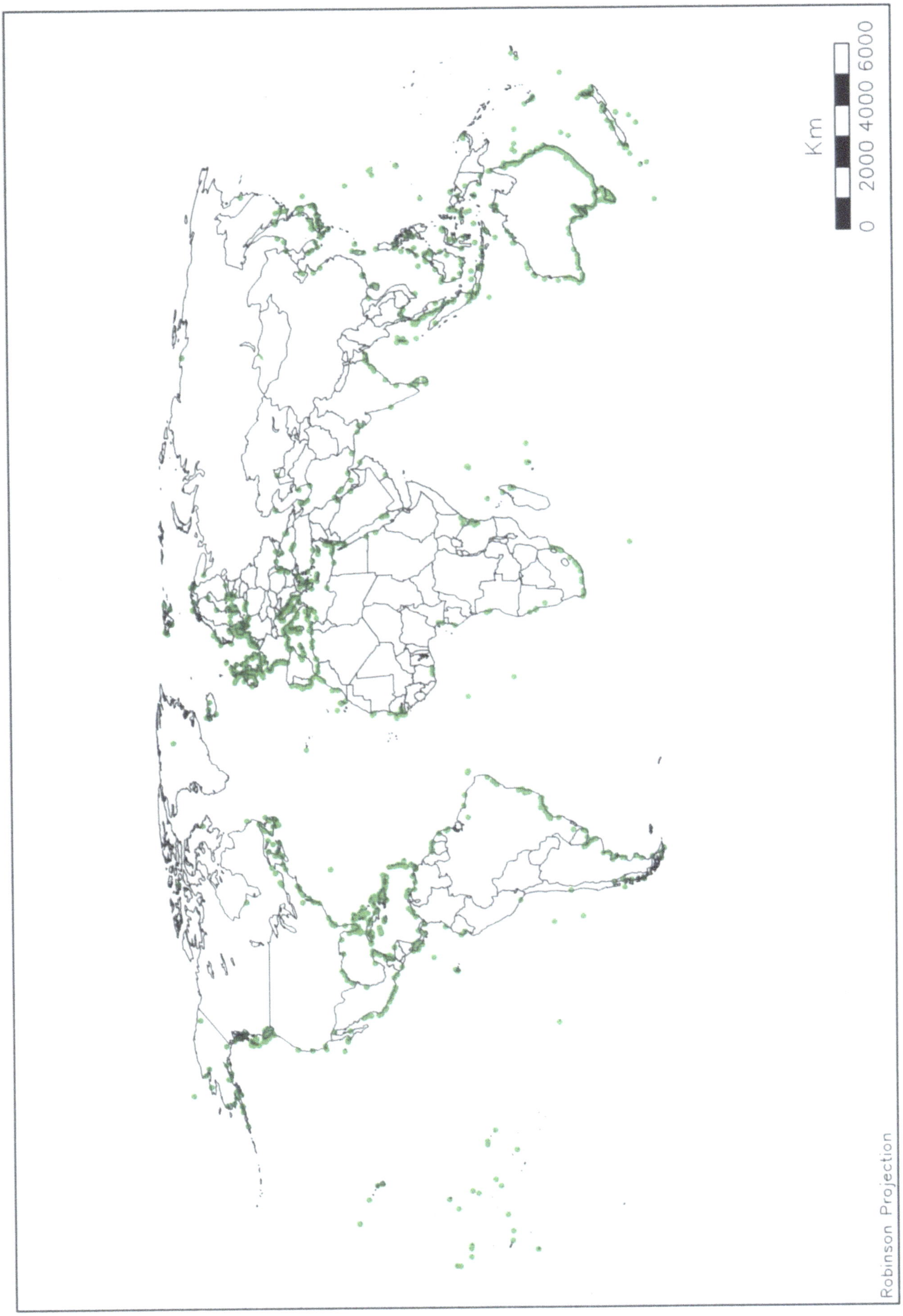

Figure 29.7 Mountainous sites

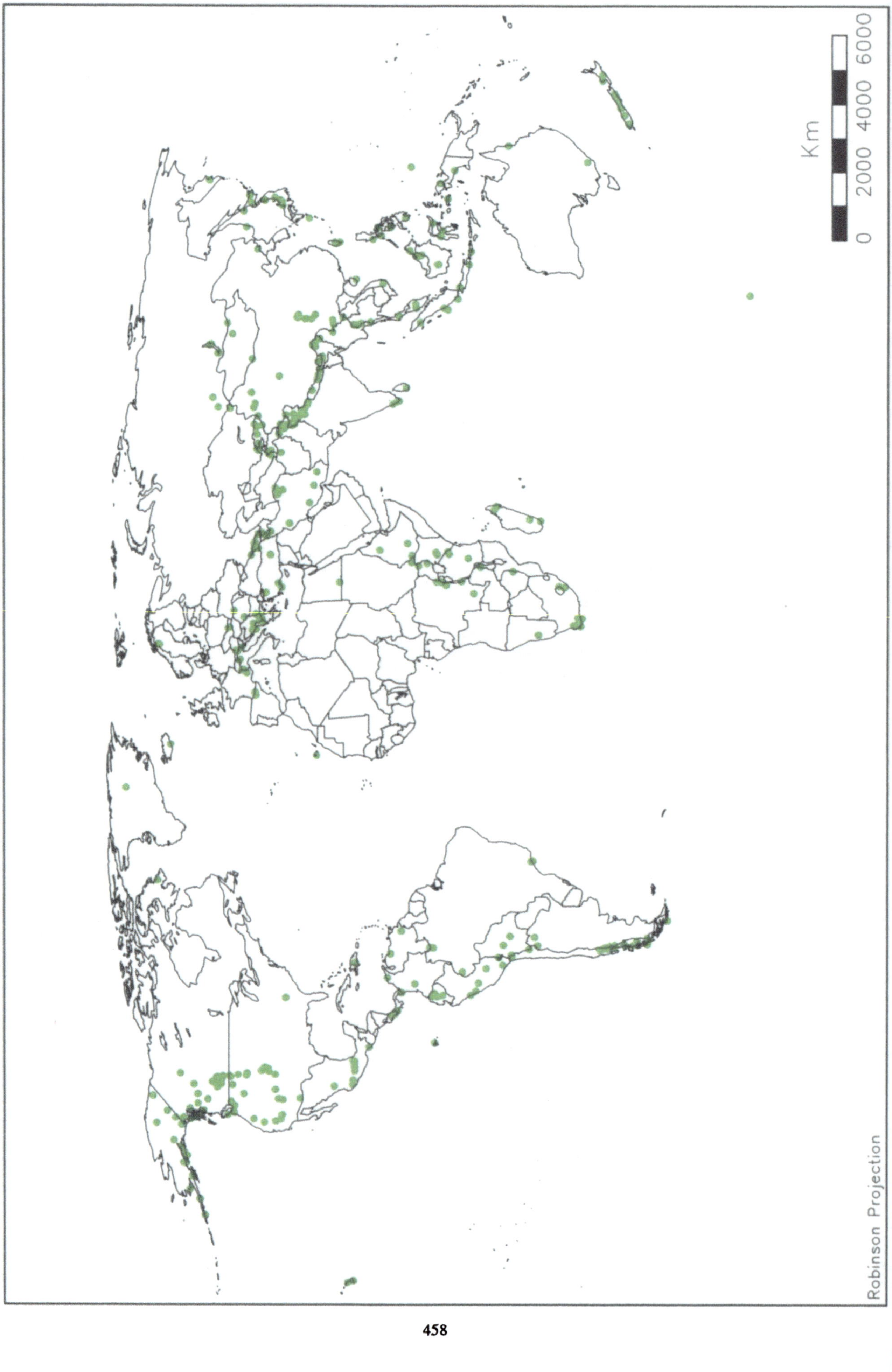

Tables 29.4 and 29.5 seem to suggest a correlation between population pressure on an ecosystem or its economic importance and the extent of its coverage. Thus, regions such as mixed mountain systems or mixed island systems, which are frequently not intensively developed, both have extensive coverage by protected areas; temperate grasslands, however, which are typically heavily used by man are poorly represented. These results illustrate the fact that socio-economic and political factors, not conservation priorities, are often the most important considerations in the establishing and siting of protected areas. Thus, in many countries protected areas are established in those regions which are the least economically valuable and with less regard to ensuring a balanced representation of the country's ecosystems. This raises a number of concerns about the ability of protected area networks alone to protect biological diversity adequately and comprehensively.

Table 29.6 shows the present state of each country's protected area network and gives figures for each IUCN Category. This table shows that, despite a global coverage of protected areas of 5.19%, there is considerable variation between continents. To simplify comparison, Categories I-V have been divided into two groups: *totally protected areas* with no extractive use (Categories I, II and III), and *partially protected areas* with local sustainable extractive use (Categories IV and V). Substantial variations can be seen between countries. Fig. 29.8 illustrates the percentage of area which is protected on a country by country basis. The protected area network for most countries covers less than 5% of the surface area. The map in Fig. 29.9 shows the period during which the greatest growth occurred for each county.

Management and funding

Any analysis of protected areas and the role they play in conservation of biodiversity is of limited value unless there is some assessment as to whether the protected areas are managed properly. Developing objective indicators to measure the degree of implementation is difficult, as proper management of a protected area is dependent on so many factors. This is an area in which IUCN and WCMC are working in an effort to develop reliable indicators by which management can be accurately assessed.

At one level, effective management requires there to be the necessary political will. One indicator of this is the promulgation of appropriate legislation. Another requirement is an administrative structure with sufficient authority and resources to manage the network adequately. Levels of funding can therefore illustrate the commitment or priority given to the establishment and management of protected areas and conservation of biodiversity in general.

WCMC is beginning to compile information on funding levels for protected areas, on a country by country basis. The information WCMC compiled to date is given in Table 29.7. It should be noted that in many instances independent verification of the levels of funding given in this table has not yet been made. Consequently, the figures provided are indicative only. Comparisons between countries at this stage could be misleading and inaccurate. These figures do show, however, that the amount of state funding devoted to protected area management in affluent countries is a different order of magnitude from that in poor countries. Thus, the annual budget for the USA of about US$2 billion dwarfs the typical budget of many less developed countries, which rarely exceeds US$500,000. Despite the limitations of the data the table indicates that many countries do devote considerable resources to protected areas management.

INTERNATIONAL PROTECTED AREA SYSTEMS

In the field of nature conservation there are two international conventions and one international programme that include provision for designation of internationally important sites in any region of the world. These are the World Heritage Convention, the Ramsar (Wetlands) Convention, and the UNESCO Man and the Biosphere (MAB) Programme. While there is a wide range of other international conventions and programmes, these cover only regions, or small groups of countries.

Both World Heritage sites and Ramsar sites must be nominated by a State that is party to the relevant convention. While there is an established review procedure for World Heritage sites (and nomination is not guarantee of listing), all nominated Ramsar sites are placed on the List of Wetlands of International Importance. Biosphere reserves are nominated by the national MAB committee of the country concerned, and are only designated following review and acceptance by the MAB Bureau.

Each Contracting Party to the Ramsar (Wetlands) Convention is obliged to nominate at least one wetland of international importance. However, a country can be party to the World Heritage Convention without having a natural site inscribed on the List, and may participate in the MAB programme without designating a biosphere reserve. See Chapter 31 for an additional view on these and other conventions.

Wetlands of International Importance (Ramsar Sites)

The Convention on Wetlands of International Importance especially as Waterfowl Habitat was signed in Ramsar (Iran) in 1971, and came into force in December 1975. This convention provides a framework for international cooperation for the conservation of wetland habitats. It places general obligations on contracting party states relating to the conservation of wetlands throughout their territories, with special obligations pertaining to those wetlands which have been designated to the 'List of Wetlands of International Importance'.

Each State Party is obliged to list at least one site. Wetlands are defined by the convention as: areas of marsh, fen, peatland or water, whether natural or artificial, permanent or temporary, with water that is static or flowing, fresh, brackish or salt, including areas of marine waters, the depth of which at low tide does not exceed six metres. Fig. 29.10 shows the parties to the Ramsar convention plus the locations of Ramsar sites around the world.

World Heritage Sites

The Convention Concerning the Protection of the World Cultural and Natural Heritage was adopted in Paris in 1972,

Table 29.6 National protection systems – distribution of land coverage by IUCN category

COUNTRY	CATEGORY I NO	CATEGORY I AREA (ha)	CATEGORY II NO	CATEGORY II AREA (ha)	CATEGORY III NO	CATEGORY III AREA (ha)	CATEGORY IV NO	CATEGORY IV AREA (ha)	CATEGORY V NO	CATEGORY V AREA (ha)	TOTALS NO	TOTALS AREA (ha)	PERCENT PROTECTED TOTALLY	PERCENT PROTECTED PARTIALLY	PERCENT PROTECTED ALL CAT.
WORLD	712	51,180,474	1,483	385,868,932	323	19,767,428	4,021	234,609,238	1,952	82,064,029	8,491	773,490,101	3.04	2.12	5.17
ASIA	169	9,750,081	239	26,570,361	8	27,200	1,651	75,878,442	113	5,975,185	2,180	118,201,269	1.31	2.95	4.26
Afghanistan	0	0	1	41,000	0	0	4	142,438	0	0	5	183,438	0.06	0.22	0.28
Bahrain	0	0	0	0	0	0	0	0	0	0	0	0	0.00	0.00	0.00
Bangladesh	0	0	0	0	0	0	6	83,332	2	13,458	8	96,790	0.00	0.67	0.67
Bhutan	0	0	1	65,800	0	0	4	840,338	0	0	5	906,138	1.41	18.03	19.44
British Indian Ocean Territory	0	0	0	0	0	0	0	0	0	0	0	0	0.00	0.00	0.00
Brunei	0	0	1	48,859	0	0	4	28,883	0	0	5	77,742	8.48	5.01	13.49
Cambodia	0	0	0	0	0	0	0	0	0	0	0	0	0.00	0.00	0.00
China	3	98,425	0	0	0	0	392	28,243,979	1	15,400	396	28,357,804	0.01	2.94	2.95
Cyprus	0	0	0	0	0	0	1	2,000	0	0	1	2,000	0.00	0.22	0.22
Hong Kong	0	0	0	0	0	0	0	0	12	37,821	12	37,821	0.00	35.61	35.61
India	2	196,043	59	4,228,908	0	0	300	9,327,006	1	18,600	362	13,770,557	1.40	2.95	4.35
Indonesia	89	6,896,141	24	6,905,069	0	0	68	5,159,643	13	270,026	194	19,230,879	7.19	2.83	10.02
Iran, Islamic Rep	18	1,904,503	7	1,075,300	2	6,150	4	1,144,918	29	3,398,105	60	7,528,976	1.81	2.76	4.57
Iraq	0	0	0	0	0	0	0	0	0	0	0	0	0.00	0.00	0.00
Israel	0	0	1	3,090	0	0	19	195,255	1	8,400	21	206,745	0.15	9.81	9.95
Japan	7	10,403	15	1,299,148	0	0	649	2,601,740	13	752,252	684	4,663,543	3.54	9.07	12.61
Jordan	1	1,200	0	0	0	0	6	79,200	1	20,000	8	100,400	0.01	1.03	1.05
Korea, Dem People's Rep	0	0	1	43,890	0	0	1	14,000	0	0	2	57,890	0.36	0.11	0.47
Korea, Rep	6	41,480	0	0	0	0	0	0	20	715,353	26	756,833	0.42	7.27	7.69
Kuwait	0	0	0	0	0	0	0	0	1	30,000	1	30,000	0.00	1.24	1.24
Laos	0	0	0	0	0	0	0	0	0	0	0	0	0.00	0.00	0.00
Lebanon	0	0	1	3,500	0	0	0	0	0	0	1	3,500	0.34	0.00	0.34
Malaysia	25	84,732	16	814,009	0	0	9	588,295	1	1,011	51	1,488,047	2.70	1.77	4.47
Maldives	0	0	0	0	0	0	0	0	0	0	0	0	0.00	0.00	0.00
Mongolia	12	224,280	3	5,943,560	0	0	0	0	0	0	15	6,167,840	3.94	0.00	3.94
Myanmar	0	0	1	160,580	0	0	0	0	1	12,691	2	173,271	0.24	0.02	0.26
Nepal	0	0	8	1,014,400	0	0	5	111,600	0	0	13	1,126,000	7.17	0.79	7.96
Oman	0	0	0	0	0	0	2	54,000	0	0	2	54,000	0.00	0.20	0.20
Pakistan	0	0	6	882,195	0	0	43	2,700,723	4	72,051	53	3,654,969	1.10	3.45	4.55
Philippines	0	0	8	204,620	6	21,050	9	324,643	4	22,553	27	572,866	0.75	1.16	1.91
Qatar	0	0	0	0	0	0	0	0	0	0	0	0	0.00	0.00	0.00
Saudi Arabia	2	260,000	0	0	0	0	7	20,487,560	1	450,000	10	21,197,560	0.11	8.72	8.83
Singapore	0	0	0	0	0	0	1	2,715	0	0	1	2,715	0.00	4.41	4.41
Sri Lanka	3	31,574	11	460,180	0	0	29	291,954	0	0	43	783,708	7.50	4.45	11.94
Syrian Arab Rep	0	0	0	0	0	0	0	0	0	0	0	0	0.00	0.00	0.00
Taiwan	0	0	2	197,490	0	0	1	47,000	2	44,087	5	288,577	5.49	2.53	8.02
Thailand	0	0	55	2,841,941	0	0	32	2,628,959	3	43,086	90	5,513,986	5.53	5.20	10.73
Turkey	1	1,300	11	194,435	0	0	3	23,150	3	50,291	18	269,176	0.25	0.09	0.35
United Arab Emirates	0	0	0	0	0	0	0	0	0	0	0	0	0.00	0.00	0.00
Viet Nam	0	0	7	142,387	0	0	52	755,111	0	0	59	897,498	0.43	2.29	2.72
Yemen Arab Rep	0	0	0	0	0	0	0	0	0	0	0	0	0.00	0.00	0.00
U.S.S.R.	153	21,981,912	22	1,926,419	0	0	34	410,731	4	55,264	213	24,374,326	1.07	0.02	1.09
EUROPE	115	2,987,617	137	4,979,706	43	345,705	712	6,338,191	1,143	27,775,724	2,150	42,426,943	0.99	7.01	8.00
Albania	0	0	6	23,000	0	0	7	21,500	0	0	13	44,500	0.80	0.75	1.55
Andorra	0	0	0	0	0	0	0	0	0	0	0	0	0.00	0.00	0.00
Austria	0	0	0	0	0	0	46	251,714	132	1,839,082	178	2,090,796	0.00	24.93	24.93
Belgium	0	0	0	0	0	0	1	3,975	1	67,854	2	71,829	0.00	2.35	2.35
Bulgaria	24	49,942	2	113,329	2	4,424	21	66,950	1	26,772	50	261,417	1.51	0.85	2.36
Czechoslovakia	5	14,069	7	274,544	1	1,517	10	24,814	42	1,743,836	65	2,058,780	2.27	13.83	16.10
Denmark	5	10,237	0	0	2	6,290	18	82,262	41	310,979	66	409,768	0.38	9.13	9.51
Faeroe Islands	0	0	0	0	0	0	0	0	0	0	0	0	0.00	0.00	0.00
Finland	16	151,170	17	354,080	0	0	0	0	2	302,000	35	807,250	1.50	0.90	2.40

Table 29.6 National protection systems – distribution of land coverage by IUCN category (continued)

COUNTRY	CATEGORY I NO	CATEGORY I AREA (ha)	CATEGORY II NO	CATEGORY II AREA (ha)	CATEGORY III NO	CATEGORY III AREA (ha)	CATEGORY IV NO	CATEGORY IV AREA (ha)	CATEGORY V NO	CATEGORY V AREA (ha)	TOTALS NO	TOTALS AREA (ha)	PERCENT PROTECTED TOTALLY	PERCENT PROTECTED PARTIALLY	PERCENT PROTECTED ALL CAT.
EUROPE (continued)															
France	4	17,054	5	261,333	0	0	39	127,616	33	4,951,558	81	5,357,561	0.51	9.34	9.85
Germany	0	0	1	13,100	0	0	68	203,014	371	5,643,339	440	5,859,453	0.04	16.38	16.42
Greece	0	0	8	60,392	2	18,000	6	11,483	5	14,678	21	104,553	0.59	0.20	0.79
Hungary	0	0	5	159,138	0	0	6	13,815	43	404,013	54	576,966	1.71	4.49	6.20
Iceland	1	270	3	180,100	5	38,604	5	51,950	8	645,000	22	915,924	2.13	6.78	8.91
Ireland	0	0	0	0	0	0	3	4,315	3	22,495	6	26,810	0.00	0.39	0.39
Italy	0	0	6	293,544	1	1,500	71	315,783	66	1,397,790	144	2,008,617	0.98	5.69	6.67
Liechtenstein	0	0	0	0	0	0	0	0	1	6,000	1	6,000	0.00	37.50	37.50
Luxembourg	0	0	0	0	0	0	0	0	0	0	0	0	0.00	0.00	0.00
Malta	0	0	0	0	0	0	0	0	0	0	0	0	0.00	0.00	0.00
Monaco	0	0	0	0	0	0	0	0	0	0	0	0	0.00	0.00	0.00
Netherlands	3	4,211	1	5,400	22	220,595	41	122,383	0	0	67	352,589	5.59	2.97	8.57
Norway	39	2,643,801	18	1,994,630	0	0	5	18,065	26	326,478	88	4,982,974	12.00	0.89	12.89
Poland	1	1,592	15	148,326	0	0	19	64,887	45	2,026,896	80	2,241,701	0.48	6.69	7.17
Portugal	0	0	1	21,100	1	2,730	12	115,936	11	416,369	25	556,135	0.26	5.81	6.07
Romania	9	51,575	1	54,400	0	0	17	618,985	13	363,678	40	1,088,638	0.45	4.14	4.58
San Marino	0	0	0	0	0	0	0	0	0	0	0	0	0.00	0.00	0.00
Spain	0	0	9	122,763	0	0	54	1,574,900	100	1,807,327	163	3,504,990	0.24	6.70	6.94
Sweden	0	0	15	495,028	0	0	157	2,171,110	23	252,414	195	2,918,552	1.10	5.39	6.49
Switzerland	1	16,887	0	0	0	0	52	268,070	59	467,935	112	752,892	0.41	17.83	18.24
United Kingdom	0	0	0	0	0	0	44	136,877	96	4,502,829	140	4,639,706	0.00	18.96	18.96
Vatican City	0	0	0	0	0	0	0	0	0	0	0	0	0.00	0.00	0.00
Yugoslavia	7	26,809	17	405,499	7	52,045	10	67,787	21	236,402	62	788,542	1.89	1.19	3.08
NORTH AND CENTRAL AMERICA	**103**	**2,733,548**	**271**	**149,596,250**	**240**	**18,035,242**	**610**	**70,829,289**	**479**	**20,944,208**	**1,703**	**262,138,537**	**7.03**	**3.79**	**10.82**
Anguilla	0	0	0	0	0	0	0	0	0	0	0	0	0.00	0.00	0.00
Antigua and Barbuda	0	0	1	4,128	0	0	0	0	0	0	1	4,128	9.34	0.00	9.34
Aruba	0	0	0	0	0	0	0	0	0	0	0	0	0.00	0.00	0.00
Bahamas	0	0	4	121,576	0	0	1	1,813	0	0	5	123,389	8.77	0.13	8.90
Barbados	0	0	0	0	0	0	0	0	0	0	0	0	0.00	0.00	0.00
Belize	0	0	1	4,144	0	0	6	113,846	0	0	7	117,990	0.18	4.96	5.14
Bermuda	0	0	0	0	0	0	1	12,000	0	0	1	12,000	**	**	**
Canada	57	501,147	80	26,309,148	1	3,090	162	18,506,762	126	4,132,136	426	49,452,283	2.70	2.28	4.98
Cayman Islands	1	1,731	0	0	0	0	1	3,310	0	0	2	5,041	6.68 *	12.78 *	19.46 *
Costa Rica	6	27,099	12	456,252	0	0	10	134,027	3	5,670	31	623,048	9.50	2.74	12.24
Cuba	8	38,328	13	390,987	1	5,300	6	151,075	4	100,310	32	686,000	3.79	2.20	5.99
Dominica	0	0	1	6,872	0	0	0	0	0	0	1	6,872	9.15	0.00	9.15
Dominican Rep	0	0	10	488,069	0	0	7	476,090	0	0	17	964,159	10.08	9.83	19.90
El Salvador	0	0	5	18,369	1	2,170	3	5,613	0	0	9	26,152	0.96	0.26	1.22
Greenland	1	1,050,000	1	97,200,000	0	0	0	0	0	0	2	98,250,000	45.16	0.00	45.16
Grenada	0	0	0	0	0	0	0	0	0	0	0	0	0.00	0.00	0.00
Guadeloupe	0	0	1	17,300	0	0	1	3,700	0	0	2	21,000	9.72	2.08	11.80
Guatemala	0	0	5	764,400	5	10,975	6	58,591	1	1,000	17	834,966	7.12	0.55	7.67
Haiti	0	0	2	7,500	0	0	0	0	1	2,200	3	9,700	0.27	0.08	0.35
Honduras	0	0	14	588,616	0	0	21	129,253	0	0	35	717,869	5.25	1.15	6.40
Jamaica	0	0	2	37,953	0	0	0	0	0	0	2	37,953	3.32	0.00	3.32
Martinique	0	0	0	0	0	0	0	0	1	70,150	1	70,150	0.00	65.01 *	65.01 *
Mexico	13	603,303	33	1,618,263	1	1,600	9	3,902,669	7	3,947,280	63	10,073,115	1.13	3.98	5.11
Montserrat	0	0	0	0	0	0	0	0	0	0	0	0	0.00	0.00	0.00
Netherlands Antilles	0	0	2	7,760	0	0	0	0	0	0	2	7,760	9.70	0.00	9.70
Nicaragua	1	295,000	3	27,300	0	0	7	40,438	0	0	11	362,738	2.18	0.27	2.45
Panama	2	6,833	11	1,188,049	0	0	2	2,258	1	129,000	16	1,326,140	15.22 *	1.67 *	16.89 *
Puerto Rico	0	0	0	0	0	0	14	28,548	0	0	14	28,548	0.00	3.19	3.19
St Kitts and Nevis	0	0	0	0	0	0	0	0	0	0	0	0	0.00	0.00	0.00
St Lucia	0	0	0	0	0	0	1	1,494	0	0	1	1,494	0.00	2.41	2.41

Table 29.6 National protection systems – distribution of land coverage by IUCN category (continued)

COUNTRY	CATEGORY I NO	CATEGORY I AREA (ha)	CATEGORY II NO	CATEGORY II AREA (ha)	CATEGORY III NO	CATEGORY III AREA (ha)	CATEGORY IV NO	CATEGORY IV AREA (ha)	CATEGORY V NO	CATEGORY V AREA (ha)	TOTALS NO	TOTALS AREA (ha)	PERCENT PROTECTED TOTALLY	PERCENT PROTECTED PARTIALLY	PERCENT PROTECTED ALL CAT.
NORTH AND CENTRAL AMERICA (continued)															
St Vincent and the Grenadines	0	0	0	0	0	0	2	8,284	0	0	2	8,284	0.00	21.30 *	21.30 *
Trinidad and Tobago	0	0	0	0	0	0	7	15,528	0	0	7	15,528	0.00	3.03	3.03
Turks and Caicos Islands	0	0	10	86,475	1	1,259	1	4,497	2	5,301	14	97,532	**	**	**
United States	14	210,107	59	20,239,010	230	18,010,848	339	47,228,820	333	12,551,161	975	98,239,946	4.10	6.38	10.49
Virgin Islands (British)	0	0	0	0	0	0	3	673	0	0	3	673	0.00	4.40	4.40
Virgin Islands (US)	0	0	1	14,079	0	0	0	0	0	0	1	14,079	40.81	0.00	40.81
SOUTH AMERICA	**58**	**8,218,628**	**220**	**50,768,119**	**23**	**1,331,650**	**180**	**27,259,814**	**118**	**19,748,725**	**599**	**107,326,936**	**3.35**	**2.61**	**5.96**
Argentina	15	1,856,873	24	2,069,135	2	26,000	64	4,013,263	10	1,430,137	115	9,395,408	1.42	1.96	3.38
Bolivia	2	136,500	7	4,096,120	0	0	16	5,598,269	2	29,876	27	9,860,765	3.85	5.12	8.98
Brazil	31	3,550,130	63	10,965,664	0	0	42	3,511,375	36	3,539,535	172	21,566,704	1.71	0.83	2.53
Chile	0	0	30	8,358,367	2	13,606	34	5,350,152	0	0	66	13,722,125	11.14	7.12	18.26
Colombia	2	1,947,000	33	7,043,790	0	0	6	57,395	0	0	41	9,048,185	7.89	0.05	7.94
Ecuador	3	642,565	6	1,952,842	0	0	2	7,994,613	7	158,367	18	10,748,387	5.62	0.35	5.98
French Guiana	0	0	0	0	0	0	0	0	0	0	0	0	0.00	0.00	0.00
Guyana	0	0	1	11,655	0	0	0	0	0	0	1	11,655	0.05	0.00	0.05
Paraguay	0	0	8	1,173,038	0	0	2	2,000	4	29,193	14	1,204,231	2.88	0.08	2.96
Peru	0	0	7	2,381,126	7	156,466	2	75,347	4	74,907	20	2,687,846	1.97	0.12	2.09
Suriname	0	0	2	86,570	0	0	11	649,400	0	0	13	735,970	0.53	3.96	4.49
Uruguay	0	0	0	0	2	15,250	1	8,000	5	8,836	8	32,086	0.08	0.09	0.17
Venezuela	5	85,560	39	12,629,812	10	1,120,328	0	0	50	14,477,874	104	28,313,574	15.17	15.87	31.04
OCEANIA	**63**	**2519627**	**372**	**65335288**	**7**	**23731**	**428**	**11543766**	**66**	**4937723**	**936**	**84360135**	**7.98**	**1.94**	**9.91**
American Samoa	0	0	1	3,725	0	0	0	0	0	0	1	3,725	18.91 *	0.00	18.91 *
Australia	18	2,075,758	356	63,152,119	1	1,540	309	11,142,616	64	4,937,415	748	81,309,448	3.99	2.09	6.09
Cook Islands	0	0	0	0	0	0	1	160	0	0	1	160	0.00	0.69	0.69
Fiji	2	5,342	0	0	0	0	0	0	0	0	2	5,342	0.29	0.00	0.29
French Polynesia	0	0	0	0	0	0	6	12,747	0	0	6	12,747	0.00	3.24	3.24
Guam	0	0	0	0	0	0	0	0	0	0	0	0	0.00	0.00	0.00
Kiribati	2	20,130	0	0	0	0	1	6,500	0	0	3	26,630	29.43 *	9.50 *	38.93 *
Marshall Is.	0	0	0	0	0	0	0	0	0	0	0	0	0.00	0.00	0.00
Micronesia	0	0	0	0	0	0	0	0	0	0	0	0	0.00	0.00	0.00
Nauru	0	0	0	0	0	0	0	0	0	0	0	0	0.00	0.00	0.00
New Caledonia	2	22,570	1	1,133	0	0	9	37,665	2	308	14	61,676	1.24	1.99	3.23
New Zealand	36	394,698	11	2,170,988	5	20,991	100	322,385	0	0	152	2,909,062	9.76	1.22	10.97
Niue	0	0	0	0	0	0	0	0	0	0	0	0	0.00	0.00	0.00
North Marianas Islands	3	1,129	0	0	0	0	0	0	0	0	3	1,129	2.40	0.00	2.40
Palau	0	0	0	0	1	1,200	0	0	0	0	1	1,200	3.29	0.00	3.29
Papua New Guinea	0	0	3	7,323	0	0	2	21,693	0	0	5	29,016	0.02	0.05	0.06
Pitcairn Island	0	0	0	0	0	0	0	0	0	0	0	0	0.00	0.00	0.00
Solomon Islands	0	0	0	0	0	0	0	0	0	0	0	0	0.00	0.00	0.00
Tokelau	0	0	0	0	0	0	0	0	0	0	0	0	0.00	0.00	0.00
Tonga	0	0	0	0	0	0	0	0	0	0	0	0	0.00	0.00	0.00
Tuvalu	0	0	0	0	0	0	0	0	0	0	0	0	0.00	0.00	0.00
Vanuatu	0	0	0	0	0	0	0	0	0	0	0	0	0.00	0.00	0.00
Wallis and Futuna Islands	0	0	0	0	0	0	0	0	0	0	0	0	0.00	0.00	0.00
Western Samoa	0	0	0	0	0	0	0	0	0	0	0	0	0.00	0.00	0.00
ANTARCTICA	**11**	**215,649**	**0**	**0**	**0**	**0**	**1**	**36,700**	**0**	**0**	**12**	**252,349**	**0.02**	**0.00**	**0.02**
Antarctica	11	215,649	0	0	0	0	0	0	0	0	11	215,649	0.02	0.00	0.02
Falkland Islands (Malvinas)	0	0	0	0	0	0	0	0	0	0	0	0	0.00	0.00	0.00
French Southern Territories	0	0	0	0	0	0	1	36,700	0	0	1	36,700	0.00	5.07	5.07
AFRICA	**40**	**2,773,412**	**222**	**86,692,789**	**2**	**3,900**	**405**	**42,312,305**	**29**	**2,627,200**	**698**	**134,409,606**	**2.99**	**1.50**	**4.49**
Algeria	4	26,200	8	12,561,150	0	0	5	31,507	1	76,438	18	12,695,295	5.28	0.05	5.33

Table 29.6 National protection systems – distribution of land coverage by IUCN category (continued)

COUNTRY	CATEGORY I NO	CATEGORY I AREA (ha)	CATEGORY II NO	CATEGORY II AREA (ha)	CATEGORY III NO	CATEGORY III AREA (ha)	CATEGORY IV NO	CATEGORY IV AREA (ha)	CATEGORY V NO	CATEGORY V AREA (ha)	TOTALS NO	TOTALS AREA (ha)	PERCENT PROTECTED TOTALLY	PERCENT PROTECTED PARTIALLY	PERCENT PROTECTED ALL CAT.
AFRICA (continued)															
Benin	0	0	2	843,500	0	0	0	0	0	0	2	843,500	7.49	0.00	7.49
Botswana	0	0	4	8,787,000	0	0	5	1,238,000	0	0	9	10,025,000	15.28	2.15	17.43
Burkina Faso	0	0	3	489,300	0	0	8	2,153,400	0	0	11	2,642,700	1.78	7.86	9.64
Burundi	0	0	0	0	0	0	0	0	3	86,735	3	86,735	0.00	3.12	3.12
Cameroon	1	1,400	6	1,030,400	0	0	6	1,002,625	0	0	13	2,034,425	2.17	2.11	4.28
Cape Verde	0	0	0	0	0	0	0	0	0	0	0	0	0.00	0.00	0.00
Central African Rep	1	86,000	4	3,102,000	0	0	7	2,668,000	0	0	12	5,856,000	5.10	4.27	9.37
Chad	0	0	2	414,000	0	0	0	0	0	0	2	414,000	0.32	0.00	0.32
Comoros	0	0	0	0	0	0	0	0	0	0	0	0	0.00	0.00	0.00
Congo	0	0	1	126,600	0	0	9	1,206,500	0	0	10	1,333,100	0.37	3.53	3.90
Cote d'Ivoire	2	128,000	8	1,762,500	0	0	2	102,350	0	0	12	1,992,850	5.86	0.32	6.18
Djibouti	0	0	1	10,000	0	0	0	0	0	0	1	10,000	0.43	0.00	0.43
Egypt	3	37,000	1	19,700	0	0	9	743,700	0	0	13	800,400	0.06	0.74	0.80
Equatorial Guinea	0	0	0	0	0	0	0	0	0	0	0	0	0.00	0.00	0.00
Ethiopia	0	0	11	2,534,100	0	0	0	0	0	0	11	2,534,100	2.48	0.00	2.48
Gabon	1	15,000	0	0	0	0	5	1,030,000	0	0	6	1,045,000	0.06	3.85	3.90
Gambia	0	0	3	18,440	0	0	0	0	0	0	3	18,440	1.72	0.00	1.72
Ghana	1	32,400	5	1,029,795	0	0	2	12,442	0	0	8	1,074,637	4.46	0.05	4.51
Guinea	2	129,170	1	38,200	0	0	0	0	0	0	3	167,370	0.68	0.00	0.68
Guinea-Bissau	0	0	0	0	0	0	0	0	0	0	0	0	0.00	0.00	0.00
Kenya	0	0	31	3,411,283	0	0	5	58,943	0	0	36	3,470,226	5.85	0.10	5.96
Lesotho	0	0	0	0	0	0	1	6,805	0	0	1	6,805	0.00	0.22	0.22
Liberia	0	0	1	130,747	0	0	0	0	0	0	1	130,747	1.17	0.00	1.17
Libya	0	0	1	35,000	0	0	2	120,000	0	0	3	155,000	0.02	0.07	0.09
Madagascar	10	568,802	6	171,307	0	0	21	375,190	0	0	37	1,115,299	1.25	0.63	1.88
Malawi	0	0	5	696,200	0	0	4	361,400	0	0	9	1,057,600	7.40	3.84	11.24
Mali	0	0	1	350,000	0	0	10	3,661,989	0	0	11	4,011,989	0.28	2.95	3.24
Mauritania	1	310,000	2	1,186,000	0	0	1	250,000	0	0	4	1,746,000	1.45	0.24	1.69
Mauritius	2	3,770	0	0	0	0	1	253	0	0	3	4,023	2.02	0.14	2.16
Mayotte	0	0	0	0	0	0	0	0	0	0	0	0	0.00	0.00	0.00
Morocco	5	55,320	0	0	0	0	3	237,000	2	69,800	10	362,120	0.12	0.67	0.79
Mozambique	0	0	0	0	0	0	1	2,000	0	0	1	2,000	0.00	0.00	0.00
Namibia	0	0	6	8,977,749	0	0	3	609,953	2	782,900	11	10,370,602	10.89	1.69	12.58
Niger	1	1,280,500	1	220,000	0	0	4	8,196,240	0	0	6	9,696,740	1.26	6.91	8.17
Nigeria	1	7,800	6	2,114,396	0	0	14	750,469	0	0	21	2,872,665	2.30	0.81	3.11
Reunion	0	0	0	0	0	0	2	5,942	0	0	2	5,942	0.00	2.37	2.37
Rwanda	0	0	2	327,000	0	0	0	0	0	0	2	327,000	12.42	0.00	12.42
Saint Helena	2	17,600	0	0	0	0	0	0	0	0	2	17,600	7.26	0.00	7.26
Sao Tome and Principe	0	0	0	0	0	0	0	0	0	0	0	0	0.00	0.00	0.00
Senegal	0	0	6	1,012,450	0	0	4	1,168,259	0	0	10	2,180,709	5.15	5.94	11.09
Seychelles	1	35,000	3	3,568	0	0	0	0	0	0	4	38,568	95.47	0.00	95.47
Sierra Leone	0	0	0	0	0	0	2	82,013	0	0	2	82,013	0.00	1.13	1.13
Somalia	0	0	0	0	0	0	1	180,000	0	0	1	180,000	0.00	0.29	0.29
South Africa	1	39,000	20	3,251,216	0	0	203	3,917,252	5	182,049	229	7,389,517	2.78	3.46	6.24
Sudan	0	0	8	8,499,000	0	0	5	742,500	1	116,000	14	9,357,500	3.39	0.34	3.73
Swaziland	0	0	0	0	0	0	4	45,920	0	0	4	45,920	0.00	2.64	2.64
Tanzania	0	0	11	3,909,975	0	0	17	9,090,000	0	0	28	12,999,975	4.16	9.67	13.83
Togo	0	0	3	357,290	0	0	8	289,616	0	0	11	646,906	6.29	5.10	11.39
Tunisia	1	450	6	44,417	0	0	0	0	0	0	7	44,867	0.27	0.00	0.27
Uganda	0	0	6	833,606	0	0	23	1,029,557	3	7,635	32	1,870,798	3.52	4.38	7.91
Western Sahara	0	0	0	0	0	0	0	0	0	0	0	0	0.00	0.00	0.00
Zaire	0	0	7	8,544,000	0	0	1	33,000	0	0	8	8,577,000	3.64	0.01	3.66
Zambia	0	0	19	6,359,000	1	1,900	0	0	0	0	20	6,360,900	8.45	0.00	8.45
Zimbabwe	0	0	10	2,701,900	1	2,000	4	18,280	10	345,643	25	3,067,823	6.93	0.93	7.86

Source: WCMC
Note: * percent protected is inflated as the protected area total includes marine areas while land area is used for percentage calculation. ** percent protected omitted as protected areas are mainly marine.

Figure 29.8 Percentage of country protected

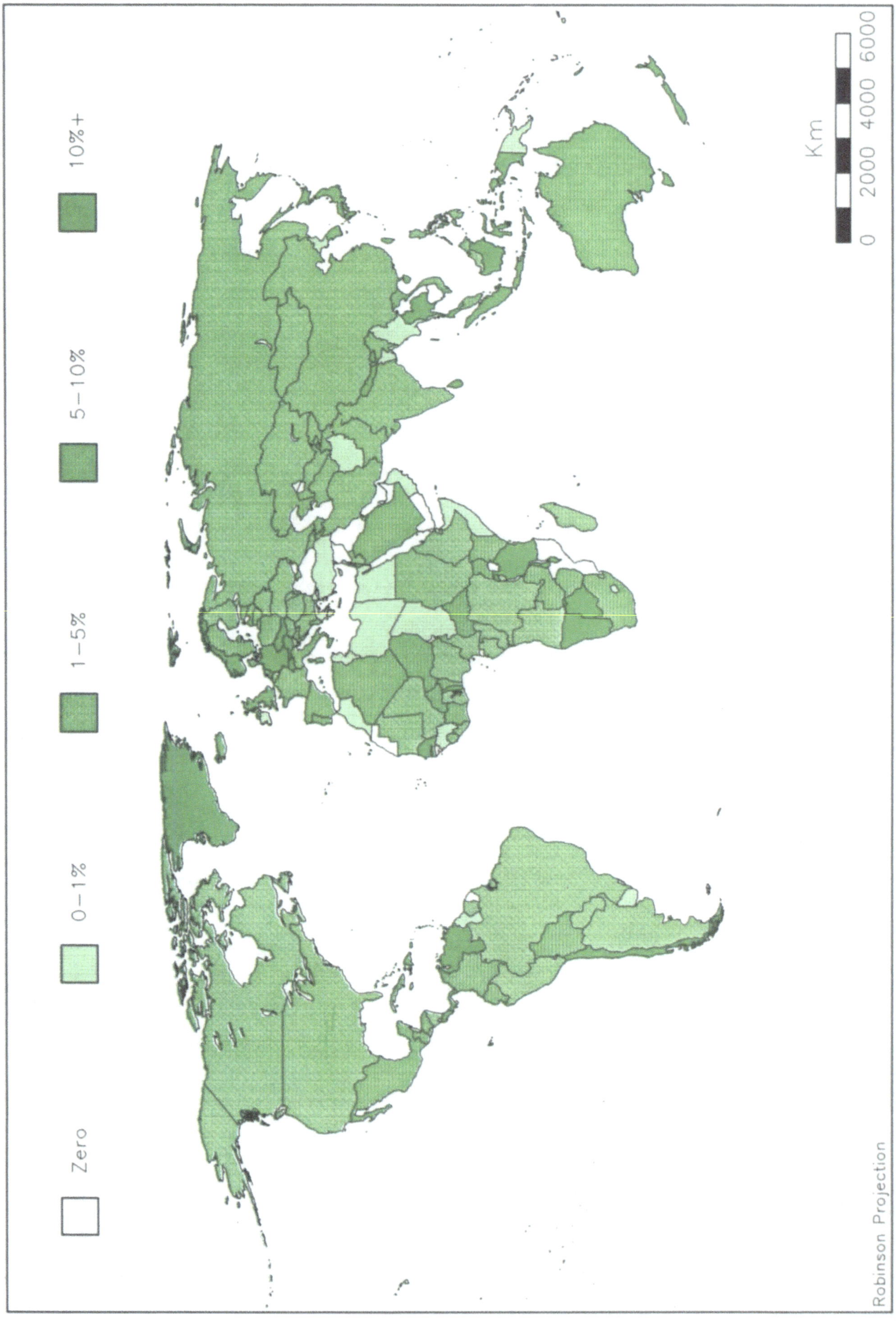

Figure 29.9 Period of greatest growth

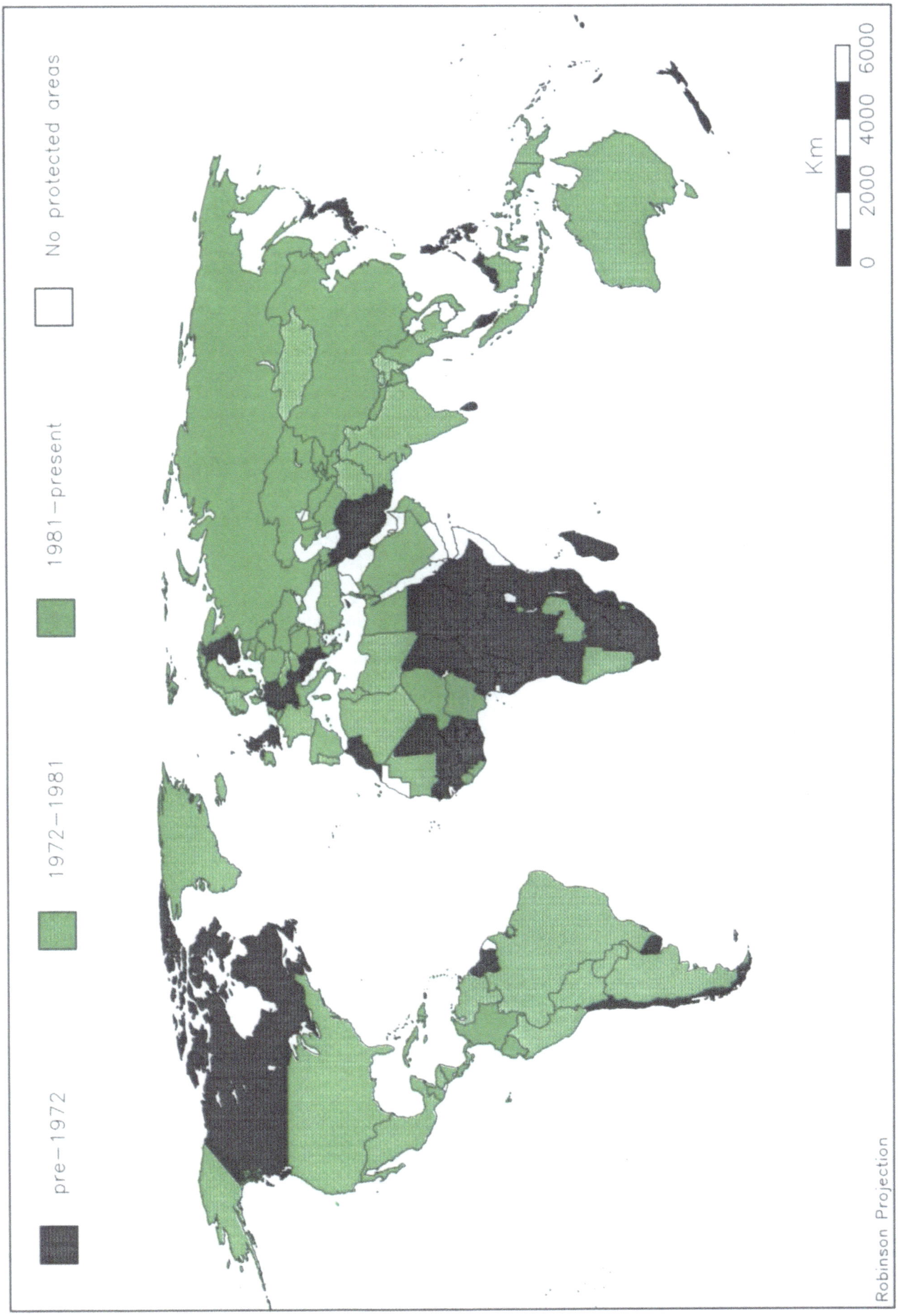

Table 29.7 Protected areas and levels of state financing

COUNTRY	AMOUNT/ CURRENCY		MILLION US$	YEAR	NOTES	SOURCE
ASIA						
Afghanistan			0.05	PRO 1991	Proposed estimate for conservation projects by the Directorate of Wildlife and National Parks	B1
Bhutan	2,128,000	Nu	0.137	1988/ 1989	Budget for Northern and Southern Wildlife circles was US$27,300 and $1,708,000 respectively	A1
Brunei	59.5 million	B$	(1.65 = 1US$)	1991	Comprises Government allocation to Forestry Department (4M), Brunei Museum (7.5M), Agriculture Department (27M), Fisheries Department (4) and District Offices (17M); proportion subsequently allocated to protected areas is unknown	A3
Cyprus	637,000[1] 138,000[2]	C£	1.28[1] 0.28[2] 0.035[3]	1991	[1]Ministry of Agriculture budget for the establishment, development and conservation of protected areas [2]Ministry of Agriculture budget for park recreational facilities [3]Running expenditure for main protected area, Lara reserve	B3
India - Project Tiger	16.2 million	Rs	(29.1 Rs = 1 US$)	1982	Allocated for 11 tiger reserves covering total area of 15,800km²	A6
Indonesia			39.8	1982-1992	Estimated government allocation for 1982-92; excludes international and private sector funding	
			6.0	1990-1991	Investment from private sector for tourism in protected areas	
Israel			14.5 28	1985 1990	Forestry department overall yearly budget, including element for forest conservation	B5
Jordan	32,326 (in addition to US$)	JD	0.12 (in addition to JD currency)	1990	[1]Funding for the National Environmental Strategy during the period up to 1990 totalled US$123,798 and JD 32,636, funded by IUCN and USAID	B6
Kuwait	2 million[1]	KD	(KD 0.294 = 1 US$)	1990[1]	[1]The government of Kuwait allocated funds for planning, research and assessment of the country's desert renewable resources	B7
	0.7-0.9 million[2]			1986[2]	[2]The original proposals for annual running budgets for the first national park with a fully functional administration and necessary maintenance	
	1 million[3]				[3]Amount allocated for all park maintenance and wardening in Kuwait	
Laos			0.01	?1991		A9
Lebanon			0.5[1]	1990	[1]FAO assistance to Department of Forests and Natural Resources institutional strengthening	B8
			0.2[2]		[2]FAO assistance for forestry activities including conservation	
Malaysia			4.98	1991		A5
Myanmar			1.4	?1991		A5
Nepal	124.3 million	NRs	2.9	1990/ 1991	Represents expenditure for total protected areas network of 10,910km², but 84% is allocated to Royal Nepal Army protection units	A7
Pakistan - Federal units	93.4 million	Rs	4.3	1990/ 1991	52.5% represents recurrent costs and 47.5% capital development costs (breakdown for provinces is available) Of Rs 2 billion allocated to forestry sub-sector in 7th Five-Year Plan (1988-1993), Rs 332 million (16.6%) is earmarked for wildlife conservation (Sheik and Jan, n.d.)	A2
Philippines	(816,200 US$)					
Saudi Arabia			c.US$ 9.06	1990-1991	NCWCD budget total per year including all aspects for protected areas, administration and management. The budget for IUCN support for two years was SAR1,444,304 (US$380,000) (IUCN 1987a)	B10
Taiwan			36.769	1989	Total budget allocation for 4 national parks covering 2,416km²	A8
Thailand			15.097	?1991		A5

Table 29.7 Protected areas and levels of state financing (continued)

COUNTRY	AMOUNT/ CURRENCY		MILLION US$	YEAR	NOTES	SOURCE
ASIA (continued)						
Turkey	300 million	TL	(6222 = 1 US$)	1985	Annual budget expended on park management by the General Directorate of Forestry	B12
Viet Nam			0.222	?1991		A5
Yemen	2 million[1]	YR	(YR 12.05= US$1) 1[2]	1988[1] 1990-92[2]	[1]Forestry (former YAR) budget towards conservation programmes including a national tree planting campaign [2]Budget for training of forest technicians	B13
Former USSR						
Former USSR	54.9 million	Rbl	32.7*	1990	Total investment in protected areas	D5
EUROPE						
Bulgaria	500 million	Leva	28.016	1989	Budget for nature protection	D2
Czechoslovakia- Czech Republic	50 million	Kcs	1.751	1991	Budget for protected areas (national parks 30.872 million, CHKOs 19.102 million)	D3
Czechoslovakia- Slovak Republic	74 million	Kcs	2.592	1991	Budget for protected areas (national parks 67.342 million, CHKOs 6.659 million)	D3
Denmark	6 billion	DKr	980.697	1988	Overall environmental expenditure. Proposals for environmental expenditure for the six year period 1989-94: Kroner 33 billion	D4
Estonia	634,000	Rbl	0.377*	1990	Total budget for protected areas	D5
France- national parks	89.939 million	FF	16.716	1989	Annual budget, broken down into capital (FF16.738 million) and current or ongoing (FF73.201 million)	D4
France- nature reserves	14.5 million	FF	2.695	1989	Annual budget, broken down into capital (FF4.5 billion) and current or ongoing (FF10 million)	D4
France	6.386 billion	FF	1186.880	1987	Total expenditure on protecting the natural heritage (including expenditure on regional parks, parks and gardens, other green spaces, centres for nature education, the improvement of the surroundings of monuments, the acquisition of green forestry spaces, forestry development and developing fishing and hunting)	D4
Germany- North-Rhine/ Westphalia	106.415 million	DM	67.634	1987	Total expenditure on promotion of nature protection and landscape preservation, including that by non-state bodies (including land acquisition, preservation and development and compensation payment). Public expenditure on nature protection areas was DM68.386 million	
Greece	200 million	Drs	1.110	1991	Maximum estimate of 'budget for protected areas' from respective bodies	D6
Iceland	134,000	£	0.242	1982	Running costs for the three national parks	D9
Italy - state	3,022 billion	Lire	2.573	1988	State environmental expenditure, divided between ongoing expenditure (Lire 677 billion) and capital expenditure (Lire 2,345 billion)	D4
Italy - regions and provinces	415 billion	Lire	348	1986	Total expenditure on nature conservation in all regions and provinces (from a total of Lire 3,026 billion on environmental expenditure)	D4
Latvia	1.188 million	Rbl	0.708*	1990	Budget for protected areas	D5
Lithuania	248,000	Rbl	0.148*	1990	Budget for protected areas	D5
Netherlands	41 million	DFl	23.110	1990	Proposed environmental spending based on the 1990 Nature Policy Plan, to be spent largely on environmental policy on specific areas, nature development, management and maintenance	D11
Norway	41 million	NKr	6.610	1991	Total budget on protected areas, divided into compensation paid for the establishment of new areas (30 million NKr) and management or ongoing expenditure (11 million NKr)	D12
Poland	27.472 billion	Zl	2.486	1990	Expenditure for national parks	D13

Table 29.7 Protected areas and levels of state financing (continued)

COUNTRY	AMOUNT/ CURRENCY		MILLION US$	YEAR	NOTES	SOURCE
EUROPE (continued)						
Portugal	3.195 billion	Esc	22.813	1991	Budget for the Portuguese Park service, of which about 64% is for capital investment and 36% for ordinary (ongoing) expenditure. The amount spent specifically on protected areas is probably about 68% of the total budget	D14
Spain - provincial	21.734 billion	Ptas	215.852	1987	Total nature conservation expenditure by the Autonomous regions, covering: protection of flora and fauna; prevention of forest fires; creation, conservation and management of forests; parks and nature reserves; game hunting and inland fishing)	D4
Spain - national	2.579 billion	Ptas	25.611	1987	National nature conservation expenditure (defined as above)	D4
Sweden	223 million	SKr	38.541	1991	Government money allotted to land purchase/compensation (SKR 140 million) and to protected areas management (SKr 83 million)	D15
UK - Northern Ireland	4 million	£	7.227	1991	Total budget of the Countryside and Wildlife Branch of the Department of the Environment for Northern Ireland (the bulk of this money is used for site protection)	D16
UK - Nature Conservancy Council	46,032,000	£	83.165	1990/ 1991	Total expenditure	D17
UK - Countryside Commission	40 million	£	72.267	1992/ 1993	Planned expenditure for all 11 national parks in England and Wales (including funding from Department of Environment, local authorities and national park authorities)	D18
UK - Gibraltar	300,000	£	0.542	1990	Expenditure on nature and landscape conservation (including public gardens) prior to the designation of the only protected area	D19
UK - Isle of Man	10,000	£	0.018	1991	Annual government expenditure on protected areas	D20
NORTH AND CENTRAL AMERICA						
Canada			282.99	1991	Canadian Parks Service and Canadian Wildlife Service	E1
Dominica	926,300	EC$	0.35	1991/ 1992	Proposed government capital expenditure on parks and protected areas	F3
Dominica	1,266,730	EC$	0.48	1991/ 1992	Recurrent expenditure for Forestry Division	F3
Guadeloupe (France)	16 million	Fr	2.97	1991	Budget for the 'Parc National de la Guadeloupe' (administrative body for Guadeloupe's PAs)	F2
Jamaica			4.89[1], 2.50[2]	1989-1990	Forest Department expenditure in the financial year, 1989/1990: [1]for recurrent (forest administration and soil conservation) [2]for capital (forestry, watershed management and conservation)	F1, Anon. 1990
Mexico			2.55	1991	All federal land management agency operations, including multiple use lands	E1
St Kitts-Nevis			0.1		Budget for Conservation Commission	
USA			1,962.70	1991	Estimates of federal government expenditures in protected areas (includes USFWS, WWF and NAWMP)	E1
SOUTH AMERICA						
Bolivia	2,964,915	Bol	0.784	1988	Budget for natural resource management as a whole not just protected area management. The CDF receives very little external funding	G1
Brazil	121,139,100	Cr$	1.211	1990	Budget for federal protected areas	G2
Ecuador	50.8 million	Sucr	0.250	1984	Budget for the national parks system	G3
Guyana	1,452,225	G$	6.201	1990-2000	Projected budget for conservation activities including the development of a protected area system, as part of the National Forestry Action Plan	G4
Peru			1.162	1990	Budget for the entire forest and wildlife department, the DGFF, not just protected areas. The distribution of funds within the DGFF is not known	G5

Table 29.7 Protected areas and levels of state financing (continued)

COUNTRY	AMOUNT/ CURRENCY		MILLION US$	YEAR	NOTES	SOURCE
SOUTH AMERICA (continued)						
Suriname	12,000	Sf	0.007	1967	Budget for the nature protection department of the Forest Service	G6
OCEANIA						
Australia	15,795,193	A$	12.22	1988-1989	Figure is for revenue, not expenditure	H1
Western Samoa	104,000	Tala	0.043257	1990	Proposed budget	I1
AFRICA						
Algeria			0.25	1990-1992	Budget allocated to a single management plan preparation by the BNEF as a trail project for future reorganisation of other existing protected areas in the country	B2
Angola			<0.02	1991	Annual personnel costs US$<20,000	K1
Benin			0.034	1991		K1
Burkina Faso			0.5	1991	Minimum possible figure Annual personnel costs US$500,000 Recurrent budget (excl. personnel) US$388	K1
Cameroon			0.409	1991	Annual personnel costs (conservators and game guards only) US$270,000 Recurrent budget (excl. personnel) US$33,000 Capital budget US$105,800	K1
Central African Rep	130 million	F CFA	0.482		The National Centre for the Protection and Management of Fauna (Centre National pour la Protection et l'Aménagement de la Faune) is a self-financing organisation, the budget being funded by hunting and ivory taxes	K4
Central African Rep	10 million	F CFA	0.037		The budget for the Ministry of Water, Forests, Hunting, Fishing and Tourism	
Chad	31 million	F CFA	0.1	1991	Budget does not include salaries	K1, K5
Congo			0.036	1991	Minimum possible figure Recurrent budget (excl. personnel) US$35,900 (1990 figure)	K1
Côte d'Ivoire			1.321	1991	Annual personnel costs US$953,571 Recurrent budget (excl. personnel) US$117,857 Capital budget US$250,000	K1
Egypt	10 million[1]	E£	(3.3 = 1 US$)	1989-1992	[1]Main government protected area initiative 'Ras Mohammed National Park project' from initial phases in 1989 to project completion. From 1989-1991 ECU 750,000 have been budgeted, funded with technical support from the EC	B4
Equatorial Guinea					The Directorate of Forestry combines four sections, one of which is the Hunting and Protected Areas Service (Servicio de Caza y Areas Protegidas), although this has no personnel, vehicles or equipment	K2
Ethiopia			0.25	1991	Annual personnel costs US$97,948 Recurrent budget (excl. personnel) US$153,623 Capital budget US$251,171	K1
Gabon			0.424	1989	Under 1987-90 development plans approximately 11% of the forestry budget (CFA 312 million) was budgeted for forest conservation (3)	K1
Gambia	370,820	D	0.041	1990/1991	Department of Wildlife Conservation (estimated)	J2
	1,708,200	D	0.19	1990/1991	Forestry Department (estimated)	
Ghana			1.05	1991	Annual personnel costs US$636,255 Recurrent budget (excl. personnel) US$55,327 Capital budget US$360,000 (1989)	K1

Table 29.7 Protected areas and levels of state financing (continued)

COUNTRY	AMOUNT/ CURRENCY		MILLION US$	YEAR	NOTES	SOURCE
AFRICA (continued)						
Kenya			18.2	1989	Annual personnel costs US$10,000,000 (1989) Recurrent budget (excl. personnel) US$84,200,000 (1989) Capital budget US$0)	K1
Malawi			0.456	1991		
Mauritius					In 1983/4 the Forest Service accounted for 0.3% of the total national budget and 13.5% of the Ministry of Agriculture, Fisheries and Natural Resources' budget	J3
Morocco	5 million	DH	0.64	1991	In 1988, 8% of the Water and Forests budget went towards parks and reserves administration and management In 1991, Water and Forests maintenance costs including salaries, equipment and information dissemination	B9
Mozambique			0.448	1986	Total annual allocation of resources available to government agencies for conservation	J4
Namibia			0.350	1991	Figure refers to Etosha NP Annual personnel costs US$280,000 Recurrent budget (excl. personnel) US$70,000	K1
Niger			1.423	1991	Annual personnel costs US$179,116 Recurrent budget (excl. personnel) US$71,326 Capital budget US$14,285	K1
Nigeria			1.66	1991		K1
Rwanda			4.73	1990	All figures for first 9 months of 1990 Annual personnel costs US$1,830,000 Recurrent budget (excl. personnel) US$2,900,000 Capital budget US$4,730,000	K1
Senegal			0.624	1991	figure refers to Niokolo Koba NP Annual personnel costs US$534,857 Recurrent budget (excl. personnel) US$89,129	K1
Sierra Leone			0.005	1991	Annual personnel costs US$4,591 Recurrent budget (excl. personnel US$388)	K1
South Africa-Natal province			0.012	1986	Total annual allocation of resources available to government agencies for conservation	J4
South Africa			3.009	1991	Figure refers to Kruger and Addo NPs; elephant budget only	K1
St Helena (UK)	7,000	£	0.012	1983/ 1984	Funding (through Project-UK) from WWF/UK, ODA, FFPS and the British Council for conservation purposes	L1
Sudan			1.0	1986	National Forestry Corporation budget is >68 million Sudanese pounds p.a. (US$12.5 million) (M.E.A.A.Ali, pers. comm., 1991)	K1
Tanzania			3.478	1991	Total earnings of the Wildlife Division for 1990/1 were TShs 591,676,500 (approx. US$2,572,506) Central government returns are approx. US$1.1 million Tanzania national Parks (TANAPA) earned >US$3.5 million in 1990/1 Ngorongoro Conservation Area Authority earned > US$1.84 million in 1990/1 (WD, 1991)	J5
Togo			0.580	1990	Annual personnel costs US$405,828 (1990) Recurrent budget (excl. personnel) US$10,521 (1990) Capital budget US$20,536 (1990)	K1
Tunisia	500,000	TD	(0.9 TD = 1 US$)	1991	[1]Maximum maintenance budget available for the key NP, Ichkeul by the Government [2]German DM 20 million made available for Ichkeul park management via the KfW Bank, Frankfurt (1991)	B11, B14
Uganda			1.846	1990		K1
Zaire			1.002	1990	Annual personnel costs US$1,000,000 (1990) Recurrent budget (excl. personnel) US$2,000 (1990)	
Zimbabwe			0.009	1986	Total annual allocation of resources available to government agencies for conservation	J4

Notes: * Converted figures area based on a commercial exchange rate of US$1 = 1.6787 roubles. The free market exchange rate of US $1 = 80 roubles gives a better approximation of purchasing power.

Figure 29.10 Wetlands of International Importance (Ramsar sites)

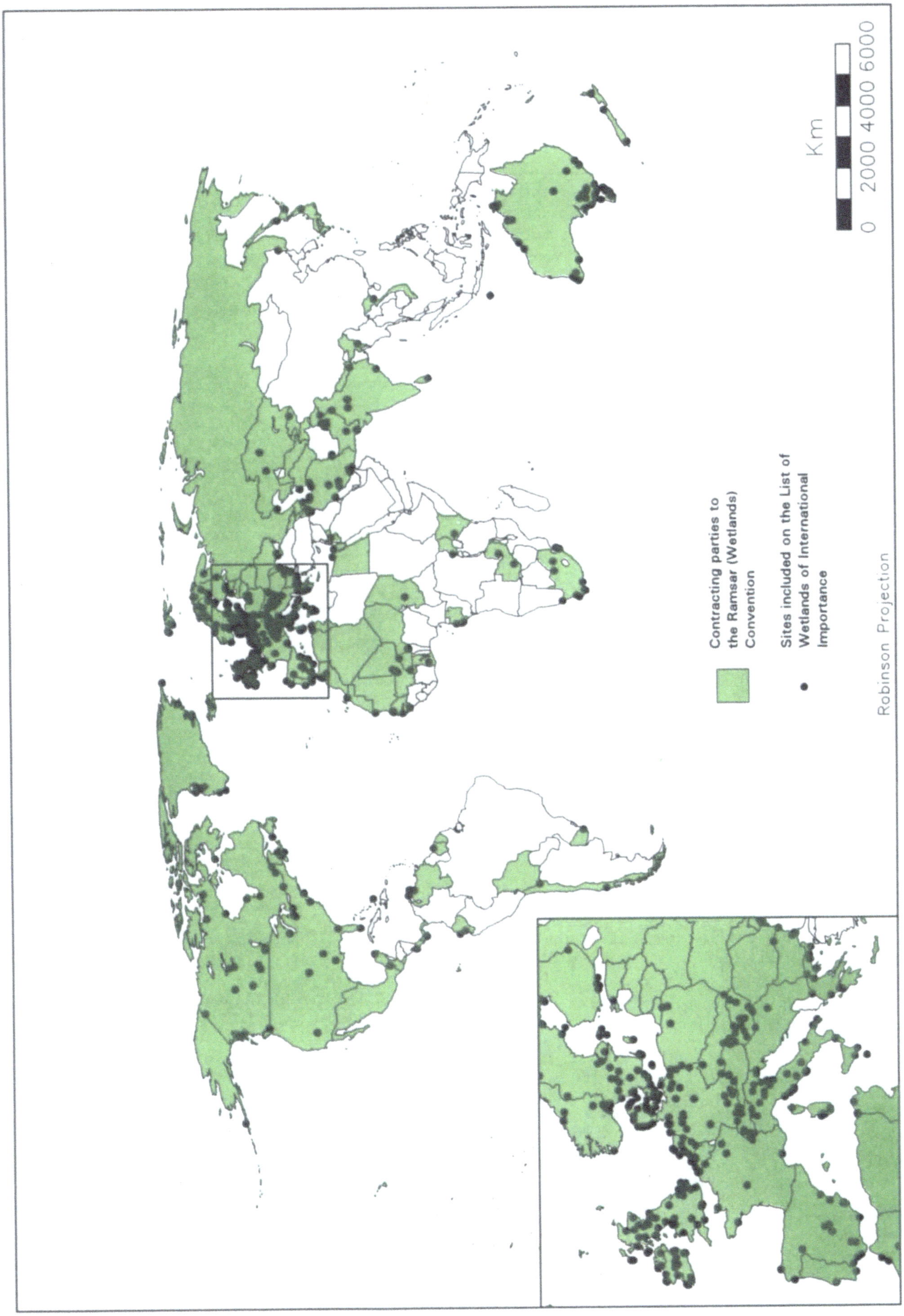

and came into force in December 1975. The convention provides for the designation of areas of 'outstanding universal value' as World Heritage Sites, with the principal aim of fostering international cooperation in safeguarding these important areas. Sites, which must be nominated by the signatory nation responsible, are evaluated for their world heritage quality before being inscribed by the international World Heritage Committee. Only natural sites, and those with mixed natural and cultural aspects are considered in this publication.

Article 2 of the World Heritage Convention considers as natural heritage: natural features consisting of physical and biological formations or groups of such formations which are of outstanding universal value from the aesthetic or scientific point of view; geological or physiographical formations and precisely delineated areas which constitute the habitat of threatened species of animals and plants of outstanding universal value from the point of view of science or conservation; and natural sites or precisely delineated areas of outstanding universal value from the point of view of science, conservation or natural beauty. Criteria for inclusion in the list are published by UNESCO. The map in Fig. 29.11 shows the location of each World Heritage Site plus the countries that are party to the convention.

Biosphere Reserves

The establishment of biosphere reserves is not covered by a specific convention, but is part of an international scientific programme, the UNESCO Man and the Biosphere (MAB) Programme. The objectives of the network of biosphere reserves, and the characteristics which biosphere reserves might display, are identified in various UNESCO-MAB documents, including the Action Plan for Biosphere Reserves.

Biosphere Reserves differ from the preceding types of site in that they are not exclusively designated to protect unique areas or important wetlands, but for a range of objectives which include research, monitoring, training and demonstration, as well as conservation. In most cases the human component is vital to the functioning of the biosphere reserve, which does not necessarily hold for either World Heritage or Ramsar sites. See Fig. 29.12 for the location of Biosphere Reserves. For this map only the green tint indicates those countries which have one or more biosphere reserves.

Table 29.8 provides summary statistics on these three international protection systems.

References

Miller, K.R. 1980. *Planning National Parks for Ecodevelopment*, Center for Strategic Wildland Management Studies, Ann Arbor.

IUCN, 1984. Categories and criteria for protected areas. In: McNeely, J.A. and Miller, K.R. (Eds), *National Parks, Conservation, and Development. The role of protected areas in sustaining society.* Smithsonian Institution Press, Washington. Pp.47-53.

IUCN/UNEP 1986. *Managing Protected Areas in the Tropics*. IUCN, Gland.

IUCN 1990. *1990 United Nations List of National Parks and Protected Areas*. IUCN, Gland. 284pp.

Udvardy, M.D.F. 1975. *A Classification of the Biogeographical Provinces of the World*. IUCN Occasional Paper No.18. Morges.

Sources for Table 29.7

A1 Blower, J.H. 1989. *Nature Conservation in Northern and Central Bhutan*. FAO, Rome. 48pp.

A2 Malik, M.M. 1990. Management status of protected areas in Pakistan. Paper presented at Regional Expert Consultation on Management of Protected Areas in the Asia-Pacific Region. FAO Regional Office for Asia and the Pacific, Bangkok, 10-14 December 1990. 40pp.

A3 Othman, M. and Ramos, V.J.A. 1991. National report on national parks and protected areas in Brunei Darussalam. Presented to 36th Working Session of IUCN CNPPA, Bangkok, 2-4 December. 66pp.

A4 Sheik, M.I. and Jan, A. undated. Role of forests and forestry in national conservation strategy of Pakistan. Draft for comment. National Conservation Strategy Secretariat, Islamabad. 86pp.

A5 World Bank 1991. Conserving biological diversity: a strategy for protected areas - Asia region. Preliminary draft. 57pp.

A6 Government of India

A7 B. Upreti, pers. comm.

A8 Taiwan Parks, 1984

A9 Information provided at the 36th IUCN CNPPA Working Session, Bangkok, 2-4 December

B1 MacPherson, N. 1991. Opportunities for improved environmental management in Afghanistan. Report of an IUCN mission under contract to the Office for the Coordination of United Nations Humanitarian and Economic Assistance Programmes relating to Afghanistan. 66pp.

B2 Anon. 1990. Algeria, watershed management and forestry project, conservation of nature. World development indicators on the environment. The World Bank, Washington DC. 10pp.

B3 Antoniou, *in litt.*, 1991

B4 EEAA 1991. Protected areas in the Arab Republic of Egypt. Paper presented at the Third Man and Biosphere Meeting on Mediterranean Biosphere Reserves and the First IUCN-CNPPA meeting for the Middle East and North Africa, 14-19 October 1991, Tunis, Tunisia. 18pp.

B5 Anon. 1990. National report on forestry in Israel. Report for the Tenth World Forestry Congress by the Land Development Authority, Kiryat-Hayim. 19pp.

B6 McEachern, J. 1990. Report on project activities, National Environment Strategy - Jordan September 1989 through January 1990. National Environment Strategy, c/o Department of Environment. IUCN Project Office, 31 January 1990.

B7 Alsdirawi, F. 1991. Protected areas in the state of Kuwait. Caracas Action Plan paper presented at the Third Man and Biosphere Meeting on Biosphere Reserves in the Mediterranean and the First IUCN-CNPPA Workshop on Protected Areas in the North Africa-Middle East Region, 14-19 October 1991, Tunis. 10pp.

B8 Child, *in litt.*, 1990

B9 Eaux et Forêts 1991. Rapport sur les aires protégées au Maroc. Paper presented at the Third Man and Biosphere Meeting on Biosphere Reserves in the Mediterranean and the First IUCN-CNPPA Workshop on Protected Areas in the North Africa-Middle East Region, 14-19 October 1991, Tunis. 18pp.

B10 Abuzinada, A.H. and Child, G. 1991 Developing a system of protected areas in Saudi Arabia. National Commission for Wildlife Conservation and Development, Riyadh. Paper presented at the Third Man and Biosphere Meeting on Mediterranean Biosphere Reserves and the First IUCN-CNPPA meeting for the Middle East and North Africa, 14-19 October 1991, Tunis. 16pp.

B11 Bel Hadj Kacem, S. 1985. La conservation de la faune et de la flore sauvages en Tunisie. Séminaire sur la conservation du patrimoine forestier national, 30-31 octobre 1985 à l'INPPSA de Sidi-Thabet.

B12 General Directorate of Forestry 1987. *Forestry in Turkey*. General Directorate of Forestry, Ministry of Agriculture, Forest and Rural Affairs, Ankara.

Figure 29.11 World Heritage Sites

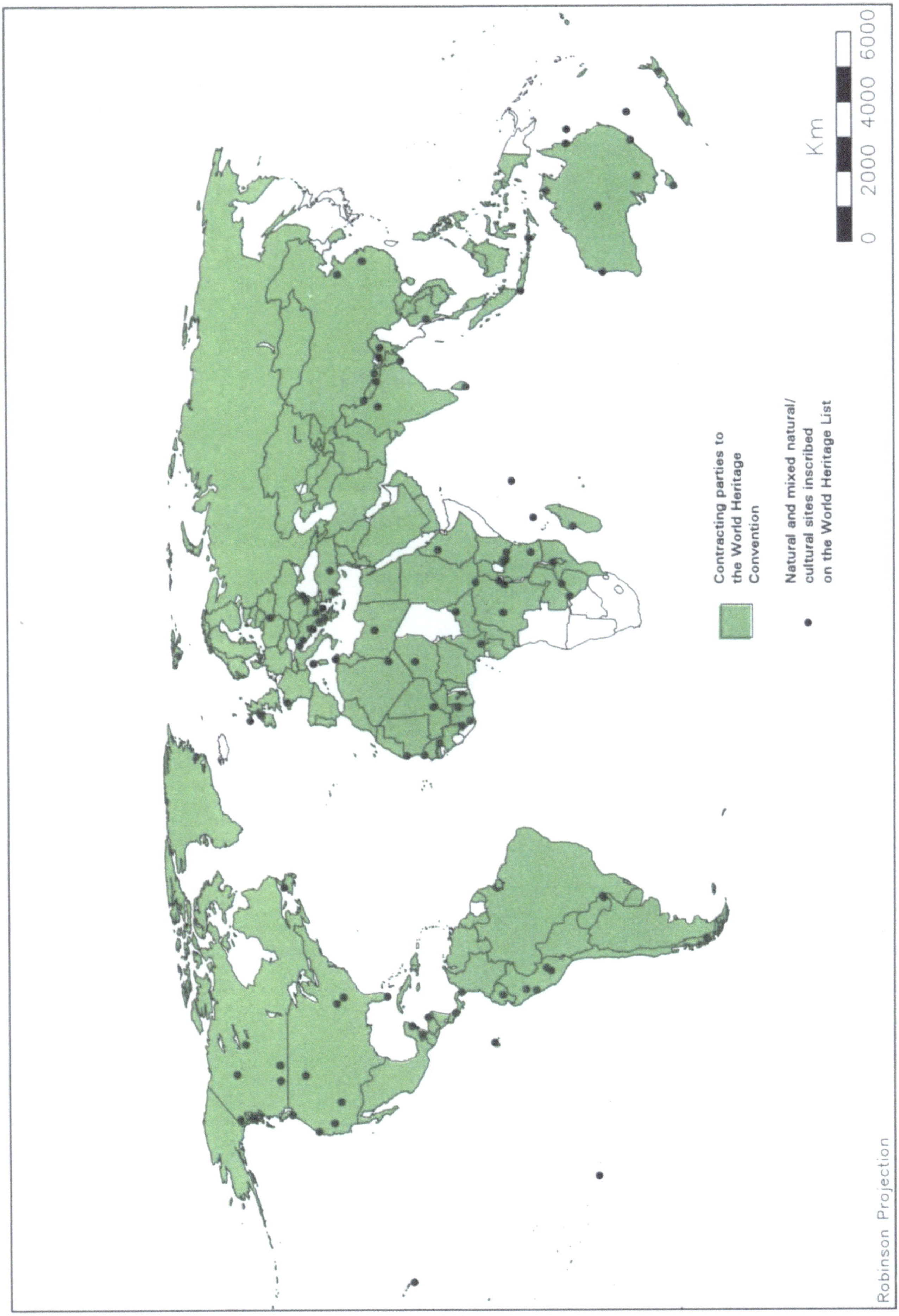

Figure 29.12 Biosphere Reserves

Table 29.8 International protection systems

	WORLD HERITAGE SITES			BIOSPHERE RESERVES		RAMSAR WETLANDS		
	DATE	NO		NO	AREA (ha)	DATE	NO	AREA (ha)
WORLD		**95**	**(8)**	**300**	**161,944,969**		**538**	**32,336,169**
ASIA		**13**	**(2)**	**38**	**12,885,459**		**40**	**1,354,493**
Afghanistan	March 1979	0		–	–	---	–	–
Bahrain	May 1991	–		–	–	---	–	–
Bangladesh	August 1983	0		–	–	---	–	–
Bhutan	---	–		–	–	---	–	–
British Indian Ocean Territory	(see UK)	0		–	–	(see UK)	0	0
Brunei	---	–		–	–	---	–	–
Cambodia	November 1991	–		–	–	---	–	–
China	December 1985	1	(1)	8	1,966,722	---	–	–
Cyprus	August 1975	0		–	–	---	–	–
Hong Kong	(see UK)	0		–	–	(see UK)	0	0
India	November 1977	5		–	–	October 1981	6	192,973
Indonesia	July 1989	2		6	1,482,400	---	–	–
Iran, Islamic Rep	February 1975	0		9	2,609,731	June 1975	18	1,087,550
Iraq	March 1974	0		–	–	---	–	–
Israel	---	–		–	–	---	–	–
Japan	---	–		4	116,000	June 1980	3	9,892
Jordan	May 1975	0		–	–	January 1977	1	7,372
Korea, Dem People's Rep	---	–		1	132,000	---	–	–
Korea, Rep	September 1988	0		1	37,430	---	–	–
Kuwait	---	–		–	–	---	–	–
Laos	March 1987	0		–	–	---	–	–
Lebanon	February 1983	0		–	–	---	–	–
Malaysia	December 1988	0		–	–	---	–	–
Maldives	May 1986	0		–	–	---	–	–
Mongolia	February 1990	0		1	5,300,000	---	–	–
Myanmar	---	–		–	–	---	–	–
Nepal	June 1978	2		–	–	December 1987	1	17,500
Oman	October 1981	0		–	–	---	–	–
Pakistan	July 1976	0		1	31,355	July 1976	9	20,990
Philippines	September 1985	0		2	1,174,345	---	–	–
Qatar	September 1984	0		–	–	---	–	–
Saudi Arabia	August 1978	0		–	–	---	–	–
Singapore	---	–		–	–	---	–	–
Sri Lanka	June 1980	1		2	9,376	June 1990	1	6,216
Syria	August 1975	0		–	–	---	–	–
Taiwan	---	–		–	–	---	–	–
Thailand	September 1987	1		3	26,100	---	–	–
Turkey	March 1983	1	(1)	–	–	---	–	–
United Arab Emirates	---	–		–	–	---	–	–
Viet Nam	October 1987	0		–	–	September 1988	1	12,000
Yemen Arab Rep	January 1984 #	0		–	–	---	–	–
USSR (former)		**0**		**20**	**10,891,366**		**12**	**2,987,185**
Byelorussian SSR	October 1988	0		1	76,201	(see USSR)	–	–
Ukrainian SSR	October 1988	0		2	120,655	(see USSR)	–	–
USSR	October 1988	0		17	10,694,510	October 1976	12	2,987,185
EUROPE		**11**	**(5)**	**91**	**4,787,243**		**328**	**3,782,517**
Albania	July 1989	0		–	–	---	–	–
Andorra	---	–		–	–	---	–	–
Austria	---	–		4	27,600	December 1982	5	102,369
Belgium	---	–		–	–	March 1986	6	9,607
Bulgaria	March 1974	2		17	39,922	September 1975	4	2,097
Czechoslovakia	---	–		6	364,170	July 1990	8	16,958
Denmark	July 1979	0		–	–	September 1977	27	734,468
Faeroe Islands	(see Denmark)	0		–	–	(see Denmark)	0	0
Finland	March 1987	0		–	–	May 1974	11	101,343
France	June 1975	1	(1)	6	575,583	October 1986	8	422,585
Germany	August 1976 *	0		9	701,849	February 1976 **	29	360,894
Greece	July 1981	0	(2)	2	8,840	August 1975	11	107,400
Hungary	July 1985	0		5	128,884	April 1979	13	110,389
Iceland	---	–		–	–	December 1977	2	57,500
Ireland	September 1991	–		2	8,808	November 1984	21	12,562
Italy	June 1978	0		3	3,798	December 1976	46	56,950
Liechtenstein	---	–		–	–	December 1991	1	90
Luxembourg	September 1983	0		–	–	---	–	–
Malta	November 1978	0		–	–	September 1988	1	11
Monaco	November 1978	0		–	–	---	–	–
Netherlands	---	–		1	260,000	May 1980	11	306,348
Norway	May 1977	0		1	1,555,000	July 1974	14	16,256
Poland	June 1976	1		4	25,836	November 1977	5	7,090
Portugal	September 1980	0		1	395	November 1980	2	30,563
Romania	May 1990	1		3	41,213	September 1991	1	647,000
San Marino	---	–		–	–	---	–	–
Spain	May 1982	1		10	537,717	May 1982	17	98,887
Sweden	January 1985	0		1	96,500	December 1974	30	382,750
Switzerland	September 1975	0		1	16,870	January 1976	8	7,049
United Kingdom	May 1984	2		13	44,258	January 1976	45	173,257
Vatican City	October 1982	0		–	–	---	–	–
Yugoslavia	May 1975	3	(2)	2	350,000	March 1977	2	18,094

Table 29.8 International protection systems (continued)

	WORLD HERITAGE SITES		BIOSPHERE RESERVES		RAMSAR WETLANDS		
	DATE	NO	NO	AREA (ha)	DATE	NO	AREA (ha)
NORTH AND CENTRAL AMERICA		**22**	**69**	**94,624,670**		**63**	**15,368,626**
Anguilla	(see UK)	0	–	–	(see UK)	0	0
Antigua and Barbuda	November 1983	0	–	–	– – –	–	–
Aruba	(see Netherlands)	0	–	–	(see Netherlands)	1	70
Bahamas	– – –	–	–	–	– – –	–	–
Barbados	– – –	–	–	–	– – –	–	–
Belize	– – –	–	–	–	– – –	–	–
Bermuda	(see UK)	0	–	–	(see UK)	0	0
Canada	July 1976	6	6	1,049,978	January 1981	30	12,937,549
Cayman Islands	(see UK)	0	–	–	(see UK)	0	0
Costa Rica	August 1977	1	2	728,955	December 1991	2	29,769
Cuba	March 1981	0	4	323,600	– – –	–	–
Dominica	– – –	–	–	–	– – –	–	–
Dominican Rep	February 1985	0	–	–	– – –	–	–
El Salvador	October 1991	–	–	–	– – –	–	–
Greenland	– – –	–	1	70,000,000	September 1977	11	1,044,500
Grenada	– – –	–	–	–	– – –	–	–
Guadeloupe	(see France)	0	–	–	(see France)	0	0
Guatemala	January 1979	1	1	1,000,000	June 1990	1	48,372
Haiti	January 1980	0	–	–	– – –	–	–
Honduras	June 1979	1	1	500,000	– – –	–	–
Jamaica	June 1983	0	–	–	– – –	–	–
Martinique	(see France)	0	–	–	(see France)	0	0
Mexico	February 1984	1	6	1,288,454	July 1986	1	47,480
Montserrat	– – –	–	–	–	– – –	–	–
Netherlands Antilles	(see Netherlands)	0	–	–	(see Netherlands)	5	2,010
Nicaragua	December 1979	0	–	–	– – –	–	–
Panama	March 1978	2	1	597,000	November 1990	1	80,765
Puerto Rico	– – –	–	2	15,346	– – –	–	–
St Kitts and Nevis	July 1986	0	–	–	– – –	–	–
St Lucia	October 1991	–	–	–	– – –	–	–
St Vincent and the Grenadines	– – –	–	–	–	– – –	–	–
Trinidad and Tobago	– – –	–	–	–	– – –	–	–
Turks and Caicos Islands	(see UK)	0	–	–	(see UK)	1	37,270
United States	December 1973	10	44	19,115,210	December 1986	10	1,140,841
Virgin Islands (British)	(see UK)	0	–	–	(see UK)	0	0
Virgin Islands (US)	(see USA)	0	1	6,127	(see USA)	0	0
SOUTH AMERICA		**9**	**26**	**13,781,071**		**7**	**322,085**
Argentina	August 1978	2	5	2,409,980	– – –	–	–
Bolivia	October 1976	0	3	435,000	June 1990	1	5,240
Brazil	September 1977	1	2	1,862,100	– – –	–	–
Chile	February 1980	0	7	2,406,633	July 1981	1	4,877
Colombia	May 1983	0	3	2,514,375	– – –	–	–
Ecuador	June 1975	2	2	1,446,244	September 1990	2	90,000
French Guiana	(see France)	0	–	–	(see France)	0	0
Guyana	June 1977	0	–	–	– – –	–	–
Paraguay	April 1988	0	–	–	– – –	–	–
Peru	February 1982	4	3	2,506,739	– – –	–	–
Suriname	– – –	–	–	–	March 1985	1	12,000
Uruguay	March 1989	0	1	200,000	May 1984	1	200,000
Venezuela	– – –	–	–	–	November 1988	1	9,968
OCEANIA		**12**	**13**	**4,745,223**		**45**	**4,515,961**
American Samoa	(see USA)	0	–	–	(see USA)	0	0
Australia	August 1974	9	12	4,743,223	May 1974	40	4,477,862
Cook Islands	(see New Zealand)	0	–	–	(see New Zealand)	0	0
Fiji	– – –	–	–	–	– – –	–	–
French Polynesia	– – –	–	1	2,000	– – –	–	–
Guam	(see USA)	0	–	–	(see USA)	0	0
Kiribati	– – –	–	–	–	– – –	–	–
Marshall Islands	– – –	–	–	–	– – –	–	–
Micronesia, Federated States of	– – –	–	–	–	– – –	–	–
Nauru	– – –	–	–	–	– – –	–	–
New Caledonia	(see France)	0	–	–	(see France)	0	0
New Zealand	November 1984	2	–	–	August 1976	5	38,099
Niue	(see New Zealand)	0	–	–	(see New Zealand)	0	0
North Marianas Islands	– – –	–	–	–	– – –	–	–
Palau	– – –	–	–	–	– – –	–	–
Papua New Guinea	– – –	–	–	–	– – –	–	–
Pitcairn Island	(see UK)	1	–	–	(see UK)	0	0
Solomon Islands	– – –	–	–	–	– – –	–	–
Tokelau	(see New Zealand)	0	–	–	(see New Zealand)	0	0
Tonga	– – –	–	–	–	– – –	–	–
Tuvalu	– – –	–	–	–	– – –	–	–
Vanuatu	– – –	–	–	–	– – –	–	–
Wallis and Futuna Islands	(see UK)	0	–	–	(see UK)	0	0
Western Samoa	– – –	–	–	–	– – –	–	–
ANTARCTICA		**0**	**0**	**0**		**0**	**0**
Antarctica	– – –	–	–	–	– – –	–	–
Falkland Islands (Malvinas)	(see UK)	0	–	–	(see UK)	0	0
French Southern Territories	(see France)	0	–	–	(see France)	0	0

Table 29.8 International protection systems (continued)

	WORLD HERITAGE SITES			BIOSPHERE RESERVES		RAMSAR WETLANDS		
	DATE	NO		NO	AREA (ha)	DATE	NO	AREA (ha)
AFRICA		28	(1)	43	20,229,937		43	4,005,302
Algeria	June 1974	1		2	7,276,438	November 1983	2	4,900
Angola	---	–		–	–	---	–	–
Benin	June 1982	0		1	880,000	---	–	–
Botswana	---	–		–	–	---	–	–
Burkina Faso	April 1987	0		1	16,300	June 1990	3	296,300
Burundi	May 1982	0		–	–	---	–	–
Cameroon	December 1982	1		3	850,000	---	–	–
Cape Verde	April 1988	0		–	–	---	–	–
Central African Rep	December 1980	1		2	1,640,200	---	–	–
Chad	---	–		–	–	June 1990	1	195,000
Comoros	---	–		–	–	---	–	–
Congo	December 1987	0		2	246,000	---	–	–
Cote d'Ivoire	January 1981	3		2	1,500,000	---	–	–
Djibouti	---	–		–	–	---	–	–
Egypt	February 1974	0		1	1,000	September 1988	2	105,700
Equatorial Guinea	---	–		–	–	---	–	–
Ethiopia	July 1977	1		–	–	---	–	–
Gabon	December 1986	0		1	15,000	December 1986	3	1,080,000
Gambia	July 1987	0		–	–	---	–	–
Ghana	July 1975	0		1	7,770	February 1988	1	7,260
Guinea	March 1979	1		2	133,300	---	–	–
Guinea–Bissau	---	–		–	–	May 1990	1	39,098
Kenya	June 1991	–		5	851,359	June 1990	1	18,800
Lesotho	---	–		–	–	---	–	–
Liberia	---	–		–	–	---	–	–
Libya	October 1978	0		–	–	---	–	–
Madagascar	July 1983	1		1	140,000	---	–	–
Malawi	January 1982	1		–	–	---	–	–
Mali	April 1977	0	(1)	1	771,000	May 1987	3	162,000
Mauritania	March 1981	1		–	–	October 1982	1	1,173,000
Mauritius	---	–		1	3,594	---	–	–
Mayotte	(see France)	0		–	–	(see France)	0	0
Morocco	October 1975	0		–	–	June 1980	4	10,580
Mozambique, People's Rep	November 1982	0		–	–	---	–	–
Namibia	---	–		–	–	---	–	–
Niger	December 1974	1		–	–	April 1987	1	220,000
Nigeria	October 1974	0		1	460	---	–	–
Reunion	(see France)	0		–	–	(see France)	0	0
Rwanda	---	–		1	15,065	---	–	–
Saint Helena	(see UK)	0		–	–	(see UK)	0	0
Sao Tome and Principe	---	–		–	–	---	–	–
Senegal	February 1976	2		3	1,093,756	July 1977	4	99,720
Seychelles	April 1980	2		–	–	---	–	–
Sierra Leone	---	–		–	–	---	–	–
Somalia	---	–		–	–	---	–	–
South Africa	---	–		–	–	March 1975	12	232,344
Sudan	June 1974	0		2	1,900,970	---	–	–
Swaziland	---	–		–	–	---	–	–
Tanzania	August 1977	4		2	2,337,600	---	–	–
Togo	---	–		–	–	---	–	–
Tunisia	March 1975	1		4	32,425	November 1980	1	12,600
Uganda	November 1987	0		1	220,000	March 1988	1	15,000
Western Sahara	---	–		–	–	---	–	–
Zaire	September 1974	4		3	297,700	---	–	–
Zambia	June 1984	1		–	–	December 1991	2	333,000
Zimbabwe	August 1982	2		–	–	---	–	–

Source: WCMC.
Notes: Dates are date of accession or ratification. The extra numbers in parenthesis in the World Heritage section, refer to mixed natural/cultural sites inscribed on the list of World Heritage on the basis of beauty resulting from the man/nature interaction, rather than natural features alone. */** The former German Democratic Republic signed the World Heritage convention in December 1988, and the Ramsar convention in July 1978. # The former People's Democratic Republic of Yemen signed the World Heritage convention in October 1980.

B13 FAO, *in litt.*, 1991

B14 Bel Hadj Kacem, S. 1991. Liste des parcs nationaux et aires protégées - Tunisie 1991. Direction Générale des Forêts, Ministère de l'Agriculture, Tunis. Paper presented to the Third Man and Biosphere Meeting on Biosphere Reserves in the Mediterranean, 14-19 October 1991, Tunis. 7pp.

D2 IUCN 1991a. *Environmental Status Report: 1990. Volume Two: Albania, Bulgaria, Romania, Yugoslavia.* IUCN East European Programme, Cambridge, UK.

D3 Kucera, B. 1991. Response to regional review questionnaire.

D4 Cutrera, A. 1991. *European Environmental Yearbook.* Institute for Environmental Studies. DocTer International UK/London. 897pp.

D5 Nikol'skii, A., Bolshova, L.I. and Karaseva, S.E. 1991. *Palaearctic-USSR Regional Protected Areas Review.* Paper proposed for IV World Parks Congress on National Parks and other Protected Areas. USSR Ministry of Natural Resources Management and Environmental Protection, Moscow.

D6 Kassioumis, K. 1991. Response to regional review questionnaire.

D9 IUCN, *in litt.*, 1982

D11 Ministry of Agriculture, Nature Management and Fisheries 1990. *Nature Policy Plan of the Netherlands.* The Hague. 103pp.

D12 Lein, B. and Nord-Varhaug, O. 1991. Response to regional review questionnaire.

D13 Oklow, C. 1991. Response to regional review questionnaire.

D14 Manners Moura, R. 1991. Response to regional review questionnaire.

D15 Larsson, T. 1991. Response to regional review questionnaire.

D16 Furphy, J.S. 1991. Response to regional review questionnaire.

D17 NCC 1991. *Seventeenth Report 1 April 1990-31 March 1991.* Nature Conservancy Council, Peterborough. 126pp.

D18 Phillips, A. 1991. Response to regional review questionnaire.

D19 Cortes, J. 1991. Response to regional review questionnaire.

D20 Pinder, N.J. 1991. Response to regional review questionnaire.

E1 Waugh and Perez Gil, 1992. Regional Review: Nearctic. Prepared for the IV World Parks Congress, Caracas, Venezuela, 10-21 February 1992.

F1 Allen, B. 1990. National park planning in Jamaica: a project in sustainable development and conservation. Paper presented to the Association of Caribbean Studies Conference on the Caribbean Environment, Santo Domingo, Dominican Republic. 22pp.

F2 Anon. 1991. Le Parc national de la Guadeloupe. Unpublished report. 5pp.

F3 IUCN 1992. Protected Areas of the World: a review of national systems. Volume 4. America. Draft.

G1 Sandoval, G.J., Reyes, J.M. and Soria, J.L. 1989. *Plan de Acción para el Desarrollo forestal 1990-1995*. Ministerio de Asuntos Campesinos y Agropecuarios, Subsecretaría de Recursos Naturales Renovables y Medio Ambiente, La Paz. 98pp.

G2 Dias, I.F.O., Gonçalves, A.R., Borges, M. and Meneses, E.O. 1991. *Sistema de Unidades de Conservação Federais do Brasil*. IBAMA-DIREC-DEUC. 11pp.

G3 Cabarle, B.J., Crespi, M., Calaway, H.D., Luzuriaga, C.C., Rose, D. and Shores, J.N. 1989. *An Assessment of Biological Diversity and Tropical Forests for Ecuador*. Prepared for US-AID/Ecuador as an Annex to the Country Development Strategy Statement 1989-1990. 110pp.

G4 GFC and CIDA 1989. *National Forestry Action Plan 1990-2000*. Guyana Forestry Commission and Canadian International Development Agency, Kingston, Georgetown. 77pp.

G5 DGFF 1991. Informe sobre progreso forestal 1988-1990 del Perú. 17th meeting of the Latin American Forestry Commission - COFLA, Venezuela, 18th-22nd February 1991. Ministerio de Agricultura, Dirección General de Forestal y Fauna, Lima. 22pp.

G6 Schultz, J.P. 1968. Nature preservation in Suriname: a review of the present situation. Suriname Forest Service, Paramaribo. 21pp.

H1 ANPWS 1989. Annual Report 1988-89. Australian National Parks and Wildlife Service. Canberra. 132pp.

I1 SPREP 1989. Country review: Western Samoa. Fourth South Pacific Conference on Nature Conservation and Protected Areas. South Pacific Commission, Noumea, New Caledonia. 12pp.

J2 Edens, J.H. 1991. *Tropical Forestry Action Plan - background paper*. MNRE, Banjul and FAO, Rome. 30pp.

J3 Forestry Service 1985. Progress report 1980-84 by the Forestry Service of the Ministry of Agriculture, Fisheries and Natural Resources. Forestry Service, Mauritius. 14 pp.

J4 IUCN/SSC 1990. *African Elephants and Rhinos Status Survey and Conservation Action Plan*. Compiled by D.H.M. Cummings, R.F Du Toit and S.N. Stuart. 72pp.

J5 WD 1991. *Elephant Conservation Plan for Tanzania*. Wildlife Division. 147pp.

K1 Data taken from the African Elephant Conservation Review 1991, produced by the African Elephant Conservation Coordinating Group AECCG. (Data refer to government expenditure on Wildlife and Protected Area Management, unless otherwise stated).

K2 MALFF 1991. *Elephant Conservation Plan. Equatorial Guinea*. Ministry of Agriculture, Livestock, Fisheries and Forestry. 44pp.

K3 McShane, T.O. and McShane-Caluzi, E. 1990. *Conservation before the Crisis: a strategy for conservation in Gabon*. WWF.

K4 MEFCPT 1986. Plan quinquennal secteur chasses et faune 1986-1990. Ministère des Eaux, Forêts, Chasses, Pêches et du Tourisme. Unpublished. 12pp.

K5 MTE 1991. *Plan de Conservation de l'Elephant au Tchad*. Ministère du Tourisme et de l'Environnement. 49pp. Ministry of Water, Forests, Hunting, Fishing and Tourism (Ministère des Eaux, Forêts, Chasses, Pêches et du Tourisme MEFCPT).

Based on text prepared by Sam Johnston, maps prepared by Joel Smith, with additional material by WCMC staff.

30. MULTILATERAL TREATIES

A multilateral treaty is an international agreement concluded between three or more states and governed by international law.

Existing international treaties which deal entirely or in part with biological diversity have evolved in an uncoordinated manner. Despite this, and the consequent gaps and duplications in overall coverage, a handful of such treaties have come to exert a very powerful effect on the conservation and management of elements of biodiversity. Perhaps foremost among these, in terms of their sophistication and global scope, are The Convention on International Trade in Endangered Species of Wild Fauna and Flora (CITES), The Convention on Wetlands of International Importance (Ramsar), and The Convention Concerning the Protection of the World Cultural and Natural Heritage (World Heritage). The Convention on the Law of the Sea (UNCLOS), which is yet to enter into force, has strong potential for enhancing marine and coastal conservation.

The names of these major treaties are indicative of their sectoral focus, and even if the many important regional and species-related treaties are also considered, it is clear that the total obligations explicit in existing treaties fall short of the demands of an adequately comprehensive system. The proposed Convention on Biological Diversity attempts to meet many of these demands, and is the first treaty planned to concentrate specifically on the conservation and use of global biodiversity (see Chapter 35).

Text

The production of a multilateral treaty usually follows several stages. The first involves negotiation of the text of the treaty. This can take many years and can require numerous meetings. The negotiation of a treaty is concluded by the adoption of the text of a treaty. This typically takes place when all the states participating in the negotiations reach agreement although the need for unanimity is not required by law. Each negotiating conference adopts its own rules concerning voting. Adoption of a treaty does not by itself create any obligations.

Consent

A treaty does not come into force until two or more States consent to be bound by the treaty. The expression of such consent is usually an entirely separate process from adoption. Consent may be expressed by "signature, exchange of instruments constituting a treaty, notification, acceptance, approval or accession or by many other means if so agreed." The permitted ways of expressing consent and becoming a party to a treaty are always outlined in the text of the treaty itself. Signature and ratification are the most frequent means of expressing consent. *Signature* refers to the signature of the diplomats negotiating the treaty and is often synonymous with the adoption of the treaty.

Ratification is the need for approval of the treaty by the head of state or the legislature. In addition to signature and ratification, a state can also become a party to a treaty by accession. Accession is the normal way that states who did not participate in the negotiations become parties to the treaty. Accession is only possible if it is provided for in the treaty; it has the same effect as signature and ratification combined.

Entry into force

The final stage in the production of a treaty is its entry into force. This usually occurs when all the negotiating states have expressed their consent to be bound by the treaty. This may be altered by agreement and it is not uncommon for the date at which a treaty enters into force to be delayed in order to give parties time to adapt themselves to its requirements. Another common variation occurs when there are a great number of states participating in the drafting of a treaty. In this case, to wait for every State to ratify the treaty before it enters into force would invariably cause excess delay, and so large multilateral treaties often enter into force when a specified number of States have ratified. However, when this specified number is reached, the treaty will only be in force between those States which have ratified it; it does not enter into force for the other States until they in turn have ratified it.

Multilateral treaty table

Table 30.3 lists all multilateral international treaties which have been adopted for the conservation of biodiversity. These treaties have here been classified into three broad groups. 'Global treaties' are ones which have no requirements as to membership and are open to any country in the world. 'Regional treaties' are ones which limit membership, normally to a certain geographical region, although in some instances other criteria are used as well. 'Species-related treaties' are ones which limit membership to those countries which have some relationship with the species which are the subject of the treaty. The scope of these treaties varies from those which, like the Antarctic Treaty, attempt to deal comprehensively with the governance of an area, to those, such as the Vicuña Treaty, which confine their scope to the conservation of one single species.

Table 30.1 represents graphically the status and membership of global and regional treaties; Table 30.2 covers species-related treaties. These tables are based on information provided to WCMC by the IUCN Environmental Law Centre (ELC) on 1 March 1992.

Table 30.1 Multilateral treaties: global and regional

Convention	in force	ASIA	Afghanistan	Bahrain	Bangladesh	Bhutan	Brunei	Cambodia	China	Cyprus	India	Indonesia	Iran, Islamic Rep	Iraq	Israel	Japan	Jordan	Korea, Dem People's Rep	Korea, Rep	Kuwait	Laos	Lebanon	Malaysia	Maldives	Mongolia	Myanmar	Nepal	Oman	Pakistan	Philippines	Qatar	Saudi Arabia	Singapore	Sri Lanka	Syria	Taiwan	Thailand	Turkey	United Arab Emirates
Regional conventions																																							
Alps (Salzburg) 1991																																							
S Pacific (Noumea) 1986	■																																						
ASEAN Agt. (Kuala Lumpur) 1985							S					S											S							S			S				S		
PA Protocol (Nairobi) • 1985	■																																						
E African Reg. (Nairobi) 1985	■																																						
SPA Protocol (Kingston) • 1990																																							
Carribean (Cartagena de Indias) 1983	■																																						
Central African (Libreville) 1983																																							
Benelux (Brussels) 1982	■																																						
Red Sea (Jeddah) 1982	■																■															■							
SE Pacific (Lima) 1981	■																																						
W & Cent. Africa (Abidjan) 1981	■																																						
SACEP (Colombo) 1981					S						S		S											S					S					S					
CCAMLR (Canberra) 1980	■										■					■			■																				
European (Bern) 1979	■									■																												■	
Amazon Pact (Brasilia) 1978	■																																						
Persian Gulf (Kuwait) 1978	■			■									■	■						■								■			■	■							■
South Pacific (Apia) 1976	■																																						
Med. Shores (Monaco) 1976	■																																						
Protocol SPA (Geneva) • 1982	■									■					■																							■	
Mediterranean (Barcelona) 1976	■									■					■							■													■			■	
Game Hunting (Yamoussoukro) 1976	■																																						
Baltic Seas (Helsinki) 1974	■																																						
Amend. (Warsaw) • 1982	■																																						
Baltic Sea and Belts (Gdansk) 1973	■																																						
Amend. 2 (Tarragona) • 1985																■																							
Amend. 1 (Tarragona) • 1985																■																							
SE Atlantic Living Resources (Rome) 1969	■													■	■	■			■																				
African (Algiers) 1968	■																																						
Protocol (Madrid) • 1991									S		S					S		S	S																				
Antarctic (Washington) 1959	■								■		■					■		■	■																				
Western Hemisphere (Washington) 1940	■																																						
Global conventions																																							
ITTA (Geneva) 1983	■								■		■	■				■			■				■				■			■							■		
UNCLOS (Montego Bay) 1982			S	■	S	S	S	S	S	■	S	■	S	■		S		S	S	■	S	S	S	S	S	S	S	■	S	■	S	S	S	S			S		S
CMS (Bonn) 1979	■										■				■														■	S		■		■					
CITES (Washington) 1973	■		■		■		■	S	■	■	■	■	■		■	■	■			S			■				■		■	■			■	■			■		■
WHC (Paris) 1972	■		■		■				■	■	■	■	■	■			■		■		■	■	■	■	■		■	■	■	■	■	■		■	■		■	■	
Wetlands (Ramsar) 1971	■										■		■			■	■										■		■					■					
High Seas (Geneva) 1958	■		■					■		■		■	S		■	■						S	■		■		■		S					S			■		
Liv.Res. High Seas (Geneva) 1958	■		S					■				S	S		S							S	■				S		S					S			■		
Plant (Rome) 1951	■			■	■			■			■	■	■	■	■	■	■		■		■	■	■					■	■	■				■			■	■	

ASIA (continued)																																									
Viet Nam						S		S																																	
Yemen																																									
USSR								S																																	
EUROPE																																									
Albania																																									
Austria								S				S																													S
Belgium								S				S		D																											
Bulgaria								S				S																													
Czechoslovakia						S		S				S																													
Denmark								S				S					D	D																							
Finland								S				S																													
France			S					S				S												S												S					S
Germany												S					D	D																							S
Greece							S	S				S																													
Hungary								S				S																													
Iceland		S	S																																						
Ireland		S	S			S		S																																	
Italy								S				S																													S
Liechtenstein								S																																	S
Luxembourg								S																																	
Malta								S																																	
Monaco								S																																	
Netherlands								S				S																								S					
Norway								S				S																													
Poland								S				S																													
Portugal								S																																	
Romania								S				S																													
San Marino																																									
Spain								S				S																													
Sweden								S				S																													
Switzerland	S							S				S																													S
United Kingdom												S																								S				S	
Vatican City			S																																						
Yugoslavia																																									S
European Community								S																											S		S	S			S
NORTH AND CENTRAL AMERICA																																									
Antigua and Barbuda																																				S					
Bahamas																																									
Barbados								S																																	
Belize																																									
Canada		S	S					S				S																													
Costa Rica		S						S																																	
Cuba		S	S							S		S																								S					
Dominica																																									
Dominican Rep								S																																	
El Salvador								S																																	
Grenada																																									
Guatemala								S																												S					
Haiti								S																																	
Honduras								S																											S						
Jamaica							S																													S					
Mexico																																				S					
Nicaragua								S																											S						
Panama		S	S					S																																	

Table 30.1 Multilateral treaties: global and regional (continued)

	N. AND CEN. AMERICA (cont.)					SOUTH AMERICA												OCEANIA																
	St Lucia	St Vincent and the Grenadines	St Kitts and Nevis	Trinidad and Tobago	United States	Argentina	Bolivia	Brazil	Chile	Colombia	Ecuador	Guyana	Paraguay	Peru	Suriname	Uruguay	Venezuela	Australia	Cook Is	Fiji	Kiribati	Marshall Is	Micronesia	Nauru	New Zealand	Niue	Palau	Papua New Guinea	Samoa	Solomon Islands	Tokelau	Tonga	Tuvalu	Vanuatu
Regional conventions																																		
Alps (Salzburg) 1991																																		
S Pacific (Noumea) 1986					■													■	■	■		■	■	S	■		S	■	■	■			S	
ASEAN Agt. (Kuala Lumpur) 1985																																		
PA Protocol (Nairobi) • 1985																																		
E African Reg. (Nairobi) 1985																																		
SPA Protocol (Kingston) • 1990	S			S	S					S							S																	
Carribean (Cartagena de Indias) 1983	■	■		■	■					■							■																	
Central African (Libreville) 1983																																		
Benelux (Brussels) 1982																																		
Red Sea (Jeddah) 1982																																		
SE Pacific (Lima) 1981									■	■	■			■																				
W & Cent. Africa (Abidjan) 1981																																		
SACEP (Colombo) 1981																																		
CCAMLR (Canberra) 1980					■	■		■	■					■		■		■							■									
European (Bern) 1979																																		
Amazon Pact (Brasilia) 1978							■	■		■	■	■		■	■		■																	
Persian Gulf (Kuwait) 1978																																		
South Pacific (Apia) 1976																		■	■	■								S	■					
Med. Shores (Monaco) 1976																																		
Protocol SPA (Geneva) • 1982																																		
Mediterranean (Barcelona) 1976																																		
Game Hunting (Yamoussoukro) 1976																																		
Baltic Seas (Helsinki) 1974																																		
Amend. (Warsaw) • 1982																																		
Baltic Sea and Belts (Gdansk) 1973																																		
Amend. 2 (Tarragona) • 1985																																		
Amend. 1 (Tarragona) • 1985																																		
SE Atlantic Living Resources (Rome) 1969																																		
African (Algiers) 1968																																		
Protocol (Madrid) • 1991					S	S		S	S	S	S			S		S		S							S			S						
Antarctic (Washington) 1959					■	■		■	■	■	■			■		■		■							■			■						
Western Hemisphere (Washington) 1940				■	■	■	S	■	■	S	■		■	■	■	■	■																	
Global conventions																																		
ITTA (Geneva) 1983				■	■		■	■		■	■			■				■										■						
UNCLOS (Montego Bay) 1982	■	S	S	■		S	S	■	S	S		S	■		S	S		S	S	■		■	■	S	S	S		S	S	S			S	S
CMS (Bonn) 1979									■				■	■	S	■		■																
CITES (Washington) 1973	■	■		■	■	■	■	■	■	■	■	■	■	■	■	■	■	■							■			■						■
WHC (Paris) 1972			■		■	■	■	■	■	■	■	■	■	■		■	■	■		■					■									
Wetlands (Ramsar) 1971					■		■		■		■				■	■	■	■							■									
High Seas (Geneva) 1958				■	■	S	S			S						S	■	■		■					S					■		■		
Liv.Res. High Seas (Geneva) 1958				■	■	S	S			■						S	■	■		■					S					■		■		
Plant (Rome) 1951			■	■	■	■	■	■	■	■	■	■	■	■	■	■	■	■							■			■		■				

AFRICA									
Algeria								S	
Angola									
Benin								S	
Botswana									
Burkina Faso								S	
Burundi								S	
Cameroon									
Cape Verde									
Central African Rep							S	S	
Chad							S	S	
Comoros								S	
Congo								S	
Côte d'Ivoire							S		
Djibouti									
Egypt									
Equatorial Guinea						S		S	
Ethiopia								S	
Gabon								S	
Gambia									
Ghana		S	S						
Guinea									
Guinea – Bissau									
Kenya									
Lesotho						S		S	
Liberia		S	S					S	
Libya								S	
Madagascar							S	S	
Malawi								S	
Mali									
Mauritania								S	
Mauritius								S	
Morocco							S	S	
Mozambique, People's Rep								S	
Namibia									
Niger								S	
Nigeria									
Rwanda								S	
Sao Tome and Principe									
Senegal									
Seychelles									
Sierra Leone								S	
Somalia								S	
South Africa								S	
Sudan									
Swaziland								S	
Tanzania									
Togo							S		
Tunisia		S	S						
Uganda							S		
Zaire									
Zambia									
Zimbabwe								S	
Number of countries in conventio	97	36	55	64	113	113	40	50	48

AFRICA																																
Algeria																																
Angola																																
Benin				S																	S											
Botswana				S																												
Burkina Faso																																
Burundi				S																												
Cameroon																									S							
Cape Verde																																
Central African Rep																																
Chad				S																												
Comoros				S																												
Congo																									S							
Côte d'Ivoire																																
Djibouti																																
Egypt																																
Equatorial Guinea																																
Ethiopia				S																												
Gabon																									S							
Gambia				S																												
Ghana																																
Guinea				S																												
Guinea – Bissau																																
Kenya																																
Lesotho				S																												
Liberia																					S											
Libya				S																												
Madagascar																												S	S			
Malawi																																
Mali																																
Mauritania				S																	S											
Mauritius				S																												
Morocco																																
Mozambique, People's Rep																																
Namibia																																
Niger																																
Nigeria																																
Rwanda																																
Sao Tome and Principe																																
Senegal																																
Seychelles																																
Sierra Leone				S																												
Somalia				S																								S	S			
South Africa																																
Sudan																									S							
Swaziland																																
Tanzania																																
Togo																																
Tunisia																																
Uganda																																
Zaire																																
Zambia																																
Zimbabwe																																
Number of countries in conventio	19	40	0	30	16	5	8	5	5	6	5	18	17	3	3	8	8	25	26	0	10	5	6	3	0	18	1	3	3	0	11	0

Source: IUCN Environmental Law Centre, Bonn.
Notes: See Table 30.3 for complete convention names. [shading] – State is a party to treaty (where treaty is not yet in force, the State has expressed its consent to become a party), S – signature only, D – terminated. * Indicates an amendment or protocol to the treaty to the left.

Table 30.2 Multilateral treaties: species–related

Treaty	in force	ASIA	Afghanistan	Bahrain	Bangladesh	Bhutan	Brunei	Cambodia	China	Cyprus	India	Indonesia	Iran, Islamic Rep	Iraq	Israel	Japan	Jordan	Korea, Dem People's Rep	Korea, Rep	Kuwait	Laos	Lebanon	Malaysia	Maldives	Mongolia	Myanmar	Nepal	Oman	Pakistan	Philippines	Qatar	Saudi Arabia	Singapore	Sri Lanka	Syria	Taiwan	Thailand	Turkey	United Arab Emirates
Others																																							
Ivory Marketing (Lilongwe) 1991																																							
Wadden Seals (Bonn) 1990	X																																						
Vicuna (Lima) 1979	X																																						
Polar Bears (Oslo) 1973	X																																						
CCAS (London) 1972	X															X																							
S Pacific Marine (Santiago) 1952	X																																						
Prot. • 1956	X								X		X					X		X	X									X		X									
IWC (Washington) 1946	X								X		X					X			X									X		X									
Plants																																							
N American Plant Protect. (Yosemite) 1976	X																																						
New Varieties (Geneva) 1961	X														X																								
Amend. • 1967	X				X			X			X	X							X		X		X			X	X		X	X				X			X		
Plant Protect (Rome) 1956	X				X			X			X	X							X		X		X			X	X		X	X				X			X		
Eur. Plant Protect. orgn. (Paris) 1951	X									X			D		X																							X	
Birds																																							
Benelux - Birds (Brussels) 1970	X																																						
Protect of Birds (Paris) 1950	X																																					X	
Fisheries																																							
Protocol I & II (Noumea) 1990																																							
Drift Net (Wellington) 1989	X																																						
S Pacific (Port Moresby) 1987																																							
E Pacific Tuna (San José) 1983																																							
N Atlantic Salmon (Reykjavik) 1982	X																																						
NEAFC (London) 1980	X																																						
S Pacific Fish (Honiara) 1979	X																																						
Nw Alantic (Ottawa) 1978	X															S																							
SE Asian Fish. (Bangkok) 1967	X															X							X							X			X				X		
N Atlantic Fish. (London) 1967	X																																						
Skagerrak Agt. (Copenhagen) 1968	X																																						
Protocol (Paris) 1984																X			X																				
Atlantic Tuna (Rio de Janeiro) 1966	X															X			X																				
Fisheries (London) 1964	X																																						
Salmon/Baltic Sea (Stockholm) 1962	X																																						
Marine Fishing (Warsaw) 10 1962	X																																						
Black Sea (Varna) 1959	X																																						
Danube (Bucharest) 1958	X																																						
Protocol (Tokyo) • 1978	X															X																							
Amend. to the Annex • 1962	X															X																							
N Pacific Fish (Tokyo) 1952	X															X																							
Protocol (Oslo) • 1959	X																																						
European crustaceans (Oslo) 1952	X																																						
Fish, Council - Med (Rome) 1949	X									X					X							X													X			X	
Inter-Am Tuna (Washington) 1949	X															X																							

ASIA (continued)																																											
Viet Nam																																											
Yemen																																											
USSR																																											
EUROPE																																											
Albania																																											
Austria																										S																	
Belgium																																											
Bulgaria																	S									S																	
Czechoslovakia																																											
Denmark																	S																										
Finland																																											
France																							S			S																	
Germany																																											
Greece																										S																	
Hungary																																											
Iceland																																											
Ireland															S																												
Italy																																											
Liechtenstein																																											
Luxembourg											S																																
Malta																																											
Monaco																										S																	
Netherlands																																											
Norway																																											
Poland															S		S																										
Portugal																	S									S																	
Romania																																											
San Marino																																											
Spain																																											
Sweden																																											
Switzerland																																											
United Kingdom																																											
Vatican City																																											
Yugoslavia																																											
European Community																																											
NORTH AND CENTRAL AMERICA																																											
Antigua and Barbuda																																											
Bahamas																																											
Barbados																																											
Belize																																			D							
Canada	D												S		S																				D							
Costa Rica	D																																										
Cuba																			S																								
Dominica																																											
Dominican Rep												S																															
El Salvador																																											
Grenada																																											
Guatemala																					S																						
Haiti																																											
Honduras																																											
Jamaica																																			D							
Mexico	D																																										
Nicaragua																																											
Panama																																			D	D						

Table 30.2 Multilateral treaties: species—related (continued)

Treaty	N. AND CEN AMERICA (cont.)	St Lucia	St Vincent and the Grenadines	St Kitts and Nevis	Trinidad and Tobago	United States	SOUTH AMERICA	Argentina	Bolivia	Brazil	Chile	Colombia	Ecuador	Guyana	Paraguay	Peru	Suriname	Uruguay	Venezuela	OCEANIA	Australia	Cook Is	Fiji	Kiribati	Marshall Is	Micronesia	Nauru	New Zealand	Niue	Palau	Papua New Guinea	Samoa	Solomon Islands	Tokelau	Tonga	Tuvalu	Vanuatu
Others																																					
Ivory Marketing (Lilongwe) 1991																																					
Wadden Seals (Bonn) 1990																																					
Vicuna (Lima) 1979									■		■		■			■																					
Polar Bears (Oslo) 1973						■																															
CCAS (London) 1972						■		■			■										■							S									
S Pacific Marine (Santiago) 1952											■		■			■																					
Prot. • 1956		■	■			■		■		■						■		■			■							■					■				
IWC (Washington) 1946		■	■			■		■		■	■					■		D			■							■					D				
Plants																																					
N American Plant Protect. (Yosemite) 1976						■																															
New Varieties (Geneva) 1961																																					
Amend. • 1967																					■		■					■	■	■	■	■	■		■		
Plant Protect (Rome) 1956																					■		■					■	■	■	■	■	■		■		
Eur. Plant Protect. orgn. (Paris) 1951																		■																			
Birds																																					
Benelux - Birds (Brussels) 1970																																					
Protect of Birds (Paris) 1950																																					
Fisheries																																					
Protocol I & II (Noumea) 1990						S					S																										
Drift Net (Wellington) 1989						S															S	■		■	S	■	S	■	S	S			S	■		S	S
S Pacific (Port Moresby) 1987						■															■	■	■	■	■	■	■		■	■	■		■				
E Pacific Tuna (San José) 1983						■																															
N Atlantic Salmon (Reykjavik) 1982						■																															
NEAFC (London) 1980																																					
S Pacific Fish (Honiara) 1979																					■	■	■	■	■	■	■	■	■	■	■	■	■		■	■	■
Nw Aiantic (Ottawa) 1978																																					
SE Asian Fish. (Bangkok) 1967																																					
N Atlantic Fish. (London) 1967						S																															
Skagerrak Agt. (Copenhagen) 1968																																					
Protocol (Paris) 1984						■				■								■	■																		
Atlantic Tuna (Rio de Janeiro) 1966						■				■								■	■																		
Fisheries (London) 1964																																					
Salmon/Baltic Sea (Stockholm) 1962																																					
Marine Fishing (Warsaw) 10 1962																																					
Black Sea (Varna) 1959																																					
Danube (Bucharest) 1958																																					
Protocol (Tokyo) • 1978						■																															
Amend. to the Annex • 1962						■																															
N Pacific Fish (Tokyo) 1952						■																															
Protocol (Oslo) • 1959																																					
European crustaceans (Oslo) 1952																																					
Fish, Council - Med (Rome) 1949																																					
Inter-Am Tuna (Washington) 1949						■							D																								■

AFRICA																																								
Algeria		■																										■												
Angola													■																											
Benin													■																											
Botswana																																								S
Burkina Faso																																								
Burundi																																								
Cameroon																																								
Cape Verde													■	■																										
Central African Rep																																								
Chad																																								
Comoros																																								
Congo																																								
Côte d'Ivoire													■																											
Djibouti																																								
Egypt		■																															D	■						
Equatorial Guinea													■	■																										
Ethiopia																																								
Gabon													■																											
Gambia																																								
Ghana													■	■																										
Guinea																																								
Guinea – Bissau																																								
Kenya																																	D	■						
Lesotho																																								
Liberia																																								
Libya		■																																						
Madagascar																																								
Malawi																																								S
Mali																																								
Mauritania																																								
Mauritius																																	■	■						
Morocco		■											■															■												
Mozambique, People's Rep																																								
Namibia																																								S
Niger																																								
Nigeria																																								
Rwanda																																								
Sao Tome and Principe													■	■																										
Senegal														■														■					■	■						
Seychelles																																	■	■						
Sierra Leone																																								
Somalia																																								
South Africa													■	■																	■		■	■		■				
Sudan																																								
Swaziland																																								
Tanzania																																								
Togo																																								
Tunisia		■																																						
Uganda																																								
Zaire																																								
Zambia																																								S
Zimbabwe																																								S
Number of countries in convent	6	20	3	3	3	3	3	0	3	5	4	12	22	17	3	13	6	7	16	10	9	4	12	5	0	10	3	31	26	25	11	3	36	42	3	13	5	4	3	0

Source: IUCN Environmental Law Centre, Bonn.
Notes: See Table 30.3 for complete convention names. ■ – State is a party to treaty (where the treaty is not yet in force, the State has expressed its consent to become a party), S – signature only, D – terminated. * Indicates an amendment or protocol to the treaty to the left.

Table 30.3 Multilateral treaties

Global conventions

1951 Plant (Rome) - International Plant Protection Convention
1958 Liv.Res. High Seas (Geneva) - Convention on Fishing and Conservation of the Living Resources of the High Seas
1958 High Seas (Geneva) - Convention on the High Seas
1971 Wetlands (Ramsar) - Convention on Wetlands of International Importance Especially as Waterfowl Habitat
1972 WHC (Paris) - Convention concerning the Protection of the World Cultural and Natural Heritage
1973 CITES (Washington) - Convention on International Trade in Endangered Species of Wild Fauna and Flora
1979 CMS (Bonn) - Convention on the Conservation of Migratory Species of Wild Animals
1982 UNCLOS (Montego Bay) - United Nations Convention on the Law of the Sea
1983 ITTA (Geneva) - International Tropical Timber Agreement

Regional conventions

1940 Western Hemisphere (Washington) - Convention on Nature Protection and Wildlife Preservation in the Western Hemisphere
1959 Antarctic (Washington) - The Antarctic Treaty
1991 Protocol (Madrid)
1968 African (Algiers) - African Convention on the Conservation of Nature and Natural Resources
1969 SE Atlantic Living Resources (Rome) - Convention on the Conservation of the Living Resources of the Southeast Atlantic
1985 Amend. 1 (Tarragona) - Amendment to Article XIII (1) of the Convention on the Conservation of the Living Resources of the Southeast Atlantic
1985 Amend. 2 (Tarragona) - Amendment for Articles VIII, XVII, XIX, and XXI of the Convention on the Conservation of the Living Resources of the Southeast Atlantic
1973 Baltic Sea and Belts (Gdansk) - Convention on Fishing and Conservation of the Living Resources in the Baltic Sea and the Belts
1982 Amend. (Warsaw) - Amendendments to the Convention on Fishing and Conservation of the Living Resources in the Baltic Sea and Belts
1974 Baltic Seas (Helsinki) - Convention on the Protection of the Marine Environment of the Baltic Sea Area
1976 Game Hunting (Yamoussoukro) - Convention on the Game Hunting Formalities Applicable to Tourists Entering Countries in the Conseil de l'Entente
1976 Mediterranean (Barcelona) - Convention for the Protection of the Mediterranean Sea against Pollution
1982 Protocol SPA (Geneva) - Protocol concerning Mediterranean Specially Protected Areas
1976 Med. Shores (Monaco) Agreement concerning the Protection of Water of the Mediterranean Shores
1976 South Pacific (Apia) - Convention on Conservation of Nature in the South Pacific
1978 Persian Gulf (Kuwait) - Kuwait Regional Convention for Cooperation on the Protection of the Marine Environment from Pollution
1978 Amazon Pact (Brasilia) - Treaty for Amazonian Cooperation
1979 European (Bern) - Convention on the Conservation of European Wildlife and Natural Habitats
1980 CCAMLR (Canberra) - Convention on the Conservation of Antarctic Marine Living Resources
1981 SACEP (Colombo) - Articles of Association of the South Asia Cooperative Environment Programme
1981 W & Cent. Africa (Abidjan) - Convention for the Cooperation in the Protection and Development of the Marine and Coastal Environment of the West and Central African Region
1981 SE Pacific (Lima) - Convention for the Protection of the Marine Environment and Coastal Area of the South-East Pacific
1982 Red Sea (Jeddah) - Regional Convention for the Conservation of the Red Sea and of the Gulf of Aden Environment
1982 Benelux (Brussels) - Benelux Convention on Nature Conservation and Landscape Protection
1983 Central African (Libreville) - Agreement for the Cooperation and Consultation Between the Central African States for the Conservation of Wild Fauna
1983 Carribean (Cartagena de Indias) - Convention for the Protection and Development of the Wider Caribbean Region
1990 SPA Protocol (Kingston) - Protocol concerning Specially Protected Areas and Wildlife to the Convention for the Protection and Development of the Marine Environment of the Wider Caribbean Region
1985 E African Reg. (Nairobi) - Convention for the Protection, Management and Development of the Marine and Coastal Environment of the Eastern African Region
1985 PA Protocol (Nairobi) - Protocol concerning Protected Areas and Wild Fauna and Flora in the Eastern African Region
1985 ASEAN Agt. (Kuala Lumpur) - ASEAN Agreement on the Conservation of Nature and Natural Resources
1986 S Pacific (Noumea) - Convention for the Protection of the Natural Resources and Environment of the South Pacific Region (SPREP Convention)
1991 Alps (Salzburg) - Convention for Protection of the Alps

Species-related conventions

Fisheries

1949 Inter-Am Tuna (Washington) - Convention for the Establishment of an Inter-American Tropical Tuna Commission
1949 Fish, Council - Med (Rome) - Agreement for the Establishment of a General Fisheries Council for the Mediterranean
1952 European Crustaceans (Oslo) - Agreement concerning Measures for the Protection of the Stocks of Deep Sea Prawns (Pandalus borealis), European Lobsters (Homarus vulgaris), Norway Lobsters (Nethrops norvegicus) and Crabs (Cancer Pagurus)
1959 Protocol (Oslo) - Protocol Amending the Agreement concerning Measures for the Protection of Stocks of Deep Sea Prawns (Pandalus borealis), European Lobsters (Homarus vulgaris), Norway Lobsters (Nethrops norvegicus) and Crabs (Cancer Pagurus)
1952 N Pacific Fish (Tokyo) - International Convention for the High Seas Fisheries of the North Pacific Ocean
1962 Amend. to the Annex - Ammendment to the Annex to the International Convention for the High Seas Fisheries of the North Pacific

Table 30.3 Multilateral treaties (continued)

Species-related conventions (continued)

1978	Protocol (Tokyo) - Protocol Amending the International Convention for the High Seas Fisheries of the North Pacific Ocean
1958	Danube (Bucharest) - Convention concerning Fishing in the Waters of the Danube
1959	Black Sea (Varna) - Convention concerning Fishing in the Black Sea
1962	Marine Fishing (Warsaw) Agreement concerning Cooperation in Marine Fishing
1962	Salmon/Baltic Sea (Stockholm) - Agreement on the Protection of the Salmon in the Baltic Sea
1964	Fisheries (London) - Fisheries Convention
1966	Atlantic Tuna (Rio de Janeiro) - International Convention for Conservation of Atlantic Tunas
1984	Protocol (Paris) - Protocol relating to Modification of the International Convention for the Conservation of Atlantic Tuna
1966	Skagerrak Agt. (Copenhagen) - Agreement on Reciprocal Access in the Skagarrak and the Kattegut
1967	N Atlantic Fish. (London) - Convention on Conduct of Fishing Operations in the North Atlantic
1967	SE Asian Fish. (Bangkok) - Agreement Establishing the Southeast Asian Fisheries Development Center
1978	NW Atlantic (Ottawa) - Convention on Future Multilateral Cooperation in the Northwest Atlantic Fisheries
1979	S Pacific Fish (Honiara) - South Pacific Forum Fisheries Agency Convention
1980	NEAFC (London) - Convention on multilateral cooperation in North-east Atlantic Fisheries
1982	N Atlantic Salmon (Reykjavik) - Convention for the Conservation of Salmon in the North Atlantic Ocean
1983	E Pacific Tuna (San José) - Eastern Pacific Ocean Tuna Fishing Agreement
1987	S Pacific (Port Moresby) - South Pacific Fisheries Treaty
1989	Drift Net (Wellington) - Convention for the Protection of Fishing with Long Driftnets in the South Pacific
1990	Protocol I & II (Noumea) - Protocol I & II to the Convention for the Prohibition of Fishing with Long Driftnets in the South Pacific
Birds	
1950	Protect of Birds (Paris) - International Convention for the Protection of Birds
1970	Benelux - Birds (Brussels) - Benelux Convention on the Hunting and Protection of Birds
Plants	
1951	Eur. Plant Protect. Orgn. (Paris) - Convention for the Establishment of the European and Mediterranean Plant Protection Organisation
1956	Plant Protect (Rome) - Plant Protection Agreement for the Asia and Pacific Region
1967	Amend. - Amendment of the Plant Protection Agreement for the Southeast Asia and Pacific Region
1961	New Varieties (Geneva) - International Convention for the Protection of New Varities of Plants (consolidated version)
1976	N.American Plant Protect. (Yosemite) - North Atlantic Plant Protection Agreement
Animals	
1946	IWC (Washington) - International Convention for the Regulation of Whaling
1956	Prot. - Protocol to the International Convention for the Regulation of Whaling
1952	S Pacific Marine (Santiago) - Treaty for the Permanent Commission on Exploitation and Conservation of Marine Resources of the South Pacific
1972	CCAS (London) - Convention for the Conservation of Antarctic Seals
1973	Polar Bears (Oslo) - Agreement on Conservation of Polar Bears
1979	Vicuña (Lima) - Convention for the Conservation and Management of the Vicuña
1990	Wadden Seals (Bonn) - Agreement on the Conservation of Seals in the Wadden Sea
1991	Ivory Marketing (Lilongwe) - Agreement for the Establishment of Southern African Centre for Ivory Marketing (SACIM)

31. INTERNATIONAL POLICY AND LEGAL ASSISTANCE

In general, conservation action takes place ultimately within a policy and legal framework established by national governments (except for those resources lying outside sovereign territory, discussed in Chapter 33). National efforts alone, however, will not ensure that biodiversity is adequately managed and additional international measures, mainly supportive in nature, are often of great importance.

Much of the world's biological diversity is found in less developed countries which rarely are able to devote adequate resources to conservation management. It is therefore desirable for more affluent countries to provide material assistance to those less affluent, and it would clearly be equitable if those who carry the burden of conservation were properly rewarded by those who benefit from it, and this requires international systems through which costs and benefits can flow freely across national boundaries. The approaches examined in this chapter begin to meet this requirement.

The chapter surveys some of the international measures which support or assist national and local efforts in biodiversity conservation through policy or legal means. The direct support of national measures by means of international development aid is the subject of the next chapter.

These measures also provide mechanisms by which the benefits of biodiversity can be registered with those that have responsibility for its care. This is especially important in the case of biodiversity, many benefits of which are global rather than national in extent. Such global benefits include, for example, the provision of migratory bird habitats (in the case of wetlands) or carbon fixing capacity (in the case of forests). Local people making decisions about local resources of this kind will place little importance on their global benefits, but if these are ignored, there is little incentive to maintain the resources in their current state. Systems that allow these benefits to be registered in the state concerned are therefore required.

The most direct means of assistance is the 'funding mechanism' approach, which provides a basis for funding domestic regimes for provision of global public goods. The best example of such a regime is the World Heritage Convention, as discussed below.

A second route to the provision of global benefits is a 'mutual listing agreement'. These international agreements confer benefits through reciprocal obligations. For example, the Ramsar agreement on wetlands provides for the listing of at least one protected wetland site by each signatory of the agreement and, in effect, the agreement acts as an international notice-board whereby each signatory agrees to confer benefits on all others (through the conservation of a global public good), and signals this agreement by recording the conserved site on the official list.

A third means of providing global public goods is to 'privatise' them by giving people the rights to compensation for benefits produced by their local resources. This can be done through the creation of internationally recognised property rights in the previously unowned resource. One example of this is the privatisation of the world's fisheries through the development of the Economic Exclusive Zone instrument in international law (see Chapter 33). In the case of biodiversity, what is required is creation of intellectual property rights in the information value of natural genetic material, or creation of internationally transferable rights in natural habitats (such as rights of exploration with regard to genetic resources).

A fourth possible mechanism by which the global benefits of a domestic resource can be registered in the state concerned is through 'regulated trading'. Certain tangible goods are closely aligned with other goods whose values are more difficult to harness. For example, many wildlife species are traded in international markets and generate substantial amounts of value. In itself, a piece of ivory or a crocodile purse does not represent a return to biodiversity, but when these goods come from natural habitats that also contain a wide variety of unused but potentially useful species, then the return from the utilised wildlife may be seen as a return to the diverse habitat. When this is the case, it is theoretically possible to subsidise biodiversity through regulated trading in wildlife products. This could become an additional role of the Convention on International Trade in Endangered Species (CITES), which is now mainly concerned with reducing the impact of trade on exploited species.

This chapter finally looks at the UNEP Regional Seas Programme, as an example of an international framework intended to promote and coordinate national actions for conservation of the coastal and marine environment in defined supra-national regions.

FUNDING

Many countries lack the resources to address properly the environmental problems with which they are confronted. At the international level, the need for financial assistance to help less developed countries tackle such problems has long been apparent; international development aid has been an important response to this need. Another way in which financial assistance has been provided is through the framework of international conventions.

Most international environmental conventions provide at least some limited assistance to the less affluent contracting parties by providing for the administration of the convention, or funding national delegates to attend the council conference or by supporting technical studies in relation to the implementation of the objectives of the conventions. This type of assistance, although important for the effectiveness of a particular convention, is of limited effect. There are, however, a number of conventions which establish a trust fund for the explicit purpose of providing material assistance to biodiversity conservation.

The World Heritage Fund

The best known international environment trust fund is the World Heritage Fund (WHF) which was established pursuant to the 1972 World Heritage Convention (WHC). The WHF grants financial assistance to protect cultural and natural heritage of outstanding universal value. The fund is administered by the World Heritage Committee (the 'Committee'), which was established within UNESCO.

The annual budget of the WHF is approximately $2.0 million (Table 31.1). This is raised by a combination of voluntary or compulsory contributions from the contracting parties. Although Article 15(2) lists a number of potential sources of funding, most important is the obligation contained in Article 16 which stipulates that contracting parties will contribute to the fund either compulsorily or voluntarily one per cent of their contribution to the regular budget of UNESCO every two years. The voluntary contributions are in effect the same as the compulsory contributions with respect to amount and timing; the distinction was made because it was felt that internal ratification procedures would be simpler in some states if contributions were technically voluntary.

The WHC does not normally allow the World Heritage Committee to accept contributions to be used only for a certain programme or project. An exception can be made, however, if the Committee has decided on the implementation of the programme or project. The Committee is composed of 21 members elected by the parties to the convention. Election of Committee members must ensure an equitable representation of the different regions and cultures of the world.

The WHF is used to provide assistance to contracting parties to the WHC. Any State Party to the WHC may request international assistance for property forming part of the world cultural or natural heritage. The request should include a description of the contemplated operation, the necessary work, the expected cost, the degree of urgency, and the reasons that the requesting State cannot meet the expenses of the project with its own resources. Before making a decision, the Committee must carry out any studies and consultations that it deems necessary.

Under the WHC, assistance may take the form of: studies, provision of experts and other staff to ensure that approved work is carried out, training of staff and specialists, supply of equipment needed by the State, loans, and non-repayable subsidies. Assistance under the WHC may also be granted to national or regional centres for the training of staff and specialists. Large-scale assistance must be preceded by scientific, economic, and technical studies. Table 31.1 outlines expenditure of the WHF with regard to each of these types of assistance.

The assistance provided by the WHF generally covers only a part of the work necessary. The State benefiting from the assistance must contribute a substantial share of the resources for its programme or project, unless its resources do not permit it to do so.

Despite the relatively small size of the fund's resources it does illustrate a number of important features. The WHC, with 117 contracting parties, is one of the most widely accepted international environmental conventions and this is vital for the success of an international convention. The WHF is an important reason for this popularity as it provides an incentive to balance the obligations of the convention. Therefore, many states can see accession to the WHC as not only satisfying a moral and political need but also as providing some material benefit.

The system of raising contributions, based on the UNESCO scale, means that both overall donors and recipients from the WHF must contribute to the fund. This requirement is important because it gives the WHF an international basis and means that it is seen to be more than just another form of aid from the developed world. It means that the World Heritage Committee retains a greater degree of control over the use of resources because it is not simply another form of multilateral aid, and it is less likely to be the subject of political manoeuvring.

The WHC itself is discussed briefly below, and its importance for protected area systems noted in Chapter 29.

The International Oil Pollution Fund

The earliest example of this class of funding mechanism is the International Maritime Organisation (IMO) Oil Pollution Fund, established in 1971 pursuant to the International Convention on the Establishment of an International Fund for Compensation for Oil Pollution Damage. This convention provides for a free-standing fund, that awards additional compensation to any person suffering oil pollution damage, to the extent that the protection offered by its companion treaty, the 1971 International Convention on Civil Liability for Oil Pollution Damage, is inadequate.

The Fund is administered by an Assembly, a Secretariat headed by a Director, and an Executive Committee. The Assembly consists of all Contracting States to the Convention. The Assembly's responsibilities include deciding how to distribute available compensation. The Assembly must meet once a year, and can hold extraordinary sessions if requested by the Executive Committee or at least one-third of the members of the Assembly.

The Executive Committee consists of one-third of members of the Assembly but of not less than seven or more than 15 members. There are 47 States which are members of the IOPC Fund with a further 14 expected to join in the near future. In electing the members of the Executive Committee the Assembly must secure an equitable geographic distribution on the basis of an adequate representation of Contracting States particularly exposed to the risks of oil pollution and of Contracting Parties having large tanker fleets. The Executive Committee must meet at least once a year. The primary responsibility of the Executive Committee is approving the settlement of claims against the IOPC Fund.

The IOPC Fund is financed by initial and annual contributions. Initial contributions are payable when a State becomes a Member of the IOPC Fund and is calculated on the basis of a fixed amount per tonne of oil received the year preceding the State's entry to the convention. Annual contributions are paid by any person who has received in the relevant calendar year more than 150,000 tonnes of

Table 31.1 World Heritage Fund accounts 1988-1991

	1988		1989		1990		1991	
	FUNDS ALLOCATED BY COMMITTEE US$	FUNDS OBLIGATED-SPENT US$	FUNDS ALLOCATED BY COMMITTEE US$	FUNDS OBLIGATED-SPENT US$	FUNDS ALLOCATED BY COMMITTEE US$	FUNDS OBLIGATED-SPENT US$	FUNDS ALLOCATED BY COMMITTEE US$	FUNDS OBLIGATED-SPENT US$
Preparatory assistance and regional studies	100,000	82,800	100,000	30,000	150,000	121,476	150,000	52,500
Technical co-operation	700,000	435,463	700,000	515,500	700,000	364,900	600,000	372,782
Training	500,000	384,430	500,000	278,500	550,000	411,500	500,000	208,185
Emergency assistance	200,000	30,000	100,000	-	100,000	41,785	100,000	2,568
Promotional activites	150,000	94,415	150,000	74,750	200,000	179,044	250,000	165,600
Advisory services	280,000	279,700	247,200	242,200	300,000	258,176	420,000	175,471
Travel for experts of LDC's of Committee	-	-	-	-	20,000	11,554	20,000	-
Temporary assistance to secretariat	260,000	260,000	210,700	210,700	135,000	135,000	190,000	190,000
TOTAL	2,190,000	1,566,808	2,007,900	1,351,650	2,155,000	1,523,435	2,230,000	1,167,106
3% contingency funds	-	-	-	-	65,000	-	70,000	-
					2,220,000		2,300,000	

Source: World Heritage Committee Annual Accounts 1988, 1989, 1990 and 1991.

crude oil in a Member State. Annual contributions are levied to meet the anticipated payments by the IOPC Fund and the administrative expenses of the Fund during the coming year. The levy of contributions is based on reports of oil receipts which are submitted by Governments of Member States. The contributions are paid by the individual contributors directly to the IOPC Fund. Governments have no responsibility for these payments.

The IOPC Fund establishes two types of accounts or funds. The first is the general fund from which are paid the administrative expenses of the funds and general claims. The other type of fund is the major claims fund which are established to meet any potential liability from major incidents, such as the sinking of an oil tanker.

In October 1991 the Assembly decided to levy annual contributions which amounted to £26.7 million. Of this, £5.0 million was for the general fund, and £6.7 and £15.0 million were for specific major claims funds. The payments made by the IOPC Fund vary considerably from year to year. As a result, the level of contributions to the Fund varies, as illustrated in Table 31.2.

Two important and unusual features of the IOPC Fund are the method of raising the funds and the system of assessing contributions. The method of fund-raising is the first example where governments have allowed an international fund to raise income directly from private individuals. This method has a number of advantages over restricting the source to the public sector. It is a more efficient in that it eliminates layers of unnecessary administration and it means that politically unpalatable choices are more easily made. It allows the fund to assume a degree of impartiality which is desirable in the situations with it is involved. The other unusual feature of this fund is the system of assessing the amount of contributions which are required from year to year. Unlike the World Heritage Convention, which is tied to the UN scale, the IOPC fund has a potentially open-ended method of calculation based upon what is required for the purposes of the convention, not what governments may be willing to give.

Table 31.2 Contributions to IOPC fund

YEAR FUND	GENERAL CLAIMS FUNDS £	MAJOR LEVY £	TOTAL £
1979	750,000	0	750 000
1980	800,000	9,200,000	10,000,000
1981	500,000	0	500,000
1982	600,000	260,000	860,000
1983	1,000,000	23,106,000	24,106,000
1984	0	0	0
1985	1,500,000	0	1,500,000
1986	1,800,000	0	1,800,000
1987	800,000	400,000	1,200,000
1988	2,900,000	90,000	2,990,000
1989	1,600,000	3,200,000	4,800,000
1990	500,000	0	500,000
1991	5,000,000	21,700,000	26,700,000

The Global Environmental Facility

Despite the advantages of private sources of income for an international convention, governments are generally reluctant to establish this type of funding mechanism. Rather, there is a strong preference for restricting the income of these funding mechanisms to public sources. An illustration of this preference of donor governments to use this 'voluntary' or public method of raising finances for international environmental funds is the Global Environmental Facility (GEF) which is examined in greater detail in Chapter 32.

The GEF has been proposed as the vehicle for funding arrangements pursuant to any new international environmental agreements. As such it would take over the role that the trust funds described above have been established for. The International Bank for Reconstruction and Development (World Bank) Draft Resolution on the GEF contains a provision that embodies this approach: "The Bank is authorised to enter into other agreements with countries party to international agreements for the protection of the global environment, international organisations and other entities in order to administer and manage financing for the purpose of, and on terms consistent with, this Resolution."

The GEF will establish a new multilateral fund under which grant or concessional loans will be given on an additional basis to developing countries to enable them to implement programmes that protect the global commons. The GEF is capitalised at $1.0 billion to spend by the end of 1993. The fund is financed by voluntary contributions mainly from the developed countries. The World Bank manages the GEF and organises project selection, appraisal and supervision, with UNDP and UNEP participation.

The GEF allocates resources to projects that have any of the following aims: protection of the ozone layer, limitation of greenhouse gas emissions, protection of biodiversity, or protection of international waters. To be eligible for funding the project must also (1) be within cost-effectiveness guidelines to be defined; and (2) provide measurable benefits to the implementing country's economy that are too low to trigger investment by the implementing country, or provide global environmental benefits that warrant modification of project design.

Projects that are economically viable on the basis of domestic benefits and costs to the implementing country are not eligible for GEF financing unless a compelling case is made that the operation would not proceed without GEF involvement.

The level of capital for the GEF is the largest ever allocated to this type of mechanism. The GEF in its short history has, however, been the subject of much controversy. The important role played by the World Bank in its administration is seen by some as compounding the problems which the GEF was established to solve.

The Wetlands Conservation Fund

Conventions which have been established for some time are now establishing funding mechanisms. One example of this is the 1971 Ramsar treaty where the conference of the contracting parties in January 1990 (pursuant to resolution C.4.3.) established a 'Wetlands Conservation Fund' to assist countries to implement the objectives of the convention (see Chapter 29 for details of Ramsar sites).

The fund established pursuant to this convention is to be operated in a similar way to the WHF. On request from a competent national authority, the fund may provide any developing country which is a Contracting Party to the Convention with financial support for wetland conservation activities in one of the following fields: improving management of sites on the Ramsar List (e.g. management plans, emergency action); designating new sites (e.g. surveys, delineation of boundaries); promoting wise use (e.g. preparing requests to development agencies, institutional development, training); regional and promotional activities (e.g. seminars, public education, information activities).

Developing countries which are not yet Contracting Parties may request a grant to support activities necessary for designating a site for the List (e.g. site identification, delineation or mapping).

Applications to the fund are reviewed by the Standing Committee and administered by the Bureau. A meeting of a sub-committee of the Standing Committee, held in Australia in December 1990, developed procedures for the operation of the fund.

By early 1991, voluntary contributions had been received from the Netherlands government and WWF, and had been promised by the governments of Austria, Switzerland, UK and the USA. Other governments which have indicated interest include: Denmark, Finland, France, Germany, Italy, Japan, Norway and Sweden.

The Kuwait Fund

The UN Kuwait Compensation Fund established pursuant to Security Council Resolution 687 (1991) (The Kuwait Fund) is an international fund which, although not established pursuant to a convention, could be an indicator of possible future developments in this type of mechanism. The fund is intended to meet compensation claims resulting from the Gulf War for, among other reasons, "environmental damage and the depletion of natural resources ...".

On 2 May 1991 the UN Secretary-General presented a report to the Security Council setting out his recommendations for the establishment and administration of the Kuwait Fund. The fund is to operate in accordance with UN Financial Rules and Regulations. It will be administered by a Commission, which will function under the authority of the Security Council and be a subsidiary organ thereof. The principal arm of the Commission will be a 15-member Governing Council, assisted by commissioners to be nominated by the Secretary-General and appointed by the Governing Council, and a secretariat.

This fund has a number of unique characteristics. Although the Secretary-General's report did not specify the size of the Kuwait Fund, it is expected that it will raise up to $35 billion over the next 10 years. This figure would make the Kuwait Fund the largest trust fund ever established.

INTERNATIONAL OBLIGATIONS: PROTECTED AREAS

The best way to ensure the fullest possible protection of biodiversity is to pursue its preservation *in situ*. This means protecting natural habitat to the extent that the integrity of all of its ecological functions are maintained.

The most important mechanism used in international treaties to protect natural habitats is the inclusion of an obligation for the parties to establish protected areas. The paragraphs below discuss international systems from this point of view; sites and coverage are discussed in Chapter 29.

These obligations facilitate the protection of natural habitats in several ways. Firstly, they are public declarations by governments committing themselves, morally if not legally, to protecting natural habitats. This public commitment may then be exploited by interested parties within a State to promote the establishment of protected areas needed to satisfy the obligations of the convention. This can be an effective means of overcoming government inertia, reluctance or opposition. A prime example of this is the use of the obligations in the World Heritage Convention by environmental pressure groups in Australia; several new national parks have been created despite strong opposition within government.

International obligations are also useful because of the clear capacity for mutual gain to be achieved by mutual obligations regarding the protection of natural habitats. Each State that undertakes to protect some parts of its diverse natural resources benefits from undertakings made by other parties. However, it is also limited for the same reason. This is because the world's diverse resources are not uniformly distributed across all nations; some have much more and others much less of the global total. Reciprocity in the declaration of equal amounts of protected areas is not a sufficient basis for ensuring full protection of the diversity that exists in those States with the greatest shares.

The development of the 'mutual listing' mechanism has evolved with the changing attitude of man towards nature. Initially, this mechanism was incorporated into conventions whose primary purpose was the protection of 'important' wildlife, by the establishment of game reserves. An early example of this is the 1909 Convention for the Preservation of Wild Animals, Birds and Fisheries in Africa which 'encouraged nature reserves'. A few decades later the intrinsic value of natural habitat itself, as something more than the producer of game, came to be recognised. One of the first treaties to incorporate this shift in emphasis to the protection of natural habitat for its own sake was the 1940 Washington Convention on Nature Protection and Wildlife Preservation in the Western Hemisphere (Western Hemisphere Convention).

This convention became a model for many subsequent treaties. Its operative language called upon the contracting parties to establish various types of protected areas, and then to list these with the Organisation of American States. The four types of protected areas defined in the convention are: National Parks, National Reserves, Nature Monuments, and Strict Wilderness Reserves. The careful definition of what constitutes a protected area and the provision of an international 'notice-board' for making these designations public are the essential ingredients of a listing regime.

Although the Western Hemisphere Convention was the first to extend protection to habitat for reasons other than game and wildlife conservation, the intended scope of the treaty remained somewhat narrow. It provided only for the protection of areas labelled of special significance because of a special animal or monument.

During the 1960s the concept of what was of special significance and therefore worthy of protection expanded to include areas of particular biological richness and diversity, even though the areas might not necessarily include any one species of special significance. This development is well illustrated by the adoption of the 1971 Convention on Wetlands of International Importance especially as Waterfowl Habitat (Ramsar). Wetlands had long been under particular threat and were generally regarded as wastelands. However, the wide range of ecosystem services rendered by these wetlands, in the maintenance of fisheries, wildlife and general services, came to be recognised and the result was a protected areas convention providing for the mutual obligation of all parties to designate protected wetlands.

As the perceived threats to protected areas have changed so has the nature of the obligation built in to establish such areas. Initially, the integrity of a protected area was believed to be safeguarded by simply ensuring that activities within the area were controlled. In the early treaties, such as the Western Hemisphere Convention, no attention is given to activities outside the protected area which may have a harmful effect on its integrity; this was remedied in later treaties. An example of this is the 1968 African Convention on the Conservation of Nature and Natural Resources (African Convention), which requires parties to the convention to establish buffer zones in order to control activities "which may have harmful consequences on the ecosystem" within the established protected areas. By 1985 when the ASEAN Agreement on the Conservation of Nature and Natural Resources (ASEAN Convention) was adopted, the establishment of buffer zones had become standard practice.

Probably the best-known example of this approach is the UNESCO Man and the Biosphere Programme (MAB). This programme was established to promote sustainable utilisation of natural resources, and to protect natural habitats from incompatible developments in the immediate vicinity. Initiated officially in 1971, MAB was a direct consequence of the Biosphere Conference of 1968 and the earlier international biological programme of the International Council of Scientific Unions. MAB became operational in 1976, and provides for the establishment of 'Biosphere Reserves' of various types throughout the world. UNESCO biosphere reserves are a special kind of protected area that rely upon zoning (i.e. designated land-uses) to safeguard biological diversity. In theory, a biosphere reserve encompasses a core zone that represents one of the earth's major ecosystems and is large enough to permit *in situ* conservation of its genetic material. These core zones are meant to be undisturbed by human activity, except for scientific research. Multi-use buffer zones are intended to surround the core, and these should be managed for the economic benefit of local populations.

Recently, the protected area approach has been extended to protect natural habitats in the international commons, including the High Seas, Antarctica and Outer Space. In these instances, parties have agreed to protect natural habitats not by establishing protected areas but by mutually agreeing to regulate or ban certain activities in the area concerned. This type of protection is illustrated by the Antarctica Treaty System where under the most recent protocol to the Antarctic Treaty the entire area is to be declared a protected area.

The extent of the obligations created in these international instruments can vary from the mandatory to the purely hortatory. Most examples are intermediate. For example, in the Ramsar Convention the obligation to protect natural habitat is relatively generalised; Article 4(1) of that treaty merely requires "each contracting party to promote the conservation of wetland and waterfowl by establishing nature reserves on wetlands". However, in order to become a party to the convention the State must nominate at least one area to be included in the list of significant wetland sites. The World Heritage Convention includes more detailed and specific obligations; Article 4 requires each contracting party to recognise the duty of identification, protection, and conservation of natural heritage as defined in the convention. It goes on to require each party to "do all it can to this end, to the utmost of its own resources and, where appropriate, with any international assistance and cooperation".

The benefits of careful construction of the language of obligation are seen when attempts are made to enforce these undertakings. The nature of the obligation created by the World Heritage Convention has been the subject of judicial consideration in a series of cases in Australia, where the High Court held that the language of Articles 4 and 5 created a binding obligation on the contracting parties to do all they can to protect sites on the World Heritage List.

The extent to which these international obligations have led to increased protection of natural habitats by means of protected area establishment is difficult to assess accurately; certainly, many such areas are now listed as World Heritage or Ramsar sites, or as Biosphere Reserves (see Chapter 29). It is clear that even though the effect of these obligations may be hard to quantify, they have been an important method of protecting the world's biological diversity.

INTELLECTUAL PROPERTY RIGHTS FOR BIOTECHNOLOGY

Intellectual Property is the term used to describe the branch of law which protects the application of thoughts, ideas and information which are of commercial value. It thus covers the law relating to patents, copyrights, trademarks, trade secrets and other similar rights (Cornish, 1989).

The development of the genetic resources of biodiversity is known as *biotechnology*. Broadly defined, biotechnology includes any technique that uses living organisms or parts of organisms to make or modify products, to improve plants or animals, or to develop microorganisms for specific uses (Congress of the United States, Office of Technology Assessment, 1990). Mankind has used forms of biotechnology since the dawn of civilisation. However, it has been the recent development of new biological techniques (e.g., recombinant DNA, cell fusion, and monoclonal antibody technology) which has raised fundamental social and moral questions and created problems in intellectual property rights.

Intellectual property protection for biotechnology is currently in a state of flux. Whilst it used to be the case that living organisms were largely excluded from protection, attitudes are now changing and increasingly biotechnology is receiving some form of protection. These changes have largely taken place in the USA and other industrialised countries, but as other countries wish to compete in the new biotechnological markets, they are likely to change their national laws in order to protect and encourage investment in biotechnology.

There is at the moment no clear international consensus on how biotechnology should be treated. Although bodies such as the World Intellectual Property Organization (WIPO, the United Nations permanent body primarily responsible for international cooperation in intellectual property), and the Organization for Economic Cooperation and Development (OECD) have conducted separate studies and produced various reports, these have only sought to make governments more aware of the potential problems and to offer some suggested solutions. In view of the highly controversial nature of providing intellectual property protection for biotechnology, it is likely that in the short term developments will be at a national and regional level.

Intellectual property protection currently available

There are currently two main systems of protection for biotechnology: rights in plant varieties, and patents. Both systems provide exclusive, time-limited rights of exploitation and are described in more detail below.

Keeping biotechnology 'secret' can also be a valuable form of protection. National treatment of trade secrets is diverse, and all attempts to harmonise trade secret laws in Europe, for example, have failed. Most jurisdictions do provide some form of protection against those who steal or use others' trade secrets unfairly. However, the problem with this form of protection is that the secret generally becomes public once the biotechnology is used commercially and thus the protection is lost.

It is conceivable that the law of copyright could afford some protection for biotechnology. Lines of genetic code are analogous to some extent with computer program code, which has now been incorporated into the copyright systems of most industrialised countries. However, this route to protection is fraught with practical and conceptual difficulties and is generally thought to be unsuitable. There is as yet no recorded case of biotechnologists claiming copyright in their inventions.

Trademarks are also unlikely to be of much use in protecting biotechnology, though they may of course prove important later in regard to marketing products, processes or services. An attempt to register the name of a plant or an animal as a trade mark is unlikely to be successful as public policy would prevent it (in England, registrations for names of varieties of roses have been removed from the Trade Mark Register for lack of distinctiveness and because of the likelihood of confusion).

Rights in plant varieties

Prior to the mid-1960s only a few countries (e.g., Germany, USA) gave any intellectual property protection to plant varieties. Because of pressure from their plant breeding industries, 10 western European countries entered into a diplomatic process in the early-1960s which eventually culminated in the formation of an International Union for the Protection of New Varieties of plants (UPOV) and the signing of a Convention (the UPOV Convention 1961). Since that time a number of other countries have become parties to the UPOV Convention (the full list of 19 parties appears in Table 31.3). Amendments were made to the UPOV Convention in 1978, principally to facilitate the entry of the USA.

The UPOV Convention requires that each member country must adopt national legislation to give at least 24 genera or species protection, in accordance with the provisions of the convention, within eight years of signing. A plant variety is protectable ("a protectable variety") under the UPOV system if it is distinct, uniform, stable (DUS) and satisfies a novelty requirement. Novelty and distinctiveness equate broadly to novelty under patent law, but are more leniently applied in comparison to the patent rule. Satisfaction of the DUS criteria is conducted by the national authority responsible, usually by growing the variety over at least two seasons. There is also an important requirement that the variety be maintained throughout the duration of protection. A country may apply the system to *all* genera or species, but there is no obligation to do so and thus the system has been extended only gradually. In addition, the UPOV Convention allows national legislation to discriminate against foreigners (including nationals of a UPOV Convention country) under the principle of reciprocity. Thus amongst the UPOV members there is still some disparity in protection.

Duration of protection depends on national legislation and on the plant species to which the variety belongs, but is generally for 20-30 years. Grant of plant variety rights confers certain exclusive rights on the holder, including the exclusive right to sell the *reproductive* material (e.g. seed, cuttings, whole plants) of the protected variety. However the rights do not extend to *consumption* material (e.g. fruit, wheat seed grown for milling flour). Essentially the exclusive rights define what others may or may not do in relation to the protected varieties.

Plant breeders were for some time dissatisfied with the protection provided by the UPOV system. This eventually resulted in a major diplomatic conference in March 1991, at which the UPOV Convention was substantially revised. The new 1991 text will provide far greater protection than is afforded at present, most notably by requiring that all member countries apply the convention to *all* genera and species, by extending the exclusive rights to include *harvested* material (e.g., fruit, wheat grown for milling into flour) and, most controversially, by allowing enforcement against farm-saved seed (where a farmer produces further seed of the protected variety from the previous year's crop). However, until the national governments ratify the new convention the system will continue to be based on the 1978 text. There will be considerable national opposition to the strengthening of plant variety rights and thus these changes may take years before they are implemented and may even be superseded by greater availability of patent protection in the meantime.

Patents for biotechnology

A patent is a grant of exclusive rights for a limited time in respect of a new and useful invention. The exact requirements for grant of a patent, the scope of protection it provides and its duration differs depending on national legislation. However, generally the invention must be of patentable subject matter, novel (new), non-obvious (inventive), of industrial application and sufficiently disclosed. A patent will provide a wide range of legal rights, including the right to possess, use, transfer by sale or gift, and to exclude others from similar rights. Duration will be for around 20 years (although for only 17 years in the USA). These rights are generally restricted to the territorial jurisdiction of the country granting the patent and thus an inventor wishing to protect his/her invention in a number of countries will need to seek separate patents in each of those countries. Whilst the majority of countries provide some form of patent protection, only a few provide patent protection for biotechnology (these include: Australia, Bulgaria, Canada, Czechoslovakia, Hungary, Romania, Japan, the Soviet Union and the parties to the European Patent Convention). The reasons for this may differ, but generally it has been because biotechnology has been thought inappropriate for patent protection, either because the system was originally designed for mechanical inventions, or for technical or practical reasons, or for one or more ethical, religious or social concerns. In all the National Patent Offices where patents are granted for biotechnology there is a considerable backlog of pending applications. Even in those countries where patent protection is provided, the type and extent of that protection is different in nearly every national system.

It has largely been the USA which has broken new ground in providing the possibility of patent protection for "anything under the sun that is made by man". Patents have been granted for plants since 1930 in the USA, under The Plant Patent Act. However, prior to 1980, the US Patent Office would not grant utility patents (separate from The Plant Patent Act) for living matter because it deemed products of nature not to be within the terms of the utility patent statute. That was until the landmark decision of the US Supreme Court in *Diamond v Chakrabarty* (from which the above quote is taken), which held that a particular genetically engineered bacterium was statutory subject matter for a utility patent. This decision has been the basis upon which patents have been granted for higher life forms. Subsequently it has been held that a utility patent may be

Table 31.3 International intellectual property treaties (party states as at 1 January 1991)

	PARIS	UPOV	MICRO	PCT	EPC
ASIA					
Bangladesh	●				
China	●				
Cyprus	●				
Indonesia	●				
Iran, Islamic Rep	●				
Iraq	●				
Israel	●	●			
Japan	●	●	●	●	
Jordan	●				
Korea, Dem People's Rep	●			●	
Korea, Rep	●		●	●	
Lebanon	●				
Malaysia	●				
Mongolia	●				
Philippines	●		●		
Sri Lanka	●			●	
Syria	●				
Turkey	●				
Viet Nam	●				
USSR					
Soviet Union[1]	●		●	●	
EUROPE					
Austria	●		●	●	●
Belgium	●	●	●	●	●
Bulgaria	●		●	●	
Czechoslovakia	●		●		
Denmark *	●	●	●	●	●
Finland	●		●	●	
France	●	●	●	●	●
Germany	●	●	●	●	●
Greece	●			●	●
Hungary	●	●	●	●	
Iceland	●				
Ireland *	●	●			
Italy *	●	●	●	●	●
Liechtenstein	●		●	●	●
Luxembourg	●			●	●
Malta	●				
Monaco	●			●	
Netherlands *	●	●	●	●	●
Norway	●		●	●	
Poland	●	●		●	
Portugal *	●				
Romania	●			●	
Spain *	●	●	●	●	●
Sweden	●	●	●	●	●
Switzerland	●	●	●	●	●
United Kingdom *	●	●	●	●	●
Vatican City	●				
Yugoslavia	●				
OCEANIA					
Australia	●	●	●	●	
New Zealand	●	●			

	PARIS	UPOV	MICRO	PCT	EPC
NORTH AND CENTRAL AMERICA					
Bahamas	●				
Barbados	●			●	
Canada	●			●	
Cuba	●				
Dominican Republic	●				
Haiti	●				
Mexico	●				
Trinidad and Tobago	●				
United States	●				
SOUTH AMERICA					
Argentina	●				
Brazil	●			●	
Suriname	●				
Uruguay	●				
AFRICA					
Algeria	●				
Benin	●			●	
Burkina Faso	●			●	
Burundi	●				
Cameroon	●			●	
Central African Rep	●			●	
Chad	●			●	
Congo	●			●	
Côte d'Ivoire	●				
Egypt	●				
Gabon	●			●	
Ghana	●				
Guinea	●				
Guinea-Bissau	●				
Kenya	●				
Lesotho	●				
Libya	●				
Madagascar	●			●	
Malawi	●			●	
Mali	●				
Mauritania	●			●	
Mauritius	●				
Morocco	●				
Niger	●				
Nigeria	●				
Rwanda	●				
Senegal	●			●	
South Africa	●	●			
Sudan	●			●	
Tanzania	●	●	●	●	
Togo	●			●	
Tunisia	●				
Uganda	●				
Zaire	●				
Zambia	●				
Zimbabwe	●				

Source: World Intellectual Property Organisation 1991.
Note: * Member States of the European Community [1] Refers to former USSR.
Key to column headings: **PARIS** - means the State is a member of the International Union for the Protection of Industrial Property (Paris Union), founded by the Paris Convention for the Protection of Industrial Property, and has ratified or acceded to at least the administrative and final provisions Articles 13 to 30) of the Stockholm Act (1967) of that Convention. **UPOV** - means the State is a party to the International Convention for the Protection of New Varieties of Plants (either the 1961 version or the revised 1978 version). **MICRO** - means the State is a party to the Budapest Treaty on the International Recognition of the Deposit of Microorganisms for the Purposes of Patent Procedure. **PCT** - means the State is a party to the Patent Cooperation Treaty. **EPC** - means the State is a party to the European Patent Convention.

granted for plants and a patent has been granted for an animal. Polyploid oysters, not naturally occurring, were held to be patentable subject matter and US Patent No.3,736,866, was issued in respect of a "transgenic non-human mammal all of whose germ cells and somatic cells contain a recombinant activated oncogene sequence introduced into the said mammal, or an ancestor of said animal, at an embryonic stage" - popularly known as the 'onco-mouse'.

Elsewhere, the treatment of applications for patents for living matter is far from certain. Whilst patents are granted in many countries for plants and microorganisms, it has been the issue of patents for animals which has been most controversial. Whilst it is not possible to summarise succinctly the position in the rest of the world, it is possible to describe the present approach of those countries which are party to the European Patent Convention (the EPC, see Table 31.3). The EPC is a regional arrangement entered into by 14 European countries for the purpose of making multiple applications for any of the member countries a great deal easier and to introduce a common system for patent protection. An application under the EPC is for a European patent, or Europatent, for short. If a Europatent is granted by the European Patent Office (EPO) it has the same effect, and is subject to the same conditions, as a national patent in each of the member countries designated in the application. In other words, through a single application a bundle of national patents can be obtained.

The EPC provides that "plant or animal varieties or essentially biological processes for the production of plants or animals" are excluded from patent protection (although the exclusion is expressly stated not to apply to microbiological processes and products). These exclusions would appear to place unequivocal prohibition on Europatents for macrobiotechnology. However, the EPO has been taking an increasingly narrow view of these exclusions, and has held that they do not exclude *all* plants and animals *per se*, but only claims for *varieties* of plants or animals and that a process is not "essentially biological" if there has been substantial interference by man.

It is also important to note that there is currently before the European Parliament of the European Community (EC) a proposal for a Council Directive for harmonisation of the legal protection provided for biotechnology in the EC. This does not propose to amend the EPC, but the present draft proposal would make even more opportunities available for patenting biotechnology and thus make the EC more attractive in terms of investment in biotechnology research.

International treaties

There are three international intellectual property treaties which are of particular importance for the protection of biotechnology: the Paris Convention for the Protection of Industrial Property (the Paris Convention); the Budapest Treaty on the International Recognition of the Deposit of Microorganisms for the Purposes of Patent Procedure (the Deposit Treaty) and the Patent Cooperation Treaty (PCT) (see Table 31.3).

The Paris Convention was originally signed in 1883 by just 11 countries, but now the majority of countries who have any form of intellectual property law are parties to it. The keystone to the convention is the principle of national treatment: an applicant from one convention country shall have the same rights in a second convention country as a national of that second country. The convention covers patents and defines them so broadly that it permits application to any of the forms of industrial patents granted under the laws of the convention countries. The most important practical result of the convention is that it is possible to claim priority from an application made in a convention country for all subsequent convention countries within 12 months of the original filing.

The Deposit Treaty, as the full title suggests, is concerned with the deposit of examples of microorganisms for the purposes of patent applications. Applications for patents for biotechnology often face considerable difficulties in describing the nature of the invention sufficiently. The Deposit Treaty is a vehicle for solving these problems, primarily through the setting up of a series of International Depository Authorities (IDA) and through the recognition by all member countries of a deposit in a single IDA.

The PCT simplifies the process of filing patent applications simultaneously in a number of countries. Under the PCT a single application may be filed in one of the official receiving offices, designating any number of PCT member countries, which can eventually result in a national patent being granted in each of the designated states (and/or a Europatent). A prior-art search is performed by the receiving office and a report sent to the applicant. The application and report are published and the application will then move on either to an international preliminary examination followed by national examination, or alternatively straight to the national examination stage. Unfortunately, the eventual outcome is not a 'world patent' and there is no harmonisation patent law under the PCT apart from the procedural aspects.

Case study: the Iguana Management Programme

The Green Iguana *Iguana iguana* of Latin America is a highly prized source of meat and eggs. Green Iguanas are arboreal herbivores which can grow up to 2m in length and can weigh as much as 6kg (about 82% of the lizard is edible). They need about half as much food as a chicken or rabbit to produce the same amount of meat. The species is now widely threatened because of excess hunting and habitat destruction.

Research into the reproductive behaviour of the Green Iguana was begun in 1983 and resulted in development of new management techniques for ranching. A 'genetic brood stock' of adult iguanas which are larger, faster growing and more productive has been developed. The research has largely been the work of the Pro Iguana Verde Fundación (formed by Dagmar Werner in 1985). The Fundación's programme for training and advice on Iguana ranching is called the Iguana Management Programme (IMP). The IMP is based in Costa Rica but it is intended to implement it throughout Latin America and possibly elsewhere.

The primary purpose of the IMP is to conserve living natural resources; its basic premise is that if farmers can raise iguanas as a food crop, the status of the wild species will be improved and forest clearance might be reduced. Farmers adopting iguana ranching would have to protect or re-establish areas of forest to provide food for stock. Research indicates that meat production per hectare by iguanas is approximately three times higher than by cattle. Income can be derived from selling iguanas and their products (meat, eggs, leather) and products from the forest.

The new technology and expertise which have been incorporated into an iguana ranching model are being applied for an industrial purpose (i.e. agriculture) and are of commercial value; they thus fall within the area of intellectual property law as applied to biotechnology. The biotechnological components of the ranching model are the genetic brood stock (the Fundación has 'bioengineered' an improved stock of Green Iguanas) and the husbandry procedures (egg laying and incubation, nutrition, disease control, release and harvesting). These are forms of 'original or traditional biotechnology', as opposed to 'new biotechnology' which is largely laboratory-based and dependent upon human manipulation of genetic material.

Intellectual property rights provide the means for compensating the Fundación for its efforts. The technologies involved in the IMP are vulnerable to piracy. Much of the work of the Fundación is contained in the genetic make-up of the Genetic Brood Stock. Once these Iguanas are transferred or sold the Fundación loses its direct control over the animals. In addition, the success of the Iguana ranching model is dependent on the expertise to use the technologies efficiently; this is information which took years to develop but which can be pirated very easily once a licence is purchased. The Fundación needs to be able to disseminate its innovations and expertise in the security of knowing that it cannot be re-sold by pirates and that there will be no reduction of the licensing potential. Only internationally recognised intellectual property law can provide these types of protection.

Because of the uncertainties of the world's intellectual laws with regard to biotechnology the availability of protection for the most important components of the IMP is questionable. At present there is widespread discrimination against the application of intellectual property rights to natural genetic materials and in favour of human-modified genetic materials. This provides no incentives for exploitation of useful genetic materials in the natural environment, even though in developing countries natural resources are obvious subjects for investment. However, one important way to limit conversion of natural resources is to ensure that fair value is paid for current uses of the existing resource base. Intellectual property rights could be a means of influencing developing countries to maintain and develop diverse resources in return for the value that these resources render to the world community.

REGULATED TRADING IN WILDLIFE PRODUCTS

Regulated trading in wildlife products has the capacity of returning benefits to the users of natural habitats. It could do this if the trade were regulated in such a way as to support prices, much as is done at present with respect to agricultural commodities, where price supports provide incentives for maintaining land in its current state, as opposed to converting it to other purposes.

At present, there is no regulated trading mechanism of exactly this nature. There are, however, a number of existing international agreements which do seek to regulate trade in wildlife products. Early examples are the Western Hemisphere Convention and the 1950 Paris International Convention for Protection of Birds. These simply outlined in broad terms an obligation to control trade in wildlife products but created little structure within which these controls could be implemented. Both conventions consequently became 'sleeping treaties'. Undoubtedly the most important and effective convention which places some control on the economic exploitation of wildlife products and thereby protects biological diversity is the Convention on International Trade in Endangered Species of Wild Fauna and Flora (CITES).

The evolution of CITES

CITES is the most widely accepted of international treaties on the conservation of natural resources. The number of Parties has been steadily increasing from the initial signing of the convention in 1973 to a total of 113 in 1992 (Fig. 31.1)

The convention attempts to prevent commercial trade in species of wildlife which are in danger of extinction and to control the trade in species which might become so if their trade was allowed to continue unchecked. It does this by means of two lists of species: Appendix I contains those species banned from international commercial trade and Appendix II, those for which trade may take place provided that export permits have been issued. Importing countries are obliged to ensure that all imports of Appendix II specimens are accompanied by correct export permits.

One of the main obligations of Parties is to submit to the Convention Secretariat annual reports of all of their trade in species included in the Appendices. The number of annual reports submitted is also shown in Fig. 31.1. These data are then compiled on a computer database and in this way it is possible to determine the global levels of trade in each species. At a fine level of resolution, the trade emanating from each range state can then be compared with what is known about the wild population in that country to enable an estimation of whether it is sustainable or whether it might be detrimental to its survival. At a coarser scale, the data can show long-term trends in trade levels or trade routes, which can be used to help in understanding and therefore controlling the trade.

The convention covers not only live animals and plants but also products and derivatives of the species listed. These range from whole skins and manufactured leather products, through ivory carvings, tortoiseshell jewellery, meat, seeds, and feathers to medicinal products extracted from plants such as ginseng. This causes problems for the implementation of the Convention because it is necessary for enforcement officers to determine not only what species the product is derived from but also whether the species is included in the Appendices. In order to minimise the problems of identification, where numerous species are very

Figure 31.1 CITES: number of Parties and annual reporting

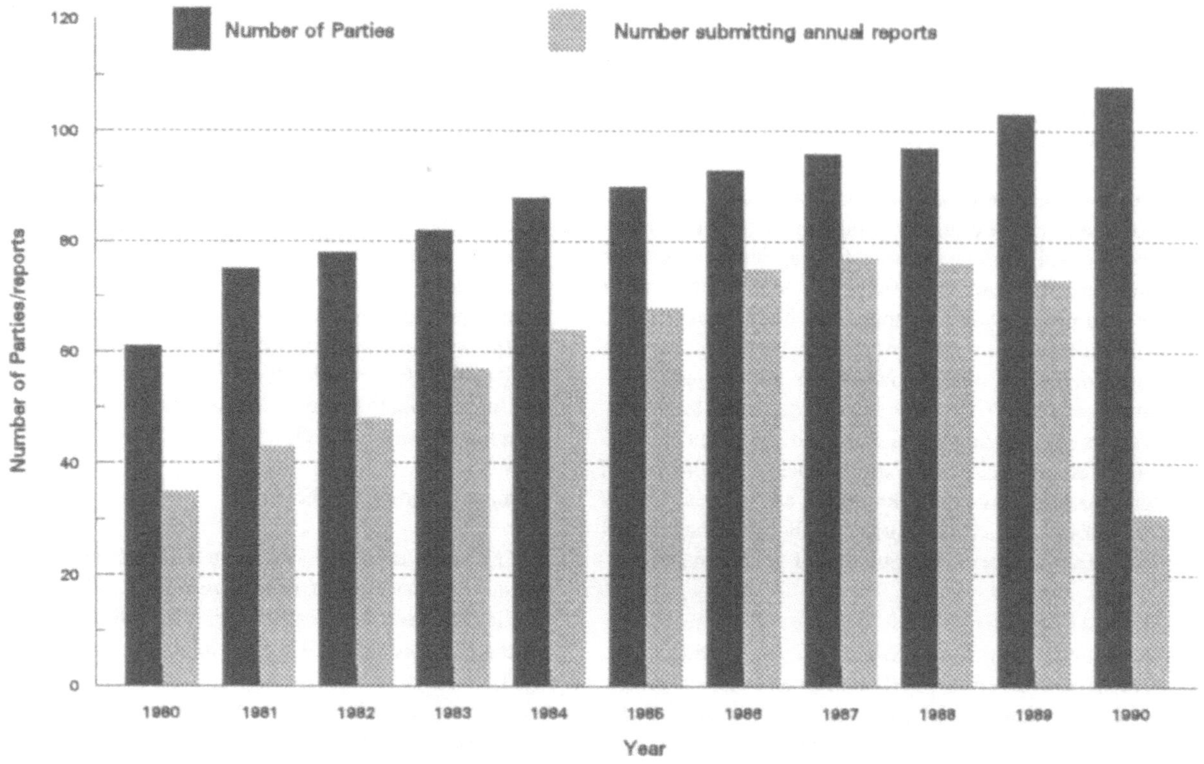

similar in appearance, the whole group of organisms may be included in Appendix II even if only a small proportion of the individual species are in trade. This provision accounts for the majority of species covered by CITES and includes examples such as: all parrots, which are mainly traded as pets; all cacti, and all orchids, which are popular in horticulture; all cats (Felidae), used mainly for the skin trade; and all primates, which are used for biomedical research and as pets.

Of the large number of international environmental conventions, CITES has probably the most detailed control structure. It was the first international wildlife treaty to provide for explicit obligations and international monitoring. As originally drafted, CITES provided little in the way of a trade regulation mechanism and was seen primarily as a protectionist measure which would essentially stop trade in endangered species. The convention is based on the premise, that where endangered status can be attributed to overuse use should be withdrawn. However, conversion of habitat rather than over-exploitation is often the primary threat. It could be argued in these cases that maximising value to local resource users, through regulated trading, is more beneficial to conservation than the elimination of that value by prohibiting trade.

Recently, the Conference of the Parties to CITES has moved toward recognition of this problem, by adopting a more flexible approach, with the attempted development of various sorts of constructive utilisation systems.

As early as 1979, the delegates from developing countries brought the anomaly of "indirect extinction in lieu of direct over-exploitation" to the attention of the Conference of the Parties. In San José, Costa Rica, they argued that there must be an economic benefit from the protected species to justify protecting their habitats from development. These concerns led to a first step towards the reform of CITES, with the adoption of Conference Resolution 3.15 at the New Delhi Conference of the Parties in 1981. This resolution provides for the transfer of certain Appendix I populations to Appendix II for the purposes of sustainable resource management. The criteria which specify how Appendix I species may be used in order to procure compensation for their habitat are known as the "ranching criteria", and each subsequent Conference of the Parties has seen a number of such proposals for review and possible acceptance. The first ranching proposal accepted involved the transfer of the Zimbabwean population of Nile crocodile to Appendix II in 1983.

Ranching proposals tend to be focused on a particular state, or operation, and do not constitute mechanisms for the control of the trade in its entirety. In essence, they continue the overall controls in effect while allowing very limited utilisation to recommence under particular conditions.

In 1983, a species-based approach was first adopted with regard to exploitation of the African Leopard. Although listed on Appendix I, it was recognised in Conference Resolution 4.13 that specimens of the leopard could be killed "to enhance the survival of the species". With this, the Conference of the Parties approved an annual quota of 460 specimens, and allocated these between the range states. In 1985 this quota was then increased to 1,140 animals, and in 1992 to 2,055.

This approach to trade management was extended in 1985 with Resolution 5.21, which provided for the systematic transfer from Appendix I to Appendix II of populations where the countries of origin agree a quota system which is sufficiently safe so as to not endanger the species. Five different species have been subject to quota systems under this Resolution: three African crocodiles, one Asian crocodile, and the Asian Bonytongue (a fish) for which Indonesia was allowed a quota of 1,250 specimens.

None of these trade control systems went further than the development of species-based quotas. In particular, no external control structure was ever implemented, this being left to the discretion of producer states. Thus, predictably, the quotas can be abused: for example, Indonesia is believed to have issued permits for about 140% of its first year's quota of Bonytongues (Anon., 1991).

The African Elephant management quota system

The third avenue of innovation under CITES, and the most concentrated attempt thus far to develop an international control structure within the system, was the creation under Resolution 5.12 of a Management Quota System for the African Elephant. This system was founded upon the ideas of controls based on management decisions taken by producer countries but enforced by consumer countries. Annual quotas were to be constructed at the outset of each year, and producer states were then to issue permits not exceeding these quotas. Then consumer states were to disallow all imports unless accompanied by a Management Quota System permit.

This did not result in an effective control system for one very important reason. The Management Quota System provided no external checks on the discretion of the producer states in determining annual quotas. There were no externally enforced incentives for sustainable use. This resulted in most states basing their annual 'management quotas' of ivory on the 'expected' confiscations from poachers. In addition, there were no disincentives for cross-border exploitation, since consumer states were allowed to import ivory unquestioningly from any exporter issuing permits.

The Management Quota System failed as a consequence of these clear inadequacies, resulting in a collapse of public confidence in the capacity for trade controls to work. This in turn resulted in the transfer of the African Elephant to Appendix I, despite the fact that there remain approximately 600,000 elephants. Each of these requires about 0.5km^2 of grazing lands and in a land where human populations are doubling every 20 years, it is difficult to maintain existing diverse resources, especially when their values are reduced (Barbier *et al.*, 1990).

Despite the difficulties experienced by CITES in achieving its aim of limiting the over-exploitation of wildlife by international trade, the convention itself has proved very popular and, with 113 signatories, it is, together with the World Heritage Convention and Ramsar, among the most significant examples of international action to preserve biodiversity.

Future trends

The 20 years in which CITES has been in existence have indicated the enormous potential that a truly effective regime could have in the effort to secure the long-term survival of significant amounts of the world's biodiversity. The importance of properly distributing the costs and benefits of this biodiversity is becoming increasingly apparent. The effort of ITTO and FAO to move world tropical timber production on to a sustainable basis is just one of many examples in which international institutions are attempting to correct previous distortions in the distribution of these costs and benefits.

REGIONAL SEAS PROGRAMME

A primary example of what can be achieved by means of international coordination of national efforts to conserve biodiversity is the UNEP Regional Seas Programme. The object of this programme, initiated in 1974, has been to develop an integrated and comprehensive approach to protect the marine environment.

Such an approach is necessary because of the nature of threats to the marine environment. Dumping from ships, land-based pollution and overfishing are among the threats which national governments acting unilaterally find difficult

Table 31.4 UNEP Regional Seas Programme: Areas and Action Plans

REGIONAL SEA AREA	ACTION PLAN ADOPTED		PUBLISHED IN RSRS*
Mediterranean	February	1975	No. 34 (1983, rev. 1985)
Gulf	April	1978	No. 35 (1983)
West/Central Africa	March	1981	No. 27 (1983)
Southeast Pacific	November	1981	No. 20 (1983)
Red Sea	February	1982	No. 81 (1986)
Caribbean	April	1981	No. 26 (1983)
Eastern Africa	June	1985	No. 61 (1985)
South Pacific	March	1982	No. 29 (1983)
East Asia	October	1981	No. 24 (1983)
South Asia	in preparation		

Note: * UNEP Regional Seas Reports and Studies.

to control. UNEP has sought to develop the necessary international cooperation through its Regional Seas Programme, which currently covers 10 different regions, consists of 24 separate international agreements and involves over 50 different countries. In a period of less than 20 years the programme has made a major impact on the conservation of the marine environment. The regions covered are illustrated in Fig. 31.2, and details of the legal instruments and the action plans developed for each are given in Tables 31.4 and 31.5.

The Mediterranean was the first region in which the programme developed a cooperative framework for environmental protection. The approach developed here has served as a blueprint for other regional plans subsequently developed by UNEP.

The first stage in this process was the development of a regional action plan. The Mediterranean Action Plan was a comprehensive interdisciplinary attempt to develop and implement substantive programmes for the protection of the marine environment. The Action Plan formed the basis of the Convention for the Protection of the Mediterranean Sea Against Pollution, otherwise known as the Barcelona Convention. This convention has four basic components:

- Environmental assessment through the MEDPOL monitoring network
- Environmental management through the 'Blue Plan' for coordinated development of the coastal regions and 'Priority Action Programmes' for cooperation in coastal settlements, agriculture, freshwater resources, soils renewable energy and tourism
- Institutional arrangements (such as the establishment of a permanent secretariat and the regular holding of conferences of the parties)
- Financial arrangements to help countries implement some requirements of the convention.

Success of this regional endeavour and of every subsequent regime developed under the auspices of this programme is entirely dependent upon the involvement of the majority of the coastal countries in the regions concerned. In order to achieve this, the Barcelona Convention was designed to be as flexible as possible. As a result of this need for flexibility the convention which the parties adopted was a *framework convention*, which outlines in broad terms what obligations the parties are willing to undertake. These basic principles are then developed into specific obligations through the adoption of *protocols* to the main convention. Another feature which provides considerable flexibility for the regulatory regime established pursuant to these programmes is the use of technical annexes, including 'black lists' and 'grey lists' for substances identified as potentially harmful to the environment. These lists may be amended through an accelerated procedure not requiring diplomatic ratification. This approach has ensured that for each region the programme has been able to enlist most if not all of the relevant coastal states.

The Barcelona Convention contains many mechanisms to foster the active cooperation of all of the contracting parties. A requirement for periodic conferences of the parties helps to retain the parties' interest and keep the convention from becoming a 'sleeping treaty'. The establishment of an active secretariat ensures that there is continuity in management. The secretariat runs numerous programmes which are designed to provide support to parties in implementing the provisions of the convention, such as: the provision of technical assistance; training programmes; financial aid; and provision of administrative support at the periodic conferences. An active administration has also been important in the dissemination of new techniques and technology amongst the contracting parties and from region to region. All of these supporting measures help develop cooperation between the contracting parties.

A measure of the success of this programme in developing cooperation and in protecting the marine environment of many of the threatened regions can be gained from comparing its development and that of the UN Convention on the Law of the Sea (UNCLOS). Both initiatives are concerned with the marine environment, both were initiated at the same time and both are international, involving a wide range of countries. The Regional Seas Programme has put in place regimes which have already had an impact on problems in the marine environment, whereas UNCLOS has yet to enter into force.

References

Anon. (TRAFFIC Japan) 1991. Asian Bonytongue exports from Indonesia. *TRAFFIC Bulletin* 12(1,2):3.

Barbier, E.B., Burgess, J.C., Swanson, T.M. and Pearce, D.W. 1990. *Elephants, Economics and Ivory*. Earthscan, London.

Cornish, W.R. 1989. *Intellectual Property: patents, copyright, trade marks and allied rights*. Sweet and Maxwell, London.

Congress of the United States, Office of Technology Assessment 1990. *New Developments in Biotechnology: patenting life*. Marcel Decker, Inc., New York.

Chapter planned by Timothy M. Swanson. Authors as follows: Funding, Sam Johnston; Regulated trading in wildlife products, Shirra Freedman; Intellectual property rights for biotechnology, Nigel Howard; International Obligations and Regional Seas programme, Sam Johnston.

Figure 31.2 UNEP Regional Seas areas

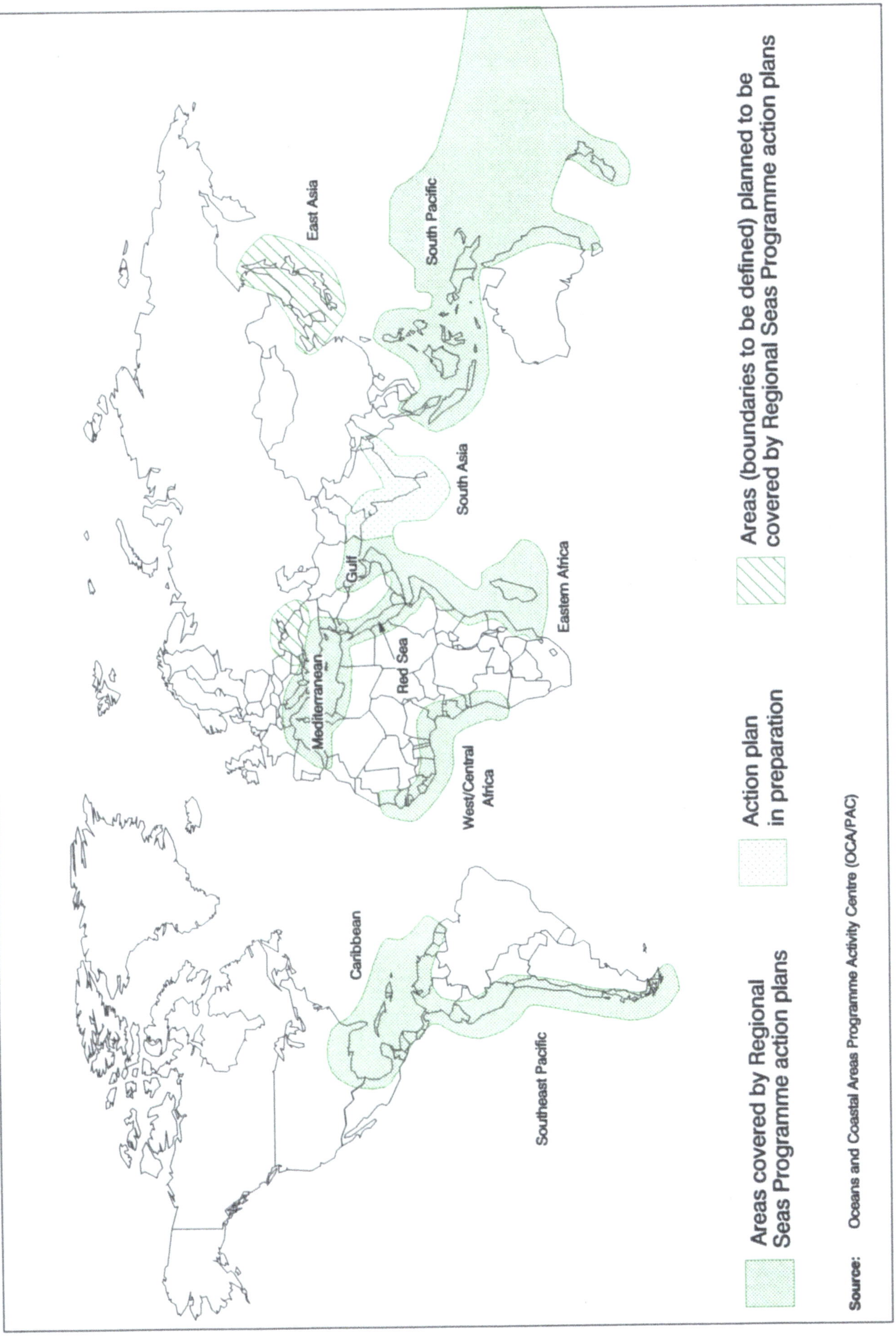

Source: Oceans and Coastal Areas Programme Activity Centre (OCA/PAC)

Table 31.5 Regional Seas conventions

MEDITERRANEAN

	BARCELONA CONVENTION[1]		DUMPING PROTOCOL[2]		EMERGENCY PROTOCOL[3]		LAND–BASED SOURCES PROTOCOL[a1]		PROTECTED AREAS PROTOCOL[a2]	
	SIGNED/ ACCEDED	IN FORCE	SIGNED/ ACCEDED	IN FORCE	SIGNED/ ACCEDED	IN FORCE	SIGNED/ ACCEDED	IN FORCE	SIGNED/ ACCEDED	IN FORCE
Algeria	16/03/81	15/04/81	16/03/81	15/04/81	16/02/81	15/04/81	02/05/83	17/06/83	16/05/85	23/03/86
Cyprus	16/02/76	19/12/79	16/02/76	19/12/79	16/02/76	19/12/79	17/05/80	27/12/87	28/06/88	28/06/88
Eygpt	16/02/76	23/09/78	16/02/76	23/09/78	16/02/76	23/09/78	18/05/83	17/06/83	16/02/83	23/03/86
France	16/02/76	10/04/78 *	16/02/76	10/04/78 *	16/02/76	10/04/78 *	17/05/80	17/06/83 *	03/04/82	02/10/86 *
Greece	16/02/76	02/02/79	16/02/76	02/02/79	16/02/76	02/02/79	17/05/80	25/02/87	03/04/82	25/02/87
Israel	16/02/76	02/04/78 *	16/02/76	31/03/84	16/02/76	02/04/78	18/05/80		04/04/82	27/11/87
Italy	16/02/76	05/03/79	16/02/76	05/03/79	16/02/76	05/03/79	17/05/80	03/08/85	03/04/82	23/03/86
Lebanon	16/02/76	12/02/78	16/02/76	12/02/78	16/02/76	12/02/78	17/05/80		n/a	n/a
Libya	31/01/77	02/03/79	31/01/77	02/03/79	31/01/77	02/03/79	17/05/80		06/06/89	06/06/89
Malta	16/02/76	12/02/78	16/02/76	12/02/78	16/02/76	12/02/78	17/05/80		03/04/82	10/02/88
Monaco	16/02/76	12/02/78	16/02/76	12/02/78	16/02/76	12/02/78	17/05/80	17/06/83	03/04/82	29/05/89
Morocco	16/02/76	15/02/80	16/02/76	15/02/80	16/02/76	15/02/80	17/05/80	11/03/87	03/04/82	22/06/90
Spain	16/02/76	12/02/78	16/02/76	12/02/78	16/02/76	12/02/78	17/05/80	05/07/84	03/04/82	21/01/88
Syria	26/12/78	25/01/79 *	26/12/78	25/01/79	26/12/78	25/01/79			n/a	n/a
Tunisia	25/05/76	12/02/78	25/05/76	12/02/78	25/05/76	12/02/78	17/05/80	17/06/83	03/04/82	23/03/86 *
Turkey	16/02/76	06/05/81	16/02/76	06/05/81	16/02/76	06/05/81	21/02/83	17/06/83	06/11/86	06/12/86
Yugoslavia	15/09/76	12/02/78	15/09/76	12/02/78	15/09/76	12/02/78			30/03/83	23/03/86
EEC	13/09/76	15/04/78	13/09/76	15/04/78	13/09/76	11/09/81	17/05/80	06/11/83	30/03/83	23/03/86

[1] Convention for the Protection of the Mediterranean Sea Against Pollution, adopted at Barcelona on 16 February 1976.
[2] Protocol for the Prevention of Pollution of the Mediterranean Sea by Dumping from Ships and Aircraft, adopted at Barcelona on the 16 February 1976.
[2] Protocol Concerning Cooperation in Combating Pollution of the Mediterranean Sea by Oil and Other Harmful Substances in Cases of Emergency, adopted at Barcelona on 16 February 1976.
[a1] Protocol for the Protection of the Mediterranean Sea Against Pollution from Land–based Sources, adopted on 17 May 1980.
[a2] Protocol Concerning Mediterranean Specially Protected Areas, adopted at Geneva on 3 April 1982.
* with reservation.

GULF

	KUWAIT CONVENTION[1]		EMERGENCY PROTOCOL[2]	
	SIGNED	IN FORCE	SIGNED	IN FORCE
Bahrain	24/04/78	01/07/79	24/04/78	01/07/79
Iran	24/04/78	01/06/80	24/04/78	01/06/80
Iraq	24/04/78	01/07/79	24/04/78	01/07/79
Kuwait	24/04/78	01/07/79	24/04/78	01/07/79
Oman	24/04/78	01/07/79	24/04/78	01/07/79
Qatar	24/04/78	01/07/79	24/04/78	01/07/79
Saudi Arabia	24/04/78	26/03/82	24/04/78	26/03/82
United Arab Emirates	24/04/78	01/03/80	24/04/78	01/03/80

[1] Kuwait Regional Convention for Co–operation on the Protection of the Marine Environment from Pollution, adopted on 23 April 1978.
[2] Protocol Concerning Regional Co–operation in Combating Pollution by Oil and Other Harmful Substances in Cases of Emergency, adopted 23 April 1978.

WEST/CENTRAL AFRICA

	ABIDJAN CONVENTION[1]		EMERGENCY PROTOCOL[2]	
	SIGNED/ ACCEDED	IN FORCE	SIGNED/ ACCEDED	IN FORCE
Benin	23/03/81		23/03/81	
Cameroon	01/03/83	05/08/84	01/03/83	05/08/84
Congo	23/03/81		23/03/81	
Côte d'Ivoire	23/03/81	05/08/84	23/03/81	05/08/84
Gabon	23/03/81		23/03/81	
Gambia	23/03/81	05/02/85	23/03/81	05/02/85
Ghana	23/03/81		23/03/81	
Guinea	23/03/81	05/08/84	23/03/81	05/08/84
Liberia	23/03/81		23/03/81	
Mauritania	22/06/81		22/06/81	
Nigeria	23/05/81	05/08/84	23/05/81	05/08/84
Senegal	23/03/81	05/08/84	23/03/81	05/08/84
Togo	23/03/81	05/08/84	23/03/81	05/08/84

[1] Convention for Co–operation in the Protection and Development of the Marine and Coastal Environment of the West and Central African Region, adopted at Abidjan on 23 March 1981.
[2] Protocol Concerning Co–operation in Combating Pollution in Cases of Emergency in the West and Central African Region, adopted at Abidjan on 23 March 1981.

SOUTH–EAST PACIFIC

	LIMA CONVENTION[1]		EMERGENCY AGREEMENT[2]		SUPPLEMENTATY PROTOCOL[3]		LAND–BASED SOURCES PROTOCOL	
	SIGNED	IN FORCE	SIGNED	IN FORCE	SIGNED	IN FORCE	SIGNED	IN FORCE
Chile	12/11/81	19/05/86	12/11/81	14/07/86	22/07/83	20/05/87	22/07/83	23/09/86
Colombia	12/11/81	19/05/86	12/11/81	14/07/86	22/07/83	20/05/87	22/07/83	23/09/86
Ecuador	12/11/81	19/05/86	12/11/81	14/07/86	22/07/83	11/01/88	22/07/83	11/01/88
Panama	12/11/81	21/09/86	12/11/81	21/09/86	22/07/83	20/05/87	22/07/83	23/09/86
Peru	12/11/81	25/02/89	12/11/81		22/07/83		22/07/83	

[1] Convention for the Protection of the Marine Environment and Coastal Area of the South–East Pacific, adopted at Lima on 12 November 1981.
[2] Agreement on Regional Co–operation in Combating Pollution of the South–East Pacific by Hydrocarbons or Other Harmful Substances in cases of Emergency, adopted at Lima on 12 November 1981.
[3] Supplementary Protocol to the Agreement on Regional Co–operation in Combating Pollution of the South–East Pacific by Hydrocarbons or Other Harmful Substances in Cases of Emergency, Adopted at Quito on 22 July 1983.
[a1] Protocol for the Protection of the South–East Pacific Against Pollution from Land–Based Sources, adopted at Quito on 22 July 1983.

Table 31.5 Regional Seas conventions (continued)

RED SEA

	JEDDAH CONVENTION[1]		EMERGENCY PROTOCOL[2]	
	SIGNED	IN FORCE	SIGNED	IN FORCE
Eygpt	21/05/90	20/08/90		
Jordan	14/02/82	07/02/89	14/02/82	
Saudi Arabia	14/02/82	20/08/82	14/02/82	20/08/82
Somalia	14/02/82	30/05/88	14/02/82	
Sudan	14/02/82	20/08/82	14/02/82	20/08/82
Yemen	14/02/82	20/08/82	14/02/82	20/08/82

1 Regional Convention for the Conservation of the Red Sea and Gulf of Aden Environment, adopted at Jeddah on 14 February 1982.
2 Protocol Concerning Regional Co-operation in Combating Pollution by Oil and Other Harmful Substances in Cases of Emergency, adopted at Jeddah on 14 February 1982.

CARIBBEAN

	CARTAGENA CONVENTION[1]		SPECIALLY PROTECTED AREAS PROTOCOL[2]		OIL SPILLS PROTOCOL[3]	
	SIGNED/ ACCEDED	IN FORCE	SIGNED/ ACCEDED	IN FORCE	SIGNED/ ACCEDED	IN FORCE
Antigua and Barbuda	11/09/86	11/10/86	18/01/90		11/09/86	11/10/86
Barbados	05/03/84	11/10/86	18/01/90		05/03/84	11/10/86
Colombia	24/03/83		18/01/90		24/03/83	
France	24/03/83	11/10/86 *	18/01/90		24/03/83	11/10/86 *
Grenada	24/03/83	16/09/87	18/01/90		24/03/83	16/09/87
Guatemala	05/07/83		18/01/90		05/07/83	
Honduras	24/03/83		18/01/90		24/03/83	
Jamaica	24/03/83	01/05/87	18/01/90		24/03/83	01/05/87
Mexico	24/03/83	11/10/86	18/01/90		24/03/83	11/10/86
Netherlands[a1]	24/03/83	11/10/86	18/01/90		24/03/83	11/10/86
Nicaragua	24/03/83		18/01/90		24/03/83	
Panama	24/03/83	06/11/87			24/03/83	06/11/87
St Lucia	24/03/83	11/10/86	18/01/90		24/03/83	11/10/86
Trinidad and Tobago	24/03/83	11/10/86	18/01/90		24/03/83	11/10/86
United Kingdom[a2]	24/03/83	11/10/86 *	18/01/90		24/03/83	11/10/86 *
United States	24/03/83	11/10/86	18/01/90		24/03/83	11/10/86
Venezuela	24/03/83	17/01/87	18/01/90		24/03/83	17/01/87
EEC	24/03/83		18/01/90			

1 Convention for the Protection and Development of the Marine Environment of the Wider Caribbean Region, adopted at Cartagena on 24 March 1983.
2 Protocol Concerning Specially Protected Areas and Wildlife to the Convention for the Protection and Development of the Marine Environment of the Wider Caribbean Region, adopted in Kingston on 17 January 1990.
3 Protocol Concerning Co-operation in Combating Oil Spills in the Wider Caribbean Region, adopted at Cartagena on 24 March 1983.
a1 On behalf of Aruba and the Netherlands Antilles Federation.
a2 On behalf of the Cayman Islands and the Turks and Caicos Islands, reserving the right to include at a future date the other territories of the United Kingdom participating in the Caribbean Action Plan (Anguilla, British Virgin Islands and Montserrat).
* with reservation.

EASTERN AFRICA

	NAIROBI CONVENTION[1]		PROTECTED AREAS PROTOCOL[2]		EMERGENCY PROTOCOL[3]	
	SIGNED	IN FORCE	SIGNED	IN FORCE	SIGNED	IN FORCE
France[a1]	21/06/85		21/06/85		21/06/85	
Kenya	11/09/90		11/09/90		11/09/90	
Madagascar	21/06/85		21/06/85		21/06/85	
Seychelles	21/06/85		21/06/85		21/06/85	
Somalia	21/06/85		21/06/85		21/06/85	
EEC	19/06/86		19/06/86		19/06/86	

1 Convention for the Protection, Management and Development of the Marine and Coastal Environment of the Eastern African Region, adopted at Nairobi on 21 June 1985.
2 Protocol Concerning Protected Areas and Wild Fauna and Flora in the Eastern African Region, adopted at Nairobi on 21 June 1985.
3 Protocol Concerning Co-operation in Combating Marine Pollution in Cases of Emergency in the Eastern African Region, adopted at Nairobi on 21 June 1985.
a1 On behalf of Réunion.

SOUTH PACIFIC

	NOUMEA CONVENTION[1]		EMERGENCY PROTOCOL[2]		DUMPING PROTOCOL[3]	
	SIGNED	IN FORCE	SIGNED	IN FORCE	SIGNED	IN FORCE
Australia	24/11/87	22/08/90	24/11/87		24/11/87	
Cook Islands	25/11/86	22/08/90	25/11/86		25/11/86	
Federated States of Micronesia	09/04/87	22/08/90	09/04/87		09/04/87	
France	25/11/86	22/08/90	25/11/86		25/11/86	
Marshall Islands	25/11/86	22/08/90	25/11/86		25/11/86	
Nauru	15/04/87		15/04/87		15/04/87	
New Zealand	25/11/86	22/08/90	25/11/86		25/11/86	
Palau	25/11/86		25/11/86		25/11/86	
Papua New Guinea	03/11/87	22/08/90	03/11/87		03/11/87	
Tuvalu	14/08/87		14/08/87		14/08/87	
United Kingdom	16/07/87		16/07/87		16/07/87	
United States	25/11/86	10/07/91	25/11/86		25/11/86	
Samoa	25/11/86	22/08/90	25/11/86		25/11/86	

1 Convention for the Protection of Natural Resources and Environment of the South Pacific Region, adopted at Noumea on 25 November 1986.
2 Protocol Concerning Co-operation in Combating Pollution Emergencies in the South Pacific Region, adopted at Noumea on 25 November 1986.
3 Protocol for the Prevention of Pollution of the South Pacific Region by Dumping, adopted at Noumea on 25 November 1986.

32. INTERNATIONAL AID

As noted in the previous chapter, national efforts alone are not sufficient, despite their fundamental importance, to ensure adequate management of biological diversity. Chapter 31 surveyed international support for national measures as provided by policy and legal assistance; this chapter will discuss the role of direct development aid.

Even though the proportion of total international aid which is specifically targeted for the conservation of biodiversity is relatively small, it nonetheless plays a very important role. The three principal kinds of assistance examined here are: international development assistance, international programmes for the conservation of tropical forests, and a form of debt purchase widely known as 'debt-for-nature' exchange.

The impact on biodiversity of international development assistance, whether intended or not, is felt through many channels and will vary in nature and extent. The following review of this international assistance attempts to assess to what extent bilateral and multilateral funds are directly targeting biodiversity conservation, and how this is incorporated into the project and loan appraisal process.

An examination of two international programmes designed to conserve tropical forests highlight some of the problems faced by any international effort to conserve biodiversity. Given the cross-sectoral nature of environmental issues and the number of different international programmes currently established, even where there is some consensus over what action is required, there still remain major problems of organisation and coordination to be overcome.

The last part of this chapter shows that there is no single and ready solution to the problem of conserving biodiversity. Several years ago 'Debt-for-Nature' exchanges were regarded as a major chance to counter many of the underlying causes of biodiversity degradation. As this mechanism has developed it has become evident that some predictions of its importance were over-optimistic.

INTERNATIONAL DEVELOPMENT ASSISTANCE

The term 'international development assistance' (IDA) (otherwise 'overseas development assistance' or ODA) is used here to refer to concessional aid provided by governments to developing countries. IDA is delivered directly by donor countries' bilateral aid or by multilateral institutions such as the United Nations and the major development funds and banks. Concessional aid includes grants and loans made at less than market interest rates, but not other types of official financial flows such as export credits, grants by private voluntary agencies or private flows at market terms. World IDA accounted for roughly half of the $110 billion in financial resources transferred to developing countries in 1989. Non-concessionary bilateral and multilateral disbursements (14%), foreign direct investment (20%) and international bank lending (7%) make up large portions of the remaining resources transfers.

Resource transfers to the developing countries are only approximately half as large in real terms as they were at their peak in 1981. At that time 38% of resource transfers consisted of private international bank lending. The drain on development finance resulting from the ensuing debt crisis of the 1980s is far from over. In 1989 developing countries paid out interest and dividends of $108 billion - a figure roughly equal to the incoming financial resources cited above. Total overseas development assistance over this period has remained fairly stable. While multilateral disbursements have been unchanging, Arab donors' contributions have fallen dramatically from close to $10 million in 1980 to less than $2 million in 1989. The group of 18 countries making up the Development Assistance Committee (DAC) of the Organization for Economic Cooperation and Development (OECD) have made up for this shortfall by steadily increasing their IDA expenditures. Commitments by DAC countries (listed in Table 32.1) have risen from an average of 62% of world IDA over the period 1980-81 to 87% in 1988-89.

The DAC average of contributing 0.35% of GNP to IDA masks considerable discrepancy between the performance of the USA and Japan, on the one hand, and Scandinavian and Dutch donors on the other. Table 32.1 clearly reveals that the average Norwegian contributes more than six times as much IDA as does the average American. The official DAC target of 0.7% of GNP, first proposed by the Pearson Commission in 1969 remains elusive for the majority of DAC members. Nonetheless, official flows of IDA remain the primary vehicle of aid transfers. In comparison, the average for aid from non-governmental organisations in 1988-89 was estimated at just 0.03% of GNP.

Two other important IDA statistics measuring aid performance are the 'grant element' and the status of 'tied aid'. In recent years only a couple of DAC countries failed to achieve the DAC target of delivering 86% of their funds on grant terms. Tying aid to purchases of goods or services from the donor country remains a more complicated and controversial topic. Tied aid benefits donors while inflating the cost of recipient country purchases by 10% to 20% (de Silva, 1982). The disparity between tied aid figures in Table 32.1 reveals the lack of coherent progress within the DAC on this issue. In 1988 approximately 30% of DAC country commitments to IDA were tied and another 7.5% partially untied.

The impact of international development assistance on biodiversity is felt through many channels: projects, sectoral and macroeconomic policy reform, and institutional and human resource initiatives. The list is long and the impacts on biodiversity, intentional and otherwise, will vary in direction and magnitude.

The following review of bilateral and multilateral aid policies on biodiversity bypasses specific negative impacts of IDA on biodiversity, which are well documented elsewhere, in favour of finding evidence of bilateral and multilateral action towards the conservation of biodiversity. A relatively new concern, such as biodiversity, is likely to be incorporated into the development assistance process either through increased funding for projects, technical cooperation and other means of addressing the issue, or by

Table 32.1 International development assistance

COUNTRY	NET AMOUNT 1988 (US$million)	AMOUNT PER CAPITA 1988 (US$)	SHARE OF GNP 1988-89 mean %	SHARE OF WORLD IDA 1988-89 mean %	TIED AID AS % OF TOTAL IDA[2]
Ireland	53.7	15.2	0.2	0.1	(3.8)
New Zealand	96.6	29.4	0.2	0.2	36.4
USA	8,749.2	35.3	0.2	15.9	37.7
Austria	298.5	39.2	0.2	0.5	68.8
UK	2,640.2	46.2	0.3	4.8	46.4
Italy	3,383.9	58.9	0.4	6.2	57.6
Australia	1,018.4	61.1	0.4	1.9	32.7
Belgium	662.9	67.0	0.4	1.2	(34.1)
Japan	9,312.5	75.8	0.3	16.9	11.5
Germany	4,946.9	80.1	0.4	9.0	32.5
Canada	2,238.3	85.8	0.5	4.1	34.5
Switzerland	609.8	91.0	0.3	1.1	15.0
Finland	643.3	129.9	0.6	1.2	47.1
France[1]	7,289.9	130.1	0.8	13.3	37.7
Netherlands	2,229.4	150.6	1.0	4.1	10.4
Denmark	950.1	185.2	0.9	1.7	15.8
Sweden	1,645.3	195.0	0.9	3.0	21.2
Norway	970.1	229.9	1.1	1.8	20.8
Total DAC	47,739.0	66.7	0.3	86.6	(30.5)
Total Non-DAC OECD	382.0	-	0.1	0.7	-
Total Arab Donors	1,900.0	-	0.7	3.5	-
Central and Eastern Europe	4,534.0	-	-	8.2	-
Total non-Arab LDC donors	425.0	-	0.0	0.8[2]	-
WORLD TOTAL	54,980.0	-	-	100.0	-

Source: DAC 1990. *Development Co-operation: efforts and policies of the members of the Development Assistance Committee*. OECD, Paris.
Notes: ( ) are DAC estimates. [1] Including aid to French possessions. [2] Figures for 1988 only.

inclusion in assessment procedures for projects that may have unintended or indirect impacts on biodiversity. For this reason the emphasis below is on reviewing whether or not, and how, biodiversity is being incorporated into the project/loan appraisal process and whether funds are targeting biodiversity directly.

Unfortunately, biodiversity issues crop up in a number of traditional sectors, such as health, energy, agriculture, mining, transport, food aid, etc., making it difficult to clearly define a biodiversity 'sector' or to sum up biodiversity expenditures. This obstacle is one shared by environmental issues in general - indeed a further confusion may occur between environmental issues and biodiversity issues broadly defined. For this reason, the approach taken below is to report on progress specific to biodiversity where such information is available. An explicit role for biodiversity in project and policy formulation, or in funding commitments, may be one indicator of the level of awareness and seriousness of commitment in donor agencies towards conserving biodiversity. The intention is not to recommend that biodiversity merits it own 'sectoral' billing, but to begin the process of evaluating the resources that development agencies are committing to conservation. When specific actions on biodiversity are not evident, progress on the general environmental front is documented as indicative of growing concern that may soon encompass biodiversity directly.

BILATERAL DEVELOPMENT ASSISTANCE

Funding

Establishing exactly what constitutes funding for biodiversity is not straightforward. Abramovitz (1991) compared the results of two studies on US-based funding for biodiversity conservation carried out by the World Resources Institute's Center for International Development and Environment. Aggregating project-level data from government sources, NGOs, foundations, universities, etc., the total investment came to $37.5 million for 1987 and $62.9 million for 1989. Listed below are the major categories of activity areas used in classifying all 1,093 projects recording for 1989:

- Research
  - Basic
    - Ecosystem
    - Species
  - Applied
    - Response to Disturbance
    - Economic Assessment
    - Social/Cultural Values
  - Systematics/Inventory
- Site or Species Management
  - Protected Areas
    - Planning and Assessment
    - Management
    - Buffer Zone Management
  - Ecosystem Restoration
  - Species Management - *in situ, ex situ*
- Policy Planning and Analysis
  - Conservation Law/Regulatory
  - Natural Resources Accounting
  - Economic Policy Analysis
  - Program/Project Planning
  - Program/Project Design
  - Statistics, Indicators
  - Environmental Impact Assessment
- Education
  - Public Awareness
  - Curriculum Development
  - Degree Training
  - Technical Training
- Institutional Support

While this study is the most comprehensive effort to date, even such a detailed survey cannot avoid the difficulties inherent in extracting biodiversity project data from programmes aligned along traditional sectoral or country boundaries. In addition, the non-equivalence of biodiversity and biological resources means that some activities will be slotted in as biodiversity funding when they have little to do with diversity *per se*. Determining whether or not a project is concerned with management of biological resources generally, or more specifically with biodiversity, is not an easy task. The difficulty with relaxing the emphasis on diversity is that the difference between funding for biodiversity and funding for the environment will become more and more blurred.

It is more difficult to categorise the importance with regard to funding levels of projects that have only a secondary or tertiary focus on biodiversity. Often, the perception that biodiversity pertains only to wild species, *ex situ* gene storage or protected areas (as in the WRI study) may also divert attention from the role of species or genetic diversity in production systems. Agriculture, aquaculture, forestry, fishery and other rural development projects may have a significant diversity component. On the other hand, projects aimed at developing sustainable resource management may involve some loss of biodiversity.

Recognising the inexactness of even the most comprehensive study, the London Environmental Economics Centre (LEEC) conducted a brief survey of DAC bilateral agencies to obtain a general impression of the visibility of biodiversity as a concern in agency funding and project appraisal. None of the agencies responding to the questionnaire currently disaggregate their expenditures to identify the amount spent on biodiversity or genetic resources. As shown in Table 32.2 roughly half of the agencies do calculate the extent of funding for the environment. Three others indicated that they will be doing so in the near future. The movement towards identification of environmental expenditures is a natural precursor to establishing similar reporting procedures for biodiversity.

Another way in which awareness regarding biodiversity may be manifested is through setting aside funds for use specifically on biodiversity. Since 1983 the US Congress has earmarked funds for biodiversity conservation in USAID's annual appropriation. Recently, a number of DAC agencies are now earmarking funds for biological diversity as described in Table 32.2.

The difficulty in interpreting such aggregate numbers is illustrated in the case of the data provided by Germany's BMZ (see Table 32.2 notes). While BMZ has earmarked DM3.5 million for biodiversity in 1991 they also reported

Table 32.2 Bilateral funding for biodiversity and the environment

DONOR COUNTRY		BILATERAL FUNDS - 1988 $US million	BIODIVERSITY FUNDS SET ASIDE	$US million	ARE EXPENDITURES ON THE ENVIRONMENT IDENTIFIED?	$US million
Australia[1]		622	A$ 4.3m in 1991/92	3	A$ 10.6m in 1990/91	9
Austria		162	none	-	yes, not available	-
Belgium		415	none	-	will identify soon	-
Canada		1,583	none	-	no	-
Denmark		478	US$ 0.115m in 1991	0	no	-
			DKr 25m in 1992-96	4		
France		5,601 [2]	plan to	-	FF 200m in 1991	35
Germany	(BMZ)	3,172 [3]	DM 3.5m in 1991	2	DM 1020m in 1990[4]	610
	(GTZ)		none	-	DM 200m in 1990	120
Ireland		22	none	-	will identify soon	-
Italy		2,408	none	-	will identify by 1992	-
Netherlands		1,552	[5]	-	US$ 135m in 1991	135
New Zealand		93	yes, not assessed	-	NZ$ 50m in 1990/91	28
Norway		570	none	-	NOK 852 in 1990[6]	130
Sweden		1,034	SKr 1.5m in 1991/92	0	no	-
Switzerland		414 [7]		-	no	-
UK		1,430	[8]	-	no[9]	-
USA[10]		6,765	US$ 9.9m in 1989	9	no	-

Source: LEEC questionnaire to DAC members. Responses received from all but Finland, Germany - KfW and Japan. EEC responses are presented under the section on multilaterals. Exchange rates used are from the rates listed in the Financial Times on 30 September 1991

Notes: [1] Directly targeted funds only. [2] Including aid to French possessions. [3] Total for all three German agencies: Bundesminister für wirtschaftliche Zusammenarbeit (BMZ), Deutsche Gesellschaft für Technische Zusammenarbeit (GTZ) and Kreditanstalt für Wiederaufbau (KfW). [4] Commitments not expenditures; does not include DM325m on tropical forests. [5] No funds are set aside, but expenditures on biodiversity in 1991 came to roughly $US 5m. [6] Includes NOK 651m on environmental integrated projects and NOK 201m on direct environmental projects. Excludes NOK 315m on population. [7] Although expenditures are not specified, the conservation of biodiversity constitutes a basic objective of the Directorate's development policy. [8] Although ODA does not set aside funds it does conduct strategic work on biodiversity conservation and finances many biodiversity projects; for example, ODA is the primary funder of this report. [9] ODA prefers to integrate environmental issues throughout its aid programme rather than treat it as a separate sector and keep separate statistics on it. [10] The figure for USAID represents actual expenditures as reported by the Biodiversity Projects Database maintained by the World Resources Institute.

forest sector expenditures of DM325 million in 1990. Setting aside money explicitly conveys BMZ's concern for biological diversity. However, the size of BMZ's existing efforts in tropical forestry indicate that the impact of the set aside funds is likely to be superseded by the funds allocated for the 'preservation and development' of tropical forests. The positive impact of the amount set aside will be small relative to the forestry expenditures if a reasonable percentage of the forestry funds goes towards preservation or sustainable use of the forests. On the other hand, if a large portion of forest sector expenditures support unsustainable logging activities, the overall negative impacts of these expenditures would overwhelm the potential benefits gained from the earmarked funds. Clearly the direct impact of donor allocations for biodiversity shown in Table 32.2 (all of them less than $10 million) will be negligible compared to the real effects of much larger sectoral expenditures on forests, agriculture, transport, etc. The significance of the allocation would be increased by using the money to sponsor innovative projects, research or institutional initiatives.

Project appraisal

Ideally, project appraisal procedures should act as a control on sectoral projects with significant impacts on biodiversity. If projects are screened for negative impacts on biodiversity then funding allocated towards biodiversity conservation will assume greater importance instead of generating suspicion that they are just compensation for the ill-effects of the remaining development portfolio. Table 32.3 reveals that most DAC agencies already undertake environmental impact assessments of project and loan proposals. Increasingly, countries are following the lead of Canada and the USA in involving host country officials and experts in the evaluation procedure.

Table 32.3 also lists a number of countries that include evaluation of project impacts on biodiversity into the appraisal process. Efforts to carry the physical data through to a complete cost-benefit analysis by monetising the impacts on the environment and biodiversity is limited. Both BMZ and the UK's Overseas Development Administration (ODA) report that assessment of the economic effects is undertaken in particular situations. Given that empirical work detailing the economic value of biodiversity, and environmental resources in general, is still an area of front-line research, it is unlikely that full cost benefit analysis is likely for each and every development project in the foreseeable future.

MULTILATERAL DEVELOPMENT ASSISTANCE

Multilateral disbursements of overseas development assistance and non-concessional finance come primarily from the World Bank Group, the regional development banks and the UN specialised agencies. Table 32.4 presents data on the net disbursements by these organisations. Most of the multilaterals belong to the Committee of International Development Institutions on the Environment (CIDIE) which coordinates multilateral activities on the environment. This section will review the specific achievements on the environment and biodiversity as reported by CIDIE members, and then look at the newest and potentially largest source of multilateral funding for biodiversity conservation: the Global Environmental Facility.

Heavily criticised for supporting, amongst other activities, large hydroelectric projects and logging schemes, the World Bank has gradually developed a policy on environmental assessment. The 1989 Environmental Assessment Operational Directive (EAOD) formalises Bank policy on environmental impact assessment (EIA). The EAOD

Table 32.3 Bilateral agency loan and project appraisal policies

COUNTRY	APPRAISAL OF IMPACTS IN PHYSICAL OR MONETARY TERMS	
	ON THE ENVIRONMENT	ON BIODIVERSITY
Australia	physical[1]	no
Austria	physical	no
Belgium	soon	no
Canada	physical	physical[2]
Denmark	physical	physical - soon
France	physical	no
Germany (BMZ)[3]	physical and monetary	physical and monetary
Germany - GTZ	physical	no
Ireland	physical	no
Italy	physical	no
Netherlands	physical	physical
New Zealand	physical	physical
Norway	physical	no
Sweden	physical	physical
Switzerland	physical	physical
UK[4]	physical and monetary	physical and monetary

Source: LEEC Questionnaire on Biodiversity to DAC members.
Notes: [1] With some attempts at monetary evaluation. [2] Impacts on wildlife flora/fauna and its habitat. [3] In physical or monetary terms 'if applicable'. [4] In physical and 'where possible' in monetary terms.

Table 32.4 Multilateral development assistance

	NET DISBURSEMENTS IN 1988 $US million CONCESSIONAL	NON-CONCESSIONAL
Major Financial Institutions		
World Bank Group		
IBRD	-	3,417
IDA	3,567	
IFC	-	356
Regional Banks/Funds		
CEC (or EEC)	2,587	56
IDB	134	1,093
Asian	707	598
African	351	625
IFAD	102	-
Other	65	374
United Nations		
UNDP	914	-
WFP	878	-
UNEP[1]	(59)	
Other UN	1,984	-
TOTALS	11,348	6,519

Sources: DAC 1990. *Development Co-operation: efforts and policies of the members of the Development Assistance Committee*. OECD, Paris; UNEP 1990. *UNEP Profile*. UNEP, Nairobi.
Notes: [1] 1989 figure showing funds sourced from regular UN budget (10%), trust funds (15%), counterpart contributions (8%) and voluntary contributions to the Environment Fund (67%). IBRD = International Bank for Reconstruction and Development; IDA = International Development Administration; IFC = International Finance Corporation; CEC = Commission of the Economic Community; IDB = Inter-American Development Bank; IFAD = International Fund for Agricultural Development; UNDP = United Nations Development Programme; WFP = World Food Programme.

separates projects by type into four categories each of which receives varying degrees of assessment (Table 32.5). Environmental projects are generally exempt from EIAs. In order to determine the nature of the EIA required, if any, World Bank projects (or components) are classified according to the likely environmental impacts of the project. EIA (or EA in the Bank's terminology) is considered a flexible procedure that is responsive to the individual project. The assessment may cover not only environmental impacts but those on health, cultural property, tribal people and the environmental impact of resettlement. EIAs should cover existing conditions, potential direct and indirect impacts, comparison of project with alternatives, compensatory measures, environmental management and training, and monitoring. Wherever possible the costs and benefits of these elements should be quantified.

As with the economic, financial, institutional and engineering analyses, the EIA is the borrower's responsibility. The Bank provides assistance in designing the terms of reference for the EIA and normally a field visit by Bank staff is suggested. The Bank usually recommends that borrowers hire experts not involved in the project to carry out the EIA. EIAs for large projects may take up to 18 months to be completed with input from the ongoing EIA occurring at relevant points in the overall project cycle. The final report is submitted to the Bank for consideration with the project or loan application. Funding for EIAs may be accomplished by a Bank loan or grant and usually comes to 5-10% of the cost of project preparation.

In its Forest Policy Paper adopted on 18 July 1991 the Bank has stated that it will not fund commercial logging in

Table 32.5 World Bank project categories and EIA requirements

CATEGORY A: Projects with diverse and significant impacts. EIA required unless directed towards rehabilitation, improved operation and maintenance, and limited upgrading of facilities.

Aquaculture/Mariculture (LS)	Mineral Development	Thermal and Hydropower
Dams and Reservoirs	Pipelines (oil, gas and water)	Tourism (LS)
Electrical Transmission (LS)	Port and Harbour Development	Transportation infrastructure
Forestry	Reclamation and New Land	Urban Development (LS)
Industrial Plants (LS) and Estates	Resettlement	Urban Water Supply/Sanitation (LS)
Irrigation and Drainage (LS)	River Basin Development	Manufacture, Transportation and
Land Clearance and Levelling	Rural Roads	Projects with serious accident risks

CATEGORY B: Projects which may have specific impacts. Limited EIA required.

Agroindustries (SS)	Mini Hydro-Power	Telecommunications
Aquaculture/Mariculture (SS)	Public Facilities	Tourism (SS)
Electrical Transmission (SS)	Renewable Energy	Urban Development (SS)
Industries (SS)	Rural Electrification	Rural Water Supply/Sanitation
Irrigation and Drainage (SS)		

CATEGORY C: Projects which normally do not have serious impacts. EIA normally unnecessary.

Education	Health	Institutional Development
Family Planning	Nutrition	Technical Assistance

CATEGORY D: Projects with a major environmental focus. EIA normally unnecessary

Source: The World Bank, 1989. Operational Directive 4.00, Annex A: Environmental Assessment.
Notes: LS = large-scale projects. SS = small-scale projects.

tropical moist forests. Full EIAs will be required for all infrastructure projects that may affect tropical moist forests or other primary forest. This is nothing new since the EAOD already lists roads, dams and mines as requiring EIAs. In the 1990 fiscal year the Bank reports that 11 free-standing environmental projects and 107 loans with environmental components were approved. The Bank's influence on conservation issues will be felt most directly through its major role in the operation of the new Global Environmental Facility and involvement in the revised Tropical Forestry Action Plan.

The Commission of the European Community (CEC) is integrating environmental considerations into their appraisal process for projects and programmes. In the future the CEC plans on incorporating into its evaluations not just physical EIA but also monetary estimation of environmental impacts. The European Development Fund (EDF) is financing $26 million worth of environmental training programmes in West Africa. The Commission does not provide its own staff with training in the environment, preferring to hire staff with the required qualifications.

Currently, the CEC does not identify its expenditures on the environment, nor are there funds specifically for biodiversity conservation. The Commission has an extensive research agenda which includes the conservation of biodiversity and tropical forests, and marine and freshwater ecology as priority themes.

In 1991 the Commission had roughly $14 million and $2.5 million for work on 'ecology and developing countries' and 'tropical forests' respectively. The Commission has a $60 million small grants facility at its disposal which frequently funds grassroots natural resource management initiatives and encourages cooperation with developed and developing country NGOs.

Over the period 1990-1995, the CEC will commit $14 billion to activities in African, Caribbean and Pacific countries. The vast majority of this assistance will be channelled through the Lome IV Convention mechanism under which international assistance is provided by the stabilisation and support of commodity prices for the raw materials of the less developed parties to the convention. Up to 75% of the programming undertaken to date has identified the environment as a key sector. A doubling of Asian and Latin American funding was accompanied by allocating $300 million over the next five years (10% of total funds) to environmental programmes.

The Inter-American Development Bank (IDB) has both a managerial-level Environmental Committee (CMA) and an Environmental Protection Division. The latter was created in 1990 under the Project Analysis Department in order to ensure that IDB operations comply with member country legislation and the guidelines on EIA developed by the Bank. EIAs are required for the proportion (about three-quarters) of IDB projects that have minor or major environmental impacts. In 1990 the costs of undertaking such project reviews were estimated to require an extra $75 million over the next three years. The Bank is also interested in increasing funding for technical assistance and NGOs, and promoting debt swaps for environmental protection.

Over the past few years the Asian Development Bank (AsDB) has upgraded its Environment Unit first to the status of a Division and most recently into the Office of the Environment. Accompanying this upgrading of status the AsDB has added five professional staff and approved a five-year programme to upgrade environmental awareness and skills among Bank staff. Guidelines for incorporating EIAs into the Bank's project cycle have been developed. In 1989, 30 loans and 43 technical assistance projects had large environmental components. The AsDB has sponsored research into the effect of projects on ecologically sensitive areas, developed guidelines for assessing the impact of projects on biodiversity and providing technical assistance funds to encourage its developing country members to conserve biodiversity. The AsDB has incorporated biodiversity conservation into its Forest Sector Policy Paper, but the tension between increasing forest production and conservation remains.

The African Development Bank (AfDB) established an Environment Unit in 1987 and in June 1990 the AfDB's Environment Policy Paper was approved by the Board of Directors. The Paper provides guidelines for environmental impact assessment of both project and non-project loans, and includes a brief on biodiversity. Review of the AfDB's 1988 loans in the transport, public utilities and agriculture sectors (67% of the Bank's loans) indicated that half of the loans would have negative environmental impacts. In 1989 just five full EIAs were conducted by the Environment Unit. During the 1987-1989 period the AfDB reports that environmentally beneficial projects and environment-linked projects more than doubled. In 1989 these projects accounted for 18% of AfDB commitments.

The traditional focus of the International Fund for Agricultural Development (IFAD) on rural poverty has recently been broadened in an effort to integrate the environmental dimension into the Fund's work programme. For this reason IFAD has rejected setting up an environmental unit and is concentrating on upgrading the knowledge and skills of existing staff. In 1991 IFAD initiated a two-year programme to develop and test the introduction of EIA into the project cycle. Preparation of guidelines for sustainable agriculture and sectoral studies of resource management is also under way.

The Nordic Investment Bank (NIB) lends roughly $1 billion per year, 20% of which goes to non-Nordic countries. In 1989-90 the NIB co-financed the Mauritius Environmental Master Plan and an afforestation project in Indonesia. Brief environmental appraisals of all NIB loans are conducted by project officers. The Nordic Environmental Finance Corporation (NEFCO) began operations in 1990 and is administered by the NIB. NEFCO provides financing to joint ventures in ex-Eastern Bloc countries that provide products beneficial to the Nordic environment.

The United Nations Development Programme (UNDP) began developing its Environmental Management Guidelines in 1989. The Guidelines provide the means for non-specialists to incorporate the principles of environmental management into their work. UNDP has increased the coverage of what it considers as 'environmental projects' from those that are purely environmental in a scientific sense to include those that encourage sustainable development and the improvement of the 'quality of human life'. Under this definition UNDP expenditures on environmental activity totalled US$600 million in 1990. In 1988 expenditures on projects characterised as biodiversity conservation projects came to $1.5 million. Adding in subtotals for plant resources and wildlife management the total spent on biodiversity increases to US$6.7 million. UNDP is currently an active partner in the Global Environmental Facility which is described below.

The World Food Programme (WFP) of the United Nations commits one-third of its 'food aid' finance to environmental project components. WFP is training its staff in incorporating environmental concerns into project design rather than approaching environmental issues by way of rigorous EIAs. A simple checklist assessment has been devised to alert staff to potential environmental risks during the project preparation and planning stages. The checklist does contain cautions regarding the loss of genetic diversity as cropping patterns change, but does not include specific diversity considerations within areas such as afforestation, road construction, or soil conservation. From time to time the Programme does undertake occasional in-depth *ex post* project and sector evaluation.

The mandate of the United Nations Environment Programme (UNEP) is to coordinate and catalyse action on the global environment. Funding for UNEP programmes comes from the United Nations, trust funds, counterpart contributions, and voluntary contributions to the Environment Fund. By far the largest source of funds is the Environment Fund category. Since contributions to the fund are voluntary they can vary substantially from year to year. Contributions decreased in real terms through the early to mid-1980s, but a recent turn around has seen real contributions increase by roughly 7% from 1987-89 and by 25% in 1990. With 1990 contributions at just over $50 million the Executive Director has called for the pace to continue in order that UNEP reach a target of $245 million by 1995. In 1990 UNEP made commitments of $3.2 million to support its efforts in biological diversity conservation. This sum includes work on genetic resources, the biodiversity convention, biotechnology transfer and cooperation with NGO conservation initiatives.

The mandate of the UN Food and Agriculture Organization (FAO) Programme to support development efforts in fisheries, forests and agriculture means that FAO policies and activities have a large impact on biodiversity. Recognising the onslaught of environmental degradation in developing countries, FAO's 1989 Governing Conference added biotechnology and the Tropical Forestry Action Plan (TFAP) as FAO priority areas. FAO Division directors meet in working groups on technical environmental matters including biological diversity. EIA procedures initiated in 1988 are now utilised for evaluating FAO field projects and projects prepared by the FAO Investment Centre.

Numerous activities undertaken or coordinated by FAO are designed to mitigate biodiversity loss. FAO is involved in revising the TFAP, assists UNESCO with the Man and the Biosphere Programme, and has cooperated with IUCN on the Caring for the Earth Strategy. Together with UNDP and the World Bank, FAO sponsors the Consultative Group on International Research (CGIAR) which supports the 13 International Agricultural Research Centers (IARCs). While Green Revolution crop research at these IARCs is criticised for leading to loss of on-farm diversity, the International Board for Plant Genetic Resources (IBPGR) - also an IARC - is heavily involved in *ex situ* preservation of genetic diversity. FAO has also formed a Commission on Plant Genetic Resources and is involved in developing Regional Animal Gene Banks.

Biodiversity and environmental initiatives undertaken to date by the major multilaterals indicate that efforts to control harmful projects and identify beneficial ones, particularly with regards to biodiversity, are still in their formative stages. Some organisations have firm guidelines in place; others are still formalising such procedures. Meanwhile other agencies prefer a less explicit approach, believing that concerns over the environment and biodiversity must become an integral part of the project cycle, instead of an extra component or evaluation tacked on to the normal appraisal process.

A key limitation in generating rapid changes surrounds the capability and availability of staff and consultants. Again the amount of effort put into retraining varies from one agency to the next, and from one discipline to the next. Technical specialists in environmental sciences capable of carrying out physical EIAs are likely to be more widely available than environmental economists. As basic and applied research into the socio-economic impacts of decrements and increments in environmental quality improves the tools of the trade, more complete cost-benefit analysis of projects will be possible.

The Global Environmental Facility

In 1989 at the annual IMF-World Bank Development Committee meetings, France suggested the creation of a global fund for encouraging developing countries to undertake environmental protection activities that provide benefits to the global community. By November 1990 agreement had been reached by 25 countries that the World Bank, UNDP and UNEP would cooperate in administering the Global Environmental Facility (GEF), a mechanism for distributing concessionary finance for the purpose of protecting the 'global commons'.

As of March 31, 1991 twenty-one countries had committed approximately US$1.4 billion to the fund over a three-year pilot stage. The Facility is accepting proposals for funding in four areas:

- protecting the ozone layer
- limiting greenhouse gas emissions
- protecting biodiversity
- protecting international waters.

The Ozone Layer Trust Fund will administer US$160 million earmarked for activities in conjunction with the Montreal Protocol. Biodiversity and the other two activity areas will receive funding from the Global Environmental Trust Fund. The GEF's mandate with respect to biodiversity is to preserve specific areas that contribute goods and services such as harvestable material for medicines or industrial products, genetic resources for food production and the regulation of climatic and rainfall patterns.

Access to GEF funds is limited to countries with GNP of less than US$4,000 in 1989 and that have UNDP programmes. In order to differentiate between projects that meet the GEF mandate and normal development projects proposals the Bank is developing criteria for GEF projects. Likely candidates for funding are those projects that do not meet overall rate of return criteria but produce global environmental benefits and funded projects that with additional investment could provide such benefits. Projects should demonstrate that funding cannot be obtained from other sources such as bilateral and internal sources and they are not economically viable by normal lending criteria. Additional qualifications of the projects are the use of appropriate technology, cost-effectiveness, merit from a global perspective, and consistency with existing

Table 32.6 Global Environment Facility (GEF) biodiversity projects

GEF investment projects

COUNTRY	PROJECT	TARGET	ASSOCIATED PROJECT	FUNDING US$millions
Congo	Congo Tropical Forest Preservation	Lowland rain forest	Free-standing	10.00
Kenya	Lower Tana River Primates	Riverine forest	IBRD US$30m	6.20
Uganda	Gorilla Reserve Bwindi Forest	Lowland and montane forest	Free-standing	4.00
Bhutan	Trust Fund for Environment Conservation	Lowland, temperate and alpine forests	Free-standing	10.00
Laos	Wildlife and Protected Areas Management	Lowland and montane forest	IBRD US$10m	5.50
Philippines	Conservation of Priority Protected Areas	10 high priority protected areas	IBRD US$158m	20.00
Algeria	El Kala National Park	Wetland	IBRD US$30m	12.00
Poland	Forest Biodiversity	Temperate/montane forest	Free-standing	4.00
North Africa	New World Screw Worm Eradication	Wildlife biodiversity	FAO Project US$56m - 1991	9.00
Brazil	National Conservation Units	25 conservation units	IBRD US$117m	30.00
Mexico	Biodiversity Conservation	20 protected areas	IBRD US$30m	30.00
			Subtotal	**140.70**

GEF technical assistance projects

COUNTRY	PROJECT DESCRIPTION	FUNDING $millions
East Africa	Support for training, research, equipment and institutional development of government, university, and NGOs working in protected area management	10.00
West/Central Africa	Establish a regional TRAFFIC office in Zaire and develop capacity to monitor both legal and illegal trade in wildlife	1.00
Viet Nam	Training/institutional development to prepare a plan for protected areas	3.00
South Pacific	Establish and manage 20 conservation areas with threatened biodiversity	8.20
Colombia	Assess diversity of the Choco Region through capacity-building research with a view to developing plans for protection and sustainable use	9.00
Guyana	Protect a large tract of rain forest, study the impact of local management	3.00
Amazon	Institutional strengthening within the eight members of the Treaty for Amazonian Cooperation	4.50
	Subtotal	**38.70**
Total GEF biodiversity funding		**179.40**

environmental conventions and national environmental strategies.

Maximum size for freestanding projects is US$10 million. Project proposals may come from bank or bilateral staff, government agencies in developing countries and NGOs. The local UNDP representative must review NGO proposals with the host government. Projects are forwarded to the World Bank Regional Environmental Division Chief for routing to the appropriate desk officer. Bank staff prepare a three to five page project summary assessing the compliance of the project with GEF criteria. If the project looks promising it is then reviewed by a technical panel which issues a Final Executive Project Summary for appraisal by the Bank and UNDP and UNEP.

For the fiscal year 1992, the first tranche of funding under the three year plan, 26 investment and technical assistance projects are slated for approval at an estimated cost of US$273 million. Table 32.6 reveals that expected funding for biodiversity projects totalled US$179 million from the first tranche. The Bank attributes this to 'unmet demand' for financing biodiversity protection. In the remaining years of the pilot programme the Bank expects to address this imbalance in the lending portfolio. The share of the initial GEF devoted to biodiversity projects is expected to reach US$400 million.

The GEF's mandate to subsidise the provision of global benefits by developing countries represents an important step forward in recognising the distribution of economic benefits provided by genetic and biological resources. However, the tone of the projects under consideration by the GEF is very 'preservationist.' As shown in Table 32.6, despite the variety of approaches employed, almost all the biodiversity projects are concerned with creating protected areas or building 'park' management capacity. This is a natural outcome of the GEF's mandate to fund projects with low rates of return and large external environmental benefits. Relatively few of the project briefs include components that diverge from pure preservation by encouraging local use and conservation of biodiversity. However, these activities are considered secondary by the GEF because of the perception that their benefits are appropriated locally not internationally. Many consider it unfortunate that GEF policy contains little overt acknowledgement that protection and sustainable use of biodiversity at the local level are often inseparable.

A portion of the GEF portfolio (perhaps through the proposed small grants window) might be allocated to research and to projects that encourage the sustainable use of local diversity with an eye towards external benefits. In this manner GEF might assist in developing or rekindling local people's respect for the benefits of biodiversity and thereby ensure its continued existence. While a pure economic analysis of the distribution of benefits from biodiversity informs the GEF biodiversity strategy, such analysis overlooks the practical problem that the generation of external benefits may not be distinct from the generation of local use benefits.

A final criticism voiced by environmental groups concerns public access throughout the GEF project cycle. The issue of confidentiality of Bank documents and public participation in project design and evaluation is a common complaint with regards to Bank projects - not just GEF biodiversity projects. An interesting conundrum of GEF financing is that since such projects are 'environmental' projects they are exempt from the EIA process under the 1989 EAOD reviewed above. The absence of a detailed EIA process exacerbates the difficulty of incorporating public participation, in both developed and developing countries, into the project selection process.

INTERNATIONAL ASSISTANCE IN FOREST MANAGEMENT

This section discusses two examples of international assistance in domestic regulation of resources.

The destruction of tropical forests has for some years been an issue of considerable concern. It is no longer perceived as simply a domestic problem for which national remedies are to be sought, but as a problem of international concern requiring an international response. A measure of the importance attached to the problem is the priority given to it in the 1992 UN Conference on Environment and Development (UNCED), where it is one of the three specific topics on the agenda (the others being biodiversity and climate change).

Widespread concern about global deforestation resulted in the genesis of two international programmes in 1983. One of these programmes, the Tropical Forestry Action Plan (TFAP), was developed by FAO with the assistance of numerous international organisations and NGOs. The second programme was the International Tropical Timber Agreement (ITTA), which was established as a type of commodities cartel between the governments of tropical timber producer and consumer countries.

THE TROPICAL FORESTRY ACTION PLAN (TFAP)

TFAP is a programme run by FAO. It is intended to provide a mechanism whereby international aid efforts could be better harmonised and coordinated, with a view to halting the destruction of tropical forests and promoting their sustainable development. The programme seeks to do this by helping countries which have tropical forests to develop national forest management strategies. These strategies are intended as the basis for increasing investment in tropical forestry, with the coordinated assistance of aid programmes from donor countries.

The plan originated from the Committee on Forest Development in the Tropics (CFDT) - a statutory body of FAO - which in October 1983 called for the establishment of *ad hoc* groups of experts to identify the main problems in tropical timber production, and the development of action programmes to address these problems at regional or global levels. The ultimate result of the various *ad hoc* meetings which followed was a five year action programme to address deforestation issues. The action programme was divided into five sections: fuel wood and agroforestry; land-use and upland watersheds; forestry management for industrial uses; conservation of tropical forest ecosystems; and strengthening institutions for research training and education. Each of the five sections contained a list of recommended actions and investments for a five year action programme.

The plan was endorsed in June 1985 by CFDT and adopted by FAO in October 1985 with the formal release of the TFAP (FAO, 1985). Initially, the means for the implementation of the recommendations were not elaborated upon to any great extent. The plan merely laid out the principles and recommendations for the guidance of development assistance agencies, in order to inform them how aid might be directed to the objective of sustainable forestry management.

The process required development of an individual national level TFAP for every country which had tropical forests. The development of these national TFAPs was a multi-stage process coordinated by the TFAP unit of FAO. The TFAP coordination unit is within the Forestry Department of FAO and is intended to receive technical support from the organisation as a whole. Funds allocated to the unit from the FAO regular programme are currently supplemented by the Multidonor Trust Fund. The process has evolved over the years and has been implemented in differing ways in different countries. The development of the procedure has been carried out by an unofficial body known as the TFAP Forestry Advisory Group which meets every six months. The General Terms of Reference for national TFAPs were outlined in the first meeting of this group in 1985 and these have been progressively expanded upon at subsequent meetings (FAO, 1989b).

Despite individual differences in the implementation of the national TFAPs, their development has involved a series of basic steps. First, the process is initiated by a request for assistance from the TFAP coordination unit in the preparation of a national forest action plan. The next stage is that FAO or another agency chosen from among the donors (the World Bank, FINNIDA, CIDA, ODA, etc.) is identified as the lead agency for the development of that national TFAP. Then, a review is prepared by the lead donor agency on the basis of existing information. The review is sometimes referred to as the 'issues paper'. This review is designed to highlight the major problems facing the forestry sector in the particular country and is used as a means of identifying sectors of intervention, terms of reference for consultants, securing participation of NGOs and local people, and as a basis for the programme and schedule for the mission. The government of the producer country concerned then has a chance to review the draft issues paper. After this, the review is finalised at a meeting of all parties involved in the forest sector (a meeting sometimes referred to as 'Roundtable I').

The forestry review mission is next set up. This usually involves foreign consultants, local government officials, and staff from the lead agency. The review mission then carries out a forestry sector review over two or three months in the country for which the national plan is being developed. The team's findings are then discussed at a meeting between representatives of the government, aid agencies, review mission team and various concerned NGOs. This meeting is commonly referred to as 'Roundtable II', and the purpose of it is to analyse from a technical point of view the various reports from the forest review mission.

The results of these discussions are then written up as the national forest action plan. This is then presented at a national planning seminar held between government officials and funding agencies to discuss effective implementation of the plan. This meeting is referred to as 'Roundtable III'.

Finally, the plan is presented to a wide range of donor countries and agencies, who can then use it as the basis for future development assistance to the particular country. The plan is not legally binding on the donor agencies or countries and they are free to use only specific parts of the plan in structuring their future assistance to the country in question. The objective of the process is to procure information that will allow donor countries to make more informed choices concerning the best uses of their aid monies, in regard to the development of the tropically forested countries.

To date, no country has completed every step of the entire process. The latest figures show that a sector review has been completed in 34 countries and was at different stages of completion in a further 51 countries. In 24 countries where the forest sector review had been completed, there had also been a Roundtable III meeting. In addition, FAO has received a further 11 requests from national governments for initiation of the TFAP process, and 10 others are expected to do so in the near future.

The effectiveness of TFAP

The number of countries involved in the TFAP process has steadily grown since its inception in 1985. At the end of March 1990, 70 countries which together include 60% of the world's remaining tropical forests have become involved in the TFAP process. However, none have yet completed the entire process as envisaged by the TFAP guidelines outlined above. The status of selected national forest plans is given in Table 32.7 and summarised in Table 32.8.

TFAP has generally been well received by most donor agencies. More than 40 aid agencies - which together account for nearly all of the official development assistance provided to the forestry sector - have collaborated to support the organisation of over 50 national forest sector reviews.

Funding commitments to the forestry sector have generally seen a dramatic increase over the last few years. Total international development assistance increased from US$603 million in 1984 to US$1,095 million in 1988 (FAO, 1989a) Recently the World Bank has committed itself to tripling investment in forestry. The ODA has also pledged £100,000,000 per year to TFAP, and USAID increased funding of forestry projects from US$50 million in 1988 to US$72 million in 1989 (Sargent, 1990).

The funds allocated to each of the five TFAP sectors are indicated in Table 32.9. By contrast, the original TFAP plan envisaged that the relative funding requirements for each sector would be: fuel and agroforestry US$1,899 million, land-use on upland watersheds US$1,231 million,

Table 32.7 Status of national TFAPs

A. Planning phase completed including Roundtable III

LATIN AMERICA/CARIBBEAN	AFRICA	ASIA/PACIFIC
Argentina (National)	Cameroon (UNDP/FAO)	Nepal (ADB)
Belize (ODA)	Equatorial Guinea	Nepal (ADB)
Bolivia(UNDP/FAO)	Ghana (FAO/WB)	Papua New Guinea (WB)
Colombia (Netherlands)	Sierra Leone((UNDP/FAO)	Philippines (ADB)
Costa Rica (Netherlands)	Sudan (WB)	Laos (UNDP/FAO)
Dominican Republic (UNDO/FAO)	Tanzania (FINNIDA)	Sri Lanka (WB)
Ecuador (National)	Zaire (CIDA)	Fiji (UNDP/FAO)
Honduras (National)		
Jamaica (UNDP/FAO)		
Panama (UNDP/FAO)		
Peru (CIDA)		
Central America (USAID)		

B. Forestry sector review completed

LATIN AMERICA/CARIBBEAN	AFRICA	ASIA/PACIFIC
Cuba (National)	Guinea (France)	Indonesia (WB/FAO)
Guatemala (USAID)	Mauritania (UNDP/FAO)	Malaysia (National)
Guyana (CIDA)	Somalia (UNDP/FAO)	Viet Nam (UNDP/FAO)
Mexico (National/FAO)		

C. Forestry sector review under way

LATIN AMERICA/CARIBBEAN	AFRICA	ASIA/PACIFIC
Chile (FAO/Netherlands)	Burkina Faso (GTZ)	Bangladesh (ADB)
Haiti (UNDL/FAO)	Burundi (UNDP/FAO)	Bhutan (ADB)
Nicaragua (SIDA)	Cape Verde (Belgium)	India
Suriname (FAO)	Central African Rep. (WB)	Pakistan (ADB)
Venezuela (National)	Congo (France)	Thailand (FINNIDA/UNDP)
	Côte d'Ivoire (FAO/WB)	Vanuatu
CARICOM (FAO/ODA)	Ethiopia (WB/UNDP)	
Antigua and Barbuda	Gabon (France)	
Barbados	Gambia	
Dominica	Guinea Bissau (WB/EEC)	
Grenada	Kenya (FINNIDA)	
Montserrat	Lesotho (UNDP/FAO)	
St Kitts and Nevis	Madagascar (UNDP/FAO)	
St Lucia	Mali (France)	
St Vincent and the Grenadines	Mozambique (FAO)	
Trinidad and Tobago	Niger (UNDP/FAO)	
	Nigeria (WB)	
	Rwanda (ACCT-Canada)	
	Senegal (UNDP/FAO)	
	Togo (UNDP/FAO)	
Amazon Pact (FAO)	Zambia (FINNIDA)	
	CILSS	
	SADCC	
	IGADD	

Table 32.7 Status of national TFAPs (continued)

D. TFAP exercise requested

LATIN AMERICA/CARIBBEAN	AFRICA	ASIA/PACIFIC
El Salvador	Angola	Myanmar
Paraguay	Liberia	Solomon Is
Uruguay	Mauritius	
	Uganda	
	Zimbabwe	

E. Preliminary contacts and inquiries

LATIN AMERICA/ CARIBBEAN	AFRICA	ASIA/PACIFIC
Brazil	Chad	China

Note: Sub-regional exercises are indicated in *italics* and are counted separately from countries (except for CARICOM, which is at the same time a subregional and multi-country exercise) where individual country issues are treated also at national level. The international core support agency is indicated within brackets. Some countries have completed their planning phase without holding a Roundtable III.

Table 32.8 Summary of national TFAPs

	COUNTRIES	SUB-REGIONAL
Total of exercises:*	86	6
Africa	37	3
Asia/Pacific	17	-
Latin America and Caribbean	32	3
Planning Phase completed	24	-
Sector Review completed	10	1
Sector Review under way	41	5
Exercise requested	11	

Note: * Inquiring countries not included

Table 32.9 Distribution of official development assistance by TFAP fields of action in 1988

FIELDS OF ACTION	DONOR COUNTRIES US$[1]	%	DEVELOPMENT BANKS US$[1]	%	UN AGENCIES US$[1]	%	TOTAL US$[1]	%
Forestry and Land-use	150.0	27.4	13.9	6.5	50	26.6	213.9	22.6
Forest-based Industries	92.6	17.0	146.4	68.9	63.8	33.9	302.8	32.0
Fuelwood and Energy	97.9	17.9	12.9	6.1	47.2	25.1	158.0	16.7
Conservation	50.3	9.2	20.0	9.4	13.2	7.0	83.5	8.8
Institutions	155.5	28.5	19.4	9.1	13.8	7.4	188.7	19.9
Subtotals	631.7[2]	100.0	212.6	100.0	188.0	100.0	1,032.3*	100.0

Source: FAO 1989c. *Review of International Cooperation in Tropical Forestry.*
Notes: [1] In millions. [2] Includes undetermined US$85.4 million, 13.5% of total, from Federal Republic of Germany.

industrial forestry US$1,640 million, ecosystem conservation US$550 million. TFAP indicated that 20% of the funding in each sector should be devoted to strengthening institutions for research, training and education. These totals represent the sum to be spent over the entire five years of the action plan.

TFAP has been an important instrument for the channelling of aid monies to conservation objectives. The extent to which it has actually generated additional developmental assistance is unclear, as is the extent to which the rate of deforestation has been slowed down by the programme. The overall trend in the problem which the TFAP was established to combat is more certain: recent FAO statistics have shown that deforestation for open and closed canopy tropical forests has increased from 11.3 million ha per year in 1980 to 19.0 million ha per year in 1990 (Collins *et al.*, 1991). No precise data are available for closed forests alone but the trend appears to be similar. As a result of this continuing rise in the rates of deforestation many NGOs have been extremely critical of the TFAP. Several reviews of the TFAP were carried out in 1990 (Colchester and Lohmann, 1990; Elliot, 1990; Winterbottom, 1990). All recommended substantial restructuring of the process. Many TFAP observers feel that the projects were too frequently developed from the top down and not from the grass roots level in areas where deforestation is occurring. This resulted in a preponderance of foreign experts involved in each project and not enough input from NGOs. Reviewers also criticised TFAP for failing to ensure that money spent in the forestry sector in developing countries is spent effectively and not with detriment to the environment. It is also claimed that even though the TFAP was established to approach the problem of deforestation on a cross-sectoral basis, it has paid scant attention to non-forest issues. There have been suggestions that TFAP be given greater independence either by removing it from the FAO altogether or by promoting the TFAP from the forest division and making it a separate division. A process of review by the architects of the TFAP was instigated at a meeting arranged by FAO in Geneva last year as a result of these criticisms. The results of this meeting at the time of writing are still being negotiated.

Despite the perceived deficiencies of the TFAP, it has acted as an important mechanism for guiding existing aid toward more effective investment in regard to the tropical forested nations.

THE INTERNATIONAL TROPICAL TIMBER AGREEMENT (ITTA)

At about the same time that the CFDT called upon the FAO to establish the TFAP, at the UN Conference on Tropical Timber an agreement on tropical timber was being negotiated and adopted. The International Tropical Timber Agreement (ITTA) came into force on 1 April 1985. The initial term of the agreement was meant to be five years, but this was extended in 1990 for a further two years. The ITTA was originally conceived as a commodities agreement and the initial version, which was drafted by the Japanese, was based upon the 1979 Rubber Agreement and the 1982 Jute Agreement. However, the final form of the adopted agreement is unlike previously negotiated commodity agreements. Typically, commodity agreements are principally concerned with price control and stabilisation, and to this end they establish buffer funds and price manipulation mechanisms. The ITTA, however, developed into a mechanism much more similar to an agreement for international development assistance.

This change in emphasis is evident in the preamble to the agreement. There it states that the parties enter into the agreement "recognising the importance of and the need for proper and effective conservation and development of tropical timber forests with a view to ensuring their optimum utilisation while maintaining the ecological balance of the regions concerned and of the biosphere". This emphasis on sustainable development is also evident in the objectives of the agreement. The ITTA's objectives include the development of the industry, but also the promotion of research and development with a view to improving forest management. The purpose of this is to "encourage members to support and develop industrial tropical timber reforestation and forest management activities" and to encourage the development of national policies aimed at sustainable utilisation and conservation of tropical forests and their genetic resources and at maintaining the ecological balance in the regions concerned.

The agreement establishes a complex institutional structure to facilitate the attainment of these objectives. The administrative structure as a whole is known as the International Tropical Timber Organization (ITTO) which is composed of the following elements:

- The International Tropical Timber Council (ITTC)
- Three permanent committees
  - Committee on Economic Information and Market Intelligence (PCM)
  - Committee on Reforestation and Forest Management (PCF)
  - Committee on Forest Industry (PCI).
- The Executive Director and staff based in Yokohama, Japan.

The ITTC acts as the principal political body of the organisation. The ITTC coordinates the work of the three permanent committees and carries out all necessary functions to fulfil the provisions of the ITTA. The Council is composed of all members of ITTO (i.e. parties to the ITTA). The ITTC is obliged to hold at least one meeting a year, although usually two are held - one in Yokohama and the other in a producer country. The voting scheme used at these meetings is quite elaborate, and is based upon achieving balanced representation between industry producers and consumers (Hypay, 1986).

Permanent committees

Most of the decisions taken by the ITTC are on the basis of the recommendations of the permanent committees who function as the operational arm of ITTO (Hypay, 1986). Participation in each of the permanent committees is open to all members of the ITTA. As the operational arm of ITTO these permanent committees play a vital and important role in the implementation of the objectives of the ITTA.

The functions of each of the three permanent committees are outlined in the ITTA. Apart from normal commodity-type functions the PCM and the PCF have functions which would normally be associated with an environmental protection organisation. For instance, the PCM functions include "making recommendations to the council on the need for and nature of appropriate studies on tropical timber including long term prospects of the international tropical timber market". The PCF is the key committee for the implementation of the environmental objectives and its functions include: reviewing assistance provided at national and international levels for reforestation and forest management; encouraging technology transfers for reforestation and forest management; setting the requirements and identifying possible sources of financing for reforestation and forest management. The most important function of the PCF is "to coordinate and harmonise these activities for cooperation in the field of reforestation and forest management with the relevant activities pursued elsewhere, such as those under FAO, UNEP, the World Bank, Regional Banks and other competent organisations". Overall, the permanent committees are orientated to assisting the general support role which ITTO plays in this area.

The work of the permanent committee

To date the vast majority of the work of ITTO has been carried out by these three permanent committees. By the end of 1990 ITTO had approved 66 projects and commissioned a further 35 pilot studies for further projects. The PCF had been delegated the largest number of these studies and consequently had the largest budget. As of 1 January 1991 this committee was engaged in 19 pre-project studies and a further 21 current projects. Its estimated budget for these studies came to over US$12.3 million of which over US$2.5 million had already been paid. The PCM was engaged in six pre-projects studies and a further eight current projects. Its estimated budget for these studies came to over US$2 million of which US$1.3 million had already been paid. The PCI was engaged in nine pre-projects and a further 19 current projects. Its estimated budget for these studies came to over US$6.4 million of which over US$2.4 million had been paid.

Table 32.10 outlines the projects which the PCF was then involved in at the end of 1990. An examination of the project descriptions show that these closely match the defined functions of the permanent committee. They also reflect the idea that this agreement is concerned with more than the narrow question of maximum price control of a particular commodity. In fact it appears from the activities of the permanent committees to date that the question of price control has not featured very largely on the agenda of the organisation. Furthermore, from Table 32.10 the support role envisaged for ITTO in pursuing these environmental objectives is evident in both the nature of the projects and the fact that in only five of the 21 current projects is ITTO the implementing or lead agency.

Apart from sponsoring these projects, the other major initiative of ITTO has been a programme to assist countries in the development of management procedures to direct timber production in tropical forests toward sustainability. The year 2000 was established as the target date for the achievement of sustainable management of tropical forests worldwide. This object has been pursued through the establishment of forestry standards for the sustainable management of natural tropical forests for timber production. These were drawn up by the PCF and adopted by the ITTC at the 7th session of the Council in May 1990. These standards contain a set of 41 principles and 36 possible actions. They cover considerations ranging from general policy to particular aspects of forestry operations. The general principles involved include the establishment of national forestry inventories and a permanent forest estate. They also recommend examination of forest lands ownership, and the establishment of separate institutions for the management of the forest estate in each country. In addition to these guidelines ITTO is presently developing another set of guidelines on biodiversity, known as 'The ITTO Guidelines on the Conservation of Biological Diversity in Tropical Production Forests'. The objective of these guidelines is to "optimise the contribution of these forests to the conservation of biological diversity that is consistent with ... the sustainable production of timber and other products". At the time of writing these guidelines, although accepted by the PCF committee, have not gained political support at the ITTC level and are unlikely to be acted upon by the ITTC before UNCED (Anon., 1991).

In addition to the development of these standards, ITTO is also generating financial assistance for producer countries, both directly from its own funds and indirectly through the solicitation of contributions from consumer countries. This assistance is usually employed to help countries in their incorporation of these guidelines into national policy and legislation.

Conclusion

From the above overview of TFAP and ITTA, it is evident there are a number of similarities between the two organisations. This is to be expected as they are both international responses to the problem of deforestation represents a significant global problem. One programme originated out of the international resource community, and the other out of the industry itself.

The stated purpose of TFAP is to harmonise and coordinate actions in the tropical forest sector so that tropical forests can be used by mankind on a sustainable basis. Similarly, the stated purpose of ITTO is "to develop proper and effective conservation and development of Tropical Forests with a view to ensuring their optimum utilisation while maintaining the ecological balance of the regions concerned and of the biosphere." ITTO states that it wants to move worldwide production to a sustainable basis by the year 2000.

Despite the fact that the objectives of the two organisations are similar, the way in which both organisations set out to achieve these objectives is ostensibly quite different; this is because of the origins of the two programmes and the difference in their legal structure.

The TFAP is technically a policy which lacks any legal content. From a legal point of view it is no more than an information document which organisations working in the area may choose to adopt to guide their decisions. The funds dispensed under the programme do not pass through the TFAP; instead, they flow directly from the donor to the recipient. At the discussions about the implementation of a national TFAP (Roundtable III), if a donor can be found for a particular aspect of the plan or the entire National TFAP, then that donor will make arrangements for the provision of funds to the recipient. Another consequence of the legal structure of the TFAP is that few of the activities involved in the development of a national TFAP are actually carried out in the name of the TFAP; they are instead coordinated and conducted in the name of the lead agency. In sum, the TFAP is only a name associated with a large number of separate agreements and relationships; it has no structure of its own.

ITTO, on the other hand, is a legally constituted organisation which does have its own legal personality,

Table 32.10 Projects of the ITTO Committee on Reforestation and Forest Management, 1990

PROJECT DESCRIPTION	AMOUNT BUDGETED US$	AMOUNT PAID US$	OUTSTANDING OBLIGATION US$	IMPLEMENTING AGENCY(IES)	DONOR(s)	AGREEMENT SIGNED
ITTC/III						
Management of natural forest in Malaysia	272,353	177,200	95,153	Forest Department Malaysia	Switzerland	08/08/88
The biology of the Okoume (*Aucoumes klaineana* Pierre) in Gabon	417,500			Tropenbos/Gabon	ITTO	
Investigation of the steps needed to rehabilitate the areas of East Kalimantan seriously affected by fire	377,600	292,000	85,600	Ministry of Forestry - Indonesia	Switzerland	01/06/88
ITTC/IV						
Rehabilitation of logged-over forests in Asia/Pacific region	240,000	240,000 [1]		JOFCA	Japan	01/12/88
Integration of forest-based development in the Western Amazon - Phase 1 - forest management to promote policies for sustainable production	1,078,000	1,036,000 [2]	42,000	FUNTAC	Japan/Switz/ Neth/WWF	22/05/88
International seminar on sustainable utilisation and conservation of tropical forests	81,032	81,032		ITTO	Japan	N.A.
ITTC/VI						
Seminar and study tour on sustainable development of tropical forests and tropical timber manufacturing industries in Japan	95,000	95,000		ITTO	Japan	N.A.
Conservation, management, utilisation, integral and sustained use of the forest of the Chimanes region department of El Beni, Bolivia	1,260,000	240,000	1,020,000	CFD/ITTO/CI	Japan/Switz/ Denmark/WWF	19/04/90
Tropical forest internship	109,500			USDA-FS, USDA-FS		
ITTC/VII						
Workshop on sustained tropical forest management with special reference to the Atlantic Fund	68,000			IBAMA & FUNATURA	Japan	10/07/90
Seminar on sustainable development of tropical forests	50,000	50,000		ITTO	Japan	N/A
Management of the Tapajos National Forest for sustainable production of industrial timber	1,513,146			IBAMA, Brazil	UK-ODA	
ITTC/VIII						
Preparation of a master land-use plan for forest areas	395,000	158,000	237,000	OMADEF, Cameroon	Japan/Norway	19/09/90

Table 32.10 . Projects of the ITTO Committee on Reforestation and Forest Management, 1990 (continued)

PROJECT DESCRIPTION	AMOUNT BUDGETED US$	AMOUNT PAID US$	OUTSTANDING OBLIGATION US$	IMPLEMENTING AGENCY(IES)	DONOR(s)	AGREEMENT SIGNED
The economic and environmental value of mangrove forests and present state of their conservation	270,000			JIAM & ISME	Japan	
Better utilisation of tropical timber resources in order to improve sustainability and reduce negative ecological impacts	500,000			FRCFFD-Germany	Germany	
Development of genetic resistance in the tropical hardwood Iroko to the damaging insect pest: *Phytolyma lata*	257,410			FPRI-Ghana	Japan/Denmark	
Project formulation workshop for establishing a network of genetic resources centres for adapting to sea level rise	100,000	68,000	32,000	CRSARD, India	Japan/UK	06/08/90
Panel discussion/seminar on promotion of a positive image of sustainable utilisation of tropical forests	100,000			ITTO	Japan	N.A.
The establishment of a demonstration plot for rehabilitation of forest affected by fire in East Kalimantan [Phase II of PD 17/87 (F)]	704,000			Ministry of Forestry-Indonesia	Japan/Norway/ Netherlands	
International network for developing human research in tropical forest management (Phase I)	500,000			ITTO	Japan/Denmark	N.A.
Sustainable forest management and human resources development in Indonesia	3,800,000			Ministry of Forestry-Indonesia	Japan	
SUB-TOTAL	12,188,541	2,437,232	1,511,753			

Source:ITTC(IX)/CRP/1
Notes: [1] $50,000 reimbursed to Pre-project subaccount; [2] $100,000 reimbursed to Pre-project subaccount.

headquarters, staff and budget. ITTO carries out many activities in its own name and raises funds and uses its resources in its own name.

An important reason for the different approaches is the different origins of the two programmes: the TFAP was developed by FAO and various NGOs active in the tropical forest sector whereas ITTO is a product of the UN Conference on Trade and Development (UNCTAD). Whilst TFAP has approached the problem from the coordination of development assistance programmes, ITTO has approached it much more from a trade point of view. Thus TFAP is meant to approach the problem in a cross-sectoral way and has sought to include as many parties as possible in the process. It has traditionally attempted to do this through the Roundtables, by increasing the importance of the forestry sector and by coordinating the various development assistance programmes that have some impact on the forest sector in the relevant country. On the other hand ITTO is not cross-sectoral and is concerned simply with the forest estate of the producing countries. This different approach is seen in the nature of the projects ITTO is involved in, which on the whole tend to be concerned with quite specific problems in tropical timber production.

Yet, despite these ostensible differences in approach, the work of the two programmes is very similar. The defined functions of the PCF are very similar to the fields of action of the TFAP. For instance the TFAP calls for "the protection and management of natural forests" along with "accelerated industrial reafforestation" and ITTO has established a permanent committee whose main purpose is to "promote better forest management and reafforestation' - the PCF. TFAP states that "financial incentives are needed to encourage investment in reafforestation and forest management" and the ITTA empowers the PCF to "identify all possible sources of financing for reforestation and forest management". Both programmes make extensive reference to promoting training, research and education in the area. Both encourage greater transfer of technology. The guidelines developed by ITTO closely resemble many national TFAPs.

This overlap between the two organisations results in some duplication of effort and illustrates a lack of direct cooperation between the TFAP and ITTO. This is perhaps one of the more easily correctable failings of international efforts to assist in the regulation of this domestic resource.

DEBT PURCHASE

The debt purchase discussed here covers a specific form of debt-equity conversion, widely termed a 'debt-for-nature' swap. Other types of debt-equity conversions such as 'debt-for-development' and 'debt-for-child' also occur. The essential aim of all these types of instruments is to convert the external debt of a developing country into a domestic obligation to support a specific programme. Table 32.11 details the debt-for-nature agreements which have been established so far. Although no two debt-for-nature swaps so far negotiated have been identical, the basic structure used in each case is similar.

The first step is that an international conservation group must raise funds in order to 'purchase' a debtor country's foreign debt. Private banks are usually reluctant to make outright donations of the debt they hold, even though in some countries like the USA such a donation is given a favourable tax treatment. Funds are usually secured from either the international conservation group's own resources or from donations from private individuals or bilateral aid agencies.

The funds raised are used to purchase the country's external debt on the secondary market at a fraction of the theoretical or face value of the debt. The 'secondary market' is a term used to describe the process whereby the original creditor of the debtor country sells on part or all of the debt to another institution. This trading can happen many times and is so prevalent that rarely will the bank who arranged the original loan retain anything but a small portion of the original debt. Table 32.11 shows that typically the country's debt has been purchased at between 15-30% of its face value. This discounting is because of a low expectation of total repayment by the debtor countries; the amount of the discount is proportional to the expectation of repayment.

Once the external debt is acquired, the environmental organisation will enter into negotiations with the debtor country to fix a favourable rate for the conversion of the external debt from the foreign currency in which the debt is denominated to the local currency of the debtor country. This rate will usually be somewhere between the local currency value of the debt and the local currency value of the price the environmental organisation paid for the debt on the secondary market. The price that is negotiated is referred to as the redemption price. Most commonly the redemption price is 100% of the face value of the acquired debt but in some instances it may be no more than the discount value of the acquired debt.

Lastly, the debtor country's government issues a financial instrument, typically a government bond, denominated in local currency in an amount equal to the redemption price of the debt. These bonds are then used to finance projects in the debtor country through local organisations.

A debt-for-nature agreement is often described as one in which all the parties involved stand to gain something.

The international conservation group is able to increase the spending power of its usually limited financial resources because of leverage provided by the difference between the redemption price and the discount rate of the purchased external debt. The group is also able by this method to influence conservation policy in countries where normally they have little impact. For the international conservation group there are also numerous collateral benefits, such as the relationship created between the parties involved (e.g. the local Ministry of Environment, Ministry of Finance, local conservation groups and creditor banks).

For the investors and the institutions who hold the country's debt there are direct benefits associated with having the extra purchases in the secondary markets, such as increasing both liquidity of the market and the price of the discounted debt.

For the debtor countries there are considerable financial benefits to be gained from the debt-for-nature agreement.

Table 32.11 Established debt-for-nature agreements

DEBTOR	DATE EST[1]	DEBT EXCH[2]	FACE VALUE[3]	2° MKT PRICE[4]	COST[5]	RDM %[6]	CNS FND[7]	CNS INV[8]	LOCAL FUND ADMIN	TERMS AND AIMS
ASIA										
Philippines	06/88	01/89 (Part I)	0.39	51.3	0.20	100	0.39	WWF-US	Haribon Foundation	For contributing to implementation of a conservation strategy, protected area management, strengthening of conservation groups, and training.
Philippines		08/90 (Part II)	0.9	48.9	0.44	100	0.90	WWF-US	Haribon Foundation	
EUROPE										
Poland	11/89	01/90	0.05	23.0	0.01	100	0.05	WWF-Int., WWF Sweden	n.a.	The exchange was set up as a first experiment exercise in what is hoped to be a large-scale Swedish project aiming at cleaning up the River Vistula. The revenue from the exchange will support the development of Biebrza National Park.
NORTH AND CENTRAL AMERICA										
Costa Rica	08/87	12/87-02/88	5.40	17.0	0.92	75	4.05	NPF	NPF in consultation with the donors	To develop three protected areas: Guanacaste, Monteverde (C.F.) and Corcovado. For Guanacaste, the programme will purchase adjacent land and for Monteverde, management support.
Costa Rica	06/88	07/88	33.00	15.1	5.00	33	9.90	Govt of the Nether-lands	Costa Rican Ministries of Natural Resources and Planning	The primary objective was reforestation operations and sustainable development activities in cooperation with various local interest groups such as peasant organisations and cooperatives.
Costa Rica	07/88	01/89	5.60	14.0	0.78	30	1.68	TNC, NPF	NPF	To help build up the National Biodiversity Institute, support NPF and the Neotropian Foundation, protection of endangered species, and management of national parks.
Costa Rica	04/89	04/89	24.50	14.3	3.5	70	17.10	Govt of Sweden	NPF	The swap proceeds will be used for expansion and protection of Guanacaste National Park.
Costa Rica	05/89	04/90 (Part Ia)	5.00	18.2	0.91	100	5.00	NPF	NPF	This three year programme (which could be classified as a debt-for-development programme) adopted in 1989 allows that US$10 million face value of debt is exchanged per year for conservation, education and micro-enterprises projects up to a total of US$45 million. Costa Rica's National Parks Foundation exchanged in April 1989 US$10.75 million face value debt for conservation bonds. These incomes will be used mainly for land compensations in La Amistad Biosphere Reserve and as a management endowment in La Amistad.
Costa Rica	05/89	04/90 (Part Ib)	5.75	18.2	1.04	80	4.60	NPF	INBIO	The income of the second tranche will support the National Biodiversity Institute of Costa Rica.
Costa Rica	01/91	01/91	0.60	60.0	0.36	90	0.54	Rain-forest Alliance MCL/TNC	n.a.	To purchase an additional 2,023ha of land for Monteverde Cloud Forest Reserve and to improve local protection.

Table 32.11 Established debt-for-nature agreements

DEBTOR	DATE EST[1]	DEBT EXCH[2]	FACE VALUE[3]	2° MKT PRICE[4]	COST[5]	RDM %[6]	CNS FND[7]	CNS INV[8]	LOCAL FUND ADMIN	TERMS AND AIMS
NORTH AND CENTRAL AMERICA (continued)										
Dominican Republic	03/90	03/90 (Part I)	0.58	20.0	0.12	100	0.58	PRCT TNC	PRONATURA Fund	The PRONATURA Fund, which consists of 11 of the Dominican Republic's conservation and development groups, reached an agreement with the Central Bank under which within four years up to US$80 million of the country's external debt may be exchanged, at 100% face value, for conservation projects. The US$582,000 (funded by the Conservation Trust of Puerto Rico) will support four conservation projects developed by the Nature Conservancy and PRONATURA, which also administrates the proceeds.
Guatemala	10/91	10/91	0.10	75.0	0.075	90	0.90	TNC	n.a.	To support Sierra de las Minas Biosphere Reserve through land acquisition and protection.
Jamaica	10/91	10/91 (Part I)	0.44	68.0	0.30	100	0.44	TNC USAID PRCT	n.a.	Programme of US$600,000 to fund and protect Montego Marine Park, Blue Mountain/John Crow Mountain.
Mexico	02/91	02/91 (Part I)	0.25	72.0	0.18	100	0.25	CI	n.a.	First tranche of a US$4 million programme to fund ecosystem conservation data centres to assess the distribution, status and conservation priority of Mexico's key species and habitats.
Mexico		08/91 (Part II)	0.25	Bank donation			0.25	CI	n.a.	Second tranche of above programme to fund communication and education campaigns at national and local levels.
SOUTH AMERICA										
Argentina	12/89	No trans-actions							Neuquén Foundation, Lorenzo Parodi Foundation	Argentina's National Development Bank (BANADE) approved that up to US$60 million in debt is exchanged for special BANADE conservation bonds, paying interest either in US dollars or local currency (depending on project needs). The bonds will benefit two national ecological groups: the Lorenzo Parodi Foundation (US$30 million and the Neuquén Foundation (US$30 million). Programmes envisaged include watershed protection and management of national parks and reserved areas.
Bolivia	07/87	07/87	0.65	15.0	0.10	39	0.25	CI	Bolivian Academy of Sciences, Ministry of Agriculture and Peasant Affairs	The debt was cancelled in exchange for full legal protection of the Beni Biosphere Reserve and a $250,000 management fund in local currency. The Bolivian government also agreed to establish four buffer zones covering 1.5 million ha in connection with the Beni Biosphere Reserve.
Ecuador	10/87	12/87 (Part I)	1.00	35.4	0.35	100	1.00	FN/ WWF-US	Fundación Natura	To support management plans for 6 protected areas in the Andes and Amazon: to develop the park infrastructure: to identify and acquire small nature reserves and fund staff and equipment and to fund species inventories.
Ecuador		04/89 (Part II)	9.00	11.9	1.07	100	9.00	FN/ WWF-US, TNC/ MBG	Fundación Natura	

Table 32.11 Established debt-for-nature agreements

DEBTOR	DATE EST[1]	DEBT EXCH[2]	FACE VALUE[3]	2° MKT PRICE[4]	COST[5]	RDM %[6]	CNS FND[7]	CNS INV[8]	LOCAL FUND ADMIN	TERMS AND AIMS
SOUTH AMERICA (continued)										
Ecuador	10/89	No transactions				50			n.a.	The government of Ecuador approved in October 1989 a US$50 million debt exchange programme. The Central Bank will allow that commercial Ecuadorian debt be exchanged (50% of face value) for monetary stabilisation bonds paying interest in local currency, to support social, cultural, environmental, and education programmes. This programme has, therefore, expanded beyond debt-for-nature into debt-for-development. The Government agreed in February 1990 that the Rotary Club may buy and exchange debt for financing a malaria eradication programme.
Peru	07/89		5.00	Bank donation		100	5.00	IFESH	IFESH	American Express donated US$5 million in Peruvian debt under a Peruvian debt-for-development law for training activities and teaching for the very poor.
AFRICA										
Madagascar	08/89	08/89 (Part I)	2.10	45.2	0.95	100	2.10	WWF-US	WWF-US in co-operation with the Government of Madagascar	This agreement aims at supporting conservation work by education, sustainable development methods and conservation, in protected areas.
Madagascar		08/90 (Part II)	0.92	0.49	0.45	100	0.92	WWF-US	WWF-US in co-operation with the Government of Madagascar	
Madagascar	05/90	01/91	0.12	50.0	0.06	100	0.12	CI	n.a.	The Government of Madagascar has agreed that Conservation International exchanges US$5 million of the nation's commercial bank debt and trade credits, at 100% face value, over the next five years. The swap proceeds may be deposited in private Malagasy banks as an endowment fund paying interest in local currency. The fund is intended to support conservation activities, including inventories of endangered species, environmental, and educational programmes. This first tranche is to provide ecosystem management programmes for 4 protected areas: Zahamena, Midongy-Sud, Manongarivo and Tsingy de Namoroka reserves.
Sudan	12/88		0.80	Bank donation			0.80	UNICEF	UNICEF	Debt-for-development swap: Midland Bank donated US$800,000 of Sudanese loans to UNICEF which has a deal with the Sudanese government that local money, in exchange for the loans, would be spent in providing water wells with hand-pumps in 10 villages.
Zambia	08/89	08/89	2.27	20.7	0.47	100	2.27	WWF Netherlands	Government of Zambia in consultation with WWF	The debts were exchanged for local currency which will be used for conservation and management of Kafue Flats and Bangweulu Basin wetlands, education programmes, support to local conservation institutions, alleviation of soil erosion, and protection of rhino and elephant populations.

Source: Primarily after Dogsé, P. and Droste, B. 1990. *Debt-for-nature exchanges and biosphere reserves: experiences and potential*. MAB Digest 6. Unesco, Paris; with additional sources.
Notes: [1] date agreement established. [2] date debt exchange executed. [3] face value of debt acquired in US$ million. [4] secondary market price, cent/US$. [5] cost, US$ million. [6] redemption price % of face value. [7] conservation funds generated in US$ million. [8] conservation investor. CI - Conservation International. FN - Fundación Natura. IFESH - International Foundation for Education and Self-Help. INBIO - National Biodiversity Institute of Costa Rica. MCL - Monteverde Conservation League. MBG - Missouri Botanical Garden. NPF - National Parks Foundation of Costa Rica. PRCT - Puerto Rican Conservation Trust. TNC - The Nature Conservancy. UNICEF - United Nations International Children's Fund. USAID - United States Agency for International Development. WWF-Int - World Wide Fund for Nature-International. WWF-US - World Wildlife Fund of the United States.
n.a. - Information not available.

Firstly, it reduces the debt owed by the country by an amount equal to the difference between the redemption price and the face value (although this can be illusory because, as noted above, countries can themselves step into the secondary market and purchase the discounted debt; therefore the redemption price becomes equal to the discount price. Most importantly, the country reduces its foreign debt by this mechanism. This in turn reduces the often crippling need to raise foreign currency to service the country's existing debt and thereby helps its external balance of payments. Another important political benefit is that the government of the debtor country will be able to control the donation made by the international conservation group, whereas if the donation had taken place directly from the international conservation group to the local conservation group the government would have less control.

Notwithstanding the potential benefits of debt-for-nature agreements, there are several problems which limit their usefulness.

Several countries, the Brazilian government foremost amongst them, have stated that the environmental conditions which are associated with such agreements are an imposition on the foreign sovereignty of the debtor nations (and contrary to the UN General Assembly Resolution 1803 on Natural Resources). This curtailment of sovereignty is supposedly manifest in two ways. Firstly, vesting in foreign creditors control over the debtor country's land and natural resources is in a sense equivalent to selling these resources to the outside interests. Secondly, it is claimed that the debt-for-nature agreements facilitate the imposition of foreign projects and values, and influence local policy in conservation projects in a manner which is more beneficial to outside interests than to local interests.

These fears seem largely unsupported by the facts of each debt-for-nature agreement established so far. None of the debt-for-nature agreements has entailed the transfer of ownership or control to foreign creditors of any sort including the international conservation group. Rather, in each instance the debt-for-nature arrangements have transferred control of the debt from foreign interests to local concerns. The involvement of the international conservation group does unquestionably to some extent impose foreign values; however, because of the need for complete cooperation of the debtor government, outside interests are subject to veto by local concerns. Most debt-for-nature agreements have been proposed by local groups within the debtor nations as a means of increasing the effectiveness of their own projects. Consequently any foreign influence exerted by the international conservation group is more in the nature of a positive exchange of ideas rather than direct imposition of inappropriate values which critics suspect.

Furthermore, debt-for-nature agreements are too small to have the consequences feared by critics. As can be seen from Table 32.11, the total amount of foreign debt which has been retired by means of debt-for-nature agreements is less than US$100 million, compared to the total debt owed by the developing world of an estimated $1.3 trillion. The insignificant size of these debt-for-nature agreements, however, does raise another point of genuine concern which is that any attention given to debt-for-nature agreements will direct attention away from solutions on a more meaningful scale to the problem of debt in less developed countries.

Finally, it is argued that the debt-for-nature mechanism, if it were to be implemented on any significant scale, would have an inflationary effect on the debtor country's economy. However, this will only be the case where the financial instruments used by the government to pay the redemption price negotiated by the international conservation group is local currency.

It should also be noted that because the attractiveness of debt-for-nature swaps for the international conservation group is the fact that the debtor countries' debt trades at a deep discount in the secondary market, this instrument cannot be used to make significant inroads into the foreign debt of a debtor country. For as soon as any significant amount of debt for a particular country starts to be purchased by parties wishing to use debt-equity instruments to fund projects in the country, this will drive the discounted price of the debt up, and it therefore becomes of diminished attraction. Indeed, as recent months have shown, the uninspiring economic situation in much of Latin America, for example, has seen a dramatic rise in the price at which debts are being traded on secondary markets.

These criticisms, however, do nothing to detract from the main importance of the mechanism. Debt-for-nature swap agreements should not be seen as a way of reducing the foreign debt of a country but rather as a means to help develop the promotion of environmental ideas and projects within a country through the local environmental programmes and groups in that country.

References

Abramovitz, J. 1991. *Investing in Biological Diversity: U.S. research and conservation efforts in developing countries.* World Resources Institute, Washington.

Anon. 1991. *Report of the Working Group on Guidelines for the Conservation of Biological Diversity in Tropical Production Forests.* International Tropical Timber Council. Eleventh Session 28 November-4 December 1991. Yokohama.

Colchester, M. and Lohmann, L. 1990. *The Tropical Forestry Action Plan: what progress?* World Rainforest Movement and The Ecologist. Penang, and Sturminster Newton, Dorset.

Collins, N.M., Sayer, J.A. and Whitmore, T.C. 1991. *The Conservation Atlas of Tropical Forests: Asia and the Pacific.* Macmillan Press, London, UK, in collaboration with IUCN, Gland, Switzerland.

DAC 1990. *Development Co-operation: efforts and policies of the members of the Development Assistance Committee.* OECD, Paris.

Dogsé, P. and Droste, B. 1990. *Debt-for-nature exchanges and biosphere reserves: experiences and potential.* MAB Digest 6. Unesco, Paris.

Elliott, C. 1990. *The Tropical Forestry Action Plan.* World Wide Fund for Nature International, Gland, Switzerland.

FAO 1985. *Tropical Forestry Action Plan.* Committee for Forest Development in the Tropics. Rome.

FAO 1989a. Committee on Forest Development in the Tropics. Ninth Session. Papers supporting the Agenda.

FAO 1989b. *Guidelines for Implementation of the Tropical Forestry Action Plan at the Country Level.* Forestry Department. Rome.

FAO 1989c. *Review of International Cooperation in Tropical Forestry.*

Hypay, T. 1986. *The International Tropical Timber Agreement. Its prospects for tropical timber trade, development and forest management.* IUCN/IIED, London. 18pp.

Sargent, C. 1990. *Defining the Issues: some thoughts and recommendations on recent critical comments of the TFAP*. IIED. 13pp.

Silva, L. de 1982. *Development Aid: a guide to facts and issues*. Third World Forum and UN NGLS, Geneva.

UNEP 1990. *UNEP Profile*. UNEP, Nairobi.

Winterbottom, R. 1990. *Taking Stock: the Tropical Forestry Action Plan after five years*. World Resources Institute, Washington, DC.

World Bank 1989. Operational Directive 4.00, Annex A: Environmental Assessment.

Authors as follows: International development assistance, Bruce Aylward (LEEC/IIED); International assistance in forest management, Sam Johnston; Debt purchase, Victoria Drake.

33. MANAGEMENT OF INTERNATIONAL RESOURCES

National boundaries do not enclose all the world's biological diversity; the high seas, the deep sea bed and Antarctica all contain natural resources, some of great interest or economic importance. Management of biodiversity in such areas can, by definition, only be achieved by means of international measures.

This chapter will discuss measures taken by the international community for biodiversity conservation in international areas, and highlight some of their strengths and weaknesses.

Use of natural resources in such areas is characterised by over-exploitation and resource depletion; this is typified by the whaling and sealing industries and several fisheries. These are classic examples of over-exploitation as a result of unrestricted access by all users, none of which has any incentive to limit extraction in the interests of long-term sustainability. With open-access resources the benefits of forbearance do not accrue to those who exercise it but rather to other users, who simply end up with a greater percentage of the market. Economically-effective use of capital also demands that exploitation occurs sooner rather than later.

A common response to these problems has been the establishment of an international commission mandated to control the use of a particular resource. The International Whaling Commission is one example, and we discuss below some other commissions concerned with international fisheries. These commissions typically use two different types of measure in an attempt to conserve their resource: setting of quotas and the setting of minimum standards which operators must adhere to.

Most of these international commissions have had only limited success in controlling over-exploitation. Some of the more common reasons for the failure of these commissions are: lack of finances, lack of political consensus, lack of scientific information about the resource and lack of power to monitor compliance with the controls established.

On the other hand, the development of the Antarctic Treaty System has been a comparatively successful venture in international resource management. The process of developing the Antarctic legal system has occurred over a period of decades, but it has been a consistent progression from a very broad and uncertain regulatory system to one that is now reasonably well-defined.

There is some tendency for change in the way ownership of international resources is perceived; they can be regarded as belonging not to those who appropriate them, but to mankind as a whole. One implication of this change is that international resources should be managed for all of mankind and not simply for those who have the ability to appropriate them. While this much is not especially contentious, many international statements have been made to the effect that biodiversity in general is the common heritage of mankind. The policy and legal implications that this view has on resources within national boundaries are complex and contentious, and have provided a major subject of debate in discussions preparatory to the proposed Biodiversity Convention.

INTERNATIONAL FISHERIES MANAGEMENT COMMISSIONS

The need for regulation

Fisheries have traditionally been regarded as common property, and so open to all without restriction. The typical pattern of use of common property resources involves increasing production beyond the point of sustainability until production declines (Hardin, 1969). This is because the economic imperatives of common property are to harvest the resource before someone else does, forgoing investment that would improve productivity of the resource.

This pattern of common property exploitation has occurred repeatedly in fisheries. The decline of stocks of herring and mackerel in the north-east Atlantic, the King Crab in the north-east Pacific, Yellow-fin Tuna in the eastern Pacific and Blue-fin Tuna in the south-west Pacific oceans are cases in point (Brown and Crutchfield, 1981). The human consequences of stock decline include smaller catch per unit effort, excess capacity of fishing equipment (large powerful vessels with sophisticated equipment) and reduced income for fishermen.

It is widely recognised that techniques must be introduced to manage stocks more efficiently, minimise costs and improve distribution of fishery benefits. Fisheries commissions have been established to perform these tasks.

Legal parameters for fisheries commissions

Overfishing became a recognised international problem initially for a few stocks in the North Sea in the late 19th century. However, international regulation is essentially a 20th century phenomenon (Underdal, 1980) and was not widespread before the first half of the 20th century.

Legal arrangements for cooperative management of fisheries can be regarded as falling into two periods, with the division between them being the rising importance of the Exclusive Economic Zone (EEZ) in the mid-1970s. This brought most traditional high seas fisheries under the jurisdiction of the coastal State, because most fish stocks are found within the 200 nautical mile (nm) EEZ. Until that time, coastal States had jurisdiction over only their territorial seas, extending 3-12nm from the coast. Beyond that, all States enjoyed open access to fisheries in the high seas. The proclamation of EEZs by coastal States extends their jurisdiction to more than half the total area of the world's seas, and the great majority of the customarily exploited living marine resources. Much variation in the catch patterns of particular countries has, therefore, been because of the new authority of coastal States over adjacent fisheries, which has enabled them to exclude or control long-range water fleets (Brown and Crutchfield, 1981).

Prior to the declaration of EEZs, there was an urgent need for cooperative management of high seas fish stocks. A range of treaties regulating the harvest of fisheries made their appearance early this century. The USA and Canada agreed on conservation treaties for halibut and salmon in the north-east Pacific, and European regional agreements were concluded for the Baltic and North seas and the northern Atlantic.

By the late 1950s fishing methods were changing rapidly. The use of sonar, mother ships with on-board processing facilities and purse seining were prominent (Knight, 1975). Fishing patterns were also altered as distant water fleets of developed countries, in particular those of Japan and the former USSR, expanded to all oceans and placed heavy pressures on fish stocks.

The 1958 Convention on Fishing and Conservation of Living Marine Resources became the first global fisheries management agreement. It placed an obligation upon its participating States to cooperate in adopting conservation measures. The intention of 'conservation of living resources' was solely to render possible the 'maximum sustainable yield' (MSY) from those resources so as to secure the maximum supply of food and goods for human consumption. The MSY approach has been criticised for failing to deal with fluctuations in stock size caused by a multitude of variables other than catch size, and because it ignores the fates of associated and dependent species.

Following the global 1958 Convention, several new regional conventions were concluded for the Atlantic, Baltic and Black seas. The next global approach to fisheries conservation to be adopted was the 1982 UN Convention on the Law of the Sea (UNCLOS). This convention included the obligation on participating states to conserve fishery resources in areas under their jurisdiction and to cooperate in their conservation beyond national jurisdiction. It also incorporated formal international recognition of the EEZ concept.

Many fish stocks are highly migratory and relatively few important commercial species remain within only one EEZ (Brown and Crutchfield, 1981). Therefore, the creation of EEZs has not eliminated the need for international institutions to facilitate cooperation in the management of fisheries but has simply redefined their role and reduced their independence. The need for international commissions is recognised in Article 63 of UNCLOS, which requires States to form regional or sub-regional organisations in order to foster cooperation between them in the conservation of the marine living resources.

UNCLOS forms a very loose and inadequate framework for decision-making by regional fisheries organisations. It requires that States maintain harvested species at population levels sufficient to produce an MSY, as qualified by relevant environmental and economic factors, and taking into account the effects on associated and dependent species with a view to preventing them from being threatened with extinction. States are also to ensure that conservation measures do not discriminate against the fishermen of any State (a constraint which may be difficult to meet in practice).

The MSY approach may be contrasted with the much wider parameters adopted in the 1980 Convention on the Conservation of Antarctic Living Marine Resources (CCALMR). It requires the "prevention of decrease in the size of harvested population to levels below those which ensure its stable recruitment", "maintenance of the ecological relationships between harvested, dependent and related populations", and "prevention of changes or minimization of risks of changes in the marine ecosystem which are not potentially reversible in two or three decades" (Article II). This approach is better suited to integrated management and conservation of ecosystems.

Decisions on the international management of fisheries are primarily concerned with the maintenance of the resource base but also need to reconcile many conflicting national and economic interests. These include maintenance of the industry's profitability and of the welfare of fishing communities, and equitable distribution of wealth between competing fishing States. The political sensitivities inherent in these interests have often resulted in international fisheries bodies being limited to very narrow and uncontentious mandates, such as the gathering and dissemination of information on fish stocks and fisheries technologies. Examples include informational bodies established by the UN Food and Agriculture Organization, such as the Indo-Pacific Fisheries Commission, the Western Central Atlantic Fisheries Commission and the General Fisheries Council for the Mediterranean.

Smaller bodies established by the participating nations themselves tend more often to have management powers. The following case studies of international fisheries commissions examine the effectiveness of a few such bodies.

European Common Fisheries

Jurisdiction

The European Community (EC) Common Fisheries Policy (CFP) evolved in the 1970s as a means of ensuring equal access by EC Member States to each other's fishing grounds. It applies in the EEZs along the North Sea and Atlantic coastlines of Member States and in certain areas of the west Atlantic, Skagerrak, Kattegat and Baltic Sea (Farnell and Elles, 1984).

Administration

The CFP is implemented by the EC. In particular, the Fisheries Council (constituted by Fisheries Ministers of the Member States) acts as the legislative organ and the Commission Directorate-General for Fisheries acts as the executive organ and enacts delegated legislation. A range of minor committees, mostly concerned with the provision of information, service these bodies. The Fisheries Council makes decisions on the long-term availability and distribution of fishery resources and on annual management matters, such as fishing restrictions, monitoring and enforcement. It votes in accordance with the usual EC procedures.

Member States can impose their own conservation measures only in relation to strictly local fisheries affecting their own citizens. They can impose unilateral conservation measures affecting fishermen from other EC States only where fishing grounds within their jurisdiction are seriously threatened and any damage would be difficult to repair. A range of other constraints are imposed upon such unilateral measures and they must be submitted to the Commission to confirm, amend or cancel. It is apparent that a great deal of exclusive management power rests with the EC.

Information

The provision of reliable information is crucial to the operation of any fishery commission. However, they are extremely difficult and expensive to obtain because of inadequate scientific knowledge of fishery dynamics. The EC therefore relies largely on a system of voluntary reporting. Fishermen must keep logbooks, and make fish landing and trans-shipment declarations. However, the voluntary reporting system is flawed as fishermen do not see it as being in their interests to make declarations which will ultimately lead to restrictions being placed upon their activities. The information provided is therefore often inaccurate. For example, fishermen landed at least 50% more cod and sole than they were permitted to in 1989. Independent research is at present conducted for the EC by the Advisory Committee on Fisheries Management of the International Council for Exploration of the Seas.

Fleet capacity containment

If the capacity of a fishing fleet does not exceed the sustainable yield of fish, no restrictions on the fishing effort of the fleet would be required in order to conserve fish. Over-capacity is, then, a root problem for fisheries management. The EC fleet capacity is currently 40% in excess of available fishing opportunities (Rose, 1991).

This problem is being dealt with by a 10-year scheme of 'Multi-Annual Guidance Programmes' (MAGPs) to restructure the EC fleet. Each Member State must present to the Commission for its approval two five-year programmes to adjust its fleet capacity, and then report back annually on implementation. Aid, in the form of a 70% reimbursement, is provided to Member States for the temporary or permanent withdrawal of vessels from service. The MAGPs also include provisions for building and modernising vessels, seeking development of aquaculture in the EC, exploratory fishing and joint ventures outside the EC and the development of markets for surplus or underfished species.

The MAGPs are not very effective as, in countries where capacity is greatly in excess of fishing opportunities, significant capacity reduction is not being achieved. As a consequence, in October 1990 the EC Commission froze MAGPs' grants for construction of new vessels in the UK, Ireland, the Netherlands and Greece. The MAGPs would be more effective if they were more tightly regulated.

The EC has recently created a central register for fishing vessels which includes details of vessel capacity. The use of a central register creates opportunities for a coordinated system of EC vessel licences entailing capacity quotas and penalty systems. Such capacity quotas are already in place in relation to Spanish and Portuguese fishermen and waters but remain politically unacceptable across the whole EC fleet. Incentives for selective fishing methods and disincentives for non-selective methods could also be introduced into the MAGPs and, ultimately, an EC licensing system (Rose, 1991).

Fishing restrictions

Restrictions on fishing effort include restrictions on access to specified areas or during certain seasons, limits on catches of particular species and undersized fish, and restrictions on the use of certain gear and on the allowable end use of some fish. No single restriction is sufficient to manage a fishery adequately, and usually a mix of measures is adopted, as is the case in the EC.

Total allowable catch

This is fixed by the Council each year, and quotas are distributed between the Member States. This is the foundation of EC fisheries conservation measures. Catch landings are monitored and reported by Member States and fishing activities are halted when the quota has been used up. States are able to trade quotas and obliged to compensate for the illegal use of another's quota. However, this is an essentially quantitative and bureaucratic method of conservation and is very inefficient. It encourages fishermen to get ahead of competitors in using up the quota and requires close monitoring and enforcement in order to prevent false understatement of landings. Fishermen are legally obliged to discard fish caught in excess of the quotas (Rose, 1991).

Minimum size restrictions are imposed on certain protected species. However, fish below the minimum size must be thrown back into the water, resulting in unnecessary wastage. The existing restrictions follow rather than prevent the catch of protected species and close surveillance is needed to enforce them.

More effective regulatory measures control fishing techniques rather than catches. For example, the EC applies area access restrictions within 12nm of the coast, and a licensing system for access to areas where there are species of special importance. These measures are sometimes supplemented by season and duration restrictions. For example, the EC has required vessels engaged in certain fisheries to lay up in port for 10 consecutive days each month (Rose, 1991). These approaches have negative economic impact, as boats, equipment and labour become under-used (Keen, 1988).

Other management measures include gear restrictions and special licensing for some fishing activities. Gear restrictions (such as prohibiting the use of guns or explosives) may be more effective in ensuring that certain fish are not landed in the first place (Rose, 1991).

Enforcement

The mix of fishing restrictions adopted by the EC requires a high degree of cooperation from fishermen or else close surveillance and firm enforcement. These are generally lacking. The Commission does not have independent monitoring powers but relies on Member States. These are obliged to report annually on their inspections at sea and in port and on the warnings, prosecutions and penalties which result. Yet national authorities often fail to ensure that conservation measures are implemented, partly because of the inadequacy of staff and facilities and also the difficulty of obtaining evidence.

Unfortunately, the CFP is failing to meet its conservation goals: 75% of fish stock within the area is exploited at unsustainably high levels (Rose, 1991). The Commissioner for Fisheries has even threatened to abandon EC control over fisheries management if Member States continue to block the Commission's management efforts. Although it is clear that the situation would be worse if the stocks were unregulated, doubts must be raised as to whether they would be better managed unilaterally under national jurisdiction.

North East Atlantic Fisheries Commission

Administration

The North East Atlantic Fisheries Commission (NEAFC) came into being in 1963, and applies to parts of the Arctic and Atlantic Oceans. It provided the framework for most international fisheries regulations in the area until 1977.

By mid-1976, it had 16 Member States, each having two commissioners, and its principal task was to recommend conservation measures to ensure rational exploitation of various stocks. Recommendations were only binding on States which did not object to them within a certain period. Application and enforcement were left to each Member State (Underdal, 1980). Clearly the participation of all significant groups was necessary for a measure to be effective; accordingly, the objection procedure had the effect of encouraging the development of the least ambitious management programme.

Catch limits

To establish a total allowable catch (TAC) required the consent of all Member States, and the approval of two-thirds of the delegations. This was not obtained until 1974 and by then some stocks were severely depleted. Until this time, the NEAFC simply made extensions and modifications to the mesh size and minimum landing size provisions. The TACs were not effective when set, as they were not adequately enforced and had to be reset each year by bargaining between the Members (Mason, 1979).

This usually resulted in the setting of such generous TACs that sacrifices were not necessary to contain the catches within them. Table 33.1 shows that in 1975 the total catch for 11 out of 15 stocks did not reach 90% of the TAC and in 1976 catches from 8 out of 15 stocks did not reach 90% of the TAC. The total catch in 1975 was only 84% of the NEAFC TAC. In many cases the TAC adopted by the NEAFC exceeded the highest TAC proposed by any party. This was because the Parties' inability to resolve arguments over individual allocations was often eventually resolved by simply raising the TAC. Table 33.2 indicates this process for 1975.

Quotas

In theory, the power to impose quotas gave the NEAFC every possible option to redistribute the TAC between Members. However, the quota did not cause redistributions significantly larger than normal fluctuations under an unregulated market. Quotas tended to be influenced by arguments based on rights (i.e. territorial or historical use of the resource) or on a concept of what is fair, taking into account needs and responsibilities. As overall economic power tended not to be a basis for argument, redistribution of quota tended to move from those with a large catch to those with smaller claims. Once a quota was agreed on, this provided a strong precedent for future decision as there was no precise or generally accepted formula for distributing TACs, which were not mentioned in the treaty. Cutbacks in quotas were usually proportional, so as to avoid redistribution. A Member usually advocated those arguments which produced the most favourable distribution for itself. Consequently, 86% of proponents' schemes gave proponents a higher proportion of the TAC than did any other proposal (Underdal, 1980).

Table 33.1 National catches as per cent of NEAFC TACs

STOCK	TOTAL CATCHES AS % OF NEAFC TACs	
	1975	1976
Arcto-Norwegian Cod	98	101
North Sea Cod	80	89
North Sea Haddock	67	100
North Sea Whiting	81	101
North Sea Plaice	86	106
North Sea Sole	146	111
North Sea Sprat	-	95
North Sea Herring	69	115
Herring w.o. Scotland	91	78
Celtic Sea Herring	55	54
Irish Sea Plaice	80	79
Irish Sea Sole	85	83
English Channel Plaice	87	77
English Channel Sole	95	116
Bristol Channel Plaice	59	47
Bristol Channel Sole	81	74
AVERAGE	84	95

Source: NEAFC reports.

Notes: In the catch statistics used as the source for this table the former USSR catches of coastal cod are included in the figures for Arcto-Norwegian Cod. For this reason the USSR quota of coastal cod has been added to its regular quota and to the TAC in the calculation of this table. The first quota regulations for the North Sea Herring fishery applied to the period July 1974 through June 1975. Available catch statistics follow calendar year.

The operation of the NEAFC can be most simply examined through case studies. The most difficult issue it faced was the management of the North Sea Herring.

In the late 1950s to early 1960s, scientists were concerned at the decline of the herring. The catch dropped from 225,000 tonnes in 1955 to 45,000 tonnes in 1963, but the scientific evidence of overfishing was not conclusive. The NEAFC was slow to respond, largely because of the self-interest shown by States such as Denmark (Underdal, 1980). It could only get agreement on more research and a study group was formed in 1969.

In 1970 conservative regulations were passed. Agreement was made possible by the continued decline of catches and stock, as it was by then clear that overfishing was a principal cause. The regulations were 'formally neutral', in that they did not directly regulate individual State quotas. They concerned limitations on mesh size and landing size,

Table 33.2 TAC decisions by NEAFC in relation to initial proposals

STOCK	FIRST QUOTA REGULATION			SECOND QUOTA REGULATION		
	Lowest Proposal	Highest Proposal	Decision	Lowest Proposal	Highest Proposal	Decision
Arcto-Norwegian Cod	510	1165	810	700/800	800/900	810
North Sea Cod	220	230	236	210	236	[1]236
North Sea Haddock	220	260	275	150	? 155	[1]206.25
North Sea Whiting	120	190	189	160	189	[1]189
North Sea Plaice	115	125/130	126	85	100/105	99.9
North Sea Sole	6	10	12.5	8	12.5	12.5
North Sea Herring	310	424	494	200	?	[2]254
Herring w.o. Scotland	156	170/200	205	-	-	[3]-
Celtic Sea Herring	25	30	32	? 19	23	25
North Sea Sprat	300	No TAC	589	-	-	-

Source: NEAFC reports and summary records, December 1973-November 1975. All figures are in 000s tonnes.
Notes: ? figure uncertain. - information not available. [1] The USSR delegation indicated indifference between the lowest and the highest TAC suggested by the LC (SR, Nov 1975: 1/8). [2] The discussion at the 13th annual meeting started with the idea of a 12 month allocation, but soon developed into a discussion on a regulation covering 18 months. For this reason some of the TAC proposals are hardly comparable. [3] In the discussion about the second quota regulation, regulations for both 1975 and 1976 were considered, the TAC for one year depending on the TAC for the other.

and season closures in 1971-1974, which had little effect, because of the exemptions from them. Depletion continued and the NEAFC requested power to limit the amount of catch and effort and to allocate quotas.

Negotiations for the introduction of explicitly distributed quotas commenced in December 1973 and were concluded in 1977. There were four quota regulations, two of which never took effect. All TACs adopted were higher than those recommended as conflict over quotas caused the TAC to be inflated to levels higher than initially agreed upon. The first TAC set in 1974 was so high that harvests did not reach it. However, recommendations for lower TACs were still rejected and by the time a lower TAC was finally imposed, it was too late. From 1977, directed fishing for herring in the North Sea was banned.

Demise of the NEAFC

By the end of 1976 it was clear that the existing voluntary regime was ineffective. The NEAFC was discredited by its failure to adopt necessary and timely conservation measures and by its inability to take necessary enforcement measures (Mason, 1979). Negotiations over the 1977 quota allocations broke down.

Following the proclamation of 200nm EEZs by most States in 1977, all North Sea fisheries became subject to the jurisdiction of coastal States. A new convention was proposed, as some kind of multilateral forum was considered desirable, but it was only to apply outside the EEZs and to generate consultation and information exchange rather than to manage stocks. In 1980 the NEAFC was resurrected to fulfil this role under the auspices of a new convention. However, no common fisheries management system has yet been put in place (Oceans Institute of Canada, 1990).

Although the situation would have been worse had it not existed, the NEAFC nevertheless presided over a decline in fish stock. It is suggested that by 1976 many stocks were further from a state of 'rational exploitation' than they were when the treaty was signed in 1959 (Underdal, 1980).

Northwest Atlantic Fisheries Organization

Administration

The Northwest Atlantic Fisheries Organization (NAFO) was formed in 1979 under the Convention on Future Multilateral Cooperation in the Northwest Atlantic Fisheries. The Regulatory Area under the convention covers only the high seas. NAFO was preceded by the International Fisheries Commission for the Northwest Atlantic (ICNAF), which was responsible for management of common fisheries prior to the declaration of EEZs in the north-west Atlantic.

NAFO is constituted by a General Council, a Scientific Council, a Fisheries Commission and a Secretariat. The Commission makes proposals for joint action by the Parties to the Convention designed to achieve the optimal utilisation of fishery resources. These measures are also to promote coordination and consistency between coastal State conservation measures in the EEZ and those taken on the high seas. The proposals are transmitted to the Parties and become binding upon Parties which do not file an objection to the proposal. That is, acceptance of management measures is voluntary.

Management measures

NAFO utilises a range of conservation measures based upon the notion of optimum yield, which is approximately 10% more conservative as a mechanism for fisheries management than maximum sustainable yield. Much of its fisheries management expertise was inherited from ICNAF, including considerable knowledge of the behaviour and status of stocks. Measures adopted included minimum mesh sizes for nets, closed areas and seasons, gear and vessel size restrictions, minimum fish size limits, TACs and national quotas for each principal commercial stock.

NAFO's conservation measures in its Regulatory Area were initially successful in the early 1980s and stocks of cod and plaice showed signs of recovery. However, the improvement was short-lived and they and other stocks have since declined or remained at low abundance. The Scientific Council has expressed concern regarding all stocks managed by NAFO on the basis of scientific findings and reduced catch per unit of effort. Cod, American Plaice, Redfish and Yellowtail Flounder are each displaying weak recruitment under fishing pressure and 1989 quotas on major stocks were reduced to approximately two-thirds of those of 1988. In recent years it has become apparent that NAFO has been ineffective in its conservation efforts (Oceans Institute of Canada, 1990).

The NAFO Joint Inspection Scheme allows Parties reciprocal rights to board and inspect vessels. However, there is no joint enforcement scheme. Breaches of the regulatory measures must therefore be conducted by the flag State.

Management crisis

The principal reasons for the recent failure of NAFO to conserve fish stocks are disunity among the Parties and continued fishing in the Regulatory Area by non-Parties. Canada, for example, is reconsidering its participation in NAFO.

Conflicts are continuing to take place between Canada and the EC, in particular with Spain and Portugal. Spain has been exceeding its quota for cod in the area designated as 3NO, where most cod is fished. Following the accession to the EC in 1985 of Spain and Portugal, the EC objected to TACs and quotas set in 1986, 1987 and 1988 for several stocks, and increased its fishing effort for them. It has used the objection procedure under the Convention to exempt itself from various conservation measures. In 1989, it set itself a quota more than 10 times that allotted to it by NAFO. While the EC has been pressing for TACs to be based upon maximum sustainable yield instead of optimum yield and against protection measures for certain stocks of cod, Canada wishes to conserve the juveniles and spawning grounds of stocks which straddle its EEZ and the NAFO Regulatory Area.

A range of countries fish in the Regulatory Area without any formal commitment to conservation of its resources. These include the USA, some Central and South American countries, Republic of Korea and also EC vessels operating under flags of convenience. Their catches are estimated to exceed any surplus available following the allocation of quotas to NAFO Parties.

North American Fisheries Commissions

International Pacific Halibut Commission

In 1923 Canada and the USA formed the International Pacific Halibut Commission to restore Pacific Halibut stocks by means of imposition of closed seasons. The Commission's conservation programme was revised in 1930, introducing an annual quota for each of four management areas and a minimum size restriction on fish landed. These measures were effective to increase stocks by 50% between 1932 and 1954. Supplementary arrangements were entered into with Japan and the former USSR during this period.

Although the quota system conserved the Pacific Halibut stock, its economic effects were in some ways detrimental. These included increased investment in gear and equipment and increased fleet capacity. Fish tended to reach the consumer in poorer condition, having been harvested earlier in the season as a result of the rush to fill the quota.

The declaration of EEZs by Canada and the USA in 1976 brought the Pacific Halibut under national jurisdiction. However, the Commission remains an example of successful international management of a shared resource. The fact that it was a bilateral rather than a multilateral organisation signals the advantages of fewer participating members. It also provides an example of effective use of the quota system, although an incidental decline in quality of fish reaching the consumer resulted.

USA administration

The USA has eight regional councils established under the US Magnusson Fisheries Conservation and Management Act. Most have adopted a species-by-species management plan. The Act restricts the use of limited access as a management tool, and rules out taxes or fees for domestic fisheries. The USA also has three marine fishery commissions for its Atlantic, Pacific and Gulf coastal states. These were formed in the 1940s to coordinate the fisheries of US states. Fish which are fully exploited and which are harvested primarily in federally controlled waters (the EEZ) must be brought under a Fisheries Management Plan.

Following proclamation of the 200nm EEZ in 1976, there was a huge increase in the number of US fishing vessels and this increased capacity led to the usual problems of overfishing, decreasing incomes and so forth (Brown and Crutchfield, 1981).

Alternative approaches

Harmonisation

States with common interests in fisheries management may choose to harmonise their fisheries laws. This has been the case in the south-west Pacific.

The South Pacific Forum Fisheries Agency (FFA), established in 1979 as an arm of the South Pacific Forum, has 13 members from the south Pacific region. In 1981 it formed the Agreement Concerning Co-operation in Management of Fisheries of Common Interest, which seeks to coordinate regional fisheries policies and to harmonise the management of fisheries, especially in the case of common stocks. To this end, it standardises licensing procedures, terms and conditions, and coordinates surveillance and enforcement functions. Although there is no limit on fishing effort, all fishing access agreements in the region must comply with a harmonised list of access conditions and a regional register of licensed fishing vessels is kept by participating local States.

In 1989, a Convention for the Prohibition of Fishing with Long Driftnets in the South Pacific was concluded to further harmonise fisheries management laws in the region. It is administered by the FFA. The convention prohibits the use of driftnets exceeding 2.5km in length by people or vessels under the jurisdiction of the Parties within the convention area, which includes both EEZs and the high seas. Parties are to take action against any fishing using driftnets in the Area by non-Parties, including prohibition of landing fish caught by driftnets in their territory.

Property rights

Traditional open access to fisheries has permitted fishermen ownership of the fishery resource on the basis of fishing effort. Fisheries commissions limit the fishing effort. An alteration of the property right may take the form of a licence to fish, within an individual quota, which can be freely traded and is itself an asset, without which the fish cannot be owned.

It has been argued that the move to a system based on full ownership with profit incentives would increase productivity and efficiency, while at the same time removing the "imperatives of the commons" (Keen, 1988). That is, the rush to exploit the resource before others would be replaced by an owner's incentive to look after the property over the longer term. Therefore, no imposition of limits upon fishing effort would be necessary.

However, this approach relies upon the licensing of only so many vessels as are required to harvest the resource. It creates a windfall for those boat-owners permitted to remain in an existing fishery and problems arise in identifying who should be given the right to remain (Keen, 1988). It is also suggested that such a scheme does not improve management of the resource because fishermen will continue to increase their harvesting capacity and compete to harvest the resource. To be successful, therefore, fishing effort must be effectively controlled, so that each fishing unit contains an optimal combination of vessels, gear and so forth. Ultimately, an independent regulatory body continues to be necessary to oversee the process of licensing and effort limitation.

Conclusions

Despite the declaration of EEZs and the introduction of new management concepts, international fisheries commissions remain necessary for the proper management of international fisheries resources. This is because fish stocks regularly cross international boundaries and fishermen habitually compete to catch them.

However, most fisheries commissions have proved to be relatively ineffective in the management of fisheries within their competence. There are several reasons for this which can be learnt from the operation of the commissions detailed above. These concern the nature of the decision-making processes involved and the mix of regulatory measures used.

- Voluntary reporting does not provide reliable information about fishery stocks; so independent research is required. This may be provided to the commission from a range of sources, including Member States.
- The process of deciding the amount of the TAC must be kept separate from decisions on the allocation of quotas. The primary decision concerning the TAC needs to be made by a scientific committee and based on biological rather than economic grounds.
- Distribution of quotas between States is best managed either by a commission with few members or in a situation where it is possible for States to engage in bargaining for an exchange of various benefits brokered by the commission.
- The quantitative approach to regulation is wasteful because it regulates a catch after it has been caught. Simpler and more enforceable restrictions should form the basis of a regulatory system. An appropriate mix of measures would centre around gear, area, season and duration restrictions, which are more amenable to enforcement within port.
- The option of a unilateral objection procedure undermines the delicate compromises which a resource distribution involves. A commission needs a strong central authority to overcome disagreements between its members.
- Enforcement of international management decisions by Member States against their own nationals tends to be lax. A commission needs direct enforcement powers against recalcitrant members and fishermen in order to ensure that its recommendations are put into practice.
- Licensing and radar surveillance are more economical and efficient systems of monitoring than the current system of inspections. Reciprocal observation arrangements, such as employed by the NAFO can supplement such a system.

The uninspiring performance of fisheries organisations to date need not be taken as conclusive of their ineffectiveness. Where the members of an international commission have the will to conserve fisheries cooperatively, measures can be designed to implement effective fisheries conservation.

ANTARCTICA: THE EVOLUTION OF AN INTERNATIONAL RESOURCE MANAGEMENT REGIME

Geography

The Antarctic region, which includes the Southern Ocean as well as the continent itself and its islands, is the largest wilderness left in the world (Laws, 1989). The region covers 13.918 million km^2, which is almost 10% of the earth's surface.

The continent is the driest, highest and coldest in the world, and is almost entirely covered by ice. During winter the sea ice rapidly increases round the continent and adds a further 20 million km^2 to the size of the ice cap. In some places the ice cap is estimated to be as much as 4.7km thick.

The Southern Oceans are some of the most turbulent in the world (Techernia and Jeannin, 1983). There are two main currents. Close to the coast a westerly current predominates while further from shore the main current is easterly; the interface between these is the Antarctic Divergence, a complex shear zone of upwelling where nutrient-rich deep water is brought to the surface, thereby providing the primary basis of the Southern Ocean food web (Deacon, 1987). Another important feature of this ocean is the Antarctic Convergence, where cold surface waters plunge beneath the warmer and less dense subtropical waters at around 50°S. The exact location of this convergence is not fixed but the pronounced changes in temperature and salinity on either side are relatively constant (Holdgate, 1984).

First human contact

The first recorded human contacts with the region were during Cook's voyages into the Southern Ocean between 1772 and 1775 (Beaglehole, 1961). Shortly after this, the huge populations of seals attracted sealers to the region, thus initiating the cycle of over-exploitation, collapse and regeneration typical of open-access resources (Bonner, 1968). Sealers at first concentrated on the islands and by 1822 (Bonner, 1968) many populations had collapsed; an estimated 1.2 million fur seals had been taken in South Georgia and one million in the South Shetland region. Sealers remained active in the region for the next 100 years and seal numbers did not recover to estimated pre-exploitation levels until recent decades (Bonner, 1982).

The Southern Oceans support many cetaceans, including the large and commercially valuable Blue, Fin and Sei Whales. Whalers lacked the technology to capture and process these whales until the 1870s, but once these difficulties were overcome, they moved into the region and operated for several decades from shore-based processing facilities on sub-Antarctic islands. Factory ships were first used in the 1925-26 season; these allowed whales to be processed at sea thereby greatly increasing the number caught (Bonner, 1980). Efforts were initially concentrated on the species with highest commercial value, such as Humpbacks, Blue, Fin and Sei Whales, but as the stocks of these declined, attention turned to other species (Bonner, 1984). By the latter half of this century only the Minke, smallest of the more common baleen whales, had not been subject to intensive commercial harvesting (Bonner, 1980).

Sovereign States of Antarctica

Seven States have made claims of territorial sovereignty in Antarctica which they have defined. These claims are based on a variety of doctrines such as: discovery, formal annexation, sector theory and occupation (Kish, 1973). In addition, both the USA and the USSR have maintained that they have a basis for such claims, although they have not made specific claims themselves and do not recognise the claims of the other States. Even though the validity of these claims may be dubious under modern principles of international law (Greig, 1988), it should be recognised that for the claimant States they are made seriously and some States may not care to relinquish them (Conforti, 1986).

The whaling treaties

The first international resource management commission which included the region within its jurisdiction was the commission established under the 1931 Convention for the Regulation of Whaling. This convention prohibited commercial whaling of two depleted species, Right Whale and Bow Whale, and banned the killing of calves and immature or female whales in the company of calves or sucklings. It further required whalers to make full use of the carcasses. However, the convention had little practical effect as several major whaling nations refused to accede to it.

The successor to this convention, the International Convention for the Regulation of Whaling, came into force in 1948 and was ratified by most major whaling nations. Its basic aim was to control whaling so as to avoid over-exploitation and to ensure conservation of the stocks (Birnie, 1985; Rosati, 1984). The convention established the International Whaling Commission (IWC) to implement the aims of the convention and regulate whaling by establishing quotas and acceptable methods of capture, and designating protected species (Smith, 1984). The schedule to the convention contains regulations governing the protection and exploitation of whales, listing protected species and setting quotas for others. The schedule may be amended by a three-quarter majority of the members at the annual general meeting of the IWC, which is composed of representatives of each contracting party to the convention.

Initially, the IWC set annual quotas based on the "Blue Whale unit", which essentially meant that the whaler could take any combination of whales of any species up to the equivalent mass of the number of Blue Whales that had been allocated. As a result, whales which were more valuable per unit weight were more heavily exploited until their stock numbers had collapsed, whereupon the next most valuable stock was exploited.

Under this regime the industry continued to grow, and numbers taken worldwide increased year after year, reaching a peak in the 1960/1961 season when approximately 64,000 whales were killed (International Whaling Commission, 1963). This mechanism of setting quotas, which proved to do little to conserve whales, was abandoned in 1972 in favour of quotas on a species basis. This new approach was then later enhanced by the "New Management Procedure" which established quotas based on a stock-by-stock approach (Birnie, 1982). This tightening in procedure was also accompanied by a reduction in the number of whales which were allowed to be harvested in any year (Birnie, 1989). As a result of this and other factors, catches declined steadily until 1982 when the IWC declared a worldwide 'pause' on commercial whaling effective from the 1985/1986 season, which is still in force.

The Antarctic Treaty

The Antarctic Treaty ('the Treaty') was adopted in 1959 and came into force in 1961. It is essentially a self-denying ordinance under which contracting parties agree: to prevent military activity in the area and to use Antarctica for peaceful purposes only; to promote international cooperation in scientific research; and to ban nuclear explosions and disposal of radioactive waste. Also within its articles, the Treaty preserves all existing rights and claims to sovereignty and the position of those who recognise no claims, and nullifies any basis of claim during its operation. The Treaty is a classic example of a 'framework convention', with the Treaty itself being quite short and general in nature, leaving matters of detail to be negotiated at a future time through recommendations or protocols. In addition, the Treaty established no independent institutional structure (Secretariat) for its implementation. Membership of the Treaty is open to all countries, with two categories of members, Consultative Parties and Non-Consultative Parties. In order to become a Consultative Party a country has to display a serious interest in Antarctica as demonstrated by substantial scientific research activity in the region. Consultative Parties have voting rights in the Antarctic Treaty system and, therefore, are responsible for the governance of the region, while Non-Consultative Parties merely have observer status at the meetings of the Parties. Details of the present membership of the Treaty are given in Table 33.3.

Periodic meetings of the Parties are held to exchange information, consult on matters of common interest pertaining to Antarctica, and formulate measures to manage and govern the region. These meetings, called Antarctic Treaty Consultative Meetings (ATCM) (Myhre, 1986), occur every two years at a conference hosted and organised by one of the Consultative Parties. In addition, special meetings (SATCM) are called from time to time to consider specific issues. Recommendations are made at meetings on a consensual basis. To date, nearly 200 recommendations have been made on a wide variety of subjects, including: protection of the environment, meteorology, telecommunications, transport and logistics, tourism and exchange of information.

In the Treaty itself there is only one brief but comprehensive reference to environmental matters: a short provision calling upon the Consultative Parties to develop measures for the "preservation and conservation of the living resources of Antarctica". From this general obligation, an elaborate management regime has been developed through additional recommendations, protocols and further conventions to provide comprehensive protection for the environment in Antarctica.

Agreed Measures for the Conservation of Antarctic Fauna and Flora

The first significant development pursuant to the general obligation to protect and conserve the Antarctic environment occurred in 1964 with the adoption of the Agreed Measures for the Conservation of Antarctic Fauna and Flora (the Agreed Measures) which represent one of the earliest examples of effective international regulation of a resource. The general intention behind this recommendation was to protect the living resources of Antarctica. In particular, it covered protection of mammal and bird life from unnecessary slaughter, and the minimisation of disturbance on land by personnel from the growing number of scientific bases.

Over and above the general provisions (which are applicable to the entire Treaty Area), the Agreed Measures allow for more stringent provisions with regard to '*Specially Protected Areas*' (SPA) and '*Specially Protected Species*'. Permits may only be granted by a contracting party for the taking of Specially Protected Species "for compelling scientific purposes" and even then they may not "jeopardise the existing natural ecosystem or survival of that species". Areas which are designated as Specially Protected Areas are protected by similarly stringent provisions (Anderson, 1968).

Criteria for review of the SPAs and for establishing *Sites of Special Scientific Interest* (SSSI) to protect sites important for research were developed later. However, the areas designated as either SPA or SSSI have been relatively small in size. A recent review by the Scientific Committee on Antarctic Research (SCAR), which acts as the scientific committee for the Agreed Measures, considered the existing SPAs and SSSIs and concluded that some areas worthy of designation remained undesignated and that the documentation of individual sites remained uneven and incomplete.

During the late 1980s moves were made to remedy these defects and, as a result, two further categories were established. The *Specially Reserved Area* is intended to protect representative examples of major geological features and those of outstanding aesthetic, scenic and wilderness value while the *Multiple Use Planning Area* (MPA) is a mechanism for controlling human activities in high-use areas, to minimise harmful environmental impacts.

The protected areas system has been rationalised under the provisions of Annex V to the Madrid Protocol, which introduced the *Antarctic Specially Protected Area* (ASPA) and the *Antarctic Specially Managed Area* (ASMA). In due course existing SPAs and SSSIs will be re-designated as ASPAs, and MPAs will be re-designated as ASMAs.

Currently there are 19 SPAs, 35 SSSIs, 1 Specially Reserved Area and 1 Multiple Use Planning Area (see Fig. 33.1 and Table 33.4).

The Convention for the Conservation of Antarctic Seals

Since the 1780s, seals had been subject to gross over-exploitation, with population collapses occurring in the 1820s and again in the 1860s. By the 1960s, seal numbers and stocks were returning to their pre-exploitation levels (Mitchell and Tinker, 1980), when Norway expressed interest in recommencing commercial exploitation. Commercial sealing has not in fact been re-established; nevertheless, Norway's actions caused considerable concern and moves were made to bring about a legal instrument to control exploitation of seals.

The conference of the Parties to the Antarctic Treaty once again provided the forum within which resource management could occur. In 1964 it was suggested that national governments should regulate pelagic sealing on a voluntary basis (Myhre, 1986), followed two years later by the adoption of Interim Guidelines for the Voluntary Regulations of Antarctic Pelagic Sealing (Recommendation IV-XXI). Finally, in 1972 Consultative Parties adopted the Convention for the Conservation of Antarctic Seals, which came into force in 1978 (Lyster, 1985) and is renewed every five years.

Table 33.3 Parties to the Antarctic Treaty

CONTRACTING PARTIES (in chronological order)

+	United Kingdom*	31 May 1960	1
+	South Africa*	21 June 1960	2
+	Belgium*	26 July 1960	3
+	Japan*	4 August 1960	4
+	USA*	18 August 1960	5
+	Norway*	24 August 1960	6
+	France*	16 September 1960	7
+	New Zealand*	1 November 1960	8
+	Commonwealth of Independent States *Φ	2 November 1960	9
+	Poland	8 June 1961 (29 July 1977)	10
+	Argentina*	23 June 1961	11
+	Australia*	23 June 1961	12
+	Chile*	23 June 1961	13
	Czechoslovakia	14 June 1962	14
	Denmark	20 May 1965	15
+	Netherlands	30 March 1967 (19 November 1990)	16
	Romania	15 September 1971	17
+	Germany, DDR#	19 November 1974 (5 October 1987)	18
+	Brazil	16 May 1975 (12 September 1983)	19
	Bulgaria	11 September 1978	20
+	Germany, BRD#	5 February 1979 (3 March 1981)	21
+	Uruguay	11 January 1980 (7 October 1985)	22
	Papua New Guinea•	16 March 1981	23
+	Italy	18 March 1981 (5 October 1987)	24
+	Peru	10 April 1981 (9 October 1989)	25
+	Spain	31 March 1982 (21 September 1988	26
+	China	8 June 1983 (7 October 1985)	27
+	India	19 August 1983 (12 September 1983)	28
	Hungary	27 January 1984	29
+	Sweden	24 April 1984 (21 September 1988)	30
+	Finland	15 May 1984 (9 October 1989)	31
	Cuba	16 August 1984	32
+	Korea, Rep	28 November 1986 (9 October 1989)	33
	Greece	8 January 1987	34
	Korea, Dem. People's Rep	21 January 1987	35
	Austria	25 August 1987	36
+	Ecuador	15 September 1987 (19 November 1990)	37
	Canada	4 May 1988	38
	Colombia	31 January 1989	39
	Switzerland	15 November 1990	40
	Guatemala	31 July 1991	41

Source: Scott Polar Research Institute, Cambridge.
Note: Made 1 December 1959; came into force 23 June 1961. The Treaty has no limits on its duration. It may be reviewed, at the request of the Consultative Party.
Key: * Original signatories; the 12 states which signed the Treaty on 1 December 1959; the dates given are those of the Deposition of the instruments of ratification, approval, or acceptance of the Treaty. + Consultative Parties; 26 states, the 12 original signatories and 14 (formerly 15#) others which achieved this status after becoming actively involved in Antarctic research (with dates in brackets). • Papua New Guinea succeeded to the Treaty after becoming independent of Australia. # The two German states unified on 3 October 1990. Thus there are now 40 member states from the 41 adherents. Φ Comprising 11 of the reassociated 15 Republics of the Soviet Union, December 1991.

The object of this convention is to "promote and achieve the objects of protection, scientific study and rational use of Antarctic Seals and to maintain a satisfactory balance with the ecological system". The convention covers six species of seal: Southern Elephant Seal, Leopard Seal, Weddell Seal, Crabeater Seal, Ross Seal and Southern Fur Seal. Like the Agreed Measures, this convention operates by a system of permits which allows the capture or killing of seals for certain purposes and, in an Annex, establishes quotas for commercial sealing. The measures outlined in the Annex also seek to control other aspects of sealing through establishing a sealing season, the areas in which sealing may take place, and the methods which may be used to capture and kill seals. In order to monitor properly the taking of seals, the convention also requires the Parties to report to SCAR all seals which have been taken in any one season and to report annually on the steps that they have taken to implement the convention. The convention also provides for the establishment of a Commission and Scientific Advisory Committee if and when commercial sealing is re-established in the region.

Figure 33.1 Protected areas in Antarctica

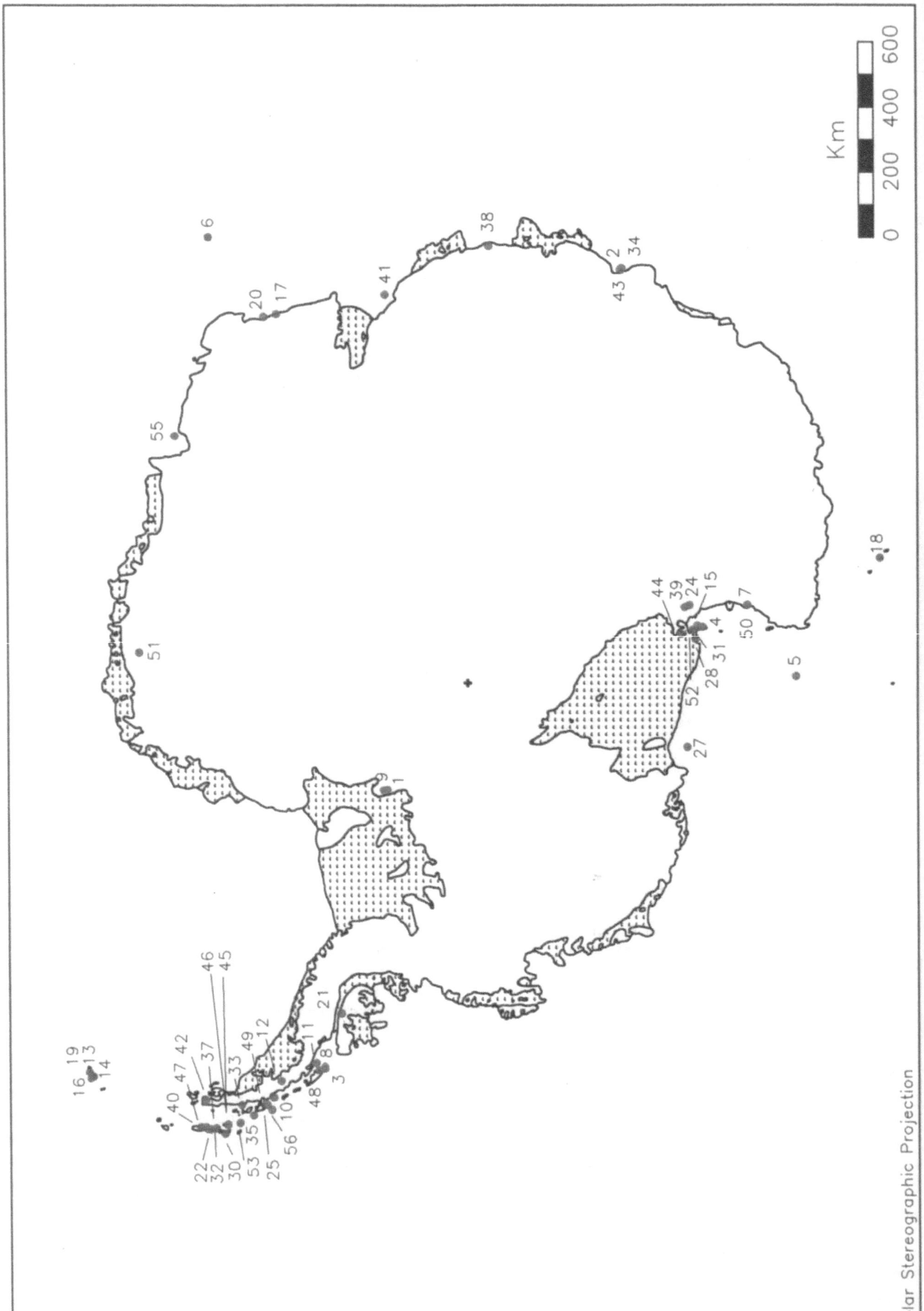

Table 33.4 Protected areas in Antarctica

	AREA (ha)	DATE ESTAB.
Specially Reserved Area		
1 North Dufek Massif	48,000	1991
Specially Protected Area		
2 Ardery Island and Odbert Island	220	1966
3 Avian Island, North-West Marguerite Bay	40	1990
4 Beaufort Island	1,865	1966
5 Cape Hallett	25	1966
6 Coppermine Peninsula	65	1970
7 Cryptogam Ridge, Mount Melbourne	60	1991
8 Dion Islands	100	1966
9 Forlidas Pond and Davis Valley ponds	600	1991
10 Green Island	25	1966
11 Lagotellerie Island	130	1985
12 Litchfield Island	250	1975
13 Lynch Island	10	1966
14 Moe Island	100	1966
15 New College Valley	10	1985
16 North Coronation Island	5,000	1985
17 Rookery Islands	65	1966
18 Sabrina Island	60	1966
19 Southern Powell & adjacent islands	610	1966
20 Taylor Rookery	30	1966
Site of Special Scientific Interest		
21 Ablation Point - Ganymede Heights, Alexander Is	18,000	1990
22 Ardley Island	300	1991
23 Arrival Heights	60	1975
24 Barwick Valley	29,120	1975
25 Biscoe Point	200	1985
26 Byers Peninsula	3,027	1975
27 Canada Glacier	100	1985
28 Cape Crozier	462	1975
29 Cape Royds	2	1975
30 Cape Shirreff	265	1989
31 Caughley Beach	25	1985
32 Chile Bay (Discovery Bay)	75	1987
33 Cierva Point	850	1985
34 Clark Peninsula	800	1985
35 East Dallman Bay	60,000	1991
36 Fildes Peninsula	154	1975
37 Harmony Point	1300	1985
38 Haswell Island	80	1975
39 Linnaeus Terrace	300	1985
40 Lions Rump	100	1991
41 Marine Plain, Mule Peninsula	2,340	1987
42 Mount Flora, Hope Bay, Antarctic Peninsula	65	1990
43 North-east Bailey Peninsula	100	1985
44 North-west White Island	1350	1985
45 Parts of Deception Island	100	1985
46 Port Foster, Deception Island	50	1987
47 Potter Peninsula	200	1985
48 Rothera Point	4	1985
49 South Bay, Doumer Island	70	1987
50 Summit of Mt Melbourne	800	1987
51 Svarthamaren	390	1987
52 Tramway Ridge	1	1985
53 Western Bransfield Strait	103,000	1991
54 Western Shore, Admiralty Bay	160,000	1979
55 Yukidori Valley	300	1987
Multiple Use Planning Area		
56 South-west Anvers Island	153,500	1991

Source: Swithinbank, C. 1991. Conservation areas of Antarctica. Unpublished contribution to the Antarctica Digital Database, in preparation. WCMC/BAS/SPRI, Cambridge.

The Convention on the Conservation of Antarctic Marine Living Resources (CCAMLR)

During the 1960s Japan and the USSR began investigating the possibility of commercial harvesting of krill, a shrimp-like crustacean. As krill play a central and vital role in the entire region's ecosystem (Auburn, 1982) the Consultative Parties decided to establish a legal regime to control this emerging industry before it developed (Barnes, 1982).

Once again the Antarctica Treaty System provided the framework and negotiations began in 1977, followed by seven official meetings and consultations on the proposed convention (Edwards and Heap, 1980). The two central concerns that shaped the resulting convention were the role played by krill in the food chain in the Antarctic region and the ever-present conflict of the territorial claims.

Krill has been the driving force behind the parties' recognition of the need to consider the Antarctic ecosystem in total. This is because krill is a key factor in the food chain in the Antarctic region; it feeds on plankton and, in turn, is the principal food for many species of birds, fish and whales (Mitchell and Sandbrook, 1980). Because the food chain in the Antarctic is simplified, with relatively few species, the mass removal of one species which is central to it is likely to have very significant effects on the whole chain. Consequently, should there be any serious disturbance to the krill, this will have serious implications for the entire Antarctic ecosystem. A further consideration is that krill has a tendency to swarm into large, densely packed formations, making it relatively easy to harvest on a large scale and particularly susceptible to over-exploitation. Therefore, even though the ostensible purpose of the negotiations was to protect krill from over-exploitation, the main concern was for the effects that overharvesting would have on other species in the Antarctic. This concern is reflected in CCAMLR in the 'ecosystem approach' adopted by the convention (Edwards and Heap, 1980).

The other issue which dominated negotiations was the legal debate arising from the conflicting territorial claims. Although these had been frozen by the Treaty in 1959, by the late 1970s the potential wealth these claims represented was closer to becoming reality; so claimants were anxious to ensure that nothing in CCAMLR would prejudice their claims or rights to exercise jurisdiction over the coastal waters. Similarly, the non-claimant States were anxious to ensure that CCAMLR did not legitimise or help legitimise the position of the claimants (Triggs, 1987).

These issues are representative of the constraints to effective regulation of international resources. Ecosystems generally overlap state boundaries, which leads to need for international environmental cooperation. It is the States' refusal to relax claims to absolute sovereignty, even in the face of obvious gains from cooperation, which leads to the difficulties of achieving effective international environmental regulation. The Antarctic system has been virtually unique in its ability to cope with both sources of conflict.

By September 1978 the key issues had been resolved and were the subject of a 'gentlemen's agreement', which lasted until the convention came into force in 1982.

The object of CCAMLR is the "conservation of Antarctic marine living resources". Because of the ecosystem approach applied to achieve this object, the convention's application extends to all areas within the Antarctic Convergence. The coverage of CCAMLR is, therefore, considerably larger than the Treaty itself. This defining of the area covered by the convention on biological grounds as compared to political ones is an innovative feature of this convention.

Within the area covered by the CCAMLR, however, there are a number of islands which are the undisputed territory of Consultative Parties, and which are not covered by the convention. The existence of these islands and the fact that they are outside the jurisdiction of CCAMLR is recognised in Article IV(2)(b). This Article provides that "nothing in this Convention and no acts or activities taking place while the present Convention is in force shall: ... be interpreted as a renunciation or diminution ... of, or as prejudicing, any right or claim or basis of claim to exercise coastal state jurisdiction under international law within the area to which this Convention applies ...". This provision deliberately does not refer to the undisputed islands, an ambiguity which allows claimant States to interpret the provision as also referring to the disputed claims within the Antarctic Treaty area (south of 60°S), whereas the non-claimant States can interpret the provision as meaning that CCAMLR applies only to where national sovereignty is generally recognised. This ambiguity therefore allows the Consultative Parties to come to agreement on the issue of conserving the marine living resources while apparently maintaining the status quo on territorial claims.

The ecosystem approach adopted in this convention means that it is unlike most other fishery agreements which set quotas based upon maximum sustainable yields (Gulland, 1968; Bean, 1983). CCAMLR sets a standard based not only on the maximum sustainable yield of the target species but also requires that equal consideration be given to the likely effects on other species and the marine ecosystem as a whole.

CCAMLR, for the first time in the Antarctic Treaty system, establishes a commission to implement its objectives, the Commission for the Conservation of Antarctic Marine Living Resources. The Commission has its headquarters in Hobart, Australia, and is composed of delegates from all Contracting Parties; it is the first permanent resource management commission established for the continent. CCAMLR also establishes a scientific body to act as a consultative body to the Commission.

To ensure that the provisions of the Convention are observed in the absence of any binding settlement procedure for disputes, the Convention adopts a number of conventional mechanisms. Each Contracting Party is required to "take appropriate measures within its competence to ensure compliance with the provisions of the Convention and with conservation measures adopted by the Convention...". To facilitate compliance with the Convention further, CCAMLR also establishes an elaborate system of observation and inspection. Contracting Parties are also required to make extensive annual reports to the Committee. A distinctive feature of the convention is the obligation on the Commission to notify other Contracting Parties of the infringements of the convention by any one Contracting Party, thereby hoping to ensure observance through public opprobrium.

Convention on the Regulation of Antarctic Mineral Resource Activities (CRAMRA)

The Antarctic region is expected to contain enormous mineral wealth but, despite numerous geological surveys, no commercially viable deposits have been discovered as yet. However, mindful of the enormous impact that a full-scale mining operation would have on the sensitive Antarctic environment, the Consultative Parties, after CCAMLR finally entered into force in 1982, began serious negotiations for a similar treaty for the regulation of the development of the mineral resources (Rich, 1981). Prior to this, a moratorium on mineral resource activity in Antarctica had been declared in 1977, and subsequently extended, dependent on the "timely conclusion of a convention on mineral resources activity".

CRAMRA was the most detailed and complex of the legal instruments making up the Antarctic Treaty System. It essentially aimed to create a regime where mineral resource activity could not take place until the proponent of such an activity could prove that the activity in question would not cause damage to the Antarctic environment. The placing of the onus of proof on the developers rather than the regulators in this convention is a major advance.

CRAMRA also provided for the establishment of five new resource management institutions. Implementation of the Convention would have been overseen by the Antarctic Minerals Resources Commission, which would have consisted of the representatives of the Consultative Parties and other nations engaged in or sponsoring mineral research. They would have been advised by a Scientific, Technical and Environmental Advisory Committee, and Regulatory Committees would have been established for each area identified by the Commission where resource activity could take place. A Secretariat would have been established to service these bodies and the special meetings of the Contracting Parties convened to discuss mineral resource issues.

In 1988, after six years of negotiation, CRAMRA was finally adopted and opened for signature. However, after considerable public lobbying, the Consultative Parties decided not to ratify the convention. In May 1989 it was declared that Australia would not ratify CRAMRA because they believed that no mining should take place at all in the region. Support for the Australian position came from the French government, and in August 1989 the two governments issued a statement to the effect that mining was incompatible with protection of the Antarctic environment and indicated that they would not ratify CRAMRA but would pursue negotiation of a comprehensive environment protection convention within the framework of the Treaty. These two countries then submitted a joint working paper proposing the preservation of Antarctica as a 'Wilderness Reserve' (Antarctic Treaty Consultative Meeting 1990a,b). Similarly, several other Consultative Parties indicated that they would not ratify CRAMRA, with some of them also submitting proposals for a comprehensive protection regime (Redgwell, 1989).

Although there has been no formal recognition by the parties that CRAMRA is defunct, it is generally accepted that CRAMRA has been overtaken by events and is no longer going to enter into force. Even so, many of the techniques developed, such as the 'onus of proof' in the environmental impact assessment being shifted from the regulator to the developer or the extensive provisions on institutional inspection to ensure compliance within the convention, are of more than passing historical interest. For not only do many of them reappear in CRAMRA's successor, the protocol for the comprehensive environment protection regime, but they are of interest in the precedent that they establish for future legislation.

The comprehensive environment protection regime

A Special Antarctic Treaty Consultative Meeting was called to consider the various proposals submitted by the Consultative Parties in relation to a comprehensive environment protection regime. The first session of SATCM was held in Chile in 1990; it agreed a draft protocol to the Treaty which formed the basis of discussion for a meeting in Madrid in 1991. At this second session a new draft protocol was agreed and recommended to governments for formal consideration. This draft protocol was generally well received except for the clause dealing with the amendment of the protocol. Eventually a compromise was reached which effectively placed a 50-year moratorium on mining in the region, after which time any Party to the Treaty could request a review. Any proposal to amend the moratorium can only become binding if it receives the approval and acceptance of three-quarters of the Consultative Parties, and only then if there is an agreed binding legal regime to regulate mineral activities. If such a modification has not entered into force within three years of the date of its adoption, any Party may withdraw from the protocol with two years' notice. With the resolution of this final problem the way was open for the 'Madrid Protocol' to be adopted in Spain, just before the XVIth ATCM in Bonn, Germany in October 1991.

The object of the protocol is to establish a "comprehensive regime for the protection of the Antarctic environment and dependent and associated ecosystems and hereby designate Antarctica as a natural reserve, devoted to peace and science". It seeks to build upon the Antarctic Treaty System by consolidating the sometimes disparate elements of the system to create a comprehensive regime. The protocol, like the Treaty, is a framework within which general obligations are agreed to, which will be translated into specific procedures and management guidelines in future annexes to the protocol. Features of this framework are:

- The environmental principles by which Parties should plan their activities in Antarctica (Article 3);
- The obligation for Parties to cooperate fully in the planning and conduct of activities in the Treaty Area (Article 6);
- A general moratorium on mineral resource activity other than for scientific research (Article 7);
- The establishment of a Committee for Environmental Protection to help Parties implement the aims of the protocol (Articles 11 and 12);
- The establishment of a system of inspections to monitor the observance by the Parties of the protocol (Article 14);
- The requirement for Parties to submit annual reports detailing the steps they have taken to implement the Protocol (Article 17);
- The establishment of Dispute Resolution Procedures (Article 18 and 19).

More detailed annexes have also been developed, covering environmental impact assessment (Annex I), conservation of flora and fauna (Annex II), waste disposal and management (Annex III), prevention of marine pollution (Annex IV) and area protection and management (Annex V), which introduces the *Antarctic Specially Protected Area* and the *Antarctic Specially Managed Area*. Under Article 9 of the Madrid Protocol the annexes form an integral part of the protocol itself, and provision is made for subsequent additional annexes to be adopted at a later date.

References

Anderson, D. 1968. The conservation of wildlife under the Antarctic Treaty. *Polar Record* 14(88):25-32.

Antarctic Treaty Consultative Meeting 1990a. Franco-Australian draft working paper on the possible components for a Comprehensive Convention for the Preservation and Protection of Antarctica. *XV ATCM/WP/3*.

Antarctic Treaty Consultative Meeting 1990b. A Joint Australian/French Proposal in the Form of a Paper including a Draft Recommendation for the ATCM XV. *XV ATCM/WP/2*.

Auburn, F.M. 1982. *Antarctic Law and Politics*. C. Hurst and Company, London.

Barnes, J. 1982. *The Emerging Convention on the Conservation of Antarctic Marine Living Resources: an attempt to meet the new realities of resource exploitation in the Southern Ocean*. Centre for Law and Social Policy, Washington D.C.

Beaglehole, J.C. (Ed.) 1961. *Journals of Captain James Cook on His Voyages of Discovery: the voyage of the Resolution and Adventure 1772-5*. Cambridge University Press. 638pp.

Bean, M. 1983. *The Evolution of National Wildlife Law*. Preager. 264pp.

Birnie, P.W. 1982. *Legal Measures for the Prevention of "Pirate" Whaling*. IUCN Environmental Policy and Law Paper No.19.

Birnie, P.W. 1985. *International Regulation of Whaling: from conservation of whaling to conservation of whales and regulation of whale-watching*, Oceana Publications.

Birnie, P.W. 1989. International legal issues in the management and protection of the whale: a review of four decades of experience. (International Law of Migratory Species). *Natural Resources Journal* 29:903-934.

Bonner, W.N. 1968. The fur seals of South Georgia. *British Antarctica Survey Scientific Reports*, No.56. 81pp.

Bonner, W.N. 1980. *Whales*. Blandford Press. 278pp.
Bonner, W.N. 1982. *Seals and Man: a study of interactions*. Washington University Press, Seattle. 170pp.
Bonner, W.N. 1984. Conservation and the Antarctic. In: Laws, R.M. (Ed.), *Antarctic Ecology*. Academic Press. Pp.821-850.
Brown, G.M. and Crutchfield, J.A. (Eds) 1981. *Economics of Ocean Resources - A Research Agenda*. Proceedings of National Workshop, Washington, 13-16 September 1981.
Conforti, B. 1986. Territorial claims in Antarctica: a modern way to deal with an old problem. (Symposium: the International Legal Regime for Antarctica). *Cornell International Law Journal* 19:249-258.
Deacon, G. 1987. *The Antarctic Circumpolar Ocean*. Cambridge University Press.
Edwards, D. and Heap, J. 1980. Convention on the conservation of Antarctic marine living resources: a commentary. *Polar Record* 20(127):354.
Farnell, J. and Elles, J. 1984. *In Search of a Common Fisheries Policy*. Gower, England.
Greig, D.W. 1988. Sovereignty, territory and the international lawyers dilemma (Antarctica). *Osgoode Hall Law Journal* 26:127-175.
Gulland, J.A. 1968. *The Concept of the Maximum Sustainable Yield and Fisheries Management*. FAO Fisheries Technical Paper No.70.
Hardin, G. 1969. The tragedy of the commons. *Science* 1243.
Holdgate, M.W. 1984. The use and abuse of polar environmental resources. *Polar Record* 22(136).
International Whaling Commission 1963. *13 International Whaling Commission Report*.
Keen, E.A. 1988. *Ownership and Productivity of Marine Resources*. McDonald, Virginia.
Kish, J. 1973. *The Law of International Spaces*.
Knight, H.G. (Ed.) 1975. *The Future of International Fisheries Management*. West Publishing Co., Minn.
Laws, R.M. (Ed.) 1989. *Antarctica: the last frontier*. Boxtree, London.
Lyster, S. 1985. *International Wildlife Law*. Grotius Publications Ltd. p.48.
Mason, C.M. (Ed.) 1979. *The Effective Management of Resources: the international politics of the North Sea*. Pinter.
Mitchell, B. and Sandbrook, 1980. *The Management of Southern Oceans*. IIED, London.
Mitchell, B. and Tinker, J. 1980. *Antarctica and its Resources*. Earthscan, London.
Myhre, J.D. 1986. *The Antarctica Treaty System: politics, law and diplomacy*. Westview Press.
Oceans Institute of Canada 1990. *Managing Fishery Resources Beyond 200 Miles: Canada's options to protect Northwest Atlantic straddling stocks*. Report prepared for the Fisheries Council of Canada.
Redgwell, C.J. 1989. Antarctica. (Current Developments: Public International Law). *International and Comparative Law Quarterly* 39:474-481.
Rich, R. 1981. A minerals regime for Antarctica. *International and Comparative Law Quarterly* 31:709-725.
Rosati, J.P. 1984. Enforcement questions of the International Whaling Commission: are exclusive economic zones the solution? *California Western International Law Journal* 14:114-147.
Rose, G.L. 1991. *Community Fisheries Management Legislation in EC Waters*. CIEL, London.
Smith, G.A. 1984. The International Whaling Commission: an analysis of the past and reflections on the future. *Natural Resources Law* 16:543-567.
Techernia, P. and Jeannin, P.F. 1983. Quelques aspects de la circulation oceanique Antarctique révélés par l'observation de la dérive d'icebergs (1972-1983). CNRS, Muséum National d'Histoire Naturelle.
Triggs, G.D. (Ed.) 1987. *The Antarctica Treaty Regime: law, environment and resources*. Cambridge University Press.
Underdal, A. 1980. *The Politics of International Fisheries Management*. Oslo.

Authors as follows: International fisheries, Greg Rose (CIEL); Antarctica, Sam Johnston.

34. CURRENT PRACTICES IN CONSERVATION

Chapters 28 to 33 have outlined the major instruments and mechanisms, both national and international, which are or can be used in the conservation of biodiversity. This chapter examines some of the most important ways in which conservation is carried out.

Actions to maintain biodiversity can be focused on three levels: ecosystem diversity, species diversity, and genetic diversity. All three are inextricably interlinked, but are not synonymous: maintenance of ecosystem diversity implies maintenance of the species (or at least the most important species) which constitute that ecosystem; however, it is perfectly feasible to maintain species independent of the ecosystems or habitats in which they normally occur. Similarly, maintenance of genetic diversity within a species self-evidently implies maintenance of that species, although the reverse does not apply, in that species can generally be maintained at far lower levels of genetic diversity than would be expected to occur under natural conditions.

However, at whatever level the problem is looked at, it is axiomatic that the maintenance of species diversity, and in particular the prevention of species extinctions, is pivotal to the conservation of biodiversity.

The preservation of species as part of a functioning, although not necessarily pristine, ecosystem is regarded as *in situ* conservation. This is, and will remain, by far the most important form of biodiversity conservation, for a variety of reasons which will be discussed below. Maintenance of species away from their normal habitat is termed *ex situ* or off-site conservation. The distinction between these forms of conservation is not absolute and becomes increasingly blurred as individual species are made the subject of complex, interventionist management strategies.

Planning for the conservation of diversity can be approached in two ways: habitat- or ecosystem-based and species-based.

Habitat or ecosystem approaches

An ecosystem approach to conservation attempts to ensure that representative samples of ecosystems or important habitat types are maintained, through the designation of a network of protected areas or through other controls on land-use. It is assumed that by so doing, the species which inhabit these ecosystems will be conserved. The principal advantage of such an approach is that it does not require detailed knowledge of the status and distribution of all species, i.e. it can be assumed to protect species for which information is not available. This applies particularly to tropical rain forests, whose diversity is at present unquantifiable because it consists largely of undescribed species. A significant proportion of these are expected to be given some measure of protection merely by the protection of large areas of habitat. Its major drawbacks are the difficulty of devising satisfactory habitat or ecosystem classifications on which to base protected area networks, and that fact that populations of particularly rare and threatened species (i.e. those in most urgent need of conservation) are likely in many cases not to be included in a network of protected areas set up on the basis of representative samples of major ecosystem types.

Species-based approaches

Species-based approaches entail the review of taxa with the aim of identifying species considered to be of high priority for conservation, most importantly threatened species and those of actual or potential resource value. Conservation or recovery plans can then be developed for these species, often entailing a combination of *in situ* and *ex situ* management. This approach is exemplified internationally by the IUCN Red Data Books (Table 34.1), which treat the status and conservation requirements of globally threatened species in detail, and the work of the IUCN Species Survival Commission (SSC).

Since its small beginnings in 1949, the SSC has grown into a large global network. In 1991 it consisted of some 95 Specialist Groups with approximately 3,500 members in 135 countries. Through its members and the work of its Specialist Groups, SSC promotes action to arrest the loss of the world's biological diversity and to restore threatened species to safe and productive population levels. The SSC is divided into Specialist Groups organised primarily on a geographical and/or taxonomic basis, although there are some 'interdisciplinary' groups. Among the existing taxon-based Groups are: Antelopes, Parrots, European Reptiles and Amphibians, Coral Reef Fish, Ants, Cycads, Carnivorous Plants, Orchids; while examples of interdisciplinary groups include Re-introductions and Ethnozoology. Membership of Specialist Groups is purely voluntary, and consists mainly of scientists and conservationists nominated by the group Chairmen, who are in turn appointed by the SSC Chairman. All appointments are ratified by Council. Some of the larger Groups (e.g. Captive Breeding) have established secretariats and employ paid staff to accomplish their core activities.

The preparation of 'Action Plans' is one of the most important activities undertaken by the SSC groups. Under the current Action Planning programme, which started in 1986, each taxon-based Specialist Group is expected to review the conservation status and needs of the species within its remit, and recommend conservation actions which will ensure their long-term survival. These recommendations may include both *in situ* measures, such as the carrying out of population surveys, gazetting of particular sites as protected areas; provision of funds or equipment to local enforcement agencies etc., and *ex situ* measures, such as the establishment of captive breeding populations. By early 1992, Action Plans covering 16 groups had been published by IUCN (Table 34.2), and many more were in preparation.

Once Action Plans are published, the Specialist Groups have a duty to promote the implementation of their recommendations by lobbying governments, conservation organisations and donors.

Table 34.1 IUCN Red Data Books

GROUP		YEAR
MAMMALS	Mammal Red Data Book: the Americas and Australasia (excluding Cetacea)	1982
	Threatened Primates of Africa	1988
	Lemurs of Madagascar and the Comoros	1990
	Dolphins, Porpoises and Whales of the World	1991
BIRDS	Threatened Birds of Africa and Related Islands*	1985
REPTILES	Amphibia-Reptilia Red Data Book: Testudines, Crocodylia, Rhynchocephalia	1982
INVERTEBRATES	Invertebrate Red Data Book	1983
	Threatened Swallowtails of the World	1985
PLANTS	Plant Red Data Book	1978

Note: * An ICBP/IUCN Red Data Book. ICBP = International Council for Bird Preservation. See references for full citations.

Table 34.2 IUCN/SSC Action Plans

GROUP		YEAR
MAMMALS	African Insectivora and Elephant Shrews	1990
	African Primates	1986
	Asian Primates	1987
	Foxes, Wolves, Jackals and Dogs	1990
	Otters	1990
	Weasels, Civets, Mongooses and their relatives	1989
	Dolphins, Porpoises and Whales	1989
	African Elephants and Rhinoceroses	1990
	The Asian Elephant	1990
	Asian Rhinoceroses	1989
	African Antelopes (3 parts)	1988-89
	The Kouprey	1988
	Rabbits, Hares and Pikas	1990
REPTILES	Tortoises and Freshwater Turtles	1989
	Crocodiles	1992
INVERTEBRATES	Swallowtail Butterflies	1991

Note: See references for full citations. Although part of the SSC network, the Bird Specialist Groups are largely coordinated by the International Council for Bird Preservation and the International Waterfowl and Wetlands Research Bureau, who are reponsible for a number of bird action plans.

The advantages of a species-based approach lie largely in its allowing resources to be directed to the most urgent cases, that is to species known to be most imminently in danger of extinction. Its disadvantages are that, given existing knowledge and resources, only a tiny proportion of the world's biota can be adequately surveyed to set priorities; even for those taxa which have been surveyed, adequate resources are available to implement recovery plans for only a small proportion of those identified as of high priority. Anthropocentric prejudice dictates that these are very largely higher vertebrates (note that no plant and only three non-mammalian animal SSC Action Plans have been produced to date). Allocation of extensive resources to the conservation of a small number of high profile or priority species may not be the most efficient use of scarce resources, particularly if conservation measures involve a large amount of *ex situ* management, because this does not benefit any other species.

Attempts to reconcile the two approaches centre on the identification of areas of high diversity and endemism (see Chapter 15), particularly of threatened species, and the use of particular species as 'flagships' to justify the preservation of areas of habitat which thereby conserve other species of lower conservation profile.

The relative merits of these various approaches in conservation planning, and the degree of intervention desirable in the management of individual species (in particular the establishment of captive-breeding groups from wild populations of critically endangered animal species) have been, and will continue to be, hotly debated.

This chapter will examine species-based approaches in more detail, comparing plants and animals and outlining *in situ* and *ex situ* approaches for the two groups.

IN SITU CONSERVATION OF THREATENED PLANT SPECIES

The most important single way that plant species can be conserved is by protection of their habitat through control of land-use. Central to this approach is the network of protected areas which nearly all countries possess (Chapter 29). However, the overall extent to which these measures actually preserve wild flora, especially in the tropics where most species occur, is not known. A survey of 25 (mostly temperate) countries revealed great variation in the extent to which listed threatened plant species occurred in protected areas.

- In New Zealand, of c. 70% of the nationally threatened species analysed, 71% are in permanent protected areas managed to benefit the biota, 18% have a low level of *in situ* protection, and 11% have no effective protection *in situ* (D. Given *in litt.*, 1990).
- In Britain, over 75% of the 317 nationally threatened plant species are represented in nature reserves (mostly county wildlife trust reserves) or in Sites of Special Scientific Interest (SSSI) (L. Farrell *in litt.*, 1990). SSSIs are in effect 'quasi protected areas': they are sites, usually on private land, designated by the government conservation service; restrictions may be put on the use of the land in return for compensation to the landowner; however, in only a few cases have management agreements on SSSIs been concluded for nationally threatened plants.
- In Spain, excluding the Canary Islands, about 35-40% of the endemic plants grow in protected areas, although many of these areas are recently designated and have no effective protection or management (C. Gómez Campo *in litt.*, 1990).
- In Bulgaria, out of 763 species listed in the National Plant Red Data Book (Velchev *et al.*, 1984), 38% occur in protected areas, and 63 (8%) of these are apparently confined to them.
- In Czechoslovakia, all 400 or so species to be included in the Red Data Book of Higher Plants (Cerovsky *et al.*, in prep.) occur in protected areas. This is partly because the occurrence of threatened species was used as a priority criterion for creating protected areas. Of the 400 species, at least 118 (30%) have a good measure of active protection.
- In Poland, out of a list of 339 threatened species of higher plants, all 308 extant are in protected areas and an estimated 30% are restricted to them.
- In Australia, almost exactly half of the 3,635 threatened species are in conservation reserves, with 179 believed confined to reserves (Table 34.3).
- In Southern Africa, figures are available which relate to the flora as a whole rather than to threatened species only. In a recent study (Siegfried, 1989), it was found that out of the 582 publicly-owned protected areas, covering 6% of the region, complete plant lists were available for 52 and partial plant lists for 153. An analysis of this and other information found 34% of Southern Africa's 23,300 known vascular plant species in reserves. Further analysis, using known plant distributions, led to a prediction that some 74% of the region's vascular plant species were represented in nature reserves. Of the rich and endangered fynbos flora, the prediction was of 99% coverage, although the author says that this is "almost certainly too high" (Siegfried, 1989). Nevertheless the region's nature reserves clearly protect a very high proportion of the flora and by implication of the threatened species.

These figures indicate that in countries where there have been long-standing programmes to identify and conserve threatened plants, some degree of success is possible. However, in other countries surveyed no threatened plants are known to be protected in conservation areas.

The information also shows that very often the large protected areas which receive the most attention, such as national parks, are not the most important sites for

Table 34.3 Threatened plant species in protected areas in Australia

ADEQUACY OF RESERVATION	CONSERVATION STATUS				
Reserved taxa	Endangered	Vulnerable	Rare	Poorly known	Total
A	4	53	218	52	327
B	38	143	45	32	258
C	11	163	625	449	1248
Total reserved	53	359	888	533	1833
All listed taxa	180	661	1173	1537	3635
% reserved	29.4	54.3	75.7	34.7	50.4

Source: J.H. Leigh *in litt.*, 18 September 1991, updating Briggs, J.D. and Leigh, J.H. 1988. *Rare or Threatened Australian Plants*. Australian National Parks and Wildlife Service.

Notes: 'Poorly known': taxa suspected, but not definitely known, to belong to the categories Extinct, Endangered, Vulnerable or Rare; other categories as used by WCMC and defined on p.234. 'Reserved taxa' column: A = 1,000 plants or more are known to occur within a conservation reserve(s); B = fewer than 1,000 plants are known to occur within conservation reserve(s); C = although recorded from a reserve, the population size is unknown.

conserving threatened or endemic plants. Smaller, less well known sites may be more important. For example in Spain, excluding the Canary Islands, no more than 5% of the nationally threatened plants occur in national parks. In the UK, the small and private nature reserves of the county wildlife trusts are widely recognised as more important for protecting threatened plants than the network of National Nature Reserves.

However, presence in a designated protected area is itself no guarantee of survival. Levels of protection are often inadequate or non-existent. Moreover, even in areas under active protection, appropriate management to maintain viable populations of threatened plant species may not be carried out for a variety of reasons - protection of a site from external disturbance may be enough in some circumstances, such as large reserves in tropical forests, but in others, such as temperate grasslands, active management may be essential.

IN SITU CONSERVATION OF CROPS AND WILD RELATIVES OF CROPS

As well as conserving threatened species, protected areas could potentially play a vital role in the conservation of plant resources which may be of more immediate benefit to mankind. Two of the most important categories of these are traditional crops and wild relatives of crops. Traditional crop material is defined here as pre-20th century varieties and 'land races' which are diverse collections of primitive types of domesticated material and related weeds.

It is widely accepted that genetic erosion of crop gene pools continues today at a rapid, albeit unquantifiable rate (Anon., 1991b). There are a number of causal agents that fuel genetic erosion and their effects have intensified over the last decade. These factors include agricultural mechanisation, spread of uniform hybrids, and habitat destruction such as forest clearing and urbanisation. The rates of genetic erosion are greatest in areas with the most fertile and most easily mechanised agricultural lands, especially near urban centres and markets. It is in impoverished and marginal areas, such as mountainous uplands, where traditional varieties are still grown and are sufficiently relied upon to allow for their careful conservation. The crop gene pools subject to active breeding programmes are among those that have been most depleted (Fowler and Mooney, 1990).

In terms of conservation requirements, crops and crop relatives can be divided into two very different groups: annuals and perennials. Seeds of annuals can be conserved *ex situ* through cold storage. However, for many perennials, *ex situ* conservation is not workable because of short-lived or recalcitrant seed, the limitations of meristem culture and reproduction requirements which are difficult to meet in field gene-banks and laboratories (Ingram, 1984). Traditional varieties and wild relatives of crops differ in their conservation requirements because wild species are evolving within natural ecosystems while land races and other primitive material are products of human practices and modified habitats.

Recent work on the structure of crop populations (Zimmerer and Douches, 1991) and on wild relatives of crops describes a fine mosaic of genotypic variation. This genetic diversity is difficult to capture and maintain even in networks of protected areas and comprehensively developed gene-bank collections (Brush, 1991). Probably much less than 10% of the alleles in the gene pools of the major crops are currently contained in protected areas. The prospects of maintenance of the genetic variation within these populations, with pressures for expanded uses of protected areas and a poor technical and institutional base for management, are doubtful, especially in view of the fact that the great majority of crop genetic resources are in the tropics and in countries with relatively ineffective programmes of protected area planning and management. Great expansion of such programmes is required in order to avoid accelerating and irreversible loss of potentially valuable genes.

Levels of effectiveness of *in situ* conservation of plant genetic resources

Virtually all protected areas and landscapes have some populations of some species of economic importance which are more or less adequately protected, at least in the short-term. However, in virtually all, substantially increased monitoring and management programmes are necessary in order to avert losses of rarer genes and other potential genetic resources. There are three general levels of *in situ* conservation of plant genetic resources which are described below.

Level I represents largely unplanned coverage through ecosystem conservation. It cannot be assumed that there will be adequate coverage within the reserve to maintain viable populations of plant species with genetic resources over the long-term.

Level II requires the planning and design of protected areas with use of distribution data for species with genetic resources. Many of the 'genetic reserves' (Jain, 1975) involve this level of conservation. Management for particular species and associated successional phases is usually necessary. Species that are monitored and managed under programmes of ecosystem coverage can also have *level II* conservation.

Level III involves site-specific monitoring, management and procurement for particular levels of conservation for specific "functional population units" (Solbrig, 1991). For this level of conservation to be attained, population viability thresholds must be set with prescriptions for maintenance of intra-specific variation and rarer alleles.

In both natural and well-protected populations, there is a constant flux of gene frequencies with some alleles becoming rare or disappearing. In protected areas, the natural and human-induced dwindling of populations can cause the narrowing of the base of variation and subsequent loss of potentially valuable genetic resources. In order to maintain rarer alleles or possible adaptive complexes associated with certain environments and selection factors, further requirements for larger and sometimes additional populations must be set.

With species for which there is no major concern for maintenance of potential genetic resources, *level I* conservation is often adequate. For *in situ* conservation of crop genetic resources, both traditional varieties and wild species, *level II* is always necessary to assure capture of desirable percentages of alleles and *level III* is necessary for long-term security as well as procurement for extended programmes of plant breeding.

The status of most populations with genetic resources is still poorly determined and the development of new theory and techniques for determining conservation requirements are necessary. There are a number of technical issues which must be addressed before protected areas can become effective for the conservation and procurement of genetic resources. Desired levels of conservation of less common genes should be determined. Access to and ease of procurement of the genetic resources of wild species must be effectively regulated and in some cases expanded. Regulation and expansion of the systems of distribution of the germplasm taken from protected populations are inevitable (Kloppenburg and Kleinman, 1988).

The text below will discuss the current status of the three levels of *in situ* conservation of crops and wild relatives of crops in terms of three categories: traditional land-tenure and farming systems; genetic reserves and other locally and nationally managed protected areas; internationally monitored protected areas.

Traditional and in-farm programmes of *in situ* conservation

Areas with traditional land tenure and farming systems provide a basis for *in situ* conservation. These areas may be designated protected areas or they may be non-protected areas where key aspects of traditional farming systems are supported and maintained.

Genetic conservation requires recognition of the interrelationships of genotypic and allelic diversity and the functioning of the agroecosystem as a whole. The existing variation has co-evolved within a mosaic of agroecosystem zones. It is especially important to know the specific environmental conditions necessary for conservation of narrowly adapted land races.

In considering the requirements for *in situ* conservation of crops, agricultural communities and systems can be placed in the following categories:

- highly traditional agricultural mosaic which has only been moderately altered through rising human population and new technologies,
- traditional agricultural mosaic under stress and with rapid rates of loss of primitive material and the diversity of selection factors,
- comparatively recent, pioneer communities with limited local knowledge and with recently introduced cultivated material.

The first generally holds the most crop genetic resources while the second has the greatest rates of genetic erosion.

In order to design and maintain effective in-farm conservation programmes, interrelationships between socio-economic systems and the structure of crop populations must be ascertained. Gender may be an important factor. Women often play key roles in farm conservation of land races, especially where they have traditionally been the selectors of seeds for planting. For example, women in a number of traditional Ethiopian societies pass knowledge of seed selection from mother to daughter (Marie Dulude, pers. comm.) and in a village in Liberia women maintain 112 varieties of rice, matched to particular micro-environments, slope categories, insolation, and soil types (Thomasson, 1991).

The traditional agroforestry system practised on Fergusson Island, Papua New Guinea, serves as an example of *in situ* conservation of traditional varieties of crops within a particularly biodiversity-rich setting (Flavelle, 1990). Because of low human populations, mountainous terrain, and distance to market, cash crop ventures have so far been unsuccessful. The system revolves around the growing of yams, principally *Dioscorea esculenta* and *D. alata*. Yams are the basis of the subsistence economy but it is their cultural importance that may ultimately prevent them from being replaced by introduced food crop species. Yam seeds are inherited through the matrilineage; they are exchanged as gifts at funeral feasts and other occasions; they are the focus of magic ritual and myths. Other tubers which farmers have grown for generations but which hold less status than yams include: sweet potato, *Ipomea* sp., *Colocasia esculenta*, and manihot, *Manihot esculenta*. Prominent food trees observed in the system include mango, *Mangifera* sp., *Citrus* sp., *Szigium* sp., *Carica* sp., banana, *Musa* sp., and *Arctocarpus* sp., and *Ficus* sp., *Cocos nucifera*, and sago, *Metroxylon* sp. Every time that a community opposes intrusion of the cash economy and environmentally damaging activities, it is a form of *in situ* conservation. Such approaches may not be permanent but can be effective over the medium- and long-term.

There is currently a very poor institutional base for more coordinated in-farm conservation programmes. The International Agricultural Research Centres (IARC) manage information on the major food crops, however, they have less expertise in integration of information about the ecosystems, societies and cultures that have created and now maintain these traditional varieties.

Almost all of the current in-farm crop conservation programmes have been initiated by small institutes and NGOs. These programmes tend to emphasise research, education, technical advice, and credit schemes that support traditional farming systems and have shown particular success where they have been able to link conservation interests with locally-driven rural development.

The Rural Advancement Fund International (RAFI) promotes the maintenance of land races and low-input farming and attempts to channel funding to local NGOs involved in crop conservation. RAFI has developed a training kit on community plant breeding, and local seed banking by maintaining living stock. In Ethiopia and Zimbabwe, the programme has been established through the umbrella organisation, Seeds for Survival. The Southeast Asia Regional Institute for Community Education (SEARICE) is working in the Mekong Delta and the

Philippines. CLADES is a South American NGO promoting sustainable agriculture, including the conservation of land races and wild crop relatives.

In Ethiopia, the Seeds for Survival programme is working with the Plant Genetic Resources Centre in Addis Ababa to reintroduce traditional varieties and land races to farmers. During droughts, farmers were forced to eat their stocks of seeds. PGRC responded by making extensive collections and is now reintroducing land races to farmers, establishing research plots, monitoring productivity, and establishing training and support programmes for farmers.

In the developed countries, there is some *in situ* conservation in historic farms and other protected cultural landscapes as well as through the networks of seed conservationists. Most efforts are organised at the grassroots level.

In order to implement long-term programmes with the dual mandate of monitoring genetic resources in agroecosystems and supporting farmer-driven development, national institutes and NGOs require greatly increased and more stable funding bases. For Latin America, Montecinos and Altieri (1991) stated that, "over 50% of the initiatives known by the authors do not have their own staff or budget, but must borrow from other projects, and strongly rely on help from farmers and other local sources. Among those that have received financial support, again over 50% have been working with less than US$5,000 per year. Over 60% of the programmes have done collection work, but do not have money to set up simple and adequate storage facilities or, even worse, to do a systematic monitoring of multiplication of material and performance tests in the field. This is, in fact, one very strong reason for diversity loss and for not seeing many breeding initiatives at the farm level."

There are major questions about the effectiveness of virtually all of the *in situ* conservation programmes for crops. At the local level, there are overlapping and jumbled objectives and many farmer organisations are more concerned with the economic benefits of producing their own seed than with conservation of genetic diversity.

Genetic reserves and other nationally or locally managed protected areas

Genetic reserves and managed protected areas often provide adequate levels of *in situ* conservation though there are often insufficient inventory data and management expertise. The institutional bases for all of the locally managed programmes of *in situ* conservation of crop relatives are remarkably weak and, after nearly a decade of negotiations, there are still no national or international frameworks funding *in situ* conservation of wild relatives of crops.

Genetic reserves (Jain, 1975) have been established for the procurement of seed and other forms of germplasm for thousands of years. In recent decades, these reserves have often been established for single and multiple crop relatives. The major problem with the current networks of genetic reserves is that the sites are often too small and do not include population levels that are adequate for maintenance of fitness and of rarer genes, nor have adequate potential for a diversity of management treatments.

In less densely populated regions with remaining forest, extractive reserves as in Amazonia could be managed for the genetic resources of wild relatives of crops. These reserves have been established for the benefit of local people to gather Brazil nuts, *Bertholletia excelsa*, and other species with commodity value. Some wild or semi-domesticated species thrive in more altered habitats. Fruit crops such as papaya, *Carica papaya*, bacuri, *Platonia insignis*, guava, *Psidium guajava*, and ciruela, *Bunchosia glandulosa*, have wild populations in more disturbed and open forest mosaics dominated by second-growth forest. Cupuaca, *Theobroma grandiflorum*, is a wild relative of cacao that is planted from seed in backyards in Amazonia, particularly in Para (Smith and Schultes, 1990).

In other types of protected area, advances in management of genetic resources have been limited. While protected area managers worldwide have become increasingly aware of issues of genetic resources and maintenance of genetic diversity, these are generally afforded lower priority than other management concerns. Moreover, traditional farming and agroforestry have tended to be suppressed within many protected areas, with a concomitant increase in rates of genetic erosion of crop plants.

There is still inadequate information on the status of crop relatives in protected areas. The level of funding for this type of highly technical and relatively expensive inventorying, monitoring and management may actually have declined, in real terms, in most of the world over the last decade. Even in developed countries attention is directed to threatened habitats and threatened species and few resources are available for monitoring populations of crop relatives, although some work has been done (e.g. on wild cranberries *Viburnum* spp. in protected areas in the mid-Atlantic States of the USA). The most effective technical linkages between programmes in the developing countries have been forged by FAO (Palmberg and Esquinas-Alcazar, 1990). However, such national programmes have tended to focus on timber species rather than food crops.

Biosphere reserves, World Heritage Sites and other internationally monitored protected areas

The internationally monitored protected areas, which are principally biosphere reserves and World Heritage Sites, hold the greatest promise for adequate inventorying, monitoring, management and procurement though respective levels of national support are too often as weak as with other protected areas.

Biosphere reserves will play an increasingly key role in the conservation and utilisation of wild relatives of crops because of institutional potentials for monitoring and international exchange of information (Ingram and Williams, 1984; Ingram, 1990b). The concept has not been effectively utilised for traditional crops. The network is very new (Batisse, 1982; Vernhes, 1989) and most reserves are still poorly inventoried.

There has been no systematic documentation of the crop varieties, land races and other primitive material in either the transition areas or in the buffer zones of biosphere reserves or within World Heritage Sites. There has been only cursory reporting of agricultural activities within and on the edge of biosphere reserves and World Heritage Sites, both natural and cultural, have rarely been inventoried for crop varieties.

The most impressive example of *in situ* conservation of crops in an internationally monitored protected area is that of Niger's Aïr and Ténéré National Nature Reserve which maintains a range of traditional and contemporary garden types and primitive, traditional and more recent crop material for the gene pools of sorghum, *Sorghum* spp., pearl millet, *Pennisetum* spp., barley, *Hordeum* sp., and wheat, *Triticum* sp. as well as for agroforestry: *Acacia* spp., *Olea* sp. and *Ziziphus* spp. (Ingram, 1990a). The area is at an ecological crossroad in terms of material and farming practices from North Africa and the Sahel. Despite the present maintenance of these gardens within the framework of protected area management, progressive desertification due largely to climatic change, could destroy prospects for long-term *in situ* conservation.

Only a very small portion of the 300 biosphere reserves (see Chapter 29) which are part of UNESCO's Man and the Biosphere Programme are in centres of crop origin or in the regions of high species diversity, such as the humid tropics. Based on cursory documentation (MAB, UNESCO files, Paris), well over 20% of the biosphere reserves have populations of the relatives of the major crops (Table 34.4); with better documentation and complete inventories of plant species in current biosphere reserves, this percentage is expected to exceed 50%.

There are, however, virtually no distribution data available and consequently no assessment of the status of populations in terms of core, buffer and transitional zones. The information on wild relatives in World Heritage Sites is even more cursory. Consequently, there is very little *level III in situ* conservation anywhere though these areas have the best institutional potentials.

The actual status of wild relatives in many biosphere reserves will remain contentious, especially with changing political and administrative contexts. For example, the status of the many crop genetic resources, such as *Malus* spp. and *Prunus* spp., which provided the original focus for a number of reserves in Soviet Asia, has not been reviewed in recent years. The Sierra de Manantlán Biosphere Reserve in Mexico was recently established for the conservation of the genetic resources of wild corn, *Zea* spp., though the requirements for *level III in situ* conservation have still barely been met.

Conclusions

Only a tiny portion, well under 10%, of the total species and allelic diversity of the major crop gene pools are currently maintained *in situ* and of this only a small portion is conserved at levels adequate to withstand threats over the long-term as well as needs for germplasm for breeding programmes. Genetic erosion continues and in many cases is accelerating both outside and within protected areas. However, there are inadequate inventory and monitoring data to determine rates. Greatly increased programmes of protected area planning, monitoring, and management are needed within the next five years in order to cause a significant reversal of the accelerating rates of genetic impoverishment.

INSTITUTIONS INVOLVED IN *EX SITU* CONSERVATION OF PLANTS

Botanic Gardens

The single most important type of institution involved in *ex situ* conservation of wild plants is the botanic garden. There are over 1,500 botanic gardens worldwide, of which about 800 are believed to be currently active in plant conservation, although this number is steadily increasing.

Botanic gardens worldwide contain well over three million accessions between them. This total illustrates the remarkable capacity of the network of botanic gardens for the cultivation of plants and their potential contribution to plant conservation. However, many botanic gardens are poorly financed and badly organised, or only weakly supported within their institutions or by their governments, and a large percentage of plants grown is of low conservation priority.

There is a considerable imbalance in the global distribution of botanic gardens. Europe has 532 botanic gardens, but Africa has only 82 and South America 66. Most tropical countries, where most of the world's flora resides, have few botanic gardens (Table 34.5) and most of these are poorly developed or funded, although most of the new botanic gardens that have been created or planned in recent years are in tropical countries, which have large floras. Indeed, many gardens in temperate countries have ceased to play any significant role in conservation or research and retain only educational significance, whereas most new tropical gardens have been created primarily as centres for plant conservation. For example, the Conservatoire et Jardin Botanique de Mascarin, Réunion Island, founded in 1987, contains over 60% of the island's rare and endangered flora.

The collections maintained in botanic gardens are very diverse. Particular groups, such as orchids, succulents, bromeliads, bulbous species and temperate trees, are particularly well represented in cultivation, as Table 34.6 shows. Collections of tropical woody species are, however, less well represented. In general, the floras of tropical and sub-tropical continental countries are less commonly grown than those of temperate countries and oceanic islands.

Efforts to coordinate the activities of botanic gardens at an international level are undertaken by Botanic Gardens Conservation International (BGCI). The purpose of the Secretariat is to disseminate information to promote and coordinate the *ex situ* conservation of threatened wild plants. It also provides technical guidance, data and support for botanic gardens in almost 100 countries and assists and promotes the development of botanic gardens and their plant conservation programmes. BGCI has a worldwide membership of 317 botanic gardens (Table 34.5).

Table 34.4 Biosphere Reserves with higher levels of documentation of wild relatives of crops and forage species

	NAME OF PROTECTED AREA	BIOGEOGRAPHIC PROVINCE	AREA OF RESERVE (ha)	BETTER DOCUMENTED GENERA WITH GENETIC RESOURCES
Algeria	Parc national de Tassili N'Ajjer	Sahara	7,200,000	*Olea* spp.
	El Kala PN	Mediterranean sclerophyll	76,438	*Olea* sp., *Pistacia* sp.
Argentina	Reserva Ecológica de Ñacuñán	Monte	11,900	*Prosopis* spp.
Australia	Croajingolong	Eastern sclerophyll	101,000	*Acacia* spp.
Austria	Lobau Reserve	Central European highlands	1,000	*Pyrus* sp.
	Benin Reserva de la biosphere de la Pendjari	West African woodland and savanna	880,000	*Acacia* spp.
Brazil	Vale do Ribeira and Serra DA Graciosa Biosphere Reserve	Serra Do Mar	1,615,000	*Prunus* sp.
Bulgaria	Reserve Boatine	Balkan Highlands	1,226	*Prunus* sp.
	Reserve Ouzounboudjak	Balkan Highlands	2,575	*Vaccinium* sp.
Cameroon	Parc national de Waza	West Africn woodland and savanna	170,000	*Oryza* sp., *Sorghum* sp.
	Parc national de la Bénoué	West African woodland and savanna	180,000	*Pennisetum* spp., *Prosopis* spp.
Canada	Waterton Lakes National Park	Rocky Mountains	52,597	*Allium* sp.
Central African Republic	Bamingui-Bangoran Conservation Area	West African woodland and savanna	1,622,000	*Prosopis* sp.
Chile	Parque Nacional Juan Fernandez	Southeastern Polynesian	9,290	*Rubus* spp.
	Reserva de la Biosfera La Compana-Peñuelas	Chilean Sclerophyll and Chilean Nothofagus	17,095	*Ribes* sp.
China	Changbai Mountain Biosphere Reserve	Manchu-Japanese Mixed Forest	217,235	*Panax* spp.
	Dinghu Nature Reserve	South Chinese Rainforest	1200	*Castanopsis* sp.
	Fanjings han Mountain Nature Reserve	Chinese Subtropical Forest	41,533	*Vaccinium* spp.
	Shennongjia Biosphere Reserve	Chinese Subtropical Forest	29,400	*Prunus* sp., *Brassica* sp., *Ribes* sp., *Malus* sp., *Prunus* sp., *Rubus* sp., *Vitis* spp., *Panax* sp., *Allium* spp., *Avena* spp., *Sorghum* sp. land races
Colombia	Cinturon Andino Cluster Biosphere Reserve	Northern Andean	855,000	*Juglans* sp.
	Sierra Nevada de Santa Marta	Venezuelan Dry Forest	731,250	*Acacia* spp.
Costa Rica	Reserva de la Biosfera de la Amistad	Central American	584,592	probably *Persea* spp., *Lycopersicon* spp.
	Cordillera Volcanica Central	Central American	144,393	probably *Persea* spp. and *Lycopersicon* spp.
Czechoslovakia	Palava Protected Landscape Area	Middle European Forest	8,017	*Prunus* sp., *Avena* sp., *Solanum* sp.
Denmark	Northeast Greenland National Park	Arctic desert and icecap	70,000,000	*Vaccinium* sp.
Egypt	Omayed Experimental Research Area	Sahara	1,000	*Gymnocarpus* sp.
Ecuador	Galapagos - Archipiélago de Colon	Galapagos Islands	766,514	*Lycopersicon* sp.
France	Réserve de la biosphère de PN des Cévennes	Atlantic	323,000	*Castanea* sp., *Vaccinium* sp.
Germany	Steckby-Loedderitz Forest Nature Reserves	Middle European Forest	3,500	*Pyru* sp.
	Vessertal Nature Reserve	Middle European Forest		*Vaccinium* spp.
	Middle Elbe Biosphere Reserve	Middle European Forest	17,500	*Malus* sp., *Pyrus* sp.
	Southeast Rügen Biosphere Reserve	Middle European Forest	18,640	*Allium* sp., *Malus* sp., *Fragaria* sp., *Vaccinium* spp.

Table 34.4 Biosphere Reserves with higher levels of documentation of wild relatives of crops and forage species (continued)

Greece	Gorge of Samaria National Park	Mediterranean Sclerophyll	4,840	*Olea* sp.
	Mount Olympus National Park	Mediterranean Sclerophyll	4,000	*Allium* sp.
Guatemala	Tikal National Park	Campechean	57,600	*Acacia* sp., *Lycopersicon* sp., *Persea* sp.
Indonesia	Cibodas Biosphere Reserve	Java	140,000	*Vaccinium* spp.
	Komodo Proposed National Park	Lesser Sunda Islands	70,000	*Ipomoea* spp.
	Tanjung Puting Proposed National Park	Borneo		*Durio* sp.
	Gunung Leuser Proposed National Park	Sumatra		*Musa* spp., *Citrus* spp.
	Siberut Nature Reserve	Sumatran		*Durio* spp., *Musa* spp.
Iran	Arasbaran Protected Area	Caucaso-Iranian Highlands	52,000	*Juglans* sp.
	Gano Protected Area	Anatolian-Iranian Desert	49,000	*Olea* sp., *Prunus* sp., *Pistacia* sp., *Acacia* spp.
	Hara Protected Area	Anatolian-Iranian Desert	85,686	*Acacia* sp., *Prosopis* sp.
	Miankaleh Protected Area	Caucaso-Iranian Highlands	68,800	*Punica* sp., *Rubus* sp.
	Touran Protected Area	Iranian Desert	1,000,000	*Hordeum* sp.
Italy	Foret Domaniale du Circeo	Mediterranean Sclerophyll	3,260	*Pistacia* sp.
Kenya	Mount Kulal Biosphere Reserve	Somalian/Lake Rudolf	700,000	*Olea* sp., *Acacia* sp.
	Amboseli Biosphere Reserve	Somalian Grasslands	283,200	*Acacia* spp.
Mali	Parc national de la Boucle du Baoucle	West African woodland and savanna	771,000	*Acacia* spp.
Mauritius	Macc habée-Bell Ombre Nature Reserve	Mascarene Islands	3,611	*Coffea* sp.
Mexico	Montes Azules	Campechean	331,200	*Persea* sp.
	Reserva de la Biosfera de Sian Ka'an	Campechean and Yucatecan	528,000	probably *Persea* sp.
	Reserva de la Biosfera Sierra de Manantlán	Madrean-Cordilleran	139,577	[1]
Mongolia	Great Gobi Biosphere Reserve	Gobi Desert	5,303,172	*Hordeum* sp., *Allium* sp.
North Korea	Mount Paekdu Nature Reserve and Biosphere Reserve	Manchu-Japanese Mixed Forest	60,000	*Ribes* spp., *Rubus* spp., *Prunus* sp., *Vaccinium* spp., *Viburnum* sp., *Allium* sp.
Pakistan	Lal Suhanra National Park	Thar Desert	31,355	*Acacia* spp.
Peru	Reserva del Noroeste	Equadorian Dry Forest	226,300	*Acacia* spp.
Philippines	Palawan Island Biosphere Reserve	Philippines	1,150,800	*Durio* spp.
Poland	Babia Gora National Park	Middle European Forest	1,741	*Allium* sp.
Romania	Pietrosul Mare Nature Reserve	Middle European Forest	3,068	*Allium* spp.
	Retezat National Park	Middle European Forest	20,000	*Juglans* spp.
	Rosca-Letea Reserve	Pontian Steppe	18,145	*Vitis* spp.
Senegal	Forêt classée de Samba Dia	West African woodland and savanna	756	*Acacia* spp.
	Parc national du Niokolo-Koba	West African Woodland and savanna	913,000	*Acacia* spp.
Spain	Reserva de Grazalema	Mediterranean Sclerophyll	32,210	*Ceratonia* sp.
	Reserva de la Biosfera de Doñana	Mediterranean Sclerophyll	77,260	*Olea* sp.
	Reserva de la Biosfera del Urdaibai	Iberian Highlands	22,500	*Castanea* sp.
	Parque Natural Del Montseny	Mediterranean Sclerophyll	17,372	*Prunus* sp.

Source: table compiled by G.B. Ingram, based on MAB/UNESCO files.
Notes: [1] This biosphere reserve was the first set up primarily for the protection of wild corn, teosinte, *Zea diploperennis*.

In recent years new national botanic garden organisations have been formed in many countries, such as Brazil and Australia, with regional groupings of the International Association of Botanic Gardens (IABG) existing in, for example, Europe, the Mediterranean region, Latin America and Asia.

Table 34.5 Number of botanic gardens and known cultivated accessions in botanic garden collections

COUNTRIES	NO. OF BOTANIC GARDENS	NO. OF BGCI MEMBERS	NO. OF ACCESSIONS IN CULTIVATION
ASIA			
Bangladesh	2	0	700
China	66	4	56,278
Hong Kong	4	3	1,200
India	68	7	86,259
Indonesia	5	4	69,840
Iran	3	0	250
Iraq	1	0	230
Israel	7	3	8,000
Japan	59	0	72,560
Korea, DPR	1	0	3,140
Korea, Rep	5	0	10,000
Malaysia	9	5	3,872
Mongolia	1	0	-
Myanmar	2	0	-
Nepal	1	1	2,300
Pakistan	5	0	430
Philippines	9	2	16,829
Saudi Arabia	2	0	-
Singapore	1	1	3,000
Sri Lanka	6	6	7,125
Taiwan	2	1	2,513
Thailand	5	0	2,400
Turkey	6	1	8,815
Viet Nam	3	1	-
USSR (former)			
USSR	160	1	344,744
EUROPE			
Albania	1	0	2,000
Austria	11	0	40,300
Belgium	15	3	45,783
Bulgaria	9	0	3,000
Czechoslovakia	34	1	53,817
Denmark	7	2	48,950
Finland	8	3	22,900
France	66	18	171,725
Germany	73	12	383,470
Greece	4	2	3,550
Hungary	17	0	7,350
Iceland	2	0	6,500
Ireland	8	5	36,500
Italy	48	10	118,432
Malta	1	0	8,000
Monaco	1	1	7,000
Netherlands	39	5	95,180
Norway	6	3	14,400
Poland	25	1	54,066
Portugal	6	1	13,204
(Azores)	3	2	100
(Macau)	1	0	-
(Madeira)	2	0	-
Romania	10	0	42,400
Spain	8	5	15,900
(Balearic Is)	1	0	-
(Canary Is)	3	1	10,000
Sweden	9	3	38,190
Switzerland	22	9	82,020
UK	60	31	217,341
(Gibraltar)	1	0	-
Yugoslavia	32	0	29,508
NORTH AND CENTRAL AMERICA			
Barbados	2	0	-
Belize	1	0	-
Bermuda	1	0	-
NORTH AND CENTRAL AMERICA (continued)			
Canada	18	7	67,374
Cayman Is	1	1	-
Costa Rica	2	2	4,000
Cuba	8	3	7,550
Dominica	1	1	750
Dominican Rep.	1	0	-
El Salvador	1	1	3,500
Grenada	1	0	-
Guadeloupe	2	1	100
Guatemala	1	1	700
Honduras	2	1	764
Jamaica	4	0	1,557
Martinique	3	0	200
Mexico	30	7	8,650
Nicaragua	1	0	-
Panama	1	0	-
Puerto Rico	4	0	3,150
St Vincent	1	1	-
Suriname	1	0	-
Trinidad and Tobago	1	0	-
USA	247	47	424,888
(Hawaii)	19	6	25,632
Virgin Is (British)	1	1	5,000
Virgin Is (US)	1	0	3,000
SOUTH AMERICA			
Argentina	9	1	18,687
Bolivia	3	1	-
Brazil	11	4	20,820
Chile	9	1	1,967
Colombia	13	4	3,000
Ecuador	2	1	500
French Guiana	2	0	-
Guyana	2	1	300
Paraguay	1	0	-
Peru	5	1	-
Uruguay	1	0	500
Venezuela	7	2	1,003
OCEANIA			
Australia	60	22	99,752
Fiji	1	0	-
New Zealand	17	6	28,231
Papua New Guinea	4	2	6,700
Solomon Is	1	1	-
Western Samoa	1	0	-
AFRICA			
Algeria	3	0	8,000
Angola	1	0	500
Benin	1	0	-
Burundi	1	0	-
Cameroon	2	1	-
Cape Verde	1	0	-
Côte d'Ivoire	1	0	1,200
Egypt	5	1	7,550
Ethiopia	1	0	100
Gabon	1	1	-
Ghana	3	1	1,000
Kenya	5	1	710
Libya	1	0	400
Madagascar	1	1	5,000
Malawi	3	3	130
Mauritius	2	2	880
Morocco	2	0	1,200
Mozambique	2	0	2,200

Table 34.5 Number of botanic gardens and known cultivated accessions in botanic garden collections (continued)

COUNTRIES	NO. OF BOTANIC GARDENS	NO. OF BGCI MEMBERS	NO. OF ACCESSIONS IN CULTIVATION	COUNTRIES	NO. OF BOTANIC GARDENS	NO. OF BGCI MEMBERS	NO. OF ACCESSIONS IN CULTIVATION
AFRICA (continued)				AFRICA (continued)			
Namibia	1	1	-	St Helena	1	0	-
Nigeria	5	1	-	Sudan	1	0	150
Réunion	4	2	-	Tanzania	2	1	-
Rwanda	1	0	175	Togo	1	1	200
Senegal	3	0	1,300	Tunisia	1	0	-
Seychelles	1	0	-	Uganda	2	0	3,320
Sierra Leone	1	0	-	Zaire	2	2	2,560
South Africa	17	11	27,582	Zimbabwe	4	2	3,250
				TOTAL	1555	317	3,077,643

Source: International Directory of Botanic Gardens 1990 and BGCI unpublished information.

Most botanic gardens now recognise that priority should be given to growing plant material of known wild origin and gradually many gardens are replacing or supplementing their collections with accessions from known wild sources. A shift in emphasis from botanic gardens growing wide and diverse collections of exotic species to the cultivation of the native flora of their region has gathered momentum.

Figures taken from the BGCI database on the occurrence of the rare and threatened plants in botanic gardens are given in Table 34.6. At present the database includes some 29,000 records of rare and threatened plants of around 10,000 taxa in 400 different institutions. This table indicates that some geographical regions and taxonomic groups are well represented in cultivation, such as those from China (63% of the rare and threatened plants are known to be cultivated) and Macronesia (77% cultivated) and the family Cactaceae (85% cultivated). Other regions and groups such as orchids have been less well surveyed for the database and appear less well represented in cultivation than they undoubtedly are; Table 34.8 gives some indication of the large number or orchid species in cultivation. The low overall figures for Europe and the USA are disappointing and suggest that many gardens in these regions have not yet sufficiently adapted their activities to be able to contribute significantly to native plant conservation.

The botanic gardens of China show the benefit of a national strategy for conservation of flora. Their priority is to bring into cultivation the protected species of China. Botanic gardens in each province have particular responsibility for the endangered species of that province (Table 34.7). Other gardens have specialist collections such as the Institute of Medicinal Plant Development, Chinese Academy of Medicinal Sciences, Beijing, which has collected 62 rare and threatened species of medicinal plants, and the South China Institute of Botany, Guangzhou, which has 99 of the 130 species of Chinese Magnoliaceae in cultivation (19 of which are protected), representing well over a third of the Magnoliaceae worldwide.

In many cases the genetic diversity maintained in the gardens' *ex situ* holdings is inadequate for conservation

Table 34.6 Examples of rare and threatened taxa known in cultivation in botanic garden collections

REGION OR PLANT GROUP	NO. OF RARE AND THREATENED TAXA IUCN SURVEYED	NO. KNOWN IN CULTIVATION BOTANIC GARDENS	%
Macaronesia[1]	557	419	75
China	338	211	63
New Zealand	230	129	56
South Africa	1,051	514	49
Australia	1,867	893	48
Mascarene[2]	377	160	42
Europe	1,723	558	32
USA	3,324	890	26
India	927	105	13
Cuba	874	55	4.5
Cacti	451	385	85
Cycads	105	81	77
Conifers	264	179	68
Palms	665	298	45
Ferns	600	73	12
Orchids	986	306	31

Source: BGCI database (1991), based on WCMC threatened plants database list.
Notes: [1] Canary Is, Madeira, Azores, Salvage Is, Cape Verde are part of Macaronesia but were not included in the survey.
[2] Mauritius, Rodrigues, Réunion in the Indian Ocean.

Table 34.7 Rare and endangered plants in cultivation in botanic gardens and arboreta in China arranged according to province

PROVINCE	NO. OF PROTECTED SPECIES IN PROVINCE	NO. OF PROTECTED SPECIES IN PROVINCE IN CULT.	TOTAL NO. NATIONALLY PROTECTED IN CULT.
Jiangxi	52		
Zhejiang	55	47	47
Heilongjiang	16	12	12
Jiangsu	76		
Hubei	55	38	59
Hunan	70	65	65
Guangdong	12		
Guangxi	115	67	67
Shaanxi garden 1	37	29	29
Shaanxi garden 2	37	27	27
Gansu	10		
Medicinal plants, Beijing	62		
Nanjing	17		

Source: Shan-An, He, Heywood, V.H. and Ashton, P.S. 1990. *Proceedings of the International Symposium on Botanical Gardens*. ISBG, 25-28 September 1988, Nanjing. Jiangsu Science and Technology Publishing House, Nanjing, China.
Note: 389 nationally protected species (National Environmental Protection Bureau of China and the Institute of Botany, 1987).

Table 34.8 Some important living collections of orchid species

COUNTRY	COLLECTION	NO. OF SPECIES
Australia	Canberra National Botanic Garden	800
Brazil	São Paulo Botanic Gardens	1500
Costa Rica	Lankester Botanic Gardens	2,000
Cuba	Orquideario Soroa	700
France	Jardin des Plantes, Paris	500
Germany	Heidelberg University Botanic Garden	2,000
	Palmengarten, Frankfurt	1,000
India	Orchid Research and Development, Arunachal Pradesh	400
	National Orchidarium and Experimental Garden, Yercaud	-
	Gurukula Botanical Sanctuary, Kerala	-
Indonesia	Bogor Botanic Garden	883
	Cibodas Botanic Garden	230
	Purwodadi Botanic Garden	546
	Bali Botanic Garden	459
Japan	Hiroshima Botanical Garden	3,000
Malaysia	Serdang (MARDI)	250
(Sabah)	Tenom Orchid Centre	453
(Sarawak)	Orchid Centre, Kuching	-
Mexico	Asociación Mexicana de Orquideologia	550
Papua New Guinea	Lipizauga Botanical Sanctuary, Goroka	-
UK	Royal Botanic Gardens, Kew	4,000
	Royal Botanic Garden, Edinburgh	1,500
	Glasgow Botanic Garden	1,000
USA	New York Botanical Garden	1,000
	Smithsonian National Orchid Collection, Washington	2,500
	Wheeler Orchid Collection, Bell State University	3,000
	The Marie Selby Botanical Garden, Florida	2,000

Source: Various, including J. Stewart, pers. comm. 1990. Royal Botanic Gardens, Kew, UK.

purposes, as the holdings do not constitute representative samples of the genetic variation of the species.

Frequently a potentially important species is represented by no more than an accession of only a few specimens. Furthermore, many current horticultural and management practices contribute to continuing genetic erosion even of these small samples. However, there has been recent recognition in the botanic garden community of the need for careful genetic management of their accessions to maximise genetic diversity. This has led a greater number of botanic gardens to define new procedures for maintaining their collections. Organisations such as BGCI and the Center for Plant Conservation (St Louis, USA) are publishing guidelines for *ex situ* collection management.

International Agricultural Research Centres

The International Agricultural Research Centres (IARC), supported by the Consultative Group on International Agricultural Research (CGIAR), have been active in the international coordination of activities concerned with plant

Table 34.9 Germplasm holdings of IARCs

IARC	MANDATE	NO. OF ACCESSIONS	GERMPLASM HOLDINGS	
CIAT Centro Internacional de Agricultura Cali,Columbia	*Phaseolus* bean, cassava, rice, tropical pastures	66,000	*Phaseolus vulgaris*	35,950
			other *Phaseous* spp.	5,111
			Manihot esculenta	4,600 *4,000
			Manihot (wild spp.)	32
			forage legumes	17,982
			forage grasses	2,514
CIMMYT Centro Internacional de Mejoramiento de maíz y Trigo Londres, Mexico	wheat maize triticale	70,000	maize	10,500
			wheat	60,000
CIP Centro Internacional de La Papa Lima, Peru	potato sweet potatao	12,000	potato	5,000
			potato (wild spp.)	1,500
			sweet potato	5,200
IBPGR International Board for Plant Genetic Resources Rome, Italy	to further the study, collecting, conservation, documentation, evaluation, and use of the genetic diversity of useful plants for the benefit of people throughout the world.	(189,000)[1]		
ICARDA International Centre for Agricultural Research in Dry Areas Aleppo, Syria	barley, lentil, faba bean, durum wheat, bread wheat, kabuli chickpea	87,000	cereals	49,749
			food legumes	16,890
			forages	19,952
ICRISAT International Crops Research Institute for the Semi-Arid Tropics Hyderabad, India	sorghum, millet, chickpea, pigeonpea, groundnut	96,000	sorghum	31,030
			pearl millet	19,796
			chickpea	15,564
			pigeonpea	11,040
			groundnut	12,160
			finger millet	2,848
			foxtail millet	1,404
			proso millet	831
			little millet	401
			barnyard millet	582
			kodo millet	544
IITA International Institute of Tropical Agriculture Ibaden, Nigeria	cassava, maize, plantain, cowpea, soybean, rice, yam,	36,000	sweet potato	*1,000
			plantain	*250
			cassava	*2,000
			yams	*1,000
			Musa spp	*200
			cowpeas	15,100
			rice	12,000
			Bambara groundnut	2,000
			soybean	1,500
			wild *Vigna*	810
ILCA International Livestock Centre for Africa Addis Ababa, Ethiopia	Livestock production systems in sub-Saharan Africa	9,000	grasses	1,524
			legumes	6,443
			browse species	1,429

Table 34.9 Germplasm holdings of IARCs (continued)

IARC	MANDATE	NO. OF ACCESSIONS	GERMPLASM HOLDINGS	
IRRI International Rice Research Institute Manila, Phillipines	rice	83,000	*Oryza sativa* (Asian rice)	78,420
			O.glaberrima (African rice)	2,408
			wild species and species hydrids	2,214
			genetic testers and mutants	208
			taxa in genera related to *oryza*	21
WARDA West African Rice Development Asociation Côte d'Ivoire)	rice	6,000	rice	5,600

Notes: * (in vitro). The 'germplasm holdings' column gives an approximate taxonomic breakdown of the 'number of accessions' column; additative differences will be because of rounding of figures and different data sources. Total holdings worldwide are estimated at 2.6 million, or, allowing for duplication, 1.3 million unique samples. [1] Number of samples collected by IBPGR or with IBPGR support 1974-1989. Three additional IARCs have no germplasm collections: IFPRI (International Food Policy Research Institute) Washington, DC, USA; ILRAD (International Laboratory for Research on Animal Diseases) Nairobi, Kenya; ISNAR (International Service for National Agriculture Research) The Hague, Netherlands.
Source: Germplasm data from Anon. nd. *Partners in conservation: plant genetic resources and the CGIAR system*. CGIAR/IBPGR. IBPGR sample estimate and accession numbers from van Sloten, D.H. 1990. IBPGR and the challenges of the 1990s: a personal point of view. *Diversity* 6(2):36-39.

Table 34.10 Status of crop germplasm collections

	TOTAL OF ACCESSIONS	
	CULTIVATED	WILD
MAJOR CEREALS		
Triticum spp.	567,190	11,986
Aegilops spp.	--	14,937
Oryza sativa	242,599	90,814
Zea mays	208,227	*53
Hordeum spp.	285,936	20,410
Sorghum bicolor	94,964	36,765
MILLETS		
Pennisetum glaucum	28,609	3,366
Setaria italica	17,234	--
Eleusine coracana	12,363	352
Paspalum scrobiculatum	1,515	--
ROOTS AND TUBERS		
Solanum spp.	63,731	3,626
Ipomoea batatas	20,160	1,853
Dioscorea spp.	10,493	78
Manihot esculenta	24,219	--

	TOTAL OF ACCESSIONS	
	CULTIVATED	WILD
FOOD LEGUMES		
Glycine max	129,043	10,342
Arachis hypogaea	62,981	4,769
Phaseolus vulgaris	96,341	16,994
Phaseolus lunatus	12,197	1,168
Phaseolus coccineus	3,543	807
Vigna unguiculata	31,180	52
Vigna radiata	19,103	20
Vigna subterranea	2,110	--
Cicer arietinum	49,176	150
Cajanus cajan	16,463	176
Vicia faba	20,739	117
Lens culinaris	20,252	54
Lupinus mutabilis	6,267	--
Lupinus albus	3,627	**19,368
Lupinus luteus	1,936	--
Psophocarpus tetragonolobus	5,725	408

Source: International Board for Plant Genetic Resources, 1991.
Note: Information as made available to IBPGR conservation database. * Refers to *Teosinte* spp. and *Tripsacum* spp. ** Refers to *Lupinus* spp. other than *L. mutabilis*, *L. albus* and *L. luteus*.

resources, particularly gene banks. The CGIAR was founded in 1971 and consists of a consortium of donor countries, foundations and development banks, sponsored by the World Bank, UNDP and FAO. The establishment of this international network was motivated by international concern over the problems of genetic erosion in cultivated species and the loss of related wild species of flora.

At present there are 13 IARCs supported by the CGIAR. Most of these centres have specific responsibilities in crop varietal development and germplasm conservation. A few of these centres also serve as an international base for specific crops and actively collect on a worldwide basis (see Table 34.9). The collection efforts of the CGIAR network were initially focused on crop plants and were based on the economic importance of the crop, the quality of existing collections and the degree of threat to the crop.

Perhaps the most important of these IARCs is the International Board for Plant Genetic Resources (IBPGR) in Rome, Italy. Established in 1974, the IBPGR does not store any germplasm itself but has a coordinating role in setting priorities and creating a network of national programmes and regional centres for the conservation of plant germplasm. It has provided training facilities, supported research into techniques of plant germplasm conservation, sponsored numerous collection missions and provided small amounts of financial assistance for conservation facilities in the developing world.

IBPGR has achieved many of its original objectives with regard to collection of germplasm of many of the major crops of the world. With IBPGR assistance the 13 IARCs and 227 seed banks in 99 countries now hold 90% or more of the known land races of such crops as wheat, corn, oats and potatoes. The IARCs have an estimated 465,000 accessions in storage, amounting to 35% of unduplicated world holdings (Van Sloten, 1990) (See Table 34.9). Data on accessions of cultivated and wild crops, made available to the IBPGR conservation database by national and other centres, are presented in Table 34.10. These figures, which do not claim to be comprehensive, in general show that collections have grown significantly (by 190% and 20% in the case of *Zea mays* and *Oryza sativa*) over the past decade.

The network has also successfully encouraged many national programmes and assisted in many scientific and educational programmes so that now IBPGR has links with over 500 institutes in some 106 countries.

The CGAIR network has recently been subject to controversy. Critics maintain that the organisation is guided too firmly by the industrial interests of the developed world (this controversty is not discussed here, but is well reviewed by Kloppenburg, 1988). FAO has recently renewed its efforts in this area of conservation, due in large part to the controversy surrounding the activities of the IBPGR and the international network it coordinates. The result of this renewed effort by FAO was the formation of a new commission called the Commission on Plant Genetic Resource (CPGR) and the drafting of a legal instrument known as the International Undertaking on Plant Genetic Resources (the Undertaking).

The Undertaking called for an international germplasm network to be established under the auspices of FAO. It lays out the duty of each nation to make all plant genetic material freely available and calls for the development of a procedure under which a germplasm conservation centre could be established by the FAO. It further provides that the IBPGR was to continue in its role of coordination but that it would do so under the supervision of FAO.

The CPRG meets every two years to review progress in germplasm conservation. The commission held its first meeting in Rome in 1985, where much of the discussion focused on concerns with the Undertaking and *in situ* preservation, which to a large extent had been ignored by the IBPGR.

The controversy which the CPGR and the Undertaking were established to resolve has bedevilled the meetings of the CPGR. As a result this initiative has been able to achieve very little so far. Consequently the IBPGR has remained fairly autonomous and continues to be the main body coordinating at an international level conservation efforts in this area.

TECHNIQUES FOR *EX SITU* PLANT CONSERVATION

Maintenance of *ex situ* populations of plants carried out by a variety of institutions, including botanical gardens, forestry institutes and agricultural research centres, involves three important techniques which will be outlined here. These are field gene banks, seed banks, and *in vitro* storage methods.

Field Gene Banks

A field gene bank is an area of land in which collections of growing plants have been assembled including as many individuals of one species as possible in order to maintain the widest practicable range of genetic diversity. This ensures that plant material is conserved and available for breeding, reintroduction, research and other purposes. Field banks are particularly appropriate for long-lived perennial trees and shrubs which cannot be adequately conserved in the wild and which may take decades to produce seeds; they thus have particular importance in forestry.

In the agricultural sector, field gene banks have been mainly established to provide germplasm for tropical crops, often trees, such as cocoa, rubber, coconut, mango, cassava and yam. The IBPGR has designated 23 field gene banks for 9 crops, at either a global or regional level. Field gene banks also contain wild relatives of economically important species as well as semi-domesticated minor crops and a number of unimproved wild plants of economic importance.

For example, the National Genetic Resources Center (CENARGEN), Brazil, is not only the designated field gene bank for *Citrus* and *Arachis* in Latin America, but also has tree crops, forest trees, some vegetables and forage plants which have recalcitrant seeds. At CENARGEN, five plants per accession are maintained of clonal material and 50-100 seedlings of wild species are planted.

Temperate trees important for commercial forestry are maintained in field gene banks by many national forestry institutes and departments. These generally act as seed orchards and for the assessment of the most suitable genotypes for large-scale production and planting.

Many important wild tropical timber species are maintained by tropical forestry research institutes. For example the Arboretum de Sibang, Libreville, Gabon maintains a collection of 40 tree taxa as a mature collection laid out in blocks. The Forestry Research Institute, Kepong, Malaysia maintains 722 taxa of woody species, mainly from Southeast Asia and especially of the commercially important dipterocarps.

Botanical gardens often have collections which are effectively field gene banks, in that they contain significant numbers of individuals of the same species, representing a considerable proportion of the known wild diversity, maintained for conservation purposes. Examples are the native palm collection of the Jardín Botánico Nacional de Cuba, the Universiti Kebangsaan Fernarium, Malaysia, which has a collection of 150 out of 650 native fern and the Lancetilla Botanic Garden and Experimental Station,

Honduras which maintains an extensive fruit tree collection, especially of Asiatic species (mangosteen, *Citrus*, mango and rambutan), as well as 100ha of *Swietenia macrophylla* (mahogany) and probably the best collection of coffee (*Coffea*) cultivars in Central America.

The national collections of US endangered native plants maintained by more than 20 botanic garden affiliates of the Center for Plant Conservation (CPC), St Louis, have minimum requirements for the number of individuals and populations of a species to include and guidelines for maximising their genetic diversity in cultivation. Over 372 species are maintained as part of the CPC national collection.

A number of well documented natural areas of varying sizes managed by many institutions, especially botanic gardens and forest research institutes, function as *de facto* field gene banks, combining *ex situ* and *in situ* approaches, often through enrichment planting, reintroductions and the genetic management of indigenous plant stocks in the reserves.

There are some deficiencies with field gene banks: they often take up a great deal of space; the collections are generally difficult to protect from natural disasters such as bushfires; they are susceptible to the spread of disease and may suffer from neglect during periods of institutional weakness. Nevertheless, for many species and in many situations they are the only available option for the conservation of important germplasm.

Seed banks

Seed banks are the most efficient and effective method of *ex situ* conservation for sexually reproducing plants whose seeds are suitable for long-term storage (termed orthodox seeds). Seeds are small and therefore take up little space, and with a few exceptions, every seed has a different genetic constitution, so samples include a wide range of genetic variability. At a practical level, a seed bank is dependent on secure power supplies, the need for careful monitoring, and testing of seed viability and the time-consuming regeneration if the viability falls below a certain pre-determined level and a new seed collection cannot be made.

However it is estimated that 50,000 plant species (20% of the world's total) produce seeds that do not survive low temperatures and/or dehydration. For example, many tropical species have seeds which possess no natural dormancy and die quickly if not allowed to germinate immediately. These are termed recalcitrant seeds. Species with recalcitrant seeds and those which do not readily produce seeds need to be maintained *ex situ* as growing plants in field gene banks or as living collections.

Seeds of orthodox types can be conserved for very long periods at sub-zero temperatures if previously dried to about 5-8% moisture content. Although longevity varies from taxon to taxon, seed viability in medium-term storage (0-5°C) can be 5-25 years, whereas long-term storage (-10°C to -20°C) gives viability of the seeds of perhaps a hundred years.

There are many seed banks for wild plants in botanic gardens (Table 34.11) - 528 of a total of 1,545 botanic gardens surveyed between 1985 and 1990 have developed a facility for seed storage and handling, with at least 144 of them known to have low-temperature seed storage facilities. An analysis of some selected seed bank accessions shows the extent to which seeds of wild plants are included in some of the world's non-crop plant seed banks.

Some seed banks specialise in a specific geographical area or taxonomic group. These are sometimes coordinated to make the best use of resources, as for example in Spain where three leading botanic institutions work in close collaboration. The Proyecto 'Artemis' is a seed bank of endemic taxa from the Iberian Peninsula and Macronesia held at the Dep. de Biologia Vegetal, Universidad Politécnica de Madrid, Spain. It has 1,000 of the 1,300 endemic Spanish taxa in its collections, with 1,500 accessions of known wild origin. The Jardín Botánico de Cordoba, Spain concentrates on the Andalucian flora of which 300 taxa are threatened as well as 439 endemic Iberian taxa (125 threatened) with a total of 1,498 Spanish accessions. The Jardín Botánico 'Viera y Clavijo', Gran Canaria has in its seed bank 350 of the 500 endemic species, most of which are threatened.

A good example of the organisation and coordination of a local seed bank is the programme of the Centre for Plant Conservation in the USA. This is a national network of 25 botanic gardens which together possess nearly 3,000 rare, threatened and endangered American native species (10% of the total American flora) as a cooperative and on a centrally managed basis. A back-up of stored seed for plants included in their programme is housed at the western regional station of NPGS and at NSSL, Fort Collins, Colorado.

An important example of the development of a seed bank of wild-collected, wild species of a crop relative is the collection of crucifers at the Instituto Nacional de Investigaciones Agrarias, Madrid, Spain (INIA). Over 80% of the accessions are collected directly from the wild or in some cases with intermediate multiplication at INIA. This seed bank was started to conserve the wild genotypes of *Brassica* and its allies.

A number of seed banks that specialise in forestry tree species, especially ones of actual or potential economic importance, are maintained by Forest Research Institutes and Forestry Departments in various countries but there is no comprehensive directory available of them or their collections. They vary in size from regional in scope to international.

The IBPGR and the Crop Genetic Resource Centres have developed about 60 gene banks in the last 20 years with long- or medium-term storage facilities of crop plants. Only recently, however, have they included wild material and then only of crop relatives. Wild species typically account for less than 2% of gene bank accessions. Currently only wild relatives of wheat (60 spp. or 75-80% of the total),

Table 34.11 Botanic garden seed banks

COUNTRY	1	2	3
ASIA			
Bangladesh	1		
China	35	7	7
Hong Kong	1		
India	12	2	4
Indonesia	1		
Iran	1		1
Israel	2	2	3
Japan	7	9	8
Korea, Rep	1		2
Malaysia	2	2	1
Mongolia	1		
Myanmar	2		
Pakistan	3		
Philippines	2		1
Saudi Arabia	2		
Sri Lanka	3		
Taiwan	2		1
Thailand	4		
Turkey	6		1
Viet Nam	1		
USSR (former)			
USSR	44	2	5
EUROPE			
Austria	1	2	6
Belgium	6	1	
Bulgaria	9	1	2
Czechoslovakia	18	2	4
Denmark	2	1	3
Finland	1	3	2
France	25	11	16
Germany	23	10	20
Greece	2		
Hungary	7	3	6
Ireland	2	1	3
Italy	16	3	14
Monaco	1		1
Netherlands	13	3	2
Norway	1	3	3
Poland	8	4	10
Portugal	6	1	3
Romania	3		3
Spain	10	2	6
Sweden	9	6	5
Switzerland	19	4	9
UK	5	10	10
Yugoslavia	12	1	3
NORTH AND CENTRAL AMERICA			
Barbados	1		
Belize	1		
Canada	7	4	5
Costa Rica	2		
Cuba	9		
Dominican Republic	1		
Grenada	1		
Honduras	1		
Martinique	1		
Mexico	18	2	2
Panama	1		
Puerto Rico	0	1	
St Vincent	1		
USA	53	28	22
SOUTH AMERICA			
Argentina	4		2
Bolivia	2		
Brazil	7	2	1
Chile	2	1	
Colombia	9		1
French Guiana	1		
Guyana	1		
Paraguay	1		
Peru	3		
Venezuela	3		
OCEANIA			
Australia	30	6	9
New Zealand	2		3
AFRICA			
Algeria	3		
Angola	1		
Benin	1		
Côte d'Ivoire	1		
Egypt	1		
Ghana	1		1
Kenya	1		
Libya	1		
Nigeria	1		
Réunion	1		
Senegal	1	2	
South Africa	12	2	3
Sudan	1		
Tanzania	2		1
Togo	1		
Tunisia	1		1
Uganda	2		
Zimbabwe	2		2
TOTAL	528	144	220

Source: Heywood, C.A., Heywood, V.H. and Wyse Jackson, P. 1990. *International Directory of Botanical Gardens V*, 5th edn. Koeltz Scientific Books on behalf of WWF, Botanic Gardens Conservation Secretariat and International Association of Botanical Gardens.
Note: 1 Botanic gardens that report having a seed bank. 2 Botanic gardens that report having a low temperature seed storage facility. 3 Botanic gardens that report having wild origin seeds available for distribution. Total number of botanic gardens surveyed: 1,545. Survey undertaken from 1985 to 1990.

potato (40 spp. or 70% of the total), tomato (10 spp. or 90% of the total) and to a limited extent, maize (15 spp. or 50% of the total) have been extensively collected and preserved in seed banks.

Over 200 botanic gardens distribute seed from plants of wild origin as part of their *Index Seminum*. This provides a mechanism for the distribution of seeds which is more valuable than that of unknown or garden origin, as it is most likely to have more genetic diversity and is of known origin. In another recent analysis of Seed Lists from botanic gardens 432 seed lists out of 600 (from 25 countries) included seed collected from the wild.

In vitro Storage

Another important form of preservation of wild flora which is carried out by many different types of institution is *in vitro* storage. *In vitro* (literally "in glass") storage of germplasm refers to the conservation of plants in laboratory conditions. For germplasm storage, *in vitro* plants are usually initiated from meristem tips, buds or stem tips and propagated through division in test tubes. *In vitro* methods are particularly suited to the long-term storage of propagules of species with recalcitrant seeds which cannot otherwise be maintained in a seed bank.

The plantings can be stored under various conditions but in general at low temperatures (-3°C to -12°C) to create a slow-growth situation and thus increase the storage period. *In vitro* storage is expensive and labour-intensive, as subculturing is necessary after a certain period (six months to two years, depending on the species). Theoretically, cultures can be stored indefinitely using cryogenic techniques which would reduce labour requirements. However, in practice only a small number of species have yet been successfully preserved in this way, such as *Malus domestica, Ribes* sp., *Rubus idaeus, Vaccinium corymboaum* and *Pyrus communis*. More research is needed before extensive cryobanks of wild material are established but it is a very important development for the long-term storage of species which are vegetatively propagated and those with recalcitrant seeds.

Table 34.12 gives the current estimates of wild material in tissue culture storage. The units include botanic gardens (29 units) which are the most important accounting for approximately 1,500 taxa *in vitro* storage, universities (12 units) and crop research centres and private laboratories (9 units). Around 500 taxa stored *in vitro* worldwide are considered threatened.

In vitro methods suffer the same disadvantages as seed banks in terms of the need for equipment and trained staff but techniques can be developed for local use in cooperation with units in the developed world.

Table 34.12 Wild plant material in tissue culture storage

COUNTRY	NUMBER OF UNITS	NUMBER OF TAXA
China	1 *	numerous
Hong Kong	1	numerous
India	2	few
Israel	1	few
Singapore	1	few
USSR	1	few
Belgium	2	160
Denmark	2	few
France	3	few
Germany	2	few
Poland	3	few
Spain	4 *	50
Sweden	3	few
UK	5	1,000
Canada	1	few
Costa Rica	1 *	few
Cuba	1	numerous
Mexico	1	few
USA	5 *	numerous
Brazil	3 *	numerous
Colombia	1 *	few
Peru	1 *	few
Australia	4 *	200 +
South Africa	1	few
TOTAL	50	

Source: M.F. Fay. 1991, pers. comm., Royal Botanic Gardens, Kew, UK.
Note: Includes wild species in crop germplasm collections.
* Cryopreservation capability in one unit.

Plant reintroductions

IUCN (1987) defines reintroduction as the "intentional movement of an organism into part of its native range from which it is has disappeared or become extirpated as a result of human activities or natural catastrophe". The intention is the establishment of a self-maintaining, viable population existing under the pressures of natural selection. The ultimate measure of success must be the reproduction and subsequent regeneration of the population. Reintroduction forms one strategy aimed at the conservation of a single species within the general umbrella of restoration that operates at the habitat or community level.

Plant reintroductions are a high risk strategy, indications of their long-term success are still uncertain. The intermittent nature of plant regeneration and the ability of individuals to survive long periods through vegetative or clonal growth means that for woody perennials it may be many years before regeneration is recorded.

Listed below are a number of plant reintroductions illustrating the range of plants and projects undertaken:

- *Stephanomeria malheurensis* (Compositae) Extinction in the wild in 1986 due to habitat changes resulting from alien weed invasion and associated change in fire regime at its only known site in Oregon, USA. Seed had been held for research, allowing subsequent reintroduction into original site (Parenti and Guerrant, 1990).
- *Pediocactus knowltonii* (Cactaceae) A vulnerable endemic restricted to a single site in New Mexico. A joint project between the US Fish and Wildlife Service and the State of New Mexico resulted in a second population of 150 individuals being established in 1985 using cuttings collected from the original population. Seed introduction has not proved successful to date, nor has regeneration been recorded.
- *Sophora fernandeziana* (Leguminosae) A tree native to the Chilean islands of Juan Fernandez. The Chilean Conservation and Forestry Organisation (CONAF) has planted this not only to bolster low numbers of this plant, but also because of its ecological function as a keystone resource for the endemic hummingbird *Sephanoides fernandensis*.
- *Ruizia cordata* (Sterculiaceae) A highly endangered shrub endemic to the Indian Ocean island of Reunion, in April 1989 several hundred young specimens propagated at the Brest Botanic Garden, France, were planted out on the cliffs of the Ravine de la Grande Chaloupe, Reunion. (Lesouef, 1991).

- *Gentiana nivalis* (Gentianaceae) A circumpolar plant with a restricted and diminishing distribution in Scotland. Vulnerable to grazing, trampling and possibly climate change. In 1980 seed was introduced to a site near to a visitor centre in the hope of establishing a new population that would divert damaging public attention from the original population (Whitten, 1990).
- *Trochetiopsis melanoxylon* (Sterculiaceae). Once thought extinct the world population of this St Helenan endemic is derived from only two individuals. A propagation programme on the island has resulted in several thousand plants being planted on the island (Drucker *et al.*, 1991).

A provisional survey of plant reintroductions indicates over 210 projects undertaken in over 22 countries involving 29 plant families, between 1980 and 1990. This is probably an underestimate since plant reintroductions have traditionally been poorly recorded and documented. The short post-reintroduction time for these projects and the poor level of documentation prevent an assessment of the degree of success of these projects. A review of Californian transplantation projects indicates a general trend: of the 15 projects reviewed 10 were unsuccessful, due to various combinations of poor horticultural practise, poor ecological understanding, lack of post planting maintenance and monitoring (Hall, 1987).

Projects have been undertaken in a wide variety of habitats, mostly in developed regions, with centres of activity in the USA, western Europe, South Africa and Australia. Most of the experience has been gained in temperate or Mediterranean regions where a flora, rich in endemics, coincides with an effective conservation infrastructure.

Botanic gardens and related institutes are holding an increasing number of species critically threatened or extinct in the wild, but the number of potential or required reintroductions far exceeds the ability to undertake such logistically demanding exercises. Because reintroductions are long-term projects requiring extensive monitoring and close collaboration with other agencies, they are best done by an institute with easy access to the planting site. The genetic viability of reintroductions originating from botanic garden collections should be questioned as the demographic management of cultivated plant stocks and international co-ordination of plant conservation projects are in their earliest stages. The material used for reintroduction comes from a variety of sources: a species may only exist in scattered botanic garden collections (e.g. *Sophora toromiro* from Easter Island); material may be salvaged from the existing wild populations prior to destruction of the habitat (e.g. *Penstemon barretiae* from the site of a hydro-electric project in Oregon, USA); dormant propagules may be sampled from the soil seed bank (e.g. *Iliamna corei* in Virginia, USA).

It is on the oceanic islands that reintroductions can play an important and immediate role. This has already been demonstrated on St. Helena and the Canary Islands; on the former island over 8000 plants of 14 species have been replanted (Drucker *et al.*, 1991). The island of Mauritius illustrates the scale of potential reintroductions; Mauritius has *c.* 112 threatened taxa with either less than 20 wild individuals or found in 1 or 2 localities only (Owadally *et al.*, 1991).

Plant reintroduction should not be regarded as a substitute for habitat protection. In contrast it offers a technique that can be used to upgrade the value of retained and protected habitats. Retained areas for conservation are influenced by increasing isolation and degradation. Accordingly reintroduction and associated restoration programmes are becoming accepted tools in an increasingly sophisticated conservation regime. In the tropical nations with much larger biological diversity and relatively poorly researched ecology, restoration and reintroduction will play an important future role; the work at Guanacaste, Costa Rica, and Mineracao Rio Norte bauxite mine, Brazil, are illustrating the potential of this work.

The scarcity of reports on past reintroductions and the need to record and co-ordinate projects has initiated the formation of the Reintroductions Specialist Group of the Species Survival Commission, the Plants Group will collate data on such projects and issue guidelines on procedure.

IN SITU CONSERVATION OF ANIMALS

Protected areas

Although it is widely accepted that protected areas are the single most important element in the preservation of animal species, relatively little work has been carried out to determine how effective protected area networks are in maintaining populations of species, either in particular taxonomic groups or in particular geographical areas. What work has been done, however, indicates that in many cases a surprisingly high percentage of species are represented in at least one protected area.

A recent study in southern Africa (Siegfried, 1989) found that 92% of amphibian, 92% of reptilian, 97% of avian and 93% of mammalian species native to the region were represented by breeding populations in protected areas (Table 34.13), despite the fact that such areas covered only 6% of the land area of the region. Moreover, over 50% of animal species were represented in more than ten reserves. Similarly Round (1985) found that in Thailand 508 out of 578 (88%) of the native bird species were recorded from protected areas. Of those that were not represented, 27 were mainly open country species unlikely to be adversely affected by habitat loss and a further ten were believed likely to occur in protected areas.

A more cursory survey of 12 African countries (Sayer and Stuart, 1988) found that in 11 of these, at least 75%, and generally well over 80%, of native bird species were present in protected areas; the exception was Somalia, where only 47% were present (Table 34.14).

Extrapolation from figures such as these indicates that in those parts of the world which have established protected area networks, the great majority of terrestrial species are likely to occur in at least one, even though such areas generally account for only a small proportion of the total land area. Sayer and Stuart (1988) noted that just under 10% of the remaining tropical moist forest in Africa was included in national parks and equivalent reserves, and considered it probable that up to 90% of tropical forest vertebrates on that continent would be maintained if these and a few additional critical sites were adequately protected.

Table 34.13 Breeding animal species in protected areas in Southern Africa

BIOME	PERCENT NUMBER OF SPECIES			
	AMPHIBIANS	REPTILES	BIRDS	MAMMALS
Fynbos	88.0 (22)	90.6 (77)	98.8 (259)	98.6 (73)
Forest	100.0 (13)	100.0 (21)	99.0 (310)	100.0 (37)
Nama-karoo	91.7 (11)	96.2 (75)	99.2 (250)	95.6 (66)
Succulent-karoo	72.7 (8)	75.8 (69)	98.6 (219)	93.8 (61)
Grassland	100.0 (33)	96.2 (100)	99.8 (416)	100.0 (94)
Moist savanna	100.0 (57)	94.1 (159)	98.1 (530)	96.7 (148)
Arid savanna	100.0 (52)	96.6 (171)	98.8 (513)	98.8 (169)

Source: Siegfried, W.R. 1989. Preservation of species in southern African nature reserves. In: Huntley, B.J. (Ed), *Biotic Diversity In Southern Africa*. Oxford University Press, Cape Town.
Note: Absolute numbers given in parentheses.

Table 34.14 Birds in protected areas in Africa

COUNTRY	NUMBER OF SPECIES IN PROTECTED AREAS	PERCENT OF BIRD FAUNA
Cameroon	649	76.5
Côte d'Ivoire	568	83.2
Ghana	558	77.4
Kenya	908	85.3
Malawi	485	77.7
Nigeria	719	86.5
Somalia	302	47.3
Tanzania	833	82.0
Uganda	880	89.0
Zaire	967	89.0
Zambia	637	87.5
Zimbabwe	581	91.5

Source: Sayer, J.A. and Stuart, S. 1988. Biological diversity and tropical forests. *Environmental Conservation* 15.

However, in some parts of the world, including some areas of very high diversity, the protected area network is manifestly inadequate for the protection of a significant proportion of the biota. This applies particularly in Oceania, where in many countries land tenure systems make it very difficult for significant areas of land to be set aside as protected areas. Here, more innovative approaches to land management (such as the Wildlife Management Areas of Papua New Guinea) are required.

Moreover, the occurrence of a species in a protected area is no guarantee of long-term security. Many such areas are protected in name only and subject to continuing pressures of encroachment, habitat degradation and hunting. Even areas which are adequately protected are often too small to maintain viable populations of species which live at low population density or which are nomadic or migratory.

These problems will become increasingly pressing as habitats outside protected areas become more and more altered and degraded, leaving protected areas as 'islands' of natural or semi-natural habitat. Under these circumstances, the areas themselves are likely to need more active management if they are to maintain their ecological integrity and continue to play their role in preventing the extinction of species.

Recovery plans

As noted above there are circumstances in which conventional protected areas are in themselves likely to be inadequate for the maintenance of some animal species. This applies when it is not possible to set aside large enough areas to maintain viable populations of given species or when species occur outside national boundaries (chiefly Antarctic and pelagic organisms).

Under these conditions, recovery plans for individual species may be developed which entail a wide range of actions designed to improve the status of the species concerned. The primary examples are those prepared under the US Endangered Species Act for nationally endangered species. Central to these is the concept of maintenance or restoration of 'critical habitat', deemed to be the minimum area of habitat necessary for the species to survive at an acceptable level (i.e. one at which it is no longer considered threatened). Such habitat does not necessarily have to be within a conventional protected area, and other land-uses may be allowed as long as they do not conflict with the requirements of the species concerned. Frequently, different degrees of protection are imposed on different parts of a species' habitat. Thus in Italy, the area of occurrence of the Brown Bear *Ursus arctos* in the Apennines, centred on the Abruzzo National Park, is divided into a variety of zones: in the core region, no human interference is allowed, the area being effectively a strict nature reserve devoted to the protection of the bear and other species; surrounding this is a region, within the park, where visitors are allowed but human activity is strictly limited. In the area immediately outside the park, which still constitutes important habitat for the bear, agricultural and pastoral activities are allowed but the bears are still strictly protected. Here compensation for damage to livestock and crops caused by the bears is paid in order to discourage (illegal) persecution which would otherwise be very difficult to control. In this way, populations of species can be maintained centred on protected areas which would otherwise be too small to sustain them.

As populations of individual species become smaller and more fragmented, active management is increasingly invoked in their conservation. This may involve translocation of individuals from remnant, isolated populations perceived to have no chance of long-term survival, to larger areas of suitable habitat or to supplement existing populations. As a last resort or as a precautionary measure, it may also entail taking animals into captivity for the purpose of captive breeding.

EX SITU CONSERVATION OF ANIMALS

The principal institutions holding *ex situ* populations of animal species are zoos and aquaria. At least 83 countries possess one or more zoos or aquaria (Table 34.15) but the overall geographic distribution is very uneven: 573 (or 65%) of zoos and aquaria are located in the developed world, Europe (298), the USA (160), Canadà (24), Australia (17), New Zealand (8) and Japan (66). These are mainly areas of low species richness. In contrast, those tropical regions with generally high species diversity and large numbers of threatened species have few or no zoos or aquaria i.e. Africa (32), South America (29), Central America (16), and Asia excluding China and Japan (55). Those institutions that do exist are mainly poorly developed and under-funded.

This divide is further reflected in the sizes of the zoological collections. Institutions which reported their specimen numbers to the International Zoo Yearbook (IZY) collectively held approximately 1,232,000 vertebrate specimens as of 31 December 1989 (Olney and Ellis, 1991). Nearly half (584,000) of these were fish. The numbers of other taxonomic groupings held were mammals (202,000 specimens), birds (351,000), reptiles (74,000) and amphibians (21,000). The developed countries together held 67% of mammal specimens, 57% of birds, 69% of reptiles, 81% of amphibians, and 76% of fish: in total 68% of all vertebrate specimens held.

Captive breeding - successes and shortcomings

Although most zoos have their origins as menageries for the entertainment, and to some extent education, of the public, they are increasingly turning their attention to conservation. It is argued that captive populations can play a significant role as demographic and genetic reservoirs from which infusions of 'new blood' may be obtained or new populations founded, and as last redoubts for species which have no immediate chance of survival in the wild.

Zoos undoubtedly have considerable capacity in this regard, but to date efforts have been relatively limited and the vast majority of captive specimens in the world's zoos have little importance for the conservation of species or even in maintaining genetic diversity amongst non-threatened species.

For example, although 629 mammalian species are considered to be wholly or partly threatened on a global scale (IUCN, 1990) only 20,628 specimens from 140 threatened species (Table 34.16) are held in zoos according to the 1991 Census of Rare and Threatened Mammals in Captivity (Olney and Ellis, 1991). This figure is probably an underestimate since some collections do not respond to the IZY's questionnaire. In other words, although some 15% of the world's mammal species are considered wholly or partly threatened, only some 22% of these are represented in captivity and only 10% of the global zoological capacity of around 200,000 mammal specimens consists of threatened mammal taxa. Moreover, of those threatened taxa, zoos included in the census make a significant contribution to the conservation of no more than 20 full species and perhaps a similar number of subspecies.

The situation is similar (or worse) for other taxonomic groups and is even less encouraging from the genetic perspective. Lande and Barrowclough (1987) suggest that in order to safeguard in the long term against the negative genetic effects of inbreeding, a minimum viable population of 500 individuals should be maintained. Only nine threatened mammalian taxa have captive populations exceeding 500 specimens, and only a further 14 have captive populations exceeding 250 (Table 34.16).

A criticism which is often made of captive breeding programmes is that they are a misallocation of resources. The basis for this criticism is that large amounts of money are spent on captive breeding efforts in comparison with that available for *in situ* conservation despite the fact that captive breeding is much less cost-effective than preservation *in situ*. For example, Leader-Williams (1990) calculates that the cost of keeping African elephants and black rhinos in zoos is 50 times that of protecting equivalent numbers in the wild in Zambian National Parks, where $1km^2$ of park can be adequately patrolled for the annual sum of only US$400. In addition, the maintenance of captive populations does not have the associated benefits of protecting an organism's habitat, and by logical extension thousands of other species.

Zoos are sensitive to these criticisms, and a significant number are attempting to improve their efforts in the conservation of threatened species, chiefly through improved international cooperation and clearer setting of priorities for breeding threatened species, as well as by devoting larger fractions of their budgets to field conservation. Coordination in these efforts is carried out through a variety of interconnected mechanisms, including studbooks, the IUCN/SSC Captive Breeding Specialist Group (CBSG), the International Species Inventory System (ISIS) and a number of regional cooperative captive breeding programmes.

Studbooks

In order to facilitate the success of captive-breeding programmes (e.g. to help prevent inbreeding) and to aid in the development of successful management techniques a series of studbooks have been developed. A studbook is an international register which lists all captive individuals of a taxon of conservation concern. Official studbooks are those recognised by the Species Survival Commission (SSC) of IUCN and the International Union of Directors of Zoological Gardens. They are coordinated through the International Zoo Yearbook and the CBSG. As of August 1991 there were 104 recognised International Studbooks and five International Registers (1 amphibian, 4 reptiles, 19 birds and 85 mammals, see Table 34.17), with a further three studbook applications awaiting endorsement (Olney 1991). In principle, studbooks are published every three years, and regular updates are available.

Table 34.15 Number of vertebrates held in zoos and aquaria

	NUMBER OF ZOOS AND AQUARIA	NUMBER OF VERTEBRATES HELD Mammals	Birds	Reptiles	Amphibians	Fishes	TOTAL
WORLD	**878**	**201706**	**351484**	**74416**	**20788**	**583832**	**1232226**
ASIA	**252**	**46646**	**110121**	**15957**	**3380**	**281878**	**457982**
Bahrain	1	575	600	16	18	60	1269
Brunei	1	0	0	7	0	492	499
China	131	12489	46175	1348	1121	70406	131539
Hong Kong	2	236	3002	220	0	2680	6138
India	17	5045	10605	5148	0	2607	23405
Indonesia	4	1847	3683	717	58	1808	8113
Israel	4	1721	1977	964	90	0	4752
Japan	66	12554	17653	4503	2014	195530	232254
Korea, Rep	4	1903	6070	130	0	261	8364
Kuwait	1	208	493	34	0	0	735
Malaysia	2	663	944	302	0	2336	4245
Myanmar	1	968	599	492	0	0	2059
Pakistan	2	–	–	–	–	–	–
Philippines	1	–	–	–	–	–	–
Qatar	2	900	401	71	4	0	1376
Saudi Arabia	1	700	18	0	0	0	718
Singapore	3	689	4662	279	18	1600	7248
Sri Lanka	1	741	1282	510	0	1946	4479
Taiwan	1	1118	1221	113	0	0	2452
Thailand	4	2295	7692	475	0	387	10849
Turkey	1	139	565	15	0	649	1368
United Arab Emirates	2	1855	2479	613	57	1116	6120
USSR (former)							
	40	16981	22557	3544	1714	30735	75531
EUROPE	**298**	**72558**	**115933**	**21936**	**8949**	**129463**	**348839**
Austria	5	1065	1158	603	204	2108	5138
Belgium	5	2409	2280	421	105	5531	10746
Bulgaria	2	274	490	67	1429	0	2260
Czechoslovakia	15	4311	4875	701	30	4437	14354
Denmark	7	1770	1593	424	278	4037	8102
Finland	3	727	602	36	6	16	1387
France	34	6520	11226	1538	34	428	19746
Germany	56	21850	35163	6667	2482	31112	97274
Hungary	6	1409	1645	214	9	2084	5361
Ireland	2	747	770	30	1	0	1548
Italy	21	2058	3545	578	382	2525	9088
Monaco	1	–	–	–	–	–	–
Netherlands	11	4365	7717	2287	1389	8578	24336
Norway	1	10	22	0	9	1500	1541
Poland	9	3290	3547	1659	325	8891	17712
Portugal	3	593	1400	258	0	3727	5978
Romania	1	297	572	28	0	0	897
Spain	11	2106	7185	888	42	6073	16294
Sweden	12	1775	1573	592	403	6222	10565
Switzerland	15	2625	4520	849	314	5158	13466
United Kingdom	76	14200	25779	4014	1507	19036	64536
Yugoslavia	2	157	271	82	0	18000	18510
NORTH AND CENTRAL AMERICA	**201**	**47713**	**65299**	**23700**	**6078**	**115796**	**258586**
Barbados	1	236	125	111	20	26	518
Belize	1	55	45	72	0	0	172
Bermuda	1	16	198	55	0	951	1220
Canada	24	4546	4511	1632	370	9928	20987
Cuba	4	430	1178	195	0	90	1893
Dominican Republic	1	–	–	–	–	–	–
Jamaica	1	–	–	–	–	–	–
Mexico	5	2637	7605	587	0	167	10996
Netherlands Antilles	1	–	–	–	–	–	–
Puerto Rico	1	135	272	88	0	20	515
Trinidad and Tobago	1	165	403	99	310	851	1828
United States	160	39493	50962	20861	5378	103763	220457
SOUTH AMERICA	**29**	**5403**	**13060**	**4064**	**67**	**9240**	**31834**
Argentina	3	595	1500	329	24	150	2598
Bolivia	1	358	1409	118	24	0	1909
Brazil	10	2368	6473	2054	0	550	11445
Chile	1						
Colombia	2	407	787	324	0	0	1518
Guyana	1	70	93	43	0	0	206
Peru	1	385	346	118	2	0	851
Uruguay	3	596	762	185	9	0	1552
Venezuela	7	624	1690	893	8	8540	11755
OCEANIA	**26**	**6035**	**9883**	**2244**	**210**	**4084**	**22456**
Australia	17	5181	7944	1815	163	3766	18869
New Zealand	8	854	1939	429	47	318	3587
Papua New Guinea	1	–	–	–	–	–	–
AFRICA	**32**	**6370**	**14631**	**2971**	**390**	**12636**	**36998**
Egypt	2	256	2317	36	11	0	2620
Ghana	1	52	76	55	0	0	183
Kenya	4	1562	187	14	0	0	1763
Libya	1	555	613	172	0	0	1340
Madagascar	2	0	0	43	0	0	43

Table 34.15 Number of vertebrates held in zoos and aquaria (contined)

	NUMBER OF ZOOS AND AQUARIA	NUMBER OF VERTEBRATES HELD Mammals	Birds	Reptiles	Amphibians	Fishes	TOTAL
AFRICA (continued)							
Mauritius	1	–	–	–	–	–	–
Morocco	1	672	1213	28	0	0	1913
Nigeria	3	130	130	118	0	0	378
Senegal	1	–	–	–	–	–	–
South Africa	12	2611	9237	1736	379	12636	26599
Sudan	1	–	–	–	–	–	–
Tunisia	1	332	658	740	0	0	1730
Zaire	1	–	–	–	–	–	–
Zimbabwe	1	200	200	29	0	0	429

Source: Olney, P.J.S. and Ellis, P. (Eds) 1991. *1990 International Zoo Yearbook, Vol.30.* Zoological Society of London, London.
Note: Some institutions did not report specimen numbers.

International Species Inventory System

Since 1974 these studbooks have been supplemented by the International Species Inventory System (ISIS), a global information network designed to support sound genetic and demographic management of zoological collections and enable zoos to meet their increasing conservation responsibilities. ISIS maintains a centralised computer database of census, demographic, genealogical, and laboratory data on wild animals held in captivity. In September 1991 the database contained information on over 141,480 living vertebrate specimens from 4,200 taxa, held in more than 395 zoological institutions in 39 countries, plus an even greater number of their ancestors. ISIS has very good coverage of North American zoos, coverage of Europe and Australasia is rapidly expanding, while participation by institutions in Latin America, Asia and Africa should be increased.

All participating institutions receive a 15,000+ page ISIS Species Distribution Report on microfiche every six months, and copies of annual bound 'Abstracts', one each for mammals, birds, reptiles and amphibians. These include information on sex and age distribution, births, deaths, and important trends for the species as a whole. Although historically zoos were able to draw on populations of wild animals to supply their specimens, this is often no longer feasible because of ethical and ecological considerations. Instead, greater reliance is placed on captive breeding. ISIS data indicate that 92% of new zoo mammals are now captive-bred, along with 71% of birds and a majority of reptiles and amphibians (Anon., 1991a).

By 1990, these various programmes had resulted in over 150 taxa of all classes of vertebrates as well as the invertebrate genus *Partula* (endemic land snails of Moorea, now extinct in the wild) being managed cooperatively by groups which totalled about 400 institutions - approximately half the world's zoos (Flesness and Foose, 1990). Zoos plan to expand the number of species now involved in multi-institution breeding programmes from 150 to 1,000 (Flesness and Foose, 1990).

Animal reintroduction programmes

One area where international networks and coordination of the world's zoos can play a role in *in situ* conservation is through reintroduction programmes. The artificial movement of individual animals between populations is becoming increasingly used as a conservation tool. Griffith *et al.* (1989) reported that over 700 translocations or repatriations occurred each year, mainly in the USA and Canada. These projects are frequently conducted with the support of international captive-breeding programmes at zoological gardens and aquaria, and may generate much public enthusiasm.

Reintroduction projects are not always successful. Griffith *et al.* (1989) examined the outcome of projects involving birds and mammals. Native game species constituted 90% of translocations and had a higher success rate (86% of 118 projects considered) than translocations of threatened, endangered or sensitive species (44% of 80 projects). Dodd and Siegel (1991) found an even lower overall success rate of only 19% for 25 projects involving reptiles and amphibians.

The reasons behind the high failure rates of reintroduction attempts are diverse. Most importantly, not all species lend themselves to reintroduction. In addition a variety of ecological factors can affect the success of reintroduction programmes, including the quality of the habitat in which the release occurs, whether the individuals released are wild or captive bred, and the feeding habits of adults. The design of the reintroduction is also crucial: such factors as the number, sex and social composition of individuals released, whether the release is 'hard' (no food and shelter provided on site) or 'soft', and planning for further releases after populations become established in order to provide injections of new blood, may all be important. In order to succeed reintroduction must therefore be carefully planned and executed, and monitoring should ideally continue for several generations after release.

Case study: reintroduction of Arabian Oryx to Oman

The Arabian (or White) Oryx, *Oryx leucoryx* formerly inhabited arid gravel plains and sandy deserts throughout the Arabian Peninsula and adjacent regions. By the early 1960s the species was confined to two small areas: where the borders of Saudi Arabia, Yemen and Oman meet; and in north-eastern Oman. The last wild oryx were probably killed in 1972 in the Jiddat al-Harasis of Oman (Henderson, 1974) although rumours of sightings persist.

Fortunately, significant numbers of oryx remained in captivity in the Middle East and elsewhere, notably at Phoenix in Arizona, USA, where a herd had been established in the 1960s in response to the continued depletion of the species in the wild. In 1974 the 'White

Table 34.16 Census of IUCN threatened animals held in captivity

	IUCN THREAT CATEGORY	NUMBERS HELD IN CAPTIVITY				NUMBERS OF CAPTIVE ANIMALS BRED IN CAPTIVITY			
		MALE	FEMALE	UNKNOWN	TOTAL	MALE	FEMALE	UNKNOWN	TOTAL
MAMMALS					20770				5500
Zaglossus bruijni	V	2	5	0	7	0	1	0	1
Myrmecobius fasciatus	E	5	6	0	11	2	3	0	5
Macrotis lagotis	E	16	11	0	27	16	11	0	27
Gymnobelideus leadbeateri	V	23	15	3	41	20	13	6	39
Potorous longipes	I	7	4	0	11	6	3	0	9
Bettongia penicillata	E	90	97	6	193	0	0	0	Most
Solenodon paradoxus	E	1	0	0	1	0	0	0	0
Pteropus rodricensis	E	64	70	18	152	0	0	0	Most
Macroderma gigas	V	4	9	0	13	3	6	0	9
Microcebus coquereli	V	31	29	4	64	0	0	0	Most
Lemur coronatus	E	20	17	0	37	16	13	0	29
Lemur m. macaco	V	109	97	4	210	0	0	0	Most
L. m. albifrons	R	96	79	8	183	0	0	0	Most
L. m. collaris	V	19	18	0	37	15	16	0	31
L. m. flavifrons	E	9	4	0	13	6	1	0	7
L. m. fulvus	R	60	65	25	150	0	0	0	Most
L. m. mayottensis	V	49	54	18	121	0	0	0	Most
L. m. rufus	R	33	42	3	78	0	0	0	Most
L. m. sanfordi	V	9	10	0	19	7	8	0	15
Lemur mongoz	E	38	34	0	72	0	0	0	Most
Lemur rubriventer	V	5	5	0	10	2	2	0	4
Hapalemur griseus	K	9	10	0	19	6	5	0	11
Varecia variegata	E	226	250	8	484	0	0	0	Most
Propithecus tattersalli	E	1	2	0	3	0	1	0	1
Propithecus verreauxi	V	6	8	0	14	2	3	0	5
Daubentonia madagascariensis	E	2	1	0	3	0	0	0	0
Tarsius syrichta	E	10	13	2	25	2	5	5	12
Callithrix aurita	E	5	3	1	9	4	3	1	8
Callithrix humeralifer	K	5	5	2	12	3	4	2	9
Callithrix jacchus flaviceps	E	1	0	0	1	0	0	0	0
Saguinus bicolor	E	16	18	9	43	13	8	9	30
Saguinus imperator	I	96	82	8	186	0	0	0	Most
Saguinus o. oedipus	E	561	509	95	1165	0	0	0	Most
Leontopithecus r.rosalia	E	274	285	33	592	0	0	0	Most
L. chrysomelas	E	109	86	29	224	83	64	28	175
L. chrysopygus	E	33	28	8	69	23	20	14	57
Callicebus personatus	E	2	0	0	2	0	0	0	0
Callimico goeldii	R	156	140	21	317	0	0	0	Most
Saimiri oerstedi	E	2	2	0	4	0	1	0	1
Chiropotes albinasus	V	0	2	0	2	0	2	0	2
Chiropotes s. satanas	E	0	2	0	2	0	0	0	0
Cacajao c. calvus	V	1	1	0	2	0	0	0	0
C. c. rubicundus	V	3	5	0	8	2	2	0	4
Cacajao melanocephalus	V	2	1	0	3	0	0	0	0
Alouatta fusca (=guariba)	V	5	0	0	5	0	0	0	0
Ateles belzebuth	V	31	56	5	92	14	20	4	38
Ateles fusciceps	V	29	46	0	75	20	21	0	41
Ateles paniscus	V	81	137	2	220	46	59	1	106
Brachyteles arachnoides	E	1	3	0	4	0	0	0	0
Lagothrix lagothricha	V	49	62	0	111	22	42	0	64
Macaca silenus	E	188	207	7	402	0	0	0	Most
Macaca sylvanus	V	368	483	23	874	0	0	0	0
Cercocebus t. torquatus	V	60	46	0	106	48	30	0	78
C. t. atys	V	73	140	0	213	65	121	0	186
C. t. lunulatus	V	11	17	0	28	7	10	0	17
Mandrillus leucophaeus	E	22	30	0	52	0	0	0	0
Theropithecus gelada	R	40	70	0	110	0	0	0	0
Cercopithecus diana	V	86	106	0	192	63	63	5	131
Cercopithecus hamlyni	V	20	30	2	52	13	20	2	35
Cercopithecus lhoesti	V	6	5	1	12	5	2	1	8
Cercopithecus preussi	E	0	1	0	1	0	0	0	0
Allenopithecus nigroviridis	K	28	27	4	59	8	10	4	22
Pygathrix nemaeus	E	25	29	0	54	14	16	0	30
Nasalis larvatus	V	8	10	0	18	6	7	0	13
Presbytis francoisi	E	19	33	0	52	10	12	0	22
Presbytis geei	R	11	9	17	37	0	2	0	2
Presbytis johni	E	10	10	6	26	6	1	0	7
Hylobates concolor	V	67	66	6	139	30	18	5	53
Hylobates klossi	E	2	2	0	4	0	0	0	0
Hylobates moloch	E	15	12	1	28	6	4	1	11
Hylobates pileatus	E	33	32	0	65	5	2	0	7
Pongo pygmaeus	E	299	379	2	680	205	255	2	462
Pan paniscus	V	30	36	0	66	20	19	0	39
Gorilla g. gorilla	V	278	339	0	617	146	154	0	300
G. g. graueri	E	3	3	0	6	0	2	0	2
Myrmecophaga tridactyla	V	47	51	2	100	18	17	1	36
Romerolagus diazi	E	3	1	0	4	3	1	0	4
Geocapromys browni	R	4	3	0	7	4	3	0	7
Plagiodontia aedium	R	1	2	0	3	0	1	0	1
Chrysocyon brachyurus	V	132	127	2	261	0	0	0	Most
Speothos venaticus	V	42	44	2	88	0	0	0	Most
Cuon alpinus	V	29	28	0	57	18	20	0	38
Lycaon pictus	E	146	123	0	269	0	0	0	0
Tremarctos ornatus	V	58	64	4	126	43	45	4	92
Melursus ursinus	V	50	52	4	106	26	23	1	50
Ailuropoda melanoleuca	E	6	2	1	9	2	0	1	3
Lutra l. longicaudis (+ platensis)	V	7	5	0	12	0	0	0	0
Lutra l. lutra	V	56	99	5	160	33	70	5	108
Pteronura brasiliensis	V	8	6	0	14	0	0	0	0
Cryptoprocta ferox	K	13	8	0	21	11	4	0	15

Table 34.16 Census of IUCN threatened animals held in captivity (continued)

	IUCN THREAT CATEGORY	NUMBERS HELD IN CAPTIVITY				NUMBERS OF CAPTIVE ANIMALS BRED IN CAPTIVITY			
		MALE	FEMALE	UNKNOWN	TOTAL	MALE	FEMALE	UNKNOWN	TOTAL
MAMMALS (continued)									
Hyaena brunnea	V	16	12	0	28	9	6	0	15
Felis margarita scheffeli	E	2	1	0	3	2	1	0	3
Felis marmorata	I	2	4	0	6	1	3	0	4
Felis planiceps	I	1	0	0	1	0	0	0	0
Felis temmincki	I	28	20	2	50	0	0	0	Most
Felis tigrina	V	10	7	0	17	1	1	0	2
Felis wiedi	V	39	41	1	81	12	20	1	33
Panthera leo persica	E	68	80	0	148	0	0	0	Most
Panthera pardus delacouri	T	4	4	0	8	3	4	0	7
P. p. fusca	T	10	14	0	24	9	14	0	23
P. p. japonensis	T	31	29	0	60	0	0	0	Most
P. p. kotiya	T	28	21	0	49	8	8	0	16
P. p. orientalis	T	39	28	1	68	0	0	0	Most
P. p. saxicolor	T	61	68	0	129	0	0	0	Most
Panthera t. tigris	E	72	75	0	147	0	0	0	Most
P. t. corbetti	E	5	6	0	11	1	5	0	6
P. t. altaica	E	325	387	0	712	0	0	0	Most
P. t. amoyensis	E	33	19	0	52	0	0	0	Most
P.t. sumatrae	E	77	88	0	165	0	0	0	Most
Panthera uncia	E	185	182	0	367	0	0	0	Most
Neofelis nebulosa	V	91	75	6	172	68	59	2	129
Acinonyx jubatus	V	250	275	0	525	175	170	0	345
Dugong dugon	V	1	1	0	2	0	0	0	0
Trichechus inunguis	V	1	0	0	1	0	0	0	0
Trichechus manatus	V	12	12	0	24	6	1	0	7
Equus przewalskii	Ex?	397	564	0	961	397	564	0	961
Equus hemionus onager	V	42	70	0	112	0	0	0	Most
E. h. kulan	V	93	191	0	284	0	0	0	Most
E. h. khur	E	4	7	0	11	2	2	0	4
E. h. kiang	V	30	25	0	55	30	25	0	55
Equus africanus	E	9	6	0	15	7	3	0	10
Equus grevyi	E	135	250	0	385	119	188	0	307
Equus zebra hartmannae	V	40	79	0	119	40	79	0	119
Tapirus bairdi	V	15	14	0	29	11	8	0	19
Tapirus indicus	E	58	72	1	131	41	54	1	96
Tapirus pinchaque	V	5	4	0	9	1	2	0	3
Rhinoceros unicornis	E	55	40	14	109	32	21	1	54
Dicerorhinus sumatrensis	E	4	9	0	13	1	0	0	1
Ceratotherium s. simum	E	4	5	0	9	2	3	0	5
Diceros bicornis	E	61	83	0	144	35	45	0	80
Babyrousa babyrussa	V	34	21	3	58	0	0	0	Most
Choeropsis liberiensis	V	81	115	0	196	63	87	0	150
Vicugna vicugna	V	61	62	0	123	0	0	0	Most
Muntiacus feaei	E	10	9	0	19	5	2	0	7
Dama dama mesopotamica	E	11	21	0	32	0	0	0	Most
Axis calamianensis	V	3	3	0	6	3	1	0	4
Axis kuhli	R	14	21	2	37	7	8	2	17
Cervus duvauceli	E	101	194	10	305	0	0	0	Most
Cervus elaphus bactrianus	E	16	20	0	36	0	0	0	Most
Cervus e. eldi	V	22	22	6	50	0	0	0	Most
C. e. siamensis	E	1	2	0	3	0	0	0	0
Cervus nippon taiouanus	E	104	182	32	318	104	182	32	318
Blastocerus dichotomus	V	3	10	0	13	3	5	0	8
Taurotragus derbianus	E	13	16	0	29	10	11	0	21
Bubalus (Anoa) depressicornis	E	23	20	0	43	0	0	0	Most
Bubalus (Anoa) quarlesi	E	5	3	0	8	3	2	0	5
Bos gaurus	V	80	125	0	205	0	0	0	Most
Bos javanicus	V	79	133	1	213	0	0	0	Most
Cephalophus jentinki	E	5	4	0	9	4	3	0	7
Cephalophus zebra	V	3	2	0	5	1	1	0	2
Kobus leche	V	105	236	2	343	0	0	0	Most
Oryx dammah	E	225	424	5	654	0	0	0	Most
Oryx leucoryx	E	209	269	13	491	0	0	0	Most
Addax nasomaculatus	E	179	310	7	496	0	0	0	Most
Damaliscus d. dorcas	V	25	54	0	79	0	0	0	Most
Damaliscus hunteri	V	2	1	0	3	1	1	0	2
Gazella cuvieri	E	33	72	0	105	0	0	0	Most
Gazella dama	E	40	82	0	122	0	0	0	Most
Gazella d. mhorr	E	56	86	0	142	0	0	0	All
Gazella dorcas isabella	V	3	17	1	21	3	16	1	20
G. d. massaesyla	V	2	2	0	4	2	2	0	4
G. d. osiris (incl. neglecta)	V	42	88	0	130	0	0	0	Most
G. d. saudiya	V	18	40	2	60	17	37	2	56
Gazella gazella arabica	V	74	105	18	197	0	0	0	Most
Gazella leptoceros	E	48	71	0	119	0	0	0	Most
Gazella rufifrons	V	0	2	0	2	0	0	0	0
Gazella spekei	V	15	21	0	36	0	0	0	Most
Gazella subgutturosa marica	E	165	185	7	357	0	0	0	Most
Rupicapra rupicapra ornata	V	3	13	5	21	3	11	5	19
Hemitragus hylocrius	V	12	25	0	37	12	25	0	37
Capra falconeri	V	45	60	4	109	0	0	0	Most
C. f. megaceros (incl. jerdoni)	E	7	15	0	22	0	0	0	Most
BIRDS					**16736**				**9487**
Apteryx oweni	V	2	2	1	5	0	1	1	2
Ciconia boyciana	R	10	8	0	18	3	1	0	4
Geronticus eremita	E	145	157	334	636	0	0	0	Most
Cairina scutulata	V	79	64	4	147	0	0	0	Most
Anas aucklandica chorotis	R	18	21	32	71	0	0	0	Most

Table 34.16 Census of IUCN threatened animals held in captivity (continued)

	IUCN THREAT CATEGORY	NUMBERS HELD IN CAPTIVITY				NUMBERS OF CAPTIVE ANIMALS BRED IN CAPTIVITY			
		MALE	FEMALE	UNKNOWN	TOTAL	MALE	FEMALE	UNKNOWN	TOTAL
BIRDS (continued)					**13775**				**8824**
Gymnogyps californianus	E	7	9	0	16	1	2	0	3
Haliaeetus albicilla	R	63	77	45	185	13	20	11	44
Haliaeetus pelagicus	R	12	12	7	31	2	1	0	3
Harpia harpyja	R	21	14	3	38	5	0	0	5
Macrocephalon maleo	V	1	3	2	6	2	0	0	2
Crax mitu	E	30	27	8	65	0	0	0	Most
Crax blumenbachi	E	4	4	0	8	3	3	0	6
Lophura bulweri	R	15	15	0	30	0	0	0	Most
Grus nigricollis	R	2	2	0	4	0	0	0	0
Grus monacha	R	30	47	10	87	16	31	5	52
Grus japonensis	V	113	126	28	267	76	90	14	180
Grus americana	E	23	24	8	55	16	22	8	46
Grus vipio	R	102	107	49	258	67	74	30	171
Grus leucogeranus	R	25	20	0	45	8	7	0	15
Rhynochetos jubatus	E	1	1	0	2	0	0	0	0
Nesoenas mayeri	E	37	32	26	95	0	0	0	Most
Psephotus c. chrysopterygius	R	29	25	0	54	0	0	0	Most
Aratinga guarouba	V	76	66	24	166	32	25	3	60
Rhynchopsitta pachyrhyncha	V	50	50	18	118	17	14	11	42
Amazona pretrei	V	15	9	0	24	0	1	0	1
Amazona versicolor	R	5	6	11	22	1	2	8	11
Amazona arausiaca	E	1	1	0	2	0	0	0	0
Amazona guildingi	R	15	16	1	32	4	4	0	8
Pharomachrus mocinno	V	1	1	0	2	0	0	0	0
Picathartes gymnocephalus	V	1	2	0	3	0	1	0	1
Picathartes oreas	V	0	1	0	1	0	1	0	1
Foudia flavicans	E	2	4	0	6	2	4	0	6
Leucopsar rothschildi	E	174	177	111	462	0	0	0	Most
REPTILES					**6187**				**4279**
Batagur baska	E	8	3	1	12	0	0	0	0
Clemmys muhlenbergi	R	26	26	11	63	12	9	11	32
Terrapene coahuila	V	10	12	95	117	7	9	95	111
Geochone elephantopus	V	77	51	108	236	13	15	107	135
Geochelone radiata	V	86	60	321	467	7	10	251	268
Geochelone yniphora	E	6	8	21	35	0	1	21	22
Pyxis arachnoides	I	4	2	0	6	0	0	0	0
Pseudoemydura umbrina	E	6	3	6	15	3	1	6	10
Alligator sinensis	E	20	29	31	80	8	22	27	57
Caiman latirostris	E	70	30	166	266	11	3	48	62
Melanosuchus niger	E	4	4	8	16	0	0	0	0
Crocodylus intermedium	E	2	3	6	11	0	1	0	1
Crocodylus moreleti	E	15	17	27	59	11	15	17	43
Crocodylus palustris	V	455	1881	728	3064	414	1852	647	2913
Crocodylus rhombifer	E	22	34	25	81	12	8	20	40
Gavialis gangeticus	E	21	43	192	256	3	13	157	173
Sphenodon punctatus	R	15	14	34	63	1	0	24	25
Phelsuma guentheri	E	10	24	3	37	0	0	0	Most
Brachylophus fasciatus	V	20	19	7	46	0	0	0	Most
Leiolopisma telfairi	R	7	9	2	18	0	0	0	Most
Heldoderma horridum	I	25	14	35	74	3	0	16	19
Heloderma suspectum	V	62	55	73	190	13	9	19	41
Varanus griseus caspius	V	18	11	13	42	0	0	0	0
Varanus komodoensis	R	2	6	4	12	2	0	0	2
Epicrates angulifer	I	63	64	49	176	34	34	47	115
Epicrates inornatus	E	27	24	34	85	17	12	31	60
Epicrates striatus fosteri	R	2	0	0	2	0	0	0	0
Epicrates subflavus	V	57	38	12	107	39	32	7	78
Acrantophis dumerili	K	77	95	3	175	0	0	0	Most
Acrantophis madagascariensis	K	19	27	4	50	8	17	3	28
Sanzinia madagascariensis	K	51	44	13	108	0	0	0	Most
Naja oxiana	E	19	15	10	44	2	5	3	10
Vipera lebetina schweizeri	V	3	3	0	6	3	2	0	5
Vipera raddei	I	14	20	9	43	5	6	4	15
Crotalus unicolor	R	37	39	1	77	0	0	0	Most
Crotalus willardi	T	28	14	6	48	10	2	2	14
AMPHIBIANS					**206**				**68**
Andrias davidianus	I	2	2	49	53	0	0	0	0
Andrias japonicus	R	12	8	119	139	68	0	0	68
Typhlomolge rathbuni	E	6	0	0	6	0	0	0	0
Bufo houstonensis	E	2	2	0	4	0	0	0	0
Conraua goliath	V	4	0	0	4	0	0	0	0

Sources: data from Olney, P.J.S. and Ellis, P. (Eds) 1991. *1990 International Zoo Yearbook, Vol.30.* Zoological Society of London, London. IUCN Threat Categories obtained from IUCN 1990. *1990 IUCN Red List of Threatened Animals.* IUCN, Gland.
Notes: It should be noted that in some cases the taxonomy of the International Zoo Yearbook differs from that of the IUCN Red List. The numbers given represent minimum estimates of species numbers held in captivity since private collections and other non–zoo organisations fall outside the scope of the International Zoo Yearbook. A number of species considered threatened by IUCN in their natural habitats are relatively common in captivity and breed regularly. For reasons of space, these taxa (which include chimpanzee, wolf, polar bear, jaguar, and various species of ducks, pheasants and reptiles) were either omitted altogether from the International Zoo Yearbook census, or only their total captive populations have been registered. No data were available for threatened fish.

Table 34.17 Current Studbooks and International Registers

	SPECIES	COMMON NAME	IUCN STATUS CATEGORY	STUDBOOK (S) OR REGISTER (R)
Amphibians	*Bufo lemur*	Puerto Rican Crested Toad	-	S
Reptiles	*Alligator sinensis*	Chinese Alligator	E	S
	Heloderma suspectum	Gila Monster	I	S
	Heloderma horridum	Beaded Lizard	V	S
	Crotalus unicolor	Aruba Island Rattlesnake	R	S
Birds	*Apteryx* spp.	Kiwis	V**	S
	Geronticus eremita	Red-cheeked Ibis (Waldrapp)	E	S
	Tragopan blythi	Blyth's Tragopan	I	S
	Tragopan caboti	Cabot's Tragopan	R	S
	Lophura edwardsi	Edward's Pheasant	I	R
	Crossoptilon crossoptilon	White Eared Pheasant	-	R
	Polyplectron inopinatum	Rothschild's Peacock	-	S
	Polyplectron malacense	Malayan Peacock Pheasant	R	S
	Afropavo congensis	Congo Peafowl	K*	S
	Grus monacha	Hooded Crane	R	S
	Grus japonensis	Red-crowned Crane	V	S
	Grus vipio	White-naped Crane	R	S
	Grus leucogeranus	Siberian White Crane	R	S
	Bugeranus carunculatus	Wattled Crane	K*	S
	Colomba (Nesoenas) mayeri	Mauritius Pink Pigeon	E	S
	Cyanopsitta spixii	Spix's Macaw	E	S
	Aratinga gauroba	Golden (Queen of Bavaria)	V	S
	Amazona guildingii	St Vincent Parrot	R	S
	Buceros bicornis	Great Indian Hornbill	-	S
Mammals	*Bettongia penicillata*	Brush-tailed Bettong	E	S
	Dendrolagus matschiei	Matschie's Tree Kangaroo	-	S
	Lemur m. macao	Black Lemur	V	S
	Lemur mongoz	Mongoose Lemur	E	S
	Varecia variegata	Ruffed Lemur	E	S
	Nycticebus pygmaeus	Pygmy Loris	V	S
	Cebuella pygmaea	Pygmy Marmoset	-	S
	Saguinus imperator	Emperor Tamarin	I	S
	Saguinus o. oedipus	Cottontop Tamarin	E	S
	Leontopithecus r. rosalia	Golden Lion Tamarin	E	S
	Leontopithecus chrysomelas	Golden-headed Lion Tamarin	E	S
	Leontopithecus chrysopygus	Black Lion Tamarin	E	S
	Callimico goeldii	Goeldi's Monkey	R	S
	Alouatta caraya	Black Howler Monkey	-	S
	Macaca silenus	Lion-tailed Macaque	E	S
	Mandrillus leucophaeus	Drill	E	S
	Theropithecus gelada	Gelada Baboon	R	S
	Cercopithecus d. diana	Diana Monkey	V[1]	S
	Pygathrix nemaeus	Douc Langur	E	S
	Hylobates concolor	Black Gibbon	V	S
	Hylobates moloch	Moloch Gibbon	E	S
	Hylobates pileatus	Pileated Gibbon	E	S
	Pongo pygmaeus	Orang-Utan	E	S
	Pan paniscus	Bonobo	V	S
	Gorilla gorilla	Gorilla	V	S
	Myrmecophaga tridactyla	Giant Anteater	V	S
	Dinomys branicki	Pacarana	E	S
	Canis lupus baileyi	Mexican Wolf	V	S
	Canis rufus	Red Wolf	E	S
	Chrysocyon brachyurus	Maned Wolf	V	S
	Speothos venaticus	Bush Dog	V	S
	Tremarctos ornatus	Spectacled Bear	V	S
	Ursus maritimus	Polar Bear	V	S
	Ailurus fulgens	Lesser or Red Panda	K	S
	Ailuropoda melanoleuca	Giant Panda	E	S

Table 34.17 Current Studbooks and International Registers (continued)

	SPECIES	COMMON NAME	IUCN STATUS CATEGORY	STUDBOOK (S) OR REGISTER (R)
Mammals (continued)	*Lutra l. lutra*	European Otter	V	S
	Aonyx cinerea	Oriental Small-clawed Otter	K	S
	Hyaena brunnea	Brown Hyena	V	S
	Felis margarita	Sand Cat	-	S
	Felis nigripes	Black-footed Cat	-	S
	Panthera leo persica	Asiatic Lion	E	S
	Panthera pardus sspp[2]	Leopard	T	S
	Panthera tigris sspp[2]	Tiger	E	S
	Panthera uncia	Snow Leopard	E	S
	Neofelis nebulosa	Clouded Leopard	V	S
	Acinonyx jubatus	Cheetah	V	S
	Equus przewalskii	Przewalski's Horse	Ex?	S
	Equus hemionus	Asiatic Wild Ass	V	S
	Equus africanus	African Wild Ass	E	S
	Equus grevyi	Grevy's Zebra	E	S
	Equus zebra hartmannae	Hartmann's Zebra	V	S
	Tapirus bairdi	Baird's Tapir	V	S
	Tapirus indicus	Malayan Tapir	E	S
	Rhinoceros unicornis	Indian Rhinoceros	E	S
	Dicerorhinus sumatrensis	Sumatran Rhinoceros	E	S
	Ceratotherium simum	White Rhinoceros	-	S
	Diceros bicornis	Black Rhinoceros	E	S
	Babyrousa babyrussa	Babirusa	V	S
	Choeropsis liberiensis	Pygmy Hippopotamus	V	S
	Vicugna vicugna	Vicuna	V	S
	Dama dama mesopotamica	Mesopotamian Fallow Deer	E	S
	Cervus duvauceli	Barasingha	E	S
	Cervus eldi	Eld's Deer	V	S
	Elaphurus davidianus	Pere David's Deer	E	R
	Ozotoceros bezoarticus	Pampas Deer	-	S
	Pudu pudu	Pudu	-	S
	Okapia johnstoni	Okapi	-	S
	Tragelaphus euryceros	Bongo	-	S
	Taurotragus derbianus gigas	Giant Eland	-	S
	Bubalus (Anoa) depressicornis	Lowland Anoa	E	S
	Bos gaurus	Gaur	V	S
	Bos javanicus	Banteng	V	S
	Bison bison athabascae	Wood Bison	-	R
	Bison bonasus	European Bison	V	R
	Kobus leche	Lechwe	V	S
	Oryx leucoryx	Arabian Oryx	E	S
	Addax nasomaculatus	Addax	E	S
	Gazella cuvieri	Cuvier's Gazelle	E	S
	Gazella dama mhorr	Mhorr Gazelle	E[1]	S
	Gazella dorcas neglecta	Dorcas Gazelle	V[1]	S
	Gazella leptoceros	Slender-horned Gazelle	E	S
	Capricornis crispus	Japanese Serow	-	S
	Ovibos moschatus	Musk Ox	-	S
	Budorcas taxicolor	Takin	-	S
	Ammotragus lervia sahariensis	Barbary Sheep	V[1]	S

Note: [1] category given is that of the whole species; [2] only some subspecies are included in the studbook; * taxa whose status is under review; ** category refers to *A. owenii*, the other two species are not threatened.

Oryx Project' was launched by the Sultan of Oman, with the aim of re-establishing a wild population. In 1980 the first oryx were returned to Oman for acclimatisation and eventual reintroduction at Yalooni in the Jiddat-al-Harasis. In 1982, the first herd of 10 was released from the $1km^2$ pre-release enclosure into the wild. Further releases were made in 1984, 1988 and 1989. Numbers increased steadily and by 1990 there were 109 free-ranging oryx, of which 80% were wild-born, occupying an unrestricted known range of more than $10,000km^2$ (Spalton, 1990). Numbers peaked at 126 in 1991, but a succession of severe drought years started to produce high mortality. Of the 15 calves born in 1991, 10 had died by January 1992 and 2 had been taken into captivity for hand-rearing, leaving the wild population at 115 (Spalton, pers. comm.). Further releases are planned to reinforce the wild population demographically and genetically.

For the first few years of the programme all released individuals were monitored closely by a force of locally-recruited Harasis rangers, using radio-tracking equipment and continuous surveillance from 4-wheel drive vehicles. Now that numbers have increased, only a selected 40 or so individuals are monitored. All the oryx are protected from poaching by strict legislation enforced by the rangers.

Reintroductions from Arizona were hampered by quarantine restrictions occasioned by the disease blue tongue, which is endemic in the USA but absent from Oman. Many captive oryx populations in the Middle East also suffer from tuberculosis. Proper veterinary procedures were therefore observed at all stages of the project.

It is estimated that the Yalooni area could eventually support 200-300 oryx, but competition with increasingly large herds of domestic livestock is beginning to cause problems. Agreement has therefore been reached with the local tribesmen not to graze their herds within a certain distance of the release site. Nevertheless, with continued sound management and effective protection - the keys to the success of this project so far - the future of the reintroduced Arabian Oryx at Yalooni seems now to be assured.

The Arabian Oryx reintroduction programme serves to demonstrate that such projects require the long-term commitment of substantial amounts of funding and manpower if they are to succeed. As such, they will of necessity be confined to a handful of species in the foreseeable future, and their contribution to the maintenance of biodiversity will remain very limited.

EX SITU CONSERVATION OF ANIMAL GENETIC RESOURCES

International efforts to conserve animal species and thereby preserve animal genetic resources are concerned either with domesticated or with wild species. At the international level few programmes attempt to conserve both domestic and wild species of animals and there is very little interaction between the two areas.

In the past there has been much less concern over the loss of genetic diversity in agricultural animals (see Chapter 26) than for agricultural plants. Consequently, there have only been limited attempts to conserve biological diversity in this area and no programme like the IBPGR presently exists for animals.

FAO in conjunction with UNEP launched a pilot programme in 1973 to conserve animal genetic resources. Initial efforts focused on developing a list of endangered breeds and of those with economic potential (and to this extent the remit of this programme was wider than simply focusing on agricultural animals). In 1980 the FAO and UNEP called for this programme to be extended and set out requirements for creating "a supranational infrastructure for animal breeding and genetics". These requirements covered a range of efforts to develop animal genetic resources and amongst other things included guidelines to develop: databanks for animal genetic resources which would also identify endangered breeds, gene banks to store semen and embryos of endangered breeds; training of scientists and administrators in genetic resource management.

The programme was developed further by a subsequent Expert Consultation in 1982. This resulted in the FAO launching its Animal Genetic Resources Programme in 1982. This programme was funded jointly by the FAO and UNEP and has concentrated on developing methodologies for a global programme for animal genetic resources. The work of this programme was published through the FAO Animal Production and Health Series and included studies on breed descriptors and databank methodology, the evolution of cryopreservation and *in situ* storage of animal genetic resources.

A five-year programme has recently been proposed by FAO in which a set of practical field orientated activities will be carried out. The main features of this programme are: the preparation of a global inventory of animal genetic resources; the creation of a 'World Watch List' to identify endangered breeds; breed preservation strategies and development programmes; development of gene technology to characterise animal biodiversity and development of a framework of international undertakings to guide access to and use of animal genetic resources. This programme illustrates the growing importance given to this previously neglected area of biodiversity conservation.

EX SITU CONSERVATION OF MICROBIAL DIVERSITY

Despite the important role that microbial diversity plays, its collection and management in the past has been carried out with a minimum of resources and on an *ad hoc* basis with little coordination within a country, let alone on an international scale. Collections of permanently preserved living cultures of microorganisms are the microbiologists' equivalent of botanic gardens, seed banks, zoos, and aquaria. Such collections are of especial importance to microbiologists as they are often the only readily available source of particular organisms required for research and assessment for exploitation. The reisolation or rediscovery in nature of desired species is often a matter of chance alone, and culture collections are thus the essential mechanism by which the earth's microbial diversity is made available to man.

One of the first formal attempts to coordinate the management of microbial resources on an international level was the establishment of a directory of institutions maintaining microbial culture collections. This is now carried out by the World Federation for Culture Collections through the World Data Centre under the auspices of UNESCO, WHO and CSIRO. The latest listing issued by the WDC details 345 culture collections distributed through 55 countries (Takishima *et al.*, 1989; Table 34.18). However, many of these collections maintain only a limited number of strains (mostly under 1,000) and are narrowly focused (e.g. only plant pathogenic bacteria, or *Rhizobium*, or human pathogens).

In 1975 UNEP, the International Cell Research Organisation and UNESCO jointly called for the establishment of a worldwide network of culture collections, and by 1992 there were 16 such collections (Table 34.19). These collections are known as microbiological resource centres (MIRCEN). The purpose of these MIRCENs is to develop and enhance the worldwide network of regional and inter-regional laboratories. Through this network it is hoped that a base of knowledge in microbiology will be developed to support biotechnology in the developing and the developed world. Activities of MIRCENs typically include collection, maintenance, testing and distribution of microbes, and training of personnel. Though each MIRCEN works according to its own set of priorities, they share a common goal of working together to strengthen the network and advance knowledge in the area. MIRCENs provide the incentives to develop and maintain microbial collections in support of national programmes. They also offer a framework that could provide a secure custodial system for national and international microbial resources.

Table 34.18 Numbers of collections of living cultures of microorganisms registered with the World Data Centre on microorganisms

ASIA	
China	10
Hong Kong	1
India	12
Indonesia	4
Iran	1
Israel	2
Japan	14
Jordan	1
Korea	1
Malaysia	3
Philippines	8
Singapore	2
Sri Lanka	4
Taiwan	1
Thailand	11
Turkey	1
USSR (former)	7
EUROPE	
Austria	1
Belgium	3
Bulgaria	3
Czechoslovakia	10
Denmark	1
Finland	2
France	12
Germany	14
Greece	3
Hungary	7
Ireland	2
Italy	8
Netherlands	6
Norway	2
Poland	6
EUROPE (continued)	
Portugal	1
Romania	1
Spain	2
Sweden	3
Switzerland	1
United Kingdom	26
Yugoslavia	2
NORTH AND CENTRAL AMERICA	
Canada	21
Guatemala	1
Mexico	9
USA	30
SOUTH AMERICA	
Argentina	7
Brazil	14
Chile	1
Colombia	2
Venezuela	1
OCEANIA	
Australia	50
New Zealand	10
Papua New Guinea	1
AFRICA	
Egypt	1
Kenya	1
Nigeria	2
Senegal	1
South Africa	3
Uganda	1
Zimbabwe	3

Source: Takishima, Y. *et al.* 1989. *Guide to World Data Centre on Microorganisms with a List of Culture Collections in the World.* World Data Centre on Microorganisms, Saitama.

Table 34.19 Microbial resource centres (MIRCENs) recognised by UNESCO

Biotechnology MIRCENs

Ain Shams University, Faculty of Agriculture, Shobra-Khaima, Cairo, Arab Republic of Egypt

Applied Research Division, Central American Research Institute for Industry (ICAITI), Ave, La Reforma 4-47 Zone 10, Apdo Postal 1552, Guatemala

Department of Bacteriology, Karolinska Institutet, Fack, S-10401 Stockholm, Sweden

Fermentation, Food and Waste Recycling MIRCEN, Thailand Institute of Scientific and Technological Research, 196 Phahonyothin Road, Bangken, Bangkok 9, Thailand

Fermentation Technology MIRCEN, ICME, University of Osaka, Suita-shi 656, Osaka, Japan

Institute for Biotechnological Studies, Research and Development Centre, University of Kent, Canterbury CT2 7TD, UK

Marine Biotechnology MIRCEN, Department of Microbiology, University of Maryland, College Park Campus, Maryland 207742, USA

Mycology MIRCEN, International Mycological Institute, Ferry Lane, Kew, Surrey TW9 3AF, UK

Planta Piloto de Procesos Industriales Microbiologicos (PROIMI), Avenida Belgrano y Pasaje Caseros, 4000 S.M. de Tucuman, Argentina

University of Waterloo, Ontario, Canada N2LK 3G1, and University of Guelph, Guelph, Ontario N1G 2W1, Canada

Rhizobium MIRCENs

Cell Culture and Nitrogen-Fixation Laboratory, Room 116, Building 011-A, Barc-West, Beltsville, Maryland 20705, USA

Centre National de Recherches Agronomiques, d'Institut Sénégalais de Recherches Agricoles, B.P. 51, Bambey, Senegal

Departments of Soil Sciences and Botany, University of Nairobi, P O Box 30197, Nairobi, Kenya

IPAGRO, Postal 776, 90000 Porto Alegre, Rio Grande do Sul, Brazil

NifTAL Project, College of Tropical Agriculture and Human Resources, University of Hawaii, P O Box "O", Paia, Hawaii 96779, USA

World Data Centre MIRCEN

World Data Centre on Collections of Micro-organisms, RIKEN, 2-1 Hirosawa, Wako, Saitama 351-01, Japan

References

Anon. 1991a. ISIS: fundamentals. A global zoo animal information system. In: Proceedings of the CBSG Annual Meeting, 27-29 September 1991, Singapore. Briefing Book - Part 1.

Anon. 1991b. *Oslo Plenary Session: Final Consensus Report: Global Initiative for the Security and Sustainable Use of Plant Genetic Resources*. Third Plenary Session. Keystone International Dialogue Series on Plant Genetic Resources, Keystone.

Batisse, M. 1982. The biosphere reserve: a tool for environmental conservation and management. *Environmental Conservation* 9(2):101-111.

Briggs, J.D. and Leigh, J.H. 1988. *Rare or Threatened Australian Plants*. Australian National Parks and Wildlife Service. 277pp.

Brush, S. 1991. A farmer-based approach to conserving crop germplasm. *Economic Botany* 45(2):153-165.

Dodd, C.K. and Siegel, R.A. 1991. Relocation, repatriation, and translocation of amphibians and reptiles: are they conservation strategies that work? *Herpetologica* 47(3):336-350.

Drucker, G., Oldfield, S., Pearce-Kelly, P., Clarke, D., Cronk, Q. *et al.* 1991. St Helena: document prepared for recognition of the entire island as an internationally recognised site of nature conservation importance. Unpublished document.

Flavelle, A. 1991. A traditional agro-forestry landscape on Fergusson Island, Papua New Guinea. Thesis submitted for an M.Sc. in Forestry. On file, University of British Columbia.

Flesness, N.R. and Foose, T.J. 1990. The role of captive breeding in the conservation of species. Guest essay. In: *1990 IUCN Red List of Threatened Animals*. IUCN, Gland.

Fowler, G. and Mooney, P. 1990. *Shattering: food, politics and the loss of genetic diversity*. The University of Arizona Press, Tucson.

Griffith, B., Michael Scott, J., Carpenter, J.W. and Reed, C. 1989. Translocation as a species conservation tool: status and strategy. *Science* 245:477-480.

Hall, L.A. 1987. Transplantation of sensitive plants as mitigation for environmental impacts. In: Elias, T.S. (Ed.), *Conservation and Management of Rare and Endangered Plants*. California Native Plant Society.

Henderson, D.S. 1974. Were they the last Arabian oryx? *Oryx* 12(3):347-350.

Heywood, C.A., Heywood, V.H. and Wyse Jackson, P. 1990. *International Directory of Botanical Gardens V*, 5th edn. Koeltz Scientific Books on behalf of WWF, Botanic Gardens Conservation Secretariat and International Association of Botanical Gardens.

Ingram, G.B. 1984. *In Situ Conservation of Genetic Resources of Plants: the scientific and technical basis*. FORGEN/MISC/84/1. FAO, Rome.

Ingram, G.B. 1990a. Multi-gene pool surveys in areas with rapid genetic erosion: an example from the Aïr Mountains, northern Niger. *Conservation Biology* 4(1):78-90.

Ingram, G.B. 1990b. The management of biosphere reserves for the conservation and utilization of genetic resources: the social choices. *Impact of Science on Society* 158:133-141.

Ingram, G.B. and Williams, J.T. 1984. *In situ* conservation of wild relatives of crops. In: Holden, J.H.W. and Williams, J.T. (Eds), *Crop Genetic Resources: conservation and evaluation*. George Allen and Unwin, London. Pp.163-179.

IUCN 1987. The IUCN position statement on translocation of living organisms: introductions, re-introductions and restocking. IUCN, Gland.

IUCN 1990. *1990 IUCN Red List of Threatened Animals*. IUCN, Gland.

Jain, S.K. 1975. Genetic reserves. In: Frankel, O.H. and Hawkes, J.G. (Eds), *Crop Genetic Resources for Today and Tomorrow*. Cambridge University Press, Cambridge. Pp.379-396.

Kloppenburg, J.R. 1988. *First the Seed*. Cambridge University Press, Cambridge.

Kloppenburg, J.R. and Kleinman, D.L. 1988. Seeds of controversy: national property versus common heritage. In: Kloppenburg, J.R. (Ed.), *Seeds and Sovereignty: the use and control of plant genetic resources*. Duke University Press, London. Pp.173-203.

Lande, R. and Barrowclough, G.F. 1987. Effective population size, genetic variation, and their use in population management. In: Soulé, M.E. (Ed.), *Viable Populations for Conservation*. Cambridge University Press, Cambridge. Pp.87-124.

Leader-Williams, N. 1990. Black rhinos and African elephants: lessons for conservation funding. *Oryx* 24(1):23-29.

Lesouef, J.Y. 1991. From rescue to re-introduction; the example of *Ruizia cordata*. In: *Tropical Botanic Gardens: their role in conservation and development*. Pp.217-222.

Montecinos, C. and Altieri, M. 1991. *Status and Trends in Grass-roots Crop Genetic Conservation Efforts in Latin America*. A contribution to the WRI/IUCN/UNEP Biodiversity Strategy Programme. CLADES - University of California, Berkeley.

Olney, P. 1991. Eighth Annual Survey of International Studbooks and Registers. In: Proceedings of the CBSG Annual Meeting, 27-29 September 1991, Singapore. Briefing Book - Part 1.

Olney, P.J.S. and Ellis, P. (Eds) 1991. *1990 International Zoo Yearbook Volume 30*. Zoological Society of London, London.

Owadally, A.W., Dulloo, M.E. and Strahm, W. 1991. Measures that are required to help conserve the flora of Mauritius and Rodrigues in *ex situ* collections. In: *Tropical Botanic Gardens: their role in conservation and development*. Academic Press.

Palmberg, C. and Esquinas-Alcazar, J.T. 1990. The role of the united conservation of plant genetic resources. *Forest Ecology and Management* 35:171-197.

Parenti, R.L. and Guerrant, E.O. 1990. Down but not out: reintroduction of the extirpated Malheur Wirelettuce, *Stephanomeria malheurensis*. *Endangered Species* Update 8(1):62-63.

Round, P.D. 1985. *The Status and Conservation of Resident Forest Birds in Thailand*. Association for the Conservation of Wildlife, Bangkok.

Sayer, J.A. and Stuart, S. 1988. Biological diversity and tropical forests. *Environmental Conservation* 15(3):193-194.

Shan-An, H., Heywood, V.H. and Ashton, P.S. 1990. *Proceedings of the International Symposium on Botanical Gardens*. ISBG, 25-28 September 1988, Nanjing. Jiangsu Science and Technology Publishing House, Nanjing. 676pp.

Siegfried, W.R. 1989. Preservation of species in southern African nature reserves. In: Huntley, B.J. (Ed.), *Biotic Diversity in Southern Africa*. Oxford University Press, Cape Town. Pp.186-201.

Smith, N. and Schultes, R.E. 1990. Deforestation and shrinking crop gene-pools in Amazonia. *Environmental Conservation* 17(3):227-234.

Solbrig, O.T. (Ed.) 1991. *From Genes to Ecosystems: a research agenda for biodiversity*. Report of an IUBS-SCOPE-UNESCO workshop, Harvard Forest, Petersham, Massachusetts, USA, 1991. IUBS, Cambridge, Massachusetts.

Spalton, A. 1990. Recent developments in the reintroduction of the Arabian oryx (*Oryx leucoryx*) to Oman. *Species* 15:27-29.

Takishima, Y., Shimura, J., Udagawa, Y. and Sugawara, H. 1989. *Guide to World Data Centre on Microorganisms with a List of Culture Collections in the World*. World Data Centre on Microorganisms, Saitama. 249pp.

Thomasson, G. 1991. Liberia's seeds of knowledge. *Cultural Survival Quarterly* 15(3):23-29.

Van Sloten, D.H. 1990. IBPGR and the challenges of the 1990s: a personal point of view. *Diversity* 6(2):36-39.

Velchev, V., Kozuharov, S., Bondev, I., Kuzmanov, B. and Markova, M. 1984. *Red Data Book of the People's Republic of Bulgaria. Vol. 1. Plants*. Bulgarian Academy of Sciences, Sofia. 447pp.

Vernhes, J.R. 1989. Biosphere reserves: the beginnings, the present, and the future challenges. In: *Proceedings of the Symposium on Biosphere Reserves, Fourth World Wilderness Congress*, September 14-17, 1987. Estes Park, Colorado, USA. National Park Service, US Department of the Interior, Atlanta.

Whitten, A.J. 1990. *Recovery: a proposed programme for Britain's protected species*. Nature Conservancy Council, CSD Report No.1089. NCC, Peterborough.

Zimmerer, K.S. and Douches, D.S. 1991. Geographical approaches to crop conservation: the partitioning of genetic diversity in Andean potatoes. *Economic Botany* 45(2):176-189.

IUCN Red Data Books:

Collar, N. and Stuart, S.N. 1985. *Threatened Birds of Africa and Related Islands: the ICBP/IUCN Red Data Book, Part I*. ICBP, Cambridge and IUCN, Gland. 796pp.

Collins, N.M. and Morris, M.G. 1985. *Threatened Swallowtail Butterflies of the World: the IUCN Red Data Book*. IUCN, Gland. 408pp.

Groombridge, B. 1982. *IUCN Amphibia-Reptilia Red Data Book: Testudines, Crocodylia, Rhynchocephalia*. IUCN, Gland. 556pp.

Harcourt, C. 1990. *Lemurs of Madagascar and the Comoros: the IUCN Red Data Book*. IUCN, Gland. 248pp.

Klinowska, M. 1991. *Dolphins, Porpoises and Whales of the World: the IUCN Red Data Book*. IUCN, Gland. 438pp.

Lee, P.C., Thornback, J. and Bennett, E.L. 1988. *Threatened Primates of Africa*. IUCN, Gland. 176pp.

Lucas, G. and Synge, H. 1978. *IUCN Plant Red Data Book*. IUCN, Morges. 540pp.

Thornback, J. and Jenkins, M. 1982. *IUCN Mammal Red Data Book: the Americas and Australasia (excluding Cetacea)*. IUCN, Gland. 556pp.

Wells, S.M., Pyle, R.M. and Collins, N.M. 1983. *IUCN Invertebrate Red Data Book*. IUCN, Gland. 682pp.

IUCN/SSC Action Plans:

Anon. 1989. *Tortoises and Freshwater Turtles. An Action Plan for their Conservation*. IUCN/SSC Tortoise and Freshwater Turtle Specialist Group. IUCN, Gland. 47pp.

Chapman, J.A. and Flux, J.E.C. 1990. *Rabbits, Hares and Pikas. Status Survey and Conservation Action Plan*. IUCN/SSC Lagomorph Specialist Group. IUCN, Gland. 168pp.

Cumming, D.H.M., du Toit, R.F. and Stuart, S.N. 1990. *African Elephants and Rhinos. Status Survey and Conservation Action Plan*. IUCN/SSC African Elephant and Rhino Specialist Group. IUCN, Gland. 73pp.

East, R. 1988. *Antelopes. Global Survey and Regional Action Plans. Part 1. East and Northeast Africa*. IUCN/SSC Antelope Specialist Group. IUCN, Gland. 96pp.

East, R. 1989. *Antelopes. Global Survey and Regional Action Plans. Part 2. Southern and South-central Africa*. IUCN/SSC Antelope Specialist Group. IUCN, Gland. 96pp.

East, R. 1989. *Antelopes. Global Survey and Regional Action Plans. Part 3. West and Central Africa*. IUCN/SSC Antelope Specialist Group. IUCN, Gland. 171pp.

Eudey, A.A. 1987. *Action Plan for Asian Primate Conservation: 1987-1991*. IUCN/SSC Primate Specialist Group. IUCN, Gland. 65pp.

Foster-Turley, P., Macdonald, S. and Mason, C. 1990. *Otters. An Action Plan for their Conservation*. IUCN/SSC Otter Specialist Group. IUCN, Gland. 126pp.

Ginsberg, J.R. and Macdonald, D.W. 1990. *Foxes, Wolves, Jackals and Dogs. An Action Plan for the Conservation of Canids*. IUCN/SSC Canid and Wolf Specialist Group. IUCN, Gland. 116pp.

Khan, M. 1989. *Asian Rhinos. An Action Plan for their Conservation*. IUCN/SSC Asian Rhino Specialist Group. IUCN, Gland. 23pp.

MacKinnon, J.R. and Stuart, S.N. 1988. *The Kouprey. An Action Plan for its Conservation*. IUCN/SSC Asian Wild Cattle Specialist Group. IUCN, Gland. 19pp.

New, T.R. and Collins, N.M. 1991. *Swallowtail Butterflies. An Action Plan for their Conservation*. IUCN/SSC Lepidoptera Specialist Group. IUCN, Gland. 36pp.

Nicoll, M.E. and Rathbun, G.B. 1990. *African Insectivora and Elephant Shrews. An Action Plan for their Conservation*. IUCN/SSC Insectivore, Tree-Shrew and Elephant-Shrew Specialist Group. IUCN, Gland. 53pp.

Oates, J.F. 1986. *Action Plan for African Primate Conservation: 1986-1990*. IUCN/SSC Primate Specialist Group. IUCN, Gland. 41pp.

Perrin, W.F. 1989. *Dolphins, Porpoises and Whales. An Action Plan for the Conservation of Biological Diversity: 1988-1992*, 2nd edn. IUCN/SSC Cetacean Specialist Group. IUCN, Gland. 27pp.

Santiapillai, C., Jackson, P. 1989. *The Asian Elephant. An Action Plan for its Conservation*. IUCN/SSC Asian Elephant Specialist Group. IUCN, Gland. 79pp.

Schreiber, A., Wirth, R., Riffel, M. and van Rompaey, H. 1989. *Weasels, Civets, Mongooses and their Relatives. An Action Plan for the Conservation of Mustelids and Viverrids*. IUCN/SSC Mustelid and Viverrid Specialist Group. IUCN, Gland. 99pp.

Thorbjarnarson, J. 1992. *Crocodiles. An Action Plan for their Conservation*. IUCN/SSC Crocodile Specialist Group. IUCN, Gland. 136pp.

Authors as follows: In situ *conservation of threatened plants, Hugh Synge;* In situ *conservation of crops, G. Brent Ingram and Alix Flavelle;* Ex situ *conservation of plants, E.A. Leadlay and P.S. Wyse Jackson; IARCs, Sam Johnston; Plant reintroductions, Mike Maunder. Remaining material by Martin Jenkins and WCMC staff.*

35. THE CONVENTION ON BIOLOGICAL DIVERSITY

BACKGROUND

As developing countries have come to realise the economic value of their biodiversity, so the political considerations surrounding its utilisation have become more complex. The direct commodity values of exploited biological resources, such as tropical hardwoods and fisheries, are well established, as are the indirect benefits from tourism and game viewing; what has changed is the recognition of the value of biodiversity as a genetic resource available for commercial exploitation through biotechnology.

Developing countries are now demanding a greater share of the economic benefits arising from the use of resources within their boundaries which until now have mainly accrued to the industrial countries with the technological capability to exploit them.

At the same time, the developed world has become increasingly apprehensive about the accelerating rate of loss of biodiversity and its global consequences (of tropical deforestation upon global climate change, for example). Developed countries want to see the use of biological resources placed on a sustainable basis, and are linking their overseas development assistance to this tenet, which often clashes with the sovereign rights of developing countries to manage their resources as they deem best on behalf of their citizens. In addition, the developed countries have until recently been able to exploit the genetic resources of tropical countries for agricultural and pharmacological advantage at little cost, but have now become concerned both about the continued erosion of these resources and the increasing restrictions developing countries are placing on their use.

These parallel concerns about the exploitation of biological resources expressed by both the industrialised and under-industrialised countries have led to the negotiations for a Convention on Biological Diversity.

Because of the very different interests and expectations of producer and consumer nations, these negotiations have become increasingly polarised, with little apparent willingness to compromise. On the one hand, the developed countries as consumers of biological resources are concerned about the sustainability of supplies and continued unrestricted access to genetic materials, whereas the developing countries as producer nations are more concerned with the transfer of the biotechnology to enable them to develop their resources more effectively for themselves, and the equitable distribution of benefits arising from the use of their resources. Redistribution of these benefits must provide the economic incentive to reinforce the conservation of biodiversity throughout the developing world. The obvious difficulties of achieving such redistribution have greatly retarded progress.

The effects of this increased polarity are exacerbated by basing the negotiations on the practice of unanimity rather than consensus, so that a single country with a strongly held position can insist on alternative wording being included in the text. The result is a plethora of square brackets including such superficial niceties as "[each Contracting Party will ....] or [the Contracting Parties shall ....]".

THE BIODIVERSITY CONVENTION

The origin of the convention negotiations goes back to the initial drafts prepared in 1987 by IUCN in response to a Resolution adopted at its 16th General Assembly. The IUCN prototype was a relatively simple document that focused on measures to reinforce the conservation of biodiversity *in situ* through the provision of economic incentives based on sustainable use. Its main breakthrough, apart from galvanising activity, was the recognition of the rights of the producer countries to share equally in the benefits of their resource use: new, innovative, funding mechanisms, such as import duties, trade tariffs, and royalty payments on the sale of commodities incorporating products of the biological resources of other countries were proposed.

In 1988, IUCN circulated a comprehensive draft amongst the participating countries at the UNEP Governing Council. This stimulated extensive discussion, resulting in acceptance of the need for an international convention which UNEP, as the appropriate inter-governmental agency, was instructed to pursue. The relatively narrow focus envisaged by IUCN was then expanded to include *ex situ* conservation, land races and the wild relatives of commercial crop varieties, access to technologies and scientific skills by developing countries, and the transfer of biotechnologies for developing countries to exploit their own genetic resources. Formal negotiations commenced in November 1990 with the first session of the Ad Hoc Working Group of Legal and Technical Experts, followed by five sessions of the Intergovernmental Negotiating Committee (INC) for a Convention on Biological Diversity. Each INC meeting has involved some 75 countries and lasted for eight days, representing a substantial investment of time and funding.

As with the Convention on Climate Change, the Biodiversity Convention has become a key component of the UN Conference on Environment and Development (UNCED), with the expectation of a formal instrument being ready for signature at the time of the conference in Brazil. However, with many of the most substantive issues still to be agreed, it is probable that only a framework convention will be ready for Rio, with the more contentious articles being negotiated as protocols to the convention according to a schedule to be decided at UNCED.

The effects of these delays will obviously influence the timetable for the convention coming into operation. As with other international legal instruments, countries must first sign and then subsequently ratify the convention, and it is only after the twentieth country has acceded (and this number is still subject to debate) that it comes into force. With ratification likely to be delayed until the protocols have been negotiated, it may be five years or more until the convention becomes operational. The non-governmental organisations are already voicing their concerns about the continued loss of biodiversity before the convention comes into force.

Table 35.1 Articles of the draft Convention on Biological Diversity

1.	Objectives	18.	Financial Needs and Means
2.	Use of Terms for the Purpose of this Convention	19.	Financial Mechanisms
3.	Fundamental Principles	20.	Relationship with other International Conventions
4.	General Obligations	21.	Conference of the Parties
5.	Implementation Measures	[22.	Procedures for Global Lists]
[5 bis	Identification and Monitoring]	23.	Secretariat
6.	*In situ* Conservation	24.	Science and Technology Committee
7.	*Ex situ* Conservation	25.	Reports
8.	Sustainable Use of Components of Biological Diversity	26.	Operational Cost
		27.	Settlement of Disputes
[8 bis	Incentive Measures]	28.	Adoption of Protocols
9.	Research and Training	29.	Amendment of the Convention or Protocols
10.	Public Education and Awareness	30.	Adoption and Amendment of Annexes
11.	Impact Assessment	31.	Right to Vote
[12.	Surveys and Inventories]	32.	Relationship between the Convention and its Protocols
[13.	Global Lists]		
[14.	Access to Genetic Material]	33.	Signature
[14 bis	Traditional Indigenous and Local Knowledge]	34.	Ratification, Acceptance or Approval
15.	Access to Technology	35.	Accession
[15 bis	Exchange of Information]	36.	Entry into Force
16.	Transfer of Technology	37.	Reservations
17.	Technical and Scientific Cooperation	38.	Withdrawals
17 bis	Handling of Biotechnology and Distribution of its Benefits	39.	Depository
		40.	Authentic Text

Notes: This table reflects the fourth revised draft of the Convention (February 1992). [ ] = Articles in square brackets are still subject to negotiation and may be dropped in their entirety. Most of the operational articles include disputed clauses, or text within clauses, in square brackets.

The contents of the convention

The contents of the draft convention are outlined in Table 35.1. Although the exact measures to be incorporated in the articles are still to be decided - and at the opening of the fourth session of the INC in February 1992 (only four months before UNCED) there were still over 350 disputed sections of text - certain features have already begun to emerge (at the time of writing this review the text of the convention was still under negotiation; this analysis is based on the provisions of the Fourth Revised Draft, including the compromise formulations proposed by the Executive Director of UNEP, which was presented to the Fourth Session of the INC, 6-15 February 1992).

The Objective (Article 1) of the convention is "to conserve the maximum possible biological diversity for the benefit of present and future generations and for its intrinsic value", which is to be achieved by:

- ensuring that the use of biological resources is sustainable
- providing adequate, new and additional funding for developing countries to facilitate the conservation and rational use of their resources
- taking account of the need to share costs and benefits between developed and developing countries
- providing economic and legal conditions favourable for the transfer of technology necessary to accomplish the objectives of this convention
- providing fair sharing of the benefits of research in biotechnology arising from the conservation of biological diversity.

The Fundamental Principles (Article 3) are extremely broad: they affirm that the conservation of biodiversity is a common concern of all people, but also stress the responsibility of states in exercising their sovereign rights to ensure that their biological resources are developed in a sustainable way. Emphasis is also given to the *in situ* conservation of ecosystems and natural habitats, whilst *ex situ* measures should preferably be undertaken in the country of origin. An important recognition is that lack of scientific certainty should not be used as a reason for postponing actions to avoid or minimise threats to biodiversity. Regarding sources of finance, countries benefiting most from the exploitation of biodiversity should contribute most to its conservation. The practices and experience of indigenous peoples in using biological resources should be recognised and rewarded.

The General Obligations (Article 4) still have much disputed text, but call on each Contracting Party to take all measures at its disposal, including national plans, policies and legislation, both individually and cooperatively, to conserve the maximum possible biological diversity within its national jurisdiction.

The Implementation Measures (Article 5) envisage the development of national strategies and programmes for the conservation and sustainable use of biological diversity, including the establishment of national bodies to implement the provisions of the Convention.

The next ten articles (Articles 5 bis to 13 inclusive) comprise the conservation core of the convention. They cover such issues as *in situ* and *ex situ* conservation, inventory and monitoring, research and training, public education, and sustainable use. The obligations on countries have been weakened by obstruction from the USA and some of the G77 states (the group of 77 non-aligned developing countries), but still include a number of positive features. For example, countries are called upon to: establish protected areas in locations requiring special conservation measures, including wildlife corridors; restore degraded ecosystems and habitats; eradicate alien species that threaten natural habitats; and to introduce legislation for the protection of threatened species, populations and varieties. Parties are also expected to undertake national surveys of their biodiversity and to maintain databases of their resources, linked into a global network. However, the convention makes it quite clear that the implementation of these obligations by developing countries is subject to the provision of new and additional financial and technical resources.

There then follow a series of articles (numbers 14 - 17 bis) that deal with access to genetic resources and the transfer of the technologies and scientific skills appropriate for their exploitation. The thrust of this section is that countries should refrain from imposing restrictions on the availability of wild genetic materials, such as breeders' or farmers' rights, but that preferential access to the research results or benefits arising from the use of genetic materials should be granted to the country of origin. At the same time, parties should undertake to provide, on mutually agreed terms, technologies appropriate to the conservation and sustainable use of biodiversity.

Aside from Articles 18 and 19 (see below), the remaining articles relate to the procedures for the establishment and operation of the convention and its protocols, and are therefore less contentious. This administrative machinery is essential for the development of an effective convention, and the procedures that have now been agreed are the most sophisticated yet seen in an environmental treaty.

Some contentious issues

Articles 18 and 19 cover the key issues of financial needs and mechanisms, upon which the viability of the whole convention depends. There is a general expectation that the developed countries must provide "adequate new and additional financial resources to enable developing countries to meet the agreed incremental costs to them of fulfilling their obligations under the Convention". There is also widespread acceptance for the establishment of a Biological Diversity Fund for developing countries to implement their obligations, but a difference of opinion about how this fund should be administered. One option is to create a 'window account' in the Global Environmental Facility (GEF) specifically for the convention, although the criteria for allocating these funds would be determined by the Conference of the Parties through a Science and Technology Committee, rather than by the three agencies in the GEF (see Chapter 32 for an outline of the GEF).

Whatever administrative mechanism is adopted, the purpose of this funding will be to empower developing countries to meet the scientific, economic and institutional requirements of the convention. It is not envisaged that the Biological Diversity Fund should provide the conduit for the economic incentives that developing countries may need to reinforce their conservation programmes: these must be derived from standard commercial practices arising from the use of biological resources negotiated by national governments.

In addition to the central issue of the financial provisions, the other contentious issues where substantive differences of opinion still need to be resolved include:

- the granting of access to genetic resources and the conditions pertaining to their use
- equitable distribution of benefits arising from the use of genetic resources between the exploiting country and the country of origin
- provisions for biotechnology safety relating to the introduction of genetically modified organisms
- fair and favourable conditions for access to and transfer of technology
- commercial patents and intellectual property rights relating to the transfer of biotechnological processes and genetic manipulation procedures
- the global lists of species and sites

Article 13 calls for a Global List of Biogeographic Areas of Particular Importance for the Conservation of Biological Diversity and a second Global List of Species Threatened with Extinction on a Global Level, but this proposal has run into opposition. Some developing countries are against lists because of the burden they would impose if species recovery plans and site management plans had to be implemented; also, global lists could be seen to conflict with the rights of national sovereignty if designations were imposed on countries without their agreement.

On the positive side, lists would focus world attention on the sites and species of global conservation concern, and would help identify priorities for funding. The lists would represent a tangible output from the convention and provide a vehicle for a concerted conservation effort involving the non-governmental organisations. It is no coincidence that the more effective conservation conventions, such as The World Heritage Convention (WHC), Ramsar and CITES, all have lists at their core. The non-governmental agencies are lobbying hard for the retention of lists on the grounds that their removal would greatly dilute the conservation provisions (see International obligations, Chapter 31).

The key role of an active administrative structure

With many of the substantive provisions of the Convention on Biological Diversity still undecided, the sophisticated administrative structure already agreed will be vital in developing the convention to a stage where it is a truly effective international instrument; this structure is in fact a major achievement of the negotiations to date.

The vital role that an active administration plays in developing a framework convention into an effective international instrument is illustrated by existing conventions. The success of both CITES and the WHC is largely due to these conventions having active and well-financed administrative structures. The comparative failure of the Bonn Convention or the Western Hemisphere Convention is in part attributable to the absence of such structures. The essential features include: a well-financed secretariat; an independent scientific committee; requirements for regular meetings of the parties and regular reporting by them to the secretariat; the involvement of outside parties (such as NGOs) in the regular meetings of the parties; and the obligation to establish or designate a national or local authority to deal with implementing the obligations of the convention.

The effectiveness of these measures arises from the fact that they keep the key issues in the public arena and on the political agenda, thereby working against political and administrative inertia. They also provide a catalyst for development of the broad objectives which a framework convention largely comprises into specific obligations which have some impact on the conduct of the Parties.

The Convention on Biological Diversity has many of these features. The convention establishes a Secretariat (Article 23) to arrange and coordinate meetings of the Parties; to assist the scientific committee in its work; and to maintain the global lists, if these are to be included. The functions of this body are to be carried out by an existing international organisation, to be decided at the first conference of the Parties, but which in the interim will be the responsibility of UNEP. The convention establishes a scientific committee which is called the Scientific and Technology Committee; its role will be to provide scientific and technological advice as required for the implementation of the convention. There are extensive reporting requirements and meetings of the Parties are to be held at regular intervals to be decided at the first conference of the Parties. Non-governmental organisations are eligible to attend these conferences provided they have informed the secretariat and not more than one third of the Parties object to their presence.

THE BIODIVERSITY COUNTRY STUDIES AND UNMET FINANCIAL NEEDS

The conservation element of the convention focuses initially on the gathering of information through national surveys and inventories, then moves on to address the benefits arising from the sustainable use of biodiversity. This information collecting exercise is to be undertaken by Country Studies detailing what is currently known about the status, threats, costs and benefits of biodiversity in each country (Table 35.2). The Country Studies will then form the basis for the development of the national plans for the conservation and sustainable use of biodiversity called for under the Implementation Measures.

At the same time, the INC needs to quantify what order of magnitude of new and additional financial resources will be required for the Biological Diversity Fund to finance the implementation of the measures in the Convention by developing countries. The Country Studies were therefore charged with calculating the unmet financial needs of each country undertaking a survey from which the total financial requirements of the Convention could be estimated.

With the coordination of UNEP, GEF funding, and the World Conservation Monitoring Centre playing a catalytic role, some 14 countries, of which 11 have reported, are undertaking Country Studies. A methodology for completing the studies was prepared by UNEP and has four main components:

- review of the status of the biological resources
- identification of the measures necessary for effective conservation and sustainable use of these resources
- determination of the costs and benefits of implementing these measures
- estimation of the current unmet financial needs

This process was expected to furnish many new data on the status and economics of biodiversity conservation and utilisation (Table 35.2). In practice, the methodology has proved to be over-ambitious so that even developed countries have had problems implementing it, although extensive new data on biological resources have been forthcoming. Of equal importance has been the recognition of the gaps in the information-base, particularly microorganisms, invertebrates and lower plants.

Table 35.3 shows the estimated unmet financial costs of the ten reporting countries. The substantial variation in the annual needs, ranging from US\$1,590/km^2 for Costa Rica to US\$64/km^2 for Kenya, reflects more the lack of standardisation in the estimation than real differences in financial requirements. Clearly these needs will vary significantly between countries - for example, countries with a sound infrastructure for biodiversity conservation will require fewer funds to implement measures in the convention than those with a neglected infrastructure.

Extrapolating from these figures of unmet costs, a number of estimates have been made, using different methods, to quantify the total financial resources required by all developing countries to implement the convention (Table 35.4) (UNEP, 1992). The average of these estimates is around US$20 billion/annum. Although no more than indicative of the order of magnitude, this estimate does suggest that substantial amounts of additional funds will have to be transferred to developing countries if they are to meet their obligations under the convention.

The current level of overseas development assistance available to developing countries for the conservation of biological diversity is estimated at US$228 million, of which US$170 million is derived from bilateral aid and US$58 million from multilateral sources (UNEP, 1991). A ten-fold increase in commitment from the donor countries is therefore required. Considering the current apprehensions being expressed by the developed countries about the likely levels of extra funding that the convention will need, it is politically inconceivable in the short-term that additional funding of this magnitude will be forthcoming. Although US$20 billion/annum in absolute terms is a substantial sum, representing about 27% of the total overseas aid budget, it is put into a realistic context by comparison with the $245 billion spend each year by the OECD countries on their own agricultural support programmes, which are themselves ultimately dependent upon biodiversity. With global military budgets at some US$980 billion in 1990, or US$185 for every person on the planet, a peace dividend from the cessation of the cold war of only 2% would seem a modest amount to save the diversity of life on earth.

Based on a pragmatic assessment of what developed countries are likely to find acceptable, UNEP is proposing the establishment of a roll-over mechanism based on the GEF to provide interim funding for the convention. This would accelerate once the institutional and human capacity were in place. UNEP is proposing an increase of US$500 million/annum over the next five years, raising the current flow of funding through the GEF from around US$100 million/annum to US$600 million/annum by 1997, and thereafter accelerating to US$850 million/annum by the end

Table 35.2 Information to be generated by Biodiversity Country Studies using the UNEP methodology of reporting

Annex I	Global and National Biodiversity Status A. Species diversity data B. Species ecological status over time C. Habitat/ecosystems diversity D. Habitat/ecosystem status and percent change over the past 10-20 years E. Areas of high species endemism F. Significant changes in populations of species of national importance over the past 10 years G. National parks/nature reserves/gazetted forests and other protected sites H. Additional national biotic communities/biogeographic provinces currently not protected I. Private wildlife sanctuaries J. Status of national *ex situ* conservation facilities K. Species in national *ex situ* conservation facilities
Annex II	Essential Planetary Services Provided by Major Taxonomic Groups of Organisms
Annex III	Categories of Value Assigned to Biological Diversity
Annex IV	Sites and Species of Significance for Conservation
Annex V	Measures to be Implemented to Achieve Desired Level of Conservation
Annex VI	Measures to be Undertaken for Effective Conservation and Rational Use of Biological Diversity
Annex VII	Calculating Costs and Benefits Associated with the Implementation of Identified Measures for Conservation and Sustainable Use of Biological Diversity
Annex IX	Current Multilateral, Bilateral and National Financial support for Biodiversity Conservation and Unmet Funding Needs in Respect of Identified Priority Areas
Annex X	Summary of Costs, Benefits and Unmet Needs of Biodiversity Conservation

Table 35.3 Unmet financial needs of countries to conserve their biodiversity

COUNTRY	TOTAL ANNUAL COSTS	UNMET ANNUAL COSTS	
		TOTAL	PER KM²*
	US$ million/year	US$ million/year	US$ million/year
ASIA			
Indonesia	290	231	120
Malaysia	X	X	X
Thailand	120	60	116
EUROPE			
Germany	1,200	950	2,662
Poland	800	100	320
SOUTH AMERICA			
Guyana	--	--	--
Peru	--	--	--
NORTH AND CENTRAL AMERICA			
Bahamas	110	84	**6,058
Costa Rica	100	81	1,590
Canada	2,686	986	99
OCEANIA			
Australia	--	--	--
AFRICA			
Kenya	160	37	64
Nigeria	593	325	352
Uganda	70	58	245

Source: Data derived from Country Study reports.

Notes: X relevant economic data not supplied; -- no Country Study Report submitted; * terrestrial land area only; ** excludes area of marine habitats.

Table 35.4 Estimates of the total unmet financial needs per annum of all developing countries to implement the measures in the Convention on Biological Diversity

	METHOD OF ESTIMATION	TOTAL UNMET FINANCIAL NEEDS US$ billion/annum
A.	Extrapolation on the basis of unmet needs of developing countries adjusted for biodiversity richness and country size	8.45
B.	Extrapolation on the basis of percentage GDP (0.5%) as desirable expenditure for biodiversity conservation and sustainable use	21.13 to 42.25
C.	Extrapolation on the basis of the number of sites and national protected areas as a percentage of the global total	42.0
D.	Extrapolation on the basis of species diversity in each country as a percentage of global total	0.68 to 15.8
E.	Extrapolation on the basis of categorising countries by their biodiversity richness and their *in-situ* conservation infrastructure (the WCMC method)	11.1

of the century. Although welcome, this level of funding is still far short of the minimum requirement estimated from the Country Studies.

FUTURE DATA NEEDS: NETWORKING AND GLOBAL MONITORING

The preparation of the Country Studies has proved to be a valuable mechanism for setting a country on course towards a better understanding of its biodiversity and a more rational use of its resources. The process has necessitated the establishment in each country of National Biodiversity Units (NBUs) to serve as coordinating centres for the gathering of data on the status, utilisation, and economic values of biodiversity. The concept of accounting the costs and benefits of biodiversity conservation and use has been introduced, and appreciation of the costs of inaction in terms of lost benefits if no action is taken to conserve biodiversity, has been accepted.

The Country Studies exercise and the resultant NBUs will provide a useful foundation upon which to build the human and institutional capacities for improved conservation practice. Already further studies are being planned for some of the most biologically rich countries such as Brazil, Colombia, Mexico, Madagascar, Zaire and Papua New Guinea, as well as countries such as Angola and Mozambique where the conservation infrastructure requires rebuilding. However, expanding the programme too rapidly will divert the limited GEF funds away from the priority activities of consolidating the results of the first tranche of reporting countries and of providing long-term support for those NBUs that have already proved their worth. If the programme is to advance it must first revise the methodology to produce a more robust system for estimating economic costs and benefits which can be applied realistically in developing countries. It must be accepted that quantifying the existence value of a threatened species or the service value of a wooded watershed to secure a constant water supply involves an element of subjective value judgement, but guidelines are needed so that estimates can be standardised between countries.

In addition, the NBUs should be further developed and strengthened into National Biodiversity Monitoring Centres responsible for the gathering and analysis of data at the country level. Such monitoring centres should then be linked into a global biodiversity information network which can be progressively expanded with each subsequent round of Country Studies. This proposal closely mirrors the key recommendation of the *Global Biodiversity Strategy* for the establishment of an Early Warning Network to monitor potential threats to biodiversity (Table 35.5) (WRI/IUCN/UNEP, 1992). The purpose of this Network is to provide a swift response to the emergence of new threats through the rapid mobilisation of information.

The best sources of early warning information are scientists, non-governmental organisations, and enforcement authorities working in the field. If they can be linked into an in-country network of data sources feeding their information into a National Biodiversity Monitoring Centre, which in turn is linked into a global network, then a mechanism can be developed to mobilise this information rapidly. The parameters that a national centre should monitor for early warning purposes must include not only direct threats but also political, legal and economic changes that could have indirect effects on biodiversity. A set of such parameters is presented in Table 35.6.

Incorporating biodiversity conservation into national policies and planning (Action 5 of Table 35.5) can help countries define and articulate their environment and development goals. A minimum set of biodiversity indicators that must be included within the monitoring programmes of a national data centre, and which provide the basic information needs for national and international policy-makers is presented in Table 35.7. This dataset provides a matrix combining the major conservation concerns with a working set of indicators that can be used to assess long-term trends in the conservation of biodiversity.

The limiting factor in such programmes for assessing biodiversity conservation trends and goals is the availability

Table 35.5 International actions to conserve the world's biological diversity as recommended by the Global Biodiversity Strategy

Action 1. Adopt in 1992 the International Convention on Biological Diversity

2. Adopt in the General Assembly of the United Nations, a resolution designating 1994-2003 the International Biodiversity Decade

3. Establish a mechanism, such as an International Panel on Biodiversity Conservation, preferably within the Convention on Biological Diversity, including scientists, non-governmental organisations and policy-makers to provide guidance on priorities for the protection, understanding, and sustainable and equitable use of biodiversity

4. Establish an Early Warning Network, linked to the Convention on Biological Diversity, to monitor potential threats to biodiversity and mobilize remedial action

5. Integrate biodiversity conservation into national planning processes

Source: WRI/IUCN/UNEP 1992. *Global Biodiversity Strategy*.

Table 35.6 Parameters that an Early Warning Network must monitor at the country level

1. Traditional crop or livestock varieties threatened by planned development projects or the introduction of new varieties
2. Increasing genetic uniformity of crops
3. Natural ecosystems subjected to new inappropriate management practices, human encroachment, or unsustainable exploitation
4. Protected areas in urgent need of financial, technical, or other support
5. Accelerating habitat loss
6. Evidence of the over-exploitation of species
7. Introductions of exotic species
8. Genebank facilities with germplasm at risk due to lack of funding for recurring costs
9. Climatic threats to biodiversity - including desertification, floods, drought, and global warming
10. Communities denied access to resources when protected areas are established
11. Pollutant discharges presenting immediate threats or chronic pollution that might pose longer-term threats
12. Changes to the legislation relating to land and other resource ownership that may disenfranchise local communities
13. Changes to national budgets that may affect the allocation of funds for conservation
14. Political or institutional developments that may influence the infrastructure for effecting conservation
15. Implementation of obligations undertaken through international conventions

Source: WRI/IUCN/UNEP 1992. *Global Biodiversity Strategy*.

and reliability of the data. In collaboration with WCMC, Reid *et al.* (in prep.) reviewed the availability, coverage and quality of the data needed for assessment of the indicators presented in Table 35.7. The conclusions make depressing reading: although the coverage at the country level for mammals and birds is reasonable, for most other species the data are lacking or, where available, of poor quality. Time series data are non-existent except for a few 'megafauna' species and for the land-use estimates by FAO. Of particular concern are the lack of data on genetic varieties of agricultural crops grown in developing countries, and the absence of base-line datasets for monitoring ecosystem changes.

These conclusions emphasise the urgent need to build monitoring capabilities at the country level. With the increasing precision of remote sensing techniques and the advances in information technology, particularly the application of Geographic Information Systems, the ability to develop sophisticated monitoring systems is within the reach of all countries. In an analysis of the relevance of technology transfer to the conservation of biological diversity, a recent report (Touche Ross, 1991) showed that the most appropriate technologies for the management and utilisation of biodiversity were 'soft': that is, information management, human skills and scientific knowledge rather than 'hard' involving physical plant and equipment.

Table 35.7 A minimum set of indicators for monitoring biodiversity at the country level

	INDICATOR	BIODIVERSITY CONSERVATION CONCERNS GENETIC DIVERSITY	SPECIES DIVERSITY	COMMUNITY DIVERSITY
	Wild Species and Genetic Diversity			
1.	Species richness (number, number per unit area, number per habitat type)	•	•	
2.	Species threatened with extinction (number or percent)	•	•	
3.	Species threatened with extirpation (number or percent)	•	•	
4.	Endemic species (number or percent)	•	•	
5.	Endemic species threatened with extinction (number or percent)	•	•	
6.	Species risk index	•	•	
7.	Species with stable or increasing populations (number or percent)	•	•	
8.	Species with decreasing populations (number or percent)	•	•	
9.	Threatened species in protected areas (percent)	•	•	
10.	Endemic species in protected areas (percent)	•	•	
11.	Threatened species in *ex situ* collections (percent)	•	•	
12.	Threatened species with viable *ex situ* populations (percent)	•	•	
13.	Species used by local residents (percent)	•	•	
	Community Diversity			
13.	Percent dominated by non-domesticated species		•	•
14.	Rate of change from dominance of non-domesticated species to domesticated species		•	•
15.	Percent of area dominated by non-domesticated species occurring in patches greater than 1,000km²		•	•
16.	Percent of area in strictly protected status		•	•
	Domesticated Species			
17.	Accessions of crops and livestock in *ex situ* storage (number)	•		
18.	Accessions regenerated in the past decade (percent)	•		
19.	Number of crops (livestock) grown as percent of number 30 years before	•		
20.	Number of varieties as percent of number 30 years before	•		
21.	Coefficient of kinship or parentage of crop	•		

Source: Reid, W.V., McNeely, J.A., Tunstall, D.B. and Bryant, D. (in prep.). World Resources Institute, Washington DC.

The main issue now is not how to monitor but what to monitor. A minimum set of parameters must be agreed, along the lines presented in Table 35.6 and 35.7, that provides a framework for determining conservation priorities and goals at the country level, that generates the data necessary to build biodiversity conservation into the national planning process, and that supplies the early warning information necessary for the rapid response to new threats. Such a system will require the standardisation of species names, habitat classifications and threat categories so that national data centres can be networked for the reciprocal exchange of information. As more centres are established, so the network will grow, enabling regional assessments to be made of needs, priorities and financial investments.

This need to build the information capacity as the basis for decision-making is recognised as the first prescribed action in the biodiversity proposals for Agenda 21 of UNCED. An encouraging start has been made with the Country Study Programme, which must now be expanded through the Convention on Biological Diversity to develop the human skills and monitoring capabilities of developing countries. The long-term goal must be to create a global biodiversity information network, linking the national centres and mobilising the substantial amounts of data worldwide to promote a more enlightened conservation and development practice.

References

Reid, W.V., McNeely, J.A., Tunstall, D.B. and Bryant, D. (in prep.). World Resources Institute, Washington DC.

Touche Ross 1991. *Conservation of Biological Diversity: the role of technology transfer*. A Report for the United Nations Conference on Environment and Development and the UNEP Intergovernmental Negotiating Committee for a Convention on Biological Diversity. Touche Ross and Co., London. 67pp.

UNEP 1991. *Guidelines for the Preparation of Country Studies on Costs, Benefits and Unmet Needs of Biological Diversity Conservation within the Framework of the Convention on Biological Diversity*. UNEP/Bio.Div./Guidelines. May 1991. 73pp.

UNEP 1992. *Biodiversity Country Studies: executive summary*. A Report to the Fifth Session of the Intergovernmental Negotiating Committee for a Convention on Biological Diversity. UNEP/Bio.Div./N7-INC.5/May 1992. 6pp.

WRI/IUCN/UNEP 1992. *Global Biodiversity Strategy: guidelines for action to save, study and use earth's biotic wealth sustainably and equitably*. World Resources Institute, Washington DC. 243pp.

GLOSSARY

This highly selective Glossary provides definitions of some of the less familiar or more technical terms to be found above.

Anadromous: (of fishes) those which ascend rivers from the sea in order to spawn.

Anaerobic respiration: liberation of energy by breakdown of substances not involving consumption of oxygen.

Archaean: belonging to or containing the group of rocks of the Archaeozoic era (the earlier part of the Precambrian era), which ended about 2,500 million years ago.

Back-arc basins: ocean floor spreading centres associated with subduction processes in deep ocean trenches.

Catadromous: (of fishes) those which descend rivers to the lower or estuarine reaches, or the sea, in order to spawn.

Chloroplasts: plastids containing chlorophyll, sometimes with other pigments, found in the cytoplasm of higher plant cells.

Chromophyte*: a member of the Chromophyta, a major division of the plant kingdom including most of the algae characterised by the presence of flagellae on the sexual spores and the presence of chlorophyll *a* (but not *b*).

Coccoliths: minute mainly marine protoctistan organisms (formerly treated as protozoa or algae) dating from Cambrian to modern times, with calcium carbonate ring or platelet structures, which form substantial chalk deposits when fossilised.

Conidial: used of fungi producing conidia, that is asexual spores formed by mitotic divisions; sexual stages are unknown in many such fungi, which are sometimes referred to as 'deuteromycetes' or 'imperfect fungi'.

Cyanobacteria: photosynthetic and nitrogen-fixing blue-green bacteria, formerly generally treated as blue-green algae but lacking nuclei and therefore regarded as belonging to the bacterial Kingdom.

Diploid: having two sets of chromosomes in the nucleus of each somatic cell. Characteristic of most normal eukaryotic higher organisms.

Ectomycorrhiza: mycorrhiza (*q.v.*) where the fungal mycelium is only associated with the first layer of epidermal cells on the plant root.

Endomycorrhiza: mycorrhiza (*q.v.*) where the fungal mycelium penetrates into the cortex of the plant root.

Epedaphon: inhabitants of the soil surface, e.g. most ground-beetles and scorpions.

Euedaphon: inhabitants of the mineral soil, e.g. most earthworms, all Symphyla, many mites.

Eukaryote: a cell or organism with a membrane-bounded nucleus, organelles and chromosomes with histone-coated DNA.

Germplasm: genetic material, especially its specific molecular and chemical constitution, that comprises the physical basis of the inherited qualities of an organism.

Haploid: having the number of chromosomes characteristic of the gametes for the organism (one set in most eukaryotic normal higher organisms).

Hemiedaphon: inhabitants of the litter and fermentation layer, e.g. many woodlice and millipedes.

Heterotrophic heterokonts: filamentous or unicellular organisms, the sexual spores (zoospores) of which have two hair-like appendages each of a different structure, and which also lack chlorophyll and obtain the carbohydrates they require by parasitising plants or utilising dead organic materials.

Hexapods: six-footed animals, specifically the insects (although certain primitive forms are occasionally excluded from the taxon Insecta and these groups together are included in the Hexapoda).

Mitochondria: double-membraned organelles in the cytoplasm of all eukaryotes where the respiratory cycles occur.

Mollicute: a bacterium-like organism such as *Spiroplasma*, lacking an independent wall and always occurring inside the cells of cellular organisms, particularly insects.

Morphospecies: a group of individuals which are considered to belong to the same species on morphological grounds alone.

Mycorrhiza: mutualistic symbiotic associations between fungi and the roots of green plants, occurring in about 80% of all vascular plants and also certain bryophytes; the fungi either form nets over the root surfaces or are mainly confined to a special layer within the root tissues themselves.

Ocean-floor spreading centres: these occur along the central axis of most oceans, and are the seismically active regions where new oceanic crust forms; as new crust is extruded from below, the older crust is pushed away from the axis of the ridge.

Organelles: the various inclusions in a cell which have special functions, e.g. mitochondria and chloroplasts.

Picoplankton: minute algal-like organisms with cells about 2 microns in diameter abundant in the upper layers of the world's oceans.

Plastid: a cytoplasmic, pigmented photosynthetic organelle or its non-photosynthetic derivative.

Primary productivity: the rate of transformation of chemical or solar energy to biomass. Most primary production is carried out by plants through photosynthesis but some bacteria can convert chemical energy to biomass through chemosynthesis.

Prokaryote: a cell or organism composed of cells lacking a membrane-bound nucleus, membrane-bound organelles and histone-coated DNA.

Protoctists*: eukaryotic organisms which are not plants, animals or fungi, i.e. protozoans and other unicellular organisms, algae, slime moulds, etc.

rRNA: ribosomal ribonucleic acid - the type of RNA which, together with proteins, makes up the ribosomes.

Subduction and fracture zones: where spreading ocean crust impinges against an unyielding continental margin, an 'active margin' forms, with the oceanic crust buckling downwards (subducting) and being destroyed within the hot interior of the earth; ocean trenches are formed along these margins.

Note: * Dependent on classification system used.

GPSR Compliance

The European Union's (EU) General Product Safety Regulation (GPSR) is a set of rules that requires consumer products to be safe and our obligations to ensure this.

If you have any concerns about our products, you can contact us on ProductSafety@springernature.com

In case Publisher is established outside the EU, the EU authorized representative is:

Springer Nature Customer Service Center GmbH
Europaplatz 3
69115 Heidelberg, Germany

Batch number: 10825074

Printed by Printforce, the Netherlands